Lecture Notes in Electrical Engineering

Volume 1066

The book series *Lecture Notes in Electrical Engineering* (LNEE) publishes the latest developments in Electrical Engineering—quickly, informally and in high quality. While original research reported in proceedings and monographs has traditionally formed the core of LNEE, we also encourage authors to submit books devoted to supporting student education and professional training in the various fields and applications areas of electrical engineering. The series cover classical and emerging topics concerning:

- Communication Engineering, Information Theory and Networks
- Electronics Engineering and Microelectronics
- Signal, Image and Speech Processing
- Wireless and Mobile Communication
- Circuits and Systems
- Energy Systems, Power Electronics and Electrical Machines
- Electro-optical Engineering
- Instrumentation Engineering
- Avionics Engineering
- Control Systems
- Internet-of-Things and Cybersecurity
- Biomedical Devices, MEMS and NEMS

For general information about this book series, comments or suggestions, please contact leontina.dicecco@springer.com.

To submit a proposal or request further information, please contact the Publishing Editor in your country:

China

Jasmine Dou, Editor (jasmine.dou@springer.com)

India, Japan, Rest of Asia

Swati Meherishi, Editorial Director (Swati.Meherishi@springer.com)

Southeast Asia, Australia, New Zealand

Ramesh Nath Premnath, Editor (ramesh.premnath@springernature.com)

USA, Canada

Michael Luby, Senior Editor (michael.luby@springer.com)

All other Countries

Leontina Di Cecco, Senior Editor (leontina.dicecco@springer.com)

**** This series is indexed by EI Compendex and Scopus databases. ****

Sanjay Sharma · Bidyadhar Subudhi ·
Umesh Kumar Sahu
Editors

Intelligent Control, Robotics, and Industrial Automation

Proceedings of International Conference, RCAAI 2022

Springer

Editors
Sanjay Sharma
Faculty of Science and Engineering
School of Engineering, Computing
and Mathematics
University of Plymouth
Plymouth, UK

Bidyadhar Subudhi
School of Electrical Sciences
Indian Institute of Technology Goa
Ponda, Goa, India

Umesh Kumar Sahu
Department of Mechatronics
Manipal Institute of Technology, Manipal
Academy of Higher Education
Manipal, Karnataka, India

ISSN 1876-1100 ISSN 1876-1119 (electronic)
Lecture Notes in Electrical Engineering
ISBN 978-981-99-4633-4 ISBN 978-981-99-4634-1 (eBook)
https://doi.org/10.1007/978-981-99-4634-1

This Springer imprint is published by the registered company Springer Nature Singapore Pte Ltd.
The registered company address is: 152 Beach Road, #21-01/04 Gateway East, Singapore 189721, Singapore

Paper in this product is recyclable.

Preface

Robotics, Control, Automation and Artificial Intelligence (RCAAI-2022) is the first series of interdisciplinary international conferences organized by the Department of Mechatronics, Manipal Institute of Technology (MIT), Manipal Academy of Higher Education (MAHE), Manipal, India, in collaboration with the Department of Mechanical Engineering, National Institute of Technology Karnataka (NITK), Surathkal, India. It was held between 24 and 26 November 2022 in virtual mode. The RCAAI is organized to provide a platform for researchers, industrial fraternity, and academia to deliberate and exchange problems and ideas on the latest trends in Robotics, Control, Automation, and Artificial Intelligence (AI).

The conference brought together academic and industry researchers from countries such as the United Arab Emirates, the USA, Sweden, Canada, Botswana, and many others. The conference has been sponsored by Harvesters Desk, India. The International Society of Automation (ISA) has been the Knowledge and Publicity Partner of the RCAAI.

The conference has attracted a large number of papers in varied disciplines such as intelligent control, industrial Internet of things, industrial automation, cyber-security, robotics, machine vision, circuits and sensors, etc. From a total of 302 papers received, 282 papers were sent for double-blind review after preliminary inspection and plagiarism check. Among those, 99 were accepted for presentation after peer review, making the acceptance rate of 33%. Out of 99 papers, 82 papers were presented at the conference over sixteen parallel technical sessions spanning six technical tracks and were considered for publication in this book as chapters.

The presented papers have been grouped into six parts and are included in the book *Intelligent Control, Robotics, and Industrial Automation—Proceedings of International Conference, RCAAI 2022* of the Springer.

Part One: Robotics and Intelligent Systems
Part Two: Advanced Control Techniques
Part Three: Industrial Automation, IoT and Cyber-Security
Part Four: Machine Vision and Image Processing
Part Five: Circuits, Sensors and Biomedical Systems
Part Six: Artificial Intelligence and Its Applications

The contributions presented in this volume widen the knowledge horizons in various domains of Robotics, Control, Automation, and AI. Brief of these advances are listed below but not limited to.

- Robotic system design for warehouse I-Robot, mecanum wheeled robot, tree-type robot, and legged bipedal robot further focusing on field of services, manufacturing, repair, and maintenance.
- Intelligent control techniques for trajectory tracking control of quadrotors, robot manipulators, autonomous underwater vehicles, and optimization and control of the hybrid solar photovoltaic (PV) system.
- Development of Internet of things (IoT) and AI-based automated device and their application in the field of safety monitoring, cyber-security, and health care.
- Performance improvement of in the existing disease diagnosis frameworks, to detect the defects of PV cells, lane recognition for intelligent ground vehicles using advanced vision-based algorithms.
- Next-generation micro-electromechanical systems (MEMS) sensors and micro-fabricated biosensors are presented for non-invasive medical diagnosis.
- Development of hybrid deep learning and quantum-based K-nearest neighbours algorithms for applications such as driver assistance, job scheduling, recommendation systems, fault diagnosis, optical fibre classification, and human locomotive.

We believe that the proceedings of the conference will append to the learning curves of the researchers, academicians, and industry professionals. Our sincere acknowledgement to all contributing authors for taking time and effort to send their research work and adhering to all review comments and formatting requirements, to the unanimous technical reviewers, session chairs, national and international advisory board committees, and institute committees. We also wish to place our gratitude to Springer Nature for publishing the accepted and presented papers of RCAAI-2022 event.

This event was made possible by the utmost support from Chancellor of MAHE Padmashree Awardee Dr. Ramadas M. Pai, Pro-Chancellor Dr. H. S. Ballal, Vice-Chancellor Lt. Gen. (Dr) M. D. Venkatesh, Registrar Dr. Narayana Sabhahit, Dy. Director (Research) Dr. Santhosh K V, Section Heads of Finance, and they deserve our heartfelt gratitude. Director (MIT) Cdr. Dr. Anil Rana, Director (NITK) Prof. Prasad Krishna, Joint Director (MIT) Dr. Somashekara Bhat, Associate Director QA Dr. Chandrashekhar Bhat, Dr. D. V. Kamath, the Head of the Department, Mechatronics, MIT, and Dr. Ravikiran Kadoli, the Head of the Department, Mechanical Engineering, NITK, Surathkal, deserve lots of appreciation for their constant guidance and motivation.

The organizing secretaries, Dr. Ankur Jaiswal and Dr. Arun Kumar Shettigar, deserve special recognition for his several months of untiring work towards this conference. Also, contributions from Dr. K. Abhimanyu K. Patro, Dr. Vijay Babu Koreboina, Mr. D. A. P. Prabhakar, Dr. Narendra Kumar, Dr. Ravi Pratap Singh, and Dr. Vijay Mohan helped a lot to make the conference, a success.

Our sincere gratitude to the administrating staffs of MAHE, MIT, the Department of Mechatronics, MIT, Manipal, and also the Department of Mechanical Engineering, NITK, Surathkal, for their wholehearted support for the conference. Finally, our deepest gratitude to all who contributed in various capacities, in organizing this event.

Plymouth, UK
Ponda, Goa, India
Manipal, Karnataka, India
November 2022

Dr. Sanjay Sharma
Prof. Bidyadhar Subudhi
Dr. Umesh Kumar Sahu

Contents

Editors and Contributors

About the Editors

Dr. Sanjay Sharma received a B.Tech. (Hons.) in Electrical Engineering from the Indian Institute of Technology (IIT) Kharagpur, M.Tech. degree with distinction in Control Systems (Electrical Engineering) from the Indian Institute of Technology Varanasi (IIT-BHU). He received his Ph.D. degree in Systems and Control Engineering from the University of Sheffield, the UK in 2000. He completed Postgraduate Certificate in Higher Education Teaching (PGCHET) course from Queen's University Belfast in the year 2004. Currently, he serves as Associate Professor, in Intelligent Autonomous Control Systems, School of Engineering, Computing and Mathematics (Faculty of Science and Engineering) at the University of Plymouth, United Kingdom. He was invited as one of the five International Experts along with colleagues from TAMU USA, MIT USA, UiT Norway, and IST Portugal to assist in the control systems technology of IIT Madras's initiative to open a Centre for Marine Autonomous Systems. He has published more than 50 referable journal papers and more than 50 conference papers in the area of marine, control, and AI. He has supervised 10 Ph.D. and 1 M.Phil. students. His current research activities include the use of AI techniques such as genetic algorithms, fuzzy logic, and neural networks for modeling and the design of novel control systems for industrial and marine plants, particularly for wave energy devices, robotics, automobiles, and uninhabited marine vehicles for both surface and underwater operations.

Prof. Bidyadhar Subudhi received a Bachelor of Electrical Engineering from the National Institute of Technology (NIT) Rourkela, M.Tech. degree in Control and Instrumentation from the Indian Institute of Technology (IIT) Delhi and a Ph.D. degree in Control System Engineering from the University of Sheffield, the UK in 2002. He served as a Post-Doctoral Research Fellow at the National University of Singapore (NUS). Currently, he serves as a Professor, at the School of Electrical Sciences, and Dean (R&D) at IIT Goa. He was a recipient of the prestigious Samanta Chandra Sekhar Award of the Odisha Bigyan Academy, Government of Odisha for his contribution to science and technology. He was awarded NITRAA Research Excellence Award in Electrical Sciences. He is a Fellow of the Indian National Academy of Engineering and a Fellow of IET (UK) and Fellow of Asia-Pacific Artificial Intelligence Association. He has supervised 40 Ph.D. theses. His research interests include system and control, PV and microgrid systems, and autonomous underwater vehicles. He featured in the top 2% of scientists in the field of industrial engineering and automation in 2020 as per Stanford University's study group.

Dr. Umesh Kumar Sahu received his Bachelor of Electronics and Telecommunication Engineering from Pandit Ravishankar Shukla University, India, and his M.Tech. degree in Instrumentation and Control Engineering from Chhattisgarh Swami Vivekanand Technical University, India. He received his Ph.D. degree in Robotics and Machine Vision from the National Institute of Technology (NIT) in Rourkela, India in 2020. Currently, he serves as an Assistant Professor in the Department of Mechatronics, Manipal Institute of Technology, Manipal Academy of Higher Education, India. He has published 16 papers in reputed journals and international conferences. He also has 3 IPRs (Copyrights) under literary work. He received two best paper awards in TENCON-2017 and IRIA-2021. He is a Fellow of the IETE India, and a Member of Young Professional Activity Group (YPAG) in ACDOS. He is a Member in The Robotics Society (TRS), IAENG Hong Kong, IAOE, and an Editorial Member in AJAI. He has supervised a total of 27 undergraduate/postgraduate students. His research interests include robotics and machine vision, image processing, instrumentation and control, network control systems, control of flexible robot manipulators, industrial automation, and intelligent robotic control.

Contributors

Bibhudendra Acharya Department of Electronics and Communication Engineering, National Institute of Technology Raipur, Raipur, Chhattisgarh, India

Madhughnea Sai Adabala SCOPE, Vellore Institute of Technology, Vellore, India

N. Thoufik Ahmed Vel Tech Rangarajan Dr. Sagunthala R&D Institute of Science and Technology, Chennai, India

N. Akshay East Point College of Engineering and Technology, Bengaluru, India

Annamalai Alagappan Department of Network and Infrastructure Management, Faculty of Engineering and Technology, Botho University, Gaborone, Botswana

Mrinal Anand R.V College of Engineering, Bangalore, India

A. B. Andhare Visvesvaraya National Institute of Technology, Nagpur, Maharashtra, India

Leo John Baptist Andrews Department of Information Technology, Faculty of Engineering and Technology, Botho University, Gaborone, Botswana

Sachin Angadi KLE Technological university, Hubli, India

Areefa S R Engineering College, Warangal, Telangana State, India

Sunil Arora Department of R&D, Bond Marine Consultancy, London, UK

Ramkumar Arsamy Department of Mechatronics Engineering, Kumaraguru College of Technology, Coimbatore, Tamilnadu, India

M. Arulmozhivarman Department of Electronics and Electrical Engineering, SASTRA Deemed University, Thanjavur, Tamil Nadu, India

N. K Arun Assistant Professor, NIT, Calicut, Kerala, India

C. R. Arunkumar Electrical Engineering Department, National Institute of Technology Warangal, Telangana, India

C. S. Asha Manipal Institute of Technology, Manipal Academy of Higher Education, Manipal, Karnataka, India

Vishwalata Bagal D.Y. Patil College of Engineering, Akurdi, Pune, India

T. Baiju Department of Mathematics, Manipal Institute of Technology, Manipal Academy of Higher Education, Manipal, Karnataka, India

Awadhesh Bansode E&TC Department, VIT, Pune, India

Aditya A. Bastapure Mechanical Engineering Department, Visvesvaraya National Institute of Technology, Nagpur, India

Manas Baviskar E&TC Department, VIT, Pune, India

A. Bhanuchandar Electrical Department, NIT Warangal, Warangal, Telangana, India

Atharva Bhatkar E&TC Department, VIT, Pune, India

Shripad Bhatlawande E&TC Department, VIT, Pune, India

Niraja Narayan Bhatta Department of Electronics and Communication Engineering, Amrita School of Engineering, Amrita Vishwa Vidyapeetham, Coimbatore, India

Neelanjan Bhattacharjee Department of Mechanical Engineering, University of Alberta, Edmonton, Canada

Kaustav Bhowmick PES University/ECE, Bangalore, India

Aashish Kumar Bohre National Institute of Technology (NIT), Durgapur, India

Ashwath Raj Capur Department of Mechatronics, Manipal Institute of Technology, Manipal Academy of Higher Education, Manipal, Karnataka, India

Jannu Chaitanya School of electronics engineering, VITAP University, Amaravati, India

V. Sri Pavan Ravi Chand Spacecraft Mechanisms Group, U.R.Rao Satellite Centre, ISRO, Bengaluru, India

K. Chandrakanth Electrical Engineering Department, National Institute of Technology Warangal, Telangana, India

Shashank Chandrakar Department of Electronics and Communication, National Institute of Technology, Raipur, Chhattisgarh, India

Shital S. Chiddarwar Mechanical Engineering Department, Visvesvaraya National Institute of Technology, Nagpur, India

Manish Choubisa Poornima College of Engineering, Jaipur, Rajasthan, India

Bishwaranjan Das KMC Hospital, Mangalore, Karnataka, India

Pallabi Das Centre for Data Science, JISIASR, JIS University, Kolkata, India

Sachikanta Dash CSE Department, GIET University, Gunupur, Odisha, India

Gerard Deepak Department of Computer Science and Engineering, Manipal Institute of Technology Bengaluru, Manipal Academy of Higher Education, Manipal, India

S. Deivalakshmi Department of Electronics and Communication Engineering, National Institute of Technology, Tiruchirappalli, India

Ravishankar P. Desai Department of Electrical and Electronics Engineering, BITS Pilani, K. K. Birla Goa Campus, South Goa, Goa, India

Apurba Dev Department of Applied Physics, KTH Royal Institute of Technology, Stockholm, Sweden

P. Uma Devi Department of ECE, G. Pullaiah College of Engineering and Technology, Kurnool, Andhra Pradesh, India

Siddharth Dewangan Department of Electronics and Communication, National Institute of Technology, Raipur, Chhattisgarh, India

Joshuva Arockia Dhanraj Centre for Automation and Robotics (ANRO), Department of Mechatronics Engineering, Hindustan Institute of Technology and Science, Chennai, Tamil Nadu, India

Pranav Vikirtan Dhayanithi Department of Mechanical Engineering, Amrita School of Engineering, Amrita Vishwa Vidyapeetham, Coimbatore, Tamil Nadu, India

Praneet Dighe Department of Mechatronics, Manipal Institute of Technology, Manipal Academy of Higher Education, Manipal, Karnataka, India

K. Dileep Research Scholar, NIT, Calicut, Kerala, India

C. Dineshkumar Department of Automobile Engineering, B.S. Abdur Rahman Crescent Institute of Science and Technology, Chennai, Tamil Nadu, India

Manish Dubey Poornima College of Engineering, Jaipur, Rajasthan, India

Pratik U. Durge Mechanical Engineering Department, V.N.I.T, Nagpur, India

Jeane Marina D'Souza Department of Mechatronics Engineering, Manipal Institute of Technology, Manipal Academy of Higher Education, Manipal, Karnataka, India

Muniyandy Elangovan Department of Biosciences, Saveetha School of Engineering, Chennai, India

K. V. Gangadharan Department of Mechanical Engineering, National Institute of Technology Karnataka, Surathkal, India

Hithesh Kumar Gatty Department of Mechatronics, Manipal Institute of Technology, Manipal, India;
Department of Applied Physics, KTH Royal Institute of Technology, Stockholm, Sweden;
Gatty Instruments AB, Green Innovation Park, Uppsala, Sweden

Kaustav Ghar Department of Mechanical Engineering, Jadavpur University, Kolkata, India

N. Gnanasekaran National Institute of Technology Karnataka, Surathkal, India

Rajesh Gogineni Department of ECE, Dhanekula Institute of Engineering and Technology, Vijayawada, India

Karthik Gogisetty Department of Mechatronics, Manipal Institute of Technology, Manipal Academy of Higher Education, Manipal, Karnataka, India

Gumalapuram Gopal EEE Dept, MGIT, Hyderabad, Telangana, India

G. Gopichand SCOPE, Vellore Institute of Technology, Vellore, India

Arun Goyal Mechanical Engineering Department, Visvesvaraya National Institute of Technology, Nagpur, India

Bhaskar Guin Department of Mechanical Engineering, Contai Polytechnic, Purba Medinipur, West Bengal, India

Vinodh Venkatesh Gumaste Central Manufacturing Technology Institute, Sc-C, C-SVTC, Bengaluru, India

Chirag Gupta Scientist/Engineer SD, North Eastern Space Applications Centre, Indian Space Research Organization, Shillong, India

Gurusiddayya Hiremath Department of Electronics & Communication Engineering, Sahyadri College of Engineering & Management, Mangaluru, Karnataka, India;
Visvesvaraya Technological University (VTU), Belagavi, Karnataka, India

P. S. Hiremath Master of Computer Applications, KLE Technological University, Hubballi, India

Baisravan HomChaudhuri Illinois Institute of Technology, Chicago, IL, USA

Thirunavukkarasu Indiran Department of Instrumentation and Control Engineering, Manipal Institute of Technology, Manipal Academy of Higher Education, Manipal, India

George Vadakkekara Itty Electrical and Electronics Engineering, Mar Baselios Christian College of Engineering and Technology, Peermade, India

Khaja Izharuddin Department of Electrical and Electronics Engineering, CBIT (A), Gandipet, Hyderabad, Telangana, India

Vinay Kumar Jadoun Manipal Institute of Technology, Manipal Academy of Higher Education (MAHE), Manipal, Karnataka, India

Pankaj Kumar Jain Indian Institute of Technology (BHU) Varanasi, Varanasi, Uttar Pradesh, India

Ankur Jaiswal Department of Mechatronics, Manipal Institute of Technology, Manipal Academy of Higher Education, Manipal, Karnataka, India

Pallavee Jaiswal Dr. C. V. Raman University, Kota, Bilaspur (C.G), India

P. D. Jeyakumar Department of Automobile Engineering, B.S. Abdur Rahman Crescent Institute of Science and Technology, Chennai, Tamil Nadu, India

Abhishek Jha Department of Mechanical Engineering, Amrita School of Engineering, Amrita Vishwa Vidyapeetham, Chennai, India

Ankit Jha Department of Electronics and Communication Engineering, Manipal Institute of Technology, Manipal Academy of Higher Education, Manipal, Karnataka, India

Bhaavin K. Jogeshwar Mechatronics Department, Manipal Institute of Technology, Manipal Academy of Higher Education, Manipal, Karnataka, India

Renu Jose Department of Electronics and Communication, Rajiv Gandhi Institute of Technology, Affiliated to APJ Abdul Kalam Technological University, Kottayam, Kerala, India

Ashutosh Joshi Tata Consultancy Services, Pune, Maharashtra, India

Devashish Joshi Prosirius Technologies, Indore, Madhya Pradesh, India

Sahana Kalligudd KLE Technological university, Hubli, India

Yogeesh Kamat KMC Hospital, Mangalore, Karnataka, India

Prasad V. Kane Department of Mechanical Engineering, VNIT, Nagpur, Maharashtra, India

Neeraj Kanwar Department of Electrical Engineering, Manipal University Jaipur, Jaipur, India

Abhinandan Kapoor Spacecraft Mechanisms Group, U.R.Rao Satellite Centre, ISRO, Bengaluru, India

Sivayazi Kappagantula Mechatronics Department, Manipal Institute of Technology, Manipal, Karnataka, India

Sachin Karadgi Department of Automation & Robotics, KLE Technological University, Hubballi, India

Shaik Karimulla IPCV Laboratory, Department of Electrical Engineering, National Institute of Technology Rourkela, Rourkela, Odisha, India

Brijesh Kumar Karna Edge AI Division, Reliance Jio Platforms Limited, Bangalore, India

S. Karthick School of Electrical and Electronics Engineering, VIT Bhopal University, Bhopal, Sehore, Madhyapradesh, India

Rachit Khandelwal Department of Electronics and Communication Engineering, National Institute of Technology, Tiruchirappalli, India

Hari Kishan Kondaveeti School of Computer Science Engineering, VIT-AP University, Vijayawada, Andhra Pradesh, India

Sivarama Krishna Koneru S R Engineering College, Warangal, Telangana State, India

Megha P. Krishna Department of Electronics and Communication, Rajiv Gandhi Institute of Technology, Affiliated to APJ Abdul Kalam Technological University, Kottayam, Kerala, India

Balachandar Krishnamurthy School of Mechanical Engineering, Sathyabama Institute of Science and Technology, Chennai, Tamil Nadu, India

H. N. Suresha Kumar Spacecraft Mechanisms Group, U.R.Rao Satellite Centre, ISRO, Bengaluru, India

M. Sathish Kumar Sri Ramachandra Faculty of Engineering and Technology, Sri Ramachandra Institute of Higher Education and Research, Chennai, Tamil Nadu, India

N. Charan Kumar East Point College of Engineering and Technology, Bengaluru, India

Santosh Kumar Indian Institute of Technology, Guwahati, India

V. Sampath Kumar Grant Thornton, Gaborone, Botswana

Manisha Kumari Department of Electronics and Communication Engineering, National Institute of Technology Raipur, Raipur, Chhattisgarh, India

Ciji Pearl Kurian Department of Electrical and Electronics Engineering, Manipal Institute of Technology, Manipal Academy of Higher Education, Manipal, India

Jan Linnros Department of Applied Physics, KTH Royal Institute of Technology, Stockholm, Sweden

Saketh Maddineni Department of Computer Science and Engineering, National Institute of Technology, Tadepalligudem, Andhra Pradesh, India

Golak Bihari Mahanta Department of Mechanical Engineering, Amrita School of Engineering, Amrita Vishwa Vidyapeetham, Chennai, India

G Maithreyan Department of Production Engineering, National Institute of Technology Tiruchirappalli, Tamil Nadu, Tiruchirappalli, India

I. Mamatha Department of Electrical and Electronics Engineering, Amrita School of Engineering, Amrita Vishwa Vidyapeetham, Bengaluru, Karnataka, India

Narayan S. Manjarekar Department of Electrical and Electronics Engineering, BITS Pilani, K. K. Birla Goa Campus, South Goa, Goa, India

Udaya Bhasker Manthati Electrical Engineering Department, National Institute of Technology Warangal, Telangana, India

Deepak Maretha Webkorps Services India Private Limited, Indore, Madhya Pradesh, India

Amit Mallikarjun Masuti KLE Technological university, Hubli, India

Rishikesh S. Mate Dr. Vishvanath Karad's MIT World Peace University, Pune, India

Nimmy Ann Mathew Department of Electronics and Communication, Rajiv Gandhi Institute of Technology, Affiliated to APJ Abdul Kalam Technological University, Kottayam, Kerala, India

Veena Mathew Department of Electrical and Electronics Engineering, Manipal Institute of Technology, Manipal Academy of Higher Education, Manipal, India

V. P. Meena Electrical Engineering Department, MNIT, Jaipur, India

S. J. Mija Associate Professor, NIT, Calicut, Kerala, India

Zeesha Mishra Department of Microelectronics and VLSI, UTD, Chhattisgarh Swami Vivekanand Technical University, Newai, Bhilai, Chhattisgarh, India

Nipu Modak Department of Mechanical Engineering, Jadavpur University, Kolkata, India

Piyush Modi Department of Electronics and Communication Engineering, Manipal Institute of Technology, Manipal Academy of Higher Education, Manipal, Karnataka, India

Thangamuthu Mohanraj Department of Mechanical Engineering, Amrita School of Engineering, Amrita Vishwa Vidyapeetham, Coimbatore, Tamil Nadu, India

Deeptej Sandeep More Manipal Institute of Technology, Manipal Academy of Higher Education, Manipal, Karnataka, India

Adwitiya Mukhopadhyay Department of Computer Science, Amrita School of Computing, Amrita Vishwa Vidyapeetham, Mysuru Campus, Mysuru, India

Abhishek Murali Department of Mechatronics Engineering, Kumaraguru College of Technology, Coimbatore, Tamilnadu, India

B. Muruganantham Department of Computing Technologies, SRM Institute of Science and Technology, Kattankulathur, Chennai, India

Solomon Jenoris Muthiya Department of Automobile Engineering, Dayananda Sagar College of Engineering, Bangalore, Karnataka, India

B. P. Nagaraj Spacecraft Mechanisms Group, U.R.Rao Satellite Centre, ISRO, Bengaluru, India

I. Madesh Naidu East Point College of Engineering and Technology, Bengaluru, India

Varsha Naik Dr. Vishvanath Karad's MIT World Peace University, Pune, India

Binoy B. Nair Department of Electronics and Communication Engineering, Amrita School of Engineering, Amrita Vishwa Vidyapeetham, Coimbatore, India

K. C. Deepika Nair Department of Electrical and Electronics Engineering, Amrita School of Engineering, Amrita Vishwa Vidyapeetham, Bengaluru, Karnataka, India

B. Lakshmi Narayana Spacecraft Mechanisms Group, U.R.Rao Satellite Centre, ISRO, Bengaluru, India

Ayan Naskar Department of Electronics and Communication Engineering, SRM University AP, Amaravati, Andhra Pradesh, India

Tarun Kanti Naskar Department of Mechanical Engineering, Jadavpur University, Kolkata, India

Rajashree Nayak Centre for Data Science, JISIASR, JIS University, Kolkata, India

Sasmita Padhy School of Computing Science and Engineering, VIT Bhopal University, Bhopal-Indore Highway, Kothrikalan, Sehore, Madhya Pradesh, India

Usha Padma Electronics and Telecommunication Engineering, RV College of Engineering, Bangalore, India

Kowstubha Palle Department of Electrical and Electronics Engineering, CBIT (A), Gandipet, Hyderabad, Telangana, India

Dency R. Pambhar Dr. Vishvanath Karad's MIT World Peace University, Pune, India

Ishita Pandey Department of Mechatronics, Manipal Institute of Technology, Manipal Academy of Higher Education, Manipal, Karnataka, India

Sandeep Pandey Thapar Institute of Engineering and Technology, Patiala, Punjab, India

Pawan Pandit Department of Mechanical Engineering, National Institute of Technology Karnataka, Surathkal, India

Arun P. Parameswaran Department of Electrical & Electronics Engineering, Manipal Institute of Technology, Manipal Academy of Higher Education, Manipal, Karnataka, India

Onika Parmar Department of Microelectronics and VLSI, UTD, Chhattisgarh Swami Vivekananda Technical University, Newai, Bhilai, India

Ruchi Patel Gyan Ganga Institute of Technology and Sciences, Jabalpur, Madhya Pradesh, India

A. V. Patil D.Y. Patil Institute of Engineering, Management and Research, Akurdi, Pune, India

Rakesh Patnaik CSE Department, GIET University, Gunupur, Odisha, India

Dipti Patra IPCV Laboratory, Department of Electrical Engineering, National Institute of Technology Rourkela, Rourkela, Odisha, India

R. Christu Paul Department of Automobile Engineering, Hindustan Institute of Technology and Science, Chennai, Tamil Nadu, India

Tinu Valsa Paul Research Scholar, Department of Instrumentation and Control Engineering, Manipal Institute of Technology, Manipal Academy of Higher Education, Manipal, India

R. Pavithra Department of Electrical and Electronics Engineering, Manipal Institute of Technology, Manipal Academy of Higher Education, Manipal, India

Chinmay Vilas Potphode Visvesvaraya National Institute of Technology, Nagpur, India

Kota Pragathi S R Engineering College, Warangal, Telangana State, India

Pothapragada Pranav Department of Electronics and Communication Engineering, SRM University AP, Amaravati, Andhra Pradesh, India

M. Praagna Prasad Electronics and Telecommunication Engineering, RV College of Engineering, Bangalore, India

S. V. Praveen Department of Computer Science and Engineering, National Institute of Technology, Tiruchirappalli, Tamil Nadu, India

Rayyan Muhammad Rafikh Electrical and Computer Engineering Department, Sultan Qaboos University, Muscat, Oman

V. M. Rahul National Institute of Technology Karnataka, Surathkal, India

K. Vineeth Rai Department of Electronics & Communication Engineering, Sahyadri College of Engineering & Management, Mangaluru, Karnataka, India; Visvesvaraya Technological University (VTU), Belagavi, Karnataka, India

Vijay Shankar Rai Spacecraft Mechanisms Group, U.R.Rao Satellite Centre, ISRO, Bengaluru, India

Parth P. Rainchwar Dr. Vishvanath Karad's MIT World Peace University, Pune, India

Pethuru Raj Edge AI Division, Reliance Jio Platforms Limited, Bangalore, India

Raymon Antony Raj Department of Electrical and Electronics Engineering, Kongu Engineering College, Perundurai, Erode, Tamil Nadu, India

Amit Singh Rajput Department of Microelectronics and VLSI, UTD, Chhattisgarh Swami Vivekananda Technical University, Newai, Bhilai, India

A. B. Raju KLE Technological university, Hubli, India

Y. Ramakrishna Department of ECE, Dhanekula Institute of Engineering and Technology, Vijayawada, India

M. Venkata Ramana Spacecraft Mechanisms Group, U.R.Rao Satellite Centre, ISRO, Bengaluru, India

H. B. Ramesh PES University/ECE, Bangalore, India

M. Pratham Rao Department of Electronics & Communication Engineering, Sahyadri College of Engineering & Management, Mangaluru, Karnataka, India; Visvesvaraya Technological University (VTU), Belagavi, Karnataka, India

Raffik Rasheed Department of Mechatronics Engineering, Kumaraguru College of Technology, Coimbatore, Tamilnadu, India

Premanshu Sekhara Rath CSE Department, GIET University, Gunupur, Odisha, India

Pooja Ravi Department of Computing Technologies, SRM Institute of Science and Technology, Kattankulathur, Chennai, India

B. Sandeep Reddy Indian Institute of Technology, Guwahati, India

Chandan N. Reddy East Point College of Engineering and Technology, Bengaluru, India

K. Harshavardhana Reddy East Point College of Engineering and Technology, Bengaluru, India

Ulasi Vivek Reddy School of Mechanical Engineering, Dr. Vishwanath Karad MIT-World Peace University, Pune, India

Koyyada Rishitha S R Engineering College, Warangal, Telangana State, India

Sampath Routu SCOPE, Vellore Institute of Technology, Vellore, India

Sudipta Roy Jio Institute, Navi Mumbai, Maharashtra, India

Shailu Sachan MANIT Bhopal, Bhopal, MP, India

Jambeswar Sahu School of Mechanical Engineering, Vellore Institute of Technology, Vellore, Tamilnadu, India

Umesh Kumar Sahu Department of Mechatronics, Manipal Institute of Technology, Manipal Academy of Higher Education, Manipal, Karnataka, India

Yahuti Sahu Department of Microelectronics and VLSI, UTD, Chhattisgarh Swami Vivekananda Technical University, Newai, Bhilai, India

A. Santhanavijayan Department of Computer Science and Engineering, National Institute of Technology, Tiruchirappalli, India

D. Sarathkumar Department of Electrical and Electronics Engineering, Kongu Engineering College, Perundurai, Erode, Tamil Nadu, India

Santosh Kumar Satapathy Information and Communication Technology, Pandit Deendayal Energy University, Gandhinagar, Gujarat, India

Madhu R Seervi Department of Computer Science, Amrita School of Computing, Amrita Vishwa Vidyapeetham, Mysuru Campus, Mysuru, India

A. S. Selvakumar Department of Mechanical Engineering, B.S. Abdur Rahman Crescent Institute of Science and Technology, Chennai, Tamil Nadu, India

Akash S. Shahade Visvesvaraya National Institute of Technology, Nagpur, Maharashtra, India

K. Shanavaj Department of Electronics and Communication Engineering, G. Pullaiah College of Engineering and Technology, Kurnool, Andhra Pradesh, India

P. Vivekananda Shanmuganathan Department of Mechanical Engineering, SRM University AP, Amaravati, Andhra Pradesh, India

Kirti Sharma Edge AI Division, Reliance Jio Platforms Limited, Bangalore, India

Neeraj Sharma Indian Institute of Technology (BHU) Varanasi, Varanasi, Uttar Pradesh, India

Sachin Sharma East Point College of Engineering and Technology, Bengaluru, India

Sanjay Kumar Sharma Department of Mechanical Engineering, Amity University, Raipur, India

Naveen Shenoy Department of Electronics & Communication Engineering, Sahyadri College of Engineering & Management, Mangaluru, Karnataka, India; Visvesvaraya Technological University (VTU), Belagavi, Karnataka, India

Swati Shilaskar E&TC Department, VIT, Pune, India

Sonam Shrivastava School of Electrical Engineering, Vellore Institute of Technology, Vellore, Tamilnadu, India

Aditya Shukla Department of Computing Technologies, SRM Institute of Science and Technology, Kattankulathur, Chennai, India

Shaivya Shukla Department of Microelectronics and VLSI, UTD, Chhattisgarh Swami Vivekananda Technical University, Newai, Bhilai, India

N. Shyamasunder Bhat Department of Orthopaedics, Kasturba Medical College, Manipal Academy of Higher Education, Manipal, Karnataka, India

Pulkit Singh Department of Electronics and Communication Engineering, MLR Institute of Technology, Hyderabad, Telangana, India

Ranjith Singh Department of Electronics & Communication Engineering, Sahyadri College of Engineering & Management, Mangaluru, Karnataka, India; Visvesvaraya Technological University (VTU), Belagavi, Karnataka, India

Suryabhan Singh Edge AI Division, Reliance Jio Platforms Limited, Bangalore, India

V. P. Singh Electrical Engineering Department, MNIT, Jaipur, India

Shambhavi Sinha Department of Electronics and Communication Engineering, Manipal Institute of Technology, Manipal Academy of Higher Education, Manipal, Karnataka, India

G. Sireesha Department of Electronics and Communication Engineering, G. Pullaiah College of Engineering and Technology, Kurnool, Andhra Pradesh, India

K. S. Sivanandan Department of Biomedical Engineering, Manipal Institute of Technology, Manipal Academy of Higher Education, Manipal, Karnataka, India

RajKumar Soni Department of Electrical Engineering, Manipal University Jaipur, Jaipur, India

Anoop Kumar Srivastava Spacecraft Mechanisms Group, U.R.Rao Satellite Centre, ISRO, Bengaluru, India

S. Srivatsan Sri Ramachandra Faculty of Engineering and Technology, Sri Ramachandra Institute of Higher Education and Research, Chennai, Tamil Nadu, India

R. Srividya Department of Electrical and Electronics Engineering, Manipal Institute of Technology, Manipal Academy of Higher Education, Manipal, India

Sundarraj Subaselvi M.Kumarasamy College of Engineering, Karur, Tamilnadu, India

Bidyadhar Subudhi School of Electrical Sciences, Indian Institute of Technology, Goa, India

B. Sunitha Department of ECE, G. Pullaiah College of Engineering and Technology, Kurnool, Andhra Pradesh, India

Sithara Mary Sunny Department of Electrical & Electronics Engineering, Manipal Institute of Technology, Manipal Academy of Higher Education, Manipal, Karnataka, India

R. Supriya Department of ECE, G. Pullaiah College of Engineering and Technology, Kurnool, Andhra Pradesh, India

Shilpa Suresh Manipal Institute of Technology, Manipal Academy of Higher Education, Manipal, Karnataka, India

Suraj Suresh Kumar Research Scholar, Department of Instrumentation and Control Engineering, Manipal Institute of Technology, Manipal Academy of Higher Education, Manipal, India

D. Susmitha Department of Electronics and Communication Engineering, G. Pullaiah College of Engineering and Technology, Kurnool, Andhra Pradesh, India

Debabrata Swain Computer Science and Engineering, Pandit Deendayal Energy University, Gandhinagar, Gujarat, India

Pankaj Swarnkar MANIT Bhopal, Bhopal, MP, India

Avinash A. Thakre Mechanical Engineering Department, V.N.I.T, Nagpur, India

Abhishek M. Thote School of Mechanical Engineering, Dr. Vishwanath Karad MIT-World Peace University, Pune, India

Manish Kumar Thukral Department of Electrical Engineering, Manipal University Jaipur, Jaipur, India

Dhruv Thummar Department of Mechanical Engineering, National Institute of Technology Karnataka, Surathkal, India

T. Tirupal Department of Electronics and Communication Engineering, G. Pullaiah College of Engineering and Technology, Kurnool, Andhra Pradesh, India

Swapnil M. Tripathi Visvesvaraya National Institute of Technology, Nagpur, India

Ganesha Udupa Department of Mechanical Engineering, Amrita Vishwa Vidyapeetham, Amritapuri, India

Dheeresh Upadhyay Mangalayatan University, Aligarh, Uttar Pradesh, India

Rajesh Kumar Upadhyay Mangalayatan University, Aligarh, Uttar Pradesh, India

V. Vanitha Sri Ramachandra Faculty of Engineering and Technology, Sri Ramachandra Institute of Higher Education and Research, Chennai, Tamil Nadu, India

Arpitha Varghese Department of Electrical and Electronics Engineering, Amrita School of Engineering, Amrita Vishwa Vidhyapeetham, Bangalore, India

P. Veeraswamy Department of ECE, Dhanekula Institute of Engineering and Technology, Vijayawada, India

Gonuguntla Venkatpathy National Institute of Technology Karnataka, Surathkal, India

M. Venusri Department of Electronics and Communication Engineering, G. Pullaiah College of Engineering and Technology, Kurnool, Andhra Pradesh, India

Khyati Verma Department of Mechanical Engineering, National Institute of Technology Karnataka, Surathkal, India

Nadanakumar Vinayagam Department of Automobile Engineering, Hindustan Institute of Technology and Science, Chennai, Tamil Nadu, India

Harshita Virwani Poornima College of Engineering, Jaipur, Rajasthan, India

L. Vishal Sri Ramachandra Faculty of Engineering and Technology, Sri Ramachandra Institute of Higher Education and Research, Chennai, Tamil Nadu, India

Sreeram Warrier Manipal Institute of Technology, Manipal Academy of Higher Education (MAHE), Manipal, Karnataka, India

Vidhun V. Warrier Department of Mechanical Engineering, Amrita Vishwa Vidyapeetham, Amritapuri, India

Soham M. Wattamwar Dr. Vishvanath Karad's MIT World Peace University, Pune, India

Surendra Kumar Yadav Poornima College of Engineering, Jaipur, Rajasthan, India

Umesh Kumar Yadav Electrical Engineering Department, MNIT, Jaipur, India

P. V. Yeswanth Department of Electronics and Communication Engineering, National Institute of Technology, Tiruchirappalli, India

Robotics and Intelligent Systems

Landmark Detection for Auto Landing of Quadcopter Using YOLOv5

Deeptej Sandeep More, Shilpa Suresh, Jeane Marina D'Souza, and C. S. Asha

Abstract The vision-based system is a crucial component of unmanned aerial vehicle (UAV) autonomous flying and is frequently regarded as a difficult phase to accomplish. Most UAV accidents happen while landing or due to obstacles in the path. Hence, it is considered as one of the most important to think about the auto landing of UAVs to reduce accidents. Some technologies, like GPS, frequently don't function indoors or in places where GPS transmissions aren't allowed. They can land up to a few meters but lack accuracy. A system that operates in such circumstances is required to overcome this and be far more suitable. Cameras are used to offer much information about their surroundings and may be helpful in certain circumstances. A vision-based system's accuracy can be as low as a few centimeters and better than GPS-based location estimation. This work involves designing a vision-based landing system that can recognize a marker by providing the bounding box around it. Typically, the H mark is employed in helicopter landing pads for vision-based landing systems. Here, the position is identified using the YOLOv5 algorithm. An image of A4-sized sheet and 2ft $\times$ 2ft printed with H mark is taken using quadcopter and is used for data set. The algorithm is tested to locate the marker at any orientation and scale. Thus, YOLOv5 identifies the marker at any distance, orientation, or size in the given data set and performs better than SVM-based approach. This could be further used to find the distance on the ground from the UAV center that aids in auto landing.

Keywords H marker · Unmanned aerial vehicle · YOLOv5 · Object detection

1 Introduction

Drones have gained enormous popularity in various fields, such as military, agriculture, traffic monitoring, surveying, disaster management, and robotics research. UAV Navigation and Guidance system must satisfy high performance in terms of accuracy.

D. S. More · S. Suresh · J. M. D'Souza · C. S. Asha (✉)
Manipal Institute of Technology, Manipal Academy of Higher Education, Manipal, Karnataka 576104, India
e-mail: asha.cs@manipal.edu

S. Sharma et al. (eds.), *Intelligent Control, Robotics, and Industrial Automation*, Lecture Notes in Electrical Engineering 1066,
https://doi.org/10.1007/978-981-99-4634-1_1

When size and payload capacity are considered, the selection of onboard sensors for mounting is difficult. Additionally, sensors that depend on GPS signals will suffer from poor accuracy and information loss. Visual sensors benefit from being diminutive, portable, less power hungry, cheaper, and provide more information that can be analyzed and used in real-time applications. However, it depends on lighting, viewing angle, and image resolution. The development of artificial intelligence for computer vision has made it possible to handle data in real-time with great accuracy and robustness. Parallelized architecture, efficient algorithms, and faster training and execution have all been made possible. Various researchers have suggested different algorithms to address the issues. Since they are fiducial and, therefore, simple to detect, most rely on synthetic markers like aruco and apriltag. Some of them don't consider possible UAV rotations, and such approaches often tend to be unworkable for actual flight. The current system relies on GPS, which has poor accuracy and may not function under certain conditions due to unstable signals. Some of the methods rely on the picture plane always being parallel to the ground plane, which is rarely the case in actual circumstances. Therefore, a system that functions in the real world is considered necessary. Authors in [1] proposed CAMShift and SURF features to track the helipad region. Landing is carried in three stages in [2]. GPS system is used to locate the landing area at approximately 2m altitude. The custom markers are used in the algorithm to find the center location easily. Chong Yu et al. overcome the limitations of GPS-based methods using fiducial navigation system [3]. In addition, Gazebo simulation is carried out in [4] to detect the markers. SIFT-based matching is employed in [5] for matching natural landmarks. The marker should be repeatable, reliable, invariant to scale, rotation, and translation. Authors in [6] use homography and projective transformation to find the location of target with respect to UAV. Lucas Kanade algorithm is used to track the pixels in the image followed by Kalman filter to get the accurate location. In [7], a simple marker with white squares with black background is used, while in [8] used neural network approach to detect the marker. Recent advances in the field overcomes the usage of markers to automatically detect the safe area to land the system [9]. Shaobao et al. discussed on the hovering control problem of drone on a mobile platform with 6 degrees of freedom [10]. Maravall et al. discussed a vision-based anticipatory controller for an unmanned aerial vehicle's autonomous indoor navigation (UAV) [11]. Huang et al., in their research, demonstrated a commercial quadcopter with an autonomous navigation system based on monocular vision [12]. The organization of the paper is as follows: Sect. 2 describes the object detection framework YOLOv5, Sect. 3 presents the proposed methodology, and Sect. 4 highlights the results followed by conclusion.

2 Background Work

YOLO is an acronym for You Only Look Once used to detect the object in an image. As drones fly at different altitudes, the object orientation and size vary. The issue such as motion blur is common in these videos. YOLOv5 proposed by Ultralytics

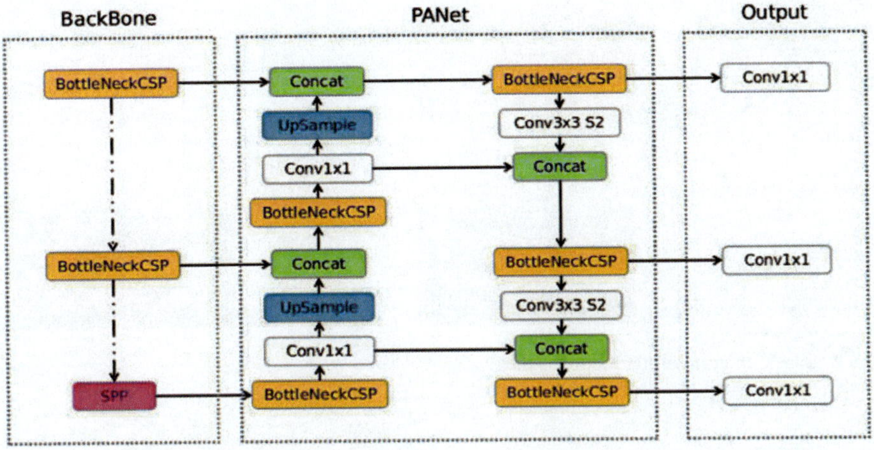

Fig. 1 Architecture of YOLOv5 (ultralytics)

[13] has various models like YOLOv5s, YOLOv5m, YOLOv5l, and YOLOv5. It uses the CSPDarknet53 architecture with SPP layer as backbone, PANNet as neck, and YOLO as detection head. The latest version YOLOv5 is known for its speed and high accuracy. The algorithm uses image as grid for classification and finally combined as a single image.

The network has three main structures; the backbone, feature pyramid, detection head. The backbone network extracts the features from different images at various scales, while feature pyramid fuses the features at different scales, passes them to the detection network.The detection network predicts the bounding box around the marker.

For this work, a small parameter size model is used. Thus having sufficient samples for training, validation and test, YOLOv5 is trained using custom data set (Fig. 1).

3 Methodology

In the proposed work, H marker is used to identify the landing location. The custom marker used in the current work is depicted in Fig 2. It is the H shape inside the circle.

The proposed method is carried out in 2 stages: The workflow of the marker detection is depicted in Fig 2. The first stage involves the training of custom marker, followed by testing using validation data.

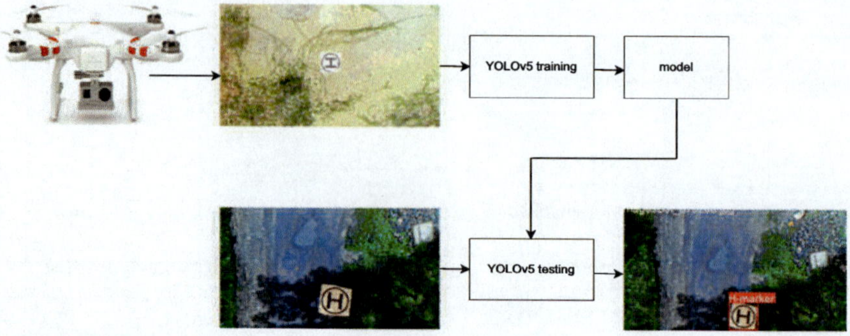

Fig. 2 Block diagram of the proposed method

3.1 Data Set

Initial phase consists of data set collection. The marker in printed in the 2 ft × 2 ft flex sheet and kept in the ground. DJI Quadcopter with fixed camera is used to capture video of the ground at various altitudes and orientation. The videos are used for generating the marker data set required for training the YOLOv5 algorithm. The UAV flying at high altitude captures images of size 1920 × 1080 pixels. Currently 50 images are used to generate positive and negative images. It is always helpful to add few background images to reduce the false positive. A Roboflow annotation platform is used to get the bounding box parameters of background and H marker. Each text file contained the bounding box (x_center, y_center, width, height). The YOLOv5 architecture is suitable for 640 × 640 pixels.

3.2 YOLOv5 Parameter Settings

The training of YOLOv5 is done using the data set captured. The image is tested using YOLOv5 to locate the marker. Since the data size is very small, the transfer learning approach is employed to train using pretrained model. The existing model is pretrained using COCO data set that contains the object detection from various scenes. The model contains backbone layer to extract the feature, while head layer computes the output predictions (Table 1).

Table 1 Parameter settings for H marker detection using YOLOv5

Parameters	Value
Batch	64
Epochs	10
Model	YOLOv5s
Optimizer	Adam
Learning rate	5e−5

4 Results and Discussion

In this section, we verify the accuracy of the proposed method in detecting the custom marker (H marker). Several metrics are adopted to measure qualitatively and quantitatively. The metrics in Fig 3 shows curves obtained for the proposed **H** marker.

4.1 Metrics

Figure 3 depicts various metrics employed in the current work. The box loss (box_loss) represents the bounding box regression loss. The obj_class denotes the confidence of object present. The cls_loss represents the cross entropy loss (classification loss). In the proposed method, there is only one class (H marker), and the classification error is zero as there are no misclassifications. Precision measures the correctness of the bounding box and measured as shown in Eq. 1, where TP describes the true positive, FP for false positive, TN for true negative.

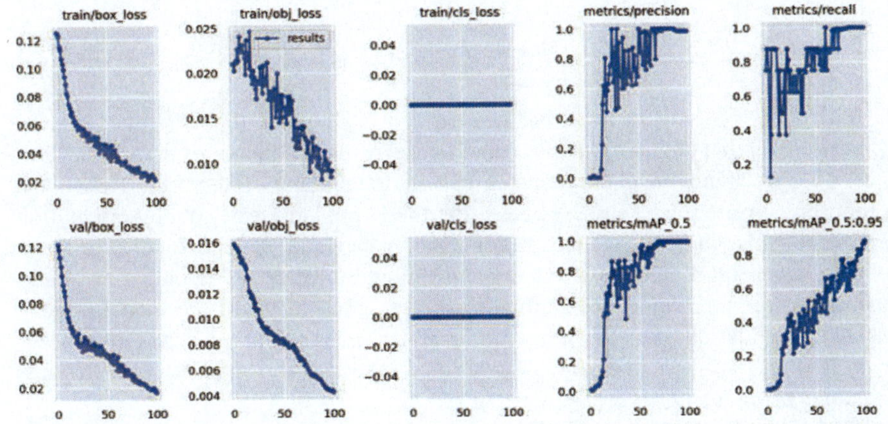

Fig. 3 Metrics used for landmark detection

Fig. 4 Ground truth bounding box for validation data

$$\text{Precision} = \frac{\text{TP}}{\text{TP} + \text{FP}} 100\% \quad (1)$$

Recall is used to measure how much of the bounding box are correctly predicted.

$$\text{Recall} = \frac{\text{TP}}{\text{TP} + \text{FN}} 100\% \quad (2)$$

mAP_0.5 provides mean Average Precision at Intersection over Union (IoU) at a threshold of 0.5, while 0.5:0.95 indicates the range from 0.5 to 0.95.

$$\text{mAP} = \frac{1}{n} \sum_{k=1}^{n} J(\text{Precision, Recall})_k \quad (3)$$

It is observed that the train/val loss curve decreases with number of iterations. Similarly, the precision, recall curve increases with the number of iterations. Hence, the proposed algorithm best suits for detecting the H marker clearly at any orientation and scale. This would help to localize the landing spot and safe auto landing process.

The data set is divided into training and validation data. The training data is augmented using horizontal flip, various scales, etc. Evaluation of the model is done using the validation data. The precision recall curve is used to infer the model accuracy. The set of sample images with ground truth label is shown in Fig 4. The results obtained using YOLOv5 are depicted in Fig. 5.

The H marker detection is carried out using Support Vector Machine (SVM)-based method for comparison purpose. This method has resulted in a lot of false

Fig. 5 Predicted bounding box for validation data

Fig. 6 Predicted bounding box using SVM for validation data

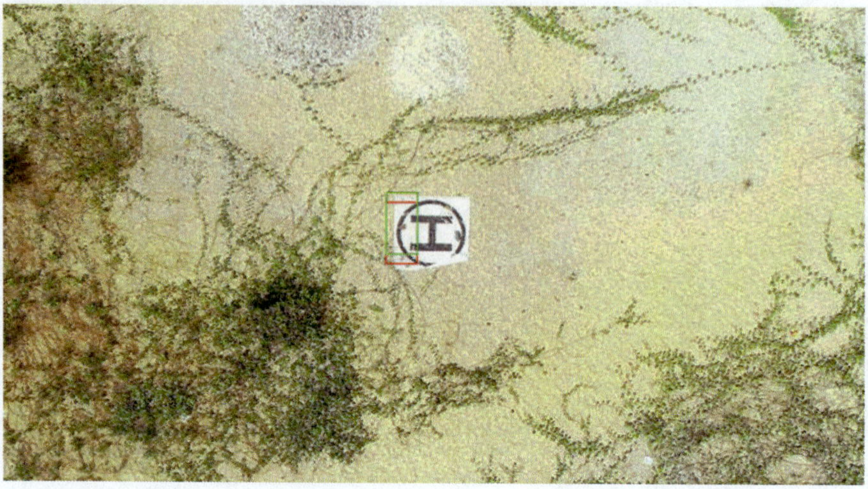

Fig. 7 Predicted bounding box for validation data

Table 2 Evaluation of SVM classifier on training data

	Precision	Recall	F1-score	Support
0	0.95	0.97	0.96	36
1	0.97	0.94	0.95	31
Accuracy			0.96	67
Macro average	0.96	0.95	0.95	67
Weighted average	0.96	0.96	0.96	67

positives that may lead to detection of many bounding boxes that are not H marker. This results in inaccurate detection of marker. The accuracy table in 2 is provided for SVM method on training data. Although SVM results in good precision, recall for training data, it fails to accurately detect the **H** marker as shown in Figs. 6 and 7. The average confidence score of SVM is 0.83 for training data and 0.65 for validation data. Hence, the proposed method using YOLOv5 finds useful in accurately detecting the marker in the captured scene.

5 Conclusion

This paper proposes H marker detection using latest YOLOv5 architecture. A set of images is used to train the YOLOv5 network. The marker is detected at all heights and orientation using the trained model. It is validated using the validations and test data set. The framework can be further extended to find the actual orientation and height

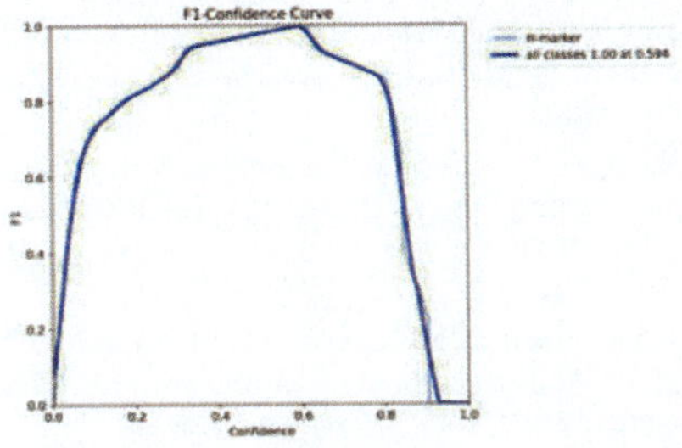

(a) F1 curve

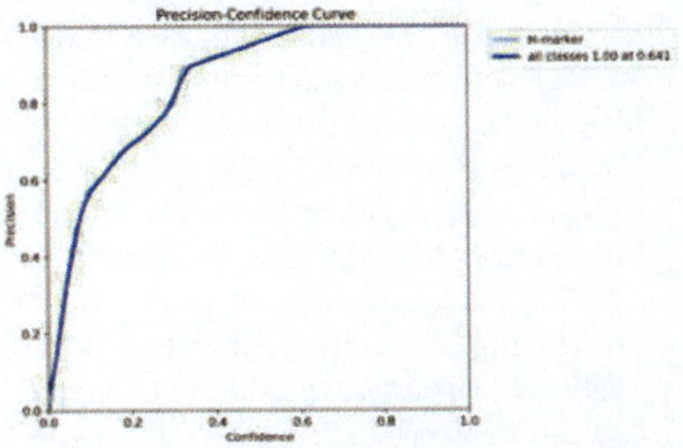

(b) Precision Curve

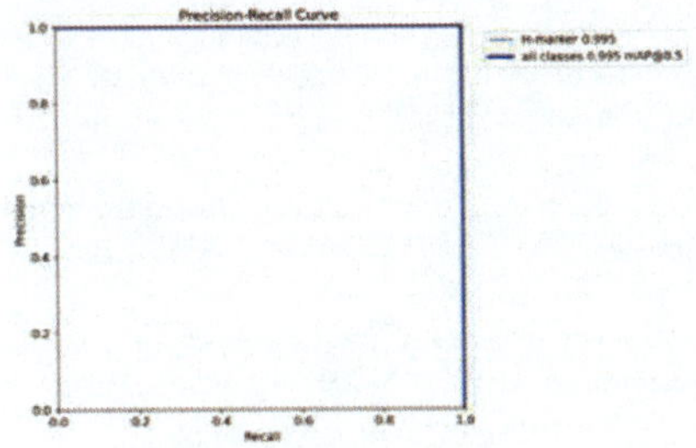

(c) Precision Recall curve

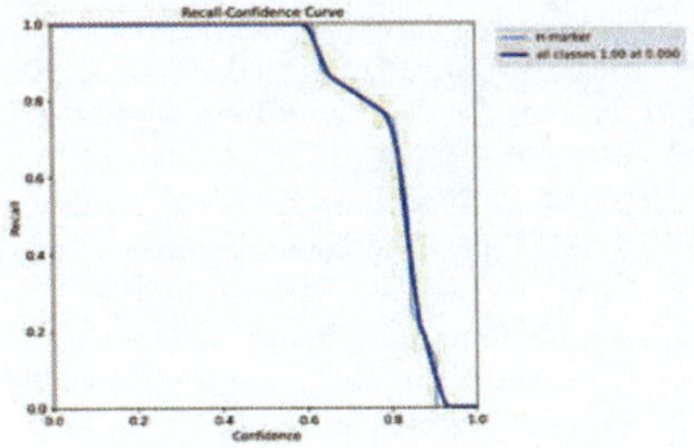

(d) Recall Curve

Fig. 8 Quantitative results of the landmark detection approach

of the quadcopter to change its speed and direction to reach the land marker automatically. Practically, the detected marker is used to retrieve the 3D vector from which UAV is able to reach with the accuracy of less than 10 cms. The current framework is successful to detect the marker on uneven surface as well, partially occluded, from the camera. It outperforms the conventional machine learning algorithms such as SVM. In addition the speed of the algorithm is suitable for real-time implementation using the portable hardware like Raspberry Pi or Jetson Nano (Fig. 8).

References

1. Zhao Y, Pei H (2012) An improved vision-based algorithm for unmanned aerial vehicles autonomous landing. Phys Procedia 33:935–941. https://doi.org/10.1016/j.phpro.2012.05.157
2. Blachut K, Szolc H, Wasala M, Kryjak T, Gorgon M (2020) A vision based hardware-software real-time control system for the autonomous landing of an uav. In: International conference on computer vision and graphics. Springer, pp 13–24. https://doi.org/10.48550/arXiv.2004.11612
3. Yu C, Cai J, Chen Q (2017) Multi-resolution visual fiducial and assistant navigation system for unmanned aerial vehicle landing. Aerosp Sci Technol 67:249–256. https://doi.org/10.1016/j.ast.2017.03.008

4. Saavedra-Ruiz M, Pinto-Vargas AM, Romero-Cano V (2021) Monocular visual autonomous landing system for quadcopter drones using software in the loop. IEEE Aerosp Electron Syst Mag 37(5):2–16. https://doi.org/10.48550/arXiv.2108.06616
5. Cesetti A, Frontoni E, Mancini A, Zingaretti P, Longhi S (2009) Vision-based autonomous navigation and landing of an unmanned aerial vehicle using natural landmarks. In: 2009 17th Mediterranean conference on control and automation. IEEE, pp 910–915. https://doi.org/10.1016/j.ast.2017.03.008
6. Mondragón IF, Campoy P, Martinez C, Olivares-Méndez MA (2010) 3d pose estimation based on planar object tracking for uavs control. In: 2010 IEEE international conference on robotics and automation. IEEE, . pp 35–41. https://doi.org/10.1109/ROBOT.2010.5509287
7. Sharp CS, Shakernia O, Sastry SS (2001) A vision system for landing an unmanned aerial vehicle. In: Proceedings 2001 ICRA. In: IEEE international conference on robotics and automation (Cat. No. 01CH37164), vol 2. IEEE, pp 1720–1727. https://doi.org/10.1109/ROBOT.2001.932859
8. Moriarty P, Sheehy R, Doody P 2017() Neural networks to aid the autonomous landing of a uav on a ship. In: 2017 28th Irish signals and systems conference (ISSC). IEEE, pp 1–4. https://doi.org/10.1109/ISSC.2017.7983613
9. Bektash O, Naundrup JJ, la Cour-Harbo A (2022) Analyzing visual imagery for emergency drone landing on unknown environments. Int J Micro Air Veh 14:17568293221106492. https://doi.org/10.1177/175682932211064
10. Li S, Durdevic P, Yang Z (2018) Hovering control for automatic landing operation of an inspection drone to a mobile platform. Ifac-papersonline 51(8):245–250. https://doi.org/10.1016/j.ifacol.2018.06.384
11. Maravall D, de Lope J, Fuentes JP (2015) Vision-based anticipatory controller for the autonomous navigation of an uav using artificial neural networks. Neurocomputing 151:101–107. https://doi.org/10.1016/j.neucom.2014.09.077
12. Huang R, Tan P, Chen BM (2015) Monocular vision-based autonomous navigation system on a toy quadcopter in unknown environments. In: 2015 International conference on unmanned aircraft systems (ICUAS). IEEE, pp 1260–1269. https://doi.org/10.1109/ICUAS.2015.7152419
13. Yolov5 (2001) https://ultralytics.com/

ROS-Based Evaluation of SLAM Algorithms and Autonomous Navigation for a Mecanum Wheeled Robot

Pranav Vikirtan Dhayanithi and Thangamuthu Mohanraj

Abstract Robot navigation is one of the most important aspects of robotics. A map of the environment and path planning based on the obstacle avoidance and optimal path algorithm are required for any robot to successfully navigate in a environment. This study investigates GMapping and Cartographer SLAM approaches and analyzes the results based on map accuracy, perception, 3D mapping, and localization. Autonomous navigation based on ROS core system is also tested for a mecanum wheel robot with respect to the performances of suitable SLAM navigated through an environment with static and dynamic obstacles.

Keywords ROS · SLAM · Mapping · Obstacle avoidance · Navigation

1 Introduction

Intelligent service robots are becoming more common as modern innovations advance. From the very first surface cleaning robot to interpersonal robots, attendant robots, academic robots, rehabilitation robots, as in [1], grocery store robots, and so on, their autonomy has been constantly improved. Autonomous mobile robots have relieved pressure on people, allowing them to be more productively used in innovation and employment creation. Smith et al. were the first to propose the concept of Robot SLAM [2]. It is a method where sensors in robot are used to understand their surroundings in a indoor surrounding to create a map of the area and determine the location. Using a collection of weighted particles, it aligns probabilistic model of the robot in a given system. However, because of the extensive computing, the approach is taking a long time to build. Murphy et al. proposed the Rao-Blackwellized particle filtering (RBPF) approach around the turn of the twentieth century [3]. Furthermore used RBPF to solve the SLAM challenge [4, 5].

P. V. Dhayanithi · T. Mohanraj (✉)
Department of Mechanical Engineering, Amrita School of Engineering, Amrita Vishwa Vidyapeetham, Coimbatore, Tamil Nadu, India
e-mail: t_mohanraj@cb.amrita.edu

S. Sharma et al. (eds.), *Intelligent Control, Robotics, and Industrial Automation*, Lecture Notes in Electrical Engineering 1066,
https://doi.org/10.1007/978-981-99-4634-1_2

The particle filtering SLAM method's computing complexity is considerably reduced by RBPF-SLAM, which advances the advancement of SLAM technology. Furthermore, the graph-based formulation is yet another simple approach frequently utilized in modern times to address problems with large maps and significant computational complexity. It entails building a network whose nodes stand for the robot's posture and whose edges stand for restrictions on those positions. The effectiveness of two widely used SLAM techniques will be compared and discussed in this work, which will also serve as a review for future SLAM system efficiency improvements. A designed mecanum wheeled robot model being simulated using Robot Operating System (ROS) with SLAM that has been selected from the comparative analysis of both SLAM algorithms. The path is assessed in a test run in a simulated space created using Gazebo Simulator. ROS creates a network linking all processes, and it is simple to finish the advancement of the robot innovation stage by constructing additional nodes or function packages similar to those in [6, 7].

2 Kinematic Model

The focus of this research is on the robot, as shown in Fig. 1, that travels through their surroundings on mecanum wheels. Four mecanum wheels on the movable base allow for free sliding at 45° angle from the direction of driving while remaining nonslip. The research will concentrate on a single mecanum wheel and create a mode. The omniwheel and mecanum wheel operate on the same theoretical foundation, but both differ in the direction in which free sliding is possible.

Initially, the defined frame is vector-fixed to the chassis, x_a and y_b. The wheel center i is at (x_i, y_i) and is in forward direction of driving, or in orientation where it does not slide when rolling and must be at angle β_i in relation to vector x_b. And the rollers, of each mecanum wheel, around the wheel rim allow for free sliding for γ_i angle in relative to the direction that is diagonal to the path of motion. For an omniwheel, γ_i is 0°, while for a mecanum wheel, γ_i is 45°. The definitions established allow for the calculation of the driving speed of wheel u_i , which would be the angular velocity of a drive system attached to the wheel.

Fig. 1 Robot model

The driving velocity plus the free-sliding velocity are added to determine the velocity distribution at the center of the wheel. This linear velocity depending on location of the wheel in the robot frame [8] and is obtained from the body twist v_b. The linear speed is then converted into a frame that is fixed to wheel. To separate the wheel speed into sliding speed as well as driving speed components, the said linear speed is as yet the vector amount of both the driving speed as well as free-sliding speed. By taking the wheel velocity and vector $[1tan\gamma_i]$ dot product, the driving factor can be calculated. Its linear velocity is then changed to a rotational velocity by dividing it by the wheel radius. The outcome would be 1×3 row vector multiplied by body twist v_i .

$$u_i = \frac{1}{r_i}[1\ \tan\gamma_i]\begin{bmatrix} \cos\beta_i & \sin\beta_i \\ -\sin\beta_i & \cos\beta_i \end{bmatrix}\begin{bmatrix} -y_i & 1 & 0 \\ x_i & 0 & 1 \end{bmatrix} v_b \tag{1}$$

To create a model for the robot, its configuration of wheel should be in such a way that the sliding directions of each wheel are not all aligned. The robot length would be in distance from mobile frame b to the mecanum wheels, with width in the distance from mobile frame b to mecanum wheels. The wheel speeds will result in sliding of the wheels with in driving direction because the matrix is not square.

$$u = \begin{bmatrix} u_1 \\ u_2 \\ u_3 \\ u_4 \end{bmatrix} = \frac{1}{r_i}\begin{bmatrix} -l-w & 1 & -1 \\ l+w & 1 & 1 \\ l+w & 1 & -1 \\ -l-w & 1 & 1 \end{bmatrix}\begin{bmatrix} 0 \\ v_b x \\ 0 \end{bmatrix} \tag{2}$$

As per Eq. (2), all wheels must move at the same speed for the body to go forward and backward in the direction. If the desired mobility is a pure rotation in robot, the wheels on the same side should revolve at the same speed. If the desired mobility is sideways in the direction, the wheels on the opposite corners should be traveling at the same speed.

3 SLAM Algorithm

The SLAM method combines mapping, sensing, kinematic model generation, multiple objects, camera depth, dynamic obstacles, loop closure, and complexity. The robot uses sensors and measurement equipment to determine its location using landmarks. An autonomous robots can move around without bumping into obstacles and decide which course is optimal, given the circumstances as in [9]. Robot position may be created, updated, and estimated using SLAM [10]. Particle filter, FastSLAM and Extended Kalman filter are some techniques utilized to resolve it. Several SLAM algorithms exist, including Gmapping, Hector SLAM, Google Cartographer, Graph SLAM, Core SLAM, etc. In this study, Google Cartographer and Gmapping, a particle filter-based SLAM [11], are employed as comparison SLAM algorithms.

3.1 Gmapping SLAM

One of the most popular SLAM algorithms utilized in the past and now is the GMapping technique, also known for being based just on particle filter approach. As a greater number of particles are needed to provide results with greater precision, SLAM approaches relying on particle filtration usually need high amounts of CPU and memory [12]. SLAM method which is based on RBPF was created as a workable solution to the SLAM issue. To reduce the amount of particles required and noticeably reduce the uncertainty around the robot position in PF prediction stage, it was further refined as reported in [13]. The main goal of the RBPF method is to estimate the probability of $p(x_{1:\,t}, m|z_{1:t}, u_{1:t-1})$, a joint posterior of a robot trajectory ($x_{1:t} = x_1, x_2, x_3, x_4, ..., x_t$) as well as the environment's grid map m, where the sensor values are shown by odometry observations by ($z_{1:t} = z_1, z_2, z_3, \ldots, z_t$) and ($u_{1:t} = u_1, u_2, u_3, \ldots, u_{t-1}$), respectively. The posterior probability uses the following factorization based on the Bayes' Rule:

$$p(x_{1:\,t}, m|z_{1:t}, u_{1:t-1}) = p(m|x_{1:\,t}, z_{1:t})p(x_{1:\,t}|z_{1:t}, u_{1:t-1}) \tag{3}$$

Because it enables predicting the robot's trajectory prior to constructing the map given that trajectory, this method is known as Rao-Blackwellization. Equation (3) may be broken down into two issues: the mapping problem [$p(m|x_{1:\,t}, z_{1:t})$] and positioning difficulty [$p(x_{1:\,t}|z_{1:t}, u_{1:t-1})$]. Additionally, RBPF method uses sample significance resampling, the most used particle filtering algorithm, and the process is broken down into four steps.

1. Sampling: The previous generation ($x_{t-1}^{(i)}$) is sampled from the proposal distribution π to produce new particles ($x_t^{(i)}$). In order to improve the sampling procedure, the present observations are employed.
2. Importance Weighting: Using the sampling procedure, the important weight $w_t^{(i)}$ of every present particle $x_t^{(i)}$ is determined (Eq. 4), where π ()is the probabilistic odometry motion model.

$$w_t^{(i)} = \frac{p(x_{1:\,t}^{(i)}, m|z_{1:t}, u_{1:t-1})}{\pi(x_{1:t}^{(i)}|z_{1:t}, u_{1:t-1})} \tag{4}$$

3. Resampling: Lower weight-scoring particles will be removed, and new particles are sampled in their stead. The number of particles is unchanged after resampling, and each particle has the same weight.
4. Update of Map: $p(m^{(i)}|x_{1:t}^{(i)}, z_{1:t})$ is used to construct the map estimate after bringing in the trajectory $x_{1:t}^{(i)}$ as well as observational$history z_{1:t}$. The computational complexity brought on by the rising length of trajectory over the period of time, as indicated in Eq. (5) [14], is reduced using the recursive weight update

formula, where η is a constant across all of the particles normalization factor determined from Bayes' rule.

$$w_t^{(i)} = w_{t-1}^{(i)} \eta \frac{p(z_t | x_{1:t}^{(i)}, z_{1:t-1}) p(x_t^{(i)} | x_{1:t-1}^{(i)}, u_{1:t-1})}{\pi(x_{1:t}^{(i)} | z_{1:t}, u_{1:t-1})} \tag{5}$$

3.2 *Google Cartographer SLAM*

Graph-based techniques are more reliable and effective in this situation than particle-based methods. In 2016, Google created the cartographer, a method based on graph optimization. The network of nodes and edges used by this graph-based method similarly reflects the postures, characteristics, and constraints identified from onboard observations [15]. The front-end, also known as Local SLAM, as well as the back-end, or the Global SLAM, of the algorithm can be seen as two separate but related methods to the SLAM difficulty. For architecture, Global and Local SLAM enhance the pose $\xi = (\xi_x, \xi_y, \xi_\theta)$ of LIDAR data, referred to as scans, which will be of an translation, (x, y), as well as ξ_θ, a rotation [16]. In this circumstance, an IMU is also necessary to measure gravity orientation. Local SLAM is defined by the three ideas listed below.

1. Scans: The submap, which would be continuously matched toward the submap coordinate frames, needs to be created, therefore scans are necessary. Equation (6) illustrates the rigid transformation T_ξ, also known as the posture ξ of the conversion of scan points from the scan frame into the submap frame, or the scan frame inside the submap frame.

$$T_\xi p = \begin{pmatrix} \cos \xi_\theta & -\sin \xi_\theta \\ \sin \xi_\theta & \cos \xi_\theta \end{pmatrix} p + \begin{pmatrix} \xi_x \\ \xi_y \end{pmatrix} \tag{6}$$

2. Submaps: Scan-matching is a nonlinear optimization method that compares each consecutive scan against a small region of the global map called a submap M. Each submap is built from a number of subsequent matched scans and therefore is represented by probability grids$M : rZ * rZ \rightarrow \left[p_{\min}\ p_{\max}\right]$ that, at a given resolution r, maps from irregular grid points. A laser scan will probably strike a grid point and find the spot occupied. Similar to how it fails, the possibility of occupancy decreases [17].
3. Ceres Scan Matching: At all times, scan-matching is used to capture the scan before inserting it into a submap. The probability there at scan sites in the submap is maximized by this posture. It is represented like a nonlinear least squares

minimization problem, similar to Eq. (7), where T_ξ conforms to scan posture by translating h_k from scan frame onto submap frame. Additionally, M_{smooth} is a component of the Bi-cubic interpolation, as a function, which was used to obtain the map value for the future scan point from the previous scan point.

$$\text{argmin}\xi \sum_{k=1}^{K} (1 - M_{\text{smooth}}(T_\xi h_k))^2 \tag{7}$$

4 Map Generation

4.1 GMapping

In this experiment, the robot was operated in indoors, as shown in Fig. 2. The results from the laser scanner are used to update the map every five seconds. The outer edges of the walls are accurately recognized by the laser scanner. GMapping method functions remarkably well in indoor settings, as seen in Fig. 3. The GMapping method depends largely on the availability and reliability of accurate odometry data and performs best in a planar environment. As a result, such a strategy cannot be used in an unstructured environment, preventing the robot or carrier from experiencing extreme pitch and roll motion.

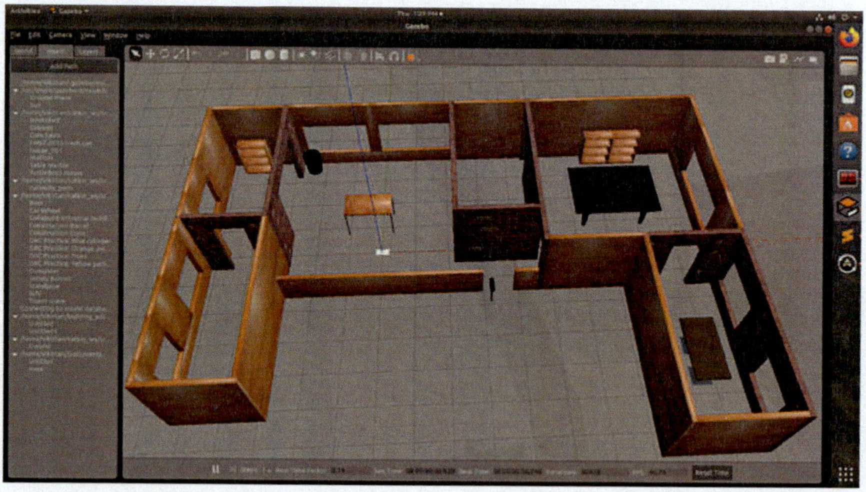

Fig. 2 Indoor environment

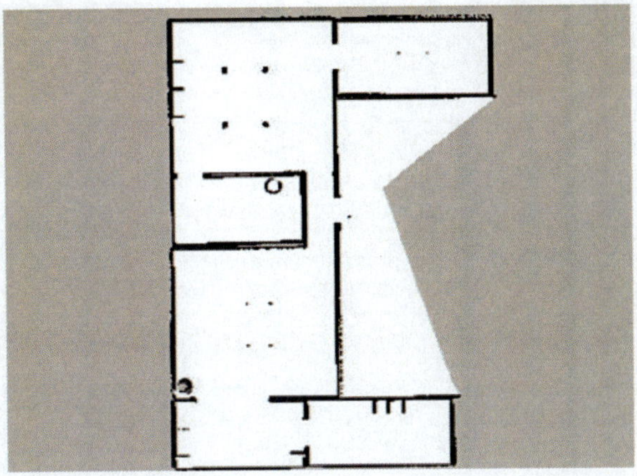

Fig. 3 GMapping-generated map

4.2 Google Cartographer

As shown in Fig 4, the IMU position and sudden turns are very sensitively displayed on the Google cartography map. The sub-maps can shift significantly as a result of even little variations between the real and modeled locations. The complexity of adjusting a cartographer is due to the abundance of choices. The delay can be decreased, especially as the values of the adjustable parameters are typically unit less scalar values which multiply against relative mistakes. But as a result shown in Fig. 5, the maps' accuracy degrades. Odometry could have a negligible influence on the final map if the scan matching effect is properly calibrated. The outcomes of the map are significantly influenced by the LIDAR's direction.

Loop closure in SLAM robotics refers to the action of the robot going back to an earlier visited place. If this happens, the robot's location will be updated using an approximation of a relative transformation between both the current and previously explored scans. Loop closure diagnostic should move faster than the installation of new scans since loops are immediately closed when a spot is examined while upgrading once every seconds.

5 2D Implementation

In the gazebo, a navigational environment is created for the mecanum wheeled robot. The robot has a camera and a LiDar plug-in and four mecanum wheels for mobility. The needed plug-ins are incorporated into the script files, and laser data produced by the lidar can be used to build the map. By adding the necessary parameters to the Gmapping SLAM from the comparative analysis that produced the best results. Using "teleop_key" package from ROS, the mecanum robot is moved to each corner of the

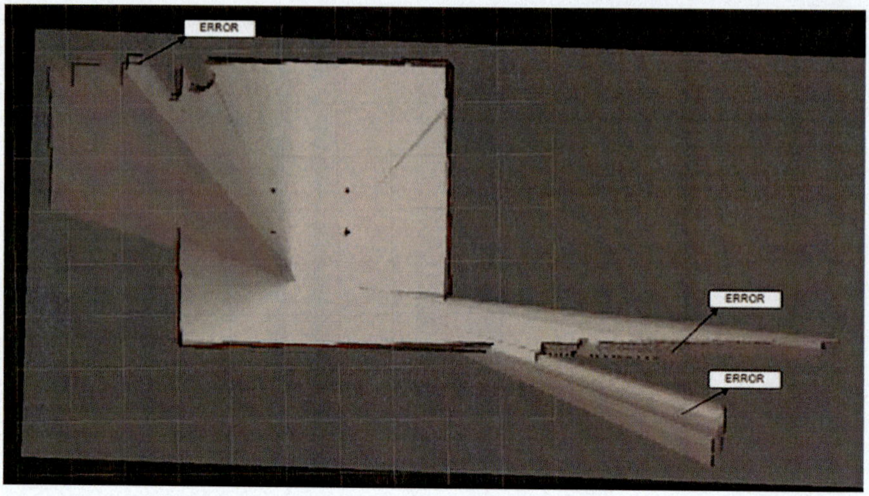

Fig. 4 Results due to sudden turn

Fig. 5 Generated Google Cartographer Map without sudden turns

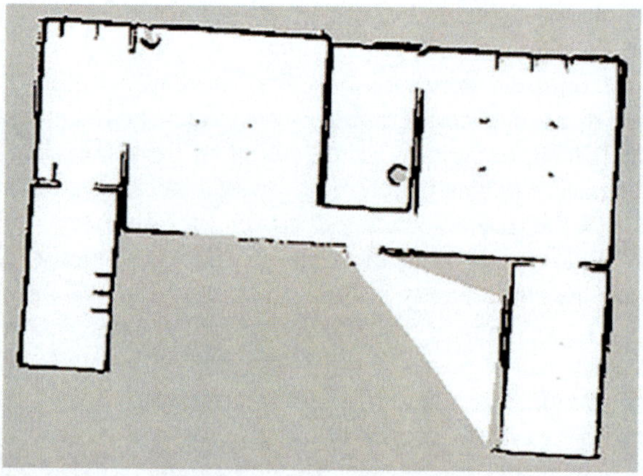

environment until a full map is created. The topics that RVIZ needed to visualize the Robot were selected and added in RViz. The lidar sensor publishes laser data in RViz as topic "/scan" which has been designated as an image acquisition topic by RViz.

Similarly, the topic "/map" is for map. The "/map server" package, which is part of ROS, saves the generated map in the system. For a robot moving in two dimensions, AMCL navigation stack packages offered a probabilistic localization system. Utilizing 2D nav goal in RViz, which assigns the robot a goal, the destination and orientation can be determined by the robot. The different topics that the different nodes have published and are subscribed to are shown with the node graph in Fig. 6. The robot constantly redirects itself in an effort to avoid and follow the new path, as in Fig. 7, even though it may not always succeed due to certain parameters.

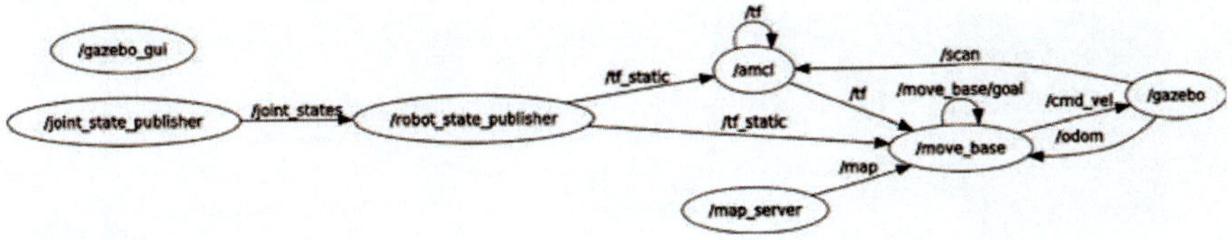

Fig. 6 Node graph

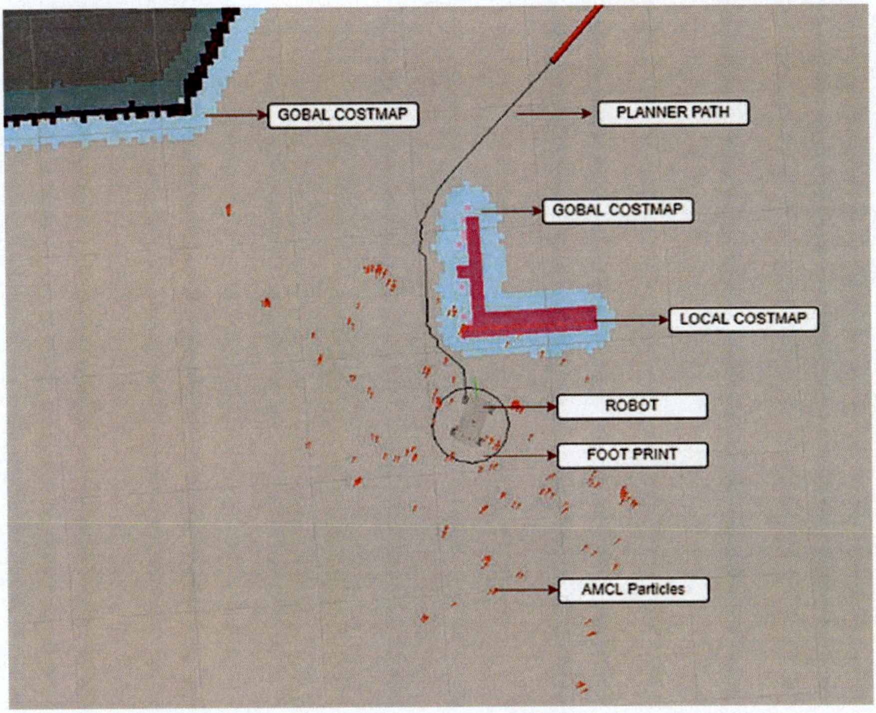

Fig. 7 Obstacle detection and path planning in the environment

6 Results and Discussion

The mobile robot was given some tasks to complete in order to test its ability to carry out the requested missions. The robot is initially given the assignment of mapping the surrounding using Gmapping, as a result of comparison of mapping conducted in previous sections. The algorithm finds the shortest path based on the generated map when any point in destination A/B/C/D, as in Fig. 8, is given as a target. Table 1 shows the time taken to reach the sections at various point locations within the chosen segments during the first test. Tables 2 and 3 display the test results when static and dynamic obstacles were introduced in the direction of the mecanum wheel robot.

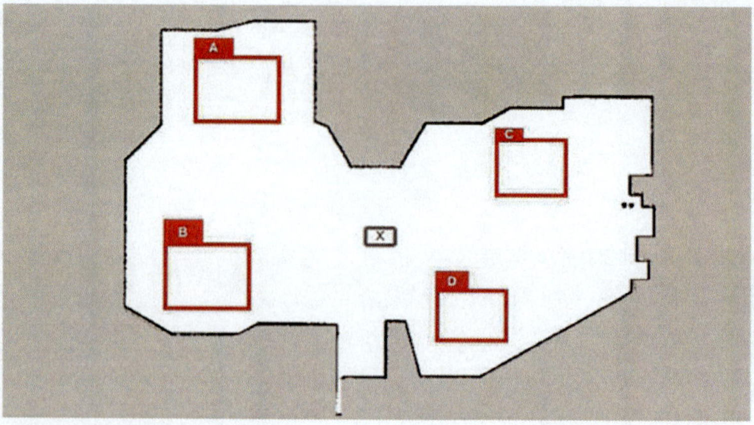

Fig. 8 Test environment with sections

Table 1 Travel time in the environment without obstacles

Trials	Source to section A	Source to section B	Source to section C	Source to section D
1	52.5	40.8	16.9	15.3
2	57.7	32.4	17.4	13.7
3	54.8	38.7	21.2	17.5
4	53.4	35.5	18.2	17
5	59.1	37.5	23.7	18.9
Average	**55.5**	**36.9**	**19.8**	**16.4**

Table 2 Travel time in the environment with static obstacles

Obstacles	Source to section A	Source to section B	Source to section C	Source to section D
1	87.8	75.7	54.8	43.8
2	111.5	108.5	93.7	77.5
3	143.9	127.4	114.9	112.3
Average	**114.4**	**103.8**	**87.8**	**77.8**

Table 3 Travel time in the environment with dynamic obstacles

Obstacles	Source to section A	Source to section B	Source to section C	Source to section D
1	96.7	85.8	61.6	53.2
2	136.2	125.4	120.7	95.3
3	160.3	143.3	159.1	146.7
Average	**131.0**	**118.1**	**113.8**	**98.4**

Obstacle avoidance and finding the optimal path were also taken into account to analysis performance of the mecanum wheel robot. The robot reacts swiftly and covers the distance in a tolerable period of time if there are no hurdles in its way. When a dynamic or static obstacle is placed in the Robot path, the laser sensor detects the obstacle and signals the Robot. By including the identified obstruction on the map, it updates it. After the map has been updated in RViz, the robot determines the next optimal route to the desired location. The robot takes longer than planned to achieve the objective with the increase in dynamic and static barriers. As in occasion of an sudden arrival of a new obstacle, the robot pauses as well as takes a long time to recalculate the new path. The search needs to be optimized to increase the efficiency.

7 Conclusion

According to the study, the key deterministic elements in the navigation and mapping process have been the sensor sensitivity, sample rate, and associated filters and algorithms. The GMapping and Google Cartographer algorithms were simulated and analyzed. The Cartographer method makes use of the idea of a submap to provide frontend matching, which compares the current frame to the submap that is being created at the moment. However, it must be modified to lower mapping mistakes as robot speed is raised. Due to the sensor sensitivity, the mapping quality of the robot has dropped nonlinearly when robot velocity was increased. Since, GMapping rely on odometry for robot localization in locations where the laser scan estimated pose is uncertain, such as wide spaces or lengthy corridors lacking features, it outperforms Cartographer in these conditions. And also, an autonomous navigation system employing ROS for the mecanum wheeled robot was examined in the paper. The map was successfully generated by the robot and made it to the designated location without colliding. Through greater software development and analysis, the existing work can increase navigational precision. Future work will focus on the whole hardware implementation and calibration for obtaining precision.

References

1. Ravuri P, Yenikapati T, Madhav B et al (2021) Design and simulation of medical assistance robot for combating covid-19. In: 2021 6th International conference on communication and electronics systems (ICCES). IEEE, pp 1548–1553
2. Smith R, Self M, Cheeseman P (1990) Estimating uncertain spatial relationships in robotics. In: Autonomous robot vehicles. Springer, pp. 167–193
3. Murphy K, Russell S (2001) Rao-blackwellised particle filtering for dynamic bayesian networks. In: Sequential Monte Carlo methods in practice. Springer, pp 499–515
4. Rodriguez-Losada D, San Segundo P, Matia F, Galan R, Jiménez A, Pedraza L (2007) Dual of the factored solution to the simultaneous localization and mapping problem. IFAC Proc Volumes 40(15):542–547

5. Montemerlo M, Thrun S, Koller D, Wegbreit B et al (2003) Fastslam 2.0: an improved particle filtering algorithm for simultaneous localization and mapping that provably converges. IJCAI 3:1151–1156
6. Nithya M, Rashmi M (2019) Gazebo-ros-simulink framework for hover control and trajectory tracking of crazyflie 2.0. In: TENCON 2019-2019 IEEE region 10 conference (TENCON). IEEE, pp 649–653
7. Mukherjee A, Adarsh S, Ramachandran K (2021) Ros-based pedestrian detection and distance estimation algorithm using stereo vision, leddar and cnn. In: Intelligent system design. Springer, pp 117–127
8. Garcia-Sillas D, Gorrostieta-Hurtado E, Vargas J, Rodríguez-Reséndiz J, Tovar S (2015) Kinematics modeling and simulation of an autonomous omni-directional mobile robot. Ingeniería e Investigación 35(2):74–79
9. Jose S, Variyar VS, Soman K (2017) Effective utilization and analysis of ros on embedded platform for implementing autonomous car vision and navigation modules. In: 2017 International conference on advances in computing, communications and informatics (ICACCI). IEEE, pp 877–882
10. Khairuddin AR, Talib MS, Haron H (2015) Review on simultaneous localization and mapping (slam). In: 2015 IEEE international conference on control system, computing and engineering (ICCSCE). IEEE, pp 85–90
11. Xuexi Z, Guokun L Genping F, Dongliang X, Shiliu L (2019) Slam algorithm analysis of mobile robot based on lidar. In: 2019 Chinese control conference (CCC). IEEE, pp 4739–4745
12. Yagfarov R, Ivanou M, Afanasyev I (2018) Map comparison of lidar-based 2d slam algorithms using precise ground truth. In: 2018 15th International conference on control, automation, robotics and vision (ICARCV). . IEEE, pp 1979–1983
13. Grisetti G, Stachniss C, Burgard W (2007) Improved techniques for grid mapping with rao-blackwellized particle filters. IEEE Trans Robot 23(1):34–46
14. Zhang X, Lai J, Xu D, Li H, Fu M (2020) 2d lidar-based slam and path planning for indoor rescue using mobile robots. J Adv Transp
15. Kohlbrecher S, Von Stryk O, Meyer J, Klingauf U (2011) A flexible and scalable slam system with full 3d motion estimation. In: 2011 IEEE international symposium on safety, security, and rescue robotics. IEEE, pp 155–160
16. Hess W, Kohler D, Rapp H, Andor D (2016) Real-time loop closure in 2d lidar slam. In: 2016 IEEE international conference on robotics and automation (ICRA). IEEE, pp 1271–1278
17. Konolige K, Grisetti G, Kümmerl R, Burgard W, Limketkai B, Vincent R (2010) Efficient sparse pose adjustment for 2d mapping. In: 2010 IEEE/RSJ international conference on intelligent robots and systems. IEEE, pp 22–29

An Autonomous Home Assistant Robot for Elderly Care

Vidhun V. Warrier and Ganesha Udupa

Abstract In this busy world, it is very difficult to manage a home that cares for everyone, especially elderly people and small kids need special attention. At home, grandparents need to take their medications and young children need to be fed at the right time. A two-wheeled robot controlled by Alexa is proposed here to address the problem. This robot will be accessible from anywhere, as long as one has access to the internet. The main aim of this robot is to deliver food and medicine to those with mobility impairment without the help of humans. If the elderly are alone in the house, using Alexa, he/she can give commands to robots and meet their needs. By connecting the robot with Alexa, it can be controlled from anywhere using the internet. Robotic operating system is used as a good platform for the navigation of this autonomous robot. Map-based navigation is used in the present robot which makes it useful in houses, hospitals, etc. The simulation results and the experiments conducted on the robot show that the navigation accuracy and performance of the robot are found to be satisfactory.

Keywords Autonomous navigation · ROS · Alexa

1 Introduction

Autonomous robots are getting very popular in today's world because of their ease of accessibility. Therefore, the scope of use for an autonomous robot is much higher than that of ordinary robots. Through vast access to so much software and hardware support, one can make an autonomous robot very easily. ROS is one such system. With the help of ROS, one can make an autonomous robot easily using a Linux-based operating system.

In this busy world, it is very difficult to manage a home that gives care to everyone. Especially when there are elderly people and small kids in the home, who always

V. V. Warrier · G. Udupa (✉)
Department of Mechanical Engineering, Amrita Vishwa Vidyapeetham, Amritapuri, India
e-mail: ganesh@am.amrita.edu

S. Sharma et al. (eds.), *Intelligent Control, Robotics, and Industrial Automation*, Lecture Notes in Electrical Engineering 1066,
https://doi.org/10.1007/978-981-99-4634-1_3

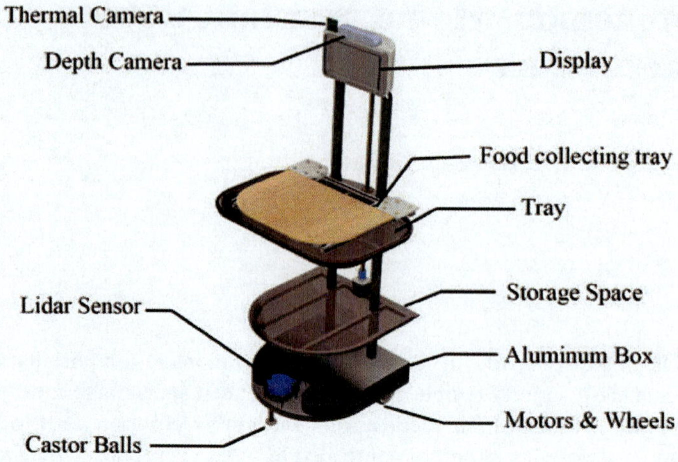

Fig. 1 Parts of the home assistant robot

need special attention. At home, there may be grandparents, who need to take their medications at right time. The robot that is presented here is able to work as an assistant for the home—especially for elderly people. The user can connect the robot with Alexa, an IoT-based voice assistant so that it can be controlled from anywhere in the world. By using ROS, a topic-based robotic middle-ware suit, the robot can achieve autonomous navigation inside the home very easily. Let us consider a family with some children and elderly people. Then managing them is a very difficult task. Also, if the other family members are not at home or work, their daily needs become difficult. Especially getting food at the right time is a problem for them. And, elderly people need their medicines at the right time. In such a scenario, it again becomes difficult to handle [1, 2].

The Care-O-Bot [3] is a robot that is similar to this robot, but they did not provide any storage space for keeping medicine and also the cost of the robot is too high. The voice-controlled robot [4] is also a similar robot that can be controlled using voice. The main limitation of the robot is that it can only control in a range limited by the microphone used.

2 Design of the Robot

The major duty of the robot is to deliver food and medicine to the elderly. So the design of the robot should help meet all these requirements. Therefore, the robot needs a tray that can be used to place food along with some extra storage spaces in case of emergencies. For moving all around the designated environment, it should be wheeled/legged. The robot box and storage space has been designed in the C shaped form so that it can be moved as close as possible to its used with the design kept in a c shaped manner, so that the robot can move closer to the elderly person and the bottom portion of it will go downside of the bed whereas the top portion with tray reaches nearer to the user and hence making it easily accessible. The robot has been provided with a depth camera for navigation and the liquid crystal display (LCD) as a user interface device at the head of the robot. The tray mounted on the two columns is used to place food and needed items. The tray can be moved up and down, which helps the robot to collect food from the source and deliver it to the user. There is also a Lidar sensor which is used for navigation. In addition it provides enough storage spaces for placing food, medicines, etc. The robot uses 2 wheels to move around. By using 2 wheels, the robot can be moved in any direction easily. Figure 1 shows the design of the robot.

2.1 Depth Camera

A depth camera is a camera that is used to find the depth of an image. Using this camera, it can improve navigation by identifying the depth and using this information to find the distance between the objects, thus aiding navigation in a 3D environment.

2.2 Display

The display used is an 8-inch touch screen for user interaction.

2.3 Thermal Camera

The thermal camera is a device used to detect the body temperature of the user. From the recent pandemic situation, it has been discovered that body temperature can be used for virus detection. For this purpose, the robot has a thermal camera that aids in diagnosing fever-like diseases.

2.4 Lidar Sensor

Lidar sensors are specific sensors that are used to detect the distance between objects at a 360-degree angle. The robot uses this sensor for navigation.

2.5 Tray and Storage Space

A Tray will be provided in the robot for placing food and medicine, which is moving up and down so that it can adjust the height with the table height in the kitchen. The robot have an extra mechanism for taking the food from the kitchen. Using this mechanism it can take food placed in a fixed place. For the movement of this mechanism and the movement of the tray,it uses 2 stepper motors and lead screws. Also, extra storage space is provided at the bottom of the robot for placing any urgent medicines or food, etc.

2.6 Motors and Wheels

The robot will be using a differential drive mechanism for the movement. It includes 2 powered wheels and a castor wheel combination.

3 Simulation of the Robot

The simulation of the robot is done using ROS. Using ROS, the robot is tested in a virtual environment. Since the robot is designed to be used in homes or hospitals, the simulation has been done in a home environment.

3.1 Autonomous Navigation of the Robot

The main aim of the simulation was to validate the navigation since this is an autonomous robot and navigation is important. The main steps of the navigation are:

1. Creating the map of the environment
2. Loading the map to the robot
3. Locating the robot in the map
4. Sharing the location of the robot where it needs to be.

3.2 Mapping of the Environment

Mapping the environment is the first step of autonomous navigation. For the navigation and map generation, it uses a Lidar sensor, which is a sensor that uses a laser beam to find the distance between

Fig. 2 Map generated after the process

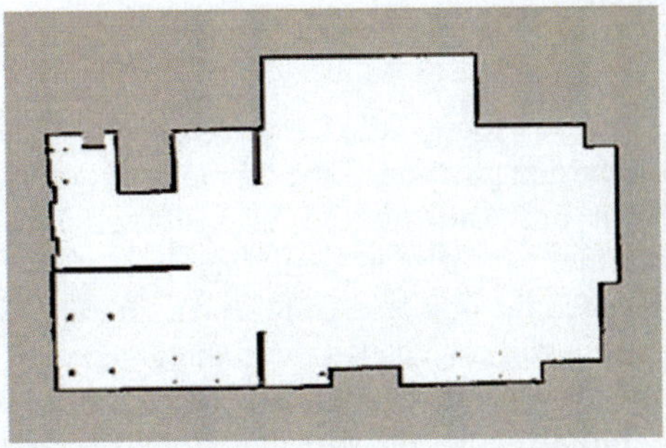

the object in front of it in 360 °C. For making the map of the environment, the robot uses lidar data, which is typically a pattern in 360 °C. Clubbing this data with the feedback from the motors and location information can generate a map of the environment. Figure 2 shows the map generated at the time of the simulation.

3.3 *Path Planning*

Path planning is the second step of navigation. Here the robot will start moving autonomously by using the map that has already been created. The user can load the map into the robot using ROS. Then the robot wants to know where it is currently, which can be entered manually or by using the sensor data.

Adaptive Monte Carlo Localization (AMCL) is the method used for the localization of the robot. After the robot knows its location, the user can send the position where it needs to go. When the robot gets the position, it will compare the map that will be created with the current map of the environment using the sensor data. Then the robot will move to the location it needs to go by comparing both of these maps. The movement of the robot will be monitored using the wheel encoders and Inertial Measurement Unit (IMU) sensors. Using the data from the sensors, the robot gets the position accurately, and this improves the path planning of the robot. Figure 3a shows the path-planning process in ROS.

(a) Path Planning in ROS (b) Environment used for Simulation

Fig. 3 Simulation in ROS

3.4 Localization of the Robot

Adaptive Monte Carlo Localization is a method used in autonomous robots for finding the current position of the robot. In an autonomous robot, it is very important that the robot must know its current position before moving to the desired location. AMCL is a method for such situations. In AMCL, the robot will first find out the distance between the object in front of it using the Lidar sensor, which is like a pattern. This pattern will be matched with the given map and find a location. Then the robot will move in a pattern and the process will be repeated till it finds out the correct location. Figure 4b shows the simulation environment used for the simulation of the robot.

3.5 Integrating Alexa

The highlight of the robot is that it can be controlled using Alexa, a voice assistant. With the help of Alexa, one can control the robot from anywhere in the world with internet access. In the scenarios discussed earlier one major problem was taking care of the elderly who are alone. In such circumstances, one can make use of Alexa and instruct the robot for that specific purpose. To integrate Alexa into a ROS-controlled robot, the robot needs:

1. An Alexa skill: It is an app for Alexa, where the user gives possible input commands. To start an Alexa skill, the user needs to say an invocation name which can be set up when the user creates a skill on the Amazon website. When the user says any commands to Alexa, it will send a respective code through the communication protocol to the robot.
2. A Server: A server is used for connecting Alexa with the robot. Then only it can transfer the data which is received at the Alexa skill.
3. A communication Protocol: It is used as a medium for communication between the robot and the Alexa skill

3.6 Communication Protocol

Message Queuing Telemetry Transport (MQTT) is a protocol that is mainly used in IoT applications to communicate between a server and a device and vice versa. Here MQTT is used for communicating with the server and the robot. The main components of MQTT are discussed below.

1. Publisher: A publisher is a data transmitter where the data is transmitted to a broker or a server. Publishers publish the message on a specific topic. Those who subscribe to the same topic will get the data. In our robot, the Alexa skill will be the publisher.

2. Broker: A broker is a server that controls the data flow between the publisher and the subscriber. In our robot, AWS IoT will be the broker.
3. Subscriber: A subscriber receives the data sent by the publisher. The subscriber needs to subscribe to a specific topic and then the data coming from the topic will be received by the subscriber. Here the robot will be the subscriber.

3.7 Server for Communication

Here, AWS IoT is used as a broker for the MQTT protocol where all the data from the publisher comes and shoots to the subscriber. AWS IoT is a good platform for working as an MQTT broker due to its high speed. It does a high-speed data transfer between the Alexa skill and the robot.

3.8 Alexa with ROS

The major purpose of Alexa integration is to control the robot using a voice assistant. When Alexa receives a command, a respective code is sent to the robot using the MQTT protocol to the robot. When the robot receives the code, a script runs in the robot to find which command is sent. When it finds the command, appropriate action will be taken. Mainly the user uses Alexa for the commands such as taking food from the kitchen. When the robot receives such a command, it processes the command and set the goal location of the robot as Kitchen/ Bedroom, which is already set by the user. The goal location can be set using a simple ROS topic.

4 Building the Robot

Building the robot by comping the physical component is the foremost step. The body of the robot is made using GI pipe, the head is 3D printed and the box at the bottom is made up of aluminum. Figure 4 shows the robot that was built.

The body of the robot is made using a 0.5-inch cast iron square pipe and the central pipe is a 1.5-inch round GI pipe. The head of the robot is 3D printed with 1.75 mm Polylactic Acid (PLA) Filament. The motherboard Raspberry Pi is placed at the head of the robot. The bottom box contains the ATmega328P IC for controlling the movement of the robot.

The tray of the robot is made with 0.5-inch cast iron square pipe and is covered using a red color acrylic sheet of 5 mm. The tray is moved using the stepper motor placed above the storage space and is connected to the tray using a lead screw of 4 mm pitch. Some bearing wheels are also added to the tray for smooth movement.

Fig. 4 Home assistant robot

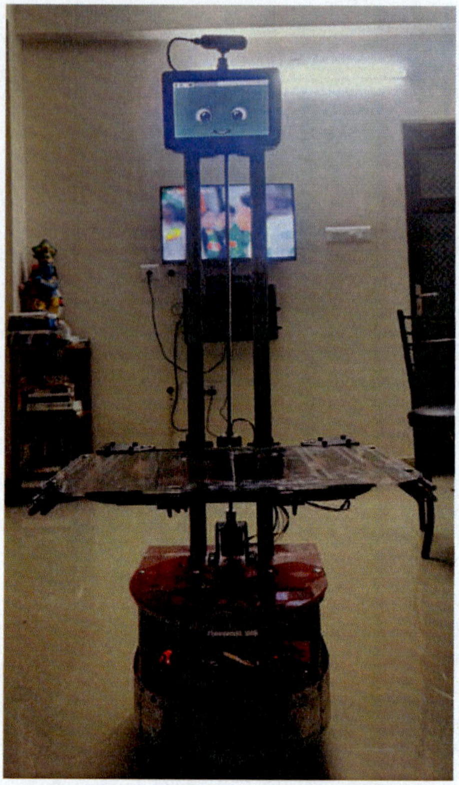

2 Johnson motor with 40 mm width wheels is used for the movement of the robot. This Johnson motor which has a torque of 100 rpm can easily control our robot weighing above 8 kg. To control the robot, L293d motor driver IC is used. The stepper motor that assists in tray movement is controlled by the A4988 stepper motor module. Figure 5 shows the block diagram of electronics in the robot.

5 Tray of the Robot

The main purpose of the robot is to serve necessary items to the user. For this purpose, it has a tray of size 30 cm × 40 cm in the robot that can be used to carry food, medicine, etc. And for taking food from the kitchen. It has a mechanism in which the robot can take the food from the kitchen. Also, the tray will move up and down, so that the user can adjust the height as per the height of the place where food is held. The endpoints of the movement are found out by using the end switches. The height of the tray can be set by the user as per their wish using the user interfaces.

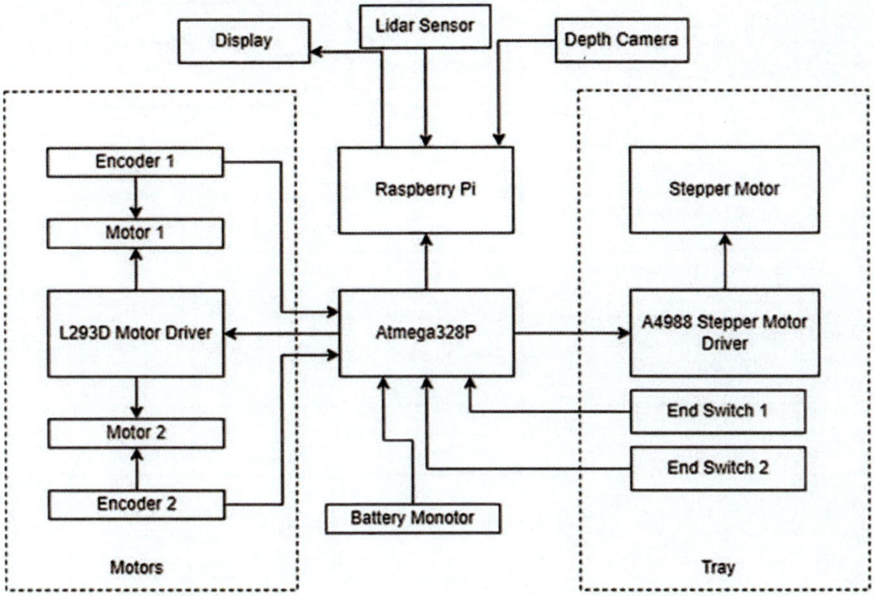

Fig. 5 Electronics in robot

Fig. 6 Tray of the robot

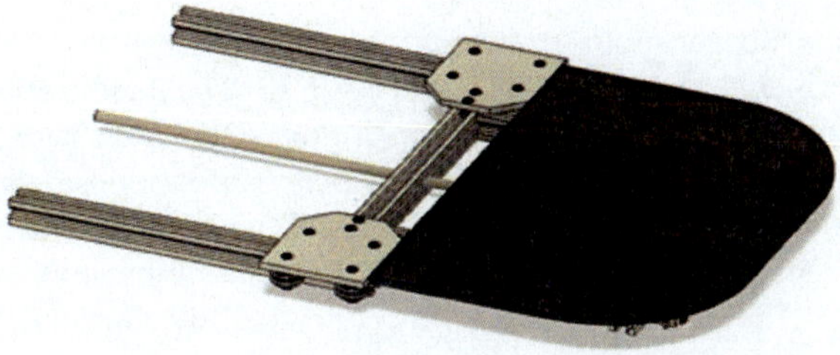

Taking food using the tray requires a mechanism that is shown in Fig. 6. It consists of a movable part, which moves forward and backward. When the robot wants to take food from a fixed place, the movable part will move forward and take the food-containing plate which can be locked by the servos in the tray. For driving the movement, a stepper motor with a lead screw is provided.

6 Experiments Conducted by the Robot

For the total control of the robot, the robot uses 3 microcontrollers, Raspberry Pi 4 Model B, ATmega328P, and Arduino Nano, where Raspberry Pi is the core part of the robot. A Raspberry Pi installed with ROS is used in the robot. For controlling

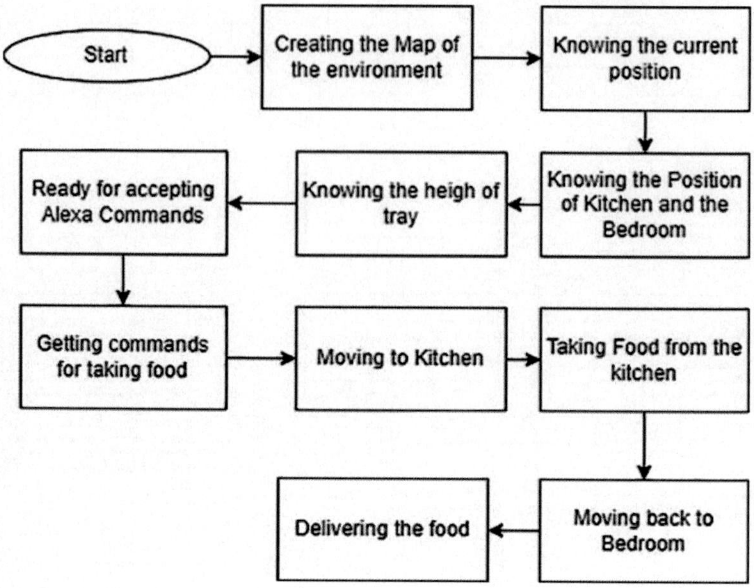

Fig. 7 Complete control process of the robot

the stepper motor driver and Johnson motor driver the robot uses ATmega328P IC, which is controlled by a Raspberry Pi. The depth camera, display, and Lidar sensor are connected to Raspberry Pi. For accuracy in navigation, it needs to implement a feedback mechanism in the robot. The robot uses motor encoders as feedback to the system, which gives the motor movement back to ROS and helps the navigation better. Figure 7 shows the block diagram of the control process of the robot.

The testing of the robot is conducted by considering there is only one patient in a room and only one kitchen. The location of the kitchen and the patient are defined after taking the map of the environment using the user interfaces. Now the robot knows where it will be and where is the location of the kitchen and the patient. When the robot gets the commands from Alexa, it will compare and start its action. The first step is to determine where it is right now. When it finds the location, the robot will start moving to the first location where the food is situated, the kitchen. When it reaches the kitchen, it takes food to the tray by using the setup. Then the robot will move toward the patient, which is defined earlier. In the testing, it is found that there are some deviations in the path that may happen due to some objects in the field. But by comparing the result with the simulation, the real robot also gives a good result.

7 Result and Discussion

The aim of this project is to serve the patient or elderly people by the autonomous robot. Using this robot, elderly people can get their food and medicine at the right time. From the simulation of the robot, it is found that the ROS used for the navigation of the robot is very efficient. Also, it can be controlled using Alexa, which is helpful because by using Alexa users can control the robot from anywhere in the world. So, if a situation like elderly is alone, his son from the office can also give commands to Alexa for food and medicine. Table 1 shows the specification of the robot after manufacturing. ROS gives almost the same results when tested in a real environment. The total weight of the robot is about 10 kg as it is made with cast iron. The good performance Johnson motor provided in the robot can easily carry this much payload. 10 Ah lithium polymer battery is used in the robot.

The simulation of the robot is carried out successfully using the ROS. It involves the autonomous navigation which is performed by mapping the environment using the lidar sensor. The 360-degree pattern of data collected by the lidar sensor for the whole area generates the surrounding map for the robot to perform its activities. Path planning of the robot has been carried out using the previously generated map of the surroundings. The robot determines its location using the Adaptive Monte Carlo Localization (AMCL) method. As per the given command, the robot moves to the destination location within the environment. The implemented navigation flowchart is shown in Fig. 8 which is self-explanatory.

The robot is designed and manufactured for the home and hospital environment. It is controlled by a voice assistant using Alexa within the home or from the office using the internet facility. Message Queuing Telemetry Transport (MQTT) is a protocol that is mainly used in IoT applications to communicate between a server and a device and vice versa. Alexa skill publishes the message data to the AWS IoT broker or server which in-turn sends it to the subscriber or the home assistant robot. The flowchart of the MQTT is shown in Fig. 9.

Table 1 Specification of the robot

Movement	Differential drive
Navigation	ROS-based navigation
Arm	6 DOF manipulator (future work)
Microcontrollers	Raspberry Pi, ATmega328P
Total weight	10 kg
Weight carrying capacity	2.5 kg
Battery	10 Ah lithium polymer
Battery life	1 hr for complete operation
Battery charging time	1 hr

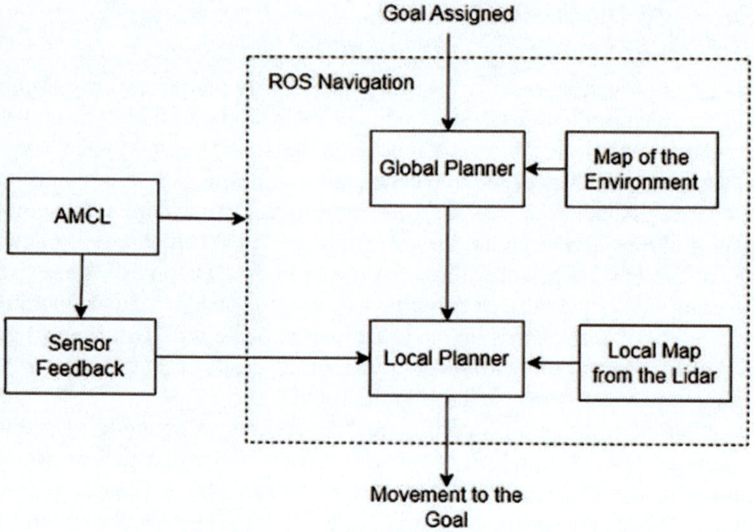

Fig. 8 Robot navigation flowchart

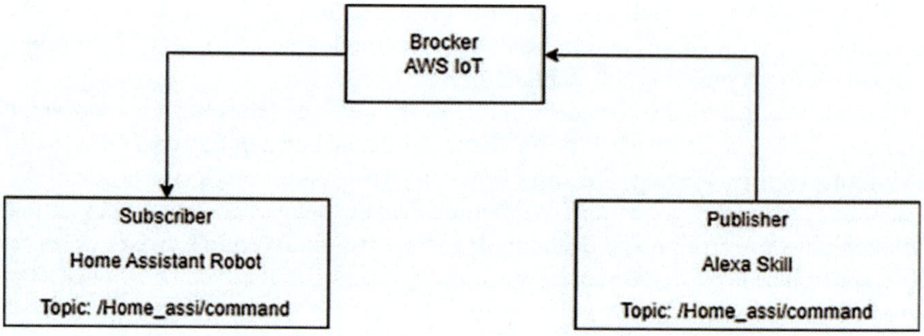

Fig. 9 MQQT flowchart

The movement of the real robot will be monitored using the wheel encoders and IMU sensors. It is found that the accuracy of reaching the destination target is within 10–20 cm.

Care-O-Bot [3] is another similar robot discussed in the introduction section. Comparing the Home Assistant Robot with Care-O-Bot, Care-O-Bot has a robotic arm for serving the food which is controlled by a separate user interface, but the present home assistant robot does not have a robotic arm currently, but work is in progress to include the arm in future. The current system has communication with a voice assistant named Alexa, which is an advantage in controlling the robot from anywhere compared to the voice-controlled robot [4].

In the future work, 2 robotic arms are planned to be integrated like the one implemented in Rover design [5] on the tray to take food from the kitchen and place it on

Fig. 10 Updated design of the robot with robotic hand

the tray and back to the user. The robotic arm consists of 6 °C of freedom so that it can move the hand very easily in the workspace. Figure 10 shows the robot with a robotic arm. Vision-based navigation [6] is another future work for the robot and is achieved using a depth camera.

8 Conclusion

An autonomous home assistant robot is implemented here. The main aim of the robot is to deliver food and medicine to the elderly. From the simulation, we have found that the navigation using ROS is a good option for navigation. The path planning, object detection, and collision avoidance in the ROS give a good result. And also navigation used on depth camera gives more accuracy in navigation and path planning. The combination of Amazon Web Service's AWS IoT platform with MQTT gives a good and faster flow of data from the Alexa skill to the robot. Alexa-based communication is a good feature because the robot can be controlled from anywhere. Using this feature, the elderly can get food or medicine from anywhere in the home, and their son children can also command Alexa to get it for them. A sliding tray has been designed and fabricated which can be moved up and down on the two columns with the help of leadscrew mechanism, for getting and carrying the food.

After testing it is observed that the robot can carry a maximum weight of 2.5 kg food/medicine in the tray and in storage spaces of about 3 kg, which is reasonable.

The total weight of the robot is about 10 kg. For the movement, it has a differential drive of 2 Johnson motors with a castor ball. The Johnson motors can withstand this much payload. All the electronic components use a 10 Ah Lithium Polymer battery, which gives a backup for 1 h if it works continuously. The robot can run at a good speed in this configuration. Also by using the differential drive system, it can move in any direction. With further development and modifications, the constructed robot has a good potential in serving the elderly people and acts as robot nurse in homes and hospitals.

Acknowledgements We would like to express our gratitude to everyone at our university, particularly our beloved chancellor, as well as all of the staff and professors, for their assistance in putting this article together.

References

1. Ribeiro T, Gonçalves F, Garcia IS, Lopes G, Ribeiro AF (2021) Charmie: a collaborative health-care and home service and assistant robot for elderly care. Appl Sci 11(16):7248. https://doi.org/10.3390/app11167248
2. Yamazaki K, Ueda R, Nozawa S, Kojima M, Okada K, Matsumoto K, Ishikawa M, Shimoyama I, Inaba M (2012) Home-assistant robot for an aging society. Proc IEEE 100(8):2429–2441. https://doi.org/10.1109/JPROC.2012.2200563
3. Hans M, Graf B, Schraft R (2002) Robotic home assistant care-o-bot: past-present-future. In: Proceedings. 11th IEEE international workshop on robot and human interactive communication. IEEE, pp 380–385. https://doi.org/10.1109/ROMAN.2002.1045652
4. Mishra A, Makula P, Kumar A, Karan K, Mittal V (2015) A voice-controlled personal assistant robot. In: 2015 International conference on industrial instrumentation and control (ICIC). IEEE, pp 523–528. https://doi.org/10.1109/IIC.2015.7150798
5. Aswath S, Ajithkumar N, Tilak CK, Saboo N, Suresh A, Kamalapuram R, Mattathil A, Anirudh H, Krishnan AB, Udupa G (2015) An intelligent rover design integrated with humanoid robot for alien planet exploration. In: Robot intelligence technology and applications, vol 3. Springer, pp 441–457. https://doi.org/10.1007/978-3-319-16841-8_41
6. Krishnan AB, Aswath S Udupa G (2014) Real time vision based soccer playing humanoid robotic platform. In: Proceedings of the 2014 international conference on interdisciplinary advances in applied computing, pp 1–8. https://doi.org/10.1145/2660859.2660966

3D Mapping Using Multi-agent Systems

Bhaavin K. Jogeshwar, Baisravan HomChaudhuri, and Sivayazi Kappagantula

Abstract Disaster response refers to the operations that must be carried out during and/or after a disaster, to safeguard people along with property. These dangers could be outdoors or indoors. 3D mapping provides extensive information, such as the site of a disaster, and aids in quick mitigation response. Mapping a region depends on parameters like an environment's area and the available resources. This paper demonstrates the relationship between the number of rooms and the number of unmanned aerial vehicles (UAVs) used to map an indoor area along with the overall time required to explore and compute it. The computational resources used include the OctoMap Library and the Frontier exploration algorithm. It was observed that increasing the number of UAVs subsequently to map an environment reduced the simulation time, only when the number of UAVs mapping the environment was less than or equal to the number of rooms in the environment.

Keywords 3D mapping · Multiple UAVs · Autonomous exploration · Indoor environment · Multiple rooms

1 Introduction

Unmanned aerial vehicles (UAVs) are aircrafts that operate without an onboard pilot. UAV technology was first employed by the government for border patrol and emergency response. The advances in drone technology were made possible by warfare. Over the years, the usage of drones has been expanded by businesses and corporations to use them for transportation of small goods, survey scans in industries,

B. K. Jogeshwar · S. Kappagantula (✉)
Mechatronics Department, Manipal Institute of Technology, Manipal, Karnataka, India
e-mail: sivayazi.k@manipal.edu

B. K. Jogeshwar
Manipal Academy of Higher Education, Manipal, Karnataka, India

B. HomChaudhuri
Illinois Institute of Technology, Chicago, IL 60616, USA

S. Sharma et al. (eds.), *Intelligent Control, Robotics, and Industrial Automation*,
Lecture Notes in Electrical Engineering 1066,
https://doi.org/10.1007/978-981-99-4634-1_4

mapping, and infrastructure, agricultural, and remote monitoring. In this paper, the terms"UAVs", "drones", and "quadrotor" are used interchangeably.

It is always a battle against time to respond to disasters. The main objective is to save people. In case of an indoor disaster, e.g., if the house is on fire, then deploying drones would be an ideal solution. Drones can generate 2D and 3D maps of the house and locate the sources of fire and those in danger. They can share the real-time footage, of what is happening in the disaster affected house, with the rescue units using onboard cameras. With the help of navigation sensors, they can find an optimal path [1] to the exit and can guide those potential victims to safety, thereby saving lives. They can do all this while simultaneously updating the situation to the emergency responders. It is evident that having multiple agents could speed up the processes, just like the "man and work" problems. However, this condition is true for certain criteria, and it is covered in this paper.

State of the art in multi-robot simulator includes ARGoS [2], the fastest general-purpose robot simulator. This effective and adaptable simulator aids in modeling intricate experiments involving several robots of various kinds in large swarms. Although, this simulator works best on mobile robots, the concept of designing a multi-robot simulator can be extended to multi-UAVs. The highly modular architecture of ARGoS can perform a 2D-dynamics simulation of 10,000 e-puck robots with a high real-time factor of 0.6. A decentralized planning approach for mobile robots was taken for the simulations done in ARGoS, which can also be used for aerial robots. In decentralized planning, each robot is a self-contained entity that reacts to its environment. For our project, we have used the distributed planning approach. The following are a few of the novelties in multi-UAV control and communication. In [3], the authors describe the multi-UAV communication using Bernoulli random variables. They also propose a centralized cooperative localization algorithm with a one-step prediction strategy and a distributed cooperative localization algorithm with a prediction compensation strategy. The work in [4] uses a Multi-UAVs cooperative control algorithm involving a search state to look for wireless signals and a relay state to transfer information and plan UAV paths.

This study highlights the link between the number of rooms and the number of unmanned aerial vehicles (UAVs) deployed to map an enclosed space, as well as the total time needed to explore and map that environment. A simulation for 3D mapping using multiple UAVs was created using Robot Operating System (ROS), Gazebo Simulator, and Rviz. Three indoor environments with different number of rooms were designed on Gazebo Sim and multiple UAVs were spawned. These UAVs explored and mapped the environments in Rviz. Numerous runs were conducted to obtain average simulation mapping times in these environments. Our results have been presented in this paper.

Task allocation among robots is an important part of coverage planning. An indoor environment's area can often be split into sections or rooms. These splits make it easier for multiple robots to pick mutually exclusive sections and collectively explore the entire environment. An efficient inter-robotic communication is implemented where they self-allocate a room. In this paper, we want to understand how proficiently can the robots communicate the auto-allocation of environment-splits, and how many of them are required to explore the environment without wasting resources. Hence, we find a relation between the number of UAVs and the number of rooms in this paper. Contributions of our paper include developing of a newer method of multi-robot communication.

We have covered the following topics. Section 2 describes the methodology. Section 3 shows the results and analysis of the study. Section 4 concludes the paper with our overall findings.

2 Methodology

We have extended the methods used in a few packages from the GitHub repositories to incorporate our methods used in this project. We first spawned a single UAV in an empty world in Gazebo Sim and tested out the OctoMap and Point Cloud Libraries and the exploration algorithm. The initial GitHub codes were then modified to spawn multiple UAVs, smooth functioning of the mapping libraries was ensured, and an approach to simultaneously sharing maps and exploring the environment was made.

2.1 Hector Quadrotor Package, Octomap and Point Cloud Libraries, and Caltech Samaritan

We installed the Hector quadrotor package [5] for simulating a UAV in an empty world. We worked with the Hector UAV from this package, as shown in Fig. 1a.

The OctoMap library [6, 7] is a free and open-source tool for creating volumetric 3D environment models using data from sensors. The OctoMap package offers a 3D occupancy grid mapping technique in C++ using data structures and mapping algorithms. The map is constructed using an octree data structure. A drone may then utilize this model data to navigate and avoid obstacles.

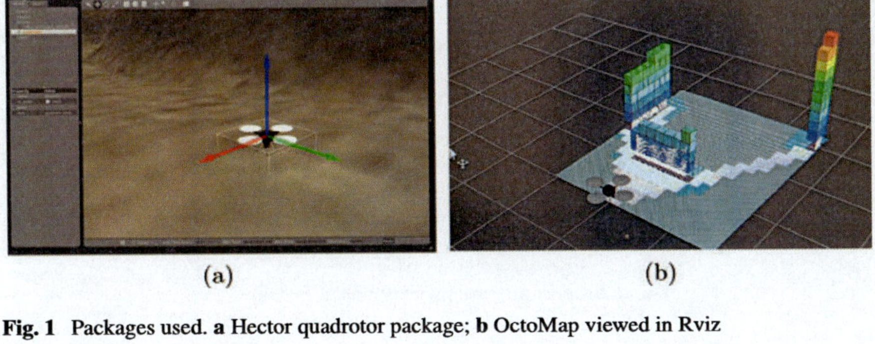

(a) (b)

Fig. 1 Packages used. **a** Hector quadrotor package; **b** OctoMap viewed in Rviz

The Point Cloud Library (PCL) [8] is an open-source library of algorithms for 3D geometry processing and tasks related to point cloud processing, both of which are essential to three-dimensional computer vision. We installed these libraries which formed the foundation of 3D mapping processes. The point clouds are formed on obstacles, and the OctoMap voxels (3D map) are formed on these point clouds. Figure 1b shows the colored voxels being formed on the white point clouds indicating obstacles.

For navigation, we used the Caltech Samaritan Package [9]. This package gave us the UAV's motion planning code during flight along with the fundamental frontier exploration algorithm. We incorporated these on the Hector UAV, and the exploration algorithm was tested.

2.2 *Frontier Exploration Algorithm and Environment*

Frontier points define the border between the explored and unexplored regions of the environment. In Fig. 2a [10], the red center dot is the robot, and the green dots are the frontiers. The black pixels are the obstacles, and the gray pixels are the unexplored regions. The robot then moves toward the green frontiers one by one, starting with the frontier that would give the UAV maximum gain.

The frontier exploration algorithm used here makes use of 4 cycles. The UAV first takes off from the ground and takes a 120° scan of the environment with a predefined range set to 15 m. It keeps a track of all the frontiers at that instant and then moves to the point of the unexplored area with a maximum potential gain, to explore the same. This cycle repeats until there are no more frontiers.

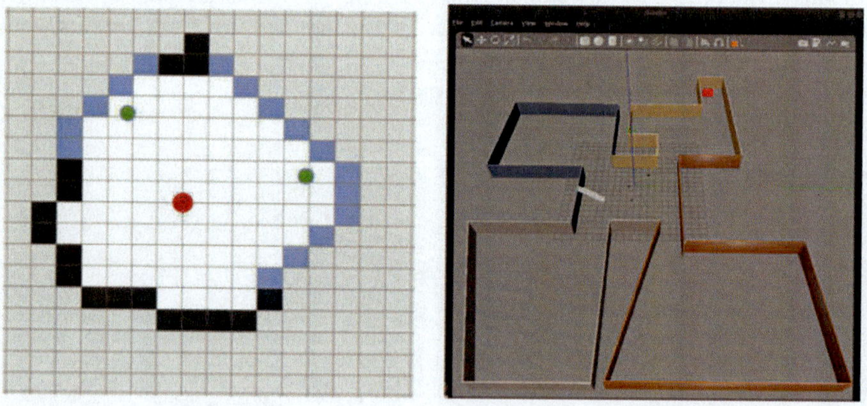

(a) (b)

Fig. 2 **a** Frontier exploration; **b** a custom 4-room environment

Different environments were tried on the Gazebo, and a custom environment was built (see Fig. 2b). We created a fire component in the top-right part of the gazebo environment to indicate an indoor disaster. UAVs were spawned in the center of the house.

2.3 Setting Up Multi-UAVs

We cloned the single UAV and spawned two UAVs in the environment and viewed them in Gazebo Sim and Rviz. Two instances of OctoMap servers and point clouds were launched, one for each UAV. In Fig. 3a, we see the top view of two UAVs in Rviz, each creating a point cloud in a world with 2 planes.

We created a 3D OctoMap of the receiving point cloud with two point cloud data streams from the UAVs. We now obtained a clean 3D map from each UAV's point cloud in Rviz (see Fig. 3b).

To cover the exploration part of the 2 UAVs, we ran the same exploration file as the single UAV's, but with two move_base nodes for the two UAVs, and tested the modified code and algorithm in simulation. There were times when we faced with a ground-mapping issue, wherein a UAV would consider the ground as an obstacle and would map it (Fig. 4). After several runs, we found out that this issue arose only sometimes and most of the time, the UAV corrected its 3D map by itself.

2.4 An Approach to Sharing Maps

Relay is a ROS node that subscribes to a topic and republishes all incoming data to a different topic. It can handle any message format. Relay is part of the topic_tools package. Figure 5 shows the flow diagram of our approach to share the map with the UAVs. We relayed the individual point clouds to a common relay topic, Point Cloud

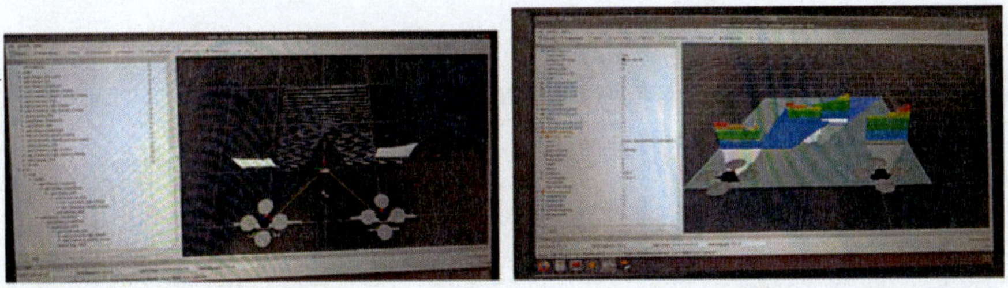

(a) (b)

Fig. 3 **a** Top view of 2 UAVs in Rviz; **b** a clean 3D map formed from each of the UAV's point cloud

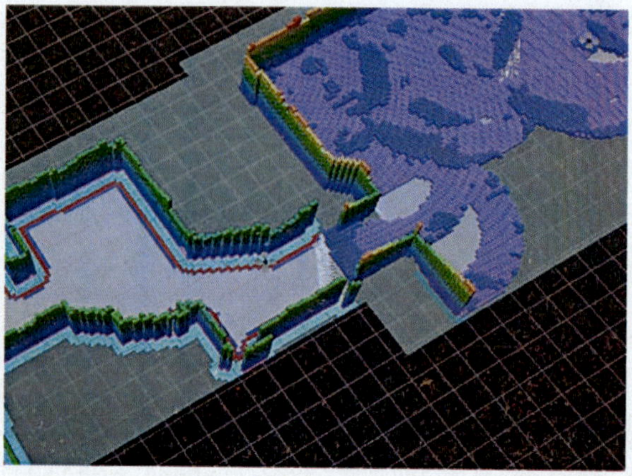

Fig. 4 Ground mapping issue: one UAV maps the ground while the second UAV ignores the ground

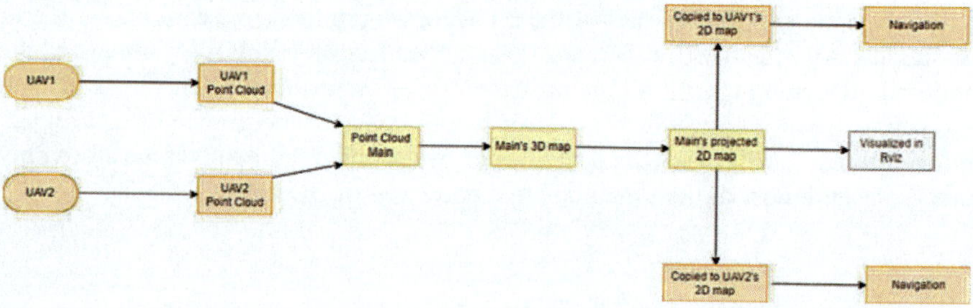

Fig. 5 A customized approach to sharing maps

Main. This helped with merging the two point clouds. We used the OctoMap library on this common relayed point cloud to obtain a 3D map. This 3D map was projected onto a 2D plane to form a 2D map which was visualized in Rviz. Lastly, the Point Cloud Main's 2D map was relayed to the other UAVs for their individual navigation.

2.5 *Need for a Multi-robot Planning Approach*

To explore the regions, the UAVs find goals with maximum gain and travel toward them. However, since both the UAVs share the same map, they both discover the one same goal with the maximum gain and travel toward that goal (see Fig. 6). This makes mapping less efficient as two UAVs map the same room. To send the multiple UAVs to different locations, we used a distributed approach to sharing maps and choosing goals. In distributed planning, the system has no single central owner, and centralization is avoided. The robots in a distributed system have equal access to data.

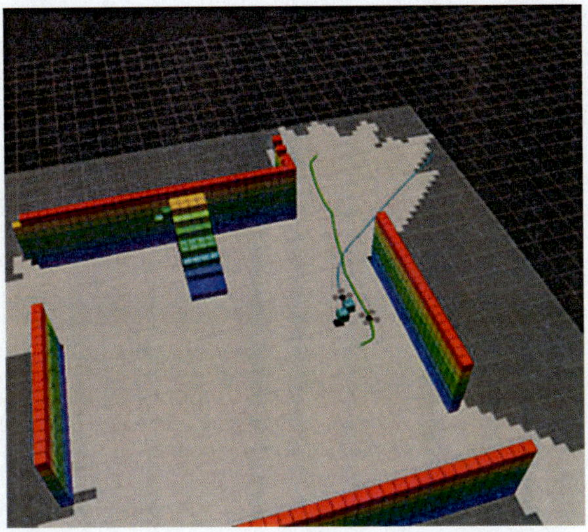

Fig. 6 Need for a multi-robot planning approach. Both UAVs go to explore the same room, which is not ideal

2.6 An Approach to Multi-UAV Exploration

We came up with a method of sending UAVs to different areas of the environment to explore. Since all the UAVs share the same map, they all have the same scores assigned to frontiers. In this method:

1. An array of the scores of each frontier was created. (e.g., [9 4 6 7 2 5 3 8 1])
2. This array was sorted in ascending order. (e.g., [1 2 3 4 5 6 7 8 9])
3. The UAVs were made to assign goals with maximum gain to all of them, i.e., the highest score of the array.
4. The UAVs were made to notify their goals to the other UAVs.
5. We also added the condition that a UAV's finalized goal should be at least 13 m away from the goals of the other UAVs. This way, all the UAVs would travel to goals having the maximum gain and would be 13 m away from other UAV's goals, thereby covering more area in less time. 13 m was chosen as the threshold distance as this value gave an optimal output compared to other values that we tried out.

Using this method, the UAVs assigned themselves a different set of goals to explore the entire environment. We then spawned three UAVs, and the code worked successfully.

In Fig. 7, a UAV's thought process at an instant is portrayed. The numbers in blue are the scores of the frontiers. In Fig. 7a, UAV1 has to find a new goal. It checks if the goal with the highest score, i.e., 6, is feasible to travel. It finds that goals with scores 6, 5, 4 are not feasible as these goals lie within a 13-meter radius from UAV3's goal. It now finds that goal with a score of 3 is feasible, as it does not lie within the threshold distance and begins to travel toward it. In Fig. 7b, a similar

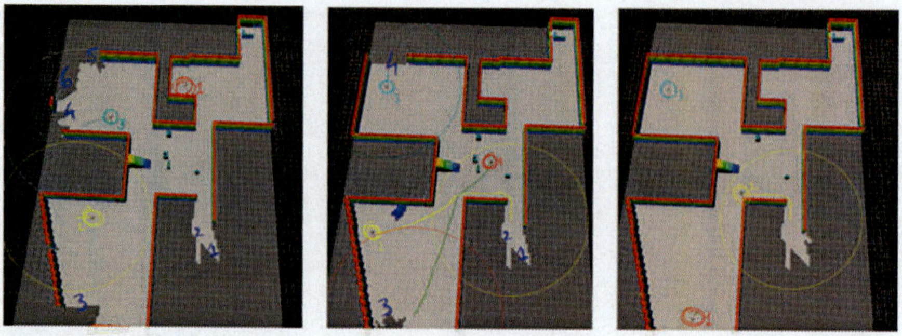

(a) (b) (c)

Fig. 7 Instance of multi-UAV exploration. The numbers in blue are the scores of the frontiers. **a** UAV1 has to find a new goal. **b** UAV2 has to find a new goal. **c** Situation that saves resources

case happens with UAV2 where it discards goals with scores 4, 3 (as they lie in the 13-meter circle) and travels to a goal with the successive score, i.e., 2. In Fig. 7c, UAVs 1 and 3 terminate their exploration as the only unexplored area lies inside the circular threshold area. This way, they save energy, power, and other resources, as traveling to a room where UAV2 would already be present to explore the room and would not be of significant advantage.

2.7 *Modified Scoring Criteria*

The code was all theoretically correct; however, there was one practical issue: one UAV abandoned the room and traveled to a goal which is closer to another UAV, while this another UAV was searching for the next goal. The one UAV does this if it finds a region of maximum gain nearer to that of another UAV. To fix this, we modified the scoring criteria and made the frontiers assign scores inversely proportional to the distance between the UAV and that frontier. This way, the UAV would prioritize exploring areas comparatively nearer to itself.

Figure 8 shows the update in the scoring pattern. Figure 8a is the initial scoring pattern: The UAV1 did not consider the distance between unexplored areas from the UAV. The UAV1 would have gone to the frontier with score 3, as shown in Fig. 7a, which could have been easily explored by UAV2, resulting in wasting time, energy, and resources. This scoring pattern is modified in Fig. 8b: for UAV1, the distance between unexplored areas from the UAV is taken into consideration, and the UAV1 now chooses to explore the frontier with score 6; thereby exploring the room which is still not explored, thus saving time for better disaster response.

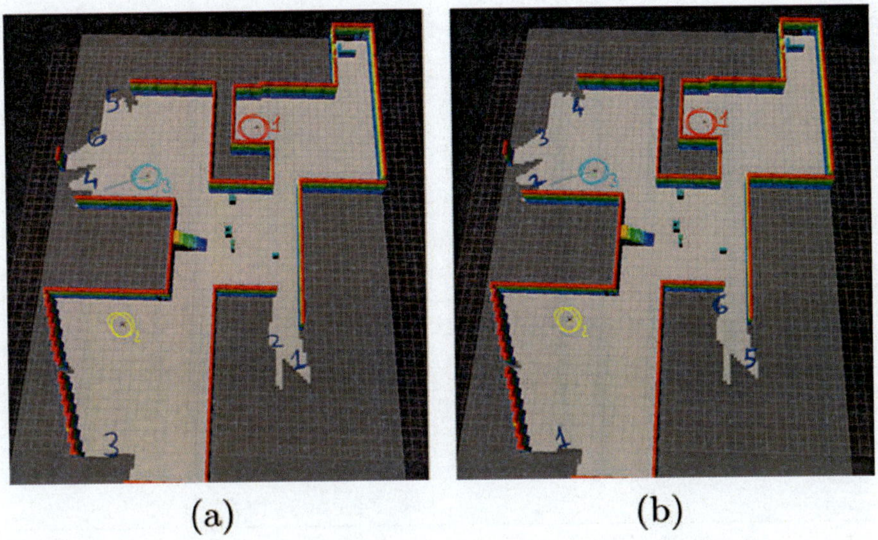

Fig. 8 Modified scoring criteria. **a** Initial scoring pattern; **b** modified scoring pattern

3 Results and Analysis

After updating the scoring criteria, numerous simulations were conducted, and data obtained was analyzed and shown in the following tables:

From the average simulation times taken from Tables 1, 2, and 3, the time difference in percentage at each instance of adding a UAV to an environment is calculated, and graphs are plotted in Fig. 9. The graphs show the time reduction (Δ) in each instance of adding a UAV in 4-room, 3-room, and 2-room environments.

From Fig. 9a, adding one UAV at each instance reduces the average total time to map the environment by nearly 30%.

Hypothesis Thus, we hypothesize that adding one UAV to any environment would reduce the time taken to explore the environment by roughly 30%.

To test out this hypothesis, we create two more environments by covering the entrances of two rooms, as shown in Fig. 10.

From Fig. 9c, we observe that adding a UAV reduces the exploration time by an estimated 30% at each instance. However, no significant difference is observed in the time taken to explore and map the environment by 4 UAVs and 3 UAVs. A similar trend is seen in Fig. 9e, where the exploration time reduces by an estimated 30%, with no significant difference in exploration time taken by 2 UAVs, 3 UAVs, and 4 UAVs.

For our observations, the maximum resource that could be tested was 4 UAVs on Ubuntu 18.04 platform running on VMware Workstation 15 Player on Dell G5 laptop with 8 GB RAM and 3GB Graphics. We did not test for resources of more than 4 UAVs.

Table 1 Times sheet of multiple UAVs mapping a 4-room environment

Number of UAVs	1 UAV	2 UAVs	3 UAVs	4 UAVs
Simulation times (min: sec)	10:00	7:42	4:30	3:07
	12:11	6:30	5:20	3:15
	10:10	7:15	5:30	3:11
	10:16	7:35	5:45	3:14
		6:00		
		5:40		
		4:50		
		5:15		
		5:55		
		3:45		
		4:07		
Average simulation time	10:39	7:15	5:08	3:11

Table 2 Times sheet of multiple UAVs mapping a 3-room environment

Number of UAVs	1 UAV	2 UAVs	3 UAVs	4 UAVs
Simulation times (min: sec)	6:41	5:15	3:38	4:06
	6:39	4:49	3:30	3:14
		4:19		
Average simulation time	6:40	4:47	3:34	3:40

Table 3 Times sheet of multiple UAVs mapping a 2-room environment

Number of UAVs	1 UAV	2 UAVs	3 UAVs	4 UAVs
Simulation times (min: sec)	4:15	3:07	3:23	3:10
	4:18	3:10	3:13	3:11
		3:11		
Average simulation time	4:16	3:09	3:18	3:10

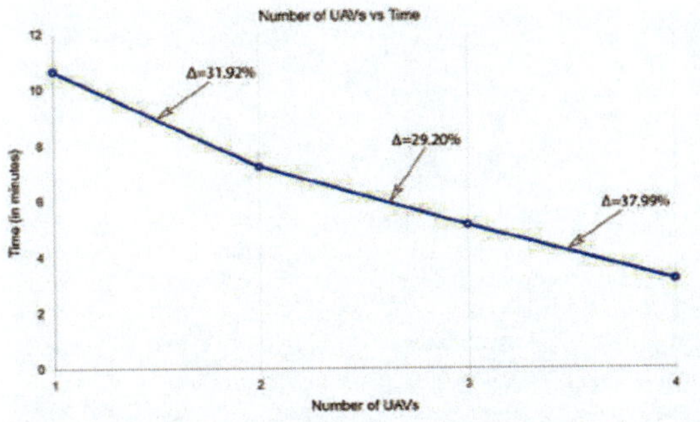

(a) For 4-room environment

For 4-room environment
Converting times to decimal:
10:39 = 10.65
7:15 = 7.25
5:08 = 5.13333
3:11 = 3.18333

Finding time difference in percentage at each instance
(10.65 - 7.25) * 100 / 10.65 = **31.92%**
(7.25 - 5.13333) * 100 / 7.25 = **29.20%**
(5.13333 - 3.18333) * 100 / 5.13333 = **37.99%**

Average time difference = (31.92 + 29.20 + 37.99) / 3
= **33.03%**

(b) Calculations for a 4-room environment

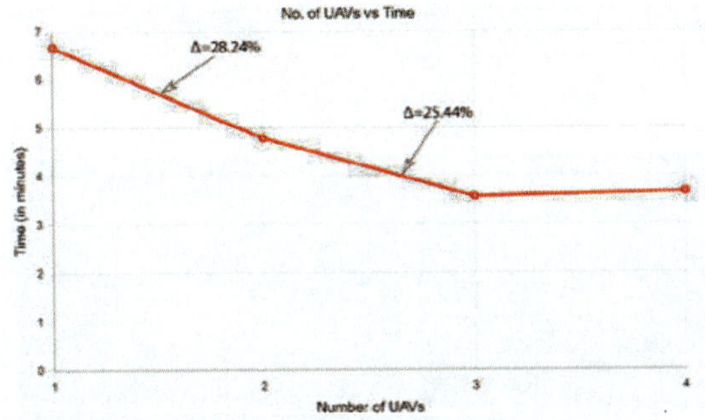

(c) For 3-room environment

For 3-room environment
Converting times to decimal:
6:40 = 6.6666
4:47 = 4.7833
3:34 = 3.5666
3:40 = 3.6666

Finding time difference in percentage at each instance
(6.6666 - 4.7833) * 100 / 6.6666 = **28.24%**
(4.7833 - 3.5666) * 100 / 4.7833 = **25.44%**

Average time difference = 28.24 + 25.44 / 2
= **26.84%**

(d) Calculations for a 3-room environment

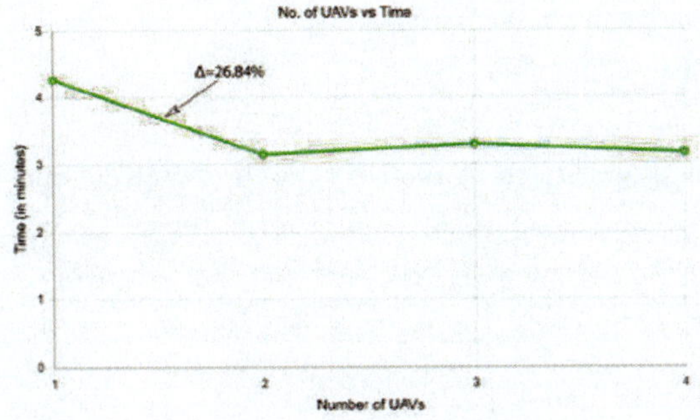

(e) For 2-room environment

For 2-room environment
Converting times to decimal:
4:16 = 4.2666
3:09 = 3.15
3:18 = 3.3
3:10 = 3.1666

Finding time difference in percentage at each instance
(4.2666 - 3.15) * 100 / 4.2666 = **26.17%**

Average time difference = **26.17%**

(f) Calculations for a 2-room environment

Fig. 9 Graphs of the number of UAVs versus time taken to map variable room environments and calculations showing the average time difference in % when a UAV is added in variable room environment

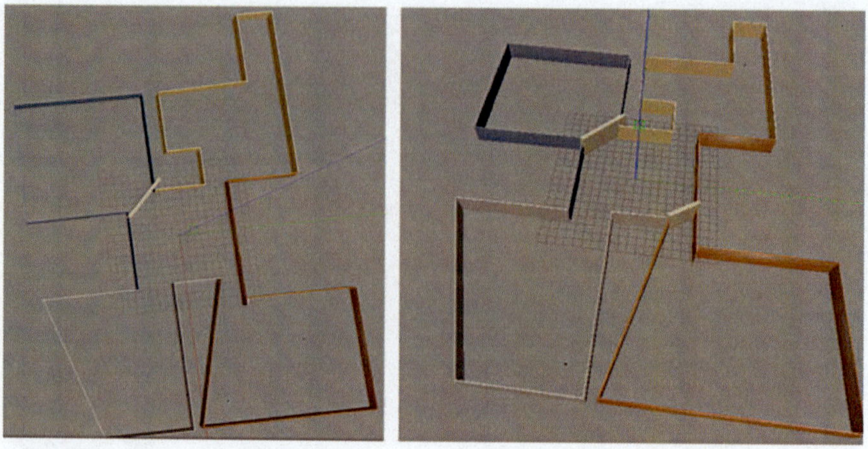

(a) A 3-room environment (b) A 2-room environment

Fig. 10 2 additional rooms created in Gazebo Sim to test our hypothesis

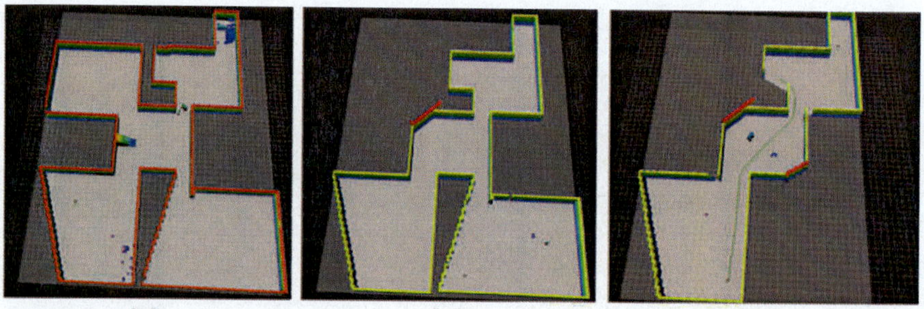

(a) (b) (c)

Fig. 11 3D maps of **a** a 4-room environment; **b** a 3-room environment; **c** a 2-room environment, generated in Rviz

Figure 11 shows the final 3D maps generated by the UAVs for the three environments in Rviz.

4 Conclusions

From our observations, we can conclude that adding one UAV at a time to explore an environment reduces the average total time to map the environment by nearly 30% at each instance. This trend of obtaining an improved disaster response is only observed in cases where the number of UAVs in an environment is lesser than or equal to the number of rooms in the environment. When the number of UAVs is equal to the number of rooms in the environment, a saturation point is reached; beyond this point, there would be no reduction in time. If the number of UAVs exceeds the number of rooms, then the average time taken to map the environment would be similar to the average time taken to map the environment by "N" UAVs, where "N" is the number of rooms in the environment. From Fig. 9, it is also evident that higher the number of rooms in an environment, the more time will be taken by a UAV to map that environment. Incorporating multiple UAVs would be a solution for faster mapping of the surroundings. This is particularly useful for mapping buildings in case of a fire evacuation.

Implementing this frontier exploration algorithm and 3D mapping tasks on multiple physical UAVs is the project's future scope. Incorporating thermal cameras on those drones and visualizing thermal data to detect fire would be more accessible when working hands-on as compared to working with this simulation.

References

1. Jogeshwar BK, Lochan K (2022) Algorithms for path planning on mobile robots. IFAC-PapersOnLine 55(1):94–100
2. Pinciroli C, Trianni V, O'Grady R, Pini G, Brutschy A, Brambilla M, Mathews N, Ferrante E, Di Caro G, Ducatelle F et al (2012) Argos: a modular, parallel, multi-engine simulator for multi-robot systems. Swarm Intell 6(4):271–295
3. Fu X, Bi H, Gao X (2017) Multi-uavs cooperative localization algorithms with communication constraints. Math Probl Eng, vol 2017
4. Xiaowei F, Xiaoguang G (2016) Multi-uavs cooperative control in communication relay. In: 2016 IEEE international conference on signal processing, communications and computing (ICSPCC). IEEE, pp 1–5
5. Meyer J, Sendobry A, Kohlbrecher S, Klingauf U, Stryk OV (2012) Comprehensive simulation of quadrotor uavs using ros and gazebo. In: International conference on simulation, modeling, and programming for autonomous robots. Springer, pp 400–411
6. Hornung A, Wurm KM, Bennewitz M, Stachniss C, Burgard W (2013) Octomap: an efficient probabilistic 3d mapping framework based on octrees. Auton Robots 34(3):189–206
7. Wurm KM, Hornung A, Bennewitz M, Stachniss C, Burgard W (2010) Octomap: A probabilistic, flexible, and compact 3d map representation for robotic systems. In: Proceedings of the ICRA 2010 workshop on best practice in 3D perception and modeling for mobile manipulation, vol 2
8. Rusu RB, Cousins S (2011) 3d is here: Point cloud library (pcl). In: 2011 IEEE international conference on robotics and automation. IEEE, pp 1–4
9. TimboKZ: Caltech samaritan. https://github.com/TimboKZ/caltech_samaritan. Last accessed 12 June 2022
10. Awabot: Frontiers. https://awabot.com/en/autonomous-exploration-method-frontiers/. Last accessed 11 May 2021

Lunokhod—A Warehouse I-Robot

Karthik Gogisetty, Ashwath Raj Capur, Ishita Pandey, and Praneet Dighe

Abstract Lunokhod is a warehouse intelligent robot conceptualized to carry out various tasks. It will be targeted to use in hazardous environments, thereby helping workers in the workspace to have a healthy atmosphere with striking use of technology to do a given task with minimum human interactions. The set environment is effectively structured by using different types of sensors and building a localized map to navigate the structured environment. Decisions are made on which nodes to navigate and who to assist depending on the real-time data. Tele-op system is set to override in case of any ethical dilemmas. The robot will be able to manoeuvre in any direction while using a standard chassis shape which will make it very easy to move around the warehouse, even in very space-constrained areas. It will be built strong enough to carry the weight of the gripper mounted on it for the pick-and-place operations.

Keywords Warehouse · OpenCV · Manipulator

K. Gogisetty (✉) · A. R. Capur
Department of Mechatronics, Manipal Institute of Technology, Manipal Academy of Higher Education, Manipal, Karnataka 576104, India
e-mail: karthikgogisetty@gmail.com

I. Pandey
Department of Computer Science and Engineering, Manipal Institute of Technology, Manipal Academy of Higher Education, Manipal, Karnataka 576104, India

P. Dighe
Department of Information and Communication Technology, Manipal Institute of Technology, Manipal Academy of Higher Education, Manipal, Karnataka 576104, India

S. Sharma et al. (eds.), *Intelligent Control, Robotics, and Industrial Automation*, Lecture Notes in Electrical Engineering 1066,
https://doi.org/10.1007/978-981-99-4634-1_5

1 Introduction

1.1 Introducing Project Lunokhod

The name 'Lunokhod' was inspired by the Soviet Lunokhod robotic lunar rovers. It is an unmanned ground vehicle with a robotic manipulator on the top for pick-and-place operations. It will be of great help in places where risk may be involved if the task was to be handled by human workforce and even in cases where the human operator would be performing mundane, repetitive tasks which don't require the degree of decision-making that human thinking is capable of. This will also ensure efficient workflow in the place of implementation. The Lunokhod robot will find multiple uses and will be able to execute tasks in warehouses or even factory assembly lines which efficiency and robustness.

The functionalities of Lunokhod have been developed while keeping in mind the robustness, agility, and efficiency necessary within the warehouse setting. Therefore, these include identification and pick-and-place ability for warehouse objects, path planning, and detection through Aruco markers and safety measures like fire detection. All of these are fully autonomous, allowing the user to remotely monitor all workings at a real-time pace, and get notified constantly through digital updates sent by Lunokhod. From a mechanical standpoint, the functionality has been decided to make Lunokhod capable to handle payloads in a warehouse setting while being able to manoeuvre effectively. To this effect, a 5 degree-of-freedom robotic arm is included. For manoeuvrability, Mecanum wheels are used which will allow superior movement in the warehouse.

From a team perspective, this project was a culmination of myriad disciplines and skills. The Aruco markers use computer vision and image processing. Path planning concepts are also included to enhance the capability of the bot while deployed in the warehouse. Sound knowledge of motors and motor drivers was also needed to select the motors. To produce the entire chassis, our team members were required to utilize their CAD skills along with elements like stress analysis to determine the structural integrity of our chassis.

The motivation for these functionalities and the entire idea of Lunokhod itself emerged during the time of the COVID pandemic, through which it became apparent that remotely monitored autonomous technologies can be highly useful in scenarios where human contact is to be minimized or there is a certain danger to humans. Although focus is to set out to design Lunokhod for warehousing applications, with more customized functionalities, it can be implemented in the healthcare sector, natural disaster settings, and several other scenarios where the environment is hazardous to human health [1].

2 Literature Overview

The most crucial part of the Lunokhod robot is its mobility with the help of motors. The paper 'Design, E.M.: Manufacturing. Drive Wheel Motor Torque Calculations' by the University of Florida gives insights about torque, force, and number of other factors required to select a motor of choice for a given application. The motor was chosen by calculating the holding torque using the gross vehicle weight and the total tractive effort. The holding torque is computed using the FOS. The overall tractive effort was determined using a MATLAB script where all forces were taken into account [2]. For a straightforward pick-and-place application, the research 'kinematic modelling and analysis of 5 DOF robotic arm' by V Deshpande and P George tries to simulate the forward and inverse Kinematics of a 5 DOF Robotic Arm. It is possible to obtain a generic D-H representation of the forward and inverse matrix. To study how the arm moves from one point in space to another, an analytical solution for the forward and inverse Kinematics of a 5 DOF robotic arm is offered [3]. Control Systems Engineering by Norman S Nise is highly renowned for its accessibility and emphasis on practical applications. It gives students and research enthusiasts an input to the design and analysis of feedback systems that enable contemporary technology. This text gives insights into computer aided design and goes beyond theory and abstract mathematics to transfer essential principles into the design of physical control systems through real-world case studies, difficult chapter problems, and in-depth explanations [4]. To provide the Lunokhod with its own sense of intelligence, object and code detection were necessary. For this, 'Learning OpenCV' by Gary Bradski and Adrian Kaehler gave a clear foundation and understanding about reading in images, contour finding and writing the information obtained from codes to separate files which can later be transferred to the controller. The Aruco library within OpenCV is then used for the detection of the markers and the QR code scanner created using the QR code detector functions in Python. It also provides an understanding of several machine learning algorithms within the computer vision context from which the linear regression model is applied upon the path learning results to predict results of a larger scale [5]. 'Practical OpenCV' by Samarth Bhatt was another useful resource as it dedicated substantial information to embedded computer vision and the practical ways of running OpenCV programmes on Raspberry Pi, allowing the integration of both concepts [6].

3 Structure of Lunokhod

3.1 Manipulator of Lunokhod

To perform any required task and enhance the usability of Lunokhod, a manipulator is assembled to aid the robot for any pick-and-place operation. It is a five degree-of-freedom robotic arm, providing flexibility in utilizing the workspace. A revolute joint

Fig. 1 Chassis and manipulator mounted

mechanically joins the first link to the base. The fixed base plate is installed on top of the chassis, and the rotating base is mounted on top of it. The material selection criteria must meet demands like high material strength and light in weight. Therefore, carbon fibre-reinforced polymer is used as the necessary material for the manipulator's body and connections. In our model, the magnetic gripper MHM-50D2-M9B has been used as the end effector for pick-and-place operations. A revolute joint allows the mount for the grippers to revolve in relation to the third link. As a result, the end effector will result to have numerous distinct orientations [3].

Figure 1 is a 3D render of the chassis, and the robotics arm is designed using Autodesk Fusion 360.

3.2 *Dimensions of Lunokhod*

Three materials—ABS plastic, carbon fibre-reinforced polymer, and aluminium—were able to bear the imposed load of 1200 N on the top frame because of the nature of the design. By dispersing the load, the vertical elements helped to prevent the platform from bowing when the load was applied. In our model, carbon fibre-reinforced polymer was selected to reduce overall model weight and reduce stress on the motor shafts. Utilizing Fusion 360 Von Mises stress analysis, with carbon fibre-reinforced polymer, it met the motor's required torque and could endure a load of 2.1×10^6 N, which is less than half the strength of ABS. The following are the dimensions of the chassis that is designed and analyzed.

Figure 2 is the outline of the dimensions of the chassis designed and drawn with the help of Autodesk Fusion 360.

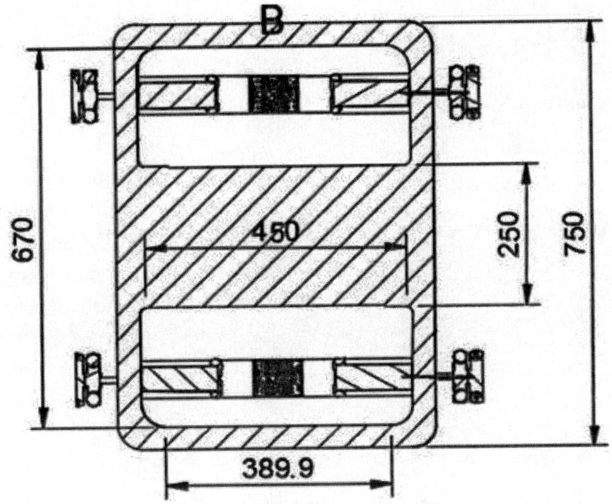

Fig. 2 Dimensions of the chassis

4 Electrical Aspects of Lunokhod

4.1 Motor Overview and Details for the Bot

DC planetary gear motors are used, namely the PG56/63ZY125-2440-55k, to meet the weight and power needs of our chassis. It is an efficient, lightweight, and small motor. It is more adaptable since a variety of reduction gear ratios are available to improve torque production. The motor weighs 2.224 kg and measures 25 × 7 × 7 cm in size. The planetary gearbox is lightweight and small. The motor has a 450 mN-m rated output torque, but as the planetary gearbox has a 55:1 reduction gear ratio, the maximum output torque is 18Nm. To produce adequate torque to support the weight of the entire chassis, including the weight of the arm assembly and the electronic components, four motors are used to reach the target torque required. It has a 1200 nN-m stall torque. The motor works with a supply dc voltage of 24 V. It has a stall current and no-load current of 22.5 A and 1.2 A, respectively. The rated power of the motor is 150 W. A magnetic field is developed by a permanent fixed magnet.

4.2 Supply, Motor Equation, and Working

A rotating armature through which a current flows, passing through this magnetic field at right angles and faces a force, $F = Bli_a(t)$. A voltage at the terminals of the conductor is developed which is $e = Blv$. The current-carrying armature is rotating in a magnetic field, its voltage is proportional to speed. Hence,

$$v_b(t) = K_b \frac{d\theta_m}{dt} \tag{1}$$

Taking Laplace transform, the following is the equation,

$$V_b(s) = K_b s \theta_m(s) \tag{2}$$

Writing a loop equation around the Laplace transformed armature circuit,

$$R_a I_a(s) + L_a s I_a(s) + V_b(s) = E_a(s) \tag{3}$$

Torque of the motor is proportional to the armature current. Hence,

$$T_m(s) = K_t I_a(s) \tag{4}$$

Substituting in the above equation, the required Laplace equation can be formulated,

$$\frac{(R_a + L_a s)\left(J_m s^2 + D_m s\right)\theta_m(s)}{K_t} = E_a(s) \tag{5}$$

Assuming that the armature inductance is small compared to its resistance, equation becomes

$$\left[\frac{R_a}{K_t}(J_m s + D_m) + K_b\right] s\theta_m(s) = E_a(s) \tag{6}$$

The desired transfer function, $\theta_m(s)/E_a(s)$ is

$$\frac{\theta_m(s)}{E_a(s)} = \frac{K_t/(R_a J_m)}{s\left[s + \frac{1}{J_m} D_m + \frac{K_t K_b}{R_a}\right]} \tag{7}$$

By solving all the equations above as per the specifics of the motor, the motor torque constant K_t is obtained as as 0.015, the back emf constant K_b as 0.0001, inertia J_m as 10 kgm^2, and damping D_m as 0 (assuming no damping system). The armature resistance R_a is 0.3 ohms which are used to check the behaviour of Lunokhod using a PID controller [2, 4].

4.3 Supply to the Electromagnet

In the model, a cylindrical electromagnet will be mounted on the end effector, and the wire is wound about the circumference of it. To obtain the required current supply to the coil, the magnetic field that need to be produced is estimated to be 10^2 T over a range of current values, the magnetic flux density is calculated. Where μ is permeability constant and n represents number of turns per unit length.

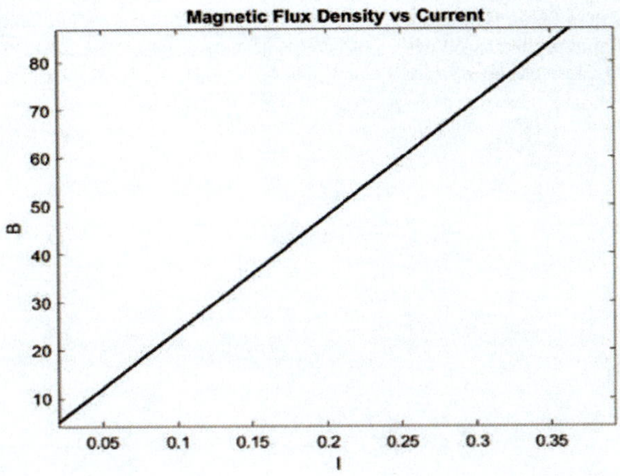

Fig. 3 Magnetic flux density versus current keeping 'μn' constant

$$B = \mu n I \tag{8}$$

The above graph, Fig. 3 of magnetic flux density vs current was plotted with the help of MATLAB by MathWorks. The force produced in newtons is also calculated using the following equation where L is the circumference of the surface area which is a circle ($2\pi r$). The choice of required current was chosen by using and plotting magnetic flux density 1.6 A was the required current. The Lunokhod end effector could take around 5000 N of maximum force which is far less than the attraction force exerted on the end effector as it is around 10^4 N as shown in Fig. 4. Using the below equations, the line graph (Fig. 4) was plotted to support the statement keeping L constant and B to be the similar above assumption.

$$B = F/IL \tag{9}$$

5 Intelligence to Lunokhod

5.1 Path Planning and Machine Learning

Path planning algorithms are used to handle a wide range of issues in a variety of sectors. It has been used to guide the robot to a specific objective, from a simple trajectory management across to the selection of an appropriate sequence of actions. The problem to find a shortest path from one vertex to another through a connected graph is of interest in multiple domains. In our model, RRT algorithm is advised to use as the algorithm, selects a random point in the environment, and connects it to the

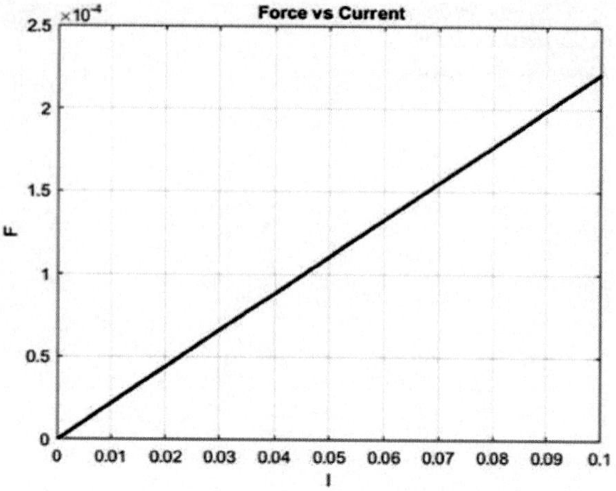

Fig. 4 Graph represents behaviour of force with respect to current

initial vertex. Subsequent random points are then connected to the closest vertex in the emerging graph. The graph is then connected to the goal node, whenever a point in the tree comes close enough given some threshold. Although RRT is generally a coverage algorithm, it can be used for path planning by maintaining the cost-to-start on each added point and biasing the selection of points to occasionally falling close to the goal. RRT can also be used to consider the non-holonomic constraints of a specific platform when generating the next random way point.

5.1.1 Path Learning

This is the additional essence of the intelligence that is provided to the Lunokhod robot. It is evident that in many path planning algorithms, the robot must explore nodes in every iteration to get the cost of the selected path. Additional intelligence is provided in our proof of concept by using machine learning to train a model for getting the cost of path to reduce the exploration time and memory. Every time it visits same or any similar environments, the cost is predicted and compared to the existing costs. The idea is to train the model with adequate number of samples with number of nodes and the cost of travelling a certain path via each node and being able to predict the total travel cost, if given any number of nodes. If it is found out to be that the new path is smaller than a cut-off cost of traversing it will be added to the stack of exploration which is used to compute the path planning algorithm excluding the large memory usage by discarding the nodes above the cut-off heuristic value [7].

5.2 *Path Traversal Using Aruco Marker*

Traversal of the robot throughout the warehouse is an essential aspect of its application, and Aruco markers can be used to make it highly efficient. These markers are fiducial square markers enclosed within a black border with an internal binary matrix, each one containing a unique identifier. Upon detection, the markers give the inputs to the camera to sense the environment in terms of angle, height, and depth. Within the warehouse context, the markers will be placed in a grid like manner over the floor, each one having a certain (x, y) co-ordinate. To initiate traversal, the robot will be fed with the co-ordinates of the respective Aruco marker it must reach and given a certain velocity, will be moved towards it. For the software aspect of this implementation, there are three major parts—the marker generation, the marker detection, and calculation of the co-ordinates of the marker. The code was implemented in Python using OpenCV libraries which can be used to do the same [5, 6].

5.3 *Simulation in Robot Operating System*

To simulate the behaviour for Lunokhod, robot operating system (ROS) is used. ROS is a flexible framework for writing robot software, comprising of a collection of programming tools, libraries, and software that aim to simplify the task of creating complex and robust robot behaviour. URDF files that are made are added with plugins to have an interactive visualization interface giving user access about various features of the robots. To support our basis and get an understanding, few inbuilt models are used and visualized the following where the LiDAR publishes a message of type LaserScan to the topic ebot/laser/scan. Data can be retrieved from this topic in the form of a list containing detected distances for each LiDAR step. The GPS publishes data to the topic /fix, with a message of type NavSatFix, providing the latitude and longitude, which must be converted to regular Cartesian co-ordinates before being used by the path planner. This conversion is done using forward equirectangular projection and is very accurate. The IMU publishes to the topic/imu, with a message of type IMU. The orientation data is stored in the form of quaternions and must be converted to the Euler angle notation before being used for calculations. Once all the data is retrieved and the required conversions are performed, the data from the sensors is used by the path planner algorithm as explained above. When a path is calculated, the required base wheel speeds are published to the topic $/cmd_v el$, which controls the motion of the base.

Simulations performed in Gazebo and Rviz are shown in the screenshots, i.e. Fig. 5.

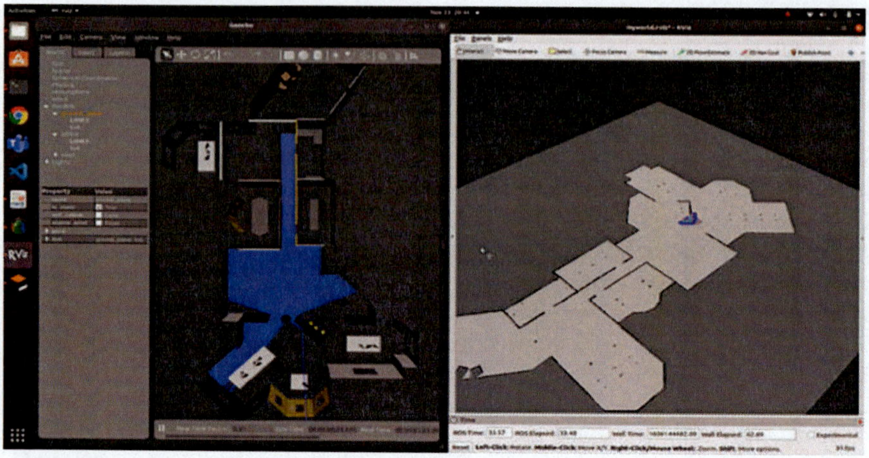

Fig. 5 A screenshot of the simulation in Gazebo (left) and RViz (right)

6 Localization and Controller

6.1 *Simultaneous Localization and Mapping (SLAM)*

Robots cannot always rely on GPS, especially when they operate indoors. GPS is not sufficiently accurate enough outdoors because precision within a few inches is required to move about safely. SLAM is a complicated multi-stage process involving sensor data alignment, motion estimation, sensor data registration, visual odometry, and map building for localization. Robots continuously do split-second gathering of sensor data on their surroundings. Camera images are taken about 90 times a second for depth-image measurements along with LiDAR images, used for precise range measurements, which are taken 20 times a second. Wheel odometry considers the rotation of a robot's wheels to help measure how far it is travelled. Inertial measurement units are also used to gauge speed and acceleration to track a robot's position. All these sensor streams are taken into consideration using sensor fusion to get a better estimate of how a robot is moving. Kalman filter algorithms and particle filter algorithms are used to fuse these sensor inputs. Bayesian filters are applied to mathematically solve where the robot is located, using the continuous stream of sensor data and motion estimates.

6.2 *Proportional Integral Derivative Controller*

Using sensor data, desired state is obtained, to reach that desired position is where control engineering plays an equal role. Proportional controller produces output proportional to the error signal sent in the feedback loop. Here, $u(t)$ is the signal that is sent to actuators and $e(t)$ is the error signal. The proportional controller is used to change the transient response as per the set goal parameter obtained from sensors.

$$u\,(t)\,\alpha e\,(t) \tag{10}$$

The integral controller produces an output that is integral of the error signal and helps to decrease the steady-state error. It sends a signal, i.e. summation of all the error values occurred up to time 't'.

$$u\,(t)\,\alpha \int e\,(t)\,\mathrm{d}t \tag{11}$$

Derivative controller produces an output which is derivative of the error signal. It is used to make the unstable control system into a stable control system by checking the slope of the error signal after every iteration.

$$ut\alpha \mathrm{d}e\,(t)\,/\mathrm{d}t \tag{12}$$

The combined error signal with their constants is fed into the model as the plants' input to reach the desired set point.

7 Results and Discussions

7.1 *Tracking of Lunokhod*

In such a system, tracking the Lunokhod is one of the most important aspects required while dealing with traversing of the robot as the user must be constantly notified with information about where the robot is. Its behaviour with respect to the occurrence of any ethical dilemmas defines the health of a working environment, making it crucial to track. As explained in the previous segment, the detected image of the QR code and Aruco markers which contain information helps in getting the location co-ordinates and the information about the object to be handled by the Lunokhod. In case of any ethical dilemma, the co-ordinates of the robot can be obtained to prevent or re-evaluate the process and maneuvre the robot while protecting factory workers. The focus is also to develop custom QR markers using Python and use it as per the workspace requirements. The major functionality of the robot takes place as the camera mounted on it detects the QR code placed on the object and uses its robot arm to pick it up. At the same time, it extracts data from the QR code and relays it to the user, notifying the details of the object that has just been picked up in real-time. The picture shown below, i.e. Fig. 6 shows how the code is detecting the Aruco markers and displaying their respective IDs.

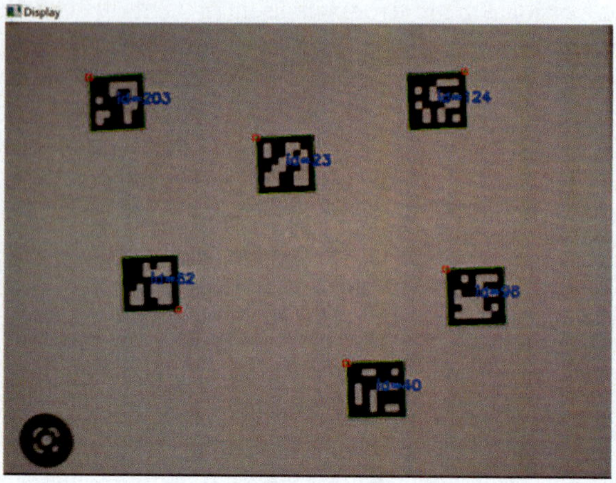

Fig. 6 Test code detecting Aruco markers and displaying their IDs

7.2 Path Learning Results

Now to explore the nodes available, a simple linear regression model was built taking a sample linear data, to conceptualize the idea stated in Sect. 1, in the model x is number of nodes in each environment and y to be the cost of travelling, given the number of nodes in a similar environment. Once the computations are done, the shortest cost of travelling a path via the existing nodes is chosen along with the exploration techniques usually done while performing any path planning algorithm. It helps reduce the exploration time exponentially as the time taken to build the model for 1000 samples was around 7.49 s which will cut short the paths that will be needed to explore by any path planning algorithm on a general basis. After the training and validation of the data set, the prediction for an evenly spaced number between 0 and 250 was performed and was observed to fit the data. As of now, the model prediction is based on linear regression for a linear data if a widespread data is given a higher order boundary condition can be kept for the model to predict the new data's traversing cost. The following graph, Fig. 7, shows the result of a linear regression model to predict the y (cost) given an x (features).

Using Aruco markers detection explained in Sect. 2 the co-ordinates of the end position to reach, a proportional integral derivative controller was implemented using the MATLAB that will actuate motors for the system to attain the desired state with minimum possible error. The analysis was first done, without using any noise parameters, and was tuned to get an idea about the ideal output behaviour. A set step response was treated as the desired state and the controller signal were generated in MATLAB Simulink as shown below in Fig. 8.

As shown the black line represents the behaviour of the Lunokhod robot whose plant equation was a DC motor, and the constants were found accordingly mentioned in the math model section of the paper. Considering that there are going to be certain external forces and noise in the signal data that is going to be sent to the plant, a

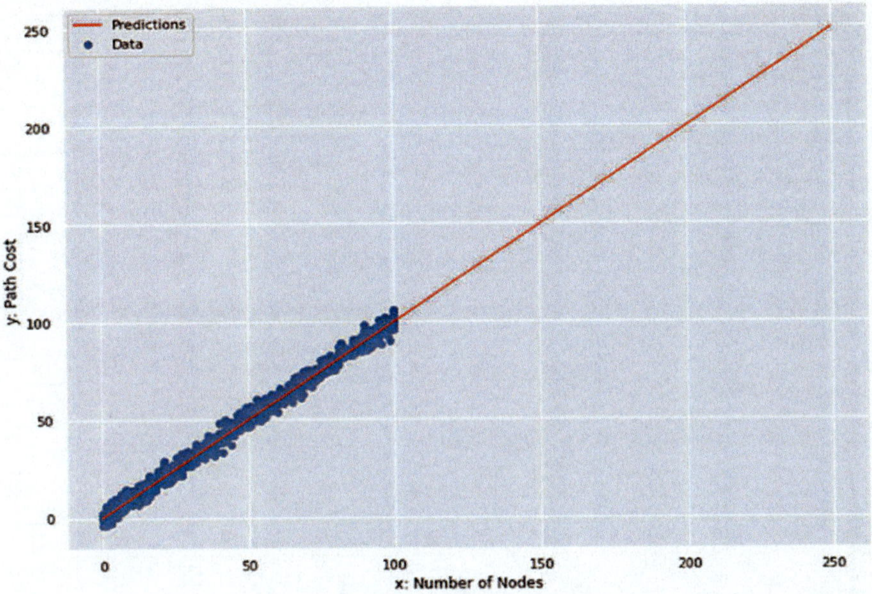

Fig. 7 Prediction versus data fed to the ML model built

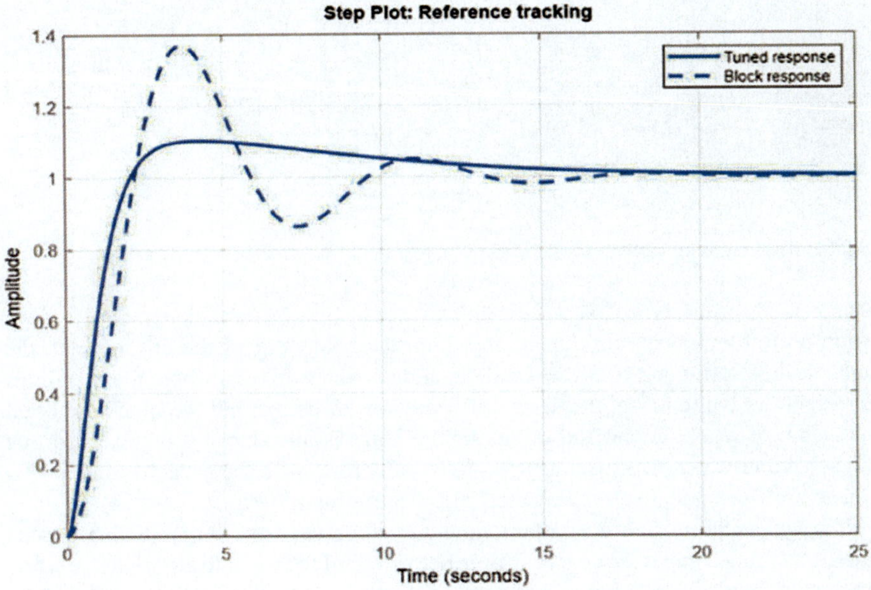

Fig. 8 Auto tuned ideal behaviour reaching a set point

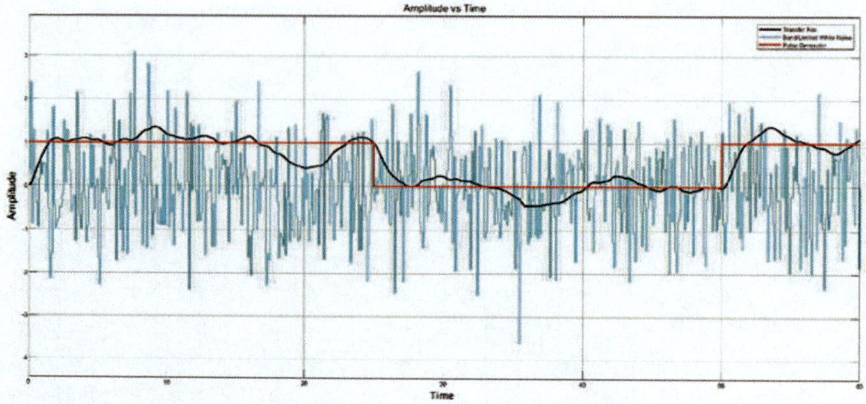

Fig. 9 Behaviour of Lunokhod when introduced noise to an ideal system

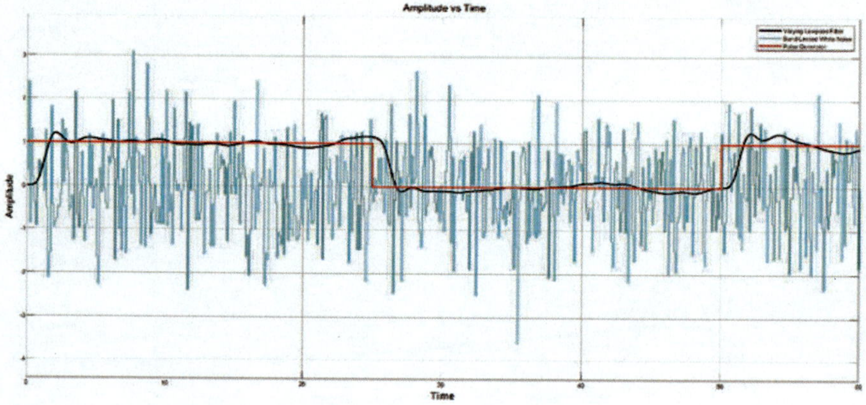

Fig. 10 PID controller with noise and filter

white noise block was introduced, and then, the behaviour of the bot to reach the same desired states and result was radical and uneven as shown where the blue lines represent the white noise, the set point behaviour to be red and the black the bots' behaviour to reach the desired point. As the noise is introduced it is observed that the behaviour was not as ideal as it is observed before, with the set auto-tunned PID constants. The same result is reflected below as a graph in Fig. 9.

To regulate the noise in the output, a low-pass filter was used, and the PID constants are slightly tuned to get the near real-time behaviour of the robot in a warehouse while traversing from one position to another as follows.

Fig. 10 shows how the noise is regulated in the output giving us the real-time behaviour of the robot. It was possible to plot this graph by performing simulations in MATLAB.

7.3 Data Visualization

Initially, the data that is available is transferred to slave micro controller from a master microcontroller using various protocols like HTTP and MQTT. The serial data that is available for the slave microcontroller will be logged using serial library of Python where the com ports data is saved in a dashboard.csv in the local computer. To make access available to everyone in the workspace, a network access storage can be used to save the data common to everyone and can be visualized using plotly and dash libraries available in Python.

8 Ethical Evaluation

8.1 Fire Detection and Evaluation

The major goal of the implementation of any robotic automation within the warehouse setup is to reduce human effort, manual errors, as well as prevent human casualties. One major source of such casualties is undetected fires which lead to large amounts of human and product damage. This can be avoided by providing fire detection-based intelligence to the Lunokhod itself. The target here would be to detect the fire and alert the operators by sending the information about the location and severity of the fire to the master microcontroller using any communication protocol. A tele-op override is also installed in case of any ethical dilemmas with respect to sensing any critical situations within the designated warehouse.

For supporting the work, a RESNET50 model was developed to detect fire and the existing ESP8266 modules help communicate remotely using their mac addresses. Figure 11 shows how the code was able to detect fire accurately, thus proving its purpose.

Fig. 11 Fire detection test result of code

9 Conclusion

Project Lunokhod implements the use of a manipulator for pick-and-place operations of the payload as well as using an autonomous guided vehicle for transporting the payload from one point in the warehouse or place of deployment to another. This ensures smooth flow in the supply management of the place. All this can be done autonomously without any external human help. The option for human overriding is always there for ethical concerns and certain critical abnormal situations that may arise. It is an ideal solution to a slow and inefficient warehousing. It is made sure the robot has safety software techniques implanted to detect any fire accidents and alarm anyone when in need. A combination of sound mechanical structure, sufficient electronics and electrical connections, and intelligence helps the Lunokhod to play a role in the upcoming technical calendar.

References

1. Lynch KM, Park FC (2017) Modern robotics. Cambridge University Press
2. Design EM (2011) Manufacturing laboratory. Drive Wheel Motor Torque Calculations. Faculty of Engineering. University of Florida, University of Florida
3. Deshpande V, George P (2014) Kinematic modelling and analysis of 5 dof robotic arm. Int J Robot Res Dev (IJRRD) 4(2):17–24
4. Nise NS (2020) Control systems engineering. Wiley
5. Kaehler A, Bradski G (2016) Learning OpenCV 3: computer vision in C++ with the OpenCV library. O'Reilly Media, Inc.
6. Brahmbhatt S (2013) Practical OpenCV. Apress
7. Buniyamin N, Ngah WW, Sariff N, Mohamad Z et al (2011) A simple local path planning algorithm for autonomous mobile robots. Int J Syst Appli Eng Dev 5(2):151–159

A Data-Driven Test Scenario Generation Framework for AD/ADAS-Enabled Vehicles

Niraja Narayan Bhatta and Binoy B. Nair

Abstract Advanced Driver Assistance System (ADAS)-enabled vehicles capable of operating at SAE J 3016—level 3 or higher extensively employ deep learning-based systems for realizing ADAS features. Such systems need to be validated with all possible scenarios that could unfold during a trip. In this paper, a novel data-driven methodology to automatically generate relevant driving scenarios based on the operational design domain (ODD) in which the vehicle being driven is proposed; variables that are necessary to evaluate the functionality of an AD/ADAS vehicle are identified and a process to quantify the relevance of each of these variables is proposed using a combination of statistical framework such as CatBoost, Weight of Evidence (WoE), and Information Value (IV). An analytical framework is then developed on the basis of analytical hierarchy process (AHP) using real-world accident scenarios data to automatically generate test scenarios. The framework proposed in this study can be adapted to generate relevant test scenarios given the variables in that ODD.

Keywords ADAS · Autonomous driving · CatBoost · Scenario priority · Machine learning · Test scenario generation · Weight of evidence and information value

1 Introduction

Global automotive industry attributes erratic judgment on the driver's part a key factor that affects safety of driver as well as passengers. Driver's inability to perceive the surrounding environment leads to erratic judgments that might lead to accidents. Systems that assist driver in the driving task by informing the driver about surrounding environment are known as an Advanced Driver Assistance System (ADAS). ADAS spans a wide range of functionalities from detecting obstacles on the road [1–3] and

N. N. Bhatta · B. B. Nair (✉)
Department of Electronics and Communication Engineering, Amrita School of Engineering, Amrita Vishwa Vidyapeetham, Coimbatore, India
e-mail: b_binoy@cb.amrita.edu

S. Sharma et al. (eds.), *Intelligent Control, Robotics, and Industrial Automation*, Lecture Notes in Electrical Engineering 1066,
https://doi.org/10.1007/978-981-99-4634-1_6

distance estimation [4] to monitoring driving-related events such as lane detection and departure [5–7] and driver attention monitoring [8, 9].

SAE standard J3016 [10] defines the different levels or degree of autonomy. Depending upon the involvement of the driving in DDT, SAE J3016 categorizes autonomous driving into five levels, starting from Level-0 (no autonomy) up to Level-5 (full autonomy with no driver input).

Testing of an AD/ADAS-enabled vehicle can be conducted in three ways: (a) testing using simulation environment, (b) testing using closed track testing, and (c) on-road testing. In simulated testing (testing in a simulation environment), maximum number of test cases with high degree of complexity and criticality can be validated in a virtual environment. In closed track testing, the actual vehicle can be subjected real-world scenarios to validate the effectiveness of the autonomous software. On-road testing provides the most realistic validation of an autonomous vehicle; however, to demonstrate the reliability of the system, the vehicle has to be driven for millions, and in some case billions of miles [11]. The alternative to on-road distance-based testing is scenario-based validation. A novel data-driven approach to automatically generate scenarios for a given ODD is presented in this work. The remainder of the paper is organized as follows: the state-of-the art is presented in Sect. 2, the problem description and the dataset employed are described in Sect. 3, the system design and implementation are presented in Sect. 4, and the conclusions are presented in Sect. 5.

2 Related Work

A system validation of AD/ADAS vehicle utilizes a pre-defined set of test cases based on a knowledge-based system constructed by experts in the field. However, having a fixed set of limited test scenarios is not sufficient to either train or test such a system. Helmer et al. defined a method to assess the safety performance of assisted driving system in [12] using virtual environment. It used the knowledge from Field Operational Tests (FOTs), naturalistic driving data along with data from driving simulator to analyze the benefits and risks in the assisted functions. Collecting FOT data involve fitting a vehicle with sensor suite and recording devices, by putting the actual vehicle on the road. Though it provides high-quality real-world data for scenario modeling, it captures bare minimal critical scenario information as well as only localized data of the geographical area where the vehicle was driven. Such limitations on the naturalistic driving data put a constraint on the scalability and generalization of the approach in using the data for scenario generation.

There are several approaches to generate data or test scenarios in simulation. Kim et al. define an approach for a test framework which generates virtual road segments automatically depending on the path and behavior of the autonomous vehicle [13]. Nalic et al. propose a software framework to test an autonomous driving system using Vissim, a traffic flow simulation software (TFSS) [14], whereas Alnaser et al. propose a novel model-based framework, using Florida Poly AV Verification Framework to assess the robustness of autonomous vehicles [15]. However, these approaches do not provide a method to define the degree relevancy of a generated scenario within the given ODD leading to validating every possible scenario. This work provides a method to generate a scenario based on real-world actual accident data as well as a relevancy score indicating the degree of relevance.

The work presented in this paper proposes a strategy to utilize real-world accident data to construct test scenario instead of depending primarily on a knowledge-based system or FOT data that lack the diversity; the method proposed in this work considers the factors contributing to an actual accident and builds the scenarios on this premise. Work presented in this paper proposes a method to quantify the relevancy of a driving scenario in order to identify if it should be considered to be a part of the scenario database; a higher relevancy value indicates a more critical scenario. Quantifying the value of a test scenario provides validity in the claim for a critical scenario.

3 Problem Description

An ODD is the constrained environment in which the vehicle is operating, thus limits the behavior of the vehicle, and thereby limits the scenarios that need to be generated. According to United Nations Economic Commission for Europe (UNECE) [16], scenarios can be identified based on: (a) analyzing driver behavior, (b) collecting prior collision data from various accident databases, (c) analyzing traffic patterns in a given ODD and/or (d) collecting and analyzing sensor data. Given a particular ODD, the large range of possible scenarios makes it challenging to identify each such scenario manually. Therefore, a better approach could be to generate the scenarios automatically. However, the challenge is that with many factors, or variables affecting a scenario, the number of generated scenarios could be in millions. Therefore, there is a need to define a process to generate scenarios with a certain degree of relevance to the ODD. In this work, we propose a data-driven methodology to determine the relevancy of a test scenario based on its contributing factors. We provide a relevancy score to each of the parameters contributing to the scenario and define the relevancy score of the generated scenarios depending on the relevancy score of the associated parameters. We referred one of largest accident datasets [17] in order to determine the relevancy score of factors contributing to a severe accident.

3.1 Dataset Description

The UK's Department of Transportation has been collecting and documenting accident data [17] in detail since the year 1979; this particular work uses the data collected and published from 1979 to 2020. The overall dataset has 86, 02, 824 records. The dataset is distributed across three different individual datasets: accident dataset, vehicle dataset, and casualty dataset. We focus on accident and vehicle datasets since the information in the casualty dataset was a result of the accident not a cause and is therefore not relevant to scenario generation process. There are a total of thirty-six features, out of which only twenty features listed below are relevant to the work presented here: (a) Accident Severity, (b) Number of vehicles, (c) Number of casualties, (d) First road class, (e) First road number, (f) Road type, (g) Speed limit, (h) Junction detail, (i) Junction control, (j) Second road class, (k) Second road number, (l) Pedestrian crossing human control, (m) Pedestrian crossing physical facilities, (n) Light conditions, (o) Weather conditions, (p) Road surface conditions, (q) Special conditions at site, (r) Carriageway hazards, (s) Urban or rural area, (t) Trunk road flag. Remaining features have no bearing on the scenario generation process and hence are not considered any further. The feature *Accident Severity* contains three responses' values: 'slight', 'severe', and 'fatal', of which 'serious' and 'fatal' are merged into one single class, 'serious'. Feature *Lighting Conditions* are re-categorized into two columns, indicating ambient lighting and street lighting. Similarly, *Weather Conditions* feature is re-categorized into weather information and high winds' information. All the different animal categories in carriageway hazards are merged into one single animal category.

4 System Design and implementation

Ulbrich et al. [18] define a scene as snapshot of the surrounding environment of the autonomous vehicle at a given instant. A scene captures all the static and dynamic elements in that environment along with the corresponding states. Ulbrich et al. [18] also defined a scenario as a consecutive sequence of several scenes. Each scenario has goal that it is supposed achieve with the given sequence of scenes. Essentially, a scenario is a sequential arranges of a group of scenes in a timeline. Menzel et al. in [19] proposed to use multiple layers of abstraction to represent a scenario with increasing level of detailing in each abstraction layer.

A typical verification and validation system will subject the autonomous vehicle to numerous scenarios in order to verify if it fulfills the requirements. In virtual validation, these scenarios are constructed using various tools.

4.1 Extracting ODD Factors from Accident Database

We analyze the UK accident database record from 1979 to 2020 [17] to prepare a list of factors affecting an ODD for AD/ADAS vehicle. Each of these factors is further analyzed to calculate the relevancy score before being considered for test scenario generation. Most of the factors derived from the UK accident database are categorical in nature. Therefore, we used a statistical analysis framework, CatBoost [20], to derive the relevancy score for each of these factors. CatBoost provides a mechanism to parse through the categorical as well as numerical data to provide importance of each factor in that dataset which provides us our relevancy score for these factors [20].

4.2 *Finding Relevancy Score of Values of Individual Factors*

Each of the contributing factors to ODD has individual relevancy scores. Weight of Evidence (WoE) and Information Values' (IVs) framework [21] has been employed here to provide evidence of how relevant each of these values is, with respect to causing a serious accident. The information value obtained for each of these values together with the relevancy score of the factors defines the relevancy score of that value within the factor (V_j). WoE and IV is primarily used because it assesses the contribution of each value [22], within a factor or variable, toward a serious or, major accident and quantifies it.

$$\mathrm{WoE}_{V_j} = \ln\left(\frac{\text{\% of min oraccidents}}{\text{\% of major accidents}}\right). \tag{1}$$

We will use the WoE value obtained for each of these individual values within a factor to calculate corresponding information value. The relevancy score for each value within a factor is the product of relevancy score of the factor and the information value of the value within the factor.

$$\text{Information Value, } \mathrm{IV}_{F_i V_j} = \sum_j \mathrm{Diff}_{V_j} \cdot \mathrm{WoE}_{V_j}, \tag{2}$$

$$\text{Relevancy score, } R_{F_i V_j} = R_{F_i} \cdot \mathrm{IV}_{F_i V_j}. \tag{3}$$

4.3 *Test Scenario Prioritization and Generation*

To constrain and control the number of test scenarios generated, Analytic Hierarchy Process (AHP) [23] has been used in this study to determine the priority of a scenario within the given ODD given the accident data. Scenario consists of individual contributing factors and their values; therefore, the overall priority of the scenario can be determined by considering the relevancy score of individual values within a contributing factor.

$$D_{F_i V_{j,k}} = \frac{R_{F_i V_j}}{R_{F_i V_k}}, \tag{4}$$

$$D_{F_i V_{k,j}} = \frac{1}{D_{F_i V_{j,k}}}. \tag{5}$$

Using the decision matrix and applying AHP, generalized priority of each value within a factor is quantified. Table 1 contains relevancy score of each factor in the dataset after applying AHP.

$$\text{Priority, } P_{F_i V_j} = \text{AHP}\left(D_{F_i V_j}\right). \tag{6}$$

5 Test Scenario Generation

To avoid generating a large number of test scenarios, laden with irrelevant and non-critical scenarios, a method is proposed to determine the priority or, importance of a test scenario within the context of the ODD, and if the given scenario has a priority value above a certain threshold, it is added to the test suite. Priority of a test scenario is dependent on the parameters constituting it and degree of contribution of those parameters in that test scenario as obtained from using AHP. Let us assume a single test case which consists of *n* different factors; the criticality of the test scenario is the sum of the importance factors of all the factors contributing to that scenario.

$$\text{Test Scenario Priority, } P_i = \sum_i \sum_j P_{(F_i)(V_j)}. \tag{7}$$

It is observed that in few of the combinations, values of certain factors do not go together in a realistic situation. For example, a scenario with weather as 'sunny' but the driving time as 'night' is impractical. To address such situations, exclusion criteria have been added in this study, as given in Table 3 by considering conditional compilation of certain word associations. If the exclusion criteria are satisfied, the scenario will not be added to the test suite. These exclusion criteria are a pre-defined word-based logical combination of certain factors which are not supposed to go together.

$$\text{TS}_{\text{Threshold}} = \{\text{TS} \in \text{TS}_{\text{All}} : P_{\text{TS}} \geq P_{\text{Threshold}}\}, \tag{8}$$

$$TS_{notExcluded} = \{TS \in TS_{Excluded} : P_{Excluded} \subset P_{All}\}, \tag{9}$$

$$TS_{Generated} = TS_{Threshold} \cap TS_{notExcluded}. \tag{10}$$

Table 1 contains the relevancy score of all the values of all the factors contributing to an accident in reference to the UK Accident Dataset [17]. Having the relevancy of each variable in an ODD provides the flexibility to pick and choose the required

Table 1 AHP relevancy score

Feature	Features and AHP relevance score
Road type	Roundabout (0.3609), one way (0.2042), SC (0.2042), slip road (0.2042), DC (0.0266)
Road speed limit	60 (0.3943), 30 (0.1991), 20 (0.1664), 70 (0.1465), 40 (0.0444), 50 (0.034), 10 (0.0153)
Junction detail	Roundabout (0.4366), no junction (0.2749), slip road (0.1256), junction with more than four arms (0.0949), T-junction (0.0283), crossroads (0.0199), private drive or entrance (0.0199)
Junction control	Automatic traffic signal (0.5114), uncontrolled (0.2553), authorized person (0.1679), no junction (0.033), stop sign (0.0324)
Pedestrian crossing (human control)	Authorized person (0.4737), control by school crossing patrol (0.4737), no pedestrian crossing (0.0526)
Pedestrian crossing (physical facility)	Non-junction pedestrian light crossing (0.2789), pedestrian phase at traffic signal junction (0.2789), no pedestrian crossing (0.2465), zebra crossing (0.0772), footbridge or subway (0.069), central refuge (0.0495)
Road surface condition	Snow (0.3868), frost or ice (0.2353), mud (0.1384), dry (0.1302), oil or diesel (0.0483), flood over 3 cm. deep (0.0451), WET or damp (0.0159)
Special condition at site	Roadworks (0.4129), non-functional traffic signal (0.2684), none (0.1292), mud (0.1075), road surface defective (0.0405), oil or diesel (0.0281), road sign or marking defective or obscured (0.0133)
Carriageway hazards	Pedestrian on road (0.4306), vehicle load on road (0.3202), other object on road (0.1134), animal on road (0.0586), none (0.0586), previous accident (0.0185)
Road location	Rural (0.6667), urban (0.3333)
Ego vehicle maneuver	Slowing or stopping (0.1723), waiting to go—held up (0.168), waiting to turn left (0.1042), going ahead left-hand bend (0.087), going ahead other (0.0701), overtaking moving vehicle—offside (0.0692) moving off (0.0601), going ahead right-hand bend (0.0538), turning right (0.0455), waiting to turn right (0.0353), reversing (0.0301), changing lane to left (0.0296), turning left (0.0197), overtaking static vehicle—offside (0.0177), U-turn (0.0153), changing lane to right (0.0101), overtaking—nearside (0.006), parked (0.006)
Other vehicle (traffic) maneuver	Going ahead other (0.1619), slowing or stopping (0.1601), waiting to go—held up (0.1523), waiting to turn left (0.0922), moving off (0.067), going ahead right-hand bend (0.0618) parked (0.0618), turning left (0.0398), changing lane to left (0.0372), going ahead left-hand bend (0.0356), waiting to turn right (0.0286), reversing (0.027), overtaking moving vehicle—offside (0.0174), overtaking—nearside (0.0168), changing lane to right (0.0165), overtaking static vehicle—offside (0.0098), turning right (0.0086), U-turn (0.0057)
Street lighting	Lights lit (0.6667), no lighting (0.3333)
Ambient lighting	Darkness (0.5), daylight (0.5)
Weather	Snowy (0.4437), fine (0.2607), rainy (0.2607), foggy (0.035)
High wind	High winds (0.8889), no high winds (0.1111)

factors based on the requirement to generate the scenarios along with corresponding test scenario relevancy score. We intend to graphically represent generated scenarios in MATLAB [24]; therefore, the features used to generate scenario, given in Table 2, are selected on the basis of possibility in modeling the scenarios in MATLAB.

After considering the exclusion criteria, the overall test scenarios are generated, a sample of which is represented in Table 4.

Table 2 Features used to generate test scenario

Feature	Feature values
Road type	Roundabout, single carriageway (SC), dual carriageway (DC)
Road speed limit	60, 70, 40, 50
Junction detail	No junction (NJ), junction with more than 4 arms (J4), T-junction (TJ), Roundabout (RA)
Road surface condition	Snow (SN), frost or ice (FI), mud, dry, oil or diesel (O), flood over 3 cm. deep (F3D), wet or damp (W)
Special condition at site	Roadworks (RW), non-functional traffic signal (SNT), none, mud, road surface defective, oil or diesel, road sign or marking defective or obscured (SDR)
Carriageway hazards	Previous accident (CHA), none, animal on road (CH), other object on road (CHO), vehicle load on road, pedestrian on road (CHP)
Ego vehicle maneuver	Slowing or stopping (ESS), overtaking moving vehicle—offside (EOO), going ahead other (EGA), going ahead left-hand bend, Waiting to turn left (EWTL), waiting to go—held up (EWH), moving off, going ahead right-hand bend, turning right (ETR), waiting to turn right (EWTR), reversing (ERV), changing lane to left (ECL), turning left, Overtaking static vehicle—offside (EOSV), U-turn (EUT), changing lane to right, overtaking—nearside, parked (EPK)
Other vehicle (traffic) maneuver	Turning right (OTR), overtaking moving vehicle—offside (OOMO), overtaking—nearside (OON), changing lane to right (OCR), overtaking static vehicle—offside (OOSO), U-turn (OUT), reversing (ORV), waiting to turn right (OWTR), going ahead left-hand bend (OGLB), changing lane to left (OCL), turning left (OTL), parked (OPK), going ahead right-hand bend (OGRB), moving off, waiting to turn left (OWTL), waiting to go—held up (OWH), slowing or stopping (OSS), going ahead other (OGA)
Street lighting	Lights lit (LL), no lighting (LN)
Ambient lighting	Darkness (DN), daylight (DL)
Weather	Snowy (S), fine (F), rainy (R), foggy (FO)
High wind	High winds (WH), no high winds (WN)

Table 3 Exclusion Conditions

Ego vehicle maneuver/other vehicle (traffic) maneuver	Ego vehicle maneuver/other vehicle (traffic) maneuver	Road surface conditions/ weather
ESS/OPK	EOO/OWH	SN/F
ESS/OGA	EOO/OWTL	FI/F
EWH/OWH	EOO/OOMO	Dry/R
EWH/EWTL	EOO/OWTR	**Road type/junction detail**
EWH/OWTR	EOO/OOSO	RA/NJ
EWH/OPK	EOO/OPK	RA/J4
EWTL/OWH	EWTR/OWH	RA/TJ
EWTL/OWTL	EWTR/OWTL	RA/PDF
EWTL/OGA	EWTR/OWTR	**Junction detail/junction control**
EWTL/OOMO	ERV/OGA	RA/NJ
EWTL/Moving off	ERV/OOSO	NJ/uncontrolled
EWTL/OGRB	ERV/OGLB	NJ/authorized person
EWTL/turning right	EOSV/slowing or stopping	J4/NJ
EWTL/OWTR	EOSV/OGLB	TJ/NJ
EWTL/OCR	EOSV/OGA	
EWTL/OON	EOSV/OOMO	
EWTL/OPK	EOSV/moving off	
EGA/OWH	EOSV/OGRB	
EGA/OWTL	EOSV/turning right	
EGA/OWTR	EOSV/OOSO	
EPK/OWH	EOSV/U-turn	
EPK/OWTL	EOSV/changing lane to right	
EPK/moving off	EOSV/OON	
EPK/OWTR		
EPK/OPK		

6 Application and Analysis

The algorithm to generate priority-based test case, using an accident dataset, is implemented using Python and MATLAB. We developed a MATLAB-based framework, using the Automated Driving Tool box, to automatically parse through the generated scenarios and create graphical scenarios. The framework is designed to adapt any MATLAB model of an ego vehicle and run the scenarios against it. Entire framework runs on a Python engine. The workflow is presented in Fig. 1. The MATLAB code to create graphical scenarios is automatically generated from the Python-based framework. Thus, we can actually generate and run graphical scenario on a vehicle

Table 4 Test scenarios with relevancy scores

Test scenarios	1	2	3	4	5	6	7	8
Road type	SC	RA	SC	SC	SC	SC	DC	RA
Road speed limit	40	60	40	40	40	40	50	30
Junction detail	TJ	RA	TJ	TJ	J4	TJ	J4	RA
Ambient lighting	DL	DN	DL	DL	DL	DL	DN	DL
Street lighting	LN	LL	LN	LN	LL	LN	LL	LN
Weather	FI	SN	FI	FI	FI	SN	FI	RA
High wind	WH	WH	WH	WH	WH	WH	WN	WN
Road surface condition	Dry	Snow	Dry	Dry	O	Snow	O	W
Sp. condition at site	None	SDR	RW	RW	RW	RW	SNT	None
Carriageway hazards	CHP	CHP	CHO	CHO	CHP	CH	CHA	None
Ego vehicle maneuver	ESS	EUT	ETR	ETR	EGA	ESS	ECL	EOO
Other vehicle (traffic) maneuver	OSS	OUT	OTL	ORV	ORV	OCL	OCR	OTR
Test relevancy score	3.28	4.57	3	2.99	3.65	3.51	2.08	2.48

algorithm with a single click. MATLAB-based scenario templates are developed and utilized to generate graphical test scenarios based on the generated test cases. Templates are basic building blocks to construct graphical test scenarios; templates include atomic units to create roads, traffic and environment. A functional interface, developed using Python, takes the deconstructed test scenarios as parameters to these templates to generate graphical interface for the generated scenarios.

Following few are some of the examples, Figs. 2, 3, and 4 of generated scenarios in graphical form. We worked with a constant velocity vehicle model to graphically generate our scenarios with two cars. The blue car represents the ego vehicle.

The framework is designed with open application programming interfaces which can be connected with other scenario rendering software such as dSpace [25] and IPG-Carmaker [26] and can be improved using deep learning techniques, as in [27].

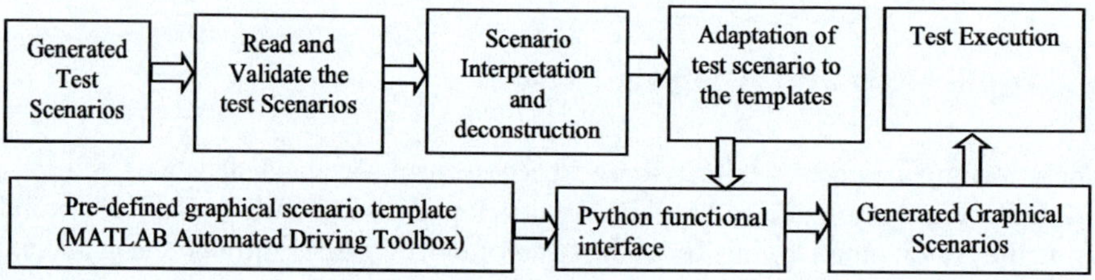

Fig. 1 Python framework to generate graphical scenarios

Fig. 2 One vehicle is crossing another: **a** vehicle approaching ego vehicle **b** vehicle passing the ego vehicle **c** vehicle successfully crosses the ego vehicle

Fig. 3 Vehicle moving ahead suddenly stops **a** both vehicles moving forward **b** vehicle in front stops in line of travel of the ego vehicle **c** possible collision with the stopped vehicle

Fig. 4 T-junction with a cyclist approaching **a** ego car moving on the carriageway **b** ego car approaching T-junction, **c** cyclist moves in front of car resulting in possible collision

7 Conclusions

A novel, data-driven scenario generation process for autonomous driving tasks, using a combination of a statistical framework, WoE and IV, along with an analytical framework, AHP, is proposed in this study. Results and corresponding analysis indicated that by considering and including the criticality factor in the scenario generation process, the overall number of generated test scenarios reduced significantly. Additionally, when higher degree of criticality is considered as a threshold, the number of scenarios generated consisted of factors carrying critical information with respect to the ODD. Scenario generation with criticality aspect not only generated a reduced set of scenarios but also scenarios which are important to the given ODD. Results and corresponding analysis indicate that by setting a minimum threshold to generate the scenario, the algorithm is forced to consider the factors with maximum contribution to the ODD, thus, generating high-priority test scenarios. Varying the priority threshold can also be used as a method to limit the number of test scenarios generated.

References

1. Tumas P, Nowosielski A, Serackis A (2020) Pedestrian detection in severe weather conditions. IEEE Access 8:62775–62784. https://doi.org/10.1109/ACCESS.2020.2982539
2. Chebrolu KNR, Kumar PN (2019) Deep learning based pedestrian detection at all light conditions. In: 2019 International conference on communication and signal processing (ICCSP). IEEE, Melmaruvathur, pp 838–842
3. Dunna S, Nair BB, Panda MK (2021) A deep learning based system for fast detection of obstacles using rear-view camera under parking scenarios. In: 2021 IEEE international power and renewable energy conference (IPRECON 2021). Kollam, pp 3–9. https://doi.org/10.1109/IPRECON52453.2021.9640804
4. Emani S, Soman KP, Sajith Variyar VV, Adarsh S (2019) Obstacle detection and distance estimation for autonomous electric vehicle using stereo vision and DNN. In: International conference on soft computing and signal processing. IEEE, Hyderabad, pp. 639–648
5. Bian Y, Ding J, Hu M, Xu Q, Wang J, Li K (2020) An advanced lane-keeping assistance system with switchable assistance modes. IEEE Trans Intell Transp Syst 21(1):385–396. https://doi.org/10.1109/TITS.2019.2892533
6. Andrade DC et al (2019) A novel strategy for road lane detection and tracking based on a vehicle's forward monocular camera. IEEE Trans Intell Transp Syst 20(4):1497–1507. https://doi.org/10.1109/TITS.2018.2856361
7. Savant KV, Meghana G, Potnuru G, Bhavana V (2022) Lane detection for autonomous cars using neural networks. In: Machine learning and autonomous systems. Springer, Singapore, pp 193–207
8. Misal S, Nair BB (2018) A machine learning based approach to driver drowsiness detection. In: International conference on information, communication and computing technology. Springer, Singapore, pp 150–159. https://doi.org/10.1007/978-981-13-5992-7_13
9. Billah T, Rahman SMM, Ahmad MO, Swamy MNS (2019) Recognizing distractions for assistive driving by tracking body parts. IEEE Trans Circuits Syst Video Technol 29(4):1048–1062. https://doi.org/10.1109/TCSVT.2018.2818407
10. Society of Automotive Engineers: Surface Vehicle Recommended Practice: J3016-Taxonomy and Definitions for Terms Related to Driving Automation Systems for On-Road Motor Vehicles, https://www.sae.org/standards/content/j3016_202104. Last Accessed 11 May 2022
11. Kalra N, Paddock SM (2016) Driving to safety: how many miles of driving would it take to demonstrate autonomous vehicle reliability? Transp Res Part A Policy Pract 94:182–193
12. Helmer T, Wang L, Kompass K, Kates R (2015) Safety performance assessment of assisted and automated driving by virtual experiments: Stochastic microscopic traffic simulation as knowledge synthesis. In: IEEE 18th international conference on intelligent transportation systems, IEEE, Gran Canaria, pp 2019–2023
13. Kim B, Masuda T, Shiraishi S (2019) Test specification and generation for connected and autonomous vehicle in virtual environments. ACM Trans Cyber Phys Syst 4(1):1–26
14. Nalic D, Pandurevic A, Eichberger A, Fellendorf M, Rogic B (2021) Software framework for testing of automated driving systems in the traffic environment of vissim. Energies 14(11):3135
15. Alnaser AJ, Akbas MI, Sargolzaei A, Razdan R (2019) Autonomous vehicles scenario testing framework and model of computation. SAE Int J Connect Autom Veh 2(4):205–218
16. UN Economic Commission for Europe: Proposal for a new UN Regulation on uniform provisions concerning the approval of vehicles with regards to Automated Lane Keeping System, https://unece.org/DAM/trans/doc/2020/wp29grva/GRVA-05-07r3e.pdf. Last Accessed 22 May 2022; 11 May 2022
17. UK Department for Transport: Road Safety Data. https://data.gov.uk/dataset/cb7ae6f0-4be6-4935-9277-47e5ce24a11f/road-safety-data. Last Accessed 23 June 2022
18. Ulbrich S, Menzel T, Reschka A, Schuldt F, Maurer M (2015) Defining and substantiating the terms scene, situation, and scenario for automated driving. In: IEEE 18th international conference on intelligent transportation systems. IEEE, Gran Canaria, pp 982–988

19. Menzel T, Bagschik G, Maurer M (2018) Scenarios for development, test and validation of automated vehicles. In: 2018 IEEE intelligent vehicles symposium (IV). IEEE, Changshu, pp 1821–1827
20. Prokhorenkova L, Gusev G, Vorobev A, Dorogush AV, Gulin A (2018) CatBoost: unbiased boosting with categorical features. Adv Neural Inf Process Syst Montreal 31
21. Weed DL (2005) Weight of evidence: a review of concept and methods. Risk Anal An Int J 25(6):1545–1557
22. Gough D (2007) Weight of evidence: a framework for the appraisal of the quality and relevance of evidence. Res Pap Educ 22(2):213–228
23. Saaty TL (2008) Decision making with the analytic hierarchy process. Int J Serv Sci 1(1):83–98
24. MathWorks Inc.: MATLAB Overview, https://in.mathworks.com/help/matlab/. Last Accessed 05 June 2022
25. dSpace: AURELION: Sensor-Realistic Simulation, https://www.dspace.com/en/inc/home/products/sw/experimentandvisualization/aurelion_sensor-realistic_sim.cfm. Last Accessed 02 Feb 2022
26. IPG-Automotive: CarMaker, https://ipg-automotive.com/en/products-solutions/software/carmaker. Last Accessed 02 Jul 2022
27. Soni RK, Nair BB (2021) Deep learning based approach to generate realistic data for ADAS applications. In: 5th International conference on computer, communication, and signal processing (ICCCSP2021), IEEE, Chennai, pp 181–185. https://doi.org/10.1109/ICCCSP52374.2021.9465529

Design of Robotic Platform for ADAS Testing

Arun Goyal, Shital S. Chiddarwar, and Aditya A. Bastapure

Abstract Car safety systems continually improve, prompting automakers to imagine a world without vehicle crashes. ADAS is one of the fast-evolving car safety systems. ADAS uses various sensor technologies to detect the environment around the vehicle and then deliver information to the driver or take action as needed. Finding a car manufacturer that is not investing heavily in ADAS technology is not easy. Traditional testing tools cannot keep up with the ADAS system. In response, new tools must be developed. A robotic platform with a height of less than 100 mm can be overrun by a testing vehicle which is one solution to avoid damaging the car. In this work, a thorough examination of existing models and requirements is carried out to develop a robotic platform. The dimensional constraint makes component sizing and selection difficult. Before being chosen and designed, each component is thoroughly examined and selected to ensure that the venue meets its requirements. Next, a CAD model is created, and an analysis is performed to validate the design. Because ADAS testing adheres to industry standards such as EuroNCAP, the robotic platform is designed to perform testing scenarios based on EuroNCAP. The framework presented in this paper shows how to design various components of the robotic platform and controller to give the robot a more precise trajectory. In addition, a vehicle safety verification and validation tool have been developed due to this method.

Keywords ADAS · V2V · Robotic platform · EuroNCAP

A. Goyal (✉) · A. A. Bastapure
Visvesvaraya National Institute of Technology, Nagpur, India
e-mail: garun2810@gmail.com

S. S. Chiddarwar
Mechanical Engineering Department, V.N.I.T, Nagpur, India

S. Sharma et al. (eds.), *Intelligent Control, Robotics, and Industrial Automation*,
Lecture Notes in Electrical Engineering 1066,
https://doi.org/10.1007/978-981-99-4634-1_7

1 Introduction

ADAS can significantly reduce the number of automotive accidents and deaths caused by human error, which is the foundation for the research and development of autonomous driving technology. Its evaluation has become one of the automotive field's research hotspots [1]. The mobile test platform vehicle can be outfitted with dummies and simulated fake cars to replicate the motion conditions of vehicles and pedestrians, among other things. While some countries have created somewhat mature test platforms, such as the full-size soft impact target vehicle produced by DRI in the USA [2], these are costly, with a long failure maintenance cycle and a high cost. As a result, independent research and the development of low-cost mobile platforms are needed.

Thanks to its unique design, the height of the platform on the robot is lower than the height of the ground plate on the testing vehicle harm which is done to the testing vehicle or the platform itself, even in the event of a collision. They are also encircled by ramps, allowing the testing vehicle to easily cross the platform even at high speeds.

Park et al. [3] designed a system to evaluate the active safety of advanced driver-assistance vehicles. To evaluate the active safety function, an unmanned robot with a height of 100 mm that was concealed beneath a target dummy car could construct vehicular pathways using a differential global positioning system. Rear-end collision tests were used to evaluate the system's performance. The target was an unmanned robot. Actual vehicle testing was done to validate the system using the Euro NCAP-established collision scenarios.

Steffan et al. [4] conducted experiments with an Ultra flat Overrunable Robot (UFO) for ADAS testing. UFO, managed by GPS, was designed such that their height was less than the ground plate of the vehicle under test. The researchers' ultimate aim was to create a robotic platform to help recreate accident scenarios. This automated platform had a driving motor system, navigation, communication, and control. They tested the platform and concluded that it was a better option for accident re-creation. Furthermore, it also suggested that this platform can be used for testing autonomous functions [5]. A few also worked on suspension system design for robotic platforms. One such article describes an automotive testing framework for ADAS reliability and stability. Its unique suspension was tested, tested, and perfected in actual automobiles under real-world conditions.

Before the actual design of the proposed robotic platform, it is essential to define design parameters. Design parameters are specific values of the features we intend to incorporate in any final design. This platform should be designed to test a vehicle of about 4500 kg, and all calculations should be performed accordingly.

For this platform, the maximum height must be less than 100 mm and maintain a ground clearance of less than 10 mm. Since this platform will be considered for vehicle testing, it is expected to have an area of 1800 * 4000 mm^2 to carry a dummy vehicle with a maximum mass of 110 kg. Additionally, this platform should reach a top speed of 80 kmph with an acceleration of 2.5 m/s^2, lateral acceleration between 0.4 and 0.5 times gravity's acceleration, and deceleration of 6 m/s^2. In addition, the design must be done such that the turning radius falls under 10 m. As far as operating conditions are concerned, it is supposed to be working within 0–50°, designed to be waterproof, and dust and mud resistant. A 4-disc hydraulic system is to be considered for the braking system.

Once designed, the platform is supposed to undergo a minimum of 20 tests per day, with the tests being EuroNCAP and Global vehicle target standards. Furthermore, the user/test authority should be flexible in creating and adding new scenarios while also being able to modify them. Moreover, if the platform is to be tested, then a remote operation with a maximum speed of up to 20 kmph is desirable. While being tested, the forum must send its health data (motor, battery, brake temperature) to the base station for continuous monitoring and diagnostics. WLAN can be used for this platform to communicate with the base station/test authority. In addition, the platform must have Bluetooth connectivity for remote data access and control. Finally, a GPS-based navigation system is mounted on the platform to locate itself on the world map.

2 Components of the Robot

Referring to prerequisites from the introduction, correct components must be selected that stand true to requirements. For any robotic platform to work, few essential elements are motors, wheels, suspension systems, positioning systems, control systems, battery systems with battery management systems (BMS), and a few complimentary systems. These systems are discussed in detail in this section. Figure 1 highlights all the vital strategies of the proposed robotic platform.

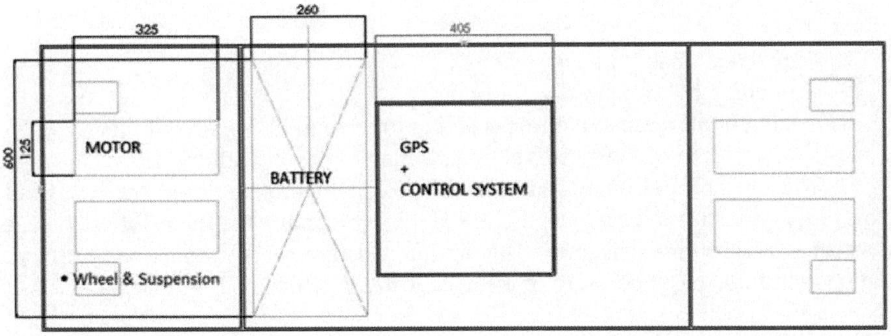

Fig. 1 Component positioning

3 Motor Selection

The electric motor is being the main driver of this robotic platform, and its selection becomes vital. The critical task is to select the appropriate engine based on the load it will carry. Cost, weight, and efficiency are key parameters affecting motor selection. The approach followed for the preference of the motor.

Vehicle dynamics such as rolling resistance, gradient resistance, aerodynamic drag, and so on must be considered when determining a vehicle's power rating. Below are sample calculations for determining a motor rating for an electric automobile with a gross weight of 240 kg.

Calculations

Given,

Payload = 120 Kg.

Velocity, $v = 80\frac{\text{Km}}{\text{h}} = 22.23\frac{\text{m}}{\text{s}}$.

Acceleration of robotic platform $a = 2.5\,\text{m/s}^2$.

Radius of wheel, $r = 50\,\text{mm} = 0.05\,\text{m}$.

Number of motors, $n = 2$.

Assuming,

Robot weight = 120 kg.

Frontal area, $A_\text{f} = 2.5\,\text{m}^2$.

Newton's second law of mechanics states that the dynamical movement of a vehicle in one coordinate axis is dictated by the sum of all forces acting on it in that same axis of direction, as expressed in the translational form.

$$\frac{\text{d}V}{\text{d}t} = \frac{\Sigma F_\text{t} - \Sigma F_\text{resistance}}{M}, \tag{1}$$

where

M—equivalent mass to be accelerated.

$\frac{\text{d}V}{\text{d}t}$—robotic platform speed $v(t)$ (m/s).

ΣF_t—sum of all the tractive forces acting to increase the platform speed.

$\Sigma F_\text{resistance}$—sum of the resistive forces acting to decrease the speed.

The vehicle's propulsion unit provides the force required to propel the car ahead. This propulsion unit force assists the vehicle in overcoming the forces of gravity, air resistance, and rolling resistance. The driving resistance forces combine to provide the traction force (F_t) necessary at the drive wheels, which is described as:

$$\Sigma F_\text{resistance} = F_\text{r} + F_\text{a} + F_\text{g}, \tag{2}$$

where

F_r—rolling resistance force,

F_g—grade resistance force.

F_a—aerodynamic drag.

$$\text{Power}_{\text{per motor}} = \frac{18.0855}{2} = 9.042753\,\text{KW}. \tag{3}$$

From Eq. (3) of the power calculation, the power requirement for the motor comes out to be 9 KW to move our platform at a speed of 80 kmph.

Many different types of motors are available in the market for robotic applications. The permanent magnet synchronous motor (PMSM) is more efficient than brushless DC motors. PMSM displays no torque ripple when the motor is commutated. It has more torque motors and, as a result, performs better. PMSM is more suitable for our application due to its efficient heat dissipation, lower noise, and smaller size. As a result, PMSM is appropriate for this robotic platform architecture.

4 Battery

Lead-acid, lithium-ion, nickel–cadmium, and nickel–metal hydride batteries are all suitable for robotic applications. One of the most important factors to consider when modelling a battery is battery type selection. This section discusses the essential characteristics of each battery type. The platform's first consideration, however, is how long it can drive on the battery.

As a result, the battery's weight and energy density should come first. A minimum of two Li-Ion batteries with capacities greater than 3 KWh are required for the system to function. Compared to other battery types, a Li-Ion battery can meet the power demands of various subsystems and drive motors. The calculations predict a total power consumption of 3 kW. As the platform is supposed to be within the height of 100 mm, a custom battery pack is suggested.

5 Suspension System

Types of Suspension Systems

Suspension systems come in various configurations, the most common of which is a spring and damper setup. The spring and damper setup are popular due to their small size and easy, quick, and direct tune-ability.

5.1 Push Rod Cantilever Suspension

In a double wishbone suspension system, a rod is used in place of the damper. Because of this, the muffler and spring are mounted inside the chassis and are managed by a bell crank that is activated by a push rod attached to the hub. The robotic

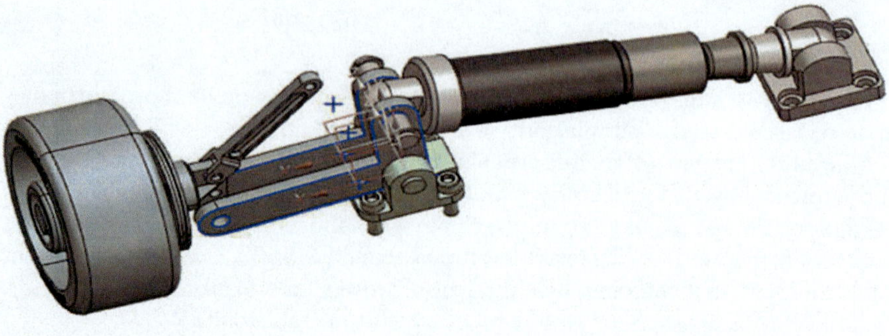

Fig. 2 Front suspension

platform design's plush-riding, a space-saving solution, is a push rod suspension in the cantilever style. The ratio of the stroke distance of the suspension and shock absorber is set to approximately 1:2, as shown in Fig. 2.

CAD model as illustrated in Fig. 3, Suspension assembly CAD model 3 primarily consists of five components. The main components are the wheel, shock absorber, clevis mount, and bell crank lever.

6 Steering Mechanism

6.1 Ackermann Steering

One of the fundamental requirements of a car's steering system is to give the steerable wheels correlated turning, making sure that the junction point of their axes lies on the rear wheel axis (point Q in Fig. 4b), which is analogous to the following relation:

$$\theta_{\mathrm{OA}}(\theta_1) = \operatorname{artan}\frac{1}{\cot(\theta_1) + {}^{1}\!/\!\left(\frac{W_b}{W_t}\right)}, \tag{4}$$

W_b and W_t are the vehicle's wheelbases and wheel track, and θ_{OA} is the ideal turning angle of the outer wheel where θ_1 is the turning angle of the inner rotation.

With the actuator piston attached, a motion analysis was done, and a displacement vs. time graph was created. Real mounting sites were analysed for the investigation as shown in Fig. 4a, and the study indicated the maximum displacement and maximum velocity (see Fig. 5) for full steering. Using the presented results' velocity, the linear actuator is selected based on stroke length and speed.

After analysing various mechanisms and motion analysis result shown in Fig. 6, a four-bar Ackermann steering is proposed. It did adhere to the principle, and as

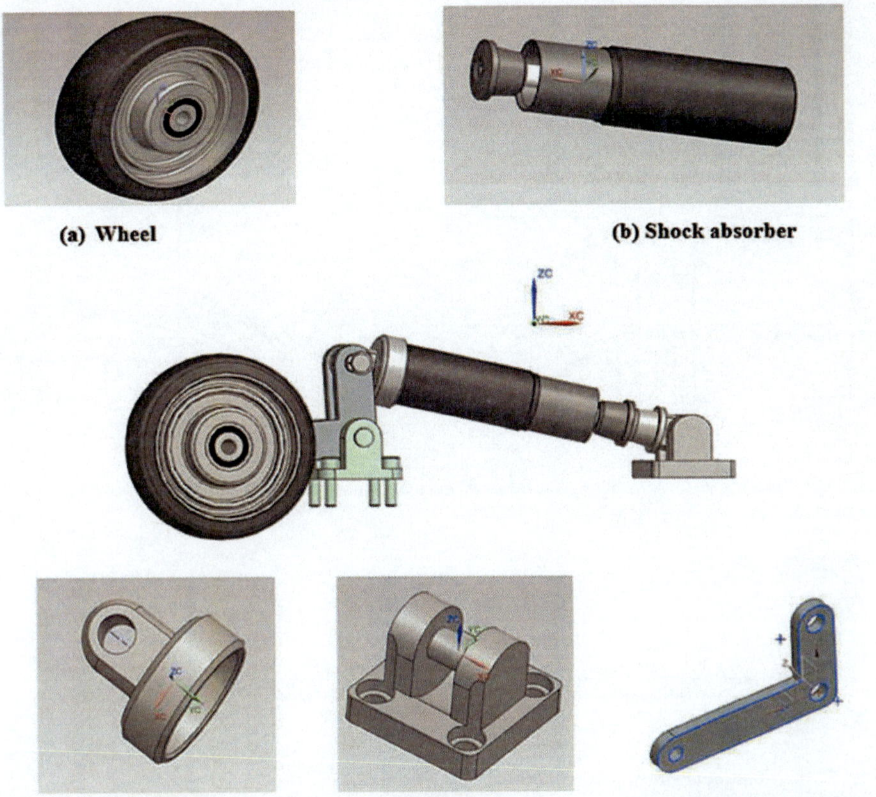

(a) Wheel **(b) Shock absorber**

(c) Clevis Mount **(d) Hinge** **(e) Bell crank lever**

Fig. 3 Rear suspension assembly

a result, a standard industrial shock absorber TA17 was chosen from the steering motion analysis results.

7 Body Frame/Chassis Design

By incorporating frame structures between the components, the body frame is designed to withstand the load of the hunter vehicle. To reduce overall weight, an arch-shaped construction is used, as shown in Fig. 7.

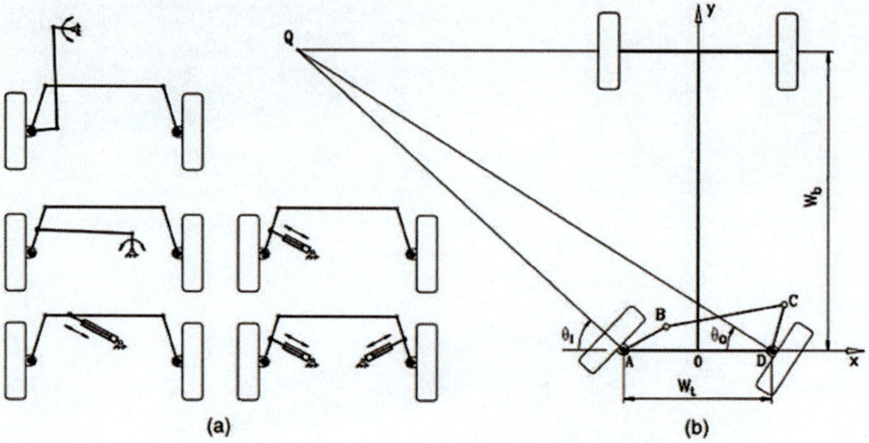

Fig. 4 **a** Steering control variations used in Ackermann linkages, **b** illustration of a four-wheeled vehicle turning

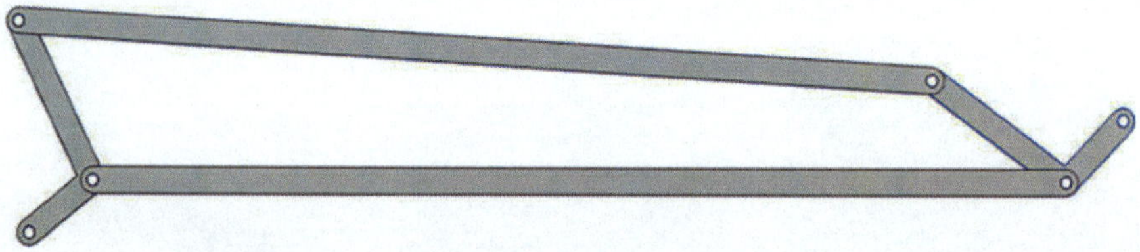

Fig. 5 Four-bar mechanism

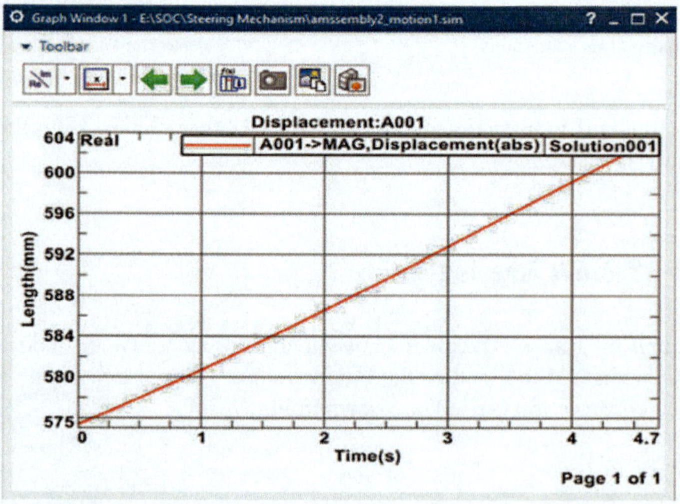

Fig. 6 Ackermann steering graph

Fig. 7 Chassis CAD model

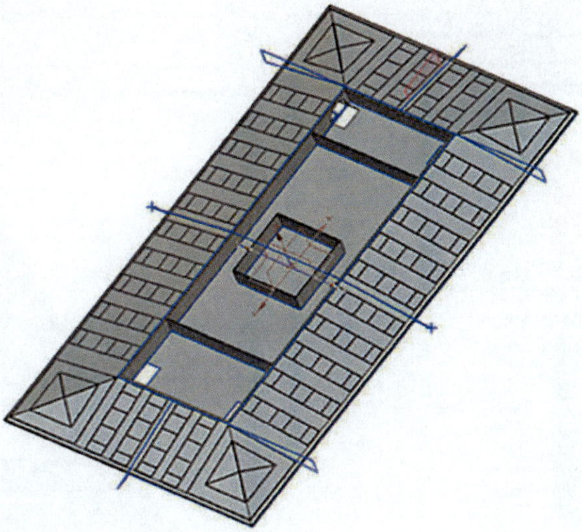

8 Result and Discussion

8.1 *Static Analysis*

After examining the chassis structure in Simcenter, the platform's maximum deformation was 0.323 mm, as shown in Fig. 8. The severity of deformation is depicted in various hues in the same graphic. The design was deemed safe to use because the most significant distortion was 0.323 mm, which was less than the allowable limit. The addition of components could explain why distortion system stiffness increases with the number of components, resulting in less deformation and stress, as shown in Fig. 9.

8.2 *Dynamic Analysis*

The maximum deflection recorded in the detailed dynamic analysis, as shown in Fig. 10, is 0.004 mm, significantly less than the static and motion analysis deflections. A reduced deflection validates the chassis design against a 1000 kg per tyre load.

The central structural part, i.e. the chassis, is designed in SiemensNX; both static and dynamic analyses were done, ensuring the design. Deformation values were observed to be within permissible bounds. As a high safety value was chosen, it is suggested that optimization can be done to lower the platform's weight.

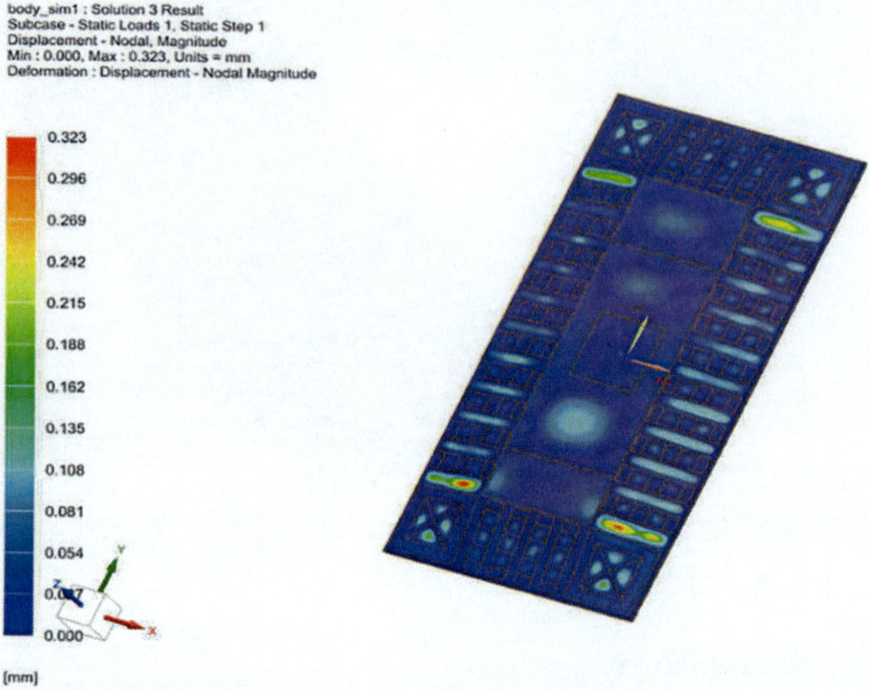

Fig. 8 Deformation analysis result

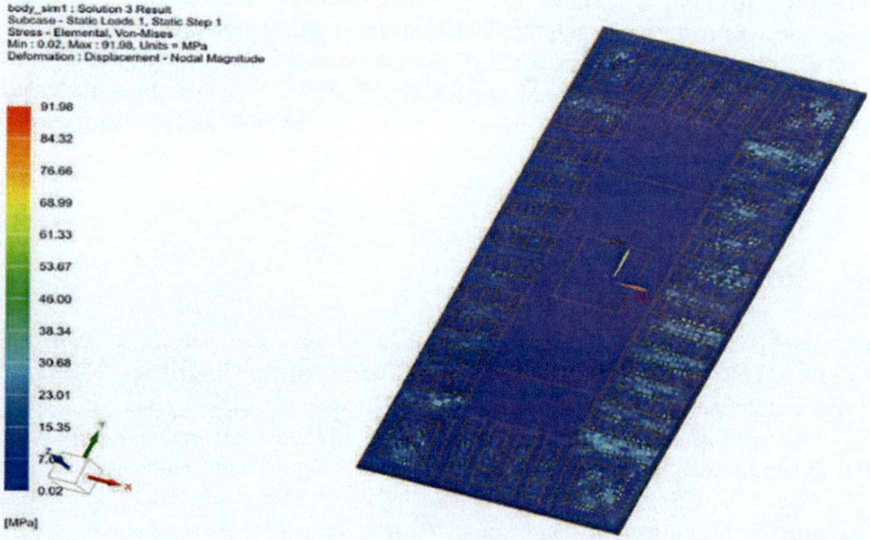

Fig. 9 Stress analysis result

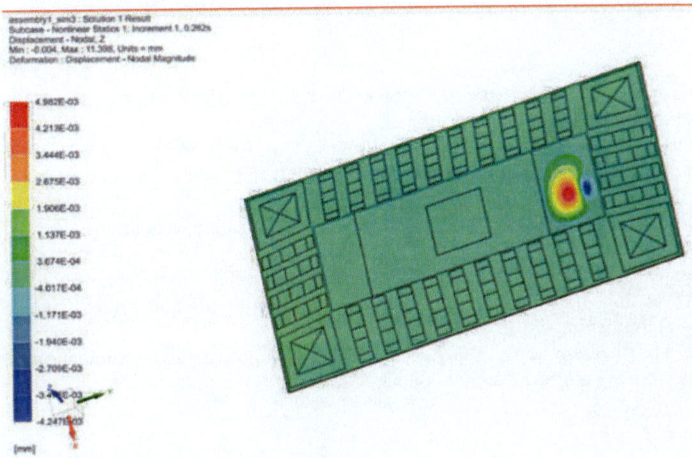

Fig. 10 Dynamic explicit analysis result

9 Conclusion

Robotic platforms have become increasingly crucial for vehicle safety testing. This work presents a design for an automated platform that meets ADAS testing requirements for four-wheel vehicles with a maximum speed of 80 kmph. Using engineering knowledge and various literature reviews, calculations are performed to select a 9KW PMSM motor and 3 KWh lithium-ion batteries to give the required power and energy to drive this automated platform at a speed of 80 kmph to function in various test cases without interruption. The suspension system is designed using a pushrod cantilever type of suspension in mind to get fitted in a given small dimension easily and to pass the forces to the chassis without affecting the transmission system. After analysing various mechanisms and motion analysis, a four-bar Ackermann steering is proposed. It did adhere to the principle, and as a result, a standard industrial shock absorber TA17 was chosen from the steering motion analysis results.

This automated robotic platform can handle force transferred by a static or moving car while taking safety into account. It is based on structural analysis of a 3D-created chassis, chassis dimensioning, and material selection (Aluminium 6061 alloy). Hence, the designed robot is suitable for performing various ADAS tests to validate vehicle safety ratings. Thus, the creation of the robotic platform is designed and validated using simulation tools from a mechanical point of view.

References

1. Kala R (2016) 4—Advanced driver assistance systems. On-Road Intell Veh. https://doi.org/10.1016/B978-0-12-803729-4.00004-0
2. Albrecht H, Barickman FS, Schnelle SC (2021) Advanced test tools for ADAS and ADS. United States. Department of Transportation. National Highway Traffic Safety Administration. Vehicle Research and Test Center
3. Park Y, Lee S, Park M, Shin J, Jeong J (2019) Target robot for active safety evaluation of ADAS vehicles. J Mech Sci Technol 33(9):4431–4438
4. Steffan H, Ellersdorfer C, Moser A, Simader J (2017) UFO: ultraflat overrunable robot for experimental ADAS testing. In: Automated driving: safer and more efficient future driving. Springer International Publishing
5. Li L, Xu L, Cui H, Abdelkareem MAA, Liu Z, Chen J (2021) Validation and optimization of suspension design for testing platform vehicle. Shock Vibrat

A Comparative Study of A*, RRT, and RRT* Algorithm for Path Planning in 2D Warehouse Configuration Space

Aditya A. Bastapure, Shital S. Chiddarwar, and Arun Goyal

Abstract Mobile robots are now being used in various interior and outdoor applications. The demands of Industry 4.0 are pushing robotic domains in the direction of decision-making. The autonomous robots should choose their direction based on their surroundings. Path planning for a mobile robot involves generation of a collision-free path from a current state to desired location while meeting particular optimization criteria through the surrounding, which includes obstacles. In this context, reliable and efficient path planning is critical. According to a thorough literature review, it was found that the A* and Rapidly Exploring Random Tree (RRT) algorithms, as well as their derivatives, are the most viable path planning algorithms for mobile robots. This paper presents a comparative study of A*, RRT, and RRT* path planning algorithms in a 2D warehouse environment. The efficiency of planners was correlated with the varied attributes of configuration spaces using several warehouse configuration spaces. The path length, runtime, and number of nodes in the path were among the performance measures. Based on the results obtained, a best-suited algorithm was selected for warehouse robot path planning.

Keywords A* · RRT · RRT*

1 Introduction

Giving mobile robots' autonomy is advantageous, and it eliminates the need for human operators, which may be favorable from both economic and safety perspectives. In most circumstances, autonomy necessitates path planners, which allow the robot to deliberate about how to go from one area to another. Mobile robots, unmanned aerial vehicles, and autonomous vehicles use path planning algorithms to

A. A. Bastapure (✉) · S. S. Chiddarwar · A. Goyal
Mechanical Engineering Department, Visvesvaraya National Institute of Technology, Nagpur, India
e-mail: aabastapure@gmail.com

S. Sharma et al. (eds.), *Intelligent Control, Robotics, and Industrial Automation*, Lecture Notes in Electrical Engineering 1066,
https://doi.org/10.1007/978-981-99-4634-1_8

find safe, efficient, collision-free, and low-cost travel paths from one point to another. Multiple literatures were studied to understand algorithms available.

Gonzalez et al. [1] presented a study in context with path and motion planning techniques applied to intelligent vehicles. Different researchers and their works were also discussed. Path planning entails determining a continuous path that will take the robot from its initial state to its final location. For doing so, algorithms like A*, Dijkstra, potential field are more common. For a single source if path is planned that goes through nodes, it is Dijkstra algorithm. A slight modification to Dijkstra is A* [2], 3. It uses heuristic function and gives priority to nodes which are better, as opposed to Dijkstra, which explores all possible nodes. These fall under Graph search-based planner family. Graph search method usually converts workspace into distinct pieces, which degrades the performance.

Md. Elbanhawi et al. [4] has explained in detail Sampling-Based Algorithms like PRM [5], RRT, and their derivative. According to a thorough literature review [6], 7, it was found that the A* and Rapidly Exploring Random Tree (RRT) algorithms, as well as their derivatives, are the most viable path planning algorithms for mobile robots.

The first algorithm is A* [3], which leverages the idea behind algorithm A* to show how a mobile robot may be mapped from one location to another using a Graphical User Interface (GUI). LaValle [8] explains that RRT is a sampling-based approach that iteratively expands by giving appropriate control parameters. The RRT builds a tree structure using numerous branches and nodes randomly selected from unoccupied space. LaValle applied RRT algorithm for problems compromising high DOF and constraints that are not holonomic. RRTs have Probabilistic Completeness, i.e., if given sufficient algorithm runtime, probability of finding the solution is one. It provides sub-optimal solutions.

A good comparison between RRT and RRT* is presented in work published by Noreen et al. [9]. These were compared based on path cost, run time by performing simulation experiments. RRT* improves initial path compared to RRT, but results in more execution time compared to simple RRT. RRT*, a variant of RRT with demonstrated asymptotically optimum properties, was developed by Frazzoli [10]. By incorporating the key components of best neighbor search and tree rewiring, RRT* increased the quality of the paths. However, the execution time and slower rate of path convergence were paid for the asymptotic optimality.

This paper presents a comparative study of A*, RRT, and RRT* path planning algorithms in a 2D warehouse environment (Scenario 1 and Scenario 2). The efficiency of planners was correlated with the varied attributes of configuration spaces using several warehouse configuration spaces. The path length, runtime, number of nodes in the path were among the performance measures. Based on the results obtained, a best-suited algorithm was selected for warehouse robot path planning.

This work can be subdivided as follows. After Sect. 1of Introduction, A*, RRT, and RRT* algorithm overview is presented in Sect. 2. The setup and environment for the simulation are explained in depth in Sect. 3. For performance analysis, Sect. 4 compiles and compares the findings from A*, RRT, and RRT* planners Conclusion and future scope are discussed in Sect. 5.

2 Algorithm Overview

The below section gives an in-depth insight of A*, RRT, and RRT* algorithms.

2.1 A* Algorithm

A* is an informed search method that uses weighted graphs to identify a path from the starting to the goal point. It uses the most effective versions of both Dijkstra's and Best-First-Search algorithms. According to Dijkstra, it chooses the section that prefers the cells nearest to the starting point. Moreover, it selects the cells closest to the goal point according to the Best-First-Search algorithm.

The algorithm selects a node by considering an f value or cost value which is the sum of the values g and h. It chooses the node with the lowest f value or cost at each stage. The g value represents the expense of moving from the initial location on the grid cell to another location, based on the path calculated to get there. Conversely, the h value is a calculated cost to move from the location on the grid cell one wants to reach to the goal point. This is described as the heuristic value, which is simply an intelligent guess. The Manhattan distance, diagonal distance, and Euclidean distance are three methods that may be used to compute the h value. In this study, the Euclidean distance was employed.

The A* employs two types of lists: open and closed. A list of unexplored nodes is in the open set, whereas a list of already explored nodes is in the closed set. In this manner, as discussed above, the node with the least f value is deleted from the open set at each step, and its neighbors' f and g values are modified correspondingly, adding them to the open set. The node that was taken out of the open set is now a part of the close set. The process stops when either the open set is empty, or the goal node's f value is less than any other node in the open set.

3 RRT Algorithm

Notations:

Let the given state space be represented by a set $Q \subset R^n n \in N$, where n describes the dimension of the given search space. The region of the state space which is filled by obstacles is denoted by $Q_{\text{obs}} \subset Q$ as well region free from obstacles is represented by $Q_{\text{free}} = Q - Q_{\text{obs}}$. $q_{\text{goal}} \subset Q_{\text{free}}$ represents goal, and $q_{\text{init}} \subset Q_{\text{free}}$ represents starting point. V is a vertex, and E is an edge. A graph, Ts, compromises of a vertex set V and an edge set E. q_{init} and q_{goal} are given as input parameters to planner. The aim is to find an obstacle-free path amidst q_{init} and $q_{\text{goal.}}$

Using random sampling approach in state space, RRT creates a tree spanning from start point q_{init} to goal point q_{goal}. As process continues, the tree eventually

Fig. 1 RRT tree expansion process [11]

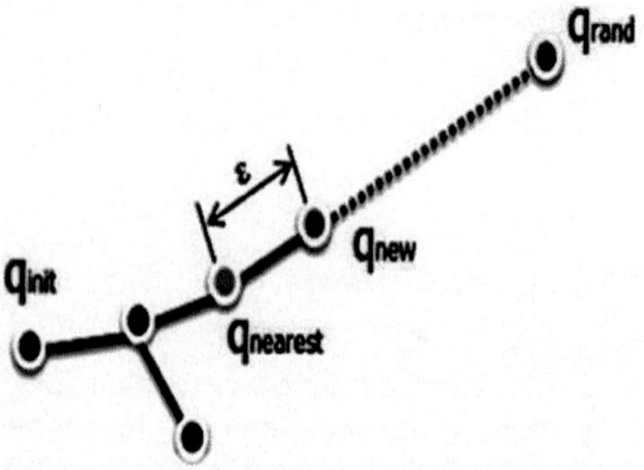

grows. From the state space (Q), a random state is chosen q_{rand} such that if it falls within obstacle-free space, a new nearest node is searched within tree. The tree then expands by establishing a connection between q_{rand} and $q_{nearest}$ providing that $q_{nearest}$ can access q_{rand} according to predefined step size. In a contradictory situation, a new node is selected with the aid of steering function. q_{new} connects with $q_{nearest}$, and thus, the tree grows. To assure obstacle-free connection, a Boolean test method is adopted amidst $q_{nearest}$ and q_{new}. The procedure continues until a certain time period has passed or a certain number of iterations have been completed. Figure 1 illustrates node expansion process and Algorithm 1 gives an idea of RRT algorithm.

A detail discussion of crucial functions is presented below.

Sample: Within the obstacle-free region (Q_{free}), a random position (q_{rand}) is generated by this function.

Nearest: Using cost function, this function helps in determining a node nearest to q_{rand}.

Steer: It calculates q_{new} at a distance ε from $q_{nearest}$ toward q_{rand}, where ε is the incremental length.

CollisionCheck: If the branch is in the obstacle-free region, CollisionCheck function returns a true value; otherwise, it returns a false value. If true, it adds the branch to Tree T (V, E), and if false, it removes it.

Near: Within the radius r of node q_{new}, it provides all the neighboring nodes in the tree.

InsertNode: This function connects node q_{min} as its parent by adding a node q_{new} to V in the tree $T = (V, E)$.

Rewire: This function checks if the cost to the nodes in q_{near} is less through q_{new} than through their previous costs, then it's parent is changed to q_{new}.

ChooseParent: From the neighboring nodes, this function chooses the optimal parent for q_{new}.

Algorithm 1. RRT algorithm [9]	**Algorithm 2**. RRT* algorithm [9]

T = (V, E) ← RRT (q_{init})
T ← InitializeTree();
T ← InsertNode(Φ, q_{init}, T);
for i = 0 to i = N **do**
q_{rand} ← Sample(i);
$q_{nearest}$ ← Nearest(T, q_{rand});
q_{new} ← Steer($q_{nearest}$, q_{rand});
if Obstaclefree($q_{nearest}$, q_{new}) **then**
T ← InsertNode($q_{nearest}$, q_{new}, T);
return T
	T = (V, E) ← RRT*(q_{init})
	T ← InitializeTree();
	T ← InsertNode(Φ, q_{init}, T);
	for i = 0 to i = N **do**
	q_{rand} ← Sample(i);
	$q_{nearest}$ ← Nearest(T, q_{rand});
	q_{new} ← Steer($q_{nearest}$, q_{rand});
	if Obstaclefree($q_{nearest}$, q_{new}) **then**
	q_{near} ← Near(T, q_{new}, |V|);
	q_{min} ← Chooseparent(q_{near}, $q_{nearest}$, q_{new});
	if Obstaclefree(q_{min}, q_{new}) **then**
	T ← InsertNode(q_{min}, q_{new}, T);
	T ← Rewire(T, q_{near}, q_{min}, q_{new});
	return T

4 RRT* Algorithm

RRT* retains all of RRT's attributes and performs analogously to RRT. It did, however, introduce two promising features: close neighbor search and rewiring tree operations. Before inserting a new node into the tree, near neighbor operations determine the best parent node. As shown in Fig. 2, the rewiring approach reconstructs the tree inside the area of radius r in order to keep the tree as cost-effective as possible between tree connections. Algorithm 2 shows the RRT* algorithm.

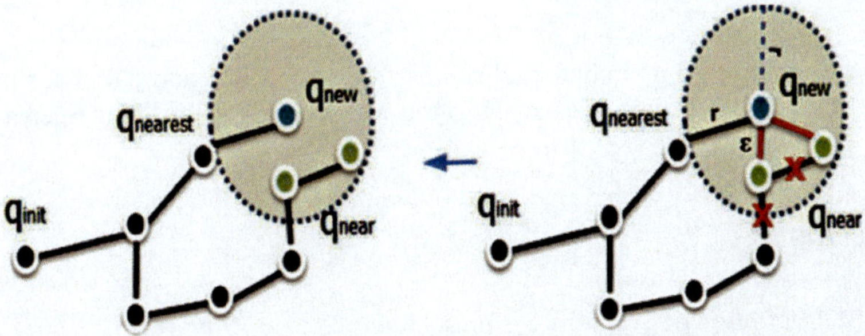

Fig. 2 RRT* tree expansion process [11]

5 Simulation Setup

This study presents an evaluation of three algorithms, A*, RRT, and RRT*. The A*, RRT, and RRT* algorithms were executed on 64-bit MATLAB R2022a version, and the results were utilized to assess their performance. The study was done on a computer running 64-bit Windows 10 Pro, with an AMD Ryzen Threadripper 2950X 16-Core CPU clocked at 3.50 GHz and 128 GB of internal RAM as the processor configuration.

The algorithms were implemented on a 2D binary map, with the white region representing free space and the rectangular black region representing an obstacle or occupied space. The two map scenarios (warehouse configuration spaces) below are taken into consideration in this study [12].

The path length, runtime, number of nodes in the path were among the performance measures. The lengths between each successive pair of nodes along the path from the starting to the goal point were added to determine the path length. Runtime is the duration needed for the path planner to generate an output path. The number of nodes is waypoints on the path generated, including start and goal points.

Five tests were performed for each scenario because the RRT and RRT* algorithms are non-deterministic in nature. The A* method is deterministic by nature and will produce results that are roughly consistent across tests. The mean and standard deviation of performance metrics were also computed for a trustworthy comparison of path planner algorithms. In both scenarios, only one test case for each planner's planned path was displayed in Figs. 3 and 4.

6 Simulation Results

This section gathered and compared results from A*, RRT, and RRT* for performance analysis. Five trials of each algorithm were run in two different Scenarios.

7 Scenario 1

Detailed description of Scenario 1 is presented in Sect. 3 (Fig. 3).

A* cells are regarded as having a size of one unit square. The A* algorithm was successfully implemented, and Table 1 shows the outcomes of five trials. Figure 5a displays one of the five paths the algorithm took. The average path length, runtime, and node count are 119.811 units, 0.6057 s, and 104, respectively (Table 2). Due to its deterministic nature, all findings were identical or nearly identical, and hence, the standard deviation is close to zero.

This work uses a maximum step size of five units for both RRT and RRT*. In addition, RRT* algorithm uses ten units as a near neighbors search radius. The data

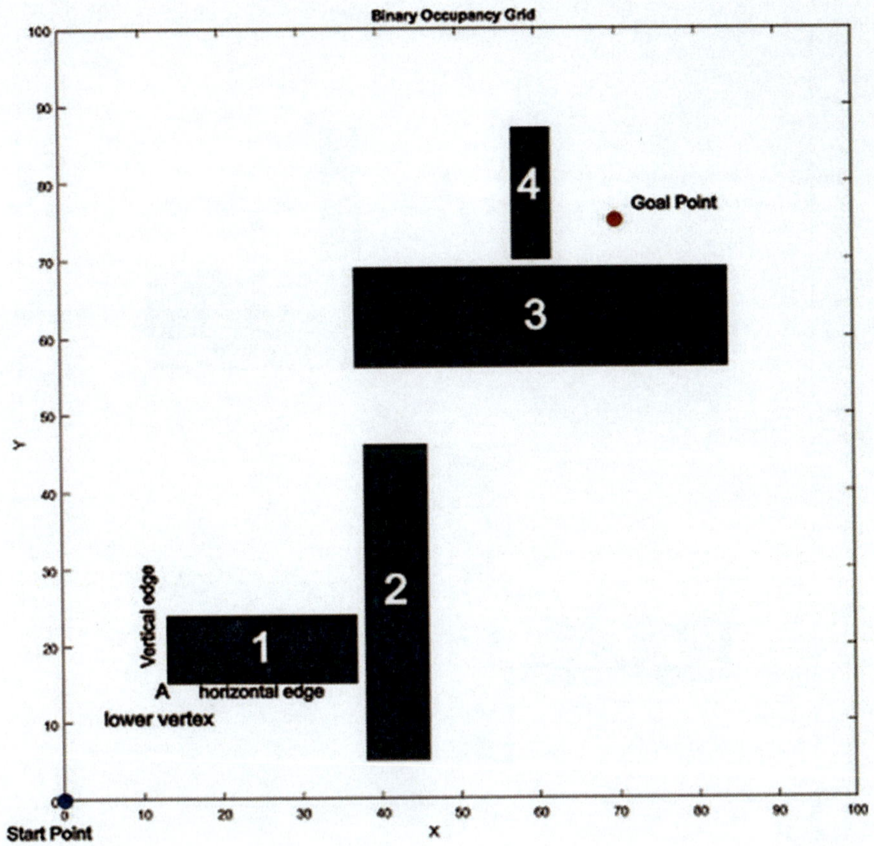

Fig. 3 Map of first warehouse scenario

obtained after implementing RRT and RRT* algorithm are stored in Table 1. The path traced by RRT and RRT* algorithms is shown in Figs. 5b and c, respectively.

For the RRT algorithm, the mean path length was 179.615 units, the runtime to obtain the path was 0.428 s, and the number of nodes in the path was 37 (Table 2). The mean path length, runtime, and the number of nodes in the path traced for the RRT* algorithm are 158.580 units, 1.414 s, and 20.2, respectively (Table 2). Due to the non-deterministic nature of RRT and RRT* planner, all results obtained were varied, and there is a significant value in standard deviation.

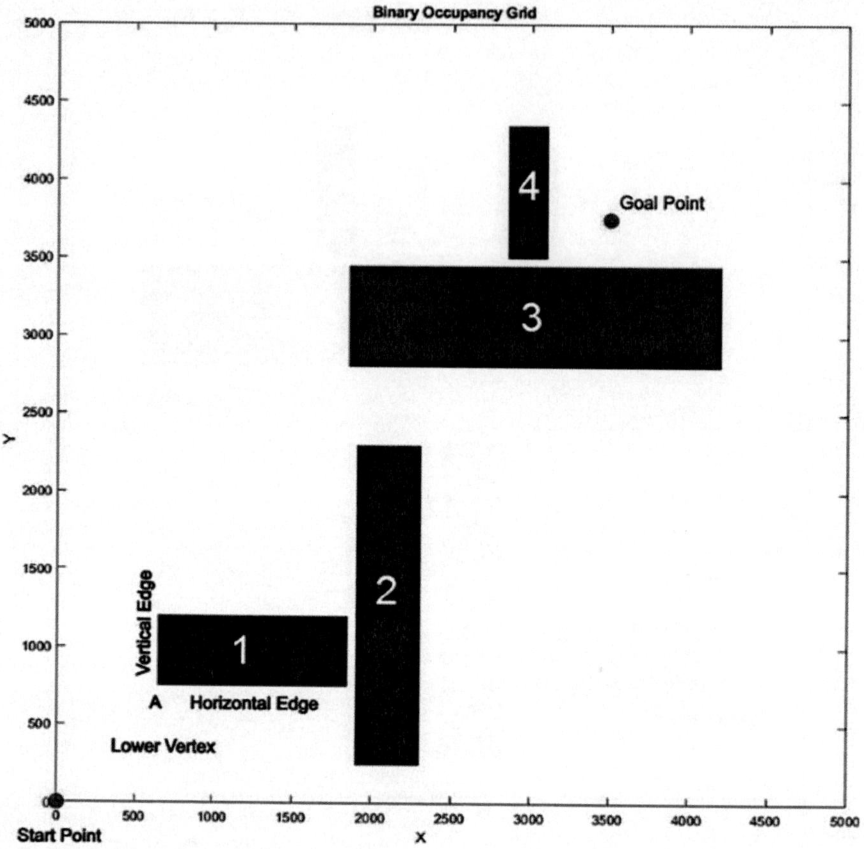

Fig. 4 Map of second warehouse scenario

Table 1 Outcome of five trials for A*, RRT, and RRT* planner in first scenario

		Number of trials				
Planner	Parameter	1	2	3	4	5
A*	Path length (units)	119.811	119.811	119.811	119.811	119.811
	Runtime (s)	0.5198	0.7368	0.5458	0.5933	0.6328
	Number of nodes in path	104	104	104	104	104
RRT	Path length (units)	170.461	154.894	210.107	195.127	167.488
	Runtime (s)	0.4397	0.4116	0.4247	0.4434	0.4208
	Number of nodes in path	35	32	43	40	35
RRT*	Path length (units)	158.423	147.071	166.471	154.628	166.309
	Runtime (s)	1.6424	1.2520	1.2353	1.5001	1.4402
	Number of nodes in path	19	19	21	20	22

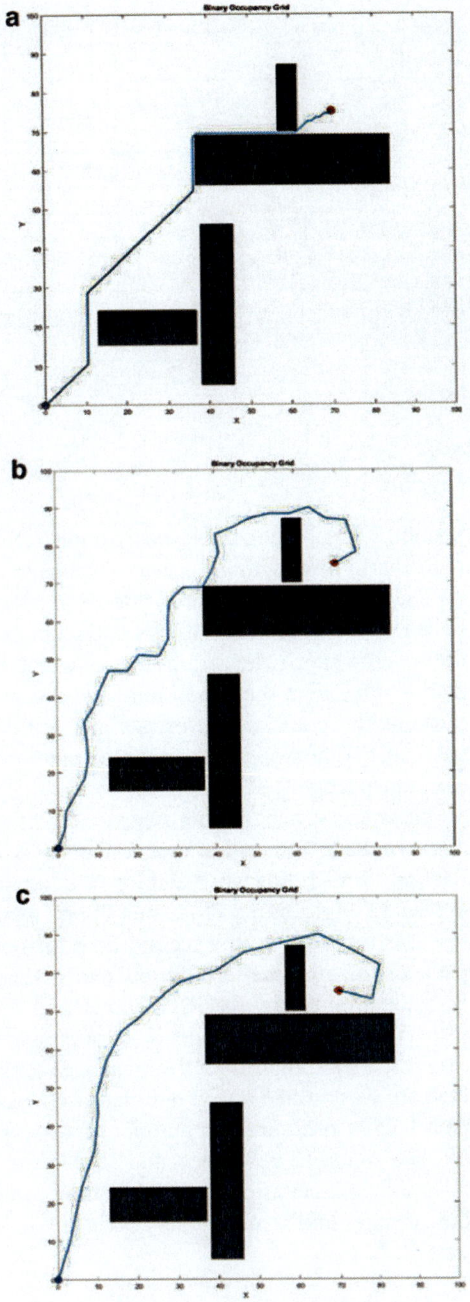

Fig. 5 Path planned in first scenario. **a** Using A* planner **b** using RRT planner **c** using RRT* planner

Table 2 Results of A*, RRT, and RRT* planner in first scenario

Planner	Parameter	Mean	Standard deviation
A*	Path length (units)	119.8112	0
	Runtime (s)	0.6057	0.0762
	Number of nodes in path	104	0
RRT	Path length (units)	179.6156	20.06198
	Runtime (s)	0.428	0.0118
	Number of nodes in path	37	3.9496
RRT*	Path length (units)	158.5809	7.3497
	Runtime (s)	1.414	0.1539
	Number of nodes in path	20.2	1.1662

7.1 Scenario 2

Section 3 (Fig. 4) contains a description of Scenario II in depth. Scenario II is similar to Scenario I but with higher dimensions.

The size of A* cells is considered to be 1 unit square. Figure 6a depicts the path traced by A* algorithm in the second scenario. The second case used five iterations of the A* algorithm. The acquired data, which included runtime, number of nodes, and path length, were then combined and included in Table 3's tabular format. Table 4 contains the calculated averages and standard deviations. The mean path length, runtime, and number of nodes in the path traced are 5990.6 units, 24.3352 secs, and 5102, respectively (Table 4).

In this work, a maximum step size of 250 units is used for both RRT and RRT*. In addition, RRT* algorithm uses 500 units as near neighbors' search radius. For both RRT and RRT*, this work employs a maximum step size of 250 units. Additionally, the RRT* algorithm's close neighbors' search radius is 500 units. The identical test was repeated in this second case for both RRT and RRT*, and the results for path length, runtime, and number of nodes in path are recorded in Table 3. The paths determined by the RRT and RRT * algorithms are shown in Figs. 6b and c, respectively. Table 4 provides a measurement of the averages and standard deviations.

For the RRT algorithm, the mean path length was 8364.86 units, runtime to obtain the path was 5.6686 s, and number of nodes in path were 35.8 (Table 4). The mean path length, runtime, and number of nodes in path traced for RRT* algorithm are 6931.32 units, 13.6892 secs, and 18.2, respectively (Table 4).

Since this scenario has a higher-dimensional complexity, it is not surprising that the averages and standard deviations are higher than in the previous scenario.

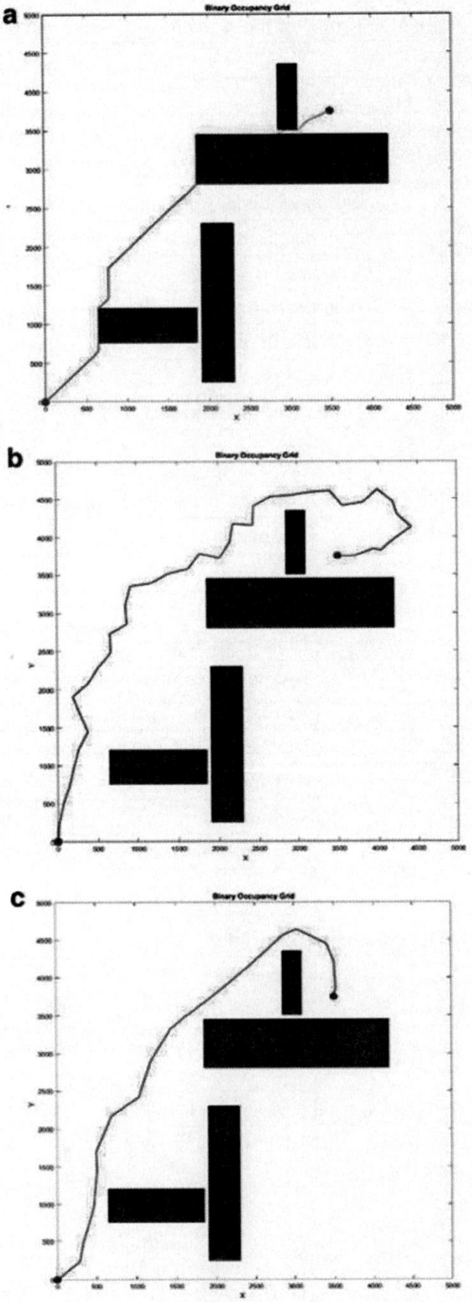

Fig. 6 Path planned in second scenario. **a** Using A* planner **b** using RRT planner **c** using RRT* planner

Table 3 Outcome of five trials for A*, RRT, and RRT* planner in second scenario

		Number of trials				
Planner	Parameter	1	2	3	4	5
A*	Path length (units)	5990.6	5990.6	5990.6	5990.6	5990.6
	Runtime (s)	25.1058	24.7922	23.6338	23.9685	24.1755
	Number of nodes in path	5102	5102	5102	5102	5102
RRT	Path length (units)	7699.2	9571.6	9777.6	7081.6	7694.3
	Runtime (s)	5.5998	5.6296	5.8151	5.5844	5.7143
	Number of nodes in path	32	40	43	31	33
RRT*	Path length (units)	6783.3	7006.9	6668.2	7059.6	7138.6
	Runtime (s)	13.6007	13.4999	13.2183	14.4340	13.6929
	Number of nodes in path	17	19	18	18	19

Table 4 Results of A*, RRT, and RRT* planner in second scenario

Planner	Parameter	Mean	Standard deviation
A*	Path length (units)	5990.6	0
	Runtime (s)	24.3352	0.5394
	Number of nodes in path	5102	0
RRT	Path length (units)	8364.86	1094.6749
	Runtime (s)	5.6686	0.0859
	Number of nodes in path	35.8	4.7917
RRT*	Path length (units)	6931.32	176.7909
	Runtime (s)	13.6892	0.4050
	Number of nodes in path	18.2	0.7483

Table 5 Summary of results

	Scenario 1			Scenario 2		
Parameter	A*	RRT	RRT*	A*	RRT	RRT*
Path length (units)	119.811	179.615	158.580	5990.6	8364.86	6931.32
Runtime (s)	0.6057	0.428	1.414	24.3352	5.6686	13.6892
Number of nodes in path	104	37	20.2	5102	35.8	18.2

8 Conclusion

This paper compared the A*, RRT, and RRT* algorithm for performance analysis in two different scenarios (warehouse configuration spaces).

In all instances, the A* planner consistently generated the shortest paths. Since the RRT* method looks for the least path parent within the near neighbors' search radius, it created shorter pathways than RRT.

The runtime for algorithms is roughly comparable in lesser dimensional maps, such as the first scenario. In contrast to RRT and RRT* planner, A* planner has a significantly longer runtime in higher dimensional spaces, such as the second scenario. The RRT* planner requires an additional step to search for the least path parent inside the near neighbors' search radius, which causes it to take almost twice as long to run as the RRT planner.

The A* method is substantially slower than the RRT family of algorithms, but the pathways identified by RRT algorithms are longer than those found by A*.

For path optimality in lower-dimensional warehouse configuration spaces (Scenario 1), the A* planner is recommended, while for runtime optimality in higher-dimensional warehouse configuration spaces (Scenario 2), the RRT family is chosen.

References

1. González D, Pérez J, Milanés V, Nashashibi F (2016) A review of motion planning techniques for automated vehicles. IEEE Trans Intell Transp Syst 17(4):1135–1145. https://doi.org/10.1109/TITS.2015.2498841
2. Foead D, Ghifari A, Kusuma MB, Hanafiah N, Gunawan E (2021) A systematic literature review of A* pathfinding. Proc Comp Sci 179:507–514. https://doi.org/10.1016/j.procs.2021.01.034
3. Loong WY, Long LZ, Hun LC (2011) A star path following mobile robot. In: 2011 4th International conference on mechatronics (ICOM). IEEE Press, pp 1–7. https://doi.org/10.1109/ICOM.2011.5937169
4. Elbanhawi M, Simic M (2014) Sampling-based robot motion planning: a review. IEEE Access 2:56–77. https://doi.org/10.1109/ACCESS.2014.2302442
5. Kavraki LE, Svestka P, Latombe JC, Overmars MH (1996) Probabilistic roadmaps for path planning in high-dimensional configuration spaces. IEEE Trans Robot Autom 12(4):566–580. https://doi.org/10.1109/70.508439
6. LaValle SM (2006) Planning algorithms. Cambridge University Press
7. Choset HM, Hutchinson S, Lynch KM, Kantor G, Burgard W, Kavraki LE, Thrun S (2005) Principles of robot motion: theory, algorithms, and implementation. MIT press
8. LaValle SM (1998) Rapidly-exploring random trees : a new tool for path planning. Annu Res Rep
9. Noreen AK, Habib Z (2016) A comparison of RRT, RRT* and RRT*—smart path planning algorithms. Int J Comp Sci Netw Secur (IJCSNS) 16(10):20–27
10. Karaman S, Walter MR, Perez A, Frazzoli E, Teller S (2011) Anytime motion planning using the RRT*. In: IEEE international conference on robotics and automation (ICRA), IEEE Press, China, pp 1478–1483. https://doi.org/10.1109/ICRA.2011.5980479
11. Improving optimality of RRT RRT. http://joonlecture.blogspot.com/2011/02/improving-optimality-of-rrt-rrt.html. Accessed 15 June 2022
12. Motion-Planning. https://in.mathworks.com/help/nav/motion-planning.html. Last Accessed 15 June 2022

Improving Payload Capacity of an ABB SCARA Robot by Optimization of Blended Polynomial Trajectories

Kaustav Ghar, Bhaskar Guin, Nipu Modak, and Tarun Kanti Naskar

Abstract SCARA robots are widely used as assembly line robots in different manufacturing industries. This paper aims at maximizing the payload capacity of robotic manipulators with the generation of optimized joint trajectories to ensure adequate motion smoothness. This would help users to operate the robot beyond its rated capacity without inducing additional stress on the actuators. The payload-carrying capacity has been maximized along with the minimization of the joint jerk and total travel time by the method of successive optimization. In this paper, the trajectories have been constructed using polynomial blending functions as they are best suited for industrial purposes. Four types of blended joint trajectories have been considered—a linear segment with parabolic blend (LSPB), a linear segment with cubic blend (LSCB), a linear segment with quintic blend (LSQB), and a linear segment with heptic blend (LSHB). It is possible to control the joint jerk by controlling the blend time of these polynomials. The jerk minimization has been done using GA with successive maximization of payload capacity considering the maximum permissible tool center point (TCP) acceleration and velocity, rated RMS torque, and force for each joint as the optimization constraints. Lagrangian dynamics have been used to establish the dynamics associated with the system. A unique concept of graded weight functions has been introduced to establish the objective function which converts a multi-objective optimization problem into a single-objective optimization problem on a priority basis. This study if implemented in industrial robotic manipulation could drastically improve performance.

Keywords Optimized joint trajectory · LSPB · LSCB · LSQB · LSHB · GA · Minimum jerk · Payload · TCP · Lagrangian dynamics

---

K. Ghar · N. Modak (✉) · T. K. Naskar
Department of Mechanical Engineering, Jadavpur University, Kolkata, India
e-mail: nmechju@gmail.com

B. Guin
Department of Mechanical Engineering, Contai Polytechnic, Purba Medinipur, West Bengal, India

S. Sharma et al. (eds.), *Intelligent Control, Robotics, and Industrial Automation*, Lecture Notes in Electrical Engineering 1066,
https://doi.org/10.1007/978-981-99-4634-1_9

1 Introduction

Modern-day industries use robots to automate the assembly line and also to manufacture components with high precision. SCARA robots are widely used in the electronics industry to handle sophisticated electronic items. To maximize productivity, the robot trajectory and payload-carrying capacity need to be optimized so that the robot can achieve its optimal performance. For manufacturing different products, robots may be required to operate at loads greater than their rated capacity. In such situations, procuring a new robot with a higher payload capacity may be uneconomical. Hence, maximizing the payloadcarrying capacity of the existing robot by optimizing its trajectory may serve the purpose with no additional investment. Using the MOPSO algorithm, trajectories have been planned for maximized payload [1] of a free-floating space manipulator. An effort to increase the capacity of carrying payload of a two-link robot has been presented in [2]. The payload has increased over its nominal value. A homogeneous payload-specific performance index has been defined for two-link and three-link RRR robotic manipulators on the basis of payload's kinetic energy [3]. For a given payload, the illustrated method optimizes the structure of the robot. By constraining of mechanism [4], the payload-carrying capacity of the endoscopic surgical robotic manipulator has been enhanced. A multi-objective two-stage planning of the trajectory of space manipulators for self-assembling heavy payloads has been proposed in [5]. The process of carrying loads has been minimized using the MOPSO algorithm. Control schemes applied to a 2-DOF robotic manipulator with variable payloads [6, 7] are effective in counteracting the model uncertainties and external disturbances. Jerk-minimized trajectories are instrumental for the cautious and delicate handling of the payload. Various techniques have been evolved till date to minimize the jerk of industrial robots. A method of jerk minimization with kinematic constraints has been described in [8]. The proposed method has fast convergence. Conventionally, trajectories are planned using classical splines and B-splines as they offer control parameters to minimize the jerk. Introducing control points in splines is effective in minimizing jerk. An innovative approach in smoothening trajectories by NURBS reparameterization has been presented in [9, 10]. The proposed methods reduce the power consumption, execution time, and jerk without changing the geometry of the trajectory. The jerk of the trajectory for a welding robot having sic-axis has been minimized using the TLBO algorithm [11]. The end-effector deviation from the weld seam path has been controlled by minimizing the jerk, resulting in reduced welding defects. A hybrid multi-objective evolutionary algorithm has been proposed for minimizing the torque rate, jerk, and travel duration [12], thus resulting in a Pareto-optimal solution set which gives the perfect trade-off among the objectives. It is evident that although lower-order polynomial trajectories are simple to implement but offer no control over jerk. In this paper, the study is focused on improving the payload-carrying capacity of robotic manipulators. Combined minimization of the jerk with payload maximization may prove to be more effective and will be explored in detail in this paper. The effect of using higher-order polynomial blends and simultaneous minimization of

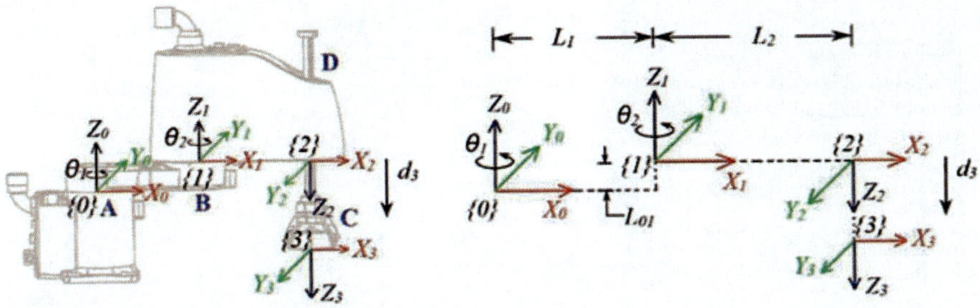

Fig. 1 Coordinate frames attached to a SCARA robot

travel time, on the payload capacity, will also be dealt with. This analysis would help industries plan optimum trajectories to improve productivity and expand the capabilities of the existing robots.

2 Mathematical Modeling

ABB SCARA robot is a four-axis RRPR robot, where rotary joint is denoted by *R* and *P* denotes prismatic (translational) joint as presented in Fig. 1. Position C is the location of end-effector with respect to base or origin located at position A. The two rotary joints are located at A and B and the prismatic joints located at D are responsible for positioning the end–effector, while the fourth joint is responsible for orienting the tool. The fourth joint responsible for the roll motion of the end-effector has been neglected in our current analysis. To develop an optimized trajectory for maximizing the payload capacity, a mathematical model of the robot needs to be developed.

3 Robot Kinematics

Robot kinematics deals with the analysis of the motion of the links and joints irrespective of the force or torque responsible for it. Forward kinematics relates the robot end-effector orientation and position for a particular value of joint actuation, while inverse kinematics is the reverse process. Using the Denavit–Hartenberg (D-H) convention, coordinate frames are attached to the ABB SCARA robot and a simplified link diagram is constructed as shown in Fig. 1.

The end-effector position and its orientation are obtained by solving the forward kinematic equations through the D-H algorithm. θ_1 and θ_2 are corresponding rotation angles of joints 1 and 2 about the z-axis, d_3 is the motion of the translational joint along the z-axis, and L_1 L_2 are the corresponding lengths of links 1 and 2. L_{01} is the offset between joints 1 and 2 measured along the z-axis. With the predefined

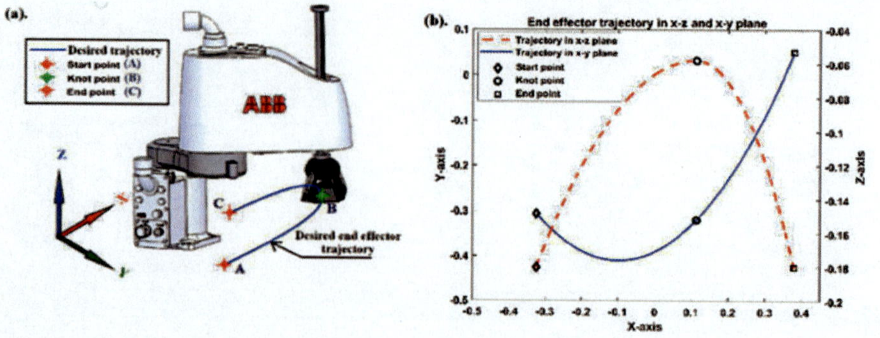

Fig. 2 **a** Path traced by end-effector, **b** end-effector trajectory in the x–z and x–y plane

range of joint actuation, the maneuverable workspace of the robotic manipulator can be obtained using the forward kinematics solution. The trajectory traced by the end-effector of the robot in Cartesian space must lie within the workspace region. The desired trajectory is defined by a set of precision points through which the end-effector must pass. Based on the trajectory under consideration, the number of precision points is chosen accordingly. An nth-order polynomial can be used to generate the path which passes through these precision points. For illustration, a trajectory is constructed which is defined by three precision points—a start point, an endpoint, and an intermediate knot point. The interpolation of these three precision points has been done using parabolic curves in x–z and x–y planes to obtain the desired trajectory of the end-effector as shown in Fig. 2b. The desired trajectory can be traced by the end-effector by providing proportionate actuation to the joints obtained by inverse kinematic analysis. The joint actuation parameters θ_1, θ_2, and d_3 for the ABB SCARA robot can be obtained as:

$$\begin{cases} \theta_2 = \pm a\cos d\left(\frac{x_e^2+y_e^2-L_1^2-L_2^2}{2L_1L_2}\right) \\ \theta_1 = a\tan d\left(\frac{-(L_2\mathrm{Sin}\theta_2)x_e+(L_1+L_2\mathrm{Cos}\theta_2)y_e}{(L_1+L_2\mathrm{Cos}\theta_2)x_e+(L_2\mathrm{Sin}\theta_2)y_e}\right) \\ d_3 = L_{01} - z_e, \end{cases} \tag{1}$$

where x_e, y_e, and z_e are the Cartesian space position coordinates of the ABB SCARA robot end-effector.

3.1 Joint Trajectory Interpolating Function

The sequence of actuation of each joint needed to trace the required trajectory of the end-effector is obtained using inverse kinematics. The direction of motion of the actuator is reversed at one or more points depending on the joint actuation sequence. These points are referred to as reversal points which divide the joint trajectory into multiple segments. While actuating a joint having a reversal point, the actuator stops upon reaching the reversal point and again starts its motion in the reverse direction. For the problem under consideration (as shown in Fig. 2a), based on the position of the reversal point, the location of knot point can be in first or second segment or coincidence might occur with reversal point. Hence, joint trajectory has a start point, a knot point, and an endpoint, with or without a reversal point. The desired trajectory is traced in Cartesian space, the trajectory of each joint needs to be interpolated with the help of some interpolation functions. Four types of blended algebraic polynomial interpolating functions have been used in this analysis. These are linear segments with parabolic blends (LSPB), linear segment with cubic blends (LSCB), linear segment with quintic blends (LSQB), and linear segment with heptic blends (LSHB). A general nth-order polynomial can be represented as:

$$f(t) = \sum_{i=0}^{n} a_i t^i. \tag{2}$$

The overall time domain is divided into three parts—[0 to t_a], [t_a to t_d], and [t_d to t_f], where t_a is the blend time for the first polynomial blend, t_d is the start time of the second polynomial blend, and t_f is the end time. A linear segment with nth-order blends can be represented as shown in Eq. 3.

$$\theta(t) = \begin{cases} f(t); & 0 \le t \le t_a \\ \sum_{i=0}^{1} a_i y^i; & t_a \le t \le t_d \\ f(t); & t_d \le t \le t_f \end{cases}. \tag{3}$$

$n = \{2,3,5,7\}$ represents LSPB, LSCB, LSQB, and LSHB joint trajectories, respectively. Incorporation of a knot point in the trajectory of each joint has been performed using the concept of a simplified robot joint trajectory [13].

4 Joint Temporal and Dynamic Behavior

Computation of joint temporal behavior is necessary to establish the dynamic model. The temporal behavior of a joint comprises velocity $\dot{\theta}$, acceleration $\ddot{\theta}$, and jerk $\dddot{\theta}$ of the joint, which are derived from angular displacements θ using the central difference method. Dynamic behavior comprises joint torque and forces required for actuation. The Lagrangian dynamic formulation can be used to establish the dynamic equation associated with each joint as a function of generalized coordinates. The dynamics of motion of a robotic manipulator can be written as:

$$M(p)\{\ddot{p}\} + H(p, \dot{p}) + F(\dot{p}) + G(p) = \tau, \tag{4}$$

where $M(p)$ is the inertia term, $H(p, \dot{p})$ is the centripetal and Coriolis term, $F(\dot{p})$ is the Coulomb friction and viscous force term, $G(p)$ is the gravity vector, and τ is generalized torque vector, respectively. The generalized displacement, generalized velocity, and generalized acceleration are represented by p, $\dot{p}$, and $\ddot{p}$, respectively. Using the inverse dynamics [14] approach, the joint forces and torques required for actuation have been computed, which are functions of joint temporal behavior.

5 Trajectory Optimization

The primary goal of this study is to maximize the payload of an industrial manipulator. Considering an arbitrary combination of blend times for the blended polynomial joint trajectory and an assumed travel time, the maximum payload capacity of the manipulator can be determined without violating the rated force/torque of the actuators. The optimization process starts with the assumption of a feasible payload, and at each successive iteration, the payload is increased/decreased proportionately till convergence is achieved. Thus, the maximum payload is obtained.

Payload maximization for minimized jerk

Minimizing the jerk often lowers the peak velocity and acceleration which may improve the payload capacity. Combined jerk minimization with payload maximization is computed using successive optimization technique where each parameter is optimized sequentially. A constant value of travel time has been assumed in the case of minimizing the joint jerk. For blended polynomials, altering the blend time alters the kinematic parameters, and hence, blend times are considered as optimization variables. The optimization constraints are the maximum velocity and acceleration of the tool center point. Minimization of jerk of a system is an optimization problem having multiple objectives that yields Pareto-optimal solutions. Hence, the conversion of the problem from a multi-objective to a single-objective optimization problem ensures fast convergence and yields a unique solution. A single objective is formulated by summing the multiple objectives that are pre-multiplied by weights that are not chosen but to prioritize the objectives. These weights are thereby termed as graded-weight functions, and the method of determining the weights is given below.

Weights coefficient selection on the basis of dependency of actuation of joint: More emphasis is given to the robotic joints having higher degree coupling, which significantly affects the overall motion of the robotic manipulator. Coefficient k_i^d has been used to designate the weights of this category, where the joint number of the robot is represented by *i*.

Weight coefficient selection on the basis of each segment actuation duration of joint trajectory: Greater joint jerk is expected for smaller travel time; hence, more emphasis is given to the joint trajectory segment having a short travel time duration. The weight coefficient of the *j*th segment of the joint trajectory is given by k_j^t.

Combined Graded-Weight coefficient (k_j^o): The product of weights based on joint actuation duration and joint dependency is chosen as the combined graded weight coefficient. The overall joint jerk which has to be minimized is given by Eq. 5.

$$\text{Overall joint jerk,}\quad J = \sum_{j=0}^{n} k_j^o * \max(\text{jerk})_j^2, \tag{5}$$

where $k_j^o = k_i^d * k_j^t$, and n—the segment number of joint trajectory of each joint. The optimization problem can be summarized as:

- *The objective function of the NLP*: minimize *overall joint jerk* (J).
- *Decision variable*: Normalized blending times t_a and t_d. The duration of blends for both acceleration and retardation is equal.
- *Restrictions*: Limiting TCP acceleration and velocity which are denoted by a_{limit} and v_{limit}, respectively.
- *Constant*: t_{traj} (total duration required for tracing the desired end-effector trajectory)

Optimization has been performed using the genetic algorithm (GA) solver in Matlab™ Optimization Toolbox which can minimize both constrained and unconstrained, highly nonlinear, multivariable problems with a single objective.

Payload maximization for minimized travel time

To improve the productivity of a manufacturing plant, the robots should complete the task in the minimum possible time. The maximum payload capacity in this situation is likely to vary from that obtainable in jerk minimization. The objective, in this case, is the simultaneous minimization of travel time and maximizing payload, which is also obtainable using successive optimization. The blend times are assumed to be constant to construct the joint trajectories. For starting the solver, a feasible travel time is assumed and subsequently modified after each iterations as long as the constraints associated are satisfied. The minimization of travel time is then followed by the maximization of payload. Thus, an optimum trajectory is obtained for which the total travel is minimum, while the payload is maximum.

Payload maximization for combined minimized jerk and travel time

The best-optimized trajectory can be obtained by ensuring combined minimization of jerk and travel time with payload maximization. The concept of simultaneous and successive optimization has to be used in this problem. To minimize the jerk and travel time simultaneously, the optimization problem in this case is similar to the case of minimization of the jerk for constant travel time. However, the total time t_{traj} required to trace the desired trajectory is no longer constant but decreased proportionately at each iteration until there occurs a violation of constraints.

6 Results and Discussion

The mathematical model has been validated by simulating a simple trajectory in MATLAB™ software. The payload gripped by the end-effector of the robot must be transported from points A to C via B. It is desired to maximize the payload capacity of the robot while ensuring the task is completed in minimum time with a minimum jerk. The manipulator chosen for the demonstration is the ABB IRB 910SC-3/0.45 robot. Within its workspace region, the end-effector trajectory has been constructed, which passes through three points of precision namely, start point A (−0.3248, −0.3059, −0.18) m, the knot or via point B (0.1131, −0.3195, −0.058) m, and endpoint C (0.3794, 0.05048, −0.18) m as presented in Fig. 3a. A parabolic curve can be constructed that passes through these precision points.

The actuation sequence of each joint is obtained by using the inverse kinematics approach and is shown in Fig. 3b. It has been found that the reversal point does not appear in θ_1, while it appears both in θ_2 and d_3. In translational joint d_3, the knot point and the reversal point coincide with each other, and hence, the reversal point can be treated as knot point and vice versa for this case. The joint trajectories are interpolated using LSPB, LSCB, LSQB, and LSHB curves with an initial assumption that the total travel time (t_{traj}) is 10 s, the normalized blend time (t_a) is 0.2, and a payload of 3 kg. The trajectories are then optimized subject to constraints. The maximum allowable TCP velocity and acceleration are considered to be 0.2 m/s and 1.5 m/s^2, respectively. The rated RMS torque of joint 1 and joint 2 is 1.4 Nm each, and the maximum force of the linear joint is 250N as per the manufacturer specifications [15].

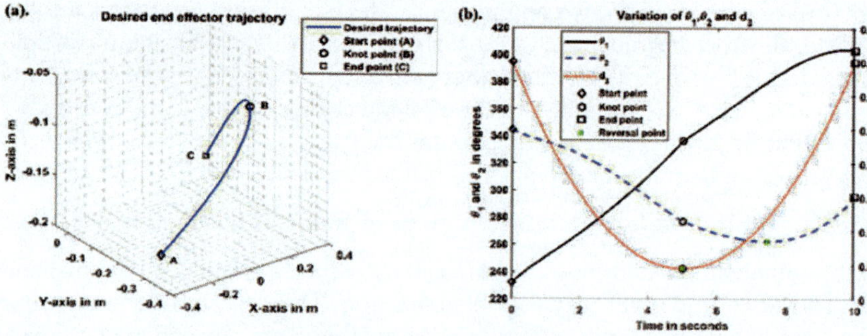

Fig. 3 **a** Required trajectory of end-effector represented in Cartesian space, **b** variation of θ_1, θ_2, and d_3

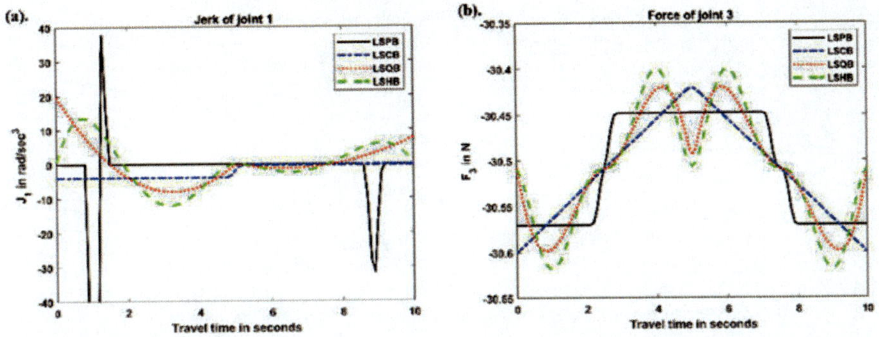

Fig. 4 **a** Jerk variation of joint 1, **b** actuator force variation for joint 3—for minimized jerk condition

7 Outcome of Jerk Minimization

The jerk experienced by rotary joint 1 and force associated with translational joint 3 for LSPB, LSCB, LSQB, and LSHB joint trajectories for jerk minimized case are shown in Fig. 4. It is observed that for the polynomial blends of higher order, the variation of joint jerk becomes more uniform since with the increase in the order of polynomial, it is possible to establish higher-order continuity at the blending points. The joint trajectory interpolated using LSHB has zero jerk at start and stop, thus imparting a smooth joint motion at the start and end. Increasing the order of the blending polynomial results in a smoother torque/force profile.

The reduction in torque/force requirement due to jerk minimization and the use of higher-order polynomial blends allows the payload to be increased.

The variation of maximum TCP jerk and the maximum payload for minimized overall jerk is plotted as a function of assumed travel time, as shown in Fig. 5. It has been observed that, for each type of blended joint trajectories, with the increase in travel time, the overall minimized jerk decreases, and the corresponding payload carrying capacity increases.

8 Outcome of Travel Time Minimization

To minimize the travel time, a constant value of blend time, $t_a = 0.2$, has been assumed to construct joint trajectories. Figures 6a and b shows the payload variation as a function of blend time for a constant travel time for LSPB and LSQB joint trajectory.

The payload becomes nearly constant on increasing the normalized blend time beyond a certain range and, increasing the travel time, significantly improves the payload capacity. Decreasing the travel time results in an increase in maximum joint torques and forces and hence decreases the payloadcarrying capacity. It has also been

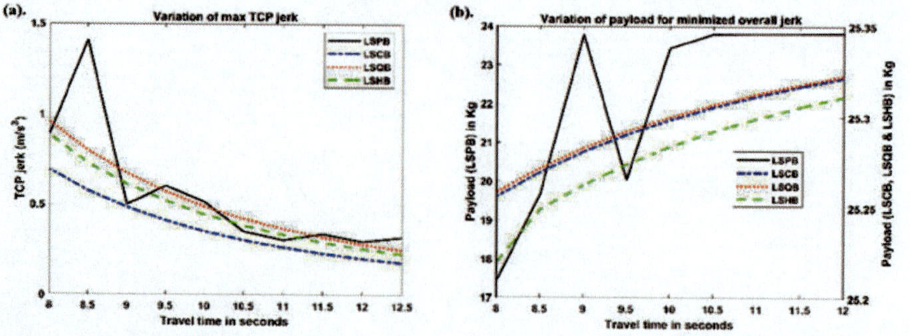

Fig. 5 **a** Maximum TCP jerk variation for minimized overall jerk and **b** maximum payload variation for minimized overall jerk

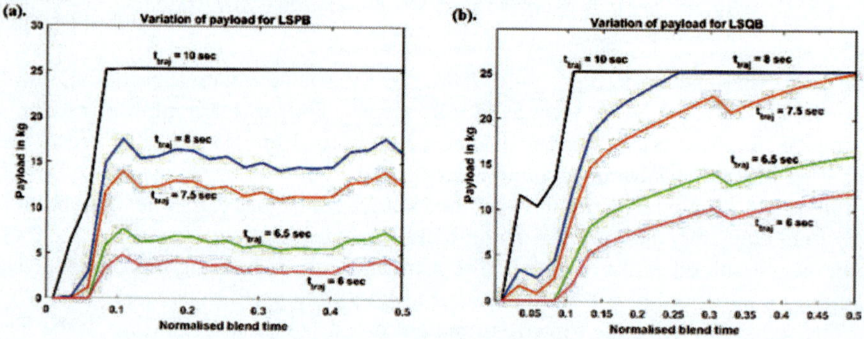

Fig. 6 Variation of payload for constant t_{traj} for **a** LSPB, **b** LSQB

observed that when t_{traj} is high, all the polynomial blended trajectories yield similar results; however, for smaller values of t_{traj}, using higher-order polynomial blends increases the maximum payload-carrying capacity.

9 Outcome of Combined Travel Time and Jerk Minimization

The jerk variation for joint 1 and force variation for joint 3 in case of combined minimization are shown in Fig. 7. It is observed that lower order polynomial blends yield lower minimum travel time, and hence, the required force for its actuation is also higher.

The variation of the payload for different polynomial blended trajectories under different optimization cases has been summarized in Table 1.

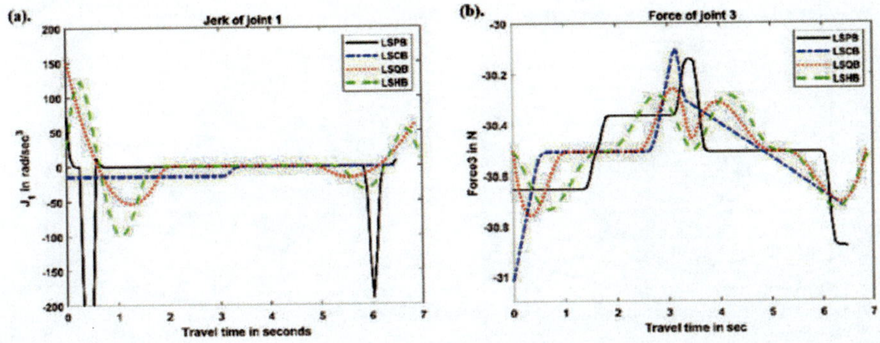

Fig. 7 **a** Jerk variation for joint 1, **b** force variation for joint 3—in the case of combined minimized jerk and travel time

Table 1 Variation of payload for different trajectory optimization cases

Case	Parameter	LSPB	LSCB	LSQB	LSHB
Assumed $t_a = 0.2$	t_{traj} (s)	10	10	10	10
	Max. payload (kg)	25.29	24.88	25.25	25.23
Minimized jerk	t_{traj} (s)	10	10	10	10
	Max. payload (kg)	25.32	25.30	25.30	25.28
Minimized travel time $t_a = 0.2$	t_{traj} (s)	8.73	6.79	6.97	7.04
	Max. payload (kg)	20.93	5.25	14.26	12.41
Minimized jerk and travel time	t_{traj} (s)	6.44	6.45	6.44	6.97
	Max. payload (kg)	5.97	17.81	10.74	12.30

It is observed that there has been a slight increase in the payload for the jerk minimized case compared to the assumed case. Similarly, when the travel time is reduced, the capacity of carrying a payload decreases. For combined minimization case, the payload further decreases than minimized travel time case. Since the possibility of further reduction of the travel time prevails in the combined minimization case, the payload-carrying capacity also decreases. The variation of maximum payload-carrying capacity for minimized travel time for the LSCB joint trajectory has been shown in Fig. 8a. The point shown by the data cursor corresponds to the maximum payload-carrying capacity for the case of combined jerk and travel time minimization. Here, t_{a4} is the normalized blend time of the first segment of the joint trajectory of d_3.

The minimum travel time variation with varying normalized blend time has been shown in Fig. 8b. The point shown by the data cursor corresponds to the condition of minimum travel time for combined minimization of jerk and travel time.

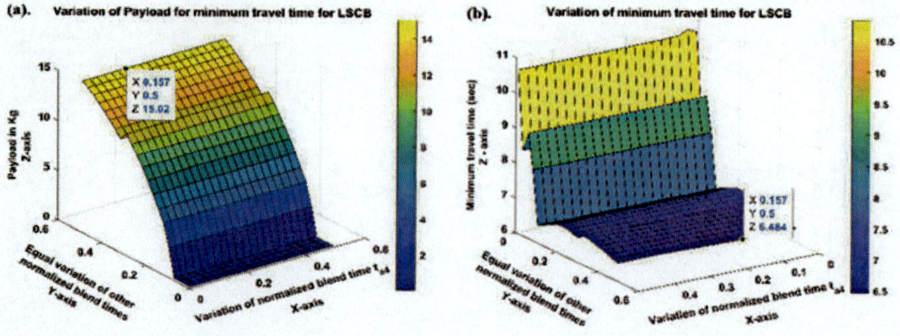

Fig. 8 Plot of **a** maximum payload, **b** minimum travel time—for LSCB joint trajectory

10 Conclusion

This paper was aimed at maximizing the payload of service robots employed in routine maneuvering tasks. This would help robots to operate beyond their rated capacity. To achieve maximum productivity, the travel time must be minimum however ensuring adequate motion smoothness to reduce stress on the actuator. For this purpose, the trajectory was optimized with respect to two mutually conflicting parameters—jerk and travel time. The variation of payload for various optimization cases reveals an interesting outcome. Minimization of jerk drastically improves the payload capacity, while minimization of travel time lowers the payload capacity due to high velocity and acceleration values. Hence, the optimum trajectory may be achieved through a trade-off between these two extremities. The novelty of this study is the use of graded-weight functions to formulate a single-objective optimization problem by combining multiple objectives on a priority basis and solving using GA. To address the industrial scenario, blended algebraic polynomials were used to interpolate the trajectories instead of complex splines. The effect of using higher-order polynomials has also been established by comparing LSPB, LSCB, LSQB, and LSHB trajectories under different optimization conditions. Higher-order polynomials are always preferred as they smoothen the kinematic profile and improve the payload capacity. Hence, this study could be effectively used in planning trajectories ensuring optimum performance of the robot in routine industrial maneuvering tasks.

References

1. Liu Y, Jia Q, Chen G, Sun H, Peng J (2015) Multi-objective trajectory planning of FFSM carrying a heavy payload. Int J Adv Rob Syst 12(9):118. https://doi.org/10.5772/61235
2. Gallant A, Gosselin C (2018) Extending the capabilities of robotic manipulators using trajectory optimization. Mech Mach Theory 121:502–514. https://doi.org/10.1016/j.mechmachtheory.2017.09.016

3. Nabavi SN, Akbarzadeh A, Enferadi J, Kardan I (2018) A homogeneous payload specific performance index for robot manipulators based on the kinetic energy. Mech Mach Theory 130:330–345. https://doi.org/10.1016/j.mechmachtheory.2018.08.007
4. Hwang M, Kwon DS (2019) Strong continuum manipulator for flexible endoscopic surgery. IEEE/ASME Trans Mechatron 24(5):2193–2203. https://doi.org/10.1109/TMECH.2019.2932378
5. Liu Y, Du Z, Wu Z, Liu F, Li X (2021) Multi objective preimpact trajectory planning of space manipulator for self-assembling a heavy payload. Int J Adv Rob Syst 18(1):1729881421990285. https://doi.org/10.1177/1729881421990285
6. Sharma R, Kumar V, Gaur P, Mittal AP (2016) An adaptive PID like controller using mix locally recurrent neural network for robotic manipulator with variable payload. ISA Trans 62:258–267. https://doi.org/10.1016/j.isatra.2016.01.016
7. Gaidhane PJ, Nigam MJ, Kumar A, Pradhan PM (2019) Design of interval type-2 fuzzy precompensated PID controller applied to two-DOF robotic manipulator with variable payload. ISA Trans 89:169–185. https://doi.org/10.1016/j.isatra.2018.12.030
8. Canali F, Guarino Lo Bianco C, Locatelli M (2014) Minimum-jerk online planning by a mathematical programming approach. Eng Optimiz 46(6):763–783. https://doi.org/10.1080/0305215X.2013.806916
9. Hashemian A, Hosseini SF, Nabavi SN (2017) Kinematically smoothing trajectories by NURBS reparameterization—an innovative approach. Adv Robot 31(23–24):1296–1312. https://doi.org/10.1080/01691864.2017.1396923
10. Wu G, Zhao W, Zhang X (2021) Optimum time-energy-jerk trajectory planning for serial robotic manipulators by reparameterized quintic NURBS curves. Proc Inst Mech Eng C J Mech Eng Sci 235(19):4382–4393. https://doi.org/10.1177/0954406220969734
11. Rout A, Dileep M, Mohanta GB, Deepak BBVL, Biswal BB (2018) Optimal time-jerk trajectory planning of 6 axis welding robot using TLBO method. Proc Comp Sci 133:537–544. https://doi.org/10.1016/j.procs.2018.07.067
12. Rout A, Mohanta GB, Gunji BM, Deepak BBVL, Biswal BB (2019) Optimal time-jerk-torque trajectory planning of industrial robot under kinematic and dynamic constraints. In: 2019 9th annual information technology, electromechanical engineering and microelectronics conference (IEMECON). IEEE, pp 36–42. 10.1109/ IEMECONX.2019.8877063
13. Williams RL (2013) Simplified robotics joint-space trajectory generation with a via point using a single polynomial. J Robot. https://doi.org/10.1155/2013/735958
14. Lauß T, Oberpeilsteiner S, Sherif K, Steiner W (2019) Inverse dynamics of an industrial robot using motion constraints. In: 2019 20th International conference on research and education in mechatronics (REM). IEEE, pp 1–7. https://doi.org/10.1109/REM.2019.8744124
15. ABB (2021) Product specification motor units and gear units. 3HAC040147-001. [Online]. Available: https://new.abb.com/products/robotics/application-equipment-and-accessories/motor-and-gear-units/motor-units

Design of a Novel Tree-Type Robot for Pipeline Repair

Santosh Kumar and B. Sandeep Reddy

Abstract In this paper, a novel design of a tree-type repair and inspection robot is presented. The modeling of the robot was done using SolidWorks software and is also presented in the paper. The robot is capable of maneuvering in horizontal pipes, vertical pipes, and flat surfaces. Taking into account the variation in types of surfaces involved when the robot moves in pipes, a special type of grip mechanism with suction cups has been proposed for better adhesion. While varying pipe orientations, the robot can be maneuvered in the pipe by simply changing the orientation of the first link (connected to the platform). Mathematical calculations have been performed to estimate the weight of the platform and links. Finite element analysis (FEA) of links and platforms is performed for validation and optimization purposes. From the simulation runs, it has been observed that the design is structurally safe and within the yield limits of the selected material. Future work involves the testing of algorithms for end-effector path following, control, and trajectory planning, among others.

Keywords Tree-type robot · In-pipe · FEA · Von Mises stress · Structural analysis

1 Introduction

Pipelines are used to transport liquids and gas, which are vital to everyday life. Pipelines often face problems owing to leakage, among others which often require repair. However, the repair of pipelines faces many problems. Firstly, identifying cracks, for instance, is difficult and this implies that very often cracks are not observed until they become large enough and are no longer able to be repaired. Secondly, repair so far has taken place manually, which requires the shutting down of the plant, which is extremely costly. The use of robots has attracted the attention of those in industry as a potential solution for such problems. Certain robots for repair purposes in industry have been built [1–3].

S. Kumar (✉) · B. S. Reddy
Indian Institute of Technology, Guwahati Assam-781039, India
e-mail: santosh_kumar@iitg.ac.in

S. Sharma et al. (eds.), *Intelligent Control, Robotics, and Industrial Automation*,
Lecture Notes in Electrical Engineering 1066,
https://doi.org/10.1007/978-981-99-4634-1_10

This paper presents a novel tree-type robot which can potentially be used for pipeline repair. The robot is an in-pipe robot, which implies that it maneuvers inside the pipeline. A tree-type robot has multiple open-loop chains in its kinematic structure. The kinematics and dynamics of such robots are comparatively more challenging than for serial robots known in the industry. One of the better methods to solve the kinematic and dynamic problems is to break the tree-type chain into multiple serial chains. Mayeda et al. [4] and Khosla [5] have derived a formula for determining the minimum set of serial link robots. The formulation proposed by Khalil et al. [6] is based on the Lagrangian. However, these methods do not give the complete minimum set of serial link robots. Moreover, Vinet et al. [7] proposed a formulation to obtain the minimum set of dynamics parameters. The minimum parameter was determined by using a recursive formulation. In this case, the tree robot could be broken up into a chain of three serial links. Therefore, this tree robot can be considered as three link serial robot.

Since the robot is in-pipe, as mentioned earlier, it is useful to look at the literature on in-pipe robots. They could be categorized into wheeled robots, [8–11] track robots, [12] inchworm robots, [13] walking robots, [14], and pig [15]-type robots depending on their traveling mechanisms. They could also be classified according to their construction single-plane [8] type with arms 180° apart, and three-plane [9, 10] four-plane types, which have arms separated by 120° and 90°, respectively. It is obvious that three-plane and four-plane [16] types of robots have more traction force and provide better stability. However, the single-plane type has a simple construction. Another parameter of the in-pipe robot classification is the pipe surface adaptation. Two adaptation techniques are used in the in-pipe robot, one active [12] and another passive [16]. If the robot adapts to the surface of the pipe with a spring, it is called passive adaptation. If robots adapt to the surface of the pipe with an actuator, it is called active adaptation. However, the active adaption technique is more effective than passive adaption given that the normal force could be controlled by the actuator.

Material choice is crucial to the parts of the robot part selection as this to a large extent determines the effectiveness of the robot performance. This paper performs a static structural analysis of the parts of the robot. The main benefit of static structural analysis (or Finite Element Analysis or FEA) is that it could be applied to any shape or dimension. This analysis can be done on homogenous and non-homogenous materials. In this paper, aluminum (5052-H34) is considered for being used for making the platform and links. Various literatures exist for FEA analysis as the methods are well known and understood (see for example, references [17, 18] and the references therein). The FEA of the platform and links have been carried out by SolidWorks software [19, 20]. The primary focus is to save the nodes that have maximum stress and maximum deformation and the element that has maximum strain. Through the simulation, it is ensured that the design is structurally safe and the results lie within the limits of the selected material. This shall serve as a basis for development of a prototype for testing various control algorithms. Furthermore, future work could be focused on developing motion planning algorithms.

The paper is organized as follows. Section 2 contains the model of a tree robot by SolidWorks modeling and its various maneuvers. The parts in the conceptual design are also listed. In Sect. 3, the mathematical calculation to find the required weight for safe design is presented. In Sect. 4, static structural analysis (FEA analysis of robotic components) is presented. Section 5 presents the conclusions and future work.

2 Model of a Tree Robot

In this paper, 5-DOF tree-type robot has been explored for inspection and repair purposes, as shown in Fig. 1. Its components are listed in Table 1. This tree-type robot consists of four links, five stepper motors for each joint, and one servomotor for the end-effector. For the locomotion of the robot, four wheels have been mounted to the platform of the robot. The two linear actuators are attached to link 1, which makes an angle of 60° to link 1, and one linear actuator is mounted on the third link. Two conditions will appear when the tree robot goes inside the pipe. One is when the tree robot performs the operation, and the other is when the robot maneuvers inside the pipe. Moreover, it is assumed only end-effector movement which will be responsible for the repair operation.

At the time of repair operation, linear actuators of the third link will be at the home position and will freely move with the third link. The linear actuators of the first link have two rubber pads mounted at the end, which work as a breaker and ensure stability at the equilibrium point of the tree robot. The plate which contains the suction cup will be operated by a solenoid. Due to the push–pull mechanism of the solenoid, the suction cup will stick to the wall of the pipe, and after the repair operation, solenoid will pull the suction cup. The plate containing the suction cup will work as a flat plate as well as an adjustable diameter plate as shown in Fig. 1e and Fig. 2. The lead screw ensures that the plate acts like a flexible plate (Fig. 2b), thereby bringing the suction cups closer to the pipe surface. The adjustable diameter of the plate as shown in Fig. 2b will ensure that the robot wheel diameter does not impair contact between the cups and the pipe surface.

The diameter of the plate can be regulated according to the pipe diameter. Figure 2a shows the shape of the suction plate when maneuvering on a flat surface, and Fig. 2b shows the shape of the plate when maneuvering inside the pipe.

2.1 Robot Maneuver in Horizontal Pipe

In the case of the horizontal pipe, the linear actuator of the first link will press the horizontal pipe, and link two, three, and four will move to perform the operation. In the rest position of the platform, three forces will act on the tree robot, making an angle of 120° to each other. The first force will act behind the platform, and the other two will act on link 1, shown in Fig. 3a. In this case, the purpose of the two linear actuators of the link first is to increase the stability during the repair operation. And the purpose of suction cups is to reduce the vibration during the repair operation. At

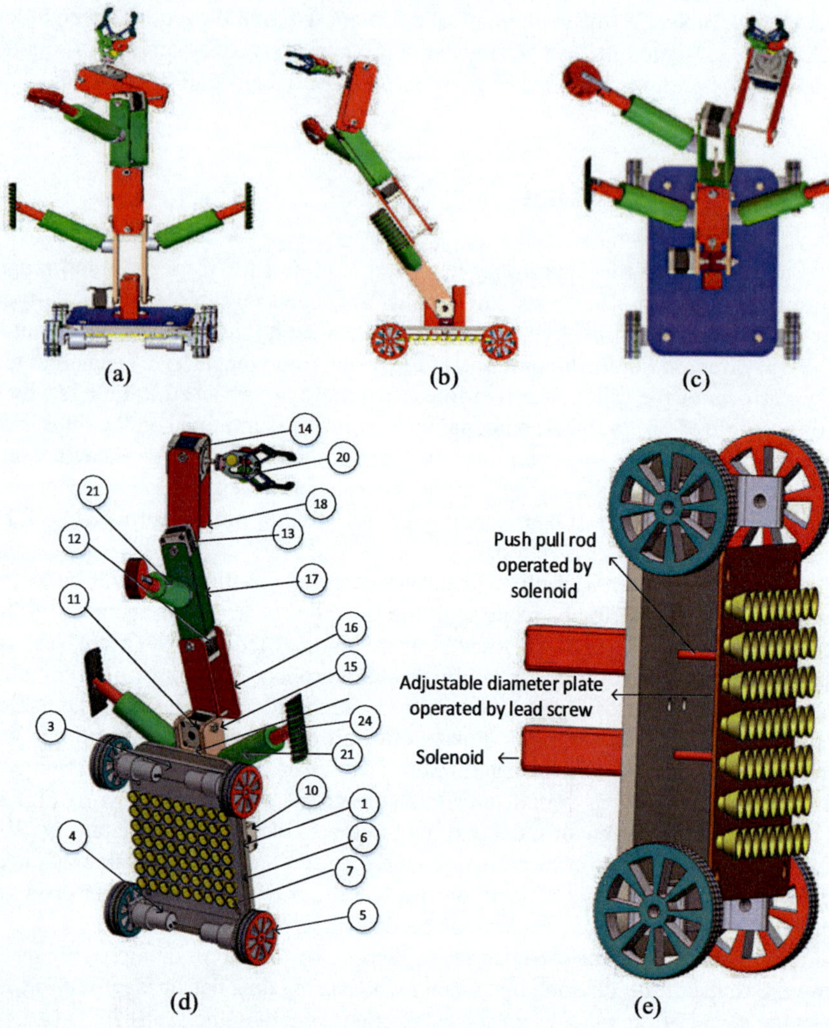

Fig. 1 CAD model of the proposed tree-type robot, **a** front view, **b** side view, **c** top view, **d** parts of the robot (listed in Table 1), **e** solenoid with push–pull mechanism of the suction plate

the time of operation platform and the first link will be at rest, and only the upper three links will move. The workspace of repair is only made with these three links.

Table 1 List of components of the robot

Sr. no.	Component	Quantity	Mass (Kg)
1	Platform	1	1.352
2	Clamp for platform and link 1	1	0.0832
3	Clamp for mounting DC gear motor	4	0.100
4	High torque DC gear motor	4	1.000
5	Robotic wheel	5	0.270
6	Thin plate of the sheet for suction cup	1	0.2368
7	Suction cup (VP15)	72	0.070
8	Push pull solenoid electromagnet	2	0.210
9	Clamp for linear actuator	3	0.075
10	Stepper motor (Nema 17 2.8 kg-cm)	1	0.220
11	Stepper motor (Nema 17 5.6 kg-cm)	1	0.380
12	Stepper motor (Nema 17 7 kg-cm)	1	0.450
13	Stepper motor (Nema 23 10 kg-cm)	1	0.650
14	Stepper motor (Nema 23 19 kg-cm)	1	1.003
15	Link 1	1	0.4296
16	Link 2	1	0.2014
17	Link 3	1	0.200
18	Link 4	1	0.2014
19	High torque metal gear standard digital servo (Stall torque)	1	0.068
20	High-quality advanced robotic gripper	1	0.048
21	Linear actuator	3	2.667
22	Wheel holder	2	0.497
23	Stepper motor holder	5	0.200
24	Rubber pad	2	0.050
25	Nuts, bolts, screws, and washers	–	0.100
		Total	10.7624

2.2 *Robot Maneuver in Vertical Pipe*

In the case of the vertical pipe, the linear actuator of the first link press the vertical pipe, and links two, three, and four will move to perform the operation, as shown in Fig. 3b. The major part of the total load is balanced by the linear actuator of the first link. And a small amount of load is balanced by the suction cup, which is mounted below the platform. The main function of suction cups is to provide stability during the repair operation.

A wheel will be mounted at the end linear actuator of the third link, which will help in the locomotion of the tree robot. When this tree robot moves, the linear actuator on the third link will support in the locomotion, and the linear actuators

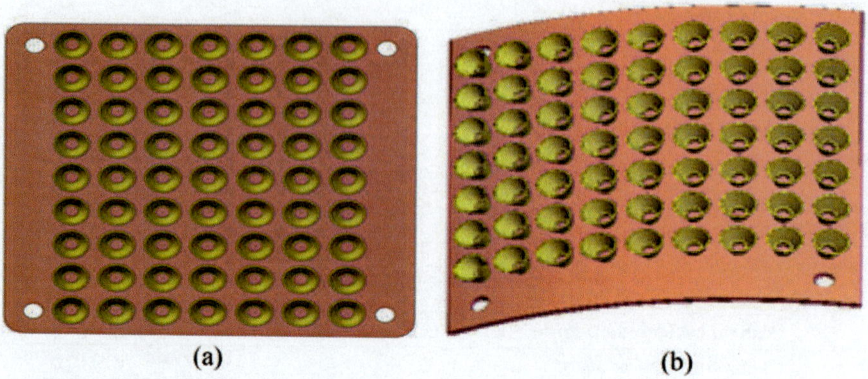

Fig. 2 **a** Plate of the suction cup when maneuvering on the flat surface **b** plate of the suction cup when maneuvering inside the pipe (figures developed in SolidWorks)

Fig. 3 **a** Robot inside an inclined horizontal pipe, **b** robot inside an inclined vertical pipe

on the first link will not work at this time. In this case, also the three link will move like a horizontal pipe, but the orientation of the link will be different from the horizontal pipe. However, the workspace coming into the picture will be the same as the horizontal pipe.

2.3 Robot Maneuver on Flat Surfaces

When a robot moves on flat surfaces, it will simply work as a 5-DOF serial robot. In this case, the linear actuator will be in the home position, i.e., it will not operate. Only suction cups will stick to the flat surface during operation, increasing the tree robot's stability. In this condition, we have to deal with a 5-DOF serial robot, the forward kinematics could be calculated by the D-H parameter, and the inverse kinematic can be calculated by Artificial neural networks (ANNs)/Adaptive Neuro-Fuzzy Inference System (ANFIS). In this case, all five angles would be involved in finding the workspace of the tree robot.

3 Computation of Safe Design Requirements

This section presents the calculation to find the required weight of the platform and links for safe design. The structure analysis is then done on the basis of the calculation of weight. In the calculation, we assume the robot maneuvers slowly and only consider the statics analysis. Dynamic calculations have not been considered in this study. Consider the free-body diagram of the robot in the case of horizontal and vertical climb, as shown in Figs. 4a and b. The coefficient of friction between the rubber tire and concrete pipe is 0.63 [22], which is considered to find the force on various links.

From Fig. 4a, considering all the weights (links, motor, and end-effector), the total forces on the platform can be calculated by the mass. Substituting the values from Table 1, we get the total weight on the platform.

Total weight on platform

$$
\begin{aligned}
&= (10.7624 - 1.3520 - 0.1000 - 1.0000 \\
&\quad - 0.2160 - 0.2368 - 0.0700) \times 9.80665 \\
&= \frac{76.3702}{2} = 38.1851\,\text{N} \sim 38.2\,\text{N}.
\end{aligned}
$$

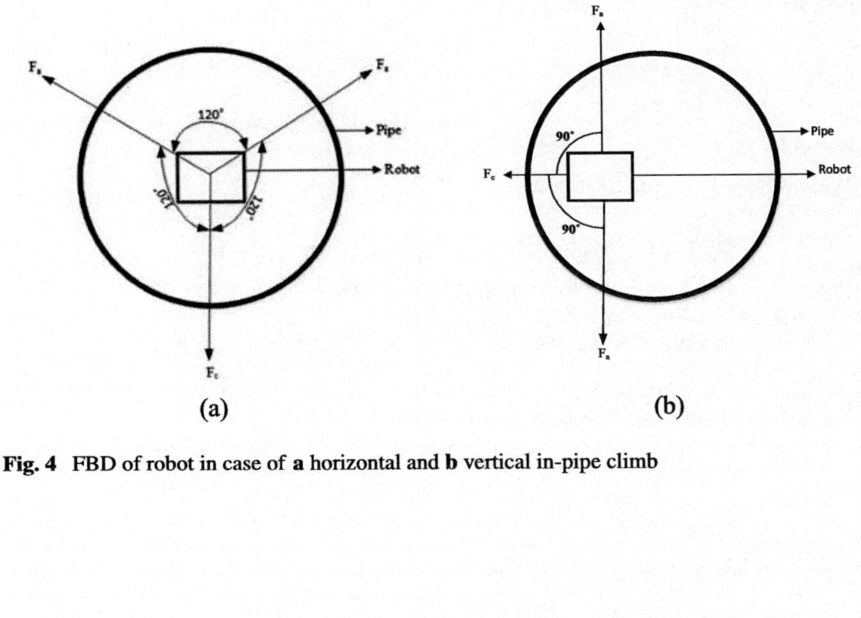

(a) (b)

Fig. 4 FBD of robot in case of **a** horizontal and **b** vertical in-pipe climb

Similarly, the total weight of the links may be calculated as below.
Total weight on link 1

$$7.7876{-}1.003 - 0.0.832 = 6.7013 \times 9.80665 = 65.7173\,\text{N} \sim 65.72\,\text{N}$$

Weight on the first part of the link 1

$$= 65.7173/2 = 32.8586\ \text{N}.$$

Weight on the upper hole of link 1

$$= (7.7876 - 0.4296 - 1.778) \times 9.8066 = 54.7208\,\text{N} \sim 54.73\,\text{N}.$$

Weight on each side of link 1

$$= 54.7208/2 = 27.3604\,\text{N} \sim 27.37\,\text{N}$$
$$F = 2 \times \mu \times N = M \times g.$$

For concrete pipe, the value of 0.63 for the coefficient of friction is obtained from [22].

$$2 \times 0.63 \times N = 10.7624 \times 9.8066$$
$$N = 83.7639\,\text{N} \sim 83.77\,\text{N}.$$

Friction force acting on each link (F_s)

$$F_s = \mu\text{N} = 0.63 \times 83.7639 = 52.7712\,\text{N} \sim 52.78\,\text{N}.$$

Force applied by the linear actuator (F_a)

$$\cos 30^\circ = F_s/F_a$$
$$F_a = \frac{F_s}{\cos 30^\circ} = 60.9349\,\text{N} \sim 60.94\,\text{N}.$$

Weight on upper hole of link 2

$$= (0.2014 + 0.450 + 0.200 + 0.889 + 0.054 + 0.025$$
$$+ 0.380 + 0.220 + 0.0680 + 0.0480) \times 9.8066$$
$$= 2.5354 \times 9.8066 = \frac{24.8636\,\text{N}}{2} = 12.4318\,\text{N} \sim 12.44\,\text{N}.$$

Weight on upper hole of link 3

$$= (0.380 + 0.2014 + 0.220 + 0.068 + 0.048) \times 9.8066$$

$$= \frac{8.9965}{2} = 4.4982\,\text{N} \sim 4.5\,\text{N}.$$

Weight on the upper hole of link 4

$$= (0.220 + 0.0680 + 0.0480) \times 9.8066$$
$$= \frac{3.2950}{2} = 1.6475\,\text{N} \sim 1.65\,\text{N}.$$

Now, the static structural analysis or finite element analysis (FEA) of the platform and links is presented. The analysis was carried out on SolidWorks software.

3.1 FEA of Platform

On the robotic platform, there are four holes, two holes for the clamp, which joint the link 1 and the platform, and two holes for the solenoid, which connects the platform to the pipe wall. The load on the hole of the platform with a clamp is 76.3702 N, and it is equally divided into 38.1851 N. The other two holes have a load of 1.05 N due to the weight of the solenoid. The platform's thickness is optimized by applying the statics load, starting from 5 mm, and the final considered is 10 mm. The total number of node and elements in the platform is found to be 22,489 and 12,364, respectively, having an element size of 9.238869 mm. Figure 5a shows the maximum value of equivalent von Mises stress which is found to be 2519 kN/m^2 (node number 21608) at the hole of filleted surfaces which is less than the yield stress of aluminum (215,000 kN/m^2). The maximum deformation of 0.005263 mm (node number 6700) is recorded near the hole, and the von Mises stress and deformation are found to be within limits, as shown in Fig. 5b. The maximum strain was found to be 1.552×10^{-5} on element 7587, as shown in Fig. 5c.

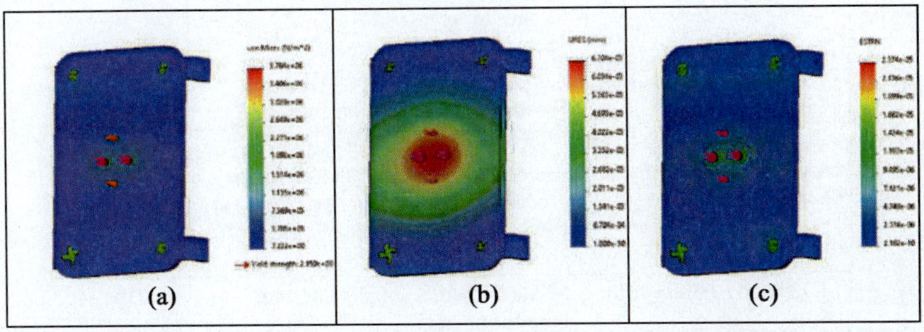

Fig. 5 **a** Stress analysis of platform, **b** deformation analysis of platform, and **c** strain analysis of platform

3.2 FEA of the Links

In the case of link 1, the force on the hole is 32.8586 N, and the load exerted by the linear actuator is 52.7712 and −83.7639 N in X and Z directional, respectively. The thickness of the link 1 is chosen from 5 mm onward, up to a value of 8 mm, while for links 2–4, the thickness is chosen from 2 mm up to a value of 5 mm. Table 2 presents a similar analysis as in Fig. 5, for each of the links. The first column shows the nodes and elements in each link. The second column shows the element size. The third column shows the maximum equivalent von Mises stress and its location, whereas the fourth column shows the maximum deformation and its location. The fifth column shows the maximum strain.

From the findings presented above, the following conclusions may be drawn:

- The maximum equivalent stress is less than yield stress of aluminum (215,000 kN/m^2) for all of the robot links as well as the robot platform. Aluminum is commonly used for manufacturing of links.
- The maximum deformation is found at the hole for the links as well as for the platform.
- The tree-type robot structure, as evident from Fig. 1, is such that the branched links are intended to hold on the pipe when the main end-effector of the robot carries out the repair. Although the structure appears to resemble a single link, when the robot is being developed, a single link shall be followed up by an appropriate end-effector to grip on the sides of the pipe. The designs for such a link plus the

Table 2 Results of FEA analysis for the robot links

Link no.	Node and elements	Elements size	Maximum equivalent von Mises stress	Maximum deformation	Maximum strain
1	13,204 and 7739	4.28328 mm	58,490 kN/m^2 (node number 13060) at joint 1 and linear actuator	1.065 mm (node number 8344) at hole of link 1	6.560×10^{-4} on element 5130
2	14,920 and 8663	3.32034 mm	208.70 kN/m^2 (node number 13873) at hole	0.0000106 mm (node number 9) at hole of link 2	2.182×10^{-6} on element 3023
3	16,896 and 9634	4.18692 mm	73,660 kN/m^2 (node number 16500) at the hole	1.523 mm (node number 12,681) at hole of link 3	8.433×10^{-4} on element 1632
4	14,920 and 8663	3.32034 mm	27.660 kN/m^2 (node number 13873) at the hole	0.000001 mm (node number 9) at hole of link 4	2.892×10^{-5} on element 3023

end-effector are already readily available in the market for gripping purposes, and hence, FEA analysis was not necessary.

From the above observations, it may be concluded that the design is safe for the link lengths chosen. To the best of the authors' knowledge, there is no work done on the design and development of tree-type kinematic chains for pipe repair operations. The direction in which this work must proceed is to perform the kinematic and dynamic analysis for such a robot along with planning and control for various repair operations like welding, crack sealing, among others.

4 Conclusions and Future Work

This paper presented the FEA analysis of a novel 5-DOF tree-type robot for inspection and repair purposes. The weight of links and platform was computed for safe design. Static analysis was presented using SolidWorks to ensure safe design. This shall ensure effective development of a prototype for testing the robot in real-world applications. The application of this robot is not limited to only the horizontal and vertical pipes and in the future could extend to angular pipes. To the best of the authors' knowledge, the work is one of the first to put forth the use of tree-chained kinematic structures for pipe repair applications. Future work involves the kinematics, dynamics, and planning for such a robot for various repair applications like welding, crack sealing, among others.

References

1. Al-Matter D, Youcef-Toumi K (2013) Pipe leakage repairing robot. Kuwait-MIT Center for Natural Resources and the Environment, 28 Aug 2013
2. Schilling K, Roth H (1999) Navigation and control for pipe inspection and repair robots. IFAC Proc Vol 32(2):8446–8449
3. Fjerdingen SA, Liljebäck P, Transeth AA (2009) A snake-like robot for internal inspection of complex pipe structures (PIKo). In: 2009 IEEE/RSJ international conference on intelligent robots and systems, IEEE, pp 5665–5671
4. Mayeda H, Yoshida K, Osuka K (1988) Base parameters of manipulator dynamic models. In: Proceedings of the IEEE international conference on robotics and automation, IEEE, pp 1367–1372
5. Khosla PK (1989) Categorization of parameters in the dynamic robot model. IEEE Trans Robot Autom 5(3):261–268
6. Khalil W, Bennis F, Gautier M (1989) Calculation of the minimum inertial parameters of tree structure robots. In: Advanced robotics. Springer, Berlin, Heidelberg, pp 189–201
7. Vinet L, Zhedanov A (2011) A 'missing' family of classical orthogonal polynomials. J Phys Math Theor 44(8):37–72. https://doi.org/10.1088/1751-8113/44/8/085201
8. Oya T, Okada T (2005) Development of a steerable, wheel-type, in-pipe robot and its path planning. Adv Robot 19(6):635–650
9. Roh SG, Choi HR (2005) Differential-drive in-pipe robot for moving inside urban gas pipelines. IEEE Trans Rob 21(1):1–17

10. Horodinca M, Dorftei I, Mignon E, Preumont A (2002) A simple architecture for in-pipe inspection robots. In: Proceedings of the international colloquium on mobile and autonomous systems, pp 61–64
11. Li P, Ma S, Li B, Wang Y (2007) Development of an adaptive mobile robot for in-Pipe inspection task. In: International conference on mechatronics and automation, pp 3622–3627
12. Park J, Hyun D, Cho W, Kim T, Yang H (2001) Normal-force control for an in-pipe robot according to the inclination of pipelines. IEEE Trans Industr Electron 58:5304–5310
13. Bertetto AM, Ruggiu M (2001) In-pipe inch-worm pneumatic flexible robot. Proc IEEE/ASME Int Conf Adv Intell Mechatron 2:1226–1231
14. Yu X, Chen Y, Chen MZ, Lam J (2015) Development of a novel in-pipe walking robot. In: 2015 IEEE international conference on information and automation, IEEE, pp 364–368
15. Mishra D, Agrawal KK, Abbas A, Srivastava R, Yadav RS (2019) Pig [pipe inspection gauge]: an artificial dustman for cross country pipelines. Proc Comput Sci 152:333–340. https://doi.org/10.1016/j.procs.2019.05.009
16. Miyagawa T, Iwatsuki N (2007) Characteristics of in-pipe mobile robot with wheel drive mechanism using planetary gears. In: Proceedings of the international conference on mechatronics and automation, pp 3646–3651
17. Razali ZB, Daud MH, Derin MAD (2015) Finite element analysis on robotic arm for waste management application. In: International conference on application and design in mechanical engineering, ICADME2015, Pulau Pinang, Malaysia
18. He DT, Guo Y (2016) Finite element analysis of humanoid robot arm. In: 2016 13th international conference on ubiquitous robots and ambient intelligence (URAI), IEEE, pp 772–776
19. Gupta M, Narayan J, Dwivedy SK (2020) Modeling of a novel lower limb exoskeleton system for paraplegic patients. In: Advances in fluid mechanics and solid mechanics 2020. Springer, Singapore, pp 199–210
20. Narayan J, Mishra S, Jaiswal G, Dwivedy SK (2020) Novel design and kinematic analysis of a 5-DOFs robotic arm with three-fingered gripper for physical therapy. Mater Today Proc 1(28):2121–2132
21. Kumar S, Reddy BS (2022) Path-planning of robot end-effector for hairline crack sealing using intelligent techniques. In: International conference on advances in mechanical engineering and material science 2022. Springer, Singapore, pp 271–282
22. Moyer RA (1934) Skidding characteristics of road surfaces. In: Highway research board proceedings, vol 13

Simulated Evaluation of Navigation System for Multi-quadrotor Coordination in Search and Rescue

Rayyan Muhammad Rafikh and Jeane Marina D'Souza

Abstract Many methods have been developed which assist in the localization of victims trapped in disaster-prone areas. Disaster management after the immediate onset of such sudden occurrences intimates readiness in technology, availability, accessibility, perception, training, evaluation, and deploy ability. This can be attained through evaluation and comparison of different techniques supplanting each other, essentially covering each aspect of the Search and Rescue operation. Developments by academia and industry have led to deep learning advancements like the use of Convolutional Neural Networks resulting in an increasing dependence of first responders on UAV technology fitted with state-of-the-art machines working with real-time information from various sensors. We have, in this paper, proposed a technique to implement a simulation involving detection of life in the immediate occurrence of disasters with the assistance of a deep learning model simultaneously deploying multi-quadrotor coordination among the vehicles with the use of an appropriate region-partitioning method to speed up the process even further. Moreover, other non-conventional techniques have also been discussed.

Keywords Disaster management · Search and rescue · Convolutional neural network · UAV · Deep learning · Multi-quadrotor coordination · Region partitioning

R. M. Rafikh
Electrical and Computer Engineering Department, Sultan Qaboos University, Muscat 123, Oman
e-mail: rayyanrafikh6@gmail.com

J. M. D'Souza (✉)
Department of Mechatronics Engineering, Manipal Institute of Technology, Manipal, Karnataka, India
e-mail: jeane.dsouza@manipal.edu

S. Sharma et al. (eds.), *Intelligent Control, Robotics, and Industrial Automation*, Lecture Notes in Electrical Engineering 1066,
https://doi.org/10.1007/978-981-99-4634-1_11

1 Introduction

A UAV is principally any aircraft that can fly autonomously or with partial remote control [1, 2]. Quadrotors, a category of aerial vehicles, have become increasingly popular in many use cases due to the numerous advantages this technology provides over its conventional alternatives [3]. These advantages include the ability to hover at place and to fly in any direction without changing heading. Furthermore, developments involving electromechanical systems and sensory techniques have made research in unmanned systems very popular. The feasibility of indoor flights has increased because of their smaller sizes and higher maneuverability. Being less complex, they have attracted applications in industries doing varying tasks such as aerial reconnaissance, search and rescue, structure inspection, and much more [4, 5].

Recently, increasing research in this area has been focused on computational methodologies which would enhance the efficacy of inter-operability among machines while they coordinate among each other to perform tasks collectively [6–13]. This has led to a glut of algorithms involving generation of waypoints for UAVs based on an appropriate division and allocation of area.

Human localization, which involves detection and tracking, has been implemented through pre-trained models developed by experts with excessive data to detect and track various classes of objects. Some of these include You Only Look Once (YOLO) and Visual Geometry Group (VGG-16). The YOLO specifically has observed developments in multiple versions, with it being updated recently to the fifth version [14, 15]. The requirements of these models are to have the output layer of the pre-trained model replaced with a custom output layer as per our target classes.

The aim of this paper is to establish an autonomous system comprising a swarm of quadrotors navigating within the open spaces of a disaster-affected area after the immediate onset of a flood, in the quest for possibility of life. The application of swarms in the search of objects of interest in water bodies has been researched thoroughly [16, 17]. This paper expounds the aforesaid concept through a realistic simulation. Simulators have since long been crucial factors in research when it comes to rapid evaluation of concepts, strategies, and algorithms, in robotics. Gazebo, one of the most eminent simulators in the field, was chosen as the simulator to proceed with for the prospects of this paper due to various reasons, one being its higher ease of integration with the framework we have used for developing algorithms for the same.

The Robot Operating System (ROS) is the leading open-source software framework used for robotics applications by industry as well as academia. Being ordered into different sets of packages, it enables ease of communication between the simulator and ROS framework. Furthermore, the robotics community, or the Open-Source Robotics Foundation in particular, provides various other resources to aid professionals in research.

Fig. 1 'Hector_quadrotor' model

2 Methodology

2.1 *'Hector_Quadrotor' Package*

The OSRF or Open-Source Robotics Foundation enables access to an substantial set of libraries and packages which can be utilized for research in integration with ROS. One such package is 'hector_quadrotor', which gives an implementation of the 'hector' quadrotor. Perception, navigation, and control modules of the quadrotors were based on and handled by the aforesaid package so that more significance can be laid on the main aim of the project.

In addition to the quadrotor model as shown in Fig. 1, it combines numerous perception, navigation, control, and guidance techniques through the use of a variety of sensors and algorithms. It considers the dynamic properties of the drone in addition to interactions with objects in its surroundings such as dynamic and static obstacles.

2.2 *Perception*

Realization of sensory data from drones was done with the help of messages encoded and published on top of different topics with separate namespaces depicting the topics published by sensors in different quadrotors. Another method of vision is by the utilization of 'rostopic echo' where the rate of display, the number of messages to be displayed, etc. can be controlled. Where the data are encoded as 'sensor_msgs.msg.Image', the built-in (into ROS) package 'image_view' carries out its role of converting it to displayable format and subsequently opening the image in a new window arising from the terminal.

2.3 Guidance

This project, focused on subsuming a hierarchy of communication between multiple quadrotors, required an ordered method for generation of waypoints and trajectories of quadrotors. Therefore, a guidance algorithm was needed that could partition a region into multiple convex non-overlapping regions simultaneously allocating each quadrotor to one of the so formed regions, thereby completing the guidance process required for autonomous multi-quadrotor coordination.

The principle requirements of the guidance module were studied to be:

1. The quadrotors must function on their own, that is, without a leader drone among them or without using a master–slave architecture.
2. The malfunction or collapse of any of the quadrotors must be handled autonomously by the functioning ones during the operation.
3. The search area of any of the drones must not overlap with that of the other drones.

The afore-mentioned objectives have been thoroughly discussed in numerous papers and comparable conclusions had been made. Moreover, present research suggests algorithms like Voronoi-based space partitioning of regions [18] and Ant Colony Optimization, in order to accomplish these objectives, and it relies on the user to select between these methods, depending on the hardware and software, as well as mission-specific needs. It is because each of the algorithms considers different variables of the operation and consumes varying amounts of computational resources.

An investigation of the mission-specific demands led us to an agreement with a few sets of points that we needed our guidance algorithm to have, which are:

1. The divisions must be convex and inside the bounds of the search area in consideration.
2. The sensor search effectiveness of drones must be considered during computation of the convex regions.
3. The return of the computation for each drone must be in the form of intermediate waypoints, while it flies from the location in the 0th iteration, that is the start position, to the location in the final iteration, that is the goal position.

This led to reroute emphasis on one algorithm, namely the Voronoi CVT. The fundamental principle behind the Voronoi algorithm is implemented in various forms, among which the CVT version matched all the afore-mentioned conditions and was hence implemented in the due course of research. It is principally a method for dividing an area into points closest to some pre-defined points and simultaneously relocating them so that they also are centroids to those assigned points. The last-mentioned feature is unique to this version of the Voronoi Partitioning Scheme, and thus, this algorithm has been implemented in unmanned aerial vehicles for the deployment of swarms to optimize parameters such as duration and accuracy.

The variables to be input to this node included the dimensions of the search area in terms of coordinates of the corners of the convex polygon of interest, the number of drones, and the sensor range.

2.4 Detection

Building on statistics of performance of different models trained on identical databases, it was established that the YOLO model assures faster and more accurate performance than similar models.

YOLO, a regression model, is faster than other deep learning methods involving classification, as it finds both the classes and bounding boxes for the entire image at once. Furthermore, high-speed detection attained with this model has made its use more promising in self-driving cars and other applications, where real-time detection is required.

Therefore, after a comprehensive study, it was found that YOLO is the most promising model for our requirement. Some of the advantageous features which led to this decision are highlighted below:

1. It enables faster detection, at approx. 45 FPS which is a necessity as a consequence of the high speed of UAVs.
2. It provides a model named as Tiny-YOLO which has a smaller, thus enabling detections at even higher framerates of 155 FPS, though with slightly lesser accuracy when compared to the original model. This results in lesser computational burden enabling drones to be employed with substantially lesser complex and more economical hardware.
3. It provides indistinguishable accuracies of detection for both real world and simulated objects. This is a necessity as this study involves drones traversing through simulated environments.

2.5 Communication Among the Nodes

The main Python node used many built-in libraries and packages of ROS. The objective of this node is to sustain integration with numerous other ones which have been launched simultaneously. Some other functions are as listed below:

1. It keeps control of each quadrotor in the Gazebo simulated world.
2. It launches the human detection node when required, publishes camera readings to the topics subscribed by the launched node, and also retrieves information from the topic that the detection node publishes to.

The control of drones, as depicted by Fig. 2, can be broken down into various parts that occur subsequently. The motors of the quadrotors are powered in succession to that of the other quadrotors in a sequence, as they consecutively takeoff one after the

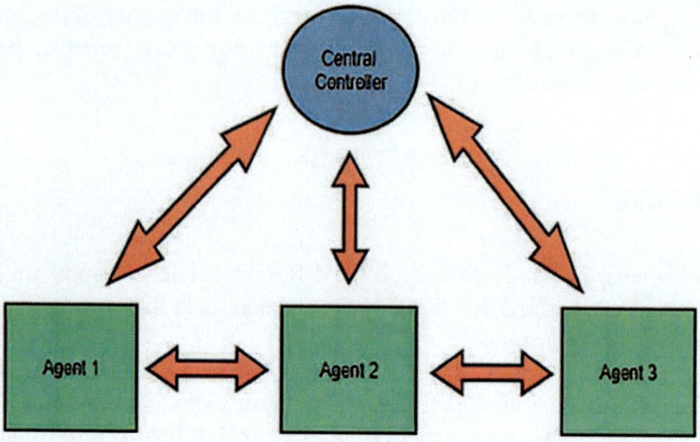

Fig. 2 Centralized control architecture for multi-robotic systems

other. This takes the drones to a pre-defined height before subsequent operation. The next control routine aims at directing the drone to a desired waypoint in the trajectory that was pre-computed by the Voronoi node as a CSV file. This step of the control strategy is implemented in steps of each waypoint as the loop iterates over each row in the afore-mentioned CSV file.

This routine executes detection of life from the camera's point of view. In this step, **the visual data** from the camera of that drone is sent to the '/camera/rgb/image_raw' topic, in which the detection node receives, processes, and sends the result as the same image but with humans detected localized in bounding boxes to the topic '/darknet_ros/detection_image' while simultaneously displaying the live detection image.

The main node then finds the message inside the topic '/darknet_ros/found_object' and appends the information of the number of humans found to a separate CSV file called the count.csv.

3 Results and Discussions

3.1 Region Partitioning Analysis

The Voronoi routine implemented through MATLAB code enabled faster computation of the partitions of separate regions for multiple (9, 11, 17, 23, 32) drones that we had used in Gazebo simulations. Furthermore, separate windows help visualize the uncertainty distribution of the search spread over the complete area and the trajectories that each drone had followed when traversing from start to current positions.

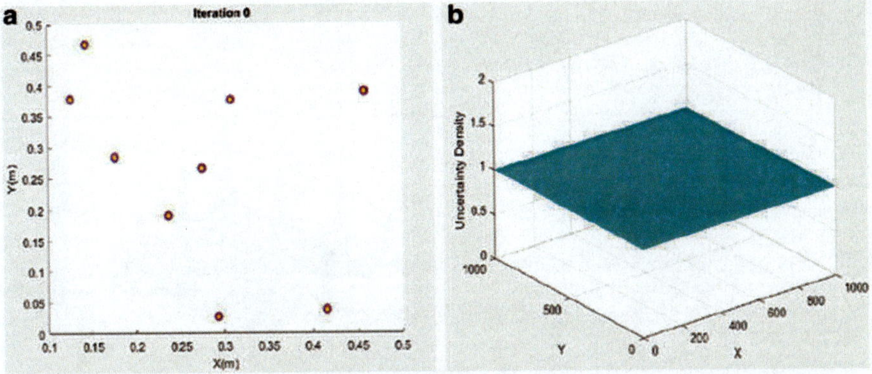

Fig. 3 **a** Initial positions and **b** corresponding total uncertainty distribution of nine quadrotors

Figure 3 denotes that the drones were initiated with an initial overall uncertainty of 1 when the positions of all the drones were close to the origin, which, in this case, is (0, 0).

Figure 4a tells us that nine drones had followed their trajectories to an updated location, with each being allocated separate, non-overlapping regions. The path followed by each of the machines is visualized with the help of curvy lines. Correspondingly, from Fig. 4b, it can be observed that the search uncertainty over the region has been reduced, resulting in nine different parabolic curves, each representing the uncertainty of search carried out by each drone.

Figure 5b denotes that as the quadrotors traverse over the region, the uncertainty in their search of the areas reduces parabolically which can be proven from the knowledge that the uncertainty distribution in regard to the search effectiveness has been modeled as a Gaussian distribution function of the sensor range, the positioning of the drones, and thus the sensors through the area grid. It can be noted from

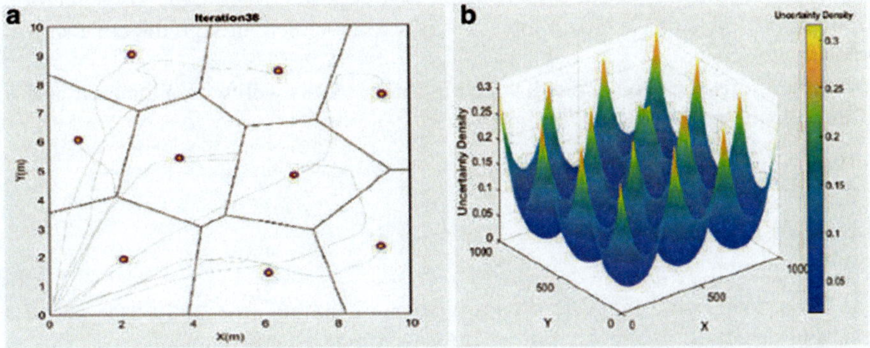

Fig. 4 **a** Intermediate positions and **b** corresponding total uncertainty distribution of nine quadrotors

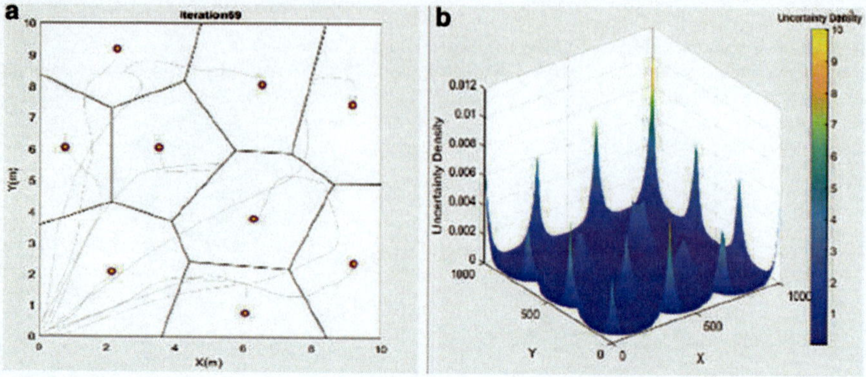

Fig. 5 **a** Final positions and **b** corresponding total uncertainty distribution of nine quadrotors

Table 1 Variation of number of iterations and cycles with number of drones

Number of drones	Cycles required	Iterations in the last cycle
9	4	29
11	3	81
17	3	142
23	2	112
32	2	120

that the calculation of new goal locations and hence new waypoints cease once the uncertainty associated with current positioning reduces to nearly zero. Likewise, these computations were carried out for eleven, seventeen, twenty-three, and thirty-two number of drones.

It can be observed from Table 1 that with an increase in the number of drones, the number of cycles required to reduce the uncertainty distribution decreases. It can also be noted that this is not a strict trend as it depends on the random generation of initial positions of the quadrotors. This is also indicated through the table shown below.

Furthermore, it can be noted that the algorithm works well within the extremities of the convex region that was assigned to it.

3.2 *Simulation Analysis*

The main node controlling the drones enables display of captured image and the resulting detection image for further observation at a later point of time. This is crucial in determining the accuracy of our model.

The detection image can be studied to the observation that the model confuses some of the objects at a distant location with humans. Moreover, it also fails to detect some humans at that very same distance from the camera.

The 'rqt' graph in Fig. 6 depicts the different active nodes and the publishing/ subscribing that takes place between these drones.

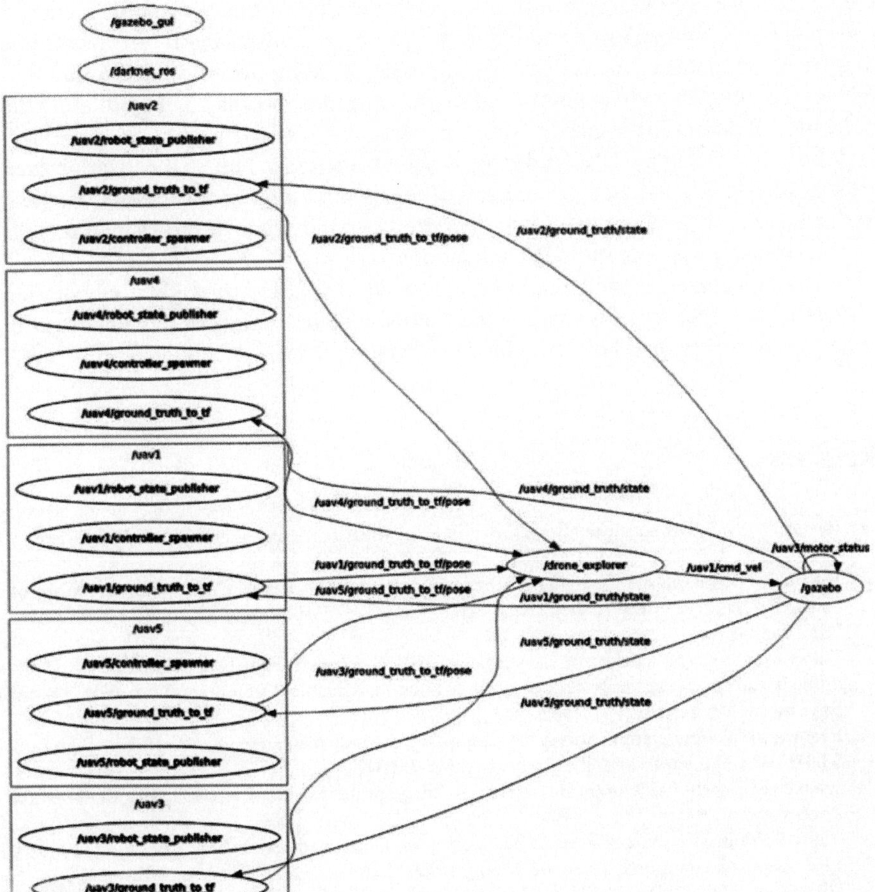

Fig. 6 'rqt' graph

4 Conclusion

The future will undoubtedly be governed by the drone business. Increasing adaptation and better perception have resulted in its existence in all industrial sectors, while also raising safety concerns. Increasing acceptance has led this technology to overcome present difficulties, resulting in rapid development. The present advancements in drone technology are to a large extent due to the accomplishments of researchers in both academia and industry. Moreover, a promising future awaits due to different modules of the technology increasingly being fueled by Artificial Intelligence, enabling them to act more and more like human beings, thereby dilating the utilization of drones in the quest to supplant humans in crisis management operations.

It was observed that YOLO-V4 model is quite efficient in human localization even with an almost forward-facing camera, but there is still a large scope for improvement in this aspect. This encourages us to dive deep into the usage of pre-trained models through transfer learning in SAR applications.

These pre-trained models could be enhanced in efficacy through augmentation of training datasets, like for say, models trained on labeled data of humans from an aerial perspective would help us achieve even better, robust detections.

References

1. Restas A (2015) Drone applications for supporting disaster management. World J Eng Technol 03:316–321. https://doi.org/10.4236/wjet.2015.33c047
2. Bucknell A, Bassindale T (2017) An investigation into the effect of surveillance drones on textile evidence at crime scenes. Sci Justice 57(5):373–375. https://doi.org/10.1016/j.scijus.2017.05.004
3. Bansod B, Singh R, Thakur R, Singhal G (2017) A comparison between satellite based and drone based remote sensing technology to achieve sustainable development: a review. J Agric Environ Int Dev (JAEID) 111:383–407
4. Restas (2015) Drone applications for supporting disaster management. World J Eng Technol 03:316–321. https://doi.org/10.4236/wjet.2015.33c047
5. Franke UE (2015) Civilian drones: fixing an image problem? ISN Blog. Int Relat Secur Netw. Retrieved 5
6. Hayat S, Yanmaz E, Muzaffar R (2016) Survey on unmanned aerial vehicle networks for civil applications: a communications viewpoint. IEEE Commun. Surv. Tutor. 18:2624–2661
7. Giyenko A, Cho YI (16–19 Oct, 2016) Intelligent UAV in smart cities using IoT. In: Proceedings of the 2016 16th international conference on control, automation and systems (ICCAS). Gyeongju, Korea, pp 207–210
8. Bupe P, Haddad R, Rios-Gutierrez F (9–12 April, 2015) Relief and emergency communication network based on an autonomous decentralized UAV clustering network. In: Proceedings of the SoutheastCon 2015. Fort Lauderdale, FL, USA, pp 1–8.
9. Giagkos A, Wilson MS, Tuci E, Charlesworth PB (7–10 June, 2016) Comparing approaches for coordination of autonomous communications UAVs. In: Proceedings of the 2016 international conference on unmanned aircraft systems (ICUAS). Arlington, VA, USA, pp 1131–1139
10. Luo C, Nightingale J, Asemota E, Grecos C (11–14 May, 2015) A UAV-Cloud System for Disaster Sensing Applications. In: Proceedings of the 2015 IEEE 81st vehicular technology conference (VTC Spring). Glasgow, UK, pp 1–5

11. Scherer J, Rinner B (21–25 Aug, 2016) Persistent multi-UAV surveillance with energy and communication constraints. In: Proceedings of the 2016 IEEE International Conference on Automation Science and Engineering (CASE). Fort Worth, TX, USA, pp 1225–1230
12. Cai C, Carter B, Srivastava M, Tsung J, Vahedi-Faridi J, Wiley C (29 April 2016) Designing a radiation sensing UAV system. In: Proceedings of the 2016 IEEE systems and information engineering design symposium (SIEDS). Charlottesville, VA, USA, pp 165–169
13. Liu Y, Lv R, Guan X, Zeng J (12–15 June, 2016) Path planning for unmanned aerial vehicle under geo-fencing and minimum safe separation constraints. In: Proceedings of the world congress on international control and automation. Guilin, China, pp 28–31
14. Handalage U, Lakshini K. Real-time object detection using YOLO: a review. https://doi.org/10.13140/RG.2.2.24367.66723
15. Ahmad T et al (2020) Object detection through modified YOLO neural network. Sci Program 2020
16. Sujit PB, Beard R (2009) Multiple UAV path planning using anytime algorithms. In: 2009 American control conference. IEEE
17. Mendonça R, Marques MM, Marques F, Lourenço A, Pinto E, Santana P, Coito F, Lobo V, Barata J (19–23 Sept, 2016) A cooperative multi-robot team for the surveillance of shipwreck survivors at sea. In: Proceedings of the OCEANS 2016 MTS/IEEE Monterey. Monterey, CA, USA, pp 1–6
18. García M, Puig D, Wu L, Solé A (2007) Voronoi-based space partitioning for coordinated multi-robot exploration. JoPha: J Pys Agents 1(1):37–44. 1. https://doi.org/10.14198/JoPha.2007.1.1.05. ISSN 1888-0258

Comparative Empirical Analysis of Biomimetic Curvy Legged Bipedal Robot with Linear Legged Bipedal Robot

Abhishek Murali, Raffik Rasheed, and Ramkumar Arsamy

Abstract Stability, desired locomotion position, and orientation are crucial in the control of bipedal robots. This paper focuses on the design and analysis of a bipedal robot with a curved knee link versus a linear knee link robot. It also focuses on its own weight balance while moving from one position to another and changing joint angles without falling. This modified design is also compared with the NAO humanoid robot in real-time. The final bipedal robot model mentioned in this paper is a biped robot with 12 degrees of freedom, with two legs coaxially connected to the hip region. The robot stands about 26 cm tall and weighs about 580 g. The design is symmetric to keep the robot's center of mass under the lower limb region. Each leg has six degrees of freedom (DOF), which are controlled by an open-loop system. Robots have undergone various design iterations for analysis. The main aim of this analysis work is to examine and compare a suitable link design that provides a stable walk for a bipedal robot. The experimental results show that the curvy link with a biomimetic design is more stable. Compared to a linear-linked robot, the biomimetic bipedal design moves and balances its weight effectively during mobility.

Keywords Bipedal robot · Joint angles · Biomimetic joints · Servo motor · DH parameters · Arduino UNO · Stability analysis · Degrees of freedom

1 Introduction

Humanoid robots are autonomous machines that resemble human anatomy and can walk, dance, speak, interact with humans, and perform various critical tasks. It has electromechanical components controlled by computer software. Mechanical design is crucial in humanoid robots. It should resemble humans and be able to move from one location to another without falling. The human leg anatomy has two links: the

A. Murali · R. Rasheed (✉) · R. Arsamy
Department of Mechatronics Engineering, Kumaraguru College of Technology, Coimbatore, Tamilnadu, India
e-mail: raffik.r.mce@kct.ac.in

S. Sharma et al. (eds.), *Intelligent Control, Robotics, and Industrial Automation*, Lecture Notes in Electrical Engineering 1066,
https://doi.org/10.1007/978-981-99-4634-1_12

Femur, which connects the knee joint and hip, and the Tibia and Fibula, which connects the foot and knee joint. The Femur bone, also known as the longest bone, handles balancing body weight and moving us. Tibia and Fibula are smaller than the Femur, but they must also withstand whole body weight in addition to the Femur's weight. A humanoid robot's main goal is to balance its own weight and move from one location to another. As a result, mimicking the design of human anatomy may improve its stability. This paper discusses a human-inspired design for a humanoid bipedal robots lower limb for stable and balanced walking. It should have its own rigid structure during the locomotion. It must be able to balance its weight during the orientation change in the swing phase. The link design is important in a robot's stance and swing phase. As a result, the replication of human link design to achieve the gait cycle. The biped is modeled as a robotic manipulator made up of a series of various links connected with suitable joints. Each link must relocate to its demanded position to achieve stable human-like biomimetic bipedal locomotion. Forward and inverse kinematics are used to correlate all joint variables to the desired end-effector position and orientation, and vice versa [1]. Human gait is the most efficient form of locomotion, so ten DOF is the minimum number for the lower extremity to achieve human-like walking. The extra four degrees of freedom are for weight balancing in roll and pitch orientation without relying on the leg DOF. To simplify kinematic computation and control, all joints are designed to intersect at a local point. The upper leg and lower leg are designed with equal length. The gap between the left and right hip joints is also significant. If the distance is too close, the legs will collide, and if it is too far, then ankle motors must work harder to swing the center of mass from side-to-side balance while walking [2]. This design has 12 DOF to facilitate locomotion and for balancing the weight of the controller and battery. The most difficult part to design the biped robot is the leg, and it has the combination of mobility and optimum weight requirements. To make the robot more stable, it should have less weight. The DOF and configuration influence leg mobility. Like other bipedal robots, the modified robot also has a flat foot, and the ankle joint servo motors control the foot's mobility [3]. As a result, the proposed generative design reduced excess materials and the links were filled with 30% infill during 3D printing.

2 Technical Details

2.1 Mechanical Design and Manufacturing

Mechanical design plays a major role in the bipedal robot. A bipedal robot should be capable of dynamic walking, stability, and push recovery [4]. Autodesk Fusion 360 Software was used to create the entire model. The Servo Motor is connected to each link via the Servo Horn. By providing opposite link support, the vibration in the robot while walking is reduced. The overall DOF of the Bipedal robot is twelve, with three axes in the hip joints, one in the knee joint, and two in the toes. 4 degrees of freedom are provided in the lateral plane which provides additional stability to the robot [5]. During the swing phase, the upper body's overall weight should support the CG on the opposite stance leg. If the CG is not perfectly balanced, it will either be stable or fall. Consider the type of manufacturing or assembling method used for making the joints during design. The length of the joint is an important parameter to consider during the design process. Because we used an SG90 motor, the torque capacity is only about 2.5 kg/cm. The distance between the joints is 40 mm from the center of the two servo shafts. The MG995 servo is used in the toe because its torque capacity is around 13 kg/cm, and because the overall weight of the robot will be concentrated only in the toe region, it is expected to supply more torque there. The SG90 Servo used in the toe could not lift the leg during the swing phase. The total length of the link was approximately 60 mm from hip to knee and 50 mm from knee to ankle. The total height of the robot is 230 mm, and no joint or part will be thicker than 5–6 mm [6]. The robot foot was designed into two parts and made flat. The length of the foot from the joint is large than the back side so that the ZMP is kept on the front side [7, 8]. The fusion deposition modeling process is used to create the parts. ABS plastic is used, and it is a stronger and more durable material for screwing. Larger parts were divided into small components, that could be fastened and easily replaced even if they broke. Using the generative design, the infill was reduced while maintaining the same strength. The servo motors are connected to their joints, the joints are connected to the servo horns to secure their position to join with the servo shaft. The toe part is a little more complicated because the total height from the heating bed is 25 mm, so the parts had to be printed separately to reduce the possibility of breakage and failure. Robots for agricultural tasks, cleaning, and health monitoring purposes were designed and developed using 3D printing techniques [9–12]. Figure 1 shows the 3D-printed parts of the bio-inspired link.

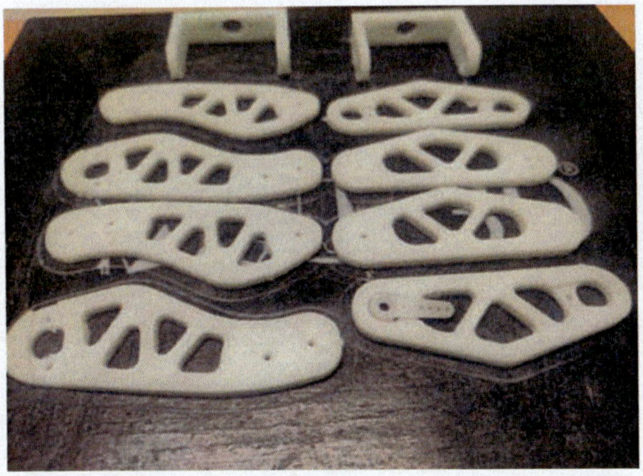

Fig. 1 Three-dimensional-printed links for the bipedal robot

2.2 Various Link Design Iterations

These are the earlier versions of the bipedal robot, with minor upgrades in each stage. The 1 DOF robot is shown in Fig. 2a which is a simple version for learning the kinematic motion of the leg mechanism. The link, joint, and link lengths were calculated and designed using human anatomy fundamentals. The metal rod of 3 mm diameter was added to provide additional stiffness and support during the walk. That resembles the muscle. The result resembled the human gait cycle. However, the Biped's stability was inadequate. It vibrates violently and is incapable of performing any other actuation other than linear walking. The 1 DOF-Biped Mechanism Prototype which is mentioned in Fig. 2b is built to study the nerve mechanism using a string. This model replicated the exact gait cycle, but the nylon string could not be able to keep its stiffness after several motions, and it needed to be tightened after a few cycles. Biped robot with 6 DOF and opposing support are mentioned in Figs. 2c and 2f. Then the bipedal robot model was created with 6 DOF by providing three servo motors in each leg. Each ankle, knee, and hip have one revolute joint. The robot's stability has improved, and its mobility was ensured with less vibration. The servo motor is not supported while actuating the knee joint, so it vibrates once more. The vibration was significantly reduced after adding the additional link support opposite to the motor. Because both links must rotate on the same axis, the opposite link is designed to be coaxial to the servo shaft. Modified 6 DOF-biped robot with no opposing support is shown in Fig. 2d and 2e. This design has no motor support, but it is unable to balance the center of mass and falls continuously during each swing phase. It also vibrates more during the stance phase. The design was improved by adding a pivot joint to support the servo motor, and this reduced vibration but increased weight. The minimum torque capability of the SG90 Servo motor was exceeded.

2.3 Bio-mimic Design

The final leg of the bipedal robot was designed with 12 DOF by reducing the link length and by mimicking the human leg design, which increased stability and reduced vibration. Figure 3 illustrates the parallels between link design and the biomimetic design of the knee joint. This design has negligible leg mass, so the body has less perturbation during the swing phase [13]. The link is built with the intention of retaining the center of mass below the knee joint, and fusion 360 is used to replicate this design. The bipedal robots center of mass is shown in Fig. 4.

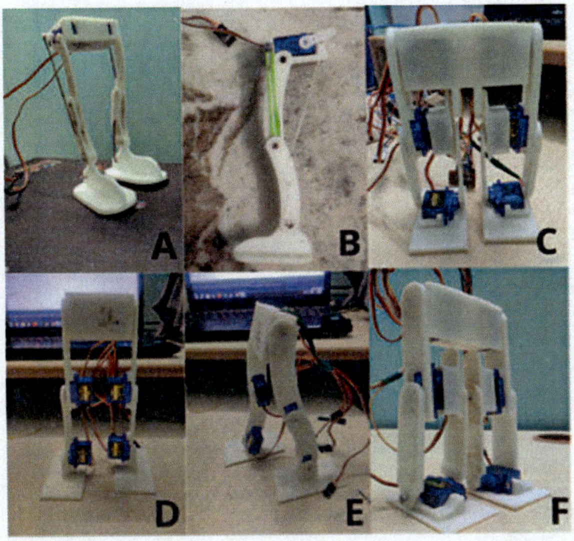

Fig. 2 **a** Two DOF bipedal robot, **b** 1 DOF-biped mechanism prototype, **c** biped robot with 6 DOF and opposing support (frontal plane), **d** DOF-biped robot with no opposing support (frontal plane), **e** DOF-biped robot with no opposing support (sagittal plane), **f** Biped robot with 6 DOF and opposing support (frontal plane)

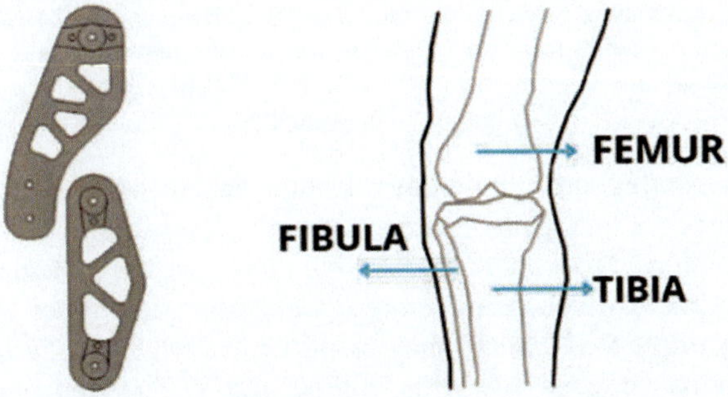

Fig. 3 Replication of human knee joint and bipedal design

2.4 Embedded Electronics

A servo motor connected to an Arduino UNO Atmega 328-p Microcontroller is used to control the robot actuators. The SG90 and MG995 servo motors are connected to a PCB board with common power pins and a separate signal pin for connecting to Arduino pins 2 through 13. It was soldered together with the USB cable and a switch to turn on and off the power bank's input voltage. The power bank provides a separate supply for the servo motor and the Arduino UNO. Because the 12 motors

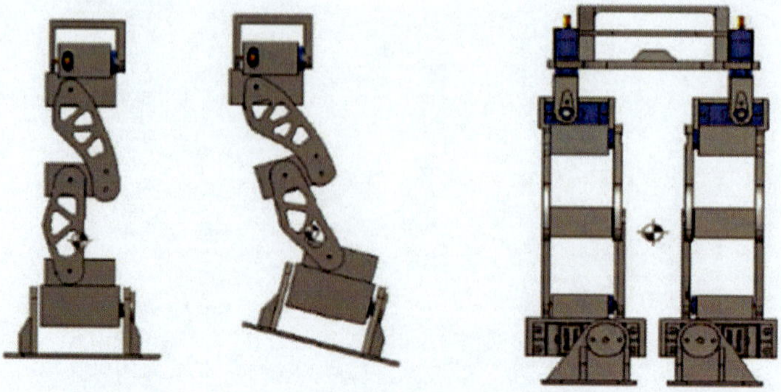

Fig. 4 CoM maintained in both the stance and swing phase and CoM of a full-assembled bipedal robot

Fig. 5 Assembly of Arduino, servo motor, and battery in the hip and connection of an Arduino and a PCB board with male-to-male header pins for connecting servo pins and a common power supply from the battery

needed to work continuously and require a huge current when it draws all at once from the Arduino, the board gets heated, and sometimes it does not even work, by plugging all the servo motor cables in the PCB as per the length and connecting the signal pins from the Arduino. The required voltage for each motor was around 5 V for the majority of the SG90 Servo motors, while the MG995 required at least 7 V for better performance, even though it worked in 5 V. The control of the robot is kept in its Hip Region which provides a downward force to keep the robot balance [14]. The SG90 did not meet the required torque of the motor because the gears were made of plastic, and at higher steps, the leg vibrates violently, which is the main disadvantage of this bipedal. Figure 5 shows the overall connection of the assembly pin connections and the servo pin connections.

2.5 DH Parameter

In 1955, Jacques Denavit and Richard Hartenberg proposed a general theory for describing an articulated sequence of joints. Four parameters describe each joint in

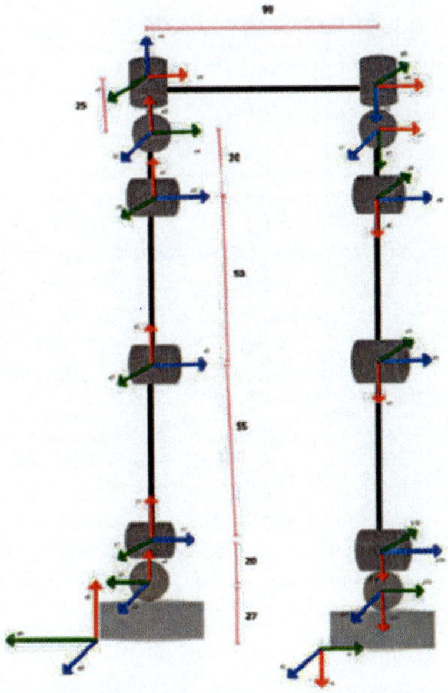

Fig. 6 DH representation of 12-axis bipedal robot

the robot. It is still a popular method for assigning reference frames in robotics. The reference frames for each joint are assigned to the 12-axis bipedal robot using the DH parameters. In Fig. 6, the blue arrow denotes the rotational axis, which is given with the Z-axis, and the red arrow is assigned by two rules: it must be perpendicular to the previous with Z-axis and must intersect it. With the right-hand coordinate frame, the Y axis is represented in green. After the frames are assigned, the DH table is built using four parameters: a—Link Length, d—Link Offset, Alpha—Twist Angle, and Theta—Joint Angle [15]. The forward kinematics of the bipedal robot were calculated using the DH table, as shown in Table 1.

2.6 *Gait Cycle*

For a perfectly smooth bipedal gait, the robot must have a controlled transition in every two phases of stance and swing [16]. In the swing phase when one leg is supported in the ground and another in the air; the distance covered in one step swing is kept minimum to stabilize the whole mass of the robot when it moves from one place to another [17]. The sagittal plane is used in robot anatomy to divide the robot longitudinally into left and right parts. The frontal plane is used in robot anatomy to divide the robot into front and back parts [18]. The robot's motion in both the sagittal and frontal planes is shown in Fig. 7.

Table 1 DH table

Link i	θ Theta (deg)	α twist angle (deg)	A link length (mm)	D displacement (mm)
Link 1	$0+\theta1$	90	27	0
Link 2	$0+\theta2$	−90	20	0
Link 3	$0+\theta3$	0	55	0
Link 4	$0+\theta4$	0	50	0
Link 5	$0+\theta5$	90	20	0
Link 6	$90+\theta6$	90	0	25
Link 7	$0+\theta7$	−90	90	0
Link 8	$0+\theta8$	90	0	25
Link 9	$90+\theta9$	−90	20	0
Link 10	$0+\theta10$	0	50	0
Link 11	$0+\theta11$	0	55	0
Link 12	$0+\theta12$	90	20	0
Link 13	0	−90	27	0

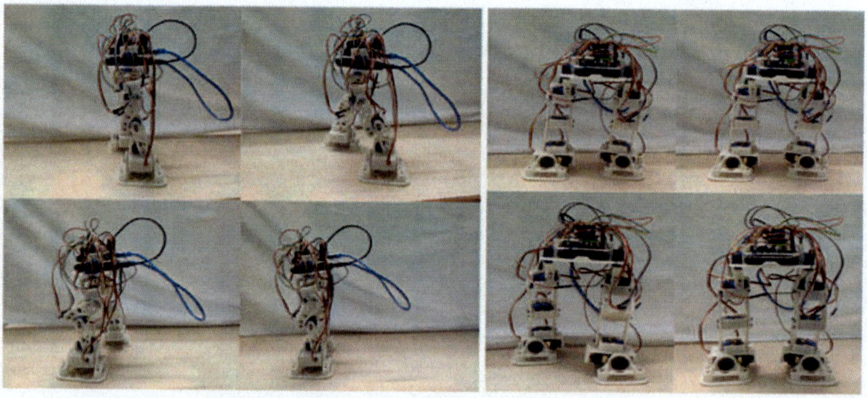

Fig. 7 Gait cycle in sagittal plane and frontal plane

2.7 *Motor Torque Specifications*

The bipedal motor is a servo motor. A servo motor is used in a closed-loop system to provide accurate angular control. It employs a potentiometer-based feedback control to calculate the amount of angular displacement and sends it to the controller and PID algorithm to move its position. It has three pins: two for power and one for a 5 V signal. SG90 servo motor and MG995 servo motor are the servo motors used in the bipedal robot. The SG90 servo motor is made of plastic and has an operating speed of 0.12 s/60°. At a maximum voltage of 6 V, it has a stall torque of about 1.5 kg cm. The motor inside the servo motor is a brushed DC motor. It measures approximately

Table 2 Servo joint details

Servo no	Name of joint
1	Left TOE-x
2	Left TOE-z
3	Left KNEE-z
4	Left HIP-z
5	Left HIP-x
6	Left HIP-y
7	Right TOE-x
8	Right TOE-z
9	Right KNEE-z
10	Right HIP-z
11	Right HIP-x
12	Right HIP-y

$23 \times 11.5 \times 24$ mm and weighs 9 g. The MG995 servo motor has dimensions of 40.4*19.9*37.5 mm and weighs 58 g. The MG995 servo's gear type is a metal gear with a Dual Ball-Bearing Horn gear spline. The servo's motor is a brushed DC motor. It has an operating voltage of 4.8 V. It has a maximum stall torque of 9.0 kg cm. Both servos are controlled by pulse width modulation command signal. The joints were directly connected to the servo horn of the servo motor which acts as a pin joint to provide sufficient motion to the links [19]. Table 2 shows the servo motor names and the order of connection in the Circuit.

2.8 *Software and Programming*

The Arduino IDE 1.8.19 software, which uses an advanced C program version was used to program the robot. The term IDE refers to the Integrated Development Environment, which is used by Arduino.cc. The bipedal was programmed using forward kinematics, which means that the end-effector was made to reach the desired position by varying the joint angles. Each joint was individually actuated by servo motors, and the joint angles were adjusted each time until it moved to the desired orientation. The 12 servo signal pins were connected from the Arduino's 2nd to 13th pins and were declared in the void setup () function, which set up the entire pin connections for giving input or output devices and declared what kind of operation was going to be undertaken. Twelve servos were declared, along with the names hip, knee, and leg, and their axis of rotation is specified as a suffix. All tasks that must be performed on a continuous basis can be programmed in the void loop () function. The program for the robot's motion was written here. Single joint is specified, and its offset is mentioned inside the for loop (). Similarly, every joint is mentioned and programmed to achieve the gait cycle.

3 Similarities Between NAO Robot and Curvy-Linked Bipedal Robot

NAO, a small humanoid robot has 25 DOF, with 12 DOF in the lower hip region. It is shown in Fig. 8. It is intended for walking, running, and dancing. It is essentially an interactive humanoid robot aimed primarily at children, researchers, and the elderly. It can recognize up to 20 languages [20]. The bipedal resembles the Nao Robot more. It is built with a region below the pelvic region and has 12 DOF. The link design is reminiscent of both the bipedal and the NAO robots. The detailed specifications of both robots are shown in Table 3.

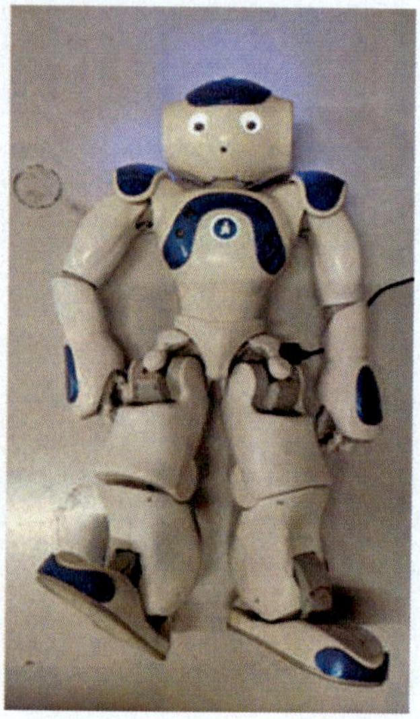

Fig. 8 Humanoid NAO robot with 25 DOF

Table 3 NAO and bipedal robot specification details

Specifications	NAO	Bipedal
Height	58 cm	26 cm
Weight	5.48 kg	580 g
Power supply	48.6 Wh lithium-ion battery	5 V lithium-ion battery
Degree of freedom	25	12
Degree of freedom (below hip)	12	12
CPU	Intel ATOM	Arduino UNO
Programing language	C++, Python, Java, C, .Net, MATLAB, Urbi	Arduino IDE

4 Results and Discussion

The experimental results show that the curved joint design provides more stability than the linear rigid joint. This modified robot was created and compared with the humanoid NAO robot's bipedal gait cycle. The links are designed like the human leg design, and all the link parameters, such as the Femur length being longer than the Tibia, are replicated in the design. By comparative empirical analysis, it is found that the curvy-linked bipedal robot has higher stability and lower vibration when compared to a rigid linked bipedal robot.

5 Conclusion

This research work assesses the viable link design for a bipedal robot by changing and varying the degree of freedom parameters and comparing the various link designs. From the observations, the bio-inspired robot model with 12 degrees of freedom has more stability and maneuverability than the linear-linked one. The complete robot was constructed from simple parts to create a bipedal robot. However, more programming changes are required in order to use inverse kinematics to control the motion. Our team is constantly working on design and programming by implementing the results of this research to build a Robust Humanoid Robot in the future.

References

1. Magsino ER (June 2019) A walking bipedal robot using a position control algorithm based on the center of mass criterion 14(11). ISSN: 1819-6608
2. Arzu RR, Russel MH, Rahman KA, Banik SC, Islam MT (May 2014) Bipedal walking robot. ICMERE-2013-PI-068
3. Kumar AS, Krishnan AG, Sridhar A, Kiruthika N, Prakash NK (July 2014) Design and fabrication of bipedal robots. IEEE 33044
4. Pratt J, Krupp B (April 2008) Design of a bipedal walking robot, 6962. https://doi.org/10.1117/12.777973
5. Medrano-Cerda GA, Akdas D (June 2018) Stabilization of a 12 degree of freedom biped robot
6. Jeon E, Jo S (2010) Human gait-based bipedal walking robot design in progress. In: ICCAS. IEEE, pp 1399–1402
7. Vukobratovic M (2004) Zero-moment point—thirty five years of its life 1(1):157–173
8. Yamaguchi J, Soga E, Inoue S, Takanishi A (1999) Development of a bipedal humanoid robot-control method of whole body cooperative dynamic biped walking. In: Proceedings of 1999 IEEE international conference on robotics and automation (Cat. No. 99CH36288C), vol 1, pp 368–374. https://doi.org/10.1109/ROBOT.1999.770006
9. Raffik R, Karthikeyan S, Abishekraj P, Manoj Guha K (2019) Design and research on amphibious robot. Int J Eng Adv Technol (IJEAT) 8(6S):560–564
10. Raffik R, Mayukha S, Hemchander J, Abishek D, Tharun R, Deepak Kumar S (2021) Autonomous weeding robot for organic farming fields. In: 2021 international conference on advancements in electrical, electronics, communication, computing and automation (ICAECA), pp 1–4. https://doi.org/10.1109/ICAECA52838.2021.9675563

11. Rakesh D, Keerthivaasan KVR, Mohan A, Samvasan P, Ganesan P, Raffik R (2021) Automated public screening and health vitals monitoring station. In: 2021 international conference on advancements in electrical, electronics, communication, computing and automation (ICAECA), pp 1–6. https://doi.org/10.1109/ICAECA52838.2021.9675550
12. An Z, Wang C, Raj B, Eswaran S, Raffik R, Debnath S, Rahin SA (2022) Application of new technology of intelligent robot plant protection in ecological agriculture. J Food Qual 2022:7, Article ID 1257015. https://doi.org/10.1155/2022/1257015
13. Kim D, Jorgensen S, Lee J, Ahn J, Luo J, Sentis L (2019) Dynamic locomotion for passive-ankle biped robots and humanoids using whole-body locomotion control. 10.1177
14. Collins S, Ruina A, Tedrake R, Wisse M (2005) Efficient bipedal robots based on passive-dynamic walkers. Science 307(5712):1082
15. Denavit J, Hartenberg RS (1955) A kinematic notation for lower-pair mechanisms based on matrices. J Appl Mech 22(2):215–221
16. Semwal VB, Katiyar SA, Chakraborty R, Nandi GC (2015) Biologically inspired push recovery capable bipedal locomotion modeling through hybrid automata. Robot Auton Syst 70:181–190
17. Dunn ER, Howe RD (1996) Foot placement and velocity control in smooth bipedal walking. In: Proceedings 1996 IEEE international conference on robotics and automation, pp 578–583
18. Fong M (2005) Mechanical design of a simple bipedal robot
19. Gini G, Scarfogliero U, Folgheraiter M (2009) New joint design to create a more natural and efficient biped 6(1):27–42
20. Ames AD, Cousineau EA, Powell MJ (17–19 April, 2012) Dynamically stable bipedal robotic walking with NAO via human-inspired hybrid zero dynamics. HSCC'12

Fractional-Order Polynomial Generation for Path Tracking of Delta Robot Through Specific Point

Dheeresh Upadhyay, Sandeep Pandey, and Rajesh Kumar Upadhyay

Abstract The tracking of the desired trajectory by the delta robot has number of applications in the industry. To accomplish accurate and fast tracking of delta robot between initial and final position through a specific point is a challenging task. In this paper, the fractional-order polynomial for Cartesian space serves as the foundation for the proposed path tracking technique. It develops different types of control strategy by using polynomials of fractional order. In first step, a new set of fractional-order polynomial coefficients is derived. Based on the generated fractional-order polynomial, a Simulink model has also been created through which the applicability of the proposed approach has been verified. Plots of the location, velocity, and acceleration profiles for step input and sinusoidal input have been made, and they have been compared.

Keywords Fractional-order polynomial · Delta robot · Path tracking

1 Introduction

Delta or parallel robots are a particular type of robot developed for manufacturing industries with more than three degrees of freedom. With three primary axes in the *x*, *y*, and *z* planes and three extra movements as pitch, roll, and yaw, the delta robot has six degrees of freedom. Due to its unique design, which offers enormous operational flexibility in manufacturing and other industries, this robot is also known as the parallel robot. This robot was initially created for the chocolate manufacturing industry, and then professor Reymond Caval created a delta robot for pick-and-place use. To do a given duty at work, the parallel robot has very high jail needs. Multiple

D. Upadhyay · R. K. Upadhyay
Mangalayatan University, Aligarh, Uttar Pradesh, India

S. Pandey (✉)
Thapar Institute of Engineering and Technology, Patiala, Punjab, India
e-mail: sandeep.pandey@thapar.edu

S. Sharma et al. (eds.), *Intelligent Control, Robotics, and Industrial Automation*,
Lecture Notes in Electrical Engineering 1066,
https://doi.org/10.1007/978-981-99-4634-1_13

parallel links are connected with the base from one end of the delta robots construction, and from the other end, these multiple links are connected with an end-effector. Through the use of forwarding dynamics and inverse dynamics analysis, the control of the delta robot requires accurate knowledge of the motion of its arms [1]. Many academics have recently shown an interest in creating kinematics theories for various delta robot versions used for various applications [2–4]. Path planning is a crucial responsibility for the robot to operate smoothly, and many approaches have been put out in accordance with polynomial theory [5–7]. There has been discussion of the joint and Cartesian polynomial schemes for delta robots in [8], where the coefficient value for an integer-order polynomial expression used in trajectory tracking was calculated. Fractional-order calculus was an emerging topic of study in many engineering and other fields, and it had several advantages. This work attempts to build a fractional-order polynomial approach for delta robot path planning. The fraction-order polynomial has been used to study the position, velocity, and acceleration of the delta robots arm. For this, the simulation model of the delta robot was initially created in the MATLAB environment. Simulink empowerment was built using the delta robot prototype created by Acrome. The proposed approach's results were then shown side by side with an existing integer-order method. By contrasting the various settings while the delta robot is in operation, the simulation results demonstrate the efficiency of fractional polynomials in Cartesian space.

2 Delta Robot Construction and Dynamics

The base of the delta robot is connected to several parallel links on one end, while the second end of each of these varied linkages is attached to an end-effector. In addition to the three fundamental movements in the x, y, and z directions, it creates an assembly to form a parallelogram and offers rotational freedom. The rotating movement of regular robots is constrained, and thus, the delta robot is a superior option for quick and flexible tasks. All of the driving actuators are positioned on the base of the delta robot as illustrated in Fig. 1 and are attached to a firm surface above or close to the working area. Due to the lack of actuators, low moment of inertia, and flexibility of the quick movement of the robotic arm, the many flexible links can be made lighter in comparison to the base. An end-effecter that is used to hold the object is attached to the opposite end of the arm.

Three identical arms are paralleled on the ACROME delta robot between the base plate and the moving end-effector plate. Three degrees of freedom for translation come from combining the restricted motion of these arms. The actuators are directly connected to the upper robot arms. On the base plate, the three actuators are fixedly spaced apart by 120 degrees. There are two parallel bars on each of the bottom arms. It affixes the traveling end-effector plate to the upper arm. For pick-and-place applications, a sphere-shaped electromagnet and a USB camera were included into the system.

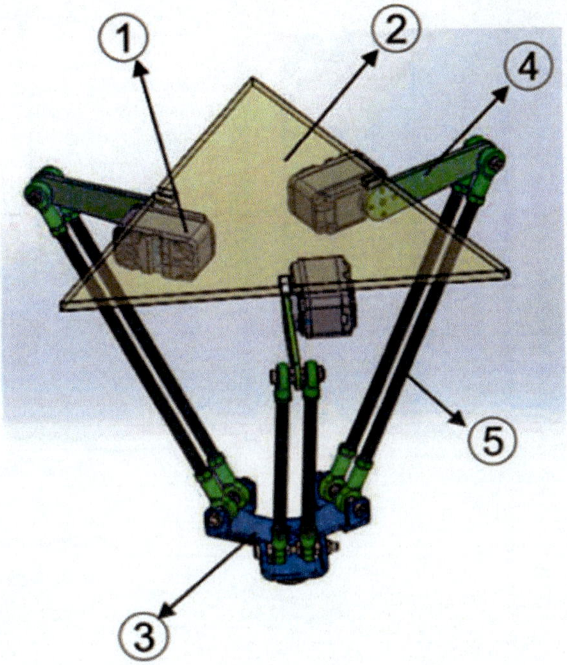

Fig. 1 General mechanical structure of delta robot: 1. Smart servo motor 2. The base plate 3. The traveling end-effector plate 4. The upper arm 5. The lower arm

2.1 Forward Kinematics

The examination of the delta robots kinematics in this part offers two types of solutions: forward and inverse kinematics. Kinematics analysis of the delta robot mechanism is merely the investigation of Cartesian space and joint space mapping. As indicated in Fig. 2, forward kinematics refers to the mapping from joint space to Cartesian space, whereas inverse kinematics refers to the mapping from Cartesian space to joint space.

The three joint angles ($\delta_1, \delta_2, \delta_3$) are inputs for forward kinematics, which is briefly mentioned above, and the Cartesian coordinates (x, y, and z) are outputs, as illustrated in Fig. 2. Servo motors are given joint angles, and then, forward kinematics equations are used to calculate the corresponding Cartesian coordinates.

Let us first provide the physical characteristics of the robot in Table 1 before building the forward kinematics of the delta robot.

The end-z-coordinate effectors will always be negative since a reference point will be selected at the triangle's base's intersection. It can be used to select the one distinct and accurate solution from the forward kinematics solution set.

The delta robots forward kinematics equations are presented in (1) and (2) [9].

$$x = \frac{a_1 z + b_1}{d}, \tag{1}$$

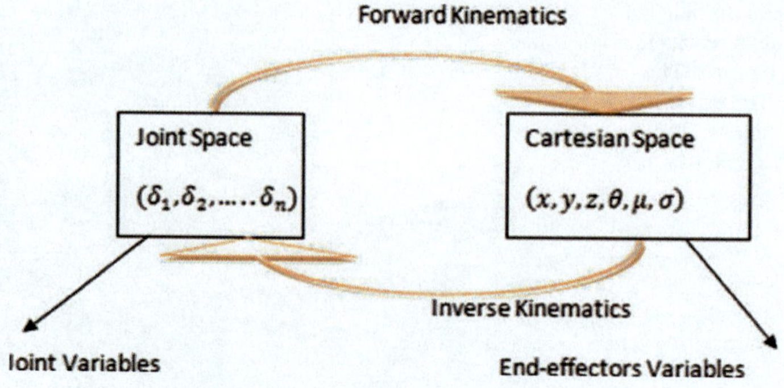

Fig. 2 Relations between forward and inverse kinematics

Table 1 Parameters' list of delta robot

Symbol	Presentation
b	Side of the base plate triangle
a	Side of the traveling plate triangle
U_f	Length of the upper arm
U_e	Length of the lower arm

$$y = \frac{a_2 z + b_2}{d}, \tag{2}$$

where the denominator parameter is defined in (3)

$$d = x_3(y_2 - y_1) - x_2(y_3 - y_1). \tag{3}$$

Finally, the value of different parameters a_1, b_1, a_2, b_2 is given in (4), (5), (6), and (7).

$$a_1 = (z_1 - z_2)(y_3 - y_1) - (z_3 - z_1)(y_2 - y_1), \tag{4}$$

$$b_1 = -0.5[(w_2 - w_1)(y_3 - y_1) - (w_3 - w_1)(y_2 - y_1)], \tag{5}$$

$$a_2 = (z_2 - z_1)x_3 - (z_3 - z_1)x_2, \tag{6}$$

$$b_2 = 0.5[(w_2 - w_1)x_3 - (w_3 - w_1)x_2]. \tag{7}$$

As was already mentioned, since the origin point is on the base plate, one can select the negative z solution. Consequently, there is a negative sign in the downward direction. The negative answer to z is provided in (8).

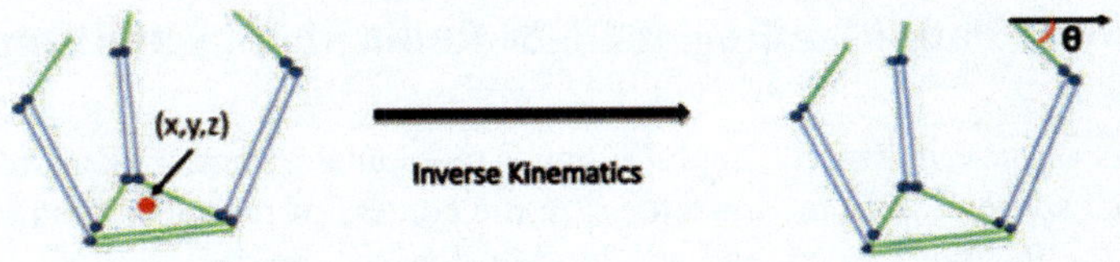

Fig. 3 Mapping from Cartesian space to joint space

$$z = \frac{-b + \sqrt{\Delta}}{2a}. \tag{8}$$

2.2 *Inverse Kinematics*

It is essential to know the angle of each joint in order to arrange motors for picking in the desired position for the end-effector. Inverse kinematics is the name given to this procedure. The mathematical equations and the inverse kinematic solutions for the delta robot are explained in the section that follows. Figure 3 depicts the schematic representation of the delta robots inverse dynamics.

Let us go back to forward kinematics section and recall the physical characteristics of the delta robot in order to build the inverse kinematics of the robot. Like the delta robots forward kinematics part, the first joint's inverse kinematics solution is provided in (10).

$$\theta_2 = \arctan\left(\frac{z_{J1}}{y_{F1} - y_{J1}}\right). \tag{10}$$

Let us rotate the coordinate system in the *XY* plane by 120 degrees and −120 degrees around the z-axis to find the final answer for the remaining angles δ_2, δ_3 which are given as in (11) and (12).

$$\mathrm{x}' = \mathrm{x}\cos\left(\frac{2\pi}{3}\right) + \mathrm{y}\sin\left(\frac{2\pi}{3}\right), \tag{11}$$

$$\mathrm{y}' = \mathrm{y}\cos\left(\frac{2\pi}{3}\right) - \mathrm{x}\sin\left(\frac{2\pi}{3}\right). \tag{12}$$

This gives the final form of inverse kinematics as shown below in (13)

$$E_0(x, y, z). \tag{13}$$

3 Proposed Path Planning of Delta Robot via Specific Point

A trajectory or path, in general, explains how a manipulator should move in a multi-dimensional space. The term "trajectory" in the context of robotics refers to a time history of the location, speed, and acceleration for each degree of freedom. Route generation, or planning the trajectory in joint space from the starting point to the finishing point, is the fundamental issue at hand. Position, velocity, and acceleration are calculated during the production of the trajectory in the most basic scenario. The trajectory points are calculated at a certain rate, known as the path-update rate, because these trajectories are computed discretely in time on computers.

Path generation in Cartesian space is defined in terms of x, y, and z coordinates' functions. Through the delta robots inverse kinematics equations, the joint angles are continually determined. The triangle formed by the base plate serves as the delta robots Cartesian reference frame. As a result, trajectories in Cartesian space change in relation to this frame of reference. There are three possible trajectories for the delta robot because each axis moves independently. The use of Cartesian space systems has benefits [9]. Every instant has a known and controllable motion between the starting and ending places via any specific point as shown in Fig. 4. Each polynomial will be evaluated over an interval starting at $t = 0$ and ending at, where $t = t_{fi}$ and $i = 1, 2$. It is simple to describe the generated path by the robot because the motion is clearly visible. There are numerous smooth functions that differ from one another between two places, in which fractional-order polynomial is one possible way to describe the trajectory. Apart from other method, fractional-order polynomial gives us a freedom to choose the fractional power of polynomial. There are some constraints which should be kept in mind to derive the desired fractional-order polynomial for a specific path generation.

Initially, four restrictions are required to produce a polynomial of fractional order. The manipulators initial position and its intended position are both restrictions.

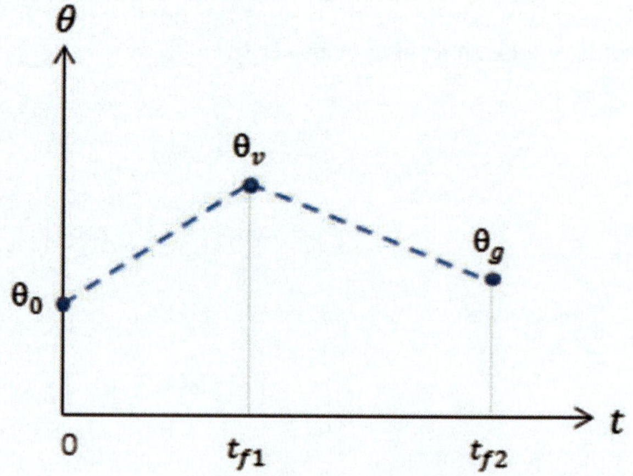

Fig. 4 Initial, via, and goal points with respect to time

$$\delta(0) = \delta_0 \tag{14}$$

and

$$\delta(t_f) = \delta_f. \tag{15}$$

However, the velocity constraints at each end are not zero, but rather, some known velocity. These two velocities imposed another two constrained which can be defined as

$$\dot{\delta}(0) = \dot{\delta}_0 \tag{16}$$

and

$$\dot{\delta}(t_f) = \dot{\delta}_f. \tag{17}$$

The proposed four equations describing this general fractional-order polynomial are:

$$\delta_0 = a_0, \tag{18}$$

$$\delta_f = a_0 + a_1 t_f + a_2 t_f^2 + a_3 t_f^{3.5}, \tag{19}$$

$$\dot{\delta}_0 = a_1, \tag{20}$$

$$\dot{\delta}_f = a_1 + 2a_2 t + 3.5 a_3 t_f^{2.5}. \tag{21}$$

The coefficient value is as follows when four constraints are applied to (19) and (21).

$$a_2 = \frac{3}{1.5t_f^2}(\delta_f - \delta_0) - \frac{\dot{\delta}_f}{1.5t_f} - \frac{2.5\dot{\delta}_0}{1.5t_f}, \tag{22}$$

$$a_3 = -\frac{4}{3.5t_f^{3.5}}(\delta_f - \delta_0) + \frac{2.33}{3.5t_f^{2.5}}(\dot{\delta}_f - \dot{\delta}_0). \tag{23}$$

With the help of the information above, the first simulation model for trajectory generation in Cartesian space using partially fractional-order polynomials has been created.

4 Simulation Results

To simulate the path tracking, one MATLAB simulation model has been developed. The same model has been used to validate the applicability of the proposed method. This model used two s-function blocks to define the desired fractional-order polynomial. The computation of the coefficient of the fractional polynomial is carried out through the derived equation as done in the above section.

In the first attempt, the initial position was defined from the origin and the final position was kept on 35, which was one kind of step signal as shown in Fig. 5. Integer-order polynomial developed in [9] was used to take the first result. Another result has been taken based on the fractional-order polynomial. These two results' comparisons show the significant difference from the integer-order polynomial and proposed fractional-order polynomials. Each polynomial acquired different paths to follow the same final position from the same initial position.

To understand the velocity and acceleration profile behavior, two separate results have been taken from the simulation model and shown in Fig. 6. The entire fractional-order polynomial shows the highest pickup in the velocity profile.

Another attempt has been made to follow the sinusoidal trajectory and the same simulation model. Figure 7 shows the trajectory tracking of all different polynomials, in which full fractional-order polynomial acquires the lowest undershoot and overshoot during the tracking operation of the same sinusoidal path. The other profile for the same signal is shown in Fig. 8 which shows the velocity. The two cases have different profiles during the path tracking operation, and fractional-order polynomial renders the fast operation and improved velocity profile.

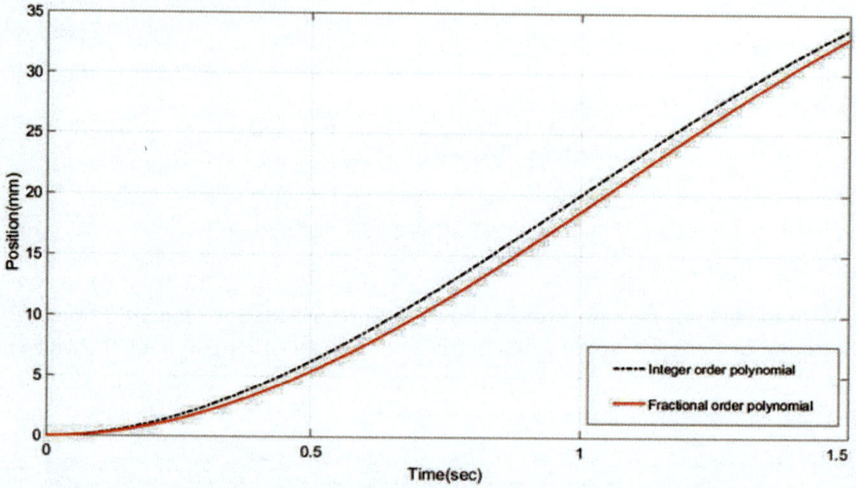

Fig. 5 Position profile comparison of different polynomial schemes for step trajectory

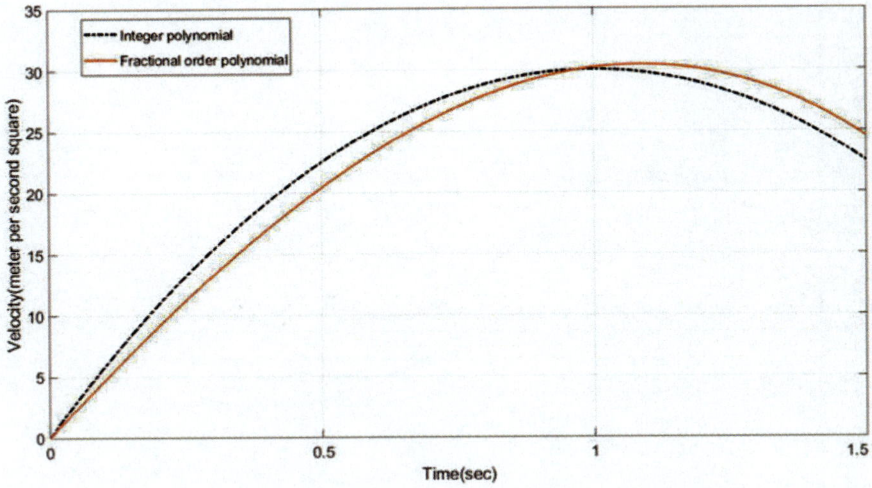

Fig. 6 Velocity profile comparison of different polynomial scheme step trajectories

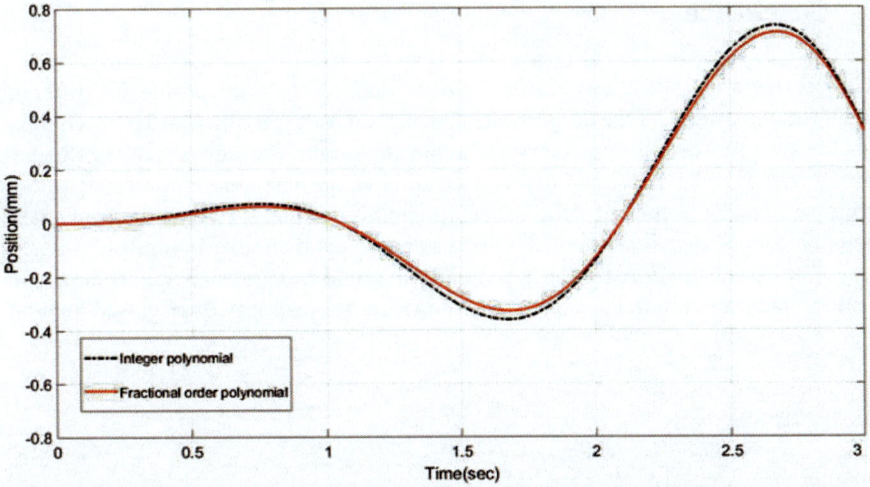

Fig. 7 Position profile comparison of different polynomial schemes for sinusoidal trajectory

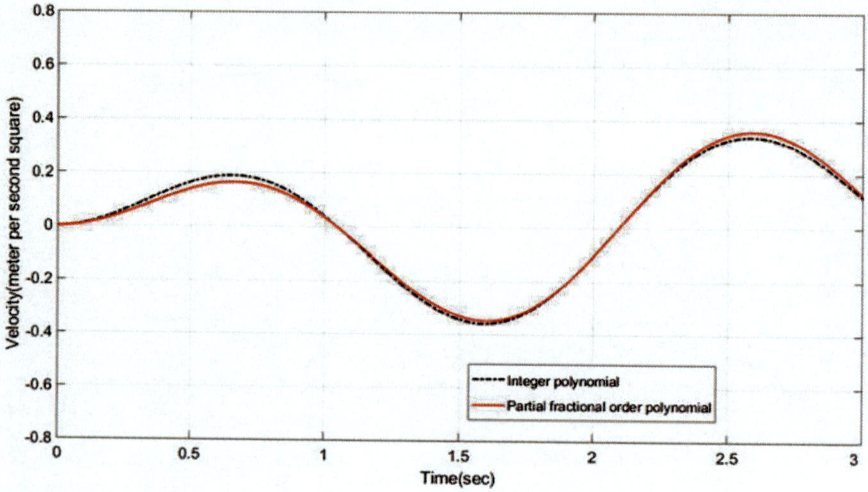

Fig. 8 Velocity profile comparison of different polynomial schemes for sinusoidal trajectory

5 Conclusion

The proposed method improves the velocity and acceleration profile for different inputs tracking used by the delta robot. The method uses fractional-order polynomial for tracking the specific trajectory via a specific point. The new set of coefficients for fractional-order polynomial was used to develop new simulation model along with parameters' values of delta robot. In addition to these, two main inputs were employed to the developed model. One is step input and another is sinusoidal trajectory. The novel fractional-order polynomial method in Cartesian space shows the faithful tracking of both the signals and shows the applicability of proposed method.

References

1. Wang J, Liu XJ (2003) Analysis of a novel cylindrical 3-DoF parallel robot. Robot Auton Syst 42(1):31–46. https://doi.org/10.1016/S0921-8890(02)00296-8
2. Kuo YL (2016) Mathematical modeling and analysis of the delta robot with flexible links. Comput Math Appl 71:1973–1989. https://doi.org/10.1016/j.camwa.2016.03.018
3. Angel V, Viola J (2018) Fractional order PID for tracking control of a parallel robotic manipulator type delta. ISA Trans 79:172–188. https://doi.org/10.1016/j.isatra.2018.04.010
4. Kuo YL, Huang PY (2017) Experimental and simulation studies of motion control of a delta robot using a model-based approach. Int J Adv Robot Syst 1–14. https://doi.org/10.1177/1729881417738
5. Ecorchard G, Maurine P (2005) Self-calibration of delta parallel robots with elastic deformation compensation. Intell Robots Syst 462–467. https://doi.org/10.1109/IROS.2005.1545024
6. Poppeova V, Uricek J, Bulej V, Sindler P (2011) Delta robots—robots for high speed manipulation. Tech Gaz 18:435–445. http://dx.doi.hrcak.srce.hr/71825

7. Brinker J, Corves B, Takeda Y (2018) Kinematic performance evaluation of high-speed delta parallel robots based Motion/force transmission indices. Mech Mach Theory 15:111–125. https://doi.org/10.1016/j.mechmachtheory.2017.11.029
8. Damic V, Cohodar M, Voloder A (2019) Modelling and path planning of delta parallel robot in virtual environment. In: Proceedings of the 29th DAAAM international symposium, DAAAM international. Vienna, Austria, pp 0149–0156. https://doi.org/10.2507/29th.daaam.proceedings.021
9. Olsson A (Feb 2009) Modeling and control of a Delta-3 robot, master thesis. Department of Automatic Control, Lund University. http://dx.doi.lup.lub.lu.se/student-papers/record/8847521

Energy-Based Approach for Robot Trajectory Selection in Task Space

Ankur Jaiswal, Abhishek Jha, Golak Bihari Mahanta, Neelanjan Bhattacharjee, and Sanjay Kumar Sharma

Abstract This work describes a simplified framework for assessment and selection of an appropriate end-effector trajectory among the multiple choices. The proposed framework is based on utilization of the kinematic and dynamic model of the robot for trajectory selection on the basis of minimum energy requirement. The effectiveness of the proposed approach is demonstrated in terms of task space trajectory planning problem for a six degrees of freedom industrial manipulator. One major concern in the trajectory planning problem is to select a best trajectory for execution by the robot, and the proposed framework resolves this issue by considering the energy requirement during execution of the trajectory.

Keywords Kinematics · Dynamics · Trajectory planning · Task space

1 Introduction

In today's competitive scenario, industries are shifting more toward robot-based automation of the various manufacturing processes. Such automation process provides many advantages like consistent quality, increased productivity, timely completion, and cost-effectiveness. A wide variety of manufacturing operations like

A. Jaiswal (✉)
Department of Mechatronics, Manipal Institute of Technology, Manipal Academy of Higher Education, Manipal, Karnataka 576104, India
e-mail: ankur.jaiswal@manipal.edu

A. Jha · G. B. Mahanta
Department of Mechanical Engineering, Amrita School of Engineering, Amrita Vishwa Vidyapeetham, Chennai, India

N. Bhattacharjee
Department of Mechanical Engineering, University of Alberta, Edmonton, Canada

S. K. Sharma
Department of Mechanical Engineering, Amity University, Raipur, India

S. Sharma et al. (eds.), *Intelligent Control, Robotics, and Industrial Automation*,
Lecture Notes in Electrical Engineering 1066,
https://doi.org/10.1007/978-981-99-4634-1_14

welding, spray painting, material handling, inspection, and assembly are well suited for robotic application [1].

To cope with the diversity of applications, industrial robots are programmed through online or offline mode. In online programming architecture, the requisite task is taught through a teach pendant, whereas in offline mode, a 3D simulation platform is used [2]. In either mode, the task is generally specified in terms of a trajectory. The trajectory planning constitutes an important aspect for the operation of any industrial robot. Its prime objective is to generate the trajectory from the specific start to goal points for completion of the desired task [3]. The end-effector of the robot is responsible for performing the desired task, and its accurate and optimal movement has to be ensured during the trajectory planning problem [4]. In trajectory planning operation, the notion of optimality varies as per the considered application and workspace condition [5, 6].

A trajectory can generally be planned either in joint or task space of the robot. In joint space scheme, trajectories are specified for each different joint of the robot. In this scheme, the definite end-effector pose is generally known at the start and end points of the trajectory. The intermediate points are estimated on the basis of a suitable polynomial scheme, where motion profile of each joint is approximated with polynomial function and resulting end-effector motions are obtained through the kinematic relations [7]. Such motion is easier to plan as it involves planning for the individual joints. However, from user's point of view, it is very difficult to visualize the trajectory due to nonlinearity and complexities involved in the kinematic relations. On the contrary, the task space trajectories are easy to specify and visualize. In this scheme, the motion profile is defined in terms of end-effector pose at each point. The joint angles corresponding to each of the specified points are obtained through the inverse kinematics. Most of the complex manufacturing applications like welding, spray painting utilize the task space trajectory planning operation [8, 9].

The trajectory planning operation gives the flexibility to define several ways to accomplish the task. By considering the nature of the application and its constraints, several trajectories can be generated for a particular task. In such a case, selection of the best trajectory for robot execution is a critical problem. Although different criteria like minimum time and minimum jerk are followed in the trajectory planning operation, but their applicability is task specific [10].

In view of above aspects, the present work aims to propose an approach for evaluation and selection of an appropriate trajectory among the multiple choices. For this purpose, a case of trajectory planning operation in the task space of the robot is considered. The proposed trajectory selection framework is based on the well-established kinematic and dynamic modeling architecture. It evaluates the different trajectories of the same task and provides a best suited trajectory for the application. This selection is made on the basis of minimum energy requirement for the robot during execution of the trajectory.

The rest of the paper is ordered as follows. The developed trajectory evaluation framework for an industrial robot is discussed in Sect. 2. Simulated experiment procedures and results are presented in Sect. 3. Concluding remarks are given in Sect. 4.

2 Framework for Trajectory Evaluation

The framework proposed for evaluation of the trajectory is based on the kinematic and dynamic modeling of the robot. These models provide all the vital information about the robot behavior during execution of a particular trajectory; therefore, it provides a reliable means to assess the trajectory for execution by the robot. The outline of the proposed framework is shown in Fig. 1.

The proposed framework is based on the utilization of the kinematic and dynamic models of the robot for assessment of the energy required for the execution of the given trajectory. The evaluation of the trajectory based on the robot parameters provides a reliable and effective way to judge suitability of the operation. The detailed methodology of the proposed approach is described in the next section.

2.1 Kinematic Modeling

The kinematic modeling of the robot provides vital information about the capabilities of the robot. Suitability of the application is also primarily dependent of the robot structure. The kinematic modeling of the robot consists of, (a) frame assignment and Denavit–Hartenberg (DH) parameter estimation, (b) forward kinematic modeling, (c) manipulator Jacobian formulation, and (d) inverse kinematic modeling

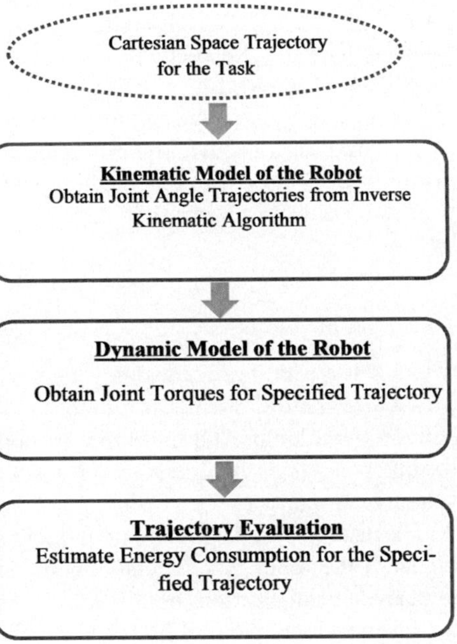

Fig. 1 Proposed framework for trajectory evaluation

[11]. The robot model considered in this work is ARISTO-XT. It possesses six degrees of freedom with vertical articulated structure. It is a popular robot training and research applications. This robot can be used for variety of applications like object manipulation, pelletizing, machine loading, and assembly operations.

The primary step in the kinematic modeling of the robot is to assign the frames to the joint and links of the manipulator and estimation of the DH parameters. The notion of DH parameters is robot configuration dependent, and with the change in home position of the robots, these parameters also tend to change. The frame assignment of the ARISTO-XT robot and the DH parameters are considered as described in [11]. After defining the DH parameters, the next step in kinematic modeling is determination of the forward kinematic equation of the robot. It describes the end-effector pose for a given set of joint variables. The forward kinematic equation for the considered robot model can be given by

$$T_0^6 = T_1^0(\theta_1)T_2^1(\theta_2)\ldots T_6^5(\theta_6). \tag{1}$$

The term T_0^6 is a 4×4 homogeneous transformation matrix, which completely defines the end-effector pose relative to the base frame.

The forward kinematic equation acts as a medium for conversion of positions from joint space to task space. While planning motion of the robot, apart from the position information, velocity information is also necessary to execute a particular task. It is essential to analyze the effect of motion of each component on the motion of the robot. Differential kinematics can be used to examine the effect of velocities of the manipulator. It describes the mapping between joint and task space velocities of the end-effector. The manipulator Jacobian is a matrix that describes this mapping. Jacobian matrix is considered to be a significant tool for robot characterization. It is mostly used for detecting singularities, examining redundancy, obtaining an inverse kinematics algorithm, and relating explicit mapping between joint torques and end-effector forces. Moreover, determination of manipulator's equation of motion and operational space control scheme design are also based on the Jacobian. The Jacobian matrix is a transformation, that maps joint velocities to Cartesian velocities, as given by

$$\dot{X} = J\dot{q}, \tag{2}$$

where X is a vector of Cartesian velocity, $\dot{q}$ is a vector of joint space velocity, and J is Jacobian matrix. Since J is a function of robot configuration, it can be calculated column-by-column. The first three rows of each column of the Jacobian represent the components of the linear velocity, while the last three rows describe the angular velocity component.

The next step in the kinematic modeling is development of the inverse kinematic model of the robot. The inverse kinematic modeling involves determination of the joint space configurations of the robot for the given end-effector pose. The solution of the inverse kinematic problem is of prime importance, as it is necessary for motion transformation from task space to the joint space of the robot. The feasibility of the inverse kinematic model allows execution of execution of the requisite motion by the robot. Because of the several complexities involved in the inverse kinematic model of the robot, a definite solution exists if the given end-effector configuration belongs to the dexterous workspace of the robot. For all the robot models, closed-form solution does not exist, or there may be multiple solutions, and even no solution. Therefore, most of the modern-day controllers rely on the numerical techniques to solve the inverse kinematic problem. The use of numerical solutions technique is advantageous because of its applicability to different kinematic structures; however, it is restricted to providing an approximate solution for a given pose.

In this work, to obtain the inverse kinematic solution of the robot model for a task space trajectory planning problem, a numerical solution technique is adopted. When the requisite path profile is specified in task space, numerical methods of inverse kinematics like Jacobian pseudoinverse and transpose approaches, damped least squares [12, 13] provide good approximation for the inverse kinematic solution. Here, Jacobian pseudoinverse method [14, 15] is adapted for obtaining the joint space configuration of the robot during a trajectory tracking operation. This method is easy to implement and provides fast convergence. The algorithm developed for inverse kinematic modeling of the robot is shown in Fig. 2.

Using the above algorithm, the joint space configurations of the robot for specified task space trajectory are obtained.

2.2 *Dynamic Modeling*

The derivation of equations of motion for the robot is part of the robot's dynamic modeling. These equations of motion serve as the basis for the design, control, and simulation of the robotic systems. The robot's dynamic modeling approach consists of two parts, (a) inverse dynamics and (b) forward dynamics. The inverse dynamics problem involves determination of the joint actuator torques/forces from a pre-specified trajectory. The forward dynamic problem consists of estimation of the joint acceleration from the pre-specified values' joint torques/forces. For this work, since the requisite trajectory is pre-defined, solution of inverse dynamic problem is considered. The most general form of the dynamic equation for an *n*-DOF robot is given by

$$\sum_{j=1}^{n} M_{i,j}(\theta)\left(\ddot{\theta}_j\right) + \sum_{j=1}^{n}\sum_{k=1}^{n} h_{ijk}\dot{\theta}_j\dot{\theta}_k + G_i = \tau_i, \quad (3)$$

where $M_{i,j}$ is the inertia matrix, h_{ijk} is the Coriolis force matrix, G_i is the gravity force matrix, $(\dot{\theta}_j, \ddot{\theta}_k)$ are the joint space velocity and acceleration, and τi represents the vector of the joint torque. The inertial parameters and the Coriolis forces are main factors affecting the system dynamics. Therefore, the accuracy of the dynamic model mostly relies of these two parameters. Using Lagrange–Euler formulation, the

Fig. 2 Inverse kinematic algorithm

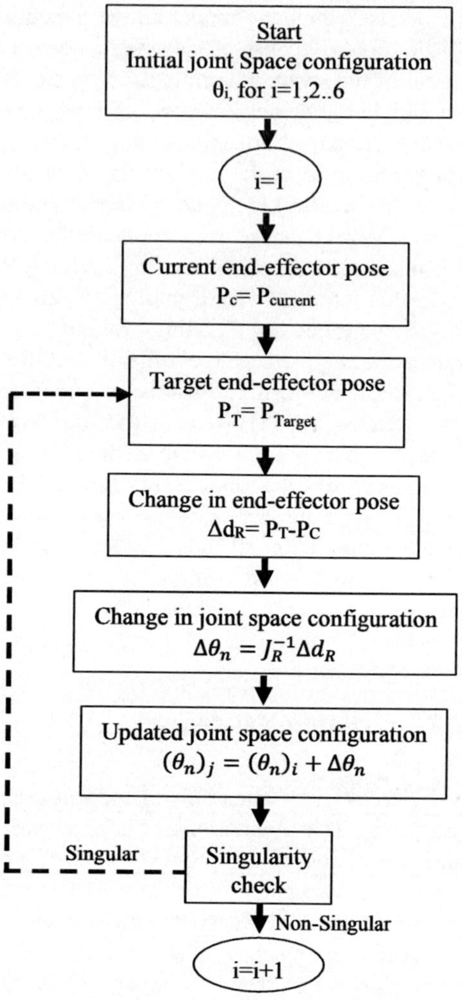

mass matrix, Coriolis component and the Gravity vector can be computed from the following relations

$$M_{ij} = \sum_{p=\max(i,j)}^{n} Tr\left[d_p I_p d_p^T\right], \tag{4}$$

$$h_{ijk} = \sum_{p=\max(i,j,k)}^{n} Tr\left[\frac{\delta(d_{pk})}{\delta q_p} I_p d_{pi}^T\right], \tag{5}$$

$$G_i = -\sum_{p=i}^{n} m_p g d_{pi} \bar{r}_p^p. \tag{6}$$

The above equations describe the dynamic model of the robot. These n-nonlinear-coupled equations are the second-order ordinary differential equations of an n-DOF robot. Since the robot under consideration is a 6-DOF robot, here the dimensionality of the matrices is six. These equations govern the dynamic behavior of the robot during motion. For the considered robot model, the different parameters for deriving the dynamic model of the robot are adapted from [11].

2.3 Trajectory Selection Framework

In order to accomplish a specified task, the end-effector of the robot moves in a particular way in the robot workspace. The execution of such task requires the robot to traverse a pre-defined or pre-planned path. The main objective of the path or trajectory planning is to describe robot motion as a time sequence of joint/end-effector locations. These locations are usually generated by the interpolation process and approximating the desired path in terms of polynomial functions. The trajectory points obtained using such a way become input to the control system of the robot.

While generating such trajectories, the consideration of the kinematic and dynamic model of the robot is essential to ensure proper execution of the trajectory by the robot. With the change in the geometrical shape of the trajectory, and application, the torque requirement for execution may change. Therefore, energy requirement by the robot for execution of a trajectory is an important factor resulting in the overall energy savings in the process. Moreover, it can also be adapted as an evaluation parameter for the assessment of a trajectory. The present work emphasizes on the utilization of the energy requirement for selection of a best trajectory for execution.

The energy consumed by the robot's joint actuators accounts for the majority of the energy usage. In general, the energy consumed by a DC motor placed at the robot joint over a time (t) can be given by [16]

$$E = \int_0^t \tau \, \dot{\theta} dt\,. \tag{7}$$

In Eq. (7), τ indicates the joint torque and $\dot{\theta}$ indicates the angular velocity. The power consumption primarily depends on the joint velocity and torque produced by each joint during the motion. In turn, the energy consumption relies on the planned trajectory. Since total energy consumption is the sum of power consumption over a time period, the minimum energy requirement reflects lower power consumption. The average power consumption for following a specified trajectory can be given by

$$P_{\text{avg}} = \frac{E_{\text{total}}}{t}, \tag{8}$$

where P_{avg} represents the average power consumption during execution of the trajectory. It must be noted that in above equations, the different mechanical and electrical factors contributing to the energy loss are not considered.

3 Simulation Results

In order to test the applicability of the proposed framework, two simulated experiments were performed. In first experiment, an end-effector trajectory consisting of a rectangular pattern was generated in the horizontal plane (XY-plane) of the robot. Such a form of trajectory is commonly used in the spray-painting operation, where the end-effector traverses the periphery of the object to paint it. In the second experiment, aforementioned trajectory was generated in the vertical plane (XZ-plane) on the robot. In each case, the objective was to select the best trajectory for the robot execution. The robot model considered for simulation is shown in Fig. 3, whereas trajectories generated for the two cases are shown in Figs. 4 and 5.

In both cases, the trajectories were obtained using the linear interpolation technique. This trajectory represents the shortest path for completing the given task, as it involves four straight-line segments. Practically, it is difficult to generate and execute a perfect straight-line path for ARISTO-XT due to the kinematic limitations. However, the adopted inverse kinematic model provided good approximation of the joint space configuration within the feasible joint limits for execution.

Both the trajectories shown in Figs. 4 and 5 can be deployed to the robot controller, as they satisfy the kinematic constraints and feasibility conditions adapted in the

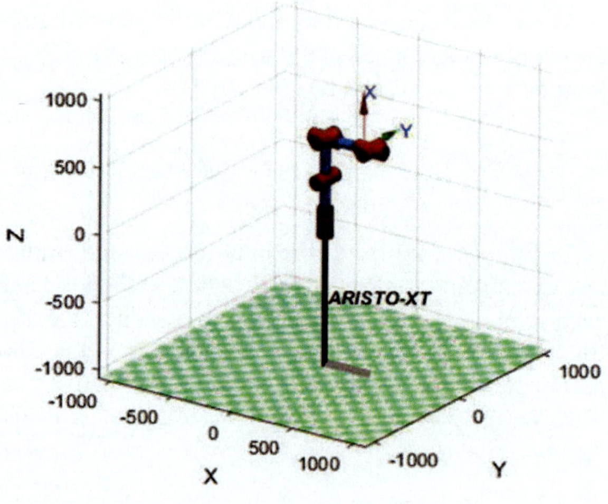

Fig. 3 ARISTO-XT robot model

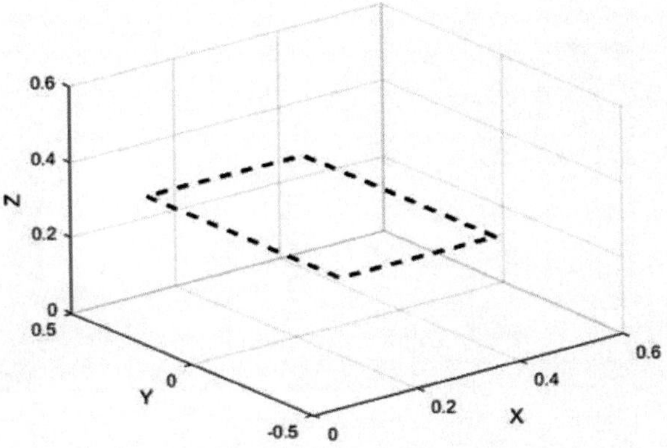

Fig. 4 Trajectory defined in horizontal plane

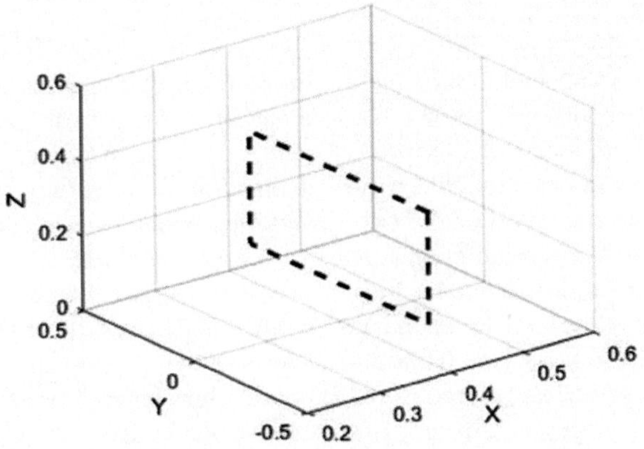

Fig. 5 Trajectory defined in the vertical plane

framework. However, selection of the best trajectory for robot execution is a critical issue, as both the trajectories can be selected. Therefore, it is necessary to further characterize the trajectories on the basis of performance measures. For this purpose, the dynamic model of the robot can be utilized. The dynamic model provides the joint torques/forces necessary for execution of such trajectories.

To evaluate the suitability of the two trajectories for robot execution, the mean joint torque values for the first three joints of the robot are compared and shown in Fig. 6.

The comparison clearly reflects that the trajectory defined in the vertical plane exerts lesser torque than the trajectory defined in the horizontal plane. Although both the trajectories are of same path length and geometrical shape, but the torque requirement is different for both. The straight-line motion is the simplest form of motion to follow but difficult for an articulated robot to achieve. The change in plane of the trajectory results in the alteration of the inverse kinematic solution, and hence, the torque requirement may vary. Another reason for variation of the joint torque in such a case is the direct dependency of the joint torque over angular acceleration. For

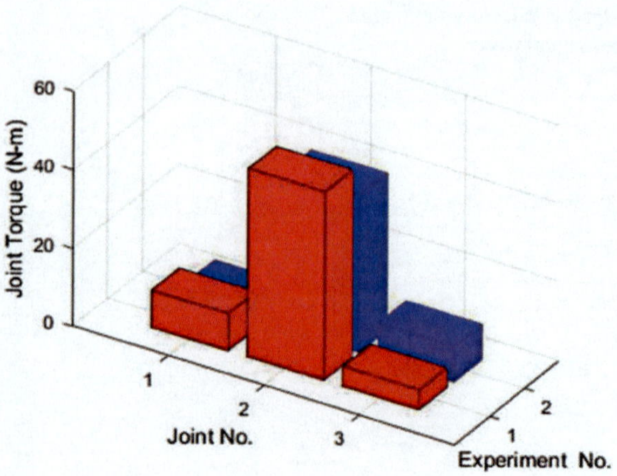

Fig. 6 Joint torque comparison

a joint actuator, the output torque required for execution of a trajectory is in direct proportion to its angular acceleration. Therefore, the velocity, acceleration, and jerk parameters of a trajectory become crucial during planning stage.

In order to make a clear distinction between two trajectories, the average power consumption of the two trajectories is compared. The comparison of the average power consumption for the two experiments clearly reflects the suitability of the trajectory defined in the vertical plane of the robot. The average power consumption for the first experiment is about 0.6031 W, whereas for second experiment, it is observed to be 0.5302 W. These values support the suitability of the trajectories defined in the vertical plane of the robot. In repetitive nature of the task, such energy saving is significant, particularly when higher number of cycles are to be executed by the robot. Moreover, the minimum power requirement by a trajectory makes it suitable for the given task execution.

4 Conclusions

In this work, a trajectory evaluation framework is discussed. It makes use of the kinematic and dynamic model of the robot to assess a task space trajectory on the basis of energy requirement for execution. The framework provides a reliable means to judge the suitability of the trajectory for a particular operation. The use of kinematic and dynamic considerations of the robot for evaluation of the trajectory makes it more realistic and efficient. The proposed trajectory evaluation framework does not rely on the nature of the trajectory and the considered application. The method adopted is helpful to select suitable trajectories for different applications, where the form or shape of the trajectory is not fixed.

References

1. Pan Z, Polden J, Larkin N, Van Duin S, Norrish J (2012) Recent progress on programming methods for industrial robots. Robot ComputIntegr Manuf 28(2):87–94
2. Mittal RK, Nagrath IJ (2003) Robotics and control. Tata McGraw-Hill, New Delhi
3. Jha A, Chiddarwar SS, Alakshendra V, Andulkar MV (2017) Kinematics-based approach for robot programming via human arm motion. J Braz Soc Mech Sci Eng 39(7):2659–2675
4. Siciliano B, Khatib O, Kroger T (eds) (2008) Springer handbook of robotics. Springer, Berlin
5. Ata AA (2007) Optimal trajectory planning of manipulators: a review. J Eng Sci Technol 2(1):32–54
6. Pragnavi RSD, Maurya A, Rao BN, Krishnan A, Agarwal S, Menon M (2019) Simple and coverage path planning for robots: a survey. In: International conference on inventive computation technologies. Springer, Cham, pp 392–403
7. Xing ZK, Xueqing L (2005) Trajectory planning of posture adjustment of welding mobile robot during auto-searching weld line [J]. Chin J Mech Eng 5
8. Stone HW (2012) Kinematic modeling, identification, and control of robotic manipulators. Springer Science & Business Media
9. Jha A, Chiddarwar SS (2017) Robot programming by demonstration using teleoperation through imitation. Ind Robot: Int J
10. Piazzi A, Visioli A (2000) Global minimum-jerk trajectory planning of robot manipulators. IEEE Trans Industr Electron 47(1):140–149
11. Jazar RN (2010) Theory of applied robotics: kinematics, dynamics, and control. Springer Science & Business Media
12. Rajeevlochana CG, Saha SK, Robo A (12 Feb, 2011) 3D model based robotic learning software. In: International conference on multi body dynamics, pp 3–13
13. Dulęba I, Opałka M (2013) A comparison of Jacobian-based methods of inverse kinematics for serial robot manipulators. Int J Appl Math Comput Sci 23(2)
14. Buss SR (2004) Introduction to inverse kinematics with jacobian transpose, pseudoinverse and damped least squares methods. IEEE J Robot Autom 17:1–19
15. Buss SR, Kim JS (2005) Selectively damped least squares for inverse kinematics. J Graph Tools 10(3):37–49
16. Stilman M (2010) Global manipulation planning in robot joint space with task constraints. IEEE Trans Rob 26(3):576–584

A Novel Collision Avoidance System for Two-Wheeler Vehicles with an Automatic Gradual Brake Mechanism

Ulasi Vivek Reddy and Abhishek M. Thote

Abstract The aim of this research study is to minimize road injuries and accidents of two-wheeler vehicles. In this research study, a novel collision avoidance system (CAS) was developed to maintain the safe distance of the two-wheeler vehicles at high speeds, i.e., generally more than 40 km/hr. In this CAS, an ultrasonic sensor and Arduino UNO (microcontroller) program in connection with a speedometer of vehicle were used to detect the safe or unsafe distance with respect to front vehicle to avoid collision. Additionally, an automatic gradual brake mechanism was developed in this study and placed on to the rear tyre of the two-wheeler vehicle. This system applies gradual brakes if it detects unsafe distance with respect to front vehicle. As soon as safe distance is achieved, this gradual braking is removed. The purpose of gradual braking was only to maintain a safe distance with respect to front vehicle without skidding and not to stop the vehicle. For this purpose, 2-s rule was used to compute the safe distance between two vehicles as per speed of the vehicle in which this system is installed. This developed CAS alerts the driver with buzzer and also shows the distance of front vehicle on screen in cm. During practical testing of this system on two-wheeler vehicle, it was observed that, the ultrasonic sensor detected the obstacle and applied automatic gradual braking successfully. Thus, it is expected that, this CAS device will play a key role to avoid road accidents of two-wheeler vehicles on high-speed roads.

Keywords Collision avoidance system (CAS) · Two-wheeler vehicles · An automatic gradual brake mechanism · Arduino Uno · Ultrasonic sensor · 2-s rule

---

U. V. Reddy · A. M. Thote (✉)
School of Mechanical Engineering, Dr. Vishwanath Karad MIT-World Peace University, Pune, India
e-mail: abhi.thote8@gmail.com

S. Sharma et al. (eds.), *Intelligent Control, Robotics, and Industrial Automation*,
Lecture Notes in Electrical Engineering 1066,
https://doi.org/10.1007/978-981-99-4634-1_15

1 Introduction

Collision avoidance system (CAS) uses small radar detectors in front of the vehicle which will continually convey speedy explosions of high recurrence radar waves. These waves bounce and return to the sensor. The separate unit is connected with a sensor which helps to calculate the position of another car, distance, speed and also relative velocity. The system will help the driver to limit their speed which automatically helps in reducing accidents. CAS uses radar and different sensors like laser, ultrasonic and cameras to recognize and avoid upcoming accidents. It also helps to alert the driver by ringing the alarm and also apply a gentle break if the drivers fail to do.

CAS is being used in a wide range of different areas and under very different circumstances, i.e., automotive collision avoidance, aerospace application, marine application and industrial robot manipulators [1, 2]. CAS also helps to reduce accidents and limits the number of deaths that occur due to fatal accidents in adverse weather by giving warning and applying automatic breaks. The complete system is safer, efficient, affordable and easily applicable.

While riding any vehicle, safe distance needs to be maintained with respect to front vehicle. It is general observation that the two-wheeler riders do not maintain safe distance. Additionally, two-wheelers are not safe when compared with cars. Hence, maintaining this safe distance is specially required at high speeds, i.e., generally more than 40 km/hr as it may cause severe injury or death of the rider.

Many researchers have developed collision avoidance system for cars [3–8]. Very few studies have developed collision avoidance system for two-wheeler vehicles. In the research study of Islam et al. [9], an alarm was developed to alert the rider with the help of ultrasonic sensor and android application. In other research studies [10–13], novel intelligent helmet was designed to alert the driver about prospective collision. So, there is a lack of full-proof system which can detect the unsafe distance before crash or collision and apply gradual braking only to maintain safe distance in two-wheeler vehicles.

The aim of this research study is to minimize road injuries and accidents of two-wheeler vehicles. In this research study, a novel collision avoidance system (CAS) was developed to maintain the safe distance of the two-wheeler vehicles at high speeds, i.e., generally more than 40 km/hr. In this CAS, an ultrasonic sensor was used to detect the safe or unsafe distance with respect to front vehicle to avoid collision. Additionally, an automatic gradual brake mechanism was developed in this study and placed on to the rear tyre of the two-wheeler vehicle. This system applies gradual brakes if CAS detects unsafe distance with respect to front vehicle. As soon as safe distance is achieved, this gradual braking is removed. The purpose of gradual braking is only to maintain a safe distance with respect to front vehicle without skidding and not to stop the vehicle. For this purpose, 2-s rule was used to compute the safe distance between two vehicles as per speed of the vehicle in which this system is installed. For this purpose, the Arduino UNO (microcontroller) program was used in this system, and it was connected to the speedometer. This developed CAS alerts the driver with buzzer and shows the distance from the front vehicle on screen in cm. Finally, practical testing of two-wheeler mounted with CAS was carried out to check application of automatic gradual brake owing to detection of obstacle.

2 Materials and Methods

To determine the safe distance for two-wheeler vehicle, 2-s rule was used [14–17]. This rule states that the rider should be 2-s away from any vehicle in front of his or her vehicle. This safe distance depends on rider's speed. So, it means that, this safe distance is equal to the distance covered by the rider in 2-s with a speed at that moment of time. This rule is widely accepted all over the world.

If the speed of the vehicle is '*v*' km/hr, then, as per 2-s rule, the safe distance (*D*) of front vehicle with respect to back vehicle is as follows:

$$D = 2v \times \frac{5}{18}\mathrm{m}$$

$$D = \frac{5v}{9}\mathrm{m} \tag{1}$$

Based on this rule, collision avoidance system (CAS) was developed for cars in previous research studies. This system detects an unsafe distance of front vehicle with respect to vehicle with CAS while driving with the help of distance measurement ultrasonic sensor. As soon as unsafe distance is detected, an automatic gradual brake mechanism is applied till the safe distance is maintained.

For this purpose, architecture of CAS for two-wheeler vehicle was prepared. Figure 1 shows the block diagram of architecture of developed collision avoidance system (CAS) for two-wheeler vehicle. The ultrasonic sensor was used to detect any vehicle in front of the two-wheeler vehicle with this system. It measures the distance of front vehicle. As soon as the two-wheeler vehicle detects unsafe distance as per 2-s rule, the Arduino Uno (8-bit AVR microcontroller) is activated and shows the distance of front vehicle in the display panel. The Arduino Uno is powered by 12 V DC battery. Also, it rings the buzzer and glows red LED light for warning to the driver. Moreover, it establishes connection between 6 V DC battery and DC wiper motor (12 V, 55 RPM) with the help of dual acting channel relay. Owing to this, DC wiper motor starts and with the help of 2-link mechanism, it pulls the brake cam to apply the brake gradually till the safe distance is detected by the ultrasonic sensor. This system ensures that vehicle will not stop owing to this gradual automatic braking.

The developed complete CAS was divided into two parts:

- Sensing and processing unit
- Automatic gradual brake mechanism.

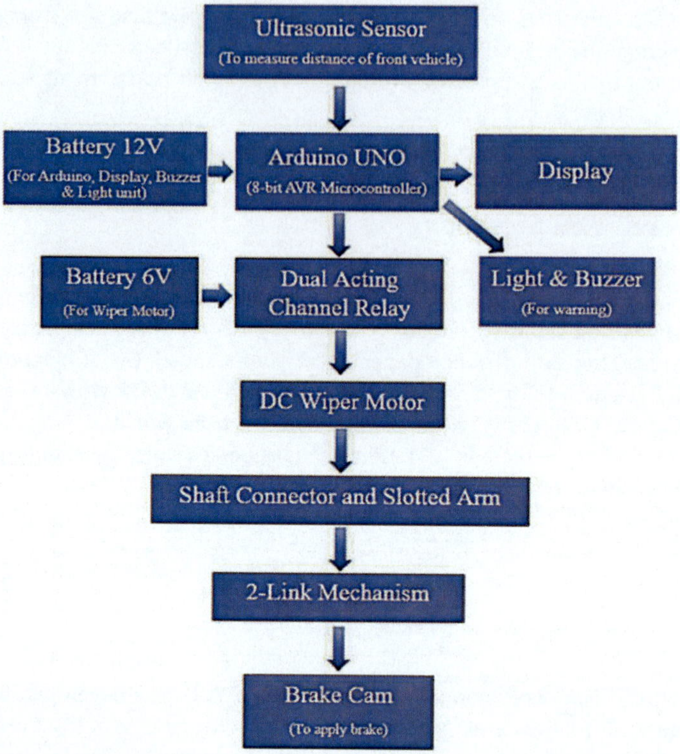

Fig. 1 Block diagram of architecture of developed collision avoidance system (CAS) for two-wheeler vehicle

2.1 *Sensing and Processing Unit*

Figure 2 shows sensing and processing unit of developed collision avoidance system (CAS). It consists of ultrasonic sensor, Arduino Uno (8-bit AVR microcontroller), distance display screen, buzzer, LED light and 12 V DC battery. The role of this unit is to measure the distance of front vehicle in real time and sending signals for an automatic gradual brake mechanism. Distance information sensed by ultrasonic sensor is sent to the Arduino Uno (microcontroller) which is powered by 12 V DC battery. This microcontroller is connected with distance display screen, buzzer and LED light as shown in Fig. 2. Dual acting channel relay is kept inbuilt in this unit to actuate automatic gradual brake mechanism.

Figure 3 shows the photograph of ultrasonic sensor used in this system, i.e., ultrasonic ranging module HC-SR04.

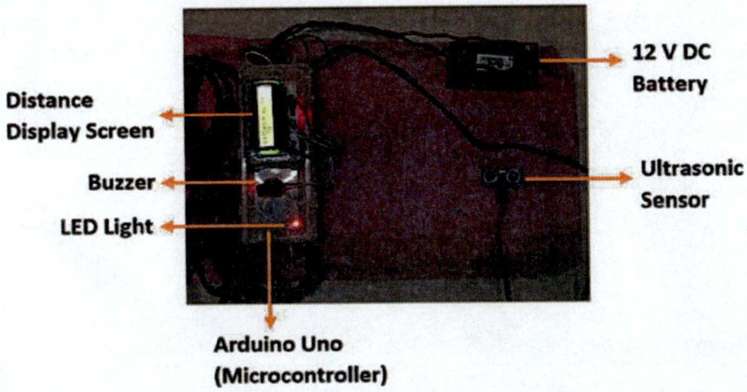

Fig. 2 Sensing and processing unit of developed collision avoidance system (CAS)

Fig. 3 Ultrasonic sensor used in CAS for two-wheeler vehicleultrasonic ranging module HC-SR04

2.2 *Automatic Gradual Brake Mechanism*

Figure 4 shows computer aided design (CAD) models of different parts of an automatic gradual brake mechanism of developed collision avoidance system (CAS). It shows different parts such as fixed arm, link 1, link 2, slotted arm, shaft connector and wiper motor. Autodesk fusion 360 software was used for CAD modeling purpose.

Figure 5 shows assembly of an automatic gradual brake mechanism of developed collision avoidance system (CAS). For the assembly of parts of an automatic gradual brake mechanism, fixed arm is connected to the footrest. DC wiper motor is attached to this fixed arm for support. The motor shaft is connected to the shaft connector. Slotted arm is present between shaft connector and two-link mechanism. Two links

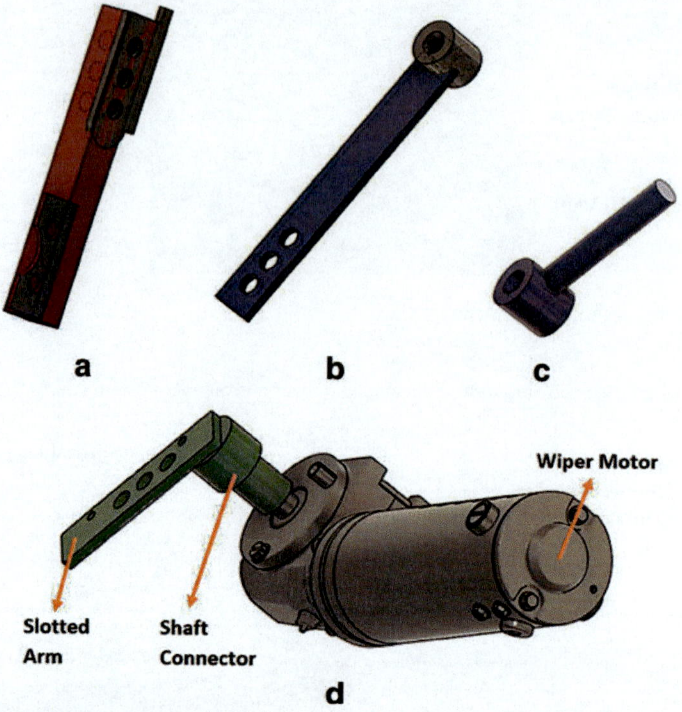

Fig. 4 Different parts of an automatic gradual brake mechanism of developed collision avoidance system (CAS), **a** Fixed arm, **b** Link 1, **c** Link 2, **d** Slotted arm, shaft connector and wiper motor

were used in this brake mechanism. Link 1 is connected to the slotted arm at one end and link 2 at another end. The link 2 is further connected to the brake cam. The wiper motor is connected to a 6 V battery, and the battery is directly connected to the CAS with electric wires. Whenever CAS is activated, it sends electric signals to start the wiper motor. This motor pulls the company fitted original brake cam gradually with the attached two links. In this way, gradual braking is applied automatically. It reduces the speed of two-wheeler vehicle. As soon as safe distance is detected by the CAS, it stops sending electric signals to the electric wiper motor. This will release the slightly pulled brake cam. In this way, gradual braking is automatically removed, and vehicle continues its running without automatic gradual braking till the unsafe distance with respect to front vehicle is detected as per 2-s rule. In this way, 2-s rule will help to advance the technology of the two-wheeler automobile sector. This brake mechanism can be manufactured as an accessory device and can be sold in all the cities and towns. Thus, the company can gain a good market share.

Figure 6 shows actual photograph of an automatic gradual brake mechanism of developed collision avoidance system (CAS) fitted on the rear brake cam of two-wheeler vehicle.

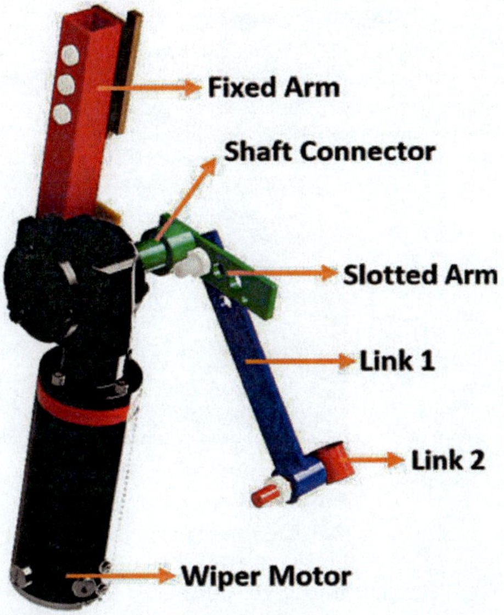

Fig. 5 Assembly of an automatic gradual brake mechanism of developed collision avoidance system (CAS)

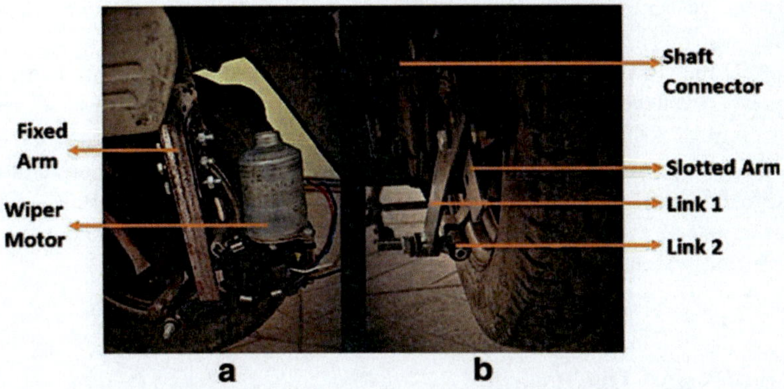

Fig. 6 Actual photograph of an automatic gradual brake mechanism of developed collision avoidance system (CAS) fitted on the rear brake cam of two-wheeler vehicle

2.3 Complete Setup of Collision Avoidance System (CAS)

Figure 7 shows the complete setup of collision avoidance system (CAS) mounted on the two-wheeler vehicle in the mechanical workshop of Dr. Vishwanath Karad, MIT-World Peace University, Pune, India. Ultrasonic sensor was placed on the front side below the headlight. Arduino Uno (microcontroller) and batteries were kept below the handles. These can be fixed at this position also at a particular height. This is the placement of parts of sensing and processing unit. An automatic gradual brake

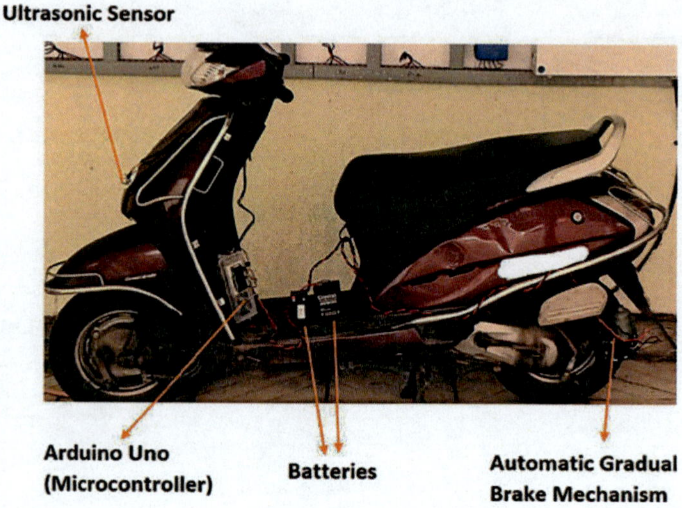

Fig. 7 Complete setup of collision avoidance system (CAS) mounted on the two-wheeler vehicle

mechanism was fixed near the rear wheel with its link 2 part attached with brake cam of wheel.

In brief, the flowchart of program of developed CAS is represented in Fig. 8. The data of front vehicle distance in real time is measured by ultrasonic sensor. The program of Arduino Uno (8-bit AVR microcontroller) determines whether the distance is safe or not as per 2-s rule. If the safe distance is detected, there will be no actuation of dual acting channel relay. Hence, an automatic gradual brake mechanism will not be active. However, if unsafe distance is detected, dual acting channel relay gets activated which will activate an automatic gradual brake mechanism.

3 Results and Discussion

The developed CAS mounted on two-wheeler vehicle was tested in a stationary condition in the mechanical workshop of Dr. Vishwanath Karad, MIT-World Peace University, Pune, India. The vehicle was kept stationary on its main stand. The accelerator was raised to rotate the rear wheel as shown in Fig. 9.

In place of another vehicle in front of this vehicle, a cardboard type of obstacle was held, and it was slowly brought close to the vehicle so that developed CAS should detect the obstacle and apply gradual brakes.

It was tested at different speeds, i.e., 5, 10, 15 and 20 km/hr. During testing, for a particular value of speed, the distance of obstacle displayed on the screen of microcontroller from the moment of activation of an automatic gradual brake mechanism was noted. This helped to check that whether the developed system

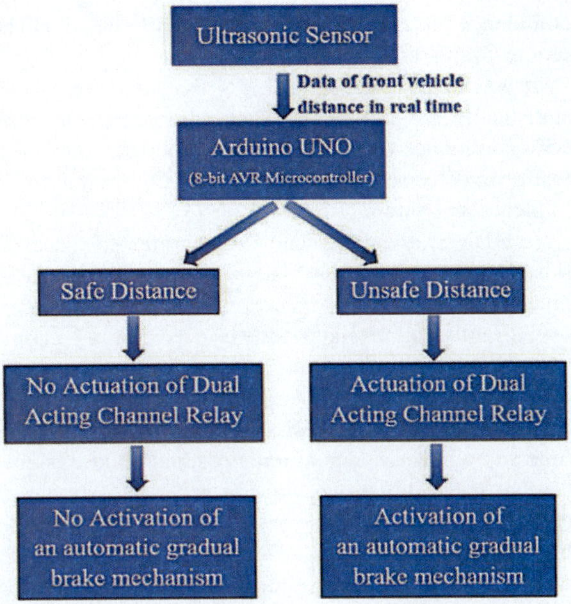

Fig. 8 Flowchart of program of developed collision avoidance system (CAS)

Fig. 9 Testing of developed CAS mounted on two-wheeler vehicle in a stationary condition

was able to follow or not 2-s rule for gradual braking purpose. Total 10 readings of distance of obstacle (D) from the moment of activation of an automatic gradual brake mechanism were noted for a particular speed value and their average value was calculated for each considered value of speed as given in Table 1. This average value of distance (D) was compared with a theoretical value as computed from Eq. (1) with the help of one sample t-test. For this purpose, two-tailed p value was computed, and confidence interval was determined. This statistical calculation was performed with online GraphPad software [18].

It was observed from Table 1 that, the average value of '*D*' was found to be statistically not significant when compared with computed theoretical value with 95% confidence interval. Hence, the performance of developed system (CAS) with respect to detection of unsafe distance (*D*) was not statistically significant with respect to theoretical value.

In this way, the developed CAS was tested successfully. Thus, it is expected that, this CAS device will play a key role to avoid road accidents of two-wheeler vehicles on high-speed roads.

Currently, this research study is in developing phase. In future prospective research study, this CAS system can be modified to activate it automatically at high speed, i.e., more than 40 km/hr. Additionally, as per current speed of two-wheeler vehicle, the future prospective CAS system will calculate the required safe distance in real time and will activate automatic gradual brake system if unsafe distance is detected for that speed of the two-wheeler vehicle. Also, it will be tested thoroughly during actual riding on the high-speed roads with proper approval of respective road and transportation authorities. After successful practical testing in future, it can be implemented in two-wheeler vehicles as an accessory device with an official approval from road and transportation authorities.

Table 1 Comparison of distance of obstacle (*D*) from the moment of activation of an automatic gradual brake mechanism with a theoretical value during testing

Speed in km/hr		5 km/hr	10 km/hr	15 km/hr	20 km/hr
Distance (*D*) recorded in meters	[1]	2.60	5.28	7.7	10.71
	[2]	3.00	5.72	9.22	11.4
	[3]	2.65	5.16	7.82	10.37
	[4]	2.95	5.84	8.68	11.54
	[5]	2.76	5.08	8.20	10.06
	[6]	2.79	6.06	8.55	11.91
	[7]	2.44	5.42	8.30	10.49
	[8]	3.18	5.54	8.40	11.17
	[9]	2.74	5.30	8.05	10.91
	[10]	2.81	5.68	8.73	11.2
Average value of '*D*' in meters		2.79	5.51	8.37	10.98
Theoretical value of '*D*' in meters		2.78	5.55	8.33	11.11
Two-tailed p value		0.8618	0.6840	0.8135	0.4799
Confidence interval (%)		95	95	95	95
Statistical significance		No	No	No	No

4 Conclusion

The portable collision avoidance system (CAS) for two-wheeler vehicles at high speed was developed in this research study, which can detect safe or unsafe distance as per 2-s rule and applies automatic gradual brake to maintain the safe distance. It can be fitted as an accessory device in the two-wheeler vehicle and can gain a good market share. This is the novelty of this research study. The developed complete CAS was divided into two parts, i.e., sensing and processing unit and automatic gradual brake mechanism. Sensing and processing unit detects safe or unsafe distance in real time. If the safe distance is detected, there will be no actuation of dual acting channel relay. Hence, an automatic gradual brake mechanism will not be active. However, if unsafe distance is detected, dual acting channel relay gets activated which will activate an automatic gradual brake mechanism. The developed CAS system was tested at different speeds, i.e., 5, 10, 15 and 20 km/hr. Total 10 readings of distance of obstacle (D) from the moment of activation of an automatic gradual brake mechanism were noted for a particular speed value and their average value was calculated for each considered value of speed. This average value of distance (D) was found to be statistically not significant when compared with a theoretical value with 95% confidence interval using one sample t-test. In this way, the developed CAS was tested successfully. Thus, it is expected that, this CAS device will play a key role to avoid road accidents of two-wheeler vehicles on high-speed roads.

References

1. Zhao Z, Zhou L, Zhu Q, Luo Y (2017) A review of essential technologies for collision avoidance assistance systems. Adv Mech Eng 9(10):1–15. https://doi.org/10.1177/1687814017725246
2. Savino G, Lot R, Massaro M, Rizzi M, Symeonidis I, Will S, Brown J (2020) Active safety systems for powered two-wheelers: a systematic review. Traffic Inj Prev 21(1):78–86. https://doi.org/10.1080/15389588.2019.1700408
3. Moon S, Yi K, Kang H (2009) Multi-vehicle adaptive cruise control with collision avoidance in multiple transitions. In: IFAC proceedings volumes. Elsevier, pp 304–311. https://doi.org/10.3182/20090902-3-US-2007.0024
4. Zhang H, Liu C, Zhao W (2022) Segmented trajectory planning strategy for active collision avoidance system. Green Energy Intell Transp 1(1):100002. https://doi.org/10.1016/j.geits.2022.100002
5. Hang P, Han Y, Chen X, Zhang B (2018) Design of an active collision avoidance system for a 4WIS-4WID electric vehicle. In: IFAC-PapersOnLine. Elsevier, pp 771–777. https://doi.org/10.1016/j.ifacol.2018.10.132
6. Cho H, Kim G, Kim B (2014) Usability analysis of collision avoidance system in vehicle-to-vehicle communication environment. J Appl Math 2014(SI01):1–10. https://doi.org/10.1155/2014/951214
7. Rodríguez-Seda EJ (2021) Collision avoidance systems, automobiles. In: Vickerman RB (ed) International encyclopedia of transportation. Elsevier, Oxford, pp 173–179. https://doi.org/10.1016/B978-0-08-102671-7.10121-6
8. Muslim H, Itoh M (2019) Trust and acceptance of adaptive and conventional collision avoidance systems. In: IFAC-PapersOnLine. Elsevier, pp 55–60. https://doi.org/10.1016/j.ifacol.2019.12.086

9. Islam N, Shamsher U, Nayyeri AD, Kulesza WJ (2013) Motorbike crash avoidance system with ultrasonic sensor and android application. Can J Electr Electron Eng 4(2):40–42
10. Chang WJ, Chen LB (2019) Design and implementation of an intelligent motorcycle helmet for large vehicle approach intimation. IEEE Sens J 19(10):3882–3892. https://doi.org/10.1109/JSEN.2019.2895130
11. Chen LB, Chang WJ, Su JP, Chen YR (2018) i-Helmet: an intelligent motorcycle helmet for rear big truck/bus intimation and collision avoidance. In: 2018 IEEE international conference on consumer electronics (ICCE), pp 1–2. https://doi.org/10.1109/ICCE.2018.8326344
12. Pangestu A, Mohammed MN, Al-Zubaidi S, Bahrain SHK, Jaenul A (2021) An internet of things toward a novel smart helmet for motorcycle: review. In: AIP conference proceedings, p 50026. https://doi.org/10.1063/5.0037483
13. Mohd Rasli MKA, Madzhi NK, Johari J (2013) Smart helmet with sensors for accident prevention. In: 2013 international conference on electrical, electronics and system engineering (ICEESE), pp 21–26. https://doi.org/10.1109/ICEESE.2013.6895036
14. TWO (2) Second rule. https://ctp.gov.in/RS2SecondRule.htm. Accessed 22 Sept 2022
15. Two-second rule. https://en.wikipedia.org/wiki/Two-second_rule. Accessed 23 Sept 2022
16. Maitra B (2017) Rules For safe driving. In: Training manual for drivers. Transport Department, Government of West Bengal, India, pp 69–110
17. 2 Second rule explained. https://www.drivingtesttips.biz/2-second-rule.html. Accessed 24 Sept 2022
18. One sample t test. https://www.graphpad.com/quickcalcs/oneSampleT1/. Accessed 24 Sept 2022

Development of a Digital Twin Interface for a Collaborative Robot

Ayan Naskar, Pothapragada Pranav, and P. Vivekananda Shanmuganathan

Abstract Collaborative Robots (Cobots) constitute a new class of industrial robot manipulators that are now becoming popular in tasks that require intelligent manipulation and human–robot interaction. The characteristic difference between these robots and the conventional industrial robots is the availability of force feedback from the wrist-force sensor. This allows the robot arm to avoid obstacles, detect variation in the part locations and estimate the type of objects being handled based on the weight or contact forces. Digital twin is the representation of a real-world object in a virtual environment. This allows the synchronization and replication of the motions of the robot present in the real world with that of the counterpart in a virtual environment and vice versa. This work is an initial attempt to develop a remote interface and a digital twin for a collaborative robot (Universal robot UR5e). Interfacing of the robot through Robotics Operating System (ROS) and the use of Unity3D for the development of the digital twin are reported here.

Keywords Collaborative robots · Virtual robot · Digital twin · Human–robot interaction · ROS · Unity3D

1 Introduction

With the advent of the age of automation and big-data analysis, it is imperative to keep on top of the chain to survive the competition, and "digital twins" (DT) are an integral part of keeping costs low while testing new technologies. Digital twin may be defined as representing a process, object or event of the real world in a digital

A. Naskar · P. Pranav
Department of Electronics and Communication Engineering, SRM University AP, Amaravati, Andhra Pradesh 522240, India

P. V. Shanmuganathan (✉)
Department of Mechanical Engineering, SRM University AP, Amaravati, Andhra Pradesh 522240, India
e-mail: vivekananda.s@srmap.edu.in

S. Sharma et al. (eds.), *Intelligent Control, Robotics, and Industrial Automation*, Lecture Notes in Electrical Engineering 1066,
https://doi.org/10.1007/978-981-99-4634-1_16

environment as accurately as possible. The data used in the digital twin to reflect current conditions from the actual world is generally collected from IoT devices such as sensors and other embedded devices. Digital twin enables the simulation of the data flow within its digital ecosystem, mirroring the data flow that occurs in the real world.

With rapid advances in GPUs, accurate 3D-model rendering, physics engines, and communication technologies, digital twin has become quite popular in the research community during the past decade.

For a digital twin, each real object and the external environment elements in the physical world which can impact the targeted result will be simulated to the extent of its physics and dimensions in the virtual world. Before executing the task on a system in the real world, we can test its performance and the presence of any potential issues by performing the same task on a digital twin, thus reducing unnecessary expenses while also saving the user time.

A good digital twin can also help in controlling the robot remotely. If the task requires any real-time human input, the input can be given remotely in the digital twin, and the same task would be mapped in the real robot at the actual site as required. This provides human safety as no human is physically present at the location. However, designing and putting in all physics engines, data flow models, and communication technologies is very tedious. Fortunately, they do not need to be built from scratch, as the game engines used in making games come pre-equipped with all the required features to generate a virtual world.

1.1 Unity as the Platform for Digital Twin

In this work, we decided to use the Unity game engine and connect it to the robot using TCP/IP. Due to Unity's vast user community, it is flexible in its usage and is supported for multiple use cases besides being useful for making games.

Unity is a platform for creating and operating real-time 3D content. This platform can also be used to develop augmented reality/virtual reality (AR/VR) applications and on-screen applications, allowing it to spread across many different industries like manufacturing, engineering, construction, architecture, media, entertainment, and industrial simulation, including robotics simulation.

Simulation is a crucial part when developing a robot. As a game engine, Unity runtime is fundamentally made up of the same component as robotics simulators, namely, the rendering engine and physics engine, both of which have gone through years of optimization in Unity. The default physics engine of Unity, Nvidia PhysX, is capable of simulating the real-world physics such as gravity, rigid body collisions, mass, acceleration, drag, force, and impulse of the virtual object.

Furthermore, it is possible to create almost any kind of environment to test and train the robot in Unity. Unity has a vast developer community and an "asset store" enriched with valuable and high-quality content that can be used to boost the environment creation process for digital twins. Moreover, Unity provides multiplatform support. The well-designed scripting API of Unity allows integration with the external software such as ROS and ROS 2.

2 Related Works

Digital twins have been on the rise, especially with the rise in automation and the latest Industry 4.0 standards. In recent years, multiple studies have been conducted on building digital twins of different control environments, including environments involving Industrial robots and autonomous vehicles. Research has been conducted on human–robot collaborations in [1, 2] that allow safe association. Also, the papers mention a more efficient ability for human–robot interactions, giving the modern industries the ability to create hybrid teams.

There is also much research on improving virtual commissioning in industrial environments [3]. Virtual reality is also on the rise concerning digital twin creation and manipulation, as seen in [4–6]. They introduce synchronized control of industrial robots based on simulation in virtual reality. Garg et al. [5] emphasizes specifically on FANUC systems, while [4, 6] give a more generalized approach. Similarly, research on the robot behavior in different industrial environments such as assembly lines, for example in [7], further showcases the possible future implications of the current research.

Research is also being conducted on making digital twins using the Unity game engine, as seen in [8]. The usage of Unity in designing digital twins simplifies the process of environment designing due to their access to a vast asset store and a multitude of physics engines. Furthermore, Artificial Intelligence algorithms could be used to improve the performance of digital twins and their ability to control the robots, as seen in [9]. Collaborative robots can also be used in medical environments [10], where the possibility of safe remote surgery was explored, and digital twins with VR gear were used to simulate an actual surgery environment. Digital twins can also be used to assess and predict collision hazards [11]; this allows workers to better asses hazards in real-time and react accordingly. Algorithms could also be designed for programming industrial robots from digital twins [12, 13], where they could be used to optimize programming from their digital counterparts.

UR5e is a member of the class of commercial collaborative robots from Universal Robots (UR). The essential literature used in this paper is the UR5 user manual [14] and the UR API Reference [15]. These manuals are the basis for building any system. The user manual introduces the components of the UR5e Robot while the API reference enables us to generate commands to receive and send data effectively.

3 Methodology

In this work, we created a digital twin of the UR5e robot, a collaborative robot manufactured by Universal robots [14, 15]. Each joint of the robot is equipped with sensors that provide the joint position and velocity. An important feature of the robot is the availability of the 6-component force-torque data from the wrist-force sensor.

The modality can be divided into multiple parts that can then be connected to form a working digital twin:

1. Robot and environment creation
2. Communication with the industrial robot.

3.1 Robot and Environment Creation

Initially, the UR5e robot's 3D mesh was imported from the official GitHub documentation. It was imported as six different segments linked using a parent–child hierarchy. Each 3D model section has the same dimensions of the real UR5e robot that are converted into a "unity meter unit". Then, a mesh collider is assigned to each part of the robot to detect continuous real-time collision with the external environment and itself.

Then, Unity's inbuilt rigid body components were assigned to each segment, letting us add physics components to the system, such as mass and gravity.

As discussed, each part of the robot in Unity is arranged in a parent–child hierarchical configuration as shown in Fig. 1. A parent joint affects the coordinate frame of the child joint, while the child joint moves without affecting the parent's coordinate frame. This allows for a serially configured robot arm to simulate a real UR5e robot.

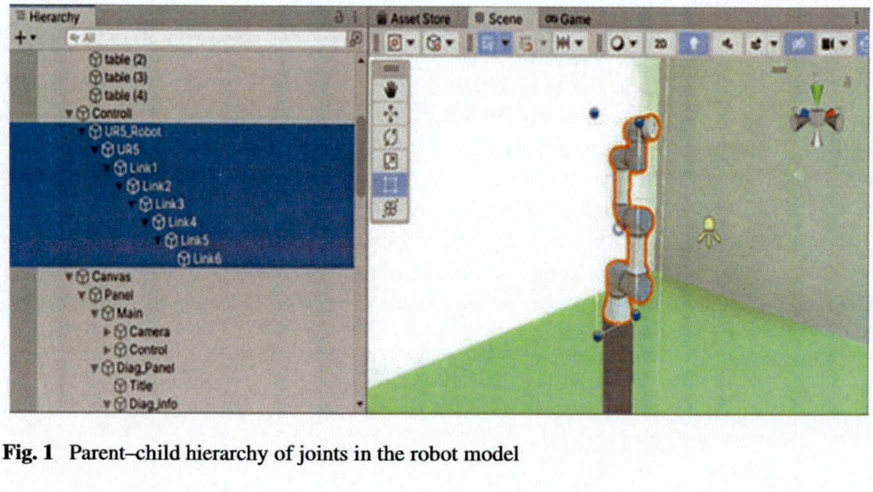

Fig. 1 Parent–child hierarchy of joints in the robot model

In the current work, the robot was made to move as per the user input as the user operates the buttons in the virtual environment. Every game object in Unity has a script attached to it, which changes the angle orientation of the game object based on the different input data. Unity provides a functionality that allows us to simulate movement similar to the real world using the rigid body physics from the physics engine.

Movement in the Robot model can be generated from one of two sources.

1. User input from pressing the virtual button
2. Change in the current position of the robot from external controls.

Unity's "transform.eulerangle()" function is used to generate the corresponding movement. This lets us rotate the game object in Unity's virtual world plane with respect to the origin coordinate (0, 0, 0) of Unity. Finally, if the input is from the virtual buttons, the angles from the aforementioned function are converted to degrees and sent to the robot using the TCP/IP connection explained below.

3.2 Communication with the Industrial Robot

The digital twin of the UR5e robot was made in Unity; hence, the program is written in C# language. "Universal Robots" provides the ability to control the UR5e robot using TCP/IP connections. The code for manipulating the UR5e robot can be divided into two parts:

1. Program that reads the current state data from the robot
2. Program that writes the commands to change the robot's pose.

Reading the state data. Universal robot provides a real-time controller data stream that can be accessed at the port number "30,003" with the corresponding read-only port at "30,013".

To read the state data of the UR robot, the stream data is generated using the Network Socket, which connects the socket to receive packets of data from PolyScope [15] about the current robot pose. This process is depicted in Fig. 2.

Initially, the virtual robot is set flat on the ground. Connecting the socket to the digital twin initializes functions that rearrange the virtual robot's joints to correspond to the UR robot's current joint configuration. Here, the end-effector position of the robot is sent through TCP/IP as a response to the established connection. This response data allows the initial pose of the virtual robot to match that of the UR robot closely.

After connecting the socket successfully, the synchronization packet is eliminated from the set, and the remaining data is divided into joint parameters to set the joint values on the digital twin. Further, based on the data transfer frequency of the robot (500 Hz for E-Series robots), the data collection thread is made to wait until the corresponding cycle (of 2 ms, in the case of the E-series robots) has ended before getting more data.

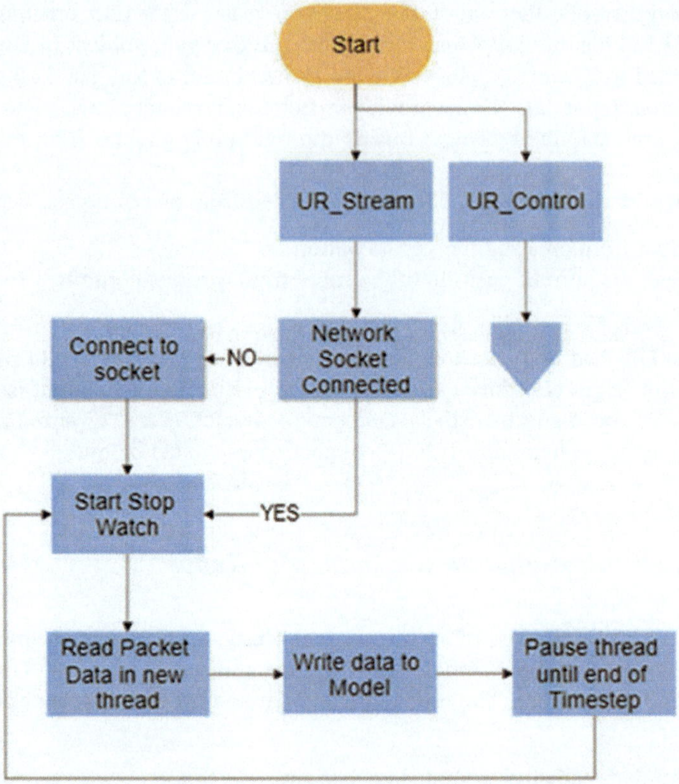

Fig. 2 Data streaming flow-chart

Sending the Control Data

Universal robots have a mode called "Remote Mode" that can be enabled on the Teach Pendant. This mode allows the robot to receive external commands from the network using TCP/IP.

Initially, a network socket is connected to the robot's IP address and the real-time port number. Once the network socket is connected, the program continuously checks for user inputs.

Currently, the robot is controlled exclusively using the "speedl" command. The script that needs to be executed by the robot is first generated based on the user input, and once done, it is converted into UTF-8 bytes that are then sent to the robot using the "networkstream.write()" function, executing the code. The complete process is depicted in Fig. 3.

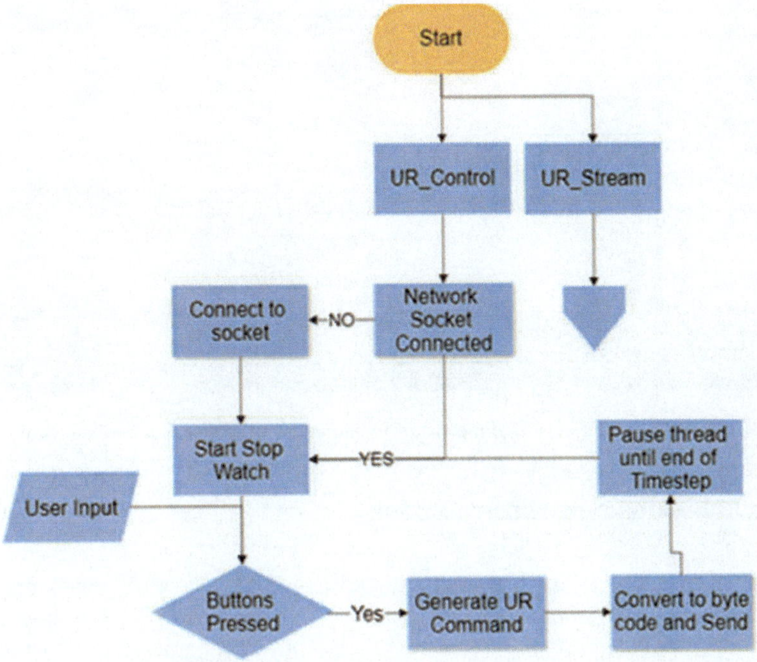

Fig. 3 Robot control flow-chart

4 Results and Discussion

A preliminary implementation of the digital twin for a UR5e robot has been demonstrated successfully in this work. The robot is completely controlled from the digital twin, and any changes to the state of the actual robot is reflected in the digital twin as well. The digital twin is capable of executing all commands the actual robot would execute in the real world.

A screenshot of the rendering on the digital twin and the real robot environment are shown in Fig. 4. Though a perfectly realistic graphical replication of the environment was not attempted due to the preliminary stage of the work, a reasonable recreation of the actual robot's environment has been achieved in the digital twin.

The pose of the virtual robot and its interaction with its recreated environment has been simulated using rigid body physics provided by the Unity engine. The twin follows the original robot closely, thanks to the real-time transmission of joint angle data. The movement between discrete data points is smoothed using the "transform.eulerAngles" function in Unity, presenting a model that does not snap into each position when supplied with new data.

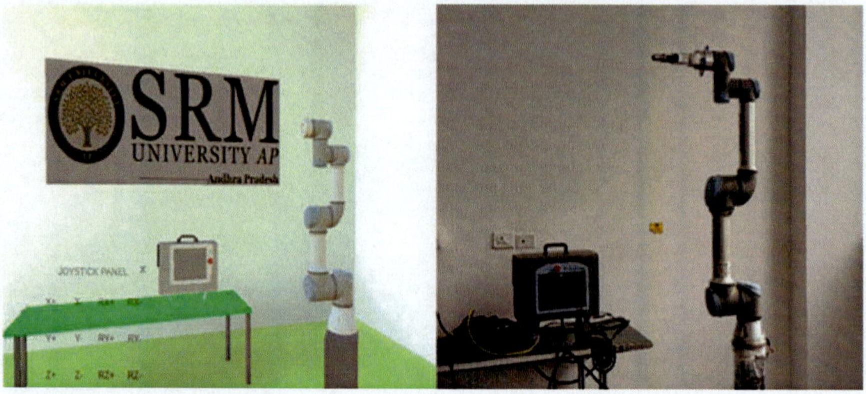

Fig. 4 A side-by-side comparison of the virtual environment with the actual environment

5 Conclusions and Future Scope

This paper documents the creation of a digital twin that can control and imitate the state of an actual UR5e industrial robot. This work demonstrates the potential for further improvement due to the recent advances in physics engines, the advent of data science allowing for better computation of robot kinematics and the flexibility of the Unity environment. Additionally, multiple systems could be added to this modality.

It is proposed to experiment further on the suitability of the robot for teleoperation and robotic surgery. The work reported here are preliminary attempts mainly toward developing a computational and control interface to the robot beyond the features of the teach pendant and the controller provided by the manufacturer. Future use cases of this DT modality could include the ability to connect multiple subsystems and using this digital twin to create custom algorithms to govern the robot's movement without first testing on the actual robot.

Furthermore, additional virtual objects may be included in the model's workspace. A virtual boundary gets created as the model will work around these objects, with the result that the robot in the real workspace respects the boundary as the motion constraints of the virtual robot apply to the real robot as well.

Unity's built-in Reinforcement Learning (RL) ToolKit can help generate inverse kinematics solutions for the robot arm with higher speed and less complexity while also working around the problem of singularities. The RL toolkit may be useful for better understanding of the robot environment using the data from multiple sensors connected in the real world.

References

1. Malik AA, Brem A (2021) Digital twins for collaborative robots: a case study in human-robot interaction. Robot Comput Integr Manuf 68:102092
2. Gallala A, Kumar AA, Hichri BP (2022) Digital twin for human–robot interactions by means of industry 4.0 enabling technologies. Sens 22(13):4950
3. Lechler T, Fischer E, Metzner M, Mayr A, Franke J (2019) Virtual commissioning—Scientific review and exploratory use cases in advanced production systems. Procedia CIRP 81:1125–1130
4. Kuts V, Otto T, Tähemaa T, Bondarenko Y (2019) Digital twin based synchronised control and simulation of the industrial robotic cell using virtual reality. J Mach Eng 19(1):128–144
5. Garg G, Kuts V, Anbarjafari G (2021) Digital twin for FANUC robots: industrial robot programming and simulation using virtual reality. Sustain 13(18):10336
6. Burghardt A, Szybicki D, Gierlak P, Kurc K, Pietruś P, Cygan R (2020) Programming of industrial robots using virtual reality and digital twins. Appl Sci 10(2):486
7. Kousi N, Gkournelos C, Aivaliotis S, Giannoulis C, Michalos G, Makris S (2019) Digital twin for adaptation of robots behavior in flexible robotic assembly lines. Procedia Manuf 28:121–126
8. Wang Z, Han K, Tiwari P (2021) Digital twin simulation of connected and automated vehicles with the unity game engine. In: 2021 IEEE 1st international conference on digital twins and parallel intelligence, DTPI. IEEE, pp 1–4
9. Matulis M, Harvey C (2021) A robot arm digital twin utilising reinforcement learning. Comput Graphics 95:106–114
10. Laaki H, Miche Y, Tammi K (2019) Prototyping a digital twin for real time remote control over mobile networks: application of remote surgery. IEEE Access 7:20325–20336
11. Messi L, Naticchia B, Carbonari A, Ridolfi L, Di Giuda GM (2020) Development of a digital twin model for real-time assessment of collision hazards. In: creative construction e-conference. Budapest University of Technology and Economics, pp 14–19
12. Bansal R, Khanesar MA, Branson D (2019) Ant colony optimization algorithm for industrial robot programming in a digital twin. In: 25th International conference on automation and computing, ICAC. IEEE, pp 1–5
13. Chen C, Liu Q, Lou P, Deng W, Hu J (2020) Digital twin system of object location and grasp robot. In: 5th International conference on mechanical, control and computer engineering, ICMCCE. IEEE, pp 65–68
14. Universal robots e-series user manual, UR5e, software version 5.12. Universal robots. https://s3-eu-west-1.amazonaws.com/ur-support-site/165908/99404_UR5e_User_Manual_en_Global.pdf. Accessed 20 Oct 2022
15. The URScript programming language for e-series, software version 5.12. Universal robots. https://www.universal-robots.com/download/manuals-e-series/script/script-manual-e-series-sw-512/. Accessed 20 Oct 2022

Design and Analysis of Combined Braking System Using Delay Valve for Automobiles

C. Dineshkumar, A. S. Selvakumar, P. D. Jeyakumar, Solomon Jenoris Muthiya, Nadanakumar Vinayagam, Balachandar Krishnamurthy, Joshuva Arockia Dhanraj, and R. Christu Paul

Abstract There are many different models and technical configurations of motorcycle braking systems, ranging from traditional drum brakes to the most recent brake systems. Additionally, the pressure of each brake circuit can be actively integrated into the system without the rider's input, ensuring that the system operates properly under all circumstances. In this study, a motorbike combined braking system (CBS) incorporating a delay valve is designed and analyzed in real time. Siemens NX is used to assist in developing and analysis. The flow rate, mass, and brake pad dimensions will not vary. The amount of pressure in the master cylinder increases when using brakes. The fluid will begin to flow along the brake lines as a result of the pressure. The delay valve in the brake line is in the front brake.

C. Dineshkumar · P. D. Jeyakumar
Department of Automobile Engineering, B.S. Abdur Rahman Crescent Institute of Science and Technology, Chennai, Tamil Nadu, India

A. S. Selvakumar
Department of Mechanical Engineering, B.S. Abdur Rahman Crescent Institute of Science and Technology, Chennai, Tamil Nadu, India

S. J. Muthiya (✉)
Department of Automobile Engineering, Dayananda Sagar College of Engineering, Bangalore, Karnataka, India
e-mail: jenoris.555@gmail.com

N. Vinayagam · R. C. Paul
Department of Automobile Engineering, Hindustan Institute of Technology and Science, Chennai, Tamil Nadu, India

B. Krishnamurthy
School of Mechanical Engineering, Sathyabama Institute of Science and Technology, Chennai, Tamil Nadu, India

J. A. Dhanraj
Centre for Automation and Robotics (ANRO), Department of Mechatronics Engineering, Hindustan Institute of Technology and Science, Chennai, Tamil Nadu, India

S. Sharma et al. (eds.), *Intelligent Control, Robotics, and Industrial Automation*, Lecture Notes in Electrical Engineering 1066,
https://doi.org/10.1007/978-981-99-4634-1_17

Keywords Brakes · Delay valve · Two-wheeler

1 Introduction

A brake system's key role is to slow down moving vehicles. A good braking system gives the driver a variety of braking options. The vehicle's handling and safety both are controlled by the braking system. More than anything else, a vehicle's braking system needs to be dependable. Brakes slow down moving vehicles by converting energy of the motion into heat. Friction between the braking rotor and the pads is created to achieve this. A significant amount of heat generated during this operation is dissipated to air. The proximity of such a high temperature requires that the components have great thermal stability [1–3]. The fundamental goal of designing a brake system is to fit it into the available space as compactly as feasible. When the driver of a two-wheeled vehicle needs to stop the vehicle immediately, occasionally gets confused over which brake to use and uses either the front brake, the rear brake, or both brakes. As a result, the wheels become out of balance, which leads to accidents. A single actuating braking system should be employed to avoid such situations. The rear brake pedal is the only actuation system used to activate the front and rear brakes in this new combined braking system [4–6]. The single actuation braking system's goal is to simultaneously lock the front and rear wheels without slipping or skidding. This will cause the brakes to bias toward the rear wheel by 60% and the front wheel by 40%, stalling the vehicle. Calculating the braking force, braking distance, and braking duration allows for system verification. The term "braking distance" in this context refers to the overall distance covered by a vehicle from a certain speed until it comes to a stop, and it can be calculated using the conventional experimental approach. This suggests that in order to fulfill the safe braking criteria, the vehicle components that have an impact on braking performance should be constructed accordingly [7–10]. Designers will discuss the many designs and technical approaches for motorcycle brake systems. An ABS system can be added to a standard system with two independent circuits to increase stability and safety. By connecting the front brake control and the rear caliper hydraulically, or vice versa, a combined brake system can be made to improve safety and comfort. Motorcycle manufacturers have the option to realize any combination integrated functionality characteristic thanks to the Motorcycle Integral Brake System (MIB). Additionally, each brake circuit's pressure can be actively built up without any rider input, ensuring that the system responds effectively in any riding circumstance [11, 12]. The combined braking system (CBS) described in this paper has two brake cables attached to it from left- and right-hand levers, and it also has two brake cables attached to the front and back wheels. A link mechanism is activated by the LH brake lever, and it is through this system that an appropriate force distribution between the front and rear wheels is achieved. The technology allows the front and rear wheels to have brakes operating concurrently. Only the front brake is applied when the RH lever is pulled; the rear brake does not get any force in this scenario. By preventing rear wheel locking, the CBS allows for greater deceleration. Drum brakes and disc brakes are the two basic types of brakes. Because they are more effective, disc brakes are frequently employed. Compared to drum brakes, they distribute heat better and cool down faster. When brake pads are used, brake rotors are pinched. After deciding master cylinder and brake disc size, pedal ratio, and caliper size can be varied. Due to dynamic weight transfer, more braking force is required at the rear as compared to front. A combined braking system is a system for linking front and rear brakes on a motorcycle or scooter. In this system, the rider's action of application one of the brake levers applies both front and rear brakes. The amount of each brake applied may be determined by a proportional control valve. This is distinct from integrated brakes, where applying pressure to brake pedal only includes application of some front brake. The active safety progression and its application currently minimize fatalities globally [13–16].

2 Methodology

The chassis dynamometer is used to compute and test the suggested model of the integrated brake system design. The calculated values utilized to determine how well the combination braking system performed. The proposed model is designed and constructed utilizing the theoretical values used to calculate the braking system. The made-up model that was used to analyze the performance. The proposed model's dynamic performance is calculated by taking into account the vehicle and rider's mass, as well as the velocity and stopping distance. The brake booster generates a specific amount of pressure in the master cylinder when the brake pedal is depressed.

In Fig. 1a. The rear brake lever is connected to the combine brake line and to the front disc where pressure is attained gradually to the disc. The horizontally moving piston of the master cylinder raises the pressure at which brake fluid flows. The brake calipers on the front and back wheels receive the fluid from the master cylinder's outlet valve by passing through valves with a particular range of pressure. To control the flow of braking fluid, front disc brakes use a delay valve system. It is situated between the master cylinder valve and the front disc caliper. As a result, the brake fluid is delivered to the delay valve. The calipers are shielded from brake fluid entry. By applying additional brake pressure, the delay valve raises the pressure in the master cylinder.

In Fig. 1b. Where the delay valve is connected to the front wheel or disc and to other side of master cylinder. The piston, which is lowered gradually after it is within the delay valve cylinder, is supported by a helical spring. The piston contracts as it lowers, allowing fluid to pass through the valve and into the front caliper. When a brake pad is pressed against a rotating disc by a caliper piston, a wheel comes to a stop. The compressed spring forces the piston back into position whenever the pressure lowers. A small time difference between the front and rear braking applications offers good stability and prevents skidding.

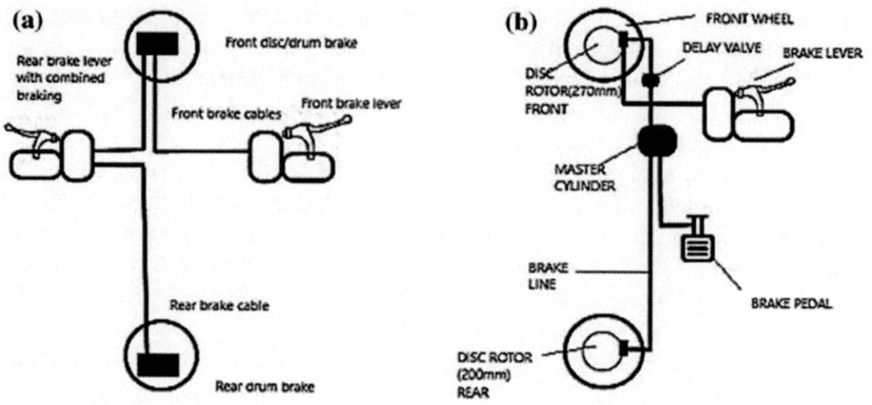

Fig. 1 **a** Typical combined braking system, **b** Proposed combined braking system with delay valve

The Combi-Brake System facilitates easier operation while braking. The front and rear brakes are simultaneously engaged when the rider depresses the rear brake lever. Therefore, you only need to utilize the rear brake lever and not both.

Additionally, the mechanism applies the proper amount of braking pressure to both wheels. As a result, it aids in safer braking and improved bike control. It allows for simple and precise braking for the rider. As a result, when the left side brake lever is applied, the system evenly distributes braking force across the front and back wheels. The mechanism effectively brakes because it applies the same amount of force to the front and back wheels. For instance, the car can be moving quickly and just use its rear brakes. Then, the conventional brakes abruptly and unevenly bring the vehicle to a stop. However, Honda automobiles have powerful front and rear brakes. Additionally, the braking distance is shortened. It enhances braking stability as well. As a result, the technology makes sure that the rider with little experience can brake with confidence. You can therefore ride with the assurance of effective brakes.

2.1 *Design Considerations*

From the Table 1. The design variables for different vehicles are shown. The design and fabrication of a combined brake system demonstration apparatus requires some considerations. These includes:

Economics and sustainability: The issue of the possibility and vitality of the design and production of the integrated brake system is addressed. There are many equipment and systems that utilize the hydraulic brake system, making the design and manufacture of the apparatus feasible.

Table 1 Design variables for different vehicles

Vehicle	Honda Activa 5G	Bajaj pulsar 150 Twin Disc	TVS apache RTR 180	Yamaha FZ25
Cubic capacity	101.19 cc air cooled	149.5 cc air cooled	177 cc air cooled	249 cc air cooled
Max.Power	8 bhp @ 7500 rpm	13.8 bhp @ 9000 rpm	17.3 bhp @ 8000 rpm	20.6 bhp @ 8500 rpm
Max.Torque	9 Nm @ 5500 rpm	13.4 Nm @ 6500 rpm	15.5 Nm @ 6000 rpm	20 Nm @ 6500 rpm
Top speed	83 kmph	110 kmph	124 kmph	134 kmph
Wheelbase	1238 mm	1320 mm	1326 mm	1360 mm
Kerb weight	109 kg	143 kg	139 kg	148 kg
Ground clearance	153 mm	165 mm	165 mm	160 mm
Gearbox	CVT	5 (S)-synchromesh	5 (S)-synchromesh	5 (S)-synchromesh
Brake-front	130 mm drum	260 mm disc	270 mm disc	282 mm disc
Brake-rear	130 mm drum	230 mm disc	200 mm disc	220 mm disc
Distribution of weight front wheel	38.15 kg	50.05 kg	48.65 kg	51.8 kg
Distribution of weight rear wheel	70.85 kg	92.95 kg	90.35 kg	96.2 kg
Horizontal distance of CG from front axle	804.70 mm	858.00 mm	861.90 mm	884.00 mm
Vertical distance of CG from front axle	433.30 mm	462.0 m	464.10 mm	476.00 mm

Expense: Since one of the goals of this research is to design and construct a low-cost integrated brake system apparatus, the proposal's expense must be kept as low as possible to maximize the likelihood that the final product would be highly marketable.

Reliability: Since one of this apparatus's key goals is to make diagnosing and maintaining the integrated braking system simpler, it should be made in such a way that it requires little to no secondary maintenance.

Security: Whether a thing is safe to use, contains only non-toxic and non-hazardous materials, or complies with norms and laws is a concern. Moving parts are confined to enclosed sections to reduce the risk of accident, even if they are exposed to allow for natural convection and user view. Hazardous and non-toxic compounds are employed. Safety equipment would be employed in its construction, and the design would place a strong emphasis on safety.

Comfort: Since this device will be used by individuals, it is essential that the dimensions, structure, and components be suitable for the consumer.

2.2 Design Input Variables

Design input variables refers to the variables taken for the design calculations. They include the speed of the vehicle, pedal ratio, master cylinder diameter, piston diameter (front & rear), and the rider and pillion weight. These values have been taken for the calculations shown in Table 2.

3 CAD Modeling 3D Model

The 3D model of the combined braking system consists of the disc rotors with calipers mounted and shown in the Fig. 2. The brake line is from the master cylinder to the brake calipers. For front wheel the flow of brake fluid is from the master cylinder to the delay valve, then to the brake caliper. In the case of rear wheels the flow of brake fluid is from the master cylinder directly to the brake caliper. The pressure in the master cylinder is increased by applying the force on the brake pedal. The disc rotors with mounted calipers make up the integrated braking system's 3D model. The brake line connects the brake calipers to the master cylinder. The master cylinder, delay valve, and brake caliper are the order in which braking fluid flows for the front wheels. When it comes to the back wheels, the brake caliper receives brake fluid directly from the master cylinder. By pressing down firmly on the brake pedal, the master cylinder's pressure is raised.

Table 2 Variables for design calculations

S. No.	Parameter	Value	Unit
1	Speed of vehicle	16.67	m/s
2	Coefficient of friction b/w tire & road	0.7	–
3	Coefficient of friction b/w brake pad & disc	0.35	–
4	Pedal ratio	2.5	–
5	Master cylinder diameter	18	mm
6	Front piston diameter	28	mm
7	Rear piston diameter	30	mm
8	Driver weight	80	kg
9	Pillion weight	80	kg

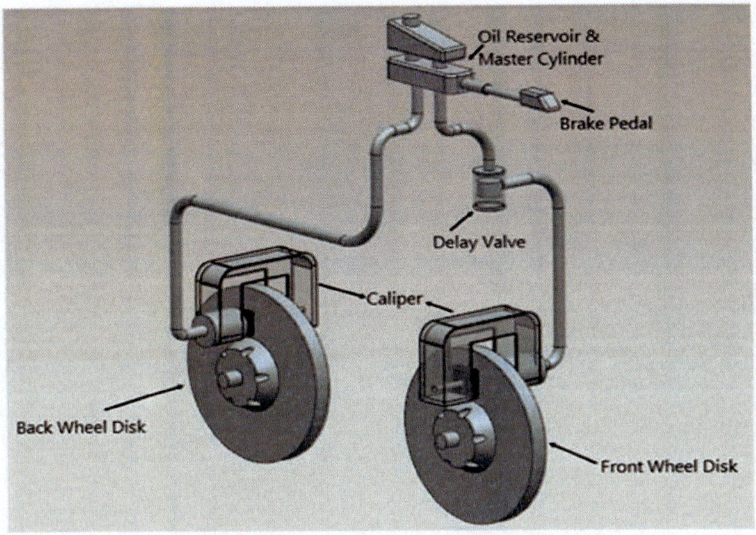

Fig. 2 3D modeling of braking system with delay valve in NX software

3.1 Fabrication Model

In the Fig. 3. Fabrication model of proposed combined braking system for two-wheeler is done in vehicle maintenance workshop at BSA crescent institute of science and technology. The brake hose is connected to the front and rear through delay valve. The both wheels lifted and placed on the stand. The motor is used for the power source for testing. The front and rear wheels are connected with the delay valve with the fluid reservoir to pressurize when depressing the brake pedal. In this model, the front brake lever operates the front calipers, which in turn triggers a secondary master cylinder to engage the front brake, while the rear brake pedal engages both the front and back brakes. By depressing the front brake lever, the pressure regulator valve receives hydraulic pressure, which is subsequently sent to the front and rear discs. The front disc is then linked to the pressure regulator valve so that it can adjust the pressure to the maximum and minimum levels needed. Motorbike design is the development, production, and assembling of a motorcycle's parts and systems to produce the motorcycle's desired functionality, price, and appearance. Current mass-produced bikes are typically built with a steel or aluminum frame, telescopic forks to support the front wheel, and disc brakes, with a few exceptions. There may be additional bodily parts that are either functionally or aesthetically designed. The swing arm-mounted rear wheel is connected to the driving shaft by a chain by a power source.

Fig. 3 Proposed model of CBS

4 Result and Discussion

The proposed model of combined braking system is tested in chassis dynamometer. After assembling the whole model is placed on the chassis frame for testing. The stopping distance is calculated for all the segment vehicles with different loads and also vehicle speed is calculated. It is evident that the braking distance grows along with the load on the vehicle. The weight on the vehicle and the vehicle speed are observed to cause the stopping distance to steadily rise. Plotted in the graph for various vehicle taken are the calculated braking distances for the corresponding flow velocities for various scenarios. It is evident that the braking distance rises along with the load on the vehicle. The weight on the vehicle and the vehicle speed are observed to cause the braking distance to steadily rise.

4.1 *Honda Activa 5G*

The following stopping distances corresponding to vehicle speeds were calculated taking into consideration the vehicle mass only.

Table 3 Stopping distances for Honda Activa 5G

S. No.	Vehicle speed (kmph)	Stopping distance (m)		
		Vehicle	Rider	Rider and pillion
1	10	0.08	0.14	0.2
2	20	0.32	0.56	0.8
3	30	0.73	1.26	1.8
4	40	1.29	2.24	3.19
5	50	2.02	3.51	4.99
6	60	2.91	5.05	7.19
7	70	3.96	6.87	9.78
8	80	5.18	8.98	12.78

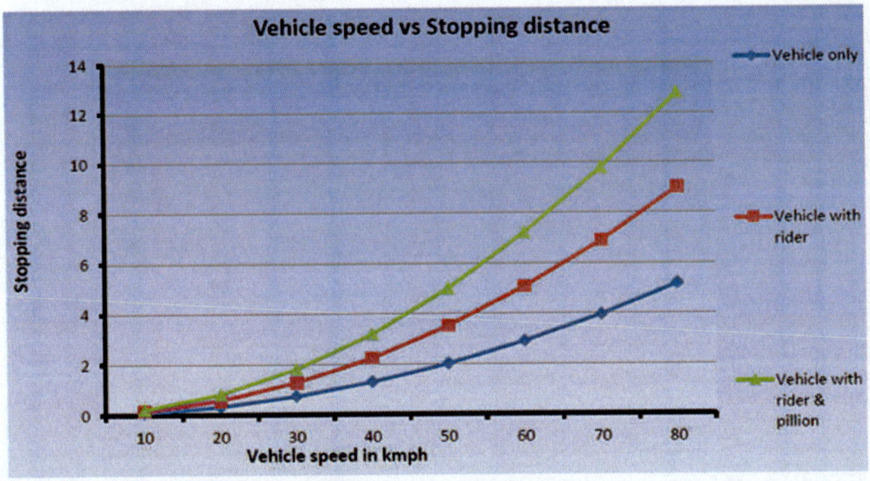

Fig. 4 Vehicle speed versus stopping distance for Honda Activa 5G

The calculated stopping distances for the corresponding vehicle speeds for different situations are plotted in the Table 3 and Fig. 4. It can be seen that as the load on the vehicle increases the stopping distance also increases. The stopping distance is seen to increase gradually with the load on the vehicle and the vehicle speed.

4.2 *Bajaj Pulsar 150 Twin Disc*

The following stopping distances corresponding to vehicle speeds were calculated taking into consideration the vehicle mass only.

The following graph depicts various stopping distances corresponding to vehicle speeds for different situations.

The calculated stopping distances for the corresponding vehicle speeds for different situations are plotted in the Table 4 and Fig. 5. It can be seen that as the load on the vehicle increases the stopping distance also increases. The stopping distance is seen to increase gradually with the load on the vehicle and the vehicle speed.

Table 4 Stopping distances for Bajaj pulsar 150 Twin disc

S. No.	Vehicle speed (kmph)	Stopping distance (m)		
		Vehicle	Rider	Rider and pillion
1	10	0.11	0.17	0.22
2	20	0.42	0.66	0.9
3	30	0.96	1.49	2.02
4	40	1.7	2.65	3.6
5	50	2.65	4.14	5.62
6	60	3.82	5.96	8.1
7	70	5.2	8.11	11.02
8	80	6.79	10.59	14.39
9	90	8.6	13.41	18.22
10	100	10.61	16.55	22.49

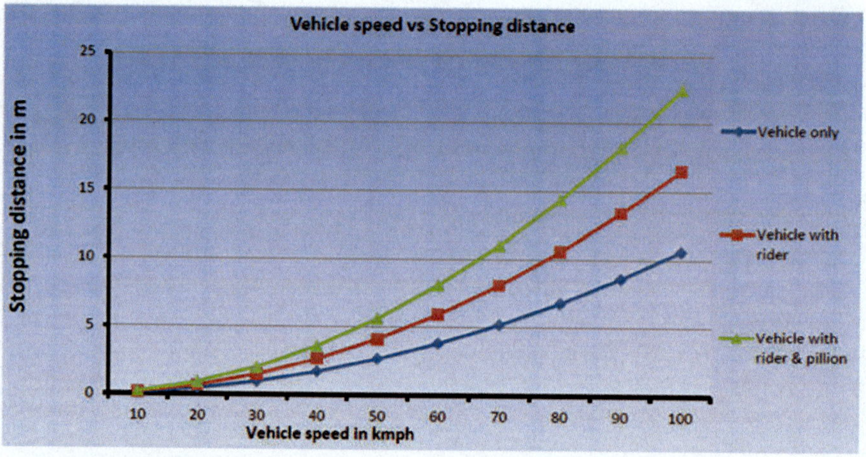

Fig. 5 Vehicle speed versus stopping distance for Bajaj Pulsar 150 Twin disc

4.3 TVS Apache 180

The following stopping distances corresponding to vehicle speeds were calculated taking into consideration the vehicle mass only.

The calculated stopping distances for the corresponding vehicle speeds for different situations are plotted in the Table 5 and Fig. 6. It can be seen that as the load on the vehicle increases the stopping distance also increases. The stopping distance is seen to increase gradually with the load on the vehicle and the vehicle speed.

Table 5 Calculated stopping distances for TVS Apache 180 vehicle only

S. No.	Vehicle speed (kmph)	Stopping distance (m)		
		Vehicle	Rider	Rider and pillion
1	10	0.1	0.16	0.22
2	20	0.41	0.65	0.89
3	30	0.93	1.46	2
4	40	1.65	2.6	3.55
5	50	2.58	4.06	5.55
6	60	3.71	5.85	7.99
7	70	5.06	7.97	10.88
8	80	6.6	10.4	14.2
9	90	8.36	13.17	17.98
10	100	10.32	16.26	22.19

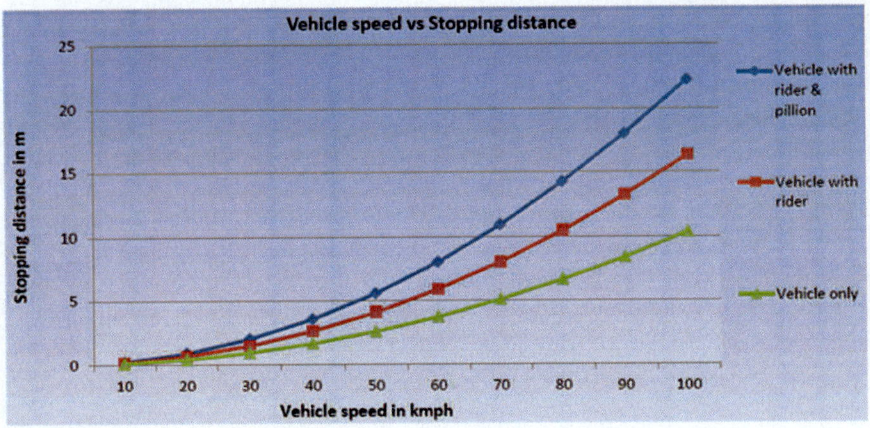

Fig. 6 Vehicle speed versus stopping distance for TVS Apache 180

Table 6 Calculated stopping distances for Yamaha FZ25 vehicle only

S. No.	Vehicle speed (kmph)	Stopping distance (m)		
		Vehicle	Rider	Rider and pillion
1	10	0.11	0.17	0.23
2	20	0.44	0.68	0.91
3	30	0.99	1.52	2.06
4	40	1.76	2.71	3.66
5	50	2.75	4.23	5.72
6	60	3.95	6.09	8.23
7	70	5.38	8.29	11.2
8	80	7.03	10.83	14.63
9	90	8.9	13.71	18.52
10	100	10.99	16.92	22.86

4.4 Yamaha FZ25

The following stopping distances corresponding to vehicle speeds were calculated taking into consideration the vehicle mass only.

The following graph depicts various stopping distances corresponding to vehicle speeds for different situations.

The calculated stopping distances for the corresponding vehicle speeds for different situations are plotted in the Table 6 and Fig. 7. It can be seen that as the load on the vehicle increases the stopping distance also increases. The stopping distance is seen to increase gradually with the load on the vehicle and the vehicle speed. The calculated stopping distances for the corresponding vehicle speeds for different situations are plotted in the Fig. 8 for various segments of vehicle taken. It can be seen that as the load on the vehicle increases the stopping distance also increases. The stopping distance is seen to increase gradually with the load on the vehicle and the vehicle speed. The following graph depicts the various stopping distances calculated for vehicles of various segments, like 110, 150, 180, and 250 cc. The calculated stopping distances for the corresponding vehicle speeds for different situations are plotted in the graph for various segments of vehicle taken. It can be seen that as the load on the vehicle increases the stopping distance also increases. The stopping distance is seen to increase gradually with the load on the vehicle and the vehicle speed.

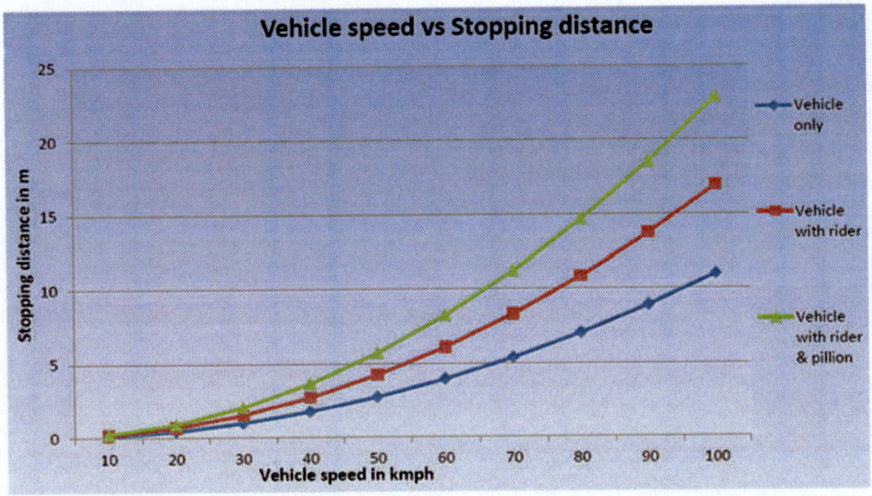

Fig. 7 Vehicle speed versus stopping distance for Yamaha FZ25

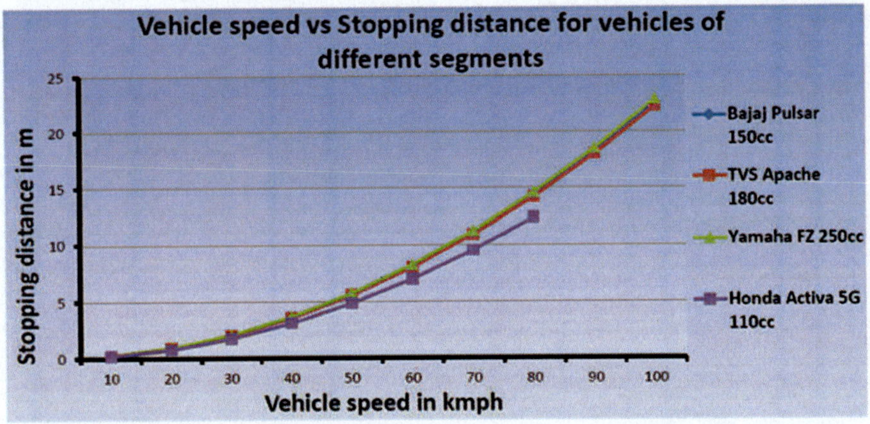

Fig. 8 Vehicle speed versus stopping distance for vehicles of different segments

5 Conclusion

Designers were able to overcome the challenge of deploying the front and rear brakes individually in sudden situations in this project. The overall flexibility of the braking system is in line with the design objectives of having a better and more responsive braking system during initial development test and after being installed on the vehicle. The brakes actuate as intended, and the second design goal of having the shortest stopping distance also was achieved. The calculations confirm that the identified braking component characteristics satisfy the design objective and are suitable for usage in motorcycles. The result provides some useful insights. It gives more sensitivity of the braking performance, comfort and is safer. In terms of stopping time is sustained, when compared to ordinary brake system it gives safer ride while applying a brake. Comparatively superior braking system that is safe for small-capacity, inexpensive motorcycles. Cost-effective safety brake system aspect comparably excellent braking system for motorcycles used for riding, cruising, and commercial use.

References

1. Indian Standard Automotive Vehicles (2018) Performance requirements and testing procedure for braking system of two and three wheeled motor vehicles. India
2. Spry SC, Girard AR (2008) Gyroscopic stabilization of unstable vehicles configurations, dynamics, and control 247–260
3. Gogoi P, Nath M, Doley BT, Boruah A, Barman HJ (2017) Design and fabrication of self balancing two wheeler vehicle using gyroscope. Int J Eng Technol. 9(3):2051–2058. https://doi.org/10.21817/ijet/2017/v9i3/1709030206/DOI
4. Colvin GR (2014) Development and validation of control moment gyroscopic stabilization. Undergraduate honors research thesis, Ohio State University, pp 1–29
5. Cossalter V, Bellati A (2012) Exploratory study of the dynamic behavior of motorcycle-rider during incipient fall events. Department of mechanical engineering via Venezia, 1. 35131 Padova (pd) University of Padova, Italy, pp 1–8
6. Khot A, Kumbhojkar N (2014) Modeling and validation of prototype of self-stabilizing motorcycle using gyroscope. Int J Adv Res Eng Technol 5(12):48–54. ISSN 0976–6480(print), ISSN 0976-6499 (online)
7. Hung JY (2000) Gyroscopic stabilization of a stationary unmanned bicycle. Auburn University, USA, pp 1–6. https://doi.org/10.1109/ACC.2014.6859392/DOI
8. Spry SC (2008) Gyroscopic stabilization of unstable vehicles configuration, dynamic, and control. USA, University of Michigan, pp 1–14. https://doi.org/10.1080/00423110801935863/DOI
9. Ksamawati WEP, Pramono AS (2018) Study of self balancing two-wheeled motor cycle with double gyroscope stabilization. Disruptive innovation in mechanical engineering for industry competitiveness, AIP conference proceedings, p 030011. https://doi.org/10.1063/1.5046246/DOI
10. Saini AJK, Bhushan B (2018) Design of combined brake system for scooter. J Emerg Technol Innovative Res 5(6):619–626. http://www.jetir.org/papers/JETIR1806684.pdf/DOI
11. Subramanian M, Muthaya J, Deepan V (2019) Health monitoring system for automobile vehicles to enhance safety. Int J Veh Struct Syst 10(6):395–398. https://doi.org/10.4273/ijvss.10.6.03/DOI
12. Murnen H, Niles A, Sigworth N (2009) System and method for providing gyroscopic stabilization to a two-wheeled vehicle, patent-PCT/US2006/02404, US
13. Hautler C. Gyroscopic device for the stabilization of laterally unstable vehicles. US, Patent No-3787066
14. Vinothkumar M (2018) Development of combined braking system for two wheeler conference.: international conference on technological advances in mechanical engineering, Chennai
15. Deepan V, Subramanian M, Dineshkumar C (2018) Motorcycle rider fatigue analyses: results of an online survey. Int J Mech Prod Eng Res Dev 8(2):509–516
16. Gupta SK, Gulhane V (2014) Design of self-balancing bicycle using object state detection. Int J Eng Res Appl 1–4

Advanced Control Techniques

Fixed-Time Information Detection-Based Secondary Control Strategy for Low Voltage Autonomous Microgrid

Sonam Shrivastava, Bidyadhar Subudhi, and Jambeswar Sahu

Abstract In this paper, information detection-based secondary control strategy is proposed for a low voltage islanded microgrid (MG) framework. The reference information is not available with all distributed generators (DG) in a clustered and meshed low voltage MG due to communication contingencies. Hence, an information detection strategy (IDS)-based control scheme is proposed to update the reference information locally at each DG terminal. Initially, in primary control layer droop boost controller is locally implemented to stabilize the MG voltage and frequency after islanding load disturbance. Then in the secondary control layer, an IDS is used to provide the reference points to every DG unit connected in the MG. Further, a consensus-based protocol is proposed to restore the voltage and frequency to their reference values in fixed time. The proposed controller is self-reliant and is independent of system parameters. The rigorous Lyapunov stability analysis is presented for guaranteed convergence. The test setup is simulated in MATLAB/Simulink to validate the effectiveness under load perturbation, DG dropout, and scalability.

Keywords Low voltage microgrid · Graph theory · Secondary control · Information detection

1 Introduction

Low voltage microgrid (MG) is a fundamental component of new generation power system that includes multiple distributed generators (DG), energy storage units, and sensitive controllable loads [1]. The on and off grid operation of MG makes it more

S. Shrivastava (✉)
School of Electrical Engineering, Vellore Institute of Technology, Vellore, Tamilnadu 632007, India
e-mail: mailtosonam24390@gmail.com

B. Subudhi
School of Electrical Sciences, Indian Institute of Technology, Goa 401403, India

J. Sahu
School of Mechanical Engineering, Vellore Institute of Technology, Vellore, Tamilnadu 632007, India

S. Sharma et al. (eds.), *Intelligent Control, Robotics, and Industrial Automation*, Lecture Notes in Electrical Engineering 1066,
https://doi.org/10.1007/978-981-99-4634-1_18

resilience and self-sufficient during the events of natural calamities and utility grid maintenance needs. To control such a complex system, a hierarchical control strategy is implemented extensively in the recent times. This system has layer-based control operation that involves primary (droop), secondary (centralized to decentralized to distributed control), and tertiary control (power flow regulation) [2].

The state-of-the-art literature includes several secondary control strategies ranging from asymptotic to fast convergence to fixed-time synchronization and event triggered ones to restores the voltage and frequency deviations caused by primary control level [3]. The conventional droop methods are designed with the assumption of dominantly inductive lines. However, while designing the droop control for high R/X line impedance, the traditional droop suffers severe power control and coupling issues that degrades system stability [4]. In this paper, the above-mentioned issue is addressed via taking a modified droop/boost primary level controller. Further, the secondary control layer has been advanced by adapting numerous controllers ranging from centralized to asymptotic with uncertain convergence time [5]. Various fixed-time control schemes with active/reactive power sharing have been presented in literature [6]. With fixed-time control, the synchronization can be achieved very fast with lesser instability and transients. The communication network is intermittent in nature, and all DGs in MG do not have access to the reference information. To address the above issue, in this paper information discovering scheme (IDS) is implemented locally with droop/boost primary control layer.

The comprehensive distributed controllers, which can address the faster convergence, while the reference information in not available is missing in literature. The primary droop/boost control layer along with IDS provides reference voltage and frequency to every DG present in the MG. It stabilizes the voltage and frequency after large signal disturbances. Further, the proposed distributed secondary control layer restores the voltage and frequency to the nominal values in fixed time.

Hence, this paper presents a fixed-time IDS-based controller for low voltage MG. The paper has following contributions.

- Fixed-time distributed IDS-based controller is designed for voltage/frequency synchronization for low voltage with high (R/X) ratio MG network.
- Proposed scheme gives access to the reference information to all DG present in low voltage network, while the convergence time is independent of system parameters.
- The upper limit on restoration time is presented with rigorous Lyapunov stability analysis.

The remaining of the paper is organized as follows:

Section 2 discusses the prilimenaries on modeling of low voltage MG network and primary droop/boost control with its communication links and graph theory. Section 3 and 4 explains the voltage and frequency restoration using IDS-based distributed controllers, respectively. Section 5 carries the result and discussions. Section 6 concludes the paper.

2 Preliminaries

2.1 Dynamic Modeling of ith-Based DG

In a MG framework, the DG unit is interpreted as a DC voltage source, VSI, LC filter, and RL output connector connected together as shown in Fig. 1. For an islanded operation, the MG works under voltage control mode; hence, voltage and frequency set points are provided by primary control layers. This dependency makes control of low voltage (LV) MG a challenging task. The conventional droop control characteristics can be given as [1]

$$\omega_i = \omega_i^d - k_{Pi}^{\omega} P_i, \quad V_{\text{odi}}^* = V_i^d - k_{Qi}^V Q_i, \tag{1}$$

where V_i^d and ω_i^d are the voltage and angular frequency references set points; V_{odi}^* is output voltage magnitude; ω_i is operating angular frequency; k_{Qi}^V and k_{Pi}^{ω} are droop coefficients for ith DG.

2.2 Droop/Boost Primary Control

Unlike the conventional droop control in (1) the line impedance in LV MG networks are highly resistive. Thus with high R/X ratio, the active and reactive power terms are proportional to voltage and frequency, respectively [7]. Hence, the modified droop/boost-based primary control equation can be derived as follows

$$\omega_i = \omega_i^d + k_{Qi}^{\omega} Q_i, \quad V_{\text{odi}}^* = V_i^d - k_{Pi}^V P_i, \tag{2}$$

where k_{Pi}^V and k_{Qi}^{ω} are droop/boost coefficients for ith DG.

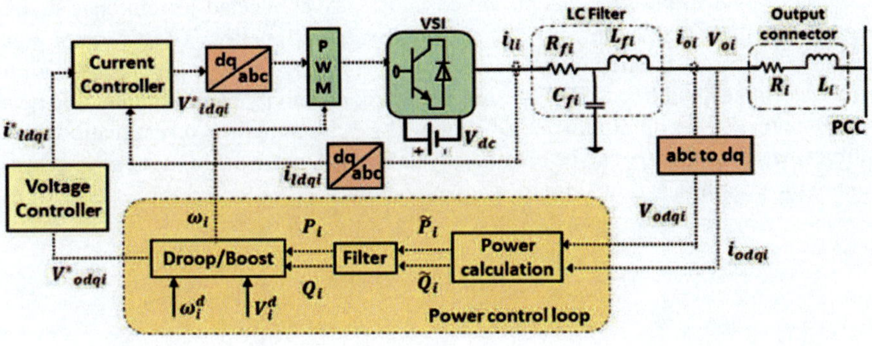

Fig. 1 Inverter-based DG model

The secondary control layer has to be designed in such a way that voltage and frequency are synchronized with in fixed time as follows

$$\omega_{i,j}(t) = \omega_r \; \forall\, i, j \in N, \; \forall t \geq t_\omega, \; V_{i,j}(t) = V_r \; \forall\, i, j \in N, \; \forall t \geq t_V. \quad (3)$$

3 Fixed-Time Frequency Control

Communication link failure between a DG unit and reference DG results in complex and challenging controller design. To resolve this drawback fixed-time information detection (ID) scheme is implemented along with primary control layer for every DG unit. This scheme identified the reference set points in fixed-time span of τ_{of}. Next, a fixed-time secondary control strategy is used to synchronize the frequency drift and achieve power allocation in fixed time τ_F.

3.1 Fixed-Time Frequency Information Detection Strategy

The direct connection to the reference node cannot be always guaranteed while using the sparse communication network for data sharing. Only few DG nodes will have access to reference information. To solve this issue and provide reference value to all the DGs present in the LV MG network, a fixed-time consensus-based ID strategy is designed as follows [8].

$$\dot{r}_i = \text{sig}\left(\sum_{j=1}^{N} a_{ij}\left(r_j - r_i\right) + b_i\left(\omega_r - r_i\right)\right)^{1/2}, \quad (4)$$

where r_i is the locally estimated value of the global reference frequency ω_r ; $b_i \geq 0$ is the communication edge weight which is directly connected to reference node. $sig(x)^\rho = \text{sgn}(x)\,|z|^\rho$ with $x \in R, \; \rho > 0$ is the signum function. The estimator presented in (4) forces the frequency values $r_i, i = 1, 2, \ldots N$ to achieve consensus with reference value ω_r in fixed-time t_{of}, subjected to connectivity of communication network and $b_i \neq 0$ for minimum one DG node. The maximum limit on estimation time of information discovery can be obtained by defining y_i as a tracking synchronization task in the continuous time domain as given below

$$y_i = \sum_{j=1}^{N} a_{ij}\left(r_j - r_i\right) + b_i\left(\omega_r - r_i\right). \quad (5)$$

The discreet time dynamics for the tracing synchronization in (5) is formulated to utilize the samples receive in discrete time interval as follows

$$\left\{ y_i(k) = \sum_{j=1}^{n} a_{ij}\left(r_j(k) - r_i(k)\right) - b_i\left(\omega_r - x_i(k)\right), \right. \tag{6}$$

where $r_i(k)$, $r_j(k)$, and $y_i(k)$ are frequencies for ith and jth distributed generator and the following estimated value at kth instant. $r_i(k+1)$ represents the ith node state at $(k+1\text{th})$ instant in discreet domain. We have assumed that the data updates are avaliable at regular periodic intervals. Further, the tracker synchronization in the (7) can be reformulated in discreet matrix version as follows

$$\mathbf{Z}(k+1) = -\left(L_G + B\right) + \left(\mathbf{r}(k) - 1_N\omega_r\right), \tag{7}$$

where $\mathbf{Z}(k+1) = [y_1(k+1),\ y_2(k+1) \ldots y_N(k+1)]^T$; $\mathbf{r}(k) = [r_1(k),\ r_2(k) \ldots r_N(k)]^T$; $B = diag(b_i)$ is a diagonal matrix; 1_N is Nth order vector matrix with all ones. Following Lemmas are utilized in construction of primary controller development based on information discovery.

Lemma 1 *[1]: If the undirected graph G is completely connected, then $(L_G + B)$ is positive definite.*

Lemma 2 *[1]: Considering a time domain continuous system $\dot{z} = f(z)$ with initial value as $f(0) = 0$. Let us assume that there reside a positive finite function $V : [0, \infty) \to [0, \infty)$, real numbers $m_1 > 0$, $0 < m_2 < 1$, such that $\dot{V} \le -m_1 V^{m_2}$. Then, V settles to zero in definite time duration*

$$T \le \frac{V(x(0))^{1-m_2}}{m_1(1-m_2)}. \tag{8}$$

Theorem 1 *With the assumption that undirected graph G is completely linked and $b_i \ne 0$, atleast for one DG node and using the distributed fixed-time information detection strategy in (4), the term r_i for all DG units achieve consensus to ω_r in fixed-time presented by (8).*

Proof Considering the discrete time domain Lyapunov candidate function as follows

$$V_1(k) = e^T(k)\left(I_N \otimes P\right)e(k), \tag{9}$$

where $e(k) = \left[e_1^T(k),\ e_2^T(k) \ldots e_N^T(k)\right]^T$ and

$$e_i(k) = r_i(k) - \frac{1}{N}\sum_{j=1}^{n} r_j(k) \tag{10}$$

is the kth error term. Now considering the values of $V_1(k)$ at indefinite sequence $k = 0, 1, \ldots$ for time t we can write

$$V_1(k+1) - V_1(k) = \sum_{p=0}^{q_k-1} \left[V_1^{p+1}(k) - V_1^{p}(k) \right] = \sum_{p=0}^{q_k-1} \left[\int_{k^p}^{k^{p+1}} \dot{V}_1(k)\, dk \right], \quad (11)$$

where $V_1^p(k)$ is the candidate Lyapunov term in discontinuous time domain, with $p = 0, 1, \ldots, q_k$, and $k = 0, 1, \ldots$.

Differentiating Lyapunov candidate function in continuous time domain

$$V_1(t) = \sum_{i=1}^{N} \int_0^{y_i} sig(z)^{1/2}\, dz \quad (12)$$

inline with the travel path of (10), between $(k+1)$ and k with an assumption that it is a continuous time function and is comparable to the divergence $V_1(k+1) - V_1(k)$ in discreet time domain.

From above discussion, differentiating (12) in time t following can be written

$$\dot{V}_1(t) = \sum_{i=1}^{N} sig(y_i)^{0.5}.\dot{y}_i = -sig\left(\mathbf{Z}^T\right)^{0.5} (L_G + B)\; sig(\mathbf{Z})^{0.5}. \quad (13)$$

From Lemma 1, $(L_G + B)$ is positive definite as far as $b_i \neq 0$ at least for one DG node. $\lambda_1(L_G + B)$ is the smallest eigen value of $(L_G + B)$ and is equal to $\lambda_2(L_G)$. For $V_1 = 0$, $\mathbf{Z} = 0$, following can be derived

$$\dot{V}_1(t) = \frac{-sig\left(\mathbf{Z}^T\right)^{0.5} (L_G + B)\, sig(\mathbf{Z})^{0.5}}{sig\left(\mathbf{Z}^T\right)^{1/2} sig(\mathbf{Z})^{0.5}} V_1^{2/3}(t) \frac{sig\left(\mathbf{Z}^T\right)^{1/2} sig(\mathbf{Z})^{1/2}}{V_1^{2/3}(t)}. \quad (14)$$

Further, $sig\left(\mathbf{Z}^T\right)^{0.5} sig(\mathbf{Z})^{0.5} = \sum_{i=1}^{N} |y_i|$, also, simplifying the 3rd term on the RHS of (14) can be modified as follows

$$\frac{sig\left(\mathbf{Z}^T\right)^{1/2} sig(\mathbf{Z})^{1/2}}{V_1^{2/3}(t)} = \frac{\sum_{i=1}^{N} |y_i|}{\left(\sum_{i=1}^{N} \frac{2}{3} |y_i|^{3/2}\right)^{2/3}}. \quad (15)$$

Further, using Lemma 3, and combining (14) and (15), we can obtain

$$\dot{V}_1(t) \le -\lambda_1(L_G + B) \left(\frac{3}{2}\right)^{2/3} V_1^{2/3}(t) \le 0. \quad (16)$$

Above equation corresponds to the negative definite continuous time function assumed in Lemma 3. Thus, the estimate of ith DG frequency r_i for $i = 1, 2, ...N$. achieves consensus in fixed time given as $t_{0f} \geq \frac{V(0)^{(1-q)}}{p(1-q)} \geq 3V_1(0)^{1/3}\lambda_1(L_G + C)$.

3.2 Secondary Frequency Restoration

Frequency restoration can be achieved using the control input derived from time domain differentiation of (2) as given below

$$\dot{\omega}_i = \dot{\omega}_i^d + k_{Qi}^{\omega}\dot{Q}_i = \varepsilon_i^{\omega},\ \dot{\omega}_i^d = \varepsilon_i^{\omega} - k_{Qi}^{\omega}\dot{Q}_i. \tag{17}$$

For balanced active power distribution among all DGs the auxiliary control input for reactive power is given as $k_{Qi}^{\omega}\dot{Q}_i = \varepsilon_i^{Q}$. The combined frequency and reactive power restoration can be given as

$$\omega_i^d = \int \left(\varepsilon_i^{\omega} - \varepsilon_i^{Q}\right) dt. \tag{18}$$

The above-mentioned auxiliary control input can be designed as

$$\begin{aligned} \varepsilon_i^{\omega} &= \alpha_{\omega}\sum_{j=1}^{N} a_{ij} sig\left(\omega_j - \omega_i\right)^f + \beta_{\omega} sig(\omega_r - \omega_i)^g, t \geq t_{0f} \\ \varepsilon_i^{Q} &= \alpha_{Q}\sum_{j=1}^{N} a_{ij} sig\left(k_{Qj}^{\omega}Q_j - k_{Qi}^{\omega}Q_i\right)^h, \end{aligned} \tag{19}$$

where α_{ω}, β_{ω}, and α_Q are the coupling gains, $0 < f < 1$, and $g = \frac{2f}{f+1}$. Define $\delta_{\omega i} = \omega_i - \omega_r$, and $\dot{\delta}_{\omega i} = \dot{\omega}_i$, and $\delta_{Qi} = k_{Qi}^{\omega}Q_i - \frac{1}{N}\sum_{i=1}^{N} k_{Qi}^{\omega}Q_i$. Since $\frac{1}{N}\sum_{i=1}^{N} k_{Qi}^{\omega}\dot{Q}_i = 0$ for an undirected network thus, $\frac{1}{N}\sum_{i=1}^{N} k_{Pi}^{\omega}Q_i$ is time invariant, and $\dot{\delta}_{Qi} = k_{Qi}^{\omega}\dot{Q}_i$. The above equation can be modified as follows

$$\begin{aligned} \dot{\delta}_{\omega i} &= \alpha_{\omega}\sum_{j=1}^{N} a_{ij} sig\left(\delta_{\omega j} - \delta_{\omega i}\right)^f + \beta_{\omega} sig(\delta_{\omega i})^g, t \geq t_{0f} \\ \dot{\delta}_{Qi} &= \alpha_{Q}\sum_{j=1}^{N} a_{ij} sig\left(\delta_{Qj} - \delta_{Qi}\right)^h. \end{aligned} \tag{20}$$

Theorem 2 *The controller presented in (19) restores the frequency in fixed time T_f and support accurate reactive power distribution in longer time frame.*

Proof Assuming the following candidate Lyapunov function as

$$V_2(t) = V_\omega(t) + V_Q(t) = \frac{1}{2}\sum_{i=1}^{N} \delta_{\omega i}^2 + \frac{1}{2}\sum_{i=1}^{N} \delta_{Qi}^2. \tag{21}$$

The time domain differentiation of (21) is obtained as follows

$$\dot{V}_2(t) = \sum_{i=1}^{N} \delta_{\omega i}\dot{\delta}_{\omega i} + \sum_{i=1}^{N} \delta_{Qi}\dot{\delta}_{Qi}. \tag{22}$$

Using (20) and (22), we can write

$$\begin{aligned}\dot{V}_2(t) = &\sum_{i=1}^{N} \delta_{\omega i}\left(\alpha_\omega \sum_{j=1}^{N} a_{ij} sig\left(\delta_{\omega j} - \delta_{\omega i}\right)^f + \beta_\omega sig(\delta_{\omega i})^g\right) \\ &+ \sum_{i=1}^{N} \delta_{Qi}\left(\alpha_Q \sum_{j=1}^{N} a_{ij} sig\left(\delta_{Qj} - \delta_{Qi}\right)^h\right),\ t \geq t_{0f}\end{aligned} \tag{23}$$

with $0 < \alpha < 1$. Using Lemma 2, above equation is modified as

$$\begin{aligned}\dot{V}_2(t) \leq &-\frac{1}{2}\left(\sum_{i=1}^{N}\sum_{j=1}^{N} (\alpha_\omega a_{ij})^{\frac{2}{1+\alpha}} \left|\delta_{\omega j} - \delta_{\omega i}\right|^{\frac{2(1+f)}{1+\alpha}} + 2\beta_\omega^{\frac{2}{1+\alpha}} \sum_{i=1}^{N} \left|\delta_{\omega i}\right|^{\frac{2(1+g)}{1+\alpha}}\right)^{\frac{1+\alpha}{2}} \\ &-\frac{1}{2}\left(\sum_{i=1}^{N}\sum_{j=1}^{N} (\alpha_Q a_{ij})^{\frac{2}{1+\alpha}} \left|\delta_{Qj} - \delta_{Qi}\right|^{\frac{2(1+h)}{1+\alpha}}\right)^{\frac{1+\alpha}{2}}.\end{aligned} \tag{24}$$

Define $\boldsymbol{\delta}_{\boldsymbol{\omega}} = [\delta_{\omega 1}, \ \ldots . \delta_{\omega N}]^T$, and $\boldsymbol{\delta}_{\mathbf{Q}} = \left[\delta_{Q1}, \ \ldots . \delta_{QN}\right]^T$, and

$$\begin{aligned}W_1(\boldsymbol{\delta}_{\boldsymbol{\omega}}) &= \sum_{i=1}^{N}\sum_{j=1}^{N} (\alpha_\omega a_{ij})^{\frac{2}{1+\alpha}} \left|\delta_{\omega j} - \delta_{\omega i}\right|^{\frac{2(1+f)}{1+\alpha}} + 2\beta_\omega^{\frac{2}{1+\alpha}} \sum_{i=1}^{N} \left|\delta_{\omega i}\right|^{\frac{2(1+g)}{1+\alpha}}, \\ W_2(\boldsymbol{\delta}_{\boldsymbol{\omega}}) &= \sum_{i=1}^{N}\sum_{j=1}^{N} (\alpha_\omega a_{ij})^{\frac{2}{1+\alpha}} \left|\delta_{\omega j} - \delta_{\omega i}\right|^2 + 2\beta_\omega^{\frac{2}{1+\alpha}} \sum_{i=1}^{N} \left|\delta_{\omega i}\right|^2 \\ Z_1(\boldsymbol{\delta}_{\mathbf{Q}}) &= \sum_{i=1}^{N}\sum_{j=1}^{N} (\alpha_Q a_{ij})^{\frac{2}{1+\alpha}} \left|\delta_{Qj} - \delta_{Qi}\right|^{\frac{2(1+h)}{1+\alpha}}, \\ Z_2(\boldsymbol{\delta}_{\mathbf{Q}}) &= \sum_{i=1}^{N}\sum_{j=1}^{N} (\alpha_Q a_{ij})^{\frac{2}{1+\alpha}} \left|\delta_{Qj} - \delta_{Qi}\right|^2\end{aligned}$$

with $W_2(\delta_\omega), Z_2(\delta_Q) \neq 0$, because if W_2 and Z_2 are zero then the connectedness of $G(A_G)$, $\delta_i = \delta_j$, for all i, j and $V_2(t) = 0$ that is conflicting . There W_2 and $Z_2 \neq 0$, and $\frac{W_1(\delta_\omega)}{W_2(\delta_\omega)} \geq k_\omega$ and $\frac{Z_1(\delta_Q)}{Z_2(\delta_Q)} \geq k_Q$, where k_ω and $k_Q > 0$ are controller gain parameters. Adapting the approach same in [8], following can be given

$$\dot{V}_2(t) \leq -\frac{1}{2}(4k_\omega\lambda_2(L_\omega))^{\frac{1+\alpha}{2}} V_\omega(t)^{\frac{1+\alpha}{2}} - \frac{1}{2}(4k_Q\lambda_2(L_Q))^{\frac{1+\alpha}{2}} V_Q(t)^{\frac{1+\alpha}{2}}, \tag{25}$$

where $\lambda_2(L_\omega)$ and $\lambda_2(L_Q)$ are the 2nd smallest eigen values of the Laplacian matrix of communication graphs G_ω and G_Q with adjacency matrices $A_\omega = [\alpha_\omega a_{ij}]^{\frac{2}{1+\alpha}}$ and $A_Q = [\alpha_Q a_{ij}]^{\frac{2}{1+\alpha}}$, respectively. Hence, by incorporating Lemma 3, $V_2(t)$ will achieve zero in fixed-time $T_f = max\{t_\omega, t_Q\}$, with

$$t_\omega \leq \frac{2V_\omega(0)^{\frac{1-\alpha}{2}}}{(1-\alpha)(4k_\omega\lambda_2(L_\omega))^{\frac{1+\alpha}{2}}},\ t_Q \leq \frac{2V_Q(0)^{\frac{1-\alpha}{2}}}{(1-\alpha)(4Q\lambda_2(L_Q))^{\frac{1+\alpha}{2}}}. \tag{26}$$

4 Secondary Voltage Restoration

A voltage disparity resulted by the primary control layer can be eliminated within fixed time and the reactive power sharing can be obtained by implementing fixed-time secondary control as discussed below. First the information detection strategy is implemented to estimate the voltage reference as follows

$$\dot{x}_i = \text{sig}\left(\sum_{j=1}^{N} a_{ij}(x_j - r_i) + b_i(V_r - x_i)\right)^{1/2}, \tag{27}$$

where x_i is locally estimated information for voltage reference V_r. The similar procedure as in *Theorem 1* is adopted to find the setting time t_{0V} such that $x_i = V_r$ as $t \geq t_{0V}$. The time domain differentiation of (2) gives

$$\dot{V}_{odi} = \dot{V}_i^d - k_{Pi}^V\dot{P}_i = \varepsilon_i^V,\ \ \dot{V}_i^d = \varepsilon_i^V + k_{Pi}^V\dot{P}_i. \tag{28}$$

The auxiliary voltage control strategy ε_i^V is proposed as follows

$$\varepsilon_i^V = \alpha_V\sum_{j=1}^{N} a_{ij}sig(V_{odj} - V_{odi})^f + \beta_V sig(V_{ref} - V_{odi})^g, t \geq t_{0V}, \tag{29}$$

where α_V and β_V are the coupling gains, $0 < f < 1$, and $g = \frac{2f}{f+1}$.

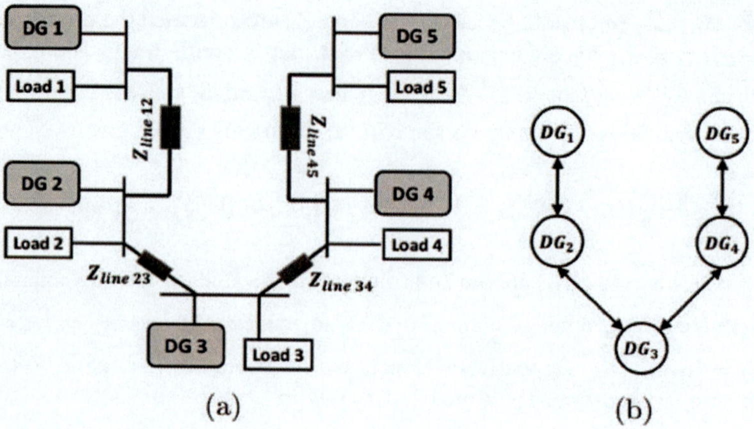

Fig. 2 **a** Islanded MG test bed **b** Communication graph

Theorem 3 *With the assumption that the communication graph is connected, the proposed controller in (29) synchronizes the voltage in fixed convergence time given as*

$$T_V \leq \frac{2V_3(0)^{\frac{1-\alpha}{2}}}{(1-\alpha)\left(4k_V\lambda_2\left(L_V\right)\right)^{\frac{1+\alpha}{2}}} \tag{30}$$

where $\lambda_2\left(L_V\right)$ *is the 2nd smallest eigen value of matrix* L_V *of graph* G_V *with* $A_V = \left[\alpha_V a_{ij}\right]^{\frac{2}{1+\alpha}}$, *and* $V_3\left(t\right)$ *represents the considered Lyapunov candidate function.*

Proof The fixed time T_V can be derived in the same manner as presented in Theorem 2.

5 Simulation Results and Discussion

A 4-bus low voltage MG (230 V, 50 Hz) test system is considered for simulation in MATLAB/Simulink to analyze the performance of the proposed controller. The test system is shown in Fig. 2. The system consists of 5 DG units with five locally connected loads as RL load-1 is $R = 300\ \Omega$, $L = 477$ mH, RL load 2 is $R = 40\ \Omega$, $L = 64$ mH, RL load 3 is $R = 50\ \Omega$, $L = 64$ mH, RL load 4 is $R = 40\ \Omega$, $L = 64$ mH, and RL load 5 is $R = 50\ \Omega$, $L = 95$ mH. The MG specifications with controller parameters are given in Tables 1 and 2, respectively.

Table 1 MG parameters

	DG 1 & DG 2 & DG 3		DG 4 & DG 5	
DGs	k_P^V	9.4×10^{-5}	k_P^V	12.5×10^{-5}
	k_Q^ω	1.3×10^{-3}	k_Q^ω	1.5×10^{-3}
	R	0.03 Ω	R	0.03 Ω
	L	0.35 mH	L	0.35 mH
	R_f	0.1 Ω	R_f	0.5 Ω
	L_f	1.35 mH	L_f	0.27 mH
	C_f	50 μF	C_f	50 μF
	K_{PV}	0.1	K_{PV}	0.05
	K_{IV}	420	K_{IV}	390
	K_{PC}	15	K_{PC}	10.5
	K_{IC}	20000	K_{IC}	16000
Lines	$Z_{Line\ 12}$	$Z_{Line\ 23}$	$Z_{Line\ 34}$	$Z_{Line\ 45}$
	$R_l = 0.35\ \Omega$	$R_l = 0.35\ \Omega$	$R_l = 0.35\ \Omega$	$R_l = 0.35\ \Omega$
	$L_l = 0.07$ mH	$L_l = 0.07$ mH	$L_l = 0.07$ mH	$L_l = 0.07$mH

Table 2 Controller parameters

Voltage controller	Frequency controller	Active power controller	Reference
$\alpha_V = 200$	$\alpha_\omega = 80$	$\alpha_P = 100$	$V_r = 380\ V$
$\beta_V = 100$	$\beta_\omega = 40$	$h = {}^1/_3$	$f_r = 50\ Hz$
$f = {}^1/_3$	$f = {}^1/_3$	$\alpha = 0.5$	
$g = {}^1/_2$	$g = {}^1/_2$		
$\alpha = 0.5$	$\alpha = 0.5$		

5.1 *Case 1: Performance Verification with Load Perturbation*

Figure 3a–b shows the state performance of voltage and frequency restoration after islanding at time instant $t = 0$ s. The presented primary droop boost and IDS strategy was active. However, the proposed secondary controller is shut off purposely. The secondary control is initiated at $t = 2$ s, which synchronizes the voltage and frequency to the reference values in fixed time. The load perturbation takes place at $t = 3$ s (RL load is connected to Load 1) and $t = 4$ s (Load of DG3 is disconnected). The results show that the synchronization takes place within 0.18 s and maintained within permissible limit with small signal perturbations.

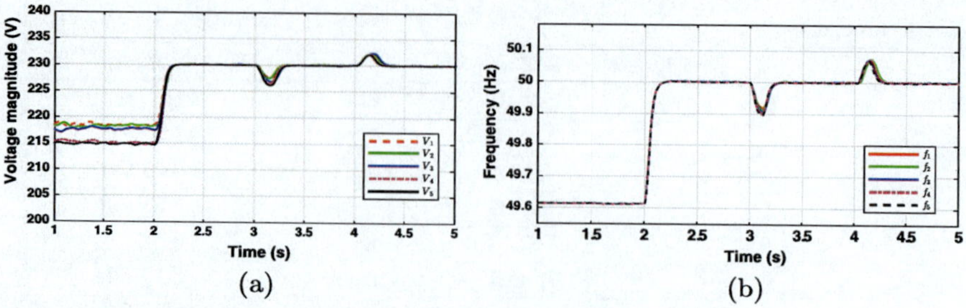

Fig. 3 Performance with load change **a** output voltage **b** frequency

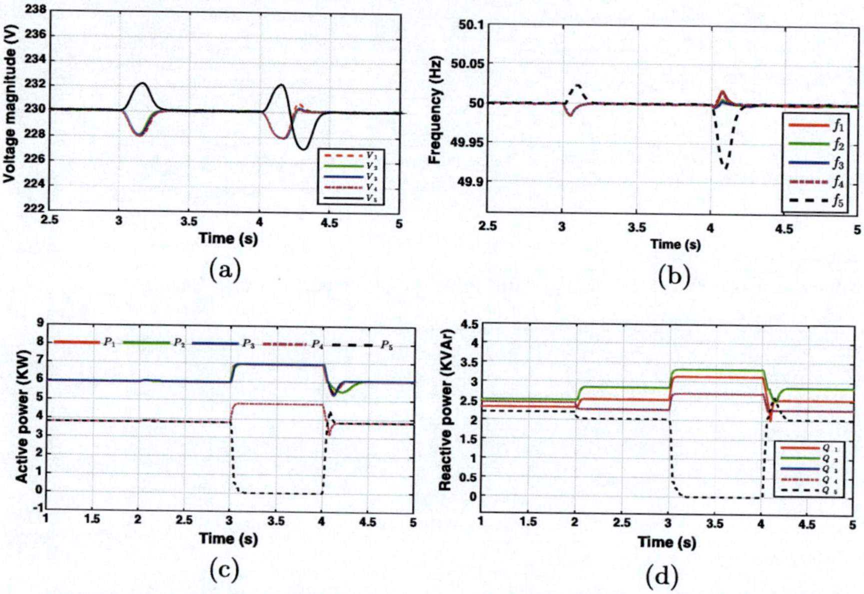

Fig. 4 Performance with loss of DG unit **a** output voltage **b** frequency **c** active power **d** reactive power

5.2 *Case 2: Performance with Loss of DG Unit*

The proposed controller is realized with plug and play scenario. DG 5 is disconnected and connected back to MG at $t = 3$s and 4s, respectively. The results envisage the efficacy of presented controller during high signal disturbance. Also, Fig. 4 shows the active and reactive power re-sharing, while DG 5 was absent from the grid.

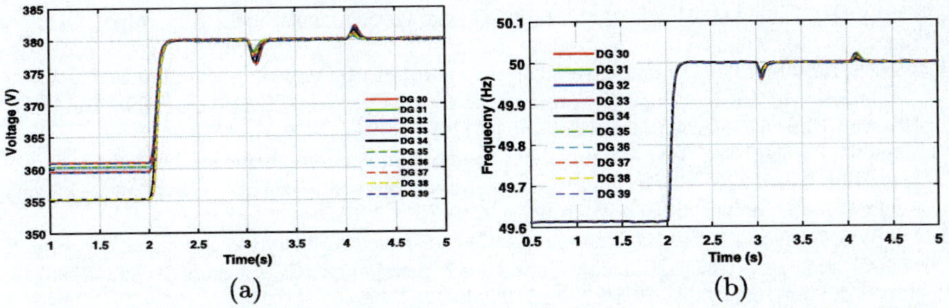

Fig. 5 Performance for IEEE 38-bus system

5.3 *Case 3: Scalability*

The scalability of the proposed IDS-based fixed-time controller is verified by testing it under 38-bus distribution network [9]. The communication graph connectivity plays an important role in the convergence time for larger grids. The upper bound on the convergence time can be optimized by setting the controller parameters accordingly. Figure 5 shows the frequency and voltage response for scalability test. The secondary controller is initiated at $t = 2$s, and it synchronizes operating frequency and voltage within permissible time limit. Hence, the proposed controller has promising application with larger systems as well.

6 Conclusion

The paper focuses on fixed-time restoration of voltage and frequency of islanded low voltage MG with communication contingency. The proposed IDS provides access of reference voltage and frequency to all DGs present in the system. The droop boost primary and consensus-based secondary control stabilizes and restores the system parameters within fixed time. The detailed Lyapunov analysis is carried out to guarantee fixed-time restoration performance of the proposed controller. The control strategy is validated via simulation carried out for a test setup in MATLAB using SimPowerSystem toolbox. The results envisage the controller performance for precipitous load perturbation, time-varying communication network, loss of DG unit, and scalability.

References

1. Zuo S, Davoudi A, Song Y, Lewis FL (2016) Distributed finite-time voltage and frequency restoration in islanded ac microgrids. IEEE Trans Ind Electron 63(10):5988–5997. https://doi.org/10.1109/TIE.2016.2577542

2. Hardy GH, Littlewood JE, Pólya G, Pólya G et al (1952) Inequalities. Cambridge university press
3. Patra S, Basu M (2022) Double-layered droop control-based frequency restoration and seamless reconnection of isolated neighboring microgrids for power sharing. IEEE J Emerg Sel Top Power Electron 10(5):6231–6242. https://doi.org/10.1109/JESTPE.2022.3197729
4. Zarei SF, Parniani M (2016) A comprehensive digital protection scheme for low-voltage microgrids with inverter-based and conventional distributed generations. IEEE Trans Power Delivery 32(1):441–452. https://doi.org/10.1109/TPWRD.2016.2566264
5. Lee WG, Nguyen TT, Yoo HJ, Kim HM (2022) Consensus-based hybrid multiagent cooperative control strategy of microgrids considering load uncertainty. IEEE Access 10:88798–88811. https://doi.org/10.1109/ACCESS.2022.3198949
6. Choi J, Habibi SI, Bidram A (2021) Distributed finite-time event-triggered frequency and voltage control of ac microgrids. IEEE Trans Power Syst 37(3):1979–1994. https://doi.org/10.1109/TPWRS.2021.3110263
7. Shrivastava S, Sahu J, Sitharthan R (2020) Noise resilient distributed voltage and frequency control for low voltage islanded microgrid. In: IOP conference series: materials science and engineering, vol 937. IOP Publishing, p 012036. https://doi.org/10.1088/1757-899X/937/1/012036
8. Wang L, Xiao F (2010) Finite-time consensus problems for networks of dynamic agents. IEEE Trans Autom Control 55(4):950–955. https://doi.org/10.1109/TAC.2010.2041610
9. Singh D, Misra RK, Singh D (2007) Effect of load models in distributed generation planning. IEEE Trans Power Syst 22(4):2204–2212. https://doi.org/10.1109/TPWRS.2007.907582

Approximation of Stand-alone Boost Converter Enabled Hybrid Solar-Photovoltaic Controller System

Umesh Kumar Yadav, V. P. Meena, Umesh Kumar Sahu, and V. P. Singh

Abstract This proposal is presenting a novel determination of reduced-order (RO) model for higher-order (HO) controller of boost converter enabled hybrid solar-photovoltaic (SPV) stand-alone system. In this work, controller of HO-SPV system is approximated to RO-SPV model by minimizing steady-state errors gain as time-moments (T-Ms) and transient-state errors gain as Markov-parameters (M-Ps) of the HO-SPV system and desired RO-SPV model. Firstly, T-Ms and M-Ps of HO-SPV controller system and RO-SPV controller model are exploited to formulate the fitness function. The minimization of formulated fitness function is done by employing grey-wolf (GW) optimization algorithm. The optimization is accomplished by satisfying the constraints of minimization of steady-state error and confirming the stability of RO model. The matching of steady-state between HO-SPV controller and RO-SPV model is accomplished by matching of first T-Ms of HO-SPV controller and RO-SPV model, whereas Hurwitz criterion is introduced to confirm the stability of RO-SPV model. The step along with impulse and Bode plots is presented to demonstrate the effectiveness and applicability of the proposed method. The comparative analysis by considering time-domain (TD) specifications and error indices are also provided and discussed for validation of proposed methodology for better understanding.

Keywords Hybrid solar-photovoltaic · Battery storage system · Boost converter · Modelling · Stability

---

U. K. Yadav (✉) · V. P. Meena · V. P. Singh
Electrical Engineering Department, MNIT, Jaipur, India
e-mail: 2019ree9085@mnit.ac.in

V. P. Meena
e-mail: vmeena1@ee.iitr.ac.in

V. P. Singh
e-mail: vinay.ee@mnit.ac.in

U. K. Sahu
Department of Mechatronics, Manipal Institute of Technology, Manipal Academy of Higher Education, Manipal 576104, Karnataka, India
e-mail: umesh.sahu@manipal.edu

S. Sharma et al. (eds.), *Intelligent Control, Robotics, and Industrial Automation*, Lecture Notes in Electrical Engineering 1066,
https://doi.org/10.1007/978-981-99-4634-1_19

1 Introduction

Due to increment in demand of electricity in today's scenario of development in industrial and transportation technologies, penetration on grid increases. The increment in demand of electricity is also a cause of environmental degradation. This degradation of environment is prime cause of increment in carbon footprint by affecting green house effect due to emission of pollutants. The technological development of renewable energy resources and limited exploitation of conventional energy can only control the increment of carbon foot print. In order to achieve the minimal pollution level, battery storage systems (BSSs) utilizing renewable energy resources is one of the finest solution. The solar-photovoltaic (SPV) power generation, wind energy (WE), hydel power, tidal energy, geothermal energy, etc., are the fast growing renewable energy resources. These energy resources are generally used with BSS as hybrid systems. The maximum part of energy generation through renewable energy resources is fulfilled by SPV system due to its advantages over other renewable energy resources.

In literature [1], Rahman et al. utilized the SPV inverter system with LCL-type grid connected hybrid system for controller design by utilizing adaptive-integral back-stepping technique. The stand-alone SPV is also utilized in [2] by Kumar and Tyagi with islanded microgrid system by incorporating pulse-width modulation-based control strategy for decentralize controlling. Similarly in [3], Rehman et al. presented frequency regulation of grid connected SPV system by controlling the synchronous generator with the help of advance genetic algorithm. Later, the Landman converter-based controller design for hybrid SPV-WE system with multi-inverter connected static compensator is presented by Vanaja et al. in [4] for proper voltage regulation of grid connected system. In addition, DC microgrid empowered SPV power generation for electric vehicle (EV) application is also utilized as EV charging station by exploiting real-time rule algorithm by Wang et al. in [5]. Further, stand-alone capacitor connected SPV system by incorporating direct sliding mode control scheme is exploited by Al-Wesabi et al. in [6] to demonstrate the control of dynamic instabilities in the system by regulating the BSS in bidirectional mode. Furthermore, in [7], Mahmud and Pota presented a designing of controller for stand-alone SPV system connected with BSS by utilizing partial feedback linearized control technique along with H_∞-based robust control scheme.

In general, SPV system is used with BSS to maintain the system reliability and to avoid system redundancy. The power flow from source to load is managed by the controller. The controllers designed for BSS enabled SPV system are of higher-order (HO), generally. The HO systems have some disadvantages like difficulty in control design and its analysis, complexity in understanding the control laws, uneconomical in designing of controllers, comparatively higher simulation time, etc. The efficient and comparative better controller for BSS enabled SPV system can be designed by exploiting the reduced-order modeling to obtain the reduced-order (RO) model for BSS enabled SPV system. In [7], controller of boost converter enabled hybrid SPV is presented. Since, the order of hybrid SPV controller system is of third order so it is desired that the HO controller of hybrid SPV must be reduced to lower order. So that, all possible disadvantages and limitations of HO system can be overcome up to some instant by proposing RO model for HO-SPV system.

This proposed work presented a determination process of second-order RO model for third-order hybrid SPV system by error minimization between time-moments (T-Ms) and Markov parameters (M-Ps) of HO-SPV system and desired RO-SPV model by utilizing grey-wolf (GW) optimization algorithm. Firstly, T-Ms and M-Ps of HO-SPV system are obtained. The unknown desired RO-SPV coefficients are determined by minimizing the weighted fitness function which is formulated by utilizing T-Ms and M-Ps of the HO-SPV system and desired RO-SPV model. The weights depicted with fitness function are considered by identifying and assigning the importance to sub-objectives. The sub-objectives are effect-on-steady-state $E(ss)$ and effect-on-transient-response $E(tr)$. The constructed fitness function is minimized with the help of GW optimization algorithm. The constraints for optimization are matching of steady-state and confirmation about stability of RO model. The matching of first T-Ms of HO-SPV system and RO-SPV model ensures the steady-state matching. The Hurwitz criterion is utilized to obtain the stable RO-SPV model. The step, impulse and Bode responses are provided to validate proposed methodology. The comparative analysis on the basis of time-domain (TD) specifications and error indices is presented in tabular form for better understanding.

This proposal is organized in different sections. In Sect. 2, mathematical representation of hybrid SPV system with its parameters is done. Problem formulation is presented in Sect. 3. Section 4 is briefly discussed about GW optimization algorithm. Order reduction of hybrid HO-SPV system and its results and discussion are described in Sect. 5. The proposed methodology is concluded and provided in Sect. 6 in addition to future scope.

2 Mathematical Representation of Hybrid Solar-Photovoltaic Stand-alone System

The mathematical representation of boost converter enabled hybrid solar-photovoltaic (SPV) stand-alone system and its controller are presented in this section. The SPV is connected to common DC link along with battery storage system (BSS). The DC-DC boost converter is connected with SPV to supply the DC load. The feasibility of the presented DC microgrid system is increased by incorporating BSS with bidirectional DC-DC boost converter. The circuit configuration of stand-alone interconnected DC microgrid system and its controller is presented in Fig. 1.

In Fig. 1, the parameters are depicted for interconnected BSS enabled SPV system. The respective voltages, currents, and parameters are presented in Fig. 1 with proper connection diagram. In SPV circuit of Fig. 1, since

$$V_{\mathrm{SPV}} = V_{cp1} = i_{\mathrm{DC}} r_p + V_{\mathrm{DC}}. \quad (1)$$

Similarly, in Fig. 1 by considering BSS circuit, the voltage equation is obtained as

$$V_B = V_{\mathrm{BSS}} - i_B(r_{b1} + r_{b2}) - V_{Cb1}. \tag{2}$$

The mathematical representation of BSS enabled interconnected SPV system can be represented as

$$\frac{dV_{Cp1}}{dt} = \frac{[i_p - i_{\mathrm{DC}}]}{C_{p1}} \tag{3}$$

$$\frac{di_{\mathrm{DC}}}{dt} = \frac{[V_{\mathrm{SPV}\text{-}} - i_{\mathrm{DC}} r_p - m_{\mathrm{SPV}} V_{\mathrm{PL}}]}{L_{p1}} \tag{4}$$

$$\frac{dV_{\mathrm{PL}}}{dt} = \frac{[m_{\mathrm{SPV}} - i_{\mathrm{PL}}]}{C_{p2}} \tag{5}$$

$$\frac{dV_{Cb1}}{dt} = \frac{[i_B - (V_{Cb1}/r_{c1})]}{C_{b1}} \tag{6}$$

$$\frac{d_{iB}}{dt} = \frac{[V_{\mathrm{BSS}} - i_B(r_{b1} + r_{b2}) - V_{cb1} - m_{\mathrm{BSS}} V_{\mathrm{BL}}]}{L_{b1}} \tag{7}$$

$$\frac{dV_{\mathrm{BL}}}{dt} = \frac{[m_{\mathrm{BSS}} i_B - i_{\mathrm{BL}}]}{C_{b2}}. \tag{8}$$

The parameters presented in (1)–(8) are depicted in Fig. 1, whereas m_{SPV} and m_{BSS} are switching inputs of DC-DC SPV converter circuit and bidirectional BSS converter circuit, respectively. The numerical values of the parameters presented in (1)-(8) are taken from [7]. The controller of the presented system is considered in Sect. 5 for proposing the reduced-order controller model for HO-SPV system.

3 Problem Description

3.1 Higher-Order (HO) System Representation

Suppose, representation of Hth order transfer function for higher-order (HO) system is done as

$$E_H(s) = \frac{\hat{F}(s)}{\hat{G}(s)} = \frac{\hat{F}_0 + \hat{F}_1 s + \hat{F}_2 s^2 + \cdots + \hat{F}_{H-1} s^{H-1}}{\hat{G}_0 + \hat{G}_1 s + \hat{G}_2 s^2 + \cdots + \hat{G}_H s^H}. \tag{9}$$

In (9), $\hat{F}_i$ for $i = 0, 1, 2, 3, 4, \ldots, (H-1)$ are coefficients of numerator and $\hat{G}_i$ for $i = 0, 1, 2, 3, 4, \ldots, H$ are coefficients of denominator.

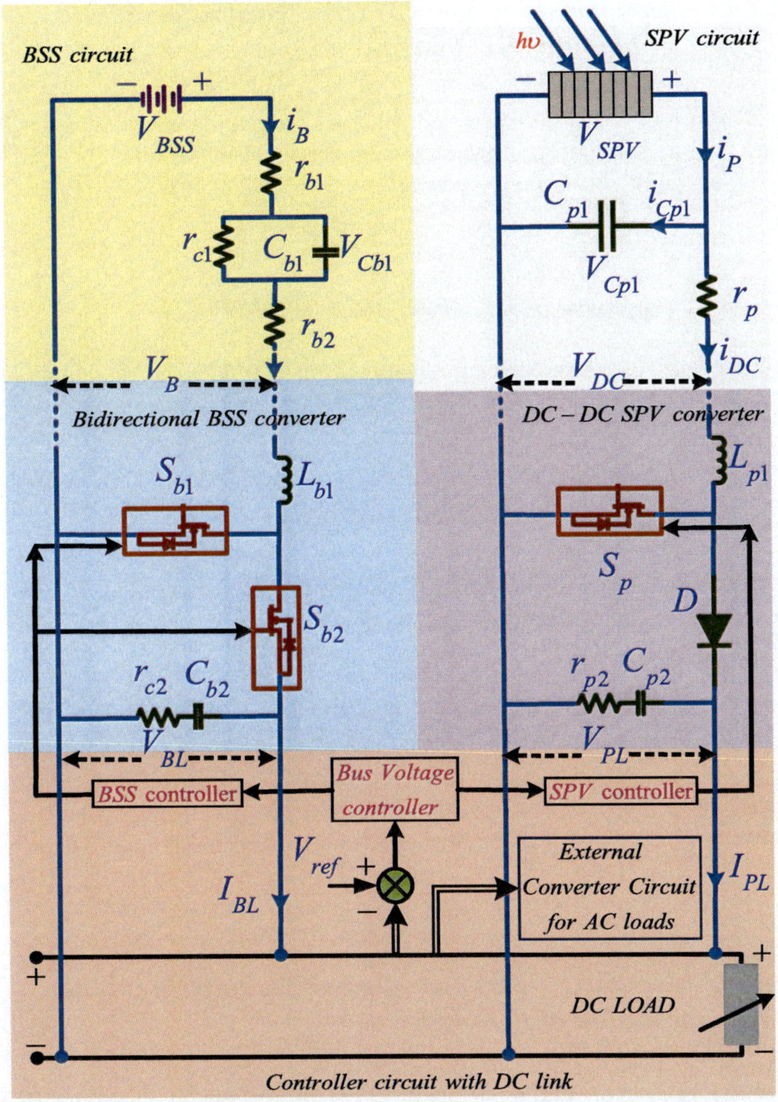

Fig. 1 Boost converter enabled solar-photovoltaic stand-alone system

3.2 Time-Moments (T-Ms) and Markov-Parameters (M-Ps) Representation of HO System

The expansions of HO system depicted in (9), by concerning time-moments (T-Ms) around $s = 0$, and by concerning Markov-parameters (M-Ps) around $s = \infty$ are done as expressed in (10) and (11), respectively.

$$E_H(s) = \hat{\tau}_0 + \hat{\tau}_1 s + \hat{\tau}_2 s^2 + \cdots + \hat{\tau}_H s^H + \cdots \tag{10}$$

$$E_H(s) = \hat{\mu}_1 s^{-1} + \hat{\mu}_2 s^{-2} + \cdots + \hat{\mu}_H s^{-H} + \cdots . \tag{11}$$

In (10), T-Ms of HO system given in (9) are $\hat{\tau}_i$ for $i = 0, 1, 2, 3, 4, \ldots$. Similarly, in (11), $\hat{\mu}_i$ for $i = 1, 2, 3, 4, \ldots$ are M-Ps of HO system represented in (9).

3.3 Reduced-Order Model Representation

The HO system of order H shown in (9) is desired to obtain the reduced-order (RO) model of order l. The order, i.e. l of RO model is such that $l < H$. Let, the lth order RO model be

$$E_l(s) = \frac{F'(s)}{G'(s)} = \frac{F'_0 + F'_1 s + F'_2 s^2 + \cdots + F'_{l-1} s^{l-1}}{G'_0 + G'_1 s + G'_2 s^2 + \cdots + G'_l s^l}, \tag{12}$$

where, F'_i for $i \in 0, 1, 2, 3, 4, \ldots, (l-1)$ and G'_i for $i \in 0, 1, 2, 3, 4, \ldots, l$ are, respectively, coefficients of numerator and denominator of RO model depicted in (12).

3.4 Time-Moments (T-Ms) and Markov-Parameters (M-Ps) Representation of RO Model

The expansions as expressed in (10) and (11) of HO system, similar expansions in terms of T-Ms and M-Ps are obtained in (13) and (14), respectively.

$$E_l(s) = \tau'_0 + \tau'_1 s + \tau'_2 s^2 + \cdots + \tau'_r s^r + \cdots \tag{13}$$

$$E_l(s) = \mu'_1 s^{-1} + \mu'_2 s^{-2} + \cdots + \mu'_r s^{-r} + \cdots . \tag{14}$$

In (13) and (14), respectively, τ'_i for $i \in 0, 1, 2, 3, 4, \ldots$ are T-Ms, and μ'_i for $i \in 1, 2, 3, 4, \ldots$ are M-Ps of the RO model depicted in (12).

3.5 Fitness Function with Its Constraints

The determination of desired RO model is accomplished by exploiting T-Ms and M-Ps of HO system shown in (9) and RO model presented in (12). The first T-Ms of HO system and RO model are exploited to ensure matching of steady-state. So, $(l-1)$ T-Ms and l M-Ps are considered and presented to frame the fitness function as

$$J=\sum_{i=1}^{l-1}\left[\omega_i^{\tau}\left(1-\frac{\tau'_i}{\hat{\tau}_i}\right)^2\right]+\sum_{j=1}^{l}\left[\omega_j^{\mu}\left(1-\frac{\mu'_i}{\hat{\mu}_j}\right)^2\right], \tag{15}$$

where, ω_i^{τ} and ω_i^{μ} are weights depicted with sub-objectives of errors between T-Ms and M-Ps of the HO system and RO model. The fitness function represented in (15) can be rearranged as given in (16).

$$J=\omega_1^{\tau}J_1^{\tau}+\omega_2^{\tau}J_2^{\tau}+\omega_3^{\tau}J_3^{\tau}+\cdots+\omega_1^{\mu}J_1^{\mu}+\omega_2^{\mu}J_2^{\mu}+\omega_3^{\mu}J_3^{\mu}+\cdots. \tag{16}$$

The fitness function formulated in (15) is optimized under constraints provided in (17) and (18) as matching of steady-state and confirming stability of RO model, respectively.

$$\hat{\tau}_0=\tau'_0, \tag{17}$$

where, first T-M of HO system is provided as $\hat{\tau}_0$ and τ'_0 is first T-M of desired RO model.

$$G'(s)\ \text{of RO model, given in (12) should be Hurwitz.} \tag{18}$$

The fitness function depicted in (15) is desired to be minimized. The minimization of fitness function shown in (15) is done by exploiting grey-wolf optimization algorithm.

4 Grey-Wolf Optimization Algorithm

The grey-wolf (GW) optimization algorithm is useful meta-heuristic-based optimization algorithm proposed by Mirjalili et al. [8]. The hunting behaviour of grey-wolves is adopted in this optimization algorithm. In grey-wolf family, generally wolves are living together, i.e. in form of pack of 5–12 members. The leader of the group is most dominant wolf in overall group and is called among all members as alpha (α). The second best wolf in grey-wolves hierarchy after alpha (α) is second dominant and second best wolf, known as beta (β). The betas are advisor of alpha (α). The betas reinforce all the directions of alpha in group. The delta (δ) wolves help upper dominant wolves for hunting. These wolves dominate omega (ω) wolves. The omegas are last ranked members in grey-wolf family. Omegas are to follow the command of other higher ranked members. The behaviour of wolves to encircle the prey and their hunting pattern are represented in (19) and (20), respectively.

$$\vec{\Delta}_w(t^+) = \vec{\Delta}_p(t) - \vec{a}\left[\left|\vec{b}.\vec{\Delta}_p(t) - \vec{\Delta}_w(t)\right|\right] \tag{19}$$

$$\vec{\Delta}_w(t^+) = (\vec{\Delta}_{w1} + \vec{\Delta}_{w2} + \vec{\Delta}_{w3})/3. \tag{20}$$

In (19), $\vec{\Delta}_w(t)$ depict current position of grey-wolf and $\vec{\Delta}_p(t)$ represent current position of targeted prey. The coefficient vectors, $\vec{a}$ and $\vec{b}$ are defined in (21) and (22).

$$\vec{a} = 2.\vec{e}.\vec{C}_1 - \vec{e} \tag{21}$$

$$\vec{b} = 2.\vec{C}_2, \tag{22}$$

where $\vec{e} \in [2, 0]$ and $\vec{C}_1$ and $\vec{C}_2 \in [0, 1]$. The grey-wolves positions in next iteration are determined with the help of (20). The positions of α along with β and δ wolves are obtained with the help of $\vec{\Delta}_{w1}$, $\vec{\Delta}_{w2}$ and $\vec{\Delta}_{w3}$ as provided in (23) (24) and (25), respectively.

$$\vec{\Delta}_{w1} = \vec{\Delta}_{w\alpha} - \vec{a_1}\left[\left|\vec{b_1}.\vec{\Delta}_{w\alpha} - \vec{\Delta}_w(t)\right|\right] \tag{23}$$

$$\vec{\Delta}_{w2} = \vec{\Delta}_{w\beta} - \vec{a_2}\left[\left|\vec{b_2}.\vec{\Delta}_{w\beta} - \vec{\Delta}_w(t)\right|\right] \tag{24}$$

$$\vec{\Delta}_{w3} = \vec{\Delta}_{w\delta} - \vec{a_3}\left[\left|\vec{b_3}.\vec{\Delta}_{w\delta} - \vec{\Delta}_w(t)\right|\right]. \tag{25}$$

5 Order Reduction of Hybrid Solar-Photovoltaic Controller System

The transfer function of higher-order (HO) hybrid solar-photovoltaic (SPV) [7] is depicted in (26).

$$E_5(s) = \left(\frac{4.078 \times 10^{07}s^2 + 2.039 \times 10^{12}s + 2.054 \times 10^{09}}{s^3 + 6.872 \times 10^{10}s^2 + 6.617 \times 10^{13}s + 3.266 \times 10^{10}}\right). \tag{26}$$

The expansion of HO-SPV system by concerning time-moments (T-Ms) is demonstrated in (27). Similarly, Markov-parameters (M-Ps) are expanded and presented in (28).

$$E_5(s) = 0.06289 - 64.98643s + 131664.05s^2 - 26.6755 \times 10^7 s^3 + 54.0452 \times 10^{10} s^4 - 1.095 \times 10^{15} s^5 + \cdots \quad (27)$$

$$E_5(s) = 40.78 \times 10^6 s^{-1} - 2.8024 \times 10^{18} s^{-2} + 1.9259 \times 10^{29} s^{-3} - 1.3235 \times 10^{40} s^{-4} + 9.09452 \times 10^{50} s^{-5} + \cdots . \quad (28)$$

Suppose, desired SPV-second-order RO model for HO-SPV system depicted in (26) is as follows:

$$E_2(s) = \frac{F_0^{'} + F_1^{'} s}{G_0^{'} + G_1^{'} s + G_2^{'} s^2}. \quad (29)$$

The expansions by concerning T-Ms and M-Ps of RO-SPV model represented in (29) can be given by (30) and (31).

$$E_2(s) = \left(\frac{F_0^{'}}{G_0^{'}}\right) + \left(\frac{F_1^{'} G_0^{'} - F_0^{'} G_1^{'}}{G_0^{'2}}\right) s + \left(\frac{F_0^{'} G_1^{'2} - F_1^{'} G_0^{'} G_1^{'} - F_0^{'} G_0^{'} G_2^{'}}{G_0^{'3}}\right) s^2 + \cdots \quad (30)$$

$$E_2(s) = \left(\frac{F_1^{'}}{G_2^{'}}\right) s^{-1} + \left(\frac{F_0^{'} G_2^{'} - F_1^{'} G_1^{'}}{G_2^{'2}}\right) s^{-2} + \left(\frac{F_1^{'} G_1^{'2} - F_0^{'} G_1^{'} G_2^{'} - F_1^{'} G_0^{'} G_2^{'}}{G_2^{'3}}\right) s^{-3} + \cdots . \quad (31)$$

The ascertainment of coefficients of desired SPV-RO model expressed in (29) is accomplished by constructing fitness function. The fitness function is formulated by employing T-Ms and M-Ps of the HO-SPV system and desired RO-SPV model. Since, desired order of SPV-RO model is $l = 2$, so at least $(2l - 1)$, i.e. $2 \times 2 - 1 = 3$ terms between T-Ms and M-Ps of HO-SPV system and RO-SPV model are matched. In fitness function, second T-M, and first and second M-Ps are considered. So, the fitness function depicted in (15) becomes

$$J = \sum_{j=1}^{1} \left[\omega_j^{\tau} \left(1 - \frac{\tau_j^{'}}{\hat{\tau}_j}\right)^2\right] + \sum_{j=1}^{2} \left[\omega_j^{\mu} \left(1 - \frac{\mu_j^{'}}{\hat{\mu}_j}\right)^2\right]. \quad (32)$$

The fitness function (32) can be rewritten as

$$J = \omega_1^{\tau} \left(1 - \frac{\tau_1^{'}}{\hat{\tau}_1}\right)^2 + \omega_1^{\mu} \left(1 - \frac{\tilde{\mu}_1}{\hat{\mu}_1}\right)^2 + \omega_2^{\mu} \left(1 - \frac{\tilde{\mu}_2}{\hat{\mu}_2}\right)^2. \quad (33)$$

Table 1 Time-domain (TD) specifications of HO-SPV controller and RO-SPV model

HO system and RO model	Time-domain specifications			
	Rise time (sec)	Settling time (sec)	Overshoot	Undershoot
HO system (26)	0.0023	0.0040	0	0
RO model (40)	3.29×10^{-3}	6.5526×10^{-3}	0	0

By utilizing T-Ms and M-Ps of HO-SPV system and RO model, fitness function shown in (33) turns out to be

$$J = \omega_1^{\tau}\left(1 + \left(\frac{F_1' G_0' - F_0' G_1'}{64.98643 G_0'^{2}}\right)\right)^2 + \omega_1^{\mu}\left(1 - \left(\frac{F_1'}{40.78 \times 10^6 G_2'}\right)\right)^2 + \omega_2^{\mu}\left(1 + \left(\frac{F_0' G_2' - F_1' G_1'}{2.8024 \times 10^{18} G_2'^{2}}\right)\right)^2. \tag{34}$$

The fitness function expressed in (34) can be demonstrated as

$$J = \omega_1^{\tau} J_1 + \omega_2^{\mu} J_2 + \omega_3^{\mu} J_3. \tag{35}$$

In (35), objectives J_1, J_2, and J_3 are formulated to minimize the errors between T-Ms and M-Ps of HO-SPV system and RO-SPV model. The weights ω_1^{τ}, ω_2^{μ} and ω_3^{μ}, associated with sub-objectives, are considered by providing equal importance to both, errors minimization between T-Ms and M-Ps. So, the weights are

$$\omega_1^{\tau} = 0.5, \quad \omega_2^{\mu} = \omega_3^{\mu} = 0.25. \tag{36}$$

By utilizing the weights from (36), the fitness function given in (34) modifies to resultant fitness function given in (37).

$$J = 0.5\left(1 + \left(\frac{F_1' G_0' - F_0' G_1'}{64.98643 G_0'^{2}}\right)\right)^2 + 0.25\left(1 - \left(\frac{F_1'}{40.78 \times 10^6 G_2'}\right)\right)^2 + 0.25\left(1 + \left(\frac{F_0' G_2' - F_1' G_1'}{2.8024 \times 10^{18} G_2'^{2}}\right)\right)^2. \tag{37}$$

The fitness function shown in (37) is optimized subjected to the constraints, provided in (17) and (18).The constraint (17) turns out to be (38) and constraint (18) becomes (39).

$$F_0' = 0.06289 G_0' \tag{38}$$

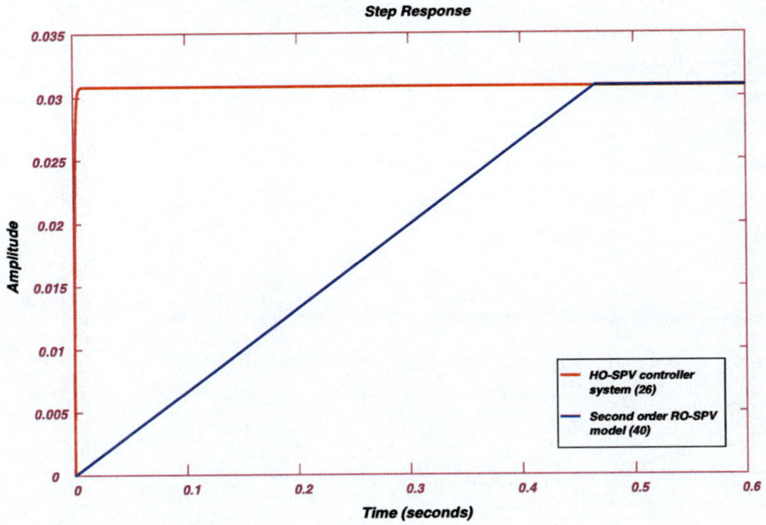

Fig. 2 Step response of HO-SPV system and RO-SPV model

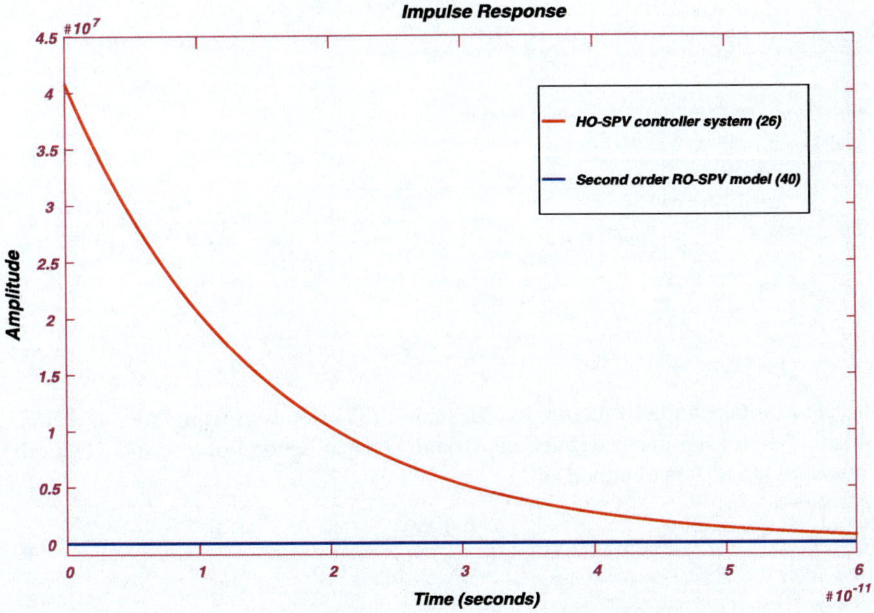

Fig. 3 Impulse response of HO-SPV system and RO-SPV model

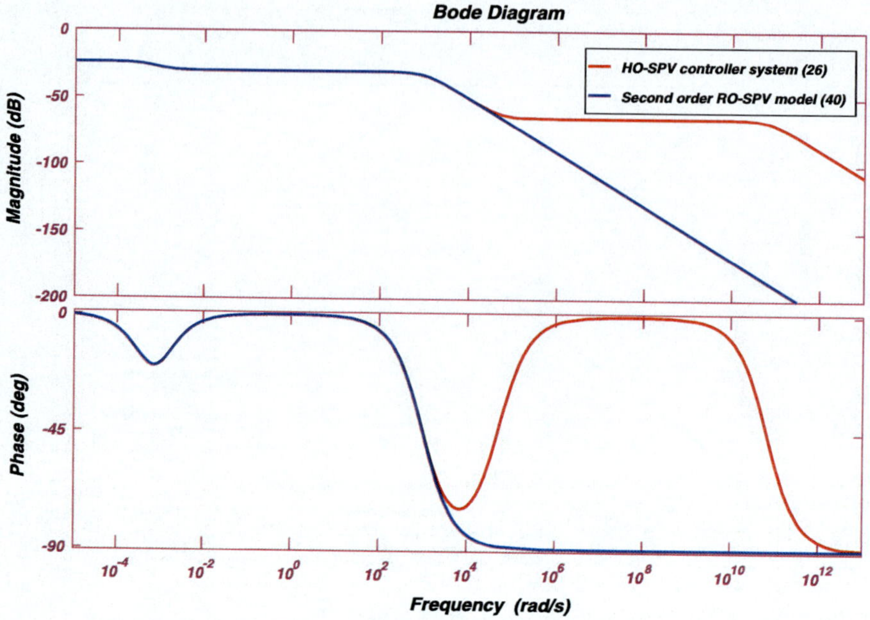

Fig. 4 Bode response of HO-SPV system and RO-SPV model

Table 2 Different performance error criterion

System (26) versus approximant (40)	Performance error criterion					
	IAE	ISE	ITAE	ITSE	IT^2AE	IT^2SE
RO model (40)	0.0005973	2.827×10^{-8}	0.003283	2.064×10^{-7}	0.02208	1453×10^6

$$G_1^{'} > 0, \ \ G_0^{'} G_1^{'} > 0 \tag{39}$$

The resultant fitness function manifested in (37) is solved using grey-wolf optimization algorithm under defined constraints. Hence, second-order SPV-RO model considered in (29) is obtained as

$$E_2(s) = \frac{0.2052 + 204s}{3.263 + 6606s + 6.819s^2}. \tag{40}$$

The second-order SPV model is obtained in (40). The step response is provided in Fig. 2. Figure 3 depicts impulse response of the HO-SPV system and RO-SPV model. The Bode plot of proposed SPV model is presented in Fig. 4 for SPV-HO system given in (26). The step response as depicted in Fig. 2 of SPV-RO model determined in (40) is closely matched with the response of HO-SPV system depicted in (26). The similar observation is also found for impulse response as provided in Fig. 3 and Bode plot demonstrated in Fig. 4 of proposed SPV-RO model given in (40) with respect to the HO-SPV system expressed in (26). The time-domain specifications as provided in Table 1 and error indices shown in Table 2, confirm the productivity and efficacy of the proposed RO-SPV model. These tabulated values are such that the time-domain specifications are closely matched with HO-SPV system and errors are considerably lesser.

Hence, the responses and plot from Figs. 2, 3, and 4 and tabulated data from Tables 1 and 2 concluded that proposed SPV-RO model depicted in (40) is comparatively well suited for the HO-SPV system given in (26). This confirms applicability and effectiveness of proposed method.

6 Conclusion

This research work provides the better approximated reduced-order solar-photovoltaic (RO-SPV) model for higher-order (HO) SPV controller system. The third-order SPV controller system is approximated to second-order RO-SPV model by exploiting the Markov-parameters (M-Ps) along with time-moments (T-Ms) of HO-SPV and RO-SPV model. The errors between M-Ps and T-Ms are minimized with the help of grey-wolf optimization algorithm for ascertainment of RO-SPV model. The superiority of proposed SPV-RO model is proved with the help of responses and tabular representations. The proposed methodology can be implemented by adopting any well suited systematic approach of determination of weights in future. Further, proposed method can also be exploited for approximation of continuous and discrete fixed coefficient and interval systems.

References

1. Rahman MM, Biswas SP, Islam MR, Rahman MA, Muttaqi KM (2021) An advanced nonlinear controller for the lcl-type three-phase grid-connected solar photovoltaic system with a dc-dc converter. IEEE Syst J. https://doi.org/10.1109/JSYST.2021.3121406
2. Kumar M, Tyagi B (2021) A robust adaptive decentralized inverter voltage control approach for solar PV and storage-based islanded microgrid. IEEE Trans Ind Appl 57(5):5356–5371. https://doi.org/10.1109/TIA.2021.3094453
3. Rehman HU, Yan X, Abdelbaky MA, Jan MU, Iqbal S (2021) An advanced virtual synchronous generator control technique for frequency regulation of grid-connected PV system. Int J Electr Power Energy Syst 125:106440. https://doi.org/10.1016/j.ijepes.2020.106440
4. Shunmugham Vanaja D, Albert JR, Stonier AA (2021) An experimental investigation on solar PV fed modular STATCOM in WECS using intelligent controller. Int Trans Electr Energy Syst 31(5):e12845. https://doi.org/10.1002/2050-7038.12845
5. Wang D, Locment F, Sechilariu M (2020) Modelling, simulation, and management strategy of an electric vehicle charging station based on a dc microgrid. Appl Sci 10(6):2053. https://doi.org/10.3390/app10062053

6. Al-Wesabi I, Fang Z, Wei Z, Dong H (2022) Direct sliding mode control for dynamic instabilities in DC-link voltage of standalone photovoltaic systems with a small capacitor. Electronics 11(1):133. https://doi.org/10.3390/electronics11010133
7. Mahmud MR, Pota H (2021) Robust partial feedback linearized controller design for standalone hybrid pv-bes system. Electronics 10(7):772. https://doi.org/10.3390/electronics10070772
8. Mirjalili S, Mirjalili SM, Lewis A (2014) Grey wolf optimizer. Advan Eng Softw 69:46–61. https://doi.org/10.1016/j.advengsoft.2013.12.007

State of the Art Sliding Mode Controller for Quadrotor Trajectory Tracking

Tinu Valsa Paul, Thirunavukkarasu Indiran, George Vadakkekara Itty, Suraj Suresh Kumar, and Chirag Gupta

Abstract The quadrotor is an under-actuated nonlinear system. The decoupling between attitude and position is done in the proposed model, and a novel homogeneous sliding mode controller is designed for the model. A six-input variable controller approach is designed for the quadrotor system, which can be used for UAV maneuvring in space where gravity is not present. The analysis of the new model is done by MATLAB simulation using the robust controller. The effectiveness of the proposed controller is clearly comprehended through the obtained results of various simulations, and a comparison with other controllers is made.

Keywords Quadrotor model · Robust control · Stability · Trajectory tracking

T. V. Paul · S. Suresh Kumar
Research Scholar, Department of Instrumentation and Control Engineering, Manipal Institute of Technology, Manipal Academy of Higher Education, Manipal 576104, India
e-mail: tinuvalsa.paul@learner.manipal.edu

S. Suresh Kumar
e-mail: suraj.suresh1@learner.manipal.edu

T. Indiran (✉)
Department of Instrumentation and Control Engineering, Manipal Institute of Technology, Manipal Academy of Higher Education, Manipal 576104, India
e-mail: it.arasu@manipal.edu
URL: http://www.itarasu.com

G. V. Itty
Electrical and Electronics Engineering, Mar Baselios Christian College of Engineering and Technology, Peermade 685531, India
e-mail: vi.george@manipal.edu

C. Gupta
Scientist/Engineer SD, North Eastern Space Applications Centre, Indian Space Research Organization, Shillong, India
e-mail: chirag.gupta@nesac.gov.in

S. Sharma et al. (eds.), *Intelligent Control, Robotics, and Industrial Automation*, Lecture Notes in Electrical Engineering 1066,
https://doi.org/10.1007/978-981-99-4634-1_20

Abbreviations

DOF Degrees of Freedom
SMC Sliding Mode Controller
UAV Unmanned Ariel Vehicle
VTOL Vertical Takeoff and Landing.

1 Introduction

Many countries have been making efforts to make advancements in space exploration over the last decade. The researchers are trying to expand the UAV field to space missions so that surveillance of the Almas, Salyut series, Skylab, Tiangong, etc. can be easily done since safety is a prime factor. For applications like disaster management and geographical monitoring in earth where gravity plays a major role, the six DOF can be easily regulated by four controls. But in future space stations, the six controller approach introduced for the SMC controllers will give ease of operation.

The quadrotor is a four-rotor VTOL that has the advantage of being flung from everywhere and the capability of hovering with an inherently unstable design in open loop operation. The problem of controlling and tracking a quadrotor is a challenging task. The proposed system can be used for maneuvring quadrotors in space by replacing the components with the compactable ones.

1.1 Motivation for the Research

Even though quadrotors flew successfully in the 1920s, automatic control was only recently developed. The quadrotor formation control [1] with a robust continuous sliding mode controller was developed, and many advancements in quadrotor control using SMC [2] have progressed. A modified integral terminal SMC [3] for position and attitude control of a quadrotor was developed with external disturbances and proved effective later in 2020. The homogeneous SMC was introduced in 2021 which is modified and implemented with simulation in this paper [4]. The paper tries to modify the existing mathematical model which uses Euler-Lagrange formulism and Newton's equations of motion. The schematic diagram of the cross configuration is shown in "Fig. 1".

1.2 Major Contributions

This paper introduces a novel system model and homogeneous SMC that can be applied to quadrotor maneuvring which ensures tracking of errors in linear positions and attitude. A comparison study was done with classical SMC and adaptive SMC. The contributions introduced are as follows:

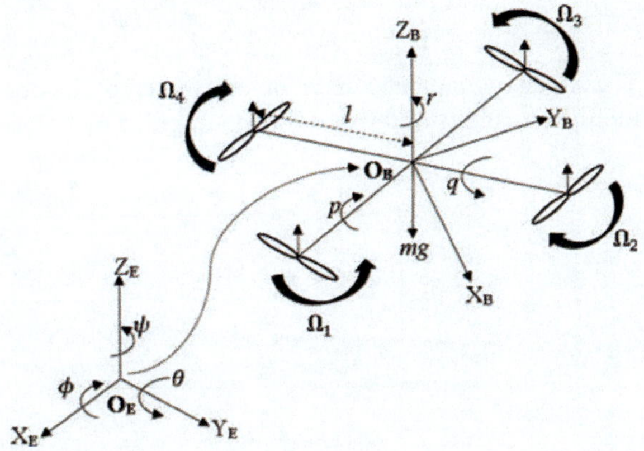

Fig. 1 Coordinate frame of the quadrotor

1. The position and attitude of the quadrotor are decoupled and separate sliding mode controllers are used for control of six DOF's so that it can be used for the application of space maneuvring.
2. A new mathematical model is developed by altering the equations of motion for the kinematic and dynamic models.
3. A novel robust homogeneous SMC is introduced which commands the position and angles of the quadortor.

1.3 Organization of the Paper

In Sect. 2, modeling of the motion equations of a quadrotor is developed. In Sects. 3 and 4, classical SMC and adaptive SMC for the developed mathematical model are discussed. In Sect. 5 a new robust SMC is introduced. In Sect. 6 simulation results are given. The paper is concluded in Sect. 7.

2 Modeling of Motion Equations of Quadrotor

The dynamic model of the quadrotor is developed with certain assumptions as follows:

Assumptions

- The structure of the quadrotor is symmetrical and rigid.
- The constraints for attitude dynamics are

$$-\pi/2 \leq (\phi, \theta) \leq \pi/2$$

$$-\pi \leq (\psi) \leq \pi$$

The standard kinematic and dynamic model of the quadrotor [5] is modified and the motion equations of the developed are given by

$$\begin{aligned}
\ddot{\phi} &= \frac{\dot{\psi}\dot{\theta}(I_y - I_z) + J_r\dot{\theta}\Omega_r - K_{ax}\dot{\phi}^2 + U_\phi}{I_x} \\
\ddot{\theta} &= \frac{\dot{\phi}\dot{\psi}(I_z - I_x) + J_r\dot{\phi}\Omega_r - K_{ay}\dot{\theta}^2 + U_\theta}{I_y} \\
\ddot{\psi} &= \frac{\dot{\theta}\dot{\phi}(I_x - I_y) - K_{az}\dot{\psi}^2 + U_\psi}{I_z} \\
\ddot{x} &= \frac{U_x(\cos\phi\sin\theta\cos\psi + \sin\phi\sin\psi)}{m} - K_{fdx}\dot{x} \\
\ddot{y} &= \frac{U_y(\cos\phi\sin\theta\cos\psi - \sin\phi\sin\psi)}{m} - K_{fdy}\dot{y} \\
\ddot{z} &= -g + \frac{U_z(\cos\phi\cos\theta)}{m} - K_{fdz}\dot{z}
\end{aligned} \tag{1}$$

For simplification, we can define,

$$\begin{aligned}
a_1 &= \frac{I_y - I_z}{I_x} & a_2 &= \frac{K_{fax}}{I_x} & a_3 &= \frac{J_r}{I_x} & a_4 &= \frac{I_z - I_x}{I_y} \\
a_5 &= \frac{K_{fay}}{I_y} & a_6 &= \frac{J_r}{I_y} & a_7 &= \frac{I_x - I_y}{I_z} & a_8 &= \frac{K_{faz}}{I_z} \\
b_1 &= \frac{1}{I_x} & b_2 &= \frac{1}{I_y} & b_3 &= \frac{1}{I_z}
\end{aligned}$$

The schematic diagram of the proposed model is depicted in "Fig. 2".

$$f(X, U) = \begin{bmatrix}
x_2 \\
a_1x_4x_6 - a_2x_2^2 - a_2x_4\Omega_r + b_1U_\phi \\
x_4 \\
a_4x_2x_6 - a_5x_4^2 - a_6x_2\Omega_r + b_2U_\theta \\
x_6 \\
a_7x_2x_4 - a_8x_6^2 + b_3U_\psi \\
x_{11} \\
\frac{-U_x}{m}(Cx_1Sx_3Cx_5 + Sx_1Sx_5) - K_{fdx}x_8 \\
x_{10} \\
\frac{-U_y}{m}(Cx_1Sx_3Cx_5 - Sx_1Sx_5) - K_{fdx}x_{10} \\
x_{12} \\
g - \frac{U_z}{m}(Cx_1Cx_3) - K_{fdx}x_{12}
\end{bmatrix} \tag{2}$$

where $C = cos(.)$ and $S = sin(.)$.

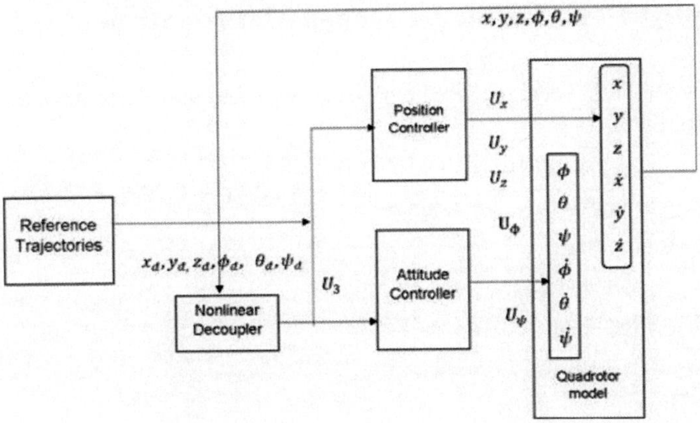

Fig. 2 Proposed control strategy of the quadrotor

3 Classical SMC for the Developed Mathematical Model

The controller tuning parameters of classical SMC, Adaptive SMC and proposed robust homogeneous SMC are given in Table 1.

The classical SMC designed for the modified Rezoug et al. [5] model is as follows:

$$U_x = \frac{m}{\cos x_7 \sin x_8 \cos x_9 + \sin x_7 \sin x_9}[-\ddot{x_d} + K_{fdx}x_4 - \lambda_x(\dot{x_d} - \dot{x}) + K_x sign\,(s_x)] \quad (3)$$

$$U_y = \frac{m}{\cos x_7 \sin x_8 \cos x_9 - \sin x_7 \cos x_9}[-\ddot{y_d} + K_{fdy}x_5 - \lambda_y(\dot{y_d} - \dot{y}) + K_y sign\,(s_y)] \quad (4)$$

$$U_z = \frac{m}{(\cos x_7 \cos x_8)}[\ddot{z_d} - g + K_{fdz}x_6 + \lambda_z(\dot{z_d} - \dot{z}) + K_z sign\,(s_z)] \quad (5)$$

$$U_\phi = \frac{1}{b_1}[\ddot{\phi}_d - a_1x_{11}x_{12} + a_2x_{10}^2 + a_3x_{11}\Omega_r + \lambda_\phi(\dot{\phi}_d - \dot{\phi}) + K_\phi sign\,(s_\phi)] \quad (6)$$

$$U_\theta = \frac{1}{b_2}[\ddot{\theta}_d - a_4x_{10}x_{12} - a_5x_{11}^2 + a_6x_{10}\Omega_r + \lambda_\theta(\dot{\theta}_d - \dot{\theta}) + K_\theta sign\,(s_\theta)] \quad (7)$$

$$U_\psi = \frac{1}{b_3}[\ddot{\psi}_d - a_7x_{10}x_{11} + a_8x_{12}^2 + \lambda_\psi(\dot{\psi}_d - \dot{\psi}) + K_\psi sign\,(s_\psi)] \quad (8)$$

The model parameters are taken from Rezoug et al. [5] and the simulation is done for the classical SMC.

4 Adaptive SMC for the Developed Mathematical Model

The adaptive SMC [2] designed in the paper is taken and applied to the mathematical model as follows:

$$U_x = \frac{m}{\cos x_7 \sin x_8 \cos x_9 + \sin x_7 \sin x_9}[\ddot{x}_d + K_{fdx}x_4 + \lambda_x e_4 + K_x sign\,(s_x)) - \hat{\Gamma}_x \\ sign\,(s_x)] \quad (9)$$

$$U_y = \frac{m}{\cos x_7 \sin x_8 \cos x_9 - \sin x_7 \cos x_9}[\ddot{y}_d + K_{fdy}x_5 + \lambda_y e_5 + K_y sign\,(s_y)) - \hat{\Gamma}_y \\ sign\,(s_y)] \quad (10)$$

$$U_z = \frac{m}{(\cos x_7 \cos x_8)}[\ddot{z}_d + K_{fdz}x_6 + g + \lambda_z e_6 + K_z sign\,(s_z) - \hat{\Gamma}_z sign\,(s_z)] \quad (11)$$

$$U_\phi = \frac{1}{b_1}[\ddot{\phi}_d - a_1 x_{11} x_{12} + a_2 x_{10}^2 + a_3 x_{11}\Omega_r + \lambda_\phi e_{10} + K_\phi sign\,(s_\phi)) - \hat{\Gamma}_\phi \\ sign\,(s_\phi)] \quad (12)$$

$$U_\theta = \frac{1}{b_2}[\ddot{\theta}_d - a_4 x_{10} x_{12} + a_5 x_{11}^2 - a_6 x_{10}\Omega_r + \lambda_\theta e_{11} + K_\theta sign\,(s_\theta)) - \hat{\Gamma}_\theta \\ sign\,(s_\theta)] \quad (13)$$

$$U_\psi = \frac{1}{b_3}[\ddot{\psi}_d - a_7 x_{10} x_{11} + a_8 x_{12}^2 + \lambda_\psi e_{12} + K_\psi sign\,(s_\psi)) - \hat{\Gamma}_\psi sign\,(s_\psi)] \quad (14)$$

where $\dot{\hat{\Gamma}} = \frac{1}{\alpha}\,|s|$ represents the adaptive law for the positions and angles, respectively.

5 Proposed Robust Homogeneous SMC for Quadrotor Model

In the paper, a modification to the homogeneous SMC [6] is done. The modified control law is comprised of two parts, i.e., an equivalent control part and a homogeneous control part.

$$U(t) = U_{eq}(t) + U_h(t) \quad (15)$$

$$U_h(t) = Ksign[s(t)] - \beta_z |s_t|^{\frac{3}{4}} (signs_t)\, t \tag{16}$$

5.1 Trajectory Control Design

The altitude errors are as follows:

$$\begin{aligned} e_{11} &= z_d - z \\ e_{12} &= \dot{z_d} - \dot{z} \end{aligned} \tag{17}$$

The sliding surface is given by

$$s_z = (\dot{z_d} - \dot{z}) + \lambda_z (z_d - z) \tag{18}$$

$$\dot{s_z} = (\ddot{z_d} - \ddot{z}) + \lambda_z (\dot{z_d} - \dot{z}) \tag{19}$$

Substituting for $\ddot{z}$ in the Eq. (1) we get

$$\dot{s_z} = \ddot{z_d} - g + \frac{1}{m}\left[-K_{fdz}\dot{z} + (\cos\phi\cos\theta)\, U_z\right] + \lambda_z(\dot{z_d} - \dot{z}) \tag{20}$$

Expanding Eq. (20), we get

$$\dot{s_z} = \ddot{z_d} - g + \frac{1}{m}(-K_{fdz}\dot{z}) + (\cos\phi\cos\theta)\, U_z \frac{1}{m} + \lambda_z e_{12} \tag{21}$$

When $\dot{s_z} = 0, U_z = U_{eqz}$ as

$$U_{eqz} = \frac{m}{(\cos\phi\cos\theta)}[\ddot{z_d} + g + \frac{K_{fdz}}{m}\dot{z} + \lambda_z(\dot{z_d} - \dot{z})] \tag{22}$$

$$U_{hz} = K_z sign[s(z)] - \beta_z |s_z|^{\frac{3}{4}} (signs_z)\, z \tag{23}$$

$$U_z = \frac{m}{\cos\phi\cos\theta}[\ddot{z_d} - g + \frac{1}{m}K_{fdz}\dot{z} + g + \lambda_z e_6 + K_z sign\,(s_z) - \beta_z |s_z|^{\frac{3}{4}} (signs_z)\, z] \tag{24}$$

The derivation of the equivalent controller for roll is described below: For the quadrotor, the error w.r.t. roll would be

$$e_7 = \phi_d - \phi$$
$$e_{10} = \dot{\phi}_d - \dot{\phi} \tag{25}$$

The sliding surface is given by

$$s_\phi = (\dot{\phi}_d - \dot{\phi}) + \lambda_\phi(\phi_d - \phi) \tag{26}$$

$$\dot{s_\phi} = (\ddot{\phi}_d - \ddot{\phi}) + \lambda_\phi(\dot{\phi}_d - \dot{\phi}) \tag{27}$$

Substituting for $\ddot{\phi}$ in the equation 27 we get

$$\dot{s_\phi} = \left[\ddot{\phi}_d - \left(\frac{I_y - I_z}{I_x}\right) x_{11}x_{12} + \frac{J_r}{I_x} x_{11}\Omega_r + \frac{L}{I_x} U_\phi\right] + \lambda_\phi(\dot{\phi}_d - \dot{\phi}) \tag{28}$$

When $\dot{s_\phi} = 0$,$U_{eq\phi}$ is obtained as

$$U_{eq\phi} = \frac{I_x}{L}[\ddot{\phi}_d - \left(\frac{I_y - I_z}{I_x}\right) x_{11}x_{12} + \frac{J_r}{I_x} x_{11}\Omega_r + \lambda_\phi(\dot{\phi}_d - \dot{\phi})] \tag{29}$$

The switching controller $U_{h\phi}$ is defined as

$$U_{h\phi} = K_\phi \operatorname{sgn}[s_\phi] - \beta_\phi |s_\phi|^{\frac{1}{2}} \left(\operatorname{sgn} s_\phi\right) \phi \tag{30}$$

The complete controller can be derived as follows:

$$U_\phi = \frac{I_x}{L}[\ddot{\phi}_d - \left(\frac{I_y - I_z}{I_x}\right) x_{11}x_{12} + \frac{J_r}{I_x} x_{11}\Omega_r + \lambda_\phi(\dot{\phi}_d - \dot{\phi}) + K_\phi \operatorname{sgn}[s_\phi]$$
$$- \beta_\phi |s_\phi|^{\frac{1}{2}} \left(\operatorname{sgn} s_\phi\right) \phi] \tag{31}$$

Similar methodologies can be used to derive the other positions and attitudes.

5.2 *Proposed Control Scheme*

The proposed Robust Homogeneous controller equations for the mathematical model are as follows:

$$U_x = p_x[-\ddot{x}_d + K_{fdx}x_4 - \lambda_x(\dot{x}_d - \dot{x}) + K_x sign\,(s_x) - \beta_x |s_x|^{\frac{3}{4}} \left(signs_x\right) x] \tag{32}$$

where $p_x = \frac{m}{\cos x_7 \sin x_8 \cos x_9 + \sin x_7 \sin x_9}$

$$U_y = p_y[-\ddot{y}_d + K_{fdy}x_5 - \lambda_y(\dot{y}_d - \dot{y}) + K_y sign\left(s_y\right) - \beta_y |s_y|^{\frac{3}{4}} \left(signs_y\right) y] \tag{33}$$

where $p_y = \frac{m}{\cos x_7 \sin x_8 \cos x_9 - \sin x_7 \cos x_9}$

$$U_z = p_z[\ddot{z_d} - g + K_{fdy}x_6 + \lambda_z(\dot{z_d} - \dot{z}) + K_z sign\,(s_z) - \beta_z|s_z|^{\frac{3}{4}}\,(signs_z)\,z] \quad (34)$$

where $p_z = \frac{m}{(\cos x_7 \cos x_8)}$

$$U_\phi = \frac{1}{b_1}[\ddot{\phi_d} - a_1x_{11}x_{12} + a_2x_{10}^2 + a_3x_{11}\Omega_r + \lambda_\phi(\dot{\phi_d} - \dot{\phi}) + K_\phi sign\,(s_\phi) - \beta_\phi|s_\phi|^{\frac{3}{4}}\,(signs_\phi)] \quad (35)$$

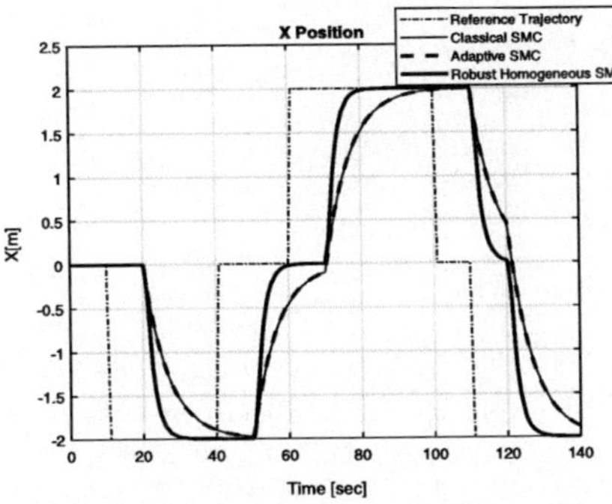

(a) Comparison of X position for the Quadrotor

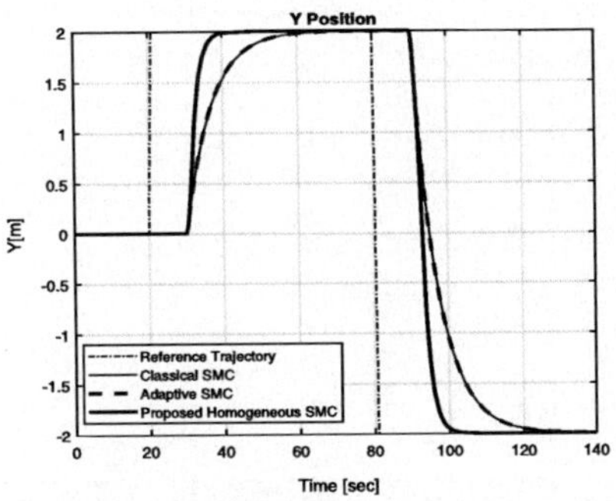

(b) Comparison of Y position for the Quadrotor

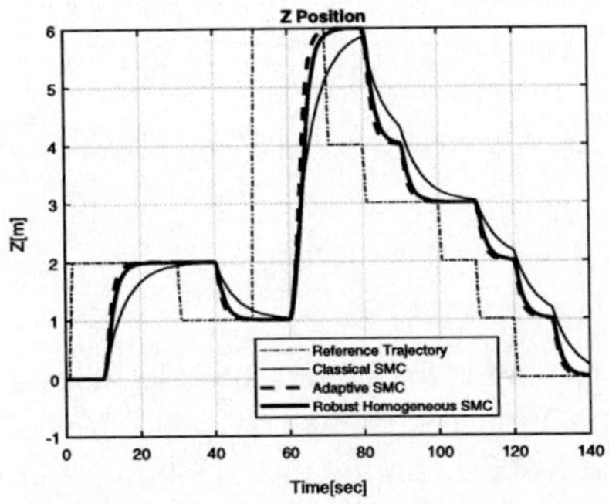

(c) Comparison of Z position for the Quadrotor

Fig. 3 Comparison of position tracking results. Under the control of the proposed methodology, better tracking is achieved.

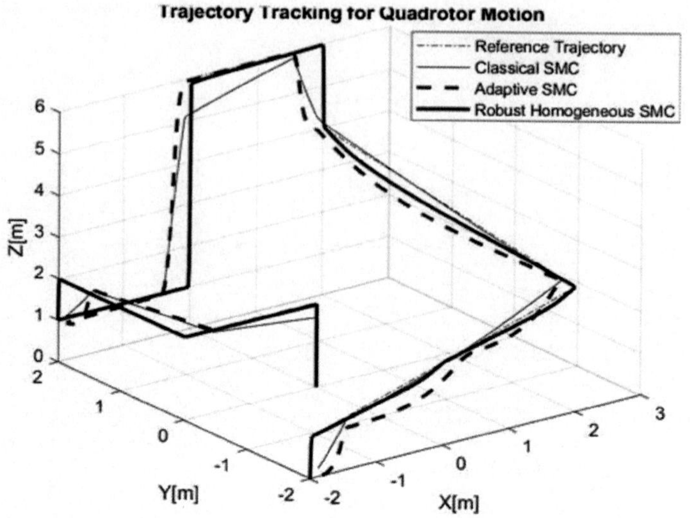

Fig. 4 3 D plot of quadrotor motion

$$U_\theta = \frac{1}{b_2}[\ddot{\theta}_d - a_4 x_{10} x_{12} + a_5 x_{11}^2 + a_6 x_{10} \Omega_r + \lambda_\theta(\dot{\theta}_d - \dot{\theta}) + K_\theta sign\,(s_\theta) - \beta_\theta |s_\theta|^{\frac{3}{4}}\,(signs_\theta)] \quad (36)$$

$$U_\psi = \frac{1}{b_3}[\ddot{\psi}_d - a_7 x_{10} x_{11} + a_8 x_{12}^2 + \lambda_\psi(\dot{\psi}_d - \dot{\psi}) + K_\psi sign\,(s_\psi) - \beta_\psi |s_\psi|^{\frac{3}{4}}\,(signs_\psi)] \quad (37)$$

6 Simulation Results

The comparison of trajectories for the quadrotor is represented in the "Fig. 3". We can infer that for the X and Y positions, the Classical SMC and Adaptive SMC are deviating and the proposed Robust Homogeneous Sliding Mode Controller is tracking properly. But in the case of Z position, the adaptive and Robust Homogeneous SMC is tracking properly where gravity plays a major role. The 3-dimensional plot of the quadrotor motion is shown in "Fig. 4". Thus, the designed controllers are able to stabilize the quadrotor UAV. The tracking error for X,Y, and Z position are illustrated in "Fig. 5" which indicates that the Robust Homogeneous SMC quickly converges to zero, but in the Z position case, the Adaptive SMC performs better than the proposed one. The tracking errors of the proposed novel Robust Homogeneous SMC converge

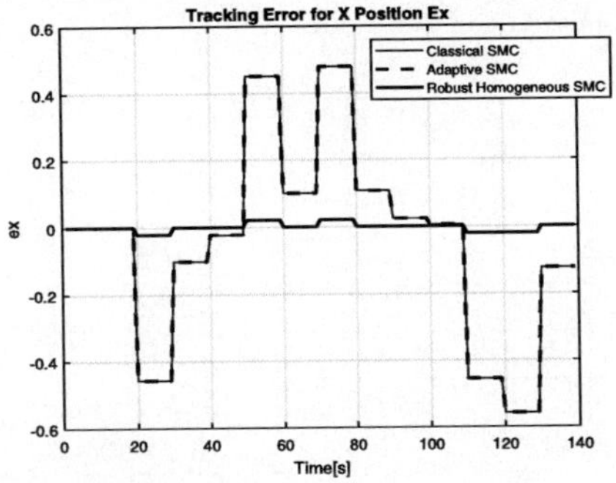

(a) Tracking Error w.r.t. X Position

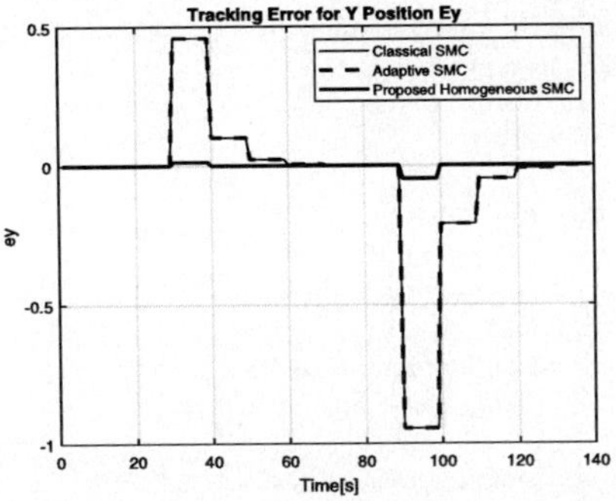

(b) Tracking Error w.r.t. Y Position

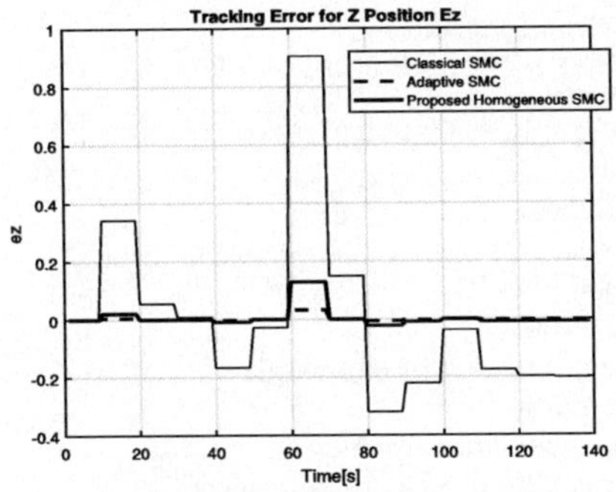

(c) Tracking Error w.r.t. Z Position

Fig. 5 Comparison of position tracking errors

faster. Thus, from the simulation results, we can infer that the overall performance of the Robust Homogeneous Sliding Mode Controller is better.

7 Conclusion

A mathematical model of the quadrotor rotational and translational is developed based on the physical laws of Lagrange-Euler and Newton-Euler. A novel Robust Homogeneous SMC technique is suggested to address the trajectory tracking control problem, and from the observed results, we can conclude that the quadrotor performs better.

Table 1 Comparison chart of parameters of the proposed nonlinear controllers

Classical SMC	K	x	y	z
		0.96	0.775	0.85
	λ	0.15	0.15	0.18
Adpative SMC	K	0.96	0.775	0.85
	λ	56799	34678	89677
	α	0.15	0.15	0.65
Proposed homogeneous SMC	K	0.6	0.65	0.55
	λ	0.5	0.55	0.5
	β	0.0009	0.0059	0.0000000049

Acronyms

b	Coefficient of Aerodynamic thrust force
C_D	Drag Coefficient
d	Coefficient of Aerodynamic moment
g	Gravitational Acceleration
K_P	Coefficient related to lift force
K	Tuning Parameter responsible for reaching phase
l	Arm Length
m	Quadrotor Mass
I_x, I_y, I_z	Inertia on x Axis, y Axis and z Axis
J_r	Rotor Inertia
K_{ax}, K_{ay}, K_{az}	Aerodynamic Friction Coefficients
$K_{fdx}, K_{fdy}, K_{fdz}$	Translational Positive Drag Coefficients
s	Sliding surface
x, y, z	X, Y and Z Positions
x_d, y_d, z_d	Reference for X Position, Y Position and Z Positions
ϕ, θ, ψ	Roll, Pitch and Yaw
ψ_d, θ_d and ϕ_d	Reference for Roll, Pitch and Yaw
U_x, U_y, U_z	Control Signals for X, Y and Z Positions
U_ϕ, U_θ, U_ψ	Control Signals for θ, ϕ and ψ
α	Zero adaptation gain depending upon tracking error
β	Homogeneous gain factor
λ	Tuning parameter design greater than zero
Ω_i	Speed of ith Rotor
$\hat{r}$	Adaptive Gain Constant
Ω_r	Relative Velocity

Acknowledgements The authors would like to thank the Indian Space Research Organization, Department of Space, Government of India for sanctioning the funds under the wide grant ID: ISRO/RES/3/822/19-20, dated August 8th, 2019.

References

1. González-Sierra J, Ríos H, Dzul A (2020) Quad-rotor robust time-varying formation control: a continuous sliding-mode control approach. Int J Control 93(7):1659–1676. https://doi.org/10.1080/00207179.2018.1526413
2. Nadda S, Swarup A (2018) On adaptive sliding mode control for improved quadrotor tracking. J Vib Control 24(14):3219–3230. https://doi.org/10.1177/1077546317703541
3. Labbadi M, Cherkaoui M (2020) Robust adaptive nonsingular fast terminal sliding-mode tracking control for an uncertain quadrotor UAV subjected to disturbances. ISA Trans 99:290–304. https://doi.org/10.1016/j.isatra.2019.10.012
4. Utkin VI (2013) Sliding modes in control and optimization. Springer Science & Business Media
5. Rezoug A, Hamerlain M, Achour Z, Tadjine M (2015) Applied of an adaptive higher order sliding mode controller to quadrotor trajectory tracking. In: 2015 IEEE international conference on control system, computing and engineering (ICCSCE). IEEE, pp 353–358. https://doi.org/10.1109/ICCSCE.2015.7482211
6. Mehta A, Bandyopadhyay B (2021) Emerging trends in sliding mode control: theory and application. Springer

Trajectory Tracking with RBF Network Estimator and Dynamic Adaptive SMC Controller for Robot Manipulator

K. Dileep, S. J. Mija, and N. K Arun

Abstract The accuracy in trajectory tracking for robotic manipulators, both in joint and cartesian space, is essential when deputed for industrial applications. The main challenge in controller design is to obtain an accurate model and to guarantee convergence even with external disturbances, friction and model uncertainties. The Radial Basis Function (RBF) neural network effectively overcomes the issue of parameter variations and model uncertainties. Moreover, RBF networks have fast learning ability and better approximation capabilities. A dynamic sliding mode controller provides fast convergence and eliminates the chattering issues associated with conventional sliding mode controllers (SMC). An adaptive dynamic SMC with RBF estimation compensates for modelling uncertainties and provides excellent trajectory tracking. The Lyapunov approach is used to show both the convergence of the RBF weight adjustment rule and the stability of the closed-loop control system. The proposed control strategies are simulated using a 2-DOF manipulator model, and the results are compared with the classical SMC.

Keywords Dynamic SMC · RBF · Robotic manipulator · Adaptive SMC

1 Introduction

The advances in automated industries have led to the widespread use of robotic manipulators for diverse applications. The ability to perform repetitive tasks with

K. Dileep (✉)
Research Scholar, NIT, Calicut, Kerala, India
e-mail: dileep_p210056ee@nitc.ac.in

S. J. Mija
Associate Professor, NIT, Calicut, Kerala, India
e-mail: mija@nitc.ac.in

N. K. Arun
Assistant Professor, NIT, Calicut, Kerala, India
e-mail: arunkm@nitc.ac.in

S. Sharma et al. (eds.), *Intelligent Control, Robotics, and Industrial Automation*, Lecture Notes in Electrical Engineering 1066,
https://doi.org/10.1007/978-981-99-4634-1_21

high precision and improved load carrying capacity makes robots very popular in industrial applications [1]. Motion control of robotic manipulators is a hot topic of intense research. Many model-based and model-free control schemes have been tried. In the model-based methods, obtaining a precise model for the robot system is very demanding because robot manipulators are extremely uncertain and greatly nonlinear due to parametric variations, frictional effects and external disturbances. To compensate for the uncertainties, advanced control techniques like neural network, adaptive control [2–4], fuzzy-neuro [5, 6] or sliding mode control [7]-based techniques need to be used. Many research contributions have already been made in the field of trajectory tracking for various applications when system uncertainties, parameter variations and external disturbances affect the model. The robot tracking problem is still a promising area of research [8].

In SMC, the system states are transferred from any initial state to the sliding surface and maintains the system after that on the switching surface using a high-frequency switching control [9]. Advanced techniques of SMC, like higher-order SMC and dynamic SMC, make use of the advantages of SMC and simultaneously reduce the chattering issues and improve the convergence rate. The dynamic SMC uses a higher-order switching function constructed using the sliding variable of the first-level sliding function. This method can sufficiently reduce the chattering phenomenon and obtain a continuous dynamic SMC rule. The dynamic sliding mode technique adds additional dynamics to the controlled system [10]. The work [11] uses a robust terminal SMC with an arctan function and a saturation function replacing the sign function. In [12], an adaptive fuzzy SMC with adaptive law was developed using the Lyapunov function. Article [13] uses finite-time stability theory and the differential inequality principle for terminal sliding control.

In the paper [14], the convergence of the reference model is achieved by a variable structure method with adaptive control using a neural network. Paper [15] presents a rehabilitation robot with a control strategy based on a neural network that adapts to the interaction between the patient and the robot. Paper [16] develops adaptive controllers to solve trajectory tracking problems for uncertain robots. An adaptive law is formulated assuming only position measurement is available. In the paper [17], an RBF network based on robust adaptive control is derived depending on the manipulator dynamics of the end effector in the workspace. RBF is an architecture that can be used to control dynamic systems [18]. The dynamic parameter matrices of a manipulator arm in the task space are derived from the RBF network, and the Lyapunov function is used to derive the RBF weight update law [19]. In [20], an NN-based approximation of the uncertain robot dynamics is performed for a group of similar robot manipulators with different tracking tasks. In article [21], trajectory tracking using adaptive RBF on a normal sliding surface is demonstrated with a robust term function for a robotic arm.

The major contributions of the paper are

- Design of a dynamic SMC for joint angle trajectory tracking for an n-DOF robot manipulator.

- Development of an RBF network with a weight adjustment law to estimate the manipulator dynamics. The network can estimate the parameter variations and uncertainties in the system design.

The paper is structured as: In Sect. 2, generalized equations for a n-DOF robot arm are developed. In Sect. 3, the dynamic sliding surface and trajectory tracking control law are developed. In Sect. 4, the RBF network for system prediction is designed; the stability and convergence are verified using the Lyapunov function. In Sect. 5, the proposed control law and estimator for a 2-DOF manipulator are simulated.

2 Dynamics of Robot Manipulator

For any robotic manipulator, the dynamics model gives the relation between the joint actuator torque (τ) to the joint angular position (q), joint velocity ($\dot{q}$) and joint acceleration ($\ddot{q}$) of an individual joint. Dynamics equations for a n-DOF robot arm derived based on the Euler-Lagrange method will be of the form:

$$\tau = D(q)\ddot{q} + H(q, \dot{q})\dot{q} + G(q) + F(\dot{q}) + d_t \tag{1}$$

with $D(q) \in \mathbb{R}^{n \times n}$ represents inertial vector, Coriolis and centripetal forces vector is represented as $H(q, \dot{q}) \in \mathbb{R}^{n \times n}$. Gravity vector is denoted as $G(q) \in \mathbb{R}^{n \times 1}$, and vector for friction is denoted as $F(\dot{q}) \in \mathbb{R}^{n \times 1}$. The unknown disturbance is represented as $d_t \in \mathbb{R}^{n \times 1}$.

These dynamics follow specific properties like:

- The symmetric, positive definite inertia matrix $D(q)$ follows the inequality

$$m_1 \|w\|^2 \leq w^T Dq(w) \leq m_2 \|w\|^2, \ \forall w \in \mathbb{R}^{n \times 1}, \tag{2}$$

  where m_1 and m_2 depend on the mass of the manipulator.
- $\dot{D}(q) - 2H(q, \dot{q})$ is a skew symmetric matrix with any vector w.

$$w^T [\dot{D}(q) - 2H(q, \dot{q})] w = 0. \tag{3}$$

- $H(q, \dot{q})$, $G(q)$ and $F(\dot{q})$ are bounded matrices. Disturbance d_t is positive and bounded, thus $\|d_t\| \leq D_o$ and $\|\dot{d}_t\| \leq D_t$.

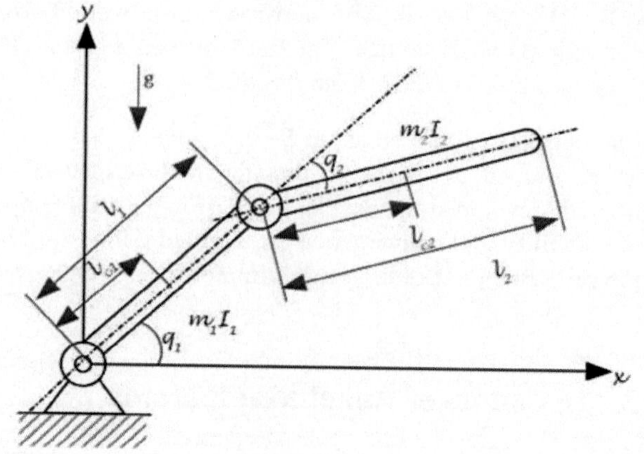

Fig. 1 2-DOF robotic manipulator

2.1 Dynamic Modelling of 2-DOF Manipulator

Figure 1 shows a 2 link manipulator with link masses m_1 and m_2 and link lengths a_1 and a_2. The dynamic equation of the form (1) is derived for the manipulator with parameter matrices as

$$D(q) = \begin{bmatrix} D_{11}(q) & D_{12}(q) \\ D_{21}(q) & D_{22}(q) \end{bmatrix} \tag{4}$$

$$D_{11} = (m_1 + m_2)a_1^2 + m_2a_2^2 + 2m_2a_1a_2\cos(q_2)$$

$$D_{12} = m_2a_2^2 + m_2a_1a_2\cos(q_2)$$

$$D_{21} = m_2a_2^2 + m_2a_1a_2\cos(q_2)$$

$$D_{22} = m_2a_2^2$$

$$H(q, \dot{q}) = \begin{bmatrix} H_{11}(q, \dot{q}) & H_{12}(q, \dot{q}) \\ H_{21}(q, \dot{q}) & H_{22}(q, \dot{q}) \end{bmatrix} \tag{5}$$

$$H_{11} = -m_2a_1a_2sin(q_2)\dot{q}_2$$

$$H_{12} = -m_2a_1a_2sin(q_2)(\dot{q}_1 + \dot{q}_2)$$

$$H_{21} = m_2a_1a_2sin(q_2)\dot{q}_1$$

$$H_{22} = 0$$

$$G(q) = \begin{bmatrix} G_1(q) \\ G_2(q) \end{bmatrix} \tag{6}$$

$$G_1 = (m_1 + m_2)a_1 g \cos(q_2) + m_2 a_2 g \cos(q_1 + q_2)$$

$$G2 = m_2 a_2 g \cos(q_1 + q_2)$$

$$F(\dot{q}) = \begin{bmatrix} F_1(\dot{q}) \\ F_2(\dot{q}) \end{bmatrix} \tag{7}$$

$$\tau = \begin{bmatrix} \tau_1 \\ \tau_2 \\ . \end{bmatrix} \tag{8}$$

The parameters for the manipulator, the controller, and the estimator are taken as follows: The mass of the links m_1 and m_2 is assumed to be 3 Kg and 2 Kg, respectively. The length of the links a_1 and a_2 is assumed to be 0.15 m and 0.15 m, respectively, and $g = 10\,\text{m/s}^2$.

2.2 Control Objective

Depending on the application for which the manipulator is used, the end effector of the robot arm must track a predefined coordinate path in the Cartesian space from an initial point to an end point. Inverse kinematics equations can be used to convert the end effector's trajectory into individual joint trajectories. Figure 2 shows the response of a 2-DOF manipulator to open-loop trajectory tracking in the presence of an external disturbance $d_t = [0.5sin(t); 0.5sin(t)]$.

Therefore, a controller must be designed that can overcome the effects of external disturbances and provide accurate trajectory tracking.

3 Dynamic SMC for Joint Trajectory Tracking

Objective of the designed dynamic SMC controller is to make the joint angle $q = [q_1, q_2, \ldots q_n]$ track the desired trajectories $q_d = [q_{d1}, q_{d2}, \ldots q_{dn}]$. The sliding function and tracking error be

$$s = \dot{e} + \lambda e \tag{9}$$

$$e = q_d - q, \tag{10}$$

where λ is positive definite hurwitz diagonal matrix. A new higher-order sliding function is designed as

$$\sigma = \dot{s} + \Lambda s. \tag{11}$$

Λ is a diagonal positive definite hurwitz matrix. By defining a suitable control signal, if σ is made zero, then $\dot{s} + \Lambda s = 0$ is asymptotically stable, therefore, $e \to 0$ and $\dot{e} \to 0$. Using Eqs. (9) and (10),

$$\dot{s} = \ddot{q}_d + \lambda \dot{e} - D(q)^{-1}(\tau - H(q, \dot{q})\dot{q} - G(q) - F(\dot{q} - d_t)). \tag{12}$$

For simplification let us take $g(q) = D(q)^{-1}$, $f(q, \dot{q}) = D(q)^{-1}(H(q, \dot{q})\dot{q} + G(q) + F(\dot{q}))$ and $D(q)^{-1}d_t = T_d$. Thus, Eq. (12) gets simplified as

$$\dot{s} = \ddot{q}_d + \lambda \dot{e} - g(q)\tau + f(q, \dot{q}) + T_d. \tag{13}$$

Differentiating Eq. (11) and using Eq. (13)

$$\dot{\sigma} = E(x) - (\dot{g}(q) + g(q)\lambda + g(q)\Lambda)\tau - g(q)\dot{\tau} + \dot{T}_d + (\lambda + \Lambda)T_d, \tag{14}$$

where

$$E(x) = (\dddot{q}_d + (\lambda + \Lambda)\ddot{q}_d + \dot{f}(q, \dot{q}) + \Lambda\lambda\dot{e}) \\ +(\lambda + \Lambda)(H(q, \dot{q})\dot{q} + G(q) + F(\dot{q})). \tag{15}$$

A dynamic control signal is chosen as

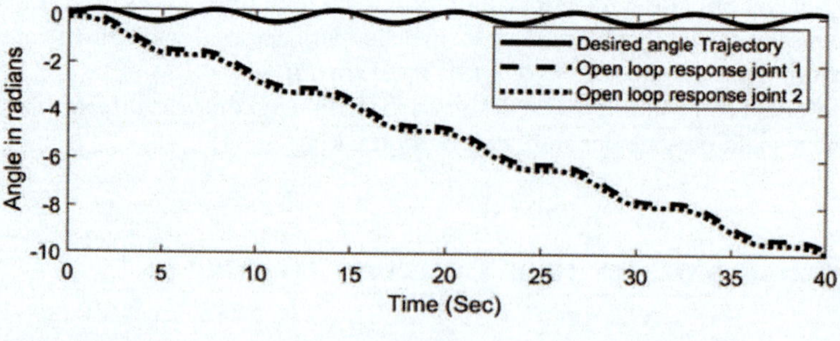

Fig. 2 Open loop joint angle trajectory tracking response of 2-DOF manipulator

$$\dot{\tau} = D(q)(E(x) - (\dot{g}(q) + g(q)\lambda + g(q)\Lambda)\tau + k_s sign(\sigma) + K\sigma). \quad (16)$$

Thus, Eq. (16) becomes

$$\dot{\sigma} = \dot{T}_d + (\lambda + \Lambda)T_d - k_s sign(\sigma) - K\sigma, \quad (17)$$

where $k_s > D_t + (\lambda + \Lambda)D_o$ and $K = diag[k_1, k_2, \ldots k_n]$ with positive elements. Now,

$$\begin{aligned}\sigma\dot{\sigma} &= \sigma(\dot{T}_d + (\lambda + \Lambda)T_d - k_s sign(\sigma) - K\sigma) \\ &\leq (D_t + (\lambda + \Lambda)D_o)\sigma - k_s|\sigma| - K\sigma^2 < 0.\end{aligned} \quad (18)$$

This ensures the possibility of a stable sliding function.

The excellence of SMC lie in its strong robustness to external disturbances and system uncertainties. However, the design of this controller requires an accurate knowledge of the state estimates and detailed dynamics. In the literature, unknown nonlinearities are effectively estimated using adaptive controllers based on neural networks due to their excellent approximation ability.

4 RBF-Based Adaptive Neural Network for State Estimation

The RBF network with the configuration shown in Fig. 3 is used to predict the uncertainties and parametric variations of the centripetal and Coriolis vectors, gravitational vector and the friction vector.

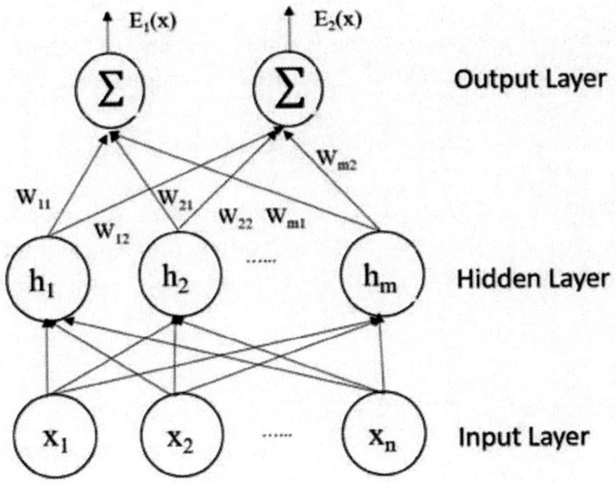

Fig. 3 RBF network structure

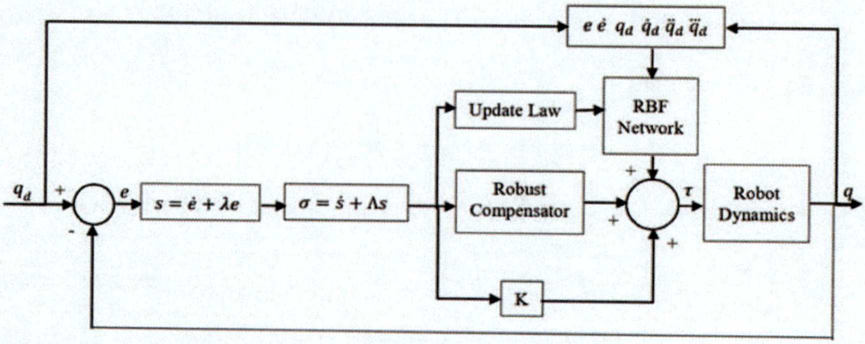

Fig. 4 Block diagram of DSMC with RBF estimator

In the first layer, the input signals $x = [e, \dot{e}, q_d, \dot{q}_d, \ddot{q}_d, \dddot{q}_d]$ are transferred to the next hidden layer with unity weights. The second layer is an array of neurons activated by a RBF. The output of the RBF function is represented as

$$h_j(x) = \exp\left(-\frac{\|x - c_j\|^2}{2d_j^2}\right) j = 1, \ldots, m, \tag{19}$$

where m denotes the count of neurons in the layer . $c_j = [c_{j1}, \ldots c_{jn}]$ is the central vector of the network, and d is the standard deviation of the jth RBF. The final layer gives the weighted sum of the RBF function from the second layer and the weights connecting the second layer to the final nodes, given by

$$f_i(x) = \sum_{j=1}^{m} W_{ij} h_j(x, c, d), \tag{20}$$

where W_{ji} represents the weight between jth hidden neuron to the ith output node. In vector form, output of the RBF neural network with Eq. (20) can be rewritten as

$$E(x) = W^T h(x) + \epsilon, \tag{21}$$

where ϵ is a small acceptable prediction error bounded by a real positive constant $\epsilon \leq \epsilon_0$ and W is the optimal weight value. Approximate output of the RBF network is given as

$$\hat{E}(x) = \hat{W}^T h(x), \tag{22}$$

where $\hat{W}^T = [\hat{W}_1^T, \hat{W}_2^T, \ldots, \hat{W}_m^T]$.

The overall control problem is now restructured with an RBF network which estimates the dynamics of manipulator as in Fig. 4. Using the estimated dynamics the control input is represented as

$$\dot{\tau} = D(q)(\hat{E}(x) - (\dot{g}(q) + g(q)\lambda + g(q)\Lambda)\tau + k_s sign(\sigma) + K\sigma). \quad (23)$$

Then, the estimated value of error will be

$$\begin{aligned}\tilde{E}(x) &= E(x) - \hat{E}(x) \\ &= W^T h(x) + \epsilon - \hat{W}^T h(x) \\ &= \tilde{W}^T h(x) + \epsilon \end{aligned} \quad (24)$$

here $\tilde{W} = W - \hat{W}$. Substituting the updated control input in Eq. (15) gives

$$\begin{aligned}\dot{\sigma} &= \tilde{E}(x) - k_s sign(\sigma) - K\sigma + \dot{T}_d + (\lambda + \Lambda)T_d \\ &= \tilde{W}^T h(x) + \epsilon - k_s sign(\sigma) - K\sigma + \dot{T}_d + (\lambda + \Lambda)T_d. \end{aligned} \quad (25)$$

4.1 Stability and Convergence Analysis

Stability of the closed-loop system and the convergence of the RBF weight adaptation function are assured using the direct Lyapunov approach. A Lyapunov candidate is taken as

$$L = \frac{1}{2}\sigma^2 + \frac{1}{2}\gamma \tilde{W}^T \tilde{W} \quad (26)$$

where γ is a positive coefficient. Now,

$$\begin{aligned}\dot{L} &= \sigma\dot{\sigma} + \gamma \tilde{W}^T \dot{\tilde{W}} \\ &= \sigma(\tilde{W}^T h(x) + \epsilon - k_s sign(\sigma) - K\sigma + \dot{T}_d + (\lambda + \Lambda)T_d) + \gamma \tilde{W}^T \dot{\tilde{W}} \\ &= \tilde{W}^T(\sigma h(x) - \gamma \dot{\tilde{W}}) - \sigma(\epsilon - k_s sign(\sigma) - K\sigma + \dot{T}_d + (\lambda + \Lambda)T_d). \end{aligned} \quad (27)$$

Let weight update rule be

$$\dot{\tilde{W}} = \frac{1}{\gamma}\sigma h(x). \quad (28)$$

Thus, $\dot{L}$ gets simplified as

$$\dot{L} = -\sigma(\epsilon - k_s sign(\sigma) - K\sigma + \dot{T}_d + (\lambda + \Lambda)T_d). \quad (29)$$

Since $k_s \geq D_t + (\lambda + \Lambda)D_o + \epsilon$, where D_t and D_o are the norm of T_d and $\dot{T}_d$. We get $\dot{L} \leq 0$. Therefore, if the control system has all the parameters bounded as per our assumptions for $t > 0$ and bounded initial conditions at $t = 0$ are ensured.

Lyapunov's direct method ensures the ultimate bound of σ. The tracking error can be minimized to the desirable range by proper choice of the control gain K.

5 Simulation Results

To asses the efficacy of the designed control technique and the estimator, simulation studies are performed for a 2-DOF manipulator designed in Sect. 2.1 on MATLAB platform. Parameters used in dynamic SMC controllers are $\lambda = diag[10, 10]$, $\Lambda = diag[10, 10]$ and $K = diag[20, 20]$. The desired trajectory is assumed to be $q_{1d} = q_{2d} = 0.25sin(t)$ in radians. The initial conditions for the joint angles in radians are $q(0) = [0.1; 0.1]$ and for the angular velocity are $\dot{q}(0) = [0; 0]$.

Variations in the payload lead to uncertainties in link mass. External disturbances may also change their magnitude. The designed adaptive controller and the RBF estimator should be able to suppress the effect of these variations. The simulation considers a mass change of the links to $m_1 = 7$ Kg and $m_2 = 5$ Kg at $t = 10$ sec and a change in disturbance amplitude to $1.5sin(t)$ at $t = 30$ s.

The simulated results of the proposed scheme with the RBF estimator are compared with that of conventional SMC [[21]]. Figure 5a shows the angle response of joints 1 and 2 for dynamic SMC and for conventional SMC. The response clearly indicates that the dynamic SMC provides faster convergence and lesser deviations. The mean squared error for the dynamic SMC is 1.48×10^{-5} radians compared to that of 6.55×10^{-5} radians for SMC with RBF estimation. Figure 5b compares torque inputs (τ) to the actuator. Dynamic SMC has lesser chattering and lower amplitude compared to SMC. The maximum torque computed for dynamic SMC is 8.58 Nm and 1.62 Nm, whereas for SMC it is 14.52 Nm and 5.48 Nm. Figure 6 gives the trajectory tracking response and actuator control input when the parametric change happens at $t = 10$ s and the amplitude of the external disturbance changes at $t = 30$ s. The adaptive controller can overcome the effects with minimum overshoots. Figure 7 compares the manipulator's actual dynamics E(x) and the estimated value of the dynamics $\hat{E}(x)$ using the RBF network estimator with the designed adaptation law. Figure 7b shows that the estimator can predict with minimal delay even when the system parameters change.

6 Conclusion

A dynamic adaptive SMC based on an RBF network is developed for tracking the trajectory of a robot manipulator in joint task space. The RBF will estimate the system dynamics based on the network inputs and the adaptation law. The control signal to dynamic SMC uses the estimated system dynamics from the RBF network. Hence, the controller can effectively overcome the variations and uncertainties in robot parameters. The chattering effect in the actuator input will be considerably reduced by the use of higher-order sliding surfaces. The robust compensation term in the control input ensures the stability and output convergence when the amplitude of external disturbances varies, and prediction error happens due to the sudden variations in the parametric values. The SMC control law and the weight updating rule are derived using the Lyapunov function, and the stability is confirmed using direct method of Lyapunov. Finally, the simulation results on a 2-DOF manipulator depict the efficacy of the developed control scheme. The results show better convergence and reduced

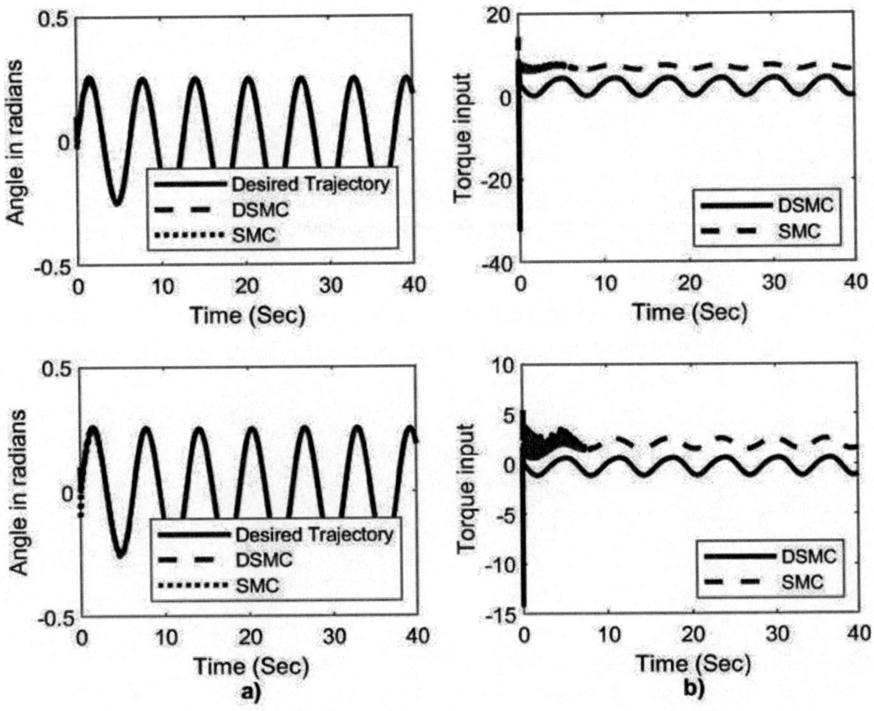

Fig. 5 **a** Trajectory tracking of joint angles **b** Control torque supplied to the joint actuators when controlled using dynamic SMC and conventional SMC

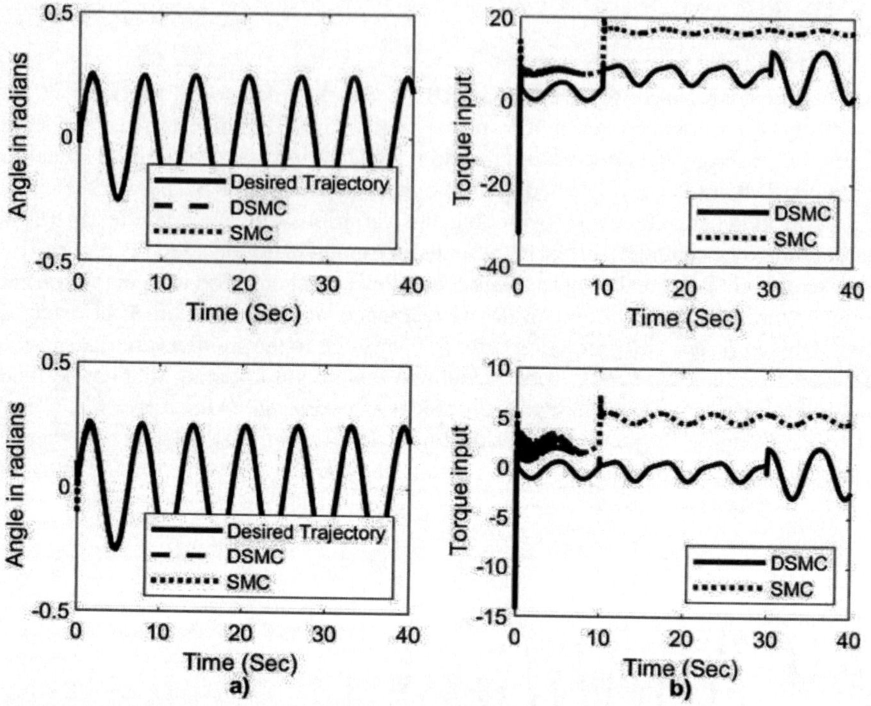

Fig. 6 **a** Angular trajectory tracking of joint 1 and 2. **b** Torque input to the actuators of joint 1 and 2, with parameter variations and change in disturbance amplitude

chattering compared to conventional SMC. Also, simulation results illustrate the ability of the designed techniques to suppress the effect of variation in payload mass and the changes in external disturbances.

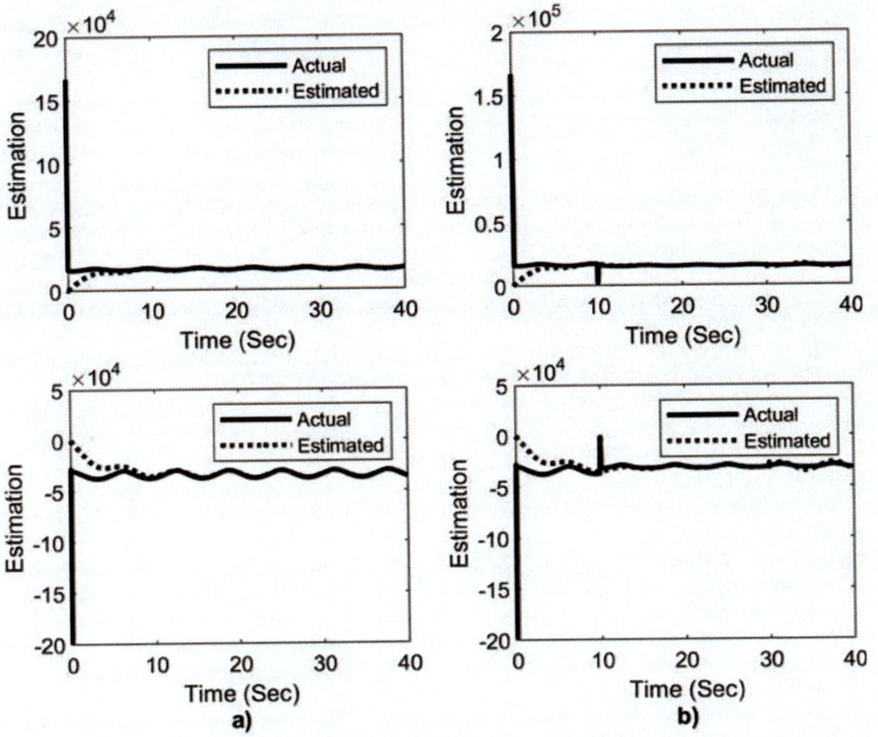

Fig. 7 RBF network predictions for system dynamics with parametric variations and magnitude change in external disturbance

References

1. Lee HW (2020) The study of mechanical arm and intelligent robot. IEEE Access 8:119624–119634. https://doi.org/10.1109/ACCESS.2020.3003807
2. Kong L, Lai Q, Ouyang Y, Li Q, Zhang S (2020) Neural learning control of a robotic manipulator with finite-time convergence in the presence of unknown backlash-like hysteresis. IEEE Trans Syst, Man, Cybern: Syst 52(3):1916–1927. https://doi.org/10.1109/TSMC.2020.3034757
3. Yang C, Huang D, He W, Cheng L (2020) Neural control of robot manipulators with trajectory tracking constraints and input saturation. IEEE Trans Neural Netw Learn Syst 32(9):4231–4242. https://doi.org/10.1109/TNNLS.2020.3017202
4. Nohooji HR (2020) Constrained neural adaptive PID control for robot manipulators. J Franklin Inst 357(7):3907–3923. https://doi.org/10.1016/j.jfranklin.2019.12.042
5. Ngo T, Wang Y, Mai TL, Nguyen MH, Chen J (2014) Robust adaptive neural-fuzzy network tracking control for robot manipulator. Int J Comput Commun Control 7(2):341–352
6. Wai RJ, Chen PC (2006) Robust neural-fuzzy-network control for robot manipulator including actuator dynamics. IEEE Trans Ind Electron 53(4):1328–1349. https://doi.org/10.1109/TIE.2006.87829
7. Islam S, Liu XP (2010) Robust sliding mode control for robot manipulators. IEEE Trans Ind Electron 58(6):2444–2453. https://doi.org/10.1109/TIE.2010.2062472

8. Xiao B, Cao L, Xu S, Liu L (2020) Robust tracking control of robot manipulators with actuator faults and joint velocity measurement uncertainty. IEEE/ASME Trans Mechatron 25(3):1354–1365. https://doi.org/10.1109/TMECH.2020.2975117
9. Hussain MA, Ho PY (2004) Adaptive sliding mode control with neural network based hybrid models. J Process Control 14(2):157–176. https://doi.org/10.1016/S0959-1524(03)00031-3
10. Koshkouei AJ, Burnham KJ, Zinober AS (2005) Dynamic sliding mode control design. IEEE Proc-Control Theory Appl 152(4):392–396. https://doi.org/10.1049/ip-cta:20055133
11. Zhai J, Xu G (2020) A novel non-singular terminal sliding mode trajectory tracking control for robotic manipulators. IEEE Trans Circ Syst II: Express Briefs 68(1):391–395. https://doi.org/10.1109/TCSII.2020.2999937
12. Guo Y, Woo PY (2003) An adaptive fuzzy sliding mode controller for robotic manipulators. IEEE Trans Syst, Man, Cybern-Part A: Syst Humans 33(2):149–159. https://doi.org/10.1109/TSMCA.2002.805804
13. Zhao D, Li S, Gao F (2009) A new terminal sliding mode control for robotic manipulators. Int J Control 82(10):1804–1813. https://doi.org/10.1080/00207170902769928
14. Yang C, Li Z, Li J (2012) Trajectory planning and optimized adaptive control for a class of wheeled inverted pendulum vehicle models. IEEE Trans Cybern 43(1):24–36. https://doi.org/10.1109/TSMCB.2012.2198813
15. He W, Ge SS, Li Y, Chew E, Ng YS (2015) Neural network control of a rehabilitation robot by state and output feedback. J Intel Robot Syst 80(1):15–31. https://doi.org/10.1007/s10846-014-0150-6
16. Colbaugh R, Glass K, Seraji H (1996) Adaptive tracking control of manipulators: theory and experiments. Robot Comput-Integr Manuf 12(3):209–216. https://doi.org/10.1016/0736-5845(96)00014-2
17. Yin X, Pan L (2018) Enhancing trajectory tracking accuracy for industrial robot with robust adaptive control. Robot Comput-Integr Manuf 51:97–102. https://doi.org/10.1016/j.rcim.2017.11.007
18. Yang ZR (2006) A novel radial basis function neural network for discriminant analysis. IEEE Trans Neural Netw 17(3):604–612. https://doi.org/10.1109/TNN.2006.873282
19. Shuzhi SG, Hang CC, Woon L (1997) Adaptive neural network control of robot manipulators in task space. IEEE Trans Ind Electron 44(6):746–752. https://doi.org/10.1109/41.649934
20. Abdelatti M, Yuan C, Zeng W, Wang C (2018) Cooperative deterministic learning control for a group of homogeneous nonlinear uncertain robot manipulators. Sci China Inform Sci 61(11):1–19. https://doi.org/10.1007/s11432-017-9363-y
21. Van Cuong P, Nan WY (2016) Adaptive trajectory tracking neural network control with robust compensator for robot manipulators. Neural Comput Appl 27(2):525–536. https://doi.org/10.1007/s00521-015-1873-4

Pitch Channel Trajectory Tracking Control of an Autonomous Underwater Vehicle

Ravishankar P. Desai and Narayan S. Manjarekar

Abstract The trajectory tracking control problem of pitch channel dynamics of an autonomous underwater vehicle (AUV) is considered. A linear extended state observer (ESO)-based sliding mode control (SMC) law is proposed under the influence of lumped disturbances. A linear ESO is designed to estimate the lumped disturbances of pitch channel dynamics in real time. A trajectory tracking control law is designed using SMC based on the estimated states. Combining linear ESO with SMC enhances the antidisturbance capability and minimizes the chattering effect of SMC. An experimentally verified pitch channel motion parameters of ODRA-I AUV is used. For robust analysis, the proposed linear ESO-based SMC law is validated under the influence of lumped disturbances. The Lyapunov theory is applied for stability analysis. The efficacy of the proposed control law is proven by comparing it with the benchmark proportional-derivative (PD) control law. The simulation results are demonstrated to support the proposed design.

Keywords Autonomous underwater vehicle · Linear extended state observer · Sliding mode control · Proportional-derivative controller · Linear system

1 Introduction

1.1 Background and Motivation

In marine technology, AUV is one of the essential solution to understand better and explore the underwater environment [1]. Due to its autonomy, AUV plays an essential role in number of potential applications such as military, navy, civil, scientific, survey, oil and gas [2]. To accomplish the potential application, AUV faces the following challenges are [1, 2]

R. P. Desai (✉) · N. S. Manjarekar
Department of Electrical and Electronics Engineering, BITS Pilani, K. K. Birla Goa Campus, NH 17 B Bypass Road, Sancoal, South Goa 403 726, Goa, India
e-mail: p20180019@goa.bits-pilani.ac.in

S. Sharma et al. (eds.), *Intelligent Control, Robotics, and Industrial Automation*, Lecture Notes in Electrical Engineering 1066,
https://doi.org/10.1007/978-981-99-4634-1_22

- System having highly nonlinear dynamics,
- Non-structural and structural uncertainties,
- Model parameters depend on underwater environmental conditions,
- The disturbances in an underwater environment.

The challenges mentioned above demand a robust control strategy. A robust controller aims to design a control law such that the controlled system guarantees the closed-loop stability and performance irrespective of uncertainties, modelling error, unmodelled dynamics in the model dynamics within a predefined limit.

Due to the high level of modelling complexity and its uncertain operational environment, it suffers modelling inaccuracies, including unmodelled dynamics, modelling errors, parameter uncertainties and external disturbances referred to as lumped uncertainty. Therefore, it is necessary to estimate and reject the effect of the lumped disturbances before it will affect the controller performance. The ESO is one class that helps to estimate the lumped disturbances by treating it as an extended state in its design [3]. The online estimation of the extended state depends on observer parameters, and selecting appropriate observer parameters is possible only when its real-time values are bounded. Furthermore, SMC is one of the promising robust controller technique for robust analysis. It is insensitive to matched uncertainties and disturbances that gives guarantees of the control-loop stability and controller performance [4]. However, switching action in control law causes chattering and bounds of uncertainties known to be before robustness. This paper synthesizes a linear ESO-based SMC law to solve the trajectory tracking control problem under lumped disturbances.

1.2 Related Work

The diving motion dynamics of REMUS 100 AUV is controlled using proportinal -derivative plus proportinal controller in [5]. ESO's basics and detailed insight refer to [3]; for SMC refer to [4]. The ESO-based SMC has been applied in many real-time applications. These applications include motors [6, 7], mobile robot [8, 9], underwater robot [10], quadrotor [11], nuclear power plant [12], pneumatic cylinder [13], etc. A speed control problem of permanent magnet synchronous motor is considered, and ESO-based SMC law is proposed in [6] to reject the external disturbances and ensure the motor's dynamic performance. Two inertia systems problem consist of the applied load and its effect on motor speed. In [7], an ESO-based adaptive integral SMC law is proposed to eliminate the load vibrations as the inertia of the load increases. The control problem based on trajectory tracking of a mobile robot is considered. A linear ESO-based adaptive SMC law is applied to the experimental set-up of a mobile robot in [8] and a nonlinear ESO-based adaptive SMC law in [9]. An adaptive multiple input multiple outputs (MIMO) ESO-based integral SMC is designed in [10] for a fully actuated underwater robot for trajectory tracking control. The control performance is validated experimentally under unmeasured velocity and unknown disturbances. The proposed controller's performance is verified under the influence of parametric variations and unknown disturbances. The attitude control problem of a quadrotor is considered, and an ESO-based integral SMC law is proposed in [11] under the influence of lumped disturbances. The control problem for power level of the nuclear power plant is considered, and an ESO-based adaptive dynamic SMC law is proposed in [12] to improve control performance in the presence of output disturbance. The tracking control method using nonlinear ESO-based integral SMC is applied on pneumatic rod cylinder servo system with varying loads and friction in [13] and verified with the experimental results.

1.3 Paper Contribution

Motivated by the proposed control law validated in the above literature and keeping in mind the AUV challenges, this paper presents a robust linear ESO-based SMC for pitch channel dynamics of an ODRA-I AUV. The paper contributions are

1. A linear ESO is designed to estimate the lumped disturbances of pitch channel dynamics of AUV in real time.
2. A SMC law is designed for trajectory tracking control under lumped disturbances.
3. The proposed linear ESO-based SMC law provide a good closed-loop control performance and faster convergence rate for reference signal tracking.
4. The closed-loop stability is proven in the sense of Lyapunov's stability theory.

1.4 Paper Summary

Section 2 presents AUV modelling for pitch channel. A linear ESO and SMC law is designed in Sect. 3. Section 4 demonstrates the simulation results. Lastly, Sect. 5 concluded the paper.

2 Autonomous Underwater Vehicle Modelling For Pitch Channel

An AUV operates in three-dimensional space and has 6-degrees of freedom. The two coordinated frames, body fixed (moving coordinated frame) and earth fixed (fixed coordinated frame) represent the motion of AUV as shown in Fig. 1. The relation between these frames forms the equations of motion stated as kinematics and dynamics. In this study, only AUV's diving motion dynamics are considered, including surge, heave and pitch, as given in Table 1. The [1] explored the detail

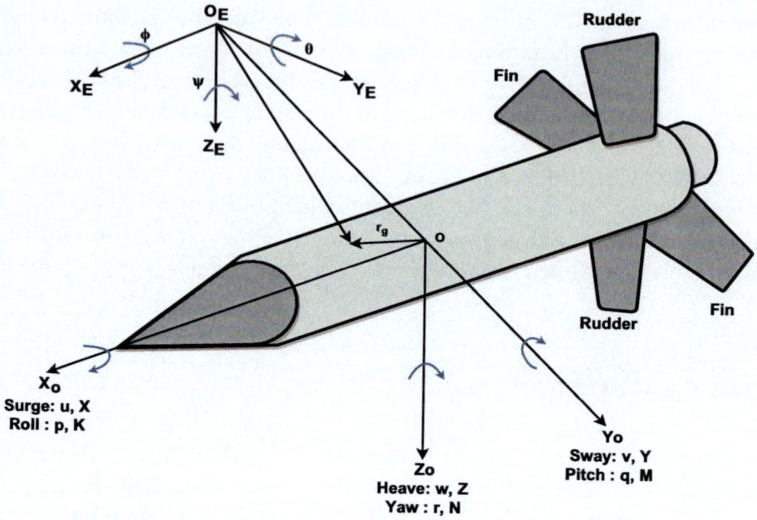

Fig. 1 AUV schematics

Table 1 Notation of AUV motion components for pitch channel dynamics [1]

DOF	Description	Earth-fixed forces and moments	Body-fixed linear and angular velocity	Body-fixed position and orientation angle
1	Surge motion	X	u	x
2	Heave motion	Z	w	z
3	Pitch rotation	M	q	θ

of mathematical modelling of an AUV. The desired depth position of the AUV is achieved by designing a robust control law on pitch channel dynamics. The pitch channel dynamics are obtained from [5] is written as

$$\begin{cases} \dot{x}_1(t) = x_2(t) \\ \dot{x}_2(t) = P_0 x_1(t) + P_1 x_2(t) + Z_0 U(t) + d(t), \end{cases} \tag{1}$$

where $x_1(t) = \theta(t)$ is the pitch angle, $x_2(t) = q(t)$ is the pitch rate, $U(t) = \delta_s(t)$ is the stern deflection, $d(t)$ is denoted as lumped disturbances, and $P_0 = \frac{M_q}{I_y - M_{\dot{q}}}$, $P_1 = \frac{M_\theta}{I_y - M_{\dot{q}}}$, $Z_0 = \frac{M\delta_s}{I_y - M_{\dot{q}}}$, where I_y is the pitch inertia, M_θ, $M_{\dot{q}}$, M_q, M_{δ_s} is the hydrostatic, added mass, combined term and fin lift, respectively.

The linear ESO and SMC law design is presented in the next section.

3 Linear Extended State Observer-Based Sliding Mode Control Law Design

The pitch channel dynamics of the vehicle get affected by lumped disturbances. Hence, there is a need to estimate it via estimator design. Based on the estimated states, a trajectory tracking control law is designed using a robust controller.

3.1 *Design of Observer Using Linear ESO*

A linear ESO is designed to estimate the lumped disturbances of pitch channel dynamics in real time. In linear ESO framework [3], an augmented variable $x_3(t) = P_0x_1(t) + P_1x_2(t) + d(t)$ is introduced in (1).

Using (1) with an augmented variable ($x_3(t)$) gives an extended state equation as

$$\begin{cases} \dot{x}_1(t) = x_2(t) \\ \dot{x}_2(t) = x_3(t) + Z_0U(t) \\ \dot{x}_3(t) = \zeta(t) \\ y(t) = x_1(t), \end{cases} \tag{2}$$

with $\zeta(t) = \frac{d}{dt}(P_0x_1(t) + P_1x_2(t) + d(t))$, i.e. rate of change of lumped disturbance.

Assumption 1 The extended state $\zeta(t)$ is unknown but bounded function.

Now, the extended state $\zeta(t)$ can be estimated using state observer. For (2), the state estimation of designed linear ESO as

$$\begin{cases} \dot{\hat{x}}_1(t) = \hat{x}_2(t) + \frac{\alpha_2}{\epsilon}(y(t) - \hat{x}_1(t)) \\ \dot{\hat{x}}_2(t) = \hat{x}_3(t) + \frac{\alpha_1}{\epsilon^2}(y(t) - \hat{x}_1(t)) - Z_0U(t) \\ \dot{\hat{x}}_3(t) = \frac{\alpha_0}{\epsilon^3}(y(t) - \hat{x}_1(t)), \end{cases} \tag{3}$$

where $\hat{x}_1(t),\ \hat{x}_2(t),\ \hat{x}_3(t)$ are the state estimates of $x_1(t),\ x_2(t),\ x_3(t)$, respectively, and $\alpha_2,\ \alpha_1,\ \alpha_0$ are the gains of the observer that need to be designed. The $\alpha_2,\ \alpha_1,\ \alpha_0$ are positive and $\epsilon << 1$ but $\neq 0$. The design goal of the observer is $\lim_{t\to\infty} \hat{x}_3(t) \to x_3(t)$, $\lim_{t\to\infty} \hat{x}_2(t) \to x_2(t)$, $\lim_{t\to\infty} \hat{x}_1(t) \to x_1(t)$. An observer error is defined as

$$\dot{\tilde{x}}(t) = \begin{bmatrix} \dot{x}_1(t) - \dot{\hat{x}}_1(t) \\ \dot{x}_2(t) - \dot{\hat{x}}_2(t) \\ \dot{x}_3(t) - \dot{\hat{x}}_3(t) \end{bmatrix} = \begin{bmatrix} \dot{\tilde{x}}_1(t) \\ \dot{\tilde{x}}_2(t) \\ \dot{\tilde{x}}_3(t) \end{bmatrix} \tag{4}$$

that satisfies

$$\begin{cases} \dot{\tilde{x}}_1(t) = -\alpha_2\tilde{x}_1(t) + \tilde{x}_2(t) \\ \dot{\tilde{x}}_2(t) = -\alpha_1\tilde{x}_1(t) + \tilde{x}_3(t) \\ \dot{\tilde{x}}_3(t) = -\alpha_0\tilde{x}_1(t) + \epsilon\zeta(t) \\ y(t) = x_1(t). \end{cases} \tag{5}$$

In the absence of lumped disturbance $d(t)$, an asymptotic error convergence is achieved by designing linear ESO gains such that

$$\dot{\tilde{x}}(t) = A\tilde{x}(t) = \begin{bmatrix} -\alpha_2 & 1 & 0 \\ -\alpha_1 & 0 & 1 \\ -\alpha_0 & 0 & 0 \end{bmatrix} \begin{bmatrix} \tilde{x}_1(t) \\ \tilde{x}_2(t) \\ \tilde{x}_3(t) \end{bmatrix} \tag{6}$$

is Hurwitz. Considering (2), matrix 'A' in (6) is Hurwitz for any positive values of $\alpha_2,\ \alpha_1,\ \alpha_0$, then the observer error $\tilde{x}(t)$ will converge to origin exponentially [14]. The following equation represents the convergence precision of observer error $\tilde{x}(t)$ is

$$\|\tilde{x}(t)\| \leq \max(\beta e^{-\gamma(t-t_0)})\|\tilde{x}(t_0)\|, \tag{7}$$

where $\beta,\ \gamma$ are positive constant. Equation (7) represents that a convergence precision is related to γ. Moreover, smaller value of ϵ in (3) represents the more rapid convergence of $\|\tilde{x}(t)\|$. Thus, the observer in (3) will improve the convergence precision of $\|\tilde{x}(t)\|$ greatly.

3.2 *Design of Controller Using SMC*

The design aim of SMC is to propose a suitable control law $U(t)$ such that output $y(t)$ tracks the desired signal asymptotically under lumped disturbances. Using [4], let the tracking error $e_1(t)$ is defined as

$$e_1(t) := x_{1d}(t) - x_1(t), \tag{8a}$$

where $x_{1d}(t)$ is the desired state. Taking time derivative of (8a)

$$\dot{e}_1(t) = \dot{x}_{1d}(t) - x_2(t) \tag{8b}$$

and the time derivative of (8b)

$$\ddot{e}_1(t) = \dot{e}_2(t) = \ddot{x}_{1d}(t) - \ddot{x}_1(t) = \ddot{x}_{1d}(t) - \dot{x}_2(t) = \ddot{x}_{1d}(t) - \zeta(t) + Z_0U(t). \tag{8c}$$

For (1), the sliding manifold is chosen as

$$S_1 = c_1e_1(t) + e_2(t) = 0, \qquad c_1 > 0. \tag{9}$$

To achieve asymptotic convergence of $\lim_{t\to\infty} e_1 = 0$, $\lim_{t\to\infty} e_2 = 0$with

$$e_1(t) = e_1(0)\exp(-c_1(t)),$$
$$e_2(t) = e_2(0)\exp(-c_1(t))$$

under an unknown disturbance, we have to drive the $S_1 \to 0$ in finite time by means of a control law $U(t)$. The reaching condition is obtained by differentiating (9) and substituting (8) as

$$\dot{S}_1 = c_1\dot{e}_1(t) + \dot{e}_2(t) = c_1\dot{e}_1(t) + \ddot{x}_d(t) - \zeta(t) + Z_0U(t). \tag{10}$$

For (10), the Lyapunov function candidate is chosen as

$$V_0 = \frac{1}{2}S_1^2. \tag{11}$$

Taking the time derivative of (11) and substituting (10), we get

$$\dot{V}_0 = S_1\dot{S}_1 = S_1[c_1\dot{e}_1(t) + \ddot{x}_d(t) - \zeta(t) + Z_0U(t)], \tag{12}$$

and to guarantee $S_1\dot{S}_1 < 0$, the SMC law is selected as

$$U(t) = -\frac{1}{Z_0}[c_1\dot{\hat{e}}_1(t) + \ddot{x}_d(t) - \hat{\zeta}(t) - \epsilon_0 \operatorname{sat}(\hat{S}_1)], \tag{13}$$

where ϵ_0 is the design parameter and its positive. Then, substituting (13) into (12), we get

$$\dot{V}_0 = [c_1\dot{\tilde{e}}_1(t) - \tilde{\zeta}(t) - \epsilon_0 \operatorname{sat}(S_1 - \tilde{S}_1)], \tag{14}$$

where $\dot{\tilde{e}}_1(t) = \dot{e}_1(t) - \dot{\hat{e}}_1(t)$, $\tilde{\zeta}(t) = \zeta(t) - \hat{\zeta}(t)$, and $\tilde{S}_1 = S_1 - \hat{S}_1$. Equation (14) satisfies

$$\dot{V}_0 = -\epsilon_0 \operatorname{sat}(S_1) < 0, \tag{15}$$

because of the convergence of the ESO is $c_1\dot{\tilde{e}}_1(t) - \tilde{\zeta}(t) - \epsilon_0 \operatorname{sat}(\tilde{S}_1)$ sufficiently small and bounded. Hence, SMC law in (13) ensures asymptotic stability of closed-loop system if $e_1 \to 0$ as $t \to \infty$.

Remark 1 A saturation function $\operatorname{sat}(S_1)$ is used inplace of discontinuous signum function $\operatorname{sign}(S_1)$ in (13) to reduce the chattering effect of $\operatorname{sign}(S_1)$. The $\operatorname{sat}(S_1)$ is defined as

$$\operatorname{sat}(S_1) := \begin{cases} +1, & S_1 > +\Delta \\ S_1/\Delta, & -\Delta \le S_1 \le \Delta \\ -1, & S_1 < -\Delta, \end{cases} \tag{16}$$

where Δ is a design constant that represents the boundary layer. The convergence result of (13) does not affect the choice of (16).

Theorem 1 *If the control law designed in (13) with observer designed in (3) are applied to pitch channel motion dynamics of an AUV in (1), then the angular velocity error as defined in (8) will asymptotically converge to the zero as* $t \to \infty$ *is guaranteed.*

Proof Based on the arguments given above.

4 Results

The simulation results of the proposed linear ESO-based SMC law under lumped disturbances are demonstrated in this section. The trajectory tracking control law proposed in (13) with the observer in (3) is applied to pitch channel motion dynamics of an AUV described in (1). Equation (16) is used to reduce the chattering phenomenon by approximating the signum function of (13). A MATLAB/Simulink (R2021a) is used for all the simulation work. The motion parameters for the ODRA-I AUV adopted in this work and pitch channel motion parameters are similar to those described in [15]. In this simulation study, the pitch angle tracking signal is chosen as $\sin(t)$, and the lumped disturbance $(d(t))$ acting on the channel is considered as $1.5 * \sin(t - \pi/2)$.

The vehicle initial states are $[\theta(0), q(0)]^T = [0, 0]^T$ and observer initial states are $[\hat{\theta}(0), \hat{q}(0)]^T = [0, 0]^T$ chosen. The following design parameters are chosen for an observer design (3), SMC law (13) and PD control law as

- ESO: $\alpha_2 = 6,\ \alpha_1 = 11,\ \alpha_0 = 6,\ \epsilon = 0.1$.
- SMC: $c_1 = 1,\ \epsilon_0 = 0.01$.
- PD control: $K_p = 0.1, K_d = 0.05$.

The controller performance is evaluated by considering the following two scenarios. The simulations are carried out by considering known pitch dynamics with lumped disturbances in the first scenario. In the second scenario, pitch dynamics are entirely unknown with lumped disturbances.

4.1 *First Scenario: In the Presence of Known States with Lumped Disturbances*

The controller performs close trajectory tracking in a known state with lumped disturbances, as shown in Fig. 2. Three subplots from Fig. 2a to c show the pitch angle, pitch rate and stern plane deflection, i.e. control signal, respectively. Figure. 2a confirms that the proposed control law rejects the lumped disturbances. Compared with

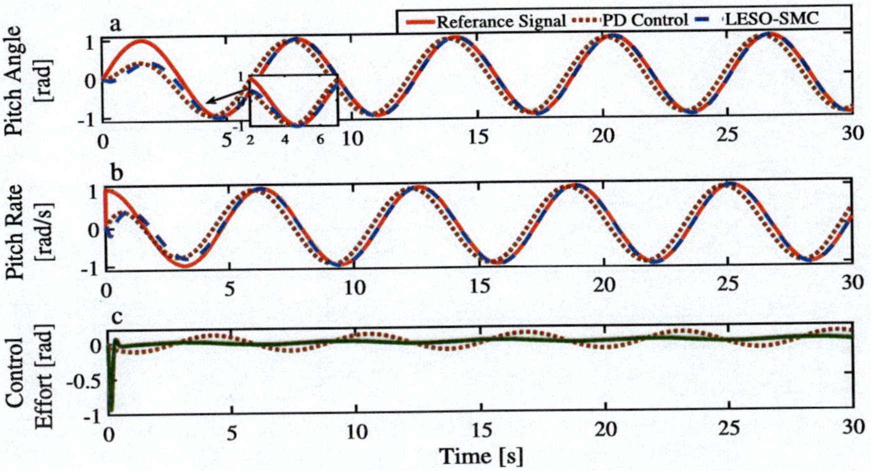

Fig. 2 Pitch channel trajectory tracking control under known state with disturbance

the PD control law, a proposed control law shows that a close trajectory tracking control performance is achieved < 4 s. The pitch tracking rate is precise to pitch rate reference signal, as seen in Fig. 2b. The control law utilizes minimum control energy, i.e. it does not violate the actuator ability (30°) as seen in Fig. 2c.

Figure 3 shows the state estimation $x1(t)$, $x2(t)$ with extended state estimation $x3(t)$. In Fig. 3, three subplots from Fig. 3a to c show the state estimation of pitch angle, pitch rate and lumped disturbances ($d(t)$), respectively. The extended state $x3(t)$ is estimated precisely with actual extended state, whereas state $x1(t)$ and $x2(t)$ are estimated accurately with actual states. The state estimation proves the proposed control law precisely estimates the extended state.

4.2 *Second Scenario: In the Presence of an Unknown States with Lumped Disturbances*

The controller performs a relative trajectory tracking in an unknown state with lumped disturbances in Fig. 4. A proposed control law shows that a relative/approximate trajectory tracking control performance is achieved < 6 s. Like the first scenario, a similar kind of analysis holds for the second scenario but a tracking error in pitch angle and pitch rate is observed in Fig. 4b and c as compared with Fig. 2b and c. It also observed a little change in the extended state estimation rate with an estimation error in Fig. 5c compared with Fig. 3c.

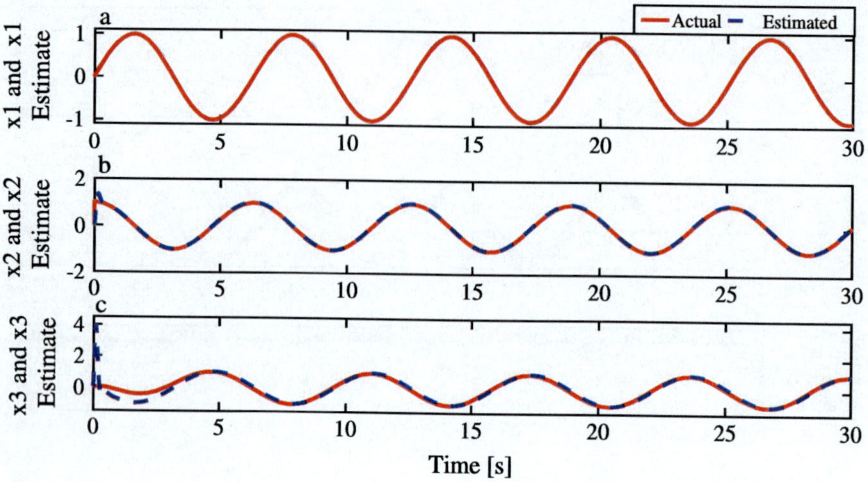

Fig. 3 State estimation of pitch channel trajectory tracking control under known state with disturbance

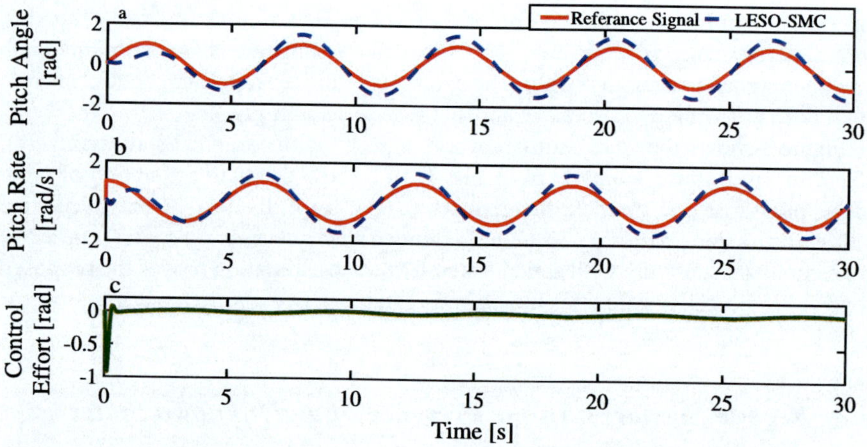

Fig. 4 Pitch channel trajectory tracking control under unknown state with disturbance

The overall simulation results of trajectory tracking prove the validity and operational capability of the pitch channel for ODRA-I AUV. Hence, it confirms the closed-loop stability and control performance of the pitch channel for ODRA-I AUV under lumped disturbances, considering known and unknown states.

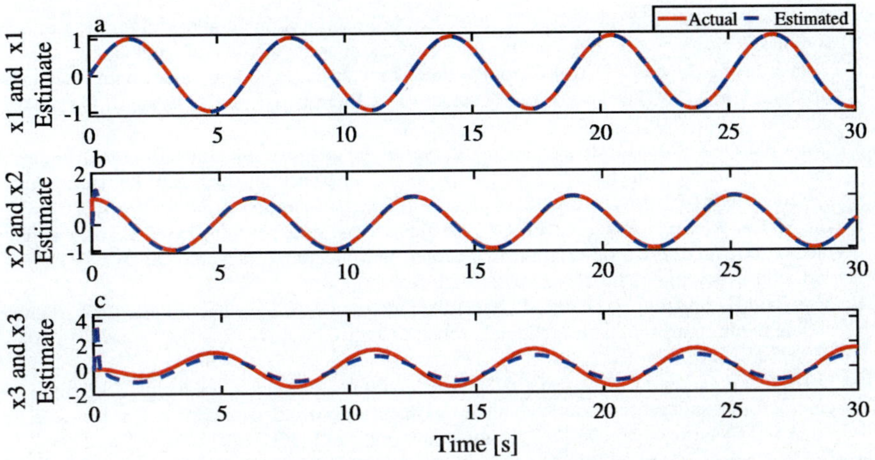

Fig. 5 State estimation of pitch channel trajectory tracking control under an unknown state with disturbance

5 Conclusion

In this paper, pitch channel dynamics of an ODRA-I AUV is considered. A linear ESO is combined with SMC to provide solution of the trajectory tracking control problem. A linear ESO is designed to estimate the lumped disturbances of pitch channel dynamics in real time. A robust trajectory tracking control law is designed using SMC based on the estimated states. The proposed linear ESO-based SMC law provides better closed-loop stability and performance under lumped disturbances with known and unknown pitch channel dynamics. The proposed control law minimizes the chattering effect and enhances the anti disturbance capability of SMC. The efficacy of the proposed linear ESO-based SMC law is validated against the benchmark controller PD control law. The simulation results are demonstrated to support the proposed control law design.

References

1. Fossen TI (1999) Guidance and control of ocean vehicles. University of Trondheim, Norway, Printed by Wiley, Chichester, England, ISBN: 0 471 94113 1, Doctors Thesis
2. Sahoo A, Dwivedy SK, Robi P (2019) Advancements in the field of autonomous underwater vehicle. Ocean Eng 181:145–160. https://doi.org/10.1016/j.oceaneng.2019.04.011
3. Guo BZ, Zhao ZL (2016) Active disturbance rejection control for nonlinear systems: an introduction. Wiley
4. Shtessel Y, Edwards C, Fridman L, Levant A et al (2014) Sliding mode control and observation, vol 10. Springer

5. Prestero TTJ (2001) Verification of a six-degree of freedom simulation model for the REMUS autonomous underwater vehicle. Ph.D. thesis, Massachusetts institute of technology
6. Qian J, Xiong A, Ma W (2016) Extended state observer-based sliding mode control with new reaching law for PMSM speed control. Math Probl Eng 2016. https://doi.org/10.1155/2016/6058981
7. Gong F, Ren X (2015) Extended state observer based adaptive integral sliding mode control for two inertia system. In: 2015 7th international conference on intelligent human-machine systems and cybernetics, vol 1. IEEE, pp 483–486. https://doi.org/10.1109/IHMSC.2015.83
8. Cui M, Liu W, Liu H, Jiang H, Wang Z (2016) Extended state observer-based adaptive sliding mode control of differential-driving mobile robot with uncertainties. Nonlinear Dyn 83(1):667–683. https://doi.org/10.1007/s11071-015-2355-z
9. Moudoud B, Aissaoui H, Diany M (2022) Extended state observer-based finite-time adaptive sliding mode control for wheeled mobile robot. J Control Decis 1–12. https://doi.org/10.1080/23307706.2021.2024458
10. Cui R, Chen L, Yang C, Chen M (2017) Extended state observer-based integral sliding mode control for an underwater robot with unknown disturbances and uncertain nonlinearities. IEEE Trans Ind Electron 64(8):6785–6795. https://doi.org/10.1109/TIE.2017.2694410
11. Cui R, Chen L, Yang C, Chen M (2017) Extended state observer-based integral sliding mode control for an underwater robot with unknown disturbances and uncertain nonlinearities. IEEE Trans Ind Electron 64(8): 6785–6795. https://doi.org/10.23919/ChiCC.2018.8484168
12. Hui J, Ge S, Ling J, Yuan J (2020) Extended state observer-based adaptive dynamic sliding mode control for power level of nuclear power plant. Ann Nucl Energy 143:107417. https://doi.org/10.1016/j.anucene.2020.107417
13. Zhao L, Zhang B, Yang H, Wang Y (2018) Observer-based integral sliding mode tracking control for a pneumatic cylinder with varying loads. IEEE Trans Syst, Man, Cybern: Syst 50(7):2650–2658. https://doi.org/10.1109/TSMC.2018.2825325
14. Khalil K (1996) Nonlinear systems, 2nd edn. Englewood Cliffs, NJ, Prentice-Hall
15. Mahapatra S, Subudhi B, Rout R, Kumar BK (2016) Nonlinear h ∞ control for an autonomous underwater vehicle in the vertical plane. IFAC-PapersOnLine 49(1):391–395. https://doi.org/10.1016/j.ifacol.2016.03.085

Simplified Current Control Method for FOC of Permanent Magnet Synchronous Motor

Amit Mallikarjun Masuti, Sachin Angadi, A. B. Raju, and Sahana Kalligudd

Abstract The vector control method is widely used to control the permanent magnet synchronus motor because of its fast response and independent control over the decoupled rotor currents. During the modeling and design phase of the vector control method designing the current controller is an essential step. The accuracy in developing the current controller will increase the effectiveness of the vector control method. This paper presents a simplified current control approach based on PMSM motor model parameters to control the rotor currents of the permanent magnet synchronous motor (PMSM) using TI's launchpad F28069M and Altair embed (VisSim) software. The effect of the implemented current controller is evaluated using experimental results. Easy to understand and less modularity are the major advantages of this current controller.

Keywords TI's launchpad F28069M · PMSM · Vector control · Current controller · Altair embed(VisSim)

1 Introduction

Due to the high emergence of electric vehicles, motors have become one of the main topics in the field of research. Synchronous motors are believed to be suitable machines to run electric vehicles. Hence for the past several years, researchers have been interested in permanent magnet synchronous motor (PMSM) for driving and automotive applications, which offers several benefits over traditional induction motors and dc motors [1]. These motors operate at high speeds, have a high torque, and don't require any maintenance [2].

There are many methods to control the PMSM motor, like the variable voltage method, variable frequency control, etc. But the accuracy that is provided by the vector control method cannot be matched to any of the other methods [3]. Hence

A. M. Masuti (✉) · S. Angadi · A. B. Raju · S. Kalligudd
KLE Technological university, Hubli, India
e-mail: amitmasuti123@gmail.com

S. Sharma et al. (eds.), *Intelligent Control, Robotics, and Industrial Automation*,
Lecture Notes in Electrical Engineering 1066,
https://doi.org/10.1007/978-981-99-4634-1_23

the vector control method is the highly preferred method in real-time applications. The vector control method is very complex because it involves two loops, mainly speed and current loops. Implementing and designing these loops is challenging and requires in-depth knowledge of rotor currents and motor parameters. Coding the vector control algorithm for a particular microcontroller or Digital signal processors (DSP)-based platform is one of the most challenging parts of its implementation. Additionally, more effort is needed to arrange the necessary components for that platform. More experience and knowledge about available microcontrollers is needed during the selection of DSPs for the implementation phase.

Texas instruments' F28069M launchpad and the DRV8301 booster pack are used in this paper to overcome the effort and difficulties in implementing the vector control algorithm [4]. F28069M launchpad is mainly designed for motor control strategies and applications in the field of power electronics [3]. Many softwares are available to program this launchpad. Altair embed (VisSim) software, along with code composer studio, explored in this paper to program the F28069M. When we dive into the vector control algorithm, the designing of the current controller will be the primary and foremost goal to achieve. Implementing the vector control method is effortless if we can control the stator currents precisely. Other difficulties exist in directly implementing the FOC algorithm for PMSM, such as calibrating the current sensors along with the encoder and adjusting the repetition gains of all three PID controllers. It has been observed that the Field oriented control's (FOC) outcomes are unpredictable without proper calibration and adjustment of the control parameters. Therefore, this study provides a relatively simple approach to create the fundamental and crucial component of the FOC algorithm.

In recent years, many studies have emerged, primarily focusing on motor control techniques [5]. And it is discovered that most vector control publications focus more on the speed loop than the current loop, with no discussion of fine-tuning of the PID current controller [6]. This paper primarily assists people who implement vector control for PMSM motors but are trapped in the current loop. There are many methods to design and model the current controller and to obtain the k_p and k_i values. In this paper, an attempt is made to explain a method that is easier than existing methods [7]. The method which requires a finding pole is more complex and requires more knowledge about the control system to understand the controller governing equations. But the method discussed in this paper is very simple and is devoid of complex equations. Minimum knowledge about the control system is sufficient to understand the modeling of the controller. It is different than the existing PI controller and will accurately control the currents and yield good results.

Implementing open loop current regulation of the permanent magnet motor is the objective of this paper [8]. A synchronous frame current controller is modeled and put into practice to control the currents. The phrase "open loop" describes an operation in which the PM machine runs without receiving any rotor position feedback; instead, the angle is determined by a frequency that is externally provided. In reality, currents are controlled in a d-q reference frame. In theory, the i_q current should produce torque, whereas the i_q current should produce flux.

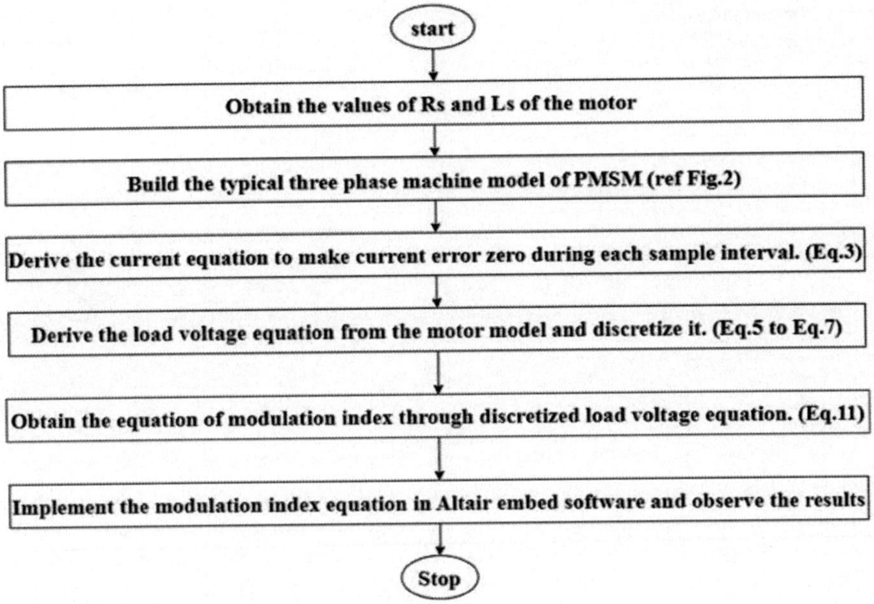

Fig. 1 Typical vector control model of PMSM motor

2 Design of Current Controller

Many steps are involved in the design process of the current controller. The flowchart illustrated in Fig. 1 gives an insight into the steps that must be followed so that we end up in designing an accurate current controller. In the design process, the motor parameters will play an important role. Because these will directly get involved in calculating the K_p and K_i values of the current controller. Hence extra care must be taken while measuring these motor parameters.

When discussing current control, it is necessary to understand the machine from the perspective of the three-phase terminals. As shown in Fig. 2, three-phase circuits that each consist of a sinusoidal EMF source e_1 linked in series with R_s and L_s stator components can be used to schematically represent the PMSM machine. The stator resistance and inductance are denoted by the parameters R_s and L_s for a permanent magnet machine, respectively. Three identically sized EMF sources with a 120 °C phase shift make up the circuit.

The Emf amplitude is given by

$$\hat{e} = \varphi P \omega_m \tag{1}$$

where ω_m , P are shaft speed and pole pair number of the machine. Basically the PMSM motor used will have non-salient pole rotor, hence flux in case of non-salient pole rotor is given by

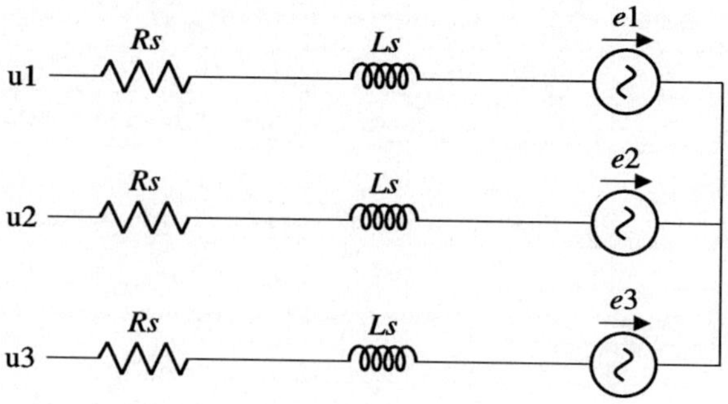

Fig. 2 Typical three-phase machine model

$$\varphi = \varphi_{\text{pm}} \tag{2}$$

The φ_{pm} is flux due to the permenant magnets. Designing the current controller is a important step, especially when using the vector control approach or other motor control strategies. The motor will draw more current and operate in an unstable manner if the current controller is improperly configured. There are three primary controllers in a control system: P, I, and D. These controllers can be used alone or in combination to control the currents. But because the output signal oscillates, we avoid employing the differentiator controller. Hence combination of proportional and integrator controller is used to control the current. The current controller is a discrete PI control module that accepts inputs for measured currents from a load represented by an inductance and resistance as well as reference currents. We are able to develop a sample controller structure that can be utilized to regulate the currents in R-L circuits in this case since the converter is a controlled voltage source for regularly sampled control systems as considered here.

It is designed to deliver average voltage per sample quantity $u(t_k)$ that corresponds to the set point $u^*(t_k)$ value that is supplied by a current control module. The developed converter is a mix of a modulator and a converter that can be represented by a voltage source. The goal of this control is to identify the average reference that will reduce the current error $i^*(t) - i(t)$ to zero, which results in the condition $i^*(t_1) - i(t_2)$ at time $t = t_1$. The current error needs to be zero during each sample interval. Hence we can write

$$i(t_{k+1}) = i^*(t_k) \tag{3}$$

for regularly sampled system with T_s as sampling time.
The modulator/converter block will satisfy the following condition

$$U^*(t_k) = \int_{t_k}^{t_{k+1}} u(\tau)\,\mathrm{d}t \tag{4}$$

u is load voltage, by observing motor model the load voltage equation can be written as

$$u = Ri + L\frac{di}{dt} \tag{5}$$

where R and L are the resistance and inductance of the motor model, respectively. From Eqs. 4 and 5

$$U^*(t_k) = \frac{R}{T_s}\int_{t_k}^{t_k+T_s} i(\tau)\,d\tau + \frac{L}{T_s}\int_{i(t_k)}^{i(t_k+T_s)} di \tag{6}$$

discretizing the Eq. 6

$$U^*(t_k) = Ri^*(t_k) + \frac{L}{T_s}\Delta i(t_k) \tag{7}$$

with $\Delta i(t_k) = i^*(t_k) + i(t_k)$ which allows this expression to be written as

$$U^*(t_k) = R\Delta i(t_k) + \frac{L}{T_s}\Delta i(t_k) + Ri(t_k) \tag{8}$$

which can also written as

$$U^*(t_k) = K_p\Delta i(t_k)(1 + \omega_i T_s) + Ri(t_k) \tag{9}$$

where $K_p = \frac{L}{T_s}$ $\omega_i = \frac{R}{L}$ are the current controller's proportional gain and bandwidth, respectively. The modulation index is defined by $m^*(t_k) = U^*(t_k)/(U_{DC}/2)$ is the modulator input in a drive. Where U_{DC} is the DC bus voltage.

$$m^*(t_k) = K_p\frac{\Delta i(t_k)}{(U_{DC}/2)}(1 + \omega_i T_s) + \frac{Ri(t_k)}{(U_{DC}/2)} \tag{10}$$

Where $m^*(t_k)$ is the value of the modulation index produced only by the integrator and sampled before the present value. A floating point-based controller was used to conduct the analysis up to this point. Scaling is necessary since a fixed point controller is commonly used in practice.

$$m^*(t_k) = (\frac{i_{fs}K_p}{u_{fs}})\frac{2\Delta i(t_k)}{(U_{DC}n)}(1 + \omega_i T_s) + mi(t_k) \tag{11}$$

Equation 11 is the designing equation of current controller. It yeilds modulation indices as the output so that the inverter will generate voltages for the motor.

3 Simulation and Experimental Results

This section discusses the experimental setup used to implement current controller along with the results. The motor's parameters are also stated because they are essential for designing controllers.

3.1 Exprimental Setup

A number of hardware components make up the experimental setup in this paper as shown in Fig. 3. The 3 ph, 28 V, 7 A, 6000 rpm PMSM is employed and has the specifications listed in Table 1. The PMSM is driven by the F28069M launchpad and the DRV8301 booster pack. To power the DRV8301 module, a DC power supply is necessary.

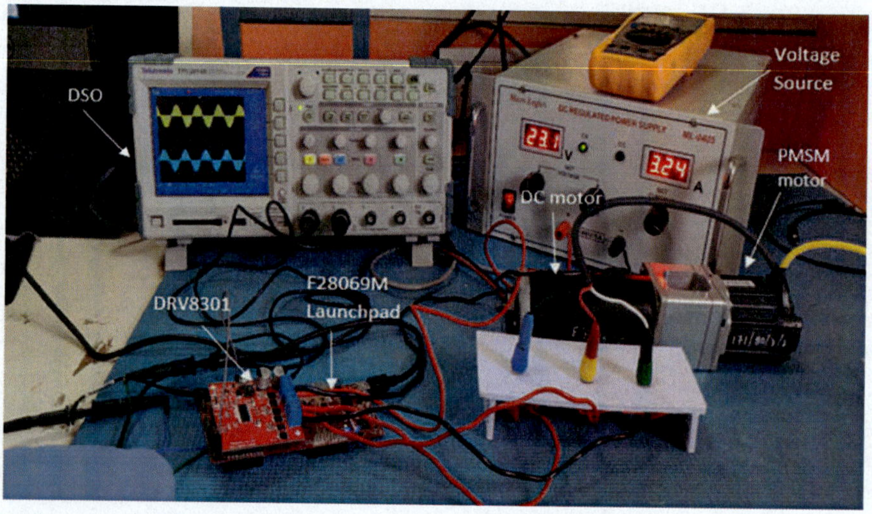

Fig. 3 Experimental setup

Table 1 TI's PMSM machine parameters

Parameter	Valaue	Units
Inductance: L_s	0.25	mH
Resistance: R_s	0.35	Ω
Magnetic flux: φ_{PM}	6.46	mWb
Pole pair: p	4	–
Inertia: J	50	µ kgm^2

Moreover, creating C code for the F28069M launchpad requires a laptop that can run code composer studio, control suite, and Altair Embed. Oscilloscopes and digital multimeters are other tools that are employed.

3.2 *Exprimental Results*

The main aim is to implement the current controller and obtain the same speed as specified in terms of frequency. Equation 11 is the equation of the current controller. It has to be used to control the currents and produce the modulation indices that are necessary to generate the voltages for the motor. So that currents of the motor are controlled by controlling the voltage. Equation 11 is the modulation index value generated by the current controller. In the present study, the two currents need to be controlled; hence this equation has to be separately implemented for d and q-axis currents so the controller will have two inputs and two outputs. The inputs are the d-axis current error Δi_d and q-axis current error Δi_q, and all other terms in equation two must be specified before simulating the controller. T_s is the time step for the simulation, in the model, we set it for 66.66us. It suggests that the discrete controller model uses a 15 kHz sampling frequency.

The simulation matches the user control parameters id = 2A, iq = 0A, and fref = 20Hz. As a result, the motor receives a 2.0 A rotating current vector at 1200 rpm. This suggests that the motor's actual shaft speed will be 300 rpm (Considering the motor's four pole pairs). Moving one level lower into the "open loop current controller" module reveals a group of modules where the designed current controller is the dominating module and all other modules, including Park's forward and reverse transformations, SVM, are involved. The reference value of i_d which will be compared with the current i_d obtained from the motor through the current sensor. Basically, we get i_a, i_b, and i_c from the launchpad with the help of current sensors. The i_a, i_b, and i_c are transformed into $i_\alpha$and i_β using Clark's transformation. These i_α and i_β are converted into the i_d and i_q so that they are compared with the given reference values and hence the error Δi_d and Δi_q are generated (Note that theta value for park's forward transformation is obtained through integrator module from user set frequency). Once the errors Δi_d and Δi_q are generated current controller will execute Eq. 11 and will provide us two outputs m_d and m_q.

Hence by considering all the values, the controller model will provide two modulation indexes as output. And these are named as m_d and m_q further these m_d and m_q are passed to the park's transformation to get rotational modulation indexes m_a and m_b finally m_A, m_B, and m_C are generated by SVM from the rotational modulation indexes. The modulation indexes created by SVM are responsible for generating the voltages for the motor through the inverter. Thetaoffset value is provided in the model to caliberate the encoder so that speed response of the motor can be visualized, and this value will varry from motor to motor.

The voltages produced by the inverter will be responsible for rotating the motor. Initially, a frequency 20 Hz was applied with i_d reference as 2A, so the speed response

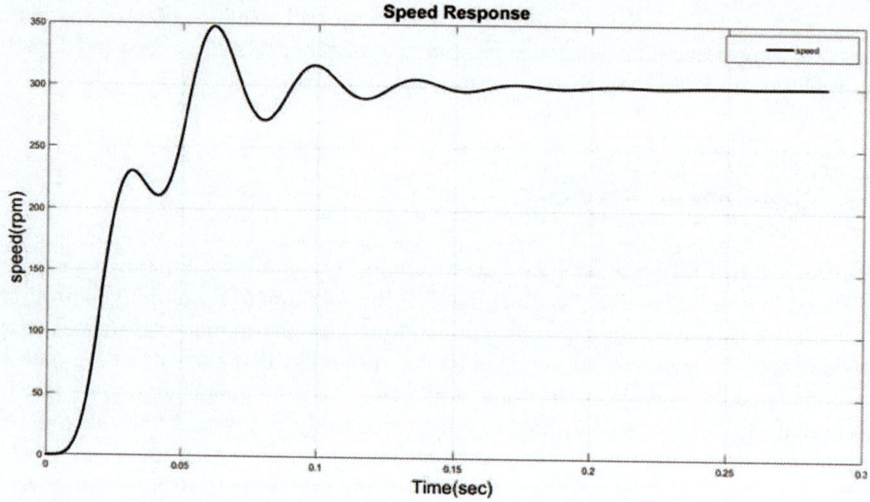

Fig. 4 Speed response of the PMSM motor

of the motor in Fig. 4 illustrates that the motor is rotating at 300 rpm. The oscillations can be seen, and after a certain period of time, output will settle for the provided reference value of speed. The speed of the motor is tracked with the help of the motor encoder, and it is displayed on the host PC using serial communication. But in the case of current waveforms monitor buffer is used to display the waveforms which are obtained through the current sensors.

The waveforms of i_α and i_β are shown in Fig. 5. These waveforms are obtained by clark's transform of i_a, i_b and i_c.

Similarly when the waveforms of i_d in Fig. 6 are observed, it can be seen that current i_d will track the given i_{dref} value, i.e., 2A. Basically, the i_d is obtained from the park's forward transformation of i_α and i_β. As stated earlier, the angle will be given from the user-specified frequency for the park's forward transformation. Hence we can say that the open loop current control has been achieved.

4 Conclusion

This paper has presented essential steps for the design of the current controller, which can be used in the implementation of the FOC of a PMSM using TI's F28069M launchpad and Altair Embed software. Details of the Simulation Platform, the hardware used, steps involved in implementation, and results are provided. By using this approach, the accuracy in controlling the currents is increased. A simple design and modeling process makes it easier than existing controlling algorithms. However, this controlling method will work fine when incorporated into the FOC algorithm and is

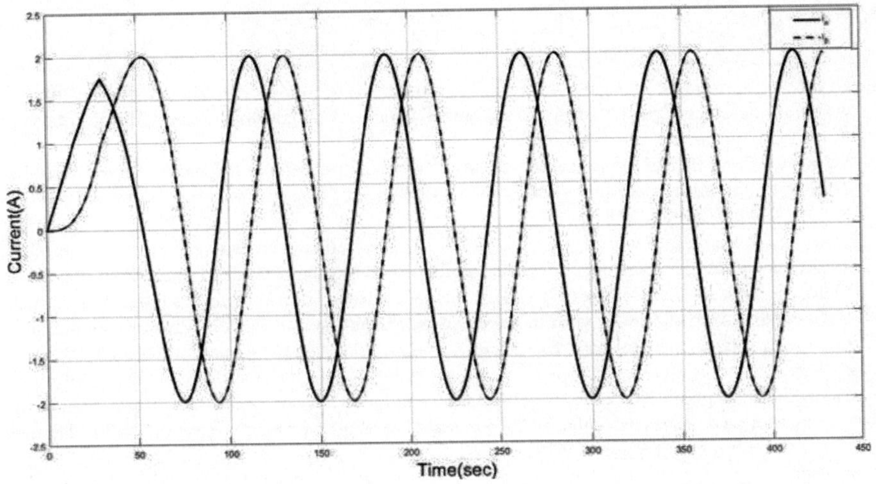

Fig. 5 Current waveforms(i_α and i_β)

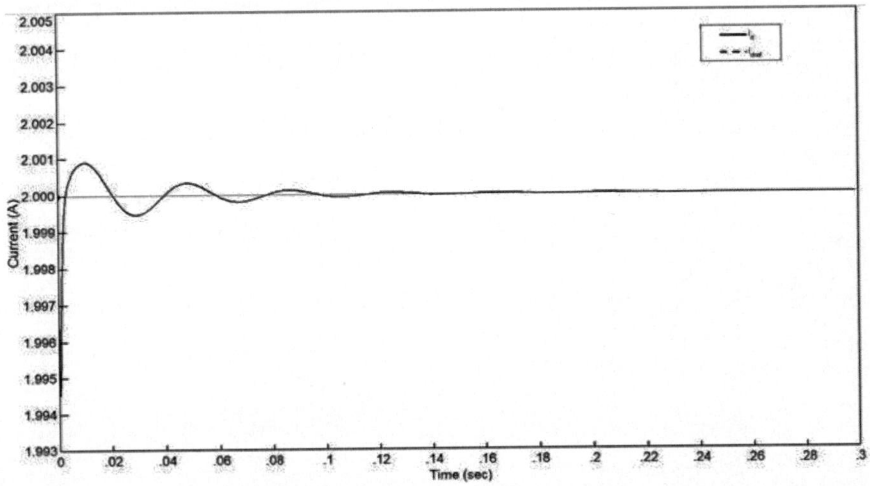

Fig. 6 Current waveforms (i_d and i_{dref})

unsuitable for any other motor speed control strategy. The current controller designed in this paper will be critical in the execution of field-oriented control of the PMSM motor and will govern the performance and efficiency of the FOC algorithm.

Acknowledgements Authors would like to thank KLE Technological University, Hubli—580 031 (INDIA) for funding this research project under Research Experience for Undergraduate (REU) scheme.

References

1. Qi F, Scharfenstein D, Weiss C, Müller C, Schwarzer U (2019) Motor handbook. Online: https://www.infineon.com/dgdl/Infineon-motorcontrol_handbook-AdditionalTechnicalInformation-v01_00-EN.pdf
2. Sinha R, Misra H (2021) Control of pmsm driven electric vehicle for indian drive cycle. In: 2021 National power electronics conference (NPEC). IEEE, pp 01–06. https://doi.org/10.1109/NPEC52100.2021.9672473
3. Rossi M, Toscani N, Mauri M, Dezza FC (2021) Introduction to microcontroller programming for power electronics control applications: coding with MATLAB® and simulink®. CRC Press
4. Mehta H, Apte A, Pawar S, Joshi V (2017) Vector control of pmsm using ti's launchpad f28069 and matlab embedded coder with incremental build approach. In: 2017 7th International conference on power systems (ICPS). IEEE, pp 771–775. https://doi.org/10.1109/ICPES.2017.8387393
5. Paitandi S, Sengupta M (2017) Design and implementation of sensorless vector control of surface mounted pmsm using back-emf estimation and pll based technique. In: 2017 National power electronics conference (NPEC). IEEE, pp 129–134. https://doi.org/10.1109/NPEC.2017.8310447
6. Jayal P, Rawat S, Bhuvaneswari G (2020) Simplified sensor based vector control of permanent magnet synchronous motor drive. In: 2020 IEEE International conference on power electronics, smart grid and renewable energy (PESGRE2020). IEEE, pp 1–6. https://doi.org/10.1109/PESGRE45664.2020.9070639
7. Sharma RK, Sanadhya V, Behera L, Bhattacharya S (2008) Vector control of a permanent magnet synchronous motor. In: 2008 Annual IEEE India conference, IEEE. vol 1, pp 81–8. https://doi.org/10.1109/INDCON.2008.4768805
8. Pulle DW, Darnell P, Veltman A (2015) Applied control of electrical drives. Switzerland, Springer pp, pp 18–31

Nonlinear Integral Sliding Mode Control for a Multivariable System, an Experimental Validation

S. Karthick

Abstract A nonlinear integral sliding mode controller design is proposed for a multivariable system which possesses inter-axis coupling effects. Real-time disturbances and model uncertainties cannot be predicted a priori in controller design. So, a linear fixed gain controller designed without the knowledge of such unknown disturbances cannot make the system to meet the desired output specifications always. As a measure to overcome this, a robust control algorithm is proposed. The proposed control algorithm is validated on a two degree of freedom helicopter system and the hardware results accentuate the merits of the proposed algorithm.

Keywords Nonlinear · SMC · Inter-axis coupling · Multivariable system

1 Introduction

In the process of modeling any real-time system, the mismatch between actual system and mathematical model is inevitable. These mismatches may be due to parasitic dynamics, unknown plant parameters and external disturbances. Several robust techniques like backstepping [1], H_∞ control [2] and sliding mode control [3–5] are developed over the years. Among them, sliding mode control (SMC) is found to be the most triumphant approach in handling parasitic dynamics and bounded uncertainties [6]. Sliding mode is revealed as a special mode in umbrella of variable structure systems. SMC has liberty in switching between structures to achieve the desired closed-loop response. This phenomenon is like combining the subsystems, and each subsystem will have a specific structured control that is going to be applicable for a specific operating region. The proper selection of sliding variables determines the performance of SMC, which is a custom-designed function [7]. The concept behind SMC is to maneuver the system trajectory toward the appropriately chosen

S. Karthick (✉)
School of Electrical and Electronics Engineering, VIT Bhopal University, Bhopal, Sehore, Madhyapradesh, India
e-mail: skpattitude@gmail.com

S. Sharma et al. (eds.), *Intelligent Control, Robotics, and Industrial Automation*, Lecture Notes in Electrical Engineering 1066,
https://doi.org/10.1007/978-981-99-4634-1_24

sliding manifold, and once it reaches the sliding manifold, it is to be maintained in the manifold with the aid of the control effort. Thereby, the system is insensitive to disturbances. In [8], adaptive SMC is implemented to perk up the performance of the maglev system and the efficiency of adaptive SMC is substantiated through simulation studies. The performance of SMC depends on the proper selection of the sliding surface. Large bodies of literatures are available for selecting the sliding surface. Linear Quadratic Regulator (LQR)-based sliding surface is designed for a rotary double-inverted pendulum to achieve the desired performance with optimal utilization of energy [9]. The constant parameters for the adaptive second-order SMC are selected through particle swarm optimization technique and the controller performance is validated through simulation studies [10]. Most of the proposed ideas available in the literature to improve the performance of SMC support only simulation studies. To date, there are still designs and implementation challenges pertain for complex systems.

The research purpose of this work is to get better tracking performance of the controller even under modeling uncertainties, worst-case disturbances and also to practically implement the proposed controller. For this purpose, a nonlinear integral sliding mode control (ISMC) using bipolar sigmoidal function is put forward. The efficiency of the proposed controller is validated through experimental studies in a 2 DoF helicopter. The 2 DoF helicopter chosen for this study is a complex benchmark apparatus with inter-axis coupling effects, which can be used to determine the performance of different controllers. The novelty lies in the practical implementation of the nonlinear ISMC for a complex multivariable system.

The research paper is arranged as follows. The mathematical model of the system is established in second section. The proposed nonlinear integral SMC and LQI controller are designed in third section. The hardware in loop simulation is executed and the outcomes are analyzed in fourth section. Fifth section represents the conclusion and future scope of the work.

2 System Model

The 2 DoF helicopter desktop system is shown in Fig. 1, which consists a Q2 data acquisition board with a maximum sampling rate of 4 kHz, two encoders to measure the yaw and pitch angles with resolution of 8192 and 4096 counts/revolution and a two-channel amplifier for power amplification. Here, front propeller serves to control the elevation about pitch axis, and rear propeller to control the sideward motion of the 2 DoF helicopter about yaw axis. The controlled variables are DC motor input voltages which are transmitted through slip ring. Yaw angle Ψ represents the motion around Z axis, and pitch angle θ represents the angular motion. The differential equations governing the laboratory 2 DoF helicopter are given as

$$(J_{\text{eq.y}} + m_{\text{heli}} l_{\text{cm}}^2)\ddot{\theta} = K_{\text{pp}} V_{\text{mp}} + K_{\text{py}} V_{\text{my}} + B_p \dot{\theta} - m_{\text{heli}} l_{\text{cm}}^2 \sin(\theta)\cos(\theta)\dot{\psi}^2$$

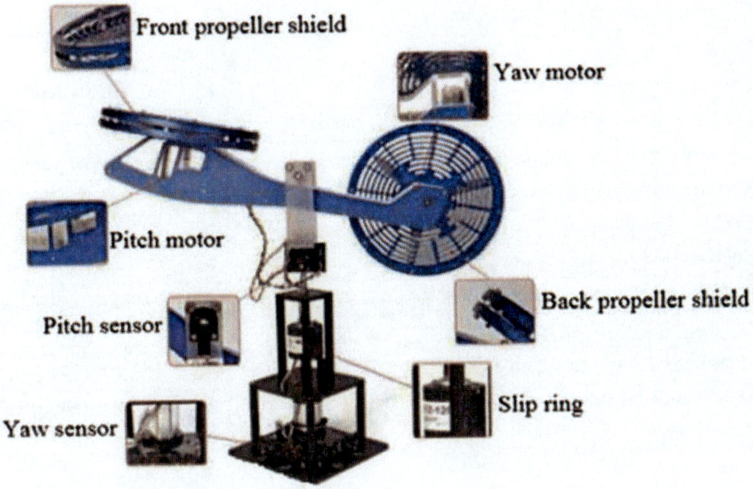

Fig. 1 2 DoF helicopter workstation

$$- m_{heli} g l_{cm} \cos(\theta), \tag{1}$$

$$\left(J_{\text{eq.y}} + m_{\text{heli}} \cos(\theta)^2 l_{\text{cm}}^2\right)\ddot{\psi} = K_{\text{yy}} V_{\text{my}} + K_{\text{yp}} V_{\text{mp}} + B_{\text{y}}\dot{\psi} + 2m_{\text{heli}} l_{\text{cm}}^2 \dot{\theta} \sin(\theta)\cos(\theta)\dot{\psi}. \tag{2}$$

State-space model and parameters for the 2 DoF helicopter are referred from [11].

$$\begin{bmatrix} \dot{\theta} \\ \dot{\psi} \\ \ddot{\theta} \\ \ddot{\psi} \\ \dot{\alpha} \\ \dot{\beta} \end{bmatrix} = \begin{bmatrix} 0 & 0 & 0 & 0 & 0 & 0 \\ 0 & 0 & 0 & 0 & 0 & 0 \\ 0 & 0 & \frac{-B_p}{J_{\text{eq}\cdot p} + m_{\text{heli}} l_{\text{cm}}^2} & 0 & 0 & 0 \\ 0 & 0 & 0 & \frac{-B_y}{J_{\text{eq}\cdot y} + m_{\text{heli}} l_{\text{cm}}^2} & 0 & 0 \\ 0 & 0 & 0 & 0 & 1 & 0 \\ 0 & 0 & 0 & 0 & 0 & 1 \end{bmatrix} \begin{bmatrix} \theta \\ \psi \\ \dot{\theta} \\ \dot{\psi} \\ \alpha \\ \beta \end{bmatrix} + \begin{bmatrix} 0 & 0 \\ 0 & 0 \\ \frac{-K_{\text{pp}}}{J_{\text{eq}\cdot p} + m_{\text{heli}} l_{\text{cm}}^2} & \frac{-K_{\text{py}}}{J_{\text{eq}\cdot p} + m_{\text{heli}} l_{\text{cm}}^2} \\ \frac{-K_{\text{yp}}}{J_{\text{eq}\cdot y} + m_{\text{heli}} l_{\text{cm}}^2} & \frac{-K_{\text{yy}}}{J_{\text{eq}\cdot y} + m_{\text{heli}} l_{\text{cm}}^2} \\ 0 & 0 \\ 0 & 0 \end{bmatrix} \begin{bmatrix} V_{\text{mp}} \\ V_{\text{my}} \end{bmatrix}, \tag{3}$$

where α and β are the integral pitch and integral yaw errors. The nominal system parameters are shown in Table 1.

Table 1 Parameters of 2 DoF helicopter

Descriptions	Symbols	Values
Yaw motor thrust force constant	K_{pp}	0.204 Nm/V
Yaw axis thrust twisting force constant from yaw motor	K_{yy}	72×10^{-3} Nm/V
Pitch axis thrust twisting force constant from yaw motor	K_{py}	6.8×10^{-3} Nm/V
Yaw axis thrust twisting force constant from pitch motor	K_{yp}	21.9×10^{-3} Nm/V
Equivalent viscous damping about pitch axis	B_p	0.545×10^{-3} N/V
Equivalent viscous damping about yaw axis	B_y	800×10^{-3} N/V
Moving mass of helicopter assembly	m_{heli}	318×10^{-3} kg
Center of mass length along helicopter body from the pitch axis	l_{cm}	0.186 m
Moment of inertia about the pitch axis	$J_{\text{eq}.p}$	0.0384 kg m^2
Moment of inertia about the yaw axis	$J_{\text{eq}.y}$	0.0432 kg m^2
Voltage for pitch motor	V_{mp}	±24 V
Voltage for yaw motor	V_{my}	±24 V

3 Problem Formulation

LQR technique is mostly preferred in aerospace applications due to its inherent robustness, stability and optimality properties. This technique can also be extended to multivariable systems.

Consider the following linear state model.

$$\begin{aligned} \dot{x}_p(t) &= Ax_p(t) + Bu(t) \\ y &= Cx_p(t) + Du(t). \end{aligned} \tag{4}$$

Let x_p (t) be the state vector, y be the output vector, $u(t)$ be the control input, and D, C, B and A are the feedforward matrices, output, input and system matrices, respectively.

State feedback control law is given by,

$$u = -k_{\text{opt}} x_p. \tag{5}$$

In LQR, optimality is achieved by minimizing the function (J).

$$J(u) = \int_0^\infty [x_p^T(t) Q_{\text{opt}} x_p(t) + u^T(t) R_{\text{opt}} u(t)]\mathrm{d}t, \tag{6}$$

where the elements of the weighing matrices Q_{opt} and R_{opt} depend upon the dimensions of the state and input variables. k_{opt} can be calculated using

$$k_{\text{opt}} = R_{\text{opt}}^{-1} B^T P, \tag{7}$$

where k_{opt} is the optimal control gain and P is the transformation matrix obtained by solving Algebraic Riccati Eq. (8)

$$A^T P + PA + Q_{\text{opt}} - PBR_{\text{opt}}^{-1} B^T P = 0. \tag{8}$$

LQR performance can be improved by augmenting an integrator

$$\dot{z} = r_p - y_p, \tag{9}$$

where r_p and y_p are being the reference and output vectors, respectively. Augmented open-loop state model of the system is given by

$$\begin{bmatrix} \dot{x}_p \\ \dot{z} \end{bmatrix} = \begin{bmatrix} A & 0 \\ -C & 0 \end{bmatrix} \begin{bmatrix} x_p \\ z \end{bmatrix} + \begin{bmatrix} B \\ 0 \end{bmatrix} u + \begin{bmatrix} 0 \\ r_p \end{bmatrix}; y = \begin{bmatrix} C & 0 \end{bmatrix} \begin{bmatrix} x_p \\ z \end{bmatrix}. \tag{10}$$

Even though LQI possesses the merit of inherent robustness in mitigating the disturbances, LQI performance starts to degrade when system experiences severe model uncertainties. As a measure to overcome this issue, an integral SMC with nonlinear sliding surface is formulated in the next section.

3.1 Integral SMC with Nonlinear Sliding Surface

Design steps involved are as follows:

Consider a sliding surface,

$$S = \begin{bmatrix} S_1 \\ S_2 \end{bmatrix} = \begin{bmatrix} e_1 + \lambda_1 \dot{e}_1 + \Lambda e_1 + \lambda_3 \int e_1 \mathrm{d}t \\ e_2 + \lambda_2 \dot{e}_2 + \Lambda e_2 + \lambda_4 \int e_2 \mathrm{d}t \end{bmatrix}, \tag{11}$$

where λ_1, λ_2 are the closed-loop Eigen values, λ_3, λ_4 are the positive constants, Λ is the uncertainty, and e_1 and e_2 are the pitch and yaw errors, respectively. If the reference trajectory is a step signal, then the derivatives can be assumed as zero.

$$S = \begin{bmatrix} (x_1 - x_{1r}) + \lambda_1 \dot{x}_1 + \Lambda e_1 + \lambda_3 \int (x_1 - x_{1r}) dt \\ (x_2 - x_{2r}) + \lambda_2 \dot{x}_2 + \Lambda e_2 + \lambda_4 \int (x_2 - x_{2r}) dt \end{bmatrix}. \tag{12}$$

Derivative of S is

$$\dot{S} = \begin{bmatrix} (\dot{x}_1 - \dot{x}_{1r}) + \Lambda \dot{e}_1 + \lambda_1 \ddot{x}_1 + \lambda_3 e_1 \\ (\dot{x}_2 - \dot{x}_{2r}) + \Lambda \dot{e}_2 + \lambda_2 \ddot{x}_2 + \lambda_4 e_2 \end{bmatrix}, \tag{13}$$

$$\dot{S} = \begin{bmatrix} 1 \; 0 \; \lambda_1 \; 0 \\ 0 \; 1 \; 0 \; \lambda_2 \end{bmatrix} \begin{bmatrix} \dot{x}_1 \\ \dot{x}_2 \\ \ddot{x}_1 \\ \ddot{x}_2 \end{bmatrix} + \begin{bmatrix} \lambda_3 & 0 \\ 0 & \lambda_4 \end{bmatrix} \begin{bmatrix} e_1 \\ e_2 \end{bmatrix} + \begin{bmatrix} \Lambda & 0 \\ 0 & \Lambda \end{bmatrix} \begin{bmatrix} \dot{e}_1 \\ \dot{e}_2 \end{bmatrix}, \quad (14)$$

$$\dot{S} = C_1(AX + Bu) + C_2E + C_3\Lambda \dot{e} \quad (15)$$

when $u = u_{eq}$, $\dot{S} = 0 \Rightarrow$

$$0 = C_1(AX + Bu_{eq}) + C_2E + C_3\Lambda \dot{e}. \quad (16)$$

Solving above Eq. (16),

$$u_{eq} = -(C_1B)^{-1}(C_1AX + C_2E + C_3\Lambda \dot{e}) \quad (17)$$

when control law is applied as, $u = u_{eq} - KSgn(S)$,
where $K = \begin{bmatrix} K_1 & 0 \\ 0 & K_2 \end{bmatrix}$ with $K_1, K_2 > 0$, then

$$\dot{S} = -K\,\text{Sgn}(S). \quad (18)$$

For $K > 0$,
$\dot{V}(S)$ is negative for all values of S. Hence, $S \to 0$ as time $t \to \infty$.

4 Results and Discussion

The hardware setup shown in Fig. 2 consists of a two-channel power amplifier, two-channel data acquisition system, and a personal computer which executes the control algorithm in MATLAB/Simulink platform.

Closed-loop response of pitch angle for ISMC is as follows. A square test signal having amplitude of $-10°$ is given as a reference to the 2 DoF laboratory helicopter, and the corresponding command tracking response of LQI is illustrated in Fig. 3. The steady-state response is oscillatory, which cannot be acceptable in real-time helicopters.

As a measure to overcome this, a first-order filter is designed. This acts as a noise rejection mechanism to mitigate the oscillations in pitch response. The noise-compensated system response is illustrated in Fig. 4.

The command tracking response of nonlinear ISMC is depicted in Fig. 5. The tracking response along with the noise rejection mechanism accentuates the efficacy of the proposed algorithm.

Fig. 2 Two DoF helicopter hardware setup

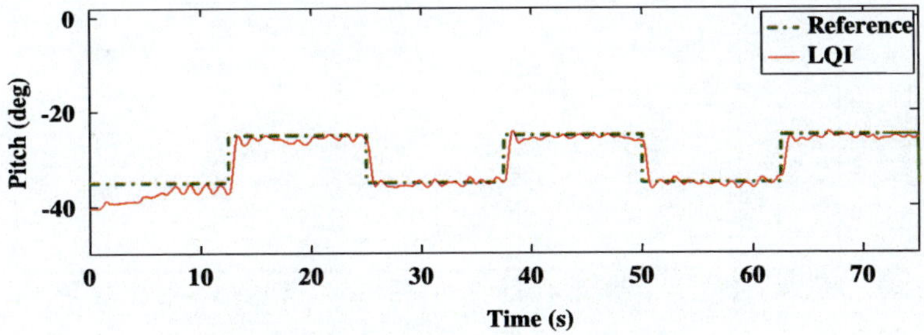

Fig. 3 LQI Command tracking response before noise compensation

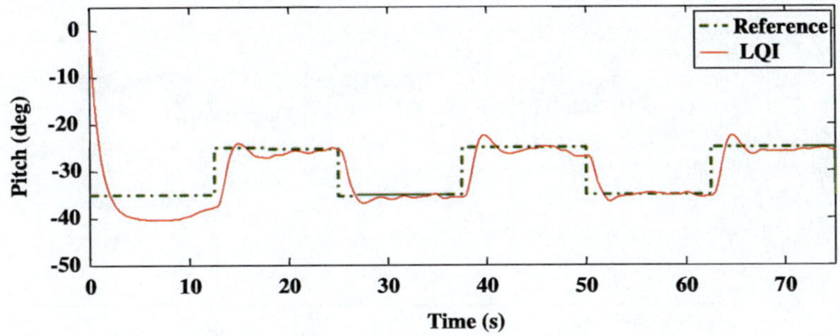

Fig. 4 LQI Command tracking response after noise compensation

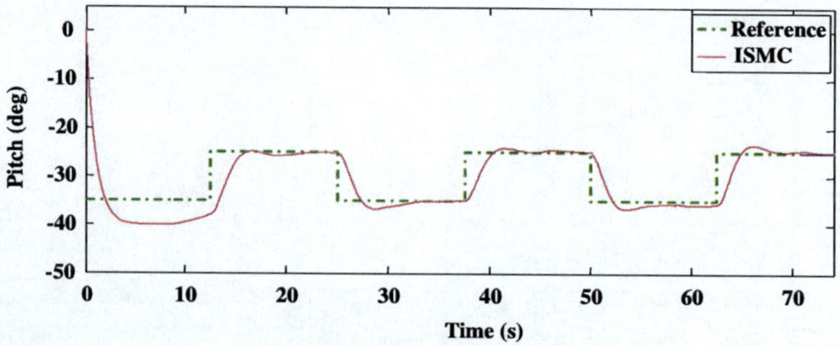

Fig. 5 ISMC Command tracking response

The pitch motor voltage profile for LQI controller and ISMC is shown in Figs. 6 and 7, respectively. From the figures, it is inferred that the voltage profile of ISMC has distortions due to chattering phenomenon in SMC.

To test the robustness properties of ISMC, a solid weight of 30 g is added at the helicopter nose, which will alter the center of gravity of the system. Moreover, this

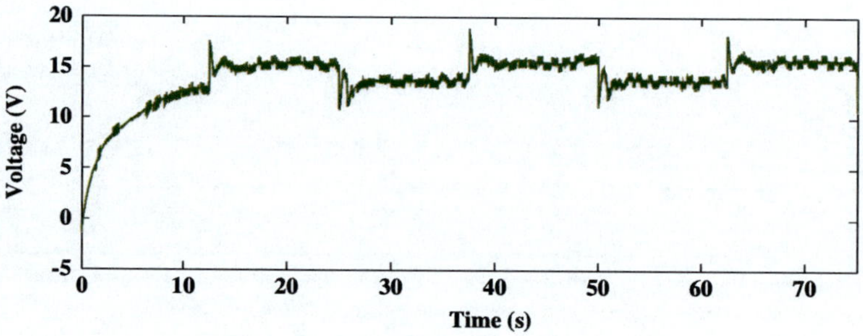

Fig. 6 LQI Pitch motor voltage during command tracking

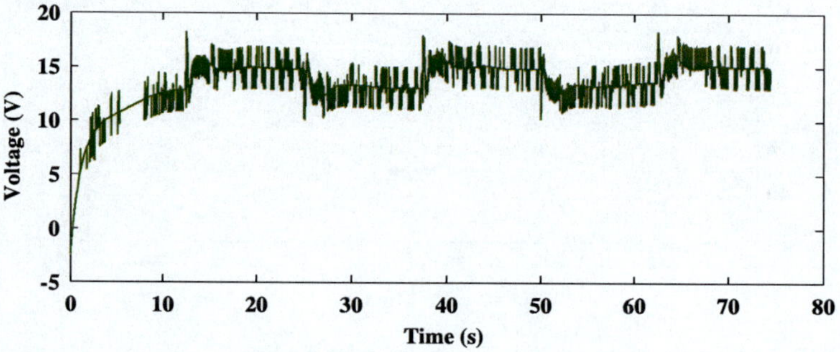

Fig. 7 ISMC Pitch motor voltage during command tracking

will induce a continuous torque in pitch motor. The efficacy of the LQI controller in handling the model uncertainty is illustrated in Fig. 8.

Efficacy of the ISMC in handling the model uncertainty is shown in Fig. 9. Quantitative comparison of the hardware results of conventional LQI and ISMC is shown in Table 2.

where t_r, t_p and t_s are the rise, peak and settling times and e_{ss} represents the steady-state error. From Figs.8 and 9, and quantitative comparison shown in Table 2, it is evident that ISMC is found to be superior in handling the uncertainties compared to LQI controller. The corresponding pitch motor voltages are shown in Figs. 10 and 11.

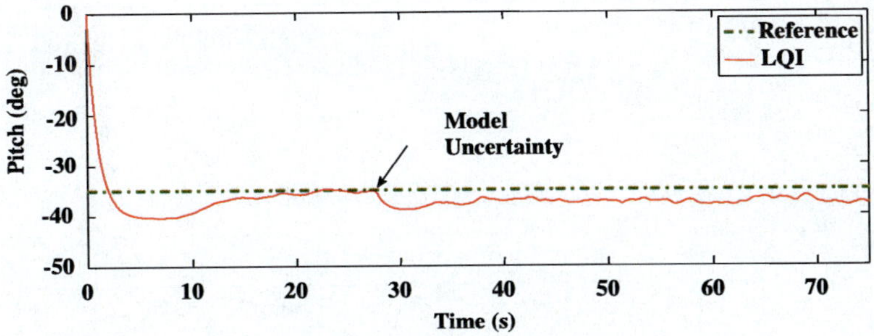

Fig. 8 LQI trajectory tracking response during model uncertainty

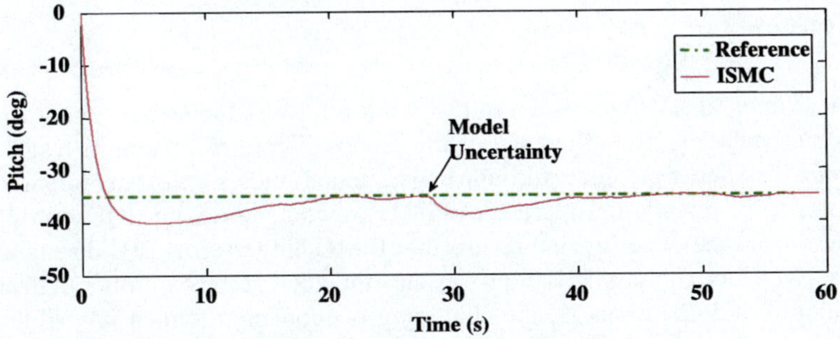

Fig. 9 ISMC trajectory tracking response during model uncertainty

Table 2 Quantitative performance comparison of hardware results

Symbol	LQI	ISMC
t_r (s)	1.67	2.78
t_p (s)	2.5	3.6
t_s (s)	6.95	3.4
e_{ss} (degrees) (during uncertainty)	2	0

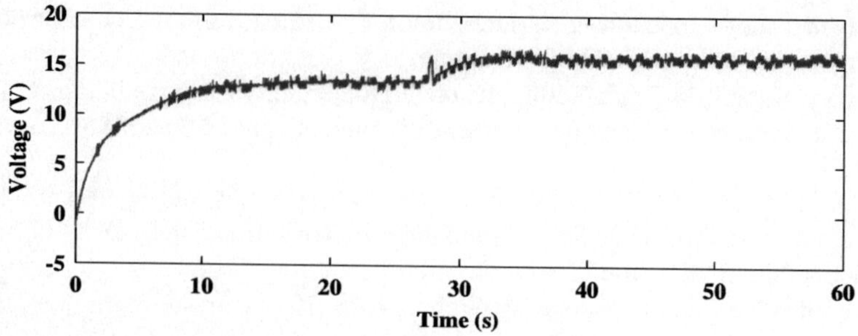

Fig. 10 LQI pitch motor voltage during uncertainty

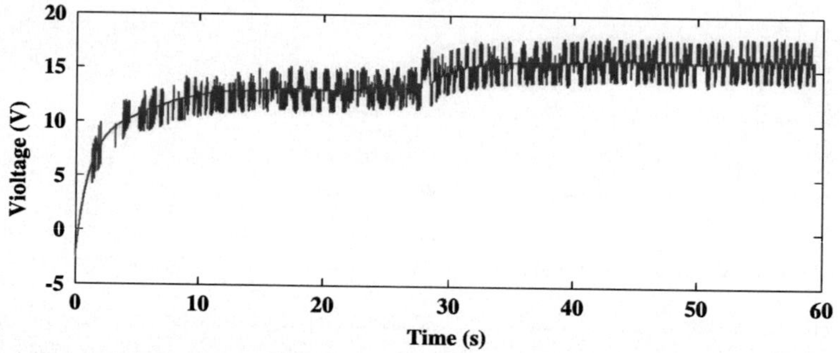

Fig. 11 ISMC pitch motor voltage during uncertainty

5 Conclusion

Nonlinear integral SMC proposed in this work improves the steady-state response during uncertainties. The efficiency of the proposed control scheme is tested on a Quanser 2 DoF helicopter plant through experimental studies. The trajectory tracking and disturbance rejection properties of ISMC are compared with LQI controller. It is evident from the experimental results that ISMC outperforms LQI during uncertainties. Even though the ISMC improves the efficiency of the controller in handling uncertainties and disturbances, the chattering phenomenon, which is visible from the pitch motor voltage profile, will decrease the life of the motor. Chattering phenomenon in SMC is still an open-ended problem. Therefore, switching intelligence can be incorporated in the control algorithm, which makes a choice between baseline LQI and robust controller according to the system needs, which will be addressed in the next work.

References

1. Sabiha AD, Kamel MA, Said E, Hussein WM (2022) ROS-based trajectory tracking control for autonomous tracked vehicle using optimized backstepping and sliding mode control. Robot Auto Syst 152
2. Kumar BA, Gayathri S, Surendhar S, Senthilrani S, Jeyabharathi R (2019) Robust H-infinity controller for two degree of freedom helicopter. In: 2019 IEEE International Conference on System, Computation, Automation and Networking (ICSCAN), pp 1–5
3. Cruz-Ortiz D, Chairez I, Poznyak A (2022) Non-singular terminal sliding-mode control for a manipulator robot using a barrier Lyapunov function. ISA Trans 121:268–283
4. Mofid O, Mobayen S, Zhang C, Esakki B (2022) Desired tracking of delayed quadrotor UAV under model uncertainty and wind disturbance using adaptive super-twisting terminal sliding mode control. ISA Trans 123:455–471
5. Wang J, Rong J, Yu L (2022) Dynamic prescribed performance sliding mode control for DC–DC buck converter system with mismatched time-varying disturbances. ISA Trans
6. Kuo T-C, Huang Y-J, Chang S-H (2008) Sliding mode control with self-tuning law for uncertain nonlinear systems. ISA Trans 47(2):171–178
7. Utkin VI (1978) Sliding modes and their applications in variable structure systems. Mir Publishers, Moscow
8. Chen Y, Cai B, Cui G (2020) The design of adaptive sliding mode controller based on RBFNN approximation for suspension control of MVAWT. In: 2020 Chinese Automation Congress (CAC), pp 1080–1084
9. Sanjeewa SDA, Parnichkun M (2022) Control of rotary double inverted pendulum system using LQR sliding surface based sliding mode controller. J Cont Dec 9(1):89–101
10. Mobayen S, Tchier F (2018) Robust global second-order sliding mode control with adaptive parameter-tuning law for perturbed dynamical systems. Trans Inst Meas Control 40(9):2855–2867
11. Subramanian R, Elumalai V (2016) Robust MRAC augmented baseline LQR for tracking control of 2 DoF helicopter. Robot Auto Syst 86:70–77

Reference Spectrum Tracking for Circadian Entrainment

Veena Mathew, R. Pavithra, Ciji Pearl Kurian, and R. Srividya

Abstract This article deals with tracking a reference spectrum, enhancing the circadian entrainment of the occupants. The choice of the reference spectrum was made with an emphasis on maintaining the CCT, maintaining a CRI value of at least 80 for the general work area, and maintaining the luminous flux. Use of the Levenberg–Marquardt (LM) algorithm results in the optimal number of LEDs needed for spectrum synthesis. As a result, this work intends to offer a solution for making a luminaire. The percentage error in CCT for the simulated spectrums was less than 0.4%. The daylight spectrum is also selected as a reference spectrum so that close characteristics of sunlight are reached, meeting human comfort needs while also delivering the advantages of sunlight. This novel method of keeping a reference spectrum to design a LED luminaire for circadian entrainment with high luminous efficacy leads to the human-centric lighting system.

Keywords Levenberg–Marquardt (LM) algorithm · Reference spectrum tracking · LED modeling · Daylight spectrum

1 Introduction

Humans require exposure to daylight and its rhythm to function normally. Light is an absolute necessity. Light is one of the primary external causes of regulating the biological clock and synchronizes with the light–dark pattern [1, 2]. Not only does the amount of light fall on the retina, but the duration and the previous light exposures play crucial roles in the hormonal suppression or secretion in the human body [3, 4]. These days, people stay more at offices or their workplaces rather than exposed to daylight. These changes in light patterns certainly affect people's mood, behavior,

V. Mathew · R. Pavithra · C. P. Kurian (✉) · R. Srividya
Department of Electrical and Electronics Engineering, Manipal Institute of Technology, Manipal Academy of Higher Education, Manipal 576104, India
e-mail: ciji.pearl@manipal.edu

S. Sharma et al. (eds.), *Intelligent Control, Robotics, and Industrial Automation*, Lecture Notes in Electrical Engineering 1066,
https://doi.org/10.1007/978-981-99-4634-1_25

and productivity [4–8]. The term "Human Centric Lighting" (HCL) refers to lighting designs in which humans are heavily involved [9].

Daylight harvesting in interior spaces helps to improve the wellness of occupants. If there is less daylight entry, it can be effectively done by implementing daylight equivalent illuminants in the interiors. Spectrally tuned Light-Emitting Diode (LED) luminaires help to improve an occupant's overall comfort in an interior space. Five channels of LEDs, including red, green, cyan, warm white, and cool white (RGCWW), were used to mimic energetic daylight [10]. Both natural sunlight and mixed white light were used to measure SPD. The color quality, luminous efficiency, and circadian tunability were improved using cyan, cool white, and warm white LEDs in color mixing. RGCWW LED color mixing was done to achieve SPD characteristics closer to daylight. According to Ohno, white LED spectra are made for general lighting with high luminous efficacy and good color rendering [11]. Supronowicz and Fryc described the spectral power distribution of LEDs using different mathematical modeling systems [12]. According to Fryc et al., who have worked with numerous LEDs, an integrating sphere is used to design a spectrally tunable light source [13]. White emitters with varying bandwidths were used to increase Color Rendering Index (CRI), it is concluded that the 60 nm emitters offered a higher CRI than the 30 nm emitters [14]. A novel spectral LED-based tunable light source can create individualized lighting solutions, particularly for recreating CIE standard illuminants [15]. Genetic algorithm has been used to maximize the spectral power distribution of an LED source over the Color-Correlated Temperature (CCT) range of 2020–7929 K [16]. The high Luminous Efficacy of Radiation (LER), the high CRI, and the high Color Quality Scale (CQS) were achieved simultaneously by using two narrowband (blue, red) and two Phosphor-Converted (PC) (yellow, green) LEDs. The authors modelled each LED individually using the Gaussian-modeling method.

This work proposed a solution for creating spectrally adjustable luminaires to enhance the mood and alertness of occupants. Here, the working algorithm follows a target/reference spectrum to obtain the minimum number of LEDs required in the luminaire design. Finally, the luminaire should be able to mimic the target spectrum. The selection of the reference spectrum is made with an emphasis on maintaining the CCT, attempting to maintain a CRI value of at least 80 for the general work area, and maintaining the luminous flux. With the help of an optimization technique, the optimal number of LEDs needed for spectrum synthesis is selected. The daylight spectrum is also used as reference spectrum selection so that close characteristics of sunlight are reached, meeting human comfort needs while also delivering the advantages of sunlight.

2 Methodology

LEDs should be chosen to fit the target spectrum's SPD with the least amount of fitness error, generate the target CCT with the lowest absolute distance (duv), and have a CRI which is at least greater than 80. This section covers the approach for

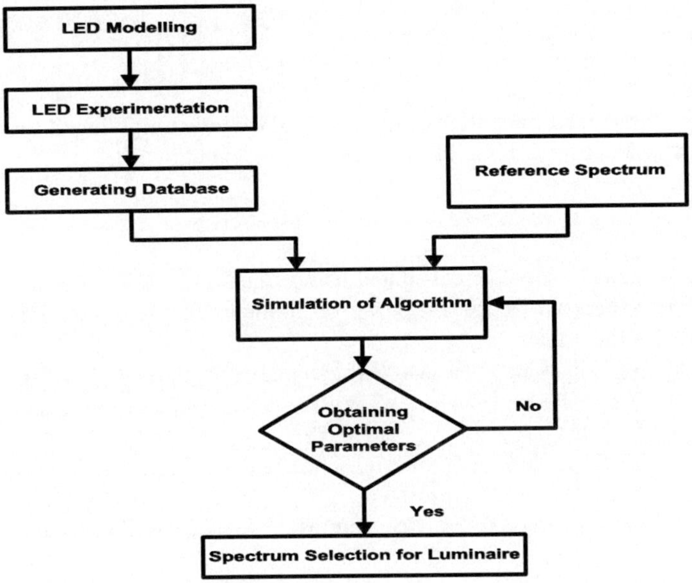

Fig. 1 Flow diagram of the methodology

modeling LEDs and the optimization algorithm for reference tracking. According to Fig. 1, LED modeling begins with selecting LEDs ranging in wavelength from 380 to 780 nm and acquiring the required set of LED data from the datasheet. Actual LEDs are purchased, and the spectrum values are measured and recorded using an integrating sphere. Finally, a database for LEDs is built. Once the reference spectrum is created, the database and reference spectra are given to the algorithm. The algorithm yields a simulated spectrum that closely resembles the reference spectrum.

2.1 Modeling of LED

A single normal or Gaussian distribution, which closely fits most LED spectra, is a simple way to model LED [11]. Peak wavelength λ0 in (nm), Full Width at Half-Maximum (FWHM) Δλ0.5 in (nm), and total radiant power Φe in (W) or total luminous flux Φv in (lumens) are the three characteristics considered when modeling polychromatic LEDs. The normalized spectrum based on the Gaussian distribution is first computed for modeling using Eq. (1).

$$\phi(\lambda, \lambda_0, \lambda_{0.5}) = \frac{g(\lambda, \lambda_0, \Delta\lambda_{0.5}) + 2 * g^5(\lambda, \lambda_0, \Delta\lambda_{0.5})}{3}. \quad (1)$$

where g is the Gaussian distribution factor, and it is calculated using Eq. (2)

$$g(\lambda, \lambda_0, \Delta\lambda_{0.5}) = \exp\left\{-\left[\frac{\lambda - \lambda_0}{\Delta\lambda_{0.5}}\right]^2\right\}. \quad (2)$$

The spectral power distribution of LED is obtained by multiplying the normalized spectrum with a conversion factor F, which is represented by Eq. (3),

$$\phi_{e,v}(\lambda) = \phi(\lambda, \lambda_0, \lambda_{0.5}) * F_{\text{LED}}. \quad (3)$$

The conversion factor F is calculated using Eqs. (4) or (5) based on available values, as the values can be radiant power or luminous flux based on the datasheet specification of the product.

If radiant power in watts is known, F is calculated using the below Eq. (4),

$$F_{\text{LED}} = \frac{\phi_e}{\int \phi(\lambda, \lambda_0, \Delta\lambda_{0.5}) \mathrm{d}\lambda}. \quad (4)$$

If the total luminous flux in lumens is known, F is calculated using Eq. (5),

$$F_{\text{LED}} = \frac{\phi_v}{K_m \int \phi(\lambda, \lambda_0, \Delta\lambda_{0.5}) V(\lambda) \mathrm{d}\lambda}. \quad (5)$$

2.2 *Design Parameters*

Visual characteristics of the light source

Color characteristics of light are generally defined in terms of CCT (K) and Duv [17, 18]. Conventionally, light sources with low CCTs (2700–3000 K) refer to warm light, and higher CCTs (4000–6500 K) refer to cool white. Mid-range CCTs are generally termed neutral white. Most commercial light sources exhibit CCT between 2700 and 6500 K. Another metric called Duv should be less than 0.005 for the CCTs to appear white. Color Rendering Index (CRI, Ra) is another metric traditionally followed to describe the rendering of objects under a test source. Lamp sources with high CRI present the illuminated object in its natural color. Reduced CRI brings dullness in the rendering of colors of illuminated objects. Generally, for office purposes, CRI greater than 80 is recommended.

Suppose $S(\lambda)$ is the spectral power distribution of the light source. In that case, $V(\lambda)$ is the spectral luminous efficiency function, and the Luminous Efficacy of Radiation (LER, lm/W) which indicates the visual stimuli per unit power is defined in Eq. (6):

$$\text{LER} = 683 \frac{\int V(\lambda) \int S(\lambda) d(\lambda)}{\int S(\lambda)}. \quad (6)$$

Non-visual characteristics of the light source

The circadian response's spectral sensitivity function is found to peak at 450–490 nm [19–21]. Recent discovery of intrinsically photosensitive retinal ganglion cells (ipRGCs) in human retina leads to various circadian quantifying metrics [22]. The "International WELL Building Institute" (IWBI) introduced their circadian metric called Equivalent Melanopic lux (EML) [23]. A minimum EML of 200 lx, including daylight, should be available at eye level. In the absence of daylight, 150 EML is sufficient in the workspace. The Lighting Research Institute (LRC) developed another metric called Circadian Stimulus (CS), which incorporates the circadian system's characteristics [24]. During the daytime, a minimum value of CS 0.3 is required for two continuous hours. CS is calculated as shown in Eq. (7), where circadian light (CL_A) is the spectrally weighted human circadian system irradiance [1, 25].

$$\boldsymbol{CS = 0.7 *} \left(1 - \frac{1}{1 + \frac{\boldsymbol{CL_A}}{355.7}^{1.1026}}\right). \tag{7}$$

Circadian Efficacy of Radiation (CER) provides the circadian effect of a spectrum under consideration. If $S(\lambda)$ is the spectral power distribution of the light source, and $C(\lambda)$, the spectral circadian efficiency function, peaks at 480 nm, CER is given by Eq. (8). CER ranges from 0 to 1. A high CER value indicates the test spectrum's high circadian potential [26].

$$\text{CER} = 683 \frac{\int C(\lambda) \int S(\lambda) d(\lambda)}{\int S(\lambda)}. \tag{8}$$

2.3 *Optimization Algorithm for Reference Tracking*

Here, the Levenberg–Marquardt (LM) algorithm is used to reproduce the reference spectrum. Levenberg first proposed the LM algorithm in 1944, and Marquardt perfected it in 1963. The LM optimization algorithm is a hybrid of the Gauss–Newton and gradient descent methods [27]. Compared to other existing algorithms, LM provides the advantages and flexibility of both Gauss–Newton and gradient descent methods. As a result, the LM algorithm is employed to solve nonlinear least-squares problems [28]. Figure 2 shows the flowchart of the algorithm. It calculates the power per LED PowerLED_i, where $i = 1, 2, \ldots$etc., and the number of LEDs required to achieve the reference spectrum. The SPD of the modelLED LEDs will be defined by $\text{SPDLED}_i(\lambda)$. The LM algorithm, as shown in Fig. 2, is implemented as follows [29]:

Step 1: Choose a starting value for each PowerLED_i.

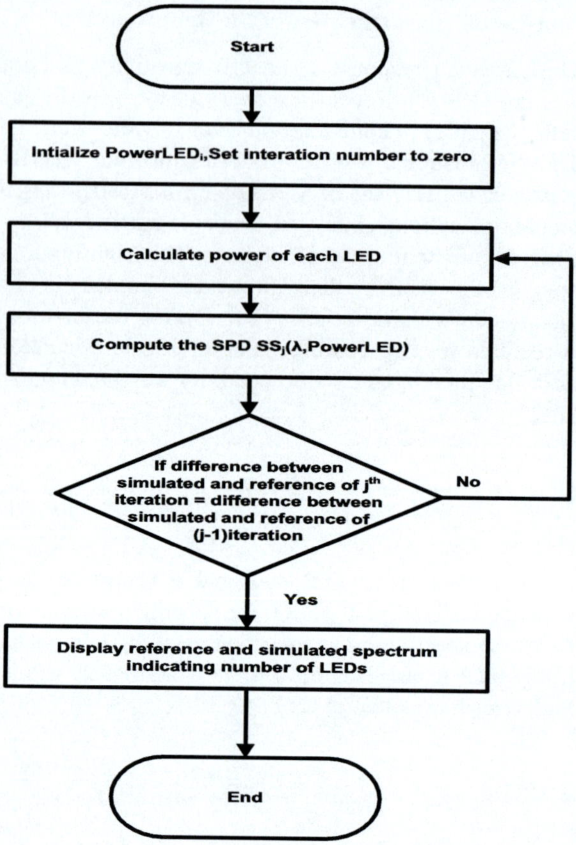

Fig. 2 Flowchart of LM algorithm

Step 2: Determine the PowerLED's jth value, where j denotes the number of iterations, with $j = 0$ being the first.

(a) For each wavelength, use Eq. (9) to calculate the difference between the simulated (I_{SS}) and target spectrums (I_{TS}).

$$R(\lambda) = I_{SS}\left(\lambda, \text{PowerLED}^{j-1}\right) - I_{TS}(\lambda). \tag{9}$$

(b) Using Eq. (10), compute the Jacobian matrix.

$$J^{j-1}_{I_{SS}(\lambda)}\left(\text{PowerLED}^{j-1}\right) = \left[\frac{\partial I_{SS}\left(\lambda, \text{Power}^{j-1}\right)}{\partial \text{Power}^{j-1}}\right]. \tag{10}$$

(c) The PowerLED's jth value is written as shown in Eq. (11).

$$\text{PowerLED}^{j} = \text{PowerLED}^{j-1} - \left[\left(J^T J\right)\alpha^{-1} J^T R\right]. \tag{11}$$

Step 3: If the error is same in current and previous iteration, as shown in Eq. (12)

$$\sum_{\lambda=380}^{780}\left|I_{SS}\left(\lambda,\text{PowerLED}^{j}\right)-I_{TS}(\lambda)\right|=\sum_{\lambda=380}^{780}\left|\begin{matrix}I_{SS}\left(\lambda,\text{PowerLED}^{(j-1)}\right)\\-I_{TS}(\lambda)\end{matrix}\right|. \tag{12}$$

The final value of the power coefficients is then represented by $\text{PowerLED}^{(j)}$, as in Eq. (13)

$$\text{(i.e.) } I_{SS}(\lambda,\text{PowerLED})=\sum_{\lambda=380}^{780}\text{PowerLED}_i\,\text{SPDLED}_i(\lambda). \tag{13}$$

The damping factor of the LM, i.e., coefficient α, is empirically determined to be 0.01. With each iteration, the LM algorithm adjusts the parameters, minimizing the squared sum of the fitness error at each wavelength [29].

3 Experimentation

3.1 LED Database from Actual LEDs

Suitable LEDs are purchased from standard manufacturers like Lumileds, OSRAM, etc. Figure 3 shows the experimental setup for the testing of purchased LEDs. Here, LEDs are mounted in the Integrating sphere (0.5 diameters). The spectral characteristic of the light source is obtained, and the data obtained from the calculation are exported to Excel® or text format and are used for further analysis. Similarly, the procedure is repeated for different LEDs.

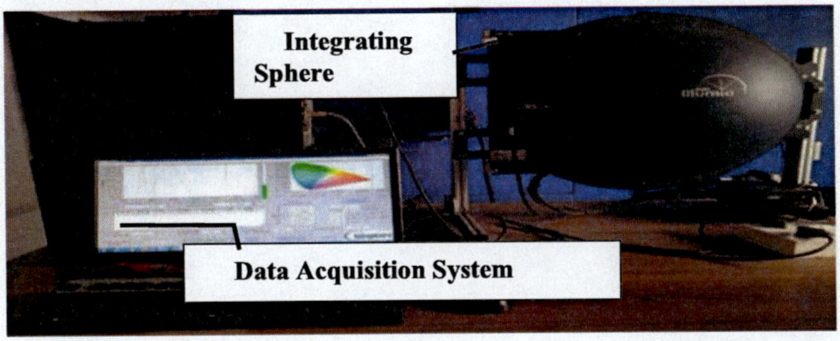

Fig. 3 Experimentation setup for LED measurement

3.2 LED Database from the Manufacturer

Initially, the LEDs ranging from 380 to 780 nm are selected from standard manufacturers. The parameters, namely: Peak wavelength in nm, Full Width at Half-Maximum (FWHM) in nm, and total radiant power in Watts or total luminous flux in lumens, Voltage in Volts, Current in mA, are required for the modeling of LEDs, values are obtained from the respective datasheet, and it is tabulated for further calculations.

3.3 Reference Spectrum from Daylight Measurement

The experiment was conducted in an area with no limitation on daylight. The experimentation was carried out from 9.00 am to 5.00 pm for every 30 min interval and was repeated for a week. For every 30 min interval, the spectrophotometer (CL500A) was placed close to the wall facing the daylight, and thus, Illuminance, CCT, and Spectral data were measured.

3.4 Selection of Spectrum for Luminaire

A solution has been presented for constructing a tunable luminaire with CCT values fixed for different day periods with varied intensity conditions. For 8.00 am to 12.00 pm, a spectrum with a blue peak at 455 nm is considered because it promotes awareness and increases activity levels [21]. A cooler tone with a CCT of 4872 K can be used for increased productivity. For 12.00 pm to 2.30 pm, the daylight spectrum is considered to provide the best coordination for the occupants, and hence, CCT of 6425 K is selected. For 2.30 pm to 5.00 pm, a spectrum with a red peak is considered. As a result, a warmer tone CCT of 2953 K is employed, as it aids humans in relaxing and preparing for sleep.

4 Results and Analysis

4.1 LED Database

Figure 4 shows LEDs with various peaks that were chosen to construct the database for the algorithm. The database includes both the modelLED LEDs and the actual ones. This database has been given as the input to the LM algorithm.

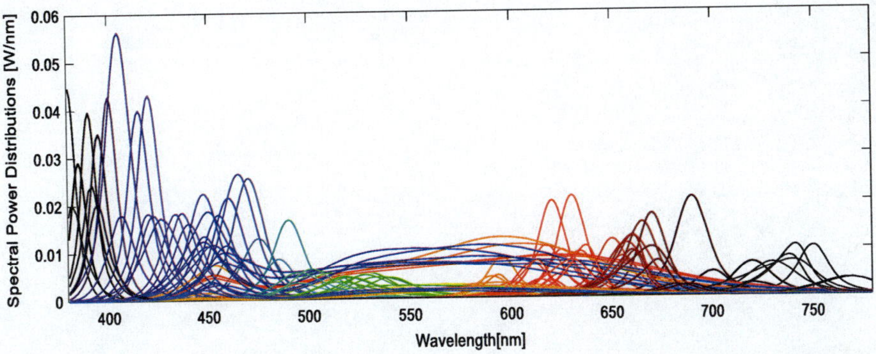

Fig. 4 Combined SPD of all the LEDs in the LED database which is formed for the algorithm

4.2 *Spectrum Tracking for Luminaire*

Figure 5 shows the reference and simulated spectrums for early morning. Similarly, Figs. 6 and 7 show the respective reference spectrums and the simulated spectrum. Table 1 gives the solution considered for the tunable luminaire design and the outcomes. EML and CS were measured for an arbitrary value of vertical eye illuminance, 250 lx. Also, the measured value of CER was high in the same period. These values for the above-mentioned circadian metrics indicate the designed luminaire's high circadian potential in the morning. It certainly helps to improve the mood and alertness of the occupants. The percentage error between actual and simulated values of CCTs 4887 K, 6425 K, and 2953 K was 0.3%, 0.06%, and 0.048%, respectively. Also, the measured LER for all the simulated spectrums was above 200, acceptable for general purposes.

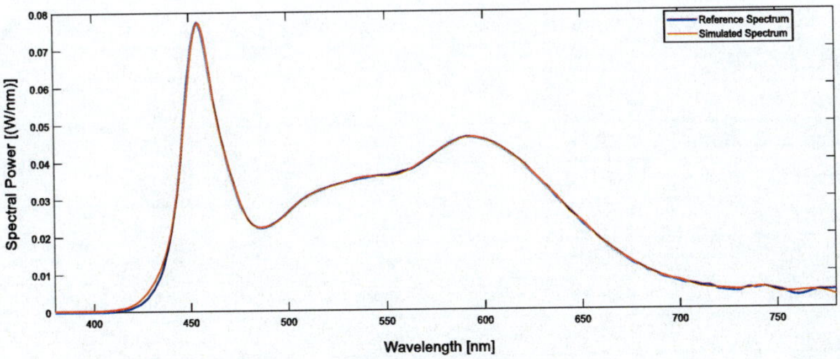

Fig. 5 Results of LM algorithm for 4887 K reference spectrum

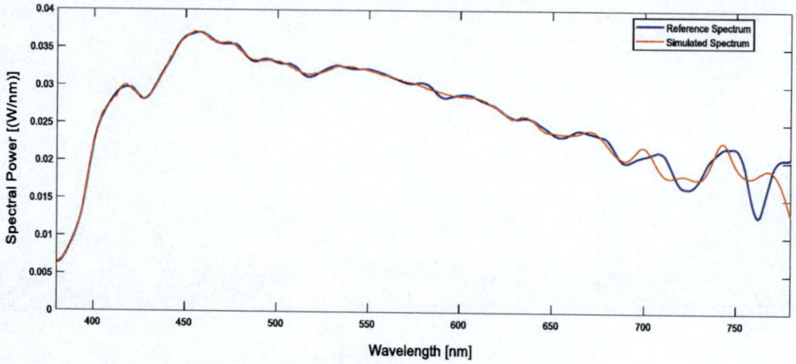

Fig. 6 Results of LM algorithm for 6425 K reference spectrum

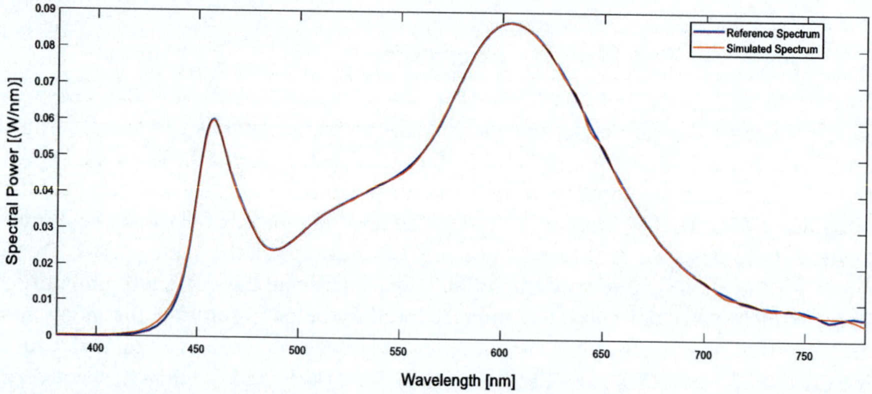

Fig. 7 Results of LM algorithm for 2953 K reference spectrum

Table 1 Outcome features of the tunable luminaire design solution

Time	6 am to 12 pm	12 pm to 2.30 pm	2.30 pm to 6 pm
Measured CCT (k)	4887	6421	2939
Measured LER	296	214	302
Measured EML (lux)	225	273	152
Measured CS	0.31	0.36	0.31
Measured CER	0.5	0.3	0.2
%error in CCT	0.308	0.062	0.48
CRI obtained	88.8797	98.83	83.4138
Duv	0.0104	0.0015	0.0093
No. of LEDs	50	52	75

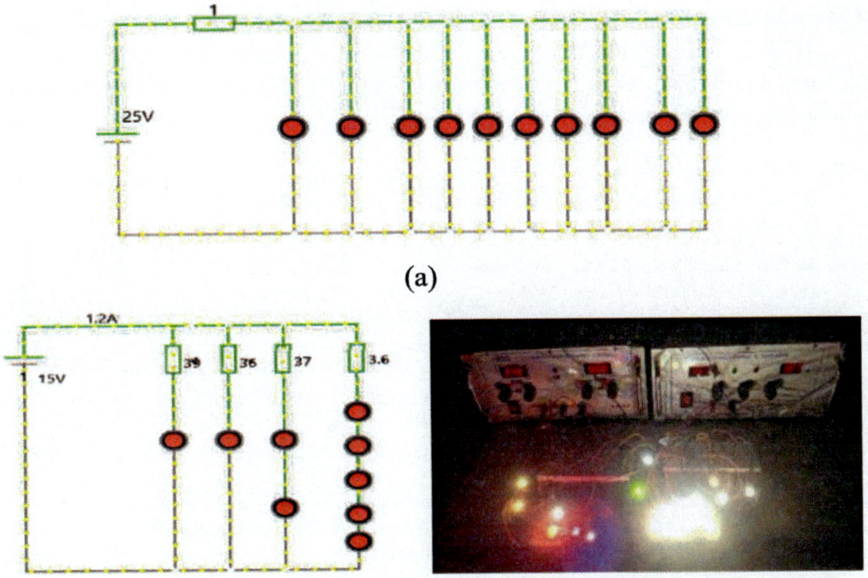

(c)

Fig. 8 Specific spectrum analysis **a** simulation circuit operating under 160 mA **b** simulation circuit operating under 350 mA **c** hardware model including all real LEDs

4.3 Hardware for a Selected Spectrum

A specific spectrum is selected and its verification is done experimentally and also simulated to check whether the circuit is working. Figure 8 shows the hardware and simulation models for the respective circuits. From this analysis, a CCT of 4317 K, CRI of 86, and luminous flux of 6065 lm are attained by considering the SPDs of all actual LEDs. The values mentioned above are obtained by inputting the Color Calculator tool with all the corresponding SPD and wavelengths of 19 LEDs.

5 Conclusions

The CCTs for the selected reference spectrums from morning to the evening were 4872 K, 6425 K, and 2953 K. The LM algorithm successfully tracked all three spectrums and delivered the optimal number of LEDs. The percentage error between actual and simulated spectrums for the three time sections was 0.3%, 0.06%, and 0.048%, respectively. During morning time, measured circadian metrics (CS, EML, and CER) values for the simulated spectrum were above the required limits for circadian entrainment. The luminous efficacy of the obtained solutions is greater than 200.

Also, a hardware implementation of a selected spectrum was demonstrated. Consequently, this optimization technique can be further used to implement spectrally tunable LED luminaires in daylight artificial-integrated systems for visual comfort and circadian performance.

Acknowledgements The study was supported by AICTE Research Promotional Scheme (RPS) India funded project, (Grant/Award Number: 8-T4/FDC/RPS (Policy-l) /2019-20).

References

1. Rea MS, Figueiro MG (2018) Light as a circadian stimulus for architectural lighting. Light Res Technol 50:497–510
2. Wright KP et al (2013) Entrainment of the human circadian clock to the natural light-dark cycle. Curr Biol 23:1554–1558
3. de Kort YAW, Smolders KCHJ (2010) Effects of dynamic lighting on office workers: first results of a field study with monthly alternating settings. Light Res Technol 42:345–360
4. Figueiro MG et al (2019) Circadian-effective light and its impact on alertness in office workers. Light Res Technol 51:171–183
5. Figueiro MG et al (2017) The impact of daytime light exposures on sleep and mood in office workers. Sleep Health 3:204–215
6. Zhu Y et al (2019) Effects of illuminance and correlated color temperature on daytime cognitive performance, subjective mood, and alertness in healthy adults. Environ Behav 51:199–230
7. Kim TW, Jeong JH, Hong SC (2015) The impact of sleep and circadian disturbance on hormones and metabolism. Int J Endocrinol
8. Figueiro MG, Rea MS (2016) Office lighting and personal light exposures in two seasons: impact on sleep and mood. Light Res Technol 48:352–364
9. Boyce P (2016) Editorial: exploring human-centric lighting. Light Res Technol 48:101
10. Nie J et al (2019) Tunable LED lighting with five channels of RGCWW for high circadian and visual performances. IEEE Photonics J 11
11. Ohno Y (2005) Spectral design considerations for white LED color rendering. Opt Eng 44
12. Supronowicz R, Fryc I (2019) The LED spectral power distribution modelLED by different functions-how spectral matching quality affected computed LED color parameters. In: 2019 2nd Balkan Junior Conference on Lighting, Balkan Light Junior 2019—Proceedings. https://doi.org/10.1109/BLJ.2019.8883564
13. Fryc I, Brown SW, Ohno Y (2005) Spectral matching with an LED-based spectrally tunable light source. In: Fifth International Conference on Solid State Lighting 5941:59411I
14. Miller ME, Gilman JM, Colombi JM (2016) A model for a two-source illuminant allowing daylight colour adjustment. Light Res Tech 48
15. Burgos-Fernández FJ et al (2016) Spectrally tunable light source based on light-emitting diodes for custom lighting solutions. Optica Applicata 46
16. Xu GQ et al (2017) Solar spectrum matching using monochromatic LEDs. Light Res Tech 49
17. Xu H, Luo MR, Rigg B (2003) Evaluation of daylight simulators. Part 1: Colorimetric and spectral variations. Coloration Techn 119:59–69
18. Colaco AM, Colaco SG, Kurian CP, Kini SG (2018) Color characterization of multicolor multichip LED luminaire for indoor. J Build Eng 18
19. Lucas RJ et al (2014) Measuring and using light in the melanopsin age. Trends Neurosci 37:1–9
20. Brainard GC et al (2001) Action spectrum for melatonin regulation in humans: evidence for a novel circadian photoreceptor. J Neurosci 21:6405–6412
21. Lockley SW, Brainard GC, Czeisler CA (2003) High sensitivity of the human circadian melatonin rhythm to resetting by short wavelength light. J Clin Endocrinol Metab 88:4502–4505

22. al Enezi J et al (2011) A ‘melanopic’ spectral efficiency function predicts the sensitivity of melanopsin photoreceptors to polychromatic lights. J Biol Rhythms 26:314–323
23. Richardson Z, The well building standard : assessment of effectiveness
24. Figueiro MG, Gonzales K, Pedler D (2016) Circadian stimulus the lighting research center proposes a metric for applying circadian light in the built environment, 31–33
25. Rea MS, Figueiro MG, Bierman A, Hamner R (2012) Modelling the spectral sensitivity of the human circadian system. Light Res Technol 44:386–396
26. Marín-Doñágueda M et al (2021) Simultaneous optimization of circadian and color performance for smart lighting systems design. Energy Build 252
27. Ranganathan A (2004) The Levenberg-Marquardt algorithm. http://www.excelsior.cs.ucsb.edu courses.cs290ipdfL.MA.pdf 142
28. Gavin HP (2019) The Levenberg-Marquardt algorithm for nonlinear least squares curve-fitting problems. Duke University
29. Kumar SR, Kurian CP, Gomes-Borges ME (2017) Multiobjective generalized extremal optimization algorithm for simulation of daylight illuminants. J Photonics Energy 7

Design and Implementation of Fuzzy Logic Based Intelligent Controller for PV System

K. Harshavardhana Reddy, Sachin Sharma, N. Charan Kumar, Chandan N. Reddy, I. Madesh Naidu, and N. Akshay

Abstract In today's world, the energy crisis is leading people to rely on renewable energy sources more than ever before. With the power-electronic technology currently in place, photovoltaic energy is one of the renewable energy source because it has given green energy and is pollution free. To increase and extract the maximum output power from the PV cell, the maximum power point technique (MPPT) is used. To get efficient maximum power, researchers are introduced wide types of algorithms. These algorithms have drawbacks such as slow and wrongly tracked. In PV systems, intelligent MPPTs are an extremely promising development. In this article, we present a method of controlling a photovoltaic system with variable insolation conditions by using MPPT and also utilize a logic controller to perform intelligent control of a converter. Here, the fuzzy intelligent controller is designed for the MPPT using MATLAB/Simulink.

Keywords Photovoltaic system · MPPT DC-DC converter · Fuzzy logic controller

1 Introduction

Power supply is a big concern in the energy sector because day to day usage is increasing and traditional energy sources aren't sufficient to supply that electricity. As a result of using fuels continuously, fossil fuel deposits have dramatically decreased and cause global warming [1]. By considering the high pollution that effects the environment with conventional sources, nowadays the researchers focus on green energy sources like solar, wind and geothermal energy.

In this paper, authors focus on solar energy for the power applications. To extract power using the solar energy, there will be a series and parallel connection of photovoltaic (PV) cells that is required. To extract the power in the photovoltaic cell, there is need to find highest track point in the panel. To track the MPPT, there are

K. H. Reddy (✉) · S. Sharma · N. C. Kumar · C. N. Reddy · I. M. Naidu · N. Akshay
East Point College of Engineering and Technology, Bengaluru, India
e-mail: kharshahari@gmail.com

S. Sharma et al. (eds.), *Intelligent Control, Robotics, and Industrial Automation*, Lecture Notes in Electrical Engineering 1066,
https://doi.org/10.1007/978-981-99-4634-1_26

various methods proposed by the researchers. To get power from the solar panels can be efficiently completed by the power tracking controller. Also MPPT can significantly improve power efficiency of the system [2]. In this, consider the Perturb & Observation (P&O) method. In these various methods of MPPT techniques, there is disadvantage like slow tracking and there will be a reduction of power.

2 Literature Survey of MPPT Techniques

PV cell typically converts between 30 and 40% of the incident irradiance into electricity. To get more output from the solar cell, MPPT-based techniques are frequently used, and boost converter is connected to get continuous voltage which is useful to all applications like connected to different loads. In boost converter, if duty cycle changes then source impedance is approximated w.r.t to load impedance [3].

Cells that produce electricity through photovoltaic methods have extremely low efficiency. To maximize efficiency, methods should be pursued to align the source with the cargo. One such method is the MPPT [2, 3]. To achieve the maximum amount of power from varying sources, this system is used. There are many PV applications that use MPPT technology, including space satellites, solar vehicles and solar water pumps.

To calculate the maximum power point tracking value, researchers are defined and implemented like P&O, etc. [3–6]. With P&O method, the algorithm alters the operating voltage within the desired direction by sampling the operating voltage, these procedures will repeat continuously until maximum point is achieved [4–9]. Despite being straightforward to implement, the main is that in the event of rapidly changing atmospheric conditions, the system may not always keep a steady maximum operating point [3, 4, 6]. The I&C algorithm will have to be operated with slope of voltage and power curves [10, 11]. These control circuits involved are expensive and complex [4–6]. To get the maximum point, fuzzy logic intelligent-based control method is used to overcome traditional methods' drawbacks. The symbolic logic controller has an efficient implementation, improved tracking performance, less oscillations and easy to implement [12]. MPPT-based fuzzy logic control method operates and controls the duty cycle in PV system by considering the curve slope of PV system that has output voltage in need [13].

3 Proposed Methodology Overview

The methodology has PV panel connected to MPPT controller and DC-DC Boost converter. The converter given has to normalize the output voltage. The DC voltage is derived by maximum power tracking from PV cell array. To track peak point, MPPT technology is used in solar PV panel. Figure 1 describes the methodology of proposed system.

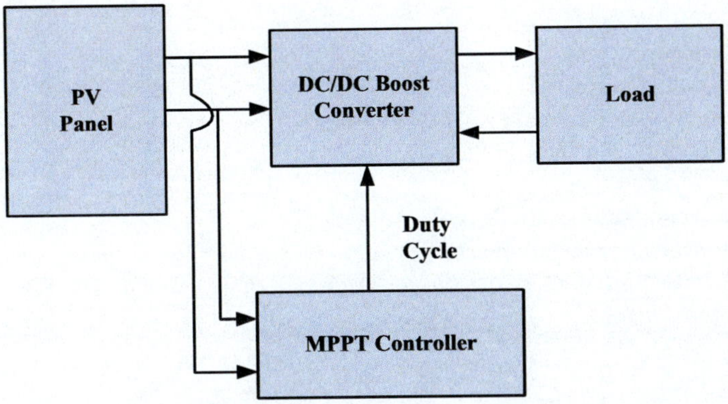

Fig. 1 Methodology of proposed system

3.1 Photovoltaic System Model

The photovoltaic (PV) cell is a semiconductor device which will generate the moving of electrons when exposed to light, so it produces the current. Figure 2, shows the electrical circuit of PV system. The conditions are taken at 25 ° and 1000 irradiance.

By considering one PV cell [14], it is often modelled by using the source of current with series connection of diode and two resistors. According to Fig. 2, the characteristic equation of current I represents in the form of current and voltages [15–18].

$$I = I_{\text{ph}} - I_s\left[\exp\left(\frac{q(V + \text{IR}_s)}{\text{KTA}}\right) - 1\right] - \frac{V + \text{IR}_s}{R_{\text{sh}}} \tag{1}$$

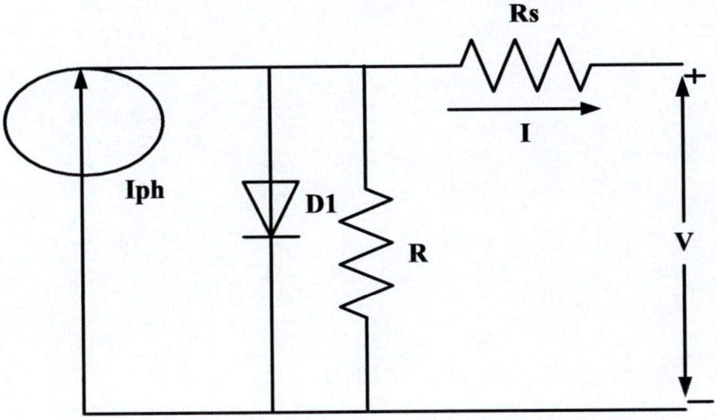

Fig. 2 Network of solar cell

I_{ph}: PV cell current,
I_s: Saturation current,
q: Charge of electrons,
k: Constant of Boltzmann (1.3810–23 J/K),
T: Temperature constant (in kelvin),
R_{sh}: Shunt resistance,
A: Ideal factor value of diode and
R_s: Resistance connected in series.

The IV and PV characteristics curves are made Figs. 3 and 4. The Eq. (2) gives the PV system current.

$$I_{ph} = \frac{G}{Gref}\left[I_{ph,ref} + C_T(T - T_r)\right] \tag{2}$$

here, terminology are given by
G: Constant (insulation) And
T: Cell temperature in Kelvin.
The I_s current is shown in Eq. (3).

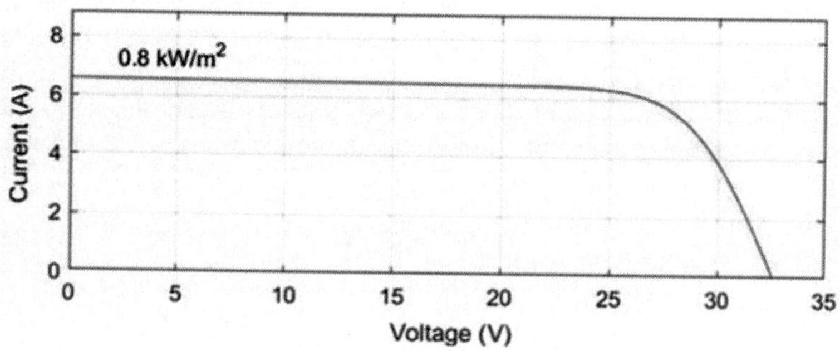

Fig. 3 PV module I V characteristics

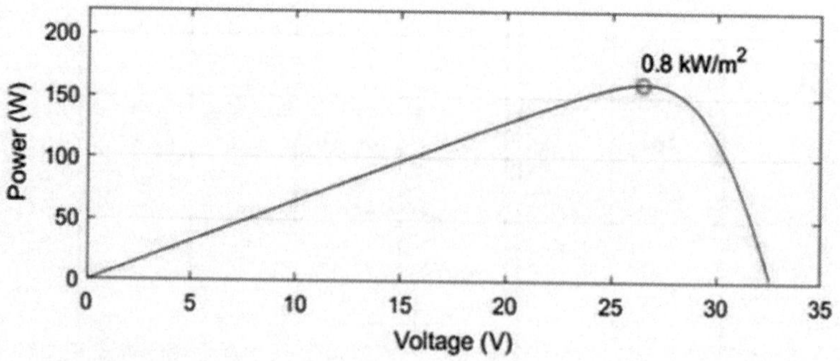

Fig. 4 PV module power—voltage characteristics

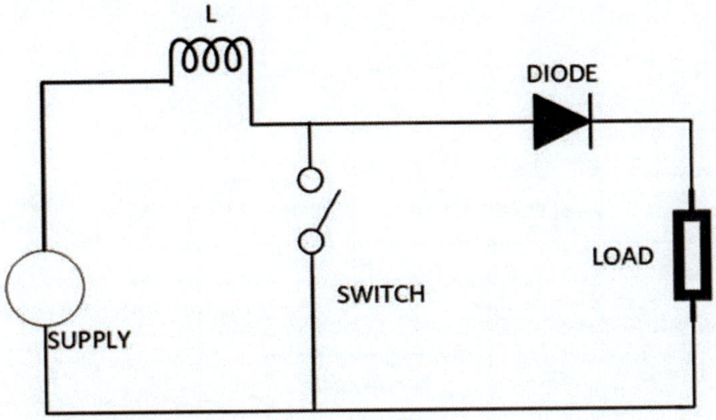

Fig. 5 Circuit diagram of boost converter

$$I_s = I_{s,\mathrm{ref}}\left[\frac{T}{T_{\mathrm{ref}}}\right]^3 \exp\left(\frac{qE_G}{\mathrm{KA}}\right)\left(\frac{1}{T_{\mathrm{ref}}} - \frac{1}{T}\right) \tag{3}$$

3.2 *Boost Converter*

Step-up converters (boost converters) step up voltage while lowering current from the input to the output. Taking into account the law of conservation of energy and assuming the converter is lossless (ideal), the input and output power should remain equal. Due to this, the output current of this converter decreases for higher output voltage. The converter can be powered by battery, solar panel, rectifier, DC generator, etc.

Figure 5 shows the circuit diagram of boost converter. The elements having in the boost converter is power supply, inductor connected with switch and diode, at last output side load is connected. By using the duty cycle with pulse width modulation, it is possible to boost the voltage of PV cell to required output that has load [20, 21].

3.3 *Fuzzy Logic-based MPPT Control Method*

The fuzzy logic-based intelligent control method is used as MPPT controller to track the maximum power point in the PV array output.

The fuzzy logic control method is used to track the maximum power point in the solar cell (PV) module. In the fuzzy logic method mapping from input to output is very convenient and it is easy to operate. The fuzzy method uses the pure mathematics, i.e. input and output variables is a degree of membership functions. Due to this

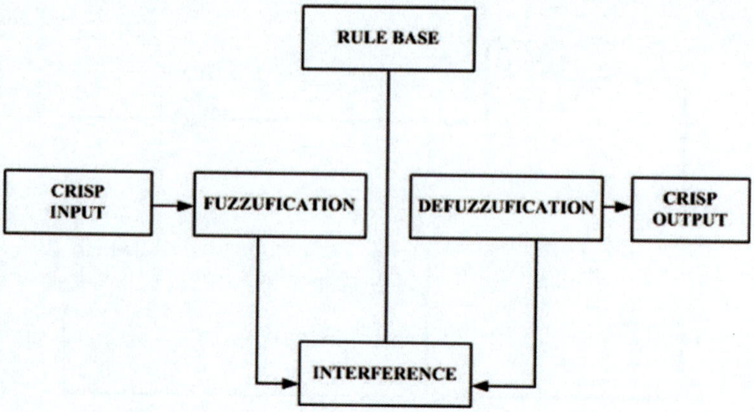

Fig. 6 Systemic diagram of fuzzy logic control method

mathematical advantage, it has been introduced in MPPT controller method. The fuzzy method is robust in nature and has simple structure. And also the designer has the entire knowledge of operation of PV module and MPPT [16]. The basic structure of proposed method is shown in Fig. 6.

The structure of fuzzy method consists of three stages, one is fuzzification, interface and rule base and defuzzification process.

In first step, fuzzification converts normal crisp value in to fuzzy input. In the second stage is interface and rule base system, here depending on rule base system, we need to give the rules to act. The sample typical rule will be like this, IF A and B is HIGH then output will be HIGH [16]. Based on these set of 64 rules, the fuzzy system interprets the input values and gives the related output values.

The final stage is defuzzification, which converts fuzzy outputs into crisps outputs. In the defuzzification, certain methods are used for conversion. Some of the methods are centroid methods, trapezoidal methods, etc. In this work, centroid method is used for defuzzification process. Here fuzzy logic is implemented, so that maximum power point can be determined quickly. In this method, fuzzy system is having two inputs and outputs. Equations (3) and (4) show the sample equations having two inputs, i.e. one is error and change in error. Equations (4) and (5) show the sample equations of error and change in error [20, 21].

$$e(k) = \left[P_{\text{pv}}(k) - P_{\text{pv}}(k-1)\right] / \left[V_{\text{pv}}(k) - V_{\text{pv}}(k-1)\right] \tag{4}$$

$$\text{ce}(k) = e(k) - e(k-1) \tag{5}$$

Figure 7 shows the fuzzy control method with inputs and outputs.

Here $P_{\text{pv}}(k)$ gives the total power of PV panel. Here, fuzzy has made to activate the duty cycle (D) which will initiate the boost converter.

Figure 8a, b shows the input membership functions.

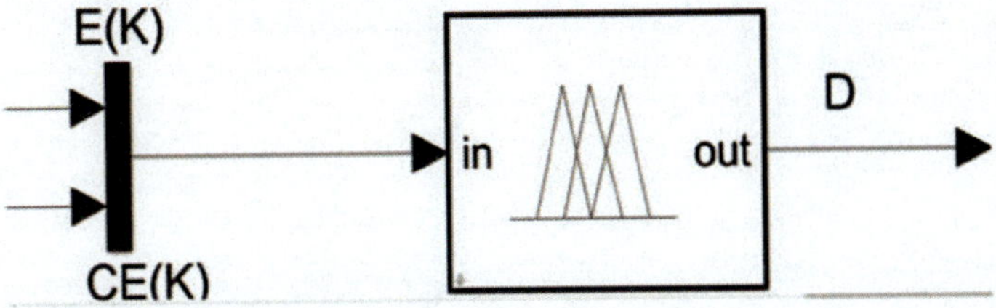

Fig. 7 Fuzzy logic control method

Figure 8c shows the output membership function which is used to design in proposed MPPT controller.

The final rule-based system is shown in Fig. 9.

4 Simulation Results

Modelling a PV module in MATLAB is done using Eq. (1). Here the solar cells are assumed to have a temperature of 25 °C with irradiance 1000 is constant with every time. Figure 10 represents a designed MATLAB PV model.

Testing of MPPT-based on fuzzy logic controller under varying solar irradiance and a constant temperature of 25 °C is done. The Simulink model consists of a PV module, a boost convertor to which a resistive load is connected. Boost convertor takes input voltage from the PV module and provides boosted voltage as the output. The Simulink model also consists of MPPT controller to track maximum power available. For this simulation, a five micro second sampling period was used [19]

The Simulation circuit of the fuzzy-based MPPT control system is shown in Fig. 11. Gate pulses to the MOSFET are provided by the PWM generator. Figure 12 shows the boosted output voltage which we obtained in MPPT algorithm by using a fuzzy system. The fuzzy logic will provide good tracking of maximum power and we get better power from solar PV cell.

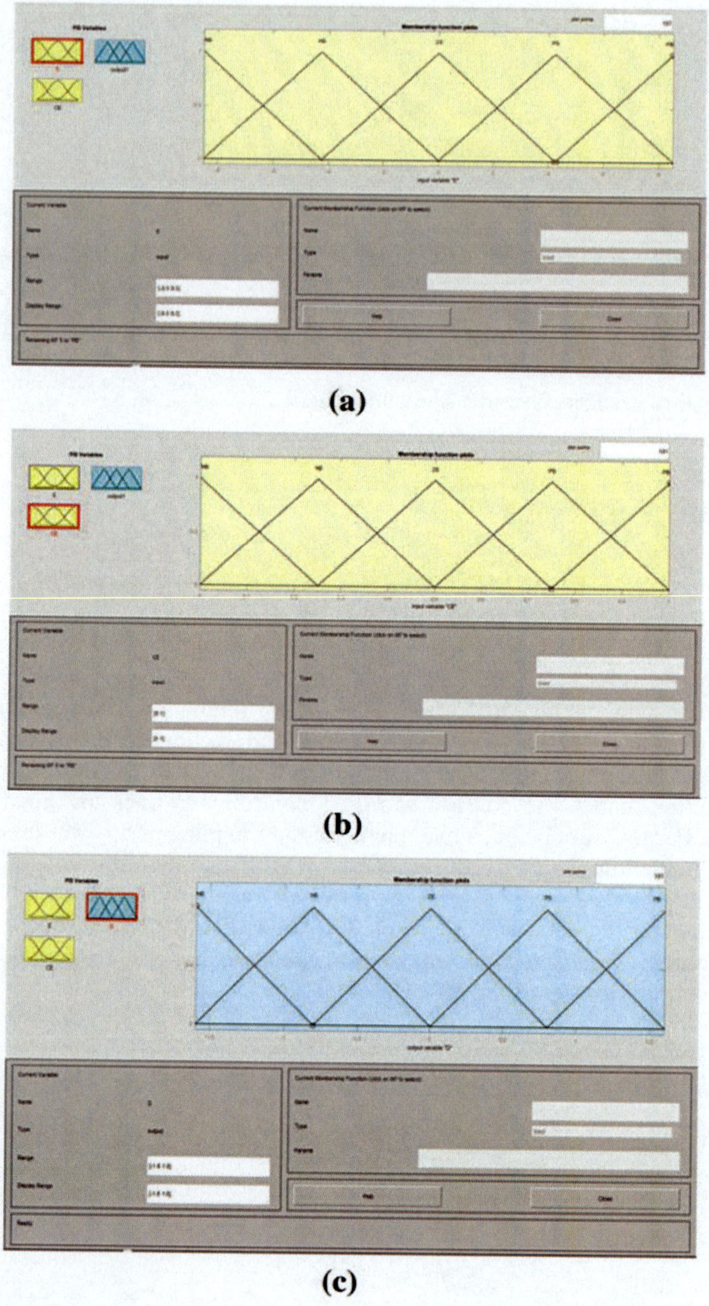

Fig. 8 **a** Error membership functions **b** Change in error membership functions **c** Duty cycle membership functions

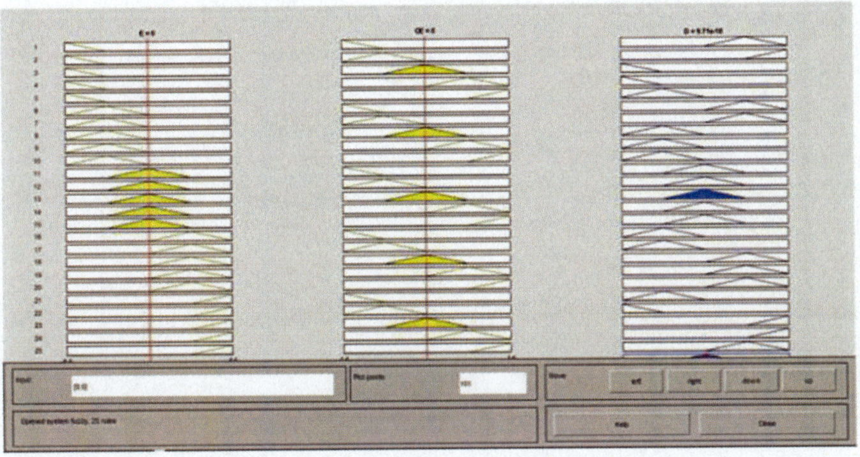

Fig. 9 Rule viewer

Fig. 10 PV model

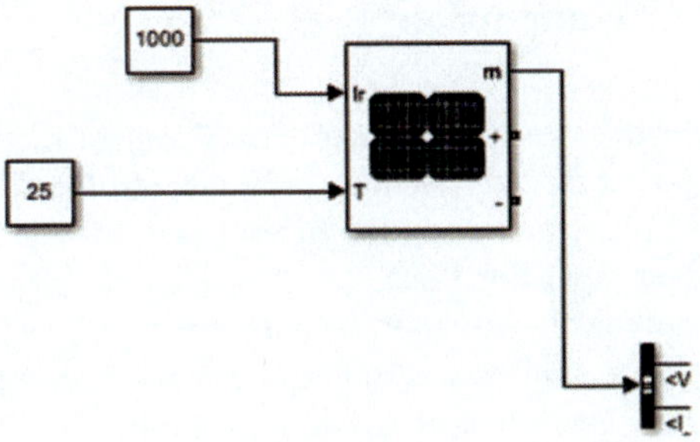

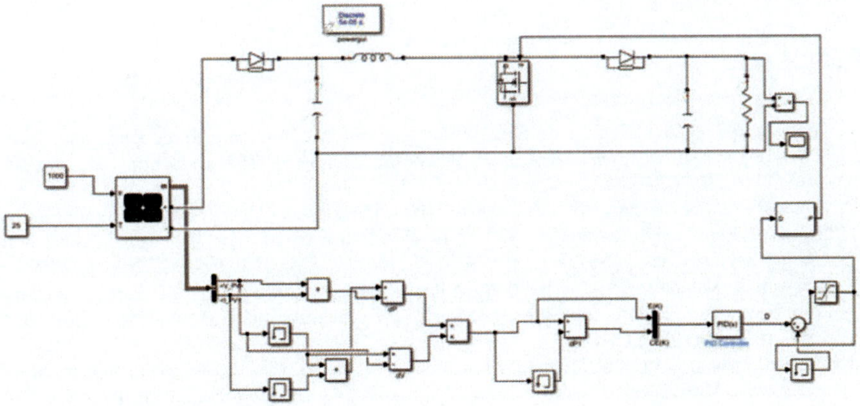

Fig. 11 Simulink circuit for the proposed method

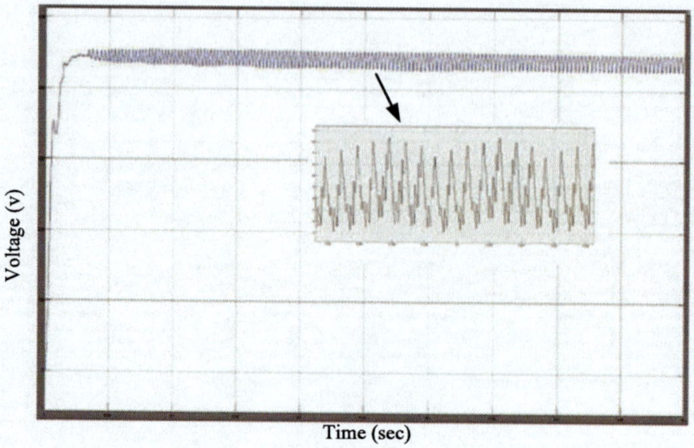

Fig. 12 Power response obtained using fuzzy logic controller

5 Conclusion

In this work, the fuzzy based control method is introduced to get the maximum power in PV system as MPPT controller. The simulation circuit is designed for the implementation of the fuzzy-based MPPT. To implement the fuzzy system, use the two input functions, i.e. error and change in error and duty cycle as an output. The duty cycle will alter the output power. The future scope of the work is to implement the neuro-fuzzy system to track the maximum power and also need to compare the fuzzy system with existing methods.

References

1. Villalva MG, Gazoli JR, Ruppert EF (2009) Comprehensive approach to modeling and simulation of photovoltaic arrays. IEEE Trans Power Electron 25(5):1198–1208. https://doi.org/10.1109/TPEL.2009.2013862
2. Alabedin AZ, El-Saadany EF, Salama MMA (2011) Maximum power point tracking for Photovoltaic systems using fuzzy logic and artificial neural networks. In: 2011 IEEE Power and Energy Society General Meeting, pp 1–9. IEEE. https://doi.org/10.1109/PES.2011.6039690
3. Rezvani A, Moghadam HM, Khalili A, Mohammadinodoushan M (2015) Comparison study of maximum power point tracker techniques for PV systems in the grid connected mode. Int J Rev Life Sci 5(10):1175–1184
4. Esram T, Chapman PL (2007) Comparison of photovoltaic array maximum power point tracking techniques. IEEE Trans Energy Convers 22(2):439–449. https://doi.org/10.1109/TEC.2006.874230
5. Jain S, Agarwal V (2007) Comparison of the performance of maximum power point tracking schemes applied to single-stage grid-connected photovoltaic systems. IET Electr Power Appl 1(5):753–762. https://doi.org/10.1049/iet-epa:20060475

6. Subudhi B, Pradhan R (2012) A comparative study on maximum power point tracking techniques for photovoltaic power systems. IEEE Trans Sustain Energy. 4(1):89–98. https://doi.org/10.1109/TSTE.2012.2202294
7. Pandey A, Dasgupta N, Mukerjee AK (2008) High-performance algorithms for drift avoidance and fast tracking in solar MPPT system. IEEE Trans Energy Convers 23(2):681–689. https://doi.org/10.1109/TEC.2007.914201
8. Berrera M, Dolara A, Faranda R, Leva S (2009) Experimental test of seven widely-adopted MPPT algorithms. In: 2009 IEEE Bucharest PowerTech, pp 1–8. IEEE. https://doi.org/10.1109/PTC.2009.528201
9. Li-Qun L, Zhi-xin W (2008) A rapid MPPT algorithm based on the research of solar cell's diode factor and reverse saturation current. WSEAS Trans Syst 7(5):568–579
10. De Cesare G, Caputo D, Nascetti A (2006) Maximum power point tracker for portable photovoltaic systems with resistive-like load. Sol Energy 80(8):982–988. https://doi.org/10.1016/j.solener.2005.07.010
11. Ankaiah B, Nageswararao J (2013) MPPT algorithm for solar photovotaic cell by incremental conductance method. Int J Inno Eng Tech (IJIET) 2(1):17–23
12. Shah N, Chudamani R (2012) Grid interactive PV system with harmonic and reactive power compensation features using a novel fuzzy logic based MPPT. In: 2012 IEEE 7th International Conference on Industrial and Information Systems (ICIIS), pp 1–6. https://doi.org/10.1109/ICIInfS.2012.6304830
13. Islam MA, Talukdar AB, Mohammad N, Khan PS (2010) Maximum power point tracking of photovoltaic arrays in Matlab using fuzzy logic controller. In: 2010 Annual IEEE India Conference (INDICON), pp 1–4. https://doi.org/10.1109/INDCON.2010.5712680
14. Sridhar R, Jeevananathan D, ThamizhSelvan N, Banerjee S (2010) Modeling of PV array and performance enhancement by MPPT algorithm. Int J Comp Appl 7(5):0975–8887
15. Said S, Massoud A, Benammar M, Ahmed S (2012) A Matlab/Simulink-based photovoltaic array model employing SimPowerSystems toolbox. J Energy Power Eng 6(12):1965–1975
16. Karthika S, Velayutham K, Rathika P, Devaraj D (2014) Fuzzy logic based maximum power point tracking designed for 10kW solar photovoltaic system with different membership functions. WASET Int J Elect Comp Eng 8(6):1022–1027
17. Salmi T, Bouzguenda M, Gastli A, Masmoudi A (2012) Matlab/simulink based modeling of photovoltaic cell. Int J Renew Ene Res 2(2):213–218
18. Revankar PS, Thosar AG, Gandhare WZ (2010) Maximum power point tracking for PV systems using MATALAB/SIMULINK. In: 2010 Second International Conference on Machine Learning and Computing, pp 8–11. IEEE. https://doi.org/10.1109/ICMLC.2010.54
19. Tsai HL, Tu CS, Su YJ (2008) Development of generalized photovoltaic model using MATLAB/SIMULINK. In: Proceedings of the world congress on Engineering and computer science, pp 1–6. ISBN: 978-988-98671-0-2
20. Ngan MS, Tan CW (2011) A study of maximum power point tracking algorithms for standalone photovoltaic systems. In: 2011 IEEE Applied Power Electronics Colloquium (IAPEC), pp 22–27. IEEE. https://doi.org/10.1109/IAPEC.2011.5779863
21. Bouchafaa F, Beriber D, Boucherit MS (2010) Modeling and simulation of a gird connected PV generation system with MPPT fuzzy logic control. In: 2010 7th International Multi-Conference on Systems, Signals and Devices, pp 1–7. IEEE. https://doi.org/10.1109/SSD.2010.5585530

Error Minimization-Based Order Diminution of Interconnected Wind-Turbine-Generator

Umesh Kumar Yadav, V. P. Meena, and V. P. Singh

Abstract In this research proposal, reduced-order (RO) model for interconnected wind-turbine-generator (WTG) system is presented by incorporating time-moments (T-Ms) as well as Markov-parameters (M-Ps) of higher-order (HO) WTG system and its RO WTG model. The T-Ms along with M-Ps of HO WTG system and its expected RO WTG model are employed for formation of fitness function. The weighted-errors as steady-state error gain, between T-Ms of the WTG model and HO WTG system are diminished to obtain the better steady-state response of RO WTG model in comparison with HO WTG system. The transient response of the model is improved by diminishing weighted errors as transient response error gain, between M-Ps of the RO WTG model and HO WTG system. The minimization of these errors are achieved by employing grey-wolf optimizer. The optimization of is done by satisfying the constraints of ensuring steady-state matching and stability of desired RO WTG model so as to obtained better RO model. The efficacy and applicability of proposed methodology are demonstrated by presenting the step response along with impulse and Bode responses. The tabulated data are also provided in terms of specifications of time-domain and error-criterion in the support of the readability of presented method.

Keywords Interconnected system · Modelling · Optimization algorithm · Reduction · Stability · Wind-turbine-generator

U. K. Yadav (✉) · V. P. Meena · V. P. Singh
Electrical Engineering Department, MNIT, Jaipur, India
e-mail: 2019ree9085@mnit.ac.in

V. P. Meena
e-mail: vmeena1@ee.iitr.ac.in

V. P. Singh
e-mail: vinay.ee@mnit.ac.in

S. Sharma et al. (eds.), *Intelligent Control, Robotics, and Industrial Automation*, Lecture Notes in Electrical Engineering 1066,
https://doi.org/10.1007/978-981-99-4634-1_27

1 Introduction

In the present era of modernization in technological development of industrial and research fields, power demand is also increasing abruptly. Due to the extensive increment in electric power requirement, exploitation of fossil fuels is also reached at threshold level. The higher exploitation of fossil fuels for industrial applications, transportation, domestic applications, etc. are main cause of pollution and degradation of green house effect. The renewable and non-conventional resources of energy are substitute of conventional one. Since, last few decades the utilization of renewable energy resources (RERs) are growing day by day. In this regard, solar-photovoltaic, wind energy, hydro power, etc. are sharing most of the percentage of energy generation throughout the world.

Among all RERs, wind energy is one of the promising, clean and freely available energy resource. But, the main limitation of wind energy is exploitation of useful power from wind due to variation in wind speed. The speed of wind-turbine is controlled by variety of speed controllers. The wind-turbine-generator (WTG) is affected by several system dynamics. All these system dynamics increase the order of the WTG system. This higher-order (HO) system needs HO controller. The HO system and controller may suffer from some limitations, due to its HO. These limitations are complication in framing control laws, design complexity, difficulty in understanding the system, uneconomical in control design, etc. The limitations of HO systems can be easily surpassed by deriving well suited reduced-order (RO) model for HO system.

In literature [1–6], several researchers exploited the order-reduction techniques (ORTs) to propose the RO model for HO systems. In [1], Mariani et al. exploited the ORT for stability analysis of micro-grid as power system applications. Similar work for DC micro-grid is also proposed by Wang et al. in [2]. The parametric averaging-based approach is applied by Zhang et al. in [3] to reduce the order of converter system. In [4], Díaz et al. utilized the ORT for power electronics applications. Later, ORT to reduce the order of HO integrated energy system is presented by Wang et al. in [5] for dynamic analysis. Further, robotic application of ORT is proposed by Sikander and Prasad in [6]. Furthermore, the parametric approximation method for permanent magnet machine is exploited by Far et al. in [7]. In some literature [8–11], several researchers utilize the ORT for interval systems also. The ORTs for reduction of WTG system are exploited in by Roy et al. [14] for single-area, two-area, and three-area interconnected systems.

In this presented work, diminution of HO interconnected WTG is accomplished with the help of time-moments (T-Ms) as well as Markov-parameters (M-Ps) of HO WTG system and its RO model. The fitness function is constructed firstly, by utilizing T-Ms and M-Ps of the system and model. The errors between RO WTG and HO WTG, T-Ms and M-Ps are desired to be minimize for ascertainment of RO WTG model for HO WTG system. The minimization is done by employing grey-wolf optimizer (GWO) algorithm. The minimization is done by satisfying matching of steady-state constraint and Hurwitz-stability criterion. In steady-state constraint, first T-M of HO WTG system and RO WTG model are matched whereas Hurwitz criterion provides the stable RO WTG model. The superiority and applicability of proposed methodology for determination of RO WTG model are proved with the help of responses and tabulated data. The step, impulse and Bode plots are provided along with specifications of time-domain and error-criterion.

This organization of sections presented in this proposal are as follows: Sect. 2 describes the interconnected WTG system with its dynamical parameters. Problem formulation is demonstrated in Sect. 3. Section 4 is briefly explains about GWO algorithm. Diminution of HO-WTG system is done in Sect. 5 by incorporating manifested results and discussion. In Sect. 6, proposed methodology is concluded by discussing the future scope of the proposal.

2 Mathematical Representation of Interconnected Wind-Turbine-Generator System

The interconnected wind-turbine-generator (WTG) dynamics and its mathematical formulation of closed-loop transfer function are presented for single-area power system model. The block diagram of interconnected single-are WTG system is combination of WTG model, governor model, machine model, load model, actuators, etc. The basic interconnected WTG block diagram is represented in Fig. 1. The mathematical representation of WTG model can be represented as

$$M_1 = \left(\frac{\kappa_{wc}}{1 + s\tau_{wc}}\right) \tag{1}$$

The transfer function of governor of WTG system can be given as

$$M_2 = \left(\frac{\kappa_{gc}}{1 + s\tau_{gc}}\right) \tag{2}$$

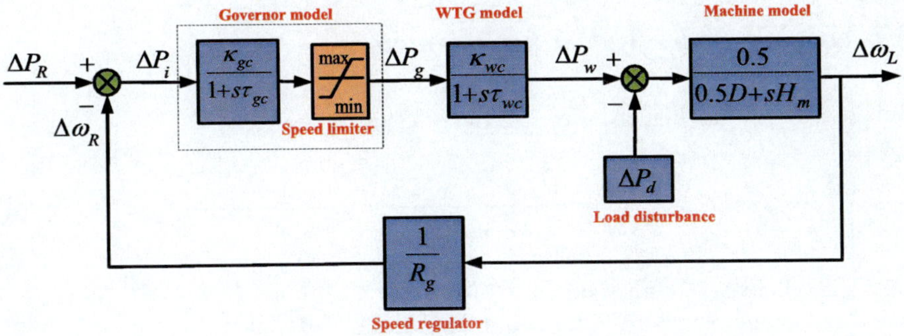

Fig. 1 Representation of interconnected wind-turbine-generator system

Table 1 Interconnected WTG system parameters

Parameter	Value
WTG time constant (τ_{wc})	1.5 s
WTG gain constant (κ_{wc})	2.0
Governor time constant (τ_{gc})	0.08 s
Governor gain constant (κ_{gc})	1.0
WTG inertia constant (H_m)	0.4 s
Load-damping coefficient (D)	1.0
Speed-regulation factor (R_g)	0.2 pu

The mathematical representation of machine and load model can be expressed as

$$M_3 = \left(\frac{0.5}{0.5D + sH_m}\right) \tag{3}$$

With the help of mathematical formulations presented in (1), (2) and (3), the overall closed-loop transfer function of interconnected WTG system can be given as

$$M_{tf} = \frac{\Delta\omega_L(s)}{-\Delta P_d(s)} = \frac{(1+s\tau_{wc})(1+s\tau_{gc})/2}{0.5/R_g + [(1+\tau_{wc})(1+s\tau_{gc})(0.5D + sHm)]} \tag{4}$$

By utilizing the WTG parameters as provided in Table 1, the third-order WTG transfer function is obtained as

$$\frac{\Delta\omega_L(s)}{-\Delta P_d(s)} = \left(\frac{0.120\,s^2 + 1.580\,s + 1.000}{0.096\,s^3 + 1.384\,s^2 + 2.380\,s + 7.670}\right) \tag{5}$$

3 Problem Formulation

3.1 Higher-Order System Representation

Suppose, H^{th} order, higher-order system (HOS) is represented by

$$G_H(s) = \frac{\hat{U}(s)}{\hat{V}(s)} = \frac{\hat{U}_0 + \hat{U}_1 s + \hat{U}_2 s^2 + \cdots + \hat{U}_{H-1} s^{H-1}}{\hat{V}_0 + \hat{V}_1 s + \hat{V}_2 s^2 + \cdots + \hat{V}_H s^H} \tag{6}$$

In (6), coefficients of numerator are $\hat{U}_i$ for $i = 0, 1, 2, 3, 4, \ldots, (H-1)$ and coefficients of denominator are $\hat{V}_i$ for $i = 0, 1, 2, 3, 4, \ldots, H$.

3.2 Time-Moments (T-Ms) and Markov-Parameters (M-Ps) Representation of HOS

The HOS depicted in (6) is expanded around $s = 0$ in terms of time-moments (T-Ms) and expanded around $s = \infty$ in terms of Markov-parameters (M-Ps) as provided in (7) and (8), respectively.

$$G_H(s) = \hat{t}_0 + \hat{t}_1 s + \hat{t}_2 s^2 + \cdots + \hat{t}_H s^H + \cdots \quad (7)$$

$$G_H(s) = \hat{m}_1 s^{-1} + \hat{m}_2 s^{-2} + \cdots + \hat{m}_H s^{-H} + \cdots \quad (8)$$

In (7), T-Ms of HOS given in (6) are $\hat{t}_i$ for $i = 0, 1, 2, 3, 4, \ldots$. Similarly, in (8), $\hat{m}_i$ for $i = 1, 2, 3, 4, \ldots$ are M-Ps of HOS represented in (6).

3.3 Reduced-Order Model Representation

The HOS shown in (6) is desired to reduce to reduced-order model (ROM) of order l. The order of ROM is such that $l < H$. Let, the lth order ROM be

$$E_l(s) = \frac{u'(s)}{v'(s)} = \frac{u'_0 + u'_1 s + u'_2 s^2 + \cdots + u'_{l-1} s^{l-1}}{v'_0 + v'_1 s + v'_2 s^2 + \cdots + v'_l s^l} \quad (9)$$

where, coefficients of numerator are u'_i for $i = 0, 1, 2, 3, 4, \ldots, (l - 1)$ and v'_i for $i = 0, 1, 2, 3, 4, \ldots, l$ are coefficients of denominator of approximated-model depicted in (9).

3.4 Time-Moments (T-Ms) and Markov-Parameters (M-Ps) Representation of ROM

The expansions as expressed in (7) and (8) of HOS, similar expansions in terms of T-Ms and M-Ps are obtained in (10) and (11), respectively.

$$E_l(s) = t'_0 + t'_1 s + t'_2 s^2 + \cdots + t'_l s^l + \cdots \quad (10)$$

$$E_l(s) = m'_1 s^{-1} + m'_2 s^{-2} + \cdots + m'_l s^{-l} + \cdots \quad (11)$$

In (10) and (11), respectively, t'_i for $i = 0, 1, 2, 3, 4, \ldots$ are T-Ms, and m'_i for $i = 1, 2, 3, 4, \ldots$ are M-Ps of (9).

3.5 Fitness Function and Constraints

The manifestation of desired ROM is accomplished by exploiting T-Ms and M-Ps of HOS shown in (6) and ROM presented in (9). The first T-Ms HOS and ROM are exploited to guarantee the steady-state matching. So, $(l-1)$ T-Ms and l M-Ps are utilized to formulate the fitness function as

$$J=\sum_{i=1}^{l-1}\left[\omega_i^t\left(1-\frac{t'_i}{\hat{t}_i}\right)^2\right]+\sum_{j=1}^{l}\left[\omega_j^m\left(1-\frac{m'_i}{\hat{m}_j}\right)^2\right] \quad (12)$$

where, ω_i^t and ω_i^m are weights, respectively, associated with T-Ms error and M-Ps error of HOS and ROM. The fitness function represented in (12) can be rearranged as given in (13).

$$J=\omega_1^t J_1^t+\omega_2^t J_2^t+\omega_3^t J_3^t+\cdots+\omega_1^m J_1^m+\omega_2^m J_2^m+\omega_3^m J_3^m+\cdots \quad (13)$$

The fitness function formulated in (12) is diminished under the constraints provided in (14) and (15). The necessary condition to determine the desired ROM is matching of first T-M of HOS and ROM to confirm zero steady-state error as

$$\hat{t}_0=t'_0 \quad (14)$$

where, $\hat{t}_0$ is first T-M of HOS and t'_0 is first T-M of desired ROM. Secondly, ensuring the stability of ROM by Hurwitz-stability criterion is also necessary to obtain stable ROM. This can be given as

$$v'(s) \text{ of ROM, given in (9) should be Hurwitz.} \quad (15)$$

The fitness function depicted in (12) is desired to be optimized. The associated weights are to be calculated in similar fashion as provided in [12]. The minimization of fitness function shown in (12) is ascertained by employing grey-wolf optimizer.

4 Grey-Wolf Optimizer

The grey-wolf optimizer (GWO) is coined by Mirjalili et al. [13]. In this algorithm, pattern and behaviour during hunting and prey tracking of grey wolves is utilized in form of mathematical representation. The wolves in a pack are known as alpha (α) (the prime member), beta (β) (second prime member), delta (δ) (helping member), and omega (ω) (least dominant member). The higher level wolf in grey-wolf hierarchy is alpha. Alpha is the leader of the group. The hunting of the prey is performed by alpha. The second and third level depends on domination level are known as beta wolf and delta wolf, respectively. Beta reinforce the command of alpha to the group and acknowledge the alpha about decisions of pack. The deltas help in hunting to alpha and betas. The members those who are caretakers, scouts, and sentinels of the pack are delta wolves. The least dominant level in grey-wolf hierarchy are omegas.

In GWO, hunting is performed by most dominant wolf, i.e. alpha. The hunting pattern with updated locations of alpha wolf, beta wolf, and delta wolf can be obtained by using encircling behaviour and hunting approach as represented in (16) and (17), respectively.

$$\vec{g}_w(t^+) = \vec{g}_p(t) - \vec{a}\left[\left|\vec{b}.\vec{g}_p(t) - \vec{g}_w(t)\right|\right] \tag{16}$$

$$\vec{g}_w(t^+) = (\vec{g}_{w1} + \vec{g}_{w2} + \vec{g}_{w3})/3 \tag{17}$$

In (16), $\vec{g}_w(t^+)$ is the updated location of the grey-wolf whereas $\vec{g}_w(t)$ is current location of grey-wolf and $\vec{g}_p(t)$ depict current location of prey, respectively. The alpha's location, beta's location and delta's location are evaluated with the help of (18), (19) and (20), respectively.

$$\vec{g}_{w1} = \vec{g}_{w\alpha} - \vec{a}_1\left[\left|\vec{b}_1.\vec{g}_{w\alpha} - \vec{g}_w(t)\right|\right] \tag{18}$$

$$\vec{g}_{w2} = \vec{g}_{w\beta} - \vec{a}_2\left[\left|\vec{b}_2.\vec{g}_{w\beta} - \vec{g}_w(t)\right|\right] \tag{19}$$

$$\vec{g}_{w3} = \vec{g}_{w\delta} - \vec{a}_3\left[\left|\vec{b}_3.\vec{g}_{w\delta} - \vec{g}_w(t)\right|\right] \tag{20}$$

The vectors utilized in (16) are such that, $\vec{a} = 2.\vec{e}.\vec{C}_1 - \vec{e}$ for $\vec{e} \in [2, 0]$, and $\vec{C}_1$ and $\vec{C}_2 \in [0, 1]$.

5 Order Reduction of Interconnected Wind-Turbine-Generator System

The transfer function of higher-order (HO) interconnected wind-turbine-generator (WTG) system [14] is depicted in (21).

$$G_3(s) = \left(\frac{0.120\,s^2 + 1.580\,s + 1.000}{0.096\,s^3 + 1.384\,s^2 + 2.380\,s + 7.670}\right) \tag{21}$$

The HO-WTG system's T-Ms are provided in (22) and, its M-Ps are depicted in (23).

$$G_5(s) = 0.130378 + 0.165541s - 0.059248s^2 \quad -0.013118s^3 + 0.012689s^4 - 0.000829s^5 + \cdots \tag{22}$$

$$G_5(s) = 1.25s^{-1} - 1.5625s^{-2} + 1.953125s^{-3} - 89.290365s^{-4} + 1363.685438s^{-5} + \cdots \tag{23}$$

Suppose, desired ROM of second-order for HOS depicted in (21) is as follows

$$G_2(s) = \frac{u_0' + u_1' s}{v_0' + v_1' s + v_2' s^2} \tag{24}$$

The ROM represented in (24) is depicted in expressions of T-Ms and M-Ps, respectively, as given in (25) and (26).

$$E_2(s) = \left(\frac{u_0'}{v_0'}\right) + \left(\frac{u_1' v_0' - u_0' v_1'}{v_0'^2}\right)s + \left(\frac{u_0' v_1'^2 - u_1' v_0' v_1' - u_0' v_0' v_2'}{v_0'^3}\right)s^2 + \cdots \tag{25}$$

$$E_2(s) = \left(\frac{u_1'}{v_2'}\right)s^{-1} + \left(\frac{u_0' v_2' - u_1' v_1'}{v_2'^2}\right)s^{-2} \quad + \left(\frac{u_1' v_1'^2 - u_0' v_1' v_2' - u_1' v_0' v_2'}{v_2'^3}\right)s^{-3} + \cdots \tag{26}$$

Now, ascertainment of unknown coefficients of ROM given in (24), are done by formulating fitness function. The fitness function is formulated by employing M-Ps of the HO-WTG system and desired ROM. Since, the desired order of ROM is $l = 2$, so at least $(2l - 1)$, i.e. $2 \times 2 - 1 = 3$ terms between T-Ms and M-Ps of the HOS and ROM are matched. In the fitness function second T-M, and first and second M-Ps are considered. So, the fitness function depicted in (12) becomes

$$J = \sum_{j=1}^{1}\left[\omega_j^t\left(1 - \frac{t_j'}{\hat{t}_j}\right)^2\right] + \sum_{j=1}^{2}\left[\omega_j^m\left(1 - \frac{m_j'}{\hat{m}_j}\right)^2\right] \tag{27}$$

The fitness function (27) can be re-framed as

$$J = \omega_1^t\left(1 - \frac{t_1'}{\hat{t}_1}\right)^2 + \omega_1^m\left(1 - \frac{\tilde{m}_1}{\hat{m}_1}\right)^2 + \omega_2^m\left(1 - \frac{\tilde{m}_2}{\hat{m}_2}\right)^2 \tag{28}$$

By utilizing the T-Ms and M-Ps of the HO WTG system and RO WTG model, the fitness function shown in (28) turns out to be

$$J = \omega_1^t\left(1 - \left(\frac{u_1^{'}v_0^{'} - u_0^{'}v_1^{'}}{0.165541{v_0^{'}}^2}\right)\right)^2 + \omega_1^m\left(1 - \frac{u_1^{'}}{1.25v_2^{'}}\right)^2 + \omega_2^m\left(1 + \left(\frac{u_0^{'}v_2^{'} - u_1^{'}v_1^{'}}{1.5625{v_2^{'}}^2}\right)\right)^2 \quad (29)$$

The fitness function depicted in (29) can be reproduced as

$$J = \omega_1^t J_1 + \omega_2^m J_3 \quad (30)$$

In (30), objectives J_1 and J_2, respectively, are formulated to minimize the errors between second T-Ms and second M-Ps of HO WTG system and desired WTG ROM. The associated weights ω_1^t and ω_2^m, with the sub-objectives are considered by providing equal importance to both, errors minimization between T-Ms and M-Ps. So, the weights ω_1^t and ω_2^m are

$$\omega_1^t = 0.5, \quad \omega_2^m = 0.5 \quad (31)$$

By utilizing the weights from (31), the fitness function given in (29) modifies to resultant fitness function given in (32).

$$J = 0.5\left(1 - \left(\frac{u_1^{'}v_0^{'} - u_0^{'}v_1^{'}}{0.165541{v_0^{'}}^2}\right)\right)^2 + 0.25\left(1 - \frac{u_1^{'}}{1.25v_2^{'}}\right)^2 + 0.25\left(1 + \left(\frac{u_0^{'}v_2^{'} - u_1^{'}v_1^{'}}{1.5625{v_2^{'}}^2}\right)\right)^2 \quad (32)$$

The fitness function depicted in (32) is optimized under the constraints, provided in (33) and (34). The constraint given in (14) is changes to (33).

$$u_0^{'} = 0.130378v_0^{'} \quad (33)$$

And, constraint given in (15), becomes

$$v_1^{'} > 0, \quad v_0^{'}v_1^{'} > 0 \quad (34)$$

The resultant fitness function represented in (32) is solved by employing GWO subjected to constraints defined in (33) and (34). Hence, desired second-order ROM represented in (24) is obtained as

$$E_2(s) = \frac{0.0252 + 0.03777s}{0.1933 + 0.05217s + 0.0285s^2} \quad (35)$$

The proposed RO-WTG model determined in (35) is compared with the models given in (36)-(43) already available in literature. These models are as follows:

$$\text{Model I}(s) = \frac{129.7s + 499.6}{s^2 + 6.835s + 16.183} \quad (36)$$

$$\text{Model II}(s) = \frac{1130s + 247.6}{s^2 + 30.25s + 94.43} \quad (37)$$

$$\text{Model III}(s) = \frac{946.3s + 227.9}{s^2 + 24.55s + 386} \quad (38)$$

$$\text{Model IV}(s) = \frac{-41.52s + 353.4}{s^2 + 1.158s + 598.6} \quad (39)$$

$$\text{Model V}(s) = \frac{266.5s + 595.6}{s^2 + 8.858s + 17.03} \quad (40)$$

$$\text{Model VI}(s) = \frac{266.5s + 595.6}{s^2 + 8.858s + 17.03} \quad (41)$$

$$\text{Model VII}(s) = \frac{266.5s + 595.6}{s^2 + 8.858s + 17.03} \quad (42)$$

$$\text{Model VIII}(s) = \frac{266.5s + 595.6}{s^2 + 8.858s + 17.03} \quad (43)$$

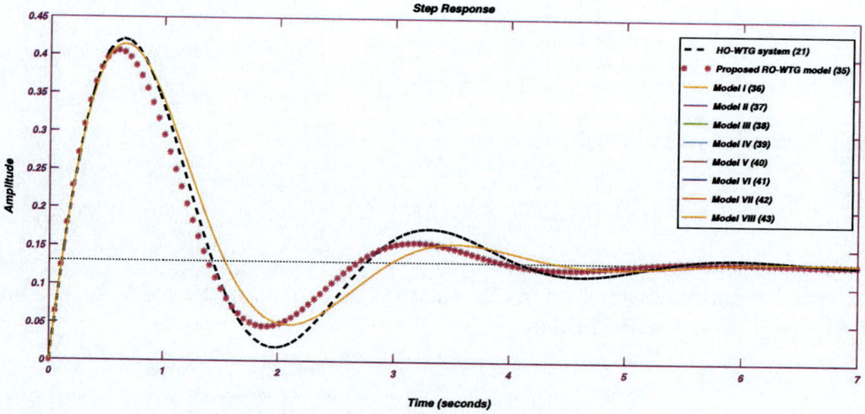

Fig. 2 Step response of HO-WTG system and RO-WTG models

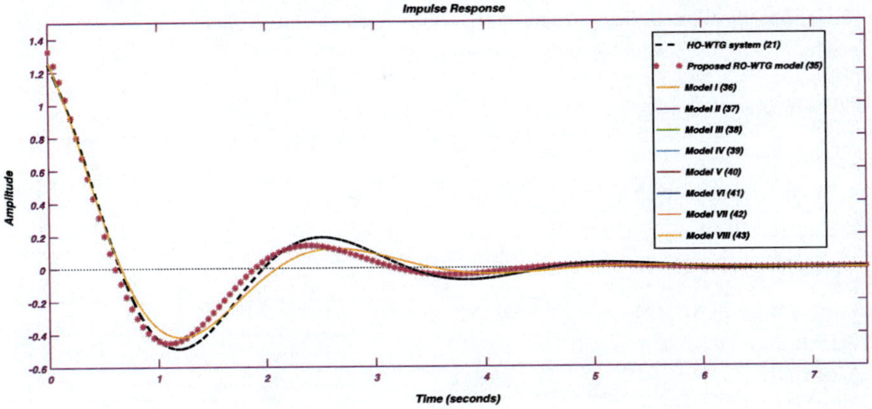

Fig. 3 Impulse response of HO-WTG system and RO-WTG models

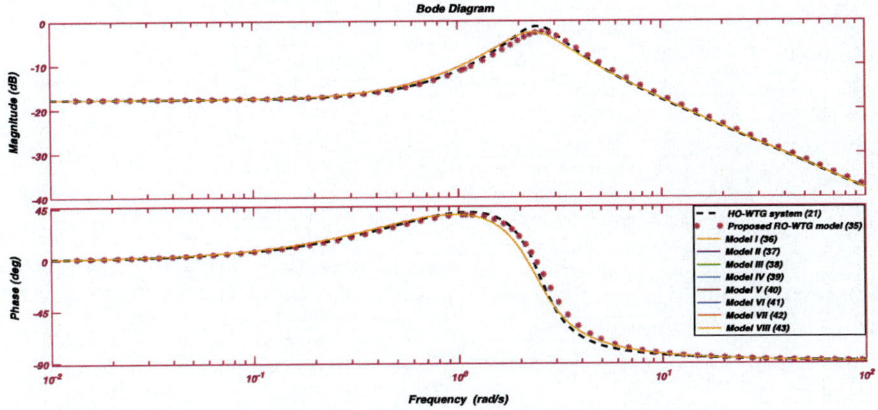

Fig. 4 Bode response of HO-WTG system and RO-WTG models

The presented results for proposed RO-WTG second-order model as derived in (35) is compared with results obtained for ROMs given in (36)-(40) derived by researchers available in literature. The responses are presented in Figs. 2, 3, and 4 for HO-WTG system, given in (21) and its RO-WTG models depicted in (35)-(43). The step response as presented in Fig. 2 of proposed RO-WTG model (35) is closely matched with HO-WTG system response and comparative better than RO-WTG models shown in (36)-(43). In similar fashion, impulse response depicted in Fig. 3 and Bode plot demonstrated in Fig. 4 of proposed RO-WTG model given in (35) is also superior than RO-WTG models depicted in (36)-(43) as presented in literature. For better understanding, comparative analysis by considering specifications of time-

Table 2 Specifications of time-domain of HO-WTG system and its ROMs

System and approximant	Specifications of Time-domain			
	Rise-time (sec)	Settling- time (s)	Overshoot	Undershoot
HO-WTG System (21)	0.0900	6.1365	223.3684	0
RO-WTG model (35)	0.0844	4.8143	211.5890	0
Model I (36)	0.0892	5.2504	218.0078	0
Model II (37)	0.0892	5.2504	218.0079	0
Model III (38)	0.0892	5.2504	218.0095	0
Model IV (39)	0.0892	5.2503	218.0041	0
Model V (40)	0.0892	5.2502	217.9965	0
Model VI (41)	0.0892	5.2501	217.9973	0
Model VII (42)	0.0892	5.2501	217.9957	0
Model VIII (43)	0.0891	5.2498	218.1898	0

Table 3 Tabular comparison of different error-criterion

System (21) versus approximant (35)	Error-criterion					
	IAE	ISE	ITAE	ITSE	IT^2AE	IT^2SE
ROM (35)	0.08502	0.001757	0.3042	0.005309	1.327	0.01817
Model I (36)	0.08988	0.002103	0.347	0.006893	1.575	0.02478
Model II (37)	0.08988	0.002103	0.347	0.006893	1.575	0.02478
Model III (38)	0.08987	0.002103	0.347	0.006892	1.575	0.02478
Model IV (39)	0.08988	0.002103	0.347	0.006893	1.575	0.02478
Model V (40)	0.08989	0.002104	0.3471	0.006895	1.575	0.02479
Model VI (41)	0.08989	0.002103	0.347	0.006894	1.575	0.02479
Model VII (42)	0.08989	0.002103	0.3471	0.006895	1.575	0.02479
Model VIII (43)	0.08975	0.0021	0.3468	0.006884	1.574	0.02476

domain and performance of error-criterion are also depicted in tabular form in Tables 2 and 3, respectively.

Thus, Figs. 2, 3 and 4, and Tables 2 and 3 confirm the utility of proposed method by providing proposed RO-WTG model as depicted in (35). The proposed RO-WTG model (35) is also better in comparison to RO-WTG models shown in (36)-(40).

6 Conclusion

This research work is dedicated to obtain the better reduced-order (RO) model for higher-order (HO) wind-turbine-generator (WTG) system by exploiting error minimization-based order diminution by employing grey-wolf optimizer. The third-order interconnected WTG system is approximated to RO-WTG model of order two. The responses and plot along with tabular data provided in previous section prove the usability and superiority of proposed error minimization-based methodology. This work also confirms the stability of proposed RO-WTG model. As a future, this proposal can be extended for ascertainment of desired RO model with the help of systematic procedure for arbitrarily taken weights associated with objectives along with diminution of interval systems.

References

1. Mariani V, Vasca F, Vasquez JC, Guerrero JM (2014) Model order reductions for stability analysis of islanded microgrids with droop control. IEEE Trans Indus Electron 62(7):4344–4354. https://doi.org/10.1109/TIE.2014.2381151
2. Wang R, Sun Q, Tu P, Xiao J, Gui Y, Wang P (2021) Reduced-order aggregate model for large-scale converters with inhomogeneous initial conditions in dc microgrids. IEEE Trans Energy Convers 36(3):2473–2484. https://doi.org/10.1109/TEC.2021.3050434
3. Zhang G, Zheng P, Yu S, Trinh H, Li Z (2021) A parameter-averaging approach to converter system order reduction. Electric Eng 103(4). https://doi.org/10.1007/s00202-020-01212-2
4. Díaz D, Meneses D, Oliver JÁ, García Ó, Alou P, Cobos JA (2009) Dynamic analysis of a boost converter with ripple cancellation network by model-reduction techniques. IEEE Trans Power Electron 24(12):2769–2775. https://doi.org/10.1109/TPEL.2009.2032187
5. Wang L, Zheng J, Li Z, Jing Z, Wu Q (2022) Order reduction method for high-order dynamic analysis of heterogeneous integrated energy systems. Appl Energy 308:118265. https://doi.org/10.1016/j.apenergy.2021.118265
6. Sikander A, Prasad R (2019) Reduced order modelling based control of two wheeled mobile robot. J Intell Manuf 30(3):1057–1067. https://doi.org/10.1007/s10845-017-1309-3
7. Far MF, Martin F, Belahcen A, Rasilo P, Awan HAA (2020) Real-time control of an ipmsm using model order reduction. IEEE Trans Indus Electron 68(3):2005–2014. https://doi.org/10.1109/TIE.2020.2973901
8. Padhy AP, Singh V, Singh VP (2021) Stable approximation of discrete interval systems. Circuits Syst Signal Process 40(10):5204–5219. https://doi.org/10.1007/s00034-021-01714-9
9. Meena VP, Singh VP, Barik L (2021) Kharitonov polynomial-based order reduction of continuous interval systems. Circuits Syst Signal Process 743–761. https://doi.org/10.1007/s00034-021-01824-4
10. Singh V, Dewangan P, Sinha S (2021) Improved approximation of siso and mimo continuous interval systems. Int J Syst Control Inf Process 3(3):246–261. https://doi.org/10.1007/s00034-020-01387-w
11. Bokam JK, Singh V (2018) Improved routh-padé approximants based on matching of markov parameters and time moments for continuous interval systems. Int J Pure Appl Math 119(12):12755–12766

12. Xiang Y, Arora JS, Rahmatalla S, Marler T, Bhatt R, Abdel-Malek K (2010) Human lifting simulation using a multi-objective optimization approach. Multibody Syst Dyn 23(4):431–451. https://doi.org/10.1007/s11044-009-9186-y
13. Mirjalili S, Mirjalili SM, Lewis A (2014) Grey wolf optimizer. Adv Eng Softw 69:46–61. https://doi.org/10.1016/j.advengsoft.2013.12.007
14. Roy R, Mukherjee V, Singh RP (2021) Harris hawks optimization algorithm for model order reduction of interconnected wind turbines. ISA Trans

Industrial Automation, IoT and Cyber-Security

FPGA Implementation of SLIM an Ultra-Lightweight Block Cipher for IoT Applications

Shashank Chandrakar, Siddharth Dewangan, Zeesha Mishra, and Bibhudendra Acharya

Abstract Increased security of radio frequency identification (RFID) systems which are resource-constrained devices is in high demand these days. Systems which have access control, payment methods and transaction banking systems are all examples of high-security applications where RFID technology are used. The attacker tries to deceive RFIDs in order to gain illegal entry to services by not paying for them or get around security measures by detection of secret password. One of the most difficult problems with RFID systems is ensuring effective security from such infringement activities. For RFID systems, lightweight cryptographic system gives security confidence. SLIM, a novel ultra-lightweight cryptography technique for RFID devices, is described in this article. As most commonly block ciphers are used cryptographic system and provides highly strong security for IoT ecosystem, SLIM which is based on Feistel structure is a block cipher of 32-bit size. The most difficult aspect of creating a lightweight block cipher is balancing cost, performance and security. SLIM, alongside all block ciphers that are symmetric, encrypts and decrypts using the same key. The suggested method performs well in both software and hardware contexts, has a small implementation footprint, a reasonable cost/security ratio for RFID devices, also is energy-saving. SLIM has shown good immunity to most successful differential and linear cryptanalysis assaults. The use of VLSI technology in the form of Xilinx software is also done in this paper. In this paper, SLIM has almost 4 times improvement in efficiency compared to other block ciphers.

Keywords Internet of Things (IoT) · RFID · Block ciphers · Lightweight cryptography · Feistel ciphers

---

S. Chandrakar · S. Dewangan · B. Acharya (✉)
Department of Electronics and Communication, National Institute of Technology, Raipur, Chhattisgarh, India
e-mail: bacharya.etc@nitrr.ac.in

Z. Mishra
Department of Microelectronics and VLSI, Chhattisgarh Swami Vivekanand Technical University, Bhilai, Chhattisgarh, India
e-mail: zmishra.phd2016.etc@nitrr.ac.in

S. Sharma et al. (eds.), *Intelligent Control, Robotics, and Industrial Automation*, Lecture Notes in Electrical Engineering 1066,
https://doi.org/10.1007/978-981-99-4634-1_28

1 Introduction

In today's world, saving life of a human is most important thing. In order to achieve this, hospitals and healthcare facilities require a system which monitors patients' conditions in real-time. One such system is RPMS which stands for Remote Patient Monitoring System. This system used Internet of Health Things (IoHT). The concept of the Internet of Health Things (IoHT) is defined as the use of IoT in the field of e-health. In this technology, the data of the patients' is stored in cloud in the real-time. Normally, there is a CCR i.e.; Central healthcare server Control. Room in which the real-time medical information of patients is transferred from the biomedical sensor devices. It usually happens in smart and active healthcare systems. But when the patient is critical, module of Global System for Mobile Communication is used for transferring medical information of patients to doctor remotely. But, as the medical information passes over internet, this info may be prone to cyber-attack as it may pass over an unauthorized network or maybe put into unauthorized cloud source. Also, IoHT devices are resource constrained as they are small in size. So, an important issue is that the existing traditional cryptography algorithms are not suitable for such resource constrained systems. Nowadays, small computation devices are becoming in demand and these devices possess high security threats. RFID is one such constrained devices in which standard cryptography algorithms don't apply. RFID systems are made up of three major components: an RFID tag, a reader, and a back-end database (server). To keep track and do identify patients, for tracking to assets and of equipment, decreasing blood control and medicine system errors, and other real-world applications for RFID devices exist in healthcare. Apart from that, it has a variety of real-world uses, including non-touch payments, passports with chips, track products, and so on. These products have an RFID tag attached to them that contains essential product information. An RFID tag, also known as a transponder, is a type of identifying device that comprises of integrated circuit (IC) and antenna. Help of IC is taken for storing and computing.

Active tags, semi-passive tags, and passive tags are three types of RFID tags based on their power source. The internal circuit and antenna of active tags are powered by a power source, which could be a battery. The RFID reader or transceiver gives energy to the RFID tags, which are passive tags with no power supply. There are now a variety of passive RFID transponders available on the market. Semi-passive tags have cell, but it solely powers the inside circuit; the antenna receives no power from it. Write-read tags and only-Read tags are the two types of tags. Write-read tag allows you for saving and alter info, whereas only-Read tags simply allow you for reading it. Security analysis has been done for IoT cipher [1].

The block-cipher has the following properties:

- The Feistel structure is used to create SLIM [2], a symmetric block cipher. This means that the keys of decryption and encryption are same.
- SLIM employs four nonlinear cipher component S-boxes which are 4 X 4 which operates on word of 16-bit in nonlinear manner.

- SLIM also has a stiffness prole against the most powerful harmful cryptanalyses "linear and differential attacks," despite its simplicity of implementation and design.
- The cipher is ideal for the Internet of Health Things and can be simply deployed with resource restricted devices like as RFID.

2 Related Work

This section covers the most up-to-date progress or works in lightweight area or algorithms specifically used to address the limits of RFID systems which has low-cost. The roots go as back as 1997 of the very first most primitive block cipher which is TEA [3]. RFID systems' proposals include commonly the hash functions for privacy and security protocols. Various recent works for lightweight block ciphers includes DESXL [4], SEA [5], KATANTAN [6], PRINT [7], KLEIN [8], CLEFINA [9], HIGHT [10], AES [11], LED [12], Hardware implementation has done for Piccolo [13] and LEA [14]. Hash functions are frequently used for privacy and security protocols of RFID systems. In comparison to block ciphers, there aren't as many ideas for lightweight stream ciphers. Some of them which have been used recently proposed are Photon [15], H-Present [16], Armadilo-C [17], DM-Present [18], Keccak [19], and Spongent [20]. Along with this hash functions there are proposed lightweight stream ciphers which includes Enocoro [21], Trivium [22]. Ref. [23] have exploited the LUT based shift register (SRL16/32) feature to perform the shift row operations. This has led to significant reduction in the resource consumption of 80 slices and double throughput on Spartan6 device. This low cost implementation and moderate throughput make solutions practical in RCE. In the article Ref. [24], Tim Good et al., presented two novel FPGA designs for AES algorithm. One design targets high throughput and the other is an extremely low area design. The comparison of different architectures, from the fastest to the smallest is also discussed. The strategies include, fully parallel loop unrolled architecture for high throughput, round-based architecture for low-area. The article Ref. [25] presented two novel FPGA designs for PRESENT algorithm. One design targets high throughput and the other is an extremely low area design. In the first architecture number of S-boxes was reduced from 16 to just 2, as S-boxes occupy large area thus reducing S-boxes number helped in reducing area footprint of the hardware. Ref. [26] utilized optimization techniques for Xilinx Spartan-3 FPGAs and applied them to lightweight cryptographic algorithms like HIGHT and PRESENT. The optimization strategies utilizing resources like LUT based 16-bit shift register (SRL16) and Distributed Random Access Memory (DRAM) in Spartan 3 FPGAs. The number of slices utilized in implementation of a shift register are dependent on the number of bits to be stored and the position of taps. The DRAMs can be utilized for implementing deeper memories with penalty on timing. The authors also suggested using ROMs for the FSM implementations for greater efficiency.

3 Theoretical Background

3.1 Processing of One Round

The internal structure of a single round can be used to determine the architecture of SLIM. The input 32-bit is split into equal two halves of sixteen-bit, called as V1 and V2. The entire each round processing, in which the input's right half V1 and Ki sub-key are changed with the help of operation XOR. The XORing method's output is sent to a substitution box, and the Sub-boxes' output is directed to a permutation block. At the end, XOR of output and the left half is performed to form next round's input of right half. The right half of the input V2 became the left half of the next round's input. Figure 1 shows SLIM encryption.

3.2 Sub-Box Layer

Designing Substitution-boxes has most difficult challenges in cryptography. As only non-linear parts of almost all latest algorithms are Sub-boxes, they can be regarded the cornerstone of all cryptosystems. As a result, the algorithm as a whole suffers from the poor design of S-boxes. Designing or choosing of Sub-boxes should done carefully. The S-box must be powerful enough to withstand differential and linear attacks while also having one of the smallest area footprints among Sub-boxes of 4-bit. The substitution layer is determined depending on cipher's differential and linear cryptanalysis.

3.3 Layer of Permutation

It is a method of rearrangement. The permutation is final step of function of SLIM in this case. The P box takes 16-bit inputs and changes/permute them according to a set of rules, resulting in a 16-bit output. The layer of permutation is determined as per the cipher's differential and linear cryptanalysis.

3.4 Key and SubKey Generation

32 sub-keys (16 bits) are required for 32 rounds and a 32-bit block, that is created with the encryption key of 80-bit. SLIM is a comparatively newer 32-bit block cipher algorithm proposed by Rabie Ai. Ramadan, Bassam Aboushoshai, Ayman El-Sayed, Ashutosh Dhari Dwivedi and Mohamed Mi. Dessouky in 2020. This block cipher depends upon Feistel structure and has a symmetric way of encryption algorithm.

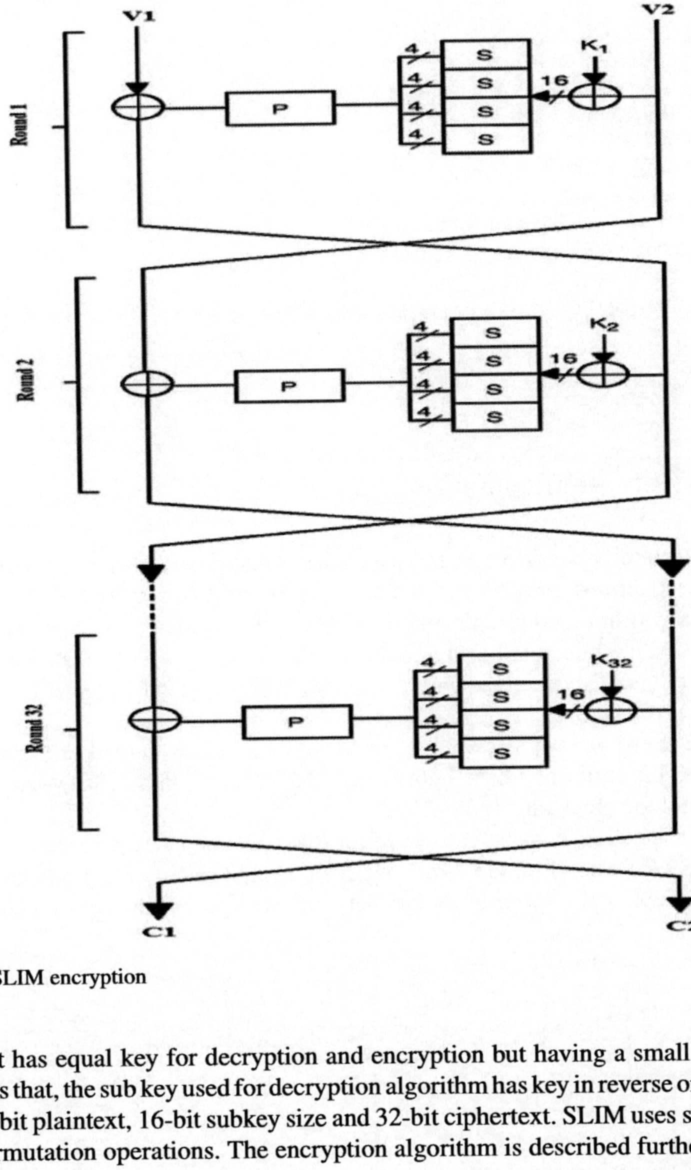

Fig. 1 SLIM encryption

Thus, it has equal key for decryption and encryption but having a small difference which is that, the sub key used for decryption algorithm has key in reverse order. SLIM has 32-bit plaintext, 16-bit subkey size and 32-bit ciphertext. SLIM uses substitution and permutation operations. The encryption algorithm is described further:

Algorithm
SLIM **Input**: V1 (16-bit), V2 (16-bit), Key (80-bit) **Output**: C (32-bit) Ciphertext

(continued)

(continued)

1. V(32) $\leftarrow$ V1(16)||V1(16)
2. K(80) $\leftarrow$ K1(16)||K2(16)||K3(16)||K4(16)||K5(16)
   (For first five sub keys)
3. for i = 0 to (round−1)
   Ki $\leftarrow$ (KeyLSB⟨⟨2 $\oplus$ KeyLSB + S − box) $\oplus$ KeyLSB⟨⟨3)
   (for next subkeys)
   $V1 = V2 \oplus P(S(K_i \oplus V1))$
   $V2 = V1 \oplus P(S(K_{i+1} \oplus V2))$
4. end for
5. C1 = V1
6. C2 = V2

4 Hardware Implementation

Use This section presents the round-based architecture of SLIM encryption algorithm. This algorithm uses 32-bit plaintext, 16-bit subkey size and 32-bit ciphertext. This algorithm uses simple Feistel structure for the encryption process. The proposed novel architecture helps in reducing the area and power consumption as it uses less number of slices. Read Only Memory (ROM) is used for implementing key generation process.

Figure 2 shows the key generation process of the cipher and the Fig. 3 shows the hardware architecture. of SLIM lightweight block cipher. In the SLIM cipher key size is 80 and the block size is 32.

The key scheduling consists of three sub-functions. When the reset is low, the input key gets selected using a single 2-to-1 MUX in the first round of key scheduling. In subsequent rounds, the selection occurs between the result of the previous round in

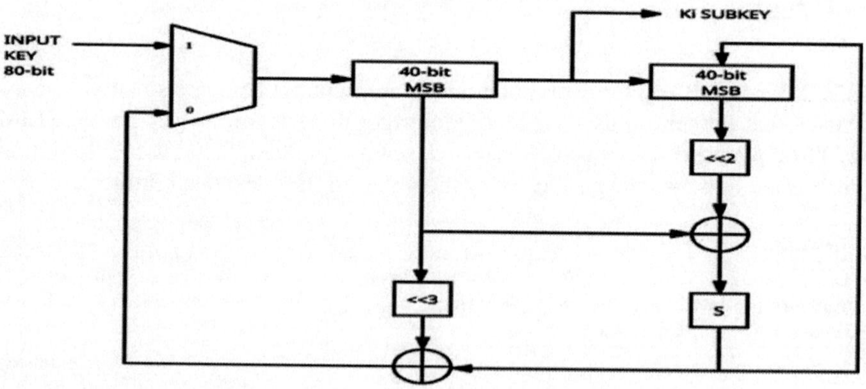

Fig. 2 Key generation process

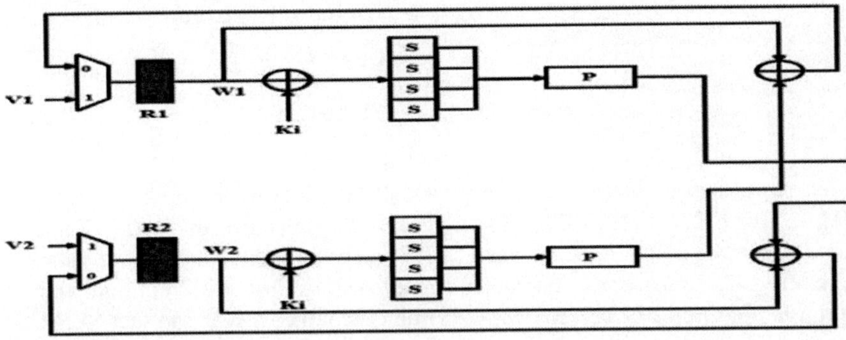

Fig. 3 Proposed hardware architecture of SLIM lightweight block cipher

the data-path and input's key. The key segments' LSBs (right-hand side) are passed in 2 steps through the left circular shifting process in SLIM, which does not require any gates. The key segments in left-hand side (MSBs) are passed through the circular shift method in left direction in three steps, and does not require any logic gates. The left side and shifted right hand side is then XORed and forwarded to the Sub-box of 4-bit. The S-box is a non-linear component which essentially has 4-AND and 4-XOR gates to reduce overhead in implementation. The Ki is passed as subkey to the encryption architecture for each round of processing.

In the implementation of encryption architecture, the encryption process accepts input in the registers R1 and R2 when it is given high reset signal. These values in the registers are fed back to the multiplexer after various operations. Firstly, these register values are XORed with the respective subkey of the current round. Then this 16-bit result is passed through 4 parallel S-boxes of 4-bit each after which its result is passed through permutation box. This output is passed in a wire W2 and XORed with wire W1 and fed back to the multiplexer.

5 Results

The architectures show the hardware implementation using Verilog HDL and FPGA. implementation uses Xilinx. ISE design suite. FPGA. Refers to a greater-speed, low-cost, flexible, and functionally programmable. logic device. that may be used for a wide variety of uses. FPGAs have a large quantity of logic cells that can be used to create any type of logic operation as sophisticated building units of digital. circuits. This paper has presented hardware implementation results and performance on FPGA (Field. Programmable Gate. Array) platform. The findings of hardware implementation and performance on the FPGA (Field. Programmable Gate. Array) platform were provided in this study. FPGA implementation can be used to realize a variety of parameters. As seen in the equations below, two of the parameters are efficiency (throughput per area) and throughput (Mbps).

$$\text{Throughput (Mbps)} = \frac{\text{Max Frequency} \times \text{Block Size}}{\text{Clock Cycle}} \tag{1}$$

$$\text{Efficiency} = \frac{\text{Throughput}}{\text{Area}} \tag{2}$$

This paper has presented synthesized results for hardware implementation of SLIM in Virtex-5 (xc5vlx20t-2ff323) platform and results are tabulated in Table 1. Table 2 shows the comparison between SLIM and other lightweight block ciphers.

From Fig. 4, compares the number of occupied slices from different light weight encryption algorithms of cryptography. While comparing it is found that SLIM 32 uses way high number of slices as compared to other algorithms that is approximately 3 times that of TEA 64 algorithm. It also surpasses XTEA 64 and XXTEA 64 in terms of slice occupied.

Figure 5 presents, the LUTs are compared among the available algorithms. In this case also, SLIM shows the highest number of LUTs among others. This factor is about 4 times the second highest algorithm that is TEA 64. SLIM 32 has 1573 LUTs in this case as compared to 355 of TEA 64 which is way more In this area SLIM remains behind the other algorithms by some margin. While the throughput of XXTEA 64 is the highest among all, SLIM's throughput is about 115 Mbps less than the second lowest algorithm while the highest one has Throughput over 700 Mbps. The maximum frequency of SLIM 32 which is 351 MHz lies at par with two

Table 1 FPGA (xc5vlx20t-2ff323) implementation results of SLIM architecture

Parameters	SLIM
LUT	1573
Slices	547
No. of free flip flop	1417
Max. frequency (MHz)	351
Cycle	33
Throughput (Mbps)	340.36
Efficiency	13.21

Table 2 Comparison between SLIM and other ciphers

Algorithm	Slices	LUT	Throughput (Mbps)	Max frequency (MHz)	Efficiency
TEA 64 [27]	166	355	517.72	266.95	1.7
XTEA 64 [27]	153	287	456.06	235.16	2.98
XXTEA 64 [27]	58	229	705.95	364.01	12.17
Unified (sel-2'b0) [28]	272	735	391.75	311	4.43
SLIM 32 *Tw	547	1573	340.36	351	13.21

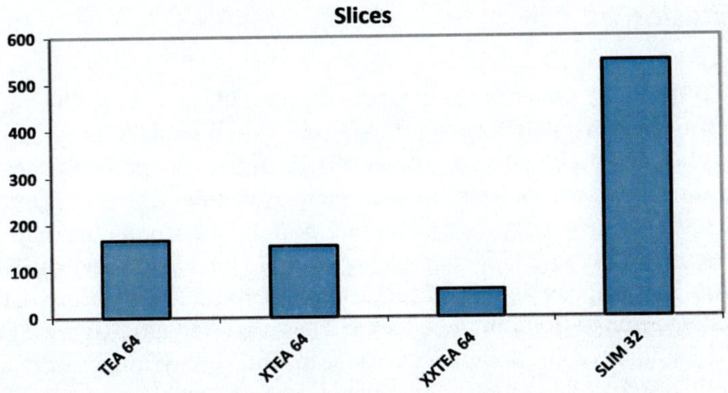

Fig. 4 Slices at maximum frequency for different ciphers

of the existing algorithms but remains behind the maximum frequency of XXTEA 64 which has about 3% better maximum frequency. The efficiency of different block cipher algorithms. In this case, SLIM 32 remains ahead of its competitors by about 8.5% than the second-best algorithm XXTEA 64.

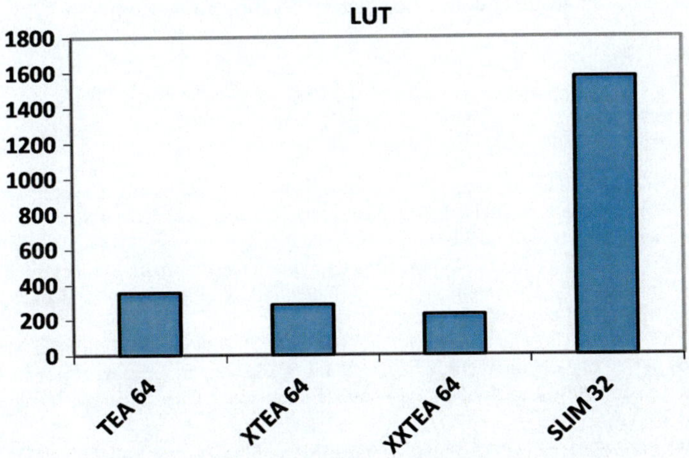

Fig. 5 Throughput at maximum frequency for different ciphers

6 Conclusion

RFIDs ecosystems are prone to many types of cyber-attacks. Providing appropriate defense against those attacks is the most important question in RFID systems. Those algorithms that are designed using modern techniques for powerful devices are unsuitable for RFIDs ecosystems because their execution will necessitate a large portion of resources, memory, computational power. The storage amount of cryptography that is usable in RFID ecosystems is limited, as is the permissible power consumption. The usage of lightweight algorithms of cryptography is one of the finest ways for safeguarding info of these contexts. The suggested ultra-lightweight. cryptography SLIM algo is appropriate for the usage in RFID ecosystems with constraints. SLIM, that is also block cypher is constructed on Feistel architecture with a 32-bit block size. To prevent unnecessary key searches, SLIM employs a long key length of 80 bits. The suggested algorithm is appropriate for wireless networks, particularly WSNs (Wireless Sensor Networks) and Internet of Things (IoT) uses, in which data packets are usually in the range of few bytes. When compared to other existing and implemented algorithms, SLIM proved to be extremely efficient as it is seen to be almost 4 times efficient than XTEA 64 algorithm and way more efficient than TEA. Also, the maximum frequency is greater than many algorithms like XXTEA, XTEA and TEA. So, it can be concluded that it has seen significant improvement in throughput per area and frequency compared to other block ciphers.

References

1. Dwivedi AD (2020) Security analysis of lightweight IoT cipher: Chaskey. Cryptography 4(3):22
2. Bassam A, Ramadan RA, Dwivedi AD, Ayman EL, Dessouky M (2020) SLIM: a lightweight block cipher for internet of health things. IEEE Access 8
3. Wheeler DJ, Needham RM (1994) Tea, a tiny encryption algorithm. In: Preneel B (ed) Proceedings 2nd International Workshop, in Lecture Notes in Computer Science, Leuven, Belgium, vol 1008. Springer, Leuven, Belgium, pp 363–366
4. Leander G, Paar C, Poschmann K, Schramm K (2007) New lightweight DES variants. In: Biryukov A (ed) Proceedings 14th International Workshop, in Lecture Notes in Computer Science, Luxembourg, Luxembourg, vol 4593. Luxembourg City, Springer, Luxembourg, pp 196–210
5. Standaert F, Piret G, Gershenfeld N, Quisquater J (2006) SEA: a scalable encryption algorithm for small embedded applications. In: Domingo-Ferrer J, Posegga J, Schreckling D (eds) Proceedings Workshop RFIP Light Weight Crypto. Lecture Notes in Computer Science, Tarragona, Spain, vol 3928. Springer, Graz, Austria, pp 222–236
6. Cannière CD, Dunkelman O, Knezevic M (2009) KATAN and KTANTAN—a family of small and efficient hardware-oriented block ciphers. In: Clavier C, Gaj K (eds) Proceedings 11th International Workshop. Lecture Notes in Computer Science, vol 5747. Springer, Lausanne, Switzerland, pp 272–288
7. Knudsen LR, Leander G, Poschmann A, Robshaw MJB (2010) Printcipher: a block cipher for IC-printing. In: Mangard S, Standaert F (eds) Proceedings 12th International Workshop. Lecture Notes in Computer Science, Santa Barbara, CA, USA, vol 6225 Springer, Santa Barbara, CA, USA, pp 16–32.

8. Gong Z, Nikova S, Law YW (2011) KLEIN: a new family of lightweight block ciphers. In: Juels A, Paar C (eds) Proceedings 7th International Workshop. Lecture Notes in Computer Science, Amherst, MA, USA, vol 7055. Springer, Amherst, MA, USA, pp 1–18
9. Sugawara T, Homma N, Aoki T, Satoh A (2008) High-performance ASIC implementations of the 128-bit block cipher CLEFIA. In: Proceedings IEEE International Symposium Circuits System, Seattle, WA, USA, May, pp 2925–2928
10. Hong D, Sung J, Hong S, Lim J, Lee S, Koo B, Lee C, Chang D, Lee J, Jeong H, Kim J, Kim, Chee S (2006) HIGHT: a new block cipher suitable for low-resource device. In: Goubin L, Matsui M (eds) Proceedings 8th International Workshop. Lecture Notes in Computer Science, Yokohama, Japan, vol 4249. Springer, Yokohama, Japan, pp 46–59
11. Daemen J, Rijmen V (2002) The sesign Rijndael: AES—the advanced encryption standard (information security and cryptography). Springer, Cham, Switzerland
12. Batina L, Das A, Ege B, Kavun EB, Mentens N, Paar C, Verbauwhede I, Yalçin T (2013) Dietary recommendations for lightweight block ciphers: power, energy and area analysis of recently developed architectures. In: Hutter M, Schmidt J (eds) Proceedings 9th International Workshop. in Lecture Notes in Computer Science, Graz, Austria, vol. 8262. Springer, Graz, Austria, Jul, pp 103–112
13. Acharya B, Mishra Z, Ramu G (2019) Hardware implementation of Piccolo Encryption Algorithm for constrained RFID application. 978-1-5386-9325-4/19. IEEE
14. Mishra S, Mishra S, Acharya B (2019) A high throughput and speed architecture of lightweight cipher LEA. In: 2019 International Conference on Computing, Communication, and Intelligent Systems (ICCCIS), October, pp 458–462
15. Guo J, Peyrin T, Poschmann A (2011) The PHOTON family of lightweight hash functions. In: Rogaway P (ed) Proceedings 31st Annual International Cryptology Conference (CRYPTO). Lecture Notes in Computer Science, vol 6841. Springer, Santa Barbara, CA, USA, Aug, pp 222–239
16. Bogdanov A, Leander G, Paar C, Poschmann A,. Robshaw MJB, Seurin Y (2008) Hash functions and RFID tags: Mind the gap. In: Oswald E, Rohatgi P (eds) Proceedings 10th International Workshop. Lecture Notes in Computer Science, vol 5154. Springer, Washington, DC, USA, Aug, pp 283–299
17. Badel S, Dagtekin N, Ouafi K, Reffé N, Sepehrdad P, Susil P, Vaudenay S (2010) ARMADILLO: a multi-purpose cryptographic primitive dedicated to hardware. In: Mangard S, Standaert F (eds) Proceedings 12th International Workshop. in Lecture Notes in Computer Science, vol 6225. Springer, Santa Barbara, CA, USA, Aug, pp 398–412
18. Poschmann AY (2009) Lightweight cryptography: cryptographic engineering for a pervasive world. Ph.D. dissertation, Dept. Elect Eng Inf Technol, Ruhr Univ, Bochum, Germany
19. Bertoni G, Daemen J, Peeters M, van Assche G (2012) Keccak sponge function family main document [Online]. Available: http://keccak.noekeon.org/Keccak-main-2.1.pdf
20. Bogdanov A, Knezevic M, Leander G, Toz D, Varici K, Verbauwhede I (2011) Spongent: a lightweight hash function. In: Preneel B, Takagi T (eds) Proceedings 13th International Workshop Cryptograph. Hardware Embedded System. Lecture Notes in Computer Science, vol 6917. Springer, Nara, Japan, Sep, pp 312–325
21. Watanabe D, Owada T, Okamoto K, Igarashi Y, Kaneko T (2010) Update on enocoro stream cipher. In: Proceedings International Symposium Inf Theory Its Application, Taichung, Taiwan, October, pp 778–783
22. Feldhofer M, Wolkerstorfer J (2008) Hardware implementation of symmetric algorithms for RFID security. Springer, Cham, Switzerland, pp 373–415
23. Chu J, Benaissa M (2012) Low area memory-free FPGA implementation of the AES algorithm. In: 22nd International Conference on Field Programmable Logic and Applications (FPL), pp 623–626
24. Good T, Benaissa T (2005) AES on FPGA from the fastest to the smallest. In: International workshop on cryptographic hardware and embedded systems, pp 427–440
25. Lara-Nino CA, Diaz-Perez A, Morales-Sandoval M (2017) Light weight hardware architectures for the present cipher in FPGA. IEEE Trans Circuits Syst I: Reg Pap 64(9):2544–2555

26. Yalla P, Kaps JP (2009) Lightweight cryptography for FPGAs. In: International Conference on Reconfigurable Computing and FPGAs, pp 225–230
27. Mishra Z, Acharya B, Efficient hardware implementation of TEA, XTEA and XXTEA lightweight ciphers for low resource IoT applications. Int J High Perform Syst Archit 10(2):80–88
28. Mishra Z, Nath PK, Acharya B (2020) High throughput unified architecture of LEA algorithm for image encryption. Microprocess Microsyst 78:103214

AR and IoT Integrated Machine Environment (AIIME)

Akash S. Shahade and A. B. Andhare

Abstract Machine maintenance accounts for 15–40% of overall manufacturing costs. Moreover, industries have progressed from breakdown type of maintenance to most recent and modern, predictive maintenance (PdM) which heavily relies on data collection. This has paved the way for IoT to be adopted for PdM. Many manufacturers are coming up with machines with built-in sensors for parameter monitoring; this makes them expensive. However, many machines are already in use, and putting sensors for data collection and applying the Internet of Things to such machines is a huge task. The concept of modular sensor systems is proposed in this article to address that. Also, with an increase in fast computing, augmented reality is getting adopted for many applications. It has the potential to enhance reality and impact many fields. It is found that less work has been done on integrating the industrial Internet of Things with augmented reality. The present work focuses on designing a modular sensor system and displaying data over a machine's digital twin in AR. Along with that, virtual step-by-step instructional guidance in augmented reality for maintenance is also adopted. This article proposes an idea of AR and IoT integrated machine environment (AIIME). AIIME is a complete machine environment where a digital twin is used to assist maintenance. Concepts and different aspects of AIIME are discussed in this article. The aim is to design a better machine environment and lay a foundation for AR industrial metaverse.

Keywords Predictive maintenance (PdM) · Internet of Things (IoT) · Augmented reality · Sensors · Cyber-physical systems

A. S. Shahade (✉) · A. B. Andhare
Visvesvaraya National Institute of Technology, Nagpur, Maharashtra, India
e-mail: akashshahade@students.vnit.ac.in

A. B. Andhare
e-mail: abandhare@mec.vnit.ac.in

S. Sharma et al. (eds.), *Intelligent Control, Robotics, and Industrial Automation*,
Lecture Notes in Electrical Engineering 1066,
https://doi.org/10.1007/978-981-99-4634-1_29

1 Introduction

Over time, there has been a divergence in maintenance practices. Some industries still use the "fail and repair" technique, while others have adopted advanced strategies such as predictive maintenance (PdM). It is based on the "monitor, learn, predict, and repair" principle. Equipment or machinery shows indicators of probable failures before eventually breaking down if appropriate measures are not taken on time. Temperature fluctuations, vibration, and noise, among several other things, can indicate that the machine needs to be maintained. Multiple strategies in the industry can cater to early warning signs of equipment degradation and offer a framework for the PdM [1]. Cost reductions, increased profit, and increased equipment availability for production are all benefits of this method, whether directly or indirectly.

Safety manuals and warnings are usually paper-based and often quickly forgotten after initial use. Also, with advances in technology, the demand for expert technicians is rising. And to cope with that is a difficult task. Massive handbooks, misplaced information, and out-of-date versions are too tough to come by and apply to operational maintenance and repair processes. The printing, supplies, and waste charges are unaccounted for. 2D diagrams are incapable of adequately displaying essential activities or being practically used in the case of complicated procedures. Real-time remote assistance is not available. All these reasons can be overcome with the correct use of the Internet of Things (IoT) and augmented reality (AR). This article addresses the method of doing it.

The Internet of Things (IoT) is a collection of physical devices linked together via electronics, software, sensors, and networking to collect and share data. The Internet of Things (IoT) allows devices to be sensed and controlled remotely using an existing network, allowing for more easy integration of the material world into computer-based systems and increased efficiency [2]. "Augmented reality" is reality that has been enhanced by interactive digital elements [3]. Simply put, it is an extension of reality. The capability to deploy virtual objects into the real world to better visualize the environment forms the core of AR. These days, frequently used AR applications depend on smartphone to represent a digitally enhanced world. People can use their phone's camera to observe the natural world on the screen and then use an AR application to alter it in various ways using digital overlays. A few examples include an overlay of digital data/ 3D CAD models, including real-time instructions, enhancing a person's appearance or surroundings with "filters" on Instagram, Snapchat, and other apps. By 2030, virtual reality and augmented reality have the potential to increase global GDP by $1.5 trillion [4]. Screens, eyewear, handheld and mobile devices, and head-mounted displays are among the devices that can display AR. This list is growing continuously.

2 Literature Review

In AR, all instructions are fed through 3D visualizations and virtual dashboards, which can easily be updated. AR can be used to self-guide the user and follow all standard protocols. This also ensures the same quality throughout. Without having to hunt, repair, and maintenance, data displays right where needed on actual items. Employees and customers can better understand procedures with 3D AR renderings of real-world things. AR can connect with an expert sitting in another part of the world. That expert can guide the user with 3D annotations by using overlays over actual machine components. During study, many fields were found where AR is playing significant role. Apart from its use in education, art, marketing [5], it has a great potential in military use [6]. Henderson et al. titled "Augmented Reality for Maintenance and Repair" [7] tried to assess the feasibility of using AR to aid military maintenance. The aim was to investigate how virtual computer graphics applied in the real world can improve technicians' productivity during training and in actual work. They examined the influence of AR on technicians' efficiency. With AR, the maintainer can access all data in the virtual world, and all instructions are needed. The article also talks about limitations such as the availability of lightweight devices complexity of the system. Augmented reality can also be used in the training of maintenance personnel. It helps them better understand procedures and indirectly boosts productivity [8] Adopting AR as a comprehensive tool for maintenance eventually minimizes downtime. With the aid of AR, technicians are better assisted in repair work, and the time required to read manuals is reduced [9]. Many researchers and organizations have realized that AR has a huge potential when applied to maintenance. Projects such as KARMA, STARMATE, ARVIKA, US Air Force's ARMAR, Airbus military's MOON project, European Aeronautic Defense and Space Company's MiRA, are a few such examples [9]. The reference [9] has also discussed use of AR for laptop maintenance. A system was designed that helps in identifying specifications of laptop components while repairing. It gives a good understanding of the current trend of AR adoption in the maintenance field. For integrating AR and IoT, it is required to study IoT also. IoT stands for the Internet of things, which connects devices. IoT is expected to generate 900 billion to 2.3 trillion USD by the year 2025 [10]. IoT forms the foundation of predictive maintenance. The institution of electrical and electronic engineers (IEEE) defines IoT as "a network of items, each embedded with sensors, which are connected to the Internet" [11]. In order to sustain in Industry 4.0 an industry must adopt to new maintenance strategy such as predictive maintenance [12]. PdM is a closed cycle with four processes, viz., capturing data, storing, analysis, and scheduling a maintenance [13]. Predictive maintenance (PdM) is based on continuous data measurement and then recognizing performance deviations from the norm. PdM focuses on maximum utilization of service life of an item [12]. Capturing data is where IoT comes into picture. Within IoT, data from any location can be sent to cloud and accessed at any location. This flexibility combined with the benefits of AR forms the core of this project.

Santi et al. [14] in their research, by using the SCOPUS database found that annually, around 2000 articles on the topics of "augmented reality" and "Industry 4.0" are published. The volume of articles containing the terms "augmented reality" AND "industry 4.0" is, however, quite small, with only roughly 100 papers per year being published [15].

From [15] it was evident that not much has been done in context to integrating IoT and AR. The primary motivation behind this project is to assess the use of augmented reality in a machine environment. AR will help eliminate the need for carrying machine manuals for maintenance and help get remote, real-time support from seniors. AR will also assist in previous data display; this would eliminate the need of having extra screens. Such assistance would help reduce stress on workers. All of these and further possibilities are discussed in this research article.

3 Proposed Solution

It was desired to have a solution that could seamlessly integrate the Internet of Things and augmented reality. This system could harness data from the machine via IoT and display the same in augmented reality. So AR and IoT integrated machine environment (AIIME) was proposed. AIIME is a machine maintenance strategy built on top of the PdM strategy. The philosophy of AIIME is simple. The entire concept is divided into three categories, mainly IoT data acquisition, control system and AR assisted machine maintenance.

The method designed for AR and IoT integration is simple and can be easily represented as a flow diagram, as shown in Fig. 1. It all starts with an industrial machine. The automatic relevant data collection system is active all year round. This is also the backbone of the AIIME strategy. Relevant data may include the machine's temperature, vibrations, motor speed, power or current consumption, coolant flow rate, etc. This data is then stored in the cloud and can be accessed in the control room. This data can then be used for visualization, i.e., real-time graphs and charts. Other than this, analytics can predict any anomaly and subsequently control actions can be taken, e.g., speed controls of motor or switching on or off the same device remotely via IoT. The technician reaches the machine to be maintained and has access to a mobile application. This application can be used to access relevant information directly from the cloud. This will help them better understand the problem. The machine will also have an AR tag, which can be scanned, and an augmented reality view of the machine is generated. This is a 3D model of a machine. The technician can interact with the model and identify which machine area is affected. This will help him correctly identify faults; this saves a lot of time. If the technician is new, s/he can be assisted with AR in quickly finding the problem, which increases efficiency. If an unexpected situation arises, an expert technician can virtually help the onsite technician with AR. The benefits are endless, which will be discussed later in the article.

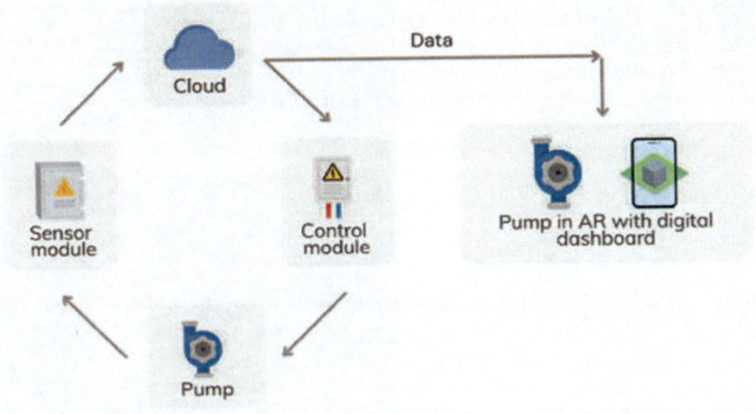

Fig. 1 Proposed solution (AIIME)

3.1 Execution

Many industries are already running on their own strategies. According to India Brand Equity Foundation's (IBEF's) report [16], there are 6.3 crore registered micro, small, and medium enterprises (MSMEs) in India as of October 21, 2021. And this is not a small number. Given that such industries operate on a small scale, there is less chance they will go for fancy maintenance strategies and intelligent machines from manufacturers. This is where AIIME stands out. AIIME is a modular design, and the entire system is focused on designing small modules for data acquisition and control. These sensor and control modules can be fitted onto existing industrial machines, which will now become intelligent machines. All relevant parameters can be measured, and further, the data can be used for failure prediction.

Benefits include affordable and quicker training, lower maintenance times, narrowing the gap in skills, remote maintenance support of field technicians, and lower disassembly time. And the same can be applied to many cases such as display of component name on machines, simulation of machine elements before/during maintenance, real-time information access, remote assistance and guide.

4 Design

4.1 IoT Data Acquisition

Data acquisition forms the backbone of predictive maintenance. Many companies are coming up with machines that have sensors built-in, which adds up to the overall cost of the device. AIIME strategy focuses on modular maintenance. This is a new

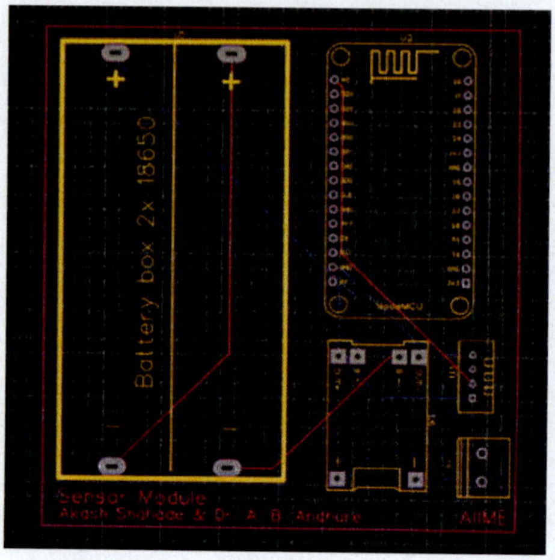

Fig. 2 Sensor module (left) and its PCB design (right)

concept where small sensor modules are created. These modules can detect temperature, pressure, vibration, and current drawn by the motor. These modules can easily be fitted onto an existing machine, and that machine is upgraded to an intelligent device. A temperature module design is shown in Fig. 2. Any IoT system works on a specific standard protocol. A detailed comparison of all IoT communication protocols were studied, and MQTT was adopted in this project. MQTT works on the publish-subscribe principle. In short, the sensor sends data to a broker, i.e., publishes the data. Any device can access this data, and such a device has to subscribe to the broker.

With the help of MQTT, a primary circuit was designed. Since this module was designed to capture temperature data, the DHT11 sensor module was considered. It was then interfaced with NodeMCU. As seen in the schematic Fig. 2, a power circuit was also designed to charge the batteries, which would make the module independent. This circuit schematic was designed using EasyEDA software.

4.2 Control System

One of the research questions was can we design a modular system that converts that machine into an intelligent machine when fitted onto an existing industrial machine? To counter this, modular maintenance or modular control system is proposed. The result is to design a control system module, a small controller to be fitted onto any machine and make it smart. By smart, we mean the device should be able to send its data to the cloud, and it is capable enough to be controlled remotely via the control room or, for that matter, from anywhere. The control module is designed based on the ESP8266 chip. The circuit schematic is shown below in Fig. 3. This schematic is then converted into PCB. And finally, it is packed into a compact enclosure. This

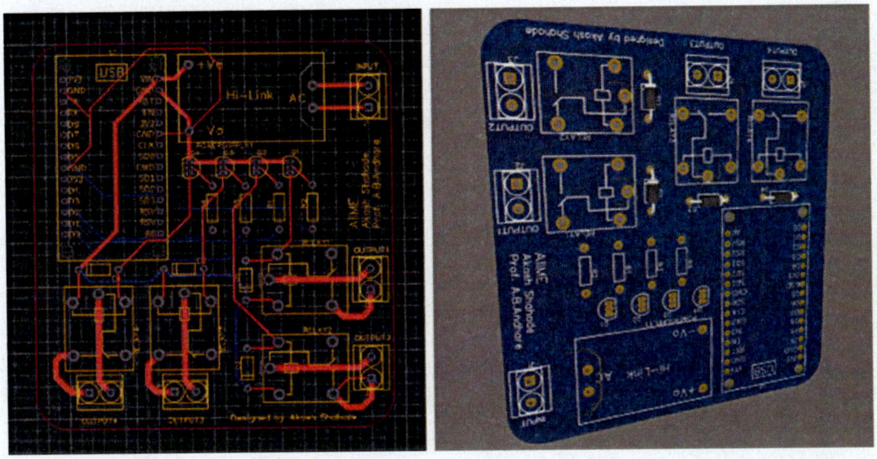

Fig. 3 PCB design for control system module

module connects to the local Internet network. A software application is present in the control room, with which this control module can be accessed. An industrial machine, such as a pump, in this case, can be connected to a control module. In the current design, four devices can be connected simultaneously.

In the above Fig. 3, the circuit schematic of the control module can be seen. NodeMCU is used as the central controller because it efficiently runs MQTT and connects to the network easily. On the right side of the schematic, four relay circuits are shown. Each relay can control a single-phase AC load, so four loads can be connected simultaneously.

4.3 Augmented Reality Assisted Maintenance

Augmented reality is an extension of our physical world. Digital elements, when added to the physical world, enhance the experience of the user. It offers better visualization of things. Slowly and steadily, with the rise in computing power, personal computing devices are flooded. In industries, mainly on the assembly lines, workers are given a set of instructions in printed format and are expected to follow these instructions. However, referring manuals may be tiresome. But if AR is used to display virtual instructions, the worker need not engage in finding the right content in manual. She/he can directly see virtual instructions floating on top of assembly line and thus productivity increases. This forms a foundation for the adoption of AR in present times. Maintenance task involves using machine manuals, papers, and previous machine data. Use of AR to eliminate the aforementioned things physically and access them virtually helps technicians visualize, de-clutter the work area, and focus more on the task. Usually, there are four significant classifications for AR

[17]. The marker-based AR approach was chosen. A marker, also known as ARTag or a Fudicial marker, is a reference with which a digital element is imposed in a real-world scenario [18]. An AR tag is a unique pattern that is placed in real space. This is then detected using detection algorithms of AR software. For this project, AR tag generator by Shawn Lehner was used for this project [19]. Two applications are taken into consideration under this project. One, when a technician needs training of a particular machine, it is a usual practice to look at a cut section of machine. So the idea is, instead of making a cut section model, can we use AR to generate a cut section? And second scenario is to develop a digital twin beside the actual machine. This digital twin, generated by AR, will be used to better visualize actual machine. A virtual dashboard will be present on top of digital twin which would display data from the actual machine. Another virtual screen to display maintenance procedures will be present. The data to be displayed can be accessed through sensor module that was previously mounted on pump. For these two applications, two different markers are generated.

Mainly three softwares were used to design the project, viz., PTC Vuforia SDK, Unity3D, and PTC Creo. Vuforia, a product by PTC is a software development kit (SDK) most widely used for developing AR programs. "Markerless" image targets, 3D model targets, and many other 2D and 3D target types are supported by the Vuforia SDK. Unity is a game development engine most widely used to develop computer and mobile games. Within Unity, a 3D environment can be designed, and 3D objects can be added. These objects can react by adding physics to them; they can be made to interact according to a particular trigger by the user or automatically by using C# scripts. Creo, as we know, is a 3D modeling software by PTC. Creo is used in this project to design and model a centrifugal pump and its components.

5 Working

Once completed, all the subsystems, i.e., IoT sensor module, IoT control module were installed on a small-sized centrifugal pump. Due to unavailability of actual pump during project development, a small-sized desert cooler pump was used.

With reference to Fig. 4, it should be noted that the sensor module and control module are fitted onto the small-sized pump, but for digital replica, a real centrifugal is considered. A modular sensor unit is designed to collect temperature data from the pump body, and this data is sent to the cloud. The sensor module is mounted on the pump body to measure skin temperature. Then, the AR application has access to this data. In the AR scene, this data is accessed and presented just above the virtual digital twin of the pump. This arrangement allows the user/technician to look at real-time values. Also, in AR, the technician has access to another virtual screen that displays maintenance procedures.

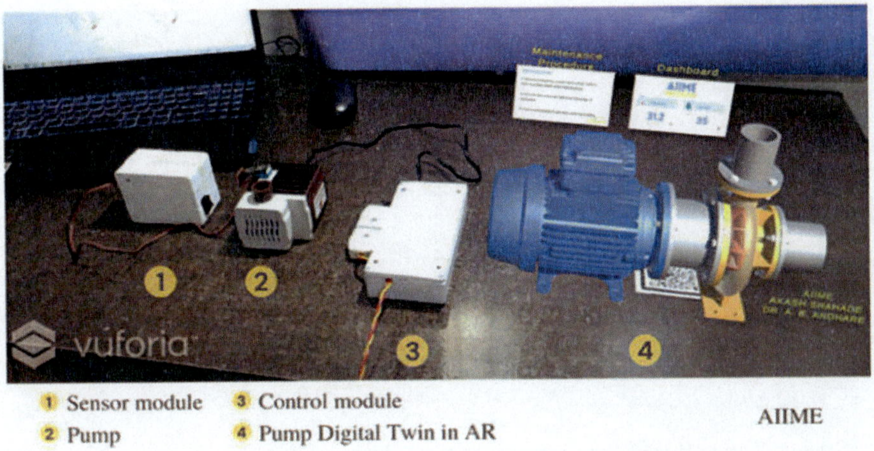

Fig. 4 AIIME setup

5.1 *Working of AR Application*

The steps necessary in augmenting reality are the same for all types of AR. A marker-based AR uses a fudicial marker as a reference and w.r.t the marker, it imposes digital content. Figure 5 explains pictorially the steps involved in creating an AR application. Steps are also discussed below.

Step 1: The software receives live camera feed.

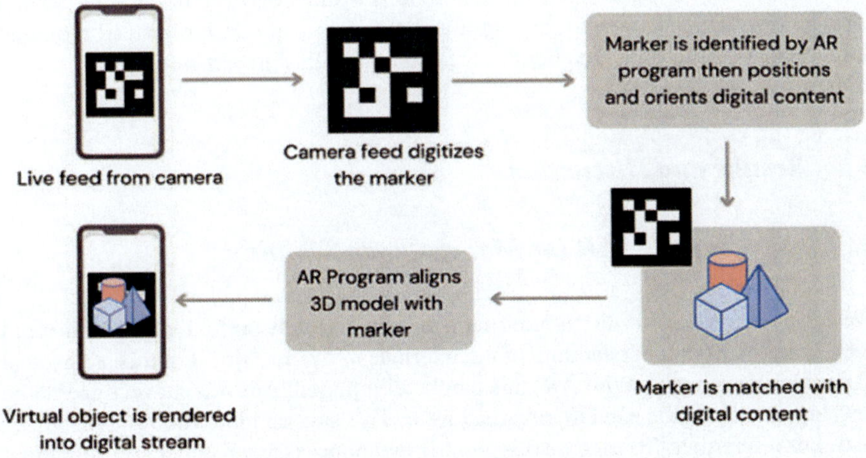

Fig. 5 Steps in developing AR application

Fig. 6 Pump cross section (left) and pump digital replica (right) in AR

Step 2: The software then takes camera feed, uses border detection, and finally identifies the marker.
Step 3: Once the marker is identified, the software then orients and positions the digital content and aligns it with the physical mark.
Step 4: The marker symbol is matched with the digital content assigned to it.
Step 5: Software then aligns 3D model with the marker.
Step 6: Once the 3D object is aligned with marker, it is then rendered and can be seen on displays, HMD, anything.

The machine temperature data is accessed in augmented reality (AR) to be displayed on a virtual screen (Fig. 6) on top of the virtual digital twin. Another screen next to the dashboard is used to display maintenance procedures. Furthermore, this arrangement helps in reducing the technician's head, and eye movement lowers maintenance time and gets real-time instructions without the need to carry and search through machine manuals. Lightweight and ergonomically designed head-mounted displays can be used for even better visualization of this project setting.

5.2 *Results and Discussion*

5.3 *Use Case 01—AR for Maintenance Training*

We are all aware that, while studying a machine, we usually prefer looking at its cross section for a better understanding of the internals of the machine. For this, a physical 3D model is prepared. With AR, this can be eliminated. Any 3D cut section can be developed without the need to manufacture it. This saves a lot of time and resources in training activities. To execute this, centrifugal pump is taken under consideration. The cut section of the centrifugal pump was first designed in PTC Creo (Fig. 7).

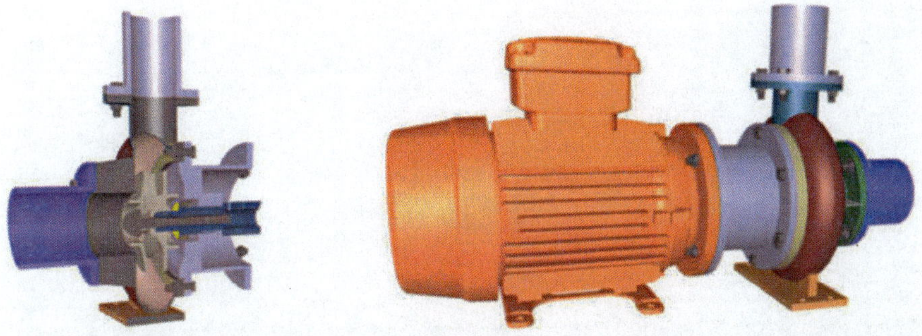

Fig. 7 3D model of pump and its cross section for AR

5.4 *Use Case 02—AR Machine Environment*

AR allows a professional to collaborate on a level that has never been feasible before. Thanks to augmented reality, the remote expert can see exactly what the user sees. As the virtual expert walks them through the process, they can carry out repairs immediately. AR allows both the user and the expert to share documentation, turning every intervention into a teaching opportunity. A shared perspective of the work environment allows for visual communication and the resolution of technical challenges. The idea is to create a digital replica of a machine and use that to guide the operator for actual work on actual machine. This digital replica can be seen in AR. To design a machine environment, a 3D model of the pump was developed in Creo (Fig. 7) and then imported into Unity. This was then linked with a proper Vuforia license. Appropriate materials were added to machine components. The MQTT client is used to display IoT data from the sensor module onto a virtual display. The data from the sensor is sent to the AR application. This is managed by writing a C# script into Unity. Once the project was ready in Unity, it was exported as an Android application. This application can be installed onto our device and was further tested. Screenshot during development can be seen in Fig. 8.

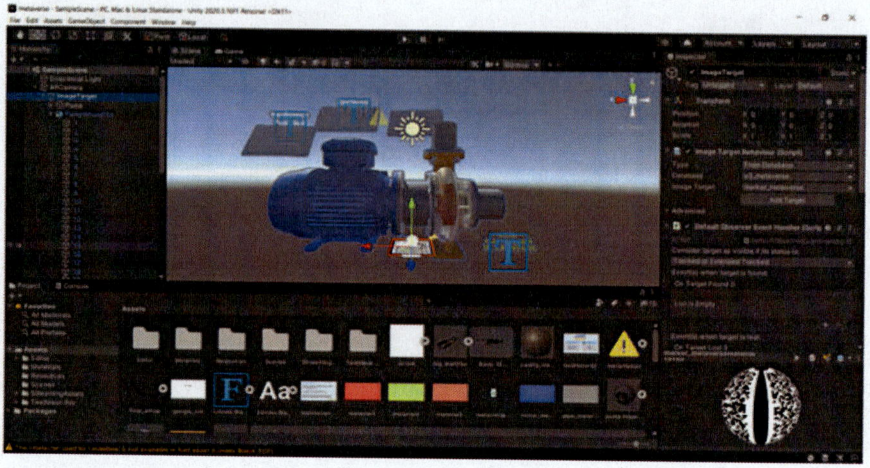

Fig. 8 Machine environment AR scene development in Unity

6 Conclusion

The motive of this project was to assess the possibility of integrating IoT with AR and, in addition, to design a system based on AR for maintenance activities. AR software can be created and deployed in the field to help with maintenance activities, reducing the amount of time spent on the job and the need for an expert to be there. Employing 3D CAD models, complicated and straightforward simulations can be created and exhibited on a mobile device, such as a tablet PC or a smartphone, or even head-mounted displays such as Microsoft HoloLens. These simulations offer the information needed to perform equipment maintenance procedures, such as the procedures to take, the appropriate tool for each step, and cautionary remarks. AIIME would be a good foundation for building an AR-based industrial metaverse. This model still needs to be tested in real environment. All the objectives were completed, and a fully functional machine environment was designed, mainly focusing on IoT and AR integration and assessing its influence on maintenance activities.

References

1. Hashemian HM, Bean WC (2011) State-of-the-art predictive maintenance techniques. IEEE Trans Instrum Meas 60:3480–3492
2. Serpanos D, Wolf M (2018) Industrial Internet of Things 5.1 introduction, 37–54
3. Kipper G (2013) Chapter 1—what us augmented reality? In: Kipper G (ed) Augmented reality. Syngress, pp 1–27. https://doi.org/10.1016/B978-1-59-749733-6.00001-2
4. PwC (2019) Seeing is believing—how virtual reality and augmented reality are transforming business and the economy. pwc.com/SeeingIsBelieving

5. Kipper G (2013) Chapter 3—the value of augmented reality. In: Kipper G (ed) Augmented Reality. Syngress, pp 51–95. https://doi.org/10.1016/B978-1-59-749733-6.00003-6
6. Wang W et al (2020) Augmented reality in maintenance training for military equipment. J Phys Conf Ser 1626:12184
7. Henderson SJ, Feiner SK (2007) Augmented reality for maintenance and repair. ARMAR
8. Simón V, Baglee D, Garfield S, Galar D, Tel (2014) The development of an advanced maintenance training programme utilizing augmented reality. https://doi.org/10.13140/2.1.5103.9685
9. Oliveira R, Farinha J, Raposo H, Pires JN (2014) Augmented reality and the future of maintenance. https://doi.org/10.14195/978-972-8954-42-0_12
10. Manyika J et al (2013) Disruptive technologies: advances that will transform life business and the global economy. www.mckinsey.com/mgi
11. Chebudie AB, Minerva R, Rotondi D (2015) Towards a definition of the Internet of Things (IoT). IEEE Internet Ini
12. Tran D, Dąbrowski K, Skrzypek K (2018) The predictive maintenance concept in the maintenance department of the "industry 4.0" production enterprise. Found Manag 10:283–292
13. Chehri A, Jeon G (2019) The industrial internet of things: examining how the IIoT will improve the predictive maintenance. In: Chen Y-W, Zimmermann A, Howlett RJ, Jain LC (eds) Innovation in medicine and healthcare systems, and multimedia. Springer, Singapore, pp 517–527. https://doi.org/10.1007/978-981-13-8566-7_47
14. Santi GM, Ceruti A, Liverani A, Osti F (2021) Innovation programs. Technologies 9:1–33
15. Santi GM, Ceruti A, Liverani A, Osti F (2021) Augmented reality in industry 4.0 and future innovation programs. Technol 9
16. Ministry of MSME Government of India (2020) Annual report. https://www.ibef.org/industry/msme.aspx
17. Kipper G (2013) Chapter 2—the types of augmented reality. In: Kipper G (ed) Augmented reality. Syngress, pp 29–50. https://doi.org/10.1016/B978-1-59-749733-6.00002-4
18. Fiala M (2005) ARTag, a fiducial marker system using digital techniques. In: 2005 IEEE Computer Society Conference on Computer Vision and Pattern Recognition (CVPR'05), vol 2, pp 590–596
19. Lehner S (2022) AR tag generator. https://shawnlehner.github.io/ARMaker/

IoT Adoption for Botswana in the Sub-Saharan Region of Africa

Leo John Baptist Andrews, Annamalai Alagappan, V. Sampath Kumar, Raymon Antony Raj, and D. Sarathkumar

Abstract The Internet of Things (IoTs) as a technology and its mainstreaming by numerous governments throughout the world have sped up the creation of numerous new IoT developments. IoTs are boosting quality of life in cities all over the world by ensuring health care, analyzing traffic, and constructing better and more efficient services for delivery in a variety of sectors, including energy, water, and sanitation. This paper seeks to comprehend some of these technologies and how they might be incorporated into Sub-Saharan Africa, with a particular emphasis on Botswana. The article makes an effort to identify key issues that must be resolved in order to achieve better IoT adoption.

Keywords IoT · Smart city · Botswana · Africa · Cyber security

1 Introduction

Researchers have had to adapt to ever-more-innovative ideas in the growing city infrastructure due to the rapid increase in population density in metropolitan regions. The consequences of climate change have accelerated the development of new technology. Around the world, 55% of people live in urban areas, which occupy 4% of

L. J. B. Andrews
Department of Information Technology, Faculty of Engineering and Technology, Botho University, Gaborone, Botswana

A. Alagappan
Department of Network and Infrastructure Management, Faculty of Engineering and Technology, Botho University, Gaborone, Botswana

V. S. Kumar
Grant Thornton, Plot 50370, Acument Park, Fair Grounds, Gaborone, Botswana

R. A. Raj · D. Sarathkumar (✉)
Department of Electrical and Electronics Engineering, Kongu Engineering College, Perundurai, Erode, Tamil Nadu, India
e-mail: sarath.eee@kongu.ac.in

S. Sharma et al. (eds.), *Intelligent Control, Robotics, and Industrial Automation*, Lecture Notes in Electrical Engineering 1066,
https://doi.org/10.1007/978-981-99-4634-1_30

the total land area and use around 67% of the energy produced. Cities are responsible for 60–70 percent of daily carbon emissions. Observers predict that rather than the national government, the world's largest cities will handle police matters as a result of expanding sustainability. By combining and connecting, contemporary and cutting-edge digital technologies like smart meters, smart homes, sensors, actuators, smartphones, and smart appliances fuel the technology-driven economies.

Cities constructed on these sophisticated, machine learning-based smart technologies are referred to as "smart cities." Smart buildings, smart health care, smart transportation, smart energy, smart education, smart technology, smart structures, smart supremacy, and smart citizens are all defined by smart parameters [1]. By enabling billions of smart devices with embedded technology to connect successfully, IoT considerably grows in scale and breadth. New challenges and opportunities result from this [2]. Many nations have long-standing development plans for their infrastructure and facilities that include IoT as a component. According to studies, by the year 2050, 70–75% of the world's estimated 6 billion people will reside in urban areas [3].

This article seeks to offer some insights into the main field of study on Internet of Technologies (IoTs) for smart cities. And how this concept can be applied to a fast-growing town in Botswana. The paper is divided into the following sections: Sect. 2 examines Internet of Technologies (IoT) for smart cities. Section 3 brings the challenges for implementation of IoT in Botswana, and Sects. 4 and 5 provide an outcomes and conclusion of the article.

2 Internet of Technologies (IoTs)—Smart Cities

The rapid development of internet communication technologies provided a tree of interlinks with each technique. It is anticipated that the total count of interlink and interconnected gadgets was far more than the amount of population [4]. The rapid stream of technological development in the application sectors has resulted in playing a fundamental role in the large-scale deployment of heterogeneous infrastructures. As a result, IoT is said to improve cities, different aspects of utility services, enhancing public transformation, reducing traffic congestion, providing better health care, and in keeping citizens safe. IoT in principle can be categorized based on their technological platform on a network basis, the ability to scale up the platform and ease of use, heterogeneity, repeatability, and involvement of end-users [5]. The improvement with IoT implementation in various sectors can render enormous benefits in the form of effective service delivery in segments, especially in municipal services, enhancing infrastructure in terms of transportation, minimizing traffic snarls, providing health care. With adequate policy changes at a national level, the demographic changes can be manifold.

2.1 *Insight of IoT into Smart Cities*

In theory, there are three layers to the IoT as (1) perception layer, (2) network layer, and (3) application layer. The perception layer is an internet-assisted layer wherever several devices with internet or intranet access can read, gather, detect, and perceive information while also interacting with one another. The network layer makes use of a variety of network communication protocols to send data to devices dependent on their communication partners' capabilities. Where the information is received and processed will depend on the application to the application layer, telecommunications technologies like 5G, 4G, also power line tools transmit the data to the higher level of long distances. Communication technologies are heavily utilized in smart cities [6].

2.2 *Parameters Defining Smart Cities*

Accordingly, IoT involves smart gadgets like mobile communication devices and various devices that aims to collaborate to achieve a joint objective with a characteristic feature with its effect on consumer's lifestyle [7]. IoT uses the internet as a source to merge the complicated things to provide ease of access and to examine the data usage design of the consumer [8]. The development of an ICT as a core creates a stronger interlink layer binding various services, thus forming an efficient base. To attain this objective, smart cities are dependent on sensor networks and connection of intelligent appliances to remotely monitor the data utilization and pattern [9].

In understanding the heterogeneous environment of smart cities, various components like participants, motivations, security policies, social impacts, environmental impacts should be taken into consideration and studies independently. Since smart cities not only portray to enhance the economics of development, attract global city investments and competitiveness but also take a holistic look at improving the lives and means of support for locals. With appropriate facilities for enjoyment, protection, reliability, it enhances the confidence in investors in infrastructure development and public facilities like transportation, energy efficiency. Smart cities also aim to boost employment opportunities and encourage competitiveness with improvement in quality and help in attaining sustainability. From the smart city interactions, various cohesions exist and there is a strong relationship with citizens, the local economy, and the local authority. The highest point of interaction lies in increased resource efficiency and sustainable mobility. In smart cities, the strategic point of interest is to ensure that the town services are more efficient while attracting investments, residents, visitors, and business community [10].

The traditional models for city development have over the years changed and aligned themselves with the global scenario attracting global investments while at the same time ensuring a balance in the local community. The new model also redefines itself and is more customer-centric requirements with the development of smart technologies [11]. This new definition of being customer-centric requires smart cities to reevaluate and adapt to the changing propositions in many sectors. A keen plan of action gives data and expresses the rationale of how it makes and gives delivery value to its clients while all the while delineating the design of business interests [12]. Figure 3 provides insight into parametric values in smart cities. The focus components defined are digital energy, smart buildings, communication technologies, healthcare, infrastructure, transportation, civic initiatives, gifted education, governance, and security. Quite some research has been undertaken and various research papers on the application of IoT on the fields are mentioned above. By using the data generated through different parameters, smart cities could be deployed. An instance, the smart energy systems will encourage a consumer to be a prosumer by application of Distributed Generation principles and will enable the consumer to be proactive in generating a substantial revenue which in turn will encourage sustainable living [13]. Likewise, applying smart technology to the construction of environmentally friendly buildings will help carbon footprint reduction. Similarly, utility corporations like water and sanitation will be more service-oriented. The biomass generated in individual homes, farms, and agriculture, which can be used for clean energy production, will aid in developing more biomass application. Smart transportation will ensure efficient transportation and will aid in improving advance technologies in the form of Hyperloop [14].

3 Challenges for Application of IoT in Botswana

The following challenges in the Botswana country is faced during the initial stage of implementation of IoT to make as smart city. Those challenges are data privacy and security issues, Big Data and Networks, Reliability, Heterogeneity, Legal and Societal Aspects, Institutional Barriers. The following Fig. 1 represents the challenges faced for implementing the IoT applications in Botswana [15].

3.1 Data Privacy and Security

IoT is a platform where data are critical. The amount and the nature of data that are transmitted to and fro in an IoT platform are inevitable to attacks. With information being crucial, the system may be subjected to numerous type of attacks like DDoS, Cross-Site Scripting, data theft, so on and so forth. The system itself may have numerous vulnerabilities that could lead to further compromise. This, therefore, warrants for efficient cyber defense mechanisms to protect citizen's data. Botswana, as a nation, has recently incorporated data privacy act and is a valuable addition in terms of providing security through governance. For effective implementation of IoT, its essential for incorporate strict data control also cyber defense mechanism. Data security platform is generally built on Confidentiality, Availability, and Integrity.

Fig. 1 Challenges for application of IoT in Botswana

The core component, however, is the privacy, trust, and data confidentiality, while Confidentiality, Integrity, and Availability form part of the CIA triad. The data privacy is critical as data comprises personal information including health care [15]. The following Fig. 2 shows a simple IoT security mechanism and security aspects of the integrated systems.

3.2 *Big Data and Networks*

IoT is a networking environment and thus communicates with numerous entities and can, therefore, pose a serious challenge in the form of processing information on a large scale [16]. The data in an IoT network can be large since as many as 5 billion devices communicate essentially in a networked environment when connected globally. This brings a challenge in form of transferring of data, storage, and database management. The primary data problem here relates speed, number, and variance. Botswana's internet platform, therefore, needs to be strengthened with stability and reliability to meet the requirements [17].

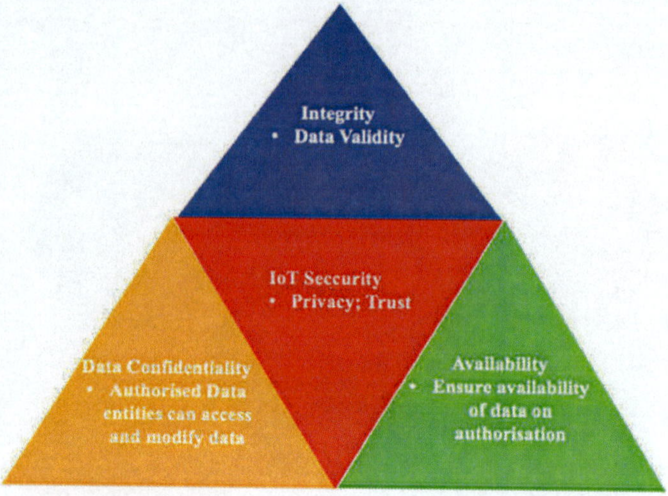

Fig. 2 IoT security aspects

3.3 *Reliability*

Due to the size of data and participation in huge numbers and complexity, IoT-based systems have some serious reliability problems. There should be a focus on this area due to various community factors that are aligned with it [18].

3.4 *Heterogeneity*

Most IoT systems are built with specifics and are interlinked and mapped to unique application environment. It is on this foundation that most have to investigate and re-examine the scenarios during practical implementation of the hardware or the software and aggregate them to form heterogeneous systems [19].

3.5 *Legal and Societal Aspects*

IoT system primarily relies on data, which in turn attracts the local and international rules and regulations in terms of data handling and compliance with the data privacy acts [20].

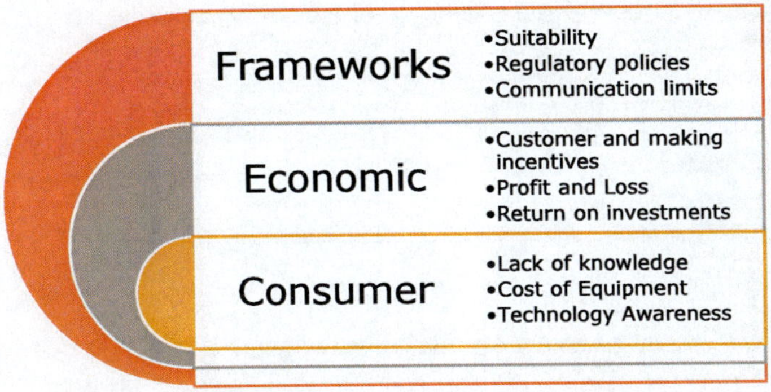

Fig. 3 Institutional barriers

3.6 *Institutional Barriers*

Institutional barriers are potential threats to the smart city environment. This is primarily due to various government regulation and policies, procedures, guidelines, communication limits, investment regulations, technology transfer and availability, lack of knowledge, etc. [21]. The barriers also include frameworks, economic, and consumer. The framework barriers include suitability, regulatory policies, and communication limits, while the economic obstacles include customer and making incentives, profit and loss, return on investments, and consumer barriers which can be in the form of lack of knowledge, cost of equipment, technology awareness, etc. The institutional barriers and its related parameters are shown in the following Fig. 3.

3.7 *Smart Parameters in Smart Cities*

The smart parameters like digital energy management, smart buildings, smart communication technology, smart health care, smart infrastructure, smart transportation and mobility, civic initiatives, intelligent education systems, smart government, smart security are considered as very important parameters for developing the smart cities. The smart parameters making in smart cities [14] are shown in the following Fig. 4.

The components of IoT infrastructure to cover smart cities are represented in the above Fig. 5. Smart cities will also encourage smart convergence and ensure smart capabilities with different participants in the system. For instance, IT participants may comprise partners in networking, digital technology, software, technology integration, network security. Similarly, telecom partners may participate in internet service providers, phones, mobile communications, and network IT services. It is to be noted that it is a congregation of companies involved in various service deliveries integrated into the technology of IoT [22]. Energy participants may transfer and share

Fig. 4 Smart parameters in smart cities

technologies in transmission and distribution, power electronics, renewable energy, substation automation. Since smart cities will comprise smart buildings, automation and building control participants will strive to achieve integration of smart grid, multi-device connectivity, demand-side management, home automation. Last but not least, the security will be smart with centralized monitoring control, firewalls and internet protocols, cloud-based services.

4 Outcomes

The Internet of Things (IoTs) as a technology and its adaptation into mainstreams by various governments across the globe have set developmental pace for many new innovations in IoT. IoTs are transforming cities across globe in the form of ensuring

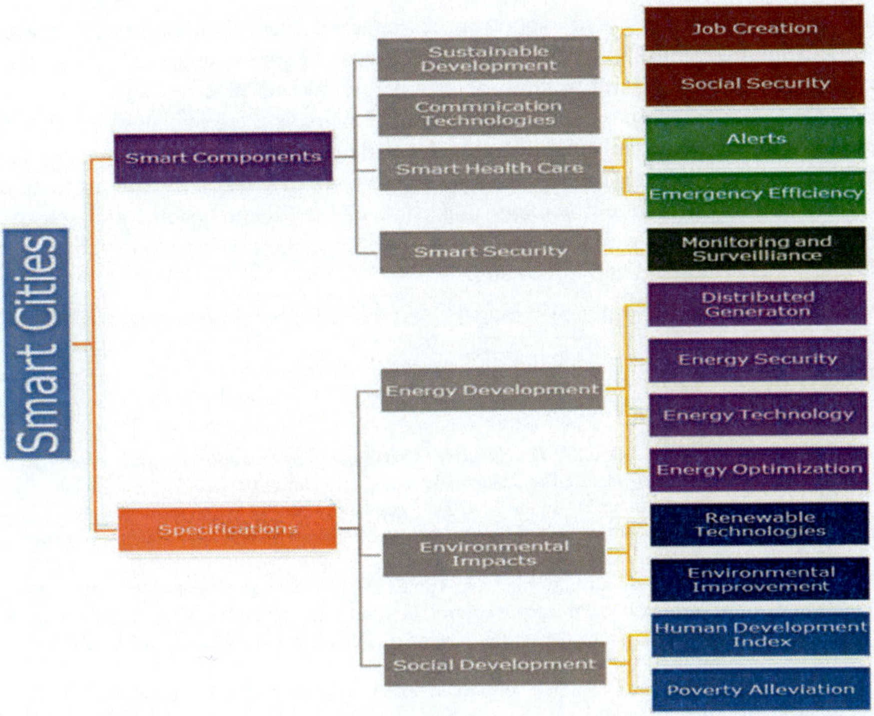

Fig. 5 Components

health care, understanding traffic, building better and efficient services in the form of delivery in various sectors like energy, water, sanitation, etc. while improving quality of life and to establish the smart city. Outcome of this paper place aims to understand some of these IoT technologies and how it can be adopted as technology into the Sub-Sharan African region with a focus on Botswana to make a smart city. The outcome of this article attempts to highlight key issues that must be resolved to achieve improved IoT adoption for Botswana's development of smart cities in Sub-Saharan Africa.

5 Remarks and Conclusions

It is important to consider how new concepts and developments, notably the Internet of Things, benefit urban areas. This article's goal was to look into the various details and features of IoT frameworks as well as the practical driving reasons for their use. Since the development of IoT infrastructures can open up a variety of chances for Botswana's communities, the greatest significant research inspirations stood first shared, and a few keys also supporting solicitations were then explained. It was demonstrated how using them might expand and enhance regular exercises. The challenges associated with putting the IoT paradigm into practice were similarly described. One of the most exciting future tendencies is the installation of an IoT platform with more established technologies. Additionally, providing a strategy for adjusting to certain substantial challenges in the form of client/occupant protection rights, etc. The IoT should actually make use of intelligent systems and sensors to provide population welfare in the form of sustainable income and development thanks to its usefulness and information.

References

1. Rathore MM, Ahmad A, Paul A, Rho S (2016) Urban planning and building smart cities based on the Internet of Things using Big Data analytics. Comp Netw, 101–108
2. Bhaskar MA, Sarathkumar D, Anand M (2014) Transient stability enhancement by using fuel cell as STATCOM. In: International Conference on Electronics and Communication Systems (ICECS), pp 1–5, IEEE, Coimbatore, India
3. Sarathkumar D, Srinivasan M, Stonier AA, Vanaja DS (2014) A brief review on optimization techniques for smart grid operation and control. In: 2021 International Conference on Advancements in Electrical, Electronics, Communication, Computing and Automation (ICAECA), pp 1–5
4. Siano P (2014) Demand response and smart grids—a survey. Renew Sustain Energy Rev 30:461–478. https://doi.org/10.1016/j.rser.2013.10.022
5. Siano P, Graditi G, Atrigna M, Piccolo A (2014) Designing and testing decision support and energy management systems for smart homes. J Ambient Intell Humaniz Comput 4(6):651–661
6. Stonier A, Yazhini M, Vanaja DS, Srinivasan M, Sarathkumar D (2021) Multi level inverter and its applications—an extensive survey. In: 2021 International Conference on Advancements in Electrical, Electronics, Communication, Computing and Automation (ICAECA), pp 1–6
7. Albert AS (2017) Development of high performance solar photovoltaic inverter with advanced modulation techniques to improve power quality. Int J Elect 104(2):174–189
8. Talari S, Shafie-khah M, Siano P, Loia V, Tommasetti A, Catalão JSP (2017) A review of smart cities based on the internet of things concept. Energies 10(4):421. https://doi.org/10.3390/en10040421
9. Sarathkumar D, Srinivasan M, Stonier AA, Samikannu R (2021) A research survey on microgrid faults and protection approaches. In: IOP Conference Series: Materials Science and Engineering 1055(012128), pp 1–15
10. Venkatachary SK, Prasad J, Samikannu R (2019) Challenges, opportunities and profitability in virtual power plant business models in Sub Saharan Africa—Botswana. Int J Energy Econ Pol 7(3):1–11
11. Sarathkumar D, Srinivasan M, Stonier AA, Samikannu R, Anand DV (2021) Design of intelligent controller for hybrid PV/wind energy based smart grid for energy management applications. In: IOP Conference Series: Materials Science and Engineering 1055 (012129), pp 1–15
12. Venkatachary SK, Prasad J, Samikannu R (2017) Economic impacts of cyber security in energy sector: a review. Int J Energy Econ Pol 7(5):250–262
13. Sarathkumar D, Srinivasan M, Stonier AA, Samikannu R, Dasari NR, Raj RA (2021) A technical review on classification of various faults in smart grid systems. In: IOP Conference Series: Materials Science and Engineering 1055(012152), pp 1–11

14. Venkatachary SK, Prasad J, Samikannu R (2017) Barriers to implementation of smart grids and virtual power plant in Sub Saharan Region—Focus Botswana. Energy Rep 4:119–128. https://doi.org/10.1016/j.egyr.2018.02.001
15. Sarathkumar D, Srinivasan M, Stonier AA, Samikannu R, Dasari NR, Raj RA (2021) A technical review on self-healing control strategy for smart grid power system. In: IOP Conference Series: Materials Science and Engineering 1055(012153), pp 1–15
16. Ye X, Huang J (2011) A framework for cloud-based smart home. In: 2011 International Conference on Computer Science and Network Technology, 2, pp 894–897. Harbin, China. https://doi.org/10.1109/ICCSNT.2011.6182105
17. Raj RA, Murugesan S (2022) Optimization of dielectric properties of Pongamia Pinnata Methyl Ester for power transformers using response surface methodology. IEEE Trans Dielect Elect Insul 29(5):1931–1939. https://doi.org/10.1109/TDEI.2022.3190257
18. Duraisamy S, Murugesan S, Palanichamy M et al (2022) Restoration of critical dielectric properties of waste/aged transformer oil using biodegradable biopolymer-activated clay composite for power and distribution transformers. Biomass Conv. Bioref. 12:4817–4833. https://doi.org/10.1007/s13399-022-03065-0
19. Raj RA, Murugesan S, Ramanujam S, Stonier AA (2022) Empirical model application to analyze reliability and hazards in Pongamia oil using breakdown voltage characteristics. IEEE Trans Dielect Elect Insul 29(5):1948–1957. https://doi.org/10.1109/TDEI.2022.3194490
20. Duraisamy S, Murugesan S, Murugan K, Raj RA (2022) Reclamation of aged transformer oil employing combined adsorbents techniques using response surface for transformer applications. IEEE Trans Dielect Elect Insul. https://doi.org/10.1109/TDEI.2022.3226162
21. Sarathkumar D, Stonier AA, Srinivasan M, Senthamil LS (2022) Review on power restoration techniques for smart power distribution systems. In: Kumar A, Srivastava SC, Singh SN (eds) Renewable energy towards smart grid. Lecture Notes in Electrical Engineering, vol 823. Springer, Singapore. https://doi.org/10.1007/978-981-16-7472-3_6
22. Sarathkumar D, Srinivasan M, Stonier AA, Kumar S, Vanaja DS (2021) A review on renewable energy based self-healing approaches for smart grid. In: 2021 International Conference on Advancements in Electrical, Electronics, Communication, Computing and Automation (ICAECA), pp 1–6. https://doi.org/10.1109/ICAECA52838.2021.9675495

Development of Affordable Smart and Secure Multilayer Locker System for Domestic Applications

Naveen Shenoy, K. Vineeth Rai, M. Pratham Rao, Ranjith Singh, and Gurusiddayya Hiremath

Abstract The security system or authorization plays an important role in preventing unauthorized users from accessing secured physical and logical locations. The most basic forms of security systems are standard door lock keys and computerized automated identification systems. The major purpose of this paper is to design and build the smart and intelligent multilayer security system for lockers. TOTP authentication, fingerprint authentication, and passcode/password authentication are all part of the smart locker system. To get access to the locker, two or more authenticators registered in the system must be present or must allow the access to unlock the locker using the system-linked mobile application. The authentication procedure begins when registered members launch the mobile application after entering the passcode/password, then pick which locker system they need to open or get access to, and then finally, the application will request fingerprint authentication to obtain the TOTP. Each user will go through the identical procedure to obtain their own TOTPs, by which they must enter into the locker system in the sequence which was assigned. Furthermore, various other security systems that may be incorporated into the system for a larger sector of security applications, such as security systems in large organizations. The goal of developing this system is to create a powerful smart intelligent multilayer locker security system that is extremely secure, low-cost, and remotely accessible.

Keywords TOTP · Multilevel · Two-factor authentication (2FA) · Security system

---

N. Shenoy · K. V. Rai · M. P. Rao · R. Singh · G. Hiremath (✉)
Department of Electronics & Communication Engineering, Sahyadri College of Engineering & Management, Mangaluru, Karnataka, India
e-mail: gurusiddayya@gmail.com

Visvesvaraya Technological University (VTU), Belagavi, Karnataka, India

S. Sharma et al. (eds.), *Intelligent Control, Robotics, and Industrial Automation*, Lecture Notes in Electrical Engineering 1066,
https://doi.org/10.1007/978-981-99-4634-1_31

1 Introduction

Normal door lock keys, as well as electronic automated identification systems, are the two most basic forms of security systems. Lockers' systems were first constructed of wood, but were later made of steel and metal. Now locker systems are developed in response to people's wants and new technology. With the introduction of new technology, the lock mechanism on a locker has developed significantly. The transition from a hefty padlock and key to an electronic system exemplifies how lockers have embraced smart technology. Smart technology enables lockers to be digital, adaptable, and equipped with a variety of functions to enhance the user experience. Smart lockers are digitally controlled storage banks that make the process of acquiring and utilizing a locker quick and easy. The technology, whether managed by a mobile phone app or a touchless kiosk, enables for automation across the whole process/workflow.

In this instance, automated password-based door locks, which are routinely used in companies and homes, are a handy alternative. The security system employs a variety of identifying methods, including Time-based One-Time Password (TOTP) authentication, fingerprint or face authentication, and passcode/password-based authentication. Time-based One-Time Password (TOTP) is a common form of two factor authentication (2FA). Unique numeric passwords are generated with a standardized algorithm that uses the current time as an input. The time-based passwords are available offline and provide user-friendly, increased account security when used as a second factor, and for organizations looking to step up their cybersecurity, they should require TOTP instead of SMS on all their IT resources, including systems, file servers, web applications, and online applications.

Under this method, only an approved set of persons has access to the locker. The fundamental security layers are TOTP and fingerprint verification. Both the hardware security system and the software security application are linked by the security system. Initially, a group of individuals are registered using the software program by entering the 16-digit seed code number, which then connects that specific mobile system to the locker system. Then, provide the necessary information, such as your phone number and email address. The fingerprint authentication methods are then validated with passcode/password credentials. The process is repeated for each member of the group, each with a unique 16-digit seed code number.

In the event of theft or illegal access to the security system, an alert system, which comprises a message, e-mail alert through GSM, and a loud alarm system, has been set up to take the appropriate steps. If someone tries to remove the whole locker, an electric shock will be delivered to the locker, rendering it untouchable, and a loud alarm will sound, alerting the entire organization and the authorities. A GPS is also inserted to track the location of the locker, and the DNO procedure will be carried out.

2 Background Study

Saleem and Alshoshan [1] proposed that multi-factor authentication is a type of account setup where a user can only visit a website after satisfactorily submitting two or more elements, or bits of proof. Because the login details used for conventional logins are not completely secure against attackers who could quickly predict them via techniques, this will be the first step in securing systems against attackers. Technologies now in operation employ extra security measures such as two-factor verification focused on a personal passcode via email or mobile devices, verification based on biometrics (fingerprints, pupil, or retinal of both the eye, and facial recognition), or identification using token handsets. But for small and midsize enterprises, these techniques require extra technology that is expensive. This paper suggests a multi-factor authentication model that integrates simplicity of use and affordability. There is no specific equipment or setup needed for the application. The user selects three images and recalls those upon enrolling since it uses visual credentials. As a result, the client only needs to select the appropriate photographs that he had envisioned in a specific arrangement throughout the registration process. The suggested approach defeats numerous security risks, including key loggers, snapshot assaults, and shoulder surfing.

Al Qahtani et al. [2] proposed the study that aims to produce a two-factor authentication (2FA) system that, in contrast to current techniques, happens with little user input. By detecting whether the multiple user gadgets were in the same geographical place, they were able to confirm that both related to the authorized source. A 2FA solution is used by numerous telecom operators to increase security, as per the report. They demonstrated how 2FA and DFA are regularly used by consumers for a variety of tasks, including transactions, finance, etc. Additionally, major corporations like Apple, Google, and Windows are progressively requiring their customers to adopt a 2FA method. Because it requires less user effort to set up the two layers of verification and to continually verify identity, the approach suggested by researchers will greatly outperform current approaches. This maintains convenience while ensuring that the accessibility to distant resources is restricted to authorized parties only. Because a CI2FA combines the simplicity of Single-Factor Authentication (SFA) with the increased security of 2FA, it will allow the 2FA procedure to be implemented in far more organizations (while minimizing user interaction). Continuous Indoor Two-Factor Authentication, or CI2FA, only needs your login and password.

Seta et al. [3] proposed a solution for a more secure authentication process by employing a two-factor authentication method, which is a time-based one-time password (TOTP) authentication which is possible using the Security Algorithm 1 plus a time-based one-time password (TOTP). In this approach, the authentication process of a webpage requires the user to provide a key or passcode in addition to their login details to access their account. Because the key can only be used once with a specified, i.e., unexpected, time restriction, time-based one-time passwords, and SHA 1 create codes that would not be identical. According to the outcomes of the testing process, SHA 1, with the feed string's length, generates a variable output text that is 160 bits in length which makes sniffing out the user id comparatively secure. Using algorithms like SHA 256, SHA 512, or AES as well as adding a protection element to become a three-factor authentication instead of relying solely on the name, account name, passcode, and pin number alone. Alternatively, add a further factor for verification, including a signature or biometric, facial recognition software, or voice recognition. Protection credentials with SHA 1 can still be cracked.

Segoro and Putro [4] suggested choosing instant messaging which is chosen for interaction in this research because of its security mechanisms, such as one-time password (OTP) codes, end-to-end encryption, and perhaps even multiple verification risk of an unauthorized party. Dual encrypting and two-factor verification, which are connected to one another, are used to maintain the safety of the instantaneous messaging service in order to avoid this identity hijacking. There are two different solution designs for each method. Its action plan aims to secure the communication process and receipt while protecting login. The QR code design is given by email, enabling login protection. When a user authenticates themselves with a biometric during both message sending and receiving, the data decoding procedure starts. For communication safety, mixed encryption employs RSA 2048 and AES 128. The applied structure has proven to lessen the effect of unauthorized parties from the various experiments.

Derhab et al. [5] proposed this paper as security experts have demonstrated that smartphone apps that employ SMS-based verification for two-factor verification are vulnerable. Consequently, they think that uploading mobile apps to the internet, which also has a lot of resources and offers a secure world, is a smart option whenever security standards and performance limits are raised. In order to achieve this, they suggested a revolutionary two-factor mutual authentication method centered on a unique method known as a "virtual chip card" as well as an offloading infrastructure for two-factor verification systems. Additionally, a method for choosing to install certain/specific authorization software and its digital card based on three factors: protection, batteryBattery life, and energy expenses. Calculate the lower bound of the smartphone platform's processing time analytically using the overall energy calculation in order to complete the installation. They look at and confirm the security features of the recommended technology. They also evaluate the two-factor mutual authentication scheme and the offloading decision-making procedure.

Gurabi et al. [6] proposed the design of an ECC-based two-factor authentication method as well as an authentication equipment component for dispersed IoT-based applications. The development of three smart card reader units for two-factor verification in IoT systems depending on multiple processing units and diverse application cases was described, along with the hardware specifications for the component's development. Additionally, they developed a customer authentication and key sharing system based on ECDH, examined it for security vulnerabilities, and assessed the techniques' resistance to various kinds of attacks. Then, they used TinyOS and the TinyECC library to create and test their verification system on IRIS sensor nodes, and they timed how quickly it ran. They also discussed how their plan may be improved in the future. In the coming future, they will examine the hardware devices in an expandable IoT environment and upgrade the authenticating system with additional capabilities, including a method for revocation of certificates.

3 Proposed System

Work is carried out on multi-layer security system. The operation of this system is to multi-layer authentication system which consists of Time-based One-Time Password (TOTP) authentication, fingerprint or face authentication, and passcode/password authentication for any locker or security system. The Android application is created using Java language in Android Studio. This application is used to get the TOTP and OTP and hence authenticate and access the locker or the security system. Furthermore, various other security systems that may be incorporated into the system for a larger sector of security applications, such as security systems in large organizations. The goal of developing this system is to create a powerful multilayer locker security system [7] that is extremely secure, low-cost, and remotely accessible. Figure 1 depicts the proposed system architecture for smart and secure multilayer locker system.

4 Methodology

There areTwo Key approaches for authentication.

4.1 OTP and TOTP-Based Authentication

Two-factor authentication (2FA) is an additional step added to the login process, such as a code delivered to your phone or a fingerprint scan, that helps validate your identity and prevents thieves from accessing your personal information. Because the criminal needs more than just your login and password credentials, 2FA provides an extra layer of security that cyber thieves cannot readily access. 2FA is a subset of multi-factor authentication, which is an electronic authentication mechanism that requires a user to authenticate their identity in numerous ways before being granted account access. Two-factor authentication gets its name from the fact that it needs a combination of two factors, whereas multi-factor authentication may require more.

In Time-based TOTP (Time-based One-Time Password) [8]. TOTP's seed is static, exactly like HOTPs; however, the movement factor is time-based rather than counter-based. A time step is the period of time that each password is valid. Timesteps are typically 30 s or 60 s long. If you have not used your password within that window, it will be invalid and you will need to get a new one to access your application.

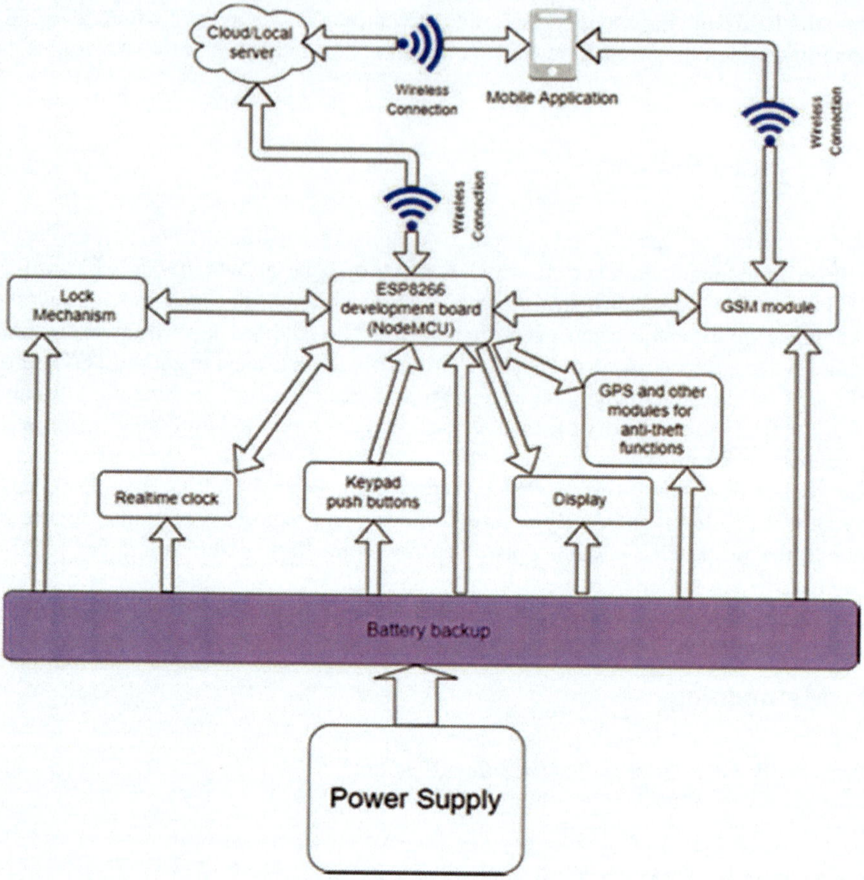

Fig. 1 Proposed system architecture

4.2 *Fingerprint-Based Authentication*

Fingerprint authentication is the process of validating a person's identification using one or more of their fingerprints. For decades, the notion has been used to a variety of initiatives, including digital identification, criminal justice, financial services, and border protection. Fingerprint authentication, often known as fingerprint scanning, is a type of biometric technology that allows users to access internet services by scanning photographs of their fingerprints. The biometric scan frequently depends on smartphone and other device native sensing technology, which has mostly supplanted software-based, third-party biometric algorithms. Some fingerprint scan systems, such as FIDO, are designed in a decentralized approach to ensure that a user's fingerprint template is secure on the user's device [9, 10]. A user's fingerprint scan is locally validated against itself, a token is delivered to the service provider, and access is permitted. Biometric authentication occurs locally, and biometric data are not kept with the service provider.

Working mechanism of the system is shown in Fig. 2. Hardware setup and authentication at first the initial setup are checked. If the initial setup is complete, then the system will proceed to the next step. If not, it will proceed to the completion of the initial setup. Initialization and cleaning of the memory to make space for relevant data storage are done first. Creation of data variable instances is done, and then, SEED is generated for TOTP authentication. To access the locker, user is asked to input the SEED and generate a TOTP in the software. Verification and storing of all the SEEDs in the safe encrypted format are done. Then OTP is sent to each user. Each user enters the TOTP and compares it with the system-generated TOTP to unlock the locker. If the TOTP does not match at the first try, an alert is sent to each user's and Do Not Open (DNO) procedure will be implemented with the electrocution of the locker casing after third try of entering each user's TOTP [11, 12]. After unlocking the locker, over a certain period of time, if the locker is not locked, then an alert will be sent to the users. Figure 3 gives anti-theft and intrusion prevention on detection of intrusion, a loud alarm is turned on, alerts will be sent to all the users, and locker will be electrocuted for a given period of time or till the registered user stops it. If theft of the locker is detected, the first step is executed, and then activating the live location using the GPS module is done.

In Figs. 4 and 5, the Initial software setup and authentication, OTP and TOTP authentication are being described, respectively. In software application, if the initial setup is complete, the software proceeds to next step. If not, completion of the initial setup is done. Signing up into the application with e-mail ID and phone number is done, which will be verified through OTP sent to the users. Creation of password and fingerprint ID for application is done next. To link the locker with the application, 16 digits' SEED code is entered. Then, enter locker details of the respective 16 digits SEED [13, 14]. User and locker details after verification are stored in the application. Now to open the application, password and fingerprint authentication is done. Then, click on generate OTP to get the OTP on your registered phone number and email ID. Enter the OTP received in your registered phone number and respective e-mail ID, and Time-based One-Time Password (TOTP) generation process starts with limited time-out period. The TOTP generated will be entered into the locker system according to the registered users, which after verification will give us the access of the locker [15, 16].

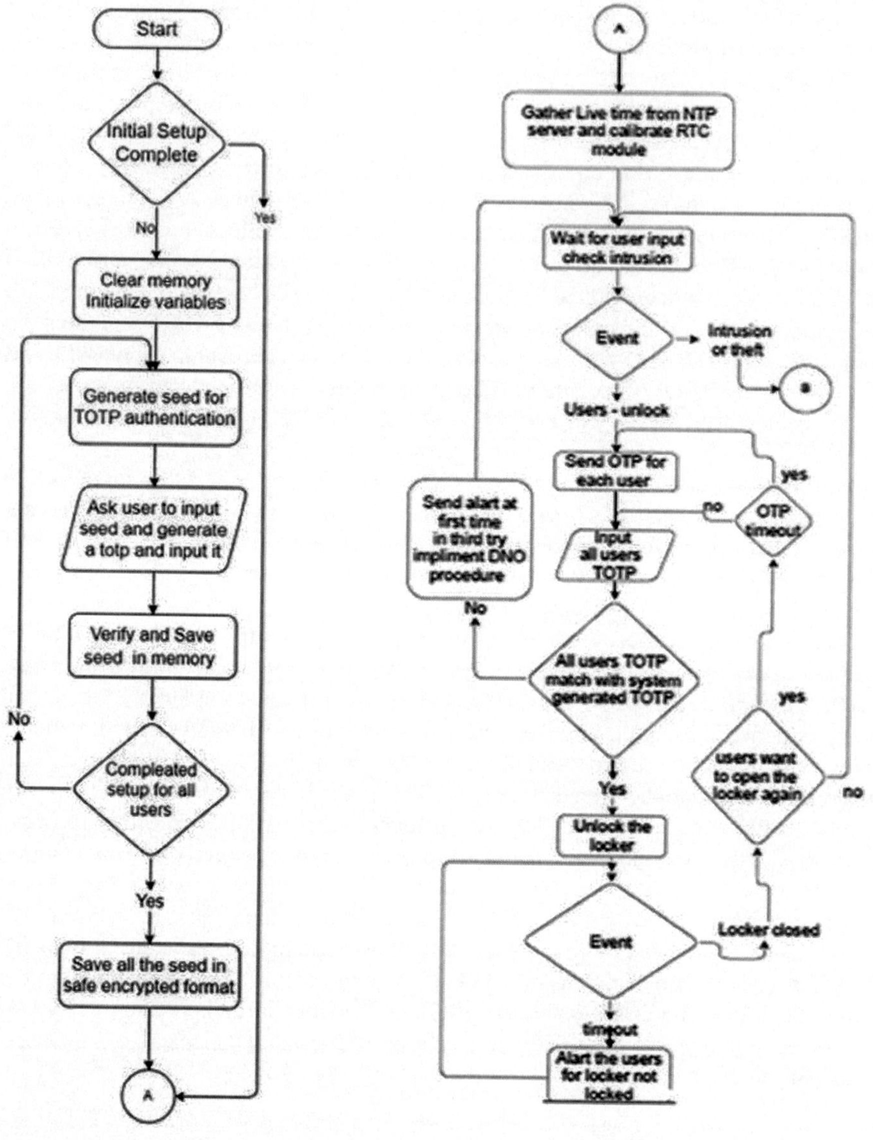

Fig. 2 Hardware setup and authentication

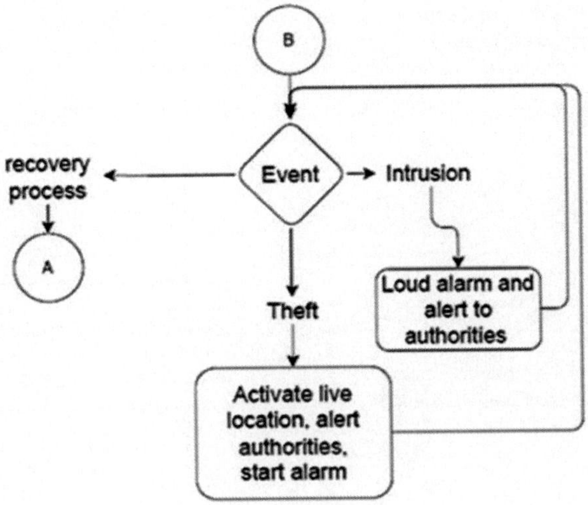

Fig. 3 Anti-theft and intrusion prevention

5 Result and Discussion

5.1 Registration and Locker Linking Process

At first, the linking of the software application and the multi-layer locker system is done. The process includes of accessing the software application after getting authenticated with fingerprint authentication of the default mobile owner. Then, the registration of each user starts with setting up of a password for password authentication and user details such as name, mobile number, and e-mail ID. Post this, each registered user must enter the 16-digit code, displayed on the screen to link the locker hardware system with software application.

5.2 OTP Authentication

In order to unlock the locker, the user must select the unlock option in the locker system. Then, each registered user will receive an OTP on his/her registered mobile number and e-mail ID, which is to be added in the OTP section present in the software application. Figure 6 shows about locker registration and linking and Fig. 7 OTP authentication process.

Fig. 4 Initial software setup and authentication

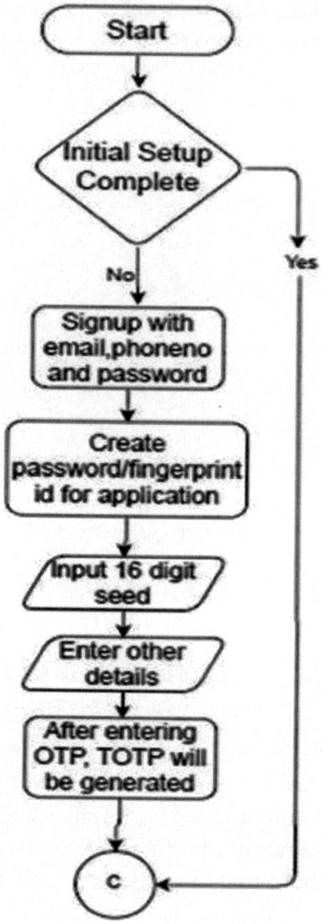

5.3 *TOTP Authentication*

After the OTP authentication, the generate TOTP button is clicked, which will generate a TOTP for limited amount of time, i.e., 60 s for each registered user. This TOTP is required to be entered into the locker system within the given period of time, according to the registered user order.

5.4 *Locker Verification and Unlocking*

The unlock option will be available to access the locker, after authentication and verification of TOTP entered by each registered user in proper order. Then, the lock

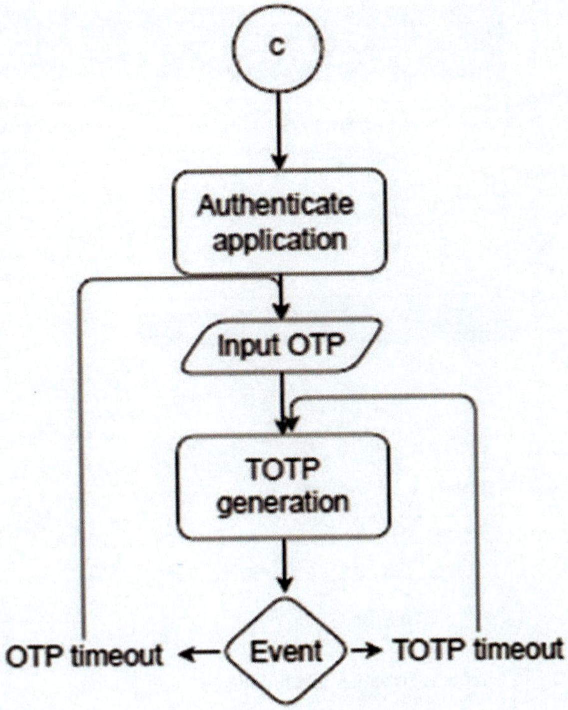

Fig. 5 OTP and TOTP authentication

option will be available to lock the locker after its usage. Figure 8 illustrates TOTP authentication process and Fig. 9 shows locker verification and unlocking process.

5.5 *Anti-Theft Mechanisms*

As shown in Fig. 10 anti-theft alert notification process, in case the user enters a wrong TOTP, a notification will be sent to each registered user in their registered e-mail ID as well as registered mobile number, as a locker alert. If a user enters more than three wrong TOTPs within given period of time, Do Not Open (DNO) feature will turn on with the electrocution of the locker for a period of time. Figure 11 shows the demo model of locket outer casing.

5.6 *Locker Outer Casing Design*

For the prototype model, the outer casing of the locker is made using half a centimeter-thick card board. The multilayer locker security system has been installed into the

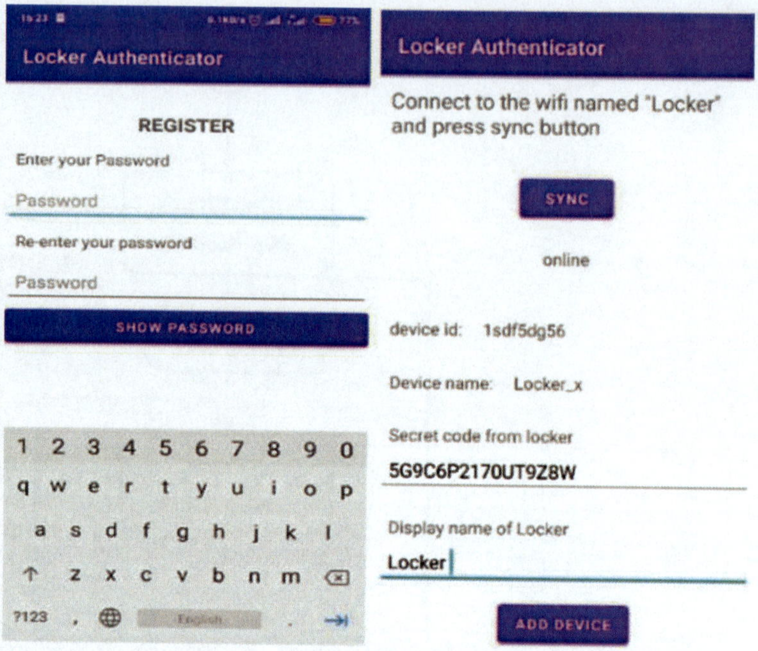

Fig. 6 Locker registration and linking

locker casing with the proper setup of authentication and anti-theft mechanism for authorization.

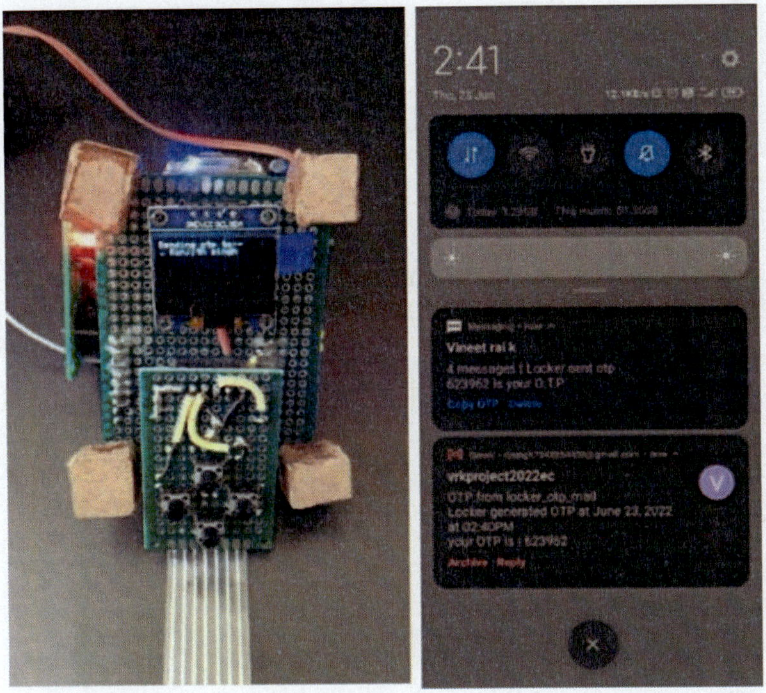

Fig. 7 OTP authentication process

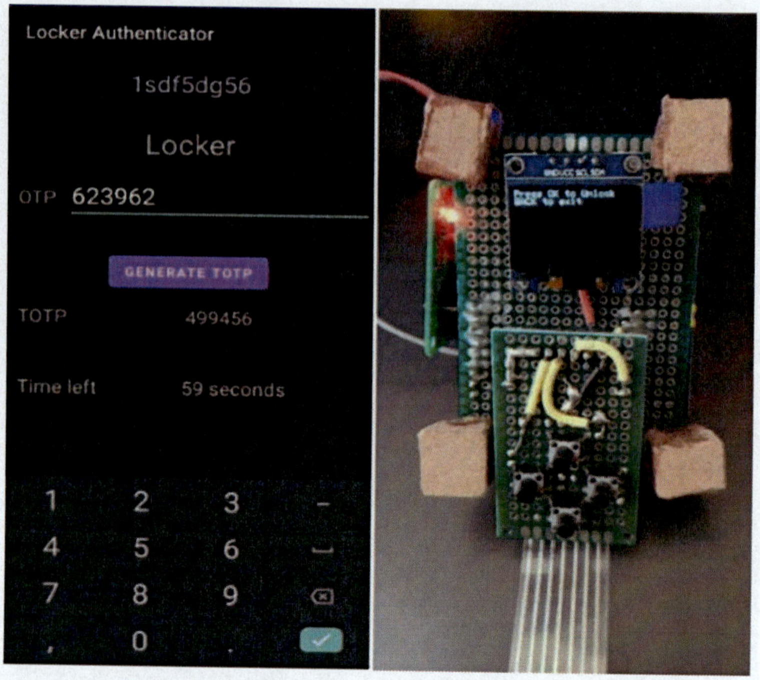

Fig. 8 TOTP authentication process

Fig. 9 Locker verification and unlocking process

Fig. 10 Anti-theft alert notification process

Fig. 11 Locker outer casing design

6 Conclusion

In today's digital environment, the locker security systems are required to be digitized with higher layers of security systems for domestic applications. The major idea behind proposed work is to create a multi-layer locker security system with anti-theft mechanism for domestic applications. As a result, the system provides multi-layer authentication systems with unique TOTP mechanism for two-factor authentication and anti-theft mechanisms, which includes remote access and locker security monitoring system. Providing high-level security for the users' lockers and door lock systems to prevent unauthorized authentication and intrusion. The proposed method will be cost effective in terms of building electronics part, mechanical part. The user will have hassle-free access toward this product. Only challenge for the user is to understand the flow of different levels open the locker system. The multi-layer locker security system can be implemented in any locker or door lock system, which is present at homes, institutes, and corporate organizations. By incorporating many levels of authentication into the system, such as OTP and TOTP authentication, password/passcode authentication, finger print authentication, and anti-theft methods such an alert notification system and alarm security system, it provides multilayer authentication. Furthermore, anti-theft mechanisms such as GPS location monitoring system, sleeping gas, and other anti-theft mechanisms can be installed. Additionally, more layers of authentication such as face lock authentication, retina scan authentication, and similar authentication methods can be set up in future.

References

1. Saleem AL, Omar B, Alshoshan AI (2021) Multi-factor authentication to systems login. In: 2021 National computing colleges conference (NCCC). https://doi.org/10.1109/NCCC49330.2021.9428806
2. Qahtani A, Ali AS, Alamleh H, Gourd J (2021) Ci2fa: continuous indoor two-factor authentication based on trilateration system. In: 2021 International conference on communication systems and networks (COMSNETS). IEEE, pp 1–5. https://doi.org/10.1109/IEMCON53756.2021.9623104
3. Seta H, Wati T, Kusuma IC (2019) Implement time based onetime password and secure hash algorithm 1 for security of website login authentication. In: 2019 International conference on informatics, multimedia, cyber and information system (ICIMCIS). IEEE, pp 115–120. https://doi.org/10.1109/ICIMCIS48181.2019.8985196
4. Segoro MB, Putro PAW (2020) Implementation of two factor authentication (2FA) and hybrid encryption to reduce the impact of account theft on android-based instant messaging (IM) applications. In: 2020 International workshop on big data and information security (IWBIS). IEEE, pp 115–120. https://doi.org/10.1109/IWBIS50925.2020.9255501
5. Derhab A, Belaoued M, Guerroumi M, Khan FA (2020) Two-factor mutual authentication offloading for mobile cloud computing. IEEE Access 8:28956–28969. https://doi.org/10.1109/ACCESS.2020.2971024
6. Gurabi MA, Alfandi O, Bochem A, Hogrefe D (2018) Hardware based two-factor user authentication for the internet of things. In: 2018 14th International wireless communications and mobile computing conference (IWCMC). IEEE, pp 1081–1086. https://doi.org/10.1109/IWCMC.2018.8450397

7. Mohammed S, Alkeelani AH (2019) Locker security system using keypad and RFID. In: 2019 International conference of computer science and renewable energies (ICCSRE). IEEE, pp 1–5. https://doi.org/10.1109/ICCSRE.2019.8807588
8. Ashraf A, Rasaily D, Dahal A (2016) Password protected lock system designed using microcontroller. Int J Eng Trends Technol 32(4):180–183
9. Srinivasan R, Mettilda T, Surendran D, Gobinath K, Sathishkumar P, Advanced locker security system. Int J Adv Res Sci Eng IJARSE 4(2):2395–6011
10. Detroja HS, Vasoya PJ, Kotadiya DD, Bambhroliya PCB (2016) GSM based bank locker security system using RFID, password and fingerprint technology. IJIRST—Int J Innov Res Sci Technol 2(11):110–115
11. Kabir AZMT, Nath ND, Akther UR, Hasan F, Alam TI (2019) Six tier multipurpose security locker system based on arduino. In: 2019 1st International conference on advances in science, engineering and robotics technology (ICASERT). IEEE, pp 1–5. https://doi.org/10.1109/ICASERT.2019.8934615
12. Komol MMR, Podder AK, Ali MN, Ansary SM (2018) RFID and finger print based dual security system: a robust secured control to access through door lock operation. Am J Embed Syst Appl 6(1):15–22
13. Chandanshive PL, Chavan VV, Jare SD, Bank locker security system based on GSM and RFID
14. Sundararaju K, Ruban Kumar S, Sathish Kumar S, Selvamani K (2020) Automatic electronic luggage locker, vol XVI, Issue VIII. ISSN:1001-1749
15. Shahane AG, Kamble SD, Apake AA, Patil OA, Khade AD, Deshmukh SD (2021) A project Phase-I report on 3 stage bank locker. Int J Rec Adv Multidisciplin Top 2(7):101–103
16. Putra IGNAW, Hananto VR, Ningsih N (2019) Design of RFID smart locker for marketplace systems, vol 04, Issue 08, pp 79–86. ISSN: 2455-4847

Machine Learning-Based Industrial Safety Monitoring System for Shop Floor Workers

Muniyandy Elangovan, N. Thoufik Ahmed, and Sunil Arora

Abstract Safety aspects are mandatory for industrial-based shop floor workers and people handling rotating machinery. Ensuring worker safety becomes tedious and has to be monitored to avoid workers, faculty, and lives. The Industry 4.0 standard provides industrial safety and production monitoring with IoT, artificial intelligence, and cloud architectures. With the help of single-board computers and camera-based sensors optimally placed in shop floor locations, workers' safety is monitored. Raspberry Pi-based computers and Jetson nano-based computers monitor safety features such as goggles, face masks, helmets, and their performance. The computing machines are trained with deep learning TensorFlow architecture, and computational time is analyzed. The machine is trained with training data sets, and incoming images from the RaspiCam sensor are validated with the data set. The Raspberry Pi and Jetson Nano provide 19 and 6 ms computing time, and the worker's acknowledgment is given through firebase cloud architecture. The round-trip time delay for sending an acknowledgment to workers is 16 ms.

Keywords IoT · Safety · Hazards · PPE · Artificial intelligence

1 Introduction

Please Health monitoring has become mandatory nowadays, which is practically tedious when patients are monitored inside hospital environments. The present situation of the COVID-19 pandemic has resulted in an increased mortality rate. Though

M. Elangovan (✉)
Department of Biosciences, Saveetha School of Engineering, Saveetha Nagar, Chennai 602105, India
e-mail: muniyandy.e@gmail.com

N. T. Ahmed
Vel Tech Rangarajan Dr. Sagunthala R&D Institute of Science and Technology, Chennai, India

S. Arora
Department of R&D, Bond Marine Consultancy, London, UK EC1V 2NX

S. Sharma et al. (eds.), *Intelligent Control, Robotics, and Industrial Automation*, Lecture Notes in Electrical Engineering 1066,
https://doi.org/10.1007/978-981-99-4634-1_32

many effective measures are under process, the self-control of every person is mandatory for a reduction in the number of COVID-positive cases and mortality rate. The government guidelines indicate that properly wearing a face mask and regularly monitoring human body temperature are essential for effective protection from the COVID-19 virus. Proper monitoring is necessary to ensure that people strictly adhere to the guidelines. Especially when workers gather in their workplaces like industries, schools, and colleges, the proper safety-ensuring measures are strictly followed, and every individual must wear necessary Personal Protective Equipment (PPE) to protect themselves and reduce the spread.

The Internet of Things (IoT)-based monitoring system is mandatory to monitor the individual for wearing a face mask. The effective computational unit, along with the camera and sensor, is necessary for monitoring the workers with physical distancing. The Raspberry Pi, along with the camera and temperature sensor, helps in serving the purpose. The Raspberry Pi is a single-board computer that is effective in computing large data. The IoT, along with machine intelligence and single-board computers, provides better monitoring of industrial events.

The Industry 4.0 standards are mainly focused on increasing system production and better remote monitoring. The existing machinery output is connected through IPV4/V6 technologies to exchange data to the cloud environment. The data are transmitted through a secure channel, and important, timely decisions can be taken through remote users. Industry 4.0 also guarantees better prognostics and better decision-making with the help of different machine learning algorithms. It also provides improved fault diagnosis and necessary safety measurement during terminal fault conditions. The present delay and jitter in the network provide limited remote operations and decision-making. It provides successful remote monitoring and prognostics to industrial machines. The rest of this paper, after the introduction, is organized as follows. Section 2 presents related works, and the proposed work is presented in Sect. 3. Results and discussion are presented in Sect. 4, and finally, Sect. 5 concludes the paper.

2 Related Works

The IoT technology has lessened human work by being capable of working on its own. The sensors and machine intelligence have made remote monitoring easy [1, 2]. Many IoT-based smart city projects have been developed [3], such as hospital care, posture recognition, industrial production monitoring, supply chain-based monitoring applications. A self-healing water quality mechanism was developed that clears its coagulation problems through a machine learning approach [4–6]. Machine learning and deep learning have a significant response in building better industrial safety and monitoring mechanisms [7–9]. The sensors are the main data-gathering units, and they act as a front end for many Industry 4.0 applications [1, 10, 11]. Deep learning-based face recognition systems have many advantages that can identify faces and their emotions even with limited quality data [12, 13]. The data set is trained through an edge computing device, and the same is used to compare the input image with the trained model. The utilization of such devices and technology for the industry can improve safety and can even save production time. The alert messages are sent through many IoT platforms, such as Raspberry Pi-based industry safety research, where the sensors are integrated and compared with the set value [14]. This article proposes a deep learning-based safety aspects' recognition system and an IoT-based care system for industrial shop floor monitoring.

3 Proposed Work

The proposed work includes monitoring of workers and industrial safety aspects on the shop floor. The proposed work includes a camera, infrared, and gas, normally COx and SOx. The sensors are connected with the help of an analog-to-digital converter. Figure 1 elucidates the architecture planned, and the same is tested with different single-board computers for checking its computational complexity.

Figure 2 depicts the procedure instructed to the single-board computer. The procedure is executed in a while loop, and the captured data are updated to the cloud environment through available IPV4/IPV6 protocols.

3.1 *Preprocessing*

Skew Correction, Linearization, and Noise Removal are the three phases in the preprocessing stage. The skewness of the acquired image is examined. With either a left or right orientation, there is a chance that images will be distorted. The image

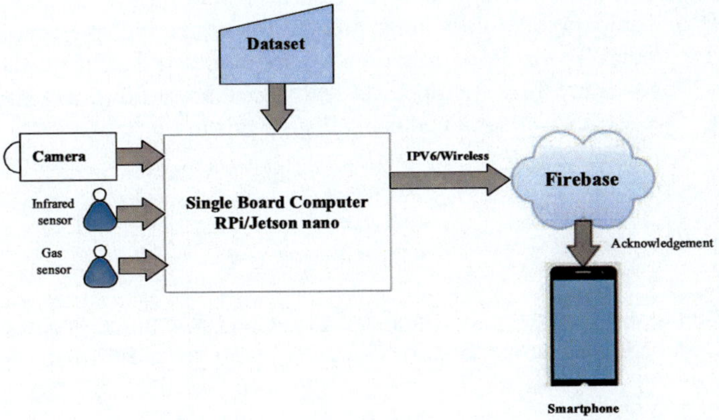

Fig. 1 Industry 4.0 testing and monitoring using single-board computers (SBCs)

Fig. 2 Industry 4.0 and approach

```
Begin Process
while
    Acquire data from the sensors camera, infrared sensors, Gas sensors
    Convert the analog sensor data to digital and compare with the set value,
    Log data in the local file and cloud
    if(set value < present sensor value)
        Alarm;
    end if
    Acquire data from the camera;
    Perform CNN based image comparison;
        perform safety requirement test from image;
        if aspects not satisfied
```

is brightened and binarized first. If an angle of orientation between 15 degrees is discovered, the skew detection function performs a basic picture rotation until the lines match the real horizontal axis, resulting in a skew-corrected image. Before processing, any noise generated when capturing or due to low page quality must be removed.

3.2 Segmentation

The noise-free image is then transferred to the segmentation step after preprocessing. It is a method of decomposing a picture of a sequence of characters into a sub-image of the individual sign (characters). Inter-line spaces are examined in the binarized image. The image is divided into sets of paragraphs across the interline gap if interline spaces are identified. The paragraph lines are examined for horizontal space intersections with the background. The width of the horizontal lines is determined using the image's histogram. The lines are then examined vertically for intersections in vertical space. The width of the words is determined using histograms in this case.

3.3 Feature Extraction

Individual picture glyphs are considered and extracted for features as part of feature extraction. To begin, the properties of a character glyph are as follows:

(1) Character height;
(2) Character width;
(3) Numbers of horizontal lines—short and long;

(4) Numbers of vertical lines—short and long;
(5) Numbers of circles;
(6) Numbers of horizontally oriented arcs;
(7) Numbers of vertically oriented arcs;
(8) Centroid of the image;
(9) Position of the various features;
(10) Pixels in the various regions.

The information is triggered via Google API to the given email ids and phone numbers, and crucial alert messages are pinged to the associated email ids and phone numbers once the sensor data have been collected and processed according to the algorithm. In the Google Firebase environment, the sensor data are also updated.

4 Result and Discussion

The proposed algorithm is tested in the Raspberry Pi 4 platform booted with Raspbian Linux. The raspiCam 8Mp camera is installed to receive data and process it with the Python package. The trained machine learning data set that can identify PPE suits and without PPE suits, Face masks, and helmets is installed in the Raspberry Pi. The TensorFlow package is used to identify the present condition of the worker. The module is interfaced with the wifi network, and information is shared through a secured channel. The entire module is powered through 5v 2 A battery source.

The screen of the Raspberry Pi can be viewed across any device within the same subnet through VNC viewer or putty. Figure 3 provides successful verification of output through the module. The system recognizes the face and verifies with the existing data set for the presence of the mask. Once the mask is not present, the system triggers an SMTP request through the internet and sends email and other necessary services.

Figure 4 provides the mask identification of the input image through machine learning approach. The data set is trained and stored inside the SBC for processing.

Figure 5 provides alert email sent to registered email id through IPV4 protocol. The email along with the photo is sent within 1.32 s of identification of no-mask condition. The module also senses toxic gases through MQ sensors, once the sensor exceeds the limit set value. There is a alert message system which was introduced to send an SMS when the temperature increases the limit. The built hardware module also shares the alert mechanisms to the users and registered email id. The gaseous sensors are analog-based and ADC1115 is used to convert its value to digital, Fig. 6.

The data are also shared in life with the firebase real-time database; a data log is created. The normal and abnormal data are shared with the Google firebase database, and the project is shared with the team people. Figure 7 shows the real-time database of the proposed hardware. The device is linked with the corresponding firebase cloud through the import firebase Python package.

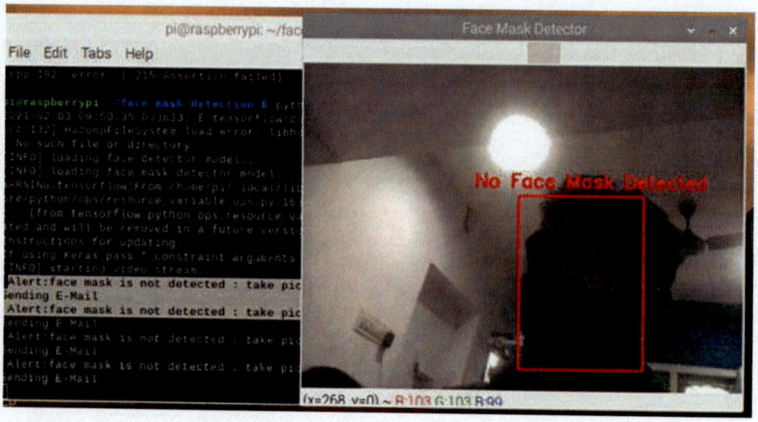

Fig. 3 Proposed model with no-mask detection

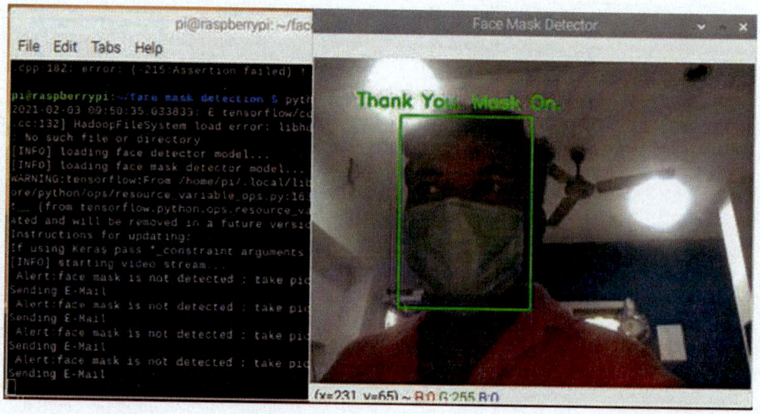

Fig. 4 Proposed model with mask identification

The same project is also realized with the NVIDIA Jetson Nano board and the processing time capabilities are tabulated. Figure 7 provides the processing capacity comparison with Raspberry Pi and Jetson Nano board.

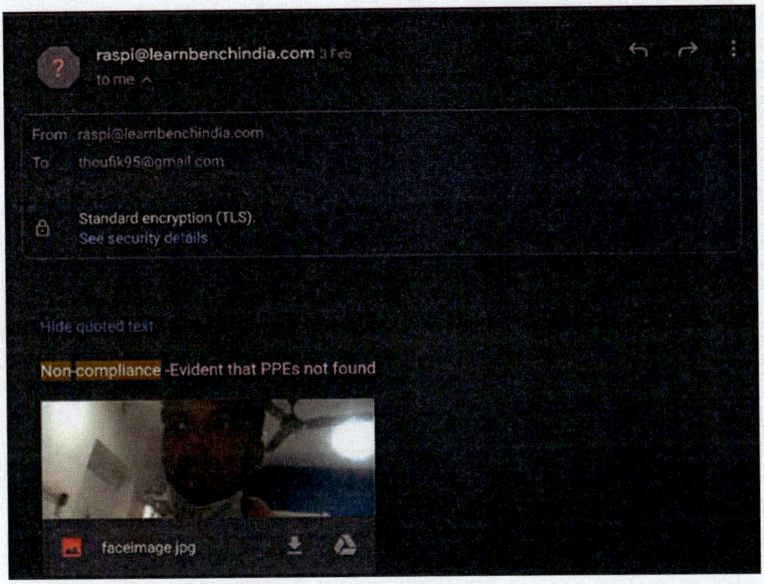

Fig. 5 SMTP-alert message on no mask

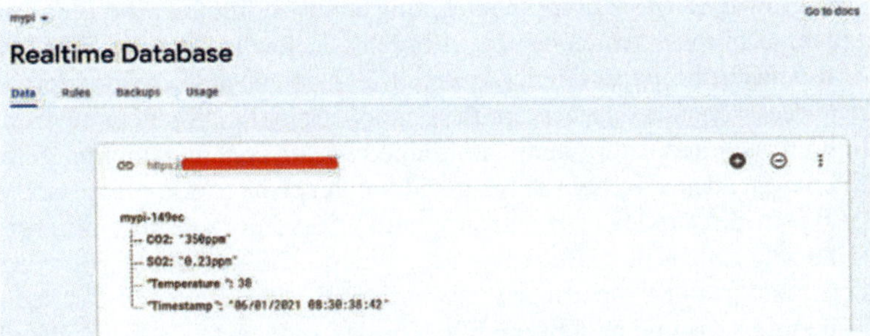

Fig. 6 Firebase real-time database connection with the hardware

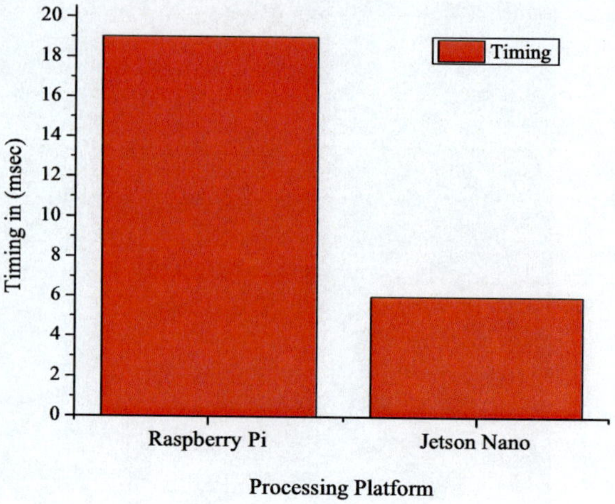

Fig. 7 Processing time comparison between Raspberry Pi and Jetson Nano boards

5 Conclusion

Technology is developing so fast and being applied to many industries for profit-making and fewer opportunities for the safety of workers in manufacturing industries. Seeing the growth of MSME companies, this safety system can be improved for the commercial purpose which can be affordable to low investment MSMEs. The proposed module in this paper efficiently identifies personal protective equipment and maintains necessary safety aspects for the employees in the shop floor environment. The minimum required safety items are trained in our system, which monitors for 24 h. The newly created system helps avoid human error and improve safety. The device also provides a quick alert response with 1.35 s from sensing and shares data within 1.89 s to upload the live data to the cloud. The module also successfully recognizes toxic gas and other necessary industrial parameters and can send alert emails. The project is operated through low power and creates 24 hours data log in the cloud environment.

References

1. Kanagachidambaresan GR, Maheswar R, Manikandan V, Ramakrishnan K (eds) (2020) Internet of Things in smart technologies for sustainable urban development. Springer Nature, Switzerland. https://doi.org/10.1007/978-3-030-34328-6
2. Prakash KB, Kanagachidambaresan GR (eds) (2021) Programming with TensorFlow: solution for edge computing applications. Springer, Switzerland. https://doi.org/10.1007/978-3-030-57077-4
3. Kanagachidambaresan GR (2021) IoT projects in smart city infrastructure. In: Role of single board computers (SBCs) in rapid IoT prototyping. Springer, Cham, pp 199–215. https://doi.org/10.1007/978-3-030-72957-8_10

4. Kanagachidambaresan GR (2021) Industry 4.0 for smart factories. In: Role of single board computers (SBCs) in rapid IoT prototyping. Springer, Cham, pp 217–227. https://doi.org/10.1007/978-3-030-72957-8_11
5. Kanagachidambaresan GR (2021) Introduction to Internet of Things and SBCs. In: Role of single board computers (SBCs) in rapid IoT prototyping. Springer, Cham, pp 1–18. https://doi.org/10.1007/978-3-030-72957-8_1
6. Kanagachidambaresan GR (2021) Sensors and SBCs for smart city infrastructure. In: Role of single board computers (SBCs) in rapid IoT prototyping. Springer, Cham, pp 47–75. https://doi.org/10.1007/978-3-030-72957-8_3
7. Bharadwaj PKB, Kanagachidambaresan GR (2021) Pattern recognition and machine learning. In: Prakash KB, Kanagachidambaresan GR (eds) Programming with TensorFlow. EAI/Springer Innovations in Communication and Computing. Springer, Cham. https://doi.org/10.1007/978-3-030-57077-4_11
8. Kanagachidambaresan GR, Prakash KB, Mahima V (2021) Programming tensor flow with single board computers. In: Prakash KB, Kanagachidambaresan GR (eds) Programming with TensorFlow. EAI/Springer Innovations in Communication and Computing. Springer, Cham (2021). https://doi.org/10.1007/978-3-030-57077-4_12
9. Kanagachidambaresan GR, Ruwali A, Banerjee D, Prakash KB (2021) Recurrent neural network. In: Prakash KB, Kanagachidambaresan GR (eds) Programming with TensorFlow. EAI/Springer Innovations in Communication and Computing. Springer, Cham. https://doi.org/10.1007/978-3-030-57077-4_7
10. Rani S, Maheswar R, Kanagachidambaresan GR, Jayarajan P (eds) (2020) Integration of WSN and IoT for smart cities. Springer, Cham
11. Maheswar R, Balasaraswathi M, Rastogi R, Sampathkumar A, Kanagachidambaresan GR (eds) (2021) Challenges and solutions for sustainable smart city development. Springer Nature. https://doi.org/10.1007/978-3-030-70183-3
12. Prasad PS, Pathak R, Gunjan VK, Ramana Rao HV (2020) Deep learning based representation for face recognition. In: Kumar A, Mozar S (eds) ICCCE 2019. Lecture Notes in Electrical Engineering, vol 570. Springer, Singapore. https://doi.org/10.1007/978-981-13-8715-9_50
13. Guo G, Zhang N (2019) A survey on deep learning based face recognition. Comput Vis Image Underst 189:102805. https://doi.org/10.1016/j.cviu.2019.102805
14. Esakki B, Ganesan S, Mathiyazhagan S, Ramasubramanian K, Gnanasekaran B, Son B, Park SW, Choi JS (2018) Design of amphibious vehicle for unmanned mission in water quality monitoring using internet of things. Sensors 18(10):3318. https://doi.org/10.3390/s18103318

Area-optimized Serial Architecture of PRINT Cipher for Resource-constrained Devices

Manisha Kumari, Pulkit Singh, and Bibhudendra Acharya

Abstract The study of lightweight symmetric cipher has gained attention because of the rising interest for security administrations in constrained computing environments, like in the Internet of Things and other resource-constrained devices. Most of the IoT devices are connected to the real world and it has a wide scope of utilization in different domains. They collect data and transfer personal information over the network, and the amount of data shared by IoT is growing at an extraordinary scale. These devices have a physical constraint in area, memory, and power. Conventional cryptography algorithms such as Data Encryption Standard (DES) and Advanced Encryption Standard (AES) provide high security to the devices and are capable of handling extensive input data. However, area required and power consumption of AES and DES ciphers are high for resource-constraint devices. AES and DES are best suited for devices where area and power are not the factor of concern. PRINT cipher is a lightweight block cipher which is a well-known encryption algorithm that consumes less hardware area due to the straightforward encryption operation. This paper discusses the FPGA implementation of the 6-bit data path serial architecture of the PRINT cipher. It is one of the block ciphers requiring minimum resources and supports block sizes of 48 and 96 bits with the key sizes of 80 and 160 bits, respectively. The proposed design is implemented in Verilog on different FPGA families. This paper presents the serial hardware architecture of PRINT cipher that supports a 48-bit data path with an 80-bit key size length. The slice count of 39 is obtained on Virtex-5, which is 44.2859% less than the slice consumed by the PRESENT architecture, and the proposed architecture is best suited for devices that require less hardware area. Hardware implementation results on different FPGA families such as Spartan-7 and vertex-5 have been discussed.

M. Kumari · B. Acharya (✉)
Department of Electronics and Communication Engineering, National Institute of Technology Raipur, Raipur, Chhattisgarh 492010, India
e-mail: bacharya.etc@nitrr.ac.in

P. Singh
Department of Electronics and Communication Engineering, MLR Institute of Technology, Hyderabad, Telangana 500043, India
e-mail: pulkit@mlrinstitutions.ac.in

S. Sharma et al. (eds.), *Intelligent Control, Robotics, and Industrial Automation*, Lecture Notes in Electrical Engineering 1066,
https://doi.org/10.1007/978-981-99-4634-1_33

Keywords IoT · Lightweight cipher · Block cipher · PRINT cipher · Cryptography · Serial architecture

1 Introduction

Information security is crucial during communication, particularly for financial and private issues. Nowadays, with the enhancement of technology and internet connectivity, computers and other electronic devices transmit millions of data every second [1]. As technology is evolving, people are becoming more dependent on it. Technologies such as RIFD tags are used in malls and offices for tracking purpose, WSN used in monitoring agriculture remotely, smart cards, etc. [2]. IoT is used in almost every field like agriculture, automobiles, industry, medical, etc. IoT devices are communicating with each other and are sharing our personal and confidential information; that is why it is crucial to secure our data so that there is no theft of personal information and invasion of privacy to something fatal [3, 4]. To provide security, cryptography is introduced. AES and DES are best suited for devices where area and power are not the factor of concern. Resource-constraint devices have small memory, require less area, consume less power, are capable of handling less input, and are cost-effective [5]. The number of resource-constrained devices, wireless networks, such as Wireless Sensor Nodes (WSNs), smart cards, Radio Frequency Identification (RFID), Internet of Things (IoT), is increasing rapidly, and these devices are broadly used in day-to-day life, which progressively leads to the requirement to provide security. Cryptography is a technique to provide end-to-end communication security. The cryptographic encryption is used to encode message (plaintext) and convert it into ciphertext. The decryption is used to decode ciphertext back into plaintext with the help of key. Conventional cryptographic algorithms such as DES and AES are unsuitable for resource-constraint devices [6]. Therefore, for low-resource devices, various lightweight cryptographic ciphers have been proposed where area, power consumption, and cost should be considered. Some examples of lightweight cryptographic algorithms are PRINT, PRESENT, PRINCE, LILLIPUT, QTL, mCRYPTON, RECTANGLE, etc. [7, 8].

Cryptography structures are based on Feistel-based network and Substitution–Permutation Network (SPN). In SPN-based ciphers, substitution and permutation layers are used. The substitution layer provides confusion, and the permutation layer provides diffusion. Diffusion means if the symbol or bit in the plaintext is changed, then the maximum or all symbols or bits in the ciphertext will also change. It is used to hide the relationship between plaintext and ciphertext [7]. Confusion means if a single bit in the key changes, then maximum or all ciphertext bits will change. It is used to hide the relationship between ciphertext and key. PRESENT, PRINT, RECTANGLE, PRINCE, Advanced Encryption Standard (AES), etc. are SPN-based ciphers, and Piccolo, LILLIPUT, SEA, MIBS, LBlock, etc. are Feistel-based ciphers.

In 2007, a new lightweight PRESENT cipher was proposed by Poshmann et al. [9]. The PRESENT is a Substitution and Permutation Network (SPN)-based lightweight cipher. The basic building block of SPN-based is S-box and P-box. S-box is the substitution box used for confusion, and P-box is used to provide diffusion. The block length of the PRESENT cipher is 64 bits. It supports the key size of 80 bits and 128 bits, and it takes 31 rounds to get the final ciphertext. It consists of a total of 16 S-box, in which each S-box takes an input of 4 bits. Later, PRESENT and HIGHT cipher optimized architecture was proposed by Yalla et al. in 2010 [10]. FPGA Spartan-3 family was used for implementation. For PRESENT cipher, the same 64-bit registers are used for both permutations and storing of data, this area is optimized, and an extra register is not required to store data. A total of 91 slices are required for this optimized hardware architecture of the PRESENT cipher. For HIGHT data path is reduced to 8 bits from a 64-bit data path. A total of 117 slices are used for this optimized hardware architecture of the HIGHT cipher. Then, two ultra-lightweight RAM-based architectures of PRINT cipher were proposed by Elif Bilge Kavun et al. in [11]. The main aim of the design is to use the existing BRAMs of FPGA for internal state storage. By this, an extra register is not required to store the output of the internal state, and this strategy optimizes the area. The slice and register count decrease by using BRAM on FPGA hardware implementation. These architectures are well suited for resource-constraint devices. Both architectures occupied 83 and 85 slices, along with a single BRAM. At 100 kHz, these architectures have 6.03 and 5.14 Kbps throughputs on the XC3S50 Xilinx Spartan family device. Then in 2012, Neil Hanley et al. proposed three different ciphers, AES, Clefia, and PRESENT hardware architecture [12]. For all the three ciphers, serial and iterative architecture is compared. The 64-bit data path is reduced to the 8-bit data path; by this, the area requirement and power consumption are also reduced. Later, the compact serial hardware architecture of the PRESENT cipher was proposed by Tay et al. in [13]. Serial architecture is best suited for compact area requirements. The Karnaugh map (K-map) technique is used to reduce the Boolean expression of S-box, which is used in the PRESENT cipher; by this, Boolean S-box is designed for compact PRESENT serial hardware architecture. A maximum frequency of 236.574 MHz with a throughput of 51.32 Mbps has been achieved with the smallest FPGA implementation on the Virtex-5 family. The total slice count of the serial architecture of PRESENT cipher on the Virtex-5 family was only 65.

1.1 Contribution

This paper proposed a 6-bit data path serial architecture of the PRINT cipher. PRINT cipher supports a 48-bit data block with an 80-bit key size and a 96-bit data block with a 160-bit key length. The proposed architecture is suitable for devices that consume less area and power.

1.2 Organization of Paper

This paper consists of a total of four sections. Section 2 gives an overview of PRINT cipher. Section 3 presents the 6-bit data path architecture of the 48-bit PRINT cipher and its hardware implementation. Section 4 depicts hardware implementation and result comparison with a different cipher on various types of FPGA families. Section 5 includes the conclusion of the work.

2 Overview of Print Cipher

PRINT cipher was first introduced by CHES in [14]. It belongs to the category of the lightweight block cipher which is used to provide security to resource-constrained devices. In block cipher, at a time, a block of plaintext is converted into ciphertext, whereas in-stream cipher, only 1 byte of plaintext is converted into ciphertext at a time [15]. It is based on the Substitution and Permutation Network (SPN). In an SPN-based network, substitution and permutation box is used to generate ciphertext from plaintext. These boxes provide confusion and diffusion in cryptography algorithms, thereby increasing cipher security. Advanced Encryption Standard (AES) uses SPN-based structure [16]. S-box provides confusion, and P-box is used to provide diffusion. Diffusion is used to hide the statistical relationship between plaintext and ciphertext. Confusion is used to hide the relationship between key and ciphertext [17].

PRINT cipher consists of the block size of 48 bits and 96 bits and is named PRINT cipher-48 and PRINT cipher-96. PRINT cipher-48 consists of 48 rounds with a key size of 80 bits. PRINT cipher-96 consists of 96 rounds with a key size of 160 bits [14]. For PRINT cipher, the number of S-box layers used is given by Eq. (1):

$$\text{Number of S - box Layer} = \frac{\text{Block size}}{3}. \tag{1}$$

So, for the 48-bit block PRINT cipher, 16 S-box layers are used, and for the 96-bit block PIRNT cipher, 32 S-box layers are used.

Process Flow of PRINT Cipher.

Input: P_T, Key
Control signal: Load, Clock
Output: C_T
Key = {K1, K0}
K1 = [79:32] and K0 = [31:0]

i. X1 ← P_T xor K1;
ii. X2 ← Player(X1);
iii. X3 ← X2 xor RC;

iv. X4 ← Kye_Based_Permutation (X3, K0);
v. X5 ← Sbox (X4).

Permutation Layer. In PRINT cipher, P-box is used to provide linear diffusion. Diffusion is used to hide the statistical relationship between plaintext and ciphertext. In simple words, if a symbol in the plaintext is changed, several or all characters of ciphertext will also change. Diffusion implies that each character in the ciphertext is dependent on some or all the symbols in the plaintext. In the permutation layer, the permutation of the input bits is performed in the following manner as Eq. (2) [14]:

$$\begin{aligned} P(x) &= 3 * x \mod d - 1 \quad \text{for } 0 \le x \le d - 2, \\ &= d - 1 \quad \text{for } x = d - 1, \end{aligned} \tag{2}$$

where x is the current position of the bit and $P(x)$ is the position of bits after permutation, and d is the cipher's block size.

Round Counter (RC). The round counter value is xored with the present state's Least significant bits (LSBs). The shift register is used to generate RC value in the following way as Eq. (3).

$$\begin{aligned} &Y = a_n \text{xor } a_{n-1} \text{xor } 1; \\ &a_i = a_{i-1} \qquad \text{for } n - 1 \ge i \ge 1, \\ &a_0 = Y \qquad \text{for } i = n. \end{aligned} \tag{3}$$

It is a linear feedback shift register that counts n-bit binary in the following manner. For 48-bit block, PRINT cipher n is 6 bits, and for 96-bit block, PRINT cipher n is 7 bits. First, linear feedback shift register (LFSR) is initialized with 000,000 or 0,000,000 on block size and then increments the record in each round [14]. Figure 1 shows the round counter of PRINT cipher.

Key Slicing. For the 48-bit block PRINT cipher, the 80-bit key is split into two, in which the first 48-bit key is used for xor operation, and another 32-bit key is generated by using a secret algorithm. The key size for the PRINT cipher depends on the block size. It is the five-third of the block size. Therefore, for a 48-bit PRINT cipher, the key size is 80 bits, and for a 96-bit block, the key size is 160 bits [14]. The key is 80 bits for a 48-bit block cipher and 160 bits for a 96-bit PRINT cipher. Let d be the block size of the cipher.

Fig. 1 Round counter (RC)

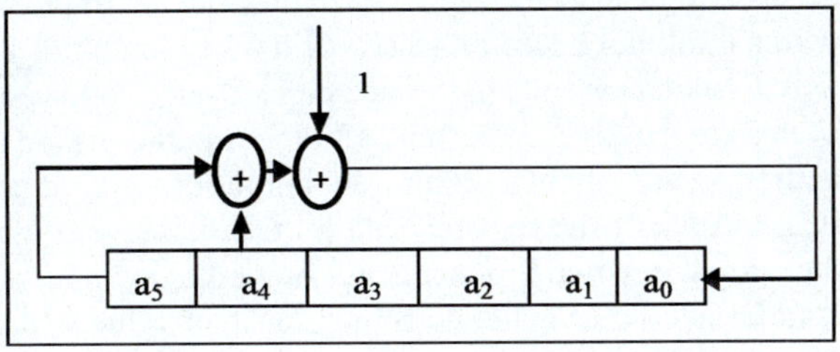

Table 1 Key-based permutation

Key $K0$	Permutation
00	E2‖E1‖E0
01	E1‖E2‖E0
10	E2‖E0‖E1
11	E0‖E1‖E2

Table 2 Substitution box

z	0	1	2	3	4	5	6	7
$S(z)$	0	1	3	6	7	4	5	2

$$\begin{aligned} \text{Key} &= \frac{5}{3} * d \\ \text{Key} &= \{K1,\ K0\}, \end{aligned} \tag{4}$$

$K1$ is given by: $\frac{2}{3} * d$, and $K0$ is given by: $\frac{1}{3} * d$.

Key-based Permutation. The lower 32-bit key is used for the key-based permutation. The permutation depends on the key. The lower 32-bit key is taken into pair of two, and permutation is performed based on this. The inputs of the S-box are first permutated based on key K2. Key K0 combines in a group of two; then based on it, permutation is performed (Table 1) [14].

S-box Layer. S-box provides confusion, and P-box is used to provide diffusion. Diffusion is used to hide the statistical relationship between plaintext and ciphertext. In simple words, if a symbol in the plaintext is changed, several or all ciphertext symbols will also change. Diffusion implies that each character in the ciphertext is dependent on some or all the symbols in the plaintext. It takes a 3-bit input and generates 3-bit output in the following way, where z is the input and $S(z)$ is the output. Table 2 depicts S-box [14].

3 Proposed Serial Architecture of Print Cipher

Figure 2 represents the 6-bit data path hardware implementation of the PRINT cipher with a total block size of 48 bits and a key length of 80 bits. In the 6-bit data path, only 6-bit data are loaded in one clock cycle. To load all 48-bit data, a total of eight cycles are required. After that, all other operations are performed. The same register is used to load input data and the output of random permutation. The shift register is used to shift the register data by 6 bits in every clock cycle, and these data are then going into two S-box; each S-box takes an input of 3 bits. Round constant (RC) counter is used to generate round constant value which is used in each round. The

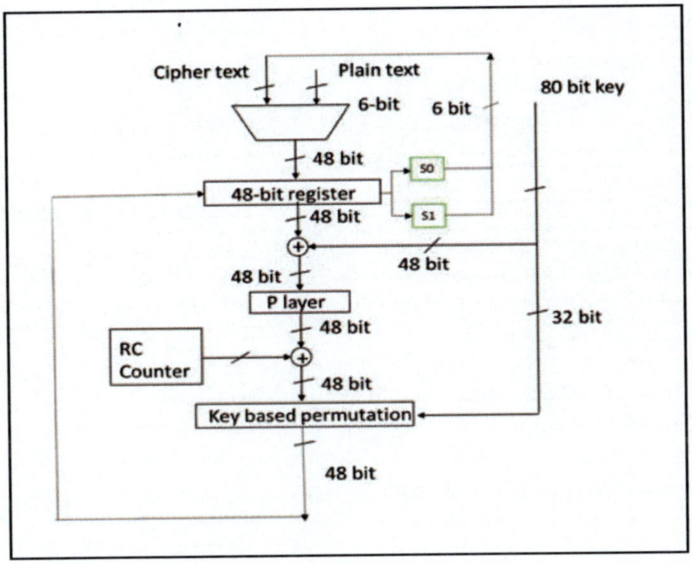

Fig. 2 Proposed 6-bit data path serial architecture of PRINT cipher

output of the S-box is used for further rounds. The latency of the serial architecture is more as compared to the round-based architecture. Figure 3 illustrates the FSM for the 6-bit data path serial architecture of the PRINT cipher. When reset is high, then FSM is at idle state. When reset becomes '0' and load_data is high, then it moves to state S0 where data are loaded and it remains at state S0 for 8 cycles, and then it moves to state S1 where key is loaded. After that it will move to state S2 where round constant starts and when start_en is active, and then it moves to state S3 where encryption process will start. Due to this, the throughput of the serial architecture is less as throughput is inversely proportional to latency.

4 FPGA Simulation Result and Result Comparison

The Xilinx Vivado and Xilinx ISE are used to implement PRINT 6-bit data path serial architecture. To explore the architecture, several FPGA (field-programmable gate array) devices such as Spartan-7 and Virtex-5 platforms are used for result comparison. The most significant attributes for evaluating the designs are power consumption and area utilization. In FPGAs, slice count, number of FFs (Flip flops), and number of LUTs count are considered in determining area utilization. Table 3 presents the result compared with other ciphers on various FPGA platforms. Slice consumed by proposed architecture is 39 and 38 on Virtex-5 and Spartan-5 platforms, respectively. Virtex-5 slice consumed by the proposed serial architecture is 44.2859% less than the slice consumed by PRESENT architecture. Table 4 depicts

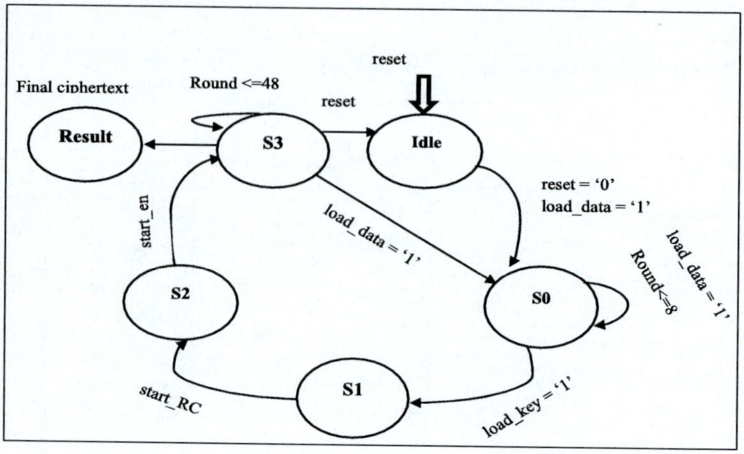

Fig. 3 FSM of controller for proposed 6-bit data path serial architecture of PRINT cipher

the area utilization, power consumption, and throughput of the 6-bit serial data path architecture on different FPGA platforms. Figure 4 describes the comparison of slice count with other lightweight ciphers. From Fig. 4, it is observed that the slice count for proposed 6-bit serial architecture of PRINT cipher is less as compared to other lightweight ciphers such as PRESENT-128, HIGHT, XTEA. Figure 5 presents the simulation output for the 6-bit data path of serial architecture on Spartan-7. From Fig. 5, it is observed that the latency of the architecture increases, due to which throughput decreases.

Table 3 Result comparison with different ciphers on various FPGA platforms

Algorithm	Devices	Block size	Slices	Max. freq. (MHz)	Throughput (Mbps)	Total power (mW)
SIMON [18]	xc3s500	–	36	136.00	3.60	–
AES [19]	xc3s50-5	128	184	45.64	36.50	–
PRESENT-128 [20]	xc5vlx50t-3ff1136	128	73	194.63	203.19	–
HIGHT [12]	xc3s50-5	–	91	163.70	65.48	–
XTEA [21]	xc5vlx85	64	66	333.00	111.00	–
XXTEA192 [22]	xc5vlx50t	192	51	257.36	588.26	–
PRESENT-80 [20]	xc5vlx50t-3ff1136	64	70	245.76	295.00	–
PRINT serial (proposed)	xc5vlx50t-2ff1136	48	39	139.76	17.47	0.124
PRINT serial (proposed)	xc7s50csga324-2	48	38	144.56	18.07	0.116

Table 4 PRINT cipher hardware metrics on FPGA in Virtex-5(xc5vlx50t-3ff1136)

Performance metrics	Parameter values
FFs	41
Slices	39
LUTs	123
Throughput (Mbps)	12.5
Dynamic power (W)	0.019
Static power (W)	0.105
Total power (W)	0.124

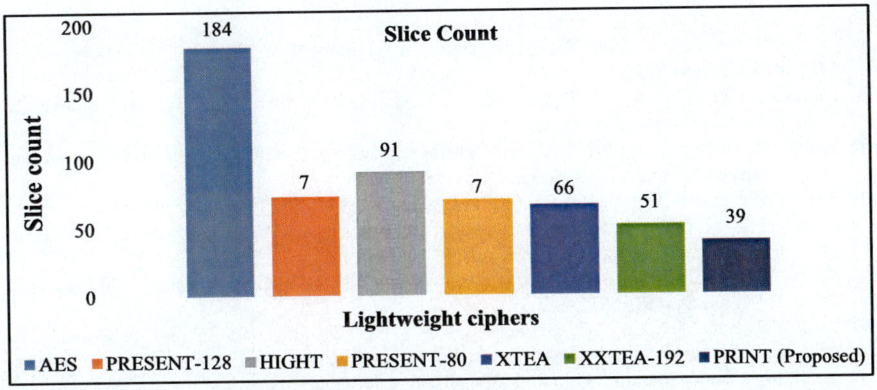

Fig. 4 Slice count comparison with different ciphers

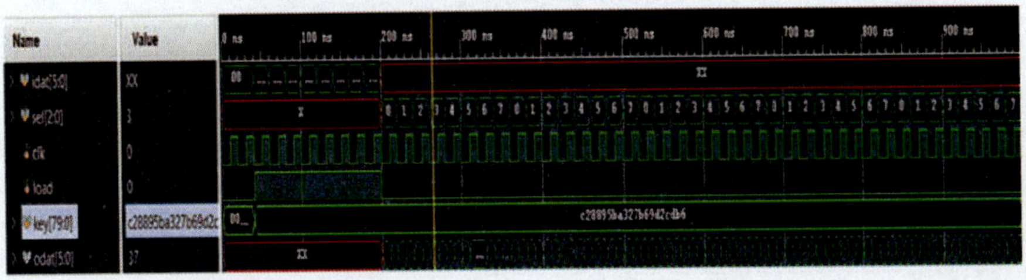

Fig. 5 Simulation result for 6-bit serial architecture of PRINT cipher on Spartan-7

5 Conclusion

This paper describes the proposed serial architecture of the PRINT cipher. The proposed architecture supports 6-bit data path input. This architecture is best suited for devices that require less area. For serial architecture, latency is more as compared to round and pipelined based architectures. Here, the number of rounds needed to get the final ciphertext is comparatively more than a round-based cipher. Still, the area requirement is less than the PRINT cipher's round-based architecture. Xilinx ISE and Xilinx Vivado are used for implementation on FPGA platform. The proposed architecture implemented on Virtex-5 requires 39 slices and 123 LUTs. The slice consumed by proposed serial architecture is 44.2859% which is less than slice consumed by PRESENT architecture. Total power dissipation of the 6-bit data path of serial architecture is significantly less. Hence, the proposed architecture is suitable for devices requiring less area and power.

References

1. Li L, Liu B, Zhou Y, Zou Y (2018) SFN: a new lightweight block cipher. Microprocess Microsyst 60:138–150
2. Hatzivasilis G, Fysarakis K, Papaefstathiou I, Manifavas C (2018) A review of lightweight block ciphers. J Cryptogr Eng 8(2):141–184
3. Mohd BJ, Hayajneh T (2018) Lightweight block ciphers for IoT: energy optimization and survivability techniques. IEEE Access 6:35966–35978
4. Singh P, Acharya B, Chaurasiya RK (2021) Lightweight cryptographic algorithms for resource-constrained IoT devices and sensor networks. In: Security and privacy issues in IoT devices and sensor networks. Academic Press, pp 153–185
5. Kassim TS, KB J (2020) ACT: An ultra light weight block cipher for Internet of Things. Int J Comput Digital Syst 9(5):921–929
6. Yang YS, Bahn JH, Lee SE, Bagherzadeh N (2009) Parallel and pipeline processing for block cipher algorithms on a network-on-chip. In: 6th International conference on information technology: new generations, pp 849–854, IEEE, Las Vegas, NV, USA
7. Kong JH, Ang LM, Seng KP (2015) A comprehensive survey of modern symmetric cryptographic solutions for resource constrained environments. J Netw Comput Appl 49:15–50
8. Singh P, Acharya B, Chaurasiya RK (2019) A comparative survey on lightweight block ciphers for resource constrained applications. Int J High Perform Syst Archit 8(4):250–270
9. Bogdanov A, Knudsen LR, Leander G, Paar C, Poschmann A, Robshaw MJ, Seurin Y, Vikkelsoe C (2007) PRESENT: an ultra-lightweight block cipher. In: International workshop on cryptographic hardware and embedded systems. Springer, Berlin, Heidelberg, pp 450–466
10. Yalla P, Kaps JP (2009) Lightweight cryptography for FPGAs. In: International conference on reconfigurable computing and FPGAs. IEEE, Cancun, Mexico, pp 225–230
11. Kavun EB, Yalcin T (2011) RAM-based ultra-lightweight FPGA implementation of PRESENT. In: International conference on reconfigurable computing and FPGAs. IEEE, Cancun, Mexico, pp 280–285
12. Hanley N, ONeill M (2012) Hardware comparison of the ISO/IEC 29192-2 block ciphers. In: IEEE computer society annual symposium on VLSI, pp 57–62. IEEE, Amherst, MA, USA (2012).
13. Tay JJ, Wong MD, Wong MM, Zhang C, Hijazin I (2015) Compact FPGA implementation of PRESENT with Boolean S-Box. In: 6th Asia symposium on quality electronic design (ASQED), IEEE, Kula Lumpur, Malaysia, pp 144–148
14. Knudsen L, Leander G, Poschmann A, Robshaw MJ (2010) PRINTcipher: a block cipher for IC-printing. In: International workshop on cryptographic hardware and embedded systems. Springer, Berlin, Heidelberg, pp 16–32
15. Zhang W (2016) A differential fault analysis attack against the light weight PRINTcipher. Chin J Electron 25(6):1019–1024
16. Blömer J, Seifert JP (2003) Fault based cryptanalysis of the advanced encryption standard (AES). In: International conference on financial cryptography. Springer, Berlin, Heidelberg, pp 162–181

17. Dalmasso L, Bruguier F, Benoit P, Torres L (2019) Evaluation of SPN-based lightweight crypto-ciphers. IEEE Access 7:10559–10567
18. Aysu A, Gulcan E, Schaumont P (2014) SIMON says: break area records of block ciphers on FPGAs. IEEE Embedded System Letter 6(2):37–40
19. Chu J, Benaissa M (2012) Low area memory-free FPGA implementation of the AES algorithm. In: 22nd International conference on field programmable logic and applications (FPL). IEEE, Oslo, Norway, pp 623–626
20. Mhaouch A, Elhamzi W, Atri M (2020) Lightweight hardware architectures for the piccolo block cipher in FPGA. In: 5th International conference on advanced technologies for signal and image processing (ATSIP). IEEE, Sousse, Tunisia, pp 1–4
21. Kaps JP (2008) Chai-tea, cryptographic hardware implementations of xtea. In: International conference on cryptology in India. Springer, Berlin, Heidelberg, pp 363–375
22. Kella C, Mishra Z, Acharya B (2021) A compact and low power architecture of XXTEA192 lightweight block cipher. In: 6th International conference on communication and electronics systems (ICCES). IEEE, Coimbatore, India, pp 972–976

Designing of IoT Device Compatible Chaos-Based Phasor Measurement Unit Data Encryption Technique

RajKumar Soni, Manish Kumar Thukral, and Neeraj Kanwar

Abstract Phasor measurement units (PMUs) have become an integral part of the power grid. Many of the crucial decisions in power grid like load shedding, early detection of fault, fault event detection, and its classification are done using PMU data. It becomes imperative to ensure integrity security of PMU data looking at its importance in making vital decisions in power grid. Although IEEE C37.118 standards provide the guidelines for phasor measurement, data format and its transmission order but no mechanism of end-to-end security is given. The presented research work focuses on designing a chaos-based cryptography technique to guarantee PMU data integrity security. For this, a logistic–sine one-dimensional (1D) chaotic map is used to encrypt the PMU data. This map is represented by simple equation which can be easily implemented in a resource constrained IoT devices. The bifurcation diagram of logistic–sine map has been simulated on MATLAB R2018b platform. It is shown that as compared to traditional maps, the logistic–sine map has better chaotic properties in terms of range of bifurcation parameter and uniformity in chaotic sequence. Later, an encryption algorithm has been proposed to encrypt PMU data using chaotic logistic–sine map and implemented on MATLAB R2018b platform. Finally, security analysis of proposed encryption algorithm has been done in terms of key space, key sensitivity, and histogram. The results exhibit the robustness of the proposed encryption technique.

Keywords Chaos theory · Bifurcation · Cryptography · One-dimensional chaotic map · Phasor measurement unit

R. Soni · M. K. Thukral · N. Kanwar (✉)
Department of Electrical Engineering, Manipal University Jaipur, Jaipur, India
e-mail: nk12.mnit@gmail.com

M. K. Thukral
e-mail: Manish.thukral@jaipur.manipal.edu

S. Sharma et al. (eds.), *Intelligent Control, Robotics, and Industrial Automation*, Lecture Notes in Electrical Engineering 1066,
https://doi.org/10.1007/978-981-99-4634-1_34

1 Introduction

Cyber-attack on electrical power system has become one of the major concerns worldwide as it causes multifaced devastating effect on a country. This is evident from famous cyber-attack on power grid on 13 July, 2019 in Manhattan, New York [1]. This resulted in complete power outage in the city. Such similar attack was launched on electrical power grid in Ukraine in year 2015 on 23rd December [1]. One of the prime methods of launching cyber-attack is by manipulating the power grid measurement data [2]. Many of the crucial decisions in power system network like load shedding, energy cut-off, and commissioning depend on these measurement data only. So, if an intruder can manipulate these measured data, then the wrong control decisions can create havoc in power system stability [3]. In this context, phasor measurement unit (PMU) which constitutes a prime device for power grid parameter measurements is most vulnerable to data integrity form of cyber-attack. This is so because PMU uses IEEE C37.118 communication frameworks which do not have significant in-build security mechanisms guidelines. Because of this PMU communication is restricted to local network only [4]. Another important point is that advanced encryption standard techniques cannot be used for encrypting PMU data as these encryption techniques are complex and hence computationally intensive. Any latency in terms of PMU data transmission because of computationally intensive encryption technique would cause havoc in taking vital decisions by grid operator. Hence, the presented work proposes chaos-based cryptography technique for ensuring PMU data integrity. This will not only ensure less time involved in PMU data encryption but also its implementation of low-cost resource constrained IoT devices is easy [5].

In chaos-based encryption technique, a pseudo-random generator is designed using simple equations which can exhibit complex dynamics like chaos. In chaotic condition, such systems can generate sequence of numbers which have random nature [6]. Chaos-based cryptography systems reported so far can be categories into two, one using 1D chaotic maps and other using multi-dimensional chaotic maps-based cryptography [7]. 1D chaotic systems have advantage of low computational load, easy implementation. At the same time, such maps have disadvantages in term of low-key space, traditional (1D) chaotic maps like logistic map, sine map have small chaotic regime and non-uniform distribution of chaotic sequences which does not fit for encryption purposes [8]. Multi-dimensional chaotic systems have large key space as well as chaotic regime but suffer from complex computation and hence difficulty to implement in hardware as compared to 1D maps. Higher more computational time involved in encryption using multi-dimensional chaotic maps make then not suitable for PMU-based real-time data transfer application [9]. In recent times, researchers have started reporting improved 1D chaotic maps in terms of key space, large chaotic region, uniformity in chaotic sequences, and hence better chaotic properties as compared their traditional counter parts [8, 10, 11]. With this research in application of 1D chaotic map in chaos-based encryption has gained momentum recently. This makes them a suitable candidate for a real time PMU data encryption which is very crucial in power system decision-making.

In presented research work, a new chaotic combination system (NCCS)-based logistic–sine map which is 1D chaotic map proposed in [10] has been used to encrypt the PMU data. This would give direct advantage of low-computational complexity involved in encryption process (as it is 1D map) at the same time providing security level as per the standard required. In this paper first, a brief discussion on PMU and the vulnerabilities associated with its data security is done. This is followed by simulation of traditional 1D logistic map and sine map bifurcation diagrams on MATLAB R2018b platform. The shortcomings of these traditional maps have been bought out through bifurcation diagram analysis. Then, it is shown how NCCS-based logistic–sine map proposed in [10] is derived from traditional maps possesses better chaotic properties. For this, bifurcation diagram of logistic–sine map has been simulated on MATLAB R2018b platform. It is shown through bifurcation diagram how logistic–sine map overcomes the shortcomings of traditional maps which are in one way seed maps. After this, an encryption algorithm has been proposed and implemented on MATLAB R2018b platform. The security analysis of the proposed encryption algorithm has been done in terms of key space, histogram, and key sensitivity.

2 Phasor Measurement Unit (PMU)

PMU is used to measure the voltage and current in terms of phasors which are represented by amplitude and angle. PMUs are installed in selected substation of the transmission system. The main feature of PMU is that these readings are time stamped with high precision as PMU gets the time clock pulse through global positioning system [12]. The measured reading by PMU is transmitted to phasor data concentrator (PDC) [13]. This data is then used for monitoring and control of electrical network in smart grid [14]. Some of the prime uses of PMU data includes making important decisions like load shedding, early detection of fault, event detection, and classification.

The PMU takes a sinusoid as input as shown in Fig. 1a. The hard-line represents the time tag or the reporting instance when the measurement of the data sample of sinusoid is taken. This reporting instance is measured with respect to the peak value of the sinusoid as shown in Fig. 1a which meets the dotted line which is synchronized with coordinated universal time (UTC). The phase difference between the peak value and the reporting instance is represented by angle Θ. The phasor corresponding to sampled date of sinusoid is shown in Fig. 1b. The PMU estimates the phasor corresponding to the sampled data of sinusoid with magnitude as root mean square value with angle using classical discrete Fourier transform [15] and gives it as output. The phasor measurement in PMU is done as per IEEE C37.118 standards.

The PMU data can be transmitted using serial communication through RS232 port. Also, the data can be communicated using transmission control protocol (TCP), user datagram protocol (UDP), or Internet protocol [16]. The PMU data format and transmission order are defined by IEEE C37.118.2 standard. The data frame and

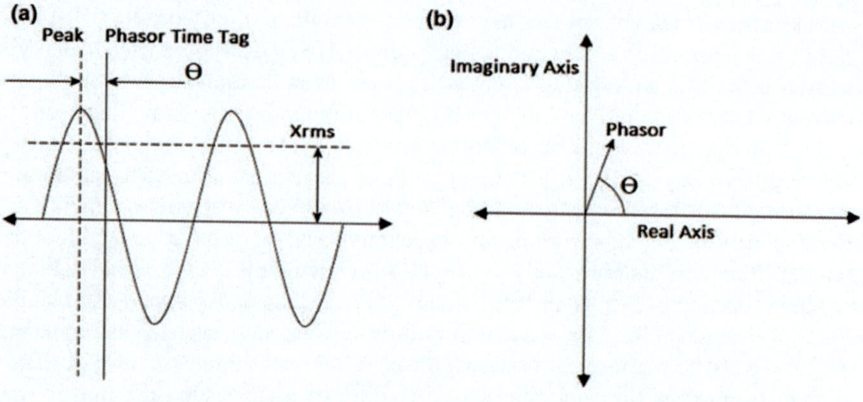

Fig. 1 **a** Representation of sinusoid. **b** The phasor corresponding to sampled data of sinusoid

transmission order are as shown in Fig. 2 where one can observe that a data file starts with 2-byte SYNC word and ends with 2-byte checksum (CHK) word. The PMU message is four types of namely command, data, configuration, and header. The SYNC word is 2 byte, and leading byte is always AA (hexadecimal). In last byte, bit 7 is default set to 0, bit 6–4 indicates whether it is a command, data, configuration, or header message. Bits 0–3 indicate whether the message transmitted is as per IEEE C37.118-2005 or IEEE C37.118.2-2011 standard. FRAMESIZE is a 2-byte word, and it signifies the number of bytes in a frame. Similarly, IDCODE word is 2-byte, and this signifies the source of the message, and each PMU has unique IDCODE. The data word length may vary depending on the phasor measurement requirements. To understand rest of words of a message one can refer to the IEEE C37.118.2-2011 standard [17]. The supervisory control and data acquisition (SCADA) system use the PMU and PDC data for monitor, control, and protection of power system components in smart grid. One of the measure problems associated with IEEE C37.118 standard is that it defines no mechanism to provide integrity security to the PMU message to be transmitted. As per this standard, the data is to be transmitted as plain text [18]. A very important issue in PMU message transmission is latency so traditional data encryption algorithms which require complex computations are not at all compatible in this case [18]. Looking into this critical gap in message integrity security, the presented research work proposes chaos-based message encryption of PMU using one-dimensional chaotic map. For this, a NCCS-based 1D logistic–sine map [10] which has been derived from seed maps, namely logistic and sine map used. In the discussion ahead, it is shown how the derived map has better chaotic properties than the seed maps. After this an encryption algorithm has been proposed to encrypt the PMU message using this chaotic map.

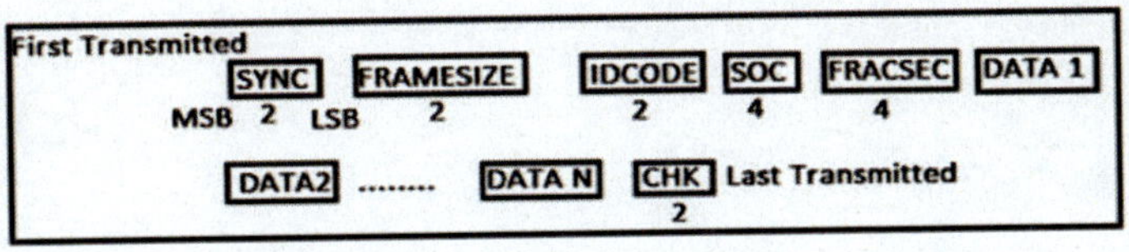

Fig. 2 PMU data file structure and transmission order [16]

3 Traditional 1D Chaotic Maps Background

In this section traditional, 1D chaotic logistic and sine map are reviewed briefly to form the basis for understanding logistic–sine map.

3.1 Logistic Map

Introduced by biologist Robert May in 1976, logistic map is known to be archetype of nonlinear dynamics. Robert May proposed mathematical model to predict population dynamics of fishes in a pond as shown in Eq. (1). Although the equation is simple, but it shows variety of complex dynamics.

$$x_{n+1} = r x_n (1 - x_n) \tag{1}$$

In Eq. (1), r is bifurcation parameter because by changing this value the dynamics of the system changes, and x_n represents the sequence generated by the map. In Fig. 3, bifurcation diagram of logistic map is shown. It is evident from Fig. 3 that the map has three regions, namely bifurcation, convergence, and chaos. The region of chaos exists in $r \in [3.57, 4]$.

3.2 Sine Map

A sine map can be mathematically represented as

$$x_{n+1} = r * \frac{\sin(pi * x_n)}{4} \tag{2}$$

In Eq. (2), r is bifurcation parameter and x_n is sequence produced by the map. The bifurcation diagram of sine map is shown in Fig. 4.

One can observe the convergence region from $r \in [1, 2.89]$, bifurcation region $r \in [2.9, 3.43]$, and chaotic region $r \in [3.44, 4]$. However, in chaotic region, there are window of stability.

Fig. 3 Bifurcation diagram of logistic map

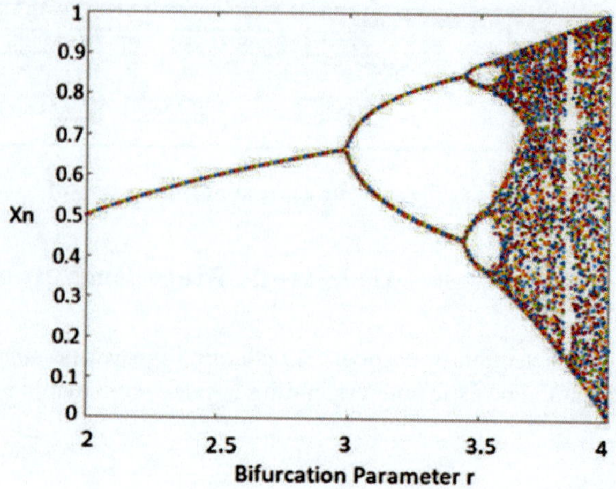

Fig. 4 Bifurcation diagram of sine map

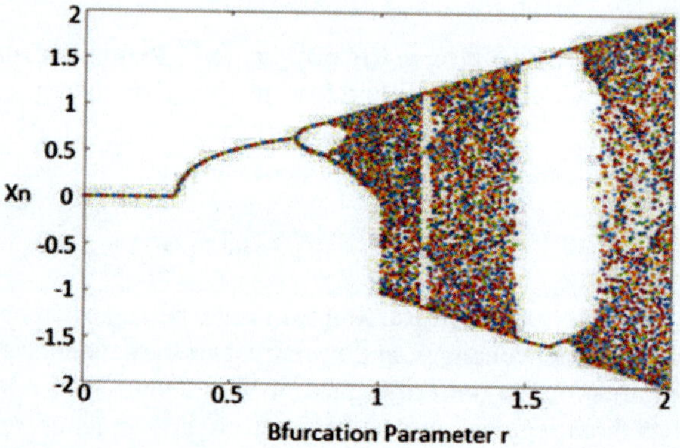

From logistic map and sine map bifurcation diagram, one can observe three important points

(i) Chaotic region is limited.
(ii) Within chaotic region, the chaotic sequence is not uniformly distributed.
(iii) Within chaotic region, there are window of stability.

These three points are limitations for a map to produce a good ergodic chaotic property to be applicable for data encryption process.

3.3 *Logistic–Sine Map*

In [10], chaotic maps have been designed by combining traditional chaotic maps. The resulting chaotic maps have been shown to be having wide chaotic regime and

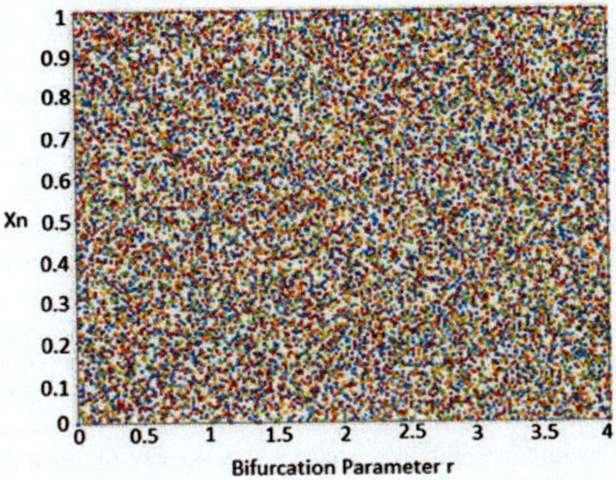

Fig. 5 Bifurcation diagram of logistic–sine map

uniformity in the sequence as compared to chaotic maps. One such map proposed is logistic–sine map. The logistic–sine map proposed in [10] is mathematically expressed as

$$x_{n+1} = \left((rx_n(1 - x_n) + r\sin(pi * x_n)) * 2\hat{\ }12\right) \mod 1 \tag{3}$$

In Eq. (3), r is known as bifurcation parameter and x_n is sequence generated by the map.

The range of bifurcation parameter is $r \in [0, 4]$. The bifurcation diagram of logistic–sine map has been shown in Fig. 5. It is evidence from the bifurcation diagram that the map is chaotic in throughout the range of bifurcation parameter i.e. $r \in [0, 4]$. Also, the chaotic sequence is uniformly distributed in the range 0–1. If compared with the traditional logistic and sine map the derived chaotic map certainly has better chaotic properties.

4 Proposed Encryption Algorithm and Implementation on MATLAB Platform

As per IEEE C37.118 standard, the PMU message is framed in form of hexadecimal number. One can refer this standard and see some examples of messages [17]. Here, the encryption algorithm of designed considering 108-character length of PMU message.

Each character is of 4 bits; hence, total size of 108-character message is 54 bytes. Although the length of message may vary depending on the data word size as shown in Fig. 2.

Algorithm steps of proposed encryption technique:

1. 108-character, 54-byte PMU data is converted from hexadecimal to binary equivalent.
2. 108 random sequences are generated by iterating the chaotic logistic–sine map. The values of generated random sequences are laying between 0 and 1.
3. Each random sequence generated is scaled by multiplying by decimal value 15. This is done because each character is of 4-bits binary equivalent, and maximum decimal equivalent of 4-bit binary is 15.
4. Round up the random sequence obtained from step 3.
5. Convert the random sequence obtained in step 3 from decimal to 4-bit binary equivalent.
6. The 4-bit binary equivalent of 108 random sequences generated in step 5 is reshaped to 432 bits using reshape function in MATLAB R2018b.
7. Bitwise XOR operation is performed of 432-bit binary sequence of PMU data sequence obtained in step 1 with the 432-bit binary sequence of random numbers obtained in step 6.

Algorithm steps for decryption technique:

1. At the receiver end, the bifurcation value r and initial value x_n of the logistic map are considered same as that used in the sending end.
2. The map is iterated to generate 108 random sequences.
3. From step 3, the random sequences generated will be having values between 0 and 1 which are scaled by multiplying with 15 decimal values, and then, each value is done round off.
4. The 108 random sequence is converted into equivalent 4-bit binary sequence and then reshaped to 432-bit sequences.
5. A receiver ends a bitwise XOR operation of 432-bit encrypted data of PMU received is done with the 432-bit random sequence obtained in step 4.
6. From step 5, the original PMU data is obtained which is then converted into hexadecimal form.

In this section, an example of 108-character PMU data encrypted using proposed algorithm is demonstrated. The encryption algorithm is implemented on MATAB R2018b platform. Consider the 108-character length data "aa01003600f14893334f003d70a4080047c3659cbfc8e73947c3631cbfc8e52b47c365574027a0d447c368643f065a850000000042a9" has been taken from [4]. For encryption, the random sequence is generated using logistic–sine chaotic map considering following parameters

Initial value of the chaotic map $x_n = 0.7$
Bifurcation parameter value $r = 3.2$.

The resulting encrypted data is as follows: "b2558ede676c60cbc83a611c8ec8e33c91b3fdb9b5133f121e0faaa9520833f9fc214110637020024ad1686785f43daf2ef5bacd8306".

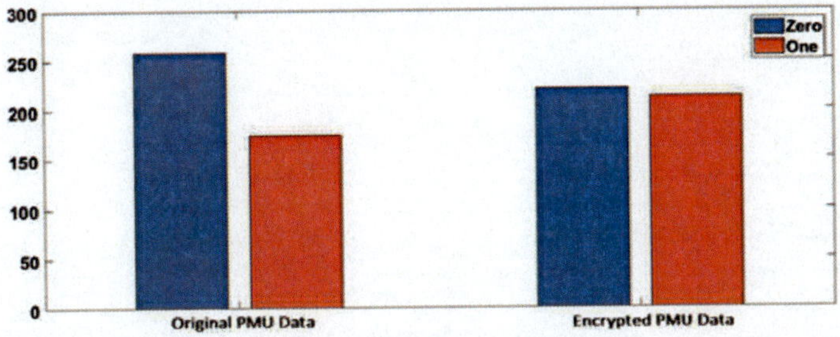

Fig. 6 Histogram distribution of original and encrypted PMU data

5 Security Analysis

To assess the robustness of the proposed encryption algorithm, various security analyzes are done.

5.1 *Histogram Analysis*

A good encryption algorithm is one which results in uniform distribution of 1 s and 0 s in encrypted data. Histogram analysis evaluates rates of 1 s and 0 s in a data. Before encrypting the PMU data, the number of zeros and ones are 258 and 174, respectively. After encrypting, the number of zeros and ones are 220 and 212, respectively. The histogram distribution of zeros and ones in original and encrypted PMU data is shown in Fig. 6.

5.2 *Key Sensitivity Analysis*

It is an important characteristic that a chaotic system must follows. A chaotic system must produce different sequences in case of slight change in initial condition in that case a chaotic system is said to be possessing good ergodic property [19, 20]. In order to exhibit key sensitivity of the logistic–sine map, the initial value of the map x_n is kept $0.7 + 10^{-15}$, and then, it is changed to $0.7 + 10^{-16}$. The value of bifurcation parameter r is kept 3.2. The map is iterated for 10 iterations only. The values of sequences x_n generated are shown in Fig. 7.

From Fig. 7, it can be observed that after two iterations, only the sequences diverge. One can observe similar phenomenon in case the initial value of the map is fixed, and bifurcation parameter value r is changed to 16 decimal places. This indicates that if an intruder must make permutation 10^32 permutations and combinations for committing brute force attack.

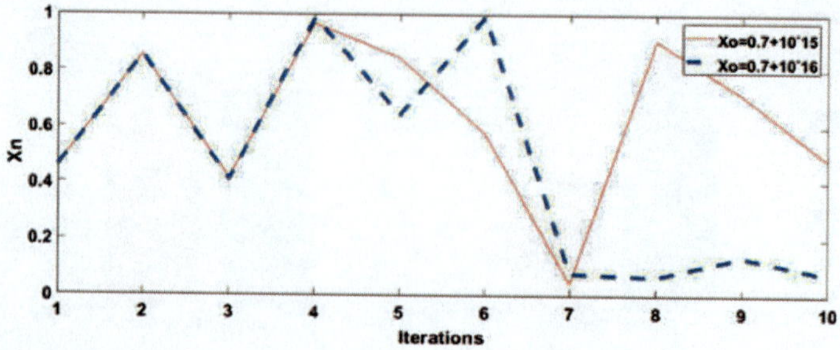

Fig. 7 Plot depicting sensitivity of logistic–sine map

5.3 *Key Space Analysis*

A large key space is an essential for a cryptosystem to counter brute force attack [11, 19]. In logistic–sine map, there are two variables, i.e., initial condition of sequence x_n and bifurcation parameter r. As evident from sensitivity analysis, an intruder would require 10^{16} keys for each variable and since there are two variables so in total, there are 10^{32} keys. If key space is more than 10^{30}, then the proposed cryptosystem can withstand brute force attack [10]. Since the key space of logistic–sine map is 10^{32}, therefore it is resistant to brute force attack.

6 Conclusions

PMU has become indispensable part of modern power grid called smart grid today as many of the vital decisions depend on PMU data. Although proliferation of PMU has been in rise since past two decades, but integrity security of its data remains a major gap. This is so because IEEE C37.118 provides no guidelines for end-to-end data security mechanism for PMU data. This research work takes into task this issue and proposes a chaos-based encryption algorithm for PMU data. The proposed mechanism not only ensures data integrity but also gives advantage of its implementation on low-cost resource constrained IoT devices. Also, the proposed encryption algorithm is compatible to the real-time data transmission constraint of PMU as the computational complexity is low.

References

1. Almasabi S et al (2022) False data injection detection for phasor measurement units. Sensors 1:1–18. https://doi.org/10.3390/s21175791

2. Khalel SI, Khuder SM (2022) Cyber-attacks risk mitigation on power system via artificial intelligence technique. In: 2022 9th International conference on electrical and electronics engineering (ICEEE). IEEE, pp 117–122. https://doi.org/10.1109/ICEEE55327.2022.9772559
3. Agrawal A, Momin DM, Syndor D, Affijulla S (2020) Impact analysis of cyber-attack under stable state of power system: voltage stability. In: 2020 IEEE region 10 symposium (TENSYMP). IEEE, pp 402–405. https://doi.org/10.1109/TENSYMP50017.2020.9230732
4. Varan M, Akgul A, Kurugollu F, Sansli A, Smith K (2021) A novel security methodology for smart grids: a case study of microcomputer-based encryption for PMU devices. Complexity 1:1–15. https://doi.org/10.1155/2021/2798534
5. Madani M, Assad SE, Dridi F, Lozi R (2022) Enhanced design and hardware implementation of a chaos-based block cipher for image protection. J Differ Equ Appl 1:1–21. https://doi.org/10.1080/10236198.2022.2069496
6. Singh AK, Chatterjee K, Singh A (2022) An image security model based on chaos and DNA cryptography for IIoT images. IEEE Trans Ind Inf 1:1–8. https://doi.org/10.1109/TII.2022.3176054
7. Yang C, Pan P, Ding Q (2022) Image encryption scheme based on mixed chaotic Bernoulli measurement matrix block compressive sensing. Entropy 1:1–23. https://doi.org/10.3390/e24020273
8. Lawnik M, Berezowski M (2022) New chaotic system: M-map and its application in chaos-Based cryptography. Symmetry 14:1–18. https://doi.org/10.3390/sym14050895
9. Naik RB, Singh U (2022) A review on applications of chaotic maps in pseudo-random number generators and encryption. Ann Data Sci 1:1–26. https://doi.org/10.1007/s40745-021-00364-7
10. Zhou W, Wang X, Wang M, Li D (2022) A new combination chaotic system and its application in a new Bit-level image encryption scheme. Opt Lasers Eng 149:1–15. https://doi.org/10.1016/j.optlaseng.2021.106782
11. Belazi A et al (2022) Improved Sine-Tangent chaotic map with application in medical images encryption. J Inf Secur Appl 66:1–24. https://doi.org/10.1016/j.jisa.2022.103131
12. Chakhchoukh Y, Lei H, Johnson BK (2020) Diagnosis of outliers and cyber-attacks in dynamic PMU-based power state estimation. IEEE Trans Power Syst 35:1188–1197. https://doi.org/10.1109/TPWRS.2019.2939192
13. Sasikala R, Meenakumari R (2021) Synchrophasor data transfer tool IEEE C37. 118.2—implementation and study of measurement standard for a futuristic substation automation. Int J Commun Syst 1:1–19. https://doi.org/10.1002/dac.4759
14. Yuvaraju V, Thangavel S (2022) Optimal phasor measurement unit placement for power system observability using teaching–learning based optimization. Int J Electr Power Energy Syst 137:1–10. https://doi.org/10.1016/j.ijepes.2021.107775
15. Luo P, Fan H, Zhang S, Hao X, Wang X (2019) A new measurement. EURASIP J Wirel Commun Netw 1:1–9. https://doi.org/10.1186/s13638-019-1492-3
16. Tu C, He X, Liu X, Li P (2018) Cyber-attacks in PMU-based power network and countermeasures. IEEE Access 6:1–10. https://ieeexplore.ieee.org/document/6111222. https://doi.org/10.1109/ACCESS.2018.2878436
17. Jolfaei A, Kant K (2017) A lightweight integrity protection scheme for fast communications in smart grid. SECRYPT 4:31–42. https://doi.org/10.5220/0006394200310042
18. Nesa N, Ghosh T, Banerjee I (2019) Design of a chaos-based encryption scheme for sensor data using a novel logarithmic chaotic map. J Inf Secur Appl 47:320–328. https://doi.org/10.1016/j.jisa.2019.05.017
19. Elghandour A, Salah A, Karawia A (2022) A new cryptographic algorithm via a two-dimensional chaotic map. Ain Shams Eng J 13:1–11. https://doi.org/10.1016/j.asej.2021.05.004
20. Alharbi AR et al (2022) A new multistage encryption scheme using linear feedback register and chaos-based quantum map. Complexity 1:1–15. https://doi.org/10.1155/2022/7047282

IoT and ML-Based Personalised Healthcare System for Heart Patients

Sreeram Warrier, Vinay Kumar Jadoun, Ankur Jaiswal, Aashish Kumar Bohre, and Pallavee Jaiswal

Abstract In this paper, a remote health monitoring system using IoT is considered. In this, a prototype was developed which keeps track of the pulse rate and body temperature of a heart patient. This prototype is to be used by the patient remotely. The sensor data is then transmitted to an IoT server using a Wi-Fi module and a microcontroller. Moreover, this live data is further recorded in a spreadsheet which can be shared with concerned authorities to track the patient's health condition. The system also allows a notification to be sent out via SMS/email when the patient's condition worsens. The second part involves the prediction of whether a patient is at risk of having a heart condition based on their medical records. It involves data analysis, feature engineering, and deploying machine learning algorithms to create a data model that predicts whether a patient is at risk with a high amount of accuracy. In addition, a hardware system is implemented and realised that could be used by hospitals to remotely track the health condition of their patients. We were also able to achieve the additional feature of notifying concerned authorities if an anomaly is detected. With the help of data analysis and machine learning tools, we were able to analyse and find the correlation between various health factors and the risk level of having a heart condition. Finally, we created a data model that was able to predict patients that are potentially at risk with a high accuracy rate that could be

S. Warrier · V. K. Jadoun (✉) · A. Jaiswal
Manipal Institute of Technology, Manipal Academy of Higher Education (MAHE), Manipal, Karnataka 576104, India
e-mail: vinay.jadoun@manipal.edu

S. Warrier
e-mail: warriersreeram@gmail.com

A. Jaiswal
e-mail: ankur.jaiswal@manipal.edu

A. K. Bohre
National Institute of Technology (NIT), Durgapur, India
e-mail: aashishkumar.bohre@ee.nitdgp.ac.in

P. Jaiswal
Dr. C. V. Raman University, Kargi Road, Kota, Bilaspur (C.G), India
e-mail: pallavee10@gmail.com

S. Sharma et al. (eds.), *Intelligent Control, Robotics, and Industrial Automation*, Lecture Notes in Electrical Engineering 1066,
https://doi.org/10.1007/978-981-99-4634-1_35

used to differentiate the level of importance of each factor. The final system that was developed during the training period was a prototype that was able to counter some of the existing challenges in the area of work we focused on. We were also able to identify newer challenges and suggest further areas of research that need to be addressed.

Keywords Internet of things · Machine learning · Data analytics · Artificial intelligence

1 Introduction

Personalised healthcare is a new form of healthcare approach which is specifically patient-oriented and aims to improve the traditional healthcare system. The focus of this style of approach is to personalise the type of diagnosis and treatment based on the complexities of each individual patient when compared with the generalised style of trial-and-error approach that was previously taken. It emphasises using tools enabled by modern technology to analyse treatments that have maximum efficacy and minimum side effects [1]. The application of the Internet of Things (IoT) and machine learning (ML) in the healthcare sector has gained a lot of traction of late. IoT-cloud-based sensor devices are used by health authorities and hospitals to remotely track patients' health conditions. This not only allows them to intervene when the onset of a condition is detected but also to create, maintain, and update the electronic health record (EHR) of each patient from the sensor data that is collected. Artificial intelligence techniques such as machine learning are then used to develop data models for further research purposes. The database of the EHRs is used to train and test ML-based data models which are further used to improve disease detection, risk analysis, and the overall clinical decision support system. In this paper, we try to set up a personalised healthcare system for heart patients with the application of IoT and ML. This work is divided into two parts based on the area of the personalised healthcare system that it touches upon (i) assistive care and (ii) monitoring and alarming. The first part utilises a cloud-based IoT server to remotely monitor and record the health status of heart patients and alert the concerned people when their condition destabilises. The second part uses machine learning algorithms to develop a data model that can analyse and predict the risk of a patient suffering from heart disease.

2 Background

Healthcare technology dedicated towards heart patients has been around for a while and is constantly evolving. The implementation of IoT and ML to create a personalised healthcare system is the newest addition to this field.

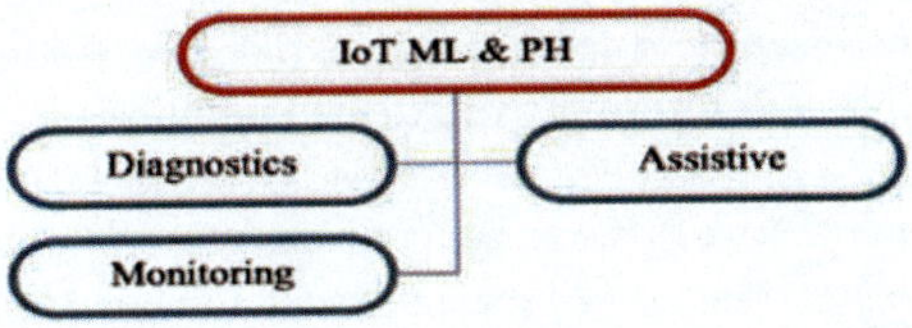

Fig. 1 Classification of IoT and ML in personalised healthcare

2.1 Classification of IoT and ML in Personalised Healthcare (PH)

The contribution of IoT and ML in personalised healthcare is classified into three areas and as shown in Fig. 1:

(1) Diagnostic care
(2) Assistive care
(3) Monitoring and alarming.

Diagnostic care refers to the usage of IoT and ML-powered technology in the diagnostic treatment of patients. Machines and tools are being invented and redesigned to support automation and intelligence to improve their efficiency and accuracy. These tools can also be used in the treatment of patients in remote areas where a medical instrument operator might not be available in emergency situations. Furthermore, ML is also used in medical research to improve the quality of the type of treatments, drugs, and medications that are produced. Assistive care is the application of IoT and ML in any form of analytical research in the medical field. A prime example of assistive care is the usage of analytics in helping health professionals during the pandemic caused by COVID-19. All mass-scale activities from allocation of essential resources to the conduction of vaccination drives were done with the support of software and systems built on the backbone of IoT and ML. It was also used for analysing the condition of patients, their reaction to different types of treatments, and even the detection of variants of the COVID-19 virus. As the name suggests, monitoring and alarming is the live tracking of the health condition and status of the patients. Remote monitoring, in particular, plays a significant role in personalised healthcare. As patients are discharged, hospitals use IoT devices to keep track of their health status by pushing the data into a cloud server. Cloud-based big data services are then used to perform intelligent predictive analysis to detect and alert the onset/worsening of any condition.

2.2 Related Works and Present-Day Scenario

A specific section of the personalised healthcare system is the one that caters to heart patients. Personalised healthcare for heart patients is an emerging area that requires a lot of attention because of the sensitivity and the growing numbers. Heart disease has become a leading cause of human fatality all over the world. The number of patients suffering from heart conditions as well as the types of variations of heart diseases has been on the rise over the last decade. It has been estimated by WHO that the worldwide heart disease rate might increase to 23% by the year 2030 [2]. While preventive measures need to be put into place, it is equally important to ensure that a support system is added to the healthcare infrastructure so that the system is equipped to combat any situation that may arise. The integration of IoT and ML in the healthcare system has opened newer avenues for professionals to venture into. One such application that has boosted the growth of personalised healthcare is remote health monitoring systems. A personalised healthcare network is built for each patient. The various sensors that are a part of this network send their data to a cloud IoT server which is further stored in a database for involved parties to access. This collected data further uses ML algorithms to analyse the data and determine the patient's risk factor and suggest further actions. This architecture is depicted in the Fig. 2 given [1]. With the growing interest in personalised healthcare for heart patients, studies are being conducted to develop remote monitoring diagnostic frameworks that detect underlying heart conditions in real time, thereby avoiding potential cardiac diseases [3]. System designs have already been proposed that use wearable sensors along with microcontrollers and Wi-Fi modules to push the sensor data that is available 24 h into an IoT server [4]. Cloud computing services that provide big data solutions for remote health monitoring are also being studied and developed [5]. Standardisation of methodologies for the risk analysis of heart patients using ML algorithms such as random forest is also being observed [6]. However, the key challenge that heart patient-based personalised healthcare systems face is the integration of these separate components to develop a sophisticated yet economical system that is reliable and can be sustainably maintained in the long run. Other issues that follow are network traffic handling, authentication issues, and data privacy.

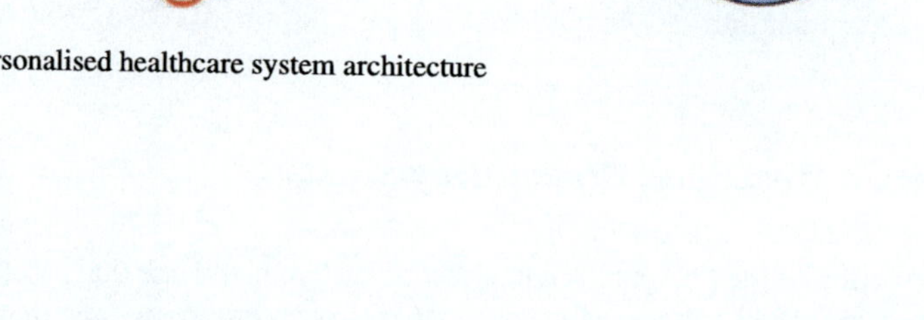

Fig. 2 Personalised healthcare system architecture

3 Methodology

As stated earlier, the work is divided into two parts based on the area of personalised healthcare that is being addressed. We shall now delve into each part in detail in the following sections.

3.1 Monitoring and Alarming Using IoT

The first part aims to create a monitoring and alarming personalised healthcare set-up by implementing a remote health monitoring system for heart patients using IoT. The system prototype feeds the body temperature, and pulse rate of the patient picked up by the sensors to a cloud IoT server using a microcontroller and a Wi-Fi module. This data is then stored in the cloud platform as well as a live google sheet that is updated with the help of an applet. The system also allows alarming by sending alerts to the concerned health authorities in case any abnormality is detected in the sensor data.

The hardware components used for the implementation of the system are:

- Arduino UNO—Microcontroller
- ESP8266—Wi-Fi module
- LM35—Temperature sensor
- Heartbeat/pulse sensor
- 16 X 2 LCD display
- LED
- Resistors.

Refer to the Fig. 3 for the block diagram of the circuit as well as the circuit set-up.

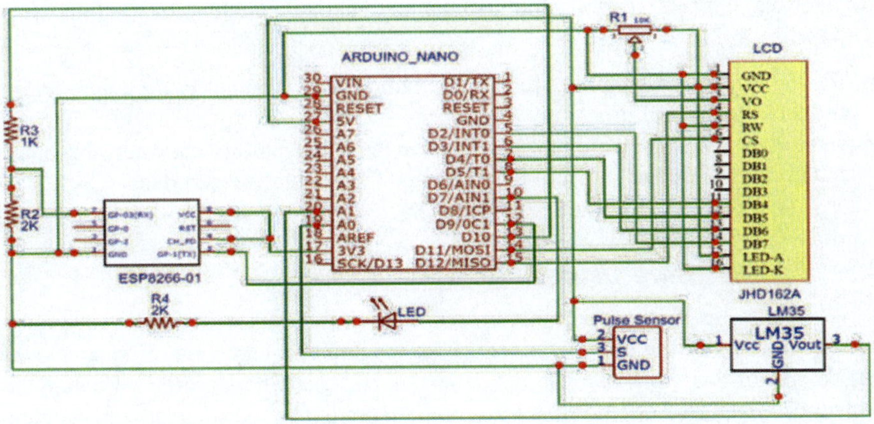

Fig. 3 Block diagram of the system circuit

To facilitate the logging and transferring of data between the sensors, Arduino UNO, ESP8266, and the IoT server, a computer programme was made and uploaded to the microcontroller using Arduino IDE. This code used for running the system can be divided into three parts:

(A) Getting the sensor readings: In this section, we use data read commands to pick up sensor readings, log them, and print them to the serial monitor. For the temperature sensor, we read the analogue value from the sensor and calibrate it to give us the temperature. For the pulse sensor, we directly read the digital values from the sensor using the "PulseSensorPlayground" library.
(B) Updating and transmitting the sensor values to the IoT server: The updated sensor readings are sent to the IoT server through API calls done by the Wi-Fi module. This is in turn configured using AT various commands. Note that the IoT server that is used for receiving the data in this work is ThingSpeak—a cloud-based IoT server.
(C) Connecting the Wi-Fi module to a wireless network: This part of the code uses the local area network details such as a Wi-Fi username and password to connect the ESP8266 to the Wi-Fi module. This configuration is done through AT commands as well.

The code also includes other snippets which perform operations such as initialising the system using a set-up function, infinitely reading the values with the help of a timer placed in a loop function, set-up of the LCD and the inclusion of the required libraries. The IoT server that is being used for this paper is a public channel made in ThingSpeak—an IoT analytics service platform that allows you to visualise and analyse live data streams in the cloud. A public channel was created in ThingSpeak for this work. The channel included live data visualisation of the pulse and temperature data along with widgets that showed the most recent value that was received by the server is given in Fig. 4.

Additionally, we also use If This Then That (IFTTT), a commercial service provider that connects various applications, to create two applets for the following purposes:

- To store the data being received by the ThingSpeak channel in a Google Sheet and update it in real time.
- To send an SMS/email alert to all people on the given mailing list when abnormal conditions are picked up from the pulse and temperature sensor data.

3.2 Assistive Care Using ML

The second part of the training comes under the assistive care area of personalised healthcare. In this work, we use data analysis and machine learning on a Jupyter Notebook with a Python interpreter to conduct a predictive risk analysis and develop a data model that can predict whether a patient is at a high or low risk of having a heart condition in the near future. The dataset we have used during the training is the heart

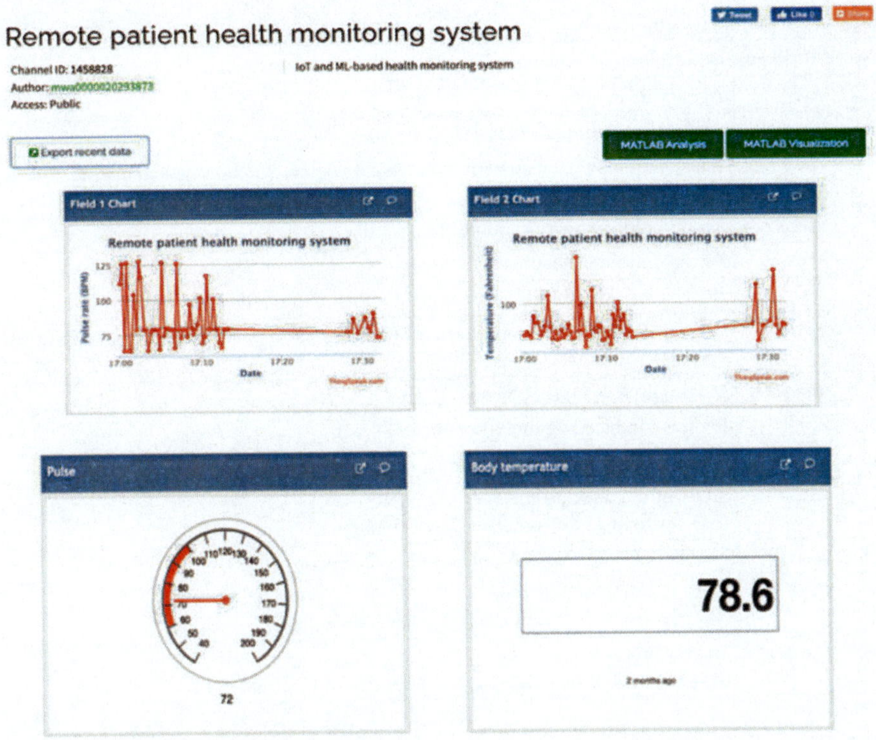

Fig. 4 ThingSpeak public channel (IoT server)

disease dataset available in the public machine learning repository of the University of California, Irvine [7]. While the original dataset contains 75 attributes, we will only be using the top 14 attributes that were identified as the most relevant by the University website upon analysing the dataset and performing feature engineering. These features along with a quick description of them can be found in the Table 1.

After importing the dataset to the Jupyter Notebook, data preprocessing had to be done to clean and wrangle the given data. This included changing the data types of certain attributes to make it more analysis friendly as well as imputing missing and erroneous data [8]. All the mentioned operations were performed on the Pandas dataframe that was made by importing the abovementioned dataset. Table 2 below contains a snippet of the final dataframe that was used.

Once the dataset was ready for further analysis, an in-depth exploratory data analysis along with data visualisation was done for each feature individually. The data visualisation was performed using the Matplotlib and Seaborn libraries of Python. We use a bar plot for discrete variables and a kernel density estimation plot for continuous variables throughout the analysis. We were also able to find Pearson's correlation coefficient between the target variable and the other attributes using the "corr" function of the Pandas library. Pearson's correlation coefficient is a stat that indicates

Table 1 Attributes description

No.	Attribute	Description
1	Age	Age of the patient
2	Sex	Sex of the patient (1 = male, 0 = female)
3	cp	Chest pain type (0 = no pain, 1 = typical angina, 2 = non typical angina, 3 = non-anginal pain)
4	trtbps	Resting blood sugar (in mm of Hg)
5	chol	Serum cholesterol level (in mg/dL)
6	fbs	Checks if fasting blood sugar is greater than 120 mg/dL (0 = false, 1 = true)
7	*Restecg*	Resting electrocardiographic results (0 = normal, 1 = ST-T wave abnormality, 2 = hypertrophy)
8	*thalachh*	Maximum heart rate achieved
9	*exng*	Exercise-induced angina (0 = no, 1 = yes)
10	*Oldpeak*	ST depression induced by exercise relative to rest (0–6)
11	*slp*	Slope of the peak exercise ST segment (2 = upsloping, 1 = flat, 0 = down sloping)
12	*caa*	Number of major vessels coloured by fluoroscopy (0–3)
13	*thall*	Thalassemia blood disorder (1 = fixed defect, 2 = normal, 3 = reversible defect)
14	*Output*	Target variable (0 = low risk of heart attack, 1 = high risk of heart attack)

Table 2 Dataset summary

Attribute	0	1	2	3	4
Age	63	37	44	56	57
Sex	4	1	0	1	0
cp	3	2	1	1	0
trtbps	145	130	130	120	120
chol	233	250	204	236	354
fbs	1	0	0	0	0
restecg	0	1	0	1	1
thalachh	150	187	172	178	163
exng	0	0	0	0	1
oldpeak	2.3	3.5	1.4	0.82	0.62
slp	0	0	2	0	0
caa	0	0	0	2	2
thall	1	2	2	1	1
Output	1	1	1		

Table 3 Pearson's correlation between each feature and the target variable

Feature	Pearson's coefficient	Feature	Pearson's coefficient	Feature	Pearson's coefficient
Age	−0.23	fbs	−0.03	oldpeak	−0.43
Sex	−0.38	restecg	0.14	slp	0.35
cp	0.43	thalschh	0.42	caa	−0.39
trtbps	−0.15	exng	−0.44	thall	−0.34
chol	−0.09				

whether there is a strong linear correlation between two variables. A value close to 0 indicates a weak correlation while a value close to + 1 or −1 indicates a strong positive or negative linear correlation, respectively, [9]. Below is the list of Pearson's correlation coefficient between each feature and the target variable given in Table 3. The next step of the process was to create a data model using machine learning. The problem statement was a classification problem that comes under supervised machine learning. Several ML algorithms such as k-nearest neighbours, decision trees, and support vector machine were tried and tested [10–12]. However, the most accurate model was made by using the random forest classifier algorithm inbuilt into the Scikit-Learn Python library [13, 14].

4 Result and Analysis

4.1 Exploratory Data Analysis

The following subsection consists of the exploratory data analysis that was performed on each feature.

A. Age

- More people lesser than age 55 have a higher risk.
- More people within the age group of 55–70 have a lower risk.
- More people greater than age 70 have a higher risk.
- There might be a nonlinear correlation between age group and heart attack risk for patients.

B. Sex

- More males are at low risk but it is comparable.
- A significant no. of women are at a high risk compared with the no. of women with low risk.
- There seems to be a correlation between sex and heart attack risk even if it isn't a causation.

C. Chest pain type

- Patients with chest pain type 1, 2, or 3 have a high chance.
- Patients with chest pain type 0, i.e. asymptomatic have a low chance.
- Linear correlation exists between the two.
- Resting blood pressure

D.

- Patients with a resting blood pressure range between 80 and 150 are observed to have a higher chance of being at high risk.
- Difficult to establish a correlation even though there might be one.

E. Cholesterol

- Density of patients with cholesterol roughly greater than 250 and lesser than 150 are more or less the same for patients with high and low chance.
- Patients within the range of 150–250 appx. are observed to have a higher chance of having a heart attack.
- Hard to predict but there is a chance of a nonlinear correlation.

F. Fasting blood sugar

- It is observed that there isn't much of a correlation between one's fasting blood sugar and heart condition risk.

G. Resting electrocardiographic results

- There seems to be a significant difference between the count low-risk and high-risk patients only for value 1. This might just be a correlation and not a causation.

H. Maximum heart rate achieved

- Patients with max heart rate lesser than 140 are more likely to have a lower risk.
- Patients with max heart rate greater than 140 are more likely to have a higher risk.
- There is a strong correlation between the max heart rate and risk of having a heart attack.

I. Exercise-induced angina

- The stark difference points out that there is a strong correlation between this feature and the target variable.

J. Exercise-induced ST depression

- There is a good correlation between these two variables.

- Lower value patients are observed to have a high risk compared with low risk.
- Higher value patients are observed to have a low risk compared with high risk.
- ST segment slope:
- For value 0: No significant difference.
- For value 1: Patients with low risk are higher in number.
- For value 2: Patients with high risk are higher in number.
- There is a correlation between these two variables.

L. No. of coloured major vessels

- As the value increases, there seems to be a higher probability of patients having a low risk compared with high risk.
- There is a linear correlation between the predictor variable and the number of major vessels.

M. Thalassemia blood disorder

- Value 1 and 3: Patients with low risk are more in number.
- Value 2: Patients with high risk are more in number.
- There seems to be some correlation between these two variables.

4.2 *ML-Based Data Model*

The random forest algorithm-based data model was able to successfully predict whether a patient was at high risk or low risk of having a heart condition with an accuracy of 88.16%. Moreover, an analysis was also performed to see the importance of each feature in the data model to check whether it concurred with our initial feature analysis (Table 4).

From this, it could be inferred that barring fasting blood sugar and resting electrocardiographic results, all other features contributed significantly towards making the final data model that was used in our paper.

Table 4 Feature importance table of the data model

Feature	%	Feature	%	Feature	%	Feature	%
age	6.78	chol	7.18	thalschh	10.29	slp	3.84
sex	5.1	fbs	0.93	exng	7.59	caa	13.19
cp	12.45	restecg	1.25	oldpeak	12.06	thal	12.63
trtbps	6.7						

5 Conclusion and Future Scope

Through the course of the project, we were able to achieve our objective and develop an IoT and ML-based personalised healthcare system for heart patients. The first part of the training realises a monitoring and alarming system through a cloud IoT-powered hardware prototype. The second part did a predictive risk analysis by deploying an ML-based data model which achieved a high accuracy rate.

5.1 IoT-based Remote Health Monitoring System for Heart Patients

The IoT-based prototype could track and record the live data of a patient such as their heartbeat rate and body temperature using a cloud IoT server. The system reads, stores, and transmits the sensor data at intervals of 30 s to the ThingSpeak public channel that acts as our cloud IoT server. Moreover, an alert is sent out if the sensor data indicates potential danger in the health condition of the patient. Although the designed system was very basic in terms of functionalities, it serves as a base or a prototype of an economical personalised healthcare system. Further studies in this field can help create a more accessible and user-friendly design with other required functionalities. A noticeable drawback is the inculcation of sensors to measure other important health parameters which can be done by further developing this prototype. Lastly, since the API call by the cloud server to fetch the sensor data from the Wi-Fi module is made periodically, there is a small delay between the recording of the measurement by the sensor and the data being shown and stored on the server. The delay depends on the time before the next periodically scheduled API call is made.

5.2 ML-based Heart Patient Risk Analysis

Assistive care-based personalised healthcare service was implemented through the data model that was trained and tested. The second part of the training required us to learn about machine learning algorithms, data analytics and data visualisation tools. It also needed us to be versatile with Python programming along with a good understanding of its various support libraries. We used these skills to do an in-depth predictive risk analysis on the given dataset on heart patients. We were able to figure out the correlation between various features and the target variable, create a data model based on the random forest classifier machine learning algorithm, and achieve an accuracy of 88%. Furthermore, we found out the five most important features that were considered while developing our data model that were: chest pain type, maximum heart rate achieved, ST depression induced by exercise, number of major coloured vessels and thalassemia blood disorder, respectively. While the data model was trained using a relatively smaller dataset, this study paves the way for further creating more enhanced and lucrative ML-based data models that can be trained and deployed on big data to achieve higher accuracy. The model can also be further designed and trained using more complex and specific classification algorithms. Other steps can also be implemented in the data modelling process such as fine-tuning the parameters of the model and checking whether the model is overfitting or under fitting. A major limitation of the model is the two parts (IoT and ML) being two separate systems and not a cohesive unit. Future studies can be conducted on trying to combine the two parts.

References

1. Ahamed F, Farid F (2018) Applying internet of things and machine-learning for personalized healthcare: issues and challenges. In: 2018 international conference on machine learning and data engineering (iCMLDE), pp 19–21. https://doi.org/10.1109/iCMLDE.2018.00014
2. Organization WH (2011) Global status report on noncommunicable diseases 2010. World Health Organization, Geneva, Switzerland
3. Abe E, Chigira H, Fujiwarai K, Yamakawa T, Kano M (2015) Heart rate monitoring by a pulse sensor embedded game controller, 1266–1269. https://doi.org/10.1109/APSIPA.2015.7415478
4. Shaikh S, Chitre V (2017) Healthcare monitoring system using IoT. In: 2017 International conference on trends in electronics and informatics (ICEI), pp 374–377. https://doi.org/10.1109/ICOEI.2017.8300952
5. Shahzad A, Lee YS, Lee M, Kim Y-G, Xiong N (2018) Real-time cloud-based health tracking and monitoring system in designed boundary for cardiology patients. J Sens 2018:15, Article ID 3202787
6. Raja MS, Anurag M, Reddy CP, Sirisala NR (2021) Machine learning based heart disease prediction system. In: 2021 international conference on computer communication and informatics (ICCCI), pp 1–5. https://doi.org/10.1109/ICCCI50826.2021.9402653
7. Heart Disease Data Set, Dua D, Graff C (2019) UCI machine learning repository [http://archive.ics.uci.edu/ml]. Irvine, CA, University of California, School of Information and Computer Science
8. Kandel S et al (2011) Research directions in data wrangling: visualizations and transformations for usable and credible data. Inf Visual 10(4):271–288
9. Benesty J et al (2009) Pearson correlation coefficient. Noise reduction in speech processing. Springer, Berlin, Heidelberg, pp 1–4
10. Peterson LE (2009) K-nearest neighbor. Scholarpedia 4(2):1883
11. Myles AJ et al (2004) An introduction to decision tree modeling. J Chemometrics: J Chemometrics Soc 18)(6):275–285.
12. Suthaharan S (2016) Support vector machine. Machine learning models and algorithms for big data classification. Springer, Boston, MA, pp 207–235
13. Rigatti SJ (2017) Random forest. J Insur Med 47(1):31–39
14. Weerts HJP, Mueller AC, Vanschoren J (2020) Importance of tuning hyperparameters of machine learning algorithms. arXiv preprint arXiv:2007.07588

Machine Vision and Image Processing

Compressively Sensed Plant Images for Diagnosis of Diseases

Nimmy Ann Mathew, Megha P. Krishna, and Renu Jose

Abstract Agriculture stands as a pivotal factor in a nation's economic advancement and the progress of human civilization. Yet, the prevalence of agricultural crop ailments exerts a detrimental influence and poses a substantial obstruction to agricultural output. Early detection of disease has become inevitable in this current scenario. Traditional practices entail long-term field monitoring and a time-consuming acquisition process. Compressive sensing, a novel sensing paradigm, acts as a great path that benefits both speed and quality of diagnosis. It allows signals to be acquired using fewer samples rate and still be able to reconstruct them without losing information. It can be leveraged to reduce the vast volume of data in data acquisition equipment. The work studies the viability of CS in the acquisition of plant images. The reconstruction performance is also evaluated in the proposed work.

Keywords Compressive sensing (cs) · Orthogonal matching pursuit (omp) · Compressive sampling matching pursuit (cosamp) · Iterative hard thresholding (iht) · Early blight · Target spot

1 Introduction

Agriculture is the base of every frugality worldwide. It is an indispensable part of profitable development. But a lot of problems are affecting agriculture day by day. Various factors contribute to the deterioration of their health, such as the over usage of pesticides, unpredictable weather patterns of our country, unhealthy soil, and

N. A. Mathew (✉) · M. P. Krishna · R. Jose
Department of Electronics and Communication, Rajiv Gandhi Institute of Technology, Affiliated to APJ Abdul Kalam Technological University, Kottayam, Kerala, India
e-mail: nimmyann29@rit.ac.in

M. P. Krishna
e-mail: meghapadmajan@gmail.com

R. Jose
e-mail: renu.jose@rit.ac.in

S. Sharma et al. (eds.), *Intelligent Control, Robotics, and Industrial Automation*, Lecture Notes in Electrical Engineering 1066,
https://doi.org/10.1007/978-981-99-4634-1_36

scarcity of clean water without the presence of chemicals. Apart from these, another grave and perhaps the most important of all that requires immediate attention is the different diseases that affect the plants. Like any living organism, diseases can greatly dwindle the health and subsequently the lifespan of a plant. It is a major concern among farmers, as it may lead to a reduction in food production and supply. If adequate precautions are not taken in this regard then it may lead to a food shortage [1–3].

Plant diseases are the cause of significant profitable losses in the husbandry business around the world. Automatic diagnosis of diseases requires long-term monitoring of farms which is a time-consuming process. Early identification of leaf infections is an integral and essential part of farming. Most diseases are recognized through human sources via visiting farms. Experts have been identifying crop diseases with their naked eyes for a long time, based on their expertise and experience. Finding an expert and contacting them is not only a time-consuming and laborious effort, but it is also a lengthy and costly process. Consequently, the whole process can take a long time, which makes it difficult to eradicate the disease and takes a long time in the case of vast areas.

To achieve good quality plant images it is required that the images be sampled at the Nyquist rate. Oftentimes, the Nyquist rate would be too high, such as digital cameras. In such cases, the conventional approach is to collect a large number of samples in the acquisition phase which are then compressed in the transmission stage. This leads to unnecessary overhead in the acquisition stage. Because most signals have a sparse characteristic, much of the information acquired would be redundant, and hence discarding samples would not lead to any visible information loss. Compressive Sensing makes use of these signal characteristics and allows data to be simultaneously compressed while sensing. This greatly improves the efficiency of the systems as it reduces the acquisition time. Consequently, by using lesser samples, CS reduces energy consumption making the system economically efficient [4].

Using this compressive sensing method the acquisition side overhead is reduced. Even if we have to transmit the data over a longer distance, we can easily do it because only a fewer samples need to be transmitted. So, this can be effectively used in a long time continuous monitoring systems. Because CS helps to reduce the time of acquisition thereby improving the life of acquisition equipment. Apart from these, having a lesser number of samples could pose a significant improvement in the battery consumption for battery devices as well as in time consumption. These have the great economic advantages of increasing the lifespan of expensive equipment and low power consumption since the acquisition time is lower [4].

The CS is also known as Magic reconstruction because of the concept of reconstruction from fewer samples. The idea of CS was initially put forward by Donoho, Candès, Romberg, and Tao in 2004 [5]. CS allows a replacement means for a lower sampling rate and its recovery if the signal is sparse or compressible representation on some suitable basis. It allows the signals to be sampled without taking into consideration its required Nyquist rate, thus reducing the data volume drastically. Sparse signals are those signals having only fewer non-zero components. Some signals have a sparse representation in their original domain, but if not, the signals can

be made sparse by transforming them into some other sparse domain. There are a few works present in the literature which studies the several domains which can be used to achieve a sparse form. Natural images are compressible in the Discrete Cosine Transform (DCT), and Wavelet basis. Once the acquisition is carried out using CS, at the receiver end, the original signal can be estimated without much information loss using nonlinear techniques [4–6]. The quality of the reconstructed output is measured using performance metrics, such as, Peak Signal to Noise Ratio (PSNR) and the Product Root Difference (PRD). Studies suggest that a reconstructed image with a PRD$\leq$9% can be considered to be of acceptable quality [7] .
CS Theory states that:

1. CS systems can effectively acquire signals
2. CS systems can effectively reconstruct signals
3. CS systems are highly under determined linear systems.

Important conditions for signal recovery are, sparsity and incoherence. Sparsity says that the signal must be sparse in some domain for it to be recovered. Incoherence means the relationships between the signals should be incoherent which means the restricted isometric property [8].

The automated classification of plant disease into multiple disease categories has gained predominance. Several works have been carried out using Machine Learning (ML) and Deep Learning (DL) algorithms on disease infected plants to achieve an accurate diagnosis [9, 10] . In [11] Support Vector Machine (SVM) techniques were used to classify disease infected plants. Their work involves noise removal methods, separation of foreground and background, feature extraction using wavelets, and finally classification using SVM. The method provided a classification accuracy of 89.6%. In [12] also SVM was used for classification, but the approach used for feature extraction was different. The authors applied K-means clustering segmentation for extracting the diseased portion from the leaf. Then the color features are extracted from the segmented portion. The accuracy obtained is 62.53%.

Another work uses a model called Genetic Algorithm and Ant Colony Optimization (GAACO) with Transductive Support Vector Machine (TSVM) for detection [13]. Here they convert the image into grayscale and the segmentation of the diseased portion is done using GAACO. Different features were extracted from them and fed to the classifier resulting in an accuracy of 91.3%. In [14] the affected part was segmented through K means and different features such as texture, color, and shape were extracted and fed to the SVM classifier, giving about 92% accuracy. In [15] color space conversion is applied and the affected portion is segmented using K means and features were extracted using Gray Level Co-occurrence Matrix (GLCM). These features were then fed to an SVM resulting in an accuracy of 90%. The authors of [16] used both SVM and KNN classifiers on features extracted using wavelet and also eighteen histogram features showing an accuracy of about 85% and 88%, respectively.

The sensing of images through a novel methodology, namely Compressive Sensing, has gained importance [17]. Many studies are being conducted to further improve the potential applications in the field. Authors of [18] propose that owing to the sparse nature of images, CS can be used for image in-painting. Their work was successful in maintaining noise and texture consistency, compared to other state-of-the-art methods.

The remaining paper is organized as follows. Section 2 deals with the system model and explains in detail the proposed methodology. Section 3 deals with the experiments and simulations that were carried out in the work. Section 4 discusses the important results and comparisons. Finally, the conclusion is given in Sect. 5 and the future scope is also discussed.

2 System Model

The proposed work consists of a CS model in the acquisition and processing stage as shown in Fig. 1. The idea is to directly acquire plant images from fields in the compressed form. Then the samples are transmitted over a long distance or uploaded to the cloud. On the receiver end, the compressed image is passed through a preprocessing block. The preprocessing block consists of a CS based image reconstruction stage. Here, the compressively sensed images are reconstructed using CS reconstruction algorithms. The reconstruction must be carried out in such a manner that the key information or the quality of the image is not compromised. The reconstruction quality is estimated using metrics, namely, Peak Signal to Noise Ratio (PSNR) and Product Root Difference (PRD).

2.1 CS for Images

2.1.1 Acquisition in CS

The acquisition method of CS can be mathematically described as

$$Y = \Phi X \tag{1}$$

where the input signal is X $\in R^n$ or C^n, of length n, the sensing matrix is of length m $\times$n, denoted by $\Phi \in R^{m\times n}$ or $C^{m\times n}$ and the compressive measurements of length m can be denoted by Y $\in R^m$ or C^m. The Compressive Sensing problem is a highly under determined system of linear equation because $m \ll n$.

A signal X is called K-sparse if it has only K non-zero coefficients.This is the basics of transform coding. Mostly, signals that are not sparse in their original domain can be made sparse in its transform domain using a sparsifying basis, Ψ of length $n \times n$. Hence, a signal X in the sparse domain can be mathematically described as:

$$X = \Psi s \tag{2}$$

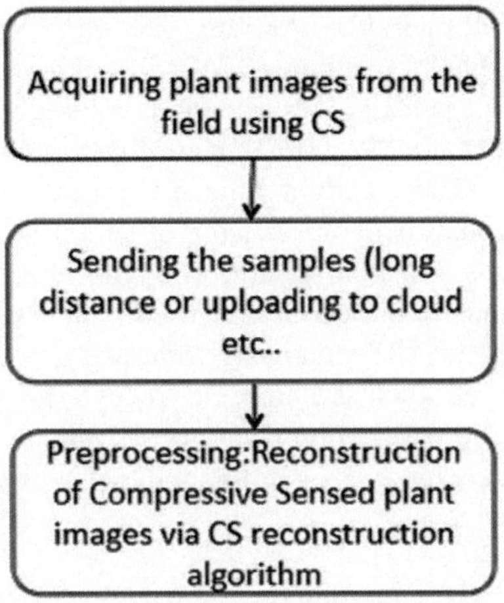

Fig. 1 Block diagram of proposed model

where s is the coefficient of the transform basis. Therefore, CS can be represented as

$$Y = \Phi X = \Phi \Psi s = \Theta s \tag{3}$$

where $\Theta = \Phi \Psi$ is an m×n matrix.

Based on related works on CS technology, the most widely used sensing matrices are random in nature, as it has proven to satisfy RIP and are also incoherent with any sparsifying basis. The elements of the matrix are mostly drawn from a Gaussian distribution density function with mean zero and variance $\frac{1}{m}$ or from Bernoulli distributions [8].

2.2 *Reconstruction Stage*

This stage works by reconstructing the image from random compressive measurements using CS reconstruction algorithms. The compressively sampled data is passed through the reconstruction block. The reconstruction is carried out using CS algorithms. The goal of the reconstruction algorithm is to obtain an output, which is the estimate of the original signal, using inputs Y and Θ. The CS reconstruction solves the inverse problem to obtain the solution to (3). Since the signal is made sparse using a transform, the output of the algorithm provides an estimate of the sparse signal, $\hat{s}$. The estimate of the original signal, $\hat{X}$, can be obtained by taking the inverse transform of $\hat{s}$. A multitude of algorithms is proposed in the literature for the recovery of signals from compressive measurements. All of these algorithms fall under six basic categories, which are the convex optimization algorithms, greedy algorithms, thresholding approach, nonconvex approach, bayesian approach, and combinatorial methods. The proposed work carries out three algorithms, namely, Orthogonal Matching Pursuit (OMP) algorithm and the Compressive Sampling Matching Pursuit (CoSaMP) algorithm, both belonging to the greedy approach and the Iterative Hard Thresholding (IHT) algorithm which belongs to the thresholding approach.

The greedy approach is an iterative process. Firstly it determines which of the columns of the sensing or sampling matrix has the maximum correlation with the measurement vector. The algorithm picks the columns in a greedy fashion, which means, once a particular column is selected it is removed from the subsequent iterations. The remaining iterations are carried out on its residue. The selected columns are known as atoms. Hence lowering the computational complexity of the algorithm. These algorithms have an advantage over convex optimization methods being that they have lower complexity, faster execution, and simpler operation.

2.2.1 Orthogonal Matching Pursuit (OMP):

It falls under the category of serial greedy approach. It is called an iterative algorithm because it finds the original image element-by-element in a step-by-step iterative manner. OMP is an effective method to find the sparse approximation and takes only K number of iterations to find the sparse approximation. OMP demands the sparsity level of the original vector. The steps involved in the algorithms are as follows:

- *Initialization*: First, all the parameters are initialized. The compressive measurements, Y is initialized to a residual vector, L. The solution vector s and the index B are both set to null. The iteration counter p is set to 1. $L_0 = y, s_0 = \emptyset$ and $B = 0$
- In each iteration, the algorithm tries to find a column of which has the highest correlation with the residual vector, L. This column is updated in the index set B. If θ_j is the jth column of , then for p=1, the index C of the chosen column can be defined as

$$C_1 = \arg\max_j |< \theta_j, L_1 >| \tag{4}$$

- Once the index of each column is obtained, the index set is augmented as

$$B_p = B_{p-1} \cup C_p \tag{5}$$

- The solution set s_p is updated using the least squares method.

$$s_p = ({\Theta_{B_p}}^T \Theta_{B_p})^{-1} {\Theta_{B_p}}^T Y \tag{6}$$

Once the solution set is updated, the approximation $\hat{Y}_p$ of the vector Y is obtained as

$$\hat{Y}_p = B_p s_p \tag{7}$$

- The final step is to update the residual so that the columns once selected will not be considered in the subsequent iterations. This can be done by subtracting the approximation vector from the measurement vector.

$$L_p = Y - B_p s_p \tag{8}$$

2.2.2 Compressive Sampling Matching Pursuit (CoSaMP)

It belongs to the parallel greedy approach. Unlike OMP, which selects a single atom during an iteration, CoSaMP tries to find K atoms or a multiple of K atoms in a single iteration. Hence it is known to have a parallel approach. Owing to this fact, this method can be considered to be more efficient as it will have a faster execution. The idea of CoSaMP is quite similar to OMP, except that during each iteration it tries to find $2K$ atoms or columns of the reconstruction matrix, Θ, that have the highest correlation with the residual vector L. Once it is obtained, it is appended with the K atoms of the previous iteration. Finally, only the best K atoms are selected after the least square solution update. The index of these columns is updated in the index set.

2.2.3 Iterative Hard Thresholding (IHT)

It belongs to the thresholding approach. The solution set s_p is updated based on a thresholding operation. In the IHT algorithm, a thresholding operator $\eta_K(.)$ is used to keep the K largest entries in the solution set and set the rest to zero. The algorithm can be explained as

$$s = \eta_K(s + \lambda\Theta^T(Y - \Theta s)) \tag{9}$$

In the above equation λ denotes the step size used in the algorithm.

3 Simulation

3.1 Dataset

The dataset is taken from the Plant Village dataset. About 3995 images have been taken for the study. These include 3 classes of leaf images having both diseases infected and healthy. The classes taken are Early Blight, Target Spot, and Healthy. Early Blight consists of 1000 images, Target Spot contains 1404, and Healthy contains 1591 images.

3.2 Tools Used

The work is simulated using MATLAB R2022 in Windows 10 Pro PC.

3.3 Acquisition and Reconstruction

All the images of the dataset, both disease infected and healthy, were compressively sensed for evaluation purposes. The reconstruction algorithm is carried out on these plant images. To obtain a better understanding of how the algorithm works, different values of m were used, i.e., the number of compressive measurements was varied. This could lead to different reconstruction quality, which in turn will provide an idea about the minimum number of samples required for optimal reconstruction. The algorithms are compared based on the time taken for reconstruction and their PSNR and PRD values.

3.4 Performance Metrics

$$\text{PSNR} = 10\log_{10}\left(\frac{\text{Maximum probable value}^2}{\text{MSE}}\right) \tag{10}$$

where MSE is the Mean Square Error.

$$\text{PRD} = \frac{\|X - \hat{X}\|_2}{\|X\|_2} \times 100 \tag{11}$$

PSNR and PRD quantifies the reconstruction quality of images.

4 Results and Discussion

The results obtained using our proposed work are discussed in this section. Initially, a suitable basis is selected in which the signals are sparse. For the proposed work, the DCT basis is selected as the sparsifying basis. Another crucial step in CS is to obtain a reconstruction matrix that is incoherent with the sparsifying basis. For evaluation, Gaussian Random matrix was selected. Once, the compressively sensed measurements are obtained, the next step is to pass the parameters on to the reconstruction algorithm. The inputs to the algorithm are the compressive measurements y, and the reconstruction matrix, Θ. The reconstruction algorithm is carried out for different

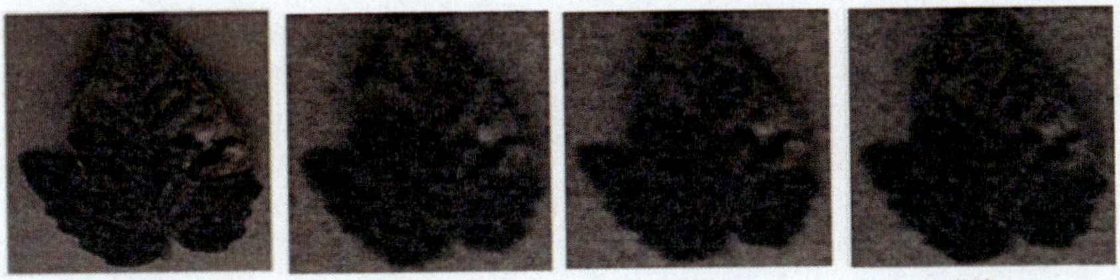

Fig. 2 **a** The original Image **b, c, d** Reconstructed Image with $m = 4500, 5000, 7000$ respectively using OMP

Table 1 Performance metrics obtained for reconstruction of plant image using OMP

No. of measurements (m)	PSNR (dB)	PRD (%)	Execution time (s)
50% (m = 5000)	25.9	0.69	233.4
70% (m = 7000)	27.5	0.64	365.1
45% (m = 4500)	25.3	0.70	213.1

Table 2 Performance metrics obtained for reconstruction of plant image using CoSaMP

No. of measurements (m)	PSNR (dB)	PRD (%)	Execution time (s)
50% (m = 5000)	25.8	0.6	201.3
70% (m = 7000)	26.0	0.62	248.6
45% (m = 4500)	25.4	0.70	155.7

Table 3 Performance metrics obtained for reconstruction of plant image using IHT

No. of measurements (m)	PSNR (dB)	PRD (%)	Execution time (s)
50% (m = 5000)	15.9	7.08	363.3
70% (m = 7000)	16.2	7.08	400.6
45% (m = 4500)	15.2	6.97	289.6

values of m, to obtain an idea as to how many measurements are required for acceptable reconstruction. The output of the algorithm is the estimate of the sparse signal from which the estimate of the original signal is obtained by taking the corresponding inverse transform. Since the visual information of the reconstructed images for both algorithms does not differ much, only the output of the OMP algorithm is shown in Fig. 2. The PSNR, PRD, and time taken for reconstruction for the algorithms are noted in Tables 1, 2 and 3.

The reconstructed output shows that as the number of measurements increases the reconstructed quality also increases. But the work shows that it is possible to reconstruct images using few samples without compromising the quality of the images.

The reconstructed images can be used further for feature extraction. The reconstructed output obtained using IHT has shown inferior quality as compared to the results obtained using OMP and CoSaMP, hence only those results will be used for future works. Gray Level Co-occurrence Matrix (GLCM) features can be used to extract important features from the images. Features such as contrast, correlation, energy, and homogeneity provide the texture information of the plant images. These features can be fed to an ML algorithm for automated disease classification. Hence, the field of disease diagnosis will extremely benefit by leveraging the ease of acquisition of CS and the efficiency of automated classification.

5 Conclusion and Future Scope

Compressive sensing is a novel sensing paradigm that allows accurate reconstruction of signals even at sampling rates much lower than the Nyquist rate. It takes advantage of the compressibility and sparsity of a signal. Traditional methods of acquiring sparse signals are rather time-consuming and also consume a lot of energy. Therefore, CS works well for sparse signals since it compresses the signals at the time of sensing, thus saving energy, time, and in turn cost of acquisition. CS is used in a wide range of image-processing applications. In the proposed work, the plant images are acquired using the CS technique. For reconstructing the original signal back from compressive measurements, reconstruction approaches, mainly algorithms from the greedy and thresholding approach have been used. The performance of the reconstruction algorithms is also noted.

The future scope of this work is to make the system more efficient by employing feature extraction, machine learning, and deep learning techniques for the classification of diseases presented in plants. In this work only leaves are taken for the study, in the future this can be tested in different parts of the plant. The reconstruction efficiency can also be further improved by using different CS approaches such as convex optimization, or any other greedy algorithms.

References

1. Li L, Zhang S, Wang B (2021) Plant disease detection and classification by deep learning-a review. IEEE Access 9:56683–56698. https://doi.org/10.1109/ACCESS.2021.3069646
2. Francis J. D, A SD, K, AB (2016) Identification of leaf diseases in pepper plants using soft computing techniques. In: Conference on emerging devices and smart systems (ICEDSS) pp 168–173. https://doi.org/10.1109/ICEDSS.2016.7587787

3. Singh V, Sharma N, Singh S, A review of imaging techniques for plant disease detection. Artific Intell Agricult. https://doi.org/10.1016/j.aiia.2020.10.002
4. Baraniuk RG (2007) Compressive sensing [lecture notes]. IEEE Signal Process Mag 24:118–121. https://doi.org/10.1109/MSP.2007.4286571
5. Donoho DL (2006) Compressed sensing. IEEE Trans Inf Theory 52:1289–1306
6. Stanković S *2015) Compressive sensing: theory, algorithms and applications. In: 4th Mediterranean conference on embedded computing (MECO) pp 4–6. https://doi.org/10.1109/MECO.2015.7181858
7. Zigel Y, Cohen A, Katz A (2000) The weighted diagnostic distortion (WDD) measure for ecg signal compression. IEEE Trans Biomed Eng 47:1422–1430
8. Candes E, Wakin M (2008) An introduction to compressive sampling. IEEE Signal Process Mag
9. Pujari JD, Yakkundimath R, Byadgi AS, Image processing based detection of fungal diseases in plants. Proc Comput Sci 47. https://doi.org/10.1016/j.procs.2015.02.137
10. Applalanaidu MV, Kumaravelan G (2021) A review of machine learning approaches in plant leaf disease detection and classification. In: Third international conference on intelligent communication technologies and virtual mobile networks (ICICV) pp 716–724. https://doi.org/10.1109/ICICV50876.2021.9388488
11. Singh K, Kumar S, Kaur P (2018) Support vector machine classifier based detection of fungal rust disease in pea plant. Int J Inf Technol
12. Kaur S, Pandey S, Goe S (2018) Semi-automatic leaf disease detection and classification system for soybean culture. IET Image Process 12:1038–1048
13. Ahmed M, Islam T, Ema R (2019) A new hybrid intelligent gaaco algorithm for automatic image segmentation and plant leaf or fruit diseases identification using tsvm classifier. In: 2nd IEEE international conference electronic computer, communication engineering, pp 1–6
14. Pinki F, Khatun N, Islam S (2018) Content based paddy leaf disease recognition and remedy prediction using support vector machine. In: 20th IEEE international conference computer science information technology, 1–5
15. MeenaPrakash R, Saraswathy G, Ramalakshmi G, Mangaleswari K, Kaviya T (2017) Detection of leaf diseases and classification using digital image processing. In: Proceedings of 2017 IEEE international conference innovation information embedded communication system, pp 1–4
16. Deshapande A, Giraddi S, Karibasappa K, Desai S (2019) Fungal disease detection in maize leaves using haar wavelet features. Information and communication technology for intelligent systems, smart innovation, systems and technologies, pp 275–286
17. Devi TG, Srinivasan A, Sudha S, Narasimhan D (2019) Web enabled paddy disease detection using compressed sensing. Math Biosci Eng 16:7719–7733. https://doi.org/10.3934/mbe.2019387
18. Shen B, Hu W, Zhang Y, Zhang YJ (2009) Image inpainting via sparse representation. In: International conference on acoustics, speech and signal processing, pp 697–200. https://doi.org/10.1109/ICASSP.2009.4959679

Comparison of Various Machine Learning and Deep Learning Classifiers for the Classification of Defective Photovoltaic Cells

Maithreyan G and Vinodh Venkatesh Gumaste

Abstract The common defects observed on the photovoltaic cells during the manufacturing process include chipping, tree crack, micro-line, soldering, and short circuits. Most of the defects mentioned above are not directly visible, making it harder for visual inspection. An appropriate method for defect classification would be by performing electroluminescence (EL) imaging, which helps reveal the defects, making it possible to visualize cracks and helps evaluate the quality of photovoltaic (PV) modules. The phenomenon where light emission occurs when current passes through PV cells is electroluminescence. The manual analysis of these electroluminescence images can be time-consuming and needs expert knowledge of various defects. This paper explains the automatic defective solar mono-cell classification task executed with different classifiers of machine learning and deep learning along with necessary image preprocessing techniques used to enhance the detection results. In comparison with the machine learning approach, deep learning offers better results on dataset of 1840 solar cell images. CNN models gave an average accuracy of 92%, and the highest accuracy of 97% was obtained with VGG16 transfer learning models after fine-tuning.

Keywords Solar cell classification · Mono-cell · Electroluminescence imaging · Machine learning · Image processing · Deep learning

G Maithreyan (✉)
Department of Production Engineering, National Institute of Technology Tiruchirappalli, Tamil Nadu, Tiruchirappalli 620015, India
e-mail: maithreyan2302@gmail.com

V. V. Gumaste
Central Manufacturing Technology Institute, Sc-C, C-SVTC, Bengaluru 560022, India

S. Sharma et al. (eds.), *Intelligent Control, Robotics, and Industrial Automation*, Lecture Notes in Electrical Engineering 1066,
https://doi.org/10.1007/978-981-99-4634-1_37

1 Introduction

Solar cells or photovoltaic cells utilize the photovoltaic effect to convert light energy into electrical energy directly. The extra energy of the excited electrons is the reason behind the potential difference, or electromotive force (e.m.f). A photovoltaic device's effectiveness depends upon the light-absorbing material used and the way of connecting it to an external circuit [1]. These solar cells are sustainable and renewable sources of energy. The carbon footprint of a solar panel is roughly 20 times lower than that of a coal-powered electrical source.

Photovoltaic modules are often shielded from environmental impacts like wind, and rain using glass lamination and an aluminum frame. External factors like heat, thermal stress, and physical collisions cause damage to these modules despite such protections. Apart from these factors, the consequences of manufacturing mistakes like poor soldering or short circuits can harm a solar cell's quality and effectiveness.

The visual inspection of defective units by direct eyes can be particularly challenging. Most of the defects can be easily viewed by the technique of electroluminescence imaging. EL imaging can be used to detect a number of different types of defects, including cracks, voids, delamination, and open circuits. EL imaging works similarly in principle to the light-emitting diodes; the current passed inside is used to produce light based on the radiative recombination of carriers. The disconnected section of the solar cell does not emit light, making some EL images darker. It is a type of non-destructive inspection methodology that provides better resolution. EL imaging is also expensive, so it may not be feasible for all companies to use this technique.

Overall, EL imaging is a promising technique for solar cell inspection. It is more sensitive than other inspection techniques. EL imaging can also be automated. This study is conducted to compare different models using EL imaging for defects detection with efficiency and also explains their benefits and drawbacks.

Figure 1 shows the common manufacturing defects in solar cells: chipping, tree cracks, micro-line, soldering, and short circuits. Mechanical deformations and damages usually cause chippings, partly or entirely broken with missing pieces. They are electrically degraded or separated parts. Tree cracks are cracks that are branched out with multiple cracks spreading across the cell. Micro-lines are cracks that are single cracks without branches. Short circuit defects in cells are disconnected areas due to the cell's interconnection degradation and appear with a completely black color because of no light emission. Soldering defects appear as patches of bright areas in a cell.

The solar panel is made up of 144 cells (monotype) arranged in 6 rows and 24 columns. Before the classification is performed, the panel is segmented into 144 cells. After segmentation, cells are used as input to the classification algorithms.

Defect detection during monitoring can be achieved using different goals. The different methods can be semantics, deep learning, computer vision, and machine learning.

Fig. 1. Classes for classification of solar cell

This paper discusses the approach of using a dataset of solar cell images for classifying defects through mathematical machine learning models and deep learning. Through this approach, overall time for detection is reduced compared with semantics, without efficiency loss, and checks the feasibility of different methods to determine defects for mono-cell. Here discussions are done using most widely and commonly machine learning models. Image processing methods were implemented to enrich the quality of image, and efficiency of detection. For better results and data cleaning, preprocessing methods are applied. Other methodologies include deep learning techniques with pretrained models like VGG16 and Inceptionv3, using transfer learning.

For better results, it's recommended to provide better lighting to the photovoltaic modules and use a camera with high spatial resolution since it increases the quality and resolution of the images.

1.1 Outline

The work will be organized as follows. Section 2 introduces the dataset and explains the image's features. The methodology implemented for performing the classification of defects will be discussed in Sect. 3. In Sect. 4, evaluations and the best results for various models are examined.

2 Images and Dataset

The electroluminescence images are extracted from photovoltaic cells of three color channels and resized to a size of 100 × 100 pixels unless otherwise stated. The image size of 224 × 224 pixels was utilized in the case of the VGG16 model in deep learning. The structure of the defects ranges from small cracks to completely dark areas depending on the type of defect present and each one affects the efficiency of the cell. Hence, the classification that is targeted is multi-class classification. Apart from multi-class classification, the binary classification of good and defect cells was performed for some models.

The dataset consists of sixmulti-class categories: good, chipping, tree crack, micro-line, soldering, and short circuits, totaling 1840 images. The bi-class categories were any two categories chosen out of six categories dataset mentioned above. The good and defects classification is done between good and all defects mentioned above. These defects negatively affect solar modules' efficiency, durability, and reliability.

Some challenges faced in detection were due to the variation in quality of image, brightness, noises and blurring. The main reasons behind these challenges are because these images are obtained by partitioning a complete solar module which is collected from a manufacturing environmental setup, where factors like the perspective of the camera, camera distortion, lighting conditions, and other factors play a vital role.

3 Methods

3.1 Preprocessing using Image processing Techniques

To overcome some of the challenges that were faced due to various factors that affected the quality of the mono-cell image, preprocessing is performed using the image processing technique. The distinct methods tested below are for noise reduction in images. OpenCV [2] is the software used for performing the above methods.

Bilateral blur filters were used to reduce the noises in places that required blurring. The below code snippet shows the values of sigmaColor and sigmaSpace that were used for blurring the images; where sigmaColor, the value of sigma in color space, is set to 75, and sigmaSpace, the value of sigma in coordinate space, is set to 75.

```
img_blur1 = cv2.bilateralFilter(img_gray, d = 7,sigmaSpace = 75,
sigmaColor =75)
```

Contouring. The grayscale images were read and inverted binary thresholding was applied. The electroluminescence images, which had some defects, showed different pixel values and some patterns in places of the defect. Considering the variations in the brightness of images, an adaptive method that captures contours based on the amount of brightness level was used. To determine the percentage defect, the contour area was calculated.

Initially, two contouring thresholds were provided based on the image's brightness level and the contours selection were made. Refer to Fig. 2 for contour selection of tree crack. The major challenge with this approach was due to noises in the image and varying quality of images; the contour selections were not accurate.

Edge detection using Sobel and Canny. Edges are defined by a difference in the values of the grayscale that characterizes an edge. When a pixel's gray level is similar to those around it, there isn't an edge there. A pixel might indicate an edge point if its neighbors have gray levels significantly different from its own.

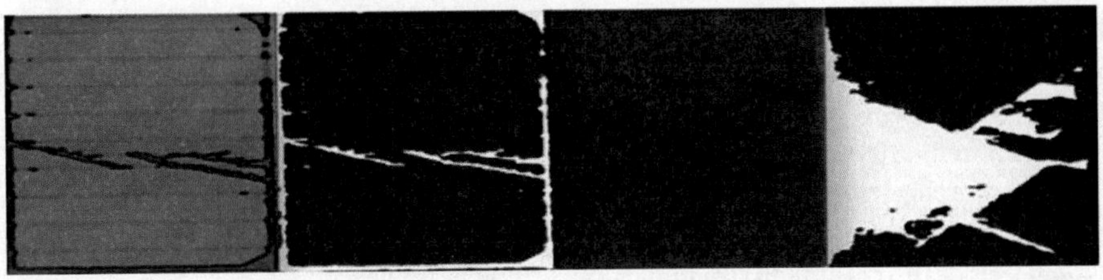

Fig. 2. Adaptive contouring based on the brightness of the images

The images were resized to 200 × 200 pixels, and kernel size was fixed to 7. Sobel edge detector uses two masks of 3 × 3 sizes. These masks calculate the gradient along the x and y-axis by moving over the images. The absolute gradient is obtained by combing both horizontal and vertical gradients (Gx and Gy). Edge magnitude is provided by:

$$|G| = \sqrt{G x^2} + \sqrt{G y^2} \quad (1)$$

As far as Canny [3] is concerned, initially, to reduce noise, images are smoothened using the Gaussian blur filter. Then, calculate partial derivatives' approximate finite-difference estimates of the gradient's magnitude and direction. Next, non-maxima suppression to the gradient magnitude is applied. To find and link edges, the double threshold technique is employed.

Compared with the Canny images obtained, Sobel made the cracks visible irrespective of the brightness or the variations in features of the images. Still, it attracted a lot of noise in the image affecting the classification process. Refer to Fig. 3 for edge detection of the tree crack image where the gradient of Sobel along x, Sobel along y, original image, Canny, and Sobel along x and y are shown, respectively.

Histogram Equalization. The cumulative distribution function (CDF) of the image is initially calculated as part of the mathematical process of histogram equalization. This function represents the probability that a given pixel will have an intensity value less than or equal to the intensity value of the current pixel. The next step is to use this function to remap the intensity values of the image so that the new image will have a flat histogram. This method improves the contrast in images, which is achieved by effectively spreading the most frequent intensity value. When close contrast values represent the data, this strategy typically boosts the overall contrast of the images [4].

Fig. 3. Edge detection for tree crack using Sobel and Canny

This enables regions with fewer local differences to acquire more distinction. This way, the image features were enhanced; it made the faint lines of crack sharper and darker. The images were converted to HSV, which was applied to adjust the image's value channel, affecting the image's brightness and contrast.

The challenge with this approach was mixing of pixels of chipping at corners of cells, making them appear like good category images and the likelihood of unwanted enhancement of features in the image causing new noises.

3.2 *Classification using Machine Learning Techniques*

This section discusses different machine learning algorithms used for supervised and unsupervised learning. The models that are discussed below are some of the most widely used algorithms and can be easily implemented by companies.

Two arrays were created to store the dataset and labels. The images are then flattened. Then they are mapped with their respective labels, split to train, and test data using the sklearn's train_test_split. In some cases, we run algorithm between a particular class and all other classes apart from it (referred to as rest). Due to relatively higher detection rate of short circuit defects compared with other classes, it is removed from classification dataset in some cases to prevent biasing of overall accuracy.

K-Nearest Neighbor (KNN) and PCA, LDA, NCA. This supervised learning algorithm worked on the concept of Euclidean distances. Here images are classified based on the nearest image in an array of samples resembling similarity to the sample image to be categorized. One of the drawbacks of using KNN is that it fails to classify accurately at higher dimensionality of the dataset, so the term coined 'curse of dimensionality' [5]. The size of the data space grows along with the number of dimensions and more data is required to maintain density. K-nearest neighbors lose all predictive potential in the absence of substantial growth in the size of the dataset. In the KNN model, scaling action was performed as a preprocessing task with Robust, Standard, and MinMax as scalers. The model was also tried out without scaling to analyze the variations in results. Figure 4 demonstrates the K-value versus error rate graph for good versus all defects. The best K-value is chosen by plotting a graph with an error rate over a range of K-values. The K-value concerning the corresponding minimum error value is preferred for better accuracy.

This model's maximum accuracy was around 56% for all classes in multi-class classification. For bi-class categorization like good versus micro-line categories, the accuracy was 77%, and for chipping versus micro-line, the accuracy was 78.84%.

Principal component analysis (PCA), neighborhood component analysis (NCA), and linear discriminant analysis (LDA) are methods of dimensionality reduction, which were used along with KNN to resolve the influence of higher dimensionality. The maximum accuracy for all classes was from PCA with around 52%. When data are subjected to PCA, a set of characteristics that account for most data variance is

```
Text(0, 0.5, 'Error Rate')
```

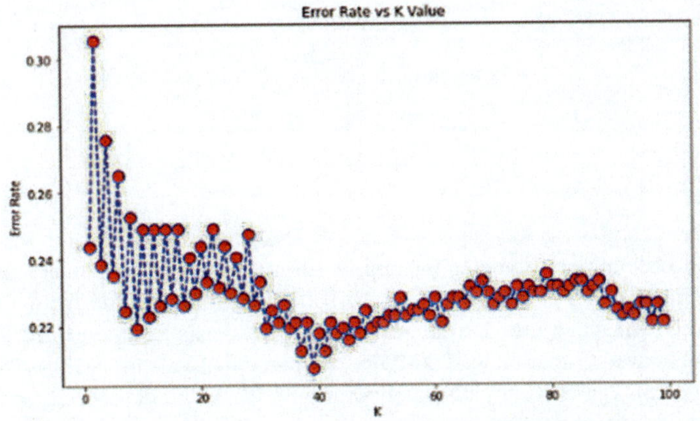

Fig. 4 K-value versus error rate graph for good versus defects, with best K-value as 39

identified. LDA identifies the qualities that significantly contribute to group diversity. LDA is a supervised method that uses known class labels, in contrast to PCA. NCA figures the most accurate feature space for the stochastic closest neighbor algorithm. Refer Fig. 5 for the clustersx formed using KNN for multi-class.

Support Vector Machine. This classifier was trained for bi-class classification. The SVM algorithm initially determines the support vectors before locating the hyperplane. These are the dataset's points that are most closely spaced from the hyperplane. The classifier then determines the appropriate hyperplane to divide the two

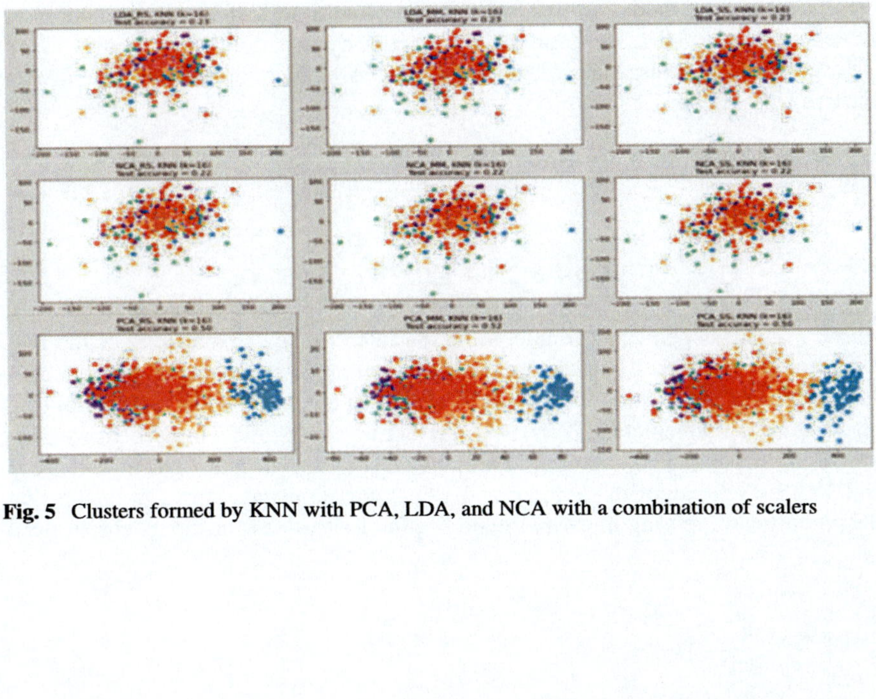

Fig. 5 Clusters formed by KNN with PCA, LDA, and NCA with a combination of scalers

Table 1 Accuracy score % of class 'A' with rest except short circuit without preprocessing

Class 'A'	Accuracy (%)
Good	92.22
Chipping	97.14
Cracks (tree crack + micro-line)	79.83
Soldering	91.68

classes after identifying the support vectors. The grid search method was used to choose the best hyperparameter by tuning it with the best C (penalty parameter of error term) from the range of [0.1,1,10,100] and gamma value from a range of [0.0001,0.001,0.1,1] and the kernel opted was radial basis function (RBF) [6]. The short circuit gave an accuracy of 99.21% compared with all other classes. Table 1 gives the percentage accuracy of a particular type versus all categories except short circuit without preprocessing with any filter.

However, comparing the false positive rate and true positive rate of 0.028 and 0.006, respectively, for chipping versus rest, the F1 score (harmonic mean of precision and recall) was 0.30 for chipping. Therefore, the classification was not proper. All other classes performed well in the confusion matrix and the accuracy score. The capacity to handle both linear and nonlinear data, the versatility in the kinds of kernels that can be utilized, and the SVM algorithm's comparatively low processing complexity are some of its benefits.

Decision Tree and Random Forest. Decision tree creates a model directly from the data. The root node of a decision tree represents the decision that needs to be made. The decision's potential outcomes are represented by splitting the root node into two or more child nodes. Random forest is an ensemble algorithm that combines the predictions of multiple decision trees. The decision tree gave an accuracy of 53% for all classes combined. Random forest contains multiple decision trees on different subsets of the given dataset. It improves predictive accuracy of that dataset by taking the average. Multi-class classification with all classes without short circuit gave an accuracy of 61.29%.

Linear and Logistic Regression. Linear regression is a mathematical process for predicting a dependent variable based on an independent variable. The basic equation is:

$$Y = b0 + b1X1 + b2X2 + \ldots + bnXn \tag{2}$$

where Y is the dependent variable, $b0$ is the intercept, $b1$ is the slope of the first independent variable, $X1$, $b2$ is the slope of the second independent variable, $X2$, and so on. The coefficients $b1$, $b2$, ..., bn are called the regression weights. Logistic regression is a mathematical process for predicting a categorical dependent variable based on one or more independent variables. The train and test splits were 70% and 30% for both algorithms. Finding the line where the data points fit the best on the plot allows applying linear regression to forecast values for inputs absent from the dataset present with the expectation that those outputs would lie on the line. The categorization of the short circuit from the rest by linear regression had a 97% accuracy rate.

Logistic regression estimates the parameters of a logistic model in regression analysis. The logistic regression for multinominal classification of all classes gave an accuracy of 52.58% for the 'blogs' solver and 52.15% for the 'saga' solver.

Naïve Bayes. This probabilistic classifier is based on probability models with high independence assumptions, yet it does not impact reality. Therefore, they are considered as naïve. This concept is inspired from the Bayes theorem for solving the probability of y given for X input features, as shown in below equation:

$$P(y|X) = (P(X|y) * P(y))/P(X) \tag{3}$$

The multi-class classification accuracy for all classes without short circuit was 42.77%.

K-Means and Transfer learning with K-Means. This unsupervised algorithm is used for clustering, which assigns k clusters to the observations. It is simple to use in classification since we may choose the number of groups, which can be equal to or greater than the number of classes. The elbow method, where the graph between the sum of squared and the number of clusters 'K' is plotted, is used to figure the best K-value.

The value of k at the elbow point [7] refer Fig. 6, i.e., the point after which the sum of squared starts decreasing linearly. The accuracy of the model in detecting for all classes except short circuit was 36.55%. This classifier's functionality is used to compare the cluster created by this classifier with the actual ones.

In transfer learning with the K-means method, initially, each image is preprocessed to match the input demanded by the transfer learning model. The particular transfer

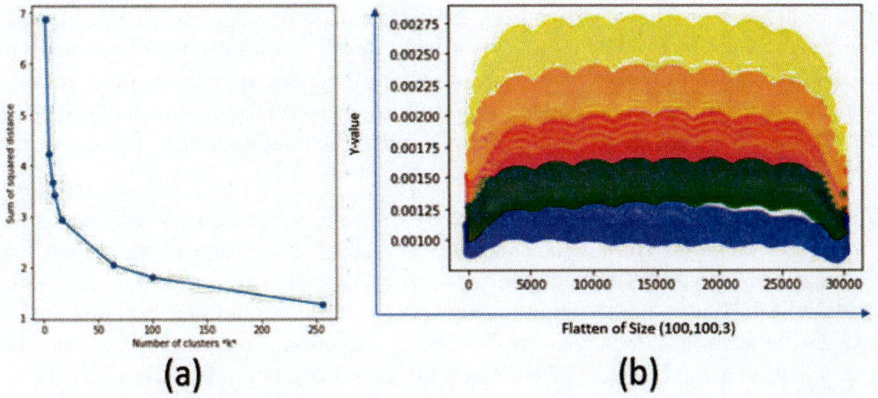

Fig. 6 **a** Elbow method for best K-value. **b** Clusters formed by K-means for all classes except short circuits (image of size (100,100,3))

learning model weights convert images to respective vectors. Later they are flattened and image weights are stored in a list. Then K-means clustering is over the list to predict the class. The images were fed at a resolution of 100 × 100 pixels and 'imagenet' weights were used. The accuracy score for VGG16 with KNN was 34.38% and with Resnet was 32.88% for all classes multi-classification without short circuit.

3.3 Classification using Deep Learning Techniques

In this paper, different CNN models like VGG16 and InceptionV3 were examined. The weights used for image classification for solar cell classification are from a pre-defined model of ImageNet, and the pooling method used is max pooling. The activation function used during the process is 'softmax' since it favors multi-class classification. As the name suggests, checkpoints in the model are used to check up on the model's parameters and help a lot during the training process by preventing the model from running multiple epochs. The setup for this technique demands complex computational hardware for faster runtime. Transfer learning [8] is performed with Inception V3, fine-tuning with VGG16. Data augmentation [9] was performed over images to improve the performance and reduce the overfitting of data and horizontal random flips and performing rotations achieve this. They are rescaled by dividing with 255 to normalize the pixel values. The deep learning is performed using Tensorflow and Keras [10] framework.

Transfer learning using Inception V3. The input images were resized to 150 × 150 pixels of three color channels. Using the flow_from_directory feature of TensorFlow, the train set and test set are linked to image location with a batch size of 20 and class mode 'categorical'. Apart from the input layers, all other layers are frozen and are made untrainable. The 'RMSprop' is used as an optimizer. The output of base model Inception V3 is flattened, followed by a dense layer to the model's output which uses 'relu' as the activation function. In addition, a dropout of 0.2 is added; this helps avoid overfitting. Then a final 'softmax' dense layer with five nodes is used to classify output—the five input categories for all classes apart from short circuit. The maximum validation accuracy for this model was 81% for the 42/50 epoch. The model was made entirely to run for 50 epochs, and the training accuracy was around 74.87% at the end of the 50 epochs.

Fine-tuning with VGG16. The input images were resized to 240 × 240 pixels of three color channels from all six classes. A total of 1450 images were used for training, 124 for validation, and 266 for testing. All layers are made trainable with a total of 14,717,766 trainable parameters [11]. A final dense layer is added to the VGG16 model's architecture using 'softmax' activation. Post employing the data augmentation using ImageDataGenerator, the model was compiled using 'Adam' as an optimizer and 'categorical_crossentropy' as a loss. This method was carried over for 15 epochs for a batch size of 32 with a learning rate of $5e^{-5}$. The dropout value was fixed as 0.1.The model gave an accuracy of 94.73% for 15 epochs. There was slight misclassification of chipping in corners of cell predicted as good and confusion between tree crack and micro-line due to almost the exact resemblance. Refer Figs. 7 and 8 for results.

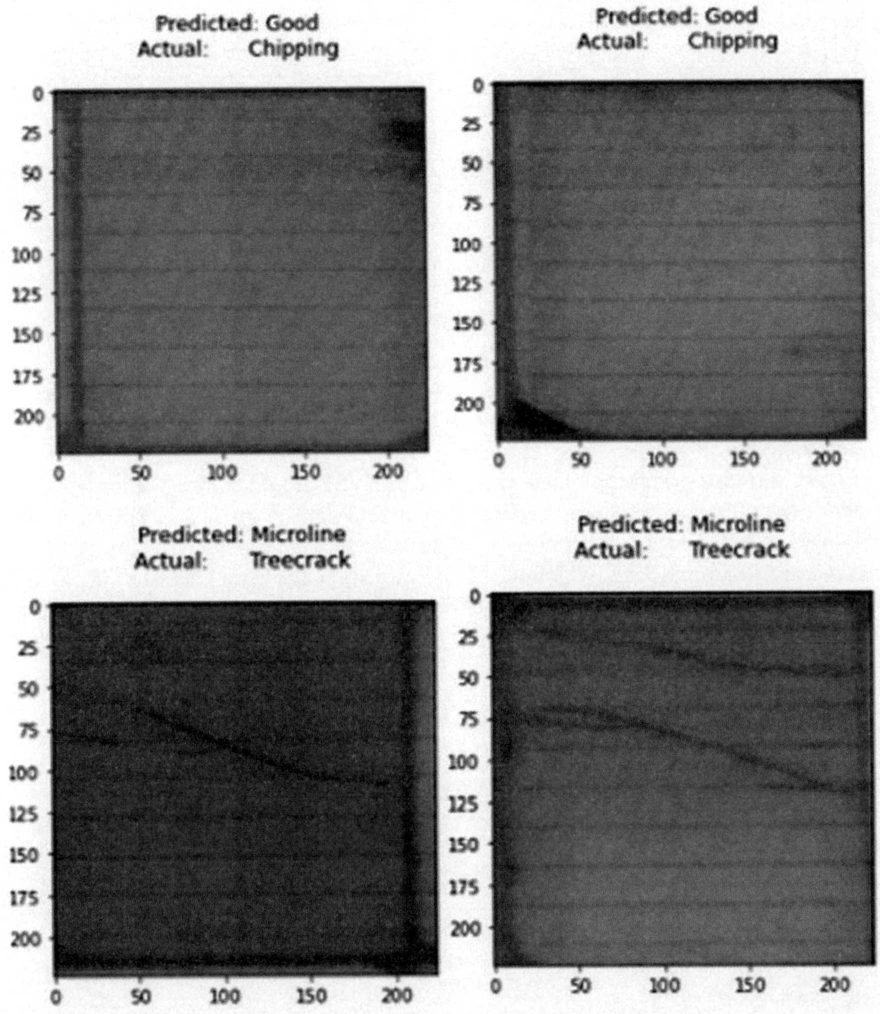

Fig. 7 Misclassification of chipping and tree crack defects (image of size (x-240, y-240))

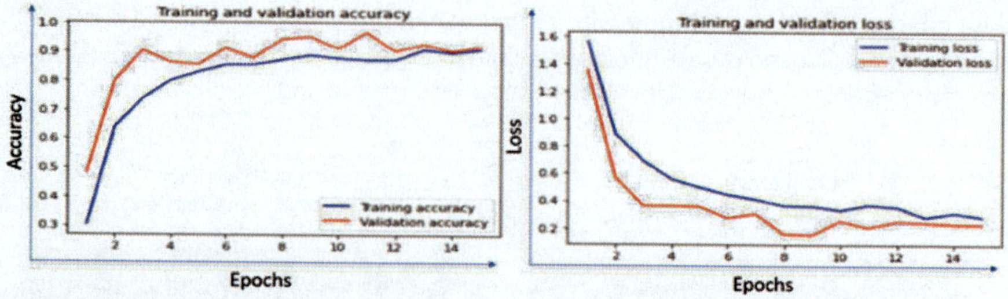

Fig. 8 Training versus validation accuracy (left) and loss graph (right)

4 Result and Discussion

4.1 Machine Learning and Image Processing Techniques

The most suitable model for bi-class classification is SVM as given in Table 2. Random forest performed better in machine learning for multi-class classification (refer Table 3).

Using the histogram equalized images with SVM, random forest improved the model's overall accuracy compared with the actual dataset by 2–3%. For example, good versus rest except short circuit for SVM using histogram equalized images gave an accuracy of around 95%, which was 2–3% more than the actual dataset's accuracy. The edge detection method using Sobel decreased the accuracy for all models due to noise creation.

Table 2 Bi-Class classification. Top five results for good versus defects

Model	Precision	Recall	F1 score	Support	Accuracy ($)	Class
SVM	0.86	0.76	0.81	80	92.36	Good
	0.94	0.97	0.95	300		Defects
Random forest	0.77	0.60	0.68	120	87.89	Good
	0.90	0.95	0.93	450		Defects
Decision tree	0.53	0.57	0.55	120	80.35	Good
	0.88	0.86	0.87	450		Defects
Logistic regression	0.42	0.54	0.48	94	79.25	Good
	0.90	0.85	0.87	446		Defects
Linear regression	0.65	0.49	0.56	159	77.22	Good
	0.81	0.89	0.85	381		Defects

Table 3 Multi-Class classification using random forest

Precision	Recall	F1 score	Support	Accuracy (%)	Class
0.64	0.84	0.73	120	62.45	Good
0.66	0.42	0.51	60		Chipping
0.61	0.65	0.63	120		Tree crack
0.72	0.67	0.69	120		Soldering
0.97	0.97	0.97	30		Short circuit
0.41	0.36	0.38	120		Micro-line

4.2 Deep Learning Techniques

The average accuracy of test data obtained after a run of 25 epochs was 96.24% as given in Table 4. The validation loss and validation accuracy % were around 0.10 and 95.97%, respectively. The accuracy % and loss of the model was 93.66% and 0.1963, respectively.

The highest accuracy obtained on the test was 97.74% (refer Fig. 9), and 96.77% on validation. There were some data imbalances initially between minority and majority of images in classes but were resolved using data augmentation.

Table 4 Multi-Class classification using VGG16

Precision	Recall	F1 score	Support	Accuracy ($)	Class
0.98	0.89	0.93	53	96.24	Chipping
0.97	0.98	0.97	59		Good
0.94	0.96	0.95	49		Micro-line
1.00	1.00	1.00	20		Short circuit
0.98	1.00	0.99	51		Soldering
0.92	0.97	0.94	34		Tree crack

```
[[48  0  0  0  0  0]
 [ 0 60  0  0  0  0]
 [ 0  0 49  0  0  1]
 [ 0  0  0 20  0  0]
 [ 0  0  0  0 52  0]
 [ 2  0  3  0  0 31]]
```

```
print(accuracy_score(y_pred,test_generator.classes))
```

```
0.9774436090225563
```

Fig. 9 Classification report for VGG16 at 25 epochs. The categories used in order: chipping, good, micro-line, short circuit, soldering, and tree crack

5 Conclusion

The challenge faced with SVM was high time consumption for computation due to high dimensional plotting of hyperplane. For multi-class classification the CNN model with VGG16 outperforms most machine learning models, despite higher computational hardware setup requirements. In future, these models can be integrated with machine vision hardware system of electroluminescence imaging for automation.

References

1. Nelson J (2003) The physics of solar cells. The physics of solar cells, by Jenny Nelson. London, Imperial College Press. 57. https://doi.org/10.1142/p276
2. Culjak I, Abram D, Pribanic T, Dzapo H, Cifrek M (2012) A brief introduction to OpenCV. In: 2012 Proceedings of the 35th international convention MIPRO, 1725–1730
3. Rashmi, Saxena R (2013) Algorithm and technique on various edge detection: a survey. Signal Image Process: Int J 4:65–75. https://doi.org/10.5121/sipij.2013.4306
4. Patel O, Maravi Y, Sharma S (2013) A comparative study of histogram equalization based image enhancement techniques for brightness preservation and contrast enhancement. Signal Image Process: Int J. 4. https://doi.org/10.5121/sipij.2013.4502
5. Kouiroukidis N, Evangelidis G (2011) The effects of dimensionality curse in high dimensional kNN search. In: Proceedings—2011 panhellenic conference on informatics, PCI 2011, pp 41–45. https://doi.org/10.1109/PCI.2011.45
6. Zhang Y (2012) Support vector machine classification algorithm and its application. In: Liu C, Wang L, Yang A (eds) Information computing and applications. ICICA 2012. Communications in computer and information science, vol 308. Springer, Berlin, Heidelberg
7. Marutho D, Hendra Handaka S, Wijaya E, Muljono (2018) The determination of cluster number at k-mean using elbow method and purity evaluation on headline news. In: 2018 international seminar on application for technology of information and communication, pp 533–538. https://doi.org/10.1109/ISEMANTIC.2018.8549751
8. Weiss K, Khoshgoftaar T, Wang DD (2016) A survey of transfer learning. J Big Data. 3. https://doi.org/10.1186/s40537-016-0043-6
9. Mikołajczyk A, Grochowski M (2018) Data augmentation for improving deep learning in image classification problem, 117–122. https://doi.org/10.1109/IIPHDW.2018.8388338
10. John Joseph FJ, Nonsiri S, Monsakul A (2021) Keras and TensorFlow: a hands-on experience. https://doi.org/10.1007/978-3-030-66519-7_4
11. Tammina S (2019) Transfer learning using VGG-16 with deep convolutional neural network for classifying images. Int J Sci Res Publ (IJSRP) 9:9420. https://doi.org/10.29322/IJSRP.9.10.2019.p9420

Four Fold Prolonged Residual Network (FFPRN) Based Super Resolution for Cherry Plant Leaf Disease Detection

P. V. Yeswanth, Rachit Khandelwal, and S. Deivalakshmi

Abstract The cherry cultivation has been threatened by crop loss due to the diseases from numerous sources such as insects, bacteria, viruses, and fungi. In order to diagnose the disease in its early stage, a novel Four Fold Prolonged Residual Network (FFPRN)-based super resolution for cherry plant leaf disease detection has been proposed in this paper. The proposed method extracts the deep features through four folds to create a high-resolution image from a given low-resolution cherry leaf image. This paper uses the plant village cherry leaf dataset to compare the performance of proposed model to Resnet50, Googlenet and Alexnet. For the super resolution factor 2 proposed model outperformed the existing Resnet50, Googlenet and Alexnet models. For super resolution factors 2,4,6 the proposed FFPRN model achieves PSNR values of 32.9305, 32.3306, 31.2962 and SSIM values of 0.8407, 0.8298, 0.8119 and classification accuracy values of 99.48, 99.08, 98.83, respectively.

Keywords Four fold prolonged residual network (FFPRN) · Prolonged residual network (PRN) · Prolonged residual block (PRB) · Cherry leaf disease detection · Super resolution

1 Introduction

Cherry is a fruit preferred by consumers with its beautiful appearance, low calorie content and positive effects on human health [1]. Due to the less yield and quality decline [2] caused by plant diseases from a variety of sources, including insects, bacteria, viruses, and fungi [3], it is necessary to employ an efficient method to diagnose the disease in early stage [4]. There are various methods for determining plant disease. In the past, farmers have examined leaves, nodes, or stems to detect the disease of plant. Manual analysis has the risk of resulting in incorrect disease

P. V. Yeswanth (✉) · R. Khandelwal · S. Deivalakshmi
Department of Electronics and Communication Engineering, National Institute of Technology, Tiruchirappalli, India
e-mail: pvy100@gmail.com

S. Sharma et al. (eds.), *Intelligent Control, Robotics, and Industrial Automation*, Lecture Notes in Electrical Engineering 1066,
https://doi.org/10.1007/978-981-99-4634-1_38

diagnosis [5] and it takes long time. Recent advances in artificial intelligence [6, 7] have led to the creation of a number of approaches that provide high accuracy and robust real-time operability.

The majority of these techniques involved classification of images, which is often divided into feature extraction and classification [8]. For disease detection in different plants, a variety of supervised machine learning techniques have been used [9]. These were less suitable for real-time use since they required a variety of user-defined attributes [10]. Due to its ability to overcome the aforementioned disadvantage, deep learning is particularly useful in this situation [11]. High resolution images were used as input in all of the aforementioned techniques. But the high cost of the required hardware makes it challenging to acquire high-resolution images. As such low-resolution images which are acquired using available resource are first converted to high-resolution images and then processed further for disease detection. In the proposed method, low resolution of input images are enhanced to generate higher resolution images, which are then used by our classification method to identify the associated disease.

2 Related Work

Crop disease diagnosis is a difficult but crucial task. Harris initially suggested employing image super-resolution [12] technique to enhance low resolution of input images for better results. Fourier transform (FT) was applied by Tsai and Huang [13] to create super-resolution image. Tom Katsaggelos [14] later used maximum likelihood to achieve this task. Then Yang et al., produced a super-resolution image using sparse representation.

2.1 Related Work on Super Resolution

Numerous techniques are proposed for using high-resolution images to better classify diseases for practical applications in agriculture. Dong et al. developed the initial super resolution method using CNN, known as super-resolution convolutional neural network (SRCNN).

To enhance efficiency, Haris et al. [15] used deep back-projection network (DBPN) method to create super-resolution image. Zhou et al. [16] used the internal layout of the high-resolution image's cross-scale features to lessen the degradation of reconstruction performance. This novel technique was called internal graph neural network (IGNN). The pixel level attention layer was used by Chen et al. to improve the performance of creating high-resolution images. Utilizing attention network, the attention layer's role is dynamically changed while actively managing attention at the pixel level. The name of this method was attention-in-attention network (AAN).

2.2 *Related Work on Plant Leaf Disease Detection*

Lately, a lot of progress has been made to use classification of leaf image datasets to diagnose disease automatically. Various methods of disease detection have been used for mango [18], tea leaf [23], olive [19], tomato [1], orange [17], rice [21], pomegranate [20], cassava [22], potato [26–28], grape [29] etc.

The accuracy and runtime of disease diagnosis were both greatly improved by Zhang et al. [24] on AI Challenger dataset. This novel technique was called fast regional convolutional neural network (FRCNN) algorithm. The attention-based approach for plant disease dataset was used by Karthik et al. [25] to detect tomato leaf disease with 98% accuracy. Later, DenseNet-121 was employed to obtain a 98.75% test accuracy.

High-resolution images were required as the input for all of the aforementioned techniques [30]. The performance of these models greatly deteriorated from the usage of lower resolution input images [31]. The proposed model uses a super-resolution block to fill this gap and produce high-resolution images for disease categorization.

3 Methodology

The four fold prolonged residual network (FFPRN) reconstructs deep features in four phases to produce a high-resolution image for the given low-resolution cherry leaf image. In this section, the proposed model FFPRN's architecture and training information is discussed.

3.1 *Four Fold Prolonged Residual Network (FFPRN) Architecture*

The FFPRN model's architecture is given in Fig. 1. I_{LR} and I_{SR} represent the low-resolution and super-resolution images, respectively. Here, the convolution block in the beginning retrieves shallow feature of the input I_{LR} image.

$$\Phi_{\text{shallow}} = K_{\text{initial}}(I_{\text{LR}}) \tag{1}$$

Φ_{shallow} stands for extracted shallow features and K_{initial} stands for the initial convolution block in the Eq. 1. Φ_{shallow} is given as input to the FFPRN to extract deep features as output.

$$\Phi_{\text{deep}} = K_{\text{FERRNI}}(\Phi_{\text{shallow}}) \tag{2}$$

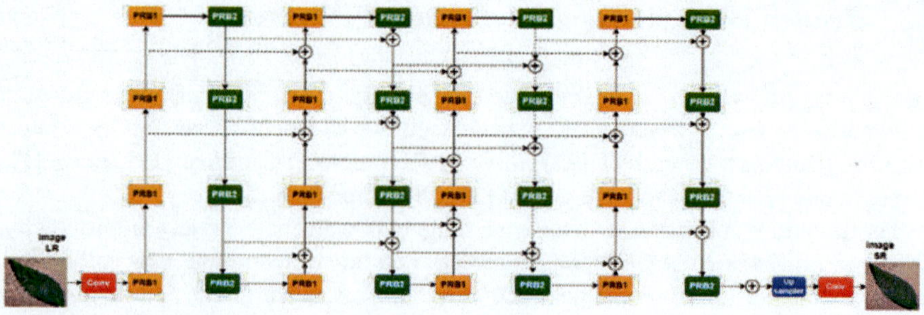

Fig. 1 Four fold prolonged residual network (FFPRN)

Φ_{deep} stands for the extracted deep features and b_{FFPRN} stands for the function of FFPRN in the Eq. 2. The extracted shallow and deep features described previously are first added ($\Phi_{\text{shallow}} + \Phi_{\text{deep}}$) and then introduced at the beginning of the upsample layer as shown in Fig. 1

$$\Phi_{\text{up}} = K_{\text{up}}\left(\Phi_{\text{shallow}} + \Phi_{\text{deep}}\right) \tag{3}$$

Φ_{up} stands for the extracted upsample features and Φ_{up} stands for the function of upsample layer in the Eq. 3. Φ_{up} is given as input to the convolution block which results in generation of the super-resolution image as output.

$$I_{\text{SR}} = K_{\text{conv}}\left(\Phi_{\text{up}}\right) = K_{\text{FFPRN}}(I_{\text{LR}}) \tag{4}$$

I_{SR} stands for the super-resolution image at the output of FFPRN, K_{conv} stands for the function of final convolution layer in the Eq. 4. K_{FFPRN} stands for the function of the entire FFPRN block, I_{LR} is given as input to FFPRN, which results in generation of I_{SR} as the output.

3.2 *FFPRN Prototype*

Figure 2 depicts model of the prolonged residual network (PRN), which consists of the prolonged residual block1 (PRB1) and prolonged residual block2 (PRB2).

The branch heading towards top extracts the deep features and utilizes the PRB1 block to upsample them, whereas the other heading towards bottom extracts the deep features and utilizes the PRB2 block to downsample them. Here, a low-resolution image (H*W*8) X is provided as input to the PRB1, which produces a resultant output of extracted features (H*W*16). As a result, every level of the channel count increases by 2, and at the top, we have an output feature size of (H*W*128). These features (H*W*128) are now sent into PRB2, which outputs features (H*W*64). We have an output feature Y (H*W*8) at the end.

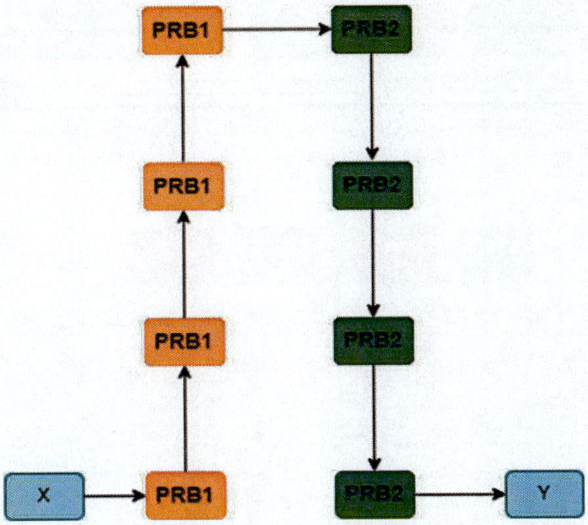

Fig. 2 Prolonged residual network (PRN)

Similar to ResNets, deep layers utilize shallow characteristics to improve performance. Figure 3 illustrates how PRB1 evaluates the difference between shallow and deep features in the bottom path of network. This difference is essentially the error, and it is further processed and upsampled to extract deep features. As depicted in Fig. 4, the only variation between PRB2 and PRB1 is the use of a convolution block rather than a deconvolution block. So, rather than upsampling the input, it downsamples it.

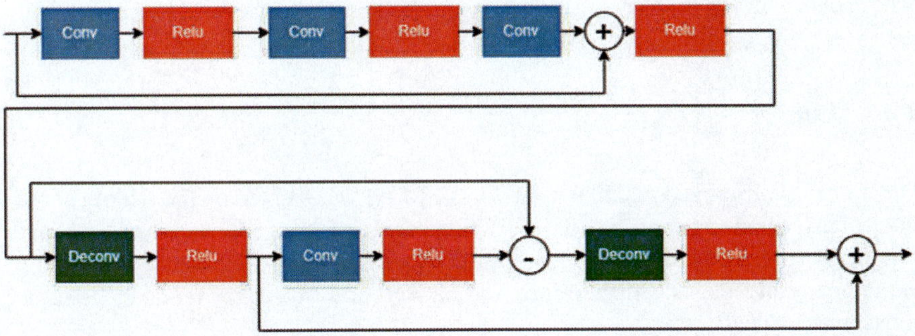

Fig. 3 Prolonged residual block 1 (PRB1)

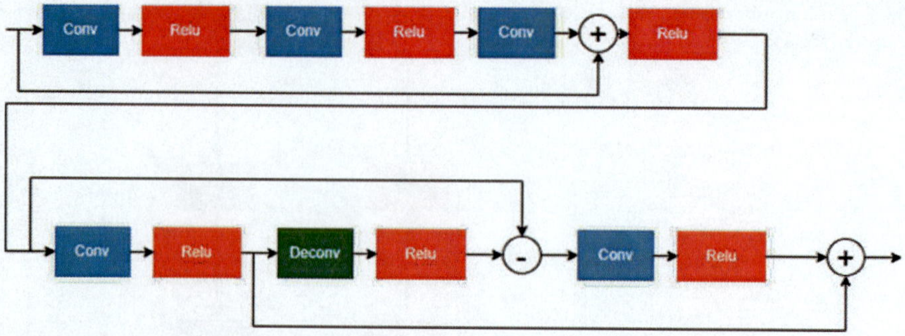

Fig. 4 Prolonged residual block 2 (PRB2)

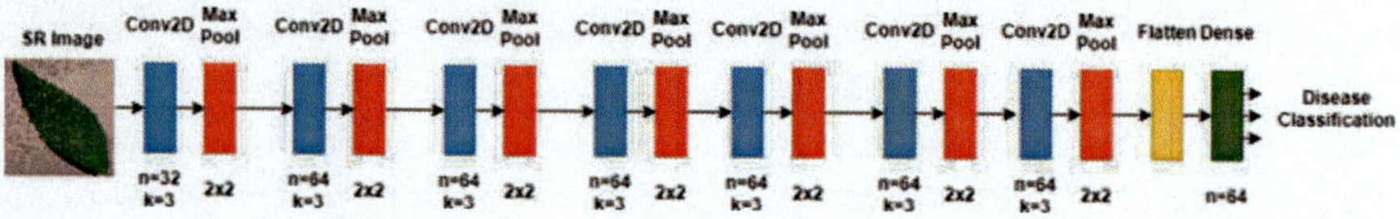

Fig. 5 Classification model

3.3 *Classification of Cherry Leaf Disease*

In the past, crop disease classification employed with high-resolution image datasets. Recently, low-resolution images are enhanced to high resolution which are then classified for disease detection. The model shown in Fig. 5 is utilized for cherry leaf disease classification.

3.4 *Dataset*

The four fold prolonged residual network (FFPRN) was trained on the widely accessible plant village cherry leaf dataset. Two kinds of images, described in Fig. 6 and Table 1, were included in this dataset. There is no performance degradation in the intensity of images. Furthermore, the background of images tends to be rather consistent.

Here, 10% of all images are randomly selected for testing, 10% are used for validation, and the remaining 80% are utilized for training.

Fig. 6 Cherry leaves images **a** Healthy leaf, **b** Powdery mildew diseased leaf

Table 1 Plant village cherry leaf dataset description

Type of leaf	No. of images
Healthy	854
Powdery disease	1052
Total images	1906

3.5 Experimentation

To conduct the experiment, all 1906 images are first downsampled using a number of different factors to produce low-resolution images, as seen in Fig. 7.

These low-resolution images are enhanced to high resolution using our FFPRN model which are then classified for disease detection. The classification is done using the classification model as shown in Fig. 5.

Here, mean square error (MSE) is employed as the loss function. During training, peak signal to noise ratio (PSNR) and structural similarity Index (SSIM) are used for evaluation by the FFPRN model.

Figure 8 depicts super-resolution model training and validation losses versus the number of epochs for downsampling factors 2, 4, 6.

Figure 9 shows the classification model training and validation accuracy versus the number of epochs for a variety of downsampling factors 2, 4, 6.

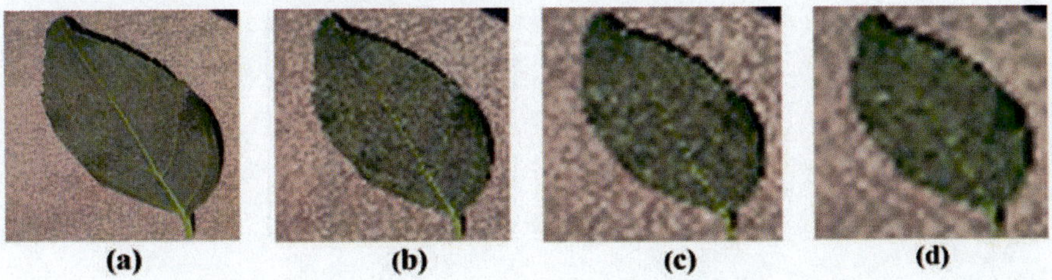

Fig. 7 **a** Original leaf image, **b** LR image (factor 2), **c** LR image (factor 4), **d** LR image (factor 6)

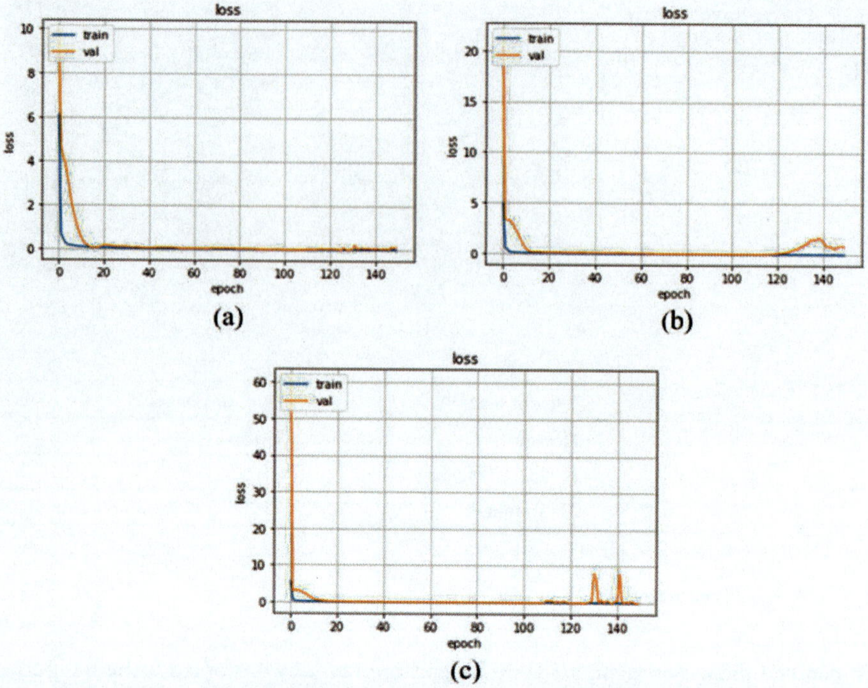

Fig. 8 FFPRN training and validation losses **a** Factor 2, **b** Factor 4, **c** Factor 6

4 Results

4.1 Performance Analysis

Tables 2 and 3 show the performance analysis of our model.

Super-resolution technique of various cherry leaf images are shown in Fig. 10. These tables show that our model has an accuracy of more than 98% in predicting diseases. Additionally the two metrics, PSNR is above 30 and SSIM above 0.8.

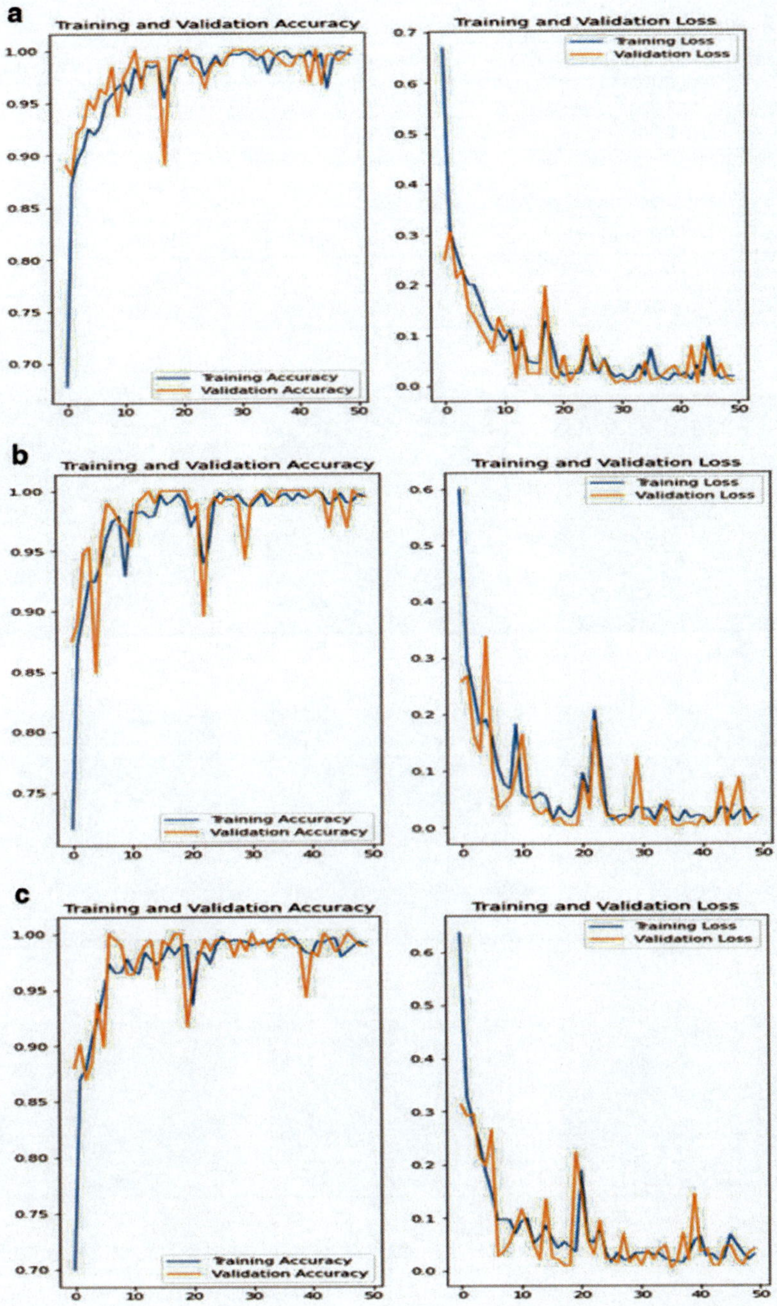

Fig. 9 Training and validation accuracies and losses versus no. of epochs. **a** Training and validation accuracies and losses versus no. of epochs (factor 2). **b** Training and validation accuracies and losses versus no. of epochs (factor 4). **c** Training and validation accuracies and losses versus no. of epochs (factor 6)

Table 2 Super-resolution model FFPRN quantitative performance analysis

Metrics	SR image (factor 2)	SR image (factor 4)	SR image (factor 6)
PSNR	32.9305	32.3306	31.2962
SSIM	0.8407	0.8298	0.8119

Table 3 Classification model performance analysis

Metrics	SR image (factor 2)	SR image (factor 4)	SR image (factor 6)
Accuracy	99.48	99.08	98.83
Loss	0.0069	0.0103	0.0405

Fig. 10 Qualitative performance analysis of FFPRN. **a** Super resolution with factor 2. **b** Super resolution with factor 4. **c** Super resolution with factor 6

Table 4 Performance comparison with pretrained model (for super resolution factor 2)

Metrics	ResNet50	AlexNet	GoogleNet	FFRN (proposed)
Accuracy	96.71	95.86	97.24	99.48

5 Conclusion

This research proposes a unique method for cherry leaf disease diagnosis using the FFPRN super-resolution model. Here model is compared with the traditional classification methods—Resnet50, Googlenet, and Alexnet as given in Table 4. For super resolution factors 2, 4, 6, the proposed FFPRN model achieves PSNR values of 32.9305, 32.3306, 31.2962, SSIM values of 0.8407, 0.8298, 0.8119, and classification accuracy values of 99.48, 99.08, 98.83, respectively. This method can be easily extended to other leaf disease detection using super-resolution image classification.

References

1. Nazki H, Yoon S, Fuentes A, Park DS (2020) Unsupervised image translation using adversarial networks for improved plant disease recognition. Comput Electron Agric 168:105117. https://doi.org/10.1016/J.COMPAG.2019.105117
2. Lacey LA, Grzywacz D, Shapiro-Ilan DI, Frutos R, Brownbridge M, Goettel MS (2015) Insect pathogens as biological control agents: back to the future. J Invertebr Pathol 132:1–41. https://doi.org/10.1016/J.JIP.2015.07.009
3. Gajjar R, Gajjar N, Thakor VJ, Patel NP, Ruparelia S (2022) Real-time detection and identification of plant leaf diseases using convolutional neural networks on an embedded platform. Visual Comput 38(8):2923–2938. https://doi.org/10.1007/S00371-021-02164-9/FIGURES/12
4. Zhang K, Zhang L, Wu Q, Identification of cherry leaf disease infected by Podosphaera Pannosa via convolutional neural network 10(2):98–110. https://services.igi-globl.com/resolvedoi/resolve.aspx? 1AD. https://doi.org/10.4018/IJAEIS.2019040105
5. Bock CH, Poole GH, Parker PE, Gottwald TR (2010) Plant disease severity estimated visually, by digital photography and image analysis, and by hyperspectral imaging 29(2):59–107 https://doi.org/10.1080/07352681003617285
6. Abiodun OI, Jantan A, Omolara AE, Dada KV, Mohamed NAE, Arshad H (2018) State-of-the-art in artificial neural network applications: a survey. Heliyon 4(11):e00938. https://doi.org/10.1016/J.HELIYON.2018.E00938
7. Patrício DI, Rieder R (2018) Computer vision and artificial intelligence in precision agriculture for grain crops: a systematic review. Comput Electron Agric 153:69–81. https://doi.org/10.1016/J.COMPAG.2018.08.001
8. Zhang T, Zheng W, Cui Z, Zong Y, Yan J, Yan K (2016) A deep neural network-driven feature learning method for multi-view facial expression recognition. IEEE Trans Multimedia 18(12):2528–2536. https://doi.org/10.1109/TMM.2016.2598092
9. Rouse MN, Nava IC, Chao S, Anderson JA, Jin Y (2012) Identification of markers linked to the race Ug99 effective stem rust resistance gene Sr28 in wheat (Triticum aestivum L.). Theor Appl Genet 125(5):877–885. https://doi.org/10.1007/S00122-012-1879-6/FIGURES/3
10. Dananjayan S, Tang Y, Zhuang J, Hou C, Luo S (2022) Assessment of state-of-the-art deep learning based citrus disease detection techniques using annotated optical leaf images. Comput Electron Agric 193:106658. https://doi.org/10.1016/J.COMPAG.2021.106658

11. Asif MKR, Rahman MA, Hena MH (2020) CNN based disease detection approach on potato leaves. In: Proceedings of the 3rd international conference on intelligent sustainable systems, ICISS 2020, 428–432. https://doi.org/10.1109/ICISS49785.2020.9316021
12. Harris JL (1964) Resolving power and decision theory*†. JOSA 54(5):606–611. https://doi.org/10.1364/JOSA.54.000606
13. Huang TS (n.d.) Advances in computer vision and image processing | Guide books. Retrieved November 21, 2022, from https://dl.acm.org/doi/abs/https://doi.org/10.5555/61892
14. Tom BC, Katsaggelos AK (1996) Reconstruction of a high-resolution image by simultaneous registration, restoration, and interpolation of low-resolution images. IEEE Int Conf Image Process 2:539–542. https://doi.org/10.1109/ICIP.1995.537535
15. Haris M, Shakhnarovich G, Ukita N (2018) Deep back-projection networks for super-resolution, pp 1664–1673
16. Zhou S, Zhang J, Zuo W, Loy CC (2020) Cross-scale internal graph neural network for image super-resolution. Adv Neural Inf Process Syst 33:3499–3509. https://github.com/sczhou/IGNN
17. Capizzi G, Sciuto Gl, Napoli C, Tramontana E, Woniak M (2016) A novel neural networks-based texture image processing algorithm for orange defects classification. Int J Comput Sci Appl 13(2):45–60. https://www.researchgate.net/publication/309769648
18. Singh UP, Chouhan SS, Jain S, Jain S (2019) Multilayer convolution neural network for the classification of mango leaves infected by anthracnose disease. IEEE Access 7:43721–43729. https://doi.org/10.1109/ACCESS.2019.2907383
19. Cruz AC, Luvisi A, de Bellis L, Ampatzidis Y (2017) X-FIDO: An effective application for detecting olive quick decline syndrome with deep learning and data fusion. Frontiers Plant Sci 8:1741. https://doi.org/10.3389/FPLS.2017.01741/BIBTEX
20. Pawar R, Jadhav A (2018) Pomogranite disease detection and classification. In: IEEE international conference on power, control, signals and instrumentation engineering, ICPCSI 2017, 2475–2479.https://doi.org/10.1109/ICPCSI.2017.8392162
21. Jiang F, Lu Y, Chen Y, Cai D, Li G (2020) Image recognition of four rice leaf diseases based on deep learning and support vector machine. Comput Electron Agric 179:105824. https://doi.org/10.1016/J.COMPAG.2020.105824
22. Ramcharan A, McCloskey P, Baranowski K, Mbilinyi N, Mrisho L, Ndalahwa M, Legg J, Hughes DP (2019) A mobile-based deep learning model for cassava disease diagnosis. Frontiers Plant Sci 10:272. https://doi.org/10.3389/FPLS.2019.00272
23. Hu G, Yang X, Zhang Y, Wan M (2019) Identification of tea leaf diseases by using an improved deep convolutional neural network. Sustain Comput: Inf Syst 24:100353. https://doi.org/10.1016/J.SUSCOM.2019.100353
24. Lu H, Yang R, Deng Z, Zhang Y, Gao G, Lan R (2021) Chinese image captioning via fuzzy attention-based DenseNet-BiLSTM. ACM Trans Multimedia Comput Commun Appl (TOMM) 17(1s). https://doi.org/10.1145/3422668
25. Karthik R, Hariharan M, Anand S, Mathikshara P, Johnson A, Menaka R (2020) Attention embedded residual CNN for disease detection in tomato leaves. Appl Soft Comput 86:105933. https://doi.org/10.1016/J.ASOC.2019.105933
26. Yeswanth PV, Khandelwal R, Deivalakshmi S (2023) Super resolution-based leaf disease detection in potato plant using broad deep residual network (BDRN). SN Comput Sci 4(2). https://doi.org/10.1007/s42979-022-01514-1
27. Yeswanth PV, Khandelwal R, Deivalakshmi S, Muttukrishnan RM, Rajarajan M, Veeravalli B, Kesswani N, Patel A (2023) Internet of things (IoT): key digital trends shaping the future. In: Proceedings of 7th International Conference on Internet of Things and Connected Technologies (ICIoTCT 2022) Two Fold Extended Residual Network Based Super Resolution for Potato Plant Leaf Disease Detection Springer Nature Singapore Singapore pp 197–209
28. Yeswanth PV, Kushal S, Tyagi G, Kumar MT, Deivalakshmi S, Ramasubramanian SP (2023) Iterative super resolution network (ISNR) for potato leaf disease detection. In: 2023 International Conference on Signal Processing, Computation, Electronics, Power and Telecommunication (IConSCEPT), Karaikal, India. pp 1–6. https://doi.org/10.1109/IConSCEPT57958.2023.10170224

29. Yeswanth PV, Deivalakshmi S, George S, Ko SB (2023) Residual skip network-based super-resolution for leaf disease detection of grape plant. Circ Syst Sig Process. https://doi.org/10.1007/s00034-023-02430-2
30. Yeswanth PV, Deivalakshmi S (2023) Extended wavelet sparse convolutional neural network (EWSCNN) for super resolution Sādhanā 48(2). https://doi.org/10.1007/s12046-023-02108-0
31. Yeswanth PV, Raviteja R, Deivalakshmi S (2023) Sovereign critique network (SCN) based super-resolution for chest X-rays images. In: 2023 International Conference on Signal Processing, Computation, Electronics, Power and Telecommunication (IConSCEPT), Karaikal, India, pp 1–5. https://doi.org/10.1109/IConSCEPT57958.2023.10170157

Real-Time Lane Recognition in Dynamic Environment for Intelligent Ground Vehicles

Shambhavi Sinha, Piyush Modi, and Ankit Jha

Abstract Robotic navigation system is a well-studied domain, and lane detection is a crucial module for autonomous vehicles. Recently, Unmanned Ground Vehicles (UGVs) have been deployed for several extensive purposes, requiring systems to be better than the traditional heuristic approach to suit specific scenarios. Although lane detecting technology has been developed for decades, many critical challenges in autonomous ground vehicles remain unresolved. This paper addresses the less pervasive but critical challenges posed in a dynamic and complex environment. Thus we chose the Intelligent Ground Vehicle Competition (IGVC) dataset used by an autonomous ground vehicle to navigate through a grass surface with white lanes and obstacles. In the research, the authors developed two core image processing algorithms Open Source Computer Vision (OpenCV), followed by two deep learning models to answer the shortcomings of the OpenCV-based solutions. The deep learning models achieve an accuracy of 98.76% (FusionNet) and 98.68% (Modified UNet) on our test dataset. They are a robust solution to erratic challenges like lighting conditions, a prevalent concern on the grassy surface.

Keywords Lane recognition · Semantic segmentation · Autonomous vehicle · CNN · Fully convoluted network · IGVC

S. Sinha (✉) · P. Modi · A. Jha
Department of Electronics and Communication Engineering, Manipal Institute of Technology, Manipal Academy of Higher Education, Manipal 576104, Karnataka, India
e-mail: sinhashambhavi1204@gmail.com

P. Modi
e-mail: modipiyush2001@gmail.com

A. Jha
e-mail: ankit.jha3@learner.manipal.edu

S. Sharma et al. (eds.), *Intelligent Control, Robotics, and Industrial Automation*, Lecture Notes in Electrical Engineering 1066,
https://doi.org/10.1007/978-981-99-4634-1_39

1 Introduction

Intelligent Ground Vehicles (IGVs) rely heavily on lane detecting technologies. They use data from Global Positioning System (GPS), ultrasonic, and other sensors to function. Autonomous vehicles have several applications, ranging from military to research to commercial. They can be used to collect data in inaccessible regions and deliver to clients. They have been widely used in the contactless delivery of medical supplies and meals with the rise of COVID-19. The Intelligent Ground Vehicle Competition (IGVC) was established to enhance engineering education in intelligent vehicles and associated technologies.This competition challenges students to develop and build an autonomous system for worldwide competition using sophisticated control theory, machine learning techniques, vehicular electronics, and mobile platform principles.

Various researchers have discovered different approaches to identify lane lines over the years. The most prevalent produces a binary picture of the lane lines and then applies the Hough transform to suit the lane [9]. Some of the earlier lane detection algorithms are based on color features and curve fitting. In general, a specific color space is chosen, followed by converting the image to the specified color space and quantizing it into a color histogram. Various color spaces, including Red Green Blue (RGB) and Hue Saturation Values (HSV), are employed to predict appropriate lane lines. However, the difficulty with using these approaches is their inability to produce excellent results in edge instances and to make judgments for new test cases.

In the past, deep learning has progressed in numerous domains by using multilayer nonlinear transformations. Several deep learning methods have been used to overcome the lane detecting problem. Start with early Convolutional Neural Network (CNN)-based techniques, end-to-end segmentation-based methods like Global Convolution Network [12], General Adversarial Network (GAN)-based approaches like EL-GAN [5], and so on.

Our primary motivation for writing this research is to propose lane recognition algorithms, ranging from straightforward OpenCV-based techniques to more intricate deep learning-based ones that can generalize better in any environment. To understand lane detection systems better, we initially implemented two OpenCV-based models. To train more data, detect curved lanes, and increase accuracy, we utilized a variety of deep learning-based models for the lane detecting method. Flaws in the model were discovered, and to address them, we implemented FusionNet (which combines Unet and FCN8) and a Modified U-Net.

2 Related Works

The author in [8] proposed a reliable lane detection technique using canny edge detection and Hough transform. Canny edge uses Gaussian blurring and extracts edge information using an intensity gradient cut off. The edge acts as input to the

Hough Line transform algorithm which plots the edge points into a polar coordinate system. Cho et al. [4] used the Hough transform with enhanced accumulator cells in a Multi-Region of Interest setting, simultaneously and efficiently recognizing lanes, hence the weaker lane recognition on the curved roads was fairly improved.

In [10] the author suggested a lane detection model based on boundary determination algorithm. The model uses a Sobel Edge detector, which determines the amplitude and direction of the local gradients. These gradients are sampled by two masks, a tracing mask and a probing mask and then a polynomial order line is fit to trace the curve as perceived by the human eye. Wang et al. [11] suggests an inverse perspective mapping method to transform binary image and then use K-means as a method to reduce interference effect and map a line on the top view of the image.

Author in [6], suggests a UNet using depth-wise separable convolutions (DSUNet) for end-to-end learning of lane detection and path prediction in autonomous driving. They also create and integrate a path prediction algorithm with CNN.

In paper [13] Lane detection is accomplished utilizing numerous continuous frames from an environment and proposes a deep hybrid architecture that combines Convolutional Neural Networks (CNN) and Recurrent Neural Networks (RNN). Liu and Yan [7] discusses a lane line segmentation technique that combines the instance segmentation network Deeplabv3 and the upgraded network Retinexnet. First, the contrast and clarity of the original images were increased using the Retinexnet network. Then, using the segmentation network Deeplabv3, lane lines were detected and segmented.

3 Methodology

A step-by-step process is employed to attain the desired objectives of detecting lanes on uneven and grass surfaces. In the following part, we will describe OpenCV-based models, followed by deep learning-based models.

3.1 OpenCV-Based Approach

OpenCV models use a series of image processing techniques applied to a single video frame, one after another, in a pipeline structure manner.

3.1.1 Dataset

This research uses two video datasets to test our OpenCV-based models. The first video is a navigation training video dataset generated by the members of RUGVED

Systems, a student robotics team, using a hand-held camera at Manipal Institute of Technology (MIT) Manipal to test the IGVC ground vehicle [3]. The second video data was collected from [2].

3.1.2 OpenCV Model-1

Lane detection on a grass lane is difficult due to the different noise sources such as grass, cones, barrels, background, and lighting conditions. We analyze different instances in this research work. The lanes to be detected are discontinuous at specific parts, straight or curved, day or night, and under any weather conditions, favorable or unfavorable.

As mentioned in Fig. 1, after reading the video, the Region of Interest (ROI) is selected, and the image is extracted within the ROI, then the image was convolved with a hybrid Gaussian blur filter using a size kernel (5,5). We used closing morphological transformation using a kernel of ones of size (5,5), used for creating smooth surface by blending narrow beaks with long thin gulfs. Morphological closing is characterized by dilation followed by erosion. The image is then converted into HSV color space. R, G, and B in RGB correlate to color luminance, hence cannot isolate color information from luminance. HSV makes it easy to work on the brightness of the image. A binary mask is obtained using lower and upper HSV bounds. The image is then subjected to canny edge detection, which extracts meaningful structural information and significantly minimizes the amount of data to be processed. Hough transformation is then applied to the image. Every edge point in the edge map is transformed by the Hough transform into all potential lines that could pass through that point. Due to the uneven nature of lane lines and to detect better lanes, we used the slope and line characteristics to interpolate the lines where the markers are not as obvious or irregular. We selected lines whose slope fell between the (-0.5,0.283) range based on the experiments, discarding all other lines, then selected lines were clustered that matched the lane mark.

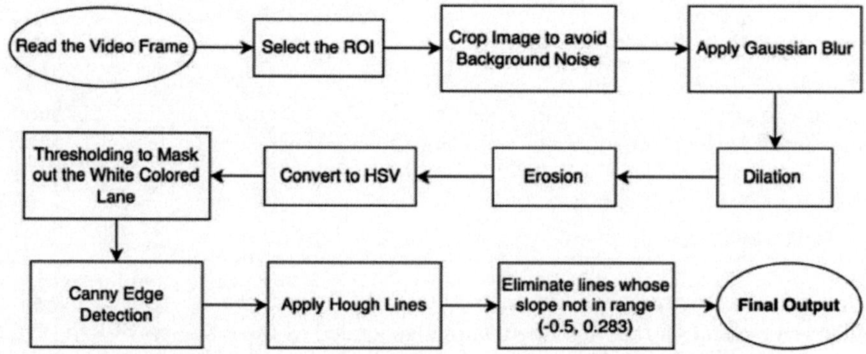

Fig. 1 Flowchart of OpenCV Model 1, based on edge detection and Hough transform

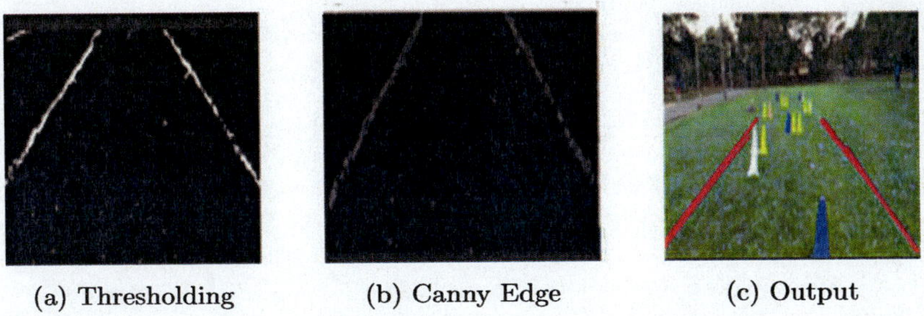

Fig. 2 Intermediate stages of OpenCV Model 1

However, the model's performance suffered during steep bends and resulted in broken lines which can be seen in Fig. 2. Taking these limitations into consideration, we created another model, OpenCV Model-2.

3.1.3 OpenCV Model-2

Our goal is to recognize white lanes on green grass. To address this issue, we examined the absorption spectra of a typical photosynthetic plant, which exhibits expected blue peaks. As a result, blue channels are used for lane detecting. The blue channel is captured and denoised with Gaussian blur.

Figure 3 shows that the model converts the feed into a blue color channel since it helps us identify green grass with a better probability. Followed by which the feed is blurred to reduce the noise in the frame. In the next step, we distinguish the white lanes from the surroundings using thresholding. Opening morphological modification is conducted, including erosion and dilation, thus reducing the noise at the boundaries and increasing the width of the object area. Canny edge detection is used to find the frame's edges. It uses vertical and horizontal gradient value pixels to detect edges. The Region of Interest is then chosen to limit undesired detection in the next phase. Finally, probabilistic Hough lines are used to detect straight lines. These lines are mixed using a weighing system to achieve the best possible contrast between the lanes, grass, and other objects in the surrounding (Fig. 4).

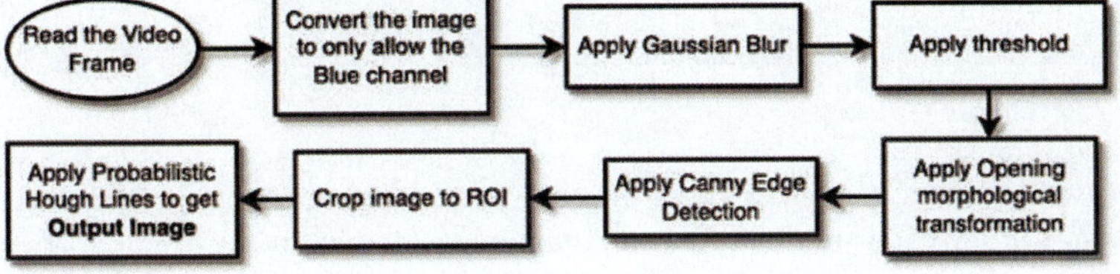

Fig. 3 Flowchart of OpenCV Model 2, based on color channel extraction

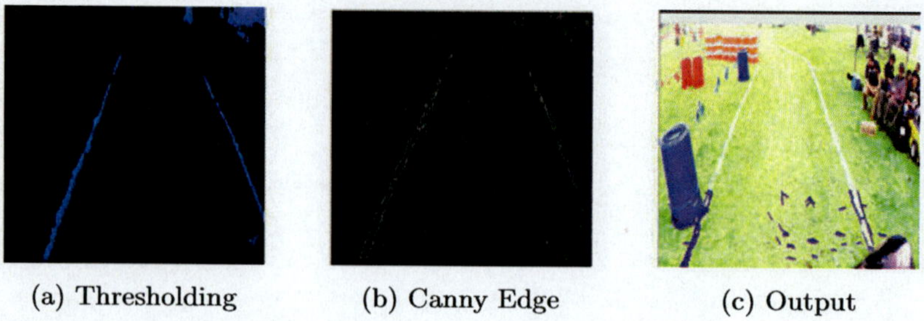

Fig. 4 Intermediate stages of OpenCV Model 2

3.2 Deep Learning Approach

During the last decade, deep convolutional neural networks have surpassed state-of-the-art approaches in numerous visual identification, classification, and segmentation applications. Convolution networks are often employed in classification tasks, where the output is a single class label; however, many visual studies, including segmentation, need localization, i.e., each pixel is allocated a class label.

Segmentation has several applications, especially in autonomous cars, biomedical image diagnosis, geo sensing, precision agriculture, business problems, and comprehension of diverse modalities such as seismic imaging, whole slide images (WSI), fundus images, and many more.

The purpose is to perform semantic image segmentation, which involves labeling each pixel of an image with the class (lane or background) to which it belongs.

UNet is a semantic segmentation encoder-decoder convolutional neural network. Its exquisite design of symmetrically growing (decoding) and contracting (encoding) sub-networks with skip connections that combine semantic (global) and appearance (local) information from both sub-networks resulted in exceptional performance in a number of contests. High-resolution characteristics from the contracting channel are concatenated with the upsampled output to localize. The primary idea is to add successive layers to contracting network, where upsampling is used instead of pooling. Based on this information, subsequent convolution layers can learn to provide a more accurate output. In this study, we describe two models inspired by the Unet architecture, Modified U-Net with cross-connection and FusionNet, with 98.79% and 98.68% accuracy, respectively.

3.2.1 Dataset

The dataset contains images of lanes on grassy land taken from a camera and annotated with the labelme module. The images in the dataset were obtained from the IGVC Course track recording from 2015 and compiled by the robotics student team (RUGVED) at Manipal Institute of Technology. The dataset can be accessed here

Table 1 Details of the dataset

Dataset	Number of images
Training	584
Validation	32
Testing	34

[1]. The lanes are faded white, with lane widths that vary from 10 to 20 ft and turn radii of at least 5 ft. Some images simulate a solid white circular depression 2 ft in diameter. The images have varying weather and lighting conditions.

The images initially acquired were in jpg format with extensive memory consumption. Therefore, to avoid such problems, we decided to convert the dataset into TFrecord. TFRecords is a TensorFlow-specific record format based on Google's protobufs. A TFRecord is the most convenient type of file format to utilize for sharing in TensorFlow. It is a binary file that contains byte-string sequences. Before converting the data into TFRecord, data must be serialized (encoded as a byte string). Wrapping data using tensorflow is the most convenient approach to serializing data in TensorFlow. It has various advantages. After performing primary data preprocessing and converting the data into TFRecords, you can test the algorithms on that dataset much faster without going to the same preprocessing step every time. It is also helpful to dynamically shuffle the dataset and change the ratio of train:test:validate from the whole dataset to avoid any biased data distribution.

For converting the dataset to the TFRecords file, each value within each observation is converted to a tf.Train.Feature containing one of the three compatible types, BytesList, FloaList, and Int64List. These are not python data types but tf. Train. Feature types store python data in compatible formats for TensorFlow operations, then we construct a dictionary from the feature name string to the encoded feature value created in step 1. As a result, the list looks like : filename pairs = [(image1, label1), (image2, label2)...]. Table 1 displays the details of the dataset used for training.

3.2.2 Modified-UNet

UNET Architecture with Cross Connection: UNET is an Image Segmentation model consisting of two paths. The first path is the encoder, which records the context of an image. The symmetric expanding channel (decoder) helps in achieving precision localization by using transposed convolutions that attempts to recover the spatial dimensions of the convolutional layer by reversing the downsampling and upsampling techniques used in it.

However, from the results, we made following improvements to the model's implementation:

For the 3*3 convolutions, we utilized a U-net model with 'SAME' padding. When the padding is set to 'SAME', the input is half-padded. Using 'SAME' guarantees

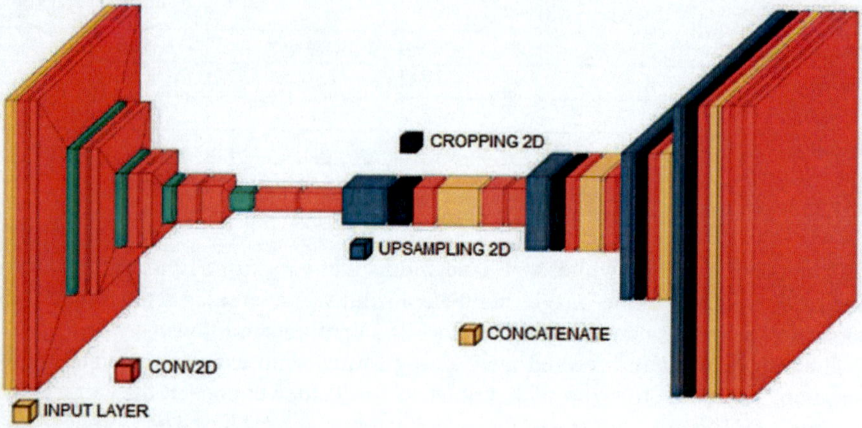

Fig. 5 Architecture of modified U-Net

that the filter is applied to all of the input components. This method of padding guarantees that no significant information is lost. In contrast, using 'VALID' padding, features on the right and bottom of the image are frequently overlooked.

U-Net consists of two 3×3 convolutions, each followed by rectified linear units (ReLU) and 2×2 max pooling with stride 2 for downsampling. With each downsampling step, we increase the number of feature channels by two. The feature map is first upsampled at the beginning of each step in the expanding path, followed by a 2×2 convolution to cut the number of feature channels in half, a concatenation with the proportionately cropped feature map from the contracting path, two 3×3 convolutions, and finally a ReLU. The loss of border pixels in each convolution necessitates cropping. A 1×1 convolution is employed at the final layer to transfer each 64- component feature vector to the appropriate number of classes. There are a total of 23 convolutional layers in the network.

Adding "skip connections" from earlier layers and combining the two feature maps during the upsampling stage provides sufficient information for the last layers to produce precise segmentation boundaries. Figure 5 shows the architecture of the model. Local predictions with nearly perfect global (spatial) structures are produced as a result of the combination of fine and coarse layers.

We utilize the kernel initializer 'he normal' for convolution operations. Kaiming Initialization, also known as He Initialization, is a neural network initialization approach that takes into consideration the nonlinearity of activation functions such as ReLU activations. It takes samples from a truncated normal distribution centered on 0 with

$$\sigma = \sqrt{(\frac{2}{\text{fan in}})} \tag{1}$$

where: σ = standarddeviation
fan in = total input units in the weight tensor

Training: The training data was scaled to 192,192 pixels, and the annotations and images were flipped at random. For training, the input picture has the dimensions 192*192*3. The activation map changes to 12*12*1024 as it approaches the bottleneck. We received network output with the dimensions 192*192*1.

We utilized the Adam optimizer with a drop of 0.9 and learning rate of 0.001. The learning rate is updated on a continuous basis using the formula

$$\alpha = \mathrm{IL} * \mathrm{drop}^{\left(\frac{1+\mathrm{epoch}}{\mathrm{epochs\ drop}}\right)} \tag{2}$$

where: α = updatedlearningrate
IL = initiallearningrate
Adam is an optimization technique which combines the features of the AdaGrad and RMSProp methods to create an optimization algorithm capable of dealing with sparse gradients on noisy issues.

We utilized binary cross entropy which calculates the difference between two distributions. It is slightly different from KL divergence as it computes the relative entropy between the probability distributions. Binary Cross-entropy is sometimes confused with logistic loss, commonly known as log loss. Log Loss represents how much the predicted probabilities deviate from the true ones used in binary cases.

$$\mathrm{Loss} = -\frac{1}{\mathrm{output\ size}} \sum_{i=1}^{\mathrm{output\ size}} \left(y_i \log(p_i) + (1 - y_i)\log(1 - p_i)\right) \tag{3}$$

where y_i is the target value and p_i is $i^t h$ probability value of the output and output size is the total number of output values in the model.

3.2.3 FusionNet

Architecture: We attempted to integrate the benefits of FCN8 and U-Net architecture into this model First, we implemented a simple U-Net and FCN architecture and observed the following:

FCN: This model performed well in bright and dim environments, although it is more sensitive to noise such as background, grass glare, and other hurdles. As a result, distinguishing lanes from other obstructions spotted along the way got harder.

UNET: Because of its backbone network, this model can effectively extract spatial data. It dealt with a lot of noise in the input image owing to the downsampling and upsampling operations. As a result, the produced image is considerably less noisy. The problem with this model is that it also fails to recognize the lanes in fluctuating

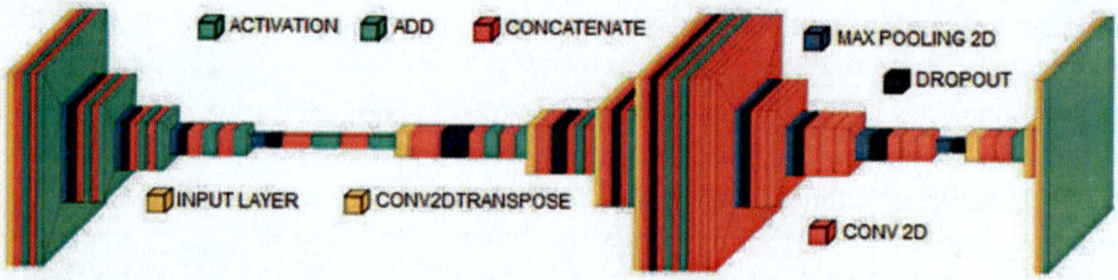

Fig. 6 Architecture of FusionNet

weather conditions, and the output appears to be significantly influenced by the lighting situation.

To get the best of both models, we created a model that integrated unet and fcn-8 architecture. We fed the image into the UNET model whose output is passed to the FCN-8 model. UNET would provide a less noisy output image, and the FCN-8 model could extract the lanes more efficiently from a less noisy image. The combined output seems to have significantly superior lane recognition than the two separately.

In the FCN section, as input image is passed through the convolution layer, its dimension changes from H $\times$ W to H/2 $\times$ W/2 and then to H/4 $\times$ W/4 in the network shown in Fig. 6. Hence, we get micro probability heatmap which is then upsampled to get final segmentation map.

Training: The model uses a Fully Convoluted Network (FCN) for the image segmentation model, with VGG-16 as its base network for feature extraction. FCN-8 allows upscaling of the output of VGG-16 to the desired resolution. FCN-8 takes the output of VGG-16 (stage 5) and upscales x2, then takes stage 4 and concatenates both. It does a second upscale x2 and concatenates with the stage 3 output. We chose FCN-8 over FCN-16 and FCN-32 as it leads to the loss of spatial information, due to which prediction is non-uniform. Hence, FCN-8 displays the best result. This integration of U-Net and FCN-8 architecture performed well in bright and dim environments despite being more sensitive to background noises, grass glare, and other hurdles. As a result, distinguishing lanes from other obstructions along the way got easier.

4 Results

Both models (Modified U-Net and FusionNet) have been trained for 20 epochs with a batch size of 8 and the loss function used was binary cross entropy. The comparison table shows that both models are subjected to different training conditions. Table 2 shows the training parameters for the models.

The performance of the models mentioned in the Table 3 on lane segmentation is expressed in terms of Accuracy score and Jaccard Index is plotted in Fig. 7.

Here, accuracy is defined as the number of pixels correctly classified as lane divided by the total number of pixel predictions.The accuracy measure is defined as

Table 2 Training parameters for the models

	FuisonNet	Modified U-Net
Basic architecture	U-Net + FCN-8	U-Net with cross-connection
Optimizer	Adam, learning rate=0.001 with a decay of $1e-6$	Adam with an initial learning rate of 0.001 with a drop of $(0.9)^{\gamma}$ where $\gamma = \frac{1+\text{epoch}}{2}$
Kernel initialization	Grator normal	He Normal
Upsampling	using convTranspose2D	using upsampling2D and conv2D

Table 3 Training Performance of the models

FusionNet	Accuracy (%)	Jaccard index
Training	98.88	0.9805
Validation	97.6	0.9800
Testing	98.76	0.9682
Modified -Unet		
Training	99.23	0.9762
Validation	98.96	0.9704
Testing	98.68	0.9666

:

$$\text{Accuracy} = \frac{\text{Total Correct Predictions}}{\text{Total Number of Predictions}} \tag{4}$$

Jaccard Index is a standard semantic image segmentation evaluation measure that computes the mean Intersection-Over-Union metric. The Jaccard index is defined for each class as follows:

$$J(G, H) = \frac{|G \cap H|}{|G \cup H|} \tag{5}$$

The models were tested on various publicly available images of the IGVC race track, which shows the following results in Figs. 8 and 9

5 Conclusion

In this paper, we tested our lane detection models on the IGVC dataset, which pushes the research to find a better lane recognition system by incorporating challenges that universal image processing techniques cannot overcome. We first discussed the classical OpenCV image processing techniques and proposed two deep learning

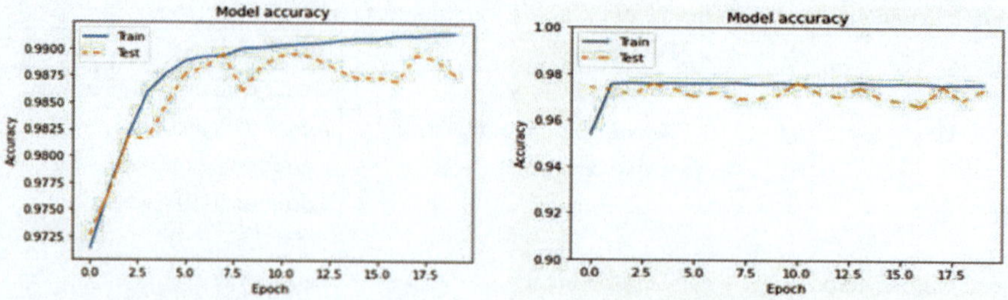

(a) Modified U-Net Accuracy (b) FusionNet Accuracy

Fig. 7 Modified UNet accuracy and FusionNet accuracy curve

(a) Input Image 1 (b) Output of FusionNet (c) Output of Modified U-Net

Fig. 8 Performance on Image 1

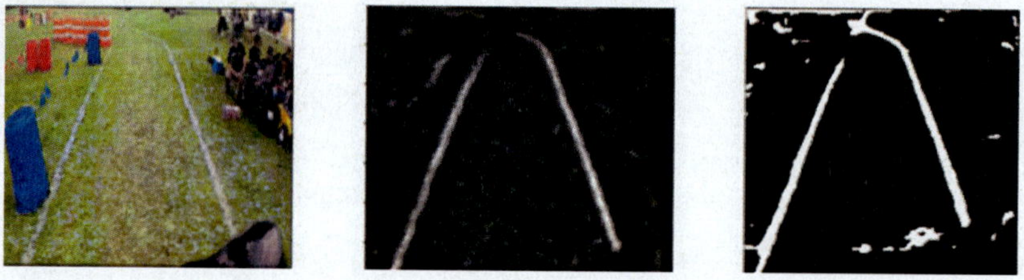

(a) Input Image 2 (b) Output of FusionNet (c) Output of Modified U-Net

Fig. 9 Performance on Image 2

models–modified UNet and FusionNet–with accuracy scores of 98.68% and 98.76%, respectively. Due to the improved computing power and GPUs, vision-based assisted driving systems have been deployed on vehicle platforms. However, there are two main challenges, the lack of generalization ability and the difficulty in deployment on mobile devices. Meta-Learning and auto-machine learning are viable explorations for a computationally intensive algorithm. In future works, we intend to make the model more efficient, lighter, and scalable so it can be deployed on embedded devices.

References

1. GitHub—Chinnu1103/Lane-Extraction-from-Grass-Surfaces: a model to extract lanes from an image using semantic segmentation. github.com. https://github.com/Chinnu1103/Lane-Extraction-from-Grass-Surfaces
2. IGVC 2015 UNSW Advanced Course GoPro—Speed Record—youtube.com. https://www.youtube.com/watch?v=A9BVr7kltl8&t=5s
3. IGVC Trial Video VGT—vimeo.com. https://vimeo.com/761864295
4. Cho JH, Jang YM, Cho SB (2014) Lane recognition algorithm using the hough transform with applied accumulator cells in multi-channel roi. In: The 18th IEEE international symposium on consumer electronics (ISCE 2014). pp 1–3. https://doi.org/10.1109/ISCE.2014.6884348
5. Ghafoorian M, Nugteren C, Baka N, Booij O, Hofmann M (2018) El-gan: embedding loss driven generative adversarial networks for lane detection. In: Proceedings of the European conference on computer vision (ECCV) workshops. https://doi.org/10.48550/arXiv.1806.05525
6. Lee DH, Liu JL (2022) End-to-end deep learning of lane detection and path prediction for real-time autonomous driving. Signal, Image Video Process 1–7. https://doi.org/10.1007/s11760-022-02222-2
7. Liu Y, Yan J (2021) Research on lane line segmentation algorithm based on deeplabv3. In: 2021 IEEE Asia-Pacific conference on image processing, electronics and computers (IPEC), IEEE, pp 787–790. https://doi.org/10.1109/IPEC51340.2021.9421285
8. Low CY, Zamzuri H, Mazlan SA (2014) Simple robust road lane detection algorithm. In: 2014 5th international conference on intelligent and advanced systems (ICIAS). pp 1–4. https://doi.org/10.1109/ICIAS.2014.6869550
9. Madan A, Jharwal D, Road lane line detection using opencv
10. Tsai SC, Huang BY, Wang YH, Lin CW, Lin CT, Tseng CS, Wang JH (2013) Novel boundary determination algorithm for lane detection. In: 2013 International conference on connected vehicles and expo (ICCVE), IEEE, pp 598–603. https://doi.org/10.1109/ICCVE.2013.6799861
11. Wang J, Mei T, Kong B, Wei H (2014) An approach of lane detection based on inverse perspective mapping. In: 17th International IEEE conference on intelligent transportation systems (ITSC), IEEE, pp 35–38. https://doi.org/10.1109/ITSC.2014.6957662
12. Zhang W, Mahale T (2018) End to end video segmentation for driving: Lane detection for autonomous car. arXiv preprint arXiv:1812.05914. https://doi.org/10.48550/arXiv.1812.05914
13. Zou Q, Jiang H, Dai Q, Yue Y, Chen L, Wang Q (2019) Robust lane detection from continuous driving scenes using deep neural networks. IEEE Trans Vehicular Technol 69(1):41–54 (2019). https://doi.org/10.48550/arXiv.1903.02193

Pansharpening of Multispectral Images Through the Inverse Problem Model with Non-convex Sparse Regularization

Rajesh Gogineni, Y. Ramakrishna, P. Veeraswamy, and Jannu Chaitanya

Abstract Pansharpening is considered as an imperative process for various remote sensing applications viz. crop monitoring, hazard monitoring, object detection and classification etc. The Pansharpening technique combines panchromatic and multispectral pictures to create a high resolution multispectral image. In this paper, the pansharpening approach and a variational optimization model are discussed. As an ill-posed inverse issue, a cost function is proposed, with three prior components, two of which are data-fidelity terms generated from the relationship between the source and output images. The third term is integrated to regularize the formulated inverse model. The eminent solver, alternating direction method of multipliers in conjunction with iterative minimization mechanism is employed to obtain the comprehensive minimum of the proposed convex cost function. The minimized solution is the required pansharpened image. The effectiveness of the suggested strategy is assessed using three different datasets and four recognized indicators. The results, both objective and subjective, show the effectiveness of the variational optimization pansharpening (VOPS) model. The merged image has greatly improved spectral and spatial properties.

Keywords Pansharpening · High resolution multispectral image · Inverse problem · Vector minmax concave · Alternating direction method of multipliers

1 Introduction

Panchromatic (PAN) and Multispectral (MS) imaging products are produced by the modern optical satellite sensors, such as IKONOS, GeoEye, worldview, and have complementary resolution properties. Due to physical and technological constraints,

R. Gogineni (✉) · Y. Ramakrishna · P. Veeraswamy
Department of ECE, Dhanekula Institute of Engineering and Technology, Vijayawada 521139, India
e-mail: rgogineni9@gmail.com

J. Chaitanya
School of electronics engineering, VITAP University, Amaravati, India

S. Sharma et al. (eds.), *Intelligent Control, Robotics, and Industrial Automation*, Lecture Notes in Electrical Engineering 1066,
https://doi.org/10.1007/978-981-99-4634-1_40

the sensors are unable to produce an entire image with a variety of properties of rich geometric information and good spectral resolution. One of the well-known methods used in Image fusion is pansharpening. In this method, a high resolution multispectral image is created by merging a monochrome image and a four band color image. The spectral characteristics of MS and PAN photos were combined to create this image. Pansharpening (PS) is a first-step technique for the majority of remote sensing applications, danger monitoring, and change detection.

The vast majority of pansharpening methods that have been presented over the past 30 years can be broadly categorized into five groups [1–3].

- component substitution (CS) techniques,
- multi resolution analysis (MRA) techniques,
- sparse represention (SR)-based techniques,
- machine learning (ML)-based techniques, and
- variational optimization (VO) techniques.

The term "classical approaches" refers to both the component substitution (CS) and multi resolution analysis (MRA) frameworks. In CS techniques, which are based on projecting the source MS image into a new transform domain, the PAN image takes the role of the intensity component. The inverse transform is then used to create the desired HRMS image. Band dependent spatial details (BDSD), partial replacement adaptive CS (PRACS), and generalized intensity-hue-saturation (GIHS) are three well-known CS approaches [4–6]. The MRA procedure involves the extraction of spatial characteristics from the PAN image and their injection into the interpolated MS image. The Additive Wavelet Luminance Proportional (AWLP) [7], Modulation Transfer Function (MTF) matched filter (MTF-GLP) [8], and High Pass Filtering (HPF) [9] are a few popular MRA approaches. The CS techniques can produce results with adequate spatial fidelity despite spectrum distortion. The merged image is tainted by spatial defects, despite the fact that MRA approaches are capable in terms of spectral quality.

The sparse representation (SR) theory is adapted for solving the pansharpening problem and became effecective in recent past. The compressive sensing theory with a dictioanry learned from the HRMS images, which are the outcomes actually, was proposed by Li and Yang [10]. The details injection mechasim of MRA-based methods is comined with the SR theory and a concept of scale invariance is introduced in, [11]. The first category of SR methods are computationally complex, since a large dictionary or multiple dictionaries need to be constructed for sparse coding of both the source images [12, 13]. Another class of SR-based PS methods uses a dictionary created from the PAN picture to achieve sparse coding of MS patches in a reduced version of the PAN image [14]. Repetitive representations show up in the fused image as a result of the patch processing utilized in SR-based techniques, and teaching the vocabulary takes time.

All remote sensing applications now use machine learning (ML), and even pansharpening of multispectral pictures has been investigated [15]. Convolutional neural networks are used in the majority of machine learning-based pansharpening algorithms (CNN). The super resolution notion in [16] uses CNN to enhance the geometric details of the MS image. A three layer CNN with different layer activations is employed for the integration of remote sensing pictures [17]. Zhang et al. [18] created the GTP-PNet, a convolutional learning network with gradient information term, for the consistency of the spatial structure. For machine learning methods to be trained, large datasets are required. Reference HRMS images are better suited for learning advantageous properties that are in fact outcomes of pansharpening.

The advancement of inverse problems and the convex optimization in image processing leads to the recent attractive development in pansharpening known as variational optimization (VO) technquies. The sensor model is used by the VO techniques for PS to modify the interaction between PAN, MS, and the HRMS. An inverse issue that is solved using some regularizers yields the required result, which is an image from the HRMS system. Three terms formed from the relationships between the input products and the target product serve as the foundation for the first VO-based technique, which was first proposed by [19]. The baysian approaches also associated with VO-based methods. A joint Gaussian model for PAN and MS image in wavelet domain is proposed in [20]. Total variation (TV) is used to regularize the incorrectly presented pansharpening problem in the variational framework Palsson et al. [21] developed. The preservation of geometric information in the combined image is changed when a regularizer based on Total Generalized Variation (TGV) is used [22]. In [23] takes use of the gradient similarities between PAN and MS pictures by employing a sparse representation method. To achieve balanced spectral quality and boost spatial information, VO techniques can be further enhanced.

A variational mechanism-based optimization model for pansharpening of multispectral images is proposed in this work. In view of various drawbacks in the existing methods a comprehensive model is developed to maintain a trade-off between the geometric details and the color information in the pansharpened image. The pansharpened image's potential spectral distortion is efficiently reduced by the SAM-based previous term. The proposed model uses an effective gradient-based regularized term to extract the necessary spatial features.
Here is a summary of the work's main contributions:

(i) The renowned remote sesning image evolution hypothesis is used to formualte the data generative terms.
(ii) It is suggested to use a unified framework to transfer to the output image all of the spectral and spatial data from the input products, including PAN and MS.
(iii) To accomplish the sparsity, the cost function is integrated with a regularizer namely, vector minmax concave (VMC) prior.
(iv) To create the necessary high resolution MS image, an optimization approach with convergence guarantees and an iterative shrinking framework is used.

The study article's remaining sections are organized as follows: Sect. 2 covers the suggested strategy and the optimization technique. The observations and analysis are depicted in Sect. 3. The convergence analysis is covered in this section as well. In Sect. 4, the essay is finished.

2 Proposed Method

A non-convex regularization prior term is used to create the variational optimization pansharpening (VOPS) model that is proposed.

For mathematical assistance, in this work the images are represented with bold and uppercase letters and the all other variables are represented in lower case. Let $\mathbf{P} \in \mathbb{R}^{m \times n}$, be the Panchromatic image, $\mathbf{M} \in \mathbb{R}^{\frac{m}{r} \times \frac{n}{r} \times b}$ be the low resolution MS image and $\mathbf{X} \in \mathbb{R}^{\frac{m}{r} \times \frac{n}{r} \times b}$ be the HRMS image. Where, $m \times n$ denotes the spatial resolution of PAN image, r denotes the spatial dimension ratio between two input images and the number of bands are represented with b.

The cost function for the proposed VOPS model is developed based on the popular remote sensing image evolution hypothesis [24]. Since there are twoinput products, i.e., PAN and MS, the cost function consists of two data generative terms.

$$\underset{X}{\operatorname{argmin}} \;\; C_1(X) + C_2(X) + \phi(X) \tag{1}$$

Where, $C_1(X)$ and $C_2(X)$ are data generative terms and $\phi(X)$, denotes the prior that regularizes the optimization problem.

The HRMS product is treated as a degraded and lower resolution variant of the MS image that was used as input. This concept results in the cost function's first term.

$$M_b = GX_b + \eta \tag{2}$$

Where, G = SH is the product of two operators: S—the downsample operator and H-is the filter that blurs the b-th band. Further, η is the Gaussin noise with zero mean and unit variance. The corresponding cost term is defined as,

$$C_1(X) = \frac{1}{2}||G_b X_b - M_b||_F^2 \tag{3}$$

The symbol $||.||_F$ represents the Frobenius norm.

The PAN image covers the entire spectral range of the MS imaging bands. This second factor in the cost function is simulated appropriately since the PAN picture is believed to be a weighted linear mixture of HRMS image bands.

$$P = \omega_b X_b + \eta \tag{4}$$

Where, ω_b represents the weights corresponding to the b-th band subjected to the constraint that $\sum_{b=1}^{N} \omega_b = 1$. The appropriate values for the weights are evaluated from the spectral response curves [25]. The equivalent cost term is expressed as

$$C_2(X) = \frac{1}{2}||P - \omega_b X_b||_F^2 \tag{5}$$

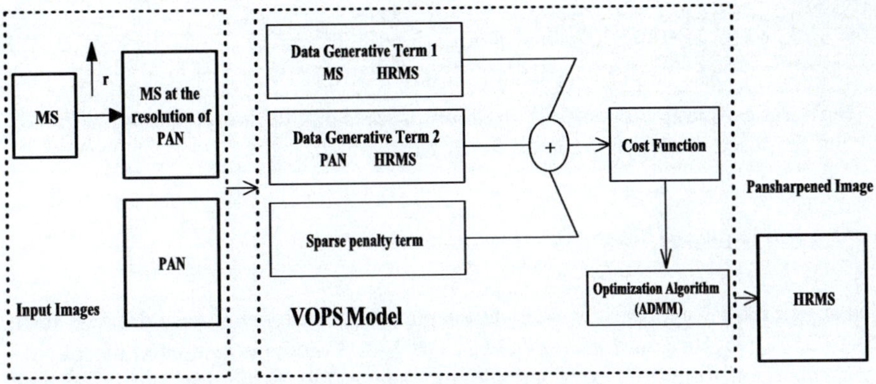

Fig. 1 The process flow diagram for the proposed VOPS model

The regularization of the ill-posed problem with vector minmax concave (VMC) penalty that turnout the global minimizer which is similar to the L0 minimization problem [26].

The comprehensive cost function for each band, b is presented as

$$C(X) = \frac{1}{2}||B_b X_b - M_b||_F^2 + \frac{1}{2}||P - \omega_b X_b||_F^2 + \beta\phi(X) \tag{6}$$

Where, $\phi(X)$ is the VMS fuction and β is a regularization parameter

The Fig. 1 presents the suggested VOPS model's process flow diagram.

2.1 *Minimizer*

Using the Alternating Direction Method of Multipliers (ADMM), the Proposed VOPS model is solved [27]. When utilizing the ADMM technique to resolve the inverse problem, which is presented as a typical regularized minimization problem, the regularization parameter is essential.

The ADMM algorithm solves the problem as

$$\min\ f_1(x_1) + f_2(x_2)\text{subject to}\,\theta_1 x_1 + \theta_2 x_2 = b \tag{7}$$

The relavent Lagrangian is defined as

$$L(x_1, x_2 : t) = f_1(x_1) + f_2(x_2) + \frac{\lambda}{2}||\theta_1 x_1 + \theta_2 x_2 - b - t||_2^2 \tag{8}$$

where t is scaled Lagrangian multiplier and β is a real variable. The estimate of the HRMS image, $\hat{X}_b$, b=1,2...N, for each band is is solved separately as

$$\hat{X}_b = \underset{X_b}{\text{argmin}} \frac{1}{2}||B_b X_b - M_b||_F^2 + \frac{1}{2}||P - \omega_b X_b||_F^2 + \beta\phi(X) \tag{9}$$

The criteria for parameter setting, iteration process and the implementation details for the proposed model are adapted from [28].

3 Experimental Results

The suggested PS approach is validated using geographically diverse datasets from IKONOS, QuickBird, and WorldView-2. The VOPS model is evaluated thoroughly by comparison to cutting-edge techniques, both at low resolution and at original resolution. Relative Dimensionless Global Error in Synthesis (ERGAS), Spectral Angle Mapper (SAM), Correlation Coefficient (CC), and Universal Quality Index (Q4) for reduced-resolution, and D_λ, D_s, and QNR for full resolution, are well-known quality measures that are used to measure the efficacy of a method. For evaluation, the set of subjective indices below is offered:

The correlation coefficient (CC), which is defined as the amount of correlation between two pictures A and B

$$\text{CC} = \frac{\sigma_{\text{A,B}}}{\sigma_{\text{A}}.\sigma_{\text{B}}} \tag{10}$$

where, $\sigma_{A,B}$, is the covarinace between A and B, and σ_A and σ_B are the positive square roots of variances of A and B, respectively.

ERGAS :

$$\text{ERGAS} = 100\frac{1}{\text{r}}\sqrt{\frac{1}{\text{K}}\sum_{\text{i}=1}^{\text{K}}\left(\frac{\text{RMSE(i)}}{\text{MEAN(i)}}\right)} \tag{11}$$

where. $\frac{l}{r}$ is the ratio of spatial dimensions of PAN and MS images and is equal to $\frac{1}{4}$ for the datasets used in evaluation and RMSE (i) stands for root mean square error of ith band.

SAM: For the given two spectral vectors, V_A ,V_B of MS images A and B,

$$\text{SAM}(\text{V}_\text{A}, \text{V}_\text{B}) = \arccos\left(\frac{\langle \text{V}_\text{A}, \text{V}_\text{B}\rangle}{\|\text{V}_\text{A}\|_2\|\text{V}_\text{B}\|_2}\right) \tag{12}$$

Q4 : is the UIQI metric's multispectral extension, which may be used to analyze images with several bands. The three components of Q4-correlation, mean bias, and contrast variation of each spectral band are used for detailed quality assessment.

The proposed method is also compared to existing visual analysis techniques from several categories, such as GIHS [4]-CS-based, AWLP [7], MTF-GLP [8]: MRA-based, SR-based [12], CNN [16]-DL-based, and variational-based techniques P+XS [19] and TGV [22]. The ideal values for the selected subjective indices are shown in Table 1.

Table 1 Ideal values of subjective indices

Quality index	CC	ERGAS	SAM	Q4	D_λ	D_s	QNR
Optimal value	1	0	0	1	0	0	1

Table 2 Subjective analysis for QuickBird imaging products

	EXP	GIHS	AWLP	MTF	P+XS	SR	CNN	TGV	Proposed
CC	0.8846	0.9347	0.9312	0.9461	0.9337	0.9414	0.9421	0.9473	0.9496
ERGAS	7.8412	4.5856	4.3126	4.2872	4.7324	4.5136	4.4618	4.2121	4.0018
SAM	5.8685	5.7261	5.2658	5.1183	5.3217	5.2845	5.0016	4.9978	4.8124
Q4	0.6984	0.8214	0.8113	0.8017	0.7967	0.8128	0.8119	0.8217	0.8275

3.1 Analysis at Reduced Resolution

The (visual and quantitative) experimental results for the QuickBird and IKONOS datasets at low resolution are presented in this section. In order to evaluate the standard Wald's technique [29], a low resolution MS image is employed as a standard against which to measure the fused result.

For QuickBird and IKONOS, the visual analyses are shown in Figs. 1 and 2, respectively. The QuickBird satellite's PAN, MS, and interpolated MS (Exp) are shown in Fig. 1a, c. The visual results make it abundantly evident that the GIHS approach creates an image with spectral distortion, which appears as a change in color in the area made up of trees (Fig. 1d). Overall, the observation of visual outcomes indicate that in the AWLP and MTF methods outcomes few geometric details surrounding the road are blurred. The outcomes of SR-based method and CNN methods appears bright in the areas of trees which indicates color distortion. The results of the suggested and TGV approaches are comparable and nearer the MS reference image. The proposed method is superior at balancing the structural and color aspects in the pansharpened image, as per the subjective indices values shown in Table 2.

In addition, reduced scale evaluation of the IKONOS dataset is performed; the visual findings are shown in Fig. 3, and key quality measures are described in detail in Table 3. Figure 3a–c, respectively, show the source pictures and the reference image. Comparatively, the edges in the fused images produced by the traditional methods are less precise. The P+XS method's results seem brighter than the typical reference image. The CNN outcome shows the preservance of comprehesive geometric information. The SR-based method outcome shows the blocking effect in the areas consists of sharp edges. The proposed method achieves considerable improvement over the other methods used in comparison.

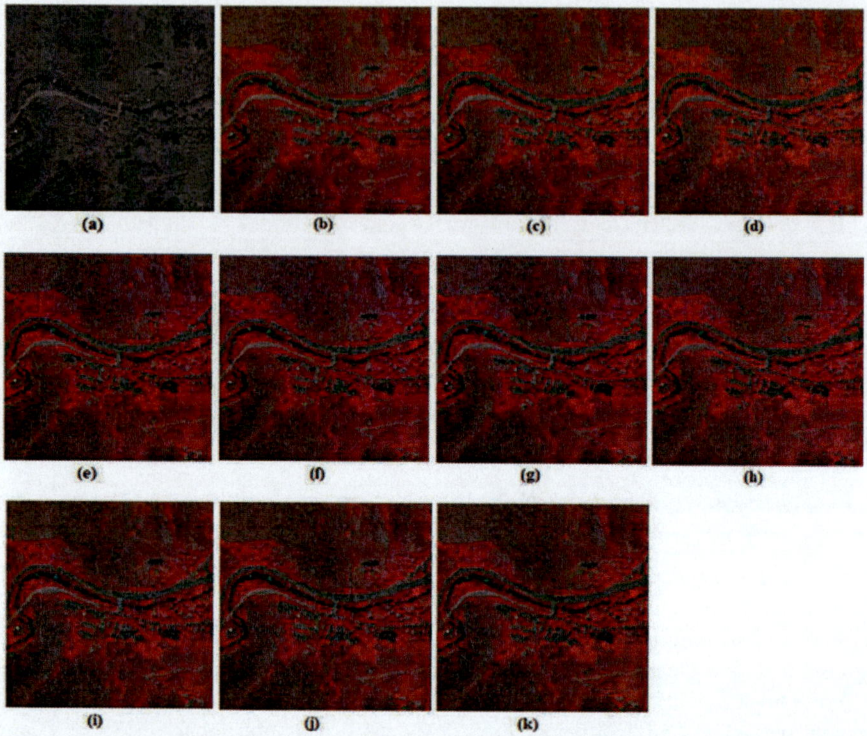

Fig. 2 Objective analysis using a lower resolution QuickBird imaging products **a** Panchromatic image **b** source MS product (input MS product) **c** Interpolated MS product (EXP) **d** GIHS **e** AWLP **f** MTF-GLP **g** P+XS **h** SR **i** CNN **j** TGV **k** Proposed approach

Table 3 Subjetive analysis for IKONOS imaging products

	EXP	GIHS	AWLP	MTF	P+XS	SR	CNN	TGV	Proposed
CC	0.8374	0.9314	0.9627	0.9326	0.9215	0.9343	0.9386	0.9415	0.9407
ERGAS	4.6843	3.9846	3.6211	3.6315	3.7796	3.6543	3.2218	3.1087	3.0105
SAM	4.7521	4.6597	4.6129	4.6356	4.7328	4.6973	4.4985	4.3682	4.1126
Q4	0.7126	0.8019	0.8132	0.8146	0.8127	0.8135	0.8271	0.8336	0.8398

3.2 Analysis at Full Resolution

The whole scale resolution evaluation is adopted from the QNR technique [30], which is predicated on the idea that the evaluation is performed without needing the reference MS image for comparison. The pansharpened image is quantitatively evaluated using the three quality metrics D_λ–spectral distortion index, D_s–spatial distortional index, and QNR–global similarity index. Figure 4 presents the visual analysis in full scale resolution.

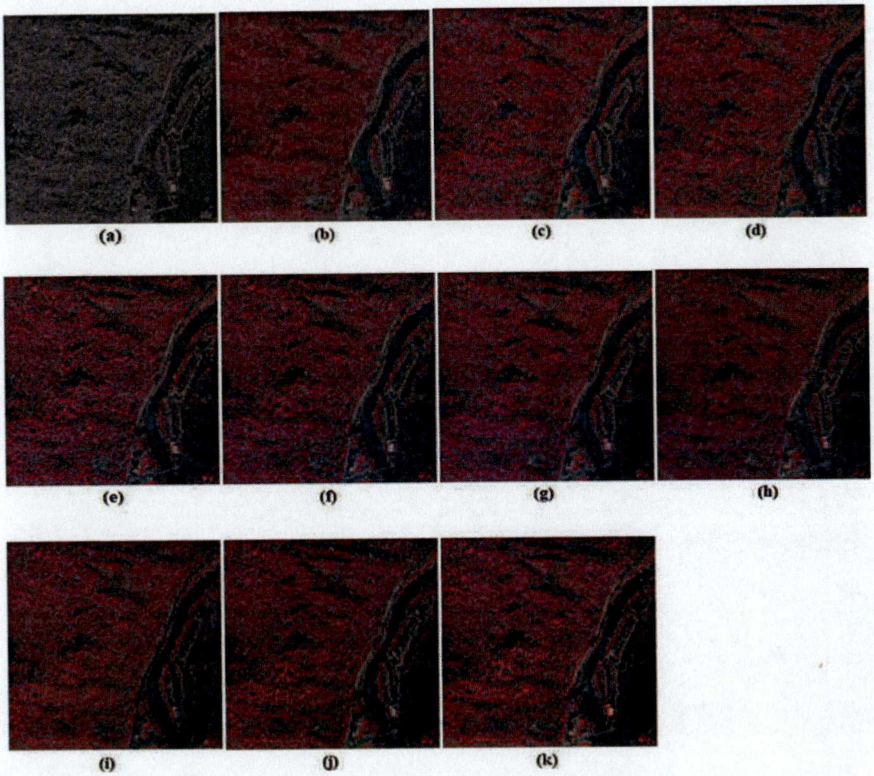

Fig. 3 Objective analysis with a lower resolution IKONOS imaging products **a** Panchromatic product **b** low resolution MS product (input MS product) **c** Interpolated MS product (EXP) **d** GIHS **e** AWLP **f** MTF-GLP **g** P+XS **h** SR **i** CNN **j** TGV **k** Proposed approach

Table 4 displays the corresponding subjective indices for all the approaches that were compared for the WorldView-2 dataset. In Fig. 4a, b, the raw images–panchromatic and interpolated MS images–are displayed (b). The fused outcome of GIHS method is spectrally different (suffer from color distortion) from the other outcomes. The presence of spatial artifacts is obvious in the AWLP result. The MTF-GLP outcome is similar to the AWLP outcome that implies the details injection process result in loss of geometric details. From the Table 4, the proposed achieves best results for all the three metrics, which approves the enhancement of spatial and spectral qualities.

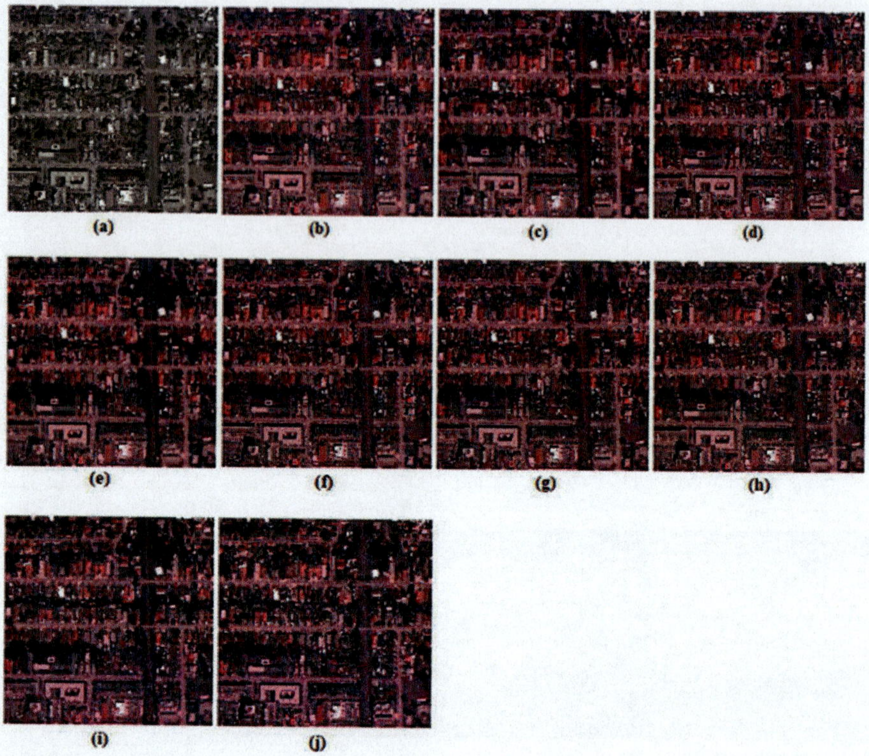

Fig. 4 Objective analysis utilizing the full resolution WorldView-2 dataset **a** Panchromatic image **b** Interpolated MS product (EXP) **c** GIHS **d** AWLP **e** MTF-GLP **f** P+XS **g** SR **h** CNN **i** TGV **j** Proposed approach

Table 4 Quantitative analysis for WorldView-2 dataset

	EXP	GIHS	AWLP	MTF	P+XS	SR	CNN	TGV	Proposed
D_λ	0.0081	0.0924	0.0965	0.1026	0.1275	0.0816	0.0643	0.0482	0.0479
D_s	0.1840	0.0931	0.1074	0.0932	0.1436	0.0839	0.0795	0.0638	0.0364
QNR	0.8094	0.8231	0.8064	0.8137	0.7472	0.8413	0.8614	0.8908	0.9174

3.3 Convergence Analysis

Iterative minimization is employed to solve the provided model, therefore the convergence in terms of iterations against relative error is assessed. The relative error is estimated as $(||\hat{X}_k - X^r||_F/||X^r||_F)$, where $\hat{X}_k$ is the estimated image after k-th iteration and X^r is the reference image used for comparison which is usually MS image. The iterations are stopped when the error between successive iterations is not varying more than the thresholdwhich is set as 0.3. For the given model the

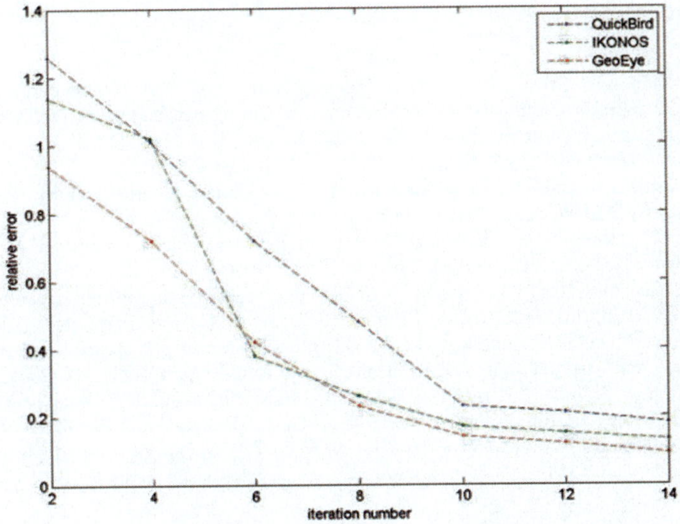

Fig. 5 Convergence analysis of the proposed method for QuickBird and IKONOS datasets

proposed method reached to the threshold value from 25 th iteration onwards. The graphical representation for the convergence analysis is displayed in Fig. 5. Further, the analysis of the effect parameters on the convergence will be carried out in future studies.

4 Conclusion

A pansharpening method based on variational optimization is suggested in this paper. Pansharpening is regarded as a preliminary step and is necessary in the majority of remote sensing applications. The proposed work exercises in a way that the fusion problem is modulated as an optimization paradigm with three fixed cost functions. Utilizing the link between low resolution and high resolution MS pictures, spectral information is added to the fused image. The HRMS image was enhanced using the spatial data from the PAN image using the second prior term. The cost function is regularized using a vector min-max concave penalty-based term. The pansharpened image is created by minimizing the optimization model suggested by using the ADMM approach. The complete results demonstrate that the proposed model is capable of balancing the geometric and color-related properties in the pansharpened image at both full resolution and decreased resolution. Additionally, by selecting reliable priors, it is feasible to enhance the pansharpened image's resolution characteristics in the context of spatial structures. The investigation of fine-tuning the cost function's parameters in relation to the quality measures will take place in subsequent studies.

References

1. Vargas-Munoz JE, Srivastava S, Tuia D, Falcao AX et al (2021) A new benchmark based on recent advances in multispectral pansharpening: revisiting pansharpening with classical and emerging pansharpening methods. IEEE Geosci Remote Sens Mag 9(1):184
2. Javan FD, Samadzadegan F, Mehravar S, Toosi A, Khatami R, Stein A (2021) A review of image fusion techniques for pan-sharpening of high-resolution satellite imagery. ISPRS J Photogrammetry Remote Sens 171:101–117
3. Yilmaz CS, Yilmaz V, Gungor O (2022) A theoretical and practical survey of image fusion methods for multispectral pansharpening. Inf Fusion 79:1–43
4. Tu TM, Huang PS, Hung CL, Chang CP (2004) A fast intensity-hue-saturation fusion technique with spectral adjustment for ikonos imagery. IEEE Geosci Remote Sens Lett 1(4):309–312
5. Garzelli A, Nencini F, Capobianco L (2007) Optimal mmse pan sharpening of very high resolution multispectral images. IEEE Trans Geosci Remote Sens 46(1):228–236
6. Choi J, Yu K, Kim Y (2010) A new adaptive component-substitution-based satellite image fusion by using partial replacement. IEEE Trans Geosci Remote Sens 49(1):295–309
7. Otazu X, González-Audícana M, Fors O, Núñez J (2005) Introduction of sensor spectral response into image fusion methods. application to wavelet-based methods. IEEE Trans Geosci Remote Sens 43(10):2376–2385
8. Aiazzi B, Alparone L, Baronti S, Garzelli A, Selva M (2006) Mtf-tailored multiscale fusion of high-resolution ms and pan imagery. Photogrammetric Eng Remote Sens 72(5):591–596
9. Witharana C, LaRue MA, Lynch HJ (2016) Benchmarking of data fusion algorithms in support of earth observation based antarctic wildlife monitoring. ISPRS J Photogrammetry Remote Sens 113:124–143
10. Li S, Yang B (2010) A new pan-sharpening method using a compressed sensing technique. IEEE Trans Geosci Remote Sens 49(2):738–746
11. Vicinanza MR, Restaino R, Vivone G, Dalla Mura M, Chanussot J (2014) A pansharpening method based on the sparse representation of injected details. IEEE Geosci Remote Sens Lett 12(1):180–184
12. Gogineni R, Chaturvedi A (2018) Sparsity inspired pan-sharpening technique using multi-scale learned dictionary. ISPRS J Photogrammetry Remote Sens 146:360–372
13. Ayas S, Gormus ET, Ekinci M (2018) An efficient pan sharpening via texture based dictionary learning and sparse representation. IEEE J Select Topics Appl Earth Observ Remote Sens 11(7):2448–2460
14. Imani M, Ghassemian H (2017) Pansharpening optimisation using multiresolution analysis and sparse representation. Int J Image Data Fusion 8(3):270–292
15. Deng LJ, Vivone G, Paoletti ME, Scarpa G, He J, Zhang Y, Chanussot J, Plaza A (2022) Machine learning in pansharpening: a benchmark, from shallow to deep networks. IEEE Geosci Remote Sens Mag 10(3):279–315
16. Zhong J, Yang B, Huang G, Zhong F, Chen Z (2016) Remote sensing image fusion with convolutional neural network. Sens Imaging 17(1):1–16
17. Scarpa G, Vitale S, Cozzolino D (2018) Target-adaptive cnn-based pansharpening. IEEE Trans Geosci Remote Sens 56(9):5443–5457
18. Zhang H, Ma J (2021) Gtp-pnet: a residual learning network based on gradient transformation prior for pansharpening. ISPRS J Photogrammetry Remote Sens 172:223–239
19. Ballester C, Caselles V, Igual L, Verdera J, Rougé B (2006) A variational model for p+ xs image fusion. Int J Comput Vision 69(1):43–58
20. Fasbender D, Radoux J, Bogaert P (2008) Bayesian data fusion for adaptable image pansharpening. IEEE Trans Geosci Remote Sens 46(6):1847–1857
21. Palsson F, Sveinsson JR, Ulfarsson MO (2013) A new pansharpening algorithm based on total variation. IEEE Geosci Remote Sens Lett 11(1):318–322
22. Liu P (2019) A new total generalized variation induced spatial difference prior model for variational pansharpening. Remote Sens Lett 10(7):659–668

23. Tian X, Chen Y, Yang C, Gao X, Ma J (2020) A variational pansharpening method based on gradient sparse representation. IEEE Signal Process Lett 27:1180–1184
24. Li S, Yin H, Fang L (2013) Remote sensing image fusion via sparse representations over learned dictionaries. IEEE Trans Geosci Remote Sens 51(9):4779–4789
25. Molina R, Vega M, Mateos J, Katsaggelos AK (2008) Variational posterior distribution approximation in bayesian super resolution reconstruction of multispectral images. Appl Comput Harmonic Anal 24(2):251–267
26. Wang S, Chen X, Dai W, Selesnick IW, Cai G, Cowen B (2018) Vector minimax concave penalty for sparse representation. Digital Signal Process 83:165–179
27. Jiao Y, Jin Q, Lu X, Wang W (2016) Alternating direction method of multipliers for linear inverse problems. SIAM J Numer Anal 54(4):2114–2137
28. Gogineni R, Chaturvedi A, BS DS, (2021) A variational pan-sharpening algorithm to enhance the spectral and spatial details. Int J Image Data Fusion 12(3):242–264
29. Wald L, Ranchin T, Mangolini M (1997) Fusion of satellite images of different spatial resolutions: assessing the quality of resulting images. Photogram Eng Remote Sens 63(6):691–699
30. Alparone L, Aiazzi B, Baronti S, Garzelli A, Nencini F, Selva M (2008) Multispectral and panchromatic data fusion assessment without reference. Photogram Eng Remote Sens 74(2):193–200

Fuzzy Set-Based Multimodal Medical Image Fusion with Teaching Learning-Based Optimization

T. Tirupal, R. Supriya, P. Uma Devi, and B. Sunitha

Abstract In clinical imaging, the fundamental step is to extract useful data from images obtained by different sources and combining to form a single image called fused image to analyse and improve diagnosis. All image fusion techniques now a day rely majorly on non-fuzzy collections. As more qualms are taken into account as compared to non-fuzzy sets, fuzzy sets are better suited for medical image processing. We provide a method for effectively combining multimodal medical images in this study. As suggested, firstly, intuitionistic fuzzy images (IFIs) are obtained considering input source images into account. The optimal membership and non-membership function values are then derived for intuitionistic fuzzy entropy (IFE). After that, the IFIs are scrutinised using the fitness function, contrast visibility (CV). Then, Teaching–Learning-Based Optimization (TLBO) is utilised to maximise the fusion coefficients that are adjusted during the teaching stage and the learning phase of TLBO, allowing the weighted coefficients to naturally adapt to the fitness function. Finally, the fused image is acquired using the optimal coefficients. On various collections of source images, simulations are run, and the output findings are correlated to the most recent fusion techniques. The superiority of the suggested system is explained and supported. Edge-based image fusion ($Q^{AB/F}$), spatial frequency (SF), entropy (E), and other objective metrics are also used to assess how superior the fused image is.

Keywords IFI · TLBO · Spatial frequency · Fuzzy sets · Image fusion

T. Tirupal (✉) · R. Supriya · P. U. Devi · B. Sunitha
Department of ECE, G. Pullaiah College of Engineering and Technology, Kurnool, Andhra Pradesh, India
e-mail: tirutalari@gmail.com

S. Sharma et al. (eds.), *Intelligent Control, Robotics, and Industrial Automation*,
Lecture Notes in Electrical Engineering 1066,
https://doi.org/10.1007/978-981-99-4634-1_41

1 Introduction

A procedure called image fusion, which combines various medical image sets, is used to improve the quality of two multimodal clinical images for medical applications [1, 2]. These image sets are used to gather clinical data that are fundamental in nature. X-rays detect fissures and abnormalities in bone locus. Dense structures like bones contained in CT images characters through fewer alterations, and fleshy lenient muscle facts are delivered by MRI, MRA successfully detects brain deserts, deepness and breadth of an infected entity are provided by VA, and feasible and metabolic data of the human brain are contained in PET and SPECT.

From this point of view, it can be stated that a single image does not produce complete meaningful information, and in the future, all relevant data should be required in a fused image, which is a single composite image [3]

All non-fuzzy set-based image fusion algorithms struggle with the poor lighting of medical images, noise, ambiguous bounds, overlapping grey levels, invisible blood veins, and difficulty differentiating image objects. They are dealt with by taking fuzzy sets into account. To reduce darkness and confusion in images, fuzzy sets are said to be the most crucial component in image processing. The fuzzy set hypothesis that Zadeh [4] predicted in 1965 cannot show weaknesses explicitly. An intuitionistic fuzzy set (IFS) with two improbability parameters—the belonging grade and the non-belonging grade—is general type of fuzzy set hypothesis suggested in 1986 by Atanassov [5]. As per the research, it offers non-membership degree with endless comprehension and successfully handles problems that arise in the present. Every step of the image processing process is fraught with uncertainty, but with IFSs, these uncertainties may be eliminated and the contrast of the image is improved.

TLBO is a widely used optimization algorithm for upgrading the fused image blocks and producing high-quality fused images. Recently created by Rao et al. [6], TLBO is a widely used improvement metric. It is an iterative learning measure that is population-based and exhibits characteristics that are almost identical to earlier evolutionary computation (EC) methods [7]. Instead of having students participate in genomic processes like selection, crossover, and mutation, TLBO examines an ideal for any student trying to grasp what the teacher is believed to be the most knowledgeable person in society to achieve ideal results. Due to its unpretentious design and high level of competition, TLBO has proven to be an excellent optimization technique used to solve real-world complications successfully [8, 9]. Generally speaking, TLBO provides superior results to other EC techniques. However, the conjunction time is consistently the most significant leap in progressive presentations. In order to form appropriate TLBO, the proposed algorithm emphasises on enlightening the conjuncture period frame deprived of negotiating the prominence of outcomes. This study suggests that despite certain ongoing technical and scientific challenges, medical image fusion has shown to be beneficial in enhancing the clinical dependability of using medical diagnosis and analysis and that it is a field of study that may become fundamental over the coming years.

2 Preliminaries

2.1 Averaging Technique

This method requires only doing mean comparisons to the pixels in each input image, as described in [10].

2.2 Discrete Wavelet Transform (DWT)

All the references [11–16] discuss the theory of wavelets and is offered as a replacement for the short-time Fourier transform and is in many ways a continuation of the Fourier hypothesis (STFT). A wavelet is a tiny wave that forms and dissipates essentially two requirements:

(1) The integral of time ought to be zero.

$$\int_{-\infty}^{\infty} \psi(t)dt = 0. \tag{1}$$

(2) The square of the wavelet that is added with respect to time should be equal to one.

$$\int_{-\infty}^{\infty} \psi^2(t)\ \mathrm{d}t = 1. \tag{2}$$

2.3 Redundant Wavelet Transform (RWT)

The discrete wave transformation (DWT), also known as Mallat's algorithm, relies on the orthogonal disintegration of the image on a wavelet foundation to prevent problem duplication in the pyramid at each stage of resolution. RWT was developed to avoid image decimation in image fusion [17] which is an image processing application.

2.4 Type-1 Fuzzy Set

The fuzzy set I [18, 19] is a well-known fuzzy set that is represented by the finite set given by

$$I = \{(y, \mu_I(y))|y \in Y\}. \quad (3)$$

2.5 Intuitionistic Fuzzy Set (IFS)

A measureable representation of an intuitionistic fuzzy set I in a limited set Y in a finite set Y is represented as an IFS [20–22].

$$I = \{(y, \mu_I(y), v_I(y), \pi_I(y))|y \in Y\}. \quad (4)$$

2.6 Type-2 Fuzzy Set

Type-2 fuzzy sets lies in an interim rather than a solitary number aimed at each component [23, 24]. The expression is given by

$$I_{\text{Type}-2} = \left\{y, \hat{\mu}_I(y)|y \in Y\right\}. \quad (5)$$

Type-2 membership is referred to as

$$\begin{aligned} \mu^{\text{upper}} &= [\mu(y)]^{\beta} \\ \mu^{\text{lower}} &= [\mu(y)]^{1/\beta}, \end{aligned} \quad (6)$$

where $0 < \beta \leq 1$. A more useful way to indicate a Type-2 fuzzy collection is

$$I_{\text{Type}\,2} = \left\{y, \mu^{\text{upper}}(y), \mu^{\text{lower}}(y)|y \in Y\right\} \quad (7)$$

and $\mu^{\text{upper}}(x) < \mu(x) < \mu^{\text{lower}}(x), \quad \mu \in [0, 1]$ where the language hedges like focus or dilatation could be used for lower and upper.

2.7 Interval-Valued Intuitionistic Fuzzy Set (IVIFS)

Atanassov and Gargov [25, 26] method increases the intuitionistic fuzzy set's ability to deal with uncertain input while also enhancing its capability to address real-world decision-making challenges.

2.8 Particle Swarm Optimization (PSO)

Kennedy et al. [27] introduced the basic improvement methods for ongoing difficulties in real-world cases. PSO is intrigued by the social interactions of bird flocks and the idea of fish teaching. In image processing, PSO is used to determine the best block size for the combined image [28].

2.9 Teaching–Learning-Based Optimization (TLBO)

There are numerous improving methods [29–31] in research for addressing various challenges, and TLBO [6] is one of them that is based on population and is used globally. It depends on the outcome of a teacher reviewing the class's students' test results. This method replicates the learning abilities of the teacher and the students in the classroom. It operates through two crucial learning hubs: cooperation with specific peers and the teacher.

The "teacher stage" and "learner stage" are the two divisions of the TLBO algorithm. While the learning stage tells what a student learns from interacting and conversing with specific peers, and, whatever studied by the student from the instructor gives the teaching stage.

Teaching Phase. The TLBO technique's global search feature depends on the teaching phase. Mean is established at this stage. A terrific educator is necessary for mean improvement. In order to achieve higher results, picking a good teacher is crucial. Assume it is the student's prior behaviour as shown by the teacher and the current grade as revealed by conversation with other students. Comparisons between the historical mean and the current mean are provided as X_{old} and X_{new}. The teaching factor, which determines how to adjust the mean value, is where represents the random value and is chosen between [0, 1].

Learner Phase. Through interaction with other students, pupils in the learning stage develop their insight. As a result, this helps with calculating the mean. If the randomly picked student has more knowledge than the person, then every understudy learns something from them. After N iterations, the algorithm terminates. The TLBO algorithm is produced by combining the teacher stage with the student stage.

2.9.1 Contrast Visibility (CV)

Contrast represents the variation of the block pixels from the block principle is known, visibility of the image is typically more reasonable than the estimation of block means and variances. According to the following, the image block is visible:

$$\text{CV} = \frac{1}{m \times n} \sum_{(i,j)\in B_k} \frac{|I(i,j) - \mu_k|}{\mu_k}. \tag{8}$$

Here, I stands for input source image, μ is mean, and $m \times n$ is block size, respectively.

2.9.2 Intuitionistic Fuzzy Entropy (IFE)

In addition to fuzzy entropy tests for blurring in a fuzzy set, which are established as in [32] and utilised to work on the suggested method, this paper provides a special target work called IFE, which is significant in image processing.

$$\text{IFE}(Z; \alpha) = \sum_{i=1}^{n} \pi_Z(y_i)\exp(1 - \pi_Z(y_i)). \tag{9}$$

Entropy (IFE) is improved by determining the extreme entropy value, which is determined using Eq. (16) for values ranging from [0.1–1].

$$\alpha_{\text{opt}} = \max(\text{IFE}(Z; \alpha)). \tag{10}$$

3 Proposed Image Fusion Using TLBO

The process shown in Fig. 1 is used to create multimodal medical images fusion. Figure 2 introduces the TLBO flowchart.

(1) Read the two multimodal medical images that were provided.
(2) Using the formula, the size image is fuzzified.

$$\mu_{Z1}(I_{\text{ij1}}) = \frac{I_{\text{ij1}} - l_{\text{min}}}{l_{\text{max}} - l_{\text{min}}}, \tag{11}$$

where it ranges between zero and L–1 (L is highest grey value).

(3) Calculate the ideal value using Eqs. (9) and (10) for entropy for the first source image, which varies for different input images.
(4) Calculate for the input source image the fuzzified IFI using the formulae below and the minimal value of to denote as:

$$\mu_{\text{YIFCS1}}(I_{\text{ij1}}; \alpha) = \mu_{Z1}(I_{\text{ij1}}) \tag{12}$$

$$\upsilon_{\text{YIFCS1}}(I_{\text{ij1}}; \alpha) = (1 - \mu_{\text{YIFCS1}}(I_{\text{ij1}}; \alpha)^{\alpha})^{1/\alpha}, \tag{13}$$

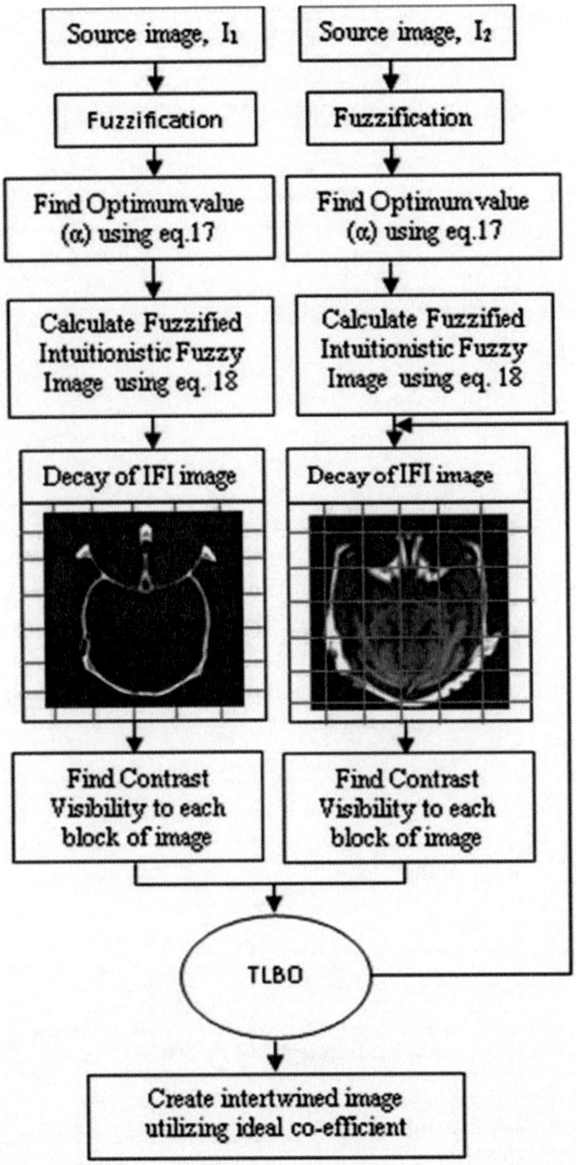

Fig. 1 Proposed technique blocks' schematic

$$\pi_{\mathrm{YIFCS1}}(I_{\mathrm{ij1}};\alpha) = 1 - \mu_{\mathrm{YIFCS1}}(I_{\mathrm{ij1}};\alpha) - \upsilon_{\mathrm{YIFCS1}}(I_{\mathrm{ij1}};\alpha), \tag{14}$$

$$I_{Y1} = \left\{(I_{\mathrm{ij1}}, \mu_{\mathrm{YIFCS1}}(I_{\mathrm{ij1}};\alpha), \upsilon_{\mathrm{YIFCS1}}(I_{\mathrm{ij1}};\alpha), \pi_{\mathrm{YIFCS1}}(I_{\mathrm{ij1}};\alpha))\right\}. \tag{15}$$

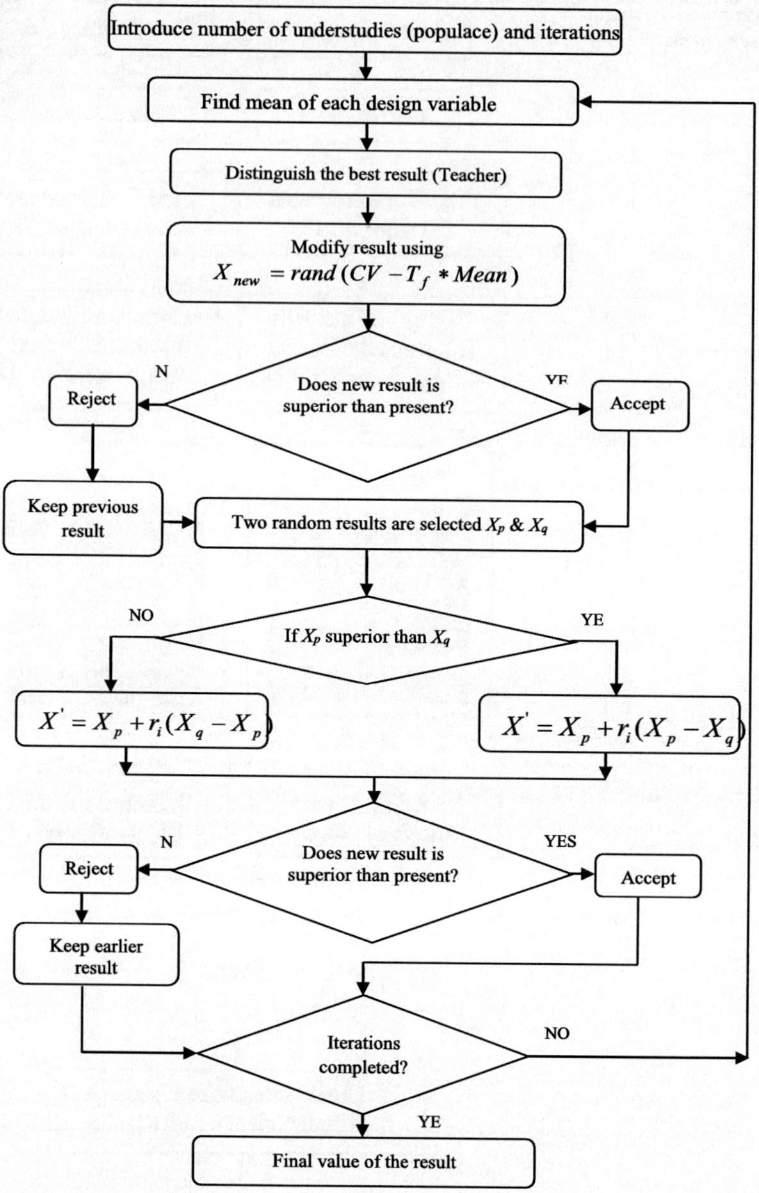

Fig. 2 TLBO algorithm flowchart

(5) For the second input image, repeat steps two through four of the aforementioned procedure to find.

$$I_{Y2} = \{(I_{ij2}, \mu_{YIFCS2}(I_{ij2}; \alpha), \upsilon_{YIFCS2}(I_{ij2}; \alpha), \pi_{YIFCS2}(I_{ij2}; \alpha))\}. \quad (16)$$

(6) Divide the two images into windows with the sizes specified for computing individual block contrast visibility of two source images separately.
(7) The TLBO technique is then used to calculate the best coefficients.
(8) Finally, using the ideal coefficients the fused image is produced.

4 Experimental Results

This section performs simulations using and comparing the existing and proposed approaches on various pairings of medical images. Two different image pairs are used in the experiments. For evaluating the quality of the images, the impressive objective performance measurements [33] PSNR, UQI, SSIM, CC, SF, E, $Q^{AB/F}$, average gradient (AG) are considered.

On several sets, overall performance of the suggested strategy is given and compared with existing methodologies. Separate figures and tables show the entire results for both subjective and objective norms. The multimodal medical image pairs come from [34, 35] and are registered.

While a magnetic resonance image provides information on soft tissue and muscles, a computed tomography image provides information on bones and rigid tissue. A single image created by fusing the two provides information that helps with diagnosis. Figure 3k represents shows that the image has more brightness and distinction than the fused images of earlier approaches. Additionally, Table 1 represents output-fused image in terms of objective criteria like PSNR (69.79 dB), which is higher for the created method than PSNR of the existing techniques. The suggested technique, as seen in the figure, provides improved execution in terms of contrast, brightness, and reflectivity of the output-fused image.

The subsequent example is shown by a T1-weighted MRI scan and an MRA image, both of which show some disease as white structures (Figs. 4a, b). The combination of these two photos provides correlated information in a single image, enhancing medical analysis. The algorithm created has a remarkable PSNR of 66.16 dB, as seen in Table 2. Figure 4k displays the algorithm's fused output image, which was produced at high spatial resolution with a pixel power of 93.38.

The third model shown in Fig. 5a,b address PET and magnetic resonance images. Table 3 shows that the PSNR of the fused output image produced by the developed approach is 65.81 dB. The fused output image provides clinicians with accurate illness diagnosis by providing substantial data on the magnitude of the disease that is not seen in other fused images.

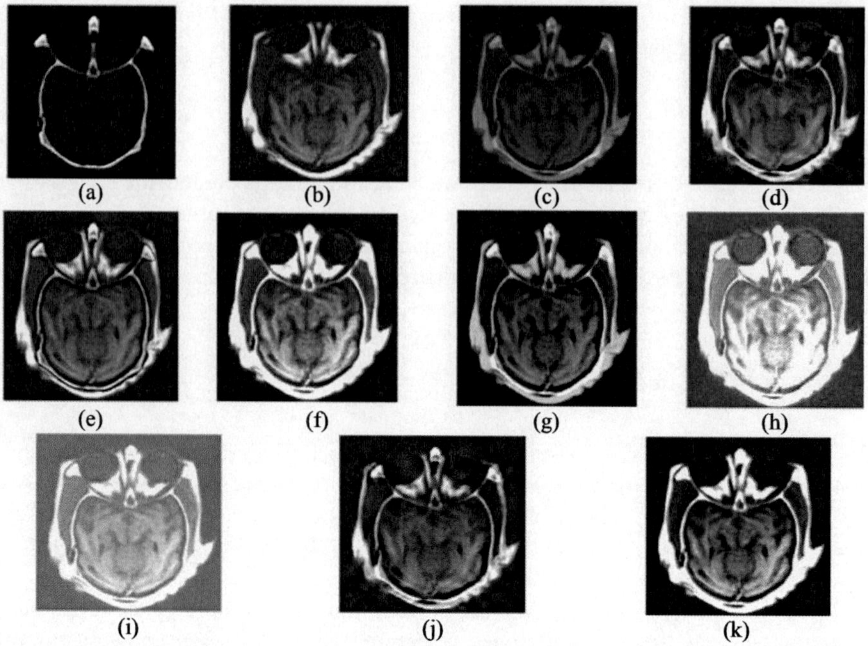

Fig. 3 Fusion yields for computed tomography as well as magnetic resonance images. **a** Computed tomography image **b** magnetic resonance image **c** fusion output by simple average **d** Fusion output by DWT **e** fusion output by RWT **f** fusion output by fuzzy set **g** fusion output by intuitionistic fuzzy set **h** fusion output by type-2 fuzzy set **I** fusion output by interval-valued intuitionistic fuzzy set **j** fusion output by PSO **k** fusion output by proposed method (TLBO)

5 Conclusion

This article uses one of the widely used optimization techniques called TLBO to demonstrate the appropriate block size for multimodal medical image fusion. The essay illustrates the fundamentals of fuzzy sets and the TLBO system. Here, the process is executed for the appropriate number of cycles to produce ideal coefficients and a fused output image. When compared to current approaches in terms of objective metrics, analysis of the results reveals that the TLBO algorithm in combination with fuzzy sets produces better results. All things considered, we can conclude that TLBO is a truly remarkable methodology for progressing the blocks that are separable and multimodal in order to provide quality perfect results in a shorter convergence time using incredibly well-known developmental strategies like PSO.

Table 1 Examination of objective parameters for various methods of image fusion for CT-MRI image pair displayed in Fig. 3

Fusion method	PSNR (dB)	UQI	SSIM	CC	AG (intensity change/ pixel)	E (bits/ pixel)	SF (Cycles/ millimetre)	$Q^{AB/F}$
Simple average [10]	63.754	0.623	0.996	0.822	9.364	5.959	24.192	0.445
DWT [11]	69.582	0.649	0.999	0.971	14.449	7.022	29.221	0.778
RWT [17]	69.651	0.872	0.999	0.972	15.991	6.822	32.345	0.834
Type-1 fuzzy set [18]	57.357	0.393	0.988	0.797	17.386	6.027	35.038	0.529
IFS [20]	65.776	0.305	0.998	0.948	14.414	5.353	30.738	0.732
Type-2 fuzzy set [24]	51.998	0.323	0.956	0.616	15.101	6.533	27.377	0.427
IVIFS [26]	56.317	0.506	0.979	0.764	10.195	6.177	20.053	0.594
PSO [28]	58.792	0.275	0.992	0.827	15.343	5.369	26.549	0.623
Proposed method (TLBO)	69.792	0.875	0.999	0.972	17.491	7.369	36.107	0.893

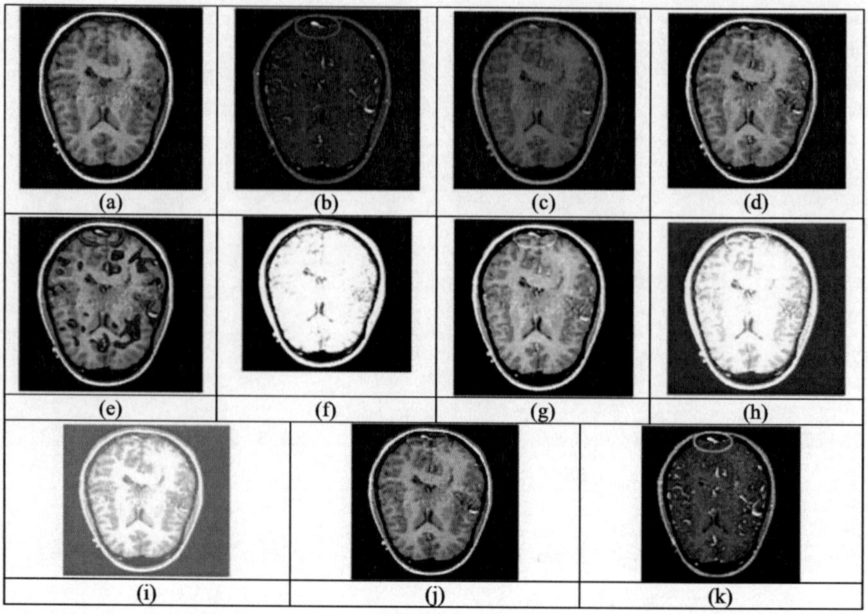

Fig. 4 Fusion outputs for MR and MRA images. **a** MR image **b** MRA image **c** fusion output by simple average **d** fusion output by DWT **e** fusion output by RWT **f** fusion output by fuzzy set **g** fusion output by INTUITIONISTIC fuzzy set **h** fusion output by type-2 fuzzy set **i** fusion output by interval-valued intuitionistic fuzzy set **j** Fusion output by PSO **k** fusion output by proposed method (TLBO)

Table 2 Examination of objective parameters for various methods of image fusion for MR-MRA image pair displayed in Fig. 4

Fusion method	PSNR (dB)	UQI	SSIM	CC	AG (intensity change/ pixel)	E (bits/ pixel)	SF (cycles/ millimetre)	$Q^{AB/F}$
Simple average [10]	63.701	0.364	0.998	0.840	11.015	5.513	25.885	0.508
DWT [11]	65.694	0.410	0.999	0.934	16.238	6.542	31.970	0.770
RWT [17]	64.194	0.342	0.998	0.899	17.489	6.206	33.831	0.760
Type-1 fuzzy set [18]	52.834	0.212	0.962	0.670	15.639	3.653	37.976	0.397
IFS [20]	61.725	0.329	0.996	0.902	17.573	5.167	35.847	0.641
Type-2 fuzzy set [24]	55.140	0.277	0.976	0.752	14.961	5.234	30.557	0.480
IVIFS [26]	55.726	0.308	0.977	0.786	10.921	5.207	21.101	0.614
PSO [28]	60.349	0.524	0.868	0.709	15.155	4.884	35.191	0.587
Proposed method (TLBO)	66.163	0.759	0.999	0.942	17.918	7.212	45.923	0.856

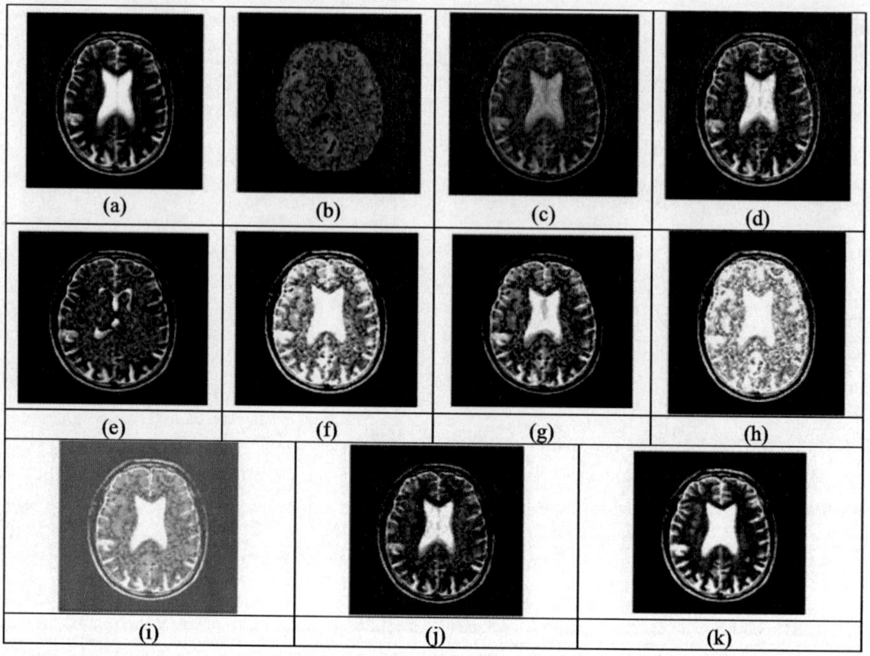

Fig. 5 Fusion outputs for MRI and PET images. **a** Magnetic resonance image **b** MRA image **c** fusion output by simple average **d** FUSION output by DWT **e** fusion output by RWT **f** fusion output by fuzzy set **g** fusion output by intuitionistic fuzzy set **h** fusion output by type-2 fuzzy set **i** fusion output by interval-valued intuitionistic fuzzy set **j** fusion output by PSO **k** fusion output by proposed method (TLBO)

Table 3 Examination of objective parameters for various methods of image fusion for MRI-PET image pair displayed in Fig. 5

Fusion method	PSNR (dB)	UQI	SSIM	CC	AG (intensity change/pixel)	E (bits/pixel)	SF (cycles/millimetre)	$Q^{AB/F}$
Simple average [10]	64.542	0.695	0.998	0.669	11.465	4.139	31.398	0.404
DWT [11]	62.795	0.175	0.997	0.715	17.586	4.865	36.352	0.376
RWT [17]	63.181	0.400	0.997	0.639	18.091	4.054	37.419	0.509
Type-1 fuzzy set [18]	53.150	0.083	0.981	0.197	23.462	3.838	47.630	0.322
IFS [20]	59.320	0.102	0.994	0.453	18.963	3.928	41.553	0.502
Type-2 fuzzy set [24]	54.823	0.566	0.983	0.323	21.877	3.954	42.797	0.322
IVIFS [26]	55.179	0.119	0.974	0.584	13.381	4.144	26.760	0.332
PSO [28]	55.122	0.177	0.713	0.333	24.435	4.116	26.707	0.399
Proposed method (TLBO)	65.813	0.717	0.988	0.881	31.599	6.230	52.983	0.751

References

1. Baum KG, Raerty K, Helguera M, Schmidt E (2007) Investigation of PET/MRI image fusion schemes for enhanced breast cancer diagnosis. In: Proceedings of IEEE seventh symposium conference on nuclear science (NSS), pp 3774–3780
2. Gholam HH, Alizad A, Fatemi M (2007) Integration of Vibro-Acoustography imaging modality with the traditional mammography. Int J Biomed Imaging,Hindawi Publishing Corporation. https://doi.org/10.1155/2007/40980
3. James AP, Dasarathy BV (2014) Medical image fusion: a survey of the state of the art. Inf Fusion 19:4–19
4. Zadeh LA (1965) Fuzzy sets. Inf Control 8:338–353
5. Atanassov KT (1986) Intuitionistic fuzzy sets. Fuzzy Sets Syst 20:87–96
6. Rao RV, Savsani VJ, Vakharia DP (2011) Teaching-learning-based optimization: a novel method for constrained mechanical design optimization problems. Comput Aided Des 43(3):303–315
7. Fogel B (1995) Evolutionary computation: towards a new philosophy of machine learning, 3rd edn. IEEE press, Wiley, Piscataway, NJ
8. Rao RV, Savsani VJ, Vakharia DP (2012) Teaching learning based optimization: an optimization method for continuous non-linear large scale problems. Inf Sci 183:1–15
9. Suresh CS (2012) Improved teaching learning optimization for global function optimization. Decis Sci Lett 2. https://doi.org/10.5267/j.dsl.2012.10.005
10. Naidu VPS, Raol JR (2008) Pixel level image fusion using wavelets and principal component analysis. Def Sci J 58(3):338–352
11. Yang Y, Park DS, Huang S, Rao N (2010) Medical image fusion via an effective wavelet based approach. EURASIP J Adv Signal Process. https://doi.org/10.1155/2010/579341

12. Prakash O, Park CM, Khare A, Jeon M, Gwak J (2019) Multiscale fusion of multimodal medical images using lifting scheme based biorthogonal wavelet transform. Optik. https://doi.org/10.1016/j.ijleo.2018.12.028
13. Vakaimalar E, Mala K, Suresh Babu R (2019) Multifocus image fusion scheme based on discrete cosine transform and spatial frequency. Multimedia Tools Appl. https://doi.org/10.1007/s11042-018-7124-9
14. Muzammil SR, Maqsood S, Haider S, Damasevicius R (2020) CSID: a novel multimodal image fusion algorithm for enhanced clinical diagnosis. Diagnostics 10. https://doi.org/10.3390/diagnostics10110904
15. Li B, Peng H, Wang J (2021) A novel fusion method based on dynamic threshold neural P systems and nonsubsampled contourlet transform for multi-modality medical images. Signal Process 178:1–13
16. Shi Z, Zhang C, Ye D, Qin P, Zhou R, Lei L (2022) MMI-fuse: multimodal brain image fusion with multiattention module. IEEE Access 10:37200–37214
17. Li X, He M, Roux M (2010) Multifocus image fusion based on redundant wavelet transform. IET Image Proc 4(4):283–293
18. Mendel JM (2001) Uncertain rule-based fuzzy logic systems introduction and new directions. Prentice-Hall, Eaglewood Cliffs, NJ
19. Gayathri K, Tirupal T (2018) Multimodal medical image fusion based on type-1 fuzzy sets. J Appl Sci Comput 5(10):1329–1341
20. Balasubramaniam P, Ananthi VP (2014) Image Fusion using Intuitionistic Fuzzy Sets. Information Fusion 20:21–30
21. Tirupal T, Mohan BC, Kumar SS (2019) Multimodal medical image fusion based on Yager's intuitionistic fuzzy sets. Iranian J Fuzzy Syst 16(1):33–48
22. Tirupal T, Mohan BC, Kumar SS (2017) Multimodal medical image fusion based on Sugeno's intuitionistic fuzzy sets. ETRI J 392:173–180
23. Karnik NN, Mendel JM, Liang Q (1999) Type-2 fuzzy logic systems. IEEE Trans Fuzzy Syst 7:643–658
24. Tirupal T, Chandra Mohan B, Srinivas Kumar S (2019) Type-2 fuzzy set based multi-modal medical image fusion. Indian Conference on Applied Mechanics (INCAM-2019), IISc BANGALORE, India
25. Atanassov K, Gargov G (1989) Interval valued intuitionistic fuzzy sets. Fuzzy Sets Syst 31(3):343–349
26. Tirupal T, Chandra Mohan B, Srinivas Kumar S (2021) Multimodal medical image fusion based on interval-valued intuitionistic fuzzy sets. In: Kumar R, Chauhan VS, Talha M, Pathak H (eds) Machines, mechansim and robotics. lecture notes in mechanical engineering 2021, Springer, pp 965–971. https://doi.org/10.1007/978-981-16-0550-5_91
27. Kennedy J, Eberhart R (1995) Particle swarm optimization. In: IEEE International conference on neural networks. Perth, Australia. https://doi.org/10.1109/ICNN.1995.488968
28. Siddiqui AB, Jaffar MA, Hussain A, Anwar MM (2011) Block-based pixel level multi-focus image fusion using particle swarm optimization. Int J Innovative Comput Inf Control 7(A):3583–3596
29. Daniel E, Anitha J, Kamaleshwaran KK, Rani I (2017) Optimum spectrum mask based medical image fusion using Gray Wolf Optimization. Biomed Signal Process Control 34:36–43
30. Daniel E (2018) Optimum wavelet based homomorphic medical image fusion using hybrid genetic–grey wolf optimization algorithm. IEEE Sens J. https://doi.org/10.1109/JSEN.2018.2822712
31. Hamid RS, Zahra T (2019) MRI and PET/SPECT image fusion at feature level using ant colony based segmentation. Biomed Signal Process Control 47:63–74
32. Chaira T (2011) A novel intuitionistic fuzzy c means clustering algorithm and its application to medical images. Appl Soft Comput 11(2):1711–1717
33. Jagalingam P, Hegde AV (2015) A review of quality metrics for fused image. In: Aquatic procedia of international conference on water resources, coastal and ocean engineering (ICWRCOE), vol 4, pp 133–142
34. Homepage, http://www.metapix.de/fusion.htm
35. Homepage, http://www.med.harvard.edu

Importance of Knee Angle and Trunk Lean in the Detection of an Abnormal Walking Pattern Using Machine Learning

Pawan Pandit, Dhruv Thummar, Khyati Verma, K. V. Gangadharan, Bishwaranjan Das, and Yogeesh Kamat

Abstract Human gait can be quantified using motion capture systems. Three-dimensional (3D) gait analysis is considered the gold standard for gait assessment. However, the process of three-dimensional analysis is cumbersome and time-consuming. It also requires complex software and a sophisticated environment. Hence, it is limited to a smaller section of the population. We, therefore, aim to develop a system that can predict abnormal walking patterns by analyzing trunk lean and knee angle information. A vision-based OpenPose algorithm was used to calculate individual trunk lean and knee angles. Web applications have been integrated with this algorithm so that any device can use it. A Miqus camera system of Qualisys 3D gait analysis system was used to validate the OpenPose algorithm. The validation method yielded an error of $\pm$ 9° in knee angle and $\pm$ 8° in trunk lean. The natural walking pattern of 100 healthy individuals was compared to simulated walking patterns in an unconstrained setting in order to develop a machine learning program. From the collected data, an RNN-based LSTM machine learning model was trained to distinguish between normal and abnormal walkings. LSTM-based models were able to distinguish between normal and abnormal gaits with an accuracy of 80%. This study shows that knee angle and trunk lean patterns collected during walking can be significant indicators of abnormal gait.

Keywords GAIT · Knee angle · Trunk lean · Machine learning · Long short-term memory · Recurrent neural network · Abnormal GAIT

P. Pandit · D. Thummar · K. Verma (✉) · K. V. Gangadharan
Department of Mechanical Engineering, National Institute of Technology Karnataka, Surathkal 575025, India
e-mail: kverma@nitk.edu.in

B. Das · Y. Kamat
KMC Hospital, Ambedkar Circle, Mangalore, Karnataka, India

S. Sharma et al. (eds.), *Intelligent Control, Robotics, and Industrial Automation*, Lecture Notes in Electrical Engineering 1066,
https://doi.org/10.1007/978-981-99-4634-1_42

1 Introduction

Human mobility is essential in keeping the joints healthy. A decline in that can result in kinematic dysfunction and reduced quality of life. Trauma and age-related degeneration are two of the most important factors that contribute to joint anomalies. Quantification of these biomechanical alterations can be done by gait analysis. This is essential for identifying factors leading to gait irregularities. This paves the way for appropriate medical treatments and rehabilitation. Three-dimensional gait analysis is considered to be the most reliable method to quantify the abnormalities with more accuracy. However, this gait analysis involves complex software and requires interpretation by experts. This makes it tedious and cumbersome.

This paved the way for the acquisition of similar parameters using the two-dimensional gait analysis method. Two types of gait approaches have attracted considerable attention: (1) sensor-based and (2) vision-based technologies.

Davis was the first to design a sensor-based system by using passive reflective markers and image processing techniques for mapping human joint motion [1]. Accuracy is one of the drawbacks of this system as reflecting markers are highly dependent on the ambient illumination conditions, and change in marker position due to body motion can produce false parameters. [2, 3]. Wearable sensors came into existence as an alternative due to their ease to use and portability nature. With recent developments in integrated circuits and microcontrollers, wearable sensor-based systems are capable of recording essential gait characteristics [4]. However, wearable devices can interfere with the subject's habitual movement. There are various non-wearable approaches for measuring gait metrics, such as pressure mats, gyroscopes, goniometers, and motion sensors [5]. Because these systems are costly, involve significant calibration, and operate in a controlled environment, they are best suited for laboratory settings. They also require maintenance in battery charging, data transferring, and cleaning.

Kinects and RGB cameras are examples of vision-based gait analysis technologies. Gait analysis with the Kinect is less expensive and provides a reasonable level of accuracy [6]. It uses infrared networks to create a 3D depth view of the physical world. Kinect lacks a depth image in bright environments [7, 8]. A study by Vishwakumar et al. found that Kinect measurements were accurate only if the knees were exposed to the camera [9]. Due to this, it is a challenge to measure the knee angle for Indian clothing using a Kinect. Camera-based devices are another substitute for vision-based technology. Gait analysis can now be performed using more detailed images captured with recent advancements in camera technology.

Furthermore, computer vision-based technologies have made RGB cameras more widely available in a number of different fields. Convolutional Neural Networks (CNNs) are widely used learning algorithms in computer vision. They are often trained using a large number of images of various individuals [10]. An integrated skeleton-like model is formed by recognizing and integrating human joints using CNN algorithms in the pose estimation of humans. Analysis of gait and its parameters was later conducted using the posture created by this algorithm [9, 11]. Deep

Neural Networks were used first time by Toshev et al. to predict human posture. This approach is well known as DeepPose: Human Pose Estimation [12]. Research on human posture estimation naturally tuned to deep learning-based systems after DeepPose demonstrated significant results. Various CNN algorithms like OpenPose, Posture Net, Blaze Pose, Deep Pose, Dense Pose, and Deep Cut are currently in use. These learning algorithms are open source and use the camera to estimate human posture. Gait parameters can be extracted with very minimal cost, time, and effort by using this algorithm.

The vision-based or sensor-based devices are primarily utilized to evaluate spatiotemporal parameters of gait. The most are Heel strike, toe-off, stride duration, and postural sway which are a few of the crucial gait characteristics. These gait metrics can help health professionals to understand an individual's altered walking pattern. However, non-unique correlations between these factors vary between different individuals. This makes it more difficult for health experts to find a generalized pattern. Hence, a gait analysis will need to be combined with machine learning (ML) to handle high-dimensional data [13, 14].

ML technology provides a solution to aid clinicians by analyzing a large number of data and assuring a qualitative comprehension of patient health records. There is a paucity of studies on the use of machine learning to assess abnormal gait. The very reliable findings obtained during this research demonstrated that the ML model could certainly be utilized for gait analysis [15]. This facilitates better patient monitoring, reduces diagnostic time, and assists clinicians in selecting appropriate treatment measures and rehabilitation. The current literature review reveals that trunk lean is not emphasized, despite its potential role in identifying abnormal gait cycles in patients with knee osteoarthritis [16, 17].

Furthermore, the majority of gait analysis research in the literature has been conducted in a controlled clinical laboratory. A small number of participants were involved. As a result, the validity of the parameters obtained during this research may be questioned as they alter the participants' natural walking [18]. In order to determine a person's natural walk, gait analysis should ideally be conducted over a longer period of walking in an unconstrained environment.

As a result, a gait analysis of around 100 healthy volunteers in natural walking conditions is conducted in this research, and a system capable of identifying abnormal gait patterns using deep learning is presented. During the gait analysis of 100 healthy volunteers in this study, only knee angle and trunk lean patterns while walking were assessed using a camera-based pose estimation algorithm to indicate their efficacy in gait analysis. The proposed classification system in the present study employs RNN approaches, particularly the LSTM network. It is used in the present study to analyze gait parameters obtained during subsequent walking sequences because it is best suitable for analyzing series data. Gait parameters recorded during successive occurrences of walking may be termed time-series data because walking is a repeating gait cycle.

Detailed information on the system architecture utilized for gait analysis of around 100 healthy volunteers and the development of an RNN-based model to distinguish between normal and abnormal walking is described in further sections.

2 System Architecture for Obtaining the Key Points of Human Posture

A vision-based posture estimation approach is utilized for gait analysis in the present work. Pose estimation-based gait analysis requires only an RGB camera and a CNN-based pose detection algorithm. The reason for using this method in the present study is that it does not require any trackers on the human body, allowing the individual to walk naturally throughout the gait analysis [10]. Human Pose Estimation is a technique for detecting and classifying human body joints. It is primarily a way of extracting a set of coordinates for each joint of the body, referred to as a key point, which can illustrate a person's posture.

There are several CNN models currently available that employ the camera to estimate human posture, which includes OpenPose, Posture net, Blaze pose, Deep Pose, DensePose, Deep cut, etc. Selecting one model over another is dependent on the application. In addition, model size, operation time, and simplicity of implementation can also be key factors to consider when selecting a model. The OpenPose CNN model developed by Cao et al. is used for posture assessment in this study because it produces more accurate results in Human Pose Estimation than other models [19]. The benefit of OpenPose is that it can identify human body joints with attire such as dhoti or saree with low error [10]. Another advantage of this method is that it is easy to implement using TensorFlow, a popular open-source ML framework.

2.1 OpenPose—Posture Estimation Algorithm

OpenPose is a live multi-person Human Pose Estimation library that has demonstrated the ability to recognize the human body, foot, palm, and face key points on single images for the first time. OpenPose supports webcams, infrared cameras, depth cameras, stereo lenses, and surveillance cameras and configured input sources such as pictures, movies, and camera streams. In terms of hardware, it supports Nvidia GPU (CUDA), AMD GPU (OpenCL), and non-GPU (CPU) computation.

The original code for OpenPose was written in C++ and Caffe. In the OpenPose working algorithm, the image is initially passed through a baseline CNN network to extract feature maps. A VGG-19 algorithm is utilized to extract features in the OpenPose algorithm. Following that, a two-branch multi-stage CNN is fed an input RGB image paired with a feature map. Two branches indicate that CNN generates two distinct outputs. The term "multi-stage" refers to the network's layers stacked one over another at every stage. The top branch predicts confidence maps for various human body parts. The bottom branch predicts affinity fields, which measure the correlation between different body parts. Predictions from the first stage and the features F obtained initially are concatenated and used to make more precise and reliable predictions in the second through sixth stages, which follow the same architecture as the first. After the six stages, the confidence maps obtained from the last stages are used to derive the position of a specific body joint through local maximums. Once the position of each body part has been determined, it must be connected to form pairs. A pair can be created between any two joints, but in order to find the correct pair, the weightage must be assigned to each pair. To assign a weightage, the line integral over the pair connecting part candidates, though associated PAFs for that pair, is calculated. The line integral will assign a score to each pair, which will be saved in a weighted bipartite graph to solve the assignment issue. The weighted bipartite graph displays all possible connections between two parts of candidates with scores assigned to each link. The links with the greatest score are chosen as the final pairs.

The final step is to transform these links or pairings into skeletons. Initially, each link is thought to belong to a separate individual. Due to that, the number of human data will be the same as the number of connections detected in the multi-person image. Following that, each human data is reviewed and compared with others; if any two humans share the pair via the same joint, these two humans are combined. This procedure is repeated until no two humans share a part. Finally, a collection of human sets is obtained, where each human contains a set of pairs, each of which contains the relative coordinate of joints. Figure 1 illustrates the 18 key points identified by the OpenPose algorithm at the end of the pipeline.

The OpenPose algorithm final output, which comprises the relative position of joints, was used to compute the knee angle trunk lean in this study. The knee angle is calculated by locating the foot joint, knee joint, and hip joint. The positions of the

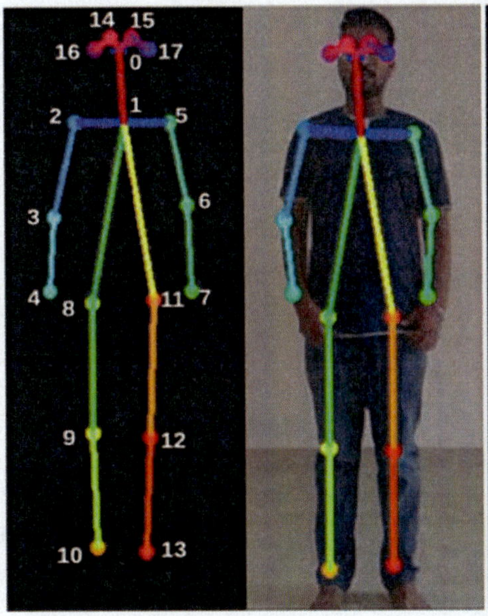

Key Point	Joint Name
0	Nose
1	Neck
2	Right Shoulder
3	Right Elbow
4	Right Wrist
5	Left Shoulder
6	Left Elbow
7	Left Wrist
8	Right Hip
9	Right Knee
10	Right Ankle
11	Left Hip
12	Left Knee
13	Left Ankle
14	Right Eye
15	Left Eye
16	Right Ear
17	Left Ear

Fig.1. Eighteen key points of the human body detected by OpenPose

Fig.2 Knee angle and trunk lean measurement

hip and shoulder joints are used to calculate the trunk lean, as shown in Fig. 2. In normal conditions, the knee angle ranges from 115 to 180 ° [20], while the lateral trunk lean ranges from 85 to 95 ° [21]. The range of angles in an abnormal situation is determined by the abnormality. In the case of knee osteoarthritis, the range of knee angle is around 130–180 ° [22], while the range of lateral trunk lean is approximately 80–100 ° [16].

To calculate the knee angle and trunk lean, cosine formula is used, which is as follows:

$$\theta = \cos^{-1} \frac{\overline{a}.\overline{b}}{\overline{a}\overline{b}}. \tag{1}$$

In Eq. 1, θ represents the angle between $\overline{a}$ and $\overline{b}$, and these vectors are calculated using joint location data obtained from OpenPose.

2.2 *Data Acquisition Platform*

After selecting OpenPose as the pose estimation algorithm, the web application is integrated with it shown in Fig. 3, to perform gait analysis on any device, regardless of operating system (Windows, Mac, or Linux). The web application is built with the Flask framework. The system is made up of two parts. The first is a front end, while the second is an analysis server. The front end is compatible with any web-enabled device, such as a laptop or smartphone. The analysis server primarily estimates coordinates of human key points using OpenPose, calculates knee angle and trunk lean, and shows it through the front end in the web application's parametric box. Furthermore, the system also produces and displays a video of the skeleton image superimposed as well as a video of solely the skeletal image on the live gait movement. It also has two-toggle buttons; that one allows users to switch between low and high-computation OpenPose models, and the second is used to record the gait analysis data. The FPS obtained with the OpenPose web application is approximately 4–5 FPS on a workstation with 8 GB ram, a 6 GB graphics card, and Windows 10 operating system. This FPS can be enhanced by employing more powerful computing resources or by not using a live web application. In the second scenario, a video of walking may be captured using a camera, and then angles can be determined separately on each frame using OpenPose such that the FPS remains the same as the camera FPS.

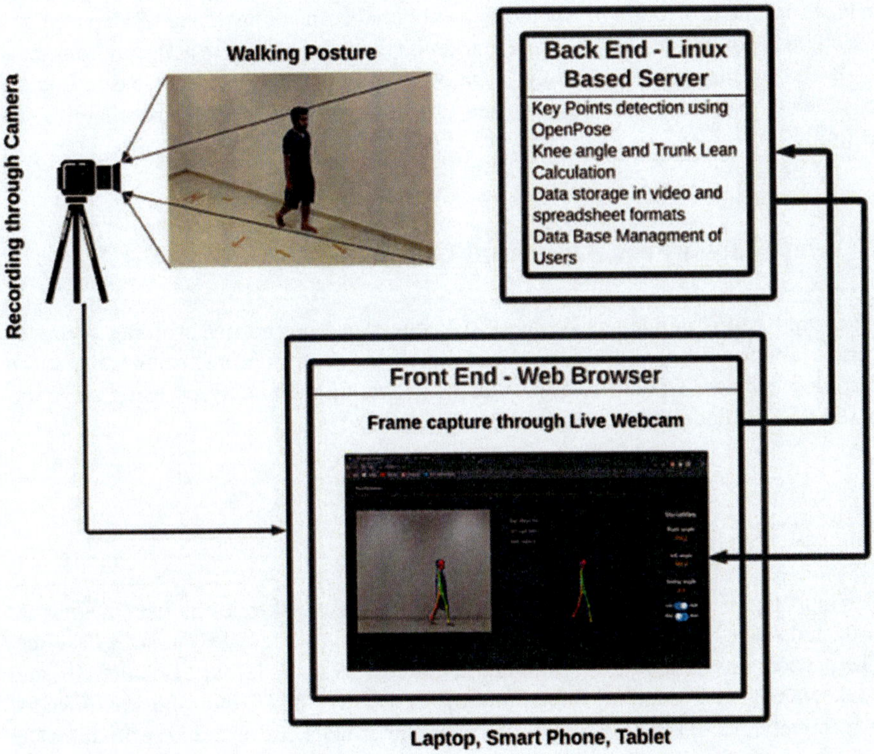

Fig.3 Web application of posture estimation

2.3 System Validation

Following the successful implementation of an OpenPose-based web application, the algorithm is tested using Qualisys, a motion capture system (Mocap) that employs reflecting markers. The present study used a 12-camera system with 55 reflecting markers mounted on the human subject according to the whole-body marker set defined in the Qualisys PAF package of Qualisys documentation [23]. Two Miqus Hybrid cameras are also used to track the subject's movement. After the motion capture system is set up and the trackers are mounted on the subject's body, the subject is asked to perform different knee and upper trunk postures, while the OpenPose-based web application and Mocap systems record the walking movement. The data acquired from both systems for knee angle and trunk lean indicate that the OpenPose approach has $\pm$ 9° error in knee angle measurement and $\pm$ 8° error in trunk lean measurement. The reason for this error might be that the angle derived from OpenPose is based on the projection of a person in a 2D image. In contrast, in a Mocap system, the angle is determined by placing a marker on the human body in a 3D environment.

Based on the validation results, it is observed that the OpenPose-based human posture estimate technique may be used for gait analysis where high precision is not required.

3 Methodology for Abnormal Gait Analysis

Abnormal gait is defined as behavior that deviates from normal walking owing to various abnormalities or disorders. In this work, the knee angle and trunk lean pattern recorded from the OpenPose algorithm while walking are analyzed using an RNN-based LSTM model to determine abnormal gait [24].

3.1 Data Collection

The first step in constructing an ML model is to collect task-relevant data. Data were collected from 100 healthy individuals (85 males and 15 females) to set a baseline. The participants' average weight, height, and age were 66.316 kg, 170.475 cm, and 22.41 years, with standard deviations of 13.473 kg, 8.907 cm, and 2.678 years, respectively. All of the participants wore regular clothes during the experiment. Data from both types of gaits must be gathered in order to distinguish between normal and altered gaits. However, collecting a large number of altered gait people data is difficult; as a result, normal people data are collected in two ways: normal walking and simulated walking as if they have a gait disorder. The walking pattern for collecting data is marked on the ground to aid volunteers in determining how far they must travel to collect their data. One camera was positioned along the walking path to capture the volunteers' walking patterns as they walked along the lines marked on the ground. The camera is linked directly to the laptop, which runs an OpenPose-based web application. Posture estimation over live walking movement captured by a camera is conducted through this application, and data are saved in video format, as well as the pattern of knee angle and trunk lean while walking is saved in spreadsheet format. This was in relation to the individual's normal walking pattern, and then the same individual was then asked to change their walking style in an abnormal way to record the data in altered gait movement. In this way, data have been collected in both a normal and altered manners.

The pattern of knee angle and trunk lean recorded during walking in spreadsheet format is used to train the LSTM model to demonstrate their importance in detecting abnormal gait. From the collected 200 spreadsheet data points, 90 data points from normal gait, and 90 data points from abnormal gait, total of 180 data points are used to train the LSTM model, and the remaining 20 data points are used to evaluate the model's accuracy.

3.2 Network Architecture (LSTM)

In the present study, an LSTM plus ANN-based model is used to distinguish between abnormal and normal gaits. An LSTM unit is made up of various gates, such as input, output, and forget. These three gates regulate the flow of information through the unit, allowing it to remember essential sequence values over long periods of time. Due to this phenomenon, LSTM can efficiently handle sequence-type data. As a result, it was used in the present study to examine the pattern of knee angle and trunk lean during walking. In contrast, ANN is used to analyze the parameters obtained from the LSTM cell. The ANN structure is made up of three layers: input, output, and one or more hidden layers. The data are passed from the input layer to the hidden layer, where it is analyzed. The hidden layer consists of a number of neurons that perform computations on the input data and tend to provide analyzed results via the output layer.

The proposed model architecture comprises five layers: an input layer, an LSTM layer, two dense layers, and an output layer. The input layer gets a 150-datapoint long sequence, including the values of two knee angles and trunk lean at each data point. This sequence is then passed through many to many LSTM cells, and analyzed parameters are collected. These parameters are then passed through a dense layer for extensive analysis. The parameters collected from the dense layer are then combined with the new four input parameters, which comprise the age, height, gender, and weight data of the individual whose walking sequence is being analyzed. Then, these parameters are further passed through a dense layer and output layer to predict the class of gait, whether it is normal or abnormal. For dense layers, the ReLU activation function is used to reduce training time [25]. For binary classification, the output layer employs the sigmoid activation function. In this study, the binary cross-entropy loss function is used to train the ML model. The equation of the binary cross-entropy loss function is as follows:

$$L = -\frac{1}{N}\sum y\log(p(y)) + (1-y)\log(p(1-y)). \quad (2)$$

In the above equation, y represents the true output, and $p(y)$ represents the predicted output obtained from the ML model. The proposed ML model architecture includes several hyperparameters, such as the number of hidden layers, the number of neurons in the hidden layers, the number of epochs, the number of cells in the LSTM layer, and the learning rate's value. The current study employs a trial-and-error methodology to identify the optimal set of hyperparameters that offers the highest accuracy when training and trained models do not overfit or underfit over training datasets in order to obtain the highest degree of accuracy for ML tasks. For each hyperparameter combination, the model is trained using the training dataset described in the previous sections using the Adam optimizer and backpropagation algorithm [26, 27]. In the present study, 80–85 hyperparameter combinations are explored. Figure 4 depicts the architecture corresponding to the optimal hyperparameter combination from the explored combinations, and Fig. 5 illustrates training over the training dataset in terms of loss value and accuracy with regard to a number of epochs.

The LSTM plus ANN model was trained for 1000 epochs, and 96 percent accuracy was achieved on the training dataset for the optimum hyperparameter combination. After training, the model is assessed over the test dataset, and an accuracy of 80% is observed. The entire training procedure is carried out in the present study using a Google Colab with TensorFlow and Keras framework.

The difference between normal and abnormal gaits can be easily and quickly determined with such great accuracy using our LSTM plus ANN-based model. OpenPose is a validated software to analyze various gait parameters. The similar studies have been undertaken using four or more cameras which require a sophisticated laboratory. We have managed to achieve comparable accuracy with only one camera and a smaller non-standardized setup. The additional benefit was provided by use of our machine learning model. The future benefit of this learning model is the potential application of such device in making gait analysis possible in doctors' outpatient clinic environment.

4 Conclusion

In the present study, an LSTM plus ANN-based ML model is used to identify abnormal gaits in real time. The proposed method uses the pattern of knee angle and trunk lean while walking to detect abnormal gaits. In the present study, OpenPose was used to detect and extract essential human body points, which were then used to measure knee angle and trunk lean during walking. The suggested technique has an accuracy of 80% for detecting abnormal gait using only knee angle and trunk lean data. According to these findings, knee angle and trunk lean patterns during walking

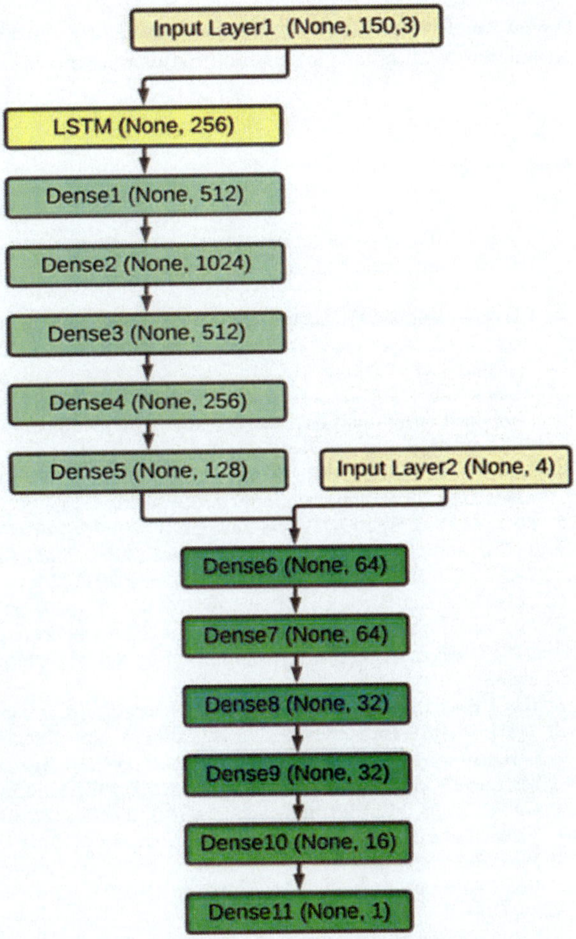

Fig.4 Architecture of LSTM plus ANN-based model

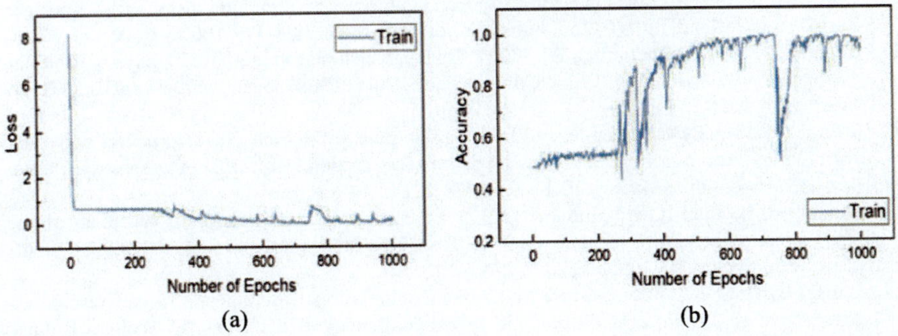

Fig.5 Graph of **a** loss versus number of epochs and **b** accuracy versus number of epochs for LSTM plus ANN-based model during the training process

can be considered essential criteria in identifying abnormal gait, and combining these parameters with other gait characteristics may aid in achieving higher accuracy.

References

1. Davis III RB, Ounpuu S, Tyburski D, Gage JR (1991) A gait analysis data collection and reduction technique. Human Movement Sci 10(5):575–87. https://doi.org/10.1016/0167-9457(91)90046-Z
2. Clayton HM (1996) Instrumentation and techniques in locomotion and lameness. Veterinary Clinics North Am: Equine Practice 12(2):337–350. https://doi.org/10.1016/S0749-0739(17)30285-7
3. Reinschmidt C, Van Den Bogert AJ, Nigg BM, Lundberg A, Murphy N (1997) Effect of skin movement on the analysis of skeletal knee joint motion during running. J Biomech 30(7):729–732. https://doi.org/10.1016/S0021-9290(97)00001-8
4. Tao W, Liu T, Zheng R, Feng H (2012) Gait analysis using wearable sensors. Sensors 12(2):2255–2283. https://doi.org/10.3390/s120202255
5. Muro-De-La-Herran A, Garcia-Zapirain B, Mendez-Zorrilla A (2014) Gait analysis methods: an overview of wearable and non-wearable systems, highlighting clinical applications. Sensors 14(2):3362–3394. https://doi.org/10.3390/s140203362
6. Gabel M, Gilad-Bachrach R, Renshaw E, Schuster A (2012) Full body gait analysis with Kinect. In: Annual international conference of the IEEE engineering in medicine and biology society 2012. IEEE, San Diego, California, USA, pp 1964–1967. https://doi.org/10.1109/EMBC.2012.6346340
7. Han J, Shao L, Xu D, Shotton J (2013) Enhanced computer vision with microsoft kinect sensor: a review. IEEE Trans Cybern 43(5):1318–1334. https://doi.org/10.1109/TCYB.2013.2265378
8. Livingston MA, Sebastian J, Ai Z, Decker JW (2012) Performance measurements for the Microsoft Kinect skeleton. IEEE Virtual Reality Workshops (VRW) 2012. IEEE, Costa Mesa, CA, USA, pp. 119–120. https://doi.org/10.1109/VR.2012.6180911
9. Viswakumar A, Rajagopalan V, Ray T, Parimi C (2019) Human gait analysis using OpenPose. In: Fifth international conference on image information processing (ICIIP) 2019. IEEE.Waknaghat, Shimla, Himanchal, pp 310–314. https://doi.org/10.1109/ICIIP47207.2019.8985781
10. Lin TY, Maire M, Belongie S, Hays J, Perona P, Ramanan D, Dollár P, Zitnick CL.: Microsoft COCO: Common objects in context. European conference on computer vision 2014. Springer. Zurich, Switzerland, pp 740–755. https://doi.org/10.1007/978-3-319-10602-1_48
11. Stenum J, Rossi C, Roemmich RT (2021) Two-dimensional video-based analysis of human gait using pose estimation. PLoS Comput Biol 17(4):e1008935. https://doi.org/10.1371/journal.pcbi.1008935
12. Toshev A, Szegedy C.: Deeppose (2014) Human pose estimation via deep neural networks. IEEE conference on computer vision and pattern recognition 2014. IEEE, Colubus, OH, USA, pp 1653–1660. https://doi.org/10.1109/CVPR.2014.214
13. Dolatabadi E, Taati B, Mihailidis A (2017) An automated classification of pathological gait using unobtrusive sensing technology. IEEE Trans Neural Syst Rehabil Eng 25(12):2336–2346. https://doi.org/10.1109/TNSRE.2017.2736939
14. Lai DTH, Begg RK, Palaniswami M (2009) Computational intelligence in gait research: a perspective on current applications and future challenges. IEEE Trans Inf Technol Biomed 13(5):687–702. https://doi.org/10.1109/TITB.2009.2022913
15. Jinnovart T, Cai X, Thonglek K (2020) Abnormal gait recognition in real-time using recurrent neural networks. In: 59th IEEE conference on decision and control (CDC) 2020. IEEE, Jeju, Korea (South), pp 972–977. https://doi.org/10.1109/CDC42340.2020.93041

16. Hunt MA, Birmingham TB, Bryant D, Jones I, Giffin JR, Jenkyn TR, Vandervoort AA (2008) Lateral trunk lean explains variation in dynamic knee joint load in patients with medial compartment knee osteoarthritis. Osteoarthritis Cartilage 16(5):591–599. https://doi.org/10.1016/j.joca.2007.10.017
17. Simic M, Hunt MA, Bennell KL, Hinman RS, Wrigley TV (2012) Trunk lean gait modification and knee joint load in people with medial knee osteoarthritis: the effect of varying trunk lean angles. Arthritis Care Res 64(10):1545–1553. https://doi.org/10.1002/acr.21724
18. Kinsella S, Moran K (2008) Gait pattern categorization of stroke participants with equinus deformity of the foot. Gait Posture 27(1):144–151. https://doi.org/10.1016/j.gaitpost.2007.03.008
19. Cao Z, Simon T, Wei SE, Sheikh Y (1017) Realtime multi-person 2d pose estimation using part affinity fields. In: IEEE conference on computer vision and pattern recognition 2017, pp 7291–7299. https://doi.org/10.48550/arXiv.1611.08050
20. Kumar N, Pankaj D, Mahajan A, Kumar A, Sohi BS (2009) Evaluation of normal gait using electro-goniometer. J Sci Indus Res 68(8):696–8. http://nopr.niscpr.res.in/handle/123456789/5302
21. Favre J, Erhart-Hledik JC, Chehab EF, Andriacchi TP (2016) General scheme to reduce the knee adduction moment by modifying a combination of gait variables. J Orthop Res 34(9):1547–1556. https://doi.org/10.1002/jor.23151
22. Favre J, Jolles BM (2016) Gait analysis of patients with knee osteoarthritis highlights a pathological mechanical pathway and provides a basis for therapeutic interventions. EFORT open reviews 1(10):368–374. https://doi.org/10.1302/2058-5241.1.000051
23. https://www.qualisys.com/ Last accesses 29 May 2022
24. Hochreiter S, Schmidhuber J (1997) Long short-term memory. Neural Comput 9(8):1735–1780. https://doi.org/10.1162/neco.1997.9.8.1735
25. Nair V, Hinton GE (2010) Rectified linear units improve restricted boltzmann machines. In: International conference on machine learning 2010. Haifa, Israel (2010). https://doi.org/10.5555/3104322.3104425
26. Kingma DP, Ba JA)(2015) A method for stochastic optimization. In: 3rd International conference for learning representation 2015 9, arXivpp1412.6980. San Diego (2015). https://doi.org/10.48550/arXiv.1412.6980
27. Rumelhart DE, Hinton GE, Williams RJ (1986) Learning representations by back-propagating errors. Nature 323(6088):533–536. https://doi.org/10.1038/323533a0

Malaria Parasite Detection and Outbreak Warning System Using Deep Learning

Areefa, Sivarama Krishna Koneru, Kota Pragathi, and Koyyada Rishitha

Abstract Malaria is an epidemic disease which causes by parasites that are spread through bites from infected Anopheles mosquitoes. Through early diagnosis and timely treatment, malaria death rates can be decreased and prevented. But the manual examination of blood smears by laboratory technicians is time-consuming and probable chances of human errors. Considering this need, in this paper, we are demonstrating a proposed system deployed over a WebApp could help laboratories to perform fast and accurate tests. For the malaria parasite detection in the blood cells, we proposed our custom CNN model and even implemented the dataset on the Transfer Learning model (VGG19) to compare the results. We got the highest accuracy of 97.74% for our CNN model which is the simplest of the current models as it prioritizes only important parameters, provides great computational efficiency, and is realistic for implementation. Additionally, a malaria outbreak warning system is proposed to warn the people in localities when there is availing risk based on climatic factors. A logistic regression model implemented by Gradient Descent Optimizer is deployed which predicts the malaria outbreak from the live weather forecast inputs from the OpenWeather API and displays the probability of the outbreak to the users with one click.

Keywords Custom CNN model · Deep learning · Web/App · Malaria

1 Introduction

Health care is a major field in which technologies must be implemented to make it more feasible and accessible for people to take preventive measures to protect themselves from epidemic diseases. Malaria, one of the world's most dangerous infectious diseases and a leading cause of public health concerns, affects more than two-thirds of the global population, killing an estimated million people each year.

Areefa (✉) · S. K. Koneru · K. Pragathi · K. Rishitha
S R Engineering College, Warangal, Telangana State, India
e-mail: areefahnk@gmail.com

S. Sharma et al. (eds.), *Intelligent Control, Robotics, and Industrial Automation*, Lecture Notes in Electrical Engineering 1066,
https://doi.org/10.1007/978-981-99-4634-1_43

These diseases are spreading rapidly in third-world countries due to a lack of medical resources. Even though there have been more cases, scientists are having a difficult time as it is time-consuming to physically check blood samples with a microscope. The examination is carried out by a trained individual who manually examines the parasite strains in the blood smear and determines whether the person is infected with malaria [1]. As a result, there is a risk that due to human errors, wrong reports may be generated, which could harm the patient's health. Hence, a computer vision system that can learn and generate accurate predictions without the intervention of laboratory staff or specialists is necessary.

In the present scenario where people are being affected by different diseases, it is very important to create a warning system for the people to warn them regarding each disease at its outbreak. Thus, with the same idea, there must be a model trained that could help the people in giving awareness about the major outbreak of malaria in their area and warn them about the risk. As we know, malaria breeds have a high capability to grow easily in humid certain temperatures. Thus, the weather conditions influence the outbreak of malaria, and whereby it is very important to tell the daily outbreak consequences to people from varying localities.

We applied deep learning techniques and created our own CNN model to detect parasite strains in blood smears (Convolutional Neural Network). We built our model by defining our custom layers based on the CNN layers. We also used the VGG19 model (Transfer Learning model) on the same dataset and recorded the accuracies. When compared to the Transfer Learning model, our custom CNN model performed well on the dataset, and we got about 98% accuracy with a smaller number of parameters. Hence, our proposed model is simple and takes about 45 min to be trained which is the least when compared to the existing model with 24 h. of training time.

We proposed to the users a malaria outbreak warning system, in which we created a web app that would run our trained model in the back end. For efficient real-time predictions of the outbreak, our proposed model is trained on logistic regression, which is implemented by the Gradient Descent Optimizer [2]. Because the optimizer is being used for implementation, it is more feasible to prediction on the live weather forecast. This paper briefly describes all the proposed models and their implementation, and the results are compared. Finally, the models we trained are hosted on Heroku, allowing users to use our web app for malaria prevention and laboratories to classify blood samples based on malaria parasite strain. The following is the rest of the article: Sect. 2 presents the literature survey of previous works, Sect. 3 outlines the proposed work's implementation, and Section summarizes the results and discussions.

2 Literature Survey

2.1 *Malaria Parasite Detection*

Starting with the work contributed by Anggraini et al. [3] where an application has been developed that could automatically apply the image segmentation methodologies over the input images such that the blood cells would be separated from the background. This created many people to consider these images and build models. Neto et al. [4] had given an image analysis approach that classifies the five phases of malaria in malaria detection. To expand the research, Rajaraman et al. [5] worked on the feature extractions that would help in building a CNN-based model for classifying the infected and uninfected blood cells. Momala [6] is a well-known scenario that has been proposed. It is a microscope-oriented system that runs as an application on a smartphone and can detect and analyze malaria strains in blood smears on slides using its phone camera. Because the proposed application has always required the microscope, we can conclude that it is not feasible for that bulky setup and is not transportable. In some scenarios, the traditional neural network methods known as multilayer perceptron (MLP) are proposed, which depict the human brain by stimulating only the connected neurons when a certain defined threshold value invokes [7]. Moreover, as we know, these use one perceptron for each corresponding input more particularly for the pixel, which generates an unreliable number of weights for large images and due to which complexity arises during training.

In many cases, the researchers used an experimental approach to identify the best and most efficient model based on the available data. Every CNN model proposed had three convolutional layers and two dense layers that were fully connected. The performance of models such as the VGG-16, Xception, ResNet-50, and AlexNet [1] was displayed and compared. Some research papers even showed their customized models, but the accuracy was lacking. Though many of the works concluded that the transfer learning models performed well and retained better results than the custom models, these required feature extraction and many numbers of model parameters to be trained that lag with time consumption [5]. All these constraints make the model more complex in size which cannot be easily deployable over any devices like IoT or smartphone applications. Some studies, such as [8], advocated simpler CNN models and even used the SGD optimizer with an adaptive learning rate approach. Though the model had good training accuracy, it failed to validate the images of blood cell smears by incorrectly predicting infected when the actual label was uninfected. This could be attributable to faster convergence with fewer experiments and a smaller set of model parameters (due to the fact that they evaluated less trainable parameters) and fewer hyperparameter updates. The classic SGD optimizer produces results by posing a saddle point problem. As a result, this optimizer might not be a good fit for the model. Therefore, there is a need for a CNN model that could consider an efficient way by taking less training time and producing accurate results without underfitting and even should not overfit. A novel solution is proposed in this paper regarding malaria parasite detection in blood smears.

2.2 *Malaria Outbreak Detection*

Few studies have used a combination of ML algorithms for warning systems for malaria outbreaks. Support vector machines and even artificial neural networks were used in the work of Sharma et al. [9]. In this basic work, Support Vector Machine (SVM) and Artificial Neural Networks (ANN) were used to predict malaria outbreaks in Maharashtra state using a dataset of outbreaks. The authors demonstrated how they were able to achieve 77 percent accuracy by directly applying SVM to the prediction of an outbreak. Noor et al. [10] used medical-based intelligence to report the locations while emphasizing the risk of an outbreak. The detection of malaria outbreaks was based on weather parameters, but because the dataset was small in most cases, it was unable to produce the efficient training seen in the current work. The authors [11] wanted to see if hybrid classification and regression models might predict disease outbreaks using datasets as some single data mining techniques were found to be inaccurate. They demonstrated that the hybridization model should solve the weaknesses of a single model by mixing many techniques such as decision trees, random forests, Nave Bayes multinomial, and simplification.

All of these studies revealed worse training accuracy, resulting in an erroneous outbreak prediction. Thus, there is a need of building the model with custom weights such that even with less dataset, it could be efficiently trained to give accurate predictions over the real-time weather forecast conditions from the API automatically. This is our proposed system, which is explained in the article that works well and accurately.

3 Proposed System

The proposed CNN model and the logistic regression with GDA models were trained and then evaluated on the Google Colab, a popular cloud-based notebook with access to high-performance graphical processing units (GPUs), such as the 16 GB NVIDIA Tesla P100 GPU with CUDA support and preinstalled libraries such as Python 3 with high-level APIs such as Keras 2.2.5 to easily access the backend frameworks such as TensorFlow 1.15.0. For developing the web app that helps users to get the prediction from our models, we used the IDE—PyCharm Community Edition and flask library for deploying the models for malaria parasite detection and malaria Outbreak prediction.

3.1 *Malaria Parasite Detection*

Dataset—Malaria Parasite Detection. For malaria parasite detection, we downloaded the dataset available at the NIH site. The National Institutes of Health has a malaria dataset with 27,558 cell pictures with an equal number of parasitized and uninfected cells as shown in Fig. 1. The red blood cells were detected and segmented using a level-set-based method [12].

Custom CNN Architecture. The proposed model is constructed using the CNN's primary building components, as seen in Fig. 2. The main blocks of CNN are:

(1) Input image, (2) convolutional layers, (3) activation function—ReLU, (4) max pooling layers, (5) flattening layer, (6) fully connected layer, and (7) output layer.

The following sections provide an overview of these blocks and more details about the layers in the custom CNN model illustrated in Fig. 2.

Input Image and Preprocessing: All the input images must be of the same size to feed a neural network. Hence after resizing, the input image size is (134, 131, 3), and the batch size is declared as 32 because of the huge dataset.

Convolutional Layers: The data from pixels is transferred into the next layers, namely the convolutional layers followed by pooling layers for the feature extraction

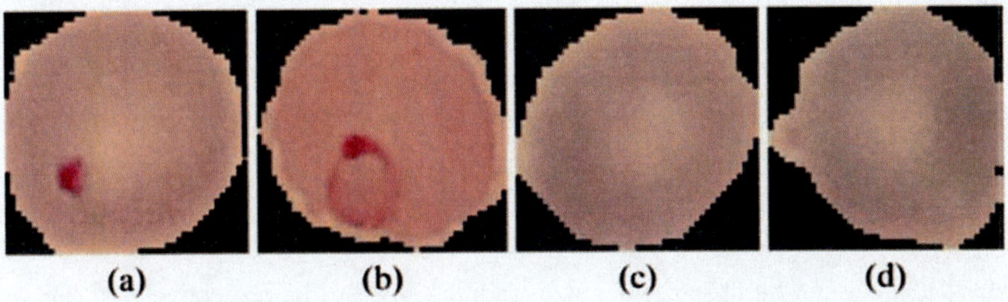

Fig. 1 Sample images of malaria **a, b** parasite, **c, d** uninfected

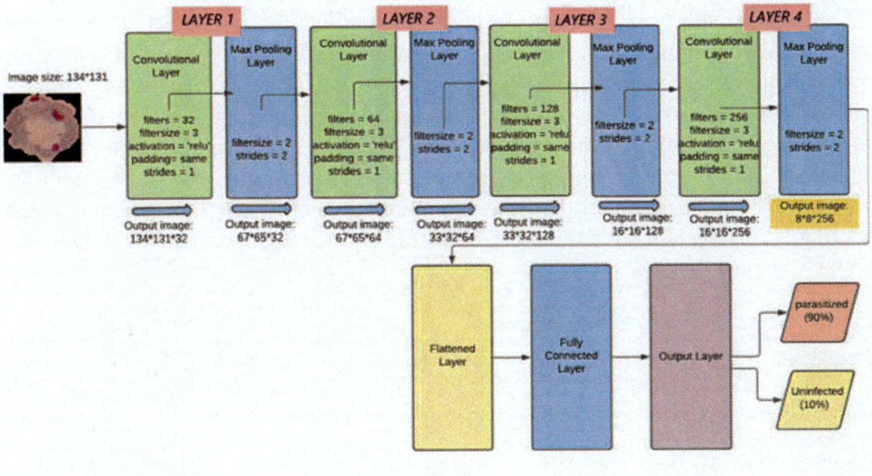

Fig. 2 Architecture of custom CNN model proposed

to grab the important data from the pixels which help in decreasing the model parameters that are needed to be given for the faster training of the model. In a convolution operation, a filter is a three-dimensional (height, width, and depth) matrix used to extract information from an image by performing convolution operations on it [13]. At every convolutional layer, we apply filters (n_f) of filter size (f) and initialized them to the random weights. We even apply the padding (p) and strides (s). These are the hyperparameters that get tuned in the propagation between the layers.

Padding: There arises the possibility of a loss of information from the edge blocks. Hence to preserve the original dimensions and not miss any important data from corners, we apply padding [13]. We give padding value as two types—same or valid. If we pad with the number on the edge, we declare padding = 'same' (choose p-value such that input shape is equal to output shape), and if we don't need padding, then simply assign it as padding = 'valid' [14].

Striding: This defines the model about how many steps it needs to take when applying the filter. When the window of the filter reaches the point where it doesn't get filled with full data (generally when reaches the last columns) then stride tells it to move 1 row below if we declare stride = 1.

The output size of the image from the convolutional layers is calculated with the help of the formula (1).

$$\text{Size of output image} = \left[\frac{n - f + 2p}{s} + 1 \times \frac{n - f + 2p}{s} + 1 \times n_c \times n_f\right], \quad (1)$$

where n, f, p, s are input size, filter size, padding, and stride, respectively. nc and nf are the number of channels and number of filters, respectively.

The values for the convolution operation parameters are listed in Table 1.

ReLU Activation Function: The ReLU activation function scales all the negative values in the pixels to 0. ReLU is used to bring nonlinearity into CNN, as most real-time systems are linear [13]. For the proposed model, the ReLU activation function performed well over all other non-linear activation functions like sigmoid, tanh.

Max Pooling Layer: This is used to minimize the feature map's dimensionality while keeping the most critical information. The maximum value in filter space is regarded as an output in max pooling and thereby highlights the most important features. The only argument is the window size, which is used to generate a max pooling layer. Because it is the most frequent, we utilize a 2 × 2 window [14]. It's

Table 1 Convolution operation parameters

Operation	Filter size	No. of filter strides	No. of filters	Padding	Output image size
1	(3, 3)	32	1	Same	(134, 131, 32)
2	(3, 3)	64	1	Same	(67, 65, 64)
3	(3, 3)	128	1	Same	(33, 32, 128)
4	(3, 3)	256	1	Same	(16, 16, 256)

not necessary to adjust the default stride length, which is 2 in our case. Thus, the filter size is 2 and the stride is 2.

The output size of the image from the max pooling layers is calculated with the help of the formula (2).

$$\text{Size of the output image} = \frac{n - f}{s} + 1, \tag{2}$$

where *n*, *f*, and *p* are the input size, filter size, and padding of the pooling layer, respectively. Every Conv2D layer is followed by the MaxPooling2D layer as shown in the architecture of the proposed model in Fig. 2.

The values for the max pooling operation parameters are listed in Table 2.

Flattening Layer: This layer will be applied after all convolution and pooling procedures have been completed. In this layer, features extracted after convolution and pooling operations are flattened. The final (and any) pooling and convolutional layer produces a three-dimensional matrix, which can be flattened by unwrapping all of its values into a vector [14].

Fully Connected Layer: Fully connected layers are the network's final layers [14]. The output from the final pooling or convolutional layer is flattened and then fed into the fully connected layer. The fully connected layer is like multilayer perceptron where the activation function is applied. As our case is about binary classification, we used the sigmoid activation function as shown in Fig. 3.

Dropout: Dropout is only applicable to input and hidden layer nodes, not output nodes [14]. The number determines the percentage of nodes (neurons) that will drop out. As shown in Fig. 3, Dropout (0.2) depicts 20% of neurons are dropped at dense layers in our model to avoid overfitting.

Output Layer: The output layer is the final layer where the final classes probabilities result. The number of nodes in this layer is equal to the number of classes in which our problem image must be classified. If the probability of class obtained < 0.5, it is considered as 0, and if the probability score is ≥ 0.5 taken as 1.

Algorithm

1. Read the dataset after downloading.
2. Improving the size of the dataset (rotation range: 20%)

Table 2 Max pooling operational parameters

Operation	Filter size	Stride at the pooling layer	No. of filters	Size of the input to the pooling layer	Output image size
1	(2, 2)	32	2	(134, 31)	(67, 65, 32)
2	(2, 2)	64	2	(67, 65)	(33, 32, 64)
3	(2, 2)	128	2	(33, 32)	(16, 16, 128)
4	(2, 2)	256	2	(16, 16)	(8, 8, 256)

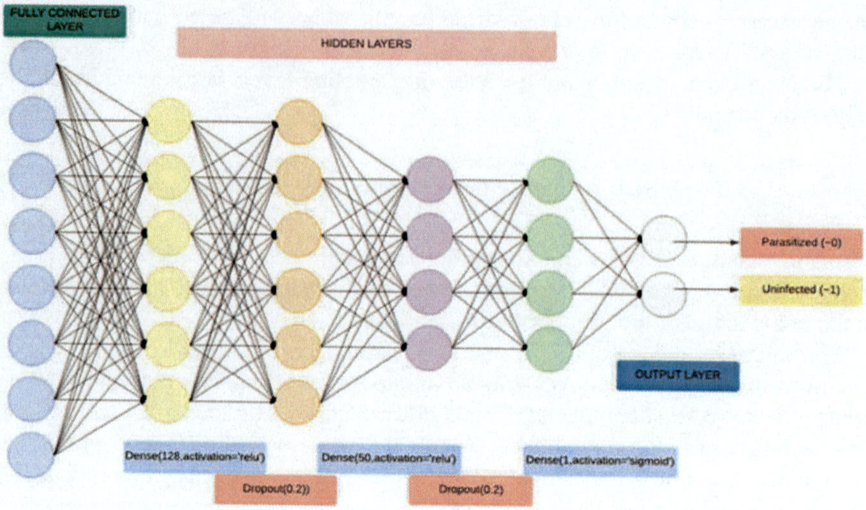

Fig. 3 Layers responsible for classification in CNN

3. Preprocess the images (resize)
4. Develop CNN model (4 Conv2D layers with 4 MaxPooling2D layers, Flattened Layer, 3 Dense Layers with Dropout (0.2), Output Layer with sigmoid activation function) – Refer Fig. 2. And Fig. 3. for layers configured.
5. Compile CNN model (optimizer = 'adam' and specify metrics)
6. Train the model
7. Test the trained model and perform Validation
8. Deploy the model

Transfer Learning—VGG19 Model. Transfer learning models are already trained in the classification of 1000's classes. We can directly use them by tuning the classes according to our needs. We just need to remove the top layer and need to perform training on the top layer according to our classification problem and the last layers including dense layers and flattened layers need to cut. VGG19 is a 19-layer deep convolutional neural network. Load a pre-trained version of the network from the ImageNet database, which has been trained on over a million images.

Algorithm

1. Read Dataset.
2. Pre-process image (resize): Resize image shape = (224, 224).
3. Import the VGG19 model and cut the 1st layer where we give the image size and remove all the last layers up to dense layers
4. Only retrain the last layers and add flattened and dense layers.

Hence, this final model is compiled, fitted, and validated after rescaling the images back. The results of validation and evaluation metrics of these two models (Custom CNN, Transfer Learning model) will be discussed in the result section.

Table 3 Malaria outbreak dataset

S. No.	maxTemp	minTemp	Avg humidity	Rainfall	Positive	pf	Outbreak
1	29	18	49.74	0	2156	112	No
2	34	23	83.27	15.22	10,717	677	Yes
3	40	23	50.74	0	1257	127	No
4	34	24	59.16	9.06	4198	211	No
5	34	27	73.23	0	11,808	712	Yes

4 Malaria Outbreak Prediction—Proposed System

Dataset. The information was gathered from a variety of sources, including the National Vector Borne Disease Control Program in Pune and Indian Meteorologic's meteorological data. The dataset has 22 records with 7 column features as given in Table 3. Prediction of Outbreak Class—Yes or No.

Methodology of Malaria Outbreak System. To predict the outbreak of malaria based on the weather condition, we applied logistic regression without using the scikit-learn library. Instead, we applied the Gradient Descent (GD) Optimizer and predicted the outbreak probability. To determine the next point, the Gradient Descent algorithm multiplies the gradient by a number (learning rate or step size) [2]. Like this, it takes the parameter such that loss reaches global minima. The proposed system in detail is depicted in Fig. 4.

5 Result and Analysis

5.1 Trained Models Performance

When we compare both models at epochs of 4, we can easily find that about 97.74% of training accuracy is obtained by our custom CNN model which is higher than the VGG19 model as shown in Figs. 6 and 8. Even validation accuracy is about 96% at the 4th epoch for the custom CNN model which is higher than that of the VGG19 model as shown in Figs. 6 and 8. The training and validation loss depicts that loss obtained is less in custom CNN when compared with VGG19 ass shown in Figs. 5 and 7.

The proposed malaria outbreak system which comprises a logistic regression model implemented using Gradient Descent Optimizer is accurate as it is trained for around 20 iterations until the Mean Squared Error is minimized to almost 0 as shown in Fig. 9.

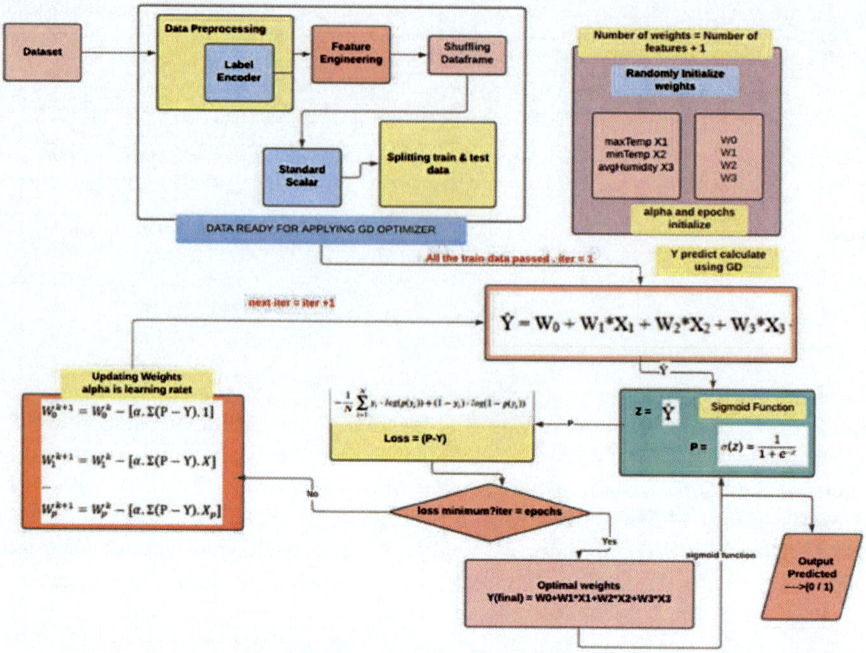

Fig. 4 Proposed methodology for implementing logistic regression with GD optimizer

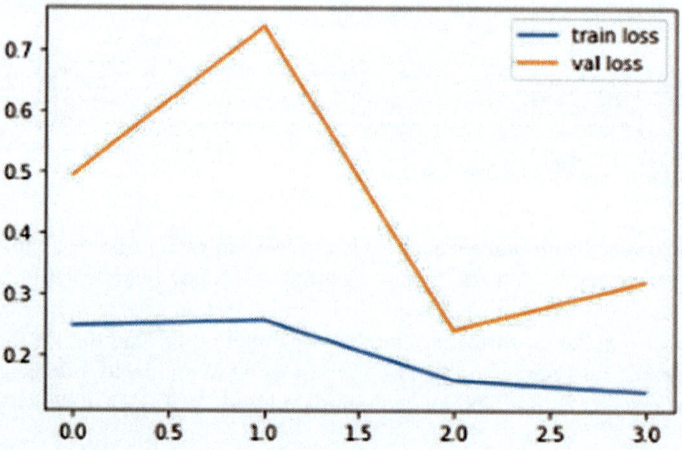

Fig. 5 Training loss and validation loss observed in VGG19

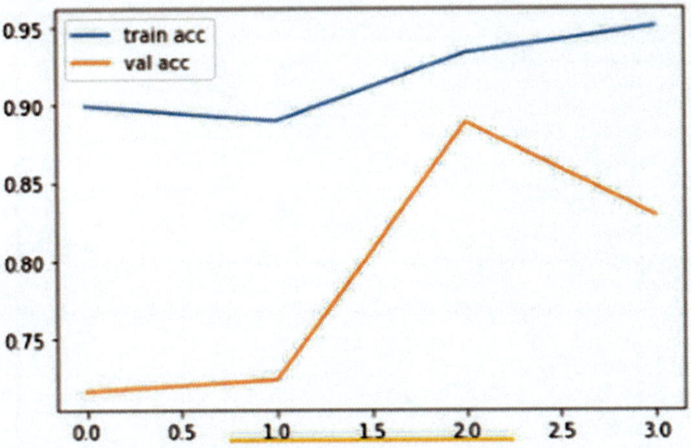

Fig. 6 Training and validation accuracy in VGG19

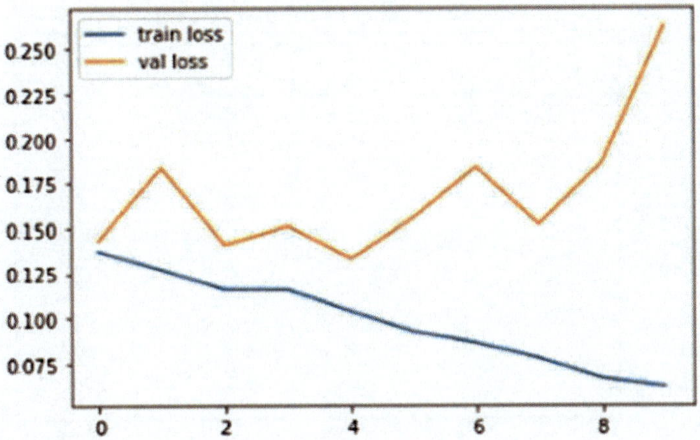

Fig. 7 Training loss and validation loss in custom CNN

6 Web-App Build and Deployed on the Cloud

Using trained models, we built a Flask App which gives inputs to the backend model for predictions. This Flask App is deployed over a cloud platform—Heroku (PaaS).

Web-App for Laboratory-Level Usage

Figures 10 and 11 depict the Web-App working on Test case 1 where the infected cell image is uploaded by user, and it is rightly classified as Infected and displayed as result to the user.

Figures 12 and 13 depict the results when the user uploads the cell image that is uninfected. Hence, the model rightly classified the cells even in this case.

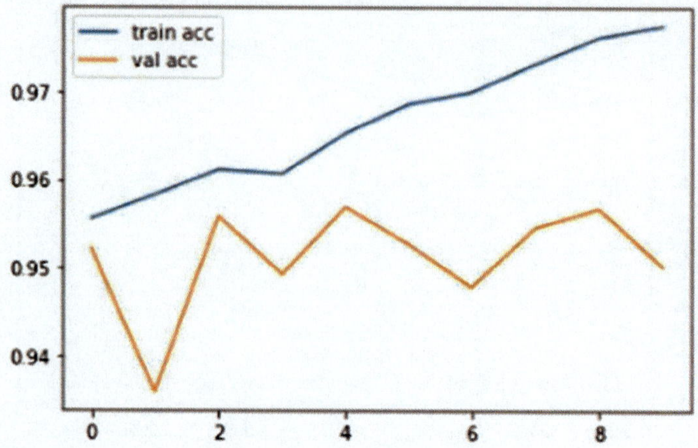

Fig. 8 Training and validation accuracy in custom CNN

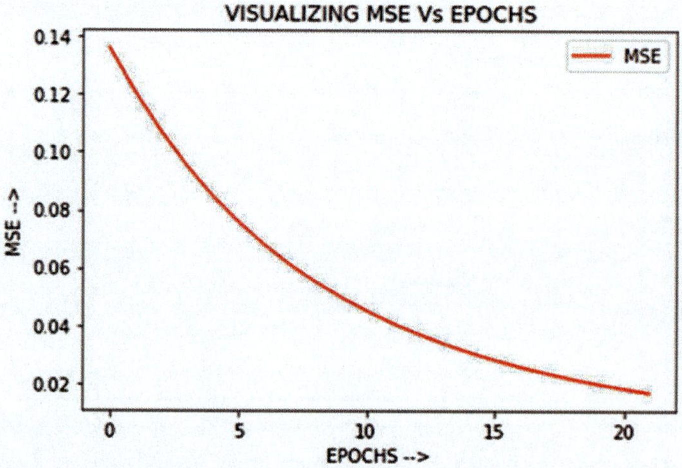

Fig. 9 Performance of logistic regression model with Gradient Descent

Web-App for Malaria Outbreak Warning System

The specialty of this WebApp is, it is connected with the OpenWeather API which helps the user to minimize the effort of manually entering weather conditions. The user just needs to select a location; thereby, automatic weather conditions would be inputted for prediction as shown in Figs. 14 and 15.

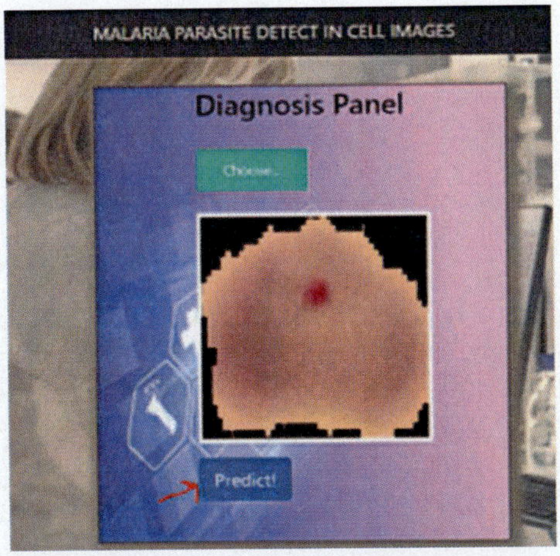

Fig. 10 User uploads cell image and clicks on Predict button—Test case 1

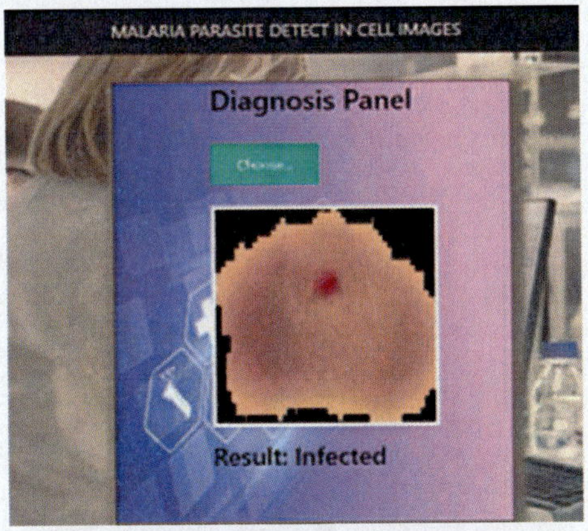

Fig. 11 Prediction is displayed as Infected for Test case 1

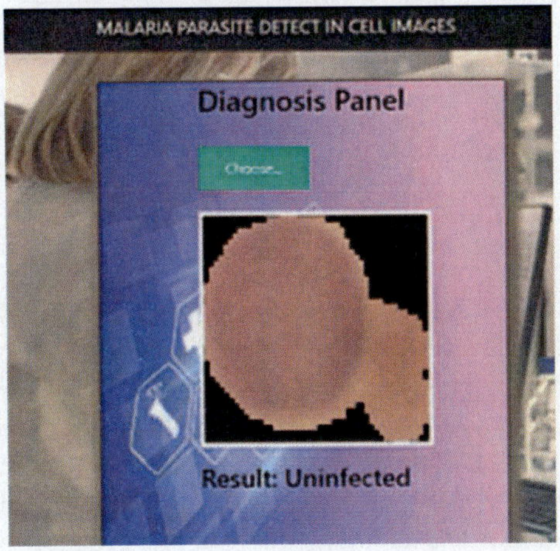

Fig. 12 Test case 2

Fig. 13 Test case 3

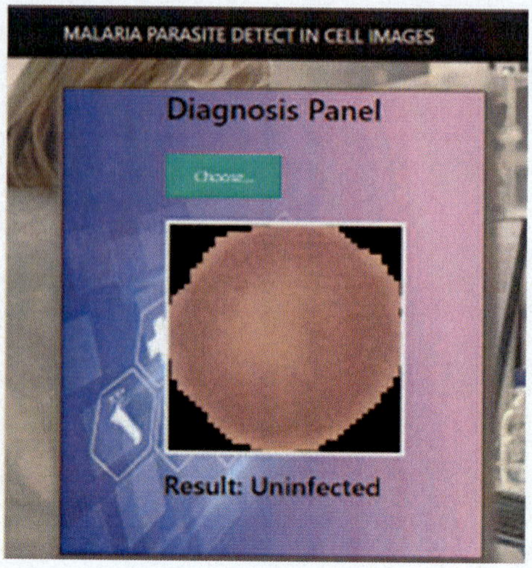

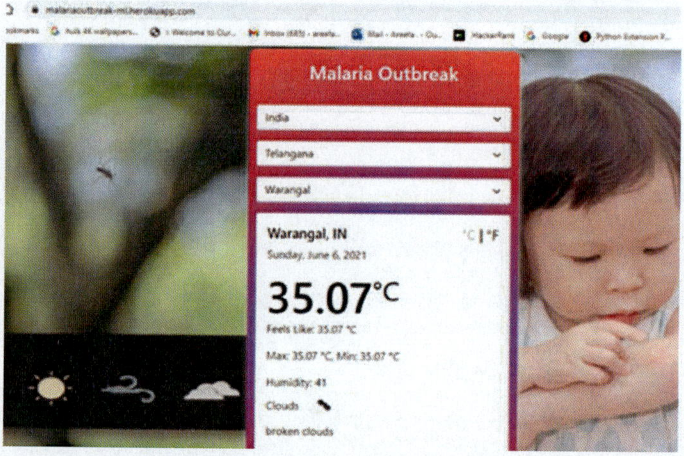

Fig. 14 User needs to just select their location; thereby, the weather conditions would be shown

7 Conclusion and Future Scope

In this paper, we implemented our custom CNN model and a Transfer Learning model (VGG19) for malaria parasite detection in blood smears at the laboratory level. Comparing the results, we got 97.74% of training accuracy and 95.03% validation accuracy with our Custom CNN model, whereas the VGG19 model showed about 94.95% training accuracy and 82.84% validation accuracy. Later, logistic regression model using a Gradient Descent Optimizer is implemented which takes the climatic conditions as input and gives the probability of the malaria outbreak in the specific

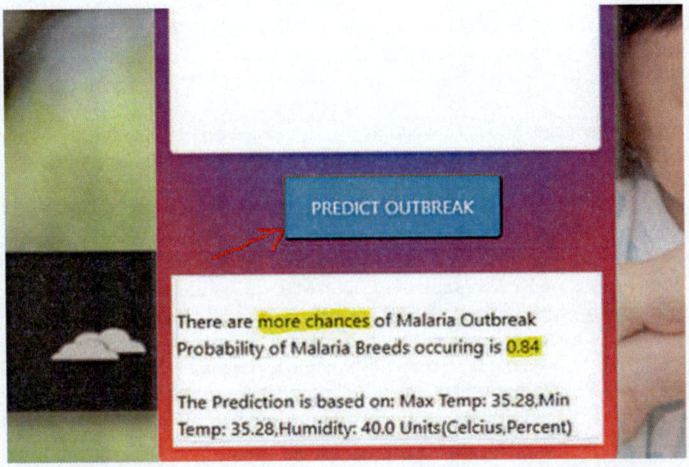

Fig. 15 Prediction is displayed when user clicks on Predict Outbreak button

locality. Finally, all these models are deployed over WebApp using flask and hosted on Heroku such that users can access to use them.

The work can be extended by experimenting changes in the custom CNN model by making trials with various optimizers as we only used ADAM optimizer for malaria parasites in blood cells. Regarding malaria outbreak system, as dataset is small, more scope to create the dataset with the help of weather API. It would help in coming up with models that help in predicting the outbreak of various diseases that depends on weather conditions. We can build a combined web app with different disease predictions that rely on the CNN model at the laboratory level.

GitHub Repository: https://github.com/Areefahnk/WebApp-Malaria-Detection-and-Outbreak-Prediction-Using-Deep-Learning.

References

1. Var E, Boray F (2018) Malaria parasite detection with deep transfer learning. In: 3rd International conference on computer science and engineering (UBMK). IEEE, Sarajevo, Bosnia and Herzegovina, pp 298–302. https://doi.org/10.1109/UBMK.2018.8566549
2. Kathuria C, Logistic regression using gradient descent optimizer in python. https://towardsdatascience.com/logistic-regression-using-gradient-descent-optimizer-in-python-485148bd3ff2. Last accessed 19 Dec 2022
3. Anggraini D, Nugroho A, Pratama C, Rozi I, Iskandar A, Hartono R (2011) Automated status identification of microscopic images obtained from malaria thin blood smears. In: ICEEI 2011 committees, pp 1–6. https://doi.org/10.1109/iceei.2011.6021762
4. Leal Neto O, Albuquerque C, Albuquerque J, Barbosa C (2014) The schisto track: a system for gathering and monitoring epidemiological surveys by connecting geographical information systems in real time. JMIR Mhealth Uhealth 2(1):e10. https://doi.org/10.2196/mhealth.2859

5. Rajaraman S, Antani S, Poostchi M, Silamut K, Hossain M, Maude R, Jaeger S, Thoma G (2018) Pre-trained convolutional neural networks as feature extractors toward improved malaria parasite detection in thin blood smear images. PeerJ 6:e4568. https://doi.org/10.7717/peerj.4568
6. Mayo Clinic Homepage. https://mayoclinic.org. Last accessed 19 Dec 2022
7. Nduati J, Introduction to neural networks. https://www.section.io/engineering-education/introduction-to-neural-networks/. Last accessed 19 Dec 2022
8. Masud M, Alhumyani H, Alshamrani S, Cheikhrouhou O, Ibrahim S, Ghulam M, Hossain M, Shorfuzzaman M (2020) Leveraging deep learning techniques for malaria parasite detection using mobile application. Wirel Commun Mob Comput. https://doi.org/10.1155/2020/8895429
9. Jha A, Vartak S, Nair K, Hingmire A (2020) Malaria outbreak prediction using machine learning. Int J Eng Res Technol (IJERT) NTASU-2020 9(03). https://doi.org/10.17577/IJERTCONV9IS03023
10. Noor A, Kinyoki D, Mundia C, Kabaria C, Mutua J, Alegana V, Fall I, Snow R (2014) The changing risk of Plasmodium falciparum malaria infection in Africa: 2000–10: a spatial and temporal analysis of transmission intensity. Lancet. Epub 383(9930):1739–1747. https://doi.org/10.1016/S0140-6736(13)62566-0
11. Gramacy R (2007) An R package for Bayesian nonstationary, semiparametric nonlinear regression and design by treed Gaussian process models. J Stat Softw 19(9):1–46. https://doi.org/10.18637/jss.v019.i09
12. Poostchi M, Silamut K, Maude R, Jaeger S, Thoma G (2018) Image analysis and machine learning for detecting malaria. Transl Res. Epub 194:36–55. https://doi.org/10.1016/j.trsl.2017.12.004
13. Mandal M, Introduction to convolutional neural networks (CNN). https://www.analyticsvidhya.com/blog/2021/05/convolutional-neural-networks-cnn/. Last accessed 19 Dec 2022
14. Dertat A, Applied deep learning—Part 4: convolutional neural networks. https://towardsdatascience.com/applied-deep-learning-part-4-convolutional-neuralnetworks-584bc134c1e2. Last accessed 19 Dec 2022

SIRR: Semantically Infused Recipe Recommendation Model Using Ontology Focused Machine Intelligence

Mrinal Anand, Gerard Deepak, and A. Santhanavijayan

Abstract In the current era, recipe recommendation is of extreme significance as there is no expert intelligence system specifically for recommendation. In the era of Web 3.0, semantically driven frameworks are mostly absent. As a result, semantically infused recipe recommendation model is highly important. This study proposes a semantically infused intelligent approach to recommend recipes using Ontology Focused Machine Intelligence to classify the documents. The semantic similarity was computed by hybridizing Normalized Compression Distance (NCD) and KL divergence under a cultural algorithm. The model uses both Recurrent Neural Network (RNN) and XGBoost at different locations to sort the data and compute semantic similarity. The model yielded an FDR of 0.04, and an accuracy of 97.09%.

Keywords NCD · KL divergence · Cultural algorithm · RNN · XGBoost · Semantic similarity

1 Introduction

A recipe is a set of instructions that illustrates how to prepare a dish. The sole purpose of a recipe is to list the right set of ingredients needed, specify their quantity and the way they are to be combined. A recommender system is a computer program that is able to provide meaningful suggestions to the user that might interest them depending upon the user's preferences and behaviour. A recommender system can be either

M. Anand
R.V College of Engineering, Bangalore, India

G. Deepak (✉)
Department of Computer Science and Engineering, Manipal Institute of Technology Bengaluru, Manipal Academy of Higher Education, Manipal, India
e-mail: gerard.deepak.christuni@gmail.com

A. Santhanavijayan
Department of Computer Science and Engineering, National Institute of Technology, Tiruchirappalli, India

S. Sharma et al. (eds.), *Intelligent Control, Robotics, and Industrial Automation*, Lecture Notes in Electrical Engineering 1066,
https://doi.org/10.1007/978-981-99-4634-1_44

based on content-based or collaborative-based filtering. A collaborative system is solely based upon past data between user and item to suggest new recommendations, whereas a content-based model takes additional information about the user such as age, gender or any other personal information into consideration to make valid and personalized recommendations. Nilesh et al. [1] recognized the large variety of cuisines available with similar ingredients and designed a technique that suggests Indian food ideas based on the items that are available.

Recipes not just give every one of the specialized and creative components vital to deliver an effective dish as a matter of course, but also help in preserving the authentic style of the dish. In this modern age, with dynamic living manners and increased workloads, diet habits of people have drastically declined which has led to a surge in cases related to heart, blood pressure, diabetes and many more. Hence, it is vital to recommend recipes to maintain health without sacrificing the taste of the dish, and this recommendation can be made effective and precise with the help of the semantic web. Vivek et al. [2] present that user profiles are used by recommender systems to assist users in finding key data from a big amount of data. Qi et al. [3] present that food is a commonly searched topic online since it is necessary for providing people with nutritional assistance, and much of the content about food found online is represented using semantically enriched markup found in ontologies.

1.1 Motivation

The Internet was designed to let computers share information among one another; it wasn't designed to teach them what the information actually means. Users want computers to recognize what is in a web page, so that they can understand what the user is interested in and suggest better response. The world is moving towards the age of the semantic web, which is a knowledge graph formed by combining linked data to work with machine comprehension and handling of content, metadata and other data objects at scale. There are just a handful of models for recipe recommendation system, and hardly any that use semantically driven approach. Hence, this approach which is socially conscious and semantically enabled is necessary.

1.2 Contribution

A semantically infused approach has been described for recipe recommendation. Tokenization, Lemmatization and Stop Word Removal are utilized for query pre-processing. The individual queries are further subjected to topic modelling using Latent Dirichlet Allocation (LDA). These enriched query topics are further subjected to semantic similarity based on crawled and classified recipes from allrecipes.com. The recipes crawled are further classified using Recurrent Neural Networks (RNN) and XGBoost at different locations. The proposed model achieved an accuracy of 97.09%, precision of 95.72%, recall of 98.45% and F-Measure of 97.07%.

2 Related Works

Nilesh et al. [1] designed a technique that suggests Indian food ideas based on the items that are available. They applied the content-based approach to recommend the recipes. Vivek et al. [2] employed two approaches: user-based and item-based method to recommend recipes. Log likelihood similarity and Tanimoto Coefficient Similarity were used to determine how similar different recipes are in the item-based method. Euclidean distance and Pearson Correlation were employed in the user-based method. Qi et al. [3] proposed ontology with the food class and the recipe class as the two fundamentals along with other supporting structures. The food class describes many food categories and divides them into several subclasses, and they also made sure to multiclass an instance of a food class into the ingredients class if used in recipes. Haussmann et al. [4] proposed an application that includes a SPARQL-based service which enables to choose cuisines depending on items that are readily accessible while taking into account restrictions like sensitivities, as well as a cognitive agent that can respond to inquiries on the knowledge graph. Ting et al. [5] presented Dietary Recommendation System (DRS) capable of determining if a user followed a healthy meal depending upon specific circumstances and recommending appropriate cuisines. Gyrard et al. [6] provide a semantic-based machine-to-machine measurement technique (M3) for automatically combining, enriching and reasoning about M2M data in order to enable cross-domain M2M applications. Gonçalves et al. [7], the Pinterest Taxonomy, which serves as the foundation of Pinterest's knowledge network, was designed as an OWL ontology. Application of semantic web technology in a menu creation system was elaborated. Bianchini et al. [8] employed a recipe collection and annotations to propose meals based on user preferences. Erp et al. [9] proposed a method for identifying recipes in digitized historical newspapers, extracting recipe tags and extracting ingredient information based on remote supervision and automatically derived lexicons. On the basis of the features associated with ontological ideas, Likavec et al. [10] offered new metrics, from which they regard two ideas to be comparable if they have the same values ascribed to some of the attributes they share in common. Chivukula et al. [11] investigate and construct an ontology that is related to the nutrition sector. The goal is to develop an ontology model in the nutrition sector that will assist users in receiving appropriate meal recommendations depending on their health issues, if there were any. Padhiar et al. [12], that offers a framework for modelling user explanations for advice on food, named Food Explanation Ontology (FEO). Bangale et al. [13] proposed an interactive online application that operates on a user device to recommend recipes. Using the idea of content-based filtering, users are given recipe suggestions based on the components they submit. Chhipa et al. [14] proposed a model that is a smartphone app that enables users to look up recipes using components they already have on hand such as fruits and vegetables and utilize Term Frequency-Inverse Document Frequency (TF-IDF) and Cosine Similarity to implement the content-based recommendation. In [17–21], several ontological models in support of the proposed framework are depicted.

The literature gap that has been identified from the related works is that there are no frameworks which are knowledge centric and ontology driven for recipe recommendation, most of them make use of static ontology, but the scale of ontology used is much light weight and lack diversity, and moreover the existing models have used ontology, where they just display the ontology itself, but in this method, there is a need for using the ontology processing, arriving at a consensus and also encompass machine intelligence which is ontology oriented artificial intelligence with machine learning. From the literature, it is also seen that most of the approaches are content-based approaches which are not knowledge centric and do not suite the needs of the semantic web. Some of them are probability-based where log likelihood similarity was alone used. Few methods used Euclidean distance and Pearson's correlation coefficient which is good as semantic similarity but they lack auxiliary knowledge and a learning model. Some of the models use recipe class, as supporting structure and several categories across classes are used, and directly they are yielded, parsing the query from standard knowledge graph also helps, they are directly displayed, but they could be further processed and ensure that a much more model can be proposed. Some of the models use machine-to-machine measurement technique which is semantically based; however, learning was lacking in these models. Some of the models used taxonomy and knowledge network, but again inferencing mechanism were quite weak, and learning methods were absent. Ontology-based models directly yield the ontology; it is not used for further estimation. Owing to the lacune and the research gap, there is a need for semantically driven model which infuses ontology for accelerating the auxiliary knowledge and reducing the cognitive gap between the external World Wide Web and the knowledge included in the framework, along with encompassing machine intelligence, by incorporating a learning scheme, along with semantically driven inferencing schemes like semantic similarity computation.

3 Proposed Architecture

Figure 1 shows the suggested ontology-driven machine learning infused model for recipe recommendation. The initial input to the recipe recommendation framework is the user query, as it is a query driven model, which is exposed to query preprocessing, which further involves Tokenization, Lemmatization and Stop Word Removal. The specific query terms are returned at the conclusion of the query preprocessing stage, which are further subjected to Latent Dirichlet Allocation (LDA) for topic modelling. Latent Dirichlet Allocation (LDA), with the prime idea being that records, is addressed as arbitrary combinations which are described by a word distribution of latent topics. Topic modelling is incorporated in order to uncover the hidden topics into the model from the World Wide Web (WWW). We choose to apply the LDA model since it is a light weight, conventional and yet efficacious model for topic modelling. After topic modelling, the upper ontology of several cuisines and ingredients is generated. Using the tool OntoCollab, this upper ontology was developed from the World Wide Web. This upper ontology is used for achieving ontology alignment between topic enriched query words with ontology for further query topic enrichment, and these enriched query topics are further subjected to semantic similarity based on crawled and classified recipes from allrecipes.com. A standard integrated dataset has been made use of which has the recipes crawled from allrecipes.com that are classified using RNN. Recurrent Neural Networks (RNN) are a type of artificial neural network that increases the network's weights in order to produce cycles in the network graph. In order for neural networks to explicitly train and use context the state is introduced to neural networks. RNN are a type of recursive networks with linear architectural variant.

Recursion promotes branching in hierarchical feature spaces, and as training goes on, the resulting network architecture replicates this. RNN has automated feature selection, and hence, this is the reason for applying RNN. Based on the classified and automatically discovered classes, the semantic similarity is computed between the ontology enriched topic modelled queries, as well the classes, which are furnished by RNN classification model, is used to compute semantic similarity. The model employs the Normalized Compression Distance (NCD) with KL divergence. For semantic similarity computation using Normalized Compression Distance (NCD), 0.75 is considered as threshold, and for KL divergence 0.2 is considered as step

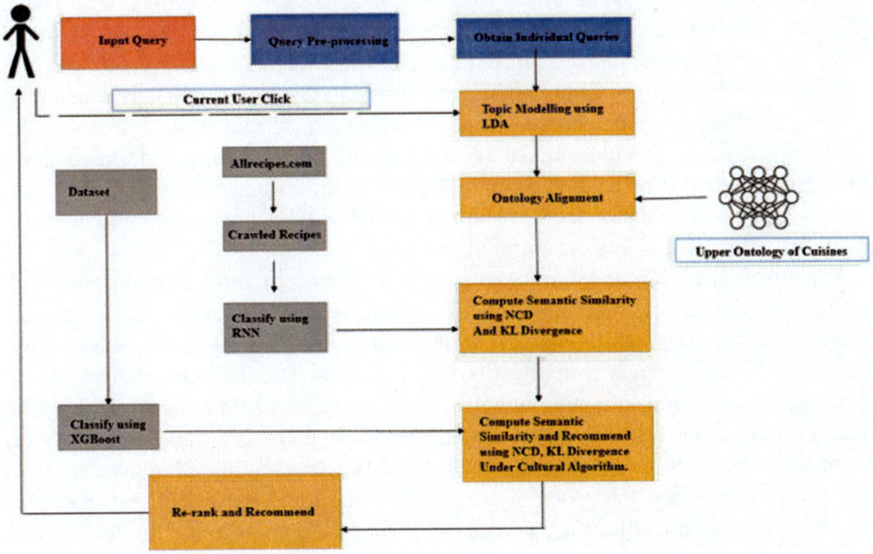

Fig. 1 Suggested ontology-driven machine learning infused model for recipe recommendation

deviation. The reason for considering 0.75 as threshold for NCD is because NCD ranges between the value 0 and 1, and most semantic similarity measures operate like probability and have variations between 0 and 1, where 0 is dissimilar and 1 being most similar, 0.75 is a very traditional threshold which is empirically decided for every semantic similarity measure, except for NPMI which is an exception, where 0.5 can be considered as threshold, as it ranges between -1 and $+1$. And 0.25 is considered as threshold for KL divergence since minimum deviation for values that cannot be normalized between 0 and 1 is 0.25; however if semantic gap has to be increased, then the step deviance too has been increased, but here only highly related entities have to be filtered out, and hence, 0.25 is considered as it ranges for one-fourth of the probability, which means that the semantic difference between the entities is not very high. At the end of semantic similarity computation phase, we have obtained aligned, matched and query relevant topics. These query relevant topics are further used to compute semantic similarity between the dataset, which is classified using XGBoost, by considering pre-processed query topics as the features.

Extreme Gradient Boosting (XGBoost) is a distributed gradient boosting library that has been tuned for higher performance and adaptability. It utilizes the Gradient Boosting framework in order to implement machine learning algorithms. It enables a parallel tree boosting (also known as GBDT, GBM) to quickly and accurately address numerous data science issues. Boosting is a sequential method built on the idea of ensemble concept. To raise the accuracy, it aggregates a cluster of ineffective learners Results that were correctly predicted were given a lesser weight than those which were incorrectly classified.

Boosting, in comparison with bagging approaches like Random Forest, which employ trees that developed to the furthest extent possible, uses trees with lesser splits. It is possible to modify parameters such as the number of trees and the depth of the tree using validation techniques like k-fold cross validation. Overfitting might happen if there are a high number of trees present. Since XGBoost is a machine learning algorithm, the pre-processed query topics are submitted as features to the XGBoost classifier. This classified dataset along with already computed semantically similar topics are further used to compute semantic similarity again by using Normalized Compression Distance (NCD), with threshold of 0.75, and KL divergence with a step deviation of 0.2 under cultural algorithm.

The Normalized Compression Distance (NCD) is a method of comparing two items, such as papers, messages, emails, music scores, languages, programmes, photographs, systems, and genomes, to mention a few. This type of measurement should not be focused on the application or be random. The Kullback–Leibler divergence score measures the difference between two probability distributions. The concept behind the KL divergence score being that there is a considerable divergence when an event from P has a high probability while having a low likelihood of occurring in Q. When the probability from Q is high and the likelihood from P is low, there is a significant divergence, although not as great as in the first instance.

In evolutionary computing, a kind known as cultural algorithms (CA), in which, on top of the population component, there is also a knowledge component termed the belief space. Cultural algorithms may be thought of as an extension of a traditional genetic algorithm in this way. The cultural algorithm's population component is similar to the genetic algorithm's population component. A population-belief space interface is necessary for cultural algorithms. The update function allows the population's finest individuals to update the belief space. The belief space's knowledge divisions can also influence the population factor through the influence function. By modifying the actions of individuals, the influence function can have an impact on the population. Cultural algorithm is a metaheuristic algorithm. The reason cultural algorithm is used in this phase and not previous phase is because, in the previous phase, the task was to enrich the number of probable topics. But here, cultural algorithm is a metaheuristic algorithm, to optimize the initial solution set to yield the best solution set. The initially yielded solution set is optimized using the cultural algorithm, in order to yield the most appropriate feasible solution. This is ranked and recommended on the basis of increasing order of NCD value and is submitted to the user, which is the first recommendation, and based on the current user clicks recorded, which are submitted to ontology alignment phase from the second half, and this search continues until there aren't any further user clicks recorded.

$$h_t = f(h_{t-1}, x_t), \tag{1}$$

$$y_t = W_{hy} h_t, \tag{2}$$

$$\mathrm{NCD}(x, y) = \frac{C(xy) - \min\{C(x), C(y)\}}{\max\{C(x), C(y)\}}, \tag{3}$$

$$F2(x) = \sigma\,(0 + 1 * h1(x) + 1 * h2(x)), \tag{4}$$

$$D_{KL}(P \| Q) = -\sum_{x \in \chi} P(x) \log\left(\frac{Q(x)}{P(x)}\right). \tag{5}$$

Here, Eq. (1) represents the current state. Equation (2) represents the output state. Equation (3) represents Normalized Compression Distance (NCD) where x and y are the two items in concern [15] and $C(x)$ is the length of x compressed by compressor C. Equation (4) represents prediction from XGBoost model. Equation (5) represents KL divergence expectation of the logarithmic difference between the probabilities P and Q, where the expectation is taken using the probabilities [16].

4 Performance Evaluation and Result

The proposed SIRR framework for semantic infused recipe recommendation is evaluated for performance, and the following is listed as potential matrix: precision, recall, accuracy, F-Measure percentages, False Discovery Rate (FDR). The first two indicate the significance of results, whereas FDR lists the number of false positive. To access effectiveness of the SIRR framework, it is baselined with RICS [1] and MLFBR [2] as the two standard baseline approaches, also decision trees, K-Means Clustering with collaborative filtering and XGBoost with LDA and Cosine Similarity are considered as potential hybrid strategies.

A recipe is a set of instructions that illustrates how to prepare a dish. The sole purpose of a recipe is to list the right set of ingredients needed, specify their quantity and the way they are to be combined.

Table 1 depicts that the proposed SIRR framework uses the highest precision of 95.72%, highest recall of 98.45%, highest accuracy of 97.09%, highest F-Measure of 97.07%, with the lowest FDR of 0.04. The experiment was carried out on 7458 queries for which ground truth was gathered over a period of 164 days among several food lovers and food vloggers. The standard formulae for these measures were utilized. In order to conduct the experimentation and performance evaluation, the baseline models and the baseline hybridization schemes were run for the exact same number of queries and exactly same environment as that of the proposed SIRR model, and the evaluations were conducted. Also, from Table 1, it is evident that RICS [1] yields 84.78% of precision, 86.42% of recall, 85.60% of accuracy, 85.59% of F-Measure, FDR of 0.15. Similarly, MLBFR [2] yields 87.45% of precision, 89.56% of recall, 88.51% of accuracy, 88.49% of F-Measure and FDR of 0.13. Decision trees yielded a precision of 84.63%, recall of 86.72%, accuracy of 85.68%, F-Measure of 85.66% and FDR of 0.15. The combination of K-Means Clustering with collaborative filtering resulted in an average precision of 82.78%, recall of 85.17%, accuracy of 83.98%, an F-Measure of 83.96% and FDR of 0.17. The integration of XGBoost, LDA and Cosine Similarity yielded a precision of 90.12%, recall of 93.28%, accuracy of 91.70%, F-Measure of 91.67% and FDR of 0.10.

The proposed SIRR framework yields highest precision, recall, accuracy, F-Measure values and lowest FDR value, since it takes two classifiers into consideration, that is, the dataset is separately classified and the knowledge resource from allrecipes.com is also classified. Most importantly, auxiliary knowledge inclusion happens using topic modelling by LDA, which uncovers all the hidden as well as

Table 1 Performance evaluation of the suggested SIRR as opposed to other methods

Search technique	Average precision %	Average Recall %	Accuracy %	F-Measure %	FDR
RICS [1]	84.78	86.42	85.60	85.59	0.15
MLBFR [2]	87.45	89.56	88.51	88.49	0.13
Decision Trees	84.63	86.72	85.68	85.66	0.15
K-Means Clustering + Collaborative Filtering	82.78	85.17	83.98	83.96	0.17
XGBoost + LDA + Cosine Similarity	90.12	93.28	91.70	91.67	0.10
Proposed SIRR	95.72	98.45	97.09	97.07	0.04

all the undiscovered topics. Topic uncovering takes place from the external environment and World Wide Web using LDA model, which leads to increase in the density of auxiliary knowledge. Also, the usage of upper ontology of cuisines ensures a diversity of cuisines, by the means of ontology alignment. Ontology incorporates static knowledge, which is highly diversified in this framework. Ontology alignment enhances density of auxiliary knowledge and enriches the entity and enhances diversity of results. Similarly, the computation of semantic similarity, using a combination of Normalized Compression Distance (NCD) and KL divergence with variable threshold and step deviation ensures a relevance computation mechanism.

Most importantly, the second phase of semantic similarity computation is done by hybridizing Normalized Compression Distance (NCD) and KL divergence but under a cultural algorithm. This is done under a cultural algorithm for it is a metaheuristic optimization scheme, which filters out the most relevant of the relevant, that is, it computes optimal solution among the sets of feasible solutions, and thereby ensuring highly relevant yet diversified results to the user, and this is the reason for high percentage of precision, recall, accuracy and F-Measure and a low FDR for the proposed SIRR model. The RICS [1] model lags mainly due to the reason that it uses ingredients and not cuisines as a feature. Using ingredients as an initial feature can lead to confusion while formulating the decision, and hence, the approach mandates proper labelling, and annotations, as well as proper knowledge bases are required, otherwise there would be lower precision, accuracy, recall and F-Measure. The usage of Pearson's correlation coefficient ensures a decent amount of relevance computation; however, the approach mandates the use of content-based filtering as well as usage of TF-IDF. But the content-based filtering isn't extremely strong, as a result the model performs above average. More importantly, external knowledge bases are incorporated only using a meal chart; however, there is a need for highly dense knowledge being adhered into the framework. As a result, due to the lack and scarcity of rich knowledge, the RICS model lags.

The MLBFR [2] model lags, although it uses a machine learning model along with log likelihood similarity, is due to the reason that it uses a naïve measure like Euclidean distance with spacing correlation. However, all the interactions and filtering techniques ensure high computation of relevance, but this approach lacks auxiliary knowledge fed into the framework. Hence, this method is not suitable for a dense semantic environment. The collaborative filtering method is used, which requires rating for every recipe on the World Wide Web. Rating for every recipe based on the user's choice is not a criterion for recommendation. The usage of decision tree alone, without the use of auxiliary knowledge also decreases the precision, accuracy, recall and F-Measure to a large extend. The incorporation of K-Means Clustering with collaborative filtering lags too, since collaborative filtering requires rating of every recipe on the World Wide Web or in the dataset. This combination gives a high lag, when compared to the proposed model. The hybridization of XGBoost with LDA and Cosine Similarity ensures slightly better performance since XGBoost is a strong classifier, and Cosine Similarity is semantically driven, and LDA uncovers the hidden topics, thereby increasing the density of knowledge to a slight extent, and hence, this hybridization has slightly higher performance relative to other hybridizations as well as baseline models.

The precision–recall curve is shown in Fig. 2, and it is obvious that the proposed framework has a superior precision vs number of recommendations curve than other methods in terms of the number of recommendations. The proposed SIRR model occupies the highest in the hierarchy, followed by combination of XGBoost and LDA and Cosine Similarity. The third framework in the hierarchy is the MLFBR [2] framework. RICS [1] model occupies the fourth spot in the list, followed by Decision trees. The lowest spot in the hierarchy is occupied by combination of K-Means Clustering and collaborative filtering. The proposed SIRR framework reveals all the hidden topics using topic modelling by LDA, which eventually leads to increase in density of auxiliary knowledge. The Ontology Alignment further enriches the results. The framework is a blend of both machine learning as well as deep learning models and hence occupies the top position in the hierarchy. The combination of XGBoost, LDA and Cosine Similarity outperforms the other models and occupies second place since, XGBoost classifies well, LDA eventually uncovers the hidden topics and Cosine Similarity is semantically determined. The MLFBR [2] model occupies third place since it uses naïve measure like Euclidean distance with spacing correlation and lacks auxiliary knowledge being fed into the system. The RICS [1] model incorporates content-based filtering along with the usage of TF-IDF, which lands it at the fourth position in the hierarchy, due to the reason that the content-based filtering is not that strong and requires high dense knowledge to be involved into the system. The combination of K-Means Clustering and collaborative filtering occupies the lowest spot in the hierarchy due to the reason that collaborative filtering requires rating for each recipe in the dataset or on the World Wide Web.

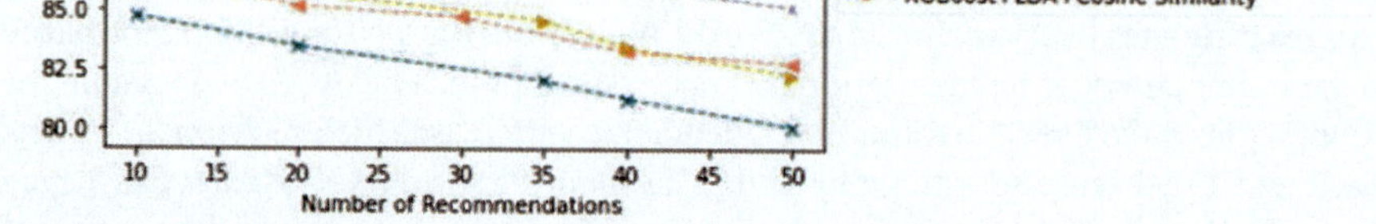

Fig. 2 Precision versus number of recommendations

5 Conclusion

A semantically integrated method is employed to create a recipe recommendation system. The proposed model deals with pre-processing user queries from allrecipes.com, and the pre-processed queries are categorized into relevance using XGBoost and RNN at various sites. The semantic similarity computation is done by hybridizing Normalized Compression Distance (NCD) and KL divergence under a cultural algorithm to ensure highly relevant and diversified results to the user. Finally, experimentation was conducted for 7458 queries over a period of 164 days among several food enthusiasts. The suggested technique achieves F-Measure of 97.07% with a low FDR of 0.04.

References

1. Nilesh N, Kumari M, Hazarika P, Raman V (2019) Recommendation of Indian cuisine recipes based on ingredients. In: 2019 IEEE 35th international conference on data engineering workshops (ICDEW), pp 96–99
2. Vivek MB, Manju N, Vijay MB (2018) Machine learning based food recipe recommendation system. In: Guru D, Vasudev T, Chethan H, Kumar Y (eds) Proceedings of international conference on cognition and recognition. Lecture notes in networks and systems, vol 14. Springer, Singapore
3. Qi M, Neeman Y, Eckhardt F, Blissett K (2018) WhatToMake: a semantic web application for recipe recommendation. Rensselaer Polytechnic Institute
4. Haussmann S, Seneviratne O, Chen Y, Ne'eman Y, Codella J, Chen CH, McGuinness DL, Zaki MJ (2019) FoodKG: a semantics-driven knowledge graph for food recommendation. In: International semantic web conference. Springer, Cham, pp 146–162
5. Ting YH, Zhao Q, Chen RC (2014) Dietary recommendation based on recipe ontology. In: 2014 IEEE 6th international conference on awareness science and technology (iCAST). IEEE, pp 1–6
6. Gyrard A, Bonnet C, Boudaoud K (2014) Enrich machine-to-machine data with semantic web technologies for cross-domain applications. In: 2014 IEEE world forum on internet of things (WF-IoT). IEEE, pp 559–564
7. Gonçalves RS, Horridge M, Li R, Liu Y, Musen MA, Nyulas CI, Obamos E, Shrouty D, Temple D (2019) Use of OWL and semantic web technologies at pinterest. In: International semantic web conference. Springer, Cham, pp 418–435
8. Bianchini D, Antonellis VD, Melchiori M (2015) A web-based application for semantic-driven food recommendation with reference prescriptions. In: International conference on web information systems engineering. Springer, Cham, pp 32–46
9. Erp MV, Wevers M, Huurdeman H (2018) Constructing a recipe web from historical newspapers. In: International semantic web conference. Springer, Cham, pp 217–232
10. Likavec S, Osborne F, Cena F (2015) Property-based semantic similarity and relatedness for improving recommendation accuracy and diversity. Int J Semant Web Inf Syst (IJSWIS) 11(4):1–40
11. Chivukula R, Lakshmi TJ, Sumalatha S, Reddy KL (2022) Ontology based food recommendation. In: IOT with smart systems. Springer, Singapore, pp 751–759
12. Padhiar I, Seneviratne O, Chari S, Gruen D, McGuinness DL (2021) Semantic modeling for food recommendation explanations. In: 2021 IEEE 37th international conference on data engineering workshops (ICDEW). IEEE, pp 13–19

13. Bangale S, Haspe A, Khemani B, Malave S (2022) Recipe recommendation system using content-based filtering. Available at SSRN 4102283
14. Chhipa S, Berwal V, Hirapure T, Banerjee S (2022) Recipe recommendation system using TF-IDF. In: ITM web of conferences, vol. 44. EDP Sciences, p 02006
15. Deepak G, Priyadarshini JS (2018) Personalized and enhanced hybridized semantic algorithm for web image retrieval incorporating ontology classification, strategic query expansion, and content-based analysis. Comput Electr Eng 72:14–25
16. Gulzar Z, Leema AA, Deepak G (2018) Pcrs: personalized course recommender system based on hybrid approach. Procedia Comput Sci 125:518–524
17. Deepak G, Gulzar Z, Leema AA (2021) An intelligent system for modeling and evaluation of domain ontologies for crystallography as a prospective domain with a focus on their retrieval. Comput Electr Eng 96:107604
18. Santhanavijayan A, Naresh Kumar D, Deepak G (2021) A semantic-aware strategy for automatic speech recognition incorporating deep learning models. In: Intelligent system design. Springer, Singapore, pp 247–254
19. Kaushik IS, Deepak G, Santhanavijayan A (2020) QuantQueryEXP: a novel strategic approach for query expansion based on quantum computing principles. J Discr Math Sci Cryptogr 23(2):573–584
20. Belu S, Coltuc D (2020) An innovative algorithm for data differencing. In: 2020 International symposium on electronics and telecommunications (ISETC). IEEE, pp 1–4
21. Csiszar I (1989) A geometric interpretation of Darroch and Ratcliff's generalized iterative scaling. Ann Stat, pp 1409–1413

Implementation of Deep Learning Models for Skin Cancer Classification

Devashish Joshi

Abstract Melanoma skin cancers are most threatening disease. Manual detection of melanomas using dermoscopic images is very time-consuming method which also demands a high level of competence. An accurate and prompt diagnosis needs the development of an intelligent classification system for the detection of skin cancer. This paper implements deep learning models for skin cancer classification and integrates features obtained from several feature extraction methods. Pre-processing, feature extraction, classification, and performance evaluation are phases of proposed approach. Any superfluous noise in the edges is removed during the pre-processing stage. The Gaussian filter method is used to improve image clarity and remove unwanted pixels. The detection of melanoma cells is based on features such as lesion segmentation and colour of images. The contour approach, contrast, Grey scale approaches, lesion segmentation using U-NET are employed for feature extraction. Deep learning-based classifiers such as ResNet50 and CNN architecture are used to classify based on extracted features. Classification techniques use these qualities to identify malignant and affected skin areas. Sensitivity, specificity, accuracy, and F-score are some of the performance measurement criteria used to evaluate the suggested approach. The classifiers are used on the HAM10000 dataset. On the HAM10000 dataset, the suggested framework outperformed existing melanoma detection systems.

Keywords Lesion segmentation · Deep learning models · Gaussian blur · Skin cancer classification

D. Joshi (✉)
Prosirius Technologies, Indore, Madhya Pradesh, India
e-mail: joshi.devashish01@gmail.com

S. Sharma et al. (eds.), *Intelligent Control, Robotics, and Industrial Automation*,
Lecture Notes in Electrical Engineering 1066,
https://doi.org/10.1007/978-981-99-4634-1_45

1 Introduction

If cancer is discovered early on, treatment options may be improved [1]. Melanoma and non-melanoma are the two most frequent types of skin cancer. Malignant melanoma has been the highest frequent form of skin cancer in Western nations and is responsible for the most skin cancer-related deaths [2]. A number of methods may be used to lower the risk of acquiring skin cancer by protecting the human body from the sun's ultraviolet rays [3]. In 2019, 96,480 people in the USA were diagnosed with melanoma, which resulted in 7230 deaths. It is possible to improve the prognosis of melanoma patients and reduce treatment expenses by up to 20 times with early diagnosis of suspicious pigmented lesions (SPLs). Considering the significance of SPL in both medical and financial terms, effective methods for identifying it are scarce.

According to skin cancer statistics, melanoma is a deadly kind of skin cancer since it is a malignant growth on the skin. Alcohol, obesity, tanning, radiation exposure, a history of impaired immune systems, and genetic illnesses all raise the risk of skin cancer. Melanoma skin cancer cases are expected to reach 91,270 new cases in 2018, with a death toll of 9320 [4]. The dermoscopy operation was previously used to conduct medical examinations of the skin and to cure skin diseases [5]. Images of the skin are taken and analysed using computer technology in dermoscopy, a form of computer-aided diagnostics Dermoscopy and the naked eye are used to assess a person's risk of contracting skin cancer. This approach has an accuracy rate of 75–80% when it comes to diagnosing skin cancer.

Computer-aided diagnosis (CAD) for melanoma skin cancer relies on the categorization of a melanoma skin lesion as benign or malignant [4, 5]. The use of computer-aided diagnostic software is currently widespread. There is a high demand for services in the healthcare industry. Developing biomedical datasets is done by saving data from people who are sick in computers. In order to develop an effective technique for early detection of skin cancer, our dataset images will be classified as benign or malignant.

2 Background

Detection and classification of melanoma have been the focus of a great deal of research in the past. 5846 clinical photographs of pigmented skin lesions from 3551 individuals were analysed by Jinnai S. and a team of researchers (2020) [6]. The pigmented skin lesions included nevus, seborrheic keratosis, senile lentigo, and hematoma/hemangioma. Photos of 666 patients were randomly selected for the test dataset, and bounding box annotations were applied to the rest of the photographs for the training dataset (4732 images, 2885 patients). By combining the training and test datasets, a faster region-based CNN (FRCNN) was developed and tested. It also resulted in the graduation of ten dermatological board-certified physicians (BCDs) and ten dermatology resident doctors (TRNs). The same tests to see how accurate they were compared to FRCNN's.

Celebi et al. (2007) [7] established a colour, texture, and form-based classification system for pigmented skin lesions based on dermoscopy images. The image has been divided into sections that show the most important clinical characteristics of the lesions in terms of colour and texture. In an optimization framework, the attributes are ranked in order to find the optimal subset. Lesion border detection is used to differentiate lesion from background skin. The boundary found is then used to determine the form attributes of lesions. At 92.34% specificity and 93.333% accuracy, the method works well on 564 samples.

Barata et al. (2014) [8] established two methods for identifying melanoma in dermoscopy images. The first system uses global features to categorise skin lesions, whereas the second system uses local characteristics and the bag-of-features classifier to categorise skin lesions. Colour features performed better than texture features when used alone to tackle the problem of skin lesion classification, according to the findings.

Soenksen et al. (2021) [9] used deep convolutional neural networks (DCNNs) to build an SPL analysis system for wide-field images and tested it on 38,283 dermatological datasets from 133 patients and public photographs. An assortment of photos taken with consumer-grade cameras was categorised by three board-certified dermatologists (15,244 nondermoscopy). Primary care physicians will be able to more rapidly and effectively identify the suspiciousness of pigmented lesions using this technique, according to the researchers, leading to better patient triage, resource use, and melanoma therapy. (Kawahara et al. (2016) [10]) Ten distinct types of skin lesions are classified using linear classifiers. A convolutional layer was added to AlexNet to replace the previous one. For feature extraction, an AlexNet was also employed.

Emara et al. (2019) [11] used Inception-v4 to classify data from the HAM10000 dataset instead of an ensemble of several sophisticated models. Reusing features from prior layers and concatenating them with high-level layers improves the classification performance of the proposed model, according to the paper. The uneven nature of the dataset necessitated the employment of a data sampling approach in this investigation.

Matsunaga et al. (2017) [12] used a CNN to categorise three forms of skin cancer based on photographs of the skin. This model won first place at the 2017 ISBI competition because to the use of data augmentation and the Keras library. Seog Han et al. [13] used the CNN approach to classify 12 unique kinds of lesions from two Korean hospitals. Kawahara et al. [14] tested CNN against Dermofit's public picture collection, and it was found to be 79.5% accurate. (Menegola et al. [15]) images from the 2017 International Skin Imaging Collaboration were used to create skin cancer categories (ISIC). Using a CNN model to classify skin lesions as benign or malignant is not new.

(Demir et al. (2019) [1]) By classifying our dataset photographs as benign or malignant, they come up with a solid way to identify skin cancer early on. In all, their dataset contains 2437 photographs for training, 660 photos for testing, and 200 photos for validation purposes. ResNet-101 and Inception-v3 deep learning architectures are used in the classification task. Differentiating characteristics of skin lesions include their structural abnormalities and colour changes. To identify skin cancer, researchers employed the nervous system to classify the lesions by their shape, colour, and tissue qualities [16, 17]. An artificial neural network classifier was used to classify skin cancer in a different study [18]. SVM and k-NN classifiers for skin cancer categorization yielded a 61% accuracy rate in skin cancer classification, according to a study [19].

3 Proposed Methodology

Dermoscopic pictures might be used to detect skin cancer, according to one study. Figure 1 depicts phases one through four. Pre-processing, feature extraction, classification and performance evaluation take place before the data can be analysed. Features are retrieved from retinal pictures after they have been normalised and scaled utilising feature extraction algorithms. Gaussian blur, region of interest (ROI), and lesion segmentation using the U-Net architecture are some of the characteristics included in these images. Using dermoscopic pictures with Gaussian blur, ROI extraction of the whole image boundary, and U-Net model extraction of lesions using lesion segmentation, this is done. For the purpose of lesion segmentation, several U-Net models are being created and put to the test. ResNet50 and a convolution neural network are now being evaluated as deep learning classification techniques (CNN). F-score is used to measure how accurate a classification model is. The next section explains the algorithm in further depth.

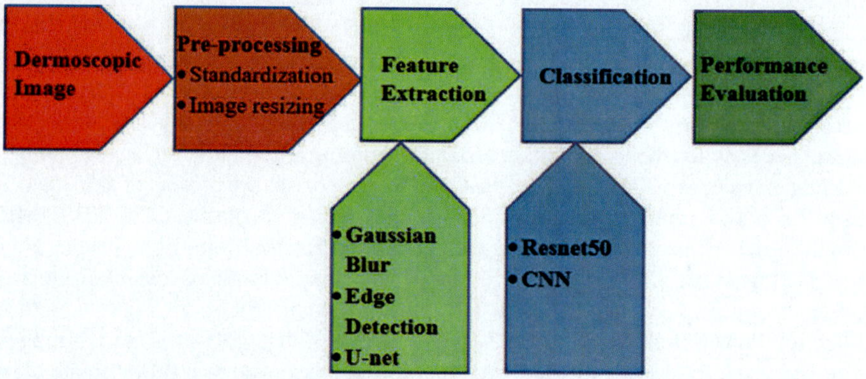

Fig. 1 Skin cancer classification proposed approach

3.1 Algorithm

The algorithm displays an explanation of each phase along with its respective module.

Phase 1: Pre-processing

The raw dermoscopic pictures were given via the HAM10000 dataset. In the collection, photographs come in a variety of shapes; in this project, images with (75, 100, 3) dimensions are selected for processing, and those lacking this shape are scaled. These suggestions include standardisation, rotating images, and enhancing the visual quality. Optical distortion, grid distortion, elastic transform, vertical flip, and horizontal flip are examples of image augmentation techniques.

Phase 2: Features Extraction

Gaussian blur, a region of interest, contrast enhancement, and lesion segmentation using U-Net topologies are some of the techniques used to extract distinct information from the pictures.

Gaussian blur: This is a technique for blurring images and eliminating high frequency components in order to extract countors from them. (75, 100, 3) is currently the established standard for further processing of images. A Gaussian blur function is provided by the Python computer vision package (cv_2), which is employed in this study. This fuzzy picture is used in the categorization process. The Gaussian kernel G, which is used to achieve Gaussian blurring, represents a circularly symmetrical bell-shaped hump. The Gaussian kernel is computed using the following function mentioned in Eq. (1).

$$G(X, Y) = \frac{1}{2\pi\sigma^2} e^{-(x^2+y^2)/2\sigma^2} \tag{1}$$

Region of Interest: Medical image analysis relies on the detection of a region of interest (ROI), defined as a bounding box enclosing the lesion in question. This technique has proven invaluable. Scooting windows are used to identify ROIs learned on skin lesion patches. The select ROI method of OpenCV is utilised to detect lesion ROIs in this research. In order to save the picked ROI points to a parameter, the picture should be provided to the select ROI function, which will then do so. The ROI parameter will store the image coordinates of the bounding box's left and right top corners. Crop the image using these points as a guide.

Lesion Segmentation using U-Net architecture: A characteristic derived from the lesion is used to segment the lesion in future. A lesion's size, shape, and thickness can be used to accurately identify melanoma cancer, since most skin problems can be identified this way. Segmentation is performed using the well-known U-Net architecture, with three different encoder and decoder layers. Each layer has a varied number of channels in it. Decoding and encoding algorithms are implemented using convolutional neural networks (CNNs). The encoding block uses the pooling layer to identify abstract characteristics by lowering position values and information. The local pixel attributes of the decoding layer are employed to ensure exact placement.

It is via the use of an up-sampling technique that data that was previously lost due to down-sampling can be recovered and included in the new feature map. By avoiding the link between encoder and decoder levels, the network can keep its low-level properties.

Model 1 has three layers of encoders, one layer of convolution, and three levels of decoders. When the maximum pooling operation for each layer is accomplished, the encoder block's size doubles in the subsequent layer. Before output, 512-dimensional convolution blocks perform batch normalisation and maximum pooling. Using maximum pooling will still preserve image texture information. Using up-sampling and image reduction, the decoder block is able to calculate the original picture's dimensions. The decoder layers convolutionally transpose and concatenate the skip connection feature vectors. Model 2 includes four encoder layers, one convolution layer, and four decoder levels to it. Model 3 consists of five encoder layers, one convolution layer, and five decoder levels. These three models are utilised to separate lesions from one another, and the resulting pictures are then used to sort the lesions into different groups. Using these five attributes, photos are then transmitted to the classifiers described in the next section.

Phase 3: Diabetes Retinopathy Classification

The terms "Malignant" and "Benign" used in the categorization of melanoma lesions imply a binary distinction. Five aspects are classified individually in order to determine which one is the most effective. This system makes use of ResNet50, InceptionV3, and CNN, three deep learning classification architectures. Classification models divide datasets into two parts: training 70% and testing 30%.

ResNet50: ResNet50's convolution block-1, which gets input from ResNet50's feature pictures, has ResNet50's ReLU activation, batch normalisation, and convolution block-1 all in one place. These strategies use mapping and filtering to determine context. ResNet50's first block, Block-1, connects to the following block in the same way. Internal adjustments have been made to 32, 64, 128, 256, 512, 1024, and 2048 types. The model incorporates two thick layers with dropout layers to avoid overfitting. One of the thick layers shows ReLU activation with penalty value 0.01 and dropouts with dropouts of 0.50. A sigmoid function is used to activate the output layer. There are two loss functions employed in this model's construction: the binary cross-entropy loss and the Adam optimizer, both with learning rates of 0.0001. For testing, epoch sizes of 32 and 40 are utilised; however, the model gets overfitting and stagnate around epoch 10.

CNN architecture: In all, the CNN model contains eight layers, which include Conv2D, MaxPooling2D, ReLU activation, dropout operations, and a completely connected layer. Node weights of 32 are used to apply the first convolution layer to a featured picture with size of 220px (264, 264, 3). The convolution layer performs striding and padding to ensure that the output picture size is constant. Using the pooling layer, the convolution layer may produce a smaller convolved feature map and save on computational resources. To do this, remove all connections between layers and work on each feature map separately. This research uses the MaxPooling2D operation and a pool size of (2, 2) to pool the most significant part of the feature map.

A dropout layer is introduced after pooling to minimise overfitting with a dropout value of (0.5), which indicates that 50% of neurons are eliminated during training and uses the ReLU activation in the convolution layer. In all, there are seven CNN layers, each with a unique amount of node weights: 64, 128, 256, 128, 32, and 16. A thick layer with ReLU activation and node weight 16 is utilised to accurately represent the vector after flattening it. Finally, this layer is shown as a completely connected layer. Figure 1 shows the output in terms of malignant and benign classifications from an extra dense layer with node weight 5 and "sigmoid" activation. Two cross-entropy losses and a learning rate of 0.0001 are utilised in the model's construction. At epoch 23, the model becomes overfit and unable to learn new information.

Phase 4: Performance Assessment

The effectiveness of the framework's classifiers is measured using a range of measures, including the following:

Accuracy (A_c): Total photos properly categorised is a measure of the classifier's overall accuracy. The formula provided in Eq. 2 can be used to determine how accurate a computation is?

Sensitivity (S_{en}): The percentage of anomalous photos that are labelled as such is the metric. Equation 3 depicts the sensitivity of a system.

Specificity (S_{pe}): The percentage of normal photos that are classified as such is the metric used. It is illustrated in Eq. 4.

$$A_c = \frac{T_N + T_P}{T_N + T_P + F_N + F_P} \tag{2}$$

$$S_{\text{en}} = \frac{T_P}{T_P + F_N} \tag{3}$$

$$S_{\text{pe}} = \frac{T_N}{T_N + F_P} \tag{4}$$

True Positive refers to the number of times a classifier accurately predicted that a sample will test positive (T_P). True Negative refers to the number of times the classifier correctly predicted the negative class as negative (T_N). False Positives are the number of times the classifier incorrectly identifies a negative class as a positive one (F_P). False Negative is a term used to describe how often a classifier has projected the positive class as the negative one (F_N). A good prediction model has a value near to 100% in these parameters.

4 Experimentation and Results Discussion

In this section, the dataset and evaluation methods are presented, and the dataset's findings are shown.

4.1 Experimental Setup and Dataset

The information was gleaned from a vast collection of high-resolution dermoscopic images. The suggested models are put to the test using the TensorFlow framework and the keras API for deep learning techniques. Many hyper parameters, including adam optimizer, binary cross-entropy loss function, 40 epochs and a batch size 32, are utilised to evaluate various models. Approximately 1e−4 to 1e−5 new things are learned per second (depending on the task). Cross-entropy loss function in Eq. (5) shows the expected probability of the point being green for all N points, where p(y) is the label and y is the probability of the point being green (1 for green points and 0 for red ones).

$$H_p(q) = -\frac{1}{N}\sum_{i=1}^{N} y_i . \log(p(y_i)) + (1 - y_i). \log(1 - p(y_i)) \quad (5)$$

4.2 Results and Discussions

Malignant and benign classes were categorised using five criteria in the proposed method. When extracting pictures, various image augmentations may be done, as seen in Figs. 2a, b, and c and 3a, b. Various classifiers have been used after receiving featured photos. Table 1 shows the results of these classifiers. The accuracy, precision, recall, and F-score of ResNet50 and CNN are analysed. CNN was beaten in every category by ResNet50. Using ResNet50, which has a 96% accuracy rate, is one of the best outcomes. One of the most accurate classifiers, ResNet50's sensitivity is 98% and its specificity is 88%, which is decent but might be better. For all classifiers, feature fusion enhanced their performance. The suggested approach, which makes use of the ResNet50 deep learning model, outclasses all other classifiers on all metrics and variables.

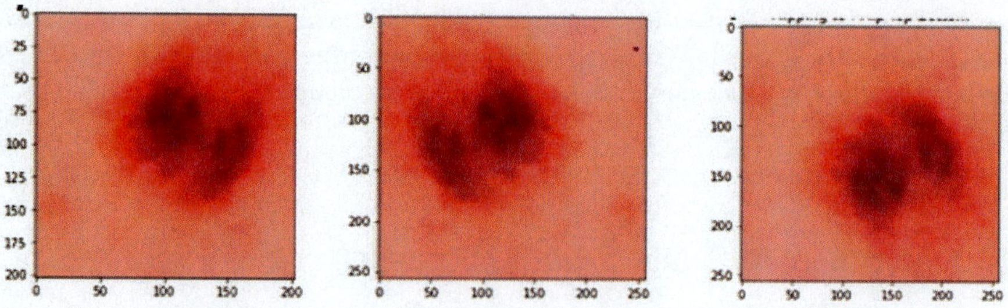

(a) (b) (c)

Fig. 2 **a** Before augmented image, **b** Images flip left to right, **c** Image flip top to bottom

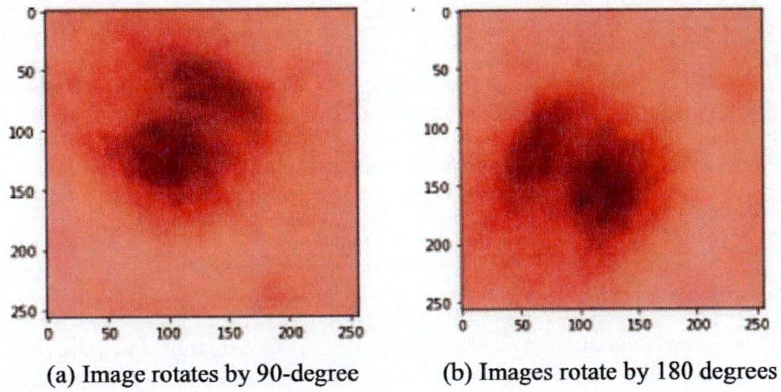

Fig. 3 **a** Image rotates by 90°, **b** Images rotate by 180°

Table 1 Performance evaluation of proposed framework based on different classifiers

Classifier	Accuracy	Specificity	Sensitivity	F-Score
ResNet50	96	88	98	94
CNN	82	62	56	59

For validation and training data, this research reveals how epoch-based classifiers fare. Figure 4a depicts the ResNet50 validation and training loss as a function of epochs. The loss diminishes with an increase in the number of epochs, and the values of the validation and training losses are identical. Figure 4b compares the validation and training accuracy of ResNet50 over epochs. After epoch 15, training accuracy remained stable for the remainder of the epochs. Validation precision was unaffected by epochs. Figure 5a depicts the validation and training losses for CNN as a function of epochs. Both validation and training loss are roughly equal until epoch 13, at which point the loss falls significantly. Cross-epoch comparison of CNN validation and training accuracy is depicted in Fig. 5b. Until epoch 11, validation and training accuracies were identical, but there was a considerable disparity in following epochs.

Figure 6 illustrates how the proposed paradigm compares to previous studies. The random forest classifier scored best when Hue-Saturation-Value was employed as a feature, according to the authors of [4]. With a 74.75% accuracy rate, they were successful. The authors in [20] used EfficientNets with a variety of input cropping methods, input resolutions, and CNN designs. They had a 74.2% accuracy rate and a 96.2% specificity rate, but their sensitivity was just 59.4%. As a consequence, Chaturvedi et al. [21] writers achieved an accuracy of 83%, which is greater than the previous findings. Using a deep convolution neural network (DCNN), the authors of [22] were able to categorise skin lesions while simultaneously dealing with issues caused by an imbalanced dataset. By 86.7%, they increased their accuracy, while also increasing the sensitivity and specificity of their results. Based on this comparison,

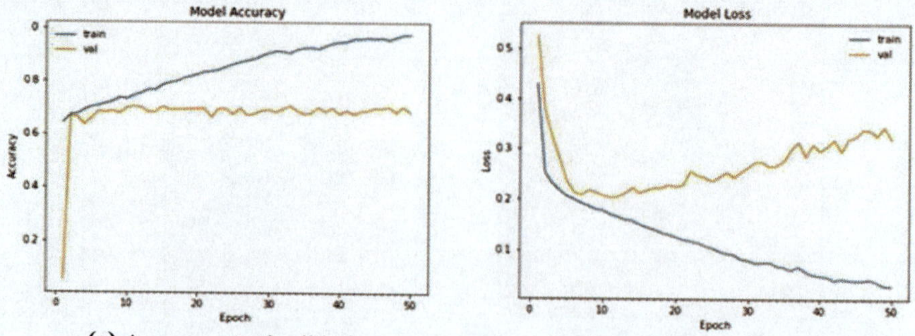

(a) Assessment of validation and training accuracy of ResNet5

(b) Assessment of validation and training loss of Resnet50

Fig. 4 **a** Assessment of validation. **b** Assessment of validation and training accuracy of ResNet5 and training loss of Resnet50

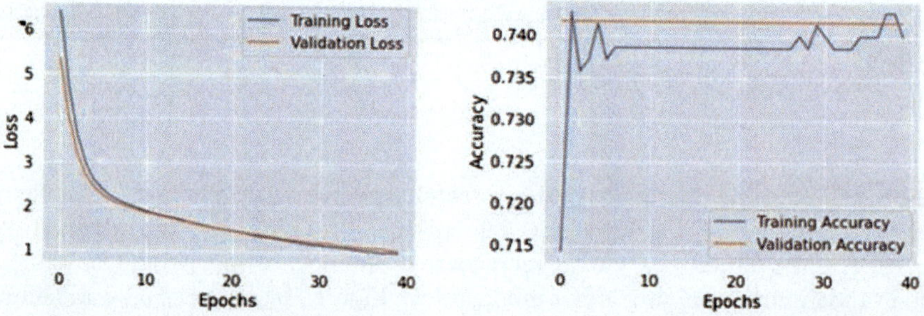

(a) Assessment of validation and training loss of CNN

(b) Assessment of validation and training accuracy of CNN

Fig. 5 **a** Assessment of validation. **b** Assessment of validation and training loss of CNN training accuracy of CNN

we can see how this earlier research stack up against the proposed ResNet50 categorization system. It was found that the proposed work had an accuracy of 94%, a sensitivity of 82%, and a specificity of 96%. Combining handmade factors like colour, edges, and the intriguing region, along with clinical features like vascular segmentation, may be more effective in the categorization of skin cancer than clinical features on their own.

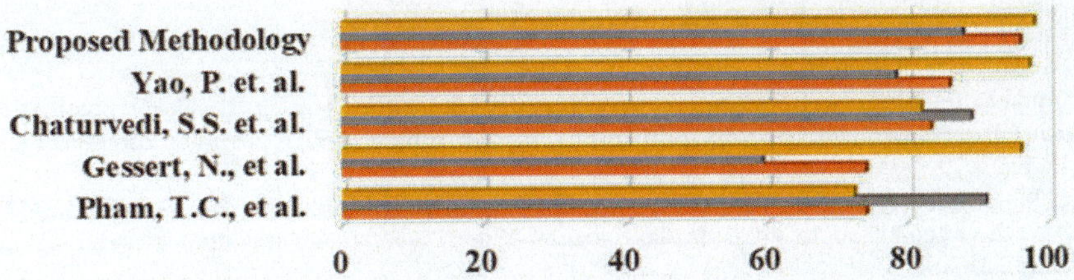

	Pham, T.C., et al.	Gessert, N., et al.	Chaturvedi, S.S. et. al.	Yao, P. et. al.	Proposed Methodology
Specificity	72.4	96	82	97.3	98
Sensitivity	91.09	59.4	89.1	78.4	88
Accuracy	74.15	74.1	83.41	86.1	96

Fig. 6 Comparison of proposed methodology with previous research

5 Conclusion

Feature extraction, classification, and performance assessment are the four parts of this research's method to skin cancer classification. Gaussian blur, ROI, and lesion segmentation utilising U-Net architecture are some of the feature extraction strategies employed in this system. For vessel segmentation, a variety of U-Net models have been developed and tested. In order to diagnose melanoma skin cancer, classification algorithms are utilised. The suggested technique makes use of already-existing deep learning architectures like ResNet50 and CNN. There are two types of images that fall into this category: malignant and benign. In order to evaluate the performance of these classification models, the accuracy, precision, recall, and F-score are all used. The feature fusion technique outperformed all others for all classifiers. In terms of classification, the ResNet50 model performs better, with a 96% accuracy. The suggested approach with the ResNet50 model beat all other classifiers on all metrics and features.

References

1. Demir A, Yilmaz F, Kose O (2019) Early detection of skin cancer using deep learning architectures: resnet-101 and inception-v3. In: 2019 medical technologies congress (TIPTEKNO). IEEE, pp 1–4
2. Schadendorf D, van Akkooi ACJ, Berking C, Griewank KG, Gutzmer R, Hauschild A, Stang A, Roesch A, Ugurel S (2018) Melanoma. The Lancet 392:971–984
3. Gandini S, Sera F, Cattaruzza MS, Pasquini P, Zanetti R, Masini C, Boyle P, Melchi CF (2005) Meta-analysis of risk factors for cutaneous melanoma: III. Family history, actinic damage and phenotypic factors. Eur J Canc 41:2040–2059
4. Pham TC, Tran GS, Nghiem TP, Doucet A, Luong CM, Hoang V-D (2019) A comparative study for classification of skin cancer. In: 2019 International conference on system science and engineering (ICSSE). IEEE, pp 267–272

5. Kittler H, Pehamberger H, Wolff K, Binder M (2002) Diagnostic accuracy of dermoscopy. Lancet Oncol 3:159–165
6. Jinnai S, Yamazaki N, Hirano Y, Sugawara Y, Ohe Y, Hamamoto R (2020) The development of a skin cancer classification system for pigmented skin lesions using deep learning. Biomolecules 10:1123
7. Celebi ME, Kingravi HA, Uddin B, Iyatomi H, Aslandogan YA, Stoecker WV, Moss RH (2007) A methodological approach to the classification of dermoscopy images. Comput Med Imaging Graph 31:362–373
8. Barata C, Ruela M, Francisco M, Mendonça T, Marques JS (2013) Two systems for the detection of melanomas in dermoscopy images using texture and color features. IEEE Syst J 8:965–979
9. Soenksen LR, Kassis T, Conover ST, Marti-Fuster B, Birkenfeld JS, Tucker-Schwartz J, Naseem A, Stavert RR, Kim CC, Senna MM, Avilés-Izquierdo J, Collins JJ Barzilay R, Gray ML (202) Using deep learning for dermatologist-level detection of suspicious pigmented skin lesions from wide-field images. Sci Trans Med 13:eabb3652
10. Kawahara, J., BenTaieb, A., Hamarneh, G.: Deep features to classify skin lesions. In: 2016 IEEE 13th international symposium on biomedical imaging (ISBI). IEEE, pp 1397–1400
11. Emara T, Afify HM, Ismail FH, Hassanien AE ()A modified inception-v4 for imbalanced skin cancer classification dataset. In: 2019 14th International conference on computer engineering and systems (ICCES). IEEE, pp 28–33
12. Matsunaga K, Hamada A, Minagawa A, Koga H (2017) Image classification of melanoma, nevus and seborrheic keratosis by deep neural network ensemble. arXiv preprint arXiv:1703.03108
13. Han SS, Kim MS, Lim W, Park GH, Park I, Chang SE (2018) Classification of the clinical images for benign and malignant cutaneous tumors using a deep learning algorithm. J Investig Dermatol 138:1529–1538
14. Kawahara J, Hamarneh G, Multi-resolution-tract CNN with hybrid pretrained and skin-lesion trained layers. In: International workshop on machine learning in medical imaging. Springer, pp 164–171
15. Menegola A, Fornaciali M, Pires R, Bittencourt FV, Avila S, Valle E (2017) Knowledge transfer for melanoma screening with deep learning. In: 2017 IEEE 14th international symposium on biomedical imaging (ISBI 2017). IEEE, pp 297–300
16. Hintz-Madsen M, Hansen LK, Larsen J, Drzewiecki KT (2001) A probabilistic neural network framework for the detection of malignant melanoma. In: Artificial neural networks in cancer diagnosis, prognosis, and patient management. CRC Press, pp. 141–184
17. Piccolo D, Ferrari A, Peris K, Daidone R, Ruggeri B Chimenti S (2002) Dermoscopic diagnosis by a trained clinician vs. a clinician with minimal dermoscopy training vs. computer-aided diagnosis of 341 pigmented skin lesions: a comparative study. Br J Dermatol 147:481–486
18. RB A, Jaleel JA, Salim S (2013) Implementation of ANN classifier using MATLAB for skin cancer detection. Academic Press
19. Mariam A, Sheha Cairo University, Mai S, Mabrouk MUST University, Amr S, Cairo University (2012) Automatic detection of melanoma skin cancer using texture analysis. Int J Comput Appl 0975-8887
20. Gessert N, Nielsen M, Shaikh M, Werner R, Schlaefer A (2020) Skin lesion classification using ensembles of multi-resolution EfficientNets with meta data. MethodsX 7:100864
21. Chaturvedi SS, Gupta K, Prasad PS, Skin lesion analyser: an efficient seven-way multi-class skin cancer classification using MobileNet. In: International conference on advanced machine learning technologies and applications. Springer, pp 165–176
22. Yao P, Shen S, Xu M, Liu P, Zhang F, Xing J, Shao P, Kaffenberger B, Xu RX (2021) Single model deep learning on imbalanced small datasets for skin lesion classification. IEEE Trans Med Imaging 41:1242–1254

Depth Estimation and Optical Flow-Based Object Tracking for Assisting Mobility of Visually Challenged

Shripad Bhatlawande, Manas Baviskar, Awadhesh Bansode, Atharva Bhatkar, and Swati Shilaskar

Abstract Vision disability is one of the most serious problems. Electronic travel aids can help visually impaired people to perform their mobility and navigation-related activities. This work presents an electronic travel aid that interprets the surrounding environment and conveys its audio representation to the user. It uses optical flow for tracking of obstacles. It computes the flow vectors of all the pixels in the frame by monitoring the current frame and the previous frame. The proposed solution implements dense optical flow which is based on Farneback's algorithm. This aid is implemented in the form of smart clothing. The weight of this wearable and portable system is 440 g. This aid accurately detects the obstacle, its position, and the distance from the user. It shortlists the priority details of the surrounding environment and translate them into simplified audio feedback. It notifies these alerts via an earphone. This aid was subjected to ten usability experiments to assess its relevance as an electronic travel aid. It consistently helped the user to understand the surrounding environment and provided an average accuracy of 81.84% for overall obstacle detection.

Keywords Aid for visually impaired · Electronic travel aid · Computer vision · Machine learning · Optical flow

S. Bhatlawande (✉) · M. Baviskar · A. Bansode · A. Bhatkar · S. Shilaskar
E&TC Department, VIT, Pune 411037, India
e-mail: shripad.bhatlawande@vit.edu

M. Baviskar
e-mail: manas.baviskar19@vit.edu

A. Bansode
e-mail: awadhesh.bansode19@vit.edu

A. Bhatkar
e-mail: atharva.bhatkar19@vit.edu

S. Shilaskar
e-mail: swati.shilaskar@vit.edu

S. Sharma et al. (eds.), *Intelligent Control, Robotics, and Industrial Automation*,
Lecture Notes in Electrical Engineering 1066,
https://doi.org/10.1007/978-981-99-4634-1_46

1 Introduction

One of the major issues that scientists face around the world is blind mobility. If we were to see in the global context, vision disability affects minimum 2200 million people globally [1]. To count avert there have been various assistive technologies ranging since the beginning of civilizations, from "low tech" to "high tech": from various types of canes and service dogs to electronic mobility aids that use ultrasonic waves and augmentative communication devices. These measures, although popular among the visually impaired population, are not as durable and cost-effective. Over the years, various kinds of commercial applications have emerged to assist the visually disabled individuals. The previous works surrounding this context include using a Smart Cane [2, 3] for detecting objects using ultrasonic sensor visually impaired. Tyflos [4] is a device consisting of conventional eyeglasses and a stereo vision module which captures the surrounding data. After capturing the data, it creates a depth map of the scanned 3D environment. A blind person is required to wear a vibratory belt on his/her abdomen which vibrates after locating obstacles. Visual White Cane [5] developed by R. Manduchi and D. Yuan is a laser-based mobility device which is an alternative to the traditional cane used by visually impaired people. The cane scans the surrounding with a laser pointer. The system also consists of a camera and a computer processor. The system thus produces a feedback which is given to the user in audio format after constantly analyzing the spatial surrounding.

Echolocation [6] uses two ultrasonic sensors mounted to regular spectacles which were used for echolocation. Using a micro-controller and an A/D converter, the sensor collects the data and converts it to a recognizable stereo sound. The varying intensities and temporal discrepancies of the reflected ultrasound waves are transferred to create a sort of localized sound pictures that show the different orientations and sizes of obstructions. However, the issue here is that it necessitated extensive training. Various GPS-powered applications [7] such as Braille Note GPS, MoBIC, and Mobile Geo have been developed and commercialized in recent years. Computer vision-based applications such as the vOICe, ENVS, and TVS are some of the most popular commercial applications that have been developed for navigation assistance. These computer vision-based modalities provide an edge over the traditional sensor technologies: sonar, infrared, and structured light. For instance, there is no limitation in observable distance and is possibly of a higher resolution. In Florida International University (FIU) [8], they proposed a project which consists of three-dimensional sounds such as head-related transfer functions (HRTF) obstacle detection system, also known as anatomical transfer functions, established on study from a multi-directional sonar system. The blueprint is made up of 2 sub-components: a sonar and compass control unit with 6 ultrasonic sensors directing in 6 directions (radial) around the user, as well as a micro-controller and a three-dimensional sound transferring system with a audio output device: such as a speaker, and a personal digital assistant consisting of a software which effectively processes data. The range unit is not ergonomically designed, and the system lacks navigation speed.

Navbelt [9] uses a mobile robot to avoid obstacles. It uses Ultrasonic sensors, a processing unit, and a head mountable system. The information from the sensors is used to produce a map of the angles and distances from this perspective to all the objects in the range. Depending on the orientation of the obstruction, different noises were created for each mode. The drawback of this solution is that it is large and requires a long period of training for the user. Navbelt has been upgraded as Guidecane [10]. A guiding cane is made up of a handle and an electronic system which has ultrasound sensors, a control movement mechanism along with a processing unit. Once the obstacle is encountered, a substituted direction is selected till the impediment is no longer in path, and the route is recommenced. The blueprint is cumbersome, hard to grasp or handle whenever it is carried by the user and has a limited region of area because tiny or aerial (stationary) objects such as sidewalks, tables are unidentified. Artificial vision system for blind (AVSB) [11] is a inexpensive navigation system for visually disabled individuals based on a micro-controller. Ultrasonic sensors have been used to estimate the distance between obstacles and guide the user down the best way. The result is voice enabled that a blind person can comprehend. This system has the disadvantage of being inflexible: visually impaired people's normal hearing is obstructed. Electro-neural vision system (ENVS) is a system which provides a virtual perception of the surrounding through electrical pulses. The stereo camera helps the system to capture the image and measure the depth which is then used to calculate the distance and converts them into electrical pulses which are then sent to the gloves. The gloves help in transmitting stimulation to the fingers through electrodes. The vOICe is also a real-time sensory system which provides a virtual experience through iterative image to sound feedbacks. It consists of a camera which is placed on the head: scans the surrounding. This frame of the video is then mapped to the audio based on two factors: height is equated to pitch, and brightness equated to intensity of the audio. Similar sensory-substitution processes have been developed in technologies such as EyeMusic and PSVA. Additionally, these measures cater to a spectrum of people who can afford it. After analyzing the economic impact of vision impairment, WHO concluded that vision impairment is a major worldwide economic distress: with worldwide yearly valuation of loss of productivity affiliated with undiagnosed presbyopia and myopia alone are approximated to a figure of US $25.4 billion and US $244 billion, respectively [1]. According to WHO, only 10% of blind population live in developed countries [1]. This highlights the need of low-cost solutions which can be affordable as well as durable.

The above-mentioned solutions face challenges such as being large and heavy which make them difficult to carry or use. Moreover, the solutions fail to consider obstacles present in the air (i.e., objects hanging in air). Additionally, the systems mentioned previously laid major emphasis on handling object detection through tactile-substitution methods which require extensive training of the user. To counter these limitations, we have proposed a system which is light-weight, efficient and takes into consideration objects in air if they are present within the region of interest. It is markedly an engrossing observation that none of the above-mentioned techniques/devices utilizes or strive to utilize of optical flow.

2 Methodology

The system detects obstacles. It conveys their distance to the user with an audio output. Figure 1 shows system block diagram. The smart clothing comprises of a chest mounted camera, a processor system, and an earphone.

As an input, the camera captures the video of the immediate surrounding. It detects the obstacles and converts their details into audio output. It uses earphone to convey the audio output to visually impaired individual.

The proposed system is implemented in 3-stages as depicted in Fig. 2, namely—(I) Optical Flow for Calculating Motion Vectors, (II) Detection of Obstacles in the Frame, (III) Object Tracking and Distance Estimation.

2.1 *Optical Flow for Calculating Motion Vectors*

Optical flow is an effective technique for tracking object movement. It works on the assumption that the pixel intensities (Z) are constant between the two frames as shown in Eq. (1). Further Taylor approximation is applied to arrive upon Eq. (2) which is finally divided by the derivative of time (dt) to yield Eq. (3):

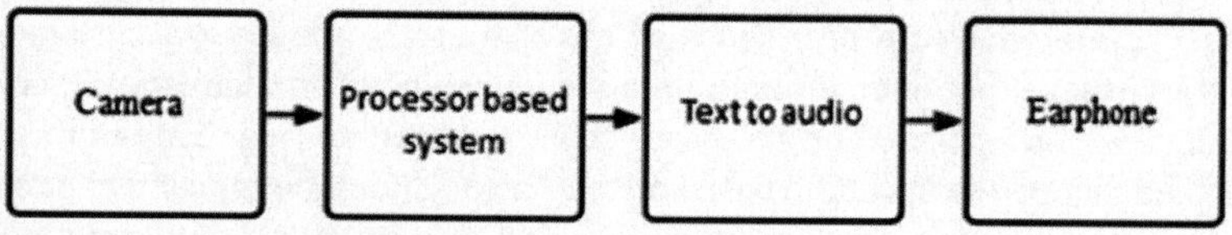

Fig. 1 Block diagram of components required

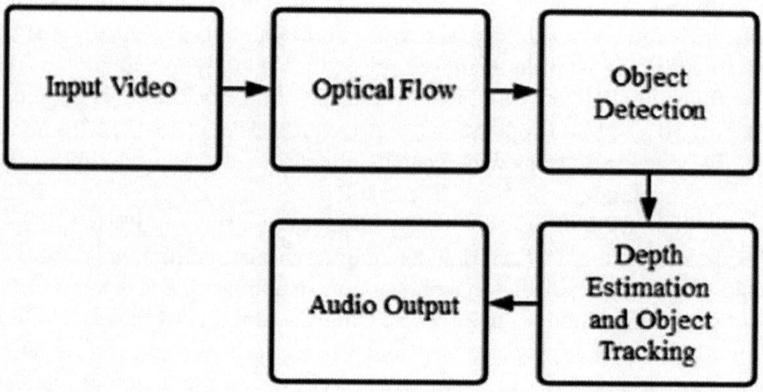

Fig. 2 Workflow of the proposed system

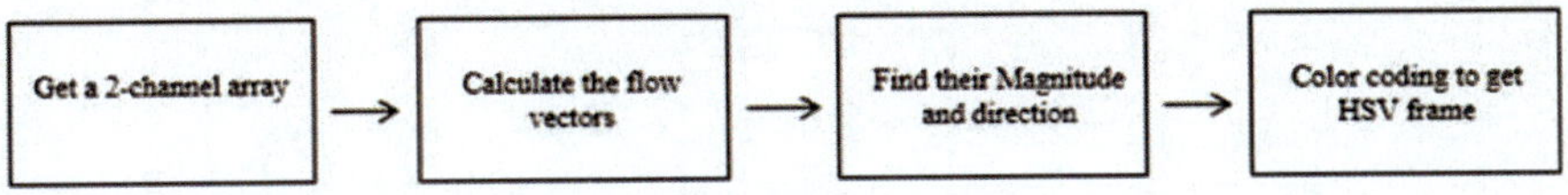

Fig. 3 Optical flow calculation for every point in frame

$$Z(x, y, t) = Z(x + \mathrm{d}x, y + \mathrm{d}y, t + \mathrm{d}t) \quad (1)$$

$$\left(\frac{\mathrm{dZ}}{\mathrm{d}x}\right)\delta x + \left(\frac{\mathrm{dZ}}{\mathrm{d}y}\right)\delta y + \left(\frac{\mathrm{dZ}}{\mathrm{d}t}\right)\delta t = 0 \quad (2)$$

$$\left(\frac{\mathrm{dZ}}{\mathrm{d}x}\right)m + \left(\frac{\mathrm{dZ}}{\mathrm{d}y}\right)n + \left(\frac{\mathrm{dZ}}{\mathrm{d}t}\right) = 0 \quad (3)$$

where:

m: (d*x*/d*t*), *n*: (d*y*/d*t*)

dZ/d*x*: Gradient of image along *x*-axis

dZ/d*y*: Gradient of image along *y*-axis

The flow vectors calculated by the algorithm are then converted to polar co-ordinates-magnitude and angle, to convert them into HSV format, with magnitude mapping hue and angle mapping value (saturation remaining constant) in HSV format. To convert them into HSV format, the angle (measured in radians) is converted to degrees using Eq. (4), and the magnitude is normalized between the pixel range—0 to 255. "MIN–MAX" normalization technique was used to normalize such that the minimum value of the sub-array considered is alpha which is the lower bound (0 in our case) and beta as the maximum value of the sub-array (255 in our case). The HSV frame is then converted to BGR format and passed for further processing to the following section. The process is depicted in Fig. 3.

$$\text{Angle in degree} = (\text{Angle in radians}) * (\text{pi}/180) \quad (4)$$

2.2 Detection of Obstacles in the Frame

The BGR formatted video is converted to grayscale format. The grayscale frame is then set up for binary thresholding technique to remove irrelevant discrepancies and noisy areas.

The threshold value considered in the project is predefined to 100 (lower limit) based on experimental trials for best accuracy of detection which is obtained after the consequent step of the project. The significance of this method in the proposed solution is: the values equal to or below the predefined threshold will be removed which will yield better detection of objects. The process is shown in Fig. 4.

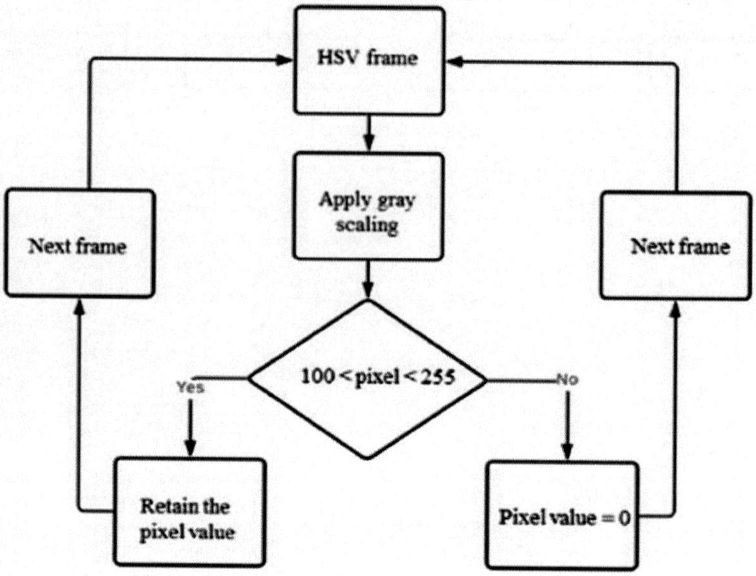

Fig. 4 Threshold-based retention of pixel values

The system then detects the contour and calculates its area. If the contour area of the object is greater than a predetermined threshold (400), it is considered as an object, else it is ignored. The predetermined threshold value was considered after numerous iterations of the system on varied scenarios. The detected object is represented on the frame using a bounding box (rectangle). The process is shown in Fig. 5. The system implements perspective transformation to change the viewpoint. This method will help us to change the problem of non-linear distance estimation to linear distance estimation. In general, the transformation to change the perspective can be expressed in Eq. (5), where transformation matrix T, the input matrix B consisting of points to be transformed and A is the resultant matrix.

$$A = T * BB \tag{5}$$

$$\begin{bmatrix} tx' \\ ty' \\ t \end{bmatrix} = \begin{bmatrix} m1\ m2\ n1 \\ m3\ m4\ n2 \\ l1\ \ l2\ \ 1 \end{bmatrix} \cdot \begin{bmatrix} x \\ y \\ 1 \end{bmatrix} \tag{6}$$

where:

t: scaling factor

m_1, m_2, m_3, m_4: define transformation such as rotation, scaling, etc.

n_1, n_2: translation vector.

l_1, l_2: projection vector.

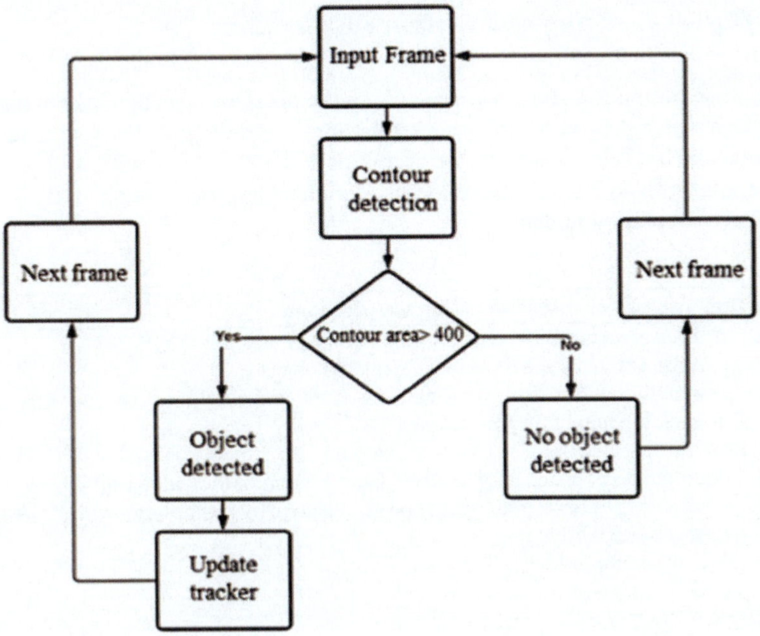

Fig. 5 Obstacle detection process

The resulting warped frame is then provided as an input for distance estimation. The input frame and warped frame are given in Fig. 6.

(a) (b)

Fig. 6 **a** Warped frame after perspective transformation, **b** original input frame

2.3 Object Tracking and Distance Estimation

For distance estimation, the system calculates the pixel location of a known distance to formulate an equation for distance estimation. According to this, the distance in estimated and if it is less than 3 m, the user is alerted of the object with an appropriate audio output with the help of text to speech module. Algorithm 1 explains the process of distance estimation in detail.

Algorithm 1: Distance estimation using warped frame.

Input : Warped objects *frame*, bounding box co-ordinates.
Output : Audio output of the distance of detected object

```
Compute the ratio of known distance on frame and pixel location: r
Compute the height of frame : ht
if object detected:
    compute pixel distance of obstacle from bottom of the video : (ht - (y+h))
    Compute the above pixel value to real distance by multiplying with r : dist
    if dist < 3.0 meters :
        Audio output : "Stop"
    end if
end if
Return dist
```

The significance of the tracker in the proposed solution is to store the co-ordinates of the bounding boxes around the obstacles detected and their count. The objective of this class is to update the central co-ordinates of every obstacle detected and remove the co-ordinates of the obstacles no longer used.

Algorithm 2 Tracking the obstacle

Input : Center co-ordinates of obstacle bounding box of 2 frames, dictionary to store current center points *cp*.
Output : New center point

```
Declare a flag same and set it to false.
for x, y in Px, Py:
Compute the Euclidean norm of (Cx-Px, Cy-Py): V
if V < 100:
Update (Px, Py) to (Cx, Cy): cp
Set same to true
if same is false:
Increase id by 1 and assign it to the new object detected.
Append new obstacle's center points to tracker dictionary: cp
end if
```

Algorithm 2 explains the tracking of the obstacle detected in detail.

3 Result

The proposed system detected the obstacles and computed the estimated distance. The system gave an appropriate audio output if the obstacle is within a predefined distance (3 m in our case). However, this predefined distance can be modified with minor changes in the system. The accuracy of the system was tested by conducting 10 experimental trials. These videos were shot in a standing, upright posture with a camera mounted, with its position around the chest of a user with height 5′9 ft. The videos were shot in a controlled environment with volunteers acting as obstacles. Extreme care has been taken to simulate the real-world scenario and safety of blindfolded volunteers. These trials, similar to [2, 4], are scientific in nature. The results obtained from these experimental trials are depicted in Table 1.

Table 1 consists of the experiment number, number of actual obstacles present in the video, number of obstacles detected by the proposed system and the accuracy computed of each of the 10 experiments. Finally, the mean accuracy is computed and presented. The result of a frame from exp.1 from Table 1 is shown in Fig. 7. The figure shows the input frame (a) which has been fed to the system proposed.

Table 1 Mean accuracy of the experimentation performance

Exp. no	ANO*	ODS*	DE*	AO*	Acc* (%)
1	1	1	ob1–6 m	No	100
2	2	2	ob1–3 m ob2–4 m	Yes No	100
3	3	2	ob1–5.3 m ob2–6 m	No No	66.67
4	5	3	ob1–4.2 m ob2–3 m ob3–5 m	No Yes No	60
5	2	2	ob1–5.6 m ob2–4.5 m	No No	100
6	4	3	ob1–4.1 m ob2–3.4 m ob3–2.9 m	No No Yes	75
7	2	1	ob1–4.7 m	No	50
8	2	2	ob1–6.7 m ob2–3 m	No Yes	100
9	2	2	ob1–5 m ob2–2.2 m	No Yes	100
10	3	2	ob1-4.4 m ob2-3.1 m ob3-3 m	No No Yes	66.67

Average Accuracy = **81.84%**

*Actual number of obstacles (ANO), obstacles detected by system (ODS), distance estimated (DE), audio output given by system (AO), accuracy (Acc), object number (ob)

Fig. 7 **a** original video frame, **b** HSV frame after implementing optical flow, **c** Object detected with estimated distance 6.0 m

Alongside are the two frames- HSV frame (b), which is the output after performing optical flow and conversion of co-ordinates to HSV format, and secondly the frame depicting the distance of the object detected (c).

As observed in Fig. 6, the vehicle is being detected with the estimated distance to be 6 m from the user. When the vehicle passes the threshold of 3 m, an appropriate audio output is given to the user.

4 Conclusion

This paper presented a technique for detection and tracking of non-stationary objects. It estimates obstacle distance from the stationary user. The solution is based on optical flow calculation applied to a stationary user video input. This method efficiently recognizes a moving target from a dynamic scene captured by a stable camera. The standalone system has a time complexity - O(n) and is implemented on Raspberry Pi board. The entire electronic system consisting of Raspberry Pi camera, battery system, alert system, and mountings weigh 440 g approximately. This solution doesn't require any machine learning technique/additional multi-sensory system to detect an obstacle. It can detect floor-level to head-level hanging obstacles in the surrounding environment. It conveys the priority details of the surrounding environment to the user via an earphone. The stability of the camera is an essential requirement for the consistent results. This has been ensured by a firm placement of the camera with flexible chest-belt system. Non-uniform illumination and occlusions can affect the performance of the aid. This aid was subjected to ten usability experiments to assess its relevance as an electronic travel aid. The existing technologies [2, 4, 11] have accuracies 82%, 75.7%, & 80%, respectively. Our system consistently helped the user to understand the surrounding environment and provided an average accuracy of 81.84% for overall obstacle detection.

Acknowledgements The authors express deep gratitude to the visually challenged participants in this study, experts and The Poona Blind Men's Association, Pune. The authors thank the Rajiv Gandhi Science and Technology Commission, Govt. of Maharashtra, Mumbai, and VIT, Pune for providing financial support (RGSTC/File-2016/DPP-158/CR-19) to conduct this research work

References s

1. https://www.who.int/news-room/fact-sheets/detail/blindness-and-visual-impairment. Ref. On 22nd Dec 2021
2. Nivedita A, Sindhuja M, Asha G, Subasree RS, Monisha S (2019) Smart cane navigation for visually impaired. Int J Innov Technol Explor Eng (IJITEE) ISSN (2019): 2278-3075
3. Narayani TL, Sivapalanirajan M, Keerthika B, Ananthi M, Arunarani M (2021) Design of smart cane with integrated camera module for visually impaired people. In: 2021 International conference on artificial intelligence and smart systems (ICAIS). IEEE, pp 999–1004
4. Bourbakis N, Keefer R, Dakopoulos D, Esposito A (2008) A multimodal interaction scheme between a blind user and the tyflos assistive prototype. In: 2008 20th IEEE international conference on tools with artificial intelligence, vol 2. IEEE, pp 487–494
5. Yuan D, Manduchi R (2004) A tool for range sensing and environment discovery for the blind. In: 2004 Conference on computer vision and pattern recognition workshop. IEEE, pp 39–39
6. Ifukube T, Sasaki T, Peng C (1991) A blind mobility aid modeled after echolocation of bats. IEEE Trans Biomed Eng 38(5):461–465
7. Pardasani A et al (2019) Smart assistive navigation de-vices for visually impaired people. In: 2019 IEEE 4th international conference on computer and communication systems (ICCCS). IEEE, pp 725–729
8. Aguerrevere D, Choudhury M, Barreto A (2004) Portable 3D sound/s onar navigation system for blind individuals. In: Presented at the 2nd LACCEI international Latin American Caribbean conference engineering technology, Miami, FL, 2–4 June 2004
9. Shoval S, Borenstein J, Koren Y (1994) Mobile robot obstacle avoidance in a computerized travel aid for the blind. In: Proceedings of 1994 IEEE robotics and automation conference, San Diego, CA, 8–13 May, pp 2023–2029
10. Ulrich I, Borenstein J (2001) The guidecane-applying mobile robot technologies to assist the visually impaired. IEEE Trans Syst Man Cybern Part A Syst Hum 31(2):131–136
11. Iqbal A, Farooq U, Mahmood H, Asad MU (2009) A low cost artificial vision system for visually impaired people. In: 2009 Second international conference on computer and electrical engineering, vol 2. IEEE, pp 474–479

Multimodal Medical Image Fusion Using the Sugeno Fuzzy Inference System

T. Tirupal, K. Shanavaj, M. Venusri, D. Susmitha, and G. Sireesha

Abstract In order to create a single image known as a fused image, a method known as multimodal medical image fusion involves extracting information from many medical images. Clinical experts frequently employ fused image analysis for the rapid identification and treatment of serious disorders. Sugeno fuzzy inference systems (SFIS), which are sophisticated fuzzy systems, are used in this work to combine multimodal medical images. The detailed information included in the input source images can be effectively transferred into the fused output image using SFIS-based fusion. By reducing the amount of memory needed to store numerous photos, Image Fusion not only offers superior information but also lowers storage costs. In comparison to current methods, the suggested work is efficient and produces better fused images. Additionally, the quality metrics Entropy (E), Mutual Information (MI), and Edge-based quality metre ($Q^{AB/F}$) are compared to the fused image. The proposed method's superiority is displayed and supported by both subjective and objective analysis.

Keywords SFIS · Entropy · Image fusion · Diagnosis · MI

1 Introduction

Gathering all the crucial data from various photos and combining it into fewer images, typically just one, is what is referred to as the image fusion process. By keeping all the data in a single input image rather of requiring several input photos, image fusion reduces the cost of storage. Image fusion's primary goal is to create images that are more acceptable and intelligible for both human and machine perception, in addition to reducing the amount of data. Numerous medical fields, such as oncology, neurology, cardiology, and radiation therapy, use image fusion [1–3], because to

T. Tirupal (✉) · K. Shanavaj · M. Venusri · D. Susmitha · G. Sireesha
Department of Electronics and Communication Engineering, G. Pullaiah College of Engineering and Technology, Kurnool, Andhra Pradesh, India
e-mail: tirutalari@gmail.com

S. Sharma et al. (eds.), *Intelligent Control, Robotics, and Industrial Automation*, Lecture Notes in Electrical Engineering 1066,
https://doi.org/10.1007/978-981-99-4634-1_47

the distinctive and enhanced representation of information. Retaining every bit of accurate and valuable information from the original photographs is the fundamental aim of image fusion without adding any artefacts. Fusion is performed on various medical image pairs [4] existing in current literature.

The literature has seen the development of numerous image fusion methods, and the fused images produced by these algorithms contain issues like contrast reduction and undesirable edges. Most often studied in image processing, the conjunctive and disjunctive characteristics of fuzzy logic have proven helpful in image fusion. Fuzzy logic for image fusion is connected as a decision administrator as well as transform administrator. Fuzzy logic-based image fusion has a few applications, including cancer treatment, brain diagnostics, image segmentation, and integration.

Fuzzy sets play a significant part in image processing, in order to eliminate the obscurity and indistinctness. Uncertainties cannot be demonstrably proven using the fuzzy set hypothesis put forth by Zadeh [5] in 1965. The intuitionistic fuzzy set (IFS), which includes the membership degree and non-membership degree, is the expanded version of the fuzzy set hypothesis proposed by Atanassov [6] in 1986. The non-membership degree offers hazy information and successfully handles urgent problems. There are several uncertainties present during every stage of the image processing process, however by using IFSs, these uncertainties can be eliminated and the contrast of the image is enriched.

2 Preliminaries

2.1 *Type-1 Fuzzy Sets*

Numerous sources of uncertainty are present in type-1 fuzzy sets [7–9], including imprecise guesses, differences between membership values and exact membership values in the data, and uncertainty in the scene, form, or other factors. Type-1 fuzzy sets cannot clearly represent such uncertainty because their membership functions are completely crisp.

2.2 *Intuitionistic Fuzzy Set (IFS)*

Atanassov [6] introduced, novel category of fuzzy set that takes into consideration both the membership and non-membership functions. Atanassov also developed a brand-new parameter known as the intuitionistic fuzzy index, or hesitation degree, which rises as a result of a lack of details or individual faults in communicating the membership degree. The uncertainties contained in multimodal medical images are removed using intuitionistic fuzzy set algorithms [10] such Sugeno's IFS [11] and

Yager's IFS [12], which also significantly boost the contrast of the combined medical images.

2.3 *Type-2 Fuzzy Set*

Type-2 fuzzy logic systems [13–15] are a special subclass of fuzzy logic systems that are proposed and presented for supervision of uncertainties in order to come up with a more reliable solution. Type-2 fuzzy sets, whose membership functions are also fuzzy restrict the effects found in medical images and exemplary uncertainties. With sharp and smooth edges, fine texture, high clarity, and the lack of artefacts, the Type-2 fuzzy method that is suggested creates fused images with increased visual quality.

2.4 *Interval Valued Intuitionistic Fuzzy Set (IVIFS)*

Atanassov and Gargov [16] created intuitionistic fuzzy set's capacity to handle uncertain input and resolve real-world decision problems is enhanced by this methodology. The additional techniques for merging multimodal medical images include fuzzy transforms [17–19] and fuzzy optimization algorithms [20, 21]. Mamdani and Sugeno-based fuzzy inference systems are a sophisticated fuzzy system that significantly contributes to the removal of the ambiguity and vagueness found in images out of all these fuzzy set techniques.

3 Sugeno Fuzzy Inference Systems (SFIS)

3.1 *Sugeno Fuzzy Inference System (SFIS)*

Takagi–Sugeno-Kang fuzzy inference is another name for the Sugeno fuzzy inference system. It uses input values with linear function and constant singleton output membership function. A Mamdani system is not as computationally efficient as a Sugeno fuzzy system. Since a Sugeno system does not compute the centroid of a two-dimensional area but rather uses a weighted average or weighted sum of a few data points. The Mamdani output membership function's centroids are matched by a constant output membership function in the Sugeno system. SFIS has good linear performance, is computationally efficient, and is well suited to mathematical analysis.

3.2 Sugeno Fuzzy Inference System

The Sugeno fuzzy inference approach is used in this paper for image fusion. The TSK fuzzy model, which was created by T, is another name for the Sugeno fuzzy model. SUGENO, M. Takagi, and K. In 1984, T. Kang. A computational overhead exists for Mamdani-type FIS. Sugeno FIS, which has a quicker processing time and works well with optimisation and adaptive approaches, is utilised to get around this. The main distinction between the two approaches is that the Sugeno fuzzy inference system's output membership function is either linear or constant. Additionally, the Sugeno technique offers more flexibility and makes it easier to integrate with MATLAB's adaptive neuro-fuzzy inference system (ANFIS) tool.

3.3 Fuzzification of Inputs and Membership Function Computation

Pixel values in the input greyscale photographs range from 0 to 255. (256 grey values). These greyscale values are broken up into a fuzzy set (B, C, G, I, W) with the following five membership functions: B = Black, C = Charcoal, G = Grey, I = Ivory, and W = White. The generated image has 256 grey levels and makes use of the same fuzzy set. In the FIS design, triangular membership function is utilised since it has a lower computing complexity than other membership functions like Gaussian, Trapezoidal, etc.

3.4 Fuzzy Rules

W1 is the input image1, W2 is the input image2, and 0 is the output in a "if–then" format for Sugeno type fuzzy model rules. Therefore, there are a total of 25 regulations, as indicated in Table 1.

Table 1 Fuzzy rules in matrix form

W2	W1				
	B	C	G	I	W
B	B	C	C	G	I
C	C	B	G	I	I
G	C	C	G	I	W
I	C	G	I	I	W
W	I	G	I	W	W

3.5 *Defuzzification*

Transferring the truth values into the output is known as defuzzification. For defuzzification, we employ the weighted average. A single column matrix representing the output of a FIS file is then transformed into an image matrix to produce the fused output image.

4 Proposed Method

Sugeno FIS information is displayed in the fuzzy inference system editor, which also offers the ability to change the amount of input and output variables as well as the fuzzy inference system's highest level properties. To work on the fuzzy inference system, it comprises of additional diverse editors. The FIS editor makes it simple to access all other editors by placing a higher value on communication with fuzzy systems that can be as flexible as possible. Figure 1 depicts the proposed method's block diagram.

4.1 *Algorithm*

The suggested fusion is predicated on fully registered input images A and B. There are various measures that must be taken, including.

Step 1: Two medical images, such as CT and MRI scans or X-rays and VA, must be used as medical images 1 and 2.

Step 2: The two photos are combined using the Sugeno fuzzy inference technique.

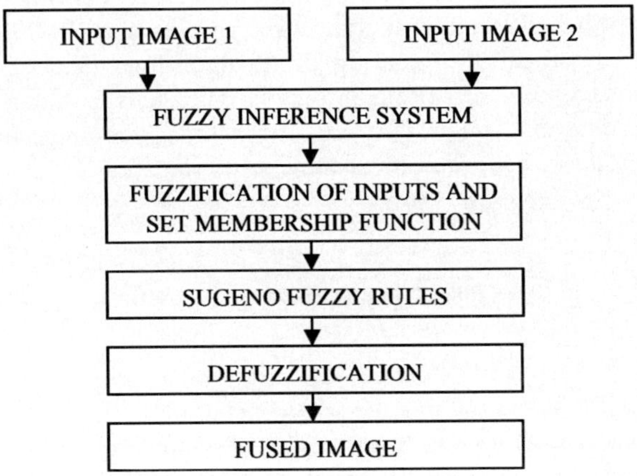

Fig. 1 Block diagram showing the suggested approach

Step 3: In the Sugeno inference system, the image is fused using the Sugeno rules and the Sugeno extension.

Step 4: Using the Sugeno extension, a fused image is produced.

Step 5: In step five, we fuse an image using the Sugeno rules; however the quality is inferior to that of the Sugeno extension.

Step 6: In step six, the input image's objective metrics are determined.

Step 7: The quality is assessed using the subjective and objective measures.

5 Experimental Results

The suggested method with different sets of medical images is used to simulate completely, and the specifics are provided in this section. Two different modalities of visuals are used in the experiments. Six brain images from MRI and CT-scans make up the input images. While the MRI provides information on muscle and soft tissue, the CT imaging provides information on bone and hard tissue. There are four such input sets, each of which comprises of a CT scan and an MRI image. One fused image is produced for each input set. The two images combined into one image provide a wealth of information that aids in more accurate diagnosis. Each image is 256 × 256 pixels in size and has a 256-level greyscale. On several pairs of multimodal medical images, the suggested algorithm's overall performance is shown and contrasted with pre-existing methods. Figure 2 and Table 2 show the total results in terms of subjective and objective standards.

5.1 *Subjective Evaluation of Results*

Visually analysing the resulting fused images allows for subjective comparison. A test is conducted on output fused images with 30 humans and asked to rank their preferences by comparing all the algorithms. Out of 30 people 28 people selected the proposed algorithm output fused image is best than the other three algorithms. It is clear that compared to existing methods, the suggested SFIS-based fusion method produces better results in subjective evaluation.

5.2 *Objective Evaluation of Results*

For objective evaluation [22, 23], which assesses the quality of the image, many performance metrics are used, including entropy, mutual information, and edge-based quality metrics. Outcomes of the recommended SFIS-based fusion approach are given in Table 2 for each of the aforementioned objective measurements. The suggested SFIS-based fusion method not only works better than the widely used

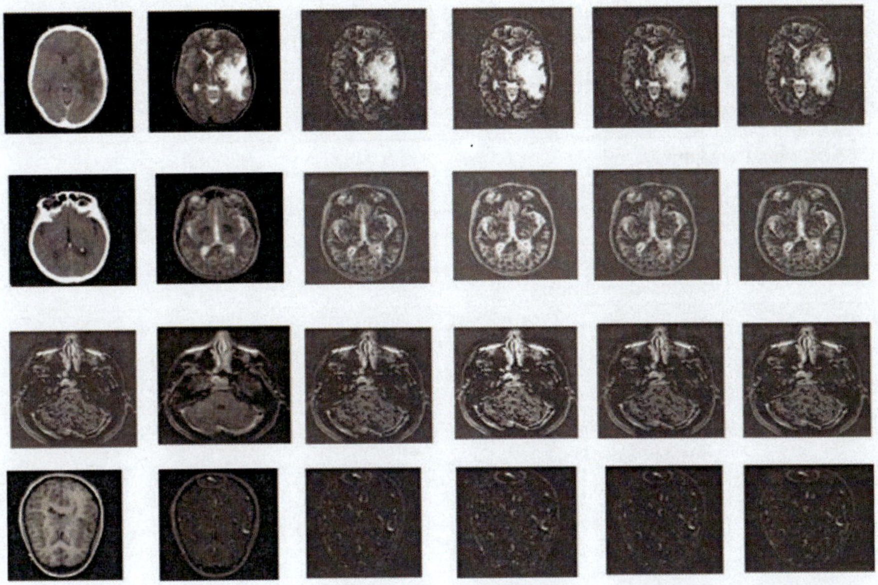

(a) Source image 1	(b) Source image 2	(c) Fused Image by Mamdani Rule	(d) Fused Image by Mamdani Extension	(e) Fused image by Sugeno Rule	(f) Fused image by Sugeno Extension

Fig. 2 Fusion yields for computed tomography as well as magnetic resonance images

spatial domain techniques, but also performs better than all the other ways, as shown by the entropy values of the fused images presented in Table 2. The tumour region is shown by an arrow in the fused image.

The fused image of first pair, i.e. CT-MRI [24] set gives good and clarity information regarding the soft tissues and hard structures. The last row of the image represents magnetic Resonance Imaging and Magnetic Resonance Angiography (MRA) image, where compared to all other fused photos, the tumour is much more visible in this one. The arrow mark represents the tumour position that is clearly mentioned in the image. Entropy is a measure of the average amount of information included in a fused image, and it is high for the image produced using the suggested method compared to all existing methods such as Mamdani rule-based methods. The suggested technique for the further sets of medical image pairs is displayed in Table 1 that have achieved better results in terms of all the objective criteria. Entropy is 6.58 for CT-MRI pair. More edges from the source images were transferred to the fused image, which is produced by 0.1981.

Table 2 Examination of objective parameters for various methods of image fusion for image pairs displayed in Fig. 2

Data set	Method	Entropy (E)	Mutual information (MI)	Edge-based quality metric ($Q^{AB/F}$)
CT-MRI pair 1	Mamdani rule	6.0781	0.2481	0.1554
	Mamdani extension	6.0852	0.2531	0.1608
	Sugeno rule	6.1781	0.3481	0.1754
	Sugeno extension (proposed method)	6.5810	0.4369	0.1982
CT-MRI pair 2	Mamdani rule	5.9550	0.0838	0.1720
	Mamdani extension	5.9650	0.0968	0.1860
	Sugeno rule	5.9820	0.1838	0.1890
	Sugeno extension (Proposed method)	6.1290	0.2170	0.1948
CT-MRI pair 3	Mamdani rule	5.6005	0.1086	0.1565
	Mamdani extension	5.7628	0.1569	0.1578
	Sugeno rule	5.9860	0.2086	0.1865
	Sugeno extension (Proposed method)	6.0005	0.2184	0.1987
MRI-MRA	Mamdani rule	12.3002	0.0673	0.0984
	Mamdani extension	12.3152	0.1024	0.0992
	Sugeno rule	12.4368	0.1173	0.1000
	Sugeno extension (Proposed method)	12.9950	0.1248	0.1368

6 Conclusion

An exceedingly important method, medical image fusion has a real significance in these kinds of applications. There are numerous ways to realise this purpose, and it is necessary to weigh them all to determine which is best for a certain committed area. Our goal is expected to be crucial for doctors who need to fuse multi-modality images because doctors can undertake image fusion from their home or medical facility. Sugeno Fuzzy Logic-based approaches for image fusion are suggested in this research. Although the proposed method is used to fuse greyscale CT and MRI images, the same techniques may also be applied in the fusion of images of other modalities (PET, X-Ray, SPECT) with their actual colour. Sugeno Fuzzy Logic has significantly less computing overhead than Mamdani FIS. The experimental results demonstrate that, in terms of luminance and contrast, the suggested Sugeno Fuzzy Logic-based method for image fusion produces visually superior images. The proposed SFIS approach is only slightly superior to the other methods, according to experimental results. This offers us the incentive to develop our methodology further in order to get better findings, which will ultimately aid in the precise identification of sickness.

References

1. Baum KG, Raerty K, Helguera M, Schmidt E (2007) Investigation of PET/MRI image fusion schemes for enhanced breast cancer diagnosis. In: Proceedings of IEEE seventh symposium conference on nuclear science (NSS), pp 3774–3780
2. Gholam HH, Alizad A, Fatemi M (2007) Integration of vibro-acoustography imaging modality with the traditional mammography. Int J Biomed Imag. Hindawi Publishing Corporation. https://doi.org/10.1155/2007/40980
3. James AP, Dasarathy BV (2014) Medical image fusion: a survey of the state of the art. Inf Fusion 19:4–19
4. Tirupal T, Chandra Mohan B, Srinivas Kumar S (2020) Multimodal medical image fusion techniques—A review. Curr Signal Transduct Ther 15(1):1–22
5. Zadeh LA (1965) Fuzzy sets. Inf Control 8:338–353
6. Atanassov KT (1986) Intuitionistic fuzzy sets. Fuzzy Sets Syst 20:87–96
7. Mendel JM (2001) Uncertain rule-based fuzzy logic systems introduction and new directions. Prentice-Hall, Eaglewood Cliffs, NJ
8. Gayathri K, Tirupal T (2018) Multimodal medical image fusion based on type-1 fuzzy sets. J Appl Sci Comput 5(10):1329–1341
9. Praneel Kumar P, Madhavi K, Tirupal T (2019) Multimodal medical image fusion based on undecimated wavelet transform and fuzzy sets. Int J Innov Technol Explor Eng 8(6):97–103
10. Balasubramaniam P, Ananthi VP (2014) Image fusion using intuitionistic fuzzy sets. Inf Fusion 20:21–30
11. Tirupal T, Chandra Mohan B, Srinivas Kumar S (2017) Multimodal medical image fusion based on Sugeno's intuitionistic fuzzy sets. ETRI J 39(2):173–180
12. Tirupal T, Chandra Mohan B, Srinivas Kumar S (2019) Multimodal Medical Image Fusion based on Yager's Intuitionistic Fuzzy Sets. Iranian J Fuzzy Syst 16(1):33–48
13. Karnik NN, Mendel JM, Liang Q (1999) Type-2 fuzzy logic systems. IEEE Trans Fuzzy Syst 7:643–658
14. Ensafi P, Tizhoosh HR (2005) Type II fuzzy image enhancement. In: Kamel M, Campilho A (eds) Lecture notes in computer sciences, vol 3656. Springer, Berlin, Germany, pp 159–166
15. Tirupal T, Chandra Mohan B, Srinivas Kumar S (2019) Type-2 fuzzy set based multimodal medical image fusion. In: Indian conference on applied mechanics (INCAM-2019), IISc Bangalore, India, 03–05 July 2019
16. Atanassov K, Gargov G (1989) Interval valued intuitionistic fuzzy sets. Fuzzy Sets Syst 31(3):343–349
17. Tirupal T, Chandra Mohan B, Srinivas Kumar S (2021) Multimodal medical image fusion based on interval-valued intuitionistic fuzzy sets. In: Kumar R, Chauhan VS, Talha M, Pathak H (eds) Machines, mechansim and robotics. Lecture notes in mechanical engineering. Springer, pp. 965–971. https://doi.org/10.1007/978-981-16-0550-5_91
18. Manchanda M, Sharma R (2016) A novel method of multimodal medical image fusion using fuzzy transform. J Vis Commun Image Represent 40:197–217
19. Nikhat Afreen H, Tirupal T (2018) Multimodal medical image fusion based on fuzzy enhancement and fuzzy transform. Int J Res 5(4):3002–3010
20. Tirupal T, Chandra Mohan B (2016) A fusion method for multimodal medical images based on fuzzy sets with teaching learning based optimization. In: International conference on advances in scientific computing, IIT Madras, Chennai, India, 28–30 Nov 2016

21. Tirupal T, Chandra Mohan B, Srinivas Kumar S (2018) Multimodal medical image fusion based on fuzzy sets with orthogonal teaching–learning-based optimization. In: Verma N, Ghosh A (eds) Computational intelligence: theories, applications and future directions—Volume II. Advances in intelligent systems and computing, vol 799, Oct 2018, Springer. https://doi.org/10.1007/978-981-13-1135-2_37
22. Homepage. http://www.metapix.de/toolbox.htm
23. Jagalingam P, Hegde AV (2015) A review of quality metrics for fused image. In: Aquatic procedia of international conference on water resources, coastal and ocean engineering (ICWRCOE), vol 4, pp 133–142 (2015)
24. Homepage. http://www.med.harvard.edu/AANLIB/home.html

Online Voting System Based on Face Recognition and QR Code Authentication

K. C. Deepika Nair and I. Mamatha

Abstract Even in this modern day of development, a significant portion of people are still unable to vote in a democracy to create a government. One reason for not voting is that the voters are not available in their constituency on the day of voting. Internet voting or online voting can enable people to cast their vote from wherever they are. In this work, an online voting system with multiple authentication processes is proposed which can allow people with Indian citizenship to vote online instead of visiting a polling place. Face recognition and QR code authentication are the verification strategies suggested to ensure voters' identity and to avoid proxying while voting. Haar cascade algorithm is used to detect human faces in the OpenCV platform and deep learning techniques are used to recognize the face. The proposed system is encoded using Python language and is implemented on Raspberry Pi 3 using a Pi camera to capture images such as faces and QR codes. For ease of interaction, two GUIs are developed: the first one for the authentication process and the second one for the voting process. Avoidance of proxy voting, multiple voting by a single voter, avoidance of visits to poll booths, and automatic counting are the major features of the proposed work.

Keywords Online voting system · Face recognition · Camera module · Raspberry Pi

1 Introduction

Elections are an integral part of life in a democratic system, and it is the government's and citizens' responsibility to ensure that they take place in a safe and secure manner. In front of the legislature, the elected candidate expresses the people's grievances and desires. In such a scenario, many qualified voters are unable to vote because they

K. C. D. Nair · I. Mamatha (✉)
Department of Electrical and Electronics Engineering, Amrita School of Engineering, Amrita Vishwa Vidyapeetham, Bengaluru, Karnataka 560035, India
e-mail: i_mamtha@blr.amrita.edu

S. Sharma et al. (eds.), *Intelligent Control, Robotics, and Industrial Automation*, Lecture Notes in Electrical Engineering 1066,
https://doi.org/10.1007/978-981-99-4634-1_48

are not present in the constituency where their names are recorded. Also, in every election method, there is a significant percentage of fraudulent or invalid voting. To overcome the problems faced by conventional voting mechanisms and to improve the percentage of voting, an online voting or Internet voting mechanism is proposed in this work. Countries like UK, Estonia, and Switzerland conduct polls as well as municipal elections with the help of Internet voting systems. This voting system is available through smartphones, tablets, computers, etc., and voting process may become easy and quick. The details are authenticated while voting to avoid proxy voting. The proposed approach uses a variety of authentication techniques to guarantee the legitimacy of the voters who are taking part in the voting process. It provides security, convenience, and accuracy. According to the survey, there is a lower prevalence of face recognition and QR code authentications together which is resolved through the proposed system. To support the concept, hardware implementation using a Raspberry Pi and an RPi camera is presented. In order to ascertain the integrity of the product, RPi camera is used for QR scanning irrespective of other mobile scanners, which are third-party devices. By doing so, an integrated model system can be formulated. The paper is organized as under Sect. 2 details the literature review carried out. Section 3 deals with proposed system. Section 4 confers the outcomes obtained, and Sect. 5 completes the work with future directions.

2 Related Works

Electronic voting machine (EVM) and Secret Ballot Voting are the existing voting systems that require a high amount of labor and are time-consuming. Once validated with the help of a voter id, individuals above the age of 18 are allowed to vote. While the Secret Ballot System is entirely a manual process, EVM is partially manual in terms of transporting the machines to different parts of the country wherever the election is taking place and the counting process. There have been few works proposed in the literature for an online voting mechanism. Several authors have done tremendous efforts in the ground of verification based on face recognition and QR Reader. A voting system based on face recognition along with a Unique ID Number and election commission ID number were proposed by Vetrivendan et al. [1, 2]. The facial image of the voters was verified by Eigen face algorithm. Every year the database has to be updated with possible deletion and addition of voters. A similar approach is used by Mahalakshmi and Patil where Haar-like features are used for face recognition [3]. A web-based E-voting system using fingerprint authentication and QR Reader is proposed by Ramakrishnan et al. where the storage of the acquired data with proper security measures has been done [4]. The authors used AES algorithm for data encryption which provides a multifactor authentication mechanism, thereby avoiding double voting. Tripathi et al. proposed a Raspberry Pi-based Electronic Voting System (EVM) with two-tier verification processes: fingerprint and face recognition. The winner of each constituency is displayed using the MySQL database [5].

Convolutional Neural Network (CNN)-based advanced voting machine is proposed by Sruthi et al. which uses image processing techniques to identify the false voter and an emergency alarm is triggered [6]. Mouli et al. proposed a smart voting machine based on three different security hierarchies such as Aadhar ID, voter ID, and face identification methods to authenticate the user. Face recognition is carried out through Local Binary Patterns Histogram algorithm [7]. A Java-based web application is developed for online voting with UID, Election ID, face and fingerprint recognition features by Deepaksa et al. [8]. The Eigen face algorithm for face recognition and AES encryption technique for data security are used. Alternately, there have been few works reported for this Internet voting mechanism on the basis of biometrics such as face, iris, and fingerprint [9–11]. The algorithm which is widely used for face recognition is Viola Jones algorithm [12, 13]. Illumination conditions greatly influence the effectiveness of face recognition algorithms. A review of the analysis of various face recognition algorithms is carried out by Lixiang et al. [14] which suggests the application of neural network and deep learning in the field of face recognition technology. Senthamizh et al. proposed a Haar cascade algorithm to detect human faces and Linear Binary Pattern Histogram (LBPH) to recognize the faces for criminal identification purposes [15, 16]. Harikrishnan et al. focused on implementing a real-time attendance and surveillance system using deep learning techniques [17]. A relative scrutiny of various procedures for face detection and human classification is carried out by Vallimeena et al. [18, 19]. CNN algorithm is found to be efficient in identifying the extent of damage in flood-hit areas by detecting the human face, its attributes, etc., within the crowd-sourced images. Hasheer et al. [20] proposed an online voting system based on QR code authentication. In the present work, an online voting system with QR code authentication and face recognition is proposed. On successful verification, the voter is allowed to cast the vote and is acknowledged for successful voting.

3 Proposed System

The functionality of online voting system is shown in Fig. 1. It is based on two authentication processes: face recognition and QR coding. Steps to follow the process:

- The user has to register the details such as the face and QR code in the database before the voting starts.
- QR code has to contain the voter details such as name, voter ID, or image ID which is the unique ID that we consider, gender, year of birth, place, district, and state.
- Each face is tagged with its corresponding QR code in the database.
- On the day of voting, the voter has to sit in front of the camera to capture his face. The face is compared with the database, and if matches, the system directs for QR code verification.

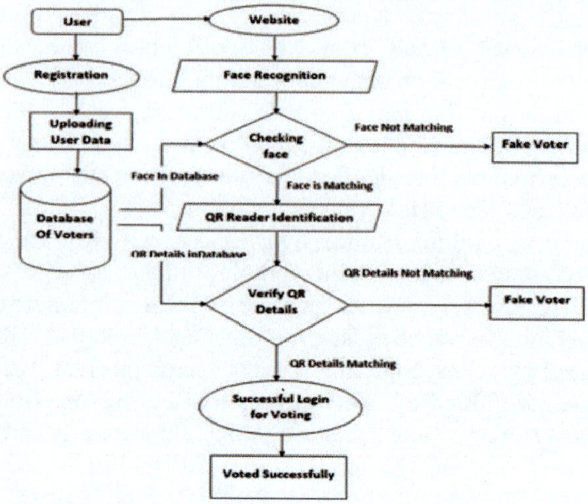

Fig. 1 Flowchart of proposed system

- Once the QR code is matched with the database with which the face is tagged, the system allows the voter to cast the vote.
- The voter casts the vote and gets an acknowledgment for the same.
- If any one of the authentication processes is failed, voter would be considered a fake voter, and multiple voting is also prohibited which is a major drawback of the conventional voting system.

3.1 Face Authentication

The two main processes followed in a face recognition system are face detection and recognition.

Face Detection. An effective object identification technique utilizing Haar feature-based cascade classifiers was introduced by Paul Viola and Michael Jones [10]. The four main concepts in the algorithm are Haar-like features, Integral image, AdaBoost technique, and Cascading classifier which help to construct an immediate face detector. In this approach, cascade function is getting trained with a huge sum of positive and negative visuals. Initially, in order to get the classifier trained, a large number of facial and non-facial images are given to the system. The following phase comprises feature extraction and the use of Haar features are shown in Fig. 2. Each feature corresponds to a value which is the total number of pixels in the white and black rectangles [2, 8].

$$\text{Feature} = \Sigma \text{ black area pixels} - \Sigma \text{ white area pixels} \quad (1)$$

The concept of integral pictures was proposed to minimize the feature calculation. Though there are any number of pixels, the sum of pixels is reduced to a

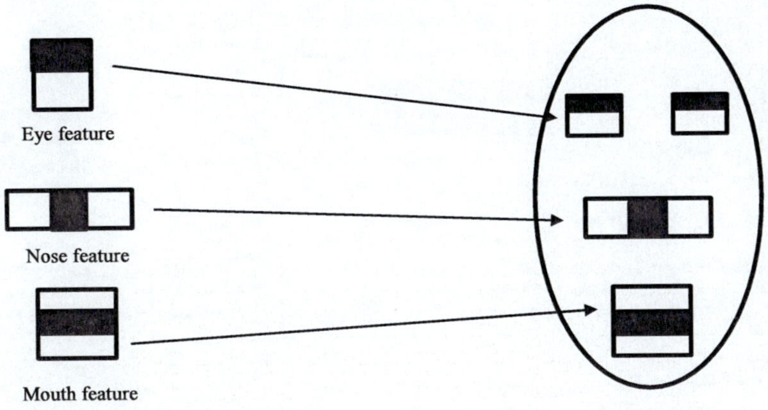

Fig. 2 Types of Haar-like features [12]

four-pixel operation, which makes the procedure quick [10] and determining the optimum threshold with the lowest error rate for classifying the faces as positive or negative. AdaBoost [8] is a good boosting technique for combining weak classifiers and lowers the training faults. The primary principle of this enhancing technique is to connect weak classifiers, which are simple classifiers. Cascade classifiers are a series of weak classifiers that work together to enhance face detection and reduce time complexity. This multi-stage classifier was created with the goal of quickly and efficiently eliminating non-face sub-windows. More non-face images are eliminated using the classifier. Following numerous steps of processing, the frequency of false positives is dramatically reduced and the face is identified as shown in Fig. 3.

Face Recognition. From the images, the features detected are extracted using face embeddings. The image of a person's face is taken as an input by a neural network and produces a vector that signifies the face's utmost vital attributes. This vector is known as embedding which is referred to as face embedding. During the training process, the network absorbs the features and gives output as comparable vectors for appearances that seems to be similar, and thus, the recognition happens. Figure 4

Fig. 3 Face detected

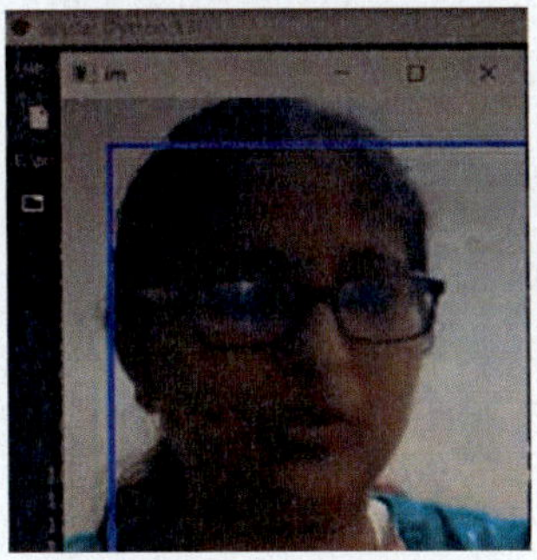

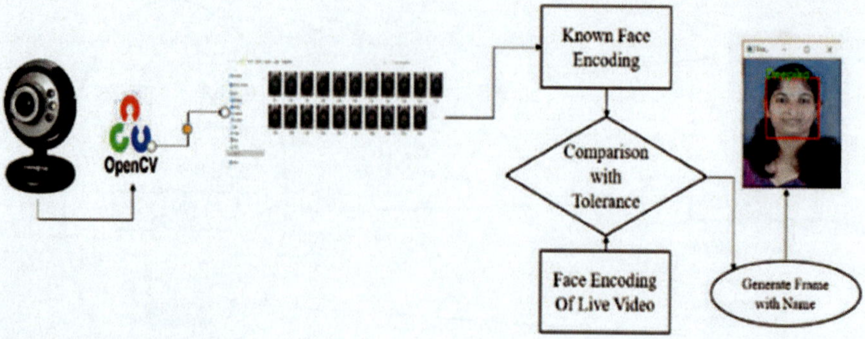

Fig. 4 Face recognized

shows the workflow of face recognition process. Once it is completed, voting process can be continued to the next level of authentication.

3.2 *QR Code Authentication*

QR code authentication uses a contactless card in the form of a QR code as shown in Fig. 5. Instead of using a typical card reader, the QR code badge is scanned using camera, and no additional hardware is required. It has superfast reading capabilities. They can be easily combined with other authentication factors for increased verification [21].

If a voter is already registered, a QR code can be generated for that voter; if the voter is new to the process, he or she should register first, and the database will generate the QR code for that voter. The voter ID number, voter name, year of birth, and address are all included in the voter information. Each pattern is encoded and rendered in QR code using black and white specific symbols for each module. Additional information may be stored in a QR code than in regular bar codes. QR codes have a unique Finder Pattern (Position Detection Pattern) [21] in each of the three corners that can be used to determine the symbol's position, size, and inclination. Once the QR code is scanned, details stored in the code are retrieved. It

Fig. 5 QR code

helps to recognize the authenticity of any voter. Once the valid QR code is identified, the voter can pass through the next authentication step, else would consider a fake voter.

4 Results and Discussion

For the proposed voting system, two Graphic User Interfaces (GUI) are developed. The former is for initial entering of details and recognition and the latter is for voting purposes. The initial GUI is shown in Fig. 6.

4.1 Hardware Implementation

The Raspberry Pi 3 [21] is a tiny computer that may be used to learn programming languages such as Python and research computing. As a desktop computer, it can be used to browse the Internet, play high-definition videos, create spreadsheets, etc. It's a card-sized computer with a regular keyboard and mouse that fits into a computer display and television. The Camera Module 2 [22] is suitable for recording high-definition video and to shoot immobile images. It is efficient in taking video modes of 1080p30, 720p60, and VGA90 and still photography. A 15 cm ribbon wire helps to connect the camera with Raspberry Pi's CSI port. The camera is compatible with the Raspberry Pi versions of 1, 2, 3, and 4. The dataset is generated by placing the RPi camera in front of the user which is in turn linked to the Pi board. Once dataset is generated, file has to be built and compared with the faces kept in the database as shown in Fig. 7. If two faces match, the face is recognized as shown in Fig. 8. Meanwhile, all the generated output is stored in the folder which is formed initially.

Figure 9 shows the QR code scanning using RPi camera. As part of this, data stored in the code would be derived. Once the QR details are captured, face recognized and fetched QR code details should be matched for logging into the voting system. After

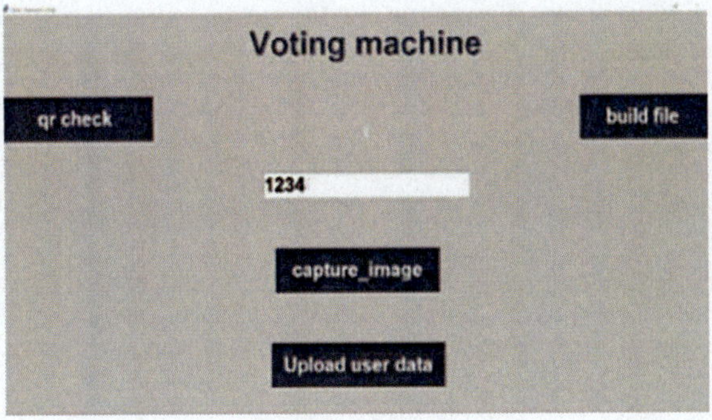

Fig. 6 GUI model

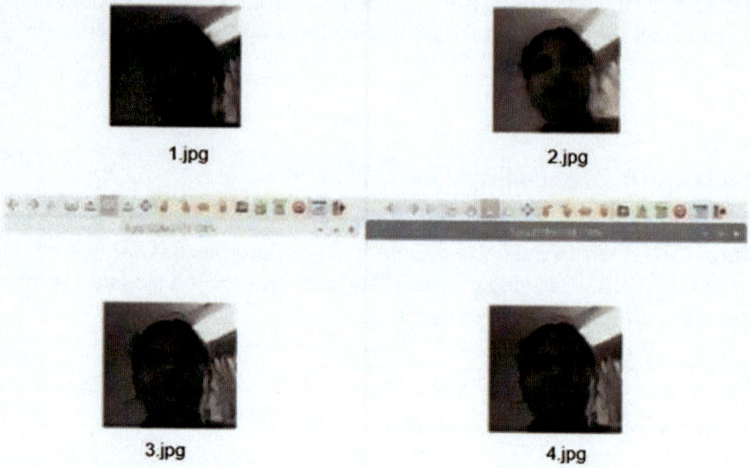

Fig. 7 Face comparing

Fig. 8 Face recognized

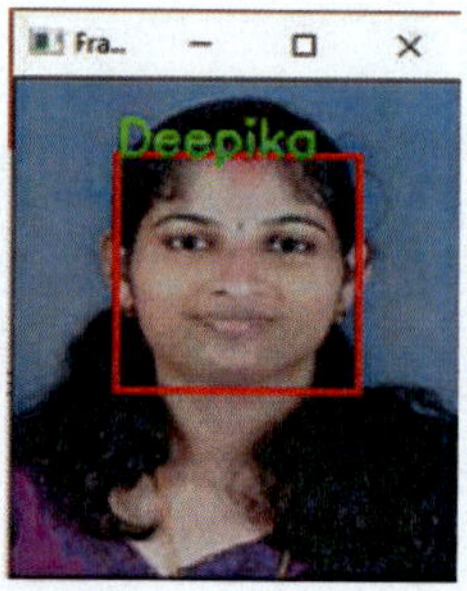

validating the details, the voter is guided to the voting GUI page as shown in Fig. 10. After successful voting, an acknowledgment for the same would be generated as shown in Fig. 11.

The results generated by the proposed system mainly depend on the approach adopted in the authentication methods and the prototype used. In the context of face

Fig. 9 Capturing QR code

Fig. 10 Voting GUI page

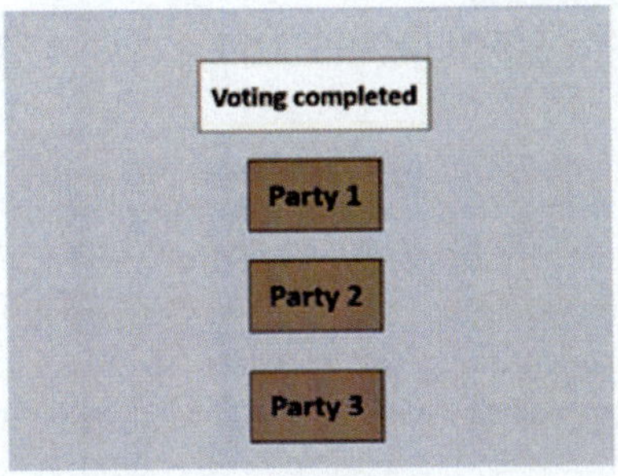

Fig. 11 Successful voting

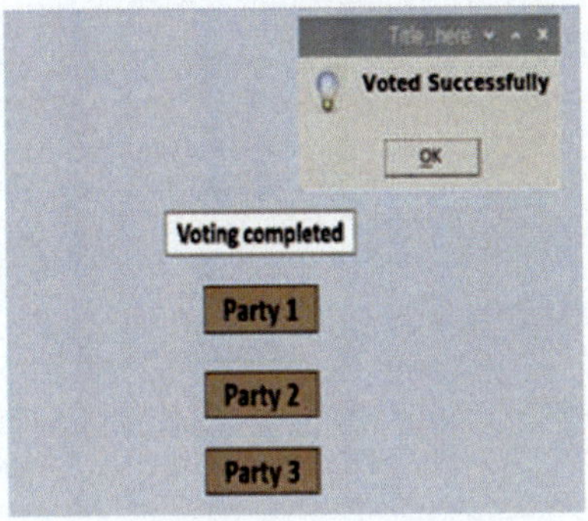

recognition, Haar cascade algorithm is very fast at computing Haar-like features due to the use of integral images. By means of AdaBoost technique, efficient feature selection is also possible. Hence, face recognition based on this algorithm can be considered as a better option for the authentication method compared to other algorithms used in other state-of-the-art methods. The hardware implementation adopted in the system is highly advantageous compared to other software systems. With a single RPi camera, both face recognition and QR code scanning have been done with high resolution. In the case of face recognition, accuracy depends on the number of faces detected from the total number of faces present in an image. The Haar cascade algorithm used in the proposed system could achieve 90–95% accuracy compared with other similar systems [1]. On the other hand, accuracy of QR code depends on the precision of data extracted from the code. Compared to mobile scanners used in other systems, RPi camera has better resolution, so that clear image of data is obtained [20]. Real-time latency of the system is influenced by the number of images taken as the dataset for the face recognition and the speed of QR scanning. Hence, the combined authentication process of the system varies from 60 to 90 s. The proposed voting system is found to be very promising, useful, and secure due to two-way authentication process, thereby eliminating false voting by all means. Face recognition technique is very beneficial in identifying fraud voters and reducing manpower. The accuracy of face authentication can be ensured again with second level security feature which is QR code authentication.

5 Conclusion and Future Work

The proposed voting systems can be used as an advancement of existing conventional voting systems like the Secret Ballot System and electronic voting machine, and there is no need of election officer or electronic voting machine, but Internet facility is required. It is not possible for all people in the country to travel to their home constituency on election day. Hence by means of this new approach, the voting percentage can be increased to a great extent. Future enhancements are required to make this system more effective and to improve its legitimacy. Face recognition and QR code approaches are explained in the current work. Together with those concepts, fingerprint recognition and OTP authentication can also be considered for the verification scenarios, thereby increasing the authenticity. These enhancements can be considered for a full-fledged online voting system with better performance in the future.

References

1. Vetrivendan L, Viswanathan R, Angelin Blessy J (2018) Smart voting system support through face recognition. Int J Eng Res Comput Sci Eng (IJERCSE) 5(4):203–207
2. Singh G, Goel AK (2020) Face detection and recognition system using digital image processing. In: International conference on innovative mechanisms for industry applications (ICIMIA), pp 348–352. https://doi.org/10.1109/ICIMIA48430.2020.9074838
3. Naik MM., Patil PN (2020) Smart voting through facial recognition. Int J Creative Res Thoughts (IJCRT) 8(5):4031–4035
4. Ramakrishnan SMK, Keerthana S, Nizamudeen VM (2020) Online voting system using Aes algorithm. Int J Adv Res Comput Sci Electron Eng IJARCSEE 9(3)
5. ripathi S, Jha A, Mishra R. Prof. NK Shinde (2020) I-Voting—raspberry pi based anywhere voting system. Int J Electron Commun Eng (IJECE) 7(4):21–23. https://doi.org/10.14445/23488549/IJECE-V7I4P105
6. Sruthi MS, Shanjai K (2021) Automatic voting system using convolutional neural network. J Phys Conf Ser (ICCCEBS) 1916(1). https://doi.org/10.1088/1742-6596/1916/1/012074
7. Mouli CC (2020) Smart voting system. Int J Innov Eng Manage Res (IJIEMR) 9(9):115–118. https://doi.org/10.2139/ssrn.3690115
8. Deepaksa Pawar D, Rajan Ahirekar M, Sanjay Lakkam S, Sagar Pachorkar Y (2020) E-smart voting system using cryptography. Int J Res Electron Comput Eng (IJRECE) 8(2)
9. Arputhamoni SJ, Saravanan AG (2021) online smart voting system using biometrics based facial and fingerprint detection on image processing and CNN. In: Third international conference on intelligent communication technologies and virtual mobile networks (ICICV), pp 1–7. https://doi.org/10.1109/ICICV50876.2021.9388405
10. Chinimilli BT, Anjali T, Kotturi A, Kaipu VR, Mandapati JV (2020) Face recognition based attendance system using haar cascade and local binary pattern histogram algorithm. In: International conference on trends in electronics and informatics (ICOEI), IEEE, vol 48184. IEEE, pp 701–704. https://doi.org/10.1109/icoei48184.2020.9143046
11. Khoirunnisaa AZ, Hakim L, Wibawa AD (2019) The biometrics system based on iris image processing: a review. In: 2nd International conference of computer and informatics engineering (IC2IE). IEEE, pp 164–169. https://doi.org/10.1109/IC2IE47452.2019.8940832

12. Al-Tuwaijari JM, Shaker SA (2020) Face detection system based viola-jones algorithm. In: 2020 6th International engineering conference sustainable technology and development (IECSTD). IEEE, pp 211–215. https://doi.org/10.1109/iec49899.2020.9122927
13. Nehru M, Padmavathi S (2017) Illumination invariant face detection using viola jones algorithm. In: 2017 4th International conference on advanced computing and communication systems (ICACCS). IEEE, pp 1–4. https://doi.org/10.1109/ICACCS.2017.8014571
14. Li L, Mu X, Li S, Peng H (2020) A review of face recognition technology. IEEE, pp 139110–139120. https://doi.org/10.1109/ACCESS.2020.3011028
15. Senthamizh SR, Sivakumar D, Sandhya JS, Ramya S, Kanaga S, Rajs S (2019) Face recognition using haar–cascade classifier for criminal identification. Int J Recent Technol Eng (IJRTE) 7(6):1871–1876
16. Purushothaman A, Palaniswamy S (2018) Pose and illumination invariant face recognition for automation of door lock system. In: 2018 Second international conference on inventive communication and computational technologies (ICICCT). IEEE, pp 1105–1108. https://doi.org/10.1109/icicct.2018.8473103
17. Harikrishnan J, Sudarsan A, Sadashiv A, Ajai RA (2019) Vision-face recognition attendance monitoring system for surveillance using deep learning technology and computer vision. In: 2019 International conference on vision towards emerging trends in communication and networking (ViTECoN). IEEE, pp 1–5. https://doi.org/10.1109/vitecon.2019.8899418
18. Vallimeena P, Gopalakrishnan U, Nair BB, Rao SN (2019) CNN algorithms for detection of human face attributes–a survey. In: 2019 International conference on intelligent computing and control systems (ICCS). IEEE, pp 576–581. https://doi.org/10.1109/iccs45141.2019.9065405
19. Neha R, Nithin S (2018) Comparative analysis of image processing algorithms for face recognition. In: 2018 International conference on inventive research in computing applications (ICIRCA). IEEE, pp 683–688. https://doi.org/10.1109/icirca.2018.8597309
20. Peerzade H, Tatipamul R, Dixit R, Pawadshetty A, Mane SM (2020) Online voting system based on Qr code. Int J Creat Res Thoughts (IJCRT) 8(4)
21. Raspberry Pi Homepage. https://www.raspberrypi.com/. Last accessed 21 June 2022
22. Raspberry Pi Homepage. https://www.raspberrypi.com/products/camera-module-v2/. Last accessed 21 June 2022

Circuits, Sensors and Biomedical Systems

Cell Balancing Techniques for Hybrid Energy Storage System in Load Support Applications

K. Chandrakanth, Udaya Bhasker Manthati, and C. R. Arunkumar

Abstract Electric vehicles are playing a major role in pushing the world toward sustainability. Battery is the essential energy storage device in EVs and supercapacitor (SC) is added to improve the battery performance by reducing the stress during the transient period. The major issue in battery/SC pack is the imbalance in cell voltage that inherently occurred during manufacturing. The cell voltage imbalance of the energy storage pack leads to a fast discharge cycle, limits the charging voltage to a lower voltage level and limits the use of total pack energy. Hence, in this work, cell balancing technique is introduced for hybrid energy storage system with battery and SC for load support applications. Initially, the work analyzes the different cell balancing techniques for independent battery and SC pack. The independent cell balancing techniques are extended for active hybrid energy storage system and tested it for different load variations. Finally, detailed simulation and comparison between different topologies are presented to validate the study.

Keywords Battery · Supercapacitor · Cell balancing · SOC · Flyback converter

1 Introduction

The electric vehicle is a promising technology to replace conventional combustion engine vehicles [1]. The major components of electric vehicles include motors, DC–DC converters, battery packs, sensors, and controllers. The battery pack is subdivided into several modules, and each module has a series and parallel connection of cells

K. Chandrakanth · U. B. Manthati · C. R. Arunkumar (✉)
Electrical Engineering Department, National Institute of Technology Warangal, Telangana 506004, India
e-mail: acr_research@student.nitw.ac.in

K. Chandrakanth
e-mail: kc921814@student.nitw.ac.in

U. B. Manthati
e-mail: ub@nitw.ac.in

S. Sharma et al. (eds.), *Intelligent Control, Robotics, and Industrial Automation*, Lecture Notes in Electrical Engineering 1066,
https://doi.org/10.1007/978-981-99-4634-1_49

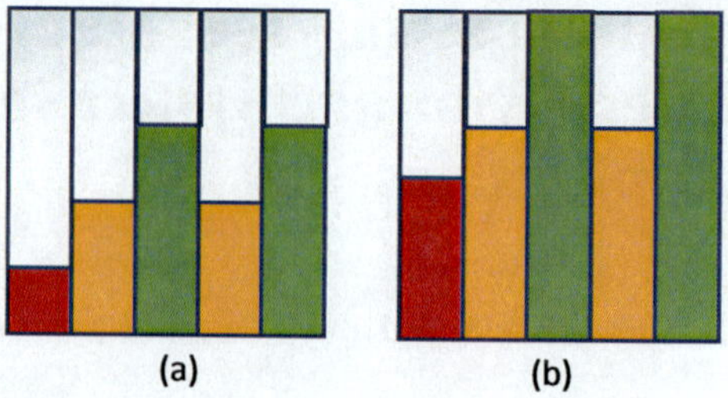

Fig. 1 **a** During discharging, **b** during charging

to get the required power [2, 3]. Unfortunately, inconsistency is inevitable for series-connected batteries due to manufacturing and usage factors [4]. In practice, battery cells are different from each other, even in the same production batch. Because of the difference, inconsistency occurs during charging and discharging due to the significant role of the battery in the electric vehicle. A battery balancing system effectively ensures that the battery's performance and lifespan are maximized, preventing the battery from over-heating, over-charging, and over-discharging [3, 5].

It will check and control the battery's charging and discharging process. Also, by balancing SOC or voltage, the maximum efficiency of the battery pack is achieved. In Fig. 1a, the discharging of cells shows that the cell having the least capacity (represented with red color) discharges faster [6]. Conversely, cells with higher energy levels compared to the lower cell remain underutilized. While cells like cell-3 and cell-5, possessing high capacities, wholly and rapidly charge during the charging process, the remaining cells may not reach full charge [7]. So, cell balancing comes into the picture in this situation. There are different kinds of cell balancing, such as SOC balancing and voltage balancing, and we will discuss them accordingly [8].

Finally, cell balancing is equalizing the voltages and state of charge among the cells. Also, imbalances must be adequately checked to minimize the effects of cell voltage fluctuations. The objective of any balancing scheme is to make the battery pack operate at its expected performance level and extend its functional capacity.

2 Cell Balancing

Generally, cell balancing is divided into two sections and they include:

- Passive cell balancing.
- Active cell balancing.

In passive cell balancing, the excess capacity of cells is dissipated so that all cells will reach the same capacity. In active cell balancing, cells with higher capacity transfer their energy to the lower capacity cells such that all cells' capacity is the same. In passive cell balancing, energy gets wasted, whereas in active cell balancing, energy is distributed equally among cells, and no energy is wasted. Since active cell balancing is highly efficient, we will go for it in implementing cell balancing to battery modules. Here will take two different active cell balancing strategies and compare their performance and other factors.

- Cell balancing using flyback converter.
- Single inductor-based cell balancing.

2.1 Using Flyback Converter

Here for 'N' cells, we will be having 'N' switches or diodes as shown in Fig. 2. Here the series cells are connected to primary such that the individual sum of the voltages of all cells is applied across the primary side and turns ratio is N:1. If the voltage of the cell is less than the secondary side voltage, then the cell will charge, but if its voltage is greater than the secondary side voltage, then it will discharge its energy to other cells and its voltage drops.

After all cells attain equilibrium, i.e., all cells attain same voltage, then the switch in the primary is to be opened. We can control MOSFET using controller, which takes all cell voltages as input and triggers if they attain equilibrium.

Here, for example, five capacitors of 5F are considered. Initially, all are having different voltages after certain time all will have same voltage as shown in Table1.

Figure 3 is the MATLAB output for five cells' active cell balancing topology; after nearly 0.15 s, the cells attained 4.18 V.

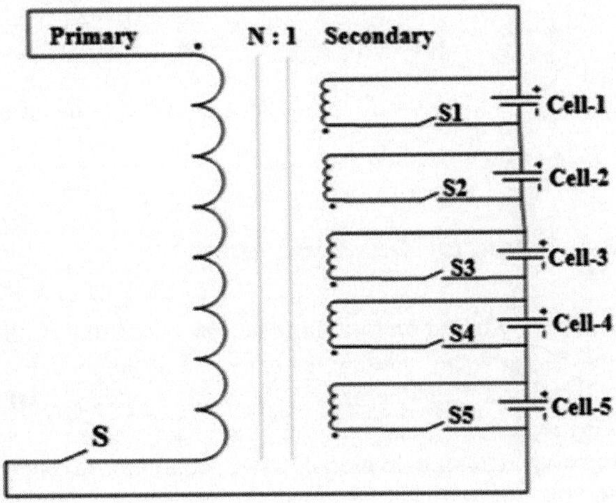

Fig. 2 Cell balancing using flyback converter

Table 1 Voltage values

S. no.	Initial voltage(V)	Final voltage(V)
Cell-1	4.4	4.17
Cell-2	4.3	4.17
Cell-3	4.2	4.17
Cell-4	4.1	4.17
Cell-5	4	4.17

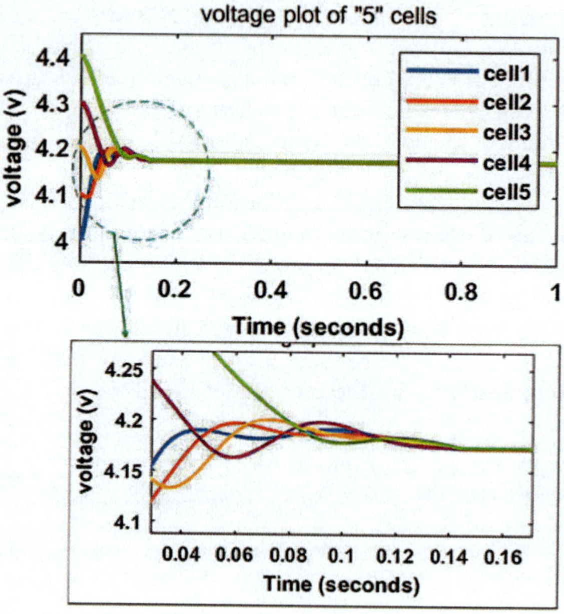

Fig. 3 Example of 5F capacitors' simulation

2.2 *Inductor-Based Balancing*

It is purely based on switching action, here, firstly cell with lowest SOC is checked, and later, the cell module is connected to the inductor and it gets charged; later, we will connect the inductor to the least SOC cell and it will get charged. From Fig. 4, we can see 'N' cells, and we are going to have '2N + 2' switches and a single inductor. Here, switches are divided into two section top layers which are referred as section-A and the bottom layer is referred as section-B. If first cell is having least SOC, then A-2 and B-1 switches are to be turned on for charging it. Before that, A-1 and B-7 switches are turned on so that inductor gets charged, and later, it is to be connected to cell-1 and similar process will go on for other cells also.

Components of MATLAB model in Fig. 5 include:

- Battery pack.

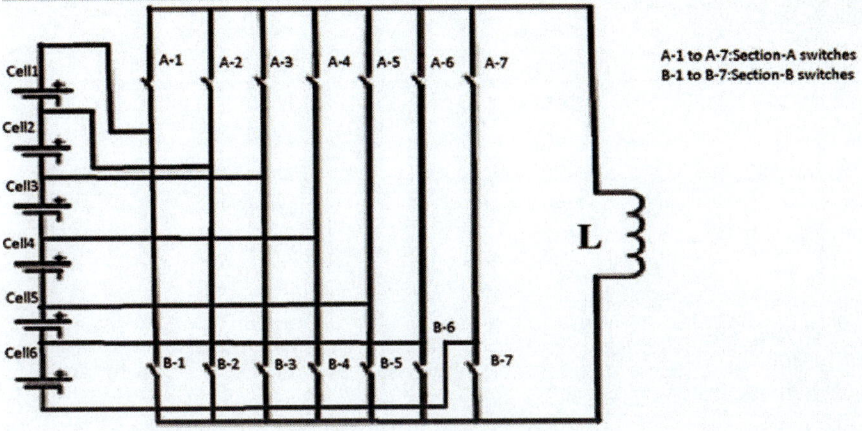

Fig. 4 Balancing circuit of single inductor-based balancing

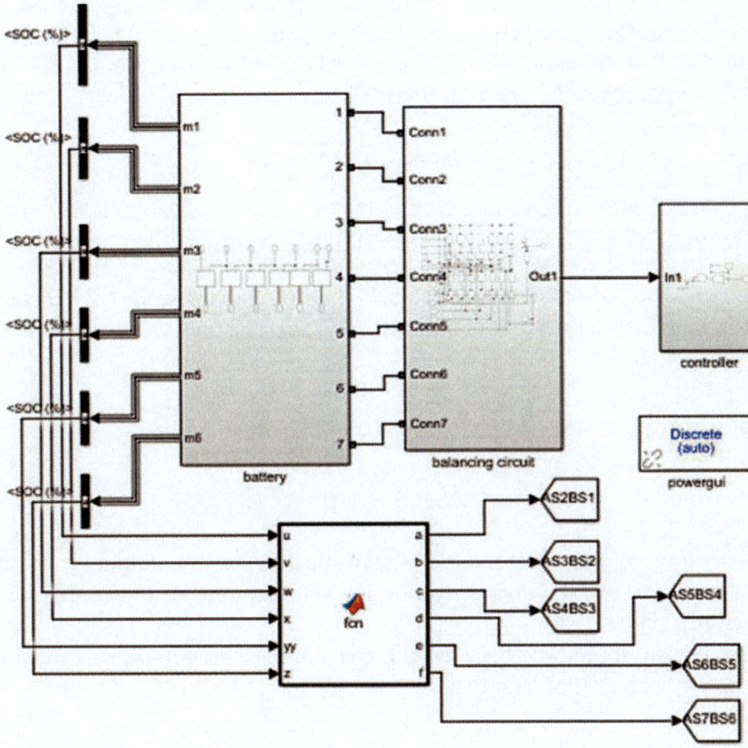

Fig. 5 MATLAB model of single inductor-based balancing

Table 2 %SOC values

S. no.	SOC (%)
CELL-1	60.01
CELL-2	60.02
CELL-3	60.03
CELL-4	60.04
CELL-5	60.05
CELL-6	60.06

- Balancing circuit.
- Controller.
- Switch controller.

Here in this study, each cell's nominal voltage is taken as 3.7 V, and the corresponding SOC is shown in Table 2.

A PI controller is used which takes inductor current and reference current as input and gives duty signal as output. During the ON period of switching cycle, i.e., 0 to $t1$, A-1 and B-7 are turned on. The current i_L of the inductor is rising faster, and at the end of t1, i_L reaches the maximum value Imax.

$$iL(t) = (v/L) * t, \tag{1}$$

$$iL(t1) <= \text{Im}. \tag{2}$$

During OFF period of switching cycle, i.e., $t1$ to $t2$, A-1 and B-7 are turned off, A-2 and B-1 are turned on, the inductor current i_L is dropping slowly, and at the time $t2$, it is equal to zero.

$$iL(t) = \text{Im} - (v5/L) * (t - t1), \tag{3}$$

$$iL(t2) <= 0. \tag{4}$$

The switching frequency of the PWM signal is 20 kHz. The output of this model is shown in Fig. 6, and the corresponding final SOC values of individual cells are mentioned in Table 2.

Clearly, it takes more time than flyback converter. Let us compare them and take one model for vehicle application.

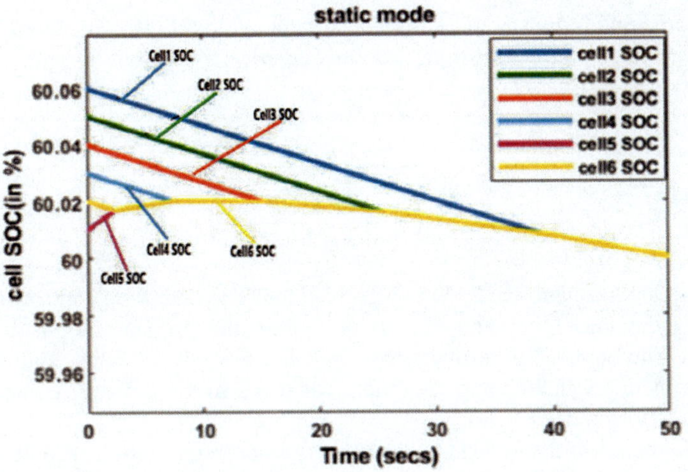

Fig. 6 Plot of %SOC versus time

Table 3 Comparison of flyback converter and single inductor balancing

Flyback converter	Single inductor balancing
1. Construction is simple and easy operation	1. Construction is complex and operation is not easy as other model
2. More space is taken	2. Since it is operating in 20 kHz, less space is taken
3. Only 'N' switches	3. '2N + 2' is required so finite switching losses are present
4. Time of operation is very less (in our model, we got in 0.15 s)	4. Because of many switches, some delay is obtained. (in our model, we got 20 s)

3 Comparison

A detailed comparison of active balancing techniques is shown in Table 3. Now, as the space taken by inductor-based balancing is cell, we will consider it for EV modeling. Now, we combine cell balancing technique for HESS in load support applications.

4 Load Support Application

In this model, we have two boost converters and they are connected to load, and SC converter is bidirectional such that it gets charged from battery during transient conditions when load drops [9, 10]. The detailed diagram is shown in Fig. 7. The converter parameters are designed based on input voltage, output voltage, inductor current ripple and output voltage ripple. The voltage ripple is considered as 1% of the output voltage, and for current, it is considered as 5% output current.

$$Lb = \left(\frac{Vo * Db}{\Delta iLB * fsw}\right), \quad (6)$$

$$\mathrm{CO} = \left(\frac{Vo * Db}{\Delta V * RL * fsw}\right). \quad (7)$$

A switching frequency of 20 kHz is considered. The dual-loop control strategy is used where the outer loop is the voltage controller, and the inner loop is the current controller. The outer PI controller gives inductor current reference and compares it with the actual inductor current. The error feeds to the inner PI controller and generates duty signals.

To tune the controllers, SISO tool is used where we consider Ziegler Nichols tuning method and get K_p and K_i values [11]. The model of the system is given in Fig. 8.

Here, transfer function of current loop is:

$$\frac{iLx(s)}{dx(s)} = \frac{(CoxVos + 2(1 - Dx)Ilx)}{\left(LxCoxs^2 + LxRLs + (1 - Dx)2\right)}. \quad (8)$$

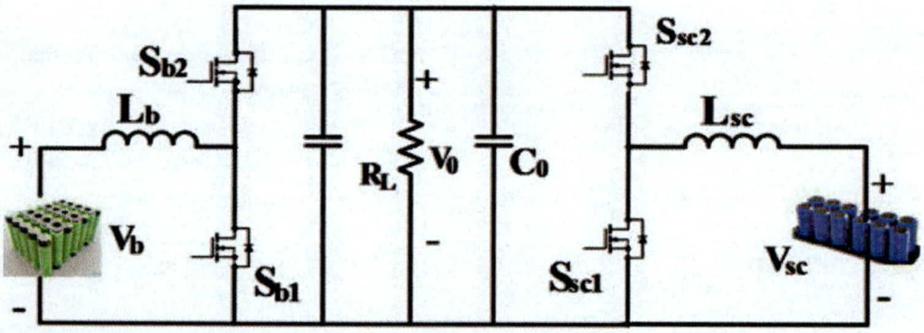

Fig. 7 Model of HESS for load support application

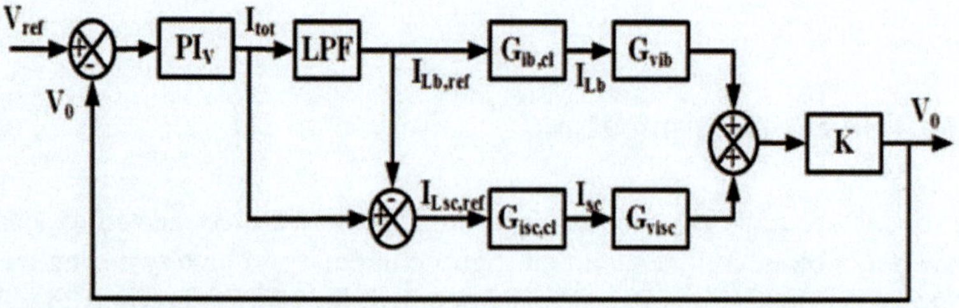

Fig. 8 Voltage and current control loops

Table 4 Components' values

Parameters	Values
C_b and C_{SC}	250 μF
L_b and L_{SC}	2.5mH
F_{SW}	20 kHz
Cell voltage of battery and SC	3.7 V. 22.2 V(pack)
Output voltage	44.4 V
K_{pv} and K_{IV}	1.5, 100
K_{PIB} and K_{PISC}	0.4, 1.5
K_{IIB} and K_{IISC}	80, 80

Transfer function for voltage loop is:

$$\frac{Vo(s)}{iLx(s)} = \frac{(1 - Dx)V0 - LxILxs}{CoxVos + 2(1 - Dx)ILx}. \tag{9}$$

Here, for time interval of 10 secs, load is varied at instant 2 s and 5 s. For normal operation condition, supercapacitor works in transient condition, and during steady state, battery operates. Table. 4 indicates the components' values that we have considered in our model.

5 Simulation Results

At three instants of time load which is being changed, we are going to add loads at instant of $t = 2$ s, and again, we repeat at $t = 5$ secs and we can see during transient conditions at $t = 0$ s; supercapacitor supplies the maximum demand and later turns to zero during steady state and for $t = 2$ s When a disturbance is created already, battery is supplying 200 W; now, supercapacitor supplies additional 100W and the same thing happens during the instant $t = 5$ s also.

From Fig. 9, we can observe that SC is supplying the transient power, and the battery is giving the steady-state power. Figure 10 shows the battery SOC balancing with during load support application. The cell with more SOC will discharge more to equalize the SOC.

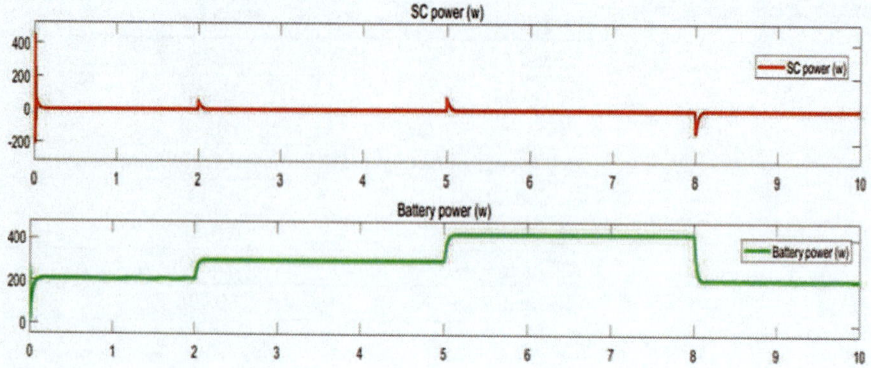

Fig. 9 Load support application

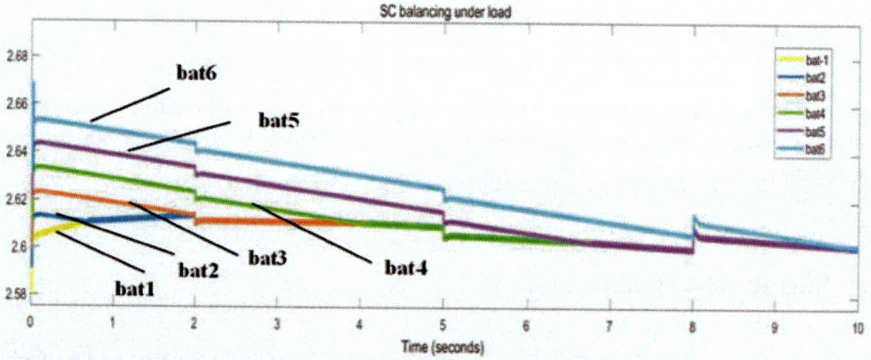

Fig. 10 Supercapacitor balancing during load support application

6 Conclusion

In this work we have discussed various active cell balancing techniques also understood how it is better compared than passive balancing system and later implemented cell balancing for both battery and super capacitor pack. In that system, we are having controller with low-pass filter to make sure that the transient condition is operated by supercapacitor and battery reference current is obtained by subtracting the supercapacitor current from the output of PI controller in voltage loop. Totally, we are having two control loops for its operation and operating switching frequency is 20 kHz. Finally, we can conclude that the balancing of cells helps in using the left or unused energy of cells; also we can charge all cells completely. Thereby, battery life will be more than the usual ones, and thereby, the batteries' performance and lifespan are maximized.

References

1. Chan CC (1993) An overview of electric vehicle technology. Proc IEEE 81(9):1202–1213
2. Vaideeswaran, V., S. Bhuvanesh, and M. Devasena. "Battery management systems for electric vehicles using lithium ion batteries." 2019 Innovations in Power and Advanced Computing Technologies (i-PACT) 1 (2019): 1–9.
3. Scarfogliero, Massimo, et al. "Lithium-ion batteries for electric vehicles: A review on aging models for vehicle-to-grid services." 2018 International Conference of Electrical and Electronic Technologies for Automotive. IEEE, 2018.
4. Balasingam B, Ahmed M, Pattipati K (2020) Battery management systems—Challenges and some solutions. Energies 13(11):2825
5. Shen J, Khaligh A (2015) A supervisory energy management control strategy in a battery/ultracapacitor hybrid energy storage system. IEEE Transactions on transportation electrification 1(3):223–231
6. Bhatt, Devang Kirtikumar, and Mohamed El Darieby. "An assessment of batteries form battery electric vehicle perspectives." 2018 IEEE International Conference on Smart Energy Grid Engineering (SEGE). IEEE, 2018.
7. Najmabadi, Armin. Evaluation of active balancing algorithms and an improved method for a deployed active battery balancer as well as physical implementation. Diss. 2013.
8. Koraddi, Shweta, et al. "Analysis of Different Cell Balancing Techniques." 2022 International Conference for Advancement in Technology (ICONAT). IEEE, 2022.
9. Arunkumar, C. R., and Udaya Bhasker Manthati. "Design and small signal modelling of battery-supercapacitor HESS for DC microgrid." In TENCON 2019–2019 IEEE Region 10 Conference (TENCON), pp. 2216–2221. IEEE, 2019.
10. Arunkumar, C. R., Udaya Bhasker Manthati, and Punna Srinivas. "Accurate modelling and analysis of Raveendran, Arunkumar Chirayarukil, and Udaya Bhasker Manthati. "Dynamic Power Sharing Assisted Hybrid Controller for PV Fed Isolated dc Microgrid." IEEE Systems Journal (2022).
11. Arunkumar, C. R., Udaya Bhasker Manthati, and Srinivas Punna. "Supercapacitor voltage based power sharing and energy management strategy for hybrid energy storage system." Journal of Energy Storage 50 (2022): 104232.

Explanable CAD System for Early Detection of Diabetic Eye Diseases: A Review

Pallabi Das and Rajashree Nayak

Abstract The most popular practice in the field of ophthalmology is to diagnose diabetic eye diseases like Diabetic Retinopathy (DR), Macular Edema (ME), Glaucoma or Age-Related Macular Degeneration (AMD) by analyzing retinography. Retinography is an imaging method used to examine the fundus of the eye and assess the severity of the disease. Manual analysis to examine and evaluate the images by trained ophthalmologists is the ultimate practice for identifying abnormalities in retinal visual impairment. However, it is time-consuming and needs expertise. Automated computer-aided diagnostic (CAD) tools with better image processing capability plays a better role to segment blood vessels and find out abnormal patterns caused by the disease. Artificial intelligence (AI)-based models develop best representation for the prediction of the disease in a self-learning technique by automatically extracting features from raw fundus images. In spite of their wide use, there is a lack of professional liability that prevents the model acceptance. It is extremely important to unfold the black box nature of the models to have a clear and detailed explanation about their learning behavior and the factors facilitating certain predictions. Here, we have written a review paper to survey the existing explainable CAD system for the early detection of diabetic eye diseases. Our aim of this paper is to identify the pros and cons of explainable AI-based CAD system to help the research community and the ophthalmologists to select intended models for clinical usage.

Keywords AMD · DR · ME · AI · CAD systems · Explainable AI

P. Das · R. Nayak (✉)
Centre for Data Science, JISIASR, JIS University, Kolkata, India
e-mail: rajashree@jisiasr.org

P. Das
e-mail: pallabi.das@gnit.ac.in

S. Sharma et al. (eds.), *Intelligent Control, Robotics, and Industrial Automation*,
Lecture Notes in Electrical Engineering 1066,
https://doi.org/10.1007/978-981-99-4634-1_50

1 Introduction

Retina is the most crucial component of the visual system in humans. It is located at the inner portion of the eye's back. Rods and cones, two types of photoreceptors, are found in the retina. Its purpose is to extract several elements that ultimately process the visual data relating both objects and motion. The visual brain cortex receives signals generated by photoreceptors for complicated processing such as texture identification, object recognition, pattern matching. Sharp vision is provided by the macula and fovea located in the retina's central region respectively. Blood is supplied to retinal tissue by a number of blood vessels. The eye's optic nerves, as well as a number of significant arteries and veins, enter and exit through this structure [1]. The structural components of the retina can be examined for an accurate pre-diagnosis of various major disorders, such as diabetic retinopathy (DR), macular edema (ME), glaucoma, or age-related macular degeneration (AMD) among many others [2]. Therefore, the automatic analysis of retinal pictures is a significant research area that has been investigated recently.

Different imaging modalities can be used to analyze the different retinal components. Retinography, also known as fundus retinal imaging, is a popular imaging technique for displaying the fundus of the eye and the structural details of the retina. Figure 1 depicts the ocular anatomy and several optical imaging modalities for diagnosing eye diseases. Using an ophthalmoscope, the fundus image captures the patient's retinal appearance. Using colored fundus images, AMD can be staged and classified. Optical coherence tomography (OCT), a more contemporary imaging technology that creates cross-sectional images of the various retinal layers. OCT can offer quantitative data on the prognostic measurement for the identification of retinal damage caused by diseases like glaucoma or macular edema. Optical coherence tomography angiography (OCTA) provides volumetric data to precisely locate structural and blood flow information. It is frequently utilized to recognize disorders of the retinal vasculature [3].

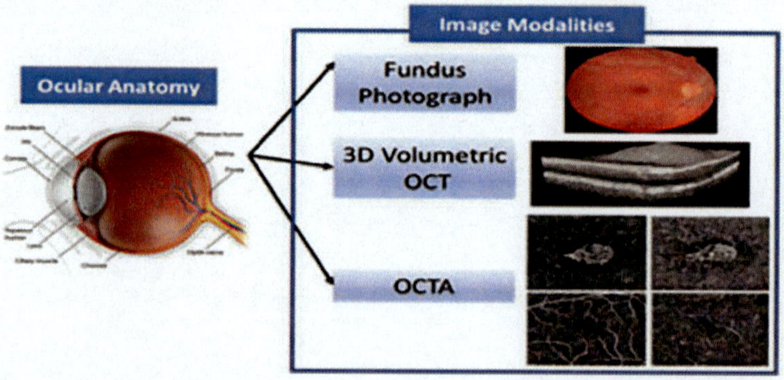

Fig. 1 Different image modalities and ocular anatomy

Worldwide, diabetes mellitus is one of the most prevalent diseases. There are two forms of diabetes mellitus: Type 1 (T1D) and Type 2 (T2D). T1D is an autoimmune disorder with a quick onset, whereas T2D is more frequent and has a slower onset. ME and DR are microvascular consequences of diabetes mellitus. When compared to T1D (20%), patients with T2D (40%) had a higher prevalence of DR [4]. According to the presence of abnormalities such as microaneurysms, exudates, intraretinal hemorrhages, and swelling of the blood vessels in the eye, Diabetic Retinopathy (DR) can be classified into four severity retinopathy grades. It is a progressive disease that can advance from mild stage to proliferative stage. There are three stages of DR: early stage, also known as non-proliferate diabetic retinopathy (NPDR) which is further divided into mild NPDR and moderate to severe NPDR, maculopathy, and progressive also known as proliferate retinopathy (PDR). The most recent and severe stage of DR is known as proliferative DR (PDR), which can cause irreversible vision loss due to retinal detachment, vitreous haemorrhage, and neovascularization on the retinal fundus [5].

Age-related macular degeneration is a significant contributor to vision loss and disability. AMD, which affects 1–3% of the population at an advanced stage, is the primary cause of retinal disorders in Western nations. A quicker illness progression is associated with a number of risk factors. There are no effective treatments for this atrophic disease at this time, despite the fact that several medications are being evaluated in clinical studies. Future progress is expected to be facilitated by improved preventative and risk factor reduction [6].

Macular edema (ME) develops when aberrant blood vessel injury in the surrounding retina causes fluid to abnormally flow into and accumulate in the macula. This may cause visual distortion, blurring and washed out look. Even irreversible eyesight loss may result from it. DR is a common reason why ME develops. ME can also develop as a result of inflammatory eye illnesses, following eye surgery, and in connection with AMD. According to experts, nearly 750,000 of the 7.7 million Americans who have DR also have diabetic macular edema (DME) [7]. According to a recent study, African Americans who are non-Hispanic are three times more likely to acquire DME. More than one-third of people with diabetes have DR, which frequently comes with vision-threatening diabetic retinopathy (severe non-proliferative or proliferative DR) or DME. According to estimates, there were 415 million people worldwide with diabetes in 2015, and that number is projected to rise to 642 by 2040. Additionally, according to Pennington and DeAngelis, 170 million people worldwide and 11 million people living in the US have AMD. These numbers are expected to rise to 196 million in 2020 and 288 million in 2040 as a result of the increased life expectancy [8]. According to scientific research, the majority of blindness caused by DR, AMD or DME can be prevented with timely treatment. Therefore, it is very important to do early screening. However, the manual technique is laborious and time-consuming, which cannot keep up with the high demand for the eye screening and may postpone the best time for treatment. The issue is made worse by a lack of equipment and a shortage of ophthalmologists in remote areas.

Methods for automatic disease detection are required because of these problems. A number of AI-enabled CAD systems have been proposed in the literature for the automatic identification of various eye illnesses as a result of the development of AI. The advantages of AI-based early detection versus manual DR detection systems include:

- Diminished the workload of ophthalmologist
- Minimized human error
- Enhanced automatic detection of lesions and abnormalities.

According on the methods employed, AI-enabled CAD systems can be roughly divided into two categories. The first one is conventional machine learning-based detection algorithms which use various image processing techniques to extract external features. The second one employs deep learning approaches by utilizing automatic feature extraction process. In [9], authors have provided a very informative review on these two types of disease detection methodologies in detail.

Despite of the success of AI-enabled CAD system in the detection process; there is lack of professional liability which prevents the model acceptance.

- Machine learning-based methodologies have a fatal flaw of not giving the reasons behind each particular prediction, which prevents its use in clinical settings.
- In contrast, deep learning-based methodologies provides no precise explanation of functionality of the deep learning model used such as considered input images, its associated classes and the grading of outputs.

However, explainability of these models is extremely important to select appropriate model for clinical use. Recently, explainable AI (Ex-AI) has grown its popularity [10]. Explainability focuses on breaking the black box of the AI model by identifying the discriminative characteristics behind the best classification accuracy of the model while justifying the diagnostic conclusion. Figure 2 depicts the overall block diagram of an AI-enabled Explainable CAD (Ex-CAD) system.

Here, we have written a review paper to survey the existing state-of-the-art explainable CAD system for the early detection of diabetic retinopathy disease. The sole aim

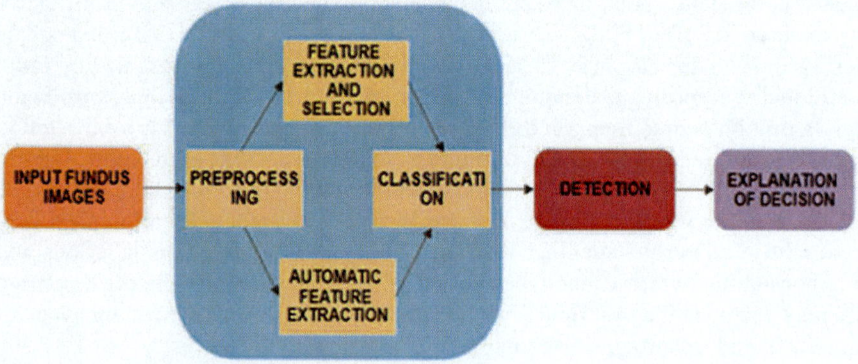

Fig. 2 Block diagram of Ex-CAD system

of this paper is to identify the pros and cons of explainable AI-based CAD system to help the research community and the ophthalmologists to select intended models for clinical usage.

Organization of the rest of the paper is as follows: Sect. 2 briefly describes several existing AI-based techniques for the detection of diabetic retinopathy. Section 3 describes explainability of these AI-based CAD system. Section 4 provides the critical findings and challenges. Concluding suggestions to deal with some of the prominent challenges in designing Ex-CAD system are described in Sect. 5.

2 Automated CAD Systems for Detection of Diabetic Eye Diseases

There are several methods for detecting DR. Researchers have experimented with blood vessels and diverse lesions including micro aneurysms, exudates, and hemorrhages to detect it. A change in the size and form of blood vessels is a reliable sign of DR. The presence of different lesions aids in the detection of DR. Authors in [11], proposed a method which integrated the multi-scale information from the SWT with multi-scale fully connected CNN. In [12], Ashraf et al. used Local Binary Patterns. In [13], Chetoui et al. used Local-Ternary-Pattern and Local- Energy-based-Shape-Histogram that outperformed LBP extracted features. In [14], authors used multi task learning technique for the grading and lesion segmentation of DR. Adaboost, Random forest, SVM, and many more ML algorithms have been proposed to enhance the effectiveness of DR detection. In [15], authors have provided a detailed reviewed on the overall techniques for DR detection including feature extraction and classification. In [1], authors have used a pre-trained model. Custom block of CNN layers are added on top of it. In [16], to categorize the retinal images, a deep NN based on Aquila optimizer was proposed. In [3], a CAD method based on optical coherence tomography (OCT) is proposed to identify early DR utilizing structural 3D retinal scans. In [17], authors suggested the Few-Shot Learning (FSL) paradigm, which efficiently trains models with a limited amount of training data. In [18], authors proposed CenterNet, on top of DenseNet-100 network. The model enabled to extract more detailed features for the proper detection and identification of DR and DME lesions.

3 Explainability of AI-Based CAD Systems for the Detection of Diabetic Eye Diseases

Explainability in machine learning aims to provide humans a deep grasp of how a model functions and makes decisions without attempting to explain every last detail of its calculations. Conceptually, explainable machine learning is comparable to how a clinician must reduce incredibly sophisticated reasoning based on years of clinical training and education into a patient-friendly explanation. Explainability is must for the better understanding and accepting prediction of certain model. Ex-AI approaches help us in providing several justifications in providing the accuracy or selecting a particular model by justifying the importance and contribution of features in obtaining the performance.

In [19], authors presented a unique explainable deep learning ensemble model to extract prominent features from multiple retinal lesions visible on fundus images. They have used gradient weighted class activation mapping. This enabled to provide adaptive explanations toward the explainability of detecting different stages of DR. In [13], improved deep CNN method was proposed. Gradient-weighted class activation mapping-based explainability technique is suggested in the method. It helped to graphically display the indicators to categorize the retinal pictures. In [20], authors developed new pathological descriptors that encode both the spatial and morphological information of lesions utilizing active neurons of the DR detector. Then, they suggested Patho-GAN, a novel network to create retinal images that are clinically realistic, to display the symptom stored in the description. In [21], for the clinical support, author offered a new interpretable CAD system. The model has utilized class enhanced Attentive Response Discovery Radiomics. The proposed system enables to yield a visual interpretation of the decision-making process. Consequently, it provides better understanding of the system performance. Proposed CLEAR-DR system addressed uninterpretability problem and provided potential breakthrough in assisting doctors. In [22], authors proposed an explainability algorithm by using DCNN based on EfficientNET B7 CNN model for recognizing various indicators of DR. In [23], authors present an Explainable Lesion Learning and Generation (ELLG) system. Their method not only offers more convincing evidences that can be explained, but it also addresses the issue of data imbalance in the diagnosis of DR, leading to greatly enhanced detection performance without adding to the inference's time complexity. In [24], author used pre-trained models such as Resnet, Inception-V3, and EfficientNet for classification. The attempt was made to shed light on the causes of misclassifications by using Explainable AI through the LIME framework. Their goal was to identify the causes of classification errors so that they can be used in the future to create models with somewhat higher accuracy than the basic model. In [25], a lightweight explainable CNN architecture was proposed to identify ME on OCT images. Substantial feature activation maps allow minimizing the number of trainable parameters. Suggested Deep OCT, model emphases on envisioning the most discriminative areas and used activation maps for minimal feature learning for the classification process. In [26], authors proposed a model for categorizing eye disorders that can be used to explain the classification of eye diseases. In [27], a platform for interpretable and extensible medical artificial intelligence has successfully been developed; it can recognize diseases, distinguish between various anatomical parts and foci, decipher diagnostic data pertinent for the diagnosis of diseases, and treatment planning. This framework can be used to identify common eye disorders in remote locations as well as in hospitals with limited medical resources to relieve burden on trained medical staff. Table 1 briefly summarizes some of the existing state-of-the-art explainable AI methodologies.

Table 1 State-of-the-art AI-enabled explainability techniques

Author	Techniques and model used	Dataset used	Observation
[19] (2021)	Gradient-weighted class activation mapping and shapely adaptive explanations created an DR identification strategy to highlight the fundus imaging	APTOS, MESSIDOR, IDRiD	Explainability technique used to regions for identification highlight the fundus imaging regions most suggestive of different DR phases has been developed. Precision: 0.970, Sensitivity: 0.980 and AUC: 0.978
[13] (2020)	Gradient-weighted class activation mapping based explainability technique	EyePACS, MESSIDOR, DIARETDB1	EyePACS- AUC: 0.986 Sensitivity: 0.958 MESSIDOR: AUC-0.963 DIARETDB1: 0.986
[20] (2022)	Retinal visuals using Patho-GAN		DR detector encodes the geographical and morphological information of lesions. Patho-GAN provides clinically realistic retinal visuals
[21] (2019)	CLass-Enhanced Attentive Response Discovery Radiomics	Kaggle diabetic retinopathy dataset	Proposed interpretable CAD system assist the clinician with an accuracy of 73.2%
[22] (2020)	EfficientNet-B7 CNN is 6.1 times faster at inference. 8.4 times smaller than the best CNN currently in use. Grad-CAM based explainability algorithm generate an improved	EyePACS APTOS 2019	Leads to higher classification visual explanation rates and attain state-of-the-art performance. AUC: 0.984, 0.990 and 0.998 for RD, DR and APTOS respectively
[23] (2020)	Explainable Lesion Learning and Generation paradigm to investigate	Kaggle, 2016 Kaggle, 2019	Gated Multilayer Saliency Map merged with other fundus pictures
[24] (2021)	LIME framework is applied to light on the reasons for misclassifications		ResNet outperforms with classification accuracy of 96.90%
[25] (2022)	Deep Learning offers computer-aided ME diagnosis for even minor diseases on OCT	Zhanglab, Singapore Eye Research (SERI)	DeepOCT focuses on visualising the most important disease on AUC: 99.20, sensitivity: 100 % and 98.40% specificity

4 Critical Findings and Challenges

Explainability is extremely important as it unfolds the black box to have a clear and detailed explanation about the learning behavior of the model and to characterize different parameters facilitating certain predictions. Lack of explainability in the system creates difficulty to the end user in

- Comprehending model functionality
- Inhibiting model acceptance for clinical use
- Finding discriminative characteristics in diagnosing eye diseases

- Understanding the reason behind the predictive conclusion of used computer-assisted systems.

In literature several contributions have been suggested regarding the interpretability and explainability of AI models for automatic detection of ophthalmology diseases. All of the studied literature demonstrates that in order to effectively streamline these techniques for practical use it is essential to

- Present the subject of interpretable AI
- Describe the key interpretability approaches
- Offer insights into the current trends and obstacles.

Explainability of AI-enabled model for detected of eye diseases is still an open challenge for research community. Traditional supervised ML requires big data sets and lengthy processing times; humans are particularly skilled at discovering patterns through exploratory learning from small samples. Large amounts of training data are frequently missing in the biomedical area, for example, with rare diseases or with human or mechanical malfunctions. Furthermore, time is a significant aspect in clinical medicine, when a doctor wants the results almost immediately or at the very least quickly. Hence, there exist several aspects to be fully explored in ML based detection methods such as

- To discuss contribution of certain features toward prediction accuracy
- Analyzing significance of Positive and negatively contributed features towards the classification performance
- Interpretability of several machine learning models in clinical uses

Interpretability of a model is directly influenced by the amount and complexity of its features. It can be challenging to comprehend a linear model with thousands of parameters or one that makes use of confusing features. Similar to this, due to their numerous model parameters, DL models are currently the most difficult to interpret machine learning models. Important research directions should be taken care while evaluating AI-based methodologies. These include

- Understanding their commonalities and differences,
- Attempting to combine them into a theoretically sound framework,
- Assessing their reliability via straightforward but efficient tests [28].

The requirements for explainable AI and trustworthy AI are intertwined. Explainable modeling can help create trustworthy AI, but the advantages of explainability still need to be demonstrated and developed in health care industry such as regulating, performing extensive (external) validation, and reporting data quality [29]. Millions of inputs can be used to train deep learning systems, and the predictions they provide can often be very accurate. The models themselves may be impenetrable to humans despite their accuracy. Analysis of proper weighting of feature vectors and establishing the linkages between those features for the classification process is much vital in the detection process. The weighting of features can fluctuate dramatically as a result of a tiny variation in a seemingly unrelated component of the data. A variety of initial conditions can also lead to the creation of various models [30].

5 Conclusion

We should first be aware of what a human can understand. Comprehension is more than just identifying, perceiving, and reproducing facts. Also it is about having an intellectual understanding of the environment in which those facts emerge. Instead, comprehension can be viewed as a link between perception and analysis. The state-of-the-art AI is still a long way off from understanding context, which is unquestionably a key sign of intelligence. Combining EX-CAD system with ontological methods is one way to close the gap between artificial intelligence and manual diagnosis. With further development, the EX-CAD system might become a reliable diagnostic assistant that addresses the accessibility issue for eye health care.

References

1. Martinez-Murcia FJ, Ortiz A, Ramírez J, Górriz JM, Cruz R (2021) Deep residual transfer learning for automatic diagnosis and grading of diabetic retinopathy. Neurocomputing 452:424–434. https://doi.org/10.1016/j.neucom.2020.04.148
2. Addo E, Bamiro OA, Siwale R (2016) Anatomy of the eye and common diseases affecting the eye. In: Ocular drug delivery: advances, challenges and applications, pp 11–25. https://doi.org/10.1007/978-3-319-47691-9_2
3. Elsharkawy M, Elrazzaz M, Ghazal M, Alhalabi M, Soliman A, Mahmoud A, El-Daydamony E, Atwan A, Thanos A, Sandhu HS et al (2021) Role of optical coherence tomography imaging in predicting progression of age-related macular disease: a survey. Diagnostics 11(12):2313. https://doi.org/10.3390/diagnostics11122313
4. Chalakkal RJ, Abdulla WH, Hong SC (2020) Fundus retinal image analyses for screening and diagnosing diabetic retinopathy, macular edema, and glaucoma disorders. In: Diabetes and fundus OCT, pp 59–111. https://doi.org/10.1016/B978-0-12-817440-1.00003-6
5. Zou B, Shan X, Zhu C, Dai Y, Yue K, Chen Y, Xiao Y, Huang J (2020) Deep learning and its application in diabetic retinopathy screening. Chin J Electron 29(6):992–1000. https://doi.org/10.1016/B978-0-12-817440-1.00003-6
6. García-Floriano A, Ferreira-Santiago Á, Camacho-Nieto O, Yáñez-Márquez C (2019) A machine learning approach to medical image classification: detecting age-related macular degeneration in fundus images. Comput Electr Eng 75:218–229. https://doi.org/10.1016/j.compeleceng.2017.11.008
7. Singh RK, Gorantla R (2020) DMENET: diabetic macular edema diagnosis using hierarchical ensemble of CNNs. Plos One 15(2):e0220677. https://doi.org/10.1371/journal.pone.0220677
8. Scuteri D, Vero A, Zito M, Naturale MD, Bagetta G, Nucci C, Tonin P, Corasaniti MT (2019) Diabetic retinopathy and age-related macular degeneration: a survey of pharmacoutilization and cost in Calabria, Italy. Neur Regener Res 14(8):1445. https://doi.org/10.4103/1673-5374.253528
9. Gangwar AK, Ravi V (2021) Diabetic retinopathy detection using transfer learning and deep learning. In: Evolution in computational intelligence, pp 679–689 (2021). https://doi.org/10.1007/978-981-15-5788-0_64
10. Oh S, Park Y, Cho KJ, Kim SJ (2021) Explainable machine learning model for glaucoma diagnosis and its interpretation. Diagnostics 11(3):510. https://doi.org/10.3390/diagnostics11030510
11. Oliveira A, Pereira S, Silva CA (2018) Retinal vessel segmentation based on fully convolutional neural networks. Expert Syst Appl 112:229–242. https://doi.org/10.1016/j.eswa.2018.06.034

12. Ashraf MN, Habib Z, Hussain M (2014) Texture feature analysis of digital fundus images for early detection of diabetic retinopathy. In: 2014 11th international conference on computer graphics, imaging and visualization, pp 57–62. https://doi.org/10.1109/CGiV.2014.29
13. Chetoui M, Akhloufi MA, Kardouchi M (2018) Diabetic retinopathy detection using machine learning and texture features. In: 2018 IEEE Canadian conference on electrical & computer engineering (CCECE), pp 1–4. https://doi.org/10.1109/CCECE.2018.8447809
14. Foo A, Hsu W, Lee ML, Lim G, Wong TY (2020) Multi-task learning for diabetic retinopathy grading and lesion segmentation. In: Proceedings of the AAAI conference on artificial intelligence, vol 34, pp 13267–13272. https://doi.org/10.1609/aaai.v34i08.7035
15. Ishtiaq U, Abdul Kareem S, Abdullah ERMF, Mujtaba G, Jahangir R, Ghafoor HY (2020) Diabetic retinopathy detection through artificial intelligent techniques: a review and open issues. Multimedia Tools Appl 79(21):15209–15252. https://doi.org/10.1007/s11042-018-7044-8
16. Gadekallu TR, Khare N, Bhattacharya S, Singh S, Maddikunta PKR, Srivastava G (2020) Deep neural networks to predict diabetic retinopathy. J Ambient Intell Hum Comput, 1–14. https://doi.org/10.1007/s12652-020-01963-7
17. Murugappan M, Prakash N, Jeya R, Mohanarathinam A, Hemalakshmi G, Mahmud M (2022) A novel few-shot classification framework for diabetic retinopathy detection and grading. Measurement 200:111485. https://doi.org/10.1016/j.measurement.2022.111485
18. Nazir T, Nawaz M, Rashid J, Mahum R, Masood M, Mehmood A, Ali F, Kim J, Kwon HY, Hussain A (2021) Detection of diabetic eye disease from retinal images using a deep learning based centernet model. Sensors 21(16):5283. https://doi.org/10.3390/s21165283
19. Shorfuzzaman M, Hossain MS, El Saddik A (2021) An explainable deep learning ensemble model for robust diagnosis of diabetic retinopathy grading. ACM Trans Multimedia Comput Commun Appl (TOMM) 17(3s):1–24. https://doi.org/10.1145/3469841
20. Niu Y, Gu L, Zhao Y, Lu F (2021) Explainable diabetic retinopathy detection and retinal image generation. IEEE J Biomed Health Inf 26(1):44–55. https://doi.org/10.1109/JBHI.2021.3110593
21. Kumar D, Taylor GW, Wong A (2019) Discovery radiomics with CLEAR-DR: interpretable computer aided diagnosis of diabetic retinopathy. IEEE Access 7:25891–25896. https://doi.org/10.1109/ACCESS.2019.2893635
22. Chetoui M, Akhloufi MA (2020) Explainable end-to-end deep learning for diabetic retinopathy detection across multiple datasets. J Med Imaging 7(4):044503. https://doi.org/10.1117/1.JMI.7.4.044503
23. Lin C, Zhu J, Shen C, Hu P, Wang Q (2020) ELLG: explainable lesion learning and generation for diabetic retinopathy detection. In: International joint conferences on artificial intelligence workshop on disease computational modeling, international joint conferences on artificial intelligence organization, pp 1–6. https://dcm-2020.github.io/5b1aa84bcd00a71de826000802315556/CameraReady/ELLG.pdf
24. Reza MT, Ahmed F, Sharar S, Rasel AA (2021) Interpretable retinal disease classification from oct images using deep neural network and explainable AI. In: 2021 international conference on electronics, communications and information technology (ICECIT), pp 1–4. https://doi.org/10.1109/ICECIT54077.2021.9641066
25. Altan G (2022) Deepoct: an explainable deep learning architecture to analyze macular edema on oct images. Int J Eng Sci Technol 34:101091. https://doi.org/10.1016/j.jestch.2021.101091
26. Kinger S, Kulkarni V (2022) Explainability of deep learning-based system in health care. In: Computational intelligence in data mining, pp 619–633. https://doi.org/10.1007/978-981-16-9447-9-47
27. Zhang K, Liu X, Liu F, He L, Zhang L, Yang Y, Li W, Wang S, Liu L, Liu Z et al (2018) An interpretable and expandable deep learning diagnostic system for multiple ocular diseases: qualitative study. J Med Internet Res 20(11):e11144. https://doi.org/10.2196/11144
28. Reyes M, Meier R, Pereira S, Silva CA, Dahlweid FM, Tengg-Kobligk Hv, Summers RM, Wiest R (2020) On the interpretability of artificial intelligence in radiology: challenges and opportunities. Radiol Artif Intell 2(3):e190043. https://doi.org/10.1148/ryai.2020190043

29. Markus AF, Kors JA, Rijnbeek PR (2021) The role of explainability in creating trustworthy artificial intelligence for health care: a comprehensive survey of the terminology, design choices, and evaluation strategies. J Biomed Inf 113:103655. https://doi.org/10.1016/j.jbi.2020.103655
30. London AJ (2019) Artificial intelligence and black-box medical decisions: accuracy versus explainability. Hastings Center Report 49(1):15–21. https://doi.org/10.1002/hast.9735

Next Generation MEMS Sensors and Systems for Non-invasive Medical Diagnosis

Hithesh Kumar Gatty

Abstract One of the popular non-invasive methods for medical diagnosis is exhaled breath. In this method, a person exhales into an instrument, such as a spirometer, that measures a component of exhaled air such as gas and/or aerosol particles. The most popular method of detection of a disease marker non-invasively is through gas detection. Gas chromatography can be used to quantify several gases that are present in the breath, and by measuring the concentration of different volatile organic compounds (VOC), a conclusion can be made about the severity of diseases such as asthma, COPD, cystic fibrosis, and even cancer. However, these measuring instruments are bulky and expensive. Therefore, there is a need for a portable measurement device that can be carried along by the patients for regular measurements. Therefore, the measurement instrument needs to be smaller and more cost-effective. To achieve this goal, we designed and developed a metal oxide gas sensor (MO_x) on a silicon-on-insulator (SOI) wafer. The sensor measures a few mm^2 in dimension and less than a millimeter in thickness. This allows us to integrate into a measurement instrument such as a spirometer, to capture and measure a gas component in the breath.

Keywords Non-invasive · MEMS · Gas sensor · Spirometer · Breath analysis

1 Introduction

In a recent study in Sweden, it is found that patients with lung inflammation (obstructive diseases) may be at risk of hospitalization due to COVID-19. According to this study, from Swedish National Airway Register (SNAR) with 271,404 patients, 1.2% with COPD and 0.5% with asthma were identified with severe COVID-19 [1]. Of hospitalized adults, 14% with asthma and 29% with COPD died. This study concludes that physicians in Sweden assess chronic obstructive pulmonary disease (COPD) to play a more significant role in the disease course of a severe COVID-19 than asthma.

H. K. Gatty (✉)
GattyInstruments AB, Green Innovation Park, Ulls väg 29C, 75651 Uppsala, Sweden
e-mail: info@gattyinstruments.com

S. Sharma et al. (eds.), *Intelligent Control, Robotics, and Industrial Automation*,
Lecture Notes in Electrical Engineering 1066,
https://doi.org/10.1007/978-981-99-4634-1_51

In contrast, for patients with severe COVID-19, comorbid asthma and COPD were not always identified.

In another recent study, the authors used electronic health records in England to estimate the excess risk of death from respiratory-related causes after accounting for other causes of death [2]. Only 14 in 100,000 people with asthma are predicted to experience a respiratory-related death up to 10 yrs. post diagnosis, where in COPD this is 98 in 100,000. The study concludes that the asthma and COPD are associated with an increased risk of respiratory-related death after accounting for other causes, whereas diagnosis of COPD carries a much higher probability. ACO (Asthma + COPD) has a lower risk compared to COPD alone, but higher risk compared to asthma alone.

To access this lung inflammation and diagnose the disease, there are several instruments in the market that are based on fractional exhaled nitric oxide (FENO), forced exhaled volume in 1 s (FEV1), nitric oxide detection, carbon monoxide detection other gas biomarkers. The most popular and ease of use instrument is a spirometer. A spirometer is a device that measures how much air a person can breathe in or out in one breath, which is a simple test used to help diagnose and monitor certain lung conditions such as resistive and obstructive lung diseases [3]. With a spirometer, several diseases can be effectively identified such as COPD, asthma, cystic fibrosis, and rheumatoid arthritis. Spirometry is also a standard test for people being considered for surgery, or to check the general health of people who have other conditions.

However, these measuring instruments are expensive, inaccurate, bulky and require calibration overtime. To address these issues, a MEMS-based sensor integration with a spirometer for measurement of flow, pressure, temperature, humidity, and other entities allows for more accurate measurement with lower cost. A hand-held monitoring device such a mobile phone when connected to a spirometer instrument allows a patient to use these instruments in bedside or home use as well. When connected to a cloud, the data could be retrieved securely and remotely by a doctor who in turn can give a prescription based on the results from the spirometer and other similar breath analyzers.

Earlier, I have fabricated an electrochemical sensor which is sensitive to nitric oxide and can potentially be used in lung inflammation detection such as asthma [4, 5].

In this work, due to the need to miniaturize a breath analyzer, we have developed a small MEMS-based metal oxide (MO_x) sensor that is used to sense gas from an exhaled breath. We propose possibility for next generation MEMS sensors with Low temperature co-fired ceramic (LTCC) packaging.

2 Materials and Methods

In this section, a large-scale method of fabricating a metal oxide (MO_x) sensor is described. In addition, how the general products around breath analyses can be used to get a beneficial reading and understand the outcome of the devices are described. A novel method of 3D printed structure for sampling the breath is achieved.

2.1 *Method*

A silicon-on-insulator wafer was used to manufacture the metal oxide gas sensors. The gas sensor consists of a heater, a sensing element, and a sensitive layer. These entities are deposited on a silicon wafer. Depending on the type of the sensitive layer, the gas can be chosen to be sensitive. These sensors are batch fabricated to get as many devices as possible in a single wafer. A review of MEMS-based gas sensors and processing is described by these authors [6].

2.2 *Fabrication Process*

To manufacture a metal oxide gas sensor, we have used a silicon-based SOI wafer with a device layer thickness of 10 μm. The fabrication process flow is shown in Fig. 1. A selection of a specific gas sensing is related to the sensitive layer deposited.

In this method of manufacturing, the SOI wafer is cleaned using RCA clean. To remove the native oxide, a mild (5%) HF etch is done to be free from native oxide and other unwanted substances. The wafer is then oxidized to achieve an oxide thickness of 3 μm. The wafer is patterned on the oxide in the handle layer to expose the silicon for further etching (Fig. 1c). A thin layer of platinum is deposited which, when patterned forms the heater layer of the MO_x sensor. The patterning of the platinum is achieved using lift-off technique. A thin layer of approximately 500–800 nm silicon nitride is deposited on the platinum layer which acts as an isolation layer. Again, a platinum layer is deposited on the nitride layer and patterned by lift-off method to realize the sensing electrode. The last step in the process is to deposit the gas sensitive layer followed by etching of the silicon in the handle layer.

Thus, a suspended membrane with heater, sensing electrode, and sensitive layer is processed on a SOI wafer. This process is scalable and a high-volume of devices can be achieved using this method.

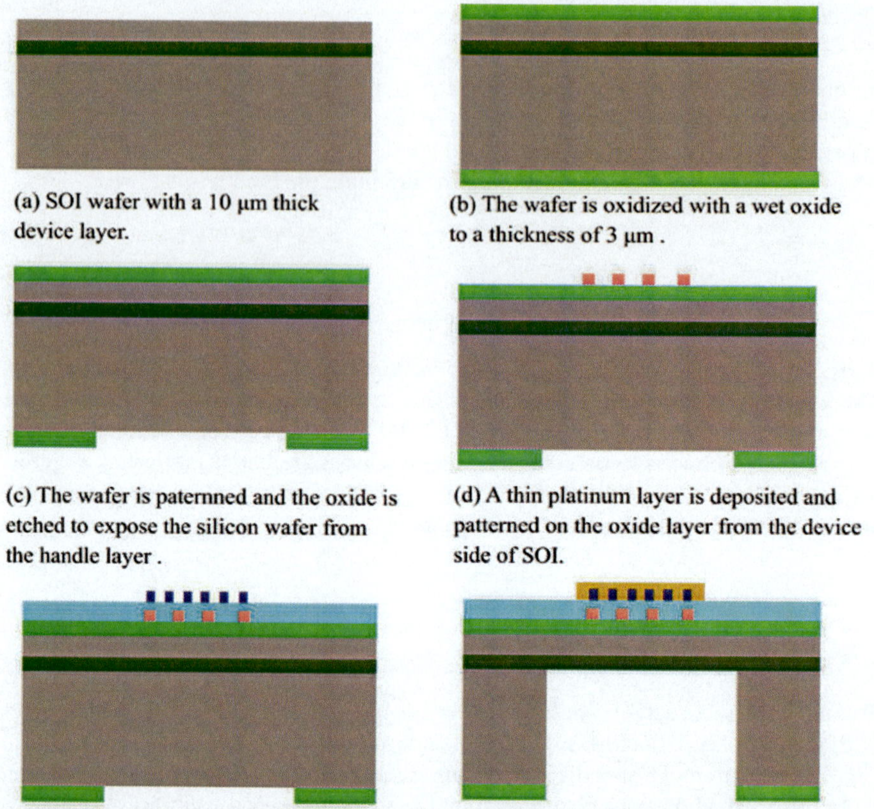

(a) SOI wafer with a 10 μm thick device layer.

(b) The wafer is oxidized with a wet oxide to a thickness of 3 μm .

(c) The wafer is paternned and the oxide is etched to expose the silicon wafer from the handle layer .

(d) A thin platinum layer is deposited and patterned on the oxide layer from the device side of SOI.

(e) A thin SiN layer is deposited to isolate from the sensing electrode. The sensing electrode is deposited above the SiN layer.

(f) Deposition of a sensitive layer and etching of handle silicon layer to achieve a miniaturized metal-oxide (MOx) gas sensor.

Fig. 1 SOI wafer is used for manufacturing the MEMS-MO_x sensor. The SOI wafer has a device layer of 10 μm thick. **a** SOI wafer is cleaned with RCA and rinsed and dried. **b** The wafer is oxidized to a thickness of 3 μm. **c** The wafer is patterned on the handle layer to expose the silicon for further etching. **d** A thin layer of platinum is deposited that forms the heater for the gas sensor. The platinum layer is patterned using lift-off technique. **e** In this step, an isolation layer of silicon nitride (SiN) layer is deposited on the heater layer. This step is followed by the deposition of platinum sensing layer. **f** In the last step, a gas sensitive layer such as SnO_2 is deposited, and the silicon is etched from the handle layer

3 Results and Discussion

In this section, the metal oxide sensor is highlighted, and a general mems fabrication is described. A parameter study of different metals is described, and an interface circuitry is mentioned.

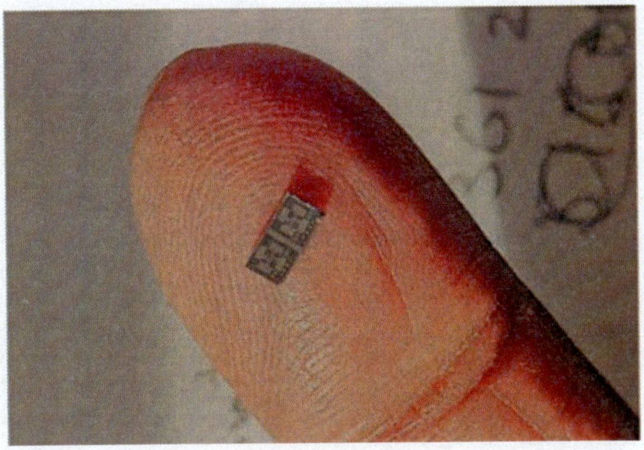

Fig. 2 Fabricated MEMS-MO_x sensor dies shown resting on a fingertip. Two sensors are shown in the picture with a size of approximately 1 mm^2 each sensing area

We have successfully designed and developed a MEMS-MO_X sensor on a SOI wafer. The wafer-level fabrication allows for several hundreds of sensors to be processed simultaneously on a single wafer. The components of the sensor such as the heater is processed using materials that can sustain high temperatures. For the sensor to work and detect gases, a high temperature between 150 and 400 °C is maintained. A photograph of two devices on a fingertip is shown in Fig. 2.

To achieve a compact formfactor of the metal oxide sensor, an ASIC integration to the sensor can be imagined. The developed sensor is compatible with CMOS manufacturing method. The process flow can be integrated into the CMOS process line. The sensing layer deposition can be a post process step in the process flow. With further optimization technique, nano-scaled electrodes can be achieve with nano-sized heater and nano-sized sensing electrodes. Furthermore, an atomic layer deposition of the sensing layer could increase the sensitivity of the sensing layer. In addition, integration of several sensing layers in a single gas sensor can be thought of to achieve multi-gas sensing module.

A batch fabricated sensors can be achieved using our method. Depending upon the wafer size, the number devices can be scaled. The measured heater resistance across the batch fabricated wafer is shown in Fig. 2. A total number of 24 heaters were measured for resistance. As can be seen, a variation in the resistance of 100–150 Ω as was observed before the heaters were annealed. After annealing the resistance increased to 60 Ω on an average.

To sustain high temperatures, we need a packaging which also allows for gas interaction with the sensitive element on the metal oxide sensor. This would allow for a packaged sensor that can be integrated on an electronic PCB with other components without the spread of heat radiation.

For next generation MEMS sensors, a "packaging cap" with the carefully designed microcavity, insulation layer, microheater, and high temperature epoxy layer is to be made as to the first step in the post-packaging process for MEMS. Low temperature co-fired ceramics (LTCC) have been employed in the creation of ceramic-based packaging of microsystems due to their excellent electrical and mechanical characteristics, as well as their ability to construct three-dimensional (3D) integrated microstructures [7]. High reliability and stability in severe environments are two significant advantages of LTCC technology. Electronic systems based on LTCC technology can operate in harsh environments with high temperatures and pressures, such as in the petroleum industry as sensors for intelligent oil and gas wells or as a dedicated sensor in a jet engine, where they must work with potentially corrosive exhaust gases in addition to high temperatures. It is often mentioned that thick-film and LTCC heaters might be used in smart planar gas sensors. The sensitivity, selectivity, and reaction time of a gas sensor are all affected by temperature. Packaging systems and components produced using LTCC technology provide stable platforms for electrochemical sensors since many of them are used at very high temperatures (up to 900 °C) [8, 9].

Modern miniaturized gas sensors with in-built heater fabricated on LTCC packaging structure along with thermal oxide layer for insulation layer on the device substrate provides additional thermal protection. Robust commercial gas sensors with chip-based integration for high temperatures are scarcely available. Historically, there have been some research efforts in the past on LTCC, LPCVD oxide-based packaging but for limited temperature ranges. There are several groups currently working on similar chip-based emitters around the world. Karlsruhe Institute of Technology, Germany, [10] has recently demonstrated the detection of H_2 gas at low temperature (120 °C) with a change in microheater pattern. We could use standard methods for miniaturized gas sensors on chip with robust packaging and thermally insulating layers to ensure long-term stability and improved energy efficiency.

To achieve the concept of next generation MEMS sensors, a packaging of LTCC layer with silicon wafer is under progress, as shown in Figs. 3 and 4. Further research work in the domain of the mems-based gas sensor is to achieve the following.

- An initial simulation of the heat spread at 400 °C shall be achieved.

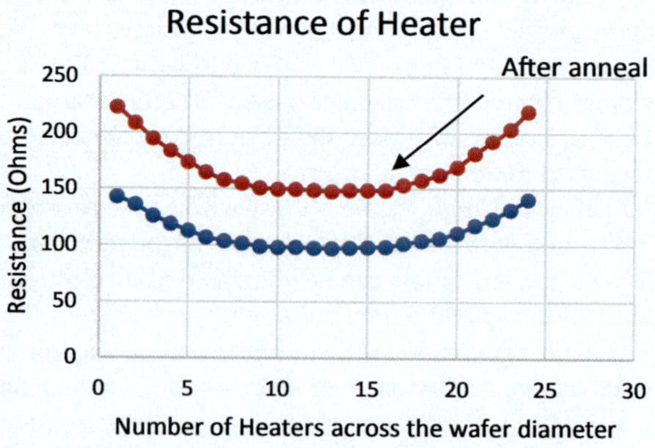

Fig. 3 Measured heater resistance across the batch fabricated wafer. A total number of 24 heaters were measured for resistance. As can be seen, a variation in the resistance of 100–150 Ω as was observed before the heaters were annealed. After annealing the resistance increased to 60 Ω on an average

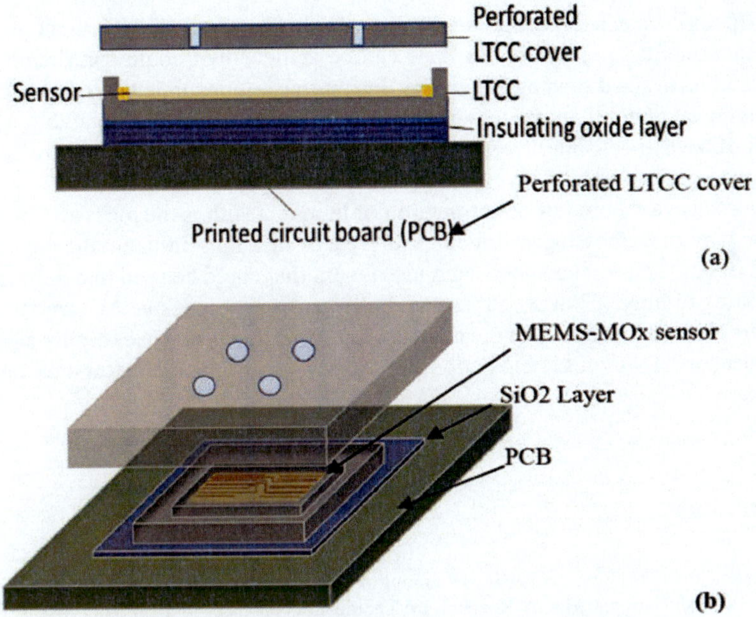

Fig. 4 **a** Possible fabrication schematic of LTCC-based packaging for the sensor. The cross-section view shows different layers that are useful in restricting the heat flow from the sensor. **b** A 3D schematic view of MEMS-MO_X packaging with LTCC capping

- After the results of the simulation, parameters leading to the thickness and materials shall be investigated.
- The MEMS-MO_X shall be integrated into the package.
- With the packaged sensor, we plan to test with a gas flow on the sensor.

Furthermore, these sensors when integrated with a breath analyzing device such as a spirometer shall achieve an increased measurement component. Hence, the breath analysis could be done reliably and more efficiently to detect and diagnose pulmonary infections. The future of exhaled breath instrument is in mobile-based data acquisition and data management. A PCB with advanced electronic circuits and algorithm can signal the concentration of the gas from the exhaled breath.

4 Conclusions

MEMS-based metal oxide sensor is an upcoming field, and the possibilities are enormous. We have designed and developed a metal oxide sensor with a small form-factor. This sensor is compatible with electronics and CMOS process. The process flow developed can be easily integrated into the CMOS process line. The sensing layer deposition can be a post process step in the process flow. With further optimization technique nano-scaled metal lines can be achieve that could form nano-sized heater and nano sized sensing electrodes. Furthermore, an atomic layer deposition of the sensing layer could increase the sensitivity of the sensing layer. In addition, integration of several sensing layers in a single gas sensor can be thought of to achieve multi-gas sensing module.

Non-invasive method of determination of human health is the most accepted and reliable way of expressing results and inferences in a lung inflammatory or health related issues. Spirometer is one such instrument that could be used to determine the functioning of lungs. However, these are bulky and very expensive. We tend to solve these problems by investing in the next generation products that uses highly sensitive mems sensors. This could help patients to monitor their lung quality at their bedside or at home.

References

1. Johanna K (2021) Severe COVID-19 among patients with asthma and COPD: a report from the Swedish National Airway Register. In: Therapeutic advances in respiratory disease
2. Gayle AV et al (2022) Respiratory-related death in individuals with incident asthma and COPD: a competing risk analysis. BMC Pulm Med
3. NHS Homepage, Spirometry—NHS. www.nhs.uk. Last accessed 30 June 2022
4. Gatty HK (2015) An amperometric nitric oxide sensor with fast response and ppb-level concentration detection relevant to asthma monitoring. Sens Act B Chem 209:639–644
5. Gatty HK (2015) A wafer-level liquid cavity integrated amperometric gas sensor with ppb-level nitric oxide gas sensitivity. JMM 25(10):105013
6. Asri M (2021) MEMS gas sensors: a review. IEEE Sens J 21(17)
7. Nowak D, Dziedzic A (2011) LTCC package for high temperature applications. Microelectron Reliab 2010(51):1241–1244
8. Akbar S, Dutta P, Lee C (2006) High-temperature ceramic gas sensors: a review. Int J Appl Ceram Tec 3:302–311
9. Benson DA, Bowman D, Filter W, Mitchell R, Perry J (1998) Presented at the ITherm'98. Sixth intersociety conference on thermal and thermomechanical phenomena in electronic systems (Cat. No.98CH36208), pp 434–441
10. Sharma A, Sharma B (2022) Influence of microheater patterns: $MoSi_2$-SnO_2 as energy-saving chemiresistors for gas sensing applications. Sens. Act B Chem 351:130901

Estimation of Breast Tumor Parameters by Random Forest Method with the Help of Temperature Data on the Surface of the Numerical Breast Model

Gonuguntla Venkatpathy, V. M. Rahul, and N. Gnanasekaran

Abstract The second most frequent reason for cancer-related fatalities in women is breast cancer. When a condition is identified early, better treatment choices are available. Different temperature patterns are seen on the breast surface due to the tumors, which change blood perfusion rate and metabolic heat production. Thermography is an infrared imaging technology for breast cancer screening that records temperature variations. The temperature dataset on the surface of the breast that corresponded to the tumor's diameter and the location was needed for the current study, but such actual data are not accessible. Thus, the modeling and development of a dataset constitute the initial component of the current study. The bio-heat transport equation is solved using COMSOL multiphysics software, and the model consists of a spherical tumor inside of a hemispherical breast model. By changing the sizes and positions of the tumor inside the breast during simulations, a reliable dataset is created. The training and testing of the dataset produced from the simulations using the random forest method make up the second portion of the current study. Breast skin temperature is used as an input in a random forest machine learning algorithm in the current work to determine the diameter and location of the tumor inside the breast. The diameter and area of the tumor location are estimated by a trained random forest algorithm.

Keywords Breast cancer · Thermography · Bio-heat transfer · Random forest · Machine learning

1 Introduction

Tumors are the result of abnormal cell growth inside the human body. All tumors are not cancerous, and there are two types of tumors: malignant (cancerous) and the other one is benign (non-cancerous). Breast cancer is a cancerous tumor present in the breast, and breast cancer is the second most typical reason for cancer-related

G. Venkatpathy · V. M. Rahul · N. Gnanasekaran (✉)
National Institute of Technology Karnataka, Surathkal 575025, India
e-mail: gnanasekaran@nitk.edu.in

S. Sharma et al. (eds.), *Intelligent Control, Robotics, and Industrial Automation*, Lecture Notes in Electrical Engineering 1066,
https://doi.org/10.1007/978-981-99-4634-1_52

deaths in women. Despite being a fatal illness, early discovery is key to curing it. Consequently, early diagnosis is crucial for effective treatment [1]. A biopsy is the only guaranteed way for the doctor to know if an area of the body has cancer in all screening tests for breast cancer. A biopsy is a procedure in which a doctor removes a small tissue sample to test in a laboratory. There are various imaging screening techniques, with mammography being the gold standard imaging approach for detecting breast malignancies. However, direct mammography is not recommended for initial screening because of its radiation risk, pain, and discomfort. Thermography thus offers details regarding the aberrant breast skin temperature distribution. One can determine the size and location of the tumor inside the breast using this temperature change [2].

Gautherie [3] conducted fourteen years of clinical studies for comprehensive research on thermography by 58,000 patients with thermal breast scans and mentioned thermography as a proper adjunct tool for screening for breast tumor prediction. He also calculated the heat conductivity of the breast and tumor tissue effectively. Feig et al. [4] compared mammography, physical examination, and thermography which were contrasted to diagnose breast cancer. The thermographic examination consisted of frontal and both oblique views and isotherms. Parisky et al. [5] conducted a digital infrared thermal imaging (DITI) study to evaluate 769 patients scheduled for breast biopsy. The infrared camera obtained multiple infrared images in rapid sequence before and during the cooling phase. The study found that thermography is a non-invasive screening tool that will complement mammography in tumor screening. Through a two-year clinical research on ninety-two women who were advised to have a breast biopsy in support of a previously detected malignancy via a mammography or ultrasound, Arora et al. [6] investigated the efficacy of DITI. In this clinical trial, ninety-two women who had breast lesions found before ultrasonography or mammography underwent DITI.

Wishart et al. [7] The results of digital infrared thermography performed on 100 patients prior to core needle biopsy were presented and evaluated using expert manual evaluation, neural network analysis, DITI, and artificial intelligence. The specificity and sensitivity of DITI in breast lesions were investigated by Kontos et al. [8]. In this study, participants are told to maintain stability and to keep their hands on top of their heads. Mammography was used to picture 90 breasts, and ultrasound was used to image 61 breasts. Thirty-nine true-negative, sixteen false positive, fifteen false negative, and five genuine positive results were obtained via thermography. The development of thermal imaging for the diagnosis of breast cancer during the past three decades was reviewed by Satish et al. [9]. They found that the depth of the tumor (20–30 mm) has a more significant influence on the surface temperature than the tumor size and that thermography needs considerable modifications to become a reliable technique for screening breast cancer. For the dataset of Wisconsin breast cancer, Hantoro [10] examined support vector machine (SVM), random forest, and multilayer perception (WBC). The multilayer perception algorithm provides the best performance, according to this study.

For the WBC dataset, Sivapriya et al. [11] examined logistic regression, SVM, random forest, and Naive Bayes. With 99.76% accuracy, they discovered that the random forest surpasses the other algorithms. The lowest error rate and maximum precision are both achieved. To make a more precise and trustworthy prediction for breast cancers, Octaviani and Rustam [12] presented categorization by a random forest applied to cancer microarray data. Additionally, the random forest provides the best accuracy for the prediction of breast cancer, with 70% and 80% of the training data accuracy being 90%.

From the literature, it is clear that the dataset used to classify or predict the tumor size and location uses datasets available from different health organizations or numerical and experimental. The breast surface temperature variation of healthy women is approximately identical for both breasts. If a tumor manifests inside the breast, it alters the thermophysical properties due to this higher surface temperature of the breast observed as compared to a healthy breast. The abnormal temperature variation on the surface of the breast with a tumor is caused by the increased blood flow and metabolic heat production inside and around the tumor. An infrared imaging method called thermography is used to record temperature changes in order to screen for breast cancer. The temperature dataset on the surface of the breast that would have identified the tumor's location and matching diameter was needed for the current study, but it was not available. As a result, a dataset created by a numerical simulation of the breast using COMSOL multiphysics software was used in the current investigation. Using the simulations dataset, a random forest machine learning algorithm is trained.

2 Materials and Methods

2.1 Numerical Model Validation

Finite volume analysis was done on the rectangular breast geometry by Das and Mishra [13]. Using the Gaussian profile features of the acquired surface temperature distribution, they concurrently predicted the tumor size and region of the site. The boundary conditions, thermophysical characteristics of the breast and tumor, and two-dimensional rectangular breast model used in this investigation are the same as those used by Das and Mishra [13] for validation. To do a steady-state bio-heat transfer analysis using COMSOL Multiphysics, the model is created and the Pennes bioheat transfer equation is solved using the finite element approach.

Figure 1 contrasts the temperature distribution found in the current investigation with that found in Das and Mishra [13]. These temperature changes fit the earlier findings quite well. It is permissible to use the model for further research after the 2D model of Das and Mishra [13] was used to validate the current model.

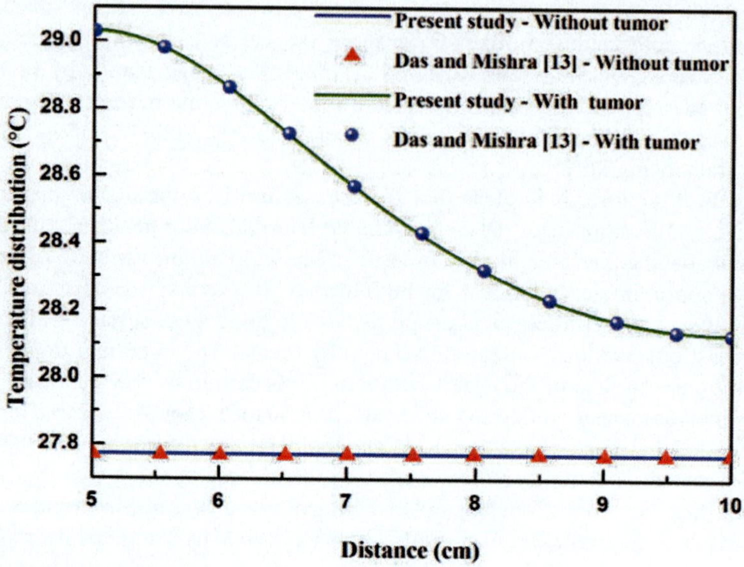

Fig. 1 Validation of present study with Das and Mishra [13]

2.2 *Numerical Simulations for Dataset*

Figure 1 represents the modeling of the hemispherical breast with spherical tumor, breast size, and tumors taken from the literature [13]. The breast diameter d is14.4 cm, the radius of the breast is denoted as r, spherical tumor diameter d_t is 1.5 cm, and the radius of spherical tumor is denoted as r_t. Properties of tissue: Tables 1 and 2 provide the thermal properties and boundary conditions needed to solve Eq. (1), respectively. As depicted in Fig. 2, the exterior skin breast surface convects heat to the surroundings, while the breast's chest wall regulates the body's core temperature.

The transfer of heat in biological tissue is governed by Pennes' bio-heat transfer equation [14].

$$\nabla \cdot (k \cdot \nabla T) + \rho_b \eta_b C_{pb}(Ta - T) + Q_m = \rho\, C_p \frac{\partial T}{\partial t}. \tag{1}$$

Subsets b, a, and m represent the blood, atrial blood temperature, and metabolic, respectively. Thermal conductivity is represented by k, density with ρ, η_b is the

Table 1 Boundary conditions

Surface of the breast	$-k\,(\nabla.T)\vert_{z=L} = h\big(T - T_f\big)$
Core body/chest wall temperature	$T_{x,y,z} = 0 = T_a = 37\ °C$

Table 2 Breast and tumor tissue properties

Tissue	k (W/m K) thermal conductivity	Q_m (W/m^3) metabolic heat generation rate	ρ (kg/m^3) density	η_b (s^{-1}) blood perfusion rate	C_p (J/kg K) specific heat
Healthy breast	0.42	450	920	0.00092	3000
Cancerous breast	0.42	29,000	920	0.009	3000
Blood	–	–	1052	–	3800

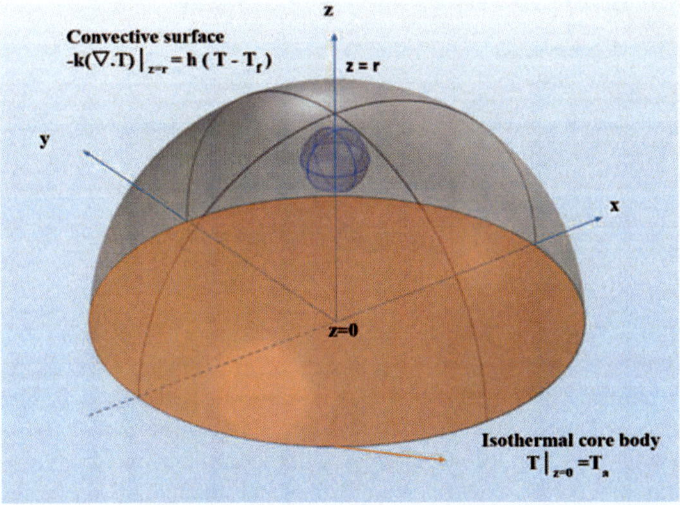

Fig. 2 Three-dimensional model of the breast with tumor

perfusion rate of the blood, and C_p is the breast tissue-specific heat. Q_m and Q_s represent the rate of metabolic heat generation and the volumetric heat source of the breast tissue.

2.3 *Element Sensitivity Study*

Element sensitivity study for 3 cm diameter of spherical tumor is located at the central axis (X = 0) and 5.65 cm from the Z-axis. A mesh ranging from 584,122 to 2,737,429 elements was considered. Among these the subset of 1,735,926 elements provide a correct temperature value with less error, as a result, this subset was utilized in all simulations, as seen in Fig. 3.

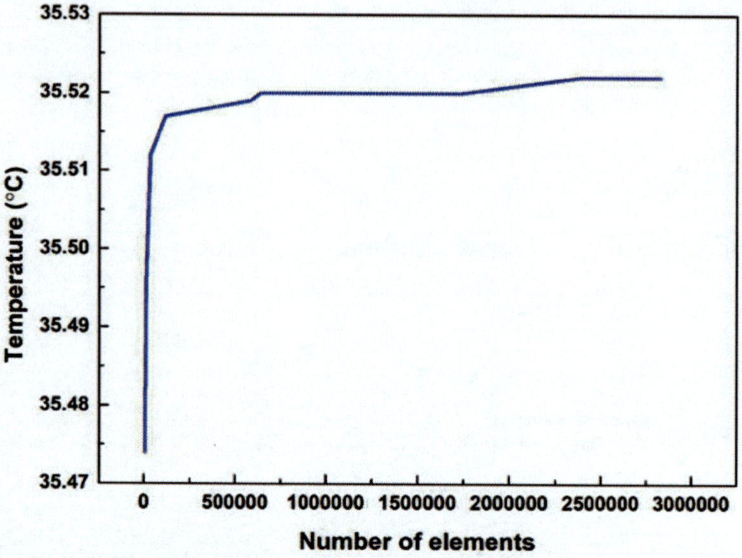

Fig. 3 Study of element independence

2.4 *Random Forest Regression Model*

Random forest is the method of combining many decision trees to predict the outcome. It reduces the dependency of the outcome based on a single decision tree and helps get more accurate data. Many decision tree models produce their output based on their constraints, and every decision tree's outcomes are analyzed [15]. The random forest model gives out the final result. This model is used for classification and regression, but the present study deals with numerical temperature data and tumor parameters; therefore, the present study used the random forest regression model [17]. The steps followed are similar to machine learning models, starting with preprocessing the simulated data and training the model by model. Fit (X, Y) is the model object, X is the input array used to predict outcomes, and Y is the actual outcome values. It uses a concept known as bagging, where the model uses bootstrapping and aggregation to distribute, as shown in Fig. 4. Bootstrapping means the samples used with the replacement for training by each decision tree in a random forest which means that a single tree may use a sample multiple times for training purposes. After the training and prediction, every decision tree's result is gathered, known as aggregation. Suppose the overfitting reduced leads to an increase in prediction accuracy.

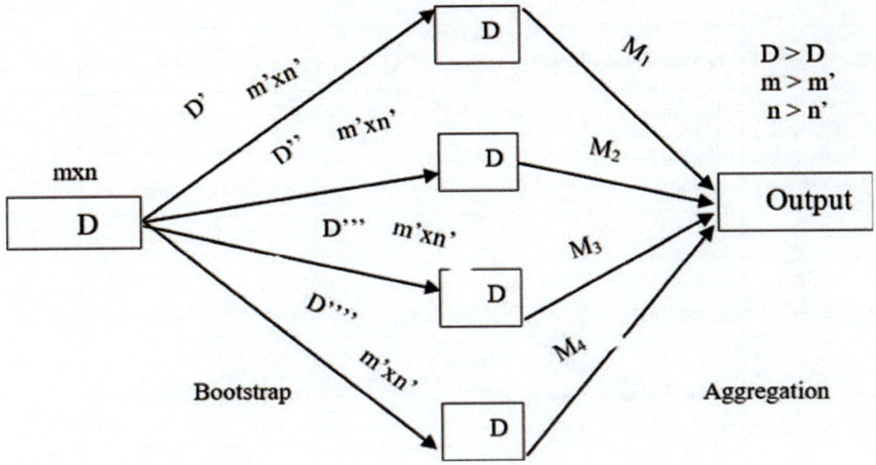

Fig. 4 Random forest [16]

3 Results and Discussion

3.1 *Training Dataset by Numerical Simulations*

The three variables that change to produce the dataset are the tumor diameter, x, and z distances. X can be anywhere between 0 and 6 cm in steps of 1 cm, and Z can be anywhere between the point at which the tumor's influence was hardly perceptible and the point at which the tumor surface met the breast surface. The diameter of a spherical tumor ranged from 3 to 0.5 cm. The temperature variation between the maximum and minimum diameter near the surface of the breast skin at the central axis is depicted in Fig. 5.

3.2 *Random Forest Predictions*

The random forest can handle data with missing values, and continuous and categorical values can be used to predict the outcome.

Input = Temperature distribution.

Output = X, Z, Tumor diameter.

Total data samples = 686.

Training data samples = 515.

Testing data samples = 171.

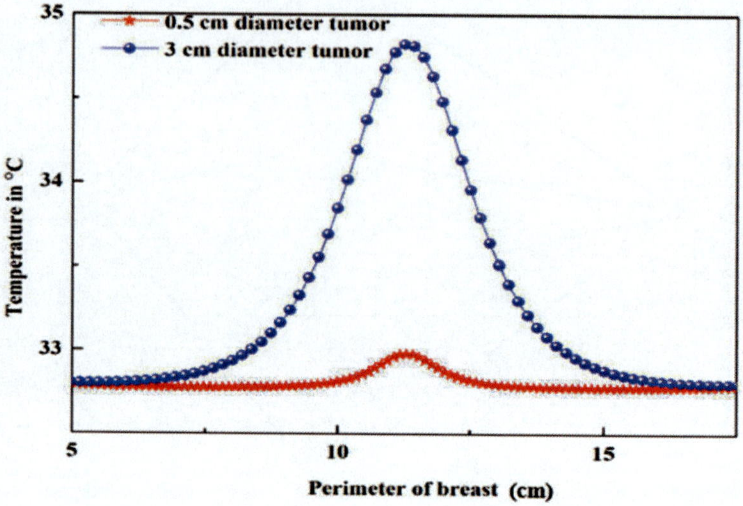

Fig. 5 Temperature variation on the surface of the breast at the central axis

Comparison of actual and predicted values of X location is shown in Fig. 6, Z location is shown in Fig. 7, and tumor diameter is shown in Fig. 8. While preparing the dataset for training, the initial dataset had many temperature values removed. The temperature near both ends of the plot had to be removed since it was similar to most trials. These values played a minimal role in determining the tumor location and size. Therefore, values from both ends are removed to ensure symmetric data, and the bulge within the range is intact, determining the diameter and tumor location.

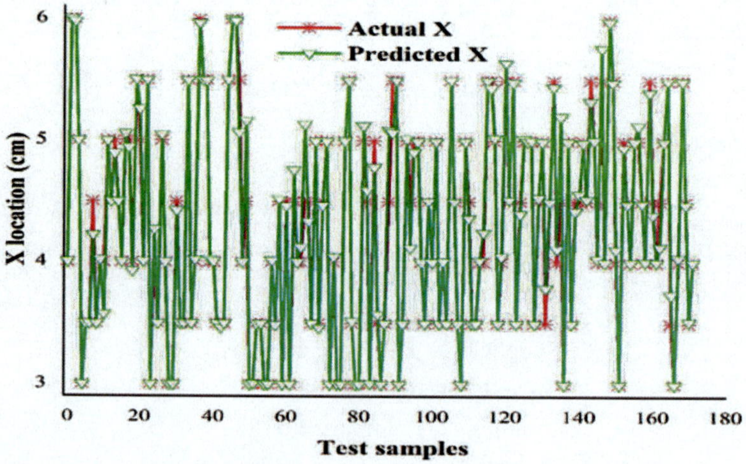

Fig. 6 Actual and predicted X location

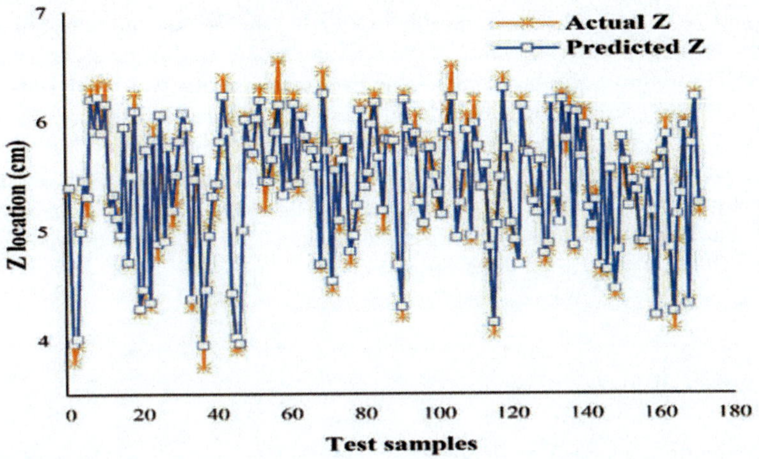

Fig. 7 Actual and predicted Z location

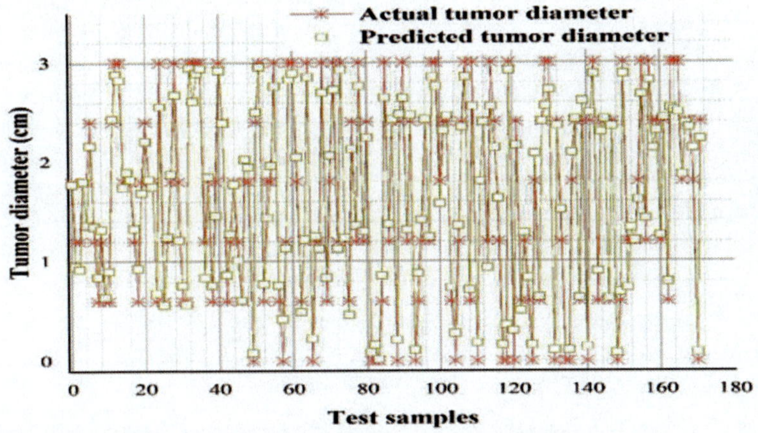

Fig. 8 Actual and predicted tumor diameters

After filtering the data, the training process initially had low accuracy. However, scaling the data between 0 and 1 improved the precision, reaching 85% to 90%. After rescaling the output values based on the variable and their original range, its accuracy increased more than 95% for location variables and crossed 90% for size variables.

The variation in the accuracy is in the order X > Z > tumor diameter because the variable X can be easily determined if its output range is known by simply figuring out where the peak of the bulge is within the temperature range. Since random forest could quickly figure it out, the accuracy of the X variable is the highest among all the variables, with 98.66% accuracy. For the variable Z, predicting its value is not as straightforward as it was for the X variable since the Z-axis goes into the plane of the paper, and the random forest should look at both X-axis and tumor diameter variable and then predict the value of the Z variable. Due to this, its accuracy is less than the X variable's accuracy, and Z-axis prediction accuracy by random forest is 98.01%. The tumor diameter depends on the X and Z variable's accuracy because the plots with significant bulge regions may have a moderate size and be very near the Z-axis. Still, if there is a comparatively small bulge, then there is a possibility that the tumor diameter may be huge and be very far along Z-axis too. Due to this, it appears smaller. Due to these interdependencies, the accuracy of the tumor diameter is at least 92.27% among all three variables.

4 Conclusion

Machine learning algorithms significantly improve the accuracy of the predictions. The present study used a simulated dataset for regression machine learning techniques to predict the diameter and tumor location. Dataset was developed by varying the spherical tumor diameters from 3 to 0.5 cm and the locations of the tumor inside the breast. Since the random forest model contains many decision trees where every decision tree's predictions are compared, and the final result depends on predictions made by all decision trees, the random forest is more accurate in predicting the values. In the present study, a random forest regression model was used to predict the location and diameter of a tumor from the numerical temperature data on the surface of the breast, and the predicted values were compared to the actual values. The random forest model had prediction accuracies of 98.66%, 98.01%, and 92.27% for X, Z location and diameter of the tumor, respectively.

References

1. Siegel RL, Miller KD, Fuchs HE, Jemal A (2021) Cancer statistics. CA Cancer J Clin 71:7–33. https://doi.org/10.3322/CAAC.21654
2. Breast Cancer Screening: Patient Version. PDQ Cancer Inf Summ, National Cancer Institute, https://www.cancer.gov/types/breast/hp/breast-screening-pdq. Accessed 13 June 2022
3. Gautherie M (1980) Thermopathology of breast cancer: measurement and analysis of in vivo temperature and blood flow. Ann N Y Acad Sci 335:383–415. https://doi.org/10.1111/J.1749-6632.1980.TB50764.X
4. Feig SA, Shaber GS, Schwartz GF, Patchefsky A, Libshitz HI, Edeiken J, Nerlinger R, Curley RF, Wallace JD (1977) Thermography, mammography, and clinical examination in breast cancer screening. Review of 16,000 studies. Radiol 122:123–127. https://doi.org/10.1148/122.1.123
5. Parisky YR, Sardi A, Hamm R (2012) Efficacy of computerized infrared imaging analysis to evaluate mammographically suspicious lesions. 180:263–269. https://doi.org/10.2214/ajr.180.1.1800263
6. Arora N, Martins D, Ruggerio D, Tousimis E, Swistel AJ, Osborne MP, Simmons RM (2008) Effectiveness of a noninvasive digital infrared thermal imaging system in the detection of breast cancer. Am J Surg 196:523–526. https://doi.org/10.1016/J.AMJSURG.2008.06.015

7. Wishart GC, Campisi M, Boswell M, Chapman D, Shackleton V, Iddles S, Hallett A, Britton PD (2010) The accuracy of digital infrared imaging for breast cancer detection in women undergoing breast biopsy. Eur J Surg Oncol 36:535–540. https://doi.org/10.1016/J.EJSO.2010.04.003
8. Kontos M, Wilson R, Fentiman I (2011) Digital infrared thermal imaging (DITI) of breast lesions: sensitivity and specificity of detection of primary breast cancers. Clin Radiol 66:536–539. https://doi.org/10.1016/J.CRAD.2011.01.009
9. Kandlikar SG, Perez-Raya I, Raghupathi PA, Gonzalez-Hernandez J-L, Dabydeen D, Medeiros L, Phatak P (2017) Infrared imaging technology for breast cancer detection ? Current status, protocols and new directions 108(B):2303–2320. https://doi.org/10.1016/j.ijheatmasstransfer.2017.01.086
10. Hantoro YN (2020) Comparative study of breast cancer diagnosis using data mining classification 9:790–795
11. Shashidhar R, Arunakumari BN, Naziya Farheen HS, Puneeth SB, Santhosh Kumar R, Roopa M (2020) Breast cancer detection using supervised machine learning algorithm. Asian J Converg Technol 6:26–31. https://doi.org/10.33130/ajct.2020v06i03.006
12. Octaviani TL, Rustam Z (2019) Random forest for breast cancer prediction. AIP Conf Proc 2168:6–11. https://doi.org/10.1063/1.5132477
13. Das K, Mishra SC (2014) Non-invasive estimation of size and location of a tumor in a human breast using a curve fitting technique. Int Commun Heat Mass Transf 56:63–70. https://doi.org/10.1016/j.icheatmasstransfer.2014.04.015
14. Penne HH (1948) Analysis of tissue and arterial blood temperatures in the resting human forearm. J Appl Physiol 1:93–122. https://doi.org/10.1152/JAPPL.1948.1.2.93
15. Random Forest—an overview ScienceDirect Topics. https://www.sciencedirect.com/topics/engineering/random-forest. Accessed 13 June 2022
16. Random Forest Regression in Python—GeeksforGeeks. https://www.geeksforgeeks.org/random-forest-regression-in-python/. Accessed 13 June 2022
17. Venkatapathy G, Mittal A, Gnanasekaran N, Desai VH (2023) Inverse estimation of breast tumor size and location with numerical thermal images of breast model using machine learning models. Heat Transf Eng 44(15):1433–1451. https://doi.org/10.1080/01457632.2022.2134081

Early Prediction of Sudden Cardiac Death Using Optimal Heart Rate Variability Features Based on Mutual Information

Shaik Karimulla and Dipti Patra

Abstract Sudden cardiac death (SCD) is a prominent cause of death around the world. SCD leads to unconsciousness and death within a few minutes. The effectiveness of available SCD prediction algorithms is minimal due to their limitations. Therefore, developing an accurate and precise methodology is essential for accurate early-stage SCD prediction, which will save many lives worldwide. In order to identify the risk of SCD in patients with congestive heart failure (CHF), atrial fibrillation (AF), coronary artery disease (CAD), ventricular tachyarrhythmia (VT), and normal sinus rhythm (NSR) subjects using heart rate variability (HRV) analysis, this study proposes a new methodology for detecting SCD 8 min before its onset. A total of 124 HRV signals of six classes acquired from the PhysioBank were used to extract 35 features for each subject in the time-domain, frequency, domain, and nonlinear methods. A mutual information-based feature selection algorithm is used to select the optimal number of features for better performance. We detected SCD 8 min before its onset using the Light gradient boosting (LightGBM) classifier, with accuracy, sensitivity, specificity, and precision of 95.52%, 94.96%, 99.08%, and 95.87%, respectively. These results indicate that the performance of the proposed method is superior to that of existing ones in terms of the number of classes, prediction time, and accuracy. The experimental results acquired by evaluating the HRV signal showed substantial results for 8 min before SCD onset, which will be helpful in automatic diagnostic systems and intensive care units (ICUs) detecting those at risk of developing SCD. Clinicians may have enough time to respond to treatment using this methodology. As a result, the proposed technique can be a valuable tool for increasing survival rates.

Keywords Artifact correction · Heart rate variability · Sudden cardiac death · LightGBM · Mutual information-based feature selection

S. Karimulla · D. Patra (✉)
IPCV Laboratory, Department of Electrical Engineering, National Institute of Technology Rourkela, Rourkela, Odisha, India
e-mail: dpatra@nitrkl.ac.in

S. Karimulla
e-mail: 520EE1006@nitrkl.ac.in

S. Sharma et al. (eds.), *Intelligent Control, Robotics, and Industrial Automation*, Lecture Notes in Electrical Engineering 1066,
https://doi.org/10.1007/978-981-99-4634-1_53

1 Introduction

Cardiovascular disease is the primary cause of death worldwide, accounting for 32% of all deaths [1]. Sudden cardiac death (SCD) is a dangerous heart condition that results in unconsciousness and death within a few minutes due to irregularities in the heart's electrical conduction system. SCD may be associated in patients with or without cardiac disease history, but patients with CAD are at higher risk of SCD [2]. The majority of SCD incidents were caused by ventricular tachyarrhythmia's such as ventricular fibrillation (VF), ventricular tachycardia (VT), and ventricular flutter (VFL). The most common underlying cause of SCD in people under the age of 50 years is coronary artery disease (CAD). While many ischemic SCDs in young and middle-aged individuals occurred without previously diagnosed CAD, many patients showed severe underlying heart disease and a previously undisclosed MI at autopsy [3]. If CAD is not promptly treated, it eventually causes a coronary artery infarction and limits the heart's capacity to deliver oxygen-rich blood to the body. This inability state condition of the heart is called congestive heart failure (CHF), which may lead to sudden cardiac arrest. According to recent research, in some instances, atrial fibrillation is associated with SCD [4]. The SCD symptoms begin one hour before its onset. After the start of sudden cardiac arrest (SCA), the survival rate is about 10.4% only due to the unavailability of early prediction methodologies [5].

Despite advances in healthcare facilities, the survival rate following SCD remains low since there are limited alternatives for treatment after it occurs. As a result, it is often desired to prevent the start of SCD by giving medical assistance and its early prediction. Defibrillator, implanted cardioverter defibrillator (ICD), and cardiopulmonary resuscitation (CPR) are the devices used to restore the normal functioning of the heart after the onset [6, 7]. According to the literature, NSR and SCD were the subjects used by many researchers in their SCD prediction studies. Few authors included three [8] and four [2] classes while developing the SCD prediction strategies. To address the shortcomings of the earlier investigations on the number of classes and prediction time, the current study included six classes of subjects for the analysis and prediction of SCD. The present work's primary vision is to develop a precise methodology for the earlier detection of SCD. The main contribution of this research can be summarized in the following points: (1) a real-time automated methodology development for SCD prediction at an early stage. (2) AF and VT subjects were included in this study for the first time in the field of SCD prediction studies because these diseases are associated with the development of SCD. (3) The proposed methodology based on HRV features extraction in the time-domain, frequency-domain, and nonlinear methods (4) A comparative analysis of features selection algorithms to identify the best combination of features (5) A new classification algorithm such as Light Gradient Boosting Machine (LightGBM) algorithm is used to classify NSR, SCD, CHF, AF CAD, and VT subjects.

Heart rate variability (HRV) is one study that monitors the heart's activity and helps distinguish abnormalities in the heart's signals. The extraction of features from

HRV signals, including time-domain, frequency-domain, and nonlinear methods, contributes to the evaluation of the cardiovascular and autonomic nervous systems (ANS). HRV features (time-domain, frequency-domain, and nonlinear methods) help to access the cardiovascular and autonomic nervous systems (ANSs) [9]. A healthy heart needs a higher HRV to respond to environmental changes and balance the two ANS branches [10]. This study used HRV analysis to develop an accurate prediction model for SCD detection. The proposed methodology is examined using four evaluation criteria: accuracy, sensitivity, specificity, and precision. The experimental results suggest that the proposed methodology can be used as an efficient tool for the early prediction and the risk of SCD in different cardiac abnormalities.

2 Methods and Materials

2.1 Data

In this study, six classes were examined for HRV-based SCD prediction. The HRV signal (RR interval) data are collected from the Physionet database, i.e., the MIT-BIH Normal Sinus Rhythm Database (NSRDB), Sudden Cardiac Death Holter Database (SDDB), MIT-BIH Atrial Fibrillation Database (AFDB), BIDMC CHF Database (CHFDB), Long-Term ST Database (LTSTDB), and the CU Ventricular Tachyarrhythmia database (CUDB). The sampling frequency of NSR subjects is 128 Hz, and the remaining are classless with a sampling frequency of 250 Hz. One hundred and twenty-four subjects were included in this study, i.e., 18 NSR, 20 SCD, 15 CHF, 23 AF, 23 CAD, and 25 VT subjects. Two channels were available for all the subjects except the VT class subject; each channel is considered a distinct observation [11]. The HRV signal of 8 minutes duration, is divided into four overlapping segments, each lasting 5 minutes. A combined total of 892 segments were extracted from 124 subjects for the purposes of feature selection and classification, as mentioned in Table 1.

Table 1 Complete description of the dataset used in the study of SCD prediction

Type	No. of subjects	Sampling frequency (Hz)	No. of channels	No. of segments	Gender	Age (range)
NSR	18	128	2	144	5 males, 13 females	20–89
SCD	20	250	2	160	10 males, 8 females, 2 unknown	30–89
CHF	15	250	2	120	11 males, 4 females	22–71
AF	23	250	2	184	Unknown	–
CAD	23	250	2	184	15 males, 3 females, 5 unknown	44–82
VT	25	250	1	100	Unknown	–

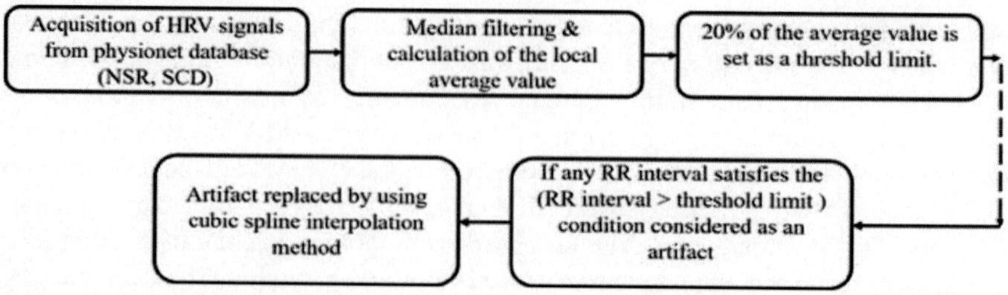

Fig. 1 Block diagram representation of artifact correction of HRV signals

2.2 Pre-Processing

The HRV signals acquired from the database may contain artifacts, including ectopic peak, extra beat, missing beats, etc. These artifacts were removed by using median filtering along with threshold value. The block diagram representation of artifact correction method is shown in Fig. 1. The pre-processing steps include:

(i) Median filtering of the HRV signal of the particular class.
(ii) Calculating the local average value of the corrected HRV signal and taking 20% of the average value as the threshold limit.
(iii) If any RR interval value > local average value + threshold limit is considered as an artifact.
(iv) If any RR interval value < local average value − threshold limit is considered as an artifact.
(v) The identified artifact is replaced by using cubic spline interpolation.

2.3 HRV Features' Extraction and Selection

In this study, the artifact-corrected HRV signals are used to extract 35 features, including twelve time-domain, sixteen frequency-domain, and seven nonlinear features [12]. The list of features is mentioned in Tables 2, 3, and 4. Feature selection (FS) is a process of choosing more relevant features and deletes irrelevant ones. FS enhances precision and reduces timing complexity. FS is one of the essential machine learning (ML) tasks due to the continual increase of datasets in recent years; it is vital to select the relevant features rather than grow all features. In machine learning, feature selection is a crucial pre-processing step. This research considers feature selection based on mutual information with a correlation coefficient (CCMI) to remove more duplicate data from the assessment criterion. In information theory, mutual information is a metric for how much information a random variable contains in the form of another random variable. Mutual information is sometimes defined as lowering the uncertainty of the original random variable when another random variable is known [13].

Table 2 Mean and SD values of time-domain HRV features for all classes

HRV feature	NSR	SCD	CHF	AF	CAD	VT
Mean RR (ms)	742.20 ± 123.21	847.04 ± 263.66	768.54 ± 133.31	728.30 ± 163.45	812.49 ± 142.83	759.31 ± 382.06
SDNN (ms)	27.05 ± 7.66	85.03 ± 54.34	60.95 ± 18.67	98.43 ± 61.18	30.36 ± 14.25	87.31 ± 57.54
Mean HR (beats/min)	82.88 ± 12.72	77.80 ± 23.93	83.79 ± 17.17	86.75 ± 20.41	76.48 ± 15.72	93.57 ± 33.19
SD HR (beats/min)	3.03 ± 0.83	11.11 ± 9.45	8.33 ± 3.60	17.73 ± 18.86	2.98 ± 1.63	42.09 ± 139.66
Min HR (beats/min)	70.38 ± 10.72	63.40 ± 18.28	71.72 ± 15.34	66.65 ± 14.85	67.76 ± 14.96	61.46 ± 32.78
Max HR (beats/min)	97.76 ± 14.27	119.64 ± 56.60	113.12 ± 22.64	126.28 ± 40.98	87.86 ± 16.39	183.37 ± 91.46
RMSSD (ms)	28.07 ± 10.13	109.69 ± 72.73	81.41 ± 33.06	132.53 ± 85.38	31.64 ± 20.40	94.34 ± 60.50
NN50 (beats)	30.69 ± 20.86	144.98 ± 118.13	99.27 ± 69.22	219.02 ± 135.71	22.45 ± 21.55	109.02 ± 97.06
pNN50 (%)	7.99 ± 6.27	38.14 ± 25.50	25.28 ± 13.62	49.81 ± 27.38	6.13 ± 5.81	33.14 ± 24.57
RR triangular index	7.16 ± 1.73	11.41 ± 7.32	8.12 ± 1.50	16.60 ± 9.76	6.83 ± 2.56	11.25 ± 9.12
TINN (ms)	140.88 ± 47.36	443.27 ± 227.83	351.13 ± 114.19	476.81 ± 249.33	195.5 ± 107.68	451.63 ± 239.55
Stress index	16.26 ± 4.11	8.23 ± 5.17	11.54 ± 7.92	7.68 ± 5.67	14.55 ± 7.00	9.98 ± 9.39

Table 3 Mean and SD values of frequency-domain HRV features of all classes

HRV feature	NSR	SCD	CHF	AF	CAD	VT
VLF (Hz)	0.037 ± 0.005	0.035 ± 0.005	0.03 ± 0.006	0.035 ± 0.004	0.035 ± 0.004	0.034 ± 0.006
LF (Hz)	0.10 ± 0.02	0.09 ± 0.03	0.08 ± 0.03	0.09 ± 0.03	0.06 ± 0.02	0.06 ± 0.03
HF (Hz)	0.21 ± 0.07	0.26 ± 0.07	0.28 ± 0.07	0.25 ± 0.06	0.26 ± 0.08	0.22 ± 0.06
VLF (ms^2)	3.35 ± 11.06	341.73 ± 606.1	209.18 ± 20.38	242.16 ± 300.17	91.16 ± 86.79	1265.57 ± 2501.46
LF (ms^2)	372.60 ± 241.2	2826 ± 4800.9	1721.98 ± 156.79	2894.29 ± 4064.63	507.7 ± 616.89	3540.33 ± 5933
HF (ms^2)	319.08 ± 349.72	3857 ± 6770.54	2386.67 ± 344.07	6845.58 ± 15,903	207.02 ± 221.78	1648.93 ± 2440
VLF (log)	0.032 ± 0.90	4.23 ± 2.08	3.35 ± 1.33	4.67 ± 1.45	3.94 ± 1.25	5.05 ± 2.59
LF (log)	5.68 ± 0.75	6.65 ± 1.82	5.47 ± 1.56	6.90 ± 1.79	5.59 ± 1.28	6.34 ± 2.59
HF (log)	5.45 ± 0.76	7.11 ± 1.72	6.02 ± 1.57	7.50 ± 1.88	4.85 ± 1.06	5.97 ± 2.18
VLF (%)	0.43 ± 1.44	4.70 ± 3.73	5.43 ± 5.52	5.53 ± 7.02	12.56 ± 7.78	15.96 ± 11.90
LF (%)	54.59 ± 16.14	37.43 ± 12.67	35.47 ± 20.49	33.74 ± 11.43	57.11 ± 17.04	45.20 ± 16.40
HF (%)	44.91 ± 16.61	57.56 ± 14.70	58.31 ± 23.41	60.48 ± 15.43	30.23 ± 17.23	38.69 ± 23.08
LF (n.u.)	54.91 ± 16.52	36.64 ± 14.08	37.98 ± 23.13	36.38 ± 14.31	65.52 ± 18.38	55.82 ± 23.50
HF (n.u.)	45.02 ± 16.52	60.09 ± 13.91	61.20 ± 22.91	63.37 ± 14.26	34.36 ± 18.30	44.03 ± 23.39
Total power (ms^2)	695.50 ± 505.65	7032.2 ± 11,607.8	4318.2 ± 432.1	9999.14 ± 19,510	806.62 ± 813.69	6457.33 ± 9933
LF/HF ratio	1.62 ± 1.22	0.77 ± 0.52	0.89 ± 1.744	0.7 ± 0.61	2.933 ± 2.35	4.05 ± 11.82

Table 4 Mean and SD values of nonlinear method HRV features for all classes

HRV feature	NSR (n = 18)	SCD (n = 20)	CHF	AF	CAD	VT
SD1 (ms)	19.88 ± 7.18	77.68 ± 51.53	57.65 ± 23.40	93.83 ± 60.47	22.40 ± 14.44	66.86 ± 42.88
SD2 (ms)	32.48 ± 8.99	90.09 ± 59.42	62.46 ± 14.07	101.76 ± 63.56	35.51 ± 16.77	101.85 ± 71.89
SD2/SD1	1.68 ± 0.35	1.17 ± 0.40	1.08 ± 0.49	1.17 ± 0.39	1.77 ± 0.64	1.58 ± 0.58
Approximate entropy (ApEn)	1.18 ± 0.09	1.00 ± 0.21	1.01 ± 0.22	1.07 ± 0.30	1.14 ± 0.12	0.74 ± 0.31
Sample entropy (SampEn)	1.74 ± 0.22	1.21 ± 0.56	1.20 ± 0.46	1.43 ± 0.62	1.55 ± 0.30	0.79 ± 0.48
Short-term fluctuations, alpha 1	1.04 ± 0.20	0.70 ± 0.21	0.62 ± 0.30	0.69 ± 0.21	1.01 ± 0.35	0.85 ± 0.31
Long-term fluctuations, alpha 2	0.14 ± 0.09	0.35 ± 0.18	0.37 ± 0.19	0.37 ± 0.16	0.47 ± 0.13	0.52 ± 0.23

2.4 Classification

The features derived from HRV signals were further reduced using feature selection tests, which improve classification accuracy with a reduced number of features. The classification was based on ten features out of 35, demonstrating a considerable variance in features across situations. This Light Gradient Boosting Machine (LightGBM) algorithm is used for disease classification because it has the following benefits: fast training, fast computational speed, excellent ability to minimize overfitting problems, suitability even for large datasets with less training time, low memory usage, produces better accuracy than other boosting methods, and higher efficiency. LightGBM can be used for classification, feature ranking, regression, and machine learning tasks [14]. LightGBM successfully classified classes of NSR, SCD, CHF, AF, CAD, and VT with an accuracy of 95.52%.

3 Results

The experimental results of HRV feature extraction from six classes from HRV signals with 892 segments are mentioned in Tables 2, 3, and 4. These tables indicate the values of each feature in terms of mean $\pm$ SD.

Table 2 lists the time-domain features extracted from various HRV signals. Six of the twelve time-domain features are chosen for classification using the mutual information 6 feature selection algorithm. Max HR (beats/min), Min HR (beats/min), TINN (ms), RMSSD (ms), pNN50 (%), and RR triangular index are the features used for classification. The box plot representation of all significant time-domain features is mentioned in Fig. 2.

Peak frequencies, absolute powers, the natural logarithm of absolute powers, relative powers, total power, and the LF/HF ratio were among the sixteen frequency-domain HRV variables extracted from HRV signals. VLF (ms^2) and VLF (%) were included in the feature selection process for classification purposes. The box plot representation of two significant features is listed in Fig. 2.

Seven features were extracted from HRV signals using nonlinear methods, and each feature's mean and SD values are listed in Table 4. Nonlinear HRV features, such as SD1 and SD2, were chosen for classification throughout the feature selection procedure. In the instance of NSR, there were substantial differences in mean and SD values compared to other classes. Box plot representation of selected nonlinear features is mentioned in Fig. 2.

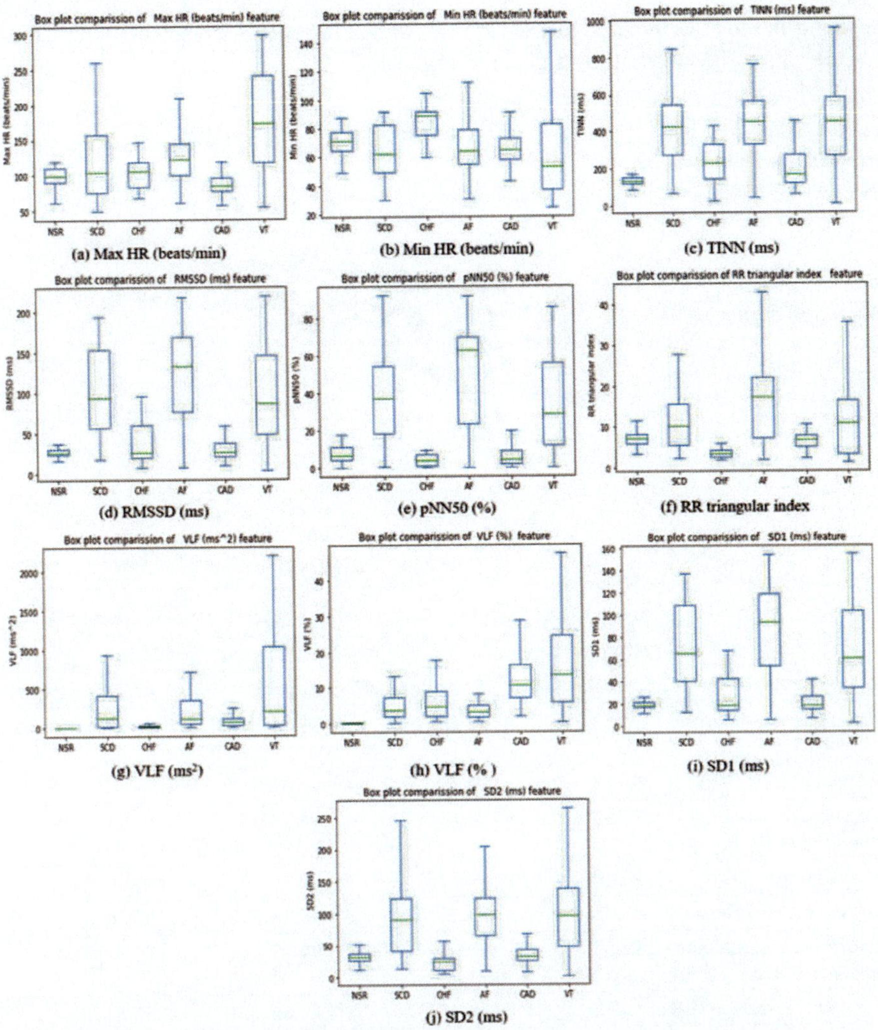

Fig. 2 Box plot representation of selected features **a** Max HR **b** Min HR **c** TINN **d** RMSSD **e** pNN50 **f** RR triangular index **g** VLF (%) **h** VLF (ms^2) **i** SD1 **j** SD2

3.1 Feature Selection and Classification Results

In order to determine which strategy provided the best HRV features for reliable SCD prediction findings, the authors used four feature selection approaches and a one-dimensionality reduction methodology (PCA) [15]. From the FS analysis of Fig. 3, it was clear that mutual information technology has produced better results with ten features. The classification results with selected features using different machine learning-based classification models are mentioned in Table 5. The confusion matrix for the LightGBM algorithm is given in Fig. 4.

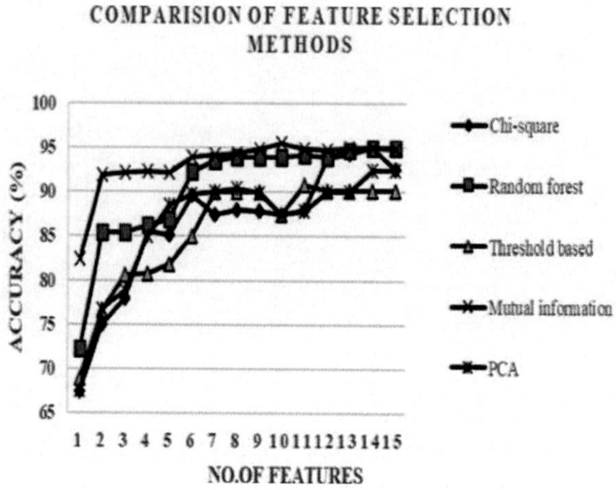

Fig. 3 Graphical representation of the performance of all FS methods with accuracies

Table 5 Comparison of performance evolution for different classification algorithms

Classification model	Accuracy (%)	Sensitivity (%)	Specificity (%)	Precision (%)
Light gradient boosting (LightGBM)	95.52	94.96	99.08	95.87
Random forest (RF)	94.63	94.50	98.49	94.83
Gradient boosting classifier (GBC)	93.95	93.65	98.78	94.21
Decision tree (DT)	90.80	90.21	97.29	91.16
K-Nearest neighbors	82.88	83.53	96.46	83.56
Quadratic discriminant analysis (QDA)	76.55	73.62	94.50	76.22
Logistic regression (LR)	68.12	66.65	93.64	66.34
Gaussian Naive Bayes (NB)	64.40	63.35	92.62	64.34
Linear discriminant analysis (LDA)	57.38	56.18	91.39	58.52
AdaBoost classifier	53.69	51.66	90.17	57.27
Support vector machine (SVM)	44.96	42.39	88.88	48.57

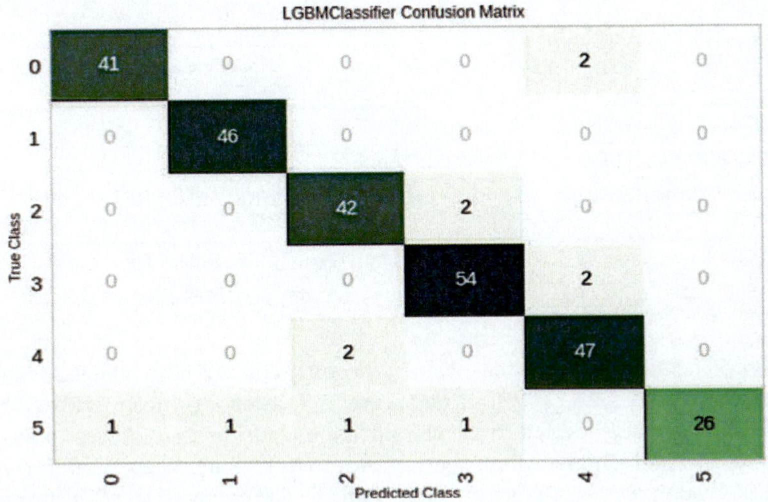

Fig. 4 Confusion matrix for LightGBM algorithm

4 Discussion

This research proposes a novel approach to detecting sudden cardiac death which is proposed in this study. The HRV signal was used to extract linear and nonlinear features. Following feature selection, which reduces the dimension of the feature space, LightGBM and RF algorithms were employed to identify healthy individuals and those at risk of SCD. In this study, six classes were examined to evaluate the risk of SCD in various cardiac scenarios. SCD is brought on by several cardiac conditions, the most frequent of which are VF and CAD. Several studies have connected CHF and AF to the development of SCD episodes. This is the first study examining six classes to identify SCD in normal and diseased circumstances. However, our research has demonstrated that a significant SCD prediction algorithm may be constructed using appropriate pre-processing, feature extraction, feature selection, and prediction methods. In comparison to prior studies, this research was able to enhance prediction time. It is appropriate that a 5-min HRV signal duration was used in this study because it is one of the most widely utilized durations in this field. After selecting the 5-min HRV signal, it has been preprocessed to remove artifacts that were present in the signal using median filtering and threshold value which is mentioned in Fig. 1. Thirty-five features were extracted from corrected HRV signals of all six classes which are reduced using the best feature selection method. Implementing a better feature selection strategy is one of the reasons for this advancement. Here in this research, five feature selection algorithms, such as Chi-square, threshold-based, random forest Random Forest-based, mutual information-based, and one-dimensionality reduction (PCA) techniques, were used to identify the best combination of features to provide more significant results. The mutual information-based algorithm has shown the best

Table 6 Comparison of the proposed method with the state-of-the-art

Author (year)	Type of signal	Prediction time (VF onset)	No. of classes	Accuracy (%)
Khazaei [16]	HRV	6 min before	NSR, SCD	95
Devi et al. [8]	HRV	10 min before	NSR, SCD, CHF	83.33
Rohila and Sharma [2]	HRV	1 h before	NSR, SCD, CHF, CAD	91.67
Present study	HRV	8 min before	NSR, SCD, CHF, AF, CAD, VT	95.52

selection of features that provides accurate results with ten best-selected features, and we received 95.52% classification accuracy. The ten significant HRV features of 892 HRV segments are fed to the classification algorithm, and it was found that the LightGBM and RF algorithms were shown to have better classification results. The classification results are mentioned in Table 5. The comparison of the proposed method with state-of-the-art is mentioned in Table 6.

5 Conclusion

The current work proposes a novel approach for determining SCD risk by comparing HRV in normal and diseased cardiac circumstances. Based on current evidence, the experiment results revealed a substantial difference in HRV between the SCD and non-SCD groups. This shows that SCD risk can be detected while other cardiac abnormalities like CHF, AF, CAD, and VT can be ruled out. Linear (time- and frequency-domain) and nonlinear features were extracted from HRV signals, and mutual information-based feature selection was used to find the best combination of features for classification. Finally, the proposed method classifies normal and abnormal conditions with 95.52%, 94.96%, 99.08%, and 95.87% accuracy, sensitivity, specificity, and precision. By predicting SCD at an early stage, the proposed method can be incorporated into automatic diagnostic equipment and continuous monitoring systems to improve survival rates. The current study can be expanded in the future to find or generate hybrid traits that can be utilized to indicate a patient's risk level.

References

1. Kaptoge S et al (2019) World health organization cardiovascular disease risk charts: revised models to estimate risk in 21 global regions. Lancet Glob Health 7(10):e1332–e1345. https://doi.org/10.1016/S2214-109X(19)30318-3
2. Rohila A, Sharma A (2020) Detection of sudden cardiac death by a comparative study of heart rate variability in normal and abnormal heart conditions. Biocybernetics Biomed Eng 40(3):1140–1154. https://doi.org/10.1016/j.bbe.2020.06.003
3. Vähätalo J et al (2021) Coronary artery disease as the cause of sudden cardiac death among victims < 50 years of age. Am J Cardiol 147:33–38. https://doi.org/10.1016/j.amjcard.2021.02.012
4. Waldmann V et al (2020) Association between atrial fibrillation and sudden cardiac death: pathophysiological and epidemiological insights. Circ Res 127(2):301–309. https://doi.org/10.1161/CIRCRESAHA.120.316756
5. Vandenberg JI, Perry MD, Hill AP (2017) Recent advances in understanding and prevention of sudden cardiac death. F1000Research 6:1–7. https://doi.org/10.12688/f1000research.11855.1
6. Hasselqvist-Ax I et al (2015) Early cardiopulmonary resuscitation in out-of-hospital cardiac arrest. N Engl J Med 372(24):2307–2315. https://doi.org/10.1056/nejmoa1405796
7. Parsi A, O'Loughlin D, Glavin M, Jones E (2020) Prediction of sudden cardiac death in implantable cardioverter defibrillators: a review and comparative study of heart rate variability features. IEEE Rev Biomed Eng 13(1):5–16. https://doi.org/10.1109/RBME.2019.2912313
8. Devi R, Tyagi HK, Kumar D (2019) A novel multi-class approach for early-stage prediction of sudden cardiac death. Biocybernetics Biomed Eng 39(3):586–598. https://doi.org/10.1016/j.bbe.2019.05.011
9. Acharya UR, Joseph KP, Kannathal N, Lim CM, Suri JS (2006) Heart rate variability: a review. Med Biol Eng Comput 44(12):1031–1051. https://doi.org/10.1007/s11517-006-0119-0
10. Robinson BF, Epstein SE, Beiser GD, Braunwald E (1966) Control of heart rate by the autonomic nervous system. Studies in man on the interrelation between baroreceptor mechanisms and exercise. Circ Res 19(2):400–411. https://doi.org/10.1161/01.RES.19.2.400
11. Holstila E, Vallittu A, Ranto S, Lahti T, Manninen A (2016) Helsinki cities as engines sustain. Compet Eur Urban Policy Pract 175–189. https://doi.org/10.4324/9781315572093-15
12. Rajendra Acharya U, Suri JS, Spaan JAE, Krishnan SM (2007) Advances in cardiac signal processing. Springer
13. Hoque N, Bhattacharyya DK, Kalita JK (2014) MIFS-ND: a mutual information-based feature selection method. Expert Syst Appl 41(14):6371–6385. https://doi.org/10.1016/j.eswa.2014.04.019.acharya
14. Ke G et al (2017) LightGBM: a highly efficient gradient boosting decision tree. Adv Neural Inf Proc Syst 2017-Decem(Nips):3147–3155
15. Araki T et al (2016) PCA-based polling strategy in machine learning framework for coronary artery disease risk assessment in intravascular ultrasound: a link between carotid and coronary grayscale plaque morphology. Comput Methods Programs Biomed 128:137–158. https://doi.org/10.1016/j.cmpb.2016.02.004
16. Khazaei M, Raeisi K, Goshvarpour A, Ahmadzadeh M (2018) Early detection of sudden cardiac death using nonlinear analysis of heart rate variability. Biocybernetics Biomed Eng 38(4):931–940. https://doi.org/10.1016/j.bbe.2018.06.003

Design and Analysis of a Multiplexer Using Domino CMOS Logic

Shaivya Shukla, Onika Parmar, Amit Singh Rajput, and Zeesha Mishra

Abstract In this study, a 2 × 1-multiplexer based on a domino logic circuit is given using three different technologies: 180 nm, 120 nm, and 90 nm. A static 2 × 1 multiplexer is compared to the proposed circuit. Digital Schematic Circuit Designing Software (DSCH2) and Microwind 3.8 were used for the schematic and stimulation, respectively. The power delay product for domino logic is 75% less than static logic in 180 nm technology, 39.28% less in 120 nm technology, and 39.33% less in 90 nm technology. As a result, the suggested 2 × 1 multiplexer with domino logic circuit outperforms static logic in terms of power dissipation and time.

Keywords Multiplexer · Static logic circuit · Domino logic circuit · Power dissipation · Power delay product

1 Introduction

The power consumption and performance of current temporal integrated circuits are critical parameters. All electronic systems are designed to be high-performing, space-efficient, and power-efficient. A multiplexer is one such electronic gadget. A multiplexer is an electrical circuit that acts as a data selector, allowing several analogue and digital input signals to be selected and forwarded to the output line. It takes 2n inputs, where n is the number of selection lines, and outputs a single line. A 2 × 1 multiplexer has two input lines and one selection line, with a single output based on the select signal line [1, 2]. A 2 × 1 multiplexers output equation is as

S. Shukla · O. Parmar (✉) · A. S. Rajput · Z. Mishra
Department of Microelectronics and VLSI, UTD, Chhattisgarh Swami Vivekananda Technical University, Newai, Bhilai, India
e-mail: faculty3_vlsi@csvtu.ac.in

A. S. Rajput
e-mail: faculty1_vlsi@csvtu.ac.in

Z. Mishra
e-mail: faculty2_vlsi@csvtu.ac.in

S. Sharma et al. (eds.), *Intelligent Control, Robotics, and Industrial Automation*, Lecture Notes in Electrical Engineering 1066,
https://doi.org/10.1007/978-981-99-4634-1_54

follows:

$$S'A + SB = Out. \tag{1}$$

Block diagram of multiplexer is depicted in Fig. 1, and it can be put together using logic diagram with two AND gates and one OR gate as shown in Fig. 2.

The truth tabled follower by multiplexer is shown in Table 1:

Different types of Complementary Metal Oxide Semiconductor (CMOS) logics can be used to create a multiplexer, and two such logics are:

(1) Static CMOS logic.
(2) Domino CMOS logic.

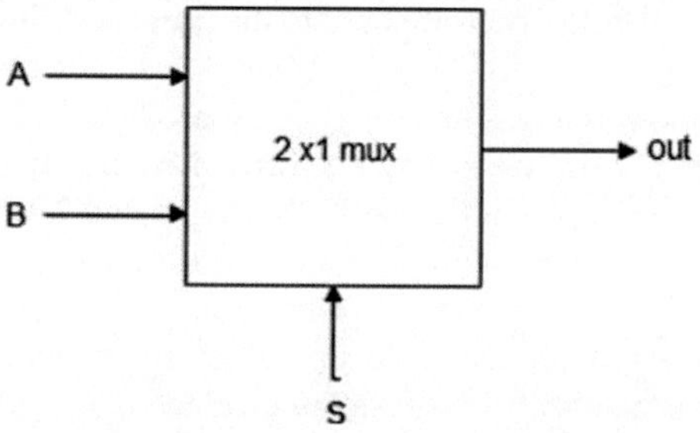

Fig. 1 Block diagram of 2 × 1 multiplexer

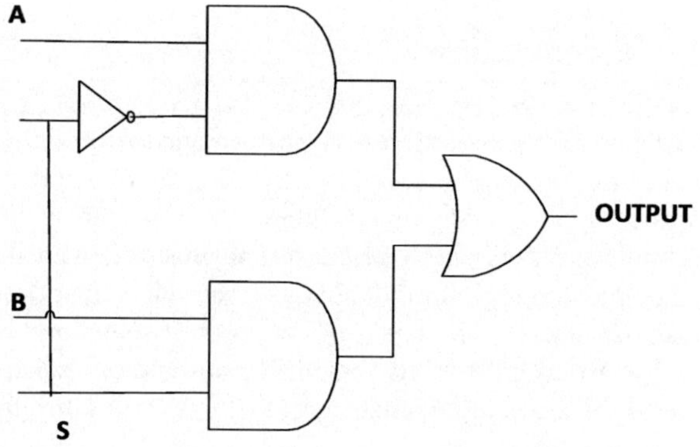

Fig. 2 Logic diagram of 2 × 1 multiplexer

Table 1 Truth table of multiplexer

Select line (S)	Input 1 (A)	Input 2 (B)	Output
0	0	0	0
0	1	0	1
1	0	1	1
1	1	1	1

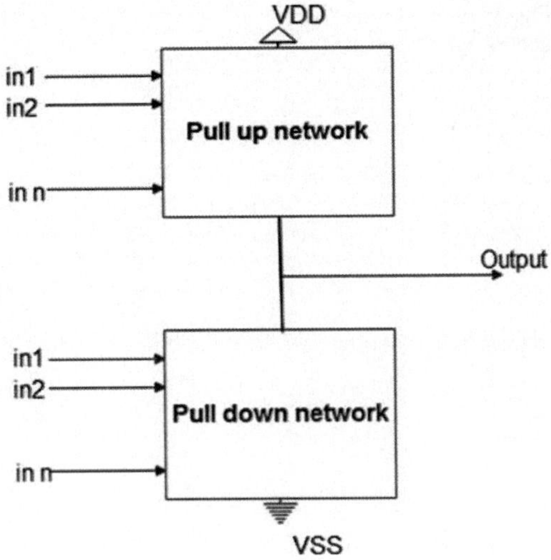

Fig. 3 Block diagram of static CMOS logic

1.1 Static CMOS Logic

The pull-up and pull-down networks are the two networks that make up static CMOS logic. When the output of a logic gate is expected to be "1," the pull-up network consists of *p* channel Metal Oxide Semiconductor (PMOS) placed together to form the connection between VDD and the output. When the output is supposed to be "0," the pull-down network consists of all the *n* channel MOS (NMOS) organised together to provide a connection between output and V_{SS}. The diagram in Fig. 3 depicts the arrangement discussed above. The static logic circuit also has good noise and leakage resistance [3].

However, the disadvantage of employing static logic is that for every *n* number of fans in a gate, 2*n* transistors are employed, resulting in a larger space requirement for static logic. This has an effect on the gate's speed.

1.2 Domino CMOS Logic

Dynamic logic provides an alternative for static logic. But, it has several disadvantages of more noise and power consumption compared to static logic. So, domino logic circuit provides a solution to this [4]. An n-type dynamic logic block is followed by a static inverter in domino CMOS logic. A dynamic logic block differs from a static logic block in that it implements a clock signal. During precharge, the n-type dynamic gate's output is charged to V_{DD} and the inverter's output is set to 0. The dynamic gate conditionally discharges during evaluation, and the inverter's output conditionally transitions from 0 to 1. The introduction of the static inverter has the

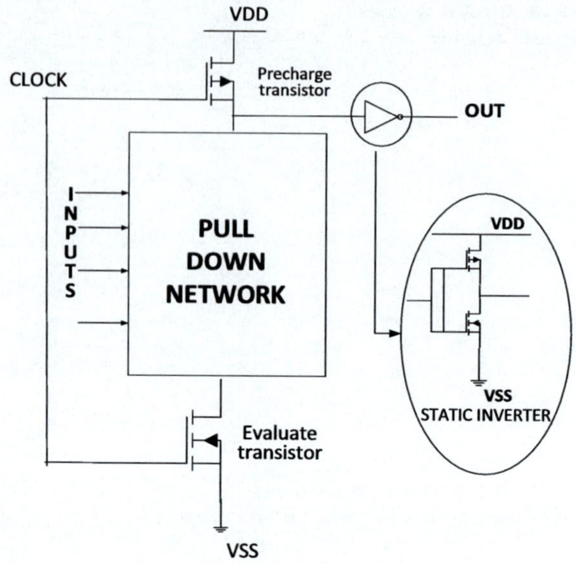

Fig. 4 Block diagram of domino CMOS logic

benefit of driving the fan out of the gate with a low impedance output, which improves noise immunity. By separating internal and load capacitance, the buffer reduces the capacitance of the dynamic output even more [5]. The block diagram of the domino CMOS logic is exhibited in Fig. 4.

The domino was invented to solve the monotonicity issue that dynamic logic circuits have. A single-phase clock system is used in the first stage of domino logic, which eliminates static power consumption. These logic circuits are glitch-free, have a rapid switching threshold, and can be cascaded.

The main disadvantage of employing domino logic circuit is that because each dynamic gate includes a static inverter, it can only be used to build non-inverting logic [6]. So this work presents a comparison between the static and domino designed multiplexers on the bases of number of transistor required, layout area and power delay. Thus, it is summed up as domino logic is a better alternative to static logic.

The paper ahead is divided into different sections previous work on the topic is discussed, then designing of multiplexer using both the logics, thereafter the result are analysed, and the work is concluded.

2 Previous Work

Domino CMOS logic refers to the combination of dynamic and static circuits. Patel C. et al. concluded in their literature that dynamic logic proves to be a better alternative of static logic. For a comparator dynamic logic gives less power delay product [7]. Domino logic is the next step in the growth of dynamic logic. A dynamic circuit is followed by a static inverter that allows diverse circuits to cascade. It was created to increase the speed of a dynamic circuit. Previous research has been done to demonstrate the benefits of domino logic [8]. The performance of the AND gate in static and domino logics is compared, and it is concluded that domino logic was created to speed up dynamic logic circuits by having a lower area than traditional CMOS. All other logic gates are also compared between static and domino logics, with the work concluding that the developed logic gate employing domino logic is superior. It provides smoother transitions and eliminates glitches. It results in a smoother transition and error-free functioning [3].

Skyler Weave and colleagues used domino logic in an Analogue-to-Digital Converter (ADC) and discovered that it is a promising option for a highly scalable and synthesizable ADC. Even in deep sub-micrometre processes, the developed ADC has the advantage of maintaining a very low design cost [9]. Using 0.25 m technology, a performance analysis of the comparator in both CMOS and high-speed domino logic is performed. Some of the comparing parameters are frequency, transistor requirement, voltage scaling, and technology. Domino logic uses less energy than CMOS logic. In addition, the maximum frequency supported by CMOS was supposed to be 1 GHz, but domino's maximum frequency was discovered to be 100 MHz Overall, it was determined that domino's logic is superior in terms of power conservation [10].

Augraha Rose V. et al. created a two-by-one multiplexer employing a variety of CMOS logic families and analysed their performance. Finally, it was deduced that domino logic is the most efficient of all because it has lower average power consumption and Power Delay Product (PDP) than all other logic families, as well as a shorter propagation time. As a result, it may be inferred that the domino logic family outperforms all other logic families [11]. In [12], the results shows that standard domino logic circuits have improvement in average speed with respect to standard static CMOS logic.

3 Designing of Multiplexer Using Static Logic

4 PMOS and 4 NMOS are required when the 2×1 multiplexer is developed using static CMOS as illustrated in Fig. 5. In order to provide input and output, two more NMOS and PMOS are necessary. As a result, 12 transistors were used in the design. One of the inputs will be mirrored in the output based on the selected input. The circuit's schematic is depicted in Fig. 5.

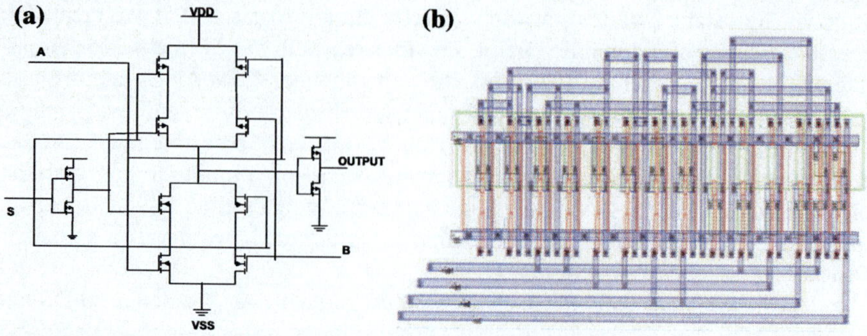

Fig. 5 **a** Schematic diagram of multiplexer using static CMOS logic **b** Layout diagram of multiplexer using static CMOS logic

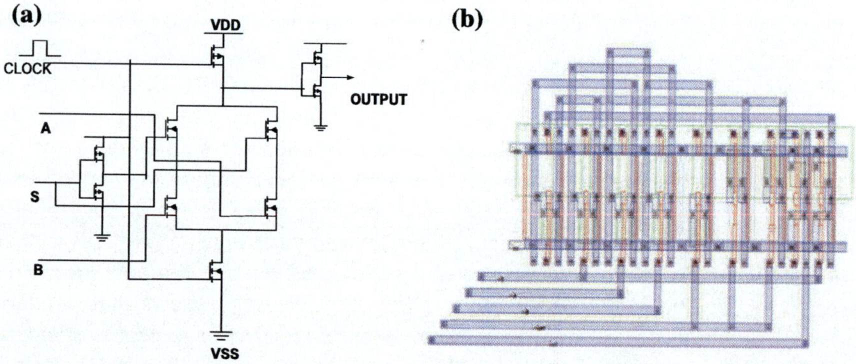

Fig. 6 **a** Schematic diagram of multiplexer using domino CMOS logic **b** Layout of multiplexer using domino CMOS logic

4 Designing of Multiplexer Using Domino Logic

The domino CMOS logic-based 2 × 1 multiplexer is shown in Fig. 6. The circuit requires 4 NMOS and 1 PMOS transistors, with each input and output inverter requiring 2 NMOS and 2 PMOS transistors. Depending on the selection line and the clock signal provided, the output will be either of the inputs.

5 Results and Discussion

Microwind 3.8 software was used to simulate the above circuit. The stimulation temperature was kept constant at room temperature. The Verilog file created by the DSCH tool is used to create the layout. At the same frequency, simulations were

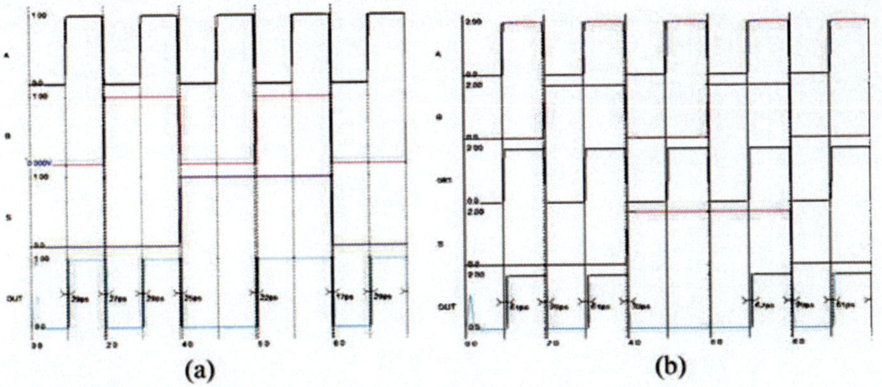

Fig. 7 **a** Output waveforms for the static logic 180 nm technology **b** Output waveform for the domino logic 180 nm technology

run on three different technologies: 180 nm, 120 nm, and 90 nm [13]. The following is the outcome of a circuit simulation in both static and domino logics for 180 nm technology:

It may be deduced from Figs. 7a and b that the output is consistent with the expected result of a 2 × 1 multiplexer for 180 nm technology. When signal *S* is logic low, the output of a multiplexer utilising static logic behaves like input A. When *S* is logic high, it i the same as *B*. When a multiplexer is built with domino logic, the output is identical to static logic except for the clock input. Using 120 nm and 90 nm technologies, similar results can be attained.

Figures 8a and b demonstrate the results of circuit simulations in both static and domino logics for 120 nm technology. According to the multiplexers input, the resultant output is as desired. The difference between the outputs of static and domino logics may also be seen because the output of domino logic is affected not only by the applied input but also by the clock.

Figures 9a and b depict circuit simulations for 90 nm technology in both static and domino logics. The resulted output is in accordance with the output equation of the multiplexers, i.e. we get output as applied. In a static logic circuit, an A input is used when the choose line is logic low, and a B input is used when the applied signal is logic high. In contrast, the output of a domino logic circuit is determined by the selection lines as well as the applied clock signal. That is, regardless of the signal and input used, the output is logic low for a low clock input. When the clock is logic high, it is determined by the signal and inputs used.

The following parameters are used to compare the static and domino logic circuits for three different technologies based on the preceding simulation results:

1. *Area.*
2. *Dissipation of power.*
3. *Delay.*
4. *Power delay product*

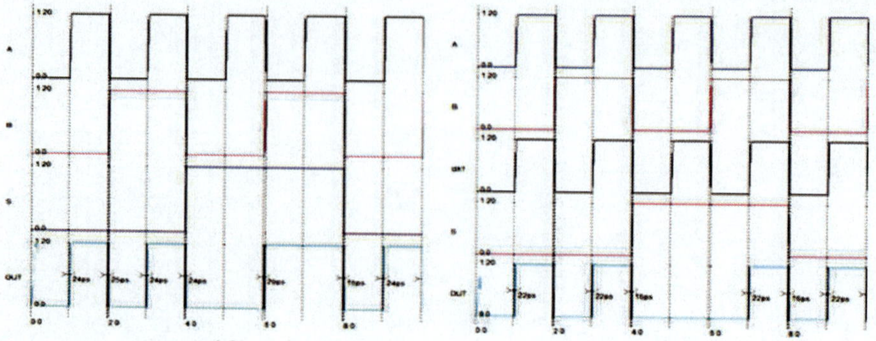

(a) (b)

Fig. 8 **a** Output waveform for static logic 120 nm technology **b** Output waveform for domino logic 120 nm technology

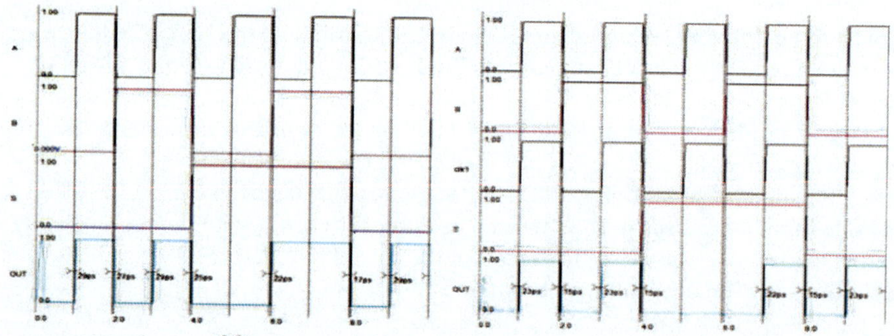

(a) (b)

Fig. 9 **a** Output waveform for static logic 90 nm technology **b** Output waveform for domino logic 90 nm technology

These variables are linked to the three most crucial components of a VLSI circuit. The requirement is to develop circuits to use less power, take up less area, and operate with less latency. As a result, we must make a trade-off between circuit area, power, and performance.

5.1 *Area (A)*

In VLSI circuits, area is one of the most important parameters to consider. The circuit is more useful if the area used is smaller. The product of the layout's height and breadth is used to compute the layout's area.

5.2 Power Dissipation (PD)

The power dissipation is calculated as the sum of static power loss (power loss due to leakage current) and the dynamic power loss (power loss during switching). The SI unit of power dissipation is in Watt (W).

$$\text{PD} = P_{\text{static}} + P_{\text{dynamic}}. \tag{2}$$

5.3 Propagation Delay (td)

Propagation delay is the time it takes for a signal to reach its final destination, and it is one of the most critical circuit performance matrices. Delay is defined as the time it takes for the output to change after the input is modified in our logic circuit propagation. The propagation delay is expressed in terms of rise and fall times as follows:

$$td = \frac{tr + tf}{2}. \tag{3}$$

The circuit's rise time is "*tr*," while its fall time is "*tf*."

5.4 Power Delay Product (PDP)

The power delay product, also known as switching energy, is a measurement of the energy expended in a CMOS circuit per switching operation. It is the total of the power consumption and the delay between the input and output.

$$\text{PDP} = \text{PD} * td. \tag{4}$$

We achieve a very profitable trade-off between the area delay and power of our planned circuit when we modify our circuit from static CMOS logic to domino logic. In every way, the developed circuit outperforms the prior static circuit.

In 180 nm technology, the percentage decrease in the power delay product of static and domino logics is 75%. The reduction in percentage in 120 nm technology is 39.28%. The reduction in percentage for 90 nm technology is 39.33%. For each technology employed, the layout area shrinks from static to domino logic.

The graph shown in Fig. 10 depicts the comparison of area for 180 nm, 120 nm, and 90 nm technology on static and domino logics:

In 180 nm technology, the percentage decrease in layout area from static to domino is 11.57%. It is 30.43% for 120 nm technology. And for 90 nm, it is 18.80%.

Fig. 10 Comparison of area

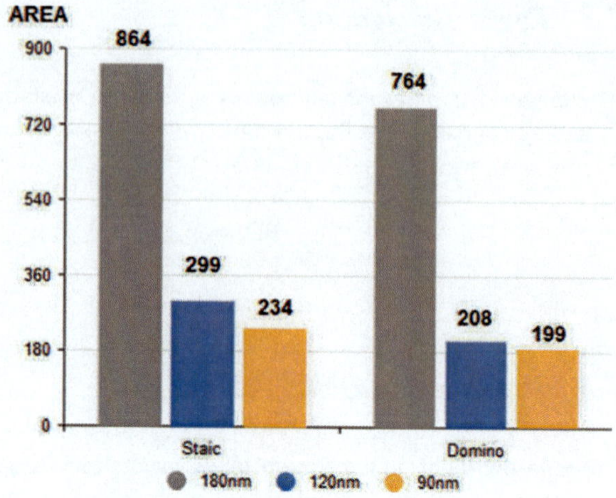

On static and domino logics, the comparison of power for 180 nm, 120 nm, and 90 nm technology is depicted in Fig. 11.

The graph in Fig. 12 depicts the comparison of delay for 180 nm, 120 nm, and 90 nm technology on static and domino logics.

The comparison of power delay product for 180 nm, 120 nm, 90 nm technology on static and domino logics is drawn as graph in Fig. 13.

Comparison of percentage decrease in power delay product for 180 nm, 120 nm, and 90 nm technology is shown in Fig. 14.

The comparison table between the needed metrics that is area, power, delay, and power delay product for several technologies that are 180 nm, 120 nm, and 90 nm is tabulated below. And it is discovered that domino logic outperformed our prior and commonly used static logic circuit in a variety of criteria.

Fig. 11 Comparison of power dissipation

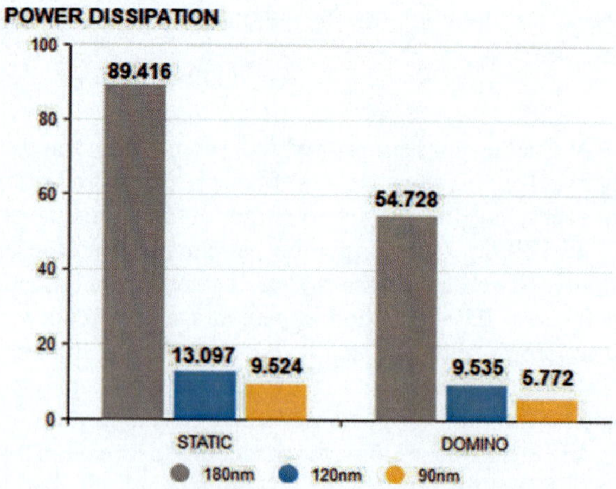

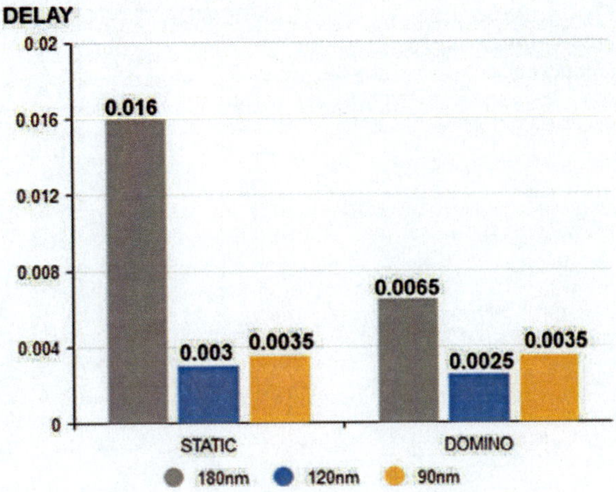

Fig. 12 Comparison of delay

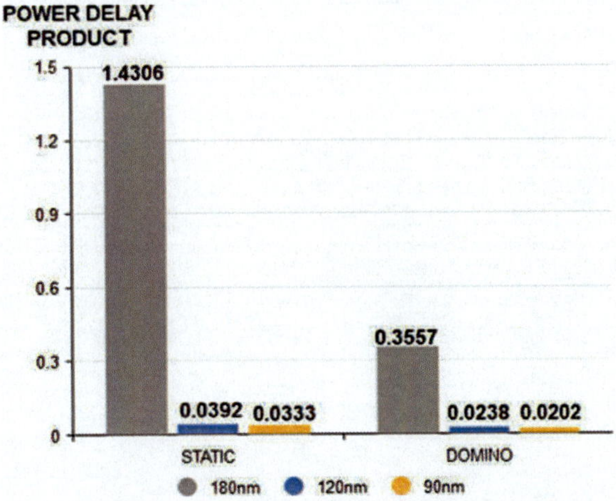

Fig. 13 Comparison of power delay product

The usage of domino logic minimises power consumption, latency, and the power delay product, as shown in Table 2. And as the technology's power delay diminishes, so does the product's power delay.

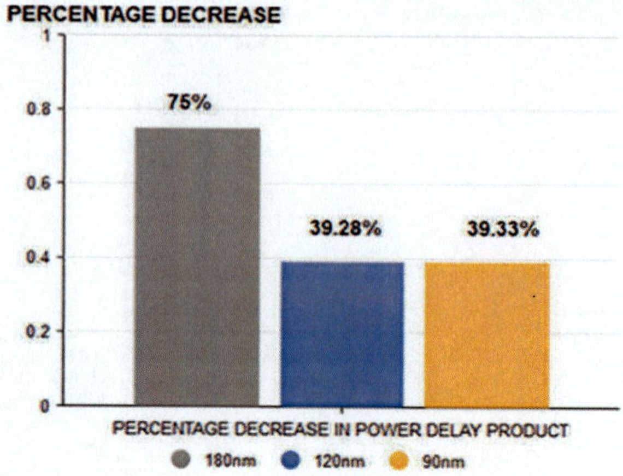

Fig. 14 Power delay product from static to domino logic

Table 2 Comparison between static and domino logic circuit parameter

Parameters	Static CMOS logic			Domino CMOS logic		
	180 nm	120 nm	90 nm	180 nm	120 nm	90 nm
Number of transistor	12	12	12	10	10	10
Area of layout	864	299	234	764	208	190
Power dissipation (μW)	89.416	13.097	9.524	54.728	9.535	5.772
Rise time (ns)	0.009	0.004	0.005	0.007	0.004	0.005
Fall time (ns)	0.007	0.002	0.002	0.006	0.001	0.002
Delay (ns)	0.016	0.003	0.0035	0.0065	0.0025	0.0035
Power delay product	1.4306	0.0392	0.0333	0.3557	0.0238	0.0202

6 Conclusion

In this study, a 2 × 1 multiplexer is developed in 180 nm, 120 nm, and 90 nm technology using static and domino logic CMOS circuits, and its performance metrics are examined. It is evident that a two-by-one multiplexer based on domino logic is more efficient. Multiplexers are used in a variety of fields, including communication systems, computer memory, satellite transmission computer systems, and telephone networks. As a result of the study presented in this work, a multiplexer based on domino logic can be employed in a variety of applications to improve performance.

References

1. Padmaja M, Prakash VNVS (2012) Design of multiplexer in multiple logic style for low power VLSI. Int J Comput Trends Technol 3(3):467–471

2. Anand KA. Fundamental of digital circuits, PHI learning private limited pp 390–393
3. Sharma A, Rao D, Mohan R (2016) Design and implementation of domino logic circuit in CMOS. J Netw Commun Emerg Technol (JNCET) 6. www.jncet.org
4. Meher P, Mahapatro KK (2013) A low power circuit technique for domino CMOS logic. IEEE, International conference on intelligent system and signal processing (ISSP)
5. Weste NHE, Harris DM (2000) CMOS VLSI design a circuit and system perspective pears on education (Asia) pvt.ltd., pp 328–342
6. Kang S-M, Leblebici Y. CMOS digital integrated circuits. McGraw hill higher education, pp 378–388
7. Patel C, Dr. Veena CS (2014) Low power comparator design base on CMOS dynamic logic circuit. In: 2nd international conference on emerging technology trends in electronics, communication and networking
8. Verma PK, Singh SK, Kumar A, Singh S (June 2012) Design and analysis of logic gate using static and domino logic technique. Int J Sci Technol Res 1(5)
9. Weaver S, Hershberg B, Maghari N, Moon U-K (Nov 2011) Domino logic based ADC for digital synthesis. IEEE Trans Circ Syst-II Express Briefs 58(11)
10. Rangari AV, Gaidhani YA (2016) Design of comparator using domino logic and CMOS logic. IEEE online international conference on green engineering and technology (IC-GET)
11. Anugraha RV, Durga DS, Avadaiammam R (19–20 May, 2017) Design and performance analysis of 2:1 multiplexer using multiple logic families at 180 nm technology. In: 2017 2nd IEEE international conference on recent trends in electronics information & communication technology (RTEICT), India
12. Thorp TJ, Yee GS, Sechen CM (Feb 2003) Design and synthesis of dynamic circuit. IEEE Trans VLSI Syst II(1)
13. Jaiswal MG, Bendre V, Sharma V (Oct 2017) Netlist optimization for CMOS place and route in microwind. Int Res J Eng Technol (IRJET) 4(10)

Single-Phase Grid-Connected 5-Level Switched Capacitor Inverter Using PLECS Tool

Khaja Izharuddin, Kowstubha Palle, A. Bhanuchandar, and Gumalapuram Gopal

Abstract In this paper, a 5-level Switched Capacitor (SC)-based grid-connected inverter (GCI) using Piecewise Linear Electrical Circuit Simulation (PLECS) tool is presented. This topology consists of six switches, 1 diode, 1 switched capacitor, and one single DC source. The SC is self-balanced based on the charging and discharging principle. This technique eliminates the usage of sensors and other additional circuitry for balancing the SC. The 1-phase grid-connected system has been subjected to a dq-frame current control technique. By incorporating an Inductor-Capacitor-Inductor (LCL) filter between the inverter and grid, a greater level of ripple attenuation capacity and active damping is achieved in the grid connection. In this control, active power is injected into the grid using the d-axis reference current, and finally, unity power factor (UPF) operation is attained by assuming that the q-axis current reference as zero. The simulation is carried out using PLECS Tool.

Keywords Switched capacitor · Grid-connected inverter · Multilevel inverter · Unity power factor

1 Introduction

In the applications of electric vehicles and grid-tied systems, the conversion of power from Direct Current (DC) to Alternating Current (AC) plays a vital role. Multilevel inverters (MLIs) are used to synthesize a staircase waveform imitating a "sine wave" [1]. The advantages of MLIs are good Total Harmonic Distortion (THD), reduced

K. Izharuddin (✉) · K. Palle
Department of Electrical and Electronics Engineering, CBIT (A), Gandipet, Hyderabad, Telangana, India
e-mail: izhar.khaja@gmail.com

A. Bhanuchandar
Electrical Department, NIT Warangal, Warangal, Telangana, India

G. Gopal
EEE Dept, MGIT, Hyderabad, Telangana, India

S. Sharma et al. (eds.), *Intelligent Control, Robotics, and Industrial Automation*, Lecture Notes in Electrical Engineering 1066,
https://doi.org/10.1007/978-981-99-4634-1_55

electromagnetic interference, decreased dv/dt stresses, small size filter requirements, etc. Voltage gain greater than unity and the self-balancing feature of capacitors are not available in the fundamental MLI topologies (such as NPC [2], FC [3], and CHB [4]). For any converter/inverter selection of number, switches play a vital role as it contributes to the efficiency of the system. The number of switches is often critical for generating a specified amount of output since each switch requires a gate driver with control and protection circuits. Depending on Reduced Switch Count (RSC) and DC voltage sources, numerous RSC-MLI topologies were introduced. To generate a 5-level output voltage with two DC sources, the no of unipolar switches required is 8, and at the same time, the blocking/standing voltage requirement is also more. This is evidenced through existing literature survey for different types of MLI topologies like multilevel DC-link inverter (MLDCL) [5], Switched Series/Parallel Source Inverter (SSPS) [6], Reverse Voltage (RV) [7], Series Connected Switched Sources (SCSS) [8], Multilevel Module (MLM) [9], Crisscross Switched Cells [10], Unity-Gain Active Neutral-Point-Clamped UG-ANPC [21].

Integration of photovoltaic (PV) systems with MLIs can boost the low input voltage as per the required load/grid voltage. In this process, several PV panels, dc-dc converters, and transformers are considered leading to the overall system size and efficiency [11]. The SC topologies are proposed to resolve the voltage imbalance between the capacitors and to lessen the requirement for stiff DC sources [12], leading to the emergence of SCMLIs.

The SC-MLI can produce the desired voltage with an increase in input voltage without any bulky transformer. To attenuate the higher frequency ripples at the grid-connected side, an Inductor-Capacitor-Inductor (LCL) filter is preferred. However, care should be taken while designing LCL values. To diminish the resonance problem of an LCL filter, active damping is introduced which leads to less power loss with better attenuation ripples in comparison with passive damping. However, the active damping method leads to more complexity in the circuit, and therefore, careful design is required to give good stability. Different PWM approaches are available as per low or high switching frequency operations [13, 14]. The Unipolar Phase Disposition-Pulse Width Modulation (UPD-PWM) technique is used in this research to manage the dq-frame current in a 5-level GCI structure. The Id ref is used in this control to add active power to the grid, and the UPF operation of the grid is accomplished by setting the q-axis current reference to zero.

In this paper, section II gives principle operation of 5L-SCMLI with GC-LCL filter, section III gives its control scheme for grid-connected operation, and section-IV briefs about simulation results. Conclusions were presented at the end.

2 Operation of 5L-SC-Based GCI Topology

The 5L-SC-based GCI topology utilized in 1-grid connected applications is shown in Fig. 1. Six unipolar switches, one diode, one SC, and a DC source make up this topology. Table.1 indicates switching patterns for the GCI's 5-level output. From the

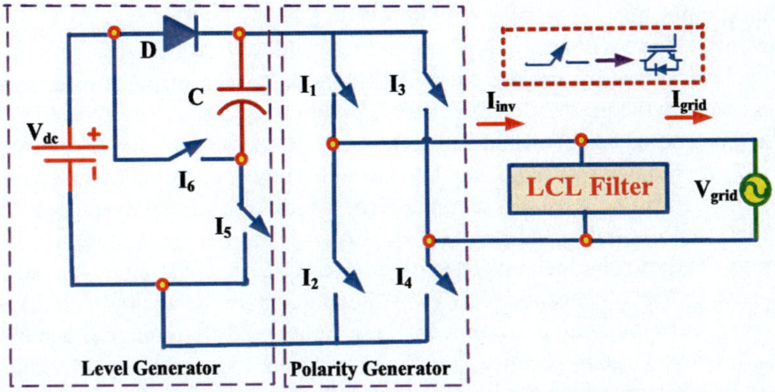

Fig. 1. 1-phase SC-based GCI

Table 1 Switching pattern for 5L-SC inverter

S. No.	ON Switches	Capacitance	V_{inv}
1	I_6, I_2, and I_3	Discharges	$-2V_{DC}$
2	I_5, I_2, and I_3	Charges	$-V_{DC}$
3	I_1, I_3 or I_2,I_4	Idle	0
4	I_5, I_1, and I_4	Charges	V_{DC}
5	I_6, I_1, and I_4	Discharges	$2V_{DC}$

switching table, it is understood that topology will give 5-level output being the states $-2V_{dc}$, V_{dc}, 0, V_{dc}, $2V_{dc}$. The basic SC unit that is provided in [15] is produced by the level generator in Fig. 1. According to this configuration, the capacitor charges to Vdc when I5 is ON and I6 is OFF, and the level generator's output voltage is V_{dc}. The capacitor discharges and the level generator's output voltage is 2Vdc when I6 is ON and I5 is OFF. The switched capacitors in the SC-MLI are charged and discharged in parallel and series configuration, respectively, with dc input supply voltage. Additionally, the system's control complexity is decreased by SC-based topologies with capacitors that proactively balance their voltages. Here, the maximum value of output voltage is $2V_{dc}$ with a boosting factor of 2 and a total standing voltage of 5.5 pu. In the end, an LCL filter is integrated with the grid.

3 Proposed Control Scheme

The suggested inverter current feedback control system for GCI is depicted in block diagram form in Fig. 2. The settings for LCL filters are established by taking into account the related design equations that are depicted in the flowchart provided in Fig. 3 [16–20] before going on to the control scheme. While designing the LCL filter,

f_r should be designed as per the equation shown in Fig. 3. To generate V_d and V_q components, a correct Phase Locked Loop (PLL) with sensed grid voltage is utilized. Grid synchronization is provided by the PLL block. By measuring inverter currents, one can compare the production of I_d and I_q components. Active and reactive power are added to grid using I_d(ref) and I_q(ref).

Now, I_d is compared with $I_{d(\text{ref})}$ (related to active power component) and similarly, I_q is compared with I_{qref} (usually taken as zero) to make the unity power factor. On the errors that occurred for I_d and I_q, the PI controller is implemented to reduce steady-state error. On these obtained errors, apply to decouple control strategy that adds V_d-(wL)I_q and V_d-(wL) I_d respectively (w is the angular frequency and L is $L_1 + L_2$) to generate the modulating β-component. Now obtained β component is a sinusoidal waveform with a large peak value, and these peaks are reduced by multiplying with proper gain P_G to produce the required modulating signal. To reduce the number of high-frequency carriers present in the RTG (Repeating Table Generation) block, the UPD-PWM approach is now applied to this modulating signal. In the UPD-PWM approach, two carrier signals and an absolute sine function (the reference signal) are compared. Whereas a traditional PD (phase disposition)-PWM requires $(K-1)$ carrier signals, we only need $(K-1)/2$ carrier signals to provide K-level output. Thus, the control scheme's complexity is thereby reduced. The necessary Gate Pulses (GP) are generated after the truncation of Aggregated Signal Generation (ASG) and Switching Table Generation (STG). UPF operation at the grid side is achieved by setting $I_{q(\text{ref})}$ equal to zero. This proposed control scheme can be used for any single-phase GCI.

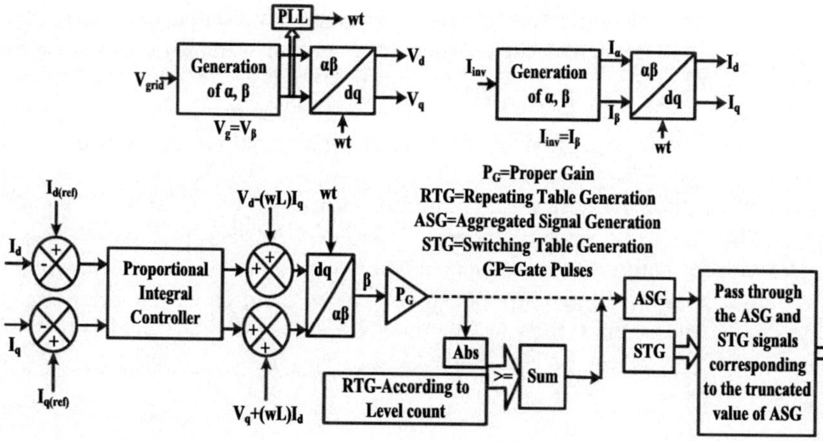

Fig. 2 Proposed dq-frame control scheme

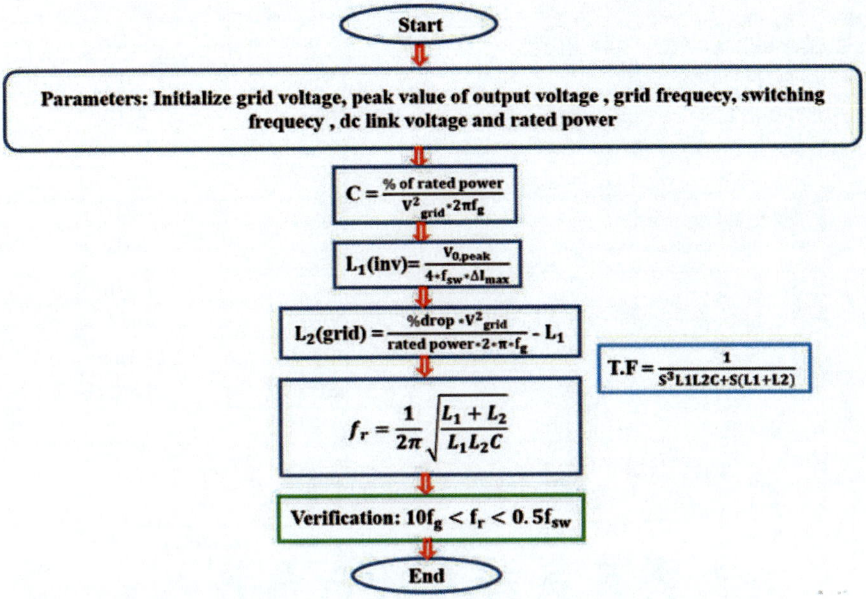

Fig. 3 Design of LCL filter flowchart [17]

4 Simulation Studies

The results acquired with the PLECS tool serve as validation for the proposed control strategy on the GCI under consideration. In grid-connected applications, simulation parameters are given in Table.2.The respective simulation results from Fig. 4 to Fig. 8 are shown below. Firstly, the capacitor is self-balanced at the V_{dc} value. Figure 4 represents the UPF operation of the grid for a step change of peak $I_{d(\text{ref})}$ from 10 to 15A. From Fig. 4 it is observed, on changing the grid current from 10 to 15A at 2.5 s, the value of grid and inverter voltages is retained the same. It is also observed that for step changes, the inverter gives 5-level output with a grid peak voltage being 325 V. Similarly, for $I_{d(\text{ref})}$=10A and $I_{d(\text{ref})} = 15$A, the peak value of grid current obtained is 9.906A and 14.882A. Therefore, even with step variations in $I_{d(\text{ref})}$ values, UPF operation at the grid side is maintained.

From Fig. 4, it is observed that inrush current of a capacitor or at the side is at 40 to 60A for a step change of $I_{d(\text{ref})=}$10A to 15A at 2.5 ms. These huge current damages the system and therefore to limit this inrush current a small value of L in micro-Henries is connected as per the method suggested in [22]. Figure 5 gives the waveforms of capacitor voltage, inverter output, and grid voltage. The injected grid current for a step change in $I_{d(\text{ref})}$ from 10 to 15 A and these step variations are provided at 0.15 ms and 3 ms. Figure 5 indicates that there is a lagging, leading, and unity between grid voltage and current during the periods (0–0.15) s, (0.15–0.3) s,

Table 2 Simulation parameters for 1-ϕ grid-connected inverter

S. No.	Parameters	Value
1	V_{grid}	230 V
2	F_{grid}	50 Hz
3	F_{sw}	10 kHz
4	f_r	1406 Hz
5	V_{dc}	200 V
6	S	2kVA
7	L_1	3.84mH
8	L_2	4mH
9	C_f	5.7μF
10	SC	2200μF

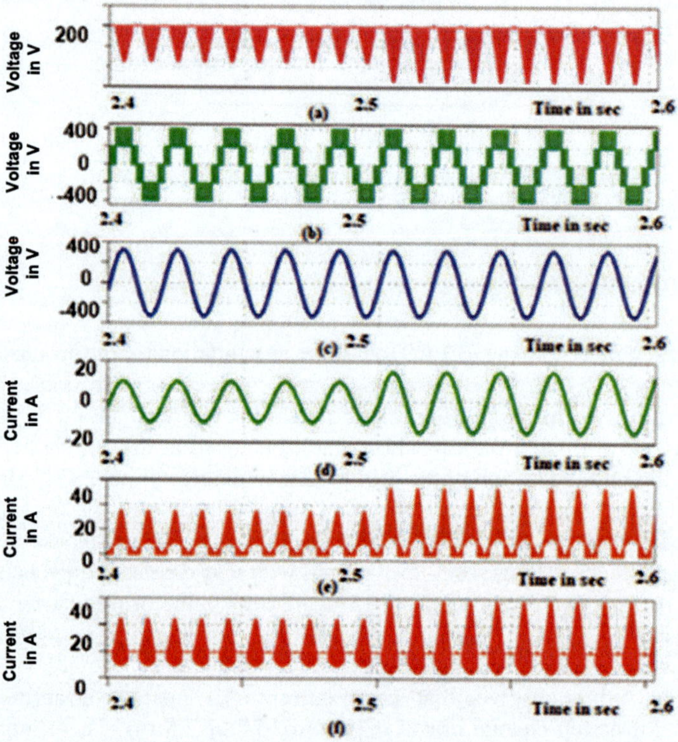

Fig. 4 UPF operation of the grid for a step change of peak $I_{d(\text{ref})}$ from 10 to 15A of **a** capacitor voltage **b** Inverter output voltage **c** grid voltage **d** grid current **e** diode current **f** capacitor current

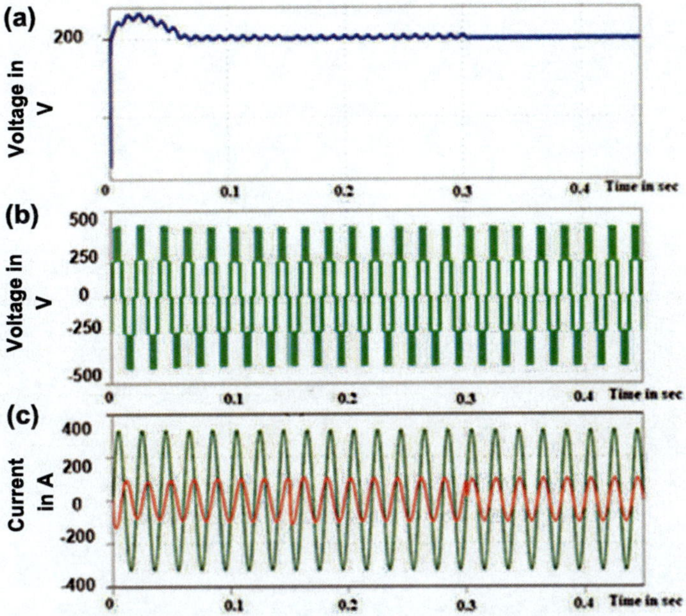

Fig. 5 **a** Capacitor voltage **b** inverter output voltage **c** waveforms of SC voltage, inverter output, and V_{grid} and injected I_{grid}—Lag, lead, and UPF modes

and (0.3–0.45) s. It is also clear that for all these changes in p.f capacitor voltage is balanced and the inverter output is 400 V ($2V_{DC}$).

Figure 6 depicts the waveforms of inverter voltage and the corresponding voltage stresses on all switches. From this Fig. 6, it is understood that the total blocking voltage of the switch is 5.5p.u which leads to cost reduction in switches with a boosting factor of 2. Figure 7 shows harmonic spectrums of grid current. THD obtained is 0.62% which is a less value on considering IEEE-519 standards.

5 Conclusions

In this paper, a single DC source, fewer semiconductor components, and SCs were used to explain the grid connection of an SC-based 5-level inverter. In general, the Residual Direct Current (RDC) circuit makes an inverter more complex in terms of protection, heat sink, and gate driver circuits with the help of any other strategies. A dq-frame current control scheme and proper switching action using the UPD-PWM technique are applied to the GCI. Here, the SC is self-balanced without the need

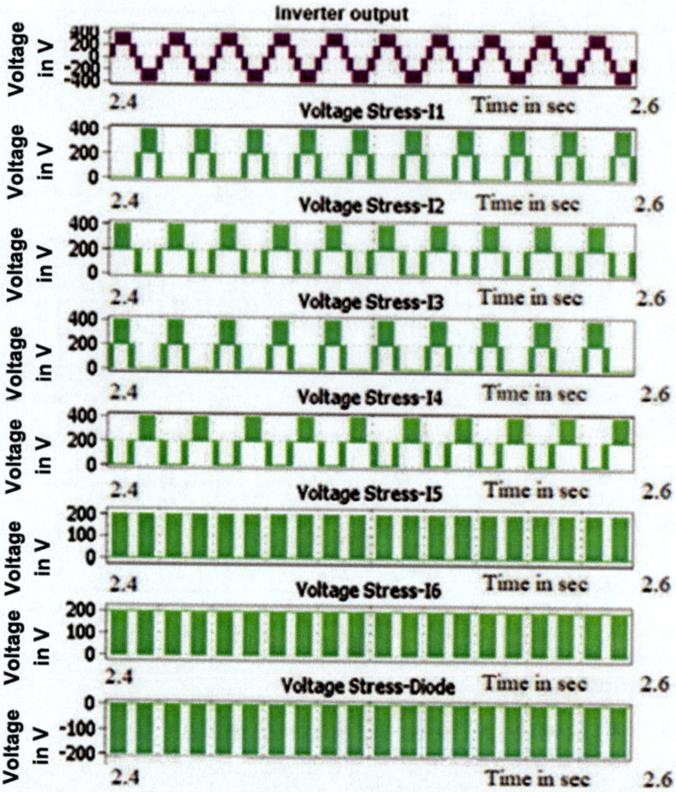

Fig. 6 Waveforms of inverter output and the corresponding voltage stresses on all switches

for any extra circuitry or independent control algorithms, and it also provides twice boosting factor. Even $I_{d(\text{ref})}$ fluctuates in stages, the respective V_{grid} and I_{grid} always permit UPF functioning on the grid side. The cases of lagging and leading power factors of the grid are also discussed. In most of the SC-based inverters, there is typically a significant inrush current at I_{dc} and at SC. In the real-time scenario, a modest inductor value estimated in micro henrys can be added to the charging loop to prevent it.

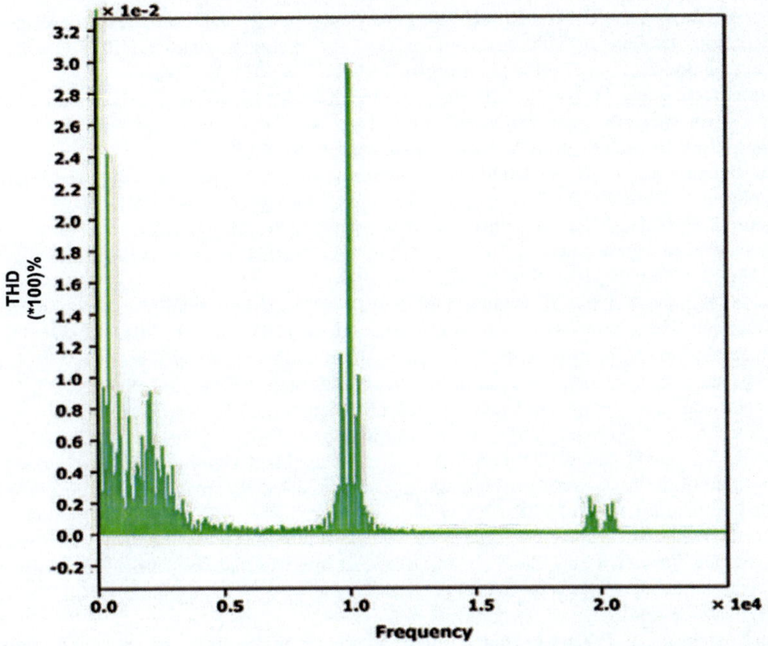

Fig. 7 Harmonic spectrum of grid current with $I_{d(\mathrm{ref})} = 10\mathrm{A}$ (UPF-Mode) and THD $= 0.62\%$

References

1. Vijeh M, Rezanejad M, Samadaei E, Bertilsson K (2019) A general review of multilevel inverters based on main submodules: structural point of view. IEEE Trans Power Electron 34(10):9479–9502. https://doi.org/10.1109/TPEL.2018.2890649
2. Nabae A, Takahashi C, Akagi H, A new neutral-point-clamped PWM inverter. IEEE Trans Industry Appl IA-17:518–523
3. Shukla A, Ghosh A, Joshi A (2007) Capacitor voltage balancing schemes in flying capacitor multilevel inverters. In: 2007 IEEE power electronics specialists conference, pp 2367–2372. https://doi.org/10.1109/PESC.2007.4342381
4. Peng FZ, Lai JS, McKeever J, VanCoevering J (1995) A multilevel voltage-source inverter with separate DC sources for static VAr generation, IAS '95. In: Conference record of the 1995 IEEE industry applications conference thirtieth IAS annual meeting, vol 3, pp 2541–2548. https://doi.org/10.1109/IAS.1995.530626
5. Su GJ (2005) Multilevel DC-link inverter. IEEE Trans Indus Appl 41(3):848–854
6. Hinago Y, Koizumi H (2010) A single-phase multilevel inverter using switched series/parallel dc voltage sources. IEEE Trans Industr Electron 57(8):2643–2650
7. Najafi E, Yatim AHM (2012) Design and implementation of a new multilevel inverter topology. IEEE Trans Industr Electron 59(11):4148–4154
8. Choi WK, Kang FS (2009) H-bridge based multilevel inverter using PWM switching function. In: 31st international conference on telecommunications energy, INTELEC 2009, pp 1–5
9. Ebrahimi J, Babaei E, Gharehpetian GB (2012) A new multilevel converter topology with a reduced number of power electronic components. IEEE Trans Indus Electron 59(2):655–667
10. Arun N, Noel MM (2018) Crisscross switched multilevel inverter using cascaded semi-half-bridge cells. IET Power Electronics 11(1):23–32

11. Trabelsi M, Vahedi H, Abu-Rub H (2021) Review on single-DC-source multilevel inverters: topologies challenges, industrial applications, and recommendations. IEEE Open J Indus Electron Soc 2:112–127. https://doi.org/10.1109/OJIES.2021.3054666
12. Kumari, Siddique M, Sarwar MD, Tariq A, Mekhilef M, Iqbal SA (2021) Recent trends and review on switched-capacitor-based single-stage boost multilevel inverter. Int Trans Electr Energ Syst 31:e12730. https://doi.org/10.1002/2050-7038.12730
13. Rodriguez J et al (2009) Multilevel converters: an enabling technology for high-power applications. Proc IEEE 97(11):1786–1817. https://doi.org/10.1109/JPROC.2009.2030235
14. Kouro S, Bernal R, Miranda H, Silva CA, Rodriguez J (2007) High-performance torque and flux control for multilevel inverter fed induction motors. IEEE Trans Power Electron 22(6):2116–2123. https://doi.org/10.1109/TPEL.2007.909189
15. Babaei E, Gowgani SS (2014) Hybrid multilevel inverter using switched capacitor units. IEEE Trans Industr Electron 61(9):4614–4621. https://doi.org/10.1109/TIE.2013.2290769
16. Bhanuchandar A, Murthy A, B K, A new single-phase five-level self-balanced and boosting grid-connected switched capacitor inverter with LCL filter. In: Panda G, Naayagi RT, Mishra S (eds) Sustainable energy and technological advancements. Advances in sustainability science and technology. Springer, Singapore. https://doi.org/10.1007/978-981-16-9033-4_11
17. Zhang N, Tang H, Yao C (2014) A Systematic method for designing a pr controller and active damping of the LCL filter for single-phase grid-connected PV inverters. Energies 7(6):3934–3954. https://doi.org/10.3390/en7063934
18. Bhanuchandar A, Murthy BK (2021) Single phase nine level switched capacitor based grid connected inverter with LCL filter. In: 2020 3rd international conference on energy, power and environment: towards clean energy technologies, pp 1–5. https://doi.org/10.1109/ICEPE50861.2021.940449
19. Ali Khan MY, Liu H, Yang Z, Yuan X (2020) A comprehensive review on grid-connected photovoltaic inverters, their modulation techniques, and control strategies. Energies 13(16):4185. https://doi.org/10.3390/en13164185
20. .Bhanuchandar A, Murthy BK (2021) Switched capacitor based 13-level boosting grid connected inverter with LCL filter. In: 2021 national power electronics conference (NPEC), 2021, pp 01–06. https://doi.org/10.1109/NPEC52100.2021.9672544
21. Lee SS, Lim CS, Siwakoti YP, Idris NRN, Alsofyani IM, Lee KB (2019) A new unity-gain 5-level active neutral-point-clamped (UG-5L-ANPC) inverter. In: 2019 IEEE conference on energy conversion (CENCON), pp 213–217. https://doi.org/10.1109/CENCON47160.2019.8974836
22. Sen P, Jha V, Sahoo AK (2021) Inrush current minimization in reduced device count multilevel inverter interfacing PV system. In: 2020 3rd international conference on energy, power and environment: towards clean energy technologies, pp 1–6. https://doi.org/10.1109/ICEPE50861.2021.9404426

Microfabricated Biosensor for Detection of Disease Biomarkers Based on Streaming Current Method

Hithesh Kumar Gatty, Jan Linnros, and Apurba Dev

Abstract A microfabricated biosensor based on the streaming current method is presented in this work. The microfabricated sensor consists of a silicon microchannel, which is enclosed with a glass capping to form a closed microchannel. The depth of the microchannel is approximately 10 μm in width and length varying from 50 to 100 μm. The silicon is etched using deep reactive ion etching (DRIE) to form a microchannel. For the capping of the channel, a glass wafer of type Borofloat is used and anodically bonded to the silicon wafer to form a closed microchannel. The microchannel is then functionalized to be specific to certain biomarkers which can be a potential biomarker for cancer, for example. The method used for detection is called the streaming current method. In this method, fluid is flown through the microchannel with high pressure close to six bars. The surface of the silicon is oxidized, which has a zeta potential of approximately 2.7. Depending on the type of fluid the charge concentration varies. By having a pressure in the channel, the charges get distributed as an anode and cathode at the inlet and outlet electrodes of the microfluidic channels. At a fixed potential, a streaming current is observed, which is proportional to the charge accumulated. The difference between the streaming current with and without the biomarker is correlated to the concentration. Hence, a biosensor based on the streaming current method can be realized, which could be used for potential cancer biomarker detection.

Keywords Microsensor · Microfluidics · Silicon · Biosensor streaming current

H. K. Gatty (✉)
Gatty Instruments AB, Ulls Väg 29C, 75651 Uppsala, Sweden
e-mail: info@gattyinstruments.com

Department of Mechatronics, Manipal Institute of Technology, Manipal, India

H. K. Gatty · J. Linnros · A. Dev
Department of Applied Physics, KTH Royal Institute of Technology, Stockholm, Sweden

S. Sharma et al. (eds.), *Intelligent Control, Robotics, and Industrial Automation*, Lecture Notes in Electrical Engineering 1066,
https://doi.org/10.1007/978-981-99-4634-1_56

1 Introduction

The importance of label-free and lab-on-a-chip compatible electrical biosensing platform is well understood and consequently has motivated the development of various miniaturized bioanalytical devices. Besides being low cost and portable, the benefit of such devices also includes the potential for massive parallelization allowing for detection/analysis of multiple target markers from a small volume of sample. Si-based devices, in this context, have a particular advantage given the well-matured silicon process technology that offers the benefit of professional mass fabrication. Such a device can also be easily integrated with robust and low-cost Si-microfluidics as well as sophisticated data handling and communication circuitry, thereby performing a broad variety of functions including automated sample preparation, sorting, and analysis [1]. Therefore, there is a general interest in Si-based bioanalytical devices.

Electrokinetic-based approaches such as streaming current/potential are among the few available methods where the transduction of biochemical response to a proportional electrical signal can be generated within a microfluidic channel. Streaming current is generated because of a pressure-driven flow in a microfluidic channel due to the convective transport of ionic charges in the electric double layer (EDL). Being sensitive to the interfacial electrical properties, the streaming current method allows for accurate surface characterization [2] as well as for precise determination of surface coverage and deposition kinetics of organic/inorganic particles [3]. Streaming current can also be easily measured by connecting two electrodes at the two ends of a microchannel and thereby offering simplicity in device architecture [4]. Various theoretical investigations have also been reported analyzing the role played by the hydrodynamic and electrostatic interaction at the solid–liquid interface, therefore, opening prospect for more advanced sensor design utilizing the streaming current principle.

Here, we report on the fabrication, characterization, and application of such an electrokinetic biochip for analysis of proteins. The sensor fabrication was done on a 4″ Si substrate accommodating 63 chips per wafer and a maximum of four devices per chip. A transparent glass substrate was finally bonded to the Si substrate allowing scope for simultaneous optical probing. Considering the challenges in interfacing such chip with external fluidic connections, particularly for high pressure application, we developed a manifold setup for easy and leak-free integration of the chips up to a pressure of six bar. The dimension of the channels was varied to characterize the design parameter in terms of detection speed and sensitivity. The capacity of the sensors for detection of both positively charged and negatively charged proteins were demonstrated by using IgG-Z domain and streptavidin–biotin model system. The improvement in sensitivity is attributed to the high quality of the surface oxide resulting in a higher density of the surface hydroxyl groups.

2 Materials and Method

The principle of detection is based on the streaming current generated due to electrolyte flow under the pressure difference across the channel as shown in Fig. 1. The streaming current is proportional to the pressure applied and is directly related to the amount of surface charges present. The streaming current is altered when a biomolecule binds to the surface of the channel, and this detection principle is used in identifying a biomolecule.

The detection principle is demonstrated using a silica capillary fiber [5]. It is found that for a given narrow channel and flow rate, a faster transport of target molecules can be achieved leading to faster response time. Aiming to take further advantage of this dimension-dependent response time, we fabricate microchannels with similar surface properties as silica fiber on a silicon wafer.

2.1 *Microchannel Processing and Assembly*

The biosensors are processed on a 525 μm thick, 100 mm diameter n-type silicon wafer. The process flow is described in Fig. 2. The wafers are cleaned in piranha solution, a mixture of concentrated H_2SO_4 and H_2O_2 (3:1) for 20 min. The piranha solution removes the organic contaminants present on the surface of the wafer. The wafer is rinsed and dried using DI water. This step is followed by IMEC-clean that consists of 0.5% HF/0.1% IPA in H20 mixture for 2 min at room temperature, followed by 10 min rinsing and drying. The two-step cleansing procedure removes the organic contaminants as well as the native SiO_2 layer. The surface of the single side polished wafer is spin-coated with a 1 μm thick photoresist (SPR 700) and lithographically patterned to expose a window for etching silicon. After exposure, the photoresist is hard baked in the hotplate at 110 °C for 60 s. The exposed silicon is

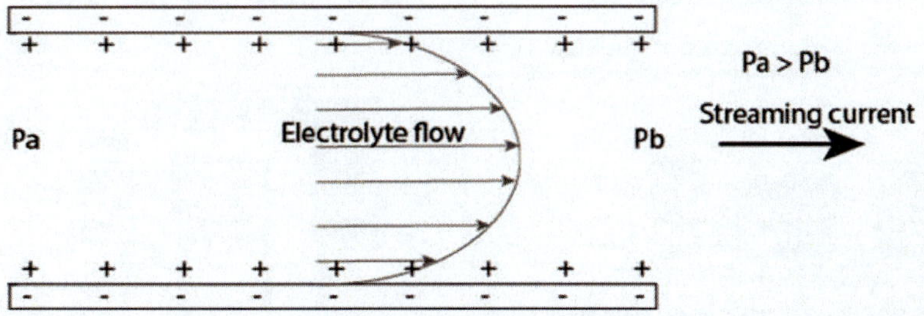

Fig. 1 Schematic of the detection principle in a microchannel. In a channel with a charged surface, a streaming current is generated when there is an electrolyte flow due to the pressure difference at the inlet and outlet ports. This streaming current is not only dependent on the surface charge but also on the size of the molecule present in the analyte. When a biomolecule binds to the surface of the channel, the streaming current is altered. This principle is used in the detection of biomolecules

etched using RIE for 4 min (STS) to achieve a depth of approx. 10 μm. The channel is used for the flow of liquid from the input to an output port. A 1 μm thick PECVD SiO_2 is deposited on the backside of the silicon wafer as shown in Fig. 2c. The SiO_2 is used as a hard mask for long duration etching of the wafer. The backside of the substrate is spin-coated and exposed to open the window on the SiO_2 layer. The photoresist is hard baked at 110 °C for 20 min in an oven. The SiO_2 is etched using oxide etcher for few minutes. The exposed silicon is etched using DRIE for 130 min at an etch rate of 4 μm/min as shown in Fig. 2d. The silicon wafer with the channel and ports is oxidized using a furnace at 1100 °C for 15 min to get a thickness of 275 nm. Thus, formed silicon oxide layer acts as the active surface of the sensor. The channel is sealed using a 250 μm thick Borofloat glass wafer using anodic bonding as shown in Fig. 2f. Thus, obtained wafer is diced into small chips of 11 mm × 7 mm dimension. Each chip contains three to four channels which act as a sensor.

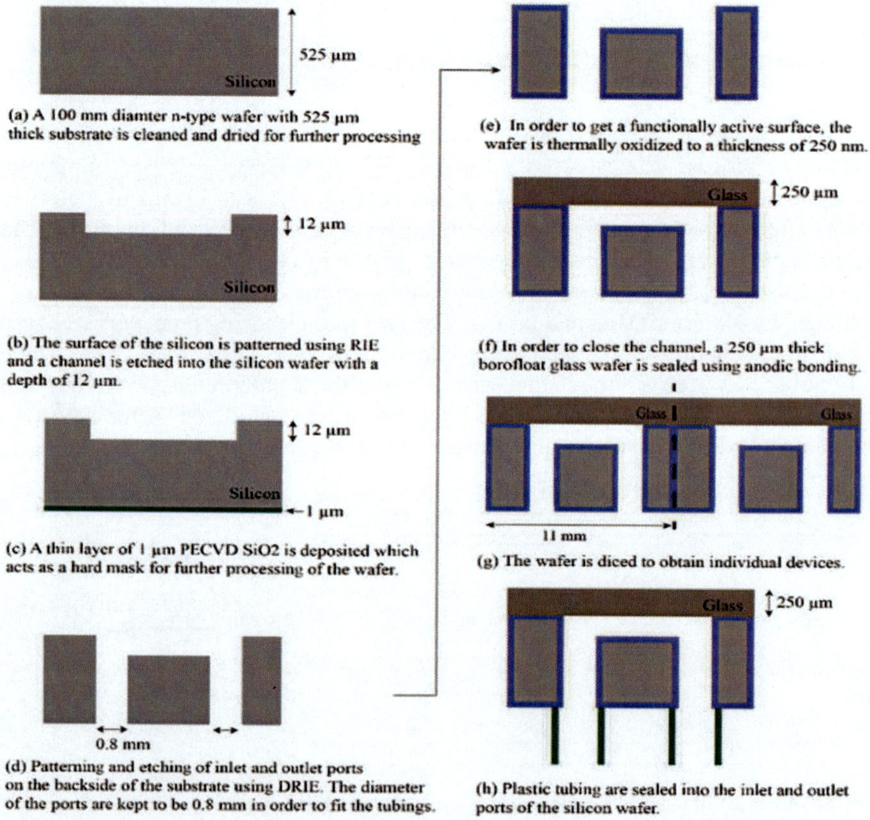

Fig. 2 **a–h** Process flow for fabricating an electrokinetic sensor on a 100 mm silicon wafer. To get an individual sensor, the wafer is diced into 11 mm × 7 mm chips. Each die contains three to four sensors

2.2 *Sensor Surface Preparation*

The sensitive silicon dioxide surface of the microchannel was prepared according to the following processes: The sensor surface was cleaned with Acetone, Isopropanol, and DIW. In addition, a TL1 cleaning with a composition of DI water, $H_2O_2$and NH_4OH (5:1:1) at 70 °C. To activate the surface of SiO2, (3-aminopropyl) triethoxysilane (APTES, 5% w/v in 95% ethanol) was injected through the channel for 10 min. Glutaraldehyde (1% w/v in 1 × PBS), which acts as a linker between the capture probes and APTES, was injected through the capillary for 1 h. The capture probes (z-domain) were on the surface for 1 h at a concentration of 50 μg/mL in 1 × PBS. Active sites that are not taking part were deactivated using Tris-Ethanolamine (0.1 M Tris buffer and 50 mM ethanolamine, pH 9.0). In between each functionalization step, the surface was cleaned with ethyl alcohol and 1 × PBS for 5 min. For all sensing measurements, the channels were treated with casein (0.05% w/v in 0.1 × PBS) for 30 min.

3 Experimental Setup

The measurement system in this study consists of a fluid flow system (Elveflow) connected to pure nitrogen gas capsule that maintains a constant pressure across the channel of the sensor as well as facilitating desired pressure pulses for the sensor characterization and biosensing experiments as shown in Fig. 3. The hydraulic pressure-driven system consists of two fluid reservoirs, one for 0.1 × PBS and the other for the desired analyte such as Avidin and Herceptin, to feed to the system. A switch is installed on the fluid path to alter the working fluid during the experiments. The chip seat platform along with the sensor and the platinum electrodes at inlet and outlet is sealed in a faraday cage to avoid interference with any external noise. A low noise Picoammeter (HP 6487, Keithley) coupled with data acquisition station are used for measurement of currents in the range of picoamps. The measurement reading is recorded using a computer with the help of a lab-view program.

4 Results and Discussions

The experiments were performed with an integrated electrode. The sensor showed a response to biomarkers and has a potential in further miniaturization for detection of biomarkers providing a complete solution for cancer biomarker detection. The

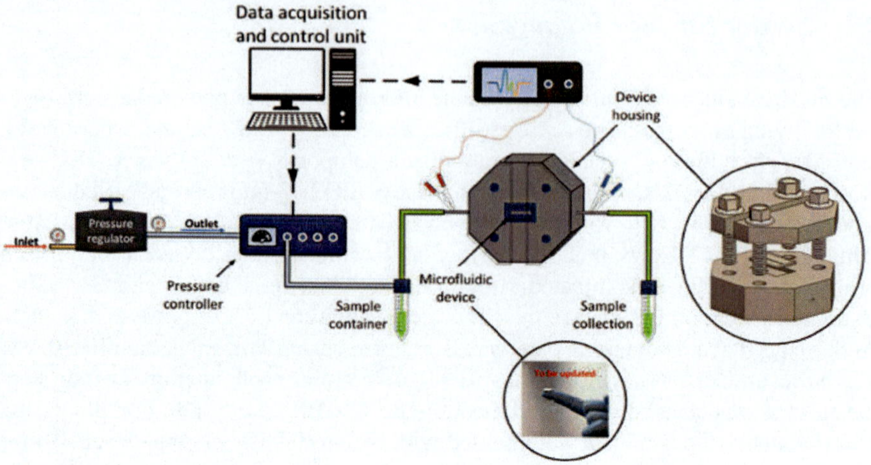

Fig. 3 Measurement system for characterizing the biosensor consisted of pressure regulator, pressure controller, sample containers and reservoirs, proper plastic tubing and microfluidic connections, chip sandwich holder and the sensor, Picoammeter (Keithley HP6487), and data acquisition system

sensor was tested for its Herceptin and Avidin concentration sensitivity, stability, and response time. All the measurements were performed with 0.1 × PBS solution. The dimension of the channel was chosen to accommodate the maximum fluid pressure from the measurement setup.

The integrated electrodes were achieved using sputtered platinum as shown in Fig. 4. The platinum electrodes are at the inlet and outlet ports of the sensor. The photograph of the fabricated chip showing three sensors on a die is shown in 4b. This sensor has a dimension of 3000 μm long rectangular microchannel, a width of 20 um, and the depth of approximately 10 μm. The assembled chip showing the PDMS cover on a circuit board is shown in Fig. 4c. The pins for electrical connections are shown are used for connecting with the external current measuring instrument.

Figure 5 presents the results of the characterization of the sensor chip with integrated electrodes. The flow characterization of the sensor is shown in Fig. 5a. As expected, a linear dependence on pressure vs the streaming current is observed as DI water is pumped through the sensor. The slope is calculated to be 3.3pA/KPa.

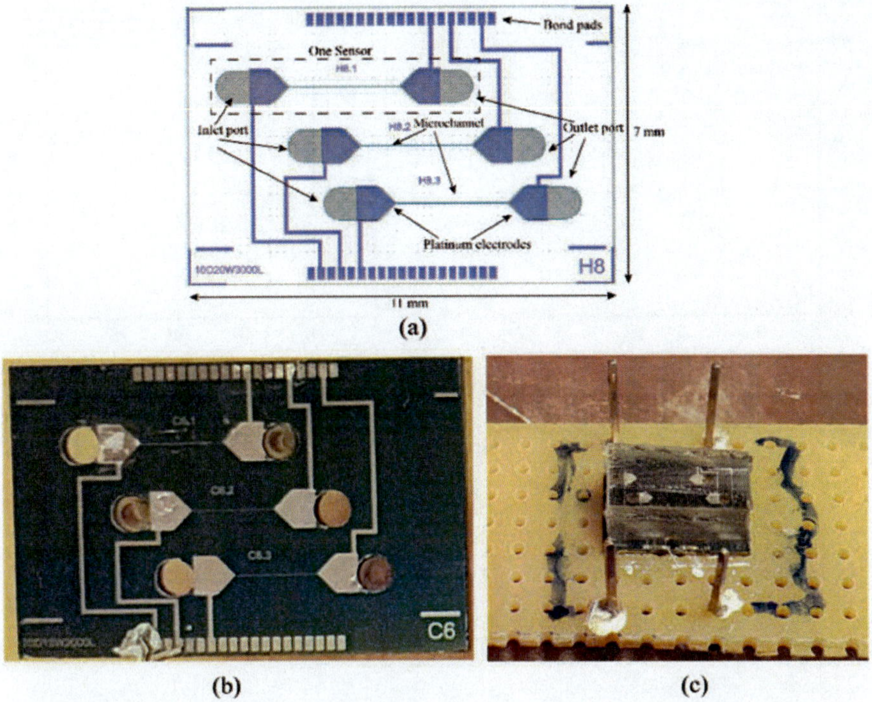

Fig. 4 Electrokinetic biosensor. **a** Schematic layout of the top view of the biosensor. There are three sensors that are processed and available for measurement. The platinum electrodes are placed at the inlet and outlet ports of the sensor. **b** Photograph of the fabricated chip showing three sensors on a die. The sensor has a dimension of 3000 um long rectangular microchannel, a width of 20 um and the depth of 9 um on the 11.5 mm $\times$ 7 mm die chip. **c** Assembled chip showing the PDMS cover on a circuit board. The pins for electrical connections are shown are used for connecting with the external current measuring instrument

When PBS 0.1 $\times$ solution is flown, the sensor response decreases to 0.85 pA/kPa. This is due to the charge screening effect on the surface of the sensor. In another measurement, the sensor was characterized with 9 nM concentration of Avidin. The response time for the biomolecules to reach an equilibrium state was approximately 45 min. When PBS 0.1 $\times$ is flown through the sensor after the equilibrium time has reached, the signal starts to decrease due to the unbinding of excessive molecules present in the vicinity of the surface. The change in the zeta potential was calculated to be approximately 30 mV. The results show a promising step toward the detection and diagnosis of biomarkers.

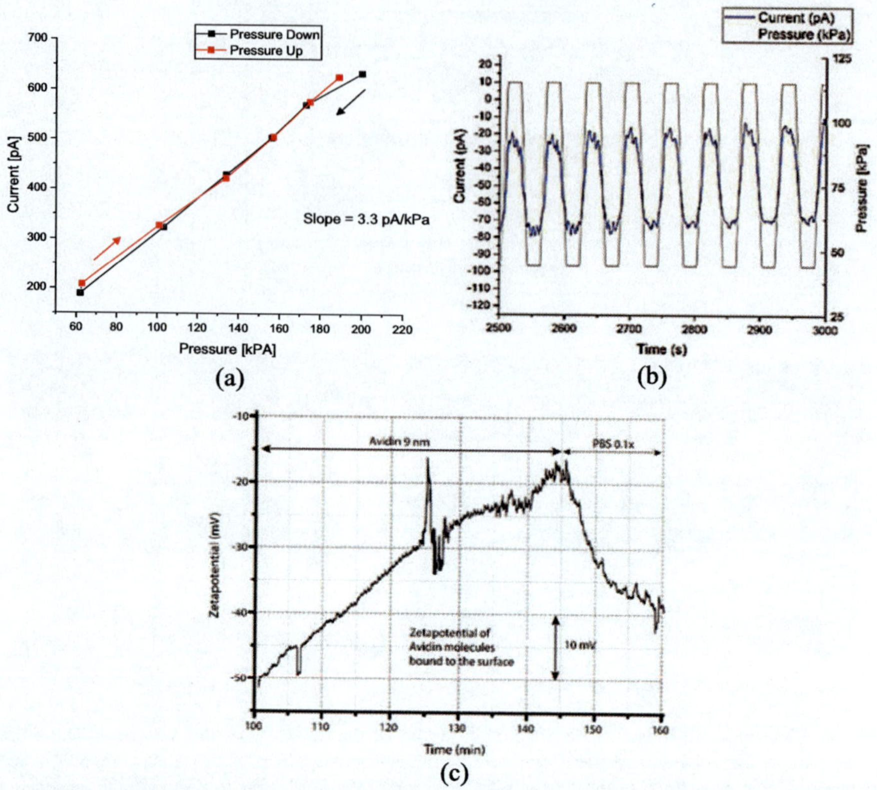

Fig. 5 Graphs showing the characterization of the sensor with integrated electrodes. **a** The flow characterization of the sensor with a slope of 3.3 pA/kPa. As expected, a linear dependence on pressure vs the streaming current is observed as DI water is pumped through the sensor at several pressures. The slope is calculated to be approx 3.3pA/kPa. **b** The graph shows the time dependence of current and pressure when a PBS 0.1 × solution is flown through the sensor, the response is calculated to be approx. 0.85 pA/KPa. **c** Graph of the zeta potential vs time for Avidin at 9 nM concentration. The equilibrium time for the biomolecules to reach an equilibrium state is approximately 45 min. When PBS 0.1 × is flown through the sensor after the equilibrium time has reached, the signal starts to decrease due to the unbinding of excessive molecules present in the vicinity of the surface. For 9 nM Avidin molecules approximately the signal is about 30 mV

5 Conclusion

In summary, we present a silicon-based novel electrokinetic biosensor for multiplexed detection of cancer biomarkers. The miniaturized electrokinetic sensor on a silicon substrate shows a higher and faster detection response and is comparably superior to commercially available silica capillaries. Novel features such as integrated electrodes, glass bonding, and PDMS bonding were successfully tested to withstand high pressures. Thus, the obtained ultra-miniaturized sensor is functionalized and can detect several biomarkers, either individually or multiplexed. A single silicon chip with a dimension of 11 mm × 7 mm contains three to four sensors and has the potential for miniaturized handheld devices. Sensors with different sizes of length and breadth were characterized. The fabricated channel can be safely used for pressures up to seven bars. Using this chip, we present the detection of Avidin and Herceptin. The silicon chip has the potential to be able to detect several protein markers and extracellular vesicles. When several sensors are used in parallel, the sensor chip can be used for multiplexed detection of markers.

References

1. Qi ZB (2013) Microfluid Nanofluid 15:361–376
2. Kirby BJ, Hasselbrink EF (2004a) Zeta potential of microfluidic substrates: 1. Theory, experimental techniques, and effects on separations. Electrophoresis 25(2):187–202
3. Downs AM, McCallum C, Pennathur S (2019) Electrophoresis 40:792–798
4. Delgado AV, Gonzalez-Caballero E, Hunter RJ, Koopal LK, Lyklema J (2005) Measurement and interpretation of electrokinetic phenomena (IUPAC technical report). Pure Appl Chem 77(10):1753–1805
5. Dev A et al (2016) Electrokinetic effect for molecular recognition: a label-free approach for real-time biosensing. Biosens Bioelectron 82:55–63

Optimised Glucose Control in Human Body using Automated Insulin Pump System

Shailu Sachan and Pankaj Swarnkar

Abstract Diabetes is a chronic condition characterised by persistently high blood glucose levels that is controllable but not curable. Controlling the blood glucose levels of diabetic patients is one of the uses of PID controllers in Biomedical Engineering. An automated insulin pump system (AIPS) has been developed to regulate glucose concentration by combining glucose monitoring, controller and insulin pump devices. This paper focuses on the design of an optimal controller for controlling glucose levels in the human body using AIPS. Traditional PID tuning methods are inefficient, time-consuming, and expensive. To address this, optimised genetic algorithm (GA) and particle swarm optimisation (PSO) algorithms for tuning AIPS controller PID gains have been provided. PSO-PID controllers outperform other PID controllers for BGC regulation because they provide superior steady-state and transient responses. The simulation study showcased that the PSO-PID controlled AIPS are more efficient in performance.

Keywords PID controller · Insulin pump system · Genetic algorithm (GA) · Particle swarm optimization (PSO)

1 Introduction

Diabetes affects the majority of the population in today's society. Diabetes mellitus (DM) or diabetes [1, 2] is a chronic condition characterised by persistently high blood glucose levels that worsen over time. It is controllable but not curable. Diabetes causes a variety of complications, including coronary heart disease, weakness, renal issues, non-traumatic amputations, blindness, etc., [3]. Insulin, which is found in the pancreas, facilitates in the transport of glucose from meals into cells where it may be utilised or stored for energy. A diabetic patient either doesn't produce enough insulin

S. Sachan (✉) · P. Swarnkar
MANIT Bhopal, Bhopal, MP, India
e-mail: sachanshailu@gmail.com

S. Sharma et al. (eds.), *Intelligent Control, Robotics, and Industrial Automation*,
Lecture Notes in Electrical Engineering 1066,
https://doi.org/10.1007/978-981-99-4634-1_57

Table 1 BGC in human body for different category while fasting

Category	BGC (in mg/dL)
Normal	70–99
Prediabetes	100–125
Diabetes	Above 126

or can't utilise insulin efficiently. There are three types of DM namely type-1 DM, type-2 DM and gestational DM (GDM).

Nonstop monitoring of glucose levels in the blood is necessary for glucose regulation. There are various methods for measuring and regulating the BGC [4, 5]. The dosage of insulin required varies as a patient's BGC changes dynamically, owing primarily to physical activity and diet. As a result, an automated insulin pump closed-loop system has been developed to regulate BGC, which will be advantageous to diabetic patients. Glucose monitoring and insulin pump devices are integrated to form an automated insulin pump system (AIPS). The BGC is measured by the sensor, and if it is above or below the set-point, the controller automatically regulate the BGC by sending a signal to the insulin pump to provide the required insulin dosage. It works similarly to the human pancreas.

Table 1 shows the BGC of different categories in the human body while fasting [5, 6]. Hypoglycaemia is low BGC, i.e. below 70 mg/dL and hyperglycaemia is high BGC. T1DM refers to the lack or absence of insulin producing β-cells. So, it is necessary to inject insulin externally to control the level of BG in TIDM. In T2DM, body is incapable of using its own insulin. Certain pregnant women have GDM, which is identical to T2DM. Prediabetes is a condition that exists between normal and diabetes and has a BGC of 100–125 mg/dL.

To regulate the insulin delivery system (IDS), several controllers have been designed. PID control, model predictive control, optimal control, robust control, fuzzy logic control are some of the methods used to design the controller [7–9]. PID controllers are utilised in IDS because they are simple to implement and inexpensive. Traditional PID tuning methods are time-consuming, costly, and ineffective [1, 10]. Optimised genetic algorithm (GA) [10–15] and particle swarm optimization (PSO) [15, 16] algorithms for tuning AIPS controller PID gains of have been provided to address this. The AIPS PSO-PID controller regulates glucose levels more quickly and efficiently.

2 Modelling of Insulin Pump System (IPS)

Diabetes characterised by persistently high blood glucose levels that worsen over time. It is controllable but not curable and causes a variety of complications. Patient's BGC changes dynamically, owing primarily to physical activity and diet. In order to improve glucose control, an artificial pancreas replicates the action of the human pancreas using closed-loop feedback. An automated closed-loop insulin pump system

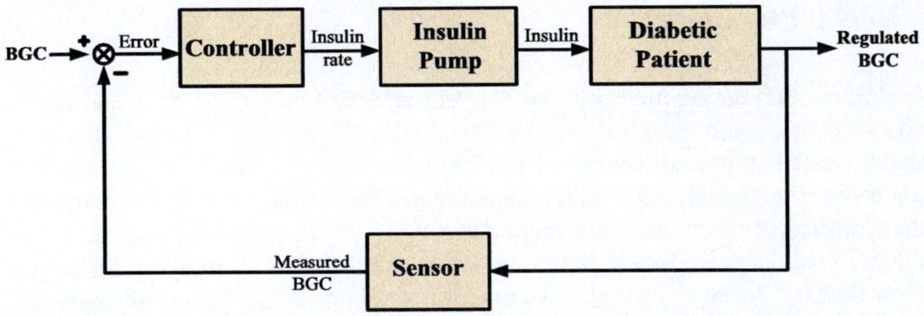

Fig. 1 AIPS

is shown in Fig. 1 has been developed to deliver required insulin dosage as a patient's BGC changes. The BGC is measured by the sensor, and if it is above or below the set-point, the controller automatically regulate the BGC by sending a signal to the insulin pump, an electro-medical device to provide the required insulin dosage. It works similarly to the human pancreas. The transfer function of insulin pump system (IPS) used for simulation study is given by the following equation [1, 2]:

$$G(s) = \frac{1}{s^3 + 6s^2 + 5s} \tag{1}$$

3 Controllers for AIPS

The various controllers for an automated insulin pump system (AIPS) to regulate BGC have been discussed below.

3.1 PID Controller

Equation (2) describes the control function of traditional and simple Proportional-Integral-Derivative (PID) controller [1, 10].

$$G_C(s) = K_P + \frac{K_I}{s} + K_D s \tag{2}$$

3.2 GA-PID Controller

Traditional PID tuning methods like ZN-PID, AMIGO-PID, CHR-PID, and ITAE-PID are time-consuming, costly and ineffective [1]. So, GA [10–15], a meta-heuristic method based on process of natural selection that has been utilised to tune the PID controller of an automated insulin pump system as shown in Fig. 2. It overcomes the shortcomings of traditional tuning methods. It will provide the optimal controller gain values to regulate glucose concentration in the human body. Integral Time Absolute Error (ITAE) is used as the objective function to minimise the error is given by:

$$J = \int_{0}^{\infty} t|e(t)|\mathrm{d}t \tag{3}$$

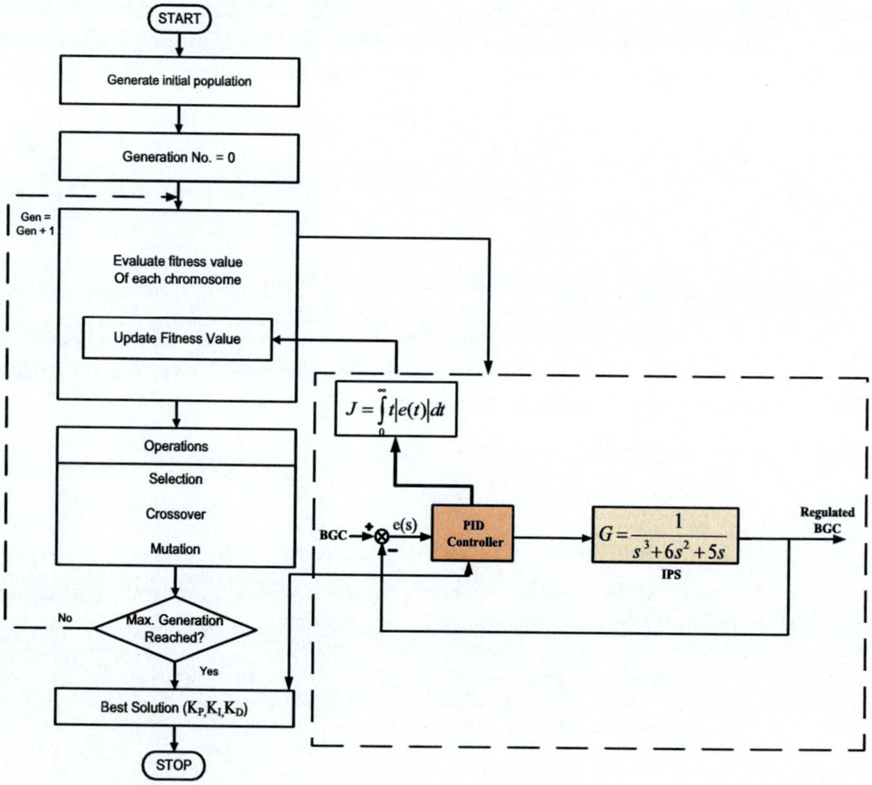

Fig. 2 Block diagram of GA-PID controlled IPS

3.3 PSO-PID Controller

Particle swarm optimisation (PSO) [15, 16] is a meta-heuristic method based on mobility and intelligence of swarm. It overcomes the shortcomings of traditional tuning methods. PSO has been utilised to tune the PID controller of an automated insulin pump system. The block diagram of PSO-PID controlled closed-loop insulin pump system is shown in Fig. 3. Integral Square Error (ISE) is used as the objective function to minimise the error is given by:

$$J = \int_{0}^{\infty} e^{2}(t)\mathrm{d}t \tag{4}$$

It will provide the optimal controller values for regulating glucose concentration in the human body.

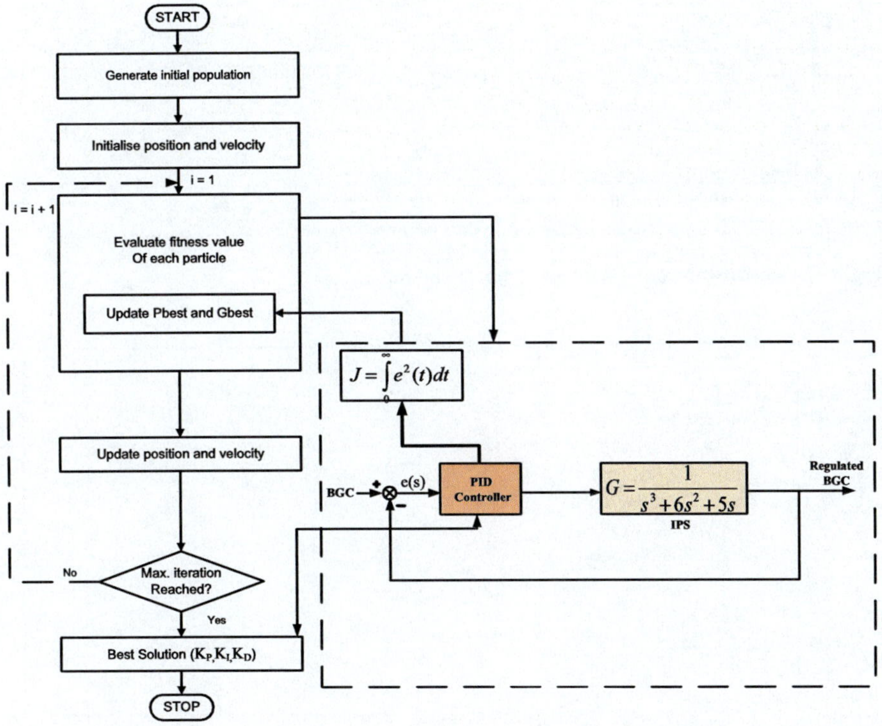

Fig. 3 Block diagram of PSO-PID controlled IPS

4 Results

The automated insulin pump closed-loop system has been developed in Simulink to regulate BGC. To control glucose concentration, various optimised techniques have been discussed in this paper. Figure 4 shows the glucose regulation of AIPS by different controllers. In addition, glucose regulation of AIPS by PSO-PID controller is more efficient and faster than other methods as the rise time, overshoot, settling time, and peak time are less. Figure 5 is the convergence curve of PSO-PID controlled AIPS providing best cost value, i.e. 0.35504.

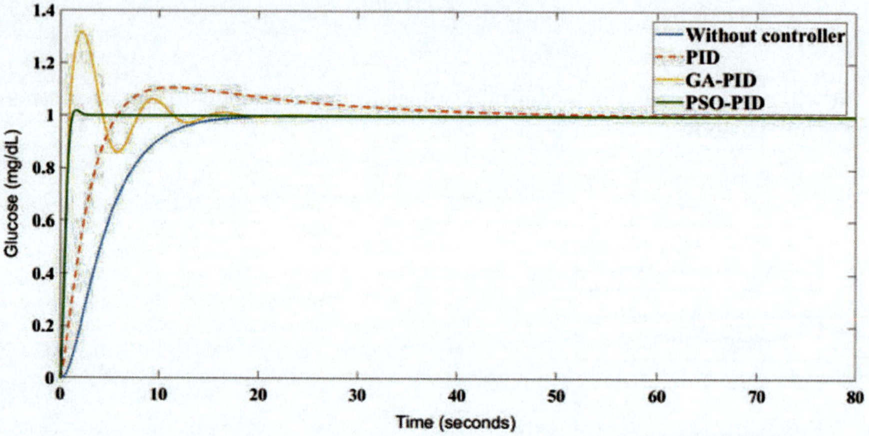

Fig. 4 Glucose regulation of AIPS by different controllers

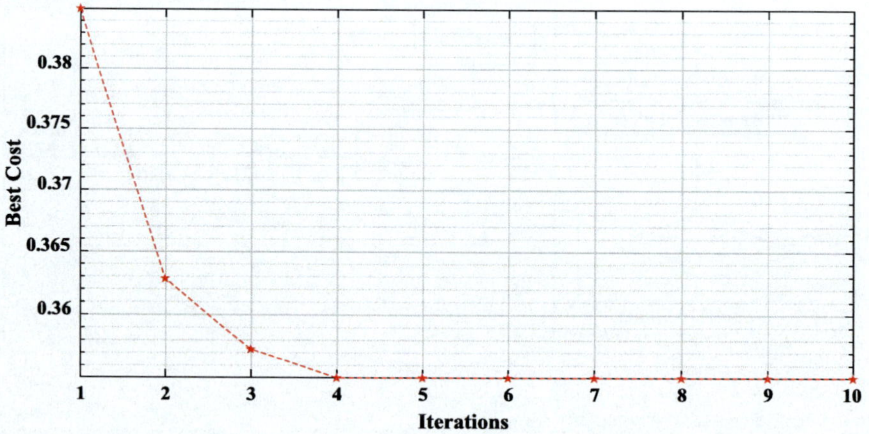

Fig. 5 Convergence curve of PSO-PID controlled AIPS

Table 2 Time response parameters of AIPS with and without controllers

	Without controller	PID	GA-PID	PSO-PID
Rise time (s)	8.3344	4.0118	0.6680	0.7693
Overshoot (%)	0	10.6033	31.8797	1.7322
Undershoot (%)	0	0	0	0
Settling time (s)	15.0125	39.1873	13.8305	1.1648
Peak time (s)	37.0932	11.3495	2.2614	1.6210

Table 3 Value of controller gains

	PID	GA-PID	PSO-PID
Kp	1.9	9.956	10
Ki	0.1	9.657	0
Kd	1.4	9.961	10

Table 2 shows the time response parameters of IDS without a controller, with a PID controller, and with GA and PSO-based PID controllers. PSO-PID controllers outperform other PID controllers for BGC regulation because they provide superior steady-state and transient responses than other controllers. The rise time, overshoot, settling time, and peak time of PSO-PID controlled AIPS have been greatly reduced, resulting in faster and more efficient performance of the system. Table 3 presents the value of PID gains for different controllers.

5 Conclusion

The regulation of glucose in human body for DM control by an automated insulin pump system using different controllers has been discussed in this paper. Traditional controllers are time-consuming, inefficient, and poor performance drawbacks that are remedied by employing a PSO-PID controller. The PSO adjusted PID controller uses integral square error as an objective function to generate optimal controller parameters. The ideally designed PSO-PID controller outperforms other PID controllers for BGC regulation because it provides superior steady-state and transient responses than other controllers. Simulation study showcased that the rise time, overshoot, settling time, and peak time of PSO-PID controlled AIPS have been greatly reduced, resulting in faster and more efficient performance of the system.

References

1. Dubey V, Goud H, Sharma PC (2021) Comparative analysis of PID tuning techniques for blood glucose level of diabetic patient. Turkish J Comput Math Educ 12(11):2948–2953
2. Sharma R, Mohanty S, Basu A (206) Improvising tuning techniques of digital PID controller for blood glucose level of diabetic patient. In: 2016 International conference on emerging trends in electrical electronics and sustainable energy systems (ICETEESES). https://doi.org/10.1109/iceteeses.2016.7581377
3. Pagkalos I, Herrero P, Toumazou C (2014) Bio-inspired glucose control in diabetes based on an analogue implementation of a β-cell model. IEEE Trans Biomed Circuits Syst 8(2). https://doi.org/10.1109/TBCAS.2014.2301377
4. Shah RB, Patel M, Maahs DM, Shah VN (2016) Insulin delivery methods: past, present and future. Int J Pharm Investig 6(1):1–9. https://doi.org/10.4103/2230-973X.176456.PMID:27014614;PMCID:PMC4787057
5. Riley L, Indicator metadata registry details. World Health Organization. Retrieved July 5, 2022, from https://www.who.int/data/gho/indicator-metadata-registry/imr-details/2380
6. Kumar A, Phadke R (2014) Design of digital PID controller for blood glucose monitoring system. Int J Eng Res Technol (IJERT) 3(12)
7. Kaveh P, Shtessel YB (2006) Blood glucose regulation in diabetics using sliding mode control techniques. In: Proceeding of the thrity-eighth Southeastern symposium on system theory. https://doi.org/10.1109/ssst.2006.1619068
8. Aiello EM, Deshpande S, Özaslan B, Wolkowicz KL, Dassau E, Pinsker JE, Doyle FJ (2021) Review of automated insulin delivery systems for individuals with type 1 diabetes: tailored solutions for subpopulations. Curr Opin Biomed Eng 19:100312. https://doi.org/10.1016/j.cobme.2021.100312
9. Al-Tabakha MM, Arida AI (2008) Recent challenges in insulin delivery systems: a review. Indian J Pharm Sci 70(3):278–286. https://doi.org/10.4103/0250-474X.42968.PMID:20046733;PMCID:PMC2792528
10. Varma A, Sachan S, Swarnkar P, Nema S (2020) Comparative analysis of conventional and meta-heuristic algorithm based control schemes for single link robotic manipulator. Intelligent Computing techniques for smart energy systems, Lecture Notes in Electrical Engineering 607. https://doi.org/10.1007/978-981-15-0214-9_6
11. Matlab Optimization Toolbox User's Guide (2017) The MathWorks, Inc., Natick, MA, USA
12. Goud H, Swarnkar P (2019) Analysis and simulation of the continuous stirred tank reactor system using genetic algorithm. Harmony search and nature inspired optimization algorithms. Adv Intell Syst Comput 741. https://doi.org/10.1007/978-981-13-0761-4_106
13. Meena DC, Devanshu A (2017) Genetic algorithm tuned PID controller for process con-trol. International conference on inventive systems and control. https://doi.org/10.1109/ICISC.2017.8068639
14. Sachan S, Goud H, Swarnkar P (2022) Performance and stability analysis of industrial robot manipulator. Intelligent computing techniques for smart energy systems (ICTSES). https://doi.org/10.1007/978-981-19-0252-9_43
15. Goud H, Swarnkar P (2019) Investigations on metaheuristic algorithm for designing adaptive PID controller for continuous stirred tank reactor. Mapan 34:113–119. https://doi.org/10.1007/s12647-018-00300-w
16. Gad AG (2022) Particle swarm optimization algorithm and its applications: a systematic review. Arch Comput Methods Eng. https://doi.org/10.1007/s11831-021-09694-4

Validation of Material Models for PDMS Material by Finite Element Analysis

Chinmay Vilas Potphode, Avinash A. Thakre, and Swapnil M. Tripathi

Abstract Generally, material constants and their corresponding stability regions of engineering materials can be obtained by well-known commercial software. Whereas for most elastomers, material constants are unavailable, this makes it difficult to carry out further mechanical studies on them. This research was aimed at performing a comparative study of various available material models and obtains the most suitable one among them, for determining hyper-elastic material constants and their stability regions from uniaxial tensile test data of polydimethylsiloxane (PDMS) material. Peel tests were conducted on a specimen of PDMS, a flexible adhesive bonded to a hard substrate, in this study. The PDMS interfacial separation from the hard substrate was detected. The PDMS was modelled as a hyper-elastic material and the interfacial characteristics were modelled with a cohesive zone model (CZM) in a finite element peeling simulation. The uniaxial tensile test experiment was used to determine the material properties of PDMS. The parameters of the cohesive zone were derived analytically from peel test data and from a parametric research. The experimental values of the peel force of the 90 degree peel test and the numerical results from the CZM/FE simulation were found to be in good agreement.

Keywords PDMS · Hyper-elastic material · Peel off · UTM

1 Introduction

Elastomers have piqued the interest of many researchers in recent decades due to their fascinating features, such as chemical stability, elasticity, and corrosion resistance. Since it is an optically transparent (see Fig. 1), biocompatible, chemically and thermally stable, highly flexible, and viscoelastic material, polydimethylsiloxane

C. V. Potphode · S. M. Tripathi
Visvesvaraya National Institute of Technology, Nagpur, India

A. A. Thakre (✉)
Mechanical Engineering Department, V.N.I.T, Nagpur, India
e-mail: aathakre@mec.vnit.ac.in

S. Sharma et al. (eds.), *Intelligent Control, Robotics, and Industrial Automation*, Lecture Notes in Electrical Engineering 1066,
https://doi.org/10.1007/978-981-99-4634-1_58

(PDMS) is one of the most researched of these materials. It has numerous applications in a variety of industries, ranging from mechanics and electronics to biomedicine. The biomedical area, mechanical sensors, flexible electronic components, and electrochemical sensors are only a few of the applications for PDMS, which is also employed in micro fluidic circuits. Sealing gaskets, micro fluidic channels, and aneurysm investigations, robotics, and tactile sensors are just a few of the applications where, polydimethylsiloxane is used. Flexible micro fluids and milli fluids, in which the interaction of fluid and structure may be studied, are a new subject that could aid a variety of fields of study, including electronics, biology, and medicine.

Various test methods for identifying material properties for use in flexible adhesive design have been developed [1]. Two volumetric processes (compression and tension) are presented, as well as three tension test methods (uniaxial, planar, and equibiaxial). Only uniaxial tension test data is required, based on comparisons of expected and experimental joint behaviour. This test, which is relatively simple to perform, can offer information on both uniaxial and volumetric tension. There is no noticeable improvement in accuracy by including the additional test data. Since flexible adhesives are compressible, volumetric factors must be included in material models. These can be determined using data from uniaxial tension tests. Investigations were carried out to study the stability of hyper-elastic materials by entering uniaxial tensile test data into FE software ABAQUS® [2]. ABAQUS contains several hyper-elastic material models, which are utilized to characterize the material characteristics of elastomers. In comparison with the stress strain data of the hyper-elastic material obtained from the uniaxial tensile test data, the material constants are also evaluated, making it easier to characterize different material properties. Peel tests were performed on specimens made up of a polyester backing membrane connected to a polyethylene substrate using an acrylic pressure-sensitive adhesive [3]. At the interfacial level, the PSA dissociated from the polyethylene substrate. The material properties of the backing membrane and the pressure-sensitive adhesive were determined using tensile testing. On specimens constructed of polyester backing membrane and an acrylic pressure-sensitive adhesive (PSA) applied to a polyethylene substrate, fixed arm peel tests were performed. The computationally predicted peel forces were found to be in good agreement with the experimentally obtained peel forces for the range of peel angles studied. A two-dimensional (2D) finite element model was created to simulate the 90-degree peeling test of hydrogels coupled to solid substrates [6]. A piece of a hydrogel strip measuring 80 mm in length and 0.86 mm

Fig. 1 PDMS strip

in thickness was removed before it was glued to a solid substrate. Under plane-strain conditions, the hydrogel strip deformation was examined. To illustrate the hydrogel's elastic properties and energy dissipation, the Ogden hyper-elastic material and Mullins effect1 were employed. The model's parameters were determined by fitting the model to experimental data from mechanical tests on the PAAm-alginate hydrogel. All of the numerical simulations were done using ABAQUS. The hydrogel and rigid backing were modelled with the CPE4R element, whereas the cohesive layer at the interface was modelled with the COH2D element. The Poisson's ratio of the hydrogel was set at 0.499 to approximate incompressibility. Using a relatively small mesh size, the adhesive contact was universally disregarded (0.1 mm). A simulation was run with a finer mesh size (0.05 mm), which produced similar peeling forces and proved the mesh insensitivity of our model. A mass scaling method was utilized in the peeling simulations to maintain the process quasi-static. From the literature, it was discovered how to synthesize polydimethylsiloxane at various concentrations and how to determine its mechanical properties using ASTM standards and a uniaxial tensile test. During the literature review, it was discovered how to perform peel off tests on hydrogel, which is also a hyper-elastic material, as well as peel off tests of various pressure-sensitive adhesives and how to perform numerical simulations with ABAQUS software, which is the most preferred software for performing simulations for the peel off test. Since polydimethylsiloxane is a pressure-sensitive adhesive or flexible adhesive that is also hyper-elastic in nature, we have a restricted database to work with, which limits our research on polydimethylsiloxane adhesion. The PDMS is manufactured with various concentrations, which necessitates further investigation: How PDMS adhesion varies with different concentrations, as well as how peel rate affects PDMS adhesive strength, and how rupture occurs. Using an experimental model, determine the strength of the adhesive junction and its separation distance between PDMS material and a hard substrate. In ABAQUS, study of numerical simulation with cohesive zone modelling was done.

2 Experimental Details

2.1 Synthesis of PDMS

The synthesis of polydimethylsiloxane is the initial stage of the experiment, polydimethylsiloxane was synthesized using two distinct processes, and it was cured at five different temperatures [4]. As a result of the five different temperatures at which it was cured, the elastic modulus of the particular concentration varied, and the same trend was observed for the various concentrations of polydimethylsiloxane. Additionally, the two distinct procedures have an impact on polydimethylsiloxane's elastic modulus value. The material is growing softer and its elastic modulus is likewise decreasing as the curing temperature rises. The basic mechanical characteristics of Sylgard 184 PDMS are temperature dependent, and this work illustrates a variety of conventional production techniques. A series of tensile, compressive, and hardness tests to ASTM D412 standards are conducted, respectively, to explore the effects of curing temperature on material properties throughout a range of curing temperatures from 25 to 200 °C. The results of these tests were used to calculate the Young's modulus, ultimate tensile strength, compressive modulus, ultimate compressive strength, and hardness ranges.

Sylgard 184 silicone elastomer consists of two-parts, a pre-polymer base (part A) and a cross-linking curing agent (part B) that cures at both room temperature (RT = 25 °C) and elevated temperatures when mixed together. To synthesize polydimethylsiloxane 20:1, 34.46 g of base was taken. To synthesize PDMS 20:1 1.71 g (approx.) of curing agent was taken and add it to the cup containing base. Mix the mixture thoroughly for 7.5 min so that it will get well mixed. The mixture was kept bubble free by putting it in vacuum at very low pressure for the 30 min. The bubble free mixture was then poured into the mould and even then if some bubbles were formed while pouring, they were removed by keeping the mould in a vacuum at very low pressure for 30 min. After the bubbles are removed, the mixture is kept in an oven for curing purpose and its curing temperature is 125 centigrade for 4.5 h. After 4.5 h, mould was kept in air for cooling and polydimethylsiloxane sheet was taken from the mould and cut into strips which were required to perform the peel test.

2.2 *Uniaxial Tensile Test*

To determine the mechanical characteristics of the materials, adequate deformation measurement techniques are needed. Through this test, the mechanical behaviour of the PDMS with respect to the stress–strain curve will be characterized, and its nonlinear and hyper-elastic nature will be confirmed. The precision of the specimen's dimensions and geometry, which in turn depend on the material's stress state—specifically, the amount of residual stresses—determines how accurate the findings of tensile testing can be. We can ensure that the geometric and dimensional tolerances comply with the ASTM D412 TYPE C standard, thanks to this assessment. Before performing the peel test on the hyper-elastic material PDMS, a uniaxial tensile test must be performed to determine the behaviour of the PDMS at room temperature, which will aid in defining the material property in the FEM Software for numerical simulation of the peel test. The current study's experimental effort comprises manufacturing and testing a hyper-elastic material, PDMS, and analyzing its uniaxial tensile test data to determine its various material model constants.

According to ASTM D412-06, the sample was made utilizing a dumbbell Die C specimen. The thickness of the specimen is (2 mm), as estimated by three measurements: two at either end and one in the middle of the reduced section. Initially, the system and testing environment were inspected for general flaws. After that, the sample ends were attached in the testing equipment and a uniaxial tensile test was performed. $L = 30$ mm is the initial gauge length used to analyze the sample's

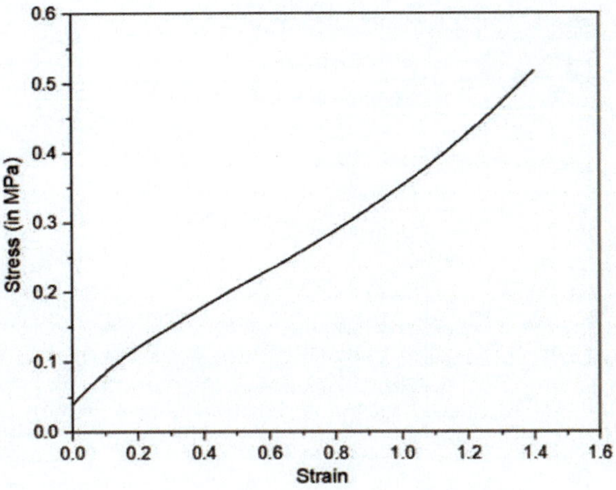

Fig. 2 Variation of stress with strain for PDMS (concentration 20:1)

longitudinal stretches (70 mm). At a grip separation speed of 5 mm/min, the load–displacement curve measurements were taken according to the test specification. During the test, the displacements and weights associated with them were manually recorded using a Sony camera to capture the computer's screen.

The testing method continued until the sample ruptured. The tests were carried out at the Tribology Laboratory of VNIT Nagpur. After obtaining axial applied loads and related longitudinal displacements from the uniaxial tensile test, stress–strain data were assessed (see Fig. 2) and listed in Table 1. To imitate its hyper-elastic nature, these results were loaded into ABAQUS as rubber test data.

3 Results and Discussions

3.1 Model Validation

When modelling hyper-elastic materials in ABAQUS the first step is to provide experimental test data as input. In order to realize the nonlinear analyses that are inherent with hyper-elastic materials, ABAQUS can construct a realistic continuous stress–strain curve via curve fitting on the test data, by providing discrete stress–strain data points. It is the obligation of the user to obtain data that are most relevant to the deformation modes that the actual material will experience. If a real-life hyper-elastic material is primarily loaded under uniaxial tensile, uniaxial tensile test data should be provided in ABAQUS for curve fitting rather than uniaxial compression data. There are seven models for PDMS 20:1, listed in Table 1, which are stable and their constant values are taken to provide material property of hyper-elastic material of different models [5].

Table 1 Properties PDMS (concentration 20:1)

Sr No.	Material models	Constants
1	Mooney Rivlin	$C_{10} = 8.6307E - 02$ $C_{01} = 3.1227E - 02$ $D_1 = 0$
2	Ogden N = 1	$\mu = 0.27146$ $\alpha = 1.854466$
3	Ogden N = 2	$\mu_1 = 8.3350E - 04$ $\alpha_1 = 8.4875$ $\mu_2 = 0.3045$ $\alpha_2 = -4.33813$ $D_1 = D_2 = 0$
4	Neo-Hookean	$C_{10} = 0.1048$ $D_1 = 0$
5	Reduce polynomial N = 2	$C_{10} = 0.10471$ $C_{20} = 3.046E - 05$ $D_1 = D_2 = 0$
6	Reduce polynomial N = 3	$C_{10} = 0.1182$ $C_{20} = -1.33288E - 02$ $C_{30} = 2.60101E - 03$ $D_1 = D_2 = D_3 = 0$
7	Arruda Boyce	$\mu = 0.20597$ $\lambda = 0.2084$

3.2 *Peel Test*

To ensure that the theoretical models were valid, a straightforward peel test experiment was carried out. According to the experimental findings, it may be possible to use theoretical models to forecast PDMS behaviour during the peeling process. With this understanding, automated PDMS membrane peeling technology might be developed in place of the current manual method [10]. The simulations can predict what will occur if a machine pulls on a PDMS membrane, shifting its tip or edge by a specific distance. After the sample has been created, a peel test is performed. Before adhering a PDMS strip to the glass substrate, the glass substrate must be washed with acetone to remove any debris, and then the PDMS strip must be glued to the glass substrate with an 8 cm bonded length. The PDMS strip was pushed 90 degrees away from the hard substrate using mechanical testing equipment (see Fig. 3). During the test, the peeling fixture kept the peeling angle at 90 degrees while the pulley was linked to the machine's crosshead.

As a result of the peel experiment on PDMS, a load vs. displacement pattern was created, which indicates the rate of rupture of PDMS at different peel rates. Experiments were carried out on PDMS 20:1 at three different speeds: $v1 = 30$ mm/min, $v2 = 60$ mm/min, and $v3 = 90$ mm/min (see Fig. 4). The load vs. displacement graph shows how different peel rates affect the rate of rupture. In a peel off test of PDMS of a specific concentration, as the peel rate increases, the interfacial toughness value increases, as does the rate of rupture. In addition, the average peel force value

Fig. 3 No force applied on PDMS

Fig. 4 Peel test of PDMS with constant velocity

Table 2 Average peel force and interfacial toughness value for different peel rate

Peel rate (mm/min)	Interfacial toughness (J/m^2)	Average peel force (N)
$v = 30$	25.08	0.627
$v = 60$	31.604	0.7901
$v = 90$	33.4	0.835

is on the rise. The interfacial toughness value, rupture rate, and average peel force for PDMS 20:1 increase as the peel rate increases as in Table 2; from $v = 30$ mm/min to $v = 60$ mm/min, the interfacial toughness value has a significant difference, and from $v = 60$ mm/min to $v = 90$ mm/min, the interfacial toughness value has a minute difference (see Fig. 5).

3.3 Numerical Validation

In numerical validation, we are trying to validate experimental results by performing finite element analysis with the help of an ABAQUS model [6] to analyze the peel test

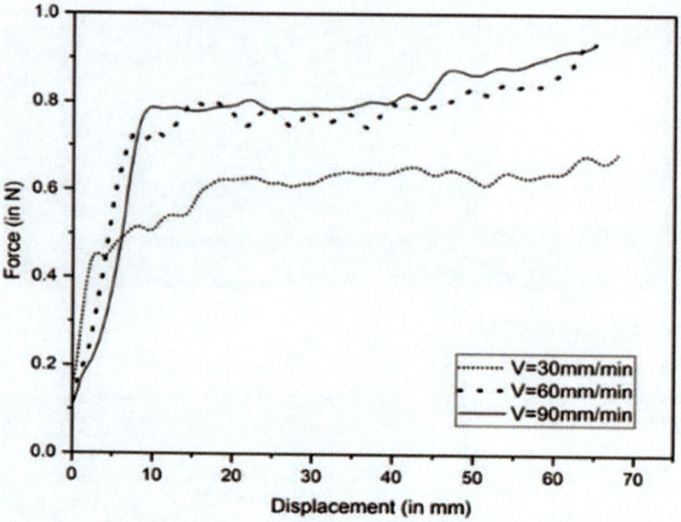

Fig. 5 PDMS 20:1 for different peel rate

numerical simulation of a hydrogel material which is also a hyper-elastic material in nature. Analyses in which different type of traction separation law is more suited in that study [8, 11], it showcases that there are many types of traction separation laws that are there such as bilinear traction separation law, trilinear traction separation law, exponential traction separation law, parabolic traction separation law, etc. [9] The model was driven by a finite element simulation to examine the mechanical processes in the tape backing and describe the peeling configuration (see Figs. 6, 7 and 8). Using finite deformation beam theory, this model may be used to calculate the cohesive zone law over a range of peel angles as well as the full-field force distribution in the process zone. A two-dimensional plane-strain simulation of the peel test was carried out using the commercial FE software ABAQUS [7]. A 2D deformable body represented the peel arm, which was then separated into the polyester backing membrane and PSA adhesive components, while an analytical rigid-body represented the polyethylene substrate in the entire assembly. As per the experimental model in which we pulled PDMS sample with a definite velocity thus maintaining 90 degree angle in peel test, to model this situation we gave two boundary conditions, one at PDMS tip from which we pulled the sample, other at glass surface from which we pulled the elastomer, both glass and PDMS sample were given the velocities, with mode 1 being the mode of contact. After the peel test, numerical validation is required to validate the experimental results, however, first it is necessary to determine which type of hyper-elastic material model will be stable for PDMS of 20:1 concentration, numerical simulation. Peel tests were performed on PDMS 20:1 at three different peel rates: $v = 30$ mm/min, $v = 60$ mm/min and $v = 90$ mm/min. Following the uniaxial tensile test on the PDMS 20:1, there are seven models in FE software that show stability to that concentration, which will be utilized to define the various hyper-elastic material models. In ABAQUS, numerical simulations of these seven alternative models are run to see if the interfacial toughness value and average peel force may be correlated with experimental values.

The experimental interfacial toughness value and average peel force for PDMS 20:1 $v = 30$ mm/min are 25.08 J/m^2 and 0.627 N, respectively, and after performing numerical simulations for seven models, the Ogden $N = 2$ model has the closest interfacial toughness and average peel force values of 23.8 J/m^2 and 0.597 N. As a result, the percentage error for Ogden N = 2 is 5.10%, which is lower than the other six models in Table 3.

The experimental interfacial toughness value and average peel force for PDMS 20:1 $v = 60$ mm/min are 31.604 J/m^2 and 0.7901 N, respectively, and after performing numerical simulations for seven models, the Ogden $N = 2$ model has the closest interfacial toughness and average peel force values of 31.88 J/m^2 and 0.797 N. As a result, the percentage error for Ogden $N = 2$ is 0.837%, which is lower than the other six models in Table 4. The experimental interfacial toughness value and average peel force for PDMS 20:1 $v = 90$ mm/min are 33.4 J/m^2 and 0.835 N, respectively, and after numerical simulation for seven models, the closest model is Ogden $N = 2$,

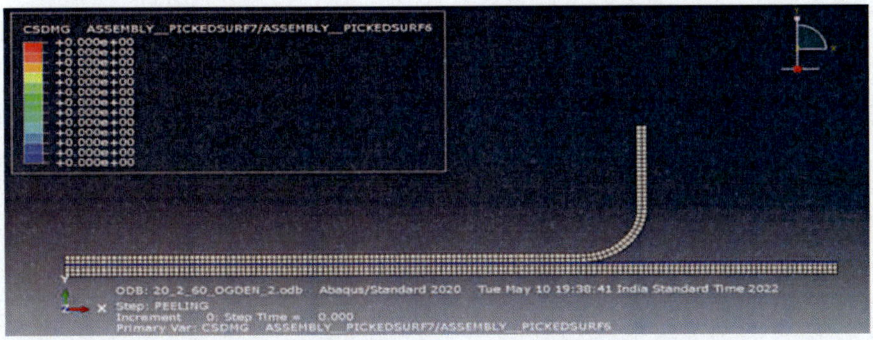

Fig. 6 PDMS 20:1 No force condition

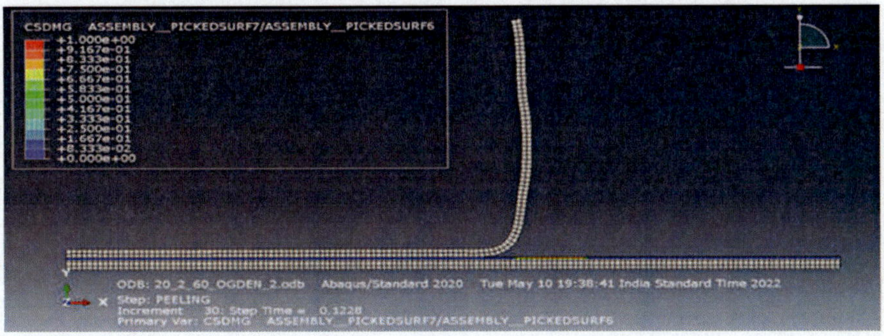

Fig. 7 PDMS 20:1 damage starts as peeling started

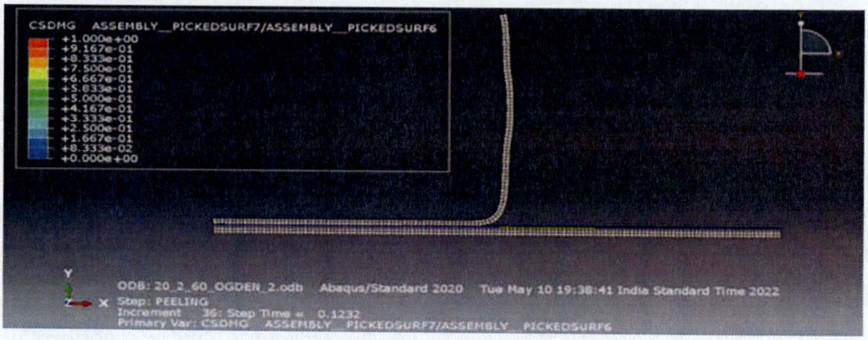

Fig. 8 PDMS 20:1 propagation of crack between interfaces

Table 3 Validation of numerical model PDMS 20:1 $v = 30$ mm/min

Models	Interfacial toughness (J/m^2)	Average peel force (N)	Percentage error (in %)
Experimental	25.08	0.627	–
Neo-Hookean	22.4	0.560	10.68
Ogden N = 1	21.8	0545	13.07
Ogden N = 2	23.8	0.597	5.10
Reduce polynomial N = 2	22.5	0.5626	10.28
Reduce polynomial N = 3	21.92	0.548	12.59
Arruda Boyce	22.04	0.551	12.12
Mooney Rivlin	22.56	0.564	10.04

with interfacial toughness and average peel force values of 29.46 J/m^2 and 0.7365 N. As a result, the percentage error for Ogden $N = 2$ is 11.74%, which is lower than the other six models in Table 5. After running simulations with seven different models for three different peel rates, the Ogden $N = 2$ model came closest to the experimental results and had the lowest % error (see Figs. 9, 10 and 11).

Table 4 Validation of numerical model PDMS 20:1 $v = 60$ mm/min

Models	Interfacial toughness (J/m^2)	Average peel force (N)	Percentage error (in %)
Experimental	31.604	0.7901	–
Neo-Hookean	28.32	0.708	10.39
Ogden N = 1	28.4	0.710	10.13
Ogden N = 2	31.88	0.797	0.837
Reduce polynomial $N = 2$	28.48	0.712	9.88
Reduce polynomial $N = 3$	28.72	0.718	9.125
Arruda Boyce	28.40	0.711	10.13
Mooney Rivlin	29	0.725	8.23

Table 5 Validation of numerical model PDMS 20:1 $v = 90$ mm/min

Models	Interfacial toughness (J/m^2)	Average peel force (N)	Percentage error (in %)
Experimental	33.4	0.835	–
Neo-Hookean	26.51	0.6628	20.62
Ogden $N = 1$	27.24	0.681	18.44
Ogden $N = 2$	29.46	0.736	11.74
Reduce polynomial $N = 2$	26.6	0.665	20.35
Reduce polynomial $N = 3$	26.8	0.671	19.76
Arruda Boyce	28.81	0.720	13.74
Mooney Rivlin	26.8	0.670	19.76

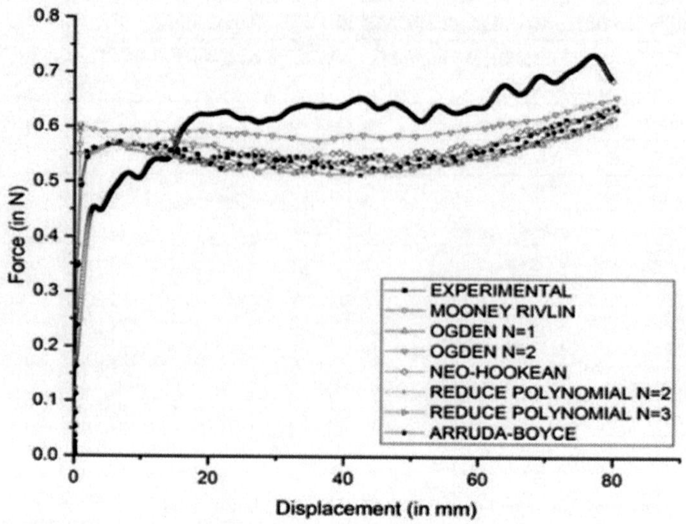

Fig. 9 Numerical simulation for PDMS 20:1v = 30 mm/min

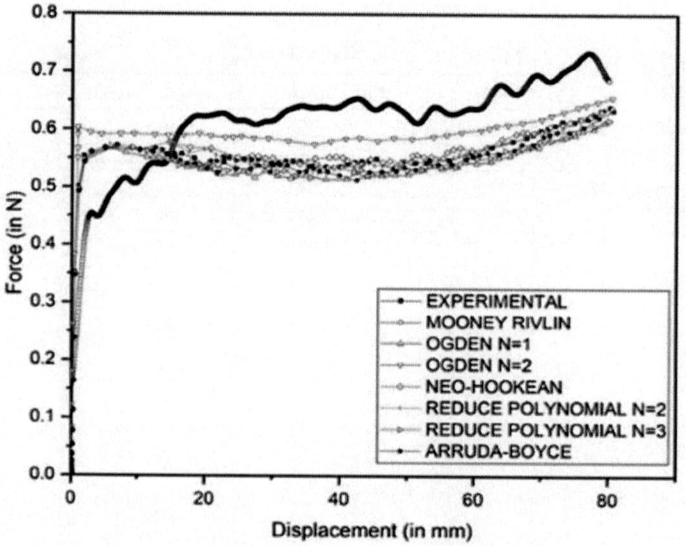

Fig. 10 Numerical simulation for PDMS 20:1 $v = 60$ mm/min

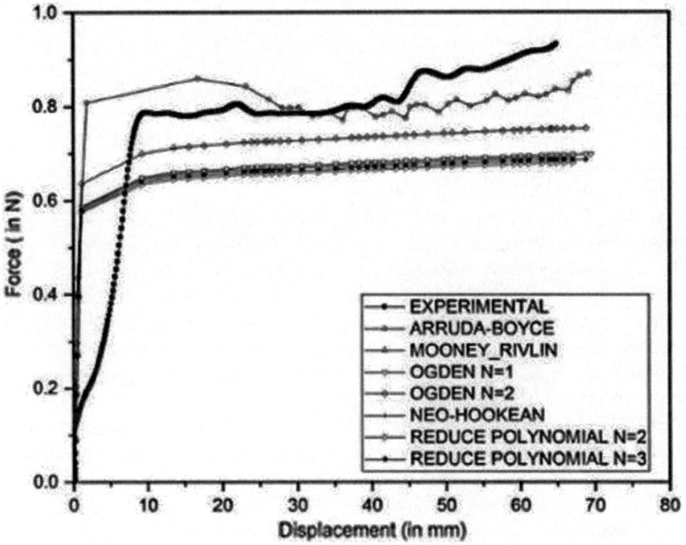

Fig. 11 Numerical simulation for PDMS 30:1 $v = 90$ mm/min

4 Conclusion

The purpose of this study was to determine the most suitable hyper-elastic material model for elastomeric materials. This study shows that the most suitable model is the one which is having least amount of error value for the average peel force and interfacial toughness value for all three peel rates. The most accurate model to our experimental model is the Ogden model. Neo-Hookean is the least suitable model for all three peel rates because it showed maximum error. In Ogden material model, strain energy density functions are used to derive stress strain relationship of the hyper-elastic material, here strain energy density is expressed in terms of principal stretches, for higher order of Ogden the material behaviour of elastomers can be described very accurately, especially for modelling rubbery and biological tissues even at higher strains. The hyper-elastic material model giving best results, i.e. the Ogden model can be effectively and efficiently employed in various experiments regarding mechanical and tribological characterization of elastomeric materials such as PDMS, numerical simulations using Ogden model can be used to predict adhesion behaviour of elastomers to best degree of certainty. FEM softwares working on this model should be preferred choice when simulations are to be run for elastomeric materials.

References

1. Bruce H, Holmqvist C (2013) Modelling adhesion in packaging materials: physical tests and virtual tests in Abaqus. Lund University, Structural Mechanics
2. Esmail, Jihan F, Mohamedmeki MZ, Ajeel AE (2020) Using the uniaxial tension test to satisfy the hyperelastic material simulation in ABAQUS. In: IOP conference series: materials science and engineering, vol 888(1). IOP Publishing
3. Mohammed IK, Charalambides MN, Kinloch AJ (2015) Modelling the interfacial peeling of pressure-sensitive adhesives. J Nonnewton Fluid Mech 222:141–150
4. Olima M (2017) Mechanical characterization of polydimethylsiloxane
5. Liravi F, Das S, Zhou C (2014) Separation force analysis based on cohesive delamination model for bottom-up stereolithography using finite element analysis. In: 2014 International solid freeform fabrication symposium. University of Texas at Austin
6. Yuk H, Zhang T, Lin S, Parada GA, Zhao X (2016) Tough bonding of hydrogels to diverse non-porous surfaces. Nat Mater 15(2):190–196
7. Zeng W, Sun W, Bowler N, Laflamme S (2015) Peel resistance of adhesive joints with elastomer–carbon black composite as surface sensing membranes. Int J Adhesion Adhesives 58:28–33
8. Zhang T, Yuk H, Lin S, Parada GA, Zhao X (2017) Tough and tunable adhesion of hydrogels: experiments and models. Acta Mech Sin 33(3):543–554
9. Kovalchick C (2011) Mechanics of peeling: cohesive zone law and stability. California Institute of Technology
10. Miao T, Tian L, Leng X, Miao Z, Xu C (2018) A comparative study of cohesive zone models for predicting delamination behaviors of arterial wall. arXiv preprint arXiv:1806.05785
11. Souza A, Marques E, Balsa C, Ribeiro J (2020) Characterization of shear strain on PDMS: numerical and experimental approaches. Appl Sci 10(9):3322

Improvement in Torque and Power Performance of an Elliptical Savonious Wind Turbine Using Numerical and Experimental Analysis

Avinash A. Thakre and Pratik U. Durge

Abstract Savonious turbines require drag force for rotation and are easy to manufacture but the torque and power performance are less when compared with lift-based turbines. There is a need to enhance the performance of savonious wind turbine. Recently elliptical blade savonious profile has better power coefficient (Cp) value than others. Hence in this research work elliptical blade of savonious wind turbine is optimized by making its hybrid profiles by adding an NACA0012 aerofoil at the end of elliptical savonious blade. Hybrid 60, Hybrid 80 and Hybrid 100 profiles were made by 60, 80 and 100% overlapping of an aerofoil on elliptical blade, respectively. 2D Numerical analysis performed using ANSYS Fluent software on three developed profiles and torque as well as power performance for input wind speed was compared with elliptical savonious blade profile. 2D analysis results show the torque performance of Hybrid 80 profile is better than elliptical savonious blade profile. Elliptical savonious turbine was manufactured along with a set of movable windshields that were 170 mm and 260 mm in length and were positioned at angles of 30°, 45°, 60° and 75°. Results are compared for an elliptical savonious rotor with and without a windshield. According to experimental findings, a windshield with a length of 170 mm has greater Cp values than a wind shield with a length of 260 mm. For a 170 mm-long windshield positioned at 45°, the elliptical savonious Cp value improves by 30.83%.

Keywords Power coefficient (Cp) · Elliptical profile · Savonious rotor · Torque · SST K-ω

A. A. Thakre (✉) · P. U. Durge
Mechanical Engineering Department, V.N.I.T, Nagpur, India
e-mail: aathakre@mec.vnit.ac.in

S. Sharma et al. (eds.), *Intelligent Control, Robotics, and Industrial Automation*, Lecture Notes in Electrical Engineering 1066,
https://doi.org/10.1007/978-981-99-4634-1_59

1 Introduction

Power generation in world is shifting toward the renewable energy with growth of 12.5% in year 2020 with renewable power capacity 3147 TerraWhr in which solar capacity and wind capacity of world is 707.5 and 733.3 TerraWhr with increase of 21.5% and 17.5%, respectively for year 2020 [1]. India has a target of achieving 175GW from renewable energy in 2022, to assist this target, there is need of enhancement of onshore renewable energy which can work on lower wind speed [2]. Major categories of wind turbines are macro and micro aeolion. Macro includes horizontal axis wind turbines (HAWT) and micro includes vertical axis wind turbine (VAWT). HAWT can't be installed everywhere because it is difficult to manufacture and noisy as well as of large size, while VAWT are simple to manufacture and can be easily installed anywhere [2]. VAWT is classified into two types, savonious type which works on drag force only and other is darrieus type which uses both drag and lift for rotation. Savonious is easy to manufacture when compared with darrieus. Also it is self-starting at low wind speed while darrieus requires initial torque for rotation at low wind speed. But the problem with savonious type turbine is, it has low Cp. Many researchers are working on increasing the power coefficient of savonious wind turbine. The blade profiles and forms of savonius wind turbines have gone through a number of modifications, with certain improvements in their power coefficient. In terms of aspect ratio (AR), higher AR models are found to be more suited at greater wind speeds, while lower AR models are shown to be more suitable at lower wind speeds. The semicircular and twisted bladed rotors' optimal AR (for CPmax) are determined to be 0.80 and 1.5–2.6, respectively. The existence of the overlap ratio improves the rotor's static torque. Based on the working conditions, the effective overlap ratio (OR) is between 0.15 and 0.25 [3]. The SST k–ω model provided greater prediction skills than the other turbulence models employed for numerical simulation [4].

The study of savonius rotor blade profiles and shapes includes Semicircular, Bach, Swinging, Benesh, Slatted, Benes, Twisted, Sistan, Zephyr, Fish-ridged Semi-elliptic, Elliptical, Slotted, Incurved, Bronzinus, Modified Bach, Roy, Airfoil shape, New elliptical, Multiple quarter Semicircular, Multiple miniature semicircular, Spline-curved and Banki wind turbine rotors in past three decades has done which made us aware about efficient elliptical profile. [5]. Alom has developed an elliptical savonious profile which has obtained the maximum Cp value of 0.33 in which optimization of the elliptical profile is done by changing the angle of cut (θ), and developed the optimum blade shape of the simple savonious wind turbine by carrying out the 2D unsteady simulations for elliptical profile at a wind speed of 6.2 m/s which shows that the maximum Cp value is obtained at the $\theta = 47.5°$, whereas the conventional semicircular profile indicates a highest power coefficient of 0.27. Hence there is increase in value of Cpmax [6]. Alom and Saha aims to conduct a systematic numerical analysis in order to determine the best OR for an elliptical shape with a sectional cut angle of 47.50°. Around the elliptical shape, a 2D unstable simulation are done with various OR ranging from 0.0 to 0.30. At

OR = 0.15, the unstable simulations reveal a peak CP of 0.34.The elliptical profile indicates a peak CP of 0.328 at OR = 0.20. When compared with OR = 0.20, a performance boost of 3.53% is gained [7]. Alom [8] has investigated numerically the effect of curtain plates placed in front of elliptical savonious profile blades. The 2D unstable numerical analysis is carried out with the help of the turbulence model SST k–ω. Nur Alom has done investigation of the influence of aerodynamic parameters on performance of elliptical profile rotor. The Cd average for an elliptical shape with curtain plates is 2.60, whereas the Cd average for an elliptical profile without curtain plates is 1.43 based on unstable numerical simulations. As a result, with the elliptical profile with curtain plates, Cd average improves by 81.81% [8]. Zhou and Rempfer has done numerical analysis on simple savonious and bach type rotor to predict the aerodynamic performance by using STARCCM+. Result shows that bach type rotor has better torque performance than simple savonious type rotor [9].

The elliptical profile developed by IIT Guwahati having Cp value of 0.33 has been numerically investigated by Alom [8] using curtain plates by 2D simulations and improved its drag coefficient. In this study, the elliptical savonious profile is optimized by adding an NACA0012 aerofoil at its end using 2D simulations and torque values are compared with elliptical savonious. Also experimental investigation is done for elliptical savonious profile using windshield at four different positions in front of rotor and power coefficient for input wind speed at optimum windshield position is obtained.

2 Simulation Details

2.1 Geometry Details of Hybrid Profiles

Figure 1 shows the sectional cut on the ellipse to obtain the different elliptical profiles. The cutting angle is the angle obtained by line intersecting with major axis OM at P. The point P is at 54% of OM from O to obtain the chord length of the blade (*d*) that influences the numerical results.

2.2 Ellipse Dimensions

OM = 0.198 m, ON = 0.132 m and θ = cut angle
Diameter of turbine = D = 0.61 m
Blade thickness = 0.0066 m
Overlap distance = 20% of chord length of the elliptical profile.

As NACA0012 has better performance when superimposed by Tartuferi for formation of SR5050 rotor, hence NACA0012 aerofoil shape with chord length of 100 mm

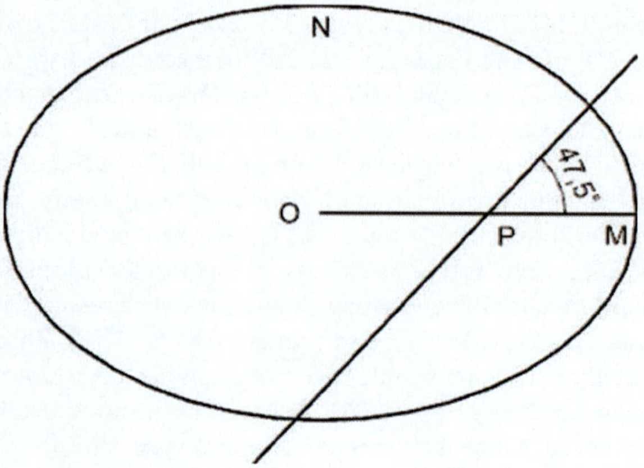

Fig. 1 Elliptical profile formation

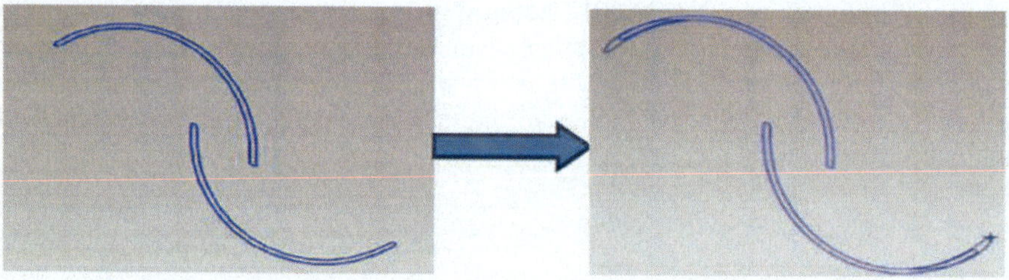

Fig. 2 Hybrid profile formation

is selected for attachment [10]. Hybrid 2D profiles are made as shown in Fig. 2, Hybrid 80 means 80 mm of chord length is overlapped with profile of elliptical savonious profile, similarly Hybrid 60 and Hybrid 100 profiles are made. Figure 3 shows the overlap percentage of NACA0012 aerofoil shape on elliptical blades savonious profile. Figure 3 shows percentage overlap of NACA0012 on elliptical profile. The dark region shows overlapped part of aerofoil.

2.3 *Mesh Generation of Hybrid Profile*

ANSYS Mesh was used to create a mesh around the rotor for two-dimensional analysis. In order to capture the boundary layers, five-layer inflation on the elliptical profiles is given and the growth rate of 1.2. Two regions are created, one is interior and other is exterior region as shown in Fig. 4a. Mesh size of interior is 15 mm and of exterior region of rotor is 25 mm. A 2D simulation was employed to achieve the same result as a 3D model because the models were simple elliptical and hybrid.

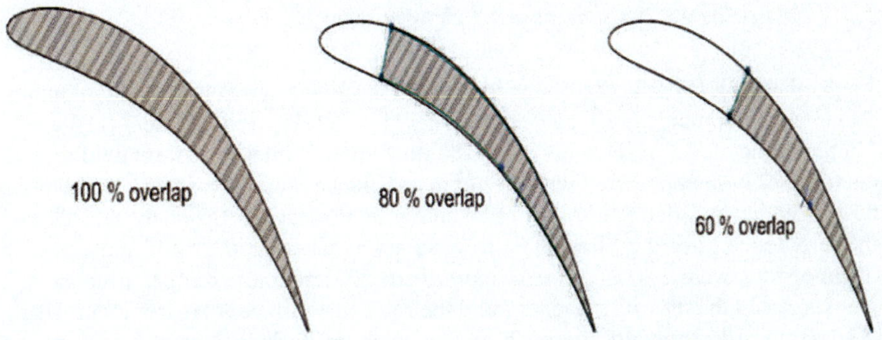

Fig. 3 Overlap of an aerofoil

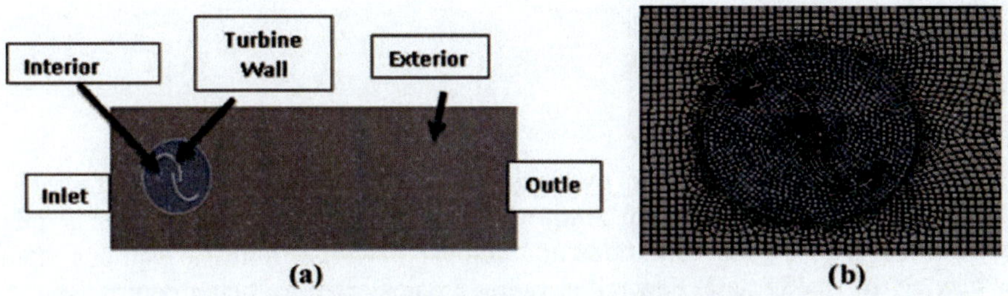

Fig. 4 **a** 2D flow analysis of elliptical savonious **b** Mesh around blade

The ANSYSFluent solver was used to produce the mesh. In order to avoid side walls effect, the distance equal to rotor diameter is kept above, below and front of turbine with computational flow domain of 5 m* 1.2 m, with the rotor 0.6 m from the inflow. Mesh around the blade is shown in Fig. 4b.

2.4 Numerical Model

The commercial software program ANSYS Fluent from ANSYS Inc. was used to conduct a systematic, iterative numerical research on different blade forms. A pressure-dependent solver and a transient solution were utilized to model air flow around the rotor, and a sliding mesh approach was employed to rotate the rotor. As K-ω turbulence model has enhanced estimation capabilities, favorable behavior in adverse pressure gradients and offers a better prediction of flow separation, the flow simulation was done using the K-ω turbulence model [8].

2.5 Boundary Conditions and Procedure

As the meshing is done, the boundary conditions are assigned to the flow domain. The inlet of domain is given as velocity inlet [$V = 3*(1-\exp(-\text{Time}/1\ [\text{s}]))$], with 5% turbulence level of intensity set at the flow domain inlet. The flow field around the turbines has been treated with *SST k*- ω turbulence model. Unsteady simulations have been carried out due to the quickening air flow at the tip of the blades and the turbulence after the blade. The time step was taken as 0.01 and the numbers of time steps were 1500. The maximum were 20 iteration taken per time step. It was supposed that the air moved around the rotor in a turbulent environment. Using the second-order upwind approach, the momentum, turbulent energy and rate of dissipation were determined. Torque, input wind speed and angular velocity of rotor is calculated from analysis.

3 Simulation Results

Initially NACA 0012 aerofoil is overlapped with elliptical savonious wind turbine with 100, 80 and 60% of overlap. And variation of torque for given flow time graphs are plotted for all three conditions and compared with elliptical savonious after achieving the steady state. Few initial values are neglected for better comparison.

3.1 60, 80 and 100% Overlap NACA0012 Aerofoil with Chord 100 mm on Elliptical Savonious

The variation of torque for given flow time is given below in graphs. These graphs are plotted after achieving the steady state, Fig. 5a–d show the variation of torque vs flow time for Hybrid 60, Hybrid80, Hybrid 100 and elliptic profiles, respectively. Hybrid 100 profile is obtained by overlapping of 100 mm chord of aerofoil on elliptical savonious profile. Similarly Hybrid 60, 80 profile is obtained by overlapping of 60 mm and 80 mm chord of aerofoil on elliptical savonious profile, respectively. For elliptical savonious rotor unnecessary drag force acting on returning blades creates the negative torque on turbine for half of the cycle, hence positive and negative torque values are obtained in torque vs. flow time graph, similar torque variation is obtained in numerical analysis of bach rotor [9].

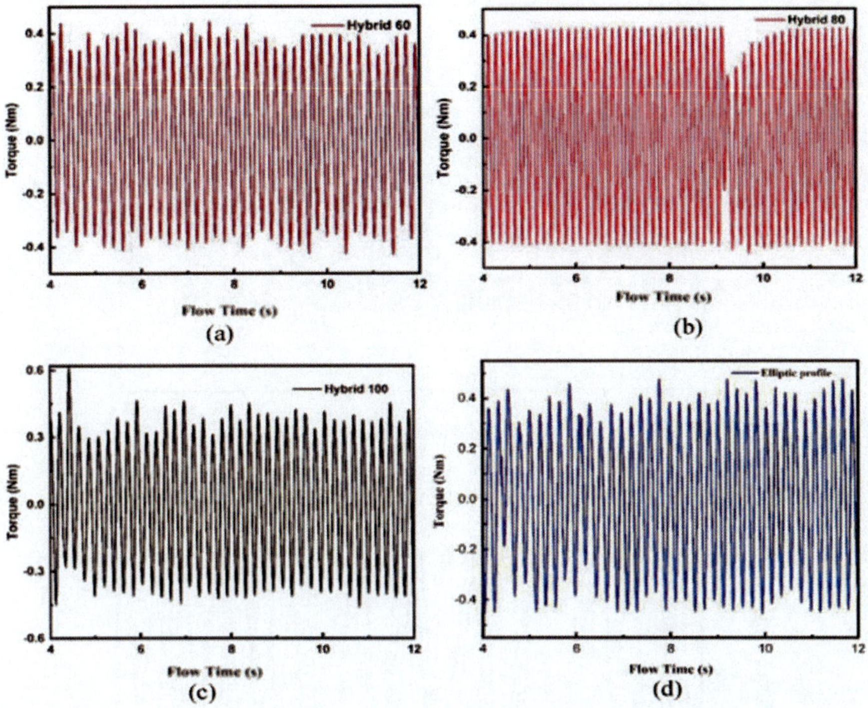

Fig. 5 Variation of torque for given flow time(s) of **a** Hybrid 60 **b** Hybrid80 **c** Hybrid 100 **d** Elliptic savonious

3.2 Comparison of All Three Conditions Hybrid 100, Hybrid 80 and Hybrid 60

All the three profile are compared, hence from Fig. 6 it is clear that 80% overlap of aerofoil on elliptical savonious gives better torque performance than other two. Figure 6a shows the variation of torque after achieving the steady state for 4–12 s. In order to interpret the results, same graph is scaled down for 2 s as shown in Fig. 6b which shows that the Hybrid 80 has higher torque performance than Hybrid 60 and 100 and torque variation is in between 0.4 and−0.4Nm.

3.3 Comparison of Hybrid 80 with Elliptical Savonious

From Fig. 6 it is clear that Hybrid 80 has better torque performance, hence it is compared with elliptical savonious profile. Figure 7a shows variation of torque for complete flow time after achieving steady state for elliptic savonious and Hybrid

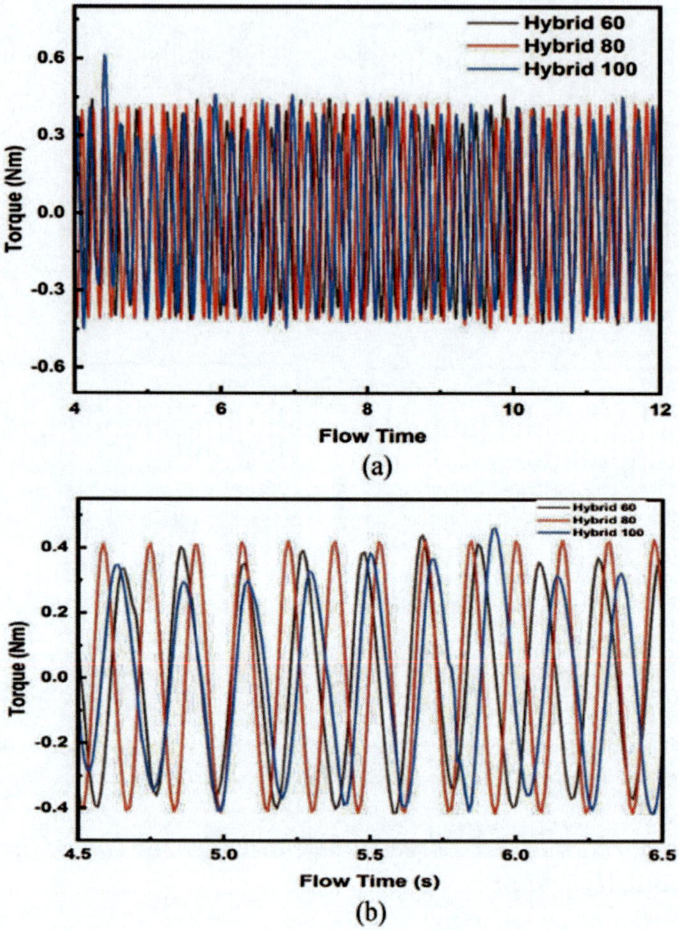

Fig. 6 Comparison of torque versus flow time of hybrid profiles **a** For large scale **b** For 2 s

80. After that the graph is scaled down to two sec for interpreting results. Figure 7b concludes that Hybrid 80 has better torque performance than elliptical savonious due to the effect of an aerofoil attached at end of rotor.

From above graphs, it is clear that the torque obtained for Hybrid 80 profile is larger than Hybrid 60 and Hybrid 100 for given flow time. Again when compared with elliptical savonious, Hybrid 80 shows the better performance for torque. If these torque values are used to calculate the power output, then following graphs are obtained for them.

Figure 8 shows the variation of power output for the given inlet wind velocity. Figure 8a shows variation of power for complete analysis and Fig. 8b is scaled down graph of Fig. 8a for inlet velocity of 2.8 to 2.96 m/s for easy comparison which shows

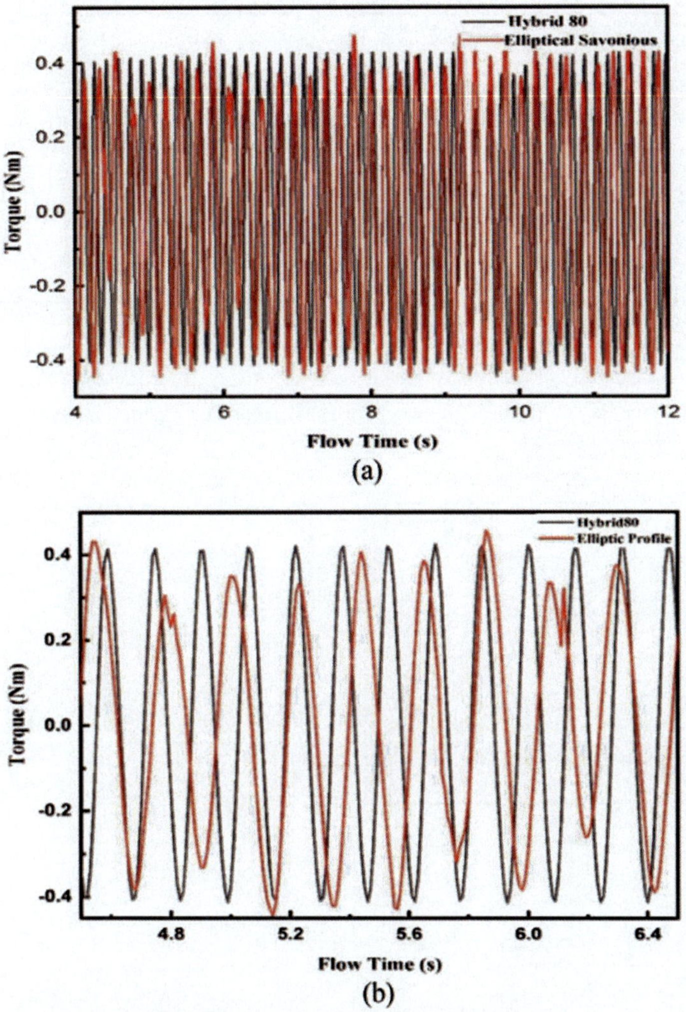

Fig. 7 Comparison of Hybrid 80 and elliptic profile **a** For complete time **b** For 2 s

that along with torque performance, power output is also better for Hybrid 80 when compared with other hybrid profiles.

From Fig. 8, it is clear that Hybrid 80 has better power performance than other Hybrid profiles. Hence it is compared with elliptical savonious profile. Figure 9a shows variation of power for given inlet velocity after achieving steady state for elliptic savonious and Hybrid 80. After that the graph is scaled down to two sec for interpreting results. From Fig. 9b we can conclude that Hybrid 80 has better power performance than elliptical savonious due to the effect of an aerofoil attached at end of rotor.

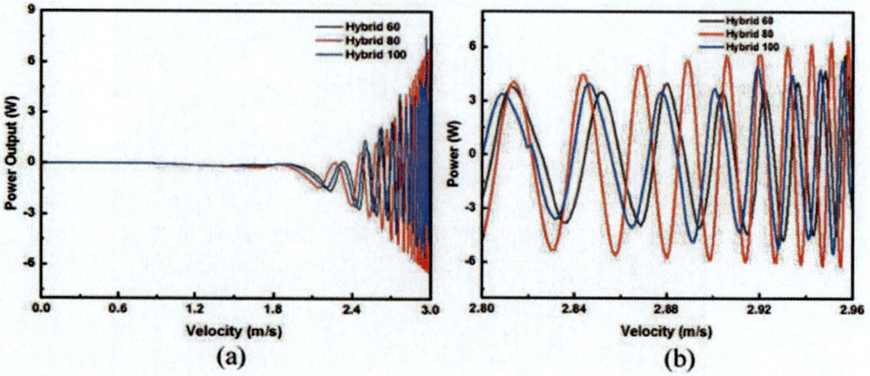

Fig. 8 Variation of power for inlet air velocity **a** For complete time period of Hybrid 60, 80 and 100 **b** Scaled to 2.8 to 2.96 s

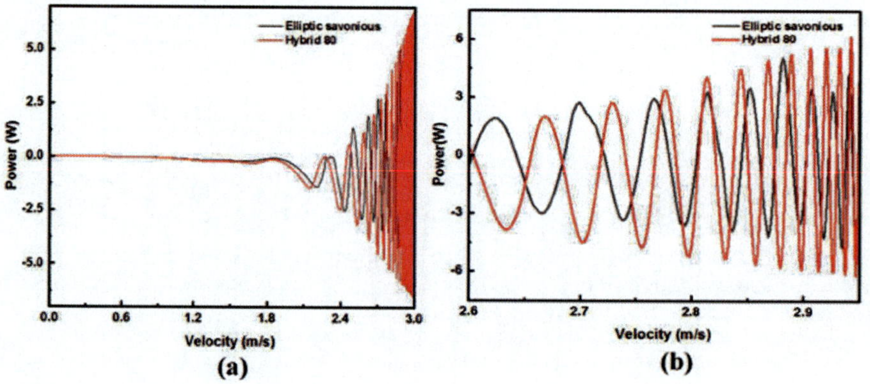

Fig. 9 Graph power versus input velocity **a** For complete time period of Hybrid 80 and elliptical savonious, **b** Scaled to 2.8 to 2.96 s

4 Experimental Details

4.1 Experimental Setup

A small prototype of elliptical savonious was manufactured having diameter of 300 mm and height of rotor as 210 mm with overlap ratio of 15% and endplates are of size 10% greater than blades diameter and aspect ratio of 0.7 as shown in Fig. 2 [7]. A stand of stainless steel was also manufactured for mounting of elliptical savonious turbine, shield and electrical motor. The windshield was linked to two plates mounted on a stand by a tiny angle plate. These angle plates have a mechanism that allows the shield to rotate at different angles of 30°, 45°, 60° and 75°as shown in Fig. 11. In order to reduce weight of turbine material two light materials, aluminum

Fig. 10 Elliptical savonious manufactured model

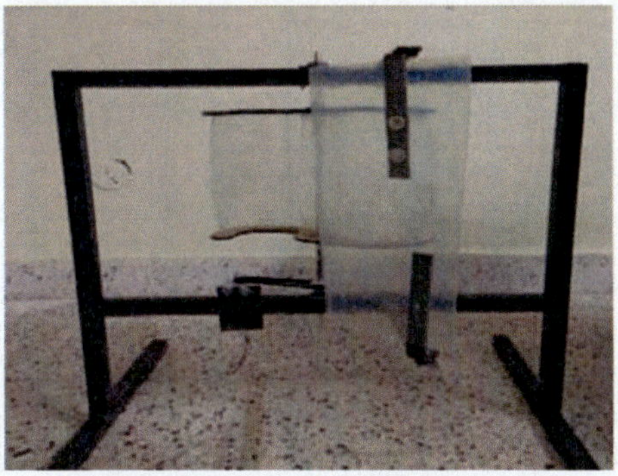

Fig. 11 Elliptical savonious with windshield

and acrylic fiber was available in market. As the density of acrylic fiber is less than aluminum, the acrylic fiber sheet of 2 mm was used as a material to manufacture the elliptical blade. And hard acrylic of 5 mm thickness is used to manufacture the end plates of rotor. The blades are fixed in the slots made by laser cutting operations at the endplates of turbine as shown in Fig. 10. Arrangement for the windshield is indicated in Fig. 11.

4.2 Setup for Voltage Measurement

A timing belt pulley was used to convert the rpm of rotor in electrical power output. Diameter of pulley attached to shaft is of 25 mm diameter and the smaller pulley of 8 mm diameter is placed at the DC motor as shown in Fig. 12. In order to convert the mechanical power into electrical power, a small DC motor of 12 V and 1800 rpm was used as a generator. And a small LED light is attached at the output of DC generator to check it is working or not.

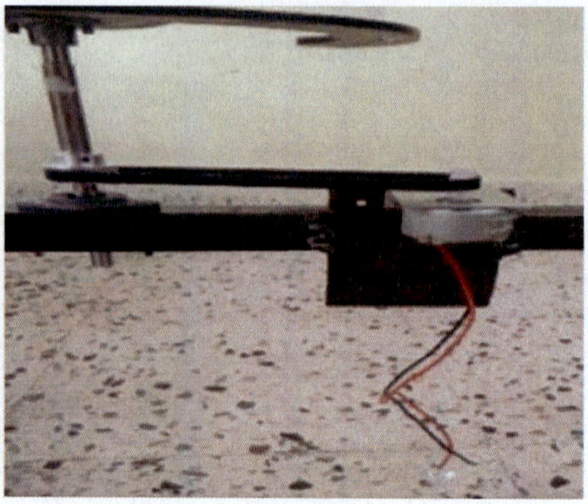

Fig. 12 Electrical output setup

4.3 *Experimental Procedure*

Initially using anemometer the flow uniformity test is done, i.e., 15–20 readings are taken at every point by varying the distance of anemometer from the fan and the points are marked for input wind speed of 4.5, 5.02, 5.56, 6.02, 6.5 m/s at distance of 40,61,88,111 and139 cm, respectively from fan [11]. Then the turbine stand with rotor is placed at every marked velocity points [12]. Digital tachometer is used to measure the rpm of the turbine for which a white color sticker is placed at the shaft of the rotor. In experiments, input parameter is wind speed and the output measuring parameter was rpm from tachometer and output voltage is measured by multimeter. Readings are taken for three conditions, i.e., elliptical savonious without windshield, elliptical savonious with windshield of length 170 mm and 260 mm placed at 30°, 45°, 60° and 75°.

5 Experimental Results

The maximum Cp value obtained was 0.1496 without using the windshield at 6 m/s of input wind speed and electrical power of 0.006612 W. Mechanical power output at same position was 5.57 W. The rpm of rotor is in the range of 164 to 259 rpm. Figure 13a shows that without using the windshield in front of elliptical savonious rotor, the Cp value increases till 6 m/s then trend changes as the input velocity increases. This shows that savonious turbine is efficient at lower wind speed only.

For elliptical savonious having windshield of length 170 mm places at position of 30°, 45°, 60° and 75° with respect to direction of windspeed, maximum Cp value of 0.179 is obtained at 5.02 m/s of input wind speed by placing windshield at position of 45° as shown in Fig. 13b. The electrical power output and mechanical power output at same position was 0.008742 and 3.918 W. In most of the cases, when the windshield is placed at 45 degrees the power coefficient values are better due to the better

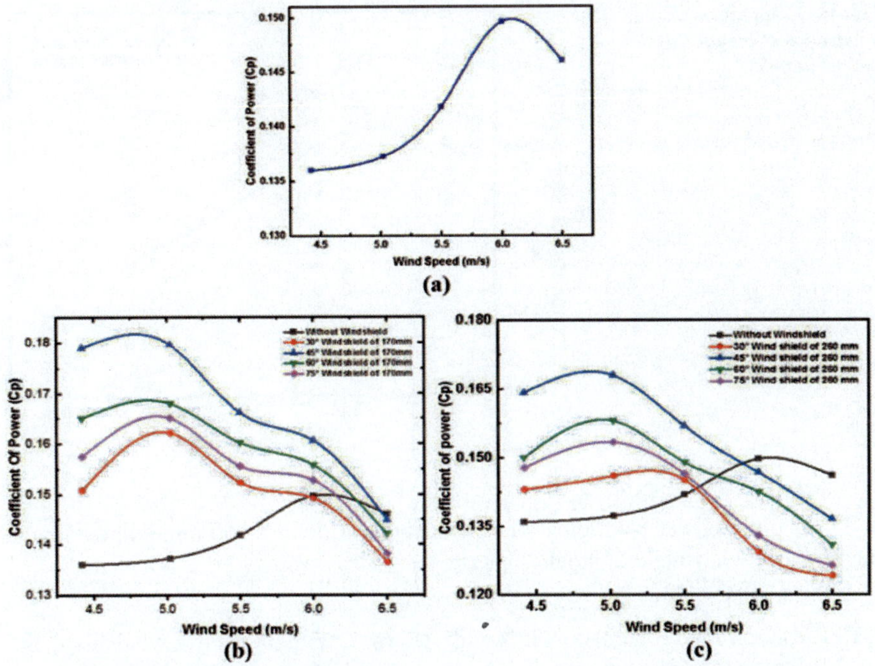

Fig. 13 **a** Elliptical savonious, **b** Elliptical savonious with windshield of length 170 mm and **c** Elliptical savonious with windshield of length 260 mm

air impact on the advancing blade and reduction of drag force on returning blade. Hence it is observed that there is an increase in Cp values for elliptical savonious with windshield when compared with without windshield. And optimum position of windshield of length 170 mm is 45°. For elliptical savonious with windshield of length 260 mm has maximum Cp value of 0.1680 and is obtained at 5.02 m/s of input wind speed by placing windshield at position of 45°. The electrical power output and mechanical power output at same position was 0.0052 and 3.713 W. Also the optimum position of windshield having length 260 mm is 45°.

Figure 13c shows variation of Cp for input wind speed in which windshield of length 260 mm is placed in front of elliptical savonious at four different angles, i.e., 30°, 45°, 60° and 75° is shown. From graph it is clear that the windshield placed at 45° is optimum when compared with other angles. Also it has maximum coefficient of power when compared with simple elliptical profile at lower velocities till 5.5 m/s. From all previous researches, it was found that savonious wind turbine has less power coefficient at higher wind velocities, this same effect is observed in Fig. 13, hence at higher wind speed it is better not to use windshield as performance using windshield decreases.

Both the windshield of length 170 and 260 mm placed at 45° with incoming air are compared as shown in Fig. 14. It shows that 170 mm length of shield placed

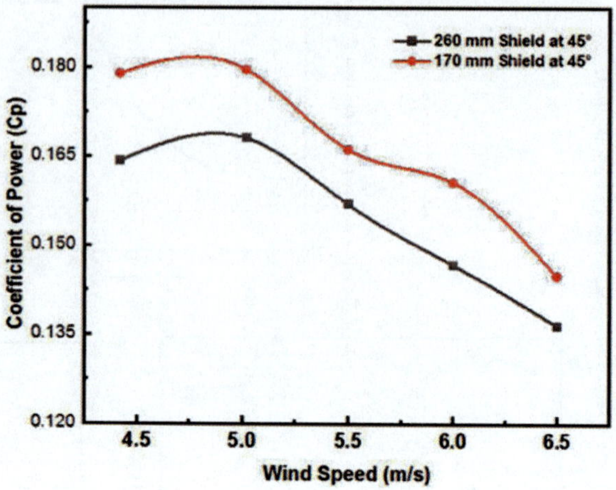

Fig. 14 Comparison of windshield of length 170 and 260 mm at 45°

at 45° as more power coefficient than windshield of length 260 mm length. Hence better to use windshield of smaller length.

Study indicated that

(1) Turbine started rotating at 1 m/s with less rpm. Without using a windshield, the mechanical power output is 5.57 W and the highest power coefficient is 0.149 at 6 m/s of input wind speed.
(2) The power output of the model is 3.91 W with a windshield of length 170 mm and a wind speed of 5.02 m/s positioned at a 45° angle, and 3W without a windshield. Additionally, the Cp value rises from 0.1372 to 0.1795, or by 30.83°.
(3) With windshield of length 260 mm at lower wind speed up to 5.5 m/s the power output with windshield is more compared with without shield, but at higher speed power output decreases when compared with elliptical savonious without shield.
(4) When elliptical savonious with shield lengths of 170 and 260 mm are compared, it is shown that the 170 mm shield has a 6.54% higher Cp value than the 260 mm length at 45° of shield.

6 Conclusion

In this study, two approaches are used to increase torque and power performance of elliptical savonious wind turbine. First by optimizing the blade by creating three hybrid profiles by adding the NACA0012 aerofoil shape at the end of elliptical savonious blade. Comparisons between the Hybrid 100, 80, and 60 profiles is done by numerical analysis which demonstrate that the Hybrid 80 has superior torque performance to the Hybrids 60 and 100. It also has superior torque and power performance compared with the elliptical savonious profile. Second approach is by introducing windshield in front of returning blades of rotor at four different angles by experimental analysis. The optimum position of windshield is obtained at 45° at shield length of 170 mm for which the elliptical savonious Cp value improves by 30.83%.

References

1. Statistical Review of World Energy: Energy Economics: Home. In: bp global. https://www.bp.com/en/global/corporate/energy-economics/statistical-review-of-world-energy.html. Accessed 21 Dec 2022
2. India plans to produce 175 GW of renewable energy by 2022 I Department of Economic and Social Affairs. In: United Nations. https://sdgs.un.org/partnerships/india-plans-produce-175-gw-renewable-energy-2022. Accessed 21 Dec 2022
3. Mahmoud NH, El-Haroun AA, Wahba E, Nasef MH (2012) An experimental study on improvement of Savonius Rotor performance. Alex Eng J 51:19–25. https://doi.org/10.1016/j.aej.2012.07.003
4. Alom N, Kumar N, Saha UK (2017) Aerodynamic performance of an elliptical-bladed Savonius rotor under the influence of number of blades and shaft, vol 2: Structures and dynamics; renewable energy (Solar, Wind); Inlets and Exhausts; Emerging Technologies (Hybrid Electric Propulsion, UAV,); GT Operation and Maintenance; Materials and Manufacturing (Including Coatings, Composites, CMCs, Additive Manufacturing); Analytics and Digital Solutions for Gas Turbines/Rotating Machinery. https://doi.org/10.1115/gtindia2017-4554
5. Alom N, Saha UK (2018) Evolution and progress in the development of Savonius wind turbine rotor blade profiles and shapes. J Solar Energy Eng 10(1115/1):4041848
6. Alom N, Kolaparthi SC, Gadde SC, Saha UK (2016) Aerodynamic design optimization of elliptical-bladed Savonius-style wind turbine by numerical simulations, vol 6. Ocean space utilization; ocean renewable energy. https://doi.org/10.1115/omae2016-55095
7. Alom N, Saha UK (2017) Arriving at the optimum overlap ratio for an elliptical-bladed Savonius rotor, vol 9. Oil and gas applications; Supercritical CO_2 power cycles; Wind energy. https://doi.org/10.1115/gt2017-64137
8. Alom N (2021) Influence of curtain plates on the aerodynamic performance of an elliptical bladed Savonius rotor (S-rotor). Energy Syst 13:265–280. https://doi.org/10.1007/s12667-021-00428-w
9. Zhou T, Rempfer D (2013) Numerical study of detailed flow field and performance of Savonius wind turbines. Renew Energy 51:373–381. https://doi.org/10.1016/j.renene.2012.09.046
10. Tartuferi M, D'Alessandro V, Montelpare S, Ricci R (2015) Enhancement of Savonius Wind rotor aerodynamic performance: a computational study of new blade shapes and curtain systems. Energy 79:371–384. https://doi.org/10.1016/j.energy.2014.11.023
11. Utomo IS, Tjahjana DD, Hadi S (2018) Experimental studies of savonius wind turbines with variations sizes and fin numbers towards performance. AIP conference proceedings. https://doi.org/10.1063/1.5024100
12. Ahmed WU, Uddin MR, Sadat QT, et al (2020) Performance assessment of a small-scale vertical axis single-stage savonius wind turbine by using artificial wind. In: 2020 IEEE Region 10 Symposium (TENSYMP). https://doi.org/10.1109/tensymp50017.2020.9230925

Self-Recurrent Neural Network-Based Event-Triggered Mobile Object Tracking Strategy for Sensor Network

Vishwalata Bagal and A. V. Patil

Abstract This paper presents an intelligent strategy to track the moving nodes in wireless sensor networks (WSN) to increase energy efficiency and tracking accuracy. The immediate location of nodes in WSN is of great importance in many applications. However, the power and cost at the various communication layers for the node localization has arisen as a critical difficulty. To improve the energy performance of the network's power-deprived sensor nodes, a distributed event-based adaptive target tracking technique is proposed. Unlike traditional target tracking systems, this research addresses the functional and parametric uncertainties introduced in the network dynamics resulted from multi-path fading, reflections, and unknown disturbances. These uncertainties are assessed using a neural network framework with a wavelet kernel function that has localized time–frequency estimation features. The usage of a wavelet neural network (WNN) to accurately identify uncertainty improves the tracking algorithm's robustness. This research derives the appropriate neural network tuning principles for training and accurate target tracking. A simulation study has been performed in the research work to ensure that the recommended approach is effective and accurate.

Keywords Neural networks (NN) · Activation function · Object detection · Wireless sensor network (WSN) · Target localization · Node tracking

1 Introduction

Because of its compact size and high processing capabilities, wireless sensor networks have emerged as a revolutionary technology that has enabled smart networking architecture. The ad hoc and transportable nature of WSNs has given

V. Bagal (✉)
D.Y. Patil College of Engineering, Akurdi, Pune, India
e-mail: vishwal.bagal@gmail.com

A. V. Patil
D.Y. Patil Institute of Engineering, Management and Research, Akurdi, Pune, India

S. Sharma et al. (eds.), *Intelligent Control, Robotics, and Industrial Automation*, Lecture Notes in Electrical Engineering 1066,
https://doi.org/10.1007/978-981-99-4634-1_60

these networks a whole new dimension. It is made up of various sensor nodes with data gathering and processing capabilities that are coupled in a meaningful way to communicate for a smart task. According to the requirements, these nodes can be joined in any topology. At the physical, data link, and network layers, multiple protocols describe the relationship between these nodes. These communication layers describe the network's parameters for frequency, bandwidth, voltage, error detection, error correction, and routing. The information given by the nodes about the sensed data is used to perform the needed action for the defined application [1–3].

One of the primary concerns in the implementation of WSN for any application is energy efficiency. The nodes are powered by batteries because the network is established on an ad hoc basis and the nodes may be stationary or mobile depending on the demand. These batteries have a finite lifespan that is determined by the node's processing and communication load. Continuous sensing of the environment in order to gather data, process the data, and transfer the data to a centralized unit necessitates a significant amount of electricity. Even a single node's energy depletion throughout the procedure could result in a hazardous or deadly condition. Many academics have offered numerous strategies to increase the energy efficiency of sensor nodes [4, 5] while keeping the relevance of sensor node lifetime in mind.

Clustering-based routing protocols, Low-energy adaptive clustering hierarchy (LEACH) routing protocols, Scalable energy-efficient clustering hierarchy (SEECH)-based routing protocols, Discontinuous transmission-based routing protocols, Event-triggered routing protocols, etc. have been proposed by many researchers [6]. The fundamental goal of these protocols was to keep sensor nodes from continuously sensing, processing, and transmitting data through a distributed architecture in terms of clusters or dataflow control. However, in all of these setups, the processing capability of each node was believed to be the same. In an effort to extend the lifetime of nodes, the processing complexity is also traded off with other characteristics. Despite the problem's intricacy, numerous researchers are still working on more efficient solutions.

The instantaneous position of the nodes, on the other hand, is critical in many applications since it reflects the location of the information source. The information contour mapping provides a very exploratory sense about the data distribution across a big area. It is crucial in the respective decision-making process as well as the sequence of subsequent significant events. Various applications of ad hoc WSN rely on the position of mobile nodes to take action. In some WSN geographic routing systems, the distance between the sensors is also crucial. Many researchers have presented solutions for target tracking of mobile nodes in WSN during the last two decades because of this element of WSN [7].

This work's main contribution is as follows:

1. An efficient and effective approach to concurrently achieving target tracking and energy efficiency for a realistic mathematical framework with uncertainties.
2. An intelligent target tracking framework that is adaptable and insensitive to network conditions.

3. To improve the estimation of functional and parametric uncertainties generated due to the parasitic effects, a proposed augmented framework that uses wavelets as the activation function in the ANN model augments the decomposition features of wavelet with the learning potential of deep learning architecture.
4. The robustness is attained without compromising the network parameters or the complexity of the calculation.

Rest of the paper is organized as follows: Detailed review of the existing literature is shown in Sect. 2 which highlights the notable solutions presented by the researchers. Section 3 discusses the problem formulation in terms of the observation model and mobility model. The mathematical model of SRWNN that uses the wavelet function as an activation function is covered in Sect. 4. In Sect. 5, a proposed event-triggered target localization model for WSN is provided. The simulation study undertaken to confirm the efficacy of the suggested technique is discussed in Sect. 6, and the paper is wrapped up in Sect. 7.

2 Literature Survey

The target tracking strategy based on received signal strength indicators (RSSI) has been the most extensively explored technique for many years. This method uses the strength of received signals from neighboring nodes to estimate the nodes' location. The tracking accuracy is greatly harmed by disturbances and uncertainty in measurements caused by environmental influences. Researchers have proposed many filters to increase tracking performance, including various configurations of Kalman filters and Bayesian filters [8]. However, due to the unknown noise distribution model, tracking accuracy is still reliant on the availability of a precise mathematical model, which is difficult to establish.

Only a few researchers, on the other hand, tackled the problem of energy efficiency and target tracking at the same time. To increase energy performance while estimating the position of mobile nodes, the sleep scheduling technique is combined with a target tracking strategy [9]. The nodes are classified into two categories in this technique: 'awake' and 'asleep,' based on their operational status on a time scale. Only the 'alive' nodes are taken into account for target tracking measurements at any given time. The sleep time distribution is calculated using a stochastic algorithm. Because sleep patterns are unpredictable, tracking accuracy cannot be guaranteed for every set of readings. The researchers also proposed transforming this task into an NP-complete optimization problem [10]. The maximization of lifetime and minimization of routing path length were evaluated as goal functions, although at the expense of tracking accuracy. Some research works [11, 12] have attempted to overcome the problem via dynamic clustering. The clustering of nodes was carried out in real time to ensure energy economy and tracking precision. However, the effectiveness of this technique was contingent on the availability of global topology data. The authors in [13] proposed an event-triggered estimation technique to provide energy-efficient target tracking for mobile nodes in a WSN environment. The estimation error or error covariance is used to make a judgment on the node's transmission and is compared to a set threshold. The authors in [14] proposed another event-triggered strategy in which measurements from different sensors are connected in bus topology of network to a central process. The effectiveness of the method was verified by examining the convergence rate of the Riccati equation in switching mode. These target tracking strategies provided an optimal solution in terms of energy efficiency and tracking accuracy, but their effectiveness was contingent on the system's accurate mathematical framework, and any deviation from the derived model had a significant impact on the overall results. Reflections, multi-path propagation, unpredictable noise distribution, ambient variables, and other factors can all contribute to these uncertain dynamics.

Various deep learning tools like Artificial Neural Network (ANN), fuzzy logic, reinforcement learning (RL), and others have recently attracted a lot of attention due to their ability to estimate unknown functional or parametric uncertainties over the last three decades. Researchers in [15] increased the network's training performance and estimate capabilities in a highly dynamic and random environment by replacing the traditional sigmoid-based activation with fast decaying wavelets. Wavelets' time–frequency localization property has been used in this wavelet neural network (WNN)-based modified estimate model. It's a feed-forward network that blends wavelet decomposition with traditional neural network learning capabilities.

We decided to pursue this research because of the limitations of existing tracking target tracking strategies in achieving higher tracking accuracy while maintaining acceptable energy efficiency in a random and uncertain environment (ill-defined mathematical model) and the potential of WNN to estimate functional and parametric uncertainties. For mobile nodes in WSN, this work proposes an intelligent adaptive moving object tracking technique. The event-triggered estimation model, in which the unknown dynamics due to parasitic effects are approximated by WNN, was implemented to meet the dual goals of tracking precision and energy efficiency.

3 Problem Statement

The mobility of sensor nodes in a WSN system has been addressed by many researchers. Traditionally, mathematical frameworks such as the random walk and pursuit mobility models, as well as singer type models, have been assumed. With the assumption that the noise distribution and disturbances are known, these models gave a state space representation in terms of the parameters of the mobile node like position, velocity, and acceleration. But, due to the randomness and uncertainties involved with the measurements, it is not possible to do so in real time. The nonlinear uncertainties in the model studied in this work are assumed to be unknown.

The node's positions in the network are represented as x and y coordinates where the network comprises of N sensor nodes whose mobility is constrained to a two-dimensional plane. A discrete time model is used to depict the mobility model as [16]:

$$\begin{aligned} x(k+1) &= Ax(k) + Bf(x(k)) + u \\ y(k) &= Cx(k) \end{aligned} \tag{1}$$

$$A = \begin{bmatrix} 0\,1\,0 \ldots 0 \\ 0\,0\,1 \ldots 0 \\ \vdots\,\vdots\,\vdots \quad \vdots \\ 0\,0\,0 \ldots 1 \\ 0\,0\,0 \ldots 0 \end{bmatrix}, B = \begin{bmatrix} 0 \\ 0 \\ 0 \\ \vdots \\ 1 \end{bmatrix} \text{ and } C = \begin{bmatrix} 1\,0\,0 \ldots 0 \end{bmatrix}$$

where $x(k) \in \Re^n$ is the state vector and $f(x(k)) : \Re^n \rightarrow \Re$ is an unknown function that shows the motion model's unknown nonlinear dynamics, which are similar to noise and external disturbances. It also contains reflection and refraction, obstructions, incorrect initial condition setup, and WSN node component aging.

The covariance of the uncertain nonlinearity is considered to be R_i. To deal with such issues, we need to create an adaptive estimator that is capable of estimating the uncertainties and variances of the model.

In terms of the sensor node, the measurement model is described as

$$z_i(k) = D_i x(k) + E_i \hat{f}(x(k)), \quad i = 1, 2, \ldots, N \tag{2}$$

where the measurement of ith node at kth sampling instant is represented by $z_i(k)$. $D_i$and E_i represent the respective measurement matrices. S_i is assumed to be the covariance of the uncertain nonlinearity. The gap between the sensor node and the target node determines the ability of ith node to track it. It's also a result of the triggering strategy. Given the distributed nature of the overall target tracking strategy, any node's capacity to take measurements in order to track a target is depicted as $\sigma_i \in \{0, 1\}$. Here $\sigma_i = 0$ represents that the node cannot take the measurements and $\sigma_i = 1$ represents its ability to do so. It is evaluated in terms of two parameters $\mu_i \in \{0, 1\}$ and $\lambda_i \in \{0, 1\}$ as $\alpha_i = \mu_i \lambda_i$. Here μ_i denotes that whether the target node lies within the sensing range of ith node or not and the value of λ_i reflects the triggering status of the corresponding node.

The goal of the proposed research is to present an optimal solution by determining the best tradeoff between estimating inaccuracy and energy efficiency. As a result, the problem is turned into an optimization problem, and a strategic utility function Q is created to assess the suggested strategy's performance. It is made up of both the estimation inaccuracy and the energy efficiency of the parameters. The estimation error is incorporated in terms of the covariance matrix $\zeta_i(k)$ and the energy of the ith node is assumed to be ε_i. The strategic utility function is derived as

$$Q_i(k) = (1 - \lambda_i(k))tr(\zeta_{i0}(k)) + \lambda_i(k)tr(\zeta_{i1}(k)) + \lambda_i(k)\varepsilon_i \quad (3)$$

where $tr(\bullet)$ represents the trace of a matrix, $\zeta_{i1}(k)$ *and* $\zeta_{i0}(k)$ represent the estimation error covariance matrix with respect to the measurement status of ith node, that whether it has taken the measurements or not, respectively. $\lambda_i(k)$ and $(1 - \lambda_i(k))$ are the measurement cost associated with $\zeta_{i1}(k)$ and $\zeta_{i0}(k)$, respectively, in the strategic utility function. The proposed work's goal is to minimize the strategic utility function, which can be accomplished by estimating states $\hat{x}(k)$ in such a way that the error tends to zero, i.e., $\tilde{x} = x - \hat{x} \approx 0$. The task also entails properly estimating the value of $f(x)$ in order to cancel out the unpredictable dynamics and achieve exact target localization.

4 Self-Recurrent Wavelet Neural Network Architecture

The activation function utilized in the relevant architecture has the possibility of identifying and estimating any machine learning framework. The sigmoid activation function is substituted by the rapidly decaying wavelet function in the hidden layer, which is a modified architecture of a typical ANN and referred as WNN. It has attracted a huge attention in the field of artificial intelligence due to its better training performance and faster convergence. Each wavelet coefficient corresponds to a hidden layer neuron in such a way that the Inverse Discrete Wavelet Transform (IDWT) for that neuron is defined locally in terms of time parameter and frequency parameter over a particular translation and dilation. WNN is discovered to be a computationally efficient estimation framework [17] since all of the wavelet coefficients are not required for the reconstruction of the original function. It also aids in the reduction of network overfitting and the improvement of convergence. Figure 1 depicts the architecture of a typical WNN.

In general, feed-forward neural network weight updates do not leverage the neural networks internal knowledge, and function approximation is highly dependent on the training data. Self-Recurrent Neural Networks (SRNN), on the other hand, which include both feed-forward and feedback connections, are widely renowned for their ability to describe and control complex processes. Because SRNN include an internal feedback loop that captures a system's dynamic response without requiring external feedback through delays, they outperform feed-forward neural networks. These networks can deal with input or output that changes over time. It results into dynamic mapping of weights over the network and shows good performance even in the uncertain environment, such as unanticipated target movement, unmodeled system dynamics, external environmental disturbance, etc. As shown in Fig. 1, the SRNN consists of a self-feedback hidden layer which enables the neurons in the hidden layer to store the network's previous knowledge and record the system's dynamic response. SRNN can now accurately estimate dynamic nonlinearities thanks to this update. As a result, SRNN is a better tool for adaptive control techniques than traditional recurrent networks.

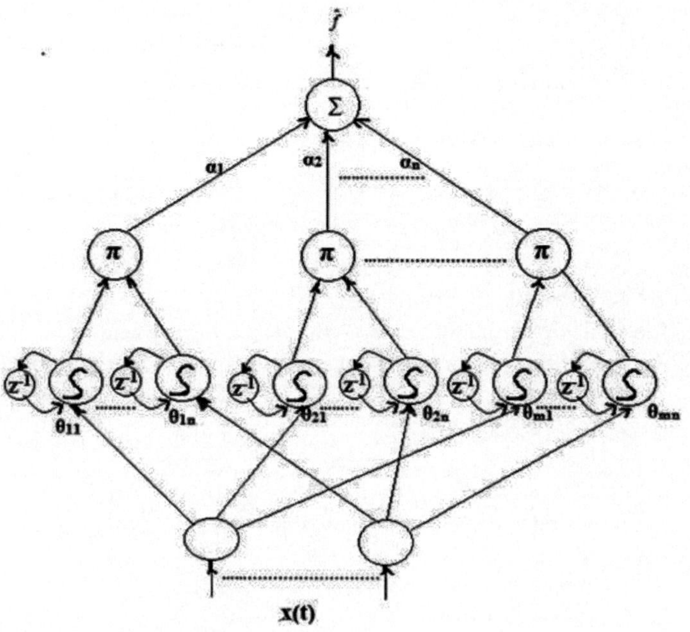

Fig. 1 Architecture of self-recurrent WNN

The network is a biased n-dimensional structure with m number of nodes whose output is shown as

$$f = \alpha^T \psi(x, \tau, \delta) + \beta^T \phi(x, \tau, \delta) \tag{4}$$

where the input vector is represented as $x = [x_1, x_2, \ldots, x_n]^T \in R^n$, and the weights for scaling and biasing are represented as $\alpha = [\alpha_1, \ldots, \alpha_m]^T \in R^m$ and $\beta = [\beta_1, \ldots, \beta_m]^T \in R^m$, respectively. $\tau = [\tau_1, \tau_2, \ldots, \tau_m]^T \in R^{mxn}$ and $\delta = [\delta_1, \delta_2, \ldots, \delta_m]^T \in R^{mxn}$ are the wavelet parameter applied with scaling (dilation) and shifting (translation), respectively. $\psi = [\psi_1, \psi_2, \ldots, \psi_m]^T \in \Re^m$ and $\phi = [\phi_1, \phi_2, \ldots, \phi_m]^T \in \Re^m$ are functions for activation and addition at hidden layer, respectively.

Considering that f^* is the best possible estimation of the uncertainty f from WNN with a small error Δ, we can assume that

$$f = f^* + \Delta = \alpha^{*T} \varphi^* + \beta^{*T} \phi^* + \Delta \tag{5}$$

where $\varphi^* = \varphi(x, \tau^*, \delta^*)$ and $\phi^* = \phi(x, \tau^*, \delta^*)$, $\alpha^*, \beta^*, \tau^*$, and δ^* are the optimal values of α, β, τ, and δ, respectively.

If estimates of $\alpha^*, \beta^*, \tau^*, \delta^*$ are assumed as $\hat{\alpha}, \hat{\beta}, \hat{\tau}, \hat{\delta}$, respectively, $\hat{\phi} = \phi\left(x, \hat{\tau}, \hat{\delta}\right)$, $\hat{\psi} = \psi\left(x, \hat{\tau}, \hat{\delta}\right)$, it can be written as

$$\hat{f} = \hat{\alpha}^T \hat{\psi} + \hat{\beta}^T \hat{\phi} \tag{6}$$

The error in estimation can deduced using (5) and (6) as

$$\begin{aligned} \tilde{f} &= f - \hat{f} \\ &= f^* - \hat{f} + \Delta \\ &= \tilde{\alpha}^T \tilde{\psi} + \hat{\alpha}^T \tilde{\varphi} + \tilde{\alpha}^T \hat{\psi} + \tilde{\beta}^T \tilde{\phi} + \hat{\beta}^T \tilde{\phi} + \tilde{\beta}^T \hat{\phi} + \Delta \end{aligned} \tag{7}$$

where $\tilde{\alpha} = \alpha^* - \hat{\alpha}$, $\tilde{\beta} = \beta^* - \hat{\beta}$, $\tilde{\psi} = \psi^* - \hat{\psi}$, $\tilde{\phi} = \phi^* - \hat{\phi}$.

To keep the value of estimation error $\tilde{f}$ within an arbitrary small limit, the dimension of network and the type of activation wavelet are selected such as the bound $\left\| \tilde{f} \right\| \leq \tilde{f}_m$ holds for all $x \in \Re$. The value of $\hat{f}(x)$ generated by the WNN is given by

$$\begin{aligned} \hat{f}(x(t)) &= \left\{ \begin{array}{l} \sum\limits_{q=M1_J}^{M2_J} \alpha^*_{J,q}(k) \phi_{J,q}(x(t)) + \\ \sum\limits_{j \geq J}^{J_N} \sum\limits_{q=M1_j}^{M2_j} \beta^*_{j,q}(t) \psi_{j,q}(x(t)) + \varepsilon(x(t)) \end{array} \right\} \\ &= \alpha^{*T} \phi(x(t)) + \beta^{*T} \psi(x(t)) + \varepsilon(x(t)) \, \forall x(t) \subset \Re^n \end{aligned} \tag{8}$$

where $N \in \Re$ and J represent the highest and lowest resolution, respectively. The number of translates at jth resolution is represented by $q = [M1_j, \cdots, M2_j] \in \Re$ and the approximation error as $\varepsilon(x(k))$

$$\varepsilon = \begin{pmatrix} \alpha^{*T} h_1 + \tilde{\alpha}^T A_1^T w^* + \tilde{\alpha}^T B_1^T c^* \\ +\beta^{*T} h_2 + \tilde{\beta}^T A_2^T w^* + \tilde{\beta}^T B_2^T c^* \end{pmatrix} \tag{9}$$

Steepest Gradient technique has been used to evaluate the optimal weights of WNN over the least square cost function:

$$J(\theta) = \frac{1}{2} \sum_{n=1}^{N} (y_p^n - y^n)^2 \tag{10}$$

where the relation between wavelets and the output governs the cost function $J(\theta)$. The tuning laws for the weight updation can be derived using backpropagation in terms of learning rate η and tuning weights w as

$$w(k+1) = w(k) + \Delta w(k) = w(k) + \eta \left(-\frac{\partial E(k)}{\partial w(k)} \right) \tag{11}$$

5 Target Tracking Model for WSN Using Self-Recurrent Wavelet Neural Network (SRWNN)

The suggested target tracking framework is for a distributed sensor network that uses an event-triggered measuring approach to save energy. This proposed technique is based on the concept that the sensor node is made active only when the target is in the detecting range, as opposed to standard time triggered measuring strategies where every sensor is kept active all of the time to measure the parameters. To propose a distributed target tracking paradigm, all of the nodes collaborate on a consensus basis. Estimation model, consensus algorithm, and event-triggered method are the three pieces of the total algorithm. The estimating model is a simple filter that is used to get a tentative first approximation of the states, i.e., $\hat{x}(k+1)$. Because the measuring environment changes over time due to the target node's mobility, the consensus algorithm provides a framework for collaboration across different nodes to follow the mobile target. The mathematical framework of decision making for the node to be triggered to take measurement is presented in the event-triggered strategy, which will aid in achieving energy efficiency. The three stages are described in full below.

5.1 Estimation Model

We present an estimating model based on the mobility model of the nodes as described in (1).

$$\begin{aligned}\hat{x}(k+1) &= A\hat{x}(k) + B\hat{f}\big(\hat{x}(t)\big) + m\big(\hat{y}(k) - C\hat{x}(k)\big)\\ \hat{y}(k) &= C\hat{x}(k)\end{aligned} \tag{12}$$

where the estimate of state vector x is represented by $\hat{x}$ and the estimator gain matrix is represented as $m = [m_1, m_2, \ldots, m_n]^T$. The most difficult aspect of state estimation of the system is determining the value of $f(x(k))$, which is unknown and indeterminate. Here, $\hat{f}(x(k))$ is the estimated value of uncertain function $f(x(k))$. Uncertainty estimation and state estimation are the two aspects of the estimation model. Noise owing to multi-path propagation, reflection, and environmental disturbances are all functions governing uncertainties in the system model, and they affect all states of the system. The suggested Wavelet neural network-based estimator provided in is used to estimate this function (8).

The following assumptions are made in this study to avoid overfitting of the proposed estimator.

(a)

$$\left\|\hat{f}(x) - \hat{f}(\hat{x})\right\| \le \gamma_1 \|\tilde{x}\| \tag{13}$$

(b)

$$(A - mC)^T P + P(A - mC) = -Q \text{ and } (PB_{21})^T = C$$
$$\text{for a symmetric positive definite matrix } Q \quad (14)$$

where state variable estimation error is shown as $\tilde{x} = x - \hat{x}$ is the.

The estimator can now be defined in terms of the uncertain term $f(\hat{x}(k))$ as

$$\begin{aligned}\hat{x}(k+1) &= A\hat{x}(k) + B\hat{f}(\hat{x}(k)) + m(\hat{y}(k) - C\hat{x}(k)) \\ \hat{y}(k) &= C\hat{x}(k)\end{aligned} \quad (15)$$

Subtracting (15) from (1), the error system can be written as

$$\begin{aligned}\tilde{x}(k+1) &= (A - mC)\hat{x}(k) + B\Big(f(\hat{x}(k)) - \hat{f}(\hat{x}(k))\Big) \\ \tilde{y}(k) &= C\tilde{x}(k)\end{aligned} \quad (16)$$

The goal has now been shifted to modifying the estimator gain values in order to reduce the values to an infinitesimally small number. The two components of the proposed estimator are combined and fine-tuned until the accuracy achieves its maximum level.

5.2 *Consensus Algorithm*

The major goal of this stage is to combine measurements from several nodes in order to estimate the target's exact location. To achieve simplicity in collaboration, it is built on the minimum trace fusion idea. Assuming that at time instant k, ith node has the prior estimation states $\hat{x}_i(k-1)$ and the respective covariance matrix $\zeta_i(k-1)$. It also receives the other new measurements, locally aggregate observed data and covariance matrix from other nodes as $\psi_i(k) = E_i^T S_i^{-1} z_i(k)$ and $\xi_i(k) = E_i^T S_i^{-1} E_i$. The weighted average strategy is used further to fuse the information as

$$\overline{x}_i(k-1) = \sum_{j \in \mathbb{N}_i} \omega_{ij}(k)\hat{x}_i(k-1) \quad (17)$$

$$\overline{\zeta}_i(k-1) = \sum_{j \in \mathbb{N}_i} \omega_{ij}(k)\hat{\xi}_i(k-1) \quad (18)$$

where $\mathbb{N}_i$ resembles to the nodes neighboring to the ith node. The weights ω_{ij} are tuned as per the minimum trace fusion principal. The convergence of the tuning laws for ω_{ij} are governed using the strategic utility function Q by minimizing $\arg\min_{j \in \mathbb{N}_i}\{Q|tr(\zeta_j(k-1))\}$.

The fused prior estimates are now utilized to estimate the posterior estimates of the state $\hat{x}_i(k)$ and its respective covariance matrix $\zeta_i(k)$ as

$$\hat{x}_i(k) = \hat{x}_i(k-1) + \zeta_i(k)\left(\sum_{j\in\mathbb{N}_i}\sigma_j(k)\big(\psi_j(k) - \xi_j\overline{x}_i(k-1)\big)\right) \tag{19}$$

$$\zeta_i(k) = \left(\overline{\zeta}_i^{-1}(k-1) + \sum_{j\in\mathbb{N}_i}\sigma_j(k)\xi_j\right)^{-1} \tag{20}$$

Also,

$$\hat{x}_i(k+1) = A\hat{x}_i(k) + B\hat{f}\big(\hat{x}_i(k)\big) + u \tag{21}$$

$$\zeta_i(k+1) = A\zeta_i(k)A^T + S_i \tag{22}$$

It results into the complete fusion of all the information at ith node using the minimum trace fusion principal.

5.3 *Event-Triggered Strategy of Measurement*

In this phase, the evaluated values of the prior estimation state and the corresponding covariance matrix are used to make a decision on whether or not to trigger the node to measure the parameters. The term 'event-triggered measurement strategy' refers to this type of decision making. The requirements for permitting the measurement to be taken by the node are as follows:

$$\lambda_i(k+1) = \begin{cases} 1\ if|\tilde{x}_i| < r_s + \upsilon_i(k+1) \text{ and} \\ \quad Q_{i1}(k+1) < Q_{i0}(k+1) \\ 0 \text{ otherwise} \end{cases} \tag{23}$$

where $\tilde{x}_i$ represents the tracking error, r_s is the sensing range, and $\upsilon_i(k+1)$ represents the estimation error. The tradeoff between the estimation accuracy and the energy efficiency is intended to be attained through the strategic utility function Q as shown in (24). However, $\lambda_j(k)$ and $(1-\lambda_j(k))$ are the measurement cost associated with $\zeta_{j1}(k)$ and $\zeta_{j0}(k)$, respectively, in the strategic utility function. Considering this the performance indices $Q_{i1}(k+1)$ and $Q_{i0}(k+1)$ may be defined as

$$\begin{aligned} Q_{i1}(k+1) &= tr(S_{i1}(k+1)) + \varepsilon_i \\ Q_{i0}(k+1) &= tr(S_{i0}(k+1)) \end{aligned} \tag{24}$$

with

$$S_{i1}(k+1) = \zeta_i(k+1) - \rho_i \zeta_i(k+1) D_i^T * \left(D_i \zeta D_i^T + R_i\right)^{-1} D_i \zeta_i(k+1) \quad (25)$$

$$S_{i0}(k+1) = \zeta_i(k+1) \quad (26)$$

Here ρ_i is the number of neighboring nodes to ith node which are communicating.

6 Simulation Analysis

The evaluation of the proposed event-triggered energy-efficient technique for target tracking in WSN environments is carried out through a simulation analysis. This study considers a WSN environment with 50 nodes distributed over a $100 \times 100\ \text{m}^2$ region. The suggested state estimator's performance is measured in terms of tracking error, WNN estimation error, and tracking output. Unlike typically reported systems, the mobility system model used in this simulation research is nonlinear. In this study, the impact of beginning conditions is also examined and comprehended. The system's tracking results are measured in terms of the error during the tracking, and that for WSN in terms of the error in estimation.

The simulation analysis here is performed over a system of the form

$$\begin{aligned} \dot{\chi}_1 &= \chi_2 \\ \dot{\chi}_2 &= \chi_3 + 0.01\chi_1\chi_2^2 + u \\ \psi &= \chi_1 \end{aligned} \quad (27)$$

The trajectory to be tracked by the proposed strategy in the analysis is

$$y_d = 40\sin(x) + 20\cos(0.5x) + 4e^{-x} \quad (28)$$

The states are assumed to possess the initial conditions as $\chi(0) = [1, 1, 1, 1, 0.6, 0.6]^T$.

Because of its efficient time–space features, the Mexican hat wavelet was chosen as the activation function in the SRWNN model. Figure 2 depicts the proposed localization strategy's tracking performance. The desired trajectory is shown in red color and the respective tracking is shown in green color in Fig. 2 which shows that the tracking is quite accurate and fast. Figure 3 demonstrates the rate of convergence of tracking error. Figure 4 shows the suggested WNN's performance in terms of estimation error, which is likewise highly encouraging. The results reveal that, even in the presence of system dynamics uncertainty and no measurements; the predicted

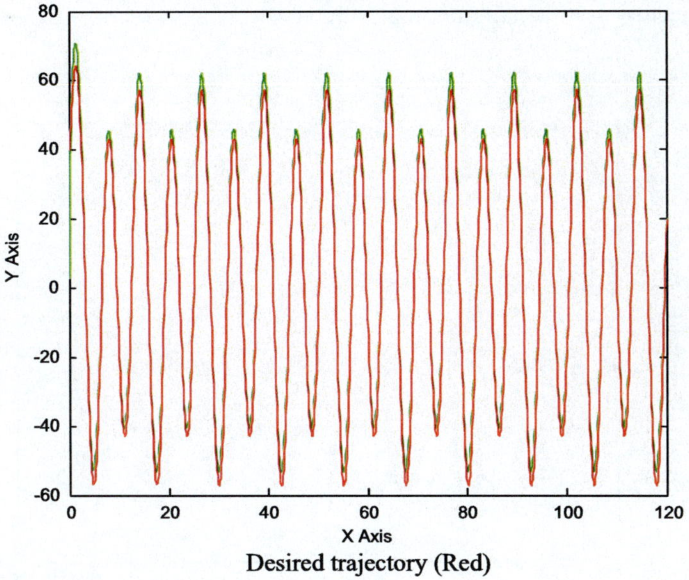

Fig. 2 Tracking performance

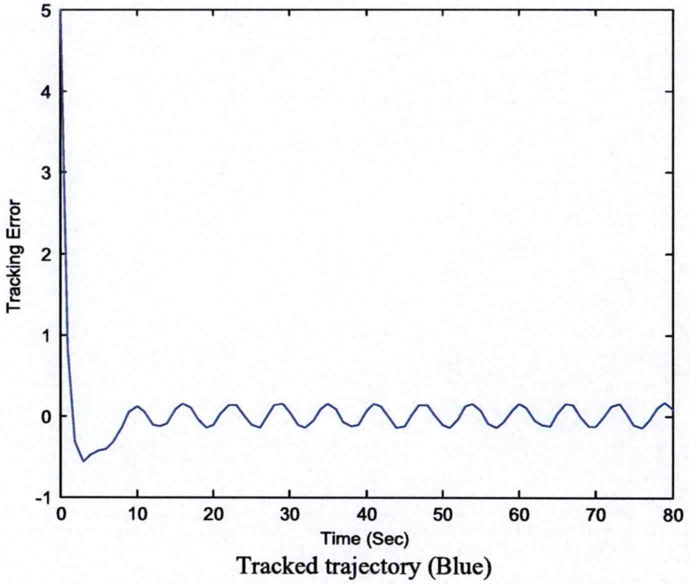

Fig. 3 Tracking error wrt time

target trajectory substantially resembles the desired trajectory of the moving node Figure 5.

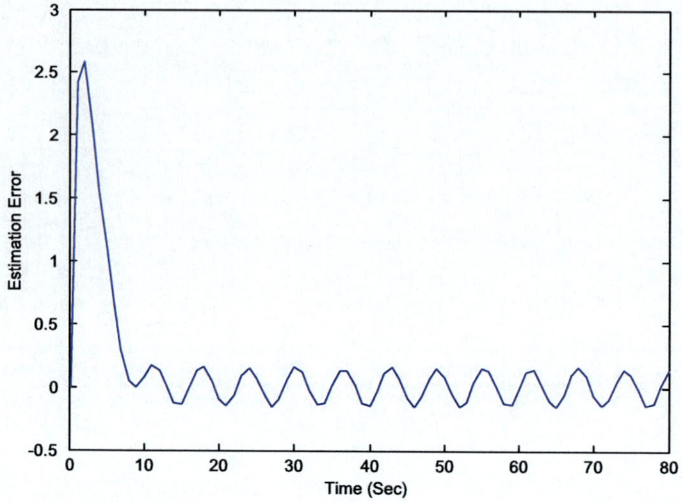

Fig. 4 Variation of estimation error

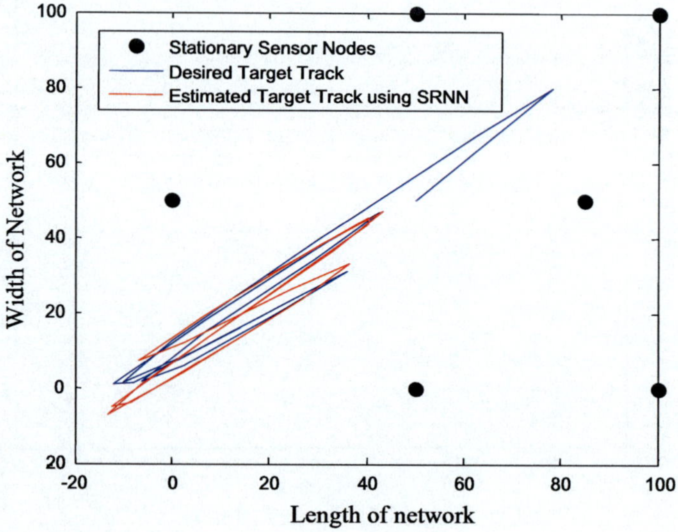

Fig. 5 Target tracking performance using SRWNN algorithm

7 Conclusion

This research presents an intelligent energy-efficient target localization approach. The moving target tracking technique is used in conjunction with the event-triggered strategy. To create a distributed sensor network with the appropriate energy efficiency, a consensus mechanism is developed. An intelligent self-recurrent framework of WNN is created and applied on the WSN environment to recognize unwanted environmental characteristics. In nature, these disruptions are thought to be nonlinear, uncertain, and unknown. Using the output of SRWNN, an adaptive estimation model is developed to determine the mobile node's current location. This study successfully presents the combination of a fundamental estimate model, a consensus-based fusion technique, and an event-triggered strategy through a detailed mathematical explanation. The simulation study is carried out to check and demonstrate the proposed work's usefulness and efficacy. In the future the target tracking can further be enhanced by using multimodal approach in the sensor network by incorporating other information like images and received signal strength, etc. Reinforcement learning can be used to enhance the performance further.

References

1. Demigha O, Hidouci WK, Ahmed T (2013) On energy efficiency in collaborative target tracking in wireless sensor network: a review. IEEE Commun Surv Tutor 15:1210–1222
2. Mahfouz S, Chehade FM, Honeine P, Farah J, Snoussi H (2014) Target tracking using machine learning and Kalman filter in wireless sensor networks. IEEE Sens J 14(10):3715–3725
3. Shah RA, Khowaja SA, Chowdhary BS (2017) Energy efficient mobile user tracking system with node activation mechanism using wireless sensor networks. In: 2017 international conference on communication, computing and digital systems (C-CODE). Islamabad, Pakistan, pp 80–85
4. Deldar F, Yaghmaee MH (2010) Energy efficient prediction-based clustering algorithm for target tracking in wireless sensor networks. In: 2010 International conference on intelligent networking and collaborative systems. Thessaloniki, Greece, pp 315–318
5. Yang WC, Fu Z, Kim JH (2007) An adaptive dynamic cluster-based protocol for target tracking in wireless sensor networks. In: Advances in data and web management, pp 157–167
6. Correal NS, Patwari N (2001) Wireless sensor networks: challenges and opportunities. In: Processing 2001 Virginia technology symposium. Wireless personal communications, pp 1–9
7. Roehrig C, Heller A, Hess D, Kuenemund F (2014) Global localization and position tracking of automatic guided vehicles using passive RFID technology. In: Processing 41st international symposium. Robot. ISR/Robot., pp 1–8
8. Qi J, Taha AF, Wang J (2017) Comparing Kalman filters and observers for power system dynamic state estimation with model uncertainty and malicious cyber attacks. arXiv preprint arXiv:1605.01030 [cs.SY]
9. Zou H, Wang H, Xie L, Jia QS (2013) An RFID indoor positioning system by using weighted path loss and extreme learning machine. In: Processing IEEE 1st international conference cyber-physics system, networks, applications (CPSNA), pp 66–71
10. Bergmann G, Molnar M, Gonczy L (2010) Optimal period length for the CGS sensor network scheduling algorithm. In: 2010 Sixth international conference on networking and services. Cancun, Mexico, pp 192–199

11. Buffi A, Nepa P, Lombardini F (2015) A phase-based technique for localization of UHF-RFID tags moving on a conveyor belt: performance analysis and test-case measurements. IEEE Sens J 15(1):387–396
12. Li S, Fang H, Chen J (2017) Energy efficient multi-target clustering algorithm for WSN-based distributed consensus filter. In: 36th Chinese control conference (CCC). Dalian, China, pp 8201–8206
13. Sijs J, Lazar M (2012) Event based state estimation with time synchronous updates. In: IEEE transactions automation control, vol 57, pp 2650–2655
14. Leong AS, Dey S, Quevedo DE (2017) Sensor scheduling in variance-based event triggered estimation with packet drops. IEEE Trans Autom Control 62:1880–1895
15. Sharma M, Verma A (2013) Wavelet reduced order observer-based adaptive tracking control for a class of uncertain delayed non-linear systems subjected to actuator saturation using actor-critic architecture. Int J Autom Control (IJAAC), Inderscience Publisher 7(4):288–303
16. Jondahale SR, Deshapande RS (2018) Modified Kalman filtering framework based real time target tracking against environmental dynamicity in wireless sensor networks. In: Ad Hoc and sensor wireless networks, vol 40, pp 119–143
17. Sharma M, Verma A (2010) Adaptive tracking control for a class of uncertain non-affine delayed systems subjected to input constraints using self recurrent wavelet neural network. In: IEEE international conference on advances in recent technologies in communication and computing, ARTCOM2010, pp 60–65

A Novel Harvesting and Data Collection Using UAV with Beamforming in Heterogeneous Wireless Sensor-Clustered Networks

Sundarraj Subaselvi

Abstract Wireless Sensor Networks (WSNs) send the sensed information to the base station by the sensor nodes. The Sensor nodes with energy constraint battery for sending and processing of data in the network which decreases the lifetime of the heterogeneous WSN. To increase the sensor nodes' lifetime in the clustered WSN, the energy harvesting and data collection with beamforming in heterogeneous WSN by Unmanned Aerial Vehicle (UAV) are proposed. The variant size of sensor nodes with different initial energies is deployed in network as heterogeneous WSN. The sensor nodes are grouped into cluster and the cluster head in each cluster is elected in the WSN. The UAV acts as a mobile sink node that collects and harvests energy from the nodes to rise the residual energy of sensor in the heterogeneous network. The aerial vehicle with multi-antenna focuses the energy signal toward the receiver single antenna sensor node instead of broadcasting the wireless signal using the Multiple Input and Single Output (MISO) beamforming technique which increases the harvesting of the heterogeneous sensor network. The projected technique efficiency is equated to the existing algorithm using UAV and proposed algorithm using mobile chargers in terms of average travel distance, residual energy, throughput, and average delay in the WSN.

Keywords Unmanned aerial vehicle · Beamforming · Wireless sensor networks

1 Introduction

The sensor nodes in Wireless Sensor Networks (WSNs) frequently broadcast information for data transfer and collection in the network. The recurrent information transfer between the sensors nodes leads to decrease the residual energy of the sensor nodes in the clustered heterogeneous WSN. Hence, various routing, clustering, and scheduling algorithms are introduced to increase the energy of the sensor in the WSN.

S. Subaselvi (✉)
M.Kumarasamy College of Engineering, Karur, Tamilnadu, India
e-mail: subaselvis.ece@mkce.ac.in

S. Sharma et al. (eds.), *Intelligent Control, Robotics, and Industrial Automation*,
Lecture Notes in Electrical Engineering 1066,
https://doi.org/10.1007/978-981-99-4634-1_61

To reduce communication distance between sensor nodes, energy-efficient recursive clustering for big data in WSN is proposed to reduce delay and energy consumption in the network [1]. The lifetime of the WSN is increased by considering the regional energy-aware clustering with isolated nodes for WSN [2].

Due to energy-constraint battery of the sensor nodes, the energy harvesting algorithms are emerged in the WSN. In harvesting algorithm, the energy-depleted nodes are harvested by the various sources of energy to decrease the consumption of energy in the network. The multi-step prediction based on adaptive dynamic programming to schedule sensors over an infinite horizon for collaborative target tracking in energy harvesting WSN is proposed to increase battery energy of sensor nodes in the WSN [3].

To predict the future energy availability, the Q-learning-based prediction technique using solar energy is considered to increase energy harvested in the WSN [4]. The static radio frequency transmitters are optimally deployed in software-defined wireless networks, and optimization problem is formulated with energy-efficient scheduling which is proposed to maximize the total energy charged in the sensor networks [5]. For energy neutral assignment for sensors when the mobile sink is used for data collection from the random deployment of sensor nodes in the WSN with reliable and stable sensor nodes, the three-stage optimization problem is proposed for determining data generation rate and optimize the data collection and aggregation by sensor nodes in the WSN to increase the network performance while performing perpetual network operation in the network [6]. To decrease consumption of energy in the sensor nodes for WSN, the multiple mobile chargers are deployed for harvesting energy in the heterogeneous clustered software-defined wireless sensor network and effective scheduling and routing algorithm for mobile chargers was proposed [7].

The UAVassisted cooperative cognitive network is proposed to analyze imperfect successive interference cancelation and increase the network performance based on throughput in the WSN. The optimization problem based on power allocation and maximizing energy efficiency in the network is formulated to maximize network throughput [8]. The harvesting and data collection using UAV and efficient routing for UAV in the network increases the lifetime of the heterogeneous WSN [9]. In half duplex and full duplex mode, data collection and energy harvesting by UAV with the energy minimization optimization problem formulation are implemented and analyzed to decrease the consumption of energy in the sensor nodes [10]. The proposed algorithm contributions are explained as follows:

1. The transfer of energy and collecting information by UAV to increase the lifetime of nodes with MISO beamforming for heterogeneous clustered WSN.
2. The sensor nodes with different size and initial energy are deployed as a heterogeneous network.
3. The random cluster formation with the election of cluster head is implemented to reduce the number of packet transmission in the WSN.
4. To transfer energy for the nodes and collecting data in a particular cluster, the UAV meets the requested cluster head to increase the remaining energy in the WSN.

5. The MISO beamforming is implemented between the sensor nodes and UAV that transmits energy in the receiver-requested node to upsurge the energy and lifetime of the heterogeneous WSN.

The proposed algorithm is projected as that Sect. 2 introduces the network model, Sect. 3 presents the projected new network method in heterogeneous clustered WSN, Sect. 4 compares the results of the proposed system model with the existing model, and Sect. 5 concludes the proposed method.

2 Network Model

In the $N \times N$ grid area, the M number of sensor nodes are not uniformly deployed with different sizes and initial energies as shown in Fig. 1. The small sensor node has less initial energy and large sensor node has high initial energy in the WSN. Once the heterogeneous sensors are randomly positioned in the sensor network, the random cluster formation method starts and cluster head in each cluster is elected based on the maximum energy among the sensor nodes in the cluster. The cluster head is reelected when the residual energy is smaller than threshold energy which is

$$T = \left\{ \frac{p}{1 - p\left(r \bmod \frac{1}{p}\right)} \times \frac{E_{\text{res}}}{E_{\text{maximum}}} \right\}, \tag{1}$$

Fig. 1 Network model

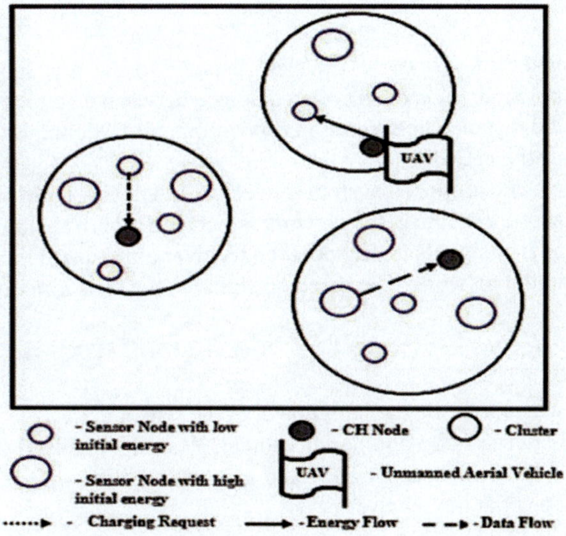

where p is the percentage of CH, r is the random number, E_{res} is the sensor nodes residual energy, and $E_{maximum}$ is the sensor nodes with maximum energy. The UAV moves around in the grid area and meets the requested cluster head in the heterogeneous WSN when the request both for harvesting and collecting data sends by the nodes in the clustered WSN.

3 Proposed Model

The proposed model in heterogeneous clustered WSN with UAV is shown in Fig. 2. Once the heterogeneous clusters are formed, the cluster head node in each cluster has elected to decrease the energy consumption of the network. The sensed information by the nodes through the CH nodes sends to the aerial vehicle. The UAV meets each CH for data collection when the energy of the sensor nodes is greater than the threshold energy. The charging request to UAV is send by the sensor nodes through maximum energy node when the preliminary energy is smaller than the onset energy. The aerial vehicle moves from the primary location to find the particular cluster head of requested sensor node for harvesting using beamforming technique and collects the data from cluster head in the network. The UAV meets the each cluster head by short-distance path planning between the current position and requested cluster head in the network. The UAV reduces the altitude at each cluster head to enhance the energy and for better data collection in the WSN. The traveled distance in the network by the UAV is

$$\mathrm{TD} = d_{(l,\mathrm{CH})} + \sum_{k=1}^{s} d_{lj}^{k} + d_{(\mathrm{CH},l)}, \tag{2}$$

where $d_{(l,\mathrm{CH})}$ is the traveled distance to the first request cluster head from initial position, d_{lj} is the traveled distance between CH node l and CH node j, and $d_{(\mathrm{CH},l)}$ is the distance between initial position and last charged cluster head that is the center of the network.

The proposed system model has UAV with multi-antenna and sensor nodes with single antenna in the network which performs transmit beamforming for transmitting energy signals in the specified receiver direction to increase the harvesting efficiency in the network. The received signal at the sensor node is

$$y_k = \sqrt{P_i}\tilde{-}_i^T W_i X_i + n_i, \tag{3}$$

where P_i is the transmit power of the sensor node, $h_\mathrm{i} = [h_{i,1}, h_{i,2}]$ is the channel vector between node and aerial vehicle, $W_i = [W_{i,1},\ W_{i,2}]$ is the beamforming vector, X_i is the transmitted signal, and n_i is the additive white Gaussian noise. The harvested power at the receiver is

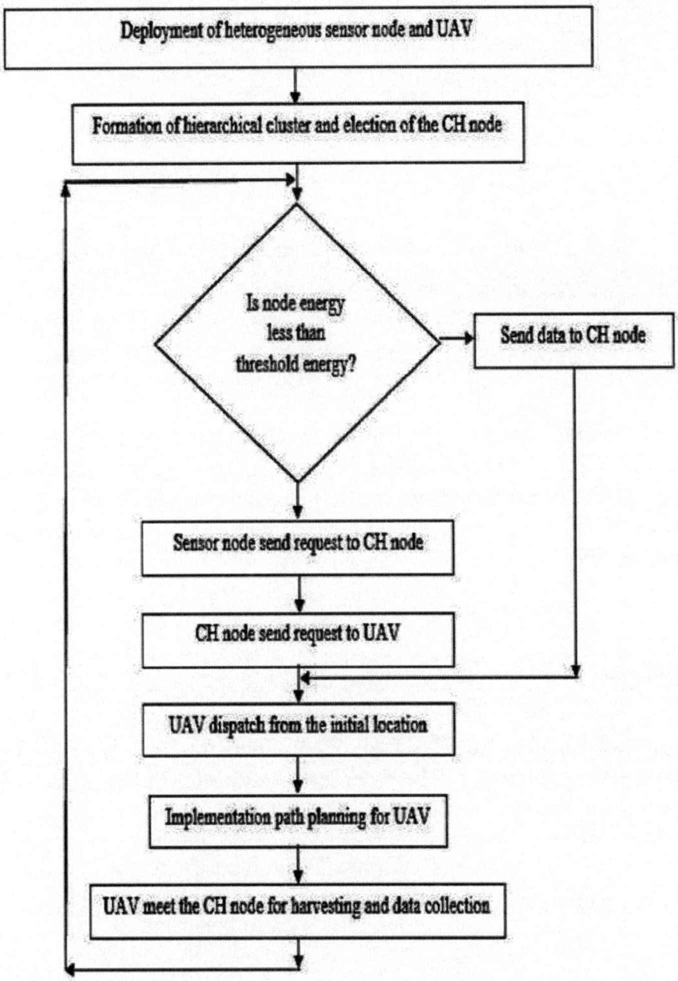

Fig. 2 Flowchart for proposed model

$$P_H = P_k \left| \eqsim_k^T W_k \right|^2. \tag{4}$$

The receiver signal-to-noise ratio is

$$SNR(\Gamma_k) = \frac{P_k \left| \eqsim_k^T W_k \right|^2}{\sigma^2}. \tag{5}$$

The harvested energy in the network is maximized by the formulation of the optimization problem as

$$\max_{W} P_H \tag{6}$$

$$\text{subject to} \quad \Gamma_k > \Upsilon_k$$

$$\sum_{k=1}^{K} W_k^2 \leq P_T,$$

where Υ_k is the signal-to-noise ratio threshold and P_T is the UAV transmit power limit. The problem resolution is given by ref. [11]

$$W_k = \frac{\tilde{\overline{\ }}^{*}_{i,i}}{\tilde{\overline{\ }}^{*}_{i,i}}. \tag{7}$$

The beamforming weight vector W_k cancels the interference from other user and increases the energy harvested for the desired user by transmitting the energy directionally to the desired receiver in the network. Further, the UAV is decreased at the altitude of 2 m [9] at each cluster head to increase the harvesting efficiency in the clustered WSN.

4 Results and Discussion

The heterogeneous clustered WSN algorithm performance is implemented and analyzed in network simulator. The heterogeneous nodes based on size and energy are deployed with the UAV as a mobile base station in the clustered WSN. The heterogeneous clustered proposed technique is equated with the existing algorithm based on UAV [9] and proposed algorithm with mobile chargers instead of UAV. The simulation parameters for the new algorithm are listed in Table 1.

Table 1 Simulation parameters

S. No.	Parameter	Value
1	Network size	1000 × 1000
2	Number of nodes	200, 250, 300, 350, 400
3	Simulation time	200, 250, 300, 350, 400
4	UAV altitude	120, 140, 160, 180, 200
5	Initial energy	0.5 and 1 J
6	Number of UAV	2
7	Variance (σ^2)	−40 dBm [11]
8	SNR threshold	20 dB [11]
9	Transmit power threshold (P_T)	30 dBm [11]

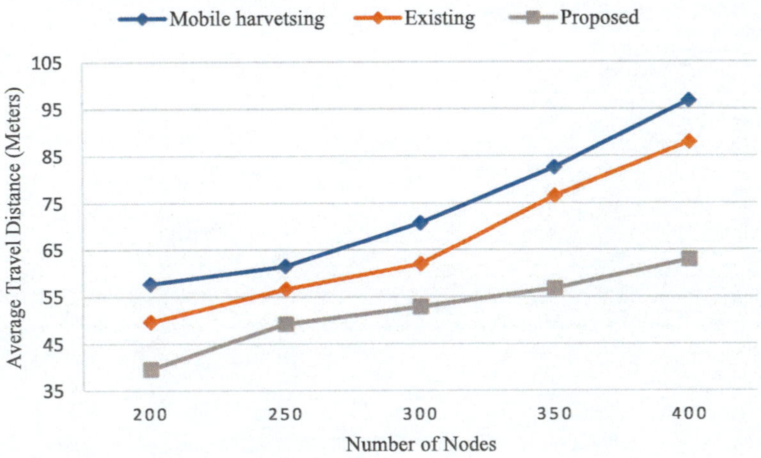

Fig. 3 Average travel distance vs. number of nodes

In Fig. 3, the average travel distance is increased as the number of sensor nodes in the WSN is increased because the data processing between the sensor nodes is increases when the number of sensor nodes is increased in the network. Hence, the UAV moves around frequently in the grid area to collect or harvest the sensor nodes in the WSN. The average distance traveled by UAV in proposed algorithm for 200 nodes is 39.53 m which is less compared to 49.53 m in existing and 57.6 m in the mobile harvesting technique. The average travel distance is higher in mobile harvesting because of less speed in the heterogeneous cluster WSN and the aerial vehicle for harvesting and data collection needs to meet each sensor node that increases the traveled distance by the UAV in the prevailing network compared to the proposed algorithm.

The average throughput is increased as the number of sensor nodes in the network is increased is shown in Fig. 4. The average throughput in proposed network with 200 nodes is 63.7 kbps which is higher compared to 48.8 kbps in existing and 42.6 kbps in the existing mobile harvesting network. The packets processing in the WSN increase the packet collision in the network which lead to decrease the throughput when the sensor nodes' count increases in the WSN. The average throughput was higher in the proposed algorithm because the UAV with high speed meets each cluster head in the WSN for energy harvesting and collecting data instead of each sensor node which reduces the control packet transmission.

The distance traveled between the UAV and sensor node is increased for collecting data in the Fig. 5. The UAV in the network takes time to reduce the altitude at each sensor node when the distance increases for energy transfer and collecting information in existing algorithm and the UAV meets each cluster head instead of each sensor node in the heterogeneous clustered algorithm which decreases the delay for data collection equated to the existing algorithm in the heterogeneous WSN. The

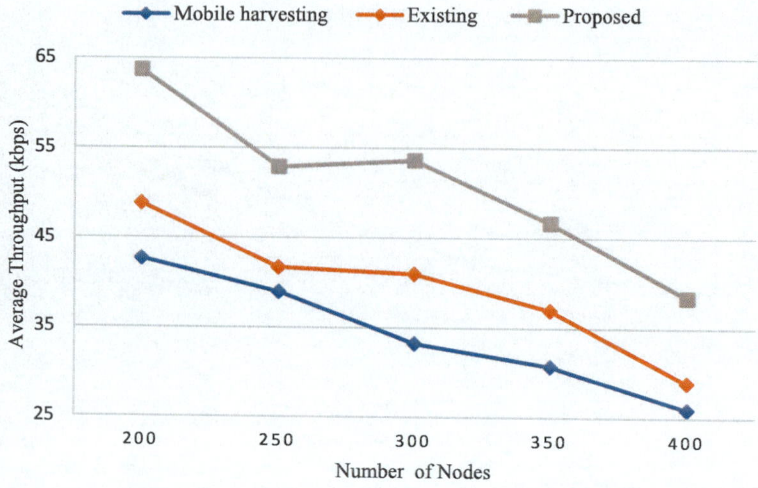

Fig. 4 Average throughput versus number of nodes

average delay for UAV at the altitude of 200 m in proposed algorithm is 0.014 s is less compared to 0.0214 s in existing algorithm.

In Fig. 6, the throughput is decreased as the UAV altitude is increased because of long-distance data transmission between the node and aerial vehicle that increase the number of packet drops in the heterogeneous WSN. The throughput for UAV at the altitude of 120 m is higher in proposed algorithm which is 57.35 kbps compared to 48.91 kbps in the existing algorithm. The high-speed data collection and harvesting by the UAV with efficient shortest path routing in the WSN instead of collecting

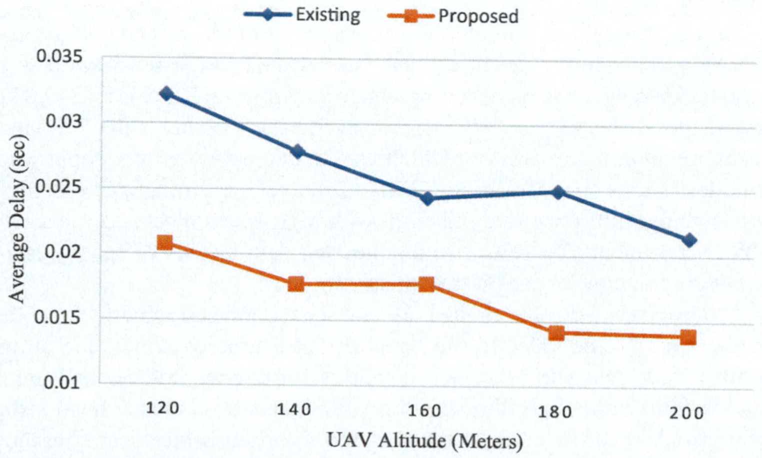

Fig. 5 Average delay versus UAV altitude

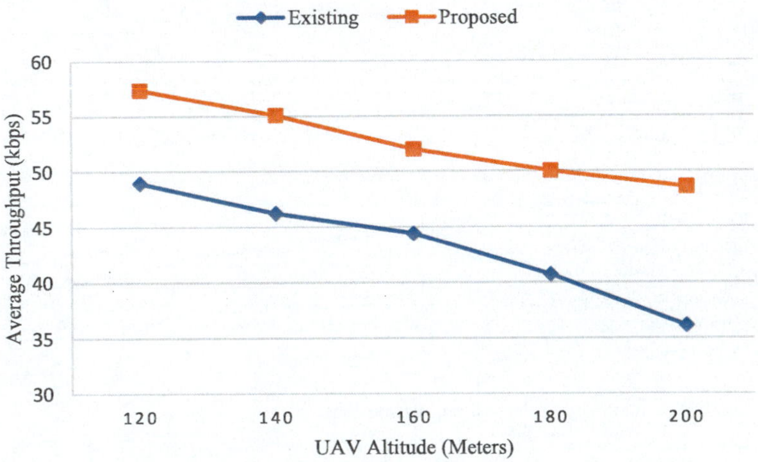

Fig. 6 Average throughput versus UAV altitude

data and harvesting at each sensor node without efficient path planning decreases the throughput in the existing technique compared to the proposed heterogeneous clustered algorithm in the WSN.

The residual energy is decreased when the simulation time increases in Fig. 7 because the information processed in the heterogeneous WSN increases which leads to frequent packet transmission of control and data packets in the clustered WSN. The average residual energy for the simulation time of 200 s is higher in proposed algorithm which is 0.83 J equated to 0.73 J in the prevailing technique and 0.55 J in movable source technique. The frequent broadcasting of packets in the WSN increases the energy depletion of the sensor nodes in the WSN. The movable harvesting algorithm residual energy is less because of increasing charging delay in the WSN. In the existing network, the UAV does not give importance to the sensor nodes with less energy which decreases the remaining energy of the existing algorithm. The UAV with MISO beamforming technique transmits energy in the particular direction which increases the residual energy in the proposed algorithm for heterogeneous WSN.

The average delay is increased when the simulation time increases because the packet drops in the network increase due to the frequent transmission of packets in Fig. 8. The charging delay for the simulation time of 400 s in proposed algorithm is 45.36 s which is less compared to 46.6 s in existing and 58.86 s in mobile harvesting algorithm. The packet drops lead to retransmission of control and data packets when the simulation time increases which increases the delay in WSN. The average delay in the mobile charging is increased due to less speed for energy transfer and collecting information in the network. In the existing algorithm, the UAV meet each sensor nodes for energy transfer and collecting information in the WSN increases the delay in the network compared to the proposed algorithm in the clustered heterogeneous WSN.

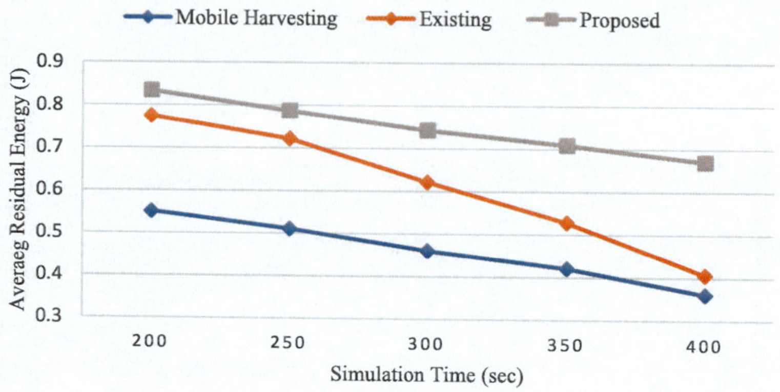

Fig. 7 Average residual energy versus simulation time

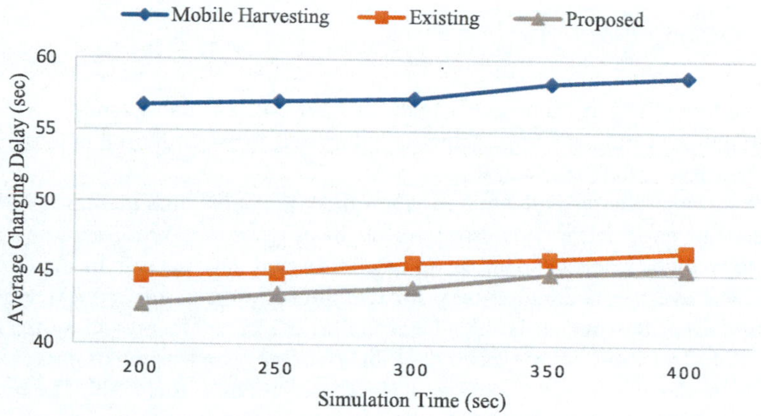

Fig. 8 Average delay versus simulation time

5 Conclusion

The proposed efficient energy harvesting and collecting data with beamforming by aerial vehicle in the heterogeneous network are implemented to increase the energy of the heterogeneous WSN. The heterogeneous sensor nodes with UAV are deployed for analyzing the efficiency of the network. The clustering process with the election of cluster head increases the residual energy of the sensor nodes by reducing the frequent transmission of packets in the network. The UAV with high speed directionally harvests energy using beamforming algorithm and meet each cluster head for data collection increases the harvesting efficiency of the heterogeneous WSN. Thus, the proposed algorithm with beamforming in heterogeneous network increases the lifetime of the WSN when compared with the proposed system with the mobile charger and the existing technique with UAV in the WSN. The future scope of the proposed algorithm is the novel scheduling algorithm that can be implemented to reduce the packet collision in the heterogeneous WSN.

References

1. Subaselvi S, Manimekalai T, Gunaseelan K (2019) Energy efficient recursive clustering and gathering in big data for wireless sensor networks. Sens Lett 17(9):680–687. https://doi.org/10.1166/sl.2019.4128
2. Leu JS, Chiang TH, Yu MC, Su KW (2015) Energy efficient clustering scheme for prolonging the lifetime of wireless sensor network with isolated nodes. IEEE Commun Lett 19(2):259–262. https://doi.org/10.1109/LCOMM.2014.2379715
3. Liu F, Jiang C, Xiao W (2021) Multistep prediction-based adaptive dynamic programming sensor scheduling approach for collaborative target tracking in energy harvesting wireless sensor networks. IEEE Trans Autom Sci Eng 18(2):693–704. https://doi.org/10.1109/TASE.2020.3019567
4. Kosunalp S (2016) A new energy prediction algorithm for energy-harvesting wireless sensor networks with Q-learning. IEEE Access 4:5755–5763. https://doi.org/10.1109/ACCESS.2016.2606541
5. Ejaz W, Naeem M, Basharat M, Anpalagan A, Kandeepan S (2016) Efficient wireless power transfer in software-defined wireless sensor networks. IEEE Sens J 16(20):7409–7420. https://doi.org/10.1109/JSEN.2016.2588282
6. Tao L, Zhang XM, Liang W (2019) Efficient algorithms for mobile sink aided data collection from dedicated and virtual aggregation nodes in energy harvesting wireless sensor networks. IEEE Trans Green Commun Network 3(4):1058–1071. https://doi.org/10.1109/TGCN.2019.2927619
7. Subaselvi S, Gunaseelan K (2022) Energy efficient mobile harvesting scheme for clustered SDWSN with beamforming technique. Intel Autom Soft Comput 34(2):1197–1213. https://doi.org/10.32604/iasc.2022.025026
8. Bhowmick A, Roy SD, Kundu S (2022) Throughput maximization of a UAV assisted CR network with NOMA-based communication and energy-harvesting. IEEE Trans Veh Technol 71(1):362–374. https://doi.org/10.1109/TVT.2021.3123183
9. Subaselvi S, Gunaseelan K (2022) Energy efficient UAV enabled harvesting with beamforming for clustered SDWSN. Computing 1–24. https://doi.org/10.1007/s00607-022-01087-0
10. Yang Z, Xu W, Shikh-Bahaei M (2020) Energy efficient UAV communication with energy harvesting. IEEE Trans Veh Technol 69(2):1913–1927. https://doi.org/10.1109/TVT.2019.2961993
11. Timotheou S, Krikidis I, Zheng G, Ottersten B (2014) Beamforming for MISO interference channels with QoS and RF energy transfer. IEEE Trans Wireless Commun 13(5):2646–2658. https://doi.org/10.1109/TWC.2014.032514.131199

Leakage Power Reduction in CMOS Inverter at 16 nm Technology

Yahuti Sahu, Amit Singh Rajput, Onika Parmar, and Zeesha Mishra

Abstract As technology reduces to nm range, switching power, short-circuit power, and total power consumption decrease, while leakage power consumption increases. There have been a few techniques reported to reduce leakage power consumption, but it remains unclear which technique is the most effective for power reduction at the 16 nm technology node. The main aim of this paper is to determine the various elements of power consumption in complementary metal–oxide–semiconductor (CMOS) circuits at various technology nodes as well as to determine the most effective method of leakage reduction at 16 nm technology. In this paper, various leakage reduction techniques are compared using a spice simulation tool and a 16 nm predictive technology model (PTM) CMOS model. As a result, it has been concluded that the best method for reducing leakage is the sleep transistor method among lector, ONOFIC, dual-stack, and sleepy keeper transistor techniques. Compared with a sleep transistor technique, lector, ONOFIC, dual-stack, and sleepy keeper techniques exhibited 151×, 156×, 2.91×, and 1.2 × times higher leakage power consumption, respectively. This study may contribute to the finding of an effective leakage reduction technique appropriate for CMOS circuit scale technology nodes.

Keywords Leakage power · LECTOR · Sleep transistor · ONOFIC · Dual-stack · Sleepy keeper techniques

Y. Sahu · A. S. Rajput · O. Parmar (✉) · Z. Mishra
Department of Microelectronics and VLSI, UTD, Chhattisgarh Swami Vivekananda Technical University, Newai, Bhilai, India
e-mail: faculty2_vlsi@csvtu.ac.in

A. S. Rajput
e-mail: 0915.ec@gmail.com

Z. Mishra
e-mail: faculty3_vlsi@csvtu.ac.in

S. Sharma et al. (eds.), *Intelligent Control, Robotics, and Industrial Automation*, Lecture Notes in Electrical Engineering 1066,
https://doi.org/10.1007/978-981-99-4634-1_62

1 Introduction

Low-power high-performance very large-scale integration (VLSI) design is growing attention because of the increasing use of various battery-operated digital devices in our daily life like laptops, smartphones, and other portable gadgets [1]. To enhance the performance of the device, the number of transistors in a unit chip area is doubled every 18 months [2]; therefore, we have to scale down the technology node to fabricate more and more transistors in a unit chip area. When technology is scaled down to the nanometre region dynamic power remains more or less constant but leakage current increases exponentially [3]. Therefore, the study of various power consumption and leakage reduction methods is important.

This work investigates static and dynamic power consumption of complementary metal oxide semiconductor (CMOS) inverter in various scaled nm (nanometre) technology node. Moreover, this work examines the suitability of various leakage reduction techniques in 16 nm technology node. We have chosen the CMOS inverter for this study because the inverter is a basic building block in various VLSI circuits.

The total power consumption in a VLSI circuit can be divided into two types; one is static power consumption, and the other is dynamic power consumption [4]. Dynamic power consumption may be divided into two-part short-circuit power consumption and switching power consumption. Short-circuit power consumption when a direct path exists between VDD and GND during the transition of input, whereas the power consumption in charging and discharging the load capacitor is recognized as switching power consumption. When no transition activity happens on the input side, i.e., the input is static, a leakage current flow from VDD to the ground produces leakage power consumption. Leakage power is dominant over the total power dissipation below the 90 nm technology node [5]. That is the reason, we examine a few leakage current reduction techniques at 16 nm technology in this paper.

This paper explores, for the first time, the comparative analysis of the ONOFIC technique [6–9], LECTOR technique [9, 10], dual-stack technique [11], sleepy keeper transistor techniques [9, 12–15], stack transistor technique, and sleep transistor technique [9, 16–18] to reduce leakage current and find their suitability at 16 nm technology node. This article is organized as follows: Sects. 2 covers CMOS inverter and static and dynamic components of power consumption. Next, Sect. 3 presents a few leakage power reduction techniques. A description of the experimental setup is presented in Sect. 4, followed by the results and discussion. Finally, in Sect. 5, the conclusion of the paper is presented.

2 CMOS Inverter

NMOS and PMOS transistors are connected at their drain and gate terminals to make a CMOS inverter, as shown in Fig. 1. The NMOS source terminal is connected to GND, while the VDD supply voltage is connected to the PMOS source terminal. A voltage V_{in} is provided to the gate terminals of NMOS and PMOS, while a voltage V_{out} is received from the drain terminals. In Fig. 1, capacitor C serves as a load capacitor to represent the parasitic capacitance of the circuit.

The NMOS and PMOS transistor change states when the input voltage varies between 0 V and VDD. When V_{in} is 0 V, then PMOS is ON, but the NMOS is OFF; therefore, the load capacitor C is charged to VDD through the PMOS transistor, and V_{out} is equal to VDD. However, when V_{in} is VDD, the PMOS transistor is turned OFF, while the NMOS transistor is turned ON. This causes load capacitor C to be discharged to GND through the NMOS transistor and V_{out} to be equal to GND. The power taken from supply VDD to charge capacitor C is considered to be switching power consumption [19]. In addition, when V_{in} transitions from 0 V to VDD or VDD to 0Va, a short-circuit current flows between VDD and ground due to partial ON of NMOS and PMOS transistors. This is referred to as a short-circuit current. When short-circuit power (when multiplied by VDD) is added to the switching power, this is considered to be the circuit's dynamic power.

When the inverter is operating in steady state mode, i.e., no transition activity occurs in the circuit, a very small magnitude current flows from VDD to GND due to the non-ideal nature of the device. This current is referred to as leakage current. Leakage current increases exponentially as technology is scaled down into nanometre technology node [20]. A leakage power can be calculated by multiplying the leakage current and the supply voltage VDD. Power consumption caused by leakage is considered static power consumption.

Power dissipation consists of three components

$$P = P_{\text{static/leakage}} + P_{\text{switching}} + P_{\text{short - circuit}} \tag{1}$$

Fig. 1 Schematic diagram of a CMOS inverter

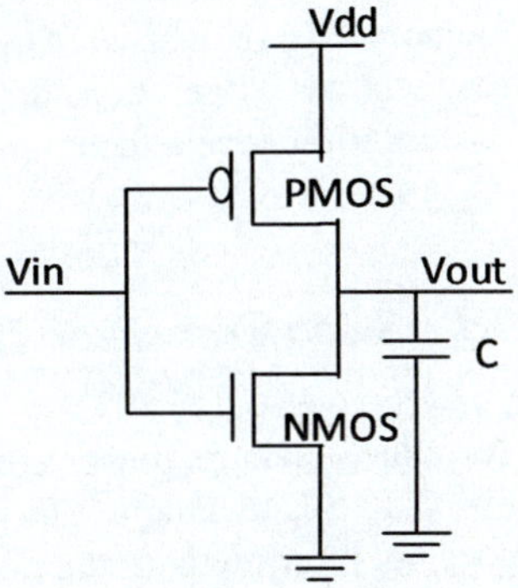

P represents total power dissipation, $P_{static/leakage}$ represents leakage power, $P_{switching}$ represents switching power, and $P_{short\text{-}circuit}$ represents short-circuit power consumption of the circuit.

$$P_{static/leakage} = I_{Leakage\ current} \times \text{VDD} \quad (2)$$

where

$$I_{leakage\ current} = I_0 e^{\frac{\text{Vgs} - (\text{Vt0} - \eta\text{Vds} - \gamma\text{Vbs})}{\text{nVT}}}\left[1 - e^{\frac{-\text{Vds}}{\text{VT}}}\right] \quad (3)$$

where $I_0 = \mu_0 C_{ox} V_T^2 e^{1.8}/L$, $V_T = k_T/q$, Vt_0 is the threshold voltage, L is the effective channel length, W is the effective transistor width, C_{ox} is the gate oxide capacitance, n is the slope coefficient, η is the drain induced barrier lowering coefficient, μ_0 is the mobility, and γ is the coefficient of the linearized body effect.

$$P_{switching} = C_L \cdot V_{DD} \cdot f \quad (1)$$

where f is the gate switching frequency and C_L is the load capacitor

$$P_{shortcircuit} = K(\text{VDD}{-}{-}2\text{V}_{th})^3 \tau \cdot f \quad (2)$$

Several parameters affect the K constant, such as transistor size and technology parameters, the supply voltage V_{DD}, the threshold voltage V_{th}, the rise and fall times of the input signal τ, and the frequency of the clock f.

3 Leakage Power Reduction Techniques

As we move toward the scale technology node, dynamic power is almost constant, since most technologies use approximately the same supply voltage, although leakage power increases exponentially [9, 16]. Leakage current is a critical issue in small-dimension high-density CMOS circuits, similar as cells used in static random-access memory (SRAM). Most of the time SRAM cells remain in the study state mode until read and write operations are performed. Therefore, in this section of the paper, we discuss some recent techniques for reducing leakage currents.

3.1 Sleep Transistor Technique

To reduce leakage power consumption, sleep transistor technique is illustrated in Fig. 2. In this technique, sleep transistors are turned on while the circuit is active, it does not interfere with the normal operation of the circuit [9, 16–18]. However, the

Fig. 2 Schematic diagram of the inverter with sleep transistor technique

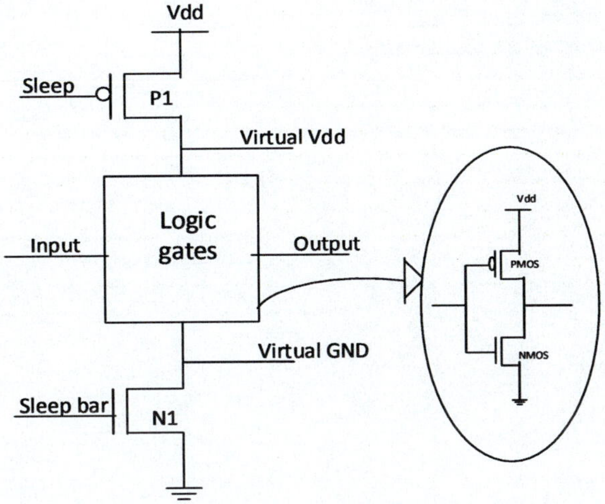

supply voltages VDD and GND are removed from the circuit, and the sleep transistors are turned off while the circuit is in standby mode [21]. To minimize leakage power, it connects the logic circuit to a lower virtual supply and virtual ground rail. Since leakage current is proportional to voltage difference, it consumes very low leakage power in the study state mode.

3.2 *Dual-Stack Technique*

Figure 3 is illustrated the dual-stack technique for a CMOS inverter [11]. In this technique, inverter made by P2 and N3 transistors is connected to VDD through a parallel network created by p1 and sleep transistor P2. Moreover, the inverter is connected to the ground through a parallel network made by N2 and sleep transistor N3. When the circuit is operating in active mode, the sleep transistors P2 and N3 are turned on, establishing a connection between VDD and ground, and the circuit shows normal operation. On the other hand, during standby mode, the sleep transistors that are connected to the N2 and P1 are turned off, which increases the resistance from VDD to GND. Therefore, it reduces leakage current during standby mode.

3.3 *Sleepy Keeper Technique*

Figure 4 illustrates a CMOS inverter with a sleepy keeper technique to reduce leakage current [9, 12–15]. High threshold voltage is used in this method. NMOS and PMOS transistors N1 and P3 are parallel connected with sleep PMOS transistor P1 and

Fig. 3 Schematic diagram of the inverter with dual-stack technique

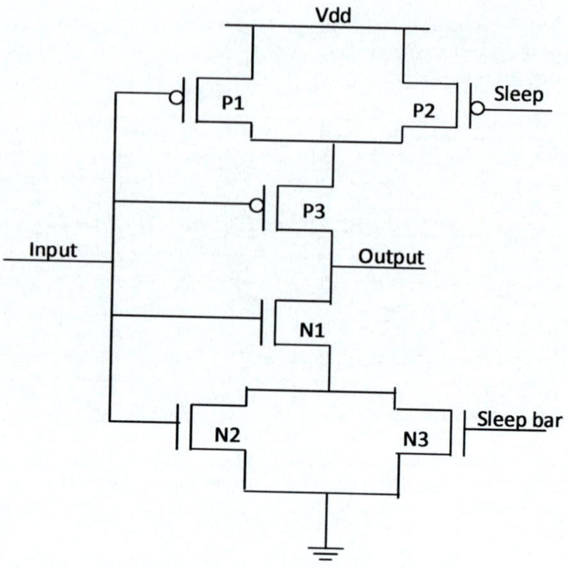

sleep_bar NMOS transistor N3. During sleep mode, sleep transistors are in the off state, and high V_t N1 and P3 are the only paths to connect the inverter (P2 and N2) to the VDD and GND since V_t of N1 and P3 is higher, resulting in a low leakage current. The inverter's output is attached to the gate of N1 and P3; therefore, N1 provides the path to load (not shown in the schematic) to link the pull-up network, and the P3 transistor provides a path to load for connecting from the pull-down network.

3.4 *Leakage Control Transistor (LECTOR) Technique*

In this technique, two extra transistors leakage control transistors (LCTs) Mt1 and Mt2 are introduced between the pull-up transistor P1 and pull-down transistor N1 as shown in Fig. 5 [9]. The gate of one LCT is controlled by the other LCT of source in such a way that one of the transistors always remains in the deep sub-threshold region [10]. When the low input signal is applied to the inverter, the transistors N1 and Mt1 are turned OFF. Therefore, we get a stacking of two series connected transistors between VDD and GND, resulting in a low leakage current. When the input signal is high, the transistors P1 and Mt2 are disabled. This increases the resistance between the supply voltage and ground path, resulting in a reduction in leakage current.

Fig. 4 Schematic diagram of the inverter with the sleepy keeper technique

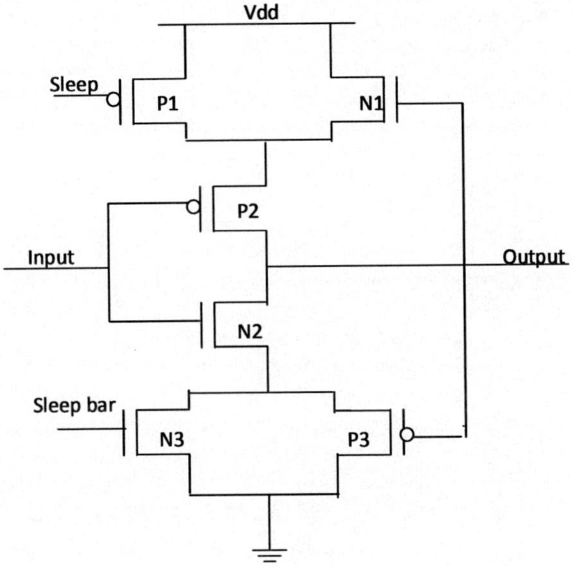

Fig. 5 Schematic diagram of the LECTOR technique

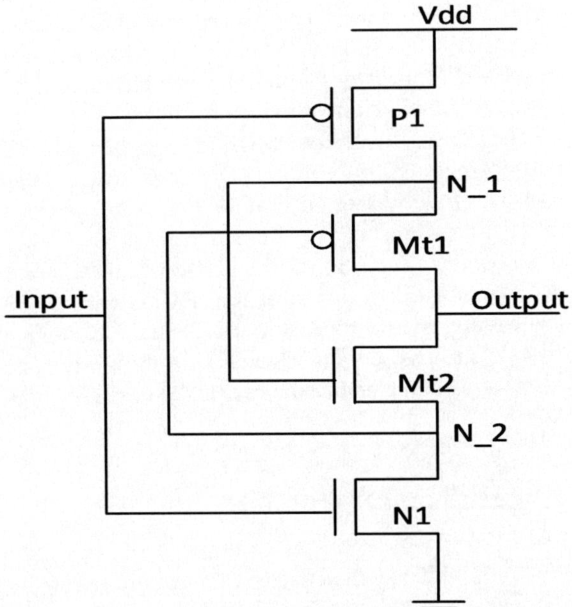

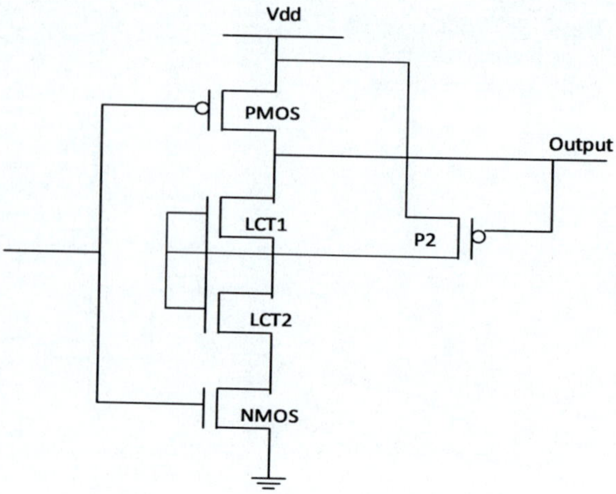

Fig. 6 Schematic diagram of the inverter with the ONOFIC technique

3.5 On Off Logic (ONOFIC) Technique

In the ONOFIC technique, two additional NMOS leakage control transistors, LCT1 and LCT2, are connected between the PMOS and NMOS transistor of the inverter as shown in Fig. 6. The gates of leakage control transistors (LCT1, LCT2) are controlled by a PMOS transistor P2, and the gate terminal of transistor P2 is controlled by the output terminal, which provides feedback to transistors LCT1 and LCT2.

The pull-up transistor PMOS is turned on when the input signal is low, and the pull-down transistor NMOS is OFF, resulting in HIGH output. Therefore, feedback transistor P2 becomes OFF, and the leakage control transistors LCT1 and LCT2 go into the sub-threshold region, resulting in negligible leakage current [6, 7]. Additionally, when input is HIGH, pull-down transistor NMOS is turned ON resulting in LOW output. In this situation, PMOS transistor P2 is turned ON, it turns ON the leakage control transistors LCT1 and LCT2 and provides a conducting path to provide strong 0 output. However, in the study state mode, LCT reduces leakage current by adding additional resistance to the inverter.

4 Result and Discussion

We followed the method used by the Author to calculate the switching, short-circuit, and leakage of the inverter [8]. This method is one of the best effective methods to calculate power consumption; therefore, it is widely adopted by the research community [8, 9]. To calculate power consumption, the inverter was simulated by spice software using the predictive technology model [22]. To measure power consumption, NMOS and PMOS length, width, and supply voltages were used as per Table 1. 10pf load capacitor was used as a load capacitor during the simulation.

Table 1 To NMOS and PMOS length, width, supply voltages, and threshold voltage

Technology (nm)	Supply (V)	Vth(V)	W/L ratio PMOS	W/L ratio NMOS
45	1.1	0.623	180 nm/45 nm	90 nm/45 nm
32	1	0.63	128 nm/32 nm	64 nm/32 nm
22	0.95	0.689	88 nm/22 nm	44 nm/22 nm
16	0.9	0.682	64 nm/16 nm	32 nm/16 nm

To determine dynamic power, a pulse is applied at the Vin terminal of the inverter using the PULSE (0 VDD 0 1n 1n 1u 2u) command and perform transient analysis using the tran command. The VDD value was selected as per the default value of the specific technology node as given in Table 1. Switching power consumption is mainly because of the charging and discharging of load capacitance. However, the rise and fall times of input waveforms that are not zero create short-circuit power, resulting in both the PMOS and NMOS being ON for a short duration. Leakage power is the power consumption caused by the leakage current flow between VDD and ground due to the non-ideal nature of the device. CMOS inverter is simulated using the parameter given in Table 1, and the result of total power consumption and its various element is presented in Table 2.

Figure 7 shows the relationship between switching power consumption and technology node. These results show that as we reduce the technology node a significant reduction in the switching power. What is noticeable here is when scale down from 45 to 22 nm or 32 nm to 16 nm switching power reduces by 0.74×. One possible explanation for this decrease is a reduction in operating supply voltage. Overall, these results indicate that switching power consumption is reduced as scale down the technology node. Similarly, Fig. 7 shows a clear trend of a reduced short-circuit power consumption as technology scales in the nanometre area. The observed decrease in short-circuit power could be interpreted as operating supply voltage reduces as technology scales down. An interesting result found that the rate of reduction in short-circuit power consumption is higher than the switching power consumption.

Leakage power is a power that flows between VDD and GND when the circuit is present in the study state mode. Figure 8 shows a leakage current of the inverter at various technology nodes. These results show that as technology scales down into the nanometre range magnitude of leakage current increases exponentially. This result is

Table 2 Result of switching, short-circuit, leakage, and total power consumption

Technology (nm)	Switching power (W)	Short-circuit power (W)	Leakage power(W)	Total power (W)
45	3.03E-06	2.83E-11	1.21E-11	3.03E-06
32	2.50E-06	3.61E-11	1.50E-11	2.50E-06
22	2.23E-06	7.48E-12	2.72E-11	2.23E-06
16	1.86E-06	1.54E-12	5.40E-11	1.86E-06

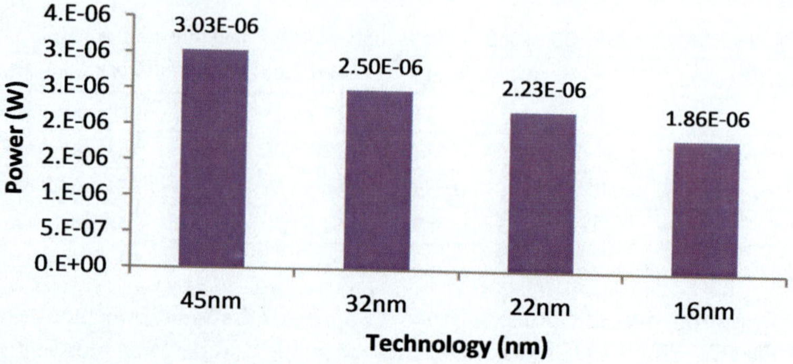

Fig. 7 Switching power with various technology node

somewhat surprising because at lower technology nodes leakage current magnitude is dominant. When we move down from 32 to 16 nm technology node magnitude of leakage current increases by 3.7×. It is concluded from Fig. 8 that problem of leakage current is critical at the lower technology node.

From the above Figs. 6, 7, 8 and Table 2, it is concluded that as technology scale down in the nanometre node magnitude of total power consumption reduces, and supply voltage reduction might be responsible for it. Switching power and short-circuit follow the same trend as total supply voltage but the value of leakage current increases exponentially as move to word the lower technology node. Therefore, in the next part of the paper, there are different leakage reduction approaches that affect the CMOS inverter's ability to reduce leakage power.

To calculate the leakage current of the inverter spice, simulation tool is used with a 16 nm PTM model [22]. We followed the method used by paper [16] to calculate the leakage current of LECTOR [9, 10], ONOFIC [6, 7], sleep transistor [9, 16–18], dual-stack transistor [11], and sleepy keeper technique-based inverter [9, 12–15]. The supply voltage of various technology nodes is different, leakage current is also

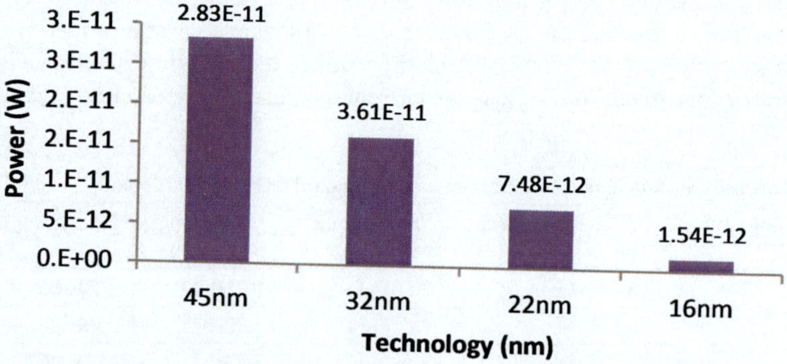

Fig. 8 Short-circuit power with various technology nodes

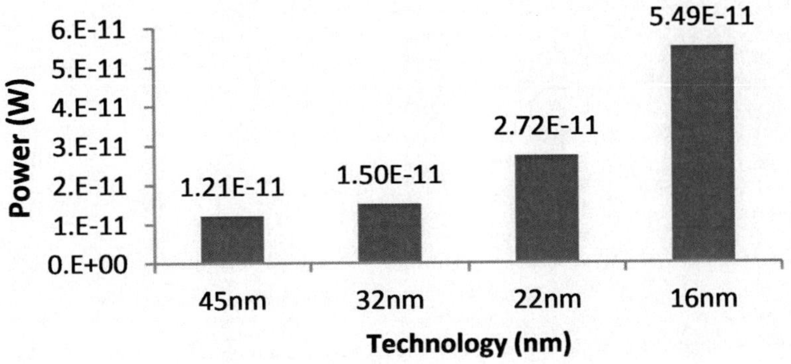

Fig. 9 Leakage power consumption with various technology nodes

Table 3 Techniques reduce leakage current of CMOS

Technique	Leakage current (A)			Av power (W)
	Vin = 0 V	Vin = VDD	Average	
LECTOR	1.12E-10	2.07E-13	5.62E-11	5.06E-11
Onofic	1.13E-10	3.13E-12	5.82E-11	5.24E-11
Sleep transistor	7.19E-13	2.75E-14	3.7E-13	3.36E-13
Dual-stack transistor	1.65E-12	5.25E-13	1.1E-12	9.79E-13
Sleepy keeper	5.93E-13	3.05E-13	4.5E-13	4.04E-13

dependent upon supply voltage; therefore, for a fair comparison between the various leakage reduction techniques, we have chosen 0.9 V for the simulation.

The same data is plotted in Fig. 9 to enhance visibility. The sleep transistor technique consumes the lowest power among the all-leakage reduction technique. Sleep transistor leakage reduction-based inverter consumes 3.36E-13 W power as leakage power consumption. However, LECTOR, ONOFIC, dual-stack transistor, and sleepy keeper technique consume 150×, 155×, 2.9×, and 1.2× higher power consumption as compared to the sleep transistor technique (Table 3 and Fig. 10).

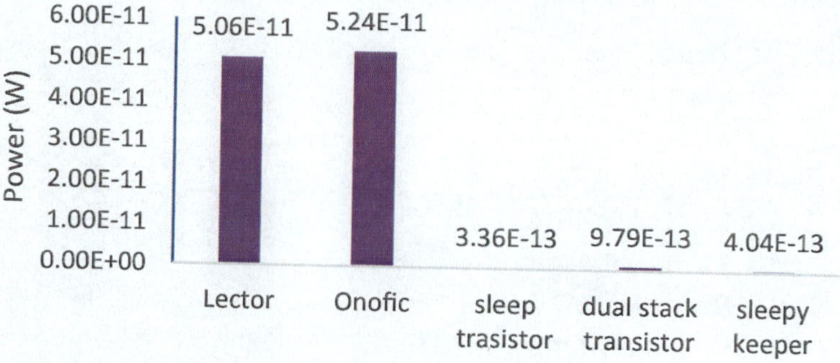

Fig. 10 Leakage power with various leakage reduction techniques

5 Conclusion

In this paper, we have compare various power consumption elements of CMOS inverters in 45, 32, 22, and 16 nm technology nodes. We also study various leakage reduction techniques in CMOS inverters at 16 nm technology node. As technology scales down to the nm range, switching power, short-circuit power, and total power consumption reduce, but leakage power increases. This study compares various leakage reduction techniques for 16 nm technology. It concludes that the sleep transistor approach is the most effective strategy for minimizing leakage power among lector, ONOFIC, dual-stack, and sleepy keeper transistor techniques. LECTOR, ONOFIC, dual-stack, and sleepy keeper techniques exhibited 151×, 156×, 2.91× , and 1.2 × higher leakage power consumption compared with a sleep transistor technique. The results of this study can help to select a leakage reduction technique that is appropriate for a CMOS circuit. As a consequence of the dominant power consumption caused by leakage currents, future studies should aim to apply the sleep transistor technique apply to memory-like circuits. Designing of multiplexer using domino logic.

References

1. Gavaskar K, Ragupathy US, Malini V (2019) Design of novel SRAM cell using hybrid VLSI techniques for low leakage and high speed in embedded memories wirel. Pers Commun 108(4). https://doi.org/10.1007/s11277-019-06523-7
2. Naffziger S et al (2021) Pioneering chiplet technology and design for the AMD EPYCTM and RyzenTM processor families: industrial product in proceedings—international symposium on computer architecture, 2021. https://doi.org/10.1109/ISCA52012.2021.00014
3. Pal S, Arif S (2015) A fully differential read-decoupled 7-T SRAM cell to reduce dynamic power consumption. ARPN J Eng Appl Sci 10(5):2142–2147
4. Roy C, Islam A. Design of low power, variation tolerant single bit line 9T SRAM cell in 16-nm technology in subthreshold region. Microelectron Reliab. https://doi.org/10.1016/j.microrel.2021.114126

5. Dadori AK, Khare K, Gupta TK, Singh RP (2017) Leakage power reduction technique by using multigate FinFET in DSM technology. Advan Intel Syst Comput. https://doi.org/10.1007/978-981-10-2750-5_25
6. Magraiya VK, Gupta TK (2019) ONOFIC-based leakage reduction technique for FinFET domino circuits Int. J Circuit Theory Appl. https://doi.org/10.1002/cta.2583
7. Chavan UJ, Patil SR (2016) High performance and low power ONOFIC approach for VLSI CMOS circuits design. In: International conference on communication and signal processing. https://doi.org/10.1109/ICCSP.2016.7754171
8. Kumar C, Mishra AS, Sharma VK (2018, 2019) Leakage power reduction in CMOS logic circuits using stack ONOFIC technique. In: Proceedings of the 2nd international conference on intelligent computing and control systems, ICICCS. https://doi.org/10.1109/ICCONS.2018.8662955
9. Raghunath A, Bathla S (2021) Analysis and comparison of leakage power reduction techniques for VLSI design. In: International conference on computer communication and informatics, ICCCI. https://doi.org/10.1109/ICCCI50826.2021.9402315
10. Saini P, Mehra R (2012) leakage power reduction in CMOS VLSI circuits. Int J Comput Appl. https://doi.org/10.5120/8778-2721
11. Banu S, Gupta S (2020) The sub-threshold leakage reduction techniques in CMOS circuits. In: Proceedings of the international conference on smart technologies in computing, electrical and electronics. https://doi.org/10.1109/ICSTCEE49637.2020.9277192
12. Deepika KG, Priyadarshini KM, David K, Raj S (2013) Sleepy keeper approach for power performance tuning in VLSI design. Int J Electron Commun Eng
13. Bhargav KN, Suresh A, Saini G (2015) Stacked keeper with body bias: a new approach to reduce leakage power for low power VLSI design. In: Proceedings of 2014 IEEE international conference on advanced communication, control and computing technologies. https://doi.org/10.1109/ICACCCT.2014.7019482
14. Pal PK, Rathore RS, Rana AK, Saini G (2010) New low-power techniques: leakage feedback with stack & sleep stack with keeper. In: 2010 international conference on computer and communication technology. https://doi.org/10.1109/ICCCT.2010.5640514
15. Agrawal R, Tomar VK (2018) Implementation and analysis of low power reduction techniques in sense amplifier. In: Proceedings of the 2nd international conference on electronics, communication and aerospace technology. https://doi.org/10.1109/ICECA.2018.8474703
16. Yousuf A, Mohamed Salih KK (2018) Comparison of sleep transistor techniques in the design of 8-bit vedic multiplier. In: 2018 international conference on emerging trends and innovations in engineering and technological research. https://doi.org/10.1109/ICETIETR.2018.8529000
17. Shi K, Howard D (2006) Challenges in sleep transistor design and implementation in low-power designs. In: Proceedings—design automation conference. https://doi.org/10.1145/1146909.1146943
18. Ruhil S, Shukla NK (2017) Leakage current optimization in 9T SRAM bit-cell with sleep transistor at 45 nm CMOS technology. In: 2017 international conference on computing and communication technologies for smart nation. https://doi.org/10.1109/IC3TSN.2017.8284487
19. Weste NHE, Harris DM (2011) CMOS VLSI design: a circuits and systems perspective, 4th edn. Addison-Wesley
20. Sachdeva A, Tomar VK, Sachdeva (2020) Design of a stable low power 11-T static random-access memory cell. J Circ, Syst Comput. https://doi.org/10.1142/S0218126620502060
21. Turi MA, Delgado-Frias JG (2020) Effective low leakage 6T and 8T FinFET SRAMs: using cells with reverse-biased FinFETs, near-threshold operation, and power gating. IEEE Trans Circ Syst II Express Briefs 67(4):765–769. https://doi.org/10.1109/TCSII.2019.2922921
22. PTM model file Online. http://ptm.asu.edu/

Artificial Intelligence and Its Applications

Hybrid Deep Learning Models for Segmentation of Atherosclerotic Plaque in B-mode Carotid Ultrasound Image

Pankaj Kumar Jain, Neeraj Sharma, and Sudipta Roy

Abstract Stroke is a fatal disease in developing countries. Every year millions of people die due to stroke and other cardiovascular diseases (CVD). Main cause of CVD and stroke is atherosclerosis disease. Modern medical treatments include automated detection of atherosclerotic plaque tissue detection in ultrasound images. Imaging plaques using ultrasound avoids radiation is inexpensive and readily available in most diagnostic centers. Image analysis can assist in stratifying stroke risk using *plaque tissue characterization* (PTC). Earlier strategies have used machine learning (ML)-based approaches but are ad-hoc, tedious, and not accurate. We presented two deep learning models for internal carotid artery (ICA) plaque segmentation. The methodology consisted of design of two deep learning models, named **UNet** and **SegNet-UNet**. Each of the models had a depth of four layers that controlled the number of parameters and computation time. We considered a database of 970 B-mode ICA ultrasound images of high-risk patients. Also, we used a K10 cross-validation on complete database. Finally, we calculated the plaque area (in mm^2) using the segmented pixels. We achieved correlation coefficient of **0.98** (p-value < 0.001) between estimated and ground truth plaque area for both models. The overall system performs segmentation of one image in fraction of a second.

Keywords Hybrid deep learning model · UNet · SegNet-UNet · Atherosclerosis

P. K. Jain (✉) · N. Sharma
Indian Institute of Technology (BHU) Varanasi, Varanasi, Uttar Pradesh, India
e-mail: pankajkrjain.rs.bme17@iitbhu.ac.in

S. Roy
Jio Institute, Navi Mumbai, Maharashtra, India

S. Sharma et al. (eds.), *Intelligent Control, Robotics, and Industrial Automation*, Lecture Notes in Electrical Engineering 1066,
https://doi.org/10.1007/978-981-99-4634-1_63

1 Introduction

Stroke and cardiovascular diseases are life threatening diseases in the modern world. A large population is affected by stroke and CVDs. Majority of people either dies or suffers by post effects of the diseases. Major causes of stroke and CVDs are atherosclerotic plaque in blood flowing arteries. Atherosclerosis disease (internal, external, or common carotid arteries) develops plaques inside lumen intima and media adventitia layers of carotid arteries [1]. These plaque when ruptures causes embolism in the brain leading to the blockage of blood supply [2]. However, unstable and prone to rupture plaque tissues are very small in size. These vulnerable plaque tissues produces 1–2% stroke rate in patients with 80% stenosis [3]. Ultrasound imaging modality proves to be best among the MRI, CT and X-ray imaging modalities, due to its non-invasive, non-radiative, no contrast agents, economic, and portable nature. Due to all these advantages, ultrasound imaging modality is being adapted in many diagnostic procedures such as abdomen, liver, thyroid, coronary, and carotid diagnosis.

Stroke risk assessment involves assessment of common and internal carotid arteries. Previously, many studies involve atherosclerotic plaque tissue characterization using automated [4, 5] and semi-automated methods [6, 7]. Although, mentioned studies are automated in nature, still they involve human interventions in their operations. These human interventions may be in terms of features extractions or selections or performance evaluations. Therefore, the main objective of this research is to propose a fully automated method of plaque tissue characterization and segmentation from ICA images. In this objective, we proposed a 'U' shaped CNN architecture having encoder and decoder and bottleneck layers in it. This architecture is a fully supervised method of image segmentation. Further, we propose a modification of the above architecture by combining two solo architectures into hybrid architecture. This hybrid architecture is SegNet-UNet and has few added advantages due to its hybrid architecture.

Common carotid artery (CCA) section was in the main focus of research for last one decade. In recent years, researchers have presented many segmentation methods for CCA plaque [8]. These methods are mainly use scale-space paradigm [9, 10], which are mainly for low to medium plaque measurements. Recently, few authors presented their research for plaque area measurements using deep learning in CCA [5, 11–13]. Further, attention-based mechanism is incorporated in DL-based segmentation by few authors in medical image segmentation [14–16]. Although the previous studies use DL methods for CCA plaque wall segmentation, they do not demonstrate the method for ICA plaque segmentation; since ICA offers challenges like sharp changes in plaque burden and consumes larger plaque area. This study is about ultrasound ICA segmentation and uses solo-DL or hybrid [17–19] DL models. Figure 1 shows the global system diagram of the plaque segmentation in B-mode ultrasound image of the carotid artery.

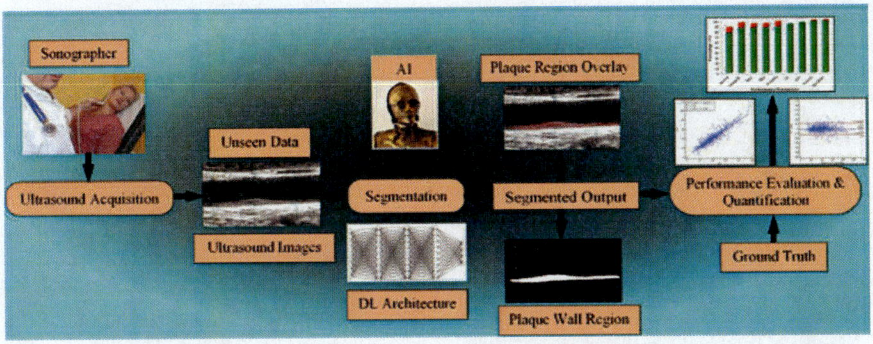

Fig. 1 Global system diagram for the plaque segmentation in carotid artery ultrasound images

2 Methodology

2.1 Patient Cohort and Data Collection

We used a cohort of 97 patients in our study. In this work, 97 US videos of ICA images were taken from Imperial College London UK and converted into still images. The male/female ratio of the database includes 47 male and 52 female patients, and the mean age of patients was 75.04 ± 9.96 years (between 45 and 96 years). The same database is used in our previously published articles [16, 18, 20].

2.2 Ground Truth Data Acquisition and Binary Mask Preparation

Ground truth binary masks are prepared by tracing the upper layer of lumen region, i.e., lumen intima (LI) and the inner layer, i.e., media adventitia (MA) borders of the ICA Far wall. These binary masks are used as ground truth (GT) for both DL models in this paper. Figure 2 shows the raw and GT binary mask images of the ICA images.

2.3 Architecture of DL Models

UNet architecture

Figure 3 below represents architecture of an encoder–decoder-based UNet model. The 'U' shaped UNet model has 4 encoder and 4 decoder stages. Both encoder and decoder stages are connected by a bottleneck layer. Each encoder stage has cascaded

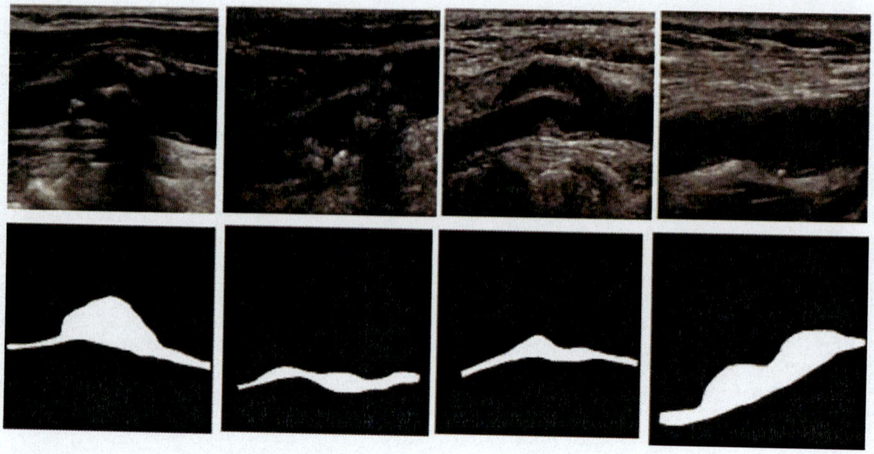

Fig. 2 Raw ICA and GT binary mask images

connections of convolution, followed by ReLU and MaxPooling layers. The convolutional and MaxPooling layers use filter size of 3 × 3 an 2 × 2, respectively. Each encoder stage MaxPooling layer down samples the image for the next stage. After encoder stages, there exists a bottleneck layer which contains finest features from the extracted by the encoder side. This bottleneck layer also known as bridge network as it stands between the encoder and decoder network. The decoder stages comprise of up-convolution followed by convolution, ReLU, and MaxPooling layer. The decoder network up-samples or resizes the image features obtained from bottleneck layer with 2 × 2 up-convolutional layer.

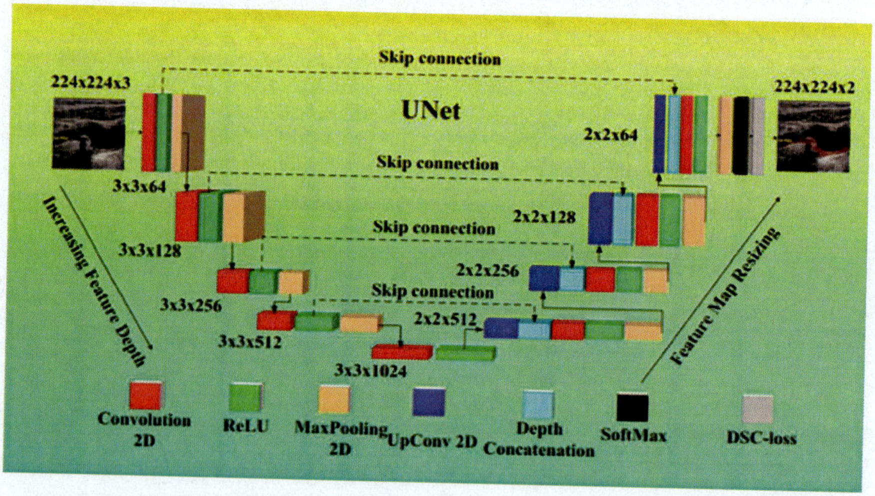

Fig. 3 A 4 stage Solo-DL UNet architecture

Table 1 Hyperparameters used for experimental setup

Learning parameter	UNet	SegNet-UNet
Optimizer	ADAM	ADAM
Training images	786	786
Validation images	87	87
Test images	97	97
Cross-validation	K10	K10
Initial learning rate	0.0001	0.0001
Max epochs	100	100
Batch size	6	6

During the feature extraction in encoder stage, some spatial features of the region of interest (ROI) are not carried to the next layer (bottom layer), which are carried by the skip connections and added to the decoder stage at the same level. Using these skip connections, the encoder stages supply the spatial features to the corresponding decoder stage. Finally, the decoder output comes out as the concatenation of the two input features: first from spatial feature information from the corresponding encoder stage and second from the adjacent bottom decoder layer. In relation to the title, this network corresponds to the solo network. Finally, after the 4th decoder stage, the optimizer (ADAM or SGDM) gradually reduces the pixel segmentation loss value, and softmax classifier classifies the unsampled data into atherosclerotic plaque (positive class) and background (negative class).

SegNet-UNet Architecture

The results of UNet architecture motivated us to develop hybrid architecture by combining the solo models. Therefore, we combined SegNet and the UNet and developed SegNet-UNet hybrid model. Further, we applied the same hyperparameter conditions defined in Table 1 and two loss functions (cross-entropy and dice similarity coefficient) to the hybrid models. Figure 4 below shows architecture of SegNet-UNet model.

2.4 Loss Function

To compare the solo and hybrid models based on various parameters, we used two majorly used loss functions, named cross-entropy and dice similarity loss functions. Both loss functions are used at the end of the DL model based on which we compared the same model twice (against the other model and against the same model with different loss function). The cross-entropy loss function is defined by the Eq. (1).

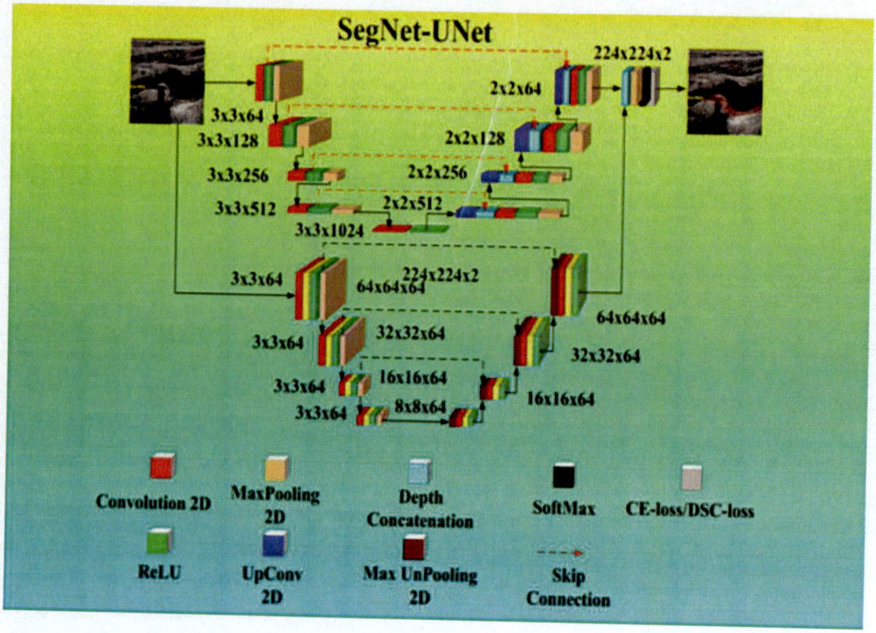

Fig. 4 4-stage hybrid DL SegNet-UNet architecture

Binary Cross-Entropy Loss Function

Equation (1) gives the binary cross-function used in the experimentation for all the models.

$$L_{CE} = -\frac{1}{N}\sum_{i=1}^{N}\left[(y_i \times \log a_i) + (1 - y_i) \times \log(1 - a_i)\right] \tag{1}$$

where 'a' is the inference values provided by the softmax layer.

Dice similarity coefficient loss function

The dice similarity coefficient loss (DSC-loss) function is calculated by using the generalized dice similarity coefficient between the GT and the predicted images. The dice similarity coefficient loss function L_{DSC} is written as Eqs. (2) and (3).

$$L_{\mathrm{DSC}} = 1 - \mathrm{DSC} \tag{2}$$

$$L_{\mathrm{DSC}} = 1 - \left[\left(\frac{2 \times \sum_{n=1}^{N} Y_n \times \hat{Y}_n}{\sum_{n=1}^{N} Y_n^2 \times \hat{Y}_n^2}\right) + \left(\frac{2 \times \sum_{n=1}^{N} (1 - Y_n) \times \left(1 - \hat{Y}_n\right)}{\sum_{n=1}^{N} \left(2 - Y_n^2 \times \hat{Y}_n^2\right)}\right)\right] \tag{3}$$

where $N =$ aggregate number of elements in the x, y-direction of the image. Ground truth and the AI-estimated images are represented by Y and $\hat{Y}$, respectively.

2.5 Cross-Validation

Cross-validation is the part of the training process to effectively utilize the database. Also, in this process, each image gets a chance to cross-validate or test by the system. In simple split process, the whole database is split into training and test subsets. However, in K cross-validation process, database is divided into K parts or subsets, where K-1 subsets are used for training, and one part or subset is used for testing. Further, the test subset is swapped from K-1 training subsets. The subset swapping process continues for K number of times, and each K subset becomes part of test process at least ones. In our case, a total of 970 images were partitioned into K = 10 subsets of 97 images. One subset of 97 images was used for the test, and the remaining 9 subsets of 873 images were used for the training process. When the training is finished, 97 test images are swapped from the remaining 873 images. In the complete process, test subsets are swapped 10 times, and all images are cross-validated.

3 Results and Performance Evaluations

Figure 5 shows the raw ICA image in the first row and first column. In the first row second column, raw image with binary mask overlay in green color is shown. In the second row, estimated plaque area in red color, its difference with GT (red and green) is shown for UNet model for both loss functions. In the third row, estimated plaque area in red color, and its difference with GT (red and green) is shown for SegNet-UNet model for both loss functions. Table 2 shows the performance of both models for CE-loss function and DSC-loss function. We evaluated the specificity, sensitivity, Matthews's correlation coefficient (MCC), accuracy, and precision for both models. Corresponding mean $\pm$ SD values of all images for UNet models (with CE-loss) are 98.56 $\pm$ 0.60, 90.32 $\pm$ 5.54, 99.23 $\pm$ 0.56, 99.23 $\pm$ 0.56, and 89.58 $\pm$ 3.69. Same mean $\pm$ SD values for SegNet-UNet models (with CE-loss) are 98.44 $\pm$ 0.70, 88.16 $\pm$ 7.00, 99.27 $\pm$ 0.45, 90.69 $\pm$ 5.14, and 88.46 $\pm$ 4.42.

After performing the segmentation task, we performed the some performance evaluation tests to validate the system performance. For this purpose, we used measured plaque area (PA) and ground truth plaque area and GT labels ('1' for PA > 40 mm^2, '0' for PA < 40 mm^2) as input variables. We performed regression analysis, receiver operating characteristics (ROC) analysis, cumulative distribution plot analysis, and paired-t-test analysis. All performance evaluation results are summarized here in brief, and their corresponding graphs and curves are shown below.

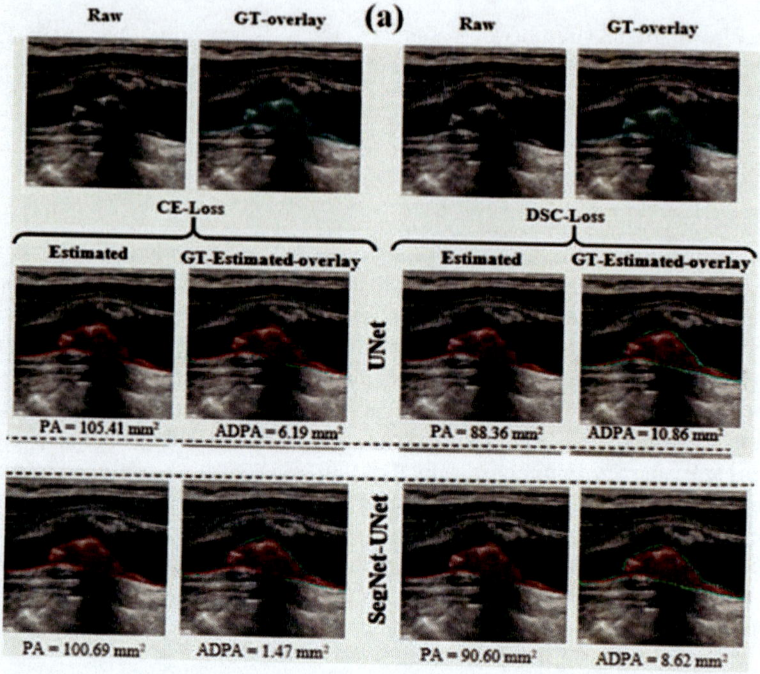

Fig. 5 Visual results of the solo and hybrid deep learning models # PA = Plaque area; ADPA = Absolute difference plaque area

Table 2 Performance parameters values (Mean ± SD) of two DL models using cross-entropy and dice loss function

Model	Spec#	Sens	MCC	Acc	Prec
UNet (CE)	99.23 ± 0.56	90.30 ± 5.50	89.58 ± 3.69	98.56 ± 0.60	99.23 ± 0.56
UNet (DSC)	98.50 ± 0.74	95.30 ± 4.47	86.50 ± 4.51	98.24 ± 0.70	86.93 ± 4.70
SegNet-UNet (CE)	99.27 ± 0.45	88.20 ± 7.00	88.46 ± 4.42	98.44 ± 0.70	90.69 ± 5.14
SegNet-UNet (DSC)	99.70 ± 0.32	78.50 ± 7.95	85.45 ± 4.78	98.13 ± 0.73	85.81 ± 5.08

#Spec Specificity; *Sens* Sensitivity; *MCC* Mathew's correlation coefficient; *Acc* Accuracy; *Prec* Precision

3.1 Regression Curve and Correlation Coefficient

Figure 6 below shows the regression curve and CC values between GT and estimated plaque area for UNet and SegNet-UNet models. For both experiments, CC value is 0.98.

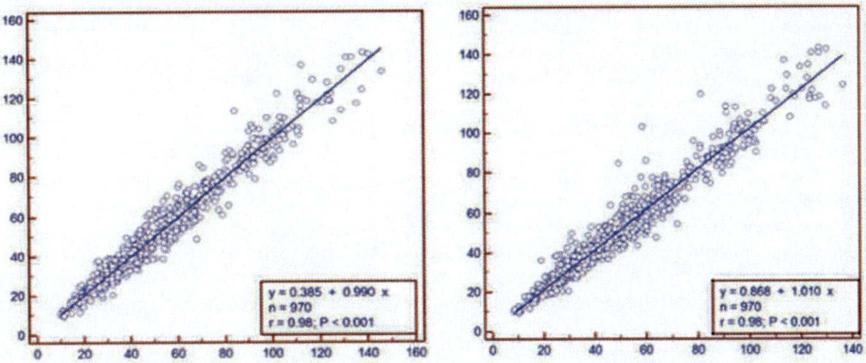

(a) (b)

Fig. 6 Regression curve and correlation coefficient for solo and hybrid models **a** UNet **b** SegNet-UNet

3.2 Receiver Operating Characteristics

Figure 7 below shows the receiver operating characteristics (ROC) curve for GT labels (1 for plaque area > 40 mm^2 and 0 for plaque area < 40 mm^2) and AI-estimated area. The area under the ROC curve is 0.910 and 0.905 for UNet and SegNet-Unet, respectively.

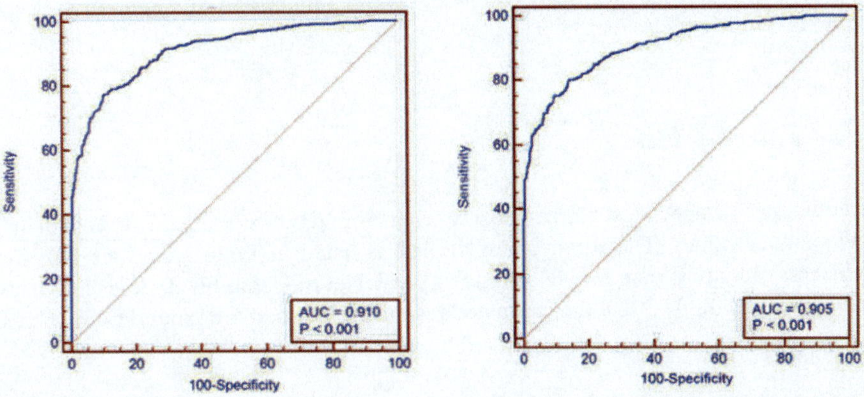

(a) (b)

Fig. 7 ROC curve and AUC values for solo and hybrid models **a** UNet **b** SegNet-UNet

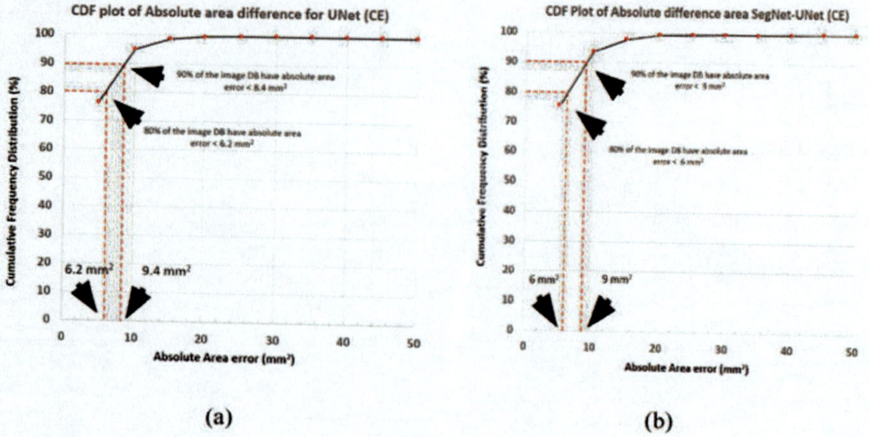

Fig. 8 Cumulative distribution plot and absolute area error for solo and hybrid models **a** UNet **b** SegNet-UNet

3.3 Cumulative Distribution Plot

Cumulative distribution plot for absolute area error vs cumulative number of images is shown in Fig. 8. As can be seen from the figure the dotted red line shows 90 and 80% of image database against the absolute area error (mm^2). It is clear that 90% of images have absolute area error < 9.4 mm^2 for UNet, and same for SegNet-UNet is 9 mm^2. Absolute area error for UNet model is < 6.2 mm^2, and for SegNet-UNet, the same is 6 mm^2.

3.4 Paired-t-Test

Another performance test paired-t-test was performed between the GT and estimated plaque area values. The summary of the test is listed in Table 3, and the box and whiskers plot are shown for UNet and SegNet-UNet model in Fig. 9. It is clear from the Fig. 9 and Table 3 that the estimated results are close to the ground truth values.

4 Benchmarking Studies

We have compared our results with some benchmark studies in past. Table 4 comprises these studies.

Table 3 Paired-t-test of GT and AI images

Properties	GT image	UNet	SegNet-UNet
Sample size	970	970	970
Arithmetic mean	47.4945	47.5650	46.1467
95% CI for the mean	45.8663 to 49.1228	45.9504 to 49.1795	44.5708 to 47.7226
Variance	667.7816	656.5795	625.5493
Standard deviation	25.8415	25.6238	25.0110
Standard error of the mean	0.8297	0.8227	0.8031

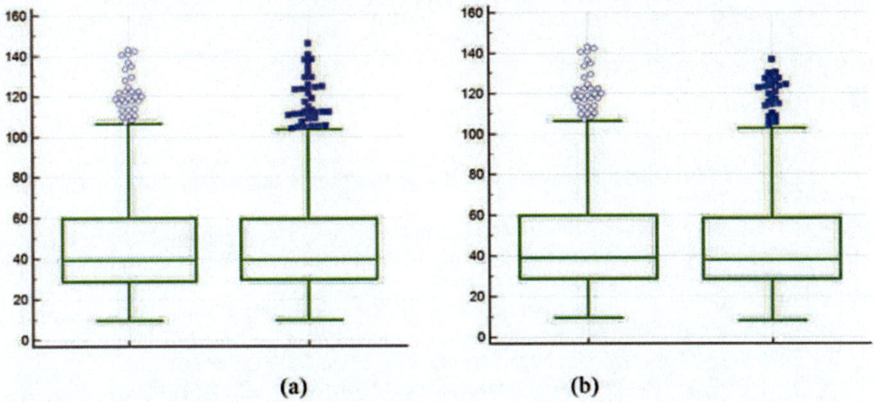

Fig. 9 Paired-t-test between GT and AI-estimated area (a) UNet (b) SegNet-UNet

Table 4 Benchmark studies in CCA and ICA plaque segmentation

Authors	Type of DL network	Measurement	Images	Results
Elisa et al. [21]	Fully connected network (FCN)	Total plaque area (TPA)	396	Model1 = 20.52 mm^2, Model2 = 19.44 mm^2
Zhou et al. [13]	UNet++	Plaque area, total plaque area (TPA)	144	Total plaque area error: 5.55 ± 4.34 mm^2
Proposed 2021	UNet (w/CE-loss) UNet (w/DSC-loss) SegNet-UNet (w/ CE-loss) SegNet-UNet (w/ DSC-loss)	Plaque area	970	Plaque area error: 3.49 mm^2 (w/CE-loss)

5 Conclusion and Future Scope

As we proposed in our hypothesis that the hybrid DL model results are better or comparable to solo-DL model, the results are in line with our hypothesis. We found that the SegNet-UNet DL model performance is better than the UNet model. The segmentation accuracy of SegNet-UNet model is 98.44%, the absolute area error is 6 mm^2 (80% of scans), and area under the ROC curve (AUC) was 0.905 ($p < 0.001$) using 10% threshold. Also, we found that the CE-loss function gave superior results compared to DSC-loss for both models. Further, the method can be extended to make faster, low memory models. Also, some other hybrid methods can be employed to enhance the model's feature extraction capability.

References

1. Libby P (2006) Inflammation and cardiovascular disease mechanisms. Am J Clin Nutr 83(2):456S-S460
2. Gupta A, Kesavabhotla K, Baradaran H, Kamel H, Pandya A, Giambrone AE et al (2015) Plaque echolucency and stroke risk in asymptomatic carotid stenosis: a systematic review and meta-analysis. Stroke 46(1):91–97
3. Virani SS, Alonso A, Aparicio HJ, Benjamin EJ, Bittencourt MS, Callaway CW et al (2021) Heart disease and stroke statistics-2021 update: a report from the American heart association. Circulation 143(8):e254–e743. https://doi.org/10.1161/cir.0000000000000950
4. Biswas M, Kuppili V, Edla DR, Suri HS, Saba L, Marinhoe RT et al (2018) Symtosis: A liver ultrasound tissue characterization and risk stratification in optimized deep learning paradigm. Comput Methods Programs Biomed 155:165–177
5. Biswas M, Saba L, Chakrabartty S, Khanna NN, Song H, Suri HS et al (2020) Two-stage artificial intelligence model for jointly measurement of atherosclerotic wall thickness and plaque burden in carotid ultrasound: a screening tool for cardiovascular/stroke risk assessment. Comput Biol Med 123:103847
6. Saba L, Jain PK, Suri HS, Ikeda N, Araki T, Singh BK et al (2017) Plaque tissue morphology-based stroke risk stratification using carotid ultrasound: a polling-based PCA learning paradigm. J Med Syst 41(6):98
7. Araki T, Ikeda N, Shukla D, Jain PK, Londhe ND, Shrivastava VK et al (2016) PCA-based polling strategy in machine learning framework for coronary artery disease risk assessment in intravascular ultrasound: a link between carotid and coronary grayscale plaque morphology. Comput Methods Programs Biomed 128:137–158
8. Molinari F, Liboni W, Giustetto P, Badalamenti S, Suri JS (2009) Automatic computer-based tracings (ACT) in longitudinal 2-D ultrasound images using different scanners. J Mech Med Biol 9(04):481–505
9. Krishna Kumar P, Araki T, Rajan J, Saba L, Lavra F, Ikeda N et al (2017) Accurate lumen diameter measurement in curved vessels in carotid ultrasound: an iterative scale-space and spatial transformation approach. Med Biol Eng Comput 55(8):1415–1434. https://doi.org/10.1007/s11517-016-1601-y
10. Molinari F, Pattichis CS, Zeng G, Saba L, Acharya UR, Sanfilippo R et al (2012) Completely automated multiresolution edge snapper–a new technique for an accurate carotid ultrasound IMT measurement: clinical validation and benchmarking on a multi-institutional database. IEEE Trans Image Process 21(3):1211–1222. https://doi.org/10.1109/tip.2011.2169270

11. Jain PK, Sharma N, Saba L, Paraskevas KI, Kalra MK, Johri A et al (2021) Unseen artificial intelligence—deep learning paradigm for segmentation of low atherosclerotic plaque in carotid ultrasound: a multicenter cardiovascular study. Diagnostics 11(12):2257
12. Jain PK, Sharma N, Saba L, Paraskevas KI, Kalra MK, Johri A et al (2021) Automated deep learning-based paradigm for high-risk plaque detection in B-mode common carotid ultrasound scans: an asymptomatic Japanese cohort study. Int Angiol: J Int Union Angiol
13. Zhou R, Guo F, Azarpazhooh R, Hashemi S, Cheng X, Spence JD et al (2021) Deep learning-based measurement of total plaque area in B-mode ultrasound images. IEEE J Biomed Health Inf. https://doi.org/10.1109/JBHI.2021.3060163
14. Oktay O, Schlemper J, Folgoc LL, Lee M, Heinrich M, Misawa K et al (2018) Attention u-net: learning where to look for the pancreas. arXiv preprint arXiv:180403999
15. Pal D, Reddy PB, Roy S (2022) Attention UW-Net: a fully connected model for automatic segmentation and annotation of chest X-ray. Comput Biol Med 106083. https://doi.org/10.1016/j.compbiomed.2022.106083
16. Jain PK, Dubey A, Saba L, Khanna NN, Laird JR, Nicolaides A et al (2022) Attention-based UNet deep learning model for plaque segmentation in carotid ultrasound for stroke risk stratification: an artificial intelligence paradigm. J Cardiovasc Dev Dis. https://doi.org/10.3390/jcdd9100326
17. Szegedy C, Liu W, Jia Y, Sermanet P, Reed S, Anguelov D et al (2015) Going deeper with convolutions. In: Proceedings of the IEEE conference on computer vision and pattern recognition, pp 1–9
18. Jain PK, Sharma N, Giannopoulos AA, Saba L, Nicolaides A, Suri JS (2021) Hybrid deep learning segmentation models for atherosclerotic plaque in internal carotid artery B-mode ultrasound. Comput Biol Med 136:104721
19. Jena B, Saxena S, Nayak GK, Saba L, Sharma N, Suri JS (2021) Artificial intelligence-based hybrid deep learning models for image classification: the first narrative review. Comput Biol Med 137:104803
20. Jain PK, Sharma N, Kalra MK, Johri A, Saba L, Suri JS (2022) Far wall plaque segmentation and area measurement in common and internal carotid artery ultrasound using U-series architectures: an unseen artificial intelligence paradigm for stroke risk assessment. Comput Biol Med 106017
21. Cuadrado-Godia E, Srivastava SK, Saba L, Araki T, Suri HS, Giannopolulos A et al (2018) Geometric total plaque area is an equally powerful phenotype compared with carotid intima-media thickness for stroke risk assessment: a deep learning approach. J Vasc Ultrasound 42(4):162–188

Real-Time GPU-Accelerated Driver Assistance System

Pooja Ravi, Aditya Shukla, and B. Muruganantham

Abstract We present an extensive driver assistance system capable of executing two essential tasks. The first module is used in assisting the driver with road safety alerts (RSA); it scans the road environment and detects any significant entities including but not limited to vehicles, pedestrians, and traffic lights. Further, alerts are issued when the estimated physical distance from detected entities is less than a set threshold. For this module, we also propose the usage of a compute-accelerated Swin Transformer model and evaluate its efficacy against other state-of-the-art models by considering relevant metrics like inference time and mAP. The second module pertains to driver alertness detection (DAD) for identifying signs of fatigue. It scans the driver's face and monitors a live video feed to ensure that the driver shows no signs of micro-sleep. When either module detects a behavioural anomaly, it will alert the driver with text-based messages and non-disruptive audio messages. We propose such a state-of-the-art safety system being integrated into the advanced driver assistance systems (ADAS's) seen in modern vehicles.

Keywords Road entity tracking · Facial behaviour monitoring · Real-time live video analysis · Object detection · ADAS

1 Introduction

Ensuring the safety and alertness of drivers is a task of utmost importance. A large number of traffic accidents that take place can be mitigated by the drivers being made alert and aware of the things happening on the road. While car accidents are often due to rule-breaking and over-speeding, more devastating accidents (caused by heavier vehicles such as trucks, trailers, and lorries) are often due to overworked and

P. Ravi (✉) · A. Shukla · B. Muruganantham
Department of Computing Technologies, SRM Institute of Science and Technology, Kattankulathur, Chennai, India
e-mail: kuchan2001@gmail.com

B. Muruganantham
e-mail: muruganb@srmist.edu.in

S. Sharma et al. (eds.), *Intelligent Control, Robotics, and Industrial Automation*, Lecture Notes in Electrical Engineering 1066,
https://doi.org/10.1007/978-981-99-4634-1_64

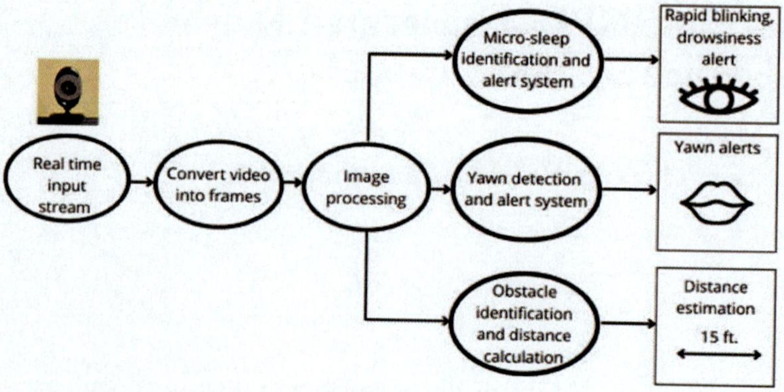

Fig. 1 Conceptual diagram for driver assistance system

sleep-deprived drivers. A system that can both alert and assist drivers on the road can help mitigate this problem. Hence, we propose a real-time driver assistance system that will equip drivers of four-wheel vehicles with some essential safety measures.

Various road entities such as crossing pedestrians, vehicles in proximity, traffic signals are potential causes of accidents if not paid due attention to. Such mishaps usually occur as a result of drowsy drivers, or due to a lack of conscious awareness of the objects moving in and around the driver's line of sight. To aid the driver in identifying such obstacles, we present an object detection mechanism with the ability to identify several classes of obstacles and possible hindrances on the road including vehicles, pedestrians, motorcyclists, bikers, as well as larger vehicles such as buses and trucks. Additionally, we compute an estimate of the distance between our vehicle and other detected cars or pedestrians. If an object comes closer than 15 ft, we propose alerting the driver via a non-disruptive audio message. This is the work of the RSA module. Such measures can help the driver foresee any plausible mishaps and regain control of the vehicle in a timely manner.

Coming inside the vehicle, the DAD module tracks the driver's behaviour and alerts them in a similar non-disruptive manner if they get drowsy. The foremost signs of drowsiness include yawning and droopy eyes.

While the RSA module scans the external environment, the DAD module keeps an eye on the internals i.e. the driver. These two modules, hence, work in tandem as seen in Fig. 1.

2 Related Works

Previous literature on this subject has introduced various algorithms for detecting drowsiness and yawns as well as for localizing & classifying driver behaviour or distraction. We have also seen various approaches explored for integrating such safety systems into ADAS's and vehicles in general.

In [1], the authors propose a drowsiness detection system on the Android interface wherein they combine the face detection and drowsiness identification module with a smartphone interface. Once the level of drowsiness is detected, they sound an alarm along with visual alerts. They detect yawns, blinks, head movements and signs of distractions. The authors of [1] also consider some special case scenarios in which the driver wears glasses or has hair covering their face. They calculate the number of correct predictions with respect to ground truth values and achieve an average hit rate of 93.37%.

The authors of [2] propose and compare two different approaches to detecting drowsiness among drivers: the first being to use a combination of recurrent and convolutional neural networks to capture the temporal aspect, and the second, to use features from images which are sent to a fuzzy classifier for obtaining predictions. As per their claims, the fuzzy system outperforms the other algorithm due to the reduced presence of false positives, helping achieve a high specificity of 93%. They use a combination of the GRU [3] and EfficientNet [4] architectures to ensure that a certain number of frames is stored in the model's memory so that accurate predictions can be obtained for the current frame. Further, a CNN system with a fuzzy triangular function is also utilized to assess blinks, yawns, and sleep. They achieve an average accuracy of 55.4% and 52% for the first and second methods respectively on the testing data.

In the work by [2], OpenCV [5] and the Dlib library [6] are employed to detect instances of drowsiness among drivers. They find the aspect ratio and establish a threshold that, if crossed, will cause an alert to be issued. Once the facial landmarks are localized successfully, their algorithm calculates an aspect ratio and sounds an alarm if the counter increases beyond a certain set limit. They also test the robustness of their proposed method and quantify how it behaves under special circumstances such as bespectacled faces and added noise.

In [7], the authors propose a CenterNet [8] based convolutional network architecture wherein they include modifications to help with optimized downsampling and spatial pyramid pooling. They obtain key point estimates and use output layers similar to [8] for producing results. The usage of atrous convolutions and lower image resolutions helps the authors of [7] to achieve computational optimality. They fulfil the object of road entity detection using their modified ASPP-CenterNet and achieve an AP of 70.6 on small objects.

Companies such as Tesla and Waymo use more sophisticated techniques such as 3D object detection, semantic segmentation, and monocular depth estimation for capturing the road environment.

While the aforementioned techniques have one specific focus, our work's objective is manifold and involves optimized computations, entity detection, and distance estimation while also ensuring the driver's alertness. Because some of the primary concerns in vehicular ADAS systems are the memory & energy footprints, we strive to achieve maximum performance with limited resources.

Fig. 2 Obstacle identification

3 Methods

We shall classify our discussions regarding the methods as per the module under which they are proposed to be implemented, namely RSA and DAD.

3.1 Road Safety Alerts (RSA)—Obstacle Localization

3.1.1 Road Entity Detection and Tracking

The task of detecting objects has been assigned to the Swin Transformer [9], a state-of-the-art architecture that excels in object detection and classification tasks. By making use of the publicly available Udacity Self-Driving Dataset (which consists of a slew of images with diverse terrains, weather conditions, and times of day), we propose the usage of the Swin [9] model to precisely detect various road entities. We further compare the performance of Swin with several variants of the popular YOLO object detection model so as to provide an idea of how our proposed detection and distance estimation methods outperform pre-existing standards.

Swin [9] is a transformer architecture that suits various computer vision tasks such as object detection, image classification and object recognition. It is a successor of the Vision Transformer [10] and its core architecture consists of various transformer blocks and their corresponding patch merging blocks.

The initial patch partition block converts the input image into 4 $\times$ 4 non-overlapping patches and the linear embedding layer converts the channels of the patches to a vector of size C. Further, the image is passed through successive patch merging and Swin Transformer blocks. The patch merging modules help combine the features and amplify the output channels.

This Swin architecture, trained on the COCO dataset [11], has been employed as the base for performing accurate object detection; an example of its performance can be seen in Fig. 2. The bounding box values predicted by Swin (and passed

Fig. 3 Masked pixel representation—Swin transformer

downstream) provide the basis for estimating the distance of the detected entities (pedestrians, trucks, other vehicles, etc.) from the driver's vehicle.

In Fig. 3, we have provided a visualization of Swin's learning process and how it localizes bounding box pixel features for further assigning its predictions.

3.1.2 Distance Estimation

Once the model has tracked the objects present in a frame, the natural next step is to develop some heuristics to sift through the objects and alert the driver to any significant entities. This may be other vehicles such as cars, trucks, bikers, or even pedestrians crossing the street.

The method used to approximate the distance of an entity from the camera is as follows: say an object is present in an image. Let its real-world width be W units and its apparent width (i.e. that in the image) be P pixels. Let us also say that the object is present at a distance of D units from the camera.

Now, we can compute a value called the perceived focal length F as follows:

$$F = \frac{P}{W} \times D \tag{1}$$

The above method leverages the property of **Triangle Similarity**.

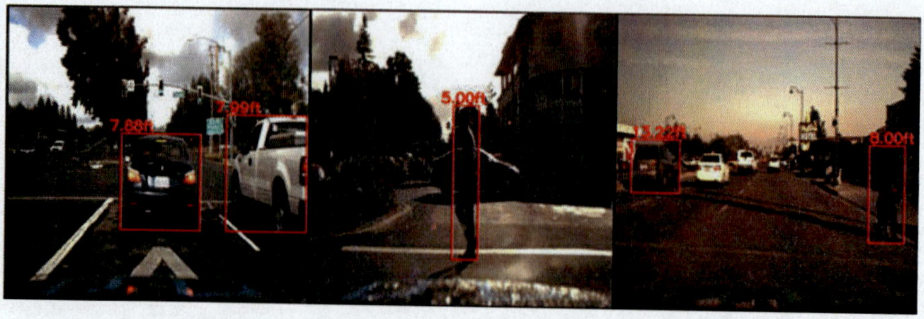

Fig. 4 Distance estimation

This value holds insight into how the size of an object scales to its size in an image. We can intuit that this has something to do with its real size, its apparent size, and how far it is from the camera. Thus, given a scenario where we do not know the 3rd factor, we can work out a guess for how far the object is if we know the first two.

Taking the example of cars, not all models and brands of cars have the same width, but we can come up with a fairly accurate ballpark: most commercial-use cars such as sedans and hatchbacks are about 5.8 ft wide. This value can serve as the real width of the object.

Now, given an image where our model has detected and put a bounding box around a car, we can very easily compute the width of that car in pixels in the image. This value can serve as the apparent width of the object.

Finally, using the below formula derived from Eq. 1, we can compute the **distance D (in ft)**, of the object from the camera to arrive at Eq. 2:

$$D = \frac{W}{P} \times F \tag{2}$$

where, **W**—real width of object (in ft), **P**—perceived width of object (in pixels), **F**—perceived focal length (computed separately beforehand).

Note that the value of F will be different for each class that our model detects. This is because each class's object has a different average length, and hence the value of F is computed manually beforehand for each class.

An example of the above method applied to an image is in Fig. 4.

3.2 *Driver Alertness Detection (DAD)—Facial Landmark Tracking*

We employ the use of facial landmarks and mathematical distance norms (such as the Euclidean distance) to track aspects like drowsiness and yawning for a driver. The requirements for the same include a functioning camera fitted inside the vehicle for

Fig. 5 Ratio of eyelid distance

collecting an input video stream of the driver's face, and a pipeline to forward this incoming data stream to a processing device that will run our proposed algorithm.

Google's open-source Mediapipe Face Detection Framework [12] helps localize facial landmark points in the form of a 3D mesh on every human subject appearing before the camera. 468 local points are identified on an individual's face and are numbered distinctly to aid with facial behaviour tracking.

The drowsiness and yawn detection algorithms work as independent modules on real-time camera input. These algorithms can be scaled across hardware regardless of the configuration and quality of equipment used.

3.2.1 Eye Motion Tracking

The *OpenCV* library [5] of Python plots the marked facial points for the upper and lower eyelids of both eyes by making use of the aforementioned Mediapipe Framework [12] for assigning 3-dimensional landmarks throughout the facial surface appearing on the frame. The tracking algorithm then converts the relevant predicted facial points to Numpy arrays by indexing the required landmarks. The arrays corresponding to the eye region are then unravelled to obtain the coordinates for the landmarks pertaining to the eyes.

Having obtained the eye-level data, an Eye Aspect Ratio (EAR) metric seen in Eq. 3 is calculated independently for each eye by using the Euclidean Distance norm as depicted in Fig. 5. The distance is calculated between the extreme coordinates of the eye, namely, the horizontal extremes across the eye and vertical extremes down the middle. This allows us to estimate the linear distance between the desired landmarks. Subsequently, we can divide the horizontal distance by the vertical distance to obtain a ratio. In this way, each time the human appearing in the frame blinks, the ratio tends towards infinity indicating that the eyes are closed.

This gives us a final heuristic in the form of a ratio metric that sheds light on whether the eye is closed or not. A set threshold of 30 frames is the criterion for ascertaining whether the driver is indeed micro-sleeping (which usually lasts up to 15 s).

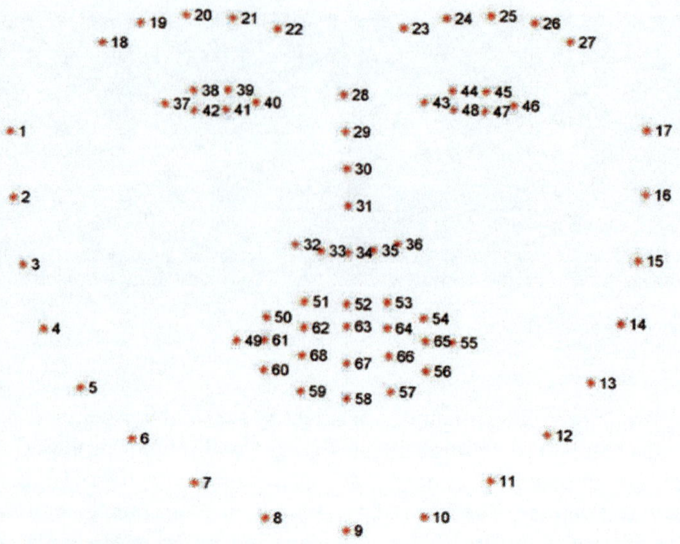

Fig. 6 Dlib facial landmarks

$$\text{Eye Aspect Ratio (EAR)} = \frac{\text{horizontal eyelid distance}}{\text{vertical eyelid distance}} \tag{3}$$

Additionally, we have included an audio alert system that will trigger an alarm upon observing recurring micro-sleep patterns so that the driver can be audibly alerted in case a visual text message is inadequate. The alarm is played at regular intervals upon detecting micro-sleep to ensure that it fulfils the role of keeping the driver attentive.

Hence, our proposed system will help detect early signs of danger, and accordingly, provide audio warnings to alert the driver. Whether the driver wishes to stop and rest or continue the journey (albeit at their own risk) is up to their own discretion.

3.2.2 Lip Tracking

Yawns are a significant and widely noticeable symptom of micro-sleep, and therefore an immediate detection and alert system will help prevent possible accidents as a consequence. The lips are detected using the Dlib facial landmark algorithm [6]. It localizes a number of landmarks across the human face, 68 to be exact. Much like in Mediapipe, we can then access data pertaining to the mouth and lips by indexing the landmark arrays relevant to that region.

The detected lips are denoted by points 49 through 68 in Fig. 6. Much like for eyes, we need a metric to approximate the distance between the upper and lower lips (which will indicate a yawn). Thus, we loop through the coordinates pertaining to the upper and lower outlines of the upper lip, as well as those for the lower lip. We then find a mean of the values for the upper lip, and similarly for the lower lip. The final metric is a modulus of the distance between these two means.

3.2.3 Euclidean Distance

This metric is used to find the distance between two given points in the coordinate plane. For the x- and y-coordinates of the points, their differences are squared and summed. The square root of this value produces the final distance between the given points as in Eq. 4.

$$d = \sqrt{(x_1 - x_2)^2 + (y_1 - y_2)^2} \tag{4}$$

4 Experiments

4.1 Dataset

The pre-trained checkpoints used by us for the Swin and YOLO models have been trained on the COCO dataset. COCO is a large-scale image dataset used for a variety of computer vision tasks such as object detection, image segmentation, and captioning. Although COCO has several classes, the ones pertaining to our interest are *car, person, bicycle, motorcycle, bus, truck, and traffic light.*

Additionally, we also tested the pre-trained model for inference on the aforementioned Udacity Self-Driving Dataset. Some of the model predictions on images from the Udacity dataset can be found in Fig. 7.

4.2 Transfer Learning

We have made use of the variant of Swin transformer pre-trained on the COCO dataset [11]. The foremost reason for adopting this technique is that COCO [11] already includes all the classes of road entities that we wish to detect, making it ideal for our use case. The other important reason why is that the results established by the pre-trained model only help us further ensure that predictions are accurate. Hence, we mitigate the possibility of putting the driver in jeopardy due to inadequate real-time performance. Thus, the process of displaying road safety alerts is accelerated by using accurate and speedy predictions, as well as instantaneous delivery of live updates.

4.3 Metrics

4.3.1 F1-Score

The $F1$-score is a highly balanced evaluation metric as it combines both the precision and recall values to obtain a single comprehensive score that can be used to evaluate any algorithm's performance while not compromising on either precision or recall.

Precision is the measure of how many frames have been correctly classified as belonging to the positive class. Recall is the measure of how many actually positive labels have also been predicted as such. $F1$-score is the harmonic mean of the precision and recall values.

$$F1\ \text{Score} = \frac{2 \times P \times R}{P + R} \tag{5}$$

4.3.2 Mean Average Precision (mAP)

The area under the precision–recall curve is termed the average precision and is used as a performance evaluation metric for object detection tasks. This metric compares the ground truth bounding box to that of the model's predictions and the returned value indicates how closely the two boxes overlap. This is calculated for each output box produced by the model for every available class.

$$AP = \sum_{n} (R_n - R_{n-1}) \times P_n \tag{6}$$

The mAP value is simply the mean of all average precision values obtained by summing and dividing by the total number of samples provided to the model.

$$\text{mAP} = \frac{1}{C} \times \sum_{c} \text{Average Precision}_c \tag{7}$$

Here, C represents the number of classes present in the image.

Specifically, we make use of the map@[0.5:0.95] which signifies a range of confidence thresholds from 0.5 (lowest) all the way to 0.95 (highest) for better analysing model performance in different scenarios.

4.3.3 Inference Times

One significant metric that we rely on to advocate the usage of our method is the inference time observed when the Swin Transformer [9] is used to obtain predictions for any image supplied.

Fig. 7 Bounding box predictions

$$\text{Inference Time} = \frac{\text{End Time} - \text{Start Time}}{\text{Number of Images}} \quad (8)$$

Here, the Start and End times correspond to the time interval for which the inference function runs.

4.4 Results

We experimented with various state-of-the-art model architectures prevalent in object detection including the aforementioned Swin Transformer [9] and other object detection models from the much acclaimed YOLO family. We made use of transfer learning by using the pre-trained weights of each of these models trained on the popular COCO dataset [11], which contains several classes of interest to us, such as cars, bikers, pedestrians, and traffic lights.

The Swin-T model has been employed for inference purposes and relevant results have been demonstrated along with corresponding inference times in Table 1 and other state-of-the-art YOLO object detection models [13] have been used for comparison purposes. Both Non-GPU and GPU-accelerated models have been considered for inference purposes.

The graph given in Fig. 8 depicts the comparison of inference times for GPU and non-GPU model performance. Upon running inference using various object detection models, we observed that the GPU-accelerated Swin Transformer provides highly optimized inference times and mAP values to account for computational efficiency while maintaining commendable mAP values.

The images displaying necessary alerts for the DAD module and aspect ratios are attached in Fig. 9 along with the frames per second used while passing them through our algorithm. They are indicative of the DAD module's performance in real-time for any surroundings regardless of the atmospheric setting. When the driver is drowsy and a certain frame threshold is crossed, the alert pops up on the screen as can be seen in Fig. 9 along with an audio alert. If and when the yawn counter registers a yawn, non-disruptive alerts are also issued.

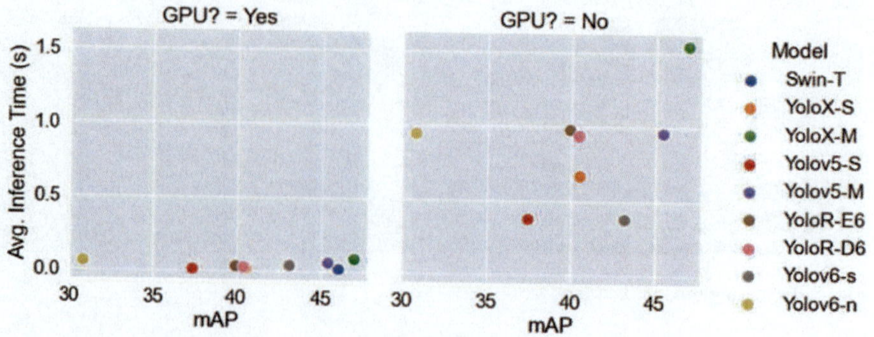

Fig. 8 Average inference time versus mAP (with and without GPU) for different models

Table 1 Inference details (confidence threshold = 75%)

Model	Architecture	No. of Parameters (millions)	mAP (0.5:0.95)	Avg. inference time per image (s)	
				With GPU	Without GPU
Swin transformer	Swin-T	28	46.0	0.0193	–
YoloX	YoloX-small	8.97	40.5	0.018	0.67
	YoloX-medium	25.3	46.9	0.087	1.55
Yolov5	Yolov5-small	7.2	37.4	0.013	0.38
	Yolov5-medium	21.2	45.4	0.064	0.96
YoloR	YoloR-E6	17.1	39.9	0.038	0.98
	YoloR-D6	21.8	40.4	0.026	0.94
YoloV6	Yolov6-s	17.2	43.1	0.042	0.38
	Yolov6-n	4.3	30.8	0.066	0.95

Fig. 9 Results of drowsiness module

Table 2 Confusion matrix—alertness detection

Module	Labels	True positive	True negative
Micro-sleep	Predicted positive	15	5
	Predicted negative	3	7
Yawns	Predicted positive	17	3
	Predicted negative	2	8

Table 3 Results—DAD module

Module	Precision	Recall	F1 score	Accuracy
Micro-sleep tracker	0.833	0.770	0.800	0.733
Yawn detection	0.850	0.895	0.872	0.833

Furthermore, we plotted the confusion matrix for the DAD module using our algorithm's predictions for 30 arbitrary frames obtained at regular intervals. Each of these frames was manually labelled to obtain the ground truth values. The confusion matrix shown in Table 2 is the depiction of how well our DAD algorithm performs in real time.

From the confusion matrix, we have also computed the precision, recall, F1-score, and accuracy for the DAD module in Table 3 as this helps us evaluate how robust our algorithm is.

5 Conclusion

The main aim of our work is to strike a necessary balance between memory efficiency, inference time, and accuracy. For road entity detection, the Swin Transformer [9] excels and outperforms all other established state-of-the-art models.

Further, we make use of the 2 modules, RSA and DAD, by effectively pooling results from both. This integration of the two modules further goes to corroborate the final decisions. This is because we are taking into account more points of reference, evaluating safety concerns from different perspectives, and finally obtaining a single-valued score for creating a more comprehensive user interface.

Such a pipeline facilitates a system that is user-friendly and minimally distracting when it comes to driver assistance.

6 Future Works

The method proposed in this paper requires certain scarcely available resources like a GPU, possible IOT integration inside the car and a good-quality camera to capture live-stream input data. This ensures the efficient usage of all the features highlighted in the paper. While the distance estimation method used by us is an approximation of the actual distance, it serves the purpose of coming up with a ballpark value so as to be able to judge whether an entity is too close. There are several other methods to perform distance estimation as well, varying in complexity and efficacy.

Technologies like IOT, V2V, and V2I networks have established a stronghold in the research surrounding driver assistance systems. The integration of such systems with the one outlined in this paper may also prove to be an exciting step in establishing a holistic ecosystem.

References

1. Galarza E, Egas F, Silva F, Velasco P, Galarza E (2018) Real time driver drowsiness detection based on driver's face image behavior using a system of human computer interaction implemented in a smartphone, pp 563–572. https://doi.org/10.1007/978-3-319-73450-7_53
2. Zaki A, Mohammed E, Aaref A (2020) Real-time driver awareness detection system. IOP Conf Ser Mater Sci Eng 745:012053. https://doi.org/10.1088/1757-899X/745/1/012053
3. Cho K, van Merrienboer B, Bahdanau D, Bengio Y (2014) On the properties of neural machine translation: encoder-decoder approaches. CoRR abs/1409.1259. http://arxiv.org/abs/1409.1259
4. Tan M, Le QV (2019) Efficientnet: rethinking model scaling for convolutional neural networks. CoRR abs/1905.11946. http://arxiv.org/abs/1905.11946
5. Bradski G (2000) The OpenCV library. Dr. Dobb's J Software Tools
6. King DE (2009) Dlib-ml: a machine learning toolkit. J Machine Learning Res 10:1755–1758
7. Li G, Xie H, Yan W, Chang Y, Qu X (2020) Detection of road objects with small appearance in images for autonomous driving in various traffic situations using a deep learning based approach. IEEE Access 8:211164–211172. https://doi.org/10.1109/ACCESS.2020.3036620
8. Zhou X, Wang D, Krähenbühl P (2019) Objects as points. CoRR abs/1904.07850. http://arxiv.org/abs/1904.07850
9. Liu Z, Lin Y, Cao Y, Hu H, Wei Y, Zhang Z, Lin S, Guo B (2021) Swin transformer: hierarchical vision transformer using shifted windows. In: International conference on computer vision (ICCV)
10. Dosovitskiy A, Beyer L, Kolesnikov A, Weissenborn D, Zhai X, Unterthiner T, Dehghani M, Minderer M, Heigold G, Gelly S, Uszkoreit J, Houlsby N (2021) An image is worth 16 × 16 words: transformers for image recognition at scale. In: International conference on learning representations. https://openreview.net/forum?id=YicbFdNTTy
11. Lin T, Maire M, Belongie SJ, Bourdev LD, Girshick RB, Hays J, Perona P, Ramanan D, Dollár P, Zitnick CL (2014) Microsoft COCO: common objects in context. CoRR abs/1405.0312. http://arxiv.org/abs/1405.0312
12. Lugaresi C, Tang J, Nash H, McClanahan C, Uboweja E, Hays M, Zhang F, Chang CL, Yong M, Lee J, Chang WT, Hua W, Georg M, Grundmann M (2019) Mediapipe: a framework for perceiving and processing reality. In: Third workshop on computer vision for AR/VR at IEEE computer vision and pattern recognition (CVPR). https://mixedreality.cs.cornell.edu/s/NewTitle_May1_MediaPipe_CVPR_CV4ARVR_Workshop_2019.pdf
13. Redmon J, Divvala SK, Girshick RB, Farhadi A (2015) You only look once: unified, real-time object detection. CoRR abs/1506.02640. http://arxiv.org/abs/1506.02640

Job Scheduling on Parallel Machines with Precedence Constraints Using Mathematical Formulation and Genetic Algorithm

Sachin Karadgi and P. S. Hiremath

Abstract Jobs need to be scheduled on identical parallel machines and are subjected to machine eligibility restrictions and various complex sequence constraints with the goal of makespan minimization. A mathematical formulation as a Mixed Integer Linear Programming (MILP) is elaborated using two decision variables. The formulation is solved using Integer Programming (IP) solver, which is NP-hard, and there are high chances of not finding an optimal/feasible solution for a bigger data set in the stipulated time. The current article proposes job scheduling using the evolutionary computing approach, specifically the Genetic Algorithm (GA). Simulation experiments are conducted for various scenarios of jobs and machines subjected to different complex constraints. The result of the proposed GA is in good agreement with the results obtained from the IP solver and is achieved at a significantly reduced computational time, especially when a larger number of jobs and machines are involved.

Keywords Mixed integer linear programming · Parallel machine · Precedence constraints · Machine eligibility restrictions · Genetic algorithm

1 Introduction

Scheduling is indispensable during the decision-making process in many industries [1]. The mathematical formulation of a scheduling scenario is not straightforward. Also, solving the mathematical formulation is much harder as it is much more difficult to obtain a feasible solution for various reasons.

Often, researchers focus on simple precedence constraints on sequencing of jobs. However, numerous other types of precedence constraints are observed in manu-

S. Karadgi (✉)
Department of Automation & Robotics, KLE Technological University, Hubballi 580031, India
e-mail: sachin.karadgi@kletech.ac.in

P. S. Hiremath
Master of Computer Applications, KLE Technological University, Hubballi 580031, India

S. Sharma et al. (eds.), *Intelligent Control, Robotics, and Industrial Automation*,
Lecture Notes in Electrical Engineering 1066,
https://doi.org/10.1007/978-981-99-4634-1_65

facturing environments. The present article primarily focuses on modeling machine eligibility restrictions and complex precedence constraints. In this scenario, multiple identical parallel machines m and numerous jobs n need to be sequenced on these machines. A machine i can perform a job j at a given time and continue to execute multiple jobs over some time. A job j can be performed on any machine i for a specific duration, known as processing time p_j, unless there is no machine eligibility restriction M_j defined for job j. Furthermore, C_j denotes the completion time of job j on any machine.

The jobs might be subjected to numerous complex precedence constraints. A processing of job k starts only after the processing of job j on the same machine, i.e., job j precedes job k ($j \prec k$). Precedence constraints are denoted by an acyclic digraph $G = (V, A)$, with V denoting job vertices, and directed edges of all precedence constraints among jobs are denoted by A [2, 3]. However, job j and job k can be allocated on any available machine for the constraint $j \prec k$ such precedence constraints denoted by $G' = (V', A')$. Similarly, the precedence constraints where job k cannot be allocated to machine i after the completion of job j but can be allocated on machine i before the beginning of job j, or job j and job k can be allocated on separate machines denoted by $G'' = (V'', A'')$. The precedence constraints, wherein, jobs j and k need to be assigned on different machines indicated by $G''' = (V''', A''')$. Finally, S is a set of jobs where a machine processes only a single job from this set.

In the current article, the mathematical formulation of a job scheduling problem is considered along with the above complex precedence constraints, and a near-optimal or optimal solution to the problem is found by employing a Genetic Algorithm (GA) approach. The article's remaining sections are arranged as follows: Sect. 2 introduces GA and presents a literature review of job scheduling. Section 3 elaborates the mathematical formulation and its solution by GA implementation. The computed experimental results of GA and their comparison with the results of a typical Integer Programming (IP) solver for different data sets and parameters are presented in Sect. 4. The article ends with Sect. 5 with analysis and points out possible future work.

2 Literature Review

Evolutionary Algorithms (EA) have been employed to solve various problems of optimization, simulation, and modeling [4]. Numerous EA techniques exist (e.g., GA, PSO—Particle Swarm Optimization). The generic idea of GA is according to the tenet of 'survival of the fittest' observed in the natural selection of surviving species during the genetic evolution of the species. This idea is imitated by generating a population of solutions or individuals having access to limited resources, with each solution competing among themselves to access the resources and thus causing natural selection based on the solution fitness [4]. The pseudocode of GA is given in Algorithm 1.

Algorithm 1 Genetic Algorithm (GA)—Pseudocode

1: Choose the solution representation;
2: Set generation counter $t = 1$ and highest allowed generation T;
3: Compute initial population, $P(t = 1)$ with population size N;
4: Evaluate objective, constraints, and allocate fitness to $P(t = 1)$;
5: **while** $t <= T$ **do**
6: Select parent population $M_p(t)$ from $P(t)$;
7: Implement crossover and mutation operations to create offspring population $Q(t)$;
8: Evaluate objective, constraints, and fitness functions for each offspring in $Q(t)$;
9: If there is no change in fitness values, then go to Step 12;
10: Select succeeding generation from $Q(t)$ and $P(t)$ to create $P(t + 1)$;
11: Increment generation counter $t = t + 1$;
12: **end while**

Finding schedules that minimize makespan is one of the most common problems in manufacturing industries. For a single-machine environment without any constraints, minimization of makespan is trivial [5]. In contrast, the makespan minimization problem in parallel machine environments subjected to constraints is NP-hard [5]. Researchers have proposed various mathematical models to minimize makespan for machines in parallel, job shops, and flow shops, and for different constraints (e.g., precedence constraints, preemption) [5].

An integer linear programming approach is suggested for scheduling n jobs to minimize the makespan, each of which must be processed on one of the available m machines [5, 6]. Further, the mathematical formulation has been improved to address precedence constraints and machine restrictions to minimize tardiness [7, 8], completion time of all jobs [9], weighted earliness-tardiness [10], and makespan [11–14].

Research has been carried out by employing GA to schedule jobs on parallel machine environments subjected to numerous constraints that optimize different objectives. For instance, GA was employed to solve parallel machine scheduling [15–18], (flexible) job shop scheduling [15, 19], unrelated parallel machine scheduling [7, 11, 20], uniform parallel machine scheduling [21, 22].

It is observed that these proposed models for job scheduling on parallel machines consider only simple precedence constraints and do not consider the several complex precedence constraints normally found in a real manufacturing environment. The present article proposes a GA-based approach for solving the job scheduling problem on identical parallel machines subjected to more complex precedence constraints and machine restrictions.

3 Proposed Methodology

A job scheduling problem as discussed in Sect. 1 can be represented as P_m |prec, M_j| $C_{\max}$. The proposed methodology for job scheduling comprises mathematical formulation as a Mixed Integer Linear Programming (MILP) with complex precedence constraints and machine eligibility restriction in Sect. 3.1 and a GA approach to solve the mathematical formulation in Sect. 3.2.

3.1 Mathematical Formulation

The following parameters, indices, and decision variables are employed in the MILP formulation:

- Indices

i Machine index
j, k Job indices

- Parameters

m Total number of machines
M Machines set, $M = \{1, 2 \ldots m\}$
n [Total number of jobs]
N' Set of jobs to be scheduled, $N' = \{1, 2 \ldots n\}$
M_j Job j's machine eligibility restrictions
p_j Time to process job j
A Precedence constraint set with jobs j and k to be allocated on the same machine
A' Precedence constraint set with jobs j and k to be allocated on any machine
A'' Precedence constraint set where job k cannot be allocated on machine i after the processing of job j, but can be allocated on machine i before the beginning of job j, or jobs j and k are allocated on separate machines
A''' Precedence constraint set with jobs j and k that need to be allocated on separate machines
S Set of jobs, where a machine processes a single job from this set
H Large positive number, say 10^6

- Continuous Variables

C_j Completion time of job j
$C_{\max}$ Makespan

- Decision Variables

$$x_{ij} = \begin{cases} 1, \text{if machine } i \text{ performs job } j \\ 0, \text{otherwise} \end{cases}$$

$$\delta_{jk} = \begin{cases} 1, \text{directly or indirectly if } j \prec k \\ 0, \text{otherwise} \end{cases}$$

The MILP formulation described below is based on the transportation model and disjunctive formulations to solve $P_m \mid \mid C_{\max}$. The formulation relies on decision variable x_{ij} to determine the allocation of job j to machine i, and δ_{jk} is employed for the sequencing of jobs j and k. The proposed mathematical formulation is as follows:

$$\text{Minimize} \quad C_{\max} \tag{1}$$

subjected to

$$\sum_{i \in M} x_{ij} = 1, \quad \forall\, j \in N' \tag{2}$$

$$C_{\max} \geq \sum_{j \in N'} x_{ij} p_j, \quad \forall\, i \in M \tag{3}$$

$$C_{\max} \geq C_j, \quad \forall\, j \in N' \tag{4}$$

$$\delta_{jk} + \delta_{kj} = 1, \quad \forall\, j, k \in N', \; j \neq k \tag{5}$$

$$H(1 - x_{ij}) + H(1 - x_{ik}) + H(1 - \delta_{jk}) + C_k \geq C_j + p_k, \quad \forall\, i \in M, \; \forall\, j, k \in N', \; j \neq k \tag{6}$$

$$H(1 - x_{ij}) + H(1 - x_{ik}) + H\delta_{jk} + C_j \geq C_k + p_j, \quad \forall\, i \in M, \forall\, j, k \in N', \; j \neq k \tag{7}$$

$$x_{ij} = x_{ik}, \forall\, i \in M, \quad \forall\, j, k \in A, \; j \neq k \tag{8}$$

$$C_k \geq C_j + p_k, \quad \forall\, j, k \in A, \; j \neq k \tag{9}$$

$$C_k \geq C_j + p_k, \quad \forall\, j, k \in A', \; j \neq k \tag{10}$$

$$H(1 - x_{ij}) + H(1 - x_{ik}) + C_j \geq C_k + p_j, \quad \forall i \in M, \; j \neq k, \quad \forall\, j, k \in A'' \tag{11}$$

$$H(1 - x_{ij}) + H(1 - x_{ik}) \geq 1, \quad \forall\, i \in M, \; j \neq k, \; \forall\, j, k \in A''' \tag{12}$$

$$\sum_{j \in N'} x_{ij} p_j \leq \; p_k + H(1 - x_{ik}), \quad \forall\, i \in M, \; \forall\, k \in S \tag{13}$$

$$C_k = p_k, \quad \forall\, k \in S \tag{14}$$

$$\sum_{i \in M_j} x_{ij} = 1, \quad \forall\, j \in N' \tag{15}$$

$$C_j \geq p_j, \quad \forall\, j \in N' \tag{16}$$

$$\delta_{jj} = 0, \quad \forall\, j \in N' \tag{17}$$

$$x_{ij} \in \{0, 1\}, \quad \forall i \in M, \ j \in N' \tag{18}$$

$$\delta_{jk} \in \{0, 1\}, \quad \forall j, k \in N' \tag{19}$$

The MILP is formulated with the makespan as objective function given in Equation (1). Constraints (2), (3), and (4) assist in determining makespan and allocation of job j to machine i without any consideration for constraints. Constraints (5), (6), and (8) support the sequencing of jobs j and k, and are not processing at the same time on a machine i.

Constraints (8) and (9) make sure that jobs j and k are assigned to same machine and follow precedence constraint $j \prec k$ with $(j, k) \in A$. Further, jobs j and k are subjected to precedence constraints $j \prec k$ with $(j, k) \in A'$, and enforced by the Constraint (10). Constraint (11) assures that a job k cannot be assigned to machine i after the completion of job j but guarantees that job k is assigned before the start of job j or on different machines with $(j, k) \in A''$. Constraint (12) enforces that jobs j and k are assigned to different machines with $(j, k) \in A'''$. Constraints (13) and (14) ensure that a single job $k \in S$ is assigned to a machine i. Constraint (15) satisfies that job j is processed on any machines in the subsets of machines as denoted by machine eligibility restriction M_j. Finally, the non-negative constraints are stated in (16)–(19).

The precedence constraints, namely, (Constraints 8–15), are the novel constraints proposed in the mathematical formulation that enhance the complexity of the optimization problem and model more realistic scenarios observed on the shop floor. Consequently, the method for obtaining an optimal solution to the proposed problem must be computationally appropriate.

3.2 Genetic Algorithm Approach

The steps in the GA approach for computing near-optimal or optimal solutions to the previously proposed mathematical formulation are explained in the paragraphs below.

The representation of a solution as a chromosome is critical. So, it is necessary to define the solution regarding its representation and length of individual genes to support encoding and decoding [7, 11, 15, 17, 20, 21, 23]. A representation was proposed with length n job genes representing both the machine allocation and sequence of jobs [16, 22]. Each gene was represented by a real number with its integer part between 1 and m representing machine assignment and fraction part between 0 and 1 representing the job sequence on the machine, which is obtained after arranging the fraction part of jobs in ascending order. An example of a gene is depicted in Fig. 1 with 9 jobs and 3 identical parallel machines. Herein, the real part is used to identify the allocated machine, and the fraction part assists in determining the job sequence on each machine after arranging the fraction part in ascending order, as depicted in Fig. 2.

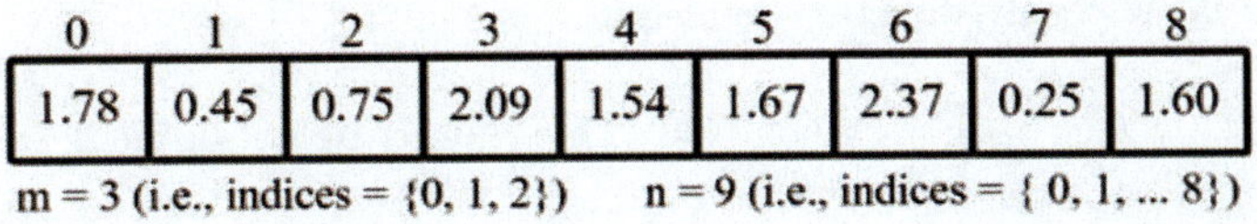

Fig. 1 Solution representation as real numbers

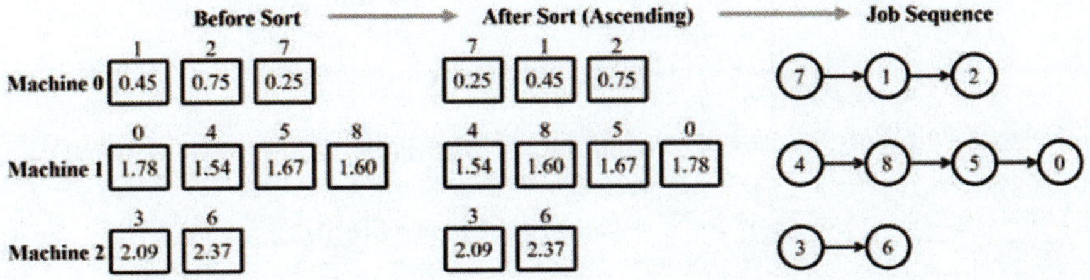

Fig. 2 Allocation and sequence of jobs for solution representation depicted in Fig. 1

An initial population of N solutions is randomly generated to enhance population diversity [7]. Here, randomness is achieved by selecting a random job from the available jobs and assigning it to a random machine. Each gene in the solution representation is filled with a real number consisting of an integer and a fraction. Further, the fraction part, up to two decimals, is randomly selected between 0 and 1. The population size N can be chosen to be large enough to arrive at the optimal solution faster, but it will incur computational costs, especially if there are many constraints [21]. The generated solution is checked for feasibility, i.e., whether it fulfills the scheduling constraints [11, 23]. Only feasible solutions are retained in the population.

The fitness function of the GA is often chosen to be the objective function, either maximization or minimization [17, 21]. There are two approaches to handling constraints [4]. First, indirect constraint handling ensures that the constraints are introduced as part of the fitness function through penalty functions [4, 21, 24]. This has the advantage of transforming the constrained problem into an equivalent unconstrained problem [25]. In contrast, direct constraint handling ensures that the constraints are handled during the execution of GA. For instance, a repair function is introduced to modify the sequence of jobs in individual solutions that violate the constraints [4, 11, 23, 26].

In the present mathematical model for job scheduling, the objective function is to minimize makespan $C_{\max}$, which can be treated as a fitness function in GA. The precedence constraints and machine eligibility restriction denoted by Constraints (8)–(15) are accounted through a combination of penalty functions and repair functions included in the objective function as in Eq. (20).

$$\text{Minimize} \quad C_{\max} + \langle \sigma \rangle + \langle \phi \rangle + \langle \varphi \rangle + \langle \gamma \rangle + \langle \theta \rangle + \langle \omega \rangle \tag{20}$$

where $\langle\cdot\rangle$ is the penalty function defined as:

$$\langle\sigma\rangle = \begin{cases} W, & \text{if Constraints (8) and (9) are not fulfilled} \\ 0, & \text{otherwise} \end{cases}$$
$$\langle\phi\rangle = \begin{cases} W, & \text{if Constraint (10) is not fulfilled} \\ 0, & \text{otherwise} \end{cases}$$
$$\langle\varphi\rangle = \begin{cases} W, & \text{if Constraint (11) is not fulfilled} \\ 0, & \text{otherwise} \end{cases}$$
$$\langle\gamma\rangle = \begin{cases} W, & \text{if Constraint (12) is not fulfilled} \\ 0, & \text{otherwise} \end{cases}$$
$$\langle\theta\rangle = \begin{cases} W, & \text{if Constraints (13) and (14) is not fulfilled} \\ 0, & \text{otherwise} \end{cases}$$
$$\langle\omega\rangle = \begin{cases} W, & \text{if Constraint (15) is not fulfilled} \\ 0, & \text{otherwise} \end{cases}$$

Constraints (12)–(15) are handled through repair functions, which are utilized after the generation of initial population, and crossover and mutation operations. Constraints (8)–(11) require penalty functions, especially exterior penalty functions [27]. Additionally, penalty functions are defined for the constraints (12)–(15) as there are chances of violating the constraints after the repair functions are carried out as dictated by the crossover and mutation operations, and also dependent on the sequence of carrying out the repair functions. Herein, the static penalty function is considered, and in general, the penalty value can vary for each of the constraints [4, 28]. A common penalty value W is chosen in the present.

At the tth iteration of GA, mating pool population $M_p(t)$ with size μ_t is selected from $P(t)$, which will undergo the crossover operation and mutation operation to generate offspring $Q(t)$ of size λ_t. There are different parent selection algorithms to select $M_p(t)$, namely, roulette wheel [4, 7, 16, 21, 26], stochastic universal sampling [4], and tournament [4, 11], among others. In the current context, a tournament selection algorithm with tournament size $k = 2$ has been employed to choose $M_p(t)$.

Crossover operation creation of offspring by exchanging genetic information between two parents selected from the mating pool $M_p(t)$ [7, 11]. There exist various ways to achieve crossover, namely, one-point [18, 20, 21, 29–31], two-point [11, 23, 31], position-based [15] and uniform crossovers [16, 31]. In this paper, a one-point crossover operation is performed to produce two offspring as depicted in Fig. 3. Randomness is introduced in parents selection, crossover point (or site) generation in $[1, n-1]$, and probability generation r_c in [0, 1]. The crossover is performed if r_c is less than the predefined crossover probability P_c. Otherwise, parents are directly copied as offspring.

The diversity of the population can be further enhanced through mutation, which is realized by slight modification in offspring [4, 11, 21, 30]. There are different ways to realize mutation: adaptive mutation operator [11], swap mutation [7, 15, 18, 20, 23, 29], among others. In the current context, a probability value r_m in [0, 1] is generated, and mutation is carried out if this value is less than the predefined mutation probability P_m. Otherwise, there is no change in offspring. For all offspring,

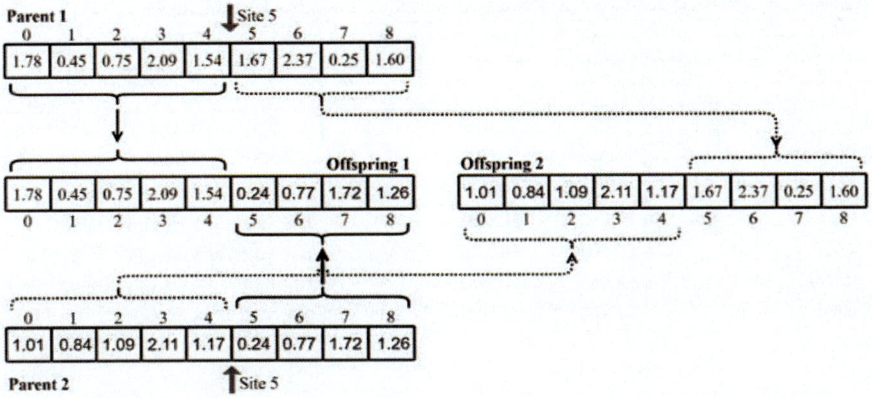

Fig. 3 Single point crossover operation

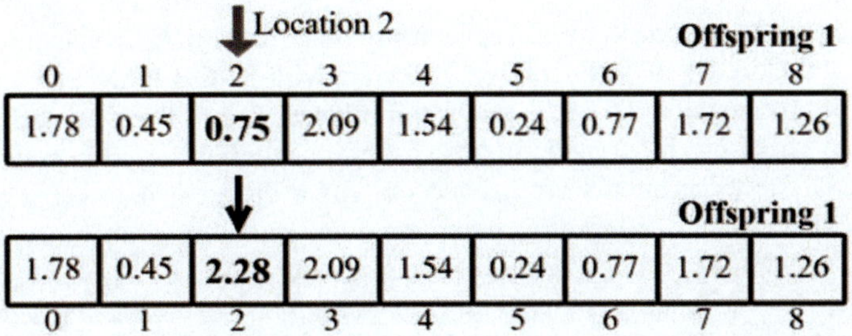

Fig. 4 Single location mutation operation

the mutation is carried out at a randomly generated gene location in $[0, n - 1]$, which involves the generation of a real part identifying the allocated machine and a fraction part determining the sequence on the machine. Figure 4 illustrates the single location mutation operation.

Survivor selection strategy is employed to select the next parent population $P(t + 1)$ from parent population $P(t)$ with size μ_t and offspring population $Q(t)$ with size λ_t [4]. Two strategies exist for survivor selection: fitness-based replacement and age-based replacement [4]. In the current context, (μ_t, λ_t) selection strategy based on age is employed. In (μ_t, λ_t) selection strategy, $\lambda_t > \mu_t$, parent population $P(t)$ is discarded and next parent population $P(t + 1)$ with size μ_t is selected from offspring $Q(t)$, which is ranked as per the fitness values [4]. The algorithm is terminated whenever there is no change in fitness values with zero penalties for a certain number of successive generations or after a fixed T generation.

Table 1 IP solver results of computational experiments for various parameters

n	m	No. of parameters	No. of binary decision variables	No. of constraints	CPU time (IP solver)	C_{max}
12	3	193	180	988	0.03 s	253
12	4	205	192	1259	0.06 s	171
12	5	217	204	1530	0.10 s	131
100	7	10,801	10,700	148,926	1.29 min	228
100	8	10,901	10,800	168,742	3.44 min	182
100	9	11,001	10,900	188,558	6.69 min	152
100	10	11,101	11,000	208,374	3.45 min	130

4 Computational Experiments

The proposed GA approach for job scheduling on parallel machines is implemented on Intel Intel Core i7-10700F CPU @ 2.90 GHz with 32 GB RAM using Microsoft Visual CSharp (version 16.11.2) programming language. The simulation experiments are carried out by considering the different numbers of jobs, machines, and constraints. The experiments are carried out for a different data set with $n = 12$ and 100. The values of algorithm parameters are mutation probability $P_m = 0.1$, crossover probability $P_c = 0.9$, survivor population ratio $(\mu, \lambda) = 0.1$, maximum generations $T = 250$, and penalty $W = 500$. A total of 7 experimental scenarios are considered to determine the optimal makespan C_{max} using IP solver (in Table 1) and 19 experimental scenarios are considered for the GA implementation (in Table 2). The GA simulation results are compared with the typical IP solver, Gurobi Optimizer [32]. It is striking that the IP solver becomes inefficient when the number of jobs and machines is larger, while GA is significantly efficient in execution time under similar scenarios.

5 Conclusions and Future Work

The current article elaborates on a mathematical formulation using MILP to schedule jobs on identical parallel machines subjected to machine eligibility restrictions and various complex precedence constraints. The problem being NP-hard, a traditional IP solver has a high chance of not finding a feasible or optimal solution in the stipulated time. This issue is addressed by choosing a GA approach to solve the formulated MILP problem. Simulation experiments are carried out by considering various scenarios of jobs and machines with complex constraints. The mathematical formulation is solved using the traditional IP solver, Gurobi Optimizer, to obtain optimal makespan. The proposed GA algorithm is implemented, and the fitness is

Table 2 GA results of computational experiments for various parameters

n	*m*	*N* (initial population in GA)	Termination similar fitness	No. of generations to reach convergence	CPU time (GA)	Fitness $C_{\max}$	Fitness gap
12	3	100	5	8	0.16 s	253	0
12	4	100	5	13	0.28 s	173	2
12	5	100	5	14	0.29 s	133	2
100	7	150	10	45	0.24 min	228	0
100	7	200	10	40	0.35 min	228	0
100	7	250	10	42	0.47 min	228	0
100	7	300	10	44	0.57 min	228	0
100	8	150	10	132	0.85 min	183	1
100	8	200	10	56	0.48 min	185	3
100	8	250	10	64	0.69 min	182	0
100	8	300	10	46	0.60 min	183	1
100	9	150	10	74	0.48 min	157	5
100	9	200	10	76	0.64 min	152	0
100	9	250	10	64	0.68 min	152	0
100	9	300	10	53	0.69 min	153	1
100	10	150	10	193	1.23 min	131	1
100	10	200	10	65	0.54 min	133	3
100	10	250	10	71	0.75 min	132	2
100	10	300	10	58	0.73 min	132	2

compared with the optimal makespan from the IP solver. The result of the proposed GA is in good agreement with the results obtained from the IP solver and is achieved at a significantly reduced computational time with a larger number of jobs and machines involved. The mathematical formulation will be generalized in future work by considering sequence-dependent setup times, uncertainty, and other parallel machine environments.

References

1. Georgiadis GP, Elekidis AP, Georgiadis MC (2019) Optimization-based scheduling for the process industries: from theory to real-life industrial applications. Processes 7(7). https://doi.org/10.3390/pr7070438
2. Skutella M, Uetz M (2005) Stochastic machine scheduling with precedence constraints. SIAM J Comput 34(4):788–802. https://doi.org/10.1137/S0097539702415007
3. Lin L, Gen M (2018) Hybrid evolutionary optimisation with learning for production scheduling: state-of-the-art survey on algorithms and applications. Int J Prod Res 56(1–2):193–223. https://doi.org/10.1080/00207543.2018.1437288
4. Eiben A, Smith J (2015) Introduction to evolutionary computing. Natural computing series. Springer Berlin Heidelberg, Berlin, Heidelberg. https://doi.org/10.1007/978-3-662-44874-8

5. Pinedo ML (2008) Scheduling: theory, algorithms, and systems. Springer. https://doi.org/10.1007/978-0-387-78935-4
6. Kurz ME, Askin RG (2001) Heuristic scheduling of parallel machines with sequence-dependent set-up times. Int J Prod Res 39(16):3747–3769. https://doi.org/10.1080/00207540110064938
7. Liu C (2013) A hybrid genetic algorithm to minimize total tardiness for unrelated parallel machine scheduling with precedence constraints. Math Probl Eng 2013:1–11. https://doi.org/10.1155/2013/537127
8. Lee JH, Yu JM, Lee DH (2013) A tabu search algorithm for unrelated parallel machine scheduling with sequence- and machine-dependent setups: minimizing total tardiness. Int J Adv Manuf Technol 69(9–12):2081–2089. https://doi.org/10.1007/s00170-013-5192-6
9. Chen N, Kang W, Kang N, Qi Y, Hu H (2022) Order processing task allocation and scheduling for e-order fulfilment. Int J Prod Res 60(13):4253–4267. https://doi.org/10.1080/00207543.2021.2018140
10. Vallada E, Ruiz R (2012) Scheduling unrelated parallel machines with sequence dependent setup times and weighted earliness-tardiness minimization. In: Just-in-time systems, pp. 67–90. No. January 2012 in Springer optimization and its applications. Springer, New York. https://doi.org/10.1007/978-1-4614-1123-9
11. Afzalirad M, Rezaeian J (2016) Resource-constrained unrelated parallel machine scheduling problem with sequence dependent setup times, precedence constraints and machine eligibility restrictions. Comput Ind Eng 98:40–52. https://doi.org/10.1016/j.cie.2016.05.020
12. Vallada E, Ruiz R (2011) A genetic algorithm for the unrelated parallel machine scheduling problem with sequence dependent setup times. Eur J Oper Res 211(3):612–622. https://doi.org/10.1016/j.ejor.2011.01.011
13. Edis EB, Ozkarahan I (2011) A combined integer/constraint programming approach to a resource-constrained parallel machine scheduling problem with machine eligibility restrictions. Eng Optim 43(2):135–157. https://doi.org/10.1080/03052151003759117
14. Gokhale R, Mathirajan M (2012) Scheduling identical parallel machines with machine eligibility restrictions to minimize total weighted flowtime in automobile gear manufacturing. Int J Adv Manuf Technol 60(9–12):1099–1110. https://doi.org/10.1007/s00170-011-3653-3
15. AK B, Koc E (2012) A guide for genetic algorithm based on parallel machine scheduling and flexible job-shop scheduling. Procedia Soc Behav Sci 62:817–823. https://doi.org/10.1016/j.sbspro.2012.09.138
16. Yeh WC, Lai PJ, Lee WC, Chuang MC (2014) Parallel-machine scheduling to minimize makespan with fuzzy processing times and learning effects. Inf Sci 269:142–158. https://doi.org/10.1016/j.ins.2013.10.023
17. Bathrinath S, Sankar SS, Ponnambalam SG, Kannan BKV (2013) Bi-objective optimization in identical parallel machine scheduling problem. In: Panigrahi BK, Suganthan PN, Das S, Dash SS (eds) Swarm, evolutionary, and memetic computing. Springer International Publishing, Cham, pp 377–388
18. Van Khanh B, Van Hop N (2021) Genetic algorithm with initial sequence for parallel machines scheduling with sequence dependent setup times based on earliness-tardiness. J Ind Prod Eng 38(1):18–28. https://doi.org/10.1080/21681015.2020.1829111
19. Guzman E, Andres B, Poler R (2022) Matheuristic algorithm for job-shop scheduling problem using a disjunctive mathematical model. Computers 11(1). https://doi.org/10.3390/computers11010001
20. Joo CM, Kim BS (2012) Non-identical parallel machine scheduling with sequence and machine dependent setup times using meta-heuristic algorithms. Ind Eng Manage Syst 11(1):114–122. https://doi.org/10.7232/iems.2012.11.1.114
21. Yeh WC, Chuang MC, Lee WC (2015) Uniform parallel machine scheduling with resource consumption constraint. Appl Math Model 39(8):2131–2138. https://doi.org/10.1016/j.apm.2014.10.012
22. Lee WC, Chuang MC, Yeh WC (2012) Uniform parallel-machine scheduling to minimize makespan with position-based learning curves. Comput Ind Eng 63(4):813–818. https://doi.org/10.1016/j.cie.2012.05.003

23. Afzalirad M, Rezaeian J (2016) A realistic variant of bi-objective unrelated parallel machine scheduling problem: NSGA-II and MOACO approaches. Applied Soft Comput 50:109–123. https://doi.org/10.1016/j.asoc.2016.10.039
24. Sawant V (2016) Genetic algorithm for resource constrained project scheduling. Int J Sci Res (IJSR) 5(6):139–146. https://doi.org/10.21275/v5i6.NOV164087
25. Sarker R, Newton C (2002) A genetic algorithm for solving economic lot size scheduling problem. Comput Ind Eng 42(2):189–198. https://doi.org/10.1016/S0360-8352(02)00027-X
26. Pongcharoen P, Hicks C, Braiden P, Stewardson D (2002) Determining optimum genetic algorithm parameters for scheduling the manufacturing and assembly of complex products. Int J Prod Econ 78(3):311–322. https://doi.org/10.1016/S0925-5273(02)00104-4
27. Coello CC, Lamont GB, van Veldhuizen DA (2007) Evolutionary algorithms for solving multi-objective problems. Genetic and evolutionary computation series. Springer US, Boston, MA (2007). https://doi.org/10.1007/978-0-387-36797-2
28. Coello CAC (1999) A survey of constraint handling techniques used with evolutionary algorithms. Tech. Rep, Laboratorio Nacional de Informática Avanzada
29. Hasani K, Kravchenko SA, Werner F (2014) Simulated annealing and genetic algorithms for the two-machine scheduling problem with a single server. Int J Prod Res 52(13):3778–3792. https://doi.org/10.1080/00207543.2013.874607
30. Xia X, Qiu H, Xu X, Zhang Y (2022) Multi-objective workflow scheduling based on genetic algorithm in cloud environment. Inf Sci
31. Hartmann S (1998) A competitive genetic algorithm for resource-constrained project scheduling. Naval Res Logist (NRL) 45(7):733–750. https://doi.org/10.1002/(SICI)1520-6750(199810)45:7<733::AID-NAV5>3.0.CO;2-C
32. Gurobi (2021) Gurobi optimization. https://www.gurobi.com/. Accessed on 22 June 2022

HyResPR: Hybridized Framework for Recommendation of Research Paper Using Semantically Driven Machine Learning Models

Saketh Maddineni, Gerard Deepak, and S. V. Praveen

Abstract The obligation for research paper recommendation is high as scientific document recommendations like research paper recommendation frameworks are not many in number. The knowledge-centric semantically inclined framework for research paper recommendation HyResPR has been proposed. HyResPR is a hybridized research paper recommendation model which is implemented on the RARD II dataset. A hybridized research paper recommendation framework (HyResPR) is proposed. It uses user queries as input to obtain the query words after preprocessing, later these query words are induvial integrated with the domain ontologies. The dataset is subjected to preprocessing to create a synthesized knowledge map with the help of category term mapping and static domain ontology alignment. Features are extracted from the synthesized knowledge map and classified using logistic regression, the extracted query words are processed along with the top 75% of the instances classified. The experimentations are conducted on the RARD II dataset which is classified using the logistic regression classifier. The SemantoSim similarity measure and Cosine similarity measures are used to compute the similarity among the extracted instances from RARD II dataset. The query words obtained from input, and relevant research papers are recommended back to user depending on the value of the semantic similarity. Auxiliary knowledge is incorporated by using static domain ontology, topic modeling has been used experimentations have been computed for 1416 queries on the RARD II dataset. The proposed HyResPR framework achieved the highest average accuracy of 96.43%, recall of 97.05%, with a least FDR value of 0.05 observed.

S. Maddineni
Department of Computer Science and Engineering, National Institute of Technology, Tadepalligudem, Andhra Pradesh, India

G. Deepak (✉)
Department of Computer Science and Engineering, Manipal Institute of Technology Bengaluru, Manipal Academy of Higher Education, Manipal, India
e-mail: gerard.deepak.christuni@gmail.com

S. V. Praveen
Department of Computer Science and Engineering, National Institute of Technology, Tiruchirappalli, Tamil Nadu, India

S. Sharma et al. (eds.), *Intelligent Control, Robotics, and Industrial Automation*, Lecture Notes in Electrical Engineering 1066,
https://doi.org/10.1007/978-981-99-4634-1_66

Keywords Hybridized recommendation · Paper recommendation · Knowledge-driven approach · Semantically inclined

1 Introduction

With the enormous growth in the research domain, data is stored digitally and scattered over the World Wide Web. Due to this overload of scientific information which is archived over the globe, recommending the relevant paper based on user queries has become a challenging and a tough task for the current architecture. With the overload of information, and booming trends in the field of research, a recommendation of a research article was required on the relevance to that user query. While retrieving the research paper recency is an important factor, the most important task is the relevance and arrangement of these research papers based on the closeness of the literature or context. We require a proper portal to recommend research papers, as existing frameworks are not semantically complied. So, with time the number of research papers increases exponentially. Today's trend becomes tomorrow's established area. Segregating the research papers based on the new trends in established areas is a tough task. As the information load on the web, handling such information is challenging because of the retrieval of research papers. Indexing of the research papers is very sparse compared to the retrieval of the normal documents as indexing is already available in the web. It is richly indexed for general topics but it isn't highly indexed for scientific-oriented topics. Moreover, we are in semantic web or web 3.0 where data is extensive, knowledge is intermediate, and wisdom is almost absent. There are needs for trends, paradigms, and methodologies for semantically driven models for research papers recommendation. A typical user must sift through a huge quantity of articles manually before finding a relevant article.

Motivation The main motivation is because of the lack of a proper framework, models, and techniques in the research paper recommendation. The need for research paper recommendations in a hierarchy of high relevance to the user query is required. In extensive age of semantic web, semantic-driven web complaint strategies are almost absent. Henceforth, semantically driven models for research recommendation are the need of the hour. Research papers are scientific papers, scientific papers are more sparsely indexed on the web when compared to general topics indexing. Retrieval of research papers requires more precision and more knowledge load. So, retrieving with more precision requires more auxiliary knowledge load and better indexing schemes. Henceforth recommending such research papers is the need of the hour.

Contribution The hybridized semantically inclined knowledge-centric framework for research paper recommendation is suggested. This architecture uses the user's query as input to obtain the query words. These extracted query words after preprocessing step are used to integrate with the static domain ontologies. Furthermore, terms and indexes are extracted. From e-books and e-resources, terms and indexes are extracted for synthesizing a knowledge map. The extracted features are classified using logistic regressor classifier, and the extracted query words are inputted along with the top 75% of the instances classified. The similarity is calculated between query words extracted from input and instances from the RARD II dataset. SemantoSim similarity measure and Cosine similarity measures are used to compute the similarity. Now, these top 75% of computed semantic similar instances are recommended back to the user, if they pass the threshold value used. The metrics namely precision, recall, accuracy, F-score, and FDR values are escalated.

Organization Rest of the paper is organized as below. A brief synopsis of the works related to our paper are provided in Sect. 2. The system architecture proposed is provided in Sect. 3. In Sect. 4 a description of the architecture implementation, results obtained, and its performance is presented. The paper is concluded in Sect. 5.

2 Related Works

Neethukrishnan and Swaraj [1] had suggested a computer science related research papers recommendation based on ontology. This approach recommends the top relevant papers based on their similarity. The similarity matching is done with the help of SVM classifier. Jomsri et al. [2] proposed a research paper recommendation using IR approach. It is a tag-based framework, it explores a group of tags while recommending a paper back to the individual user. In this approach for each individual user's profile self-defined tags are used as to demonstrate good accuracy. Xue et al. [3] have suggested a framework for personalized research paper recommendations. Their proposed methodology automatically constructs training data from the existing network of data and diverse features and recommends to users belonging to an online scholar platform. With this model, they constructed a real-time recommendation system.

Chen and Lee [4] proposed an architecture to tackle problems faced while recommending research papers on big scholar data (BSD). They built an analytic platform and developed a research paper recommendation system. This recommendation system is capable of recommending relevant research papers based on their profile matching the interest of each individual user. Haruna et al. [5] have proposed a framework using public contextual metadata. The framework customizes research documents based on user preferences, and the framework is independent of user expertise. Pan and Li [6] suggested a research paper recommender framework with the help of techniques like topic analysis to reduce the problem of cold start. For item-based recommendation system they introduced a similarity computation called thematic similarity. It is a collaborative filtering-based recommendation system. Magara et al. [7] has proposed an altimetric-based techniques and methodologies for research paper recommendation for efficient and effective way of retrieving desired research papers. Altimetric from research papers and paper ontologies are used for efficient and effective way of recommendation. Hassan et al. [8] suggested a method for evaluating models. It recommends research papers based on content. It involves reranking models based on BM25 recommendations.

Beel and Langer [9] have discussed the appropriateness of different evaluation methods in research paper. They implemented various content-based approaches for filtering for recommending research articles and performed various evaluation metrics to find the appropriateness of various metrics. In [10–16] several models have been discussed about the proposed literature. Most of the existing research paper recommendation techniques are not based on a hybrid intelligence paradigm, instead, they are based on altimetric where specific metrics are used. Some of them use collaborative filtering as a primary technique which requires rating research papers. Every research paper has to be rated, which is practically impossible to rate the research paper as rating varies with the user and their domain knowledge. Rating matrix computation for research paper recordation is not a viable scheme so collaborative filtering methods shouldn't be used. Certain metrics use content-based filtering for recommending research papers. Content-based filtering is a good option as the contents of research papers are taken into consideration. However, the existing model doesn't make use of any learning paradigms. In content-based filtering, only lightweight models are used and the inclusion of auxiliary knowledge leads to a large semantic gap.

Some of the models use static ontologies with similarity computation and training with SVM which is a decent model but however, but the strength of relevance computation mechanism, semantic similarity mechanism, the strength of incorporating ontologies, and other heterogeneous sources is weak.

Most of the techniques use standalone ML models, collaborative filtering, and content-based filtering, and some diverse features are included but these are still naive the strength of hybridization can be improved and the strength of individual models can be increased. These models lack auxiliary knowledge in static ontologies and from other heterogeneous information making the proposed model outperform the baseline models. From the existing literature, the research gap can be identified in that the existing models aren't semantically inclined and they are not knowledge-centric and they don't hybridize integrated models with a collective intelligence scheme. The standalone models lack auxiliary knowledge and they aren't knowledge-centric and the learning models incorporated are quite weak.

3 Proposed System Architecture

A semantically inclined recommendation framework for research papers has been depicted in Fig. 1. This entire framework is categorized into 2 stages. The stage one is dataset driven and the stage two is user query driven. The dataset in phase 1 is subjected to preprocessing to extract the categories. The dataset is a categorical research paper recommendation dataset. So, the categories are extracted and these categories are then further enriched with static domain ontologies pertaining to that domain by some ontology alignment. So, the domains used are energy, informatics, computational relevance, semantic intelligence, data discovery, and reasoning. These were the domains used for experimenting starting on domain ontologies that were built and subjected to ontology alignment, and ontology alignment was done using the Lin similarity measure.

Lin similarity measure is a semantic similarity measure (https://www.gabormelli.com/RKB/Node-based_Semantic_Similarity_Measure) proposed by Dekang Lin in 1998. It is a node-based measure and is based on least common subsumed (LCS) (https://www.gabormelli.com/RKB/index.php?title=least_common_subsumer&action=edit&redlink=1).

Equation (1) represents the formula for computing Lin similarity measure.

$$\mathrm{Sim}_{\mathrm{lin}}(C_1, C_2) = \frac{2 \times \mathrm{IC}(\mathrm{lcs}(c1, c2))}{\mathrm{IC}(c1) + \mathrm{IC}(c2)} \tag{1}$$

Here $IC(c)$ is Information Control measure, where $IC(c)$ is represented as

$$IC(c) = -\log\left(\frac{\mathrm{freq}(c)}{\mathrm{maxFreq}}\right) \tag{2}$$

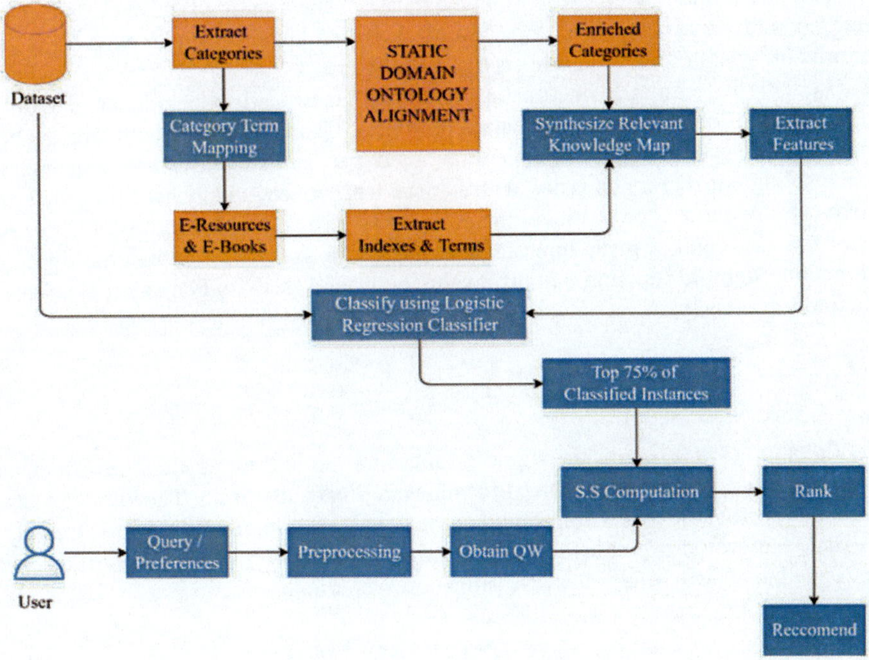

Fig. 1 Proposed semantically inclined framework

On static domain ontology alignment, we are obtaining the enriched categories and subsequently, the extracted categories are defined for subject term mapping by computing the Cosine similarity with that of keywords in the e-resources and indexes in the e-books. So, Cosine similarity with a threshold of 0.5 is considered because more entities relevant to categories need to be mapped, and all the mapped entities are extracted as extracted indexes and terms.

Cosine similarity is used to find an optimal relation between two relatable or similar documents in a quantitative way. The output is 1 if the two objects are similar to each other. Equation (3) represents the formula for computing Cosine similarity given two vectors of features, x and y.

$$\text{similarity}(x, y) = \cos(\theta) = \frac{x \cdot y}{|x||y|} \quad (3)$$

The Cosine similarity threshold is set to 0.5 at this phase, and the extracted indexes and terms which are the outcome of category term mapping and enriched categories are merged together based on common entity identification and a relevant knowledge map is synthesized. For synthesizing the relevant knowledge map the category matching is again done based on the computation of Cosine similarity with the threshold of 0.5. And from the synthesized relevant knowledge map the features are extracted randomly and sent into a logistic regression classifier for classification of the dataset. At the end of the classification of the dataset on the basis of the features extracted from the synthesized knowledge map. The classified instances are rearranged under each class, i.e., top 75% of classified instances under each class are stored in separate subspace. This is the end of phase 1.

Logistic regression is one of the supervised classification techniques generally employed to solve binary classification problems. Classification is predicting a label by identifying the category to which the object best suits, from various given paraments. Logistic regression is based on statistical approach, widely used for problems like classification, used to model the probability of a certain class. Logistic regression uses a weight of given input feature data and passes them through the sigmoid function. Sigmoid function outputs a value between 0 and 1 when any real number is fed or sent as input.

$$g(z) = \frac{1}{1 + e^{-z}} \quad (4)$$

Equation (4) represents the sigmoid function used in the logistic regressor. This function is used to map a real value to a value ranging from 0 and 1. The rule is that the value of the logistic regression should output a value belonging from 0 to 1. Usually, logistic regression is employed for predicting binary target variables. Further, it is extended to classify multi-class classification scenarios.

$$z = x_1 w_1 + x_2 w_2 + b \quad (5)$$

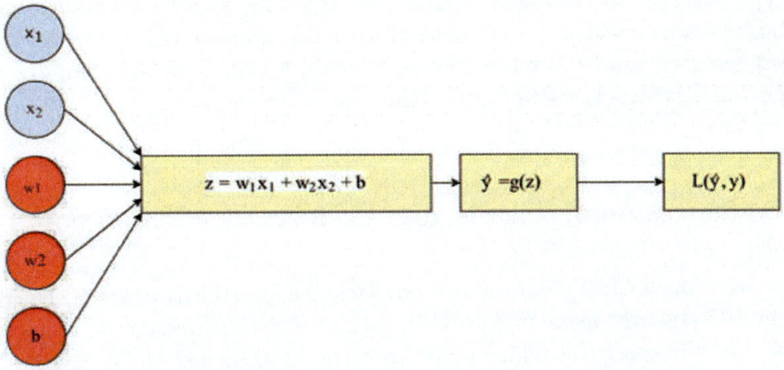

Fig. 2 Logistic regression classifier

$$\hat{y} = g(z) \tag{6}$$

In Eq. (5) x_1, x_2 are input features and $\hat{y}$ is the prediction. These are the terms used in logistic regression. Figure 2 explains the steps involved in logistic regression.

Phase 2 is driven by the user query the user submits the user query or preferences. The user query or preferences are subjected to preprocessing. Preprocessing involves stop word removal, NER, lemmatization, and tokenization. In last stage of the preprocessing phase, extraction of query words QW takes place. The top most 75% of instances classified and query words are individually subjected for semantic similarity computation (SS). Semantic similarity computations are done using two measures, i.e., Cosine similarity and SemantoSim measure. Cosine similarity with the threshold of 0.5 is maintained at this step because high relevant entities alone have to be ranked and recommended, in order to do that relevance has to be increased so that's why the threshold is increased only to the strength of the Cosine similarity measure. However, SemantoSim being a very strong semantic similarity measure is very stringent and strong. So, as a result, its relevant threshold is decreased to 0.5. The reason for using hybridized similarity measure, i.e., Cosine similarity and SemantoSim is because of ensuring a high degree of relevance, heterogeneous and diverse entities have to be recommended back to the user.

Church and Hanks proposed a semantic similarity measure called SemantoSim. SemantoSim is inspired from PointWise Mutual Information, SemantoSim measure is computed using the Eq. (7).

$$\text{SemantoSim}(x, y) = \frac{p(x, y)\log\left[p(x, y)\right] + \text{pmi}(x, y)}{\log\left[p(y, x)\right] + \left[p(x) * p(y)\right]} \tag{7}$$

At the end of this ultimately the entities which are relevant to the queries are ranked in the low–high order of SemantoSim measure, which are recommended back to the user. If the user satisfies the recommendation halts, if the user isn't satisfied the user clicks are submitted as preferences again the process continues until the entities are recommended. As the entities are recommended to the user the relevant mapped research papers to these entities which are ranked are also simultaneously recommended when the user clicks on the exact relevant terms.

4 Implementation, Results, and Performance Evaluation

For this work the RARD II dataset is used to test the model performance. The dataset contains 1474 queries formatted in XML format. The implementation was executed in an i7-1195G7 processor with a clock speed of 4.9 GHz and 16 GB RAM. Python 3.9.10 was used under Google Collaboratory IDE. For carrying out the preprocessing step Python's Natural Language Toolkit (NLTK) was employed. Domain Ontologies were formulated using web protegee but the entries for ontology models were crawled using a customized crawler. It was formalized into an ontology by using web protegee because it was lightweight and easy to handle.

Experimentations have been conducted for the HyResPR model and the baseline models on RARD II dataset—a total of 1474. The baseline models were also evaluated for the same number of queries with the same environment as the proposed HyResPR. The metrics used for performance evaluation of the proposed Hybrid Research Paper Recommendation (HyResPR) using Semantic Intelligence) approach are accuracy, recall, precision, F-score, and False Discovery Rate (FDR). From Figs. 3 and 4, we can understand that among various approaches the proposed HyResPR surpasses all the existing models in terms of accuracy, recall, precision, F-score percentage, and a low FDR rate of 1–0.95 as 0.05. So, HyResPR yielded a precision value of 0.95, a recall value of 0.97, a accuracy value of 0.96, a F-score value 0.96, and an overall FDR of 1–0.95, i.e., 0.05. To benchmark the performance of HyResPR it is contrasted with OBRR [1], TRPR [2], PPR [3], Hierarchical clustering, and Cosine similarity.

The proposed HyResPR model's performance is contrasted with various established models in Figs. 3 and 4. It is indicated that OBPR [1] yielded a precision value of 0.91, a recall value of 0.93, a accuracy value of 0.92, a F-score value of 0.92, and an FDR of 1–0.91, i.e., 0.09. The TRPR [2] architecture furnishes a precision value of 0.89, a recall value of 0.91, a accuracy value of 0.90, a F-score value of 0.90, and an overall FDR of 1–0.89, i.e., 0.11.

Similarly, PPR [3] architecture furnished a precision value of 0.88, a recall value of 0.91, a accuracy value of 0.89, a F-score value of 0.89, and an overall FDR of 1–0.89, i.e., 0.11. And, the Hierarchical Clustering and Cosine similarity combination yielded a precision value of 0.87, a recall value of 0.90, an accuracy value of 0.88, and a F-score value of 0.88, and an overall FDR of 1–0.87, i.e., 0.13. From Figs. 3 and 4 we can infer that the proposed model achieved high accuracy, precision, and recall with the lowest FPR values, the main reason behind this is due to the fact it is semantically driven, and research paper is recommended using relevant knowledge map synthesis. Their category term mapping takes place, indexes and terms extraction

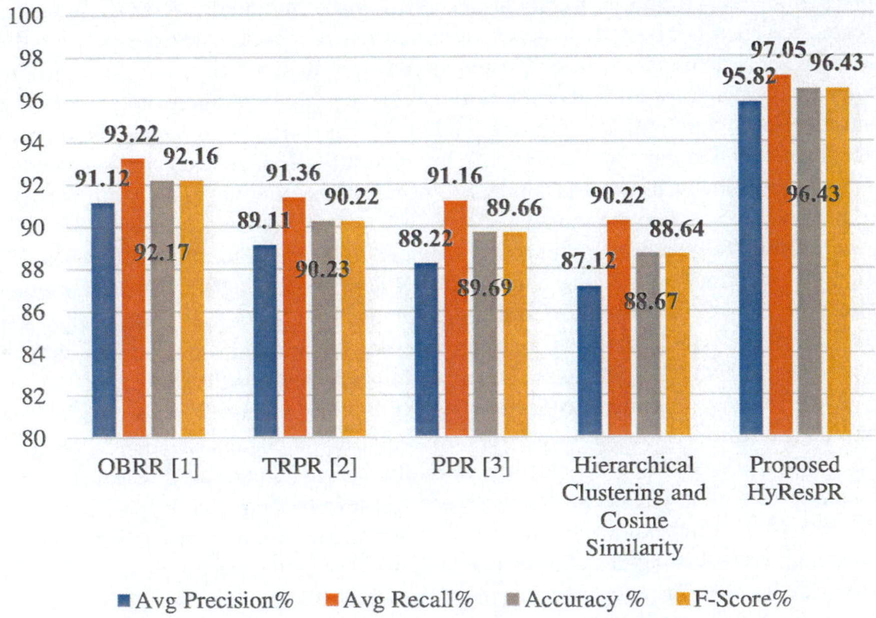

Fig. 3 Comparison of performance by proposed HyResPR with various frameworks

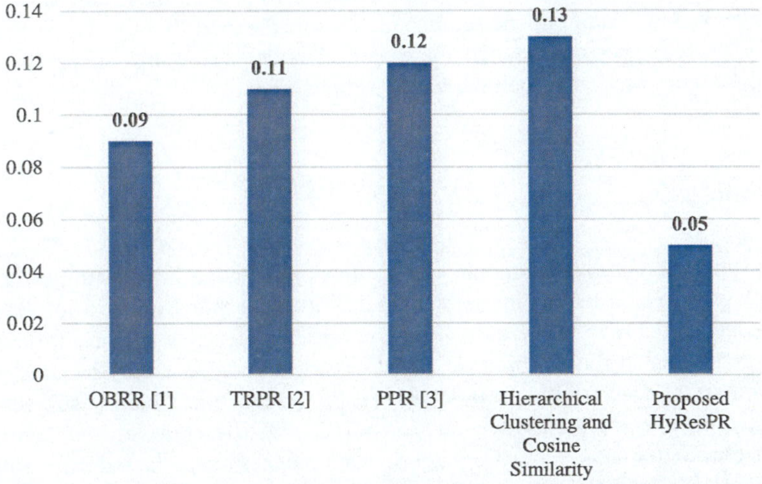

Fig. 4 Comparison of FDR value by proposed HyResPR with various frameworks

takes place apart from the alignment of static domain ontologies. Moreover, logistic regression classifier is used and semantic similarity computation takes place using the knowledge. Encompassment of Cosine similarity, Lin similarity, and SemantoSim with differential thresholds at different stages in the proposed framework ensures that strong relevant computing is present and henceforth HyResPR achieved the highest accuracy, recall, precision, F-score, and the lowest FDR value. Relevant knowledge map synthesis in the proposed framework ensures a large amount of dense knowledge to be incorporated into the model.

Precision (%) versus Number of Recommendations curve for proposed HyResPR along with various approaches namely OBRR [1], TRPR [2], PPR [3], Hierarchical clustering, and Cosine similarity is depicted in the Fig. 5. In OBRR [1], the approach uses ontology which provides a strong relevant computation mechanism but the classifier used is very light, i.e., SVM, moreover ontology used is static, therefore, it is denser in nature compared to the knowledge graph/map. Henceforth this model lags and there is absence of strong relevant computation mechanisms. The reason why the TRPR [2] model doesn't perform well although it's a tag-based research paper recommendation model because it uses user profile in the framework apart from that the lack of auxiliary knowledge and lack of strong semantic computation scheme makes sure that this tag-based model doesn't perform extensively well. The PPR [3] model doesn't perform well because the reason that although it uses personalization, and user interest ranking takes place, apart from this the entire model relies only on the dataset which employees supervised learning mechanism to rank along with the user interests. The relevant computations mechanism is not very strong in this case and the lack of auxiliary knowledge also makes sure the PPR [3] model doesn't work as it is expected to perform. Hence, the proposed model is way ahead in performance and works more accurately compared to baseline models.

5 Conclusions

In this study, a hybridized semantically inclined knowledge-centric framework for research paper recommendation is proposed. With the rapid growth of information knowledge available in the research domain, retrieving the relevant scientific documents pertaining to their submitted queries is a challenging task. The user query is passed as an input for the proposed HyPesPR framework, which is subjected to preprocessing to extract individual query words. These extracted words and enriched domain ontologies are integrated, and terms and indexes are extracted from e-books and e-resources. With the help of category term mapping and static domain ontologies a relevant knowledge map is synthesized into extra features, these extracted features are classified using a logistic regressor classifier. The instances classified are sorted in descending order and the first 75% of the instances are subjected for computing similarity. While recommending to user the top 75% of instances are submitted for similarity computation. The approach proposed accomplished the most accurate with the highest value of precision, i.e., 95.82% and with a least FDR of 0.05. The

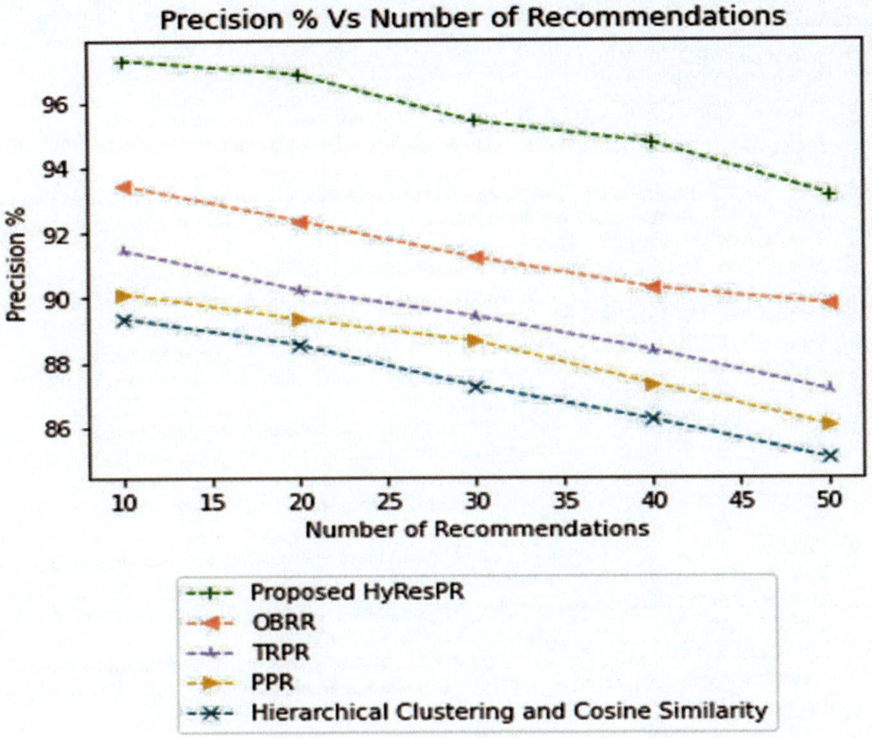

Fig. 5 Precision % versus number of recommendations

proposed framework achieved an overall accuracy way better than already existing approaches making it reliable and accurate.

References

1. Neethukrishnan KV, Swaraj KP (2017) Ontology based research paper recommendation using personal ontology similarity method. In: 2017 Second international conference on electrical, computer and communication technologies (ICECCT). IEEE, pp 1–4. https://doi.org/10.1109/ICECCT.2017.8117833
2. Jomsri P, Sanguansintukul S, Choochaiwattana W (2010) A framework for tag-based research paper recommender system: an IR approach. In: 2010 IEEE 24th International conference on advanced information networking and applications workshops. IEEE, pp 103–108. https://doi.org/10.1109/WAINA.2010.35
3. Xue H, Guo J, Lan Y, Cao L (2014) Personalized paper recommendation in online social scholar system. In: 2014 IEEE/ACM International conference on advances in social networks analysis and mining (ASONAM 2014). IEEE, pp 612–619. https://doi.org/10.1109/ASONAM.2014.6921649

4. Chen TT, Lee M (2018) Research paper recommender systems on big scholarly data. In: Knowledge management and acquisition for intelligent systems. PKAW 2018. Lecture notes in computer science, vol 11016. Springer, Cham, pp 251–260. https://doi.org/10.1007/978-3-319-97289-3_20
5. Haruna K, Ismail MA, Qazi A et al (2020) Research paper recommender system based on public contextual metadata. Scientometrics 125:101–114. https://doi.org/10.1007/s11192-020-03642-y
6. Pan C, Li W (2010) Research paper recommendation with topic analysis. In: 2010 International conference on computer design and applications. IEEE, pp V4-264–V4-268. https://doi.org/10.1109/ICCDA.2010.5541170
7. Magara MB, Ojo S, Zuva T (2017) Toward altmetric-driven research-paper recommender system framework. In: 2017 13th International conference on signal-image technology and internet-based systems (SITIS). IEEE, pp 63–68. https://doi.org/10.1109/SITIS.2017.21
8. Hassan HAM, Sansonetti G, Gasparetti F, Micarelli A, Beel J (2019) BERT, ELMo, Use and InferSent sentence encoders: the panacea for research-paper recommendation?. In: ACM RecSys 2019 Late-breaking results, pp 6–10
9. Beel J, Langer S (2015) A comparison of offline evaluations, online evaluations, and user studies in the context of research-paper recommender systems. In: Kapidakis S, Mazurek C, Werla M (eds) Research and advanced technology for digital libraries. TPDL 2015. Lecture notes in computer science, vol 9316. Springer, Cham, pp 153–168. https://doi.org/10.1007/978-3-319-24592-8_12
10. Siddiqui T, Ren X, Parameswaran A, Han J (2016) FacetGist: collective extraction of document facets in large technical corpora. In: CIKM'16: Proceedings of the 25th ACM international on conference on information and knowledge management, pp 871–880. https://doi.org/10.1145/2983323.2983828
11. Mei X, Cai X, Xu S, Li W, Pan S, Yang L (2022) Mutually reinforced network embedding: an integrated approach to research paper recommendation. Expert Syst Appl 204:117616. https://doi.org/10.1016/j.eswa.2022.117616
12. Chaudhuri A, Sarma M, Samanta D (2022) SHARE: designing multiple criteria-based personalized research paper recommendation system. Inf Sci 617:41–64. https://doi.org/10.1016/j.ins.2022.09.064
13. Gündoğan E, Kaya M (2022) A novel hybrid paper recommendation system using deep learning. Scientometrics 127:3837–3855. https://doi.org/10.1007/s11192-022-04420-8
14. Chaudhuri A, Sinhababu N, Sarma M, Samanta D (2021) Hidden features identification for designing an efficient research article recommendation system. Int J Digit Libr 22(2):233–249. https://doi.org/10.1007/s00799-021-00301-2
15. Adithya V, Deepak G (2021) HBlogRec: a hybridized cognitive knowledge scheme for blog recommendation infusing XGBoosting and semantic intelligence. In: 2021 IEEE International conference on electronics, computing and communication technologies (CONECCT). IEEE, pp 1–6. https://doi.org/10.1109/CONECCT52877.2021.9622526
16. Surya D, Deepak G, Santhanavijayan A (2021) KSTAR: a knowledge based approach for socially relevant term aggregation for web page recommendation. In: Digital technologies and applications. ICDTA 2021. Lecture notes in networks and systems, vol 211. Springer, Cham, pp 555–564. https://doi.org/10.1007/978-3-030-73882-2_50

Development of Real-Time Fault Diagnosis Technique for the Newly Manufactured Gearbox

Prasad V. Kane

Abstract For the newly manufactured gearbox, it is essential to identify the presence of defects for industries. It is found that the defects in the newly manufactured gearbox are identified by running it with a motor. Based on the sound coming from the gearbox, the inspector takes a decision about the presence of faults in the gearbox. There are many faults in the gearbox such as shaft assembly errors, profile errors, burrs, and dents on the gear teeth. This paper discusses the methodology developed for the real-time fault identification of the gearbox using the LabVIEW virtual instrumentation in real-time mode and the supervised artificial intelligence technique already trained in another programming language. The Artificial Neural Network (ANN) and Radial Basis Function Neural Network (RBFNN) are used for identifying faults. The Virtual Instrumentation (VI) is developed with different colour indicators designed to indicate a specific fault and this indicator glows when the vibration signal is acquired from the particular fault.

Keywords Gearbox · ANN · RBFNN · VI · Real-time fault detection

1 Introduction

1.1 Introduction to the Fault Identification

The survey is carried out in India's automobile plants to identify the practises followed by industries for newly manufactured gearbox fault identifications for quality inspection. It is found that the manual technique is followed in all industries. It is observed that, for ensuring identifying faults, a gearbox or transmission assembly is operated using an induction motor. The various faults which may be present at the end of the assembly line, i.e. the assembly errors like misalignment, error on the involute

P. V. Kane (✉)
Department of Mechanical Engineering, VNIT, Nagpur, Maharashtra, India
e-mail: pkane@mec.vnit.ac.in

S. Sharma et al. (eds.), *Intelligent Control, Robotics, and Industrial Automation*, Lecture Notes in Electrical Engineering 1066,
https://doi.org/10.1007/978-981-99-4634-1_67

profile of the gear teeth, burrs, and dent marks on gear teeth need to be detected. These faults get manifested as noise and vibrations. This sound emitted is listened to by the operator and based on the experience of the operator, the gear fault is identified by him. This process is found to be very subjective and tedious in nature. Hence to overcome this, a technique is developed which can be applied to fault identification. A technique is developed which works on a scientific basis to identify faults and gives output as a visual indicator. This technique can be deployed so that even the layman on the shop floor can identify faults from the coloured visual indicators.

1.2 Development of Methodology

To develop a framework to provide a solution to the problem identified, a review of the literature was carried out for identifying the different practises followed for gear fault diagnosis and which one would be suitable for the problem identified. An objective type of intelligent technique is proposed to assist the operator in identifying the recurrently occurring faults. Diagnostics of gearboxes is an interesting area of research for the last twenty years due to its complexity. Reported Literature [1–7] summarizes the different techniques in the vibration-based techniques, i.e. signal processing, feature selections, and artificial intelligence techniques. Worden et al. [8] provided an in-depth study of the literature and reported the advancements in the steps involved in vibration-based diagnostics and the further processing of the information contents of the signals and use of different classifiers. It summarizes the application of various other artificial intelligence or computing techniques for fault diagnosis of mechanical systems. The quest to propose the indices to assess and predict the health of the gearbox is found to be an area of interest for many researchers where the signals are processed in the time domain or frequency domain [9–12]. For condition monitoring, researchers proposed the features obtained from the time domain representations indicating the health of the gearbox are RMS, kurtosis, crest factor, maximum value, energy operator, sideband index, FM0, FM4, NA4, clearance fact and impulse indicator, etc. The standard deviation frequency, Shannon entropy, and spectral kurtosis, etc. are the other frequency domain indicators. A time synchronous averaging technique is applied for extracting these features in advanced and complex signal processing. However, Lebold et al. [11] reported that signal processing techniques to be applied lacks the standardization. The indicators such as FM4, FMO, and NA4 are obtained from the application of time synchronous averaging. The repeatability and reproducibility of these indicators is a matter of challenge as the methods of filtering the harmonic contents from the acquired raw signal are not standardized. Alattas and Basaleem [12] reported the results of these parameters for the fatigue testing where the faults such as dense wear, tooth breakage, single pit, and distributed pits were studied. The comprehensive review revealed that, though many techniques are reported in the literature that real-time fault diagnosis is a challenge and there is a need to pay attention. The limitation of the collection of datasets from industries is the availability of labelled datasets due to interference in the day-to-day day production process.

As a result, it was not possible to get enough vibration datasets from the industry, faults are simulated in the experimental setup in the laboratory. Laboratory experiments are conducted to simulate the faults such as an error on the involute profile of the tooth, crack at the tooth root, smashed gear tooth, and shaft misalignment error to obtain the labelled dataset. The statistical features are obtained from the signal and are made an input to the artificial intelligent technique belonging to the neural network family, i.e. ANN and RBFNN. Next section discusses the experimental setup and instrumentation used.

2 Fault Simulator Setup and Data Collection

2.1 Fault Simulator Experimental Setup

A faults simulator setup developed in-house is shown in Fig. 1 and it is having spur gear with the loading arrangement. The rotational speed of the motor can be controlled by the controller. A data acquisition (DAQ) card is used to acquire the vibration. The DC motor was operated at a speed of 720 rpm. The loading arrangement is made and the vibrations signals are acquired at no load and load condition. The gearbox is run at constant rotational speed as this is to be used for inspection of the newly manufactured gearbox which is a controlled environment. The sampling frequency of 20 kHz is selected while acquiring the vibration signal. The gear faults and misalignment errors are introduced and signals were acquired while running these gears with load and without load conditions. The statistical features were extracted which are discussed in the following section.

2.2 Feature Extraction

The features extracted are RMS value, skewness, variance, kurtosis, range, peak value, least value, etc. from the acquired time domain signal at each fault condition and healthy condition. These features contain information about the presence of a fault. Kurtosis indicates high-pitched peaks in the signal which could be generated due to the presence of a dent in the gear teeth. RMS indicates the overall condition of the gearbox. The features are not extracted from the frequency domain analysis to avoid delay involved in signal processing. These features are used as input made to the ANN and RBFNN classifier which is discussed in the following section.

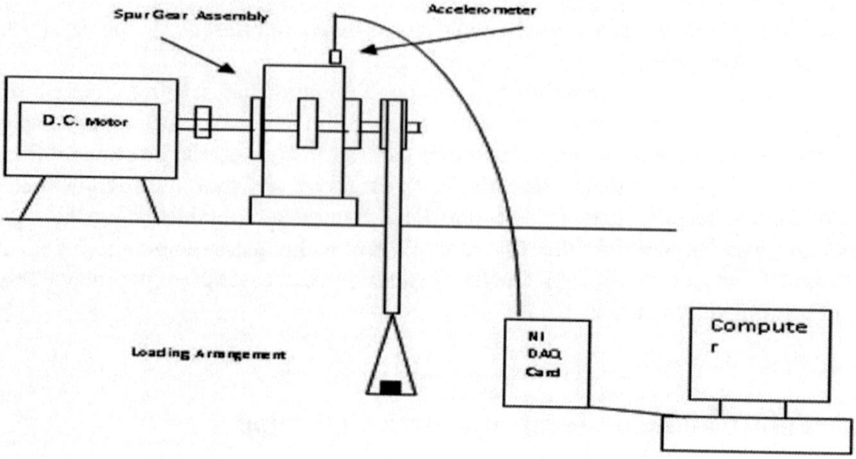

Fig. 1 Layout of experimental setup

3 Artificial Neural Network and Radial Basis Function Neural Network

The ANN and RBFN are trained by the features obtained from the vibration signal. The feed-forward back propagation neural network is having architecture with one input layer, two hidden layers, and an output layer. The input to classifiers are the eleven statistical features and the output indicates one of the five gear conditions. The ANN training algorithm selected is *trainlm*. The RBFNN is a variation of the simple feed-forward neural network. Radial basis function network works on the idea of function approximation using localized basis functions. For training and testing with RBFNN the goal and spread values are provided. The built-in function in MATLAB, i.e. *newrb* is applied which designs the radial basis network with *radbas* as a neural transfer function in the hidden layer [13]. This code automatically creates several hidden nodes based on the number of clusters using a clustering algorithm and the value of the spread. The process followed for applying the ANN and RBFNN is shown in Fig. 2.

Figures 3 and 4 depicts the ANN training regression plot and performance plot indicating satisfactory training without overfitting. Figure 5 indicates the training performance plot of RBFNN. The training and testing are carried out with 70 and 30% of the total dataset for both ANN and RBFNN.

The fault identification accuracy obtained for the training and testing of these techniques is stated in Table 1. It also specifies the Pearson's correlation coefficient (COR) and root mean square error (RMSE) for the two classifiers' output and target values. These trained ANN and RBFNN modules are invoked in the LabVIEW to design a graphical user interface (GUI) so that it can identify the gear condition based on the training provided which is discussed in the next section.

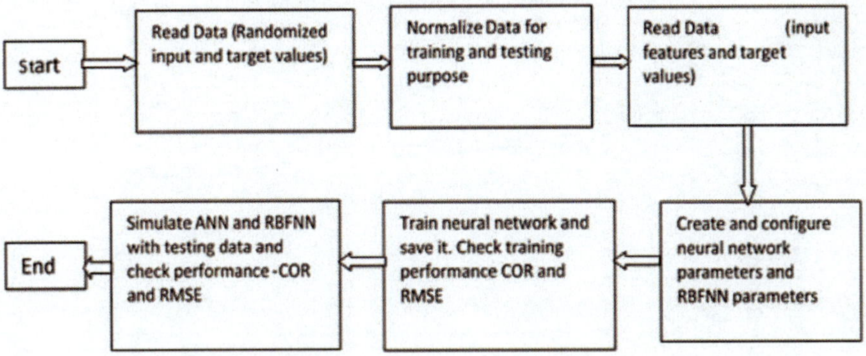

Fig. 2 Procedure to apply ANN and RBFNN for classification

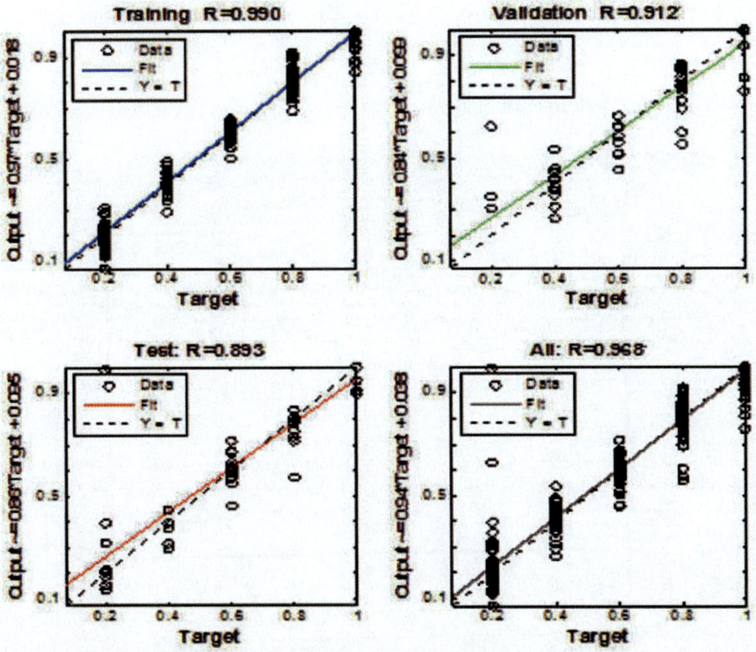

Fig. 3 ANN training regression plot

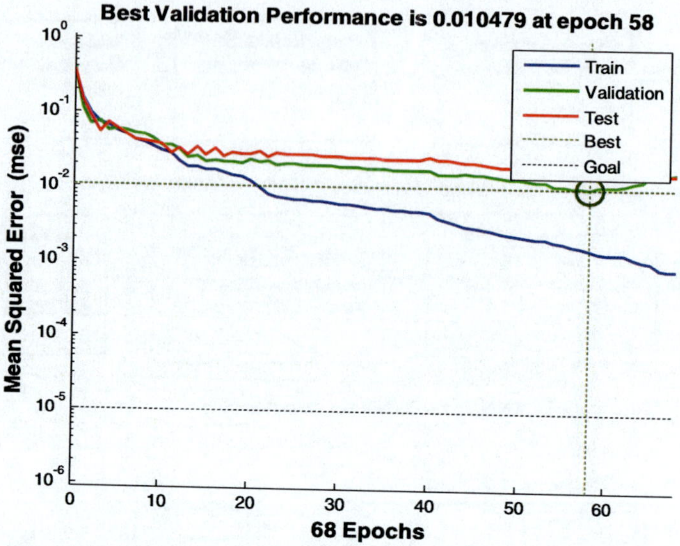

Fig. 4 ANN training performance plot

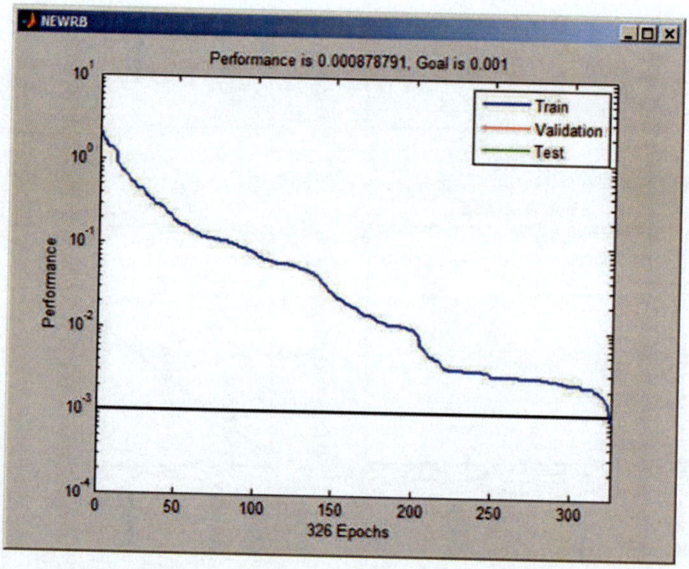

Fig. 5 Training performance plot of RBFNN

Table 1 Training and testing accuracy of ANN and RBFNN

Classifier	Accuracy (%)		Correlation coefficient		RMSE	
	Training	Testing	Training	Testing	Training	Testing
ANN	95.00	92.19	0.962	0.924	0.156	0.379
RBFNN	89.37	79.12	0.94	0.88	0.450	0.550

4 Development of Graphics User Interface for Fault Diagnosis

LabVIEW, a virtual instrumentation software, along with the NI DAQ card was used in this work for data acquisition and feature extraction. LabVIEW also provides computational abilities and virtual instrumentation (VI) is designed for online diagnosis of faults and the best diagnostic technique selected from the trained offline module in MATLAB can be used in it. The limitation of the LabVIEW software is its inability to process massive data in a limited time. Therefore, the facility of using MATLAB scripts in LabVIEW is used which helps in fast computing to develop online fault diagnosis VI which is presented in Fig. 6.

Figure 7 depicts the front end of intelligent virtual instrumentation to diagnose faults by glowing visual indicators for the particular status or condition of the gearbox. The LabVIEW module acquires the vibration signal from the gearbox and computes its statistical features and these features become input to the MATLAB script module which is invoked in LabVIEW as an intelligent diagnostic module. As per the predicted values of this module, the different coloured visual indicators glow showing the presence of faults.

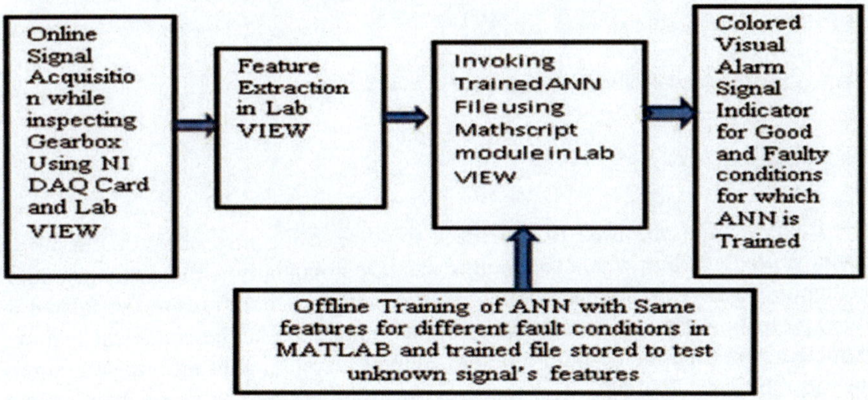

Fig. 6 Flow diagram for the development of the fault diagnosis module

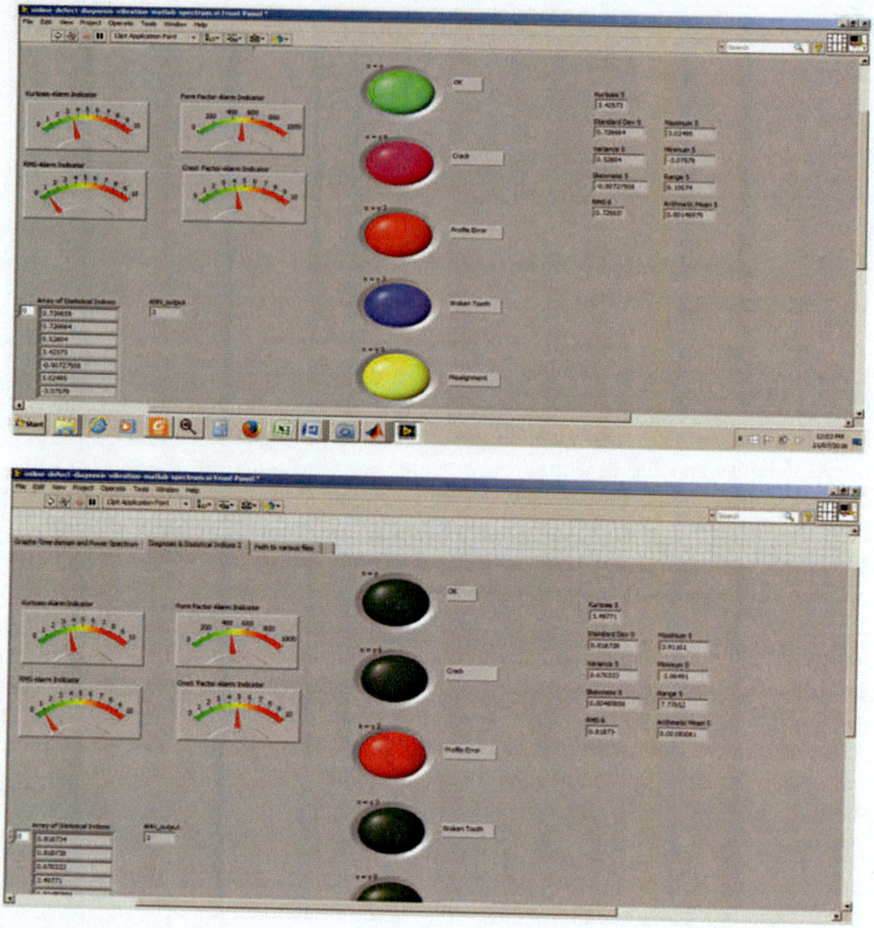

Fig. 7 GUI developed to indicate the presence of a fault in real-time

5 Conclusions

The framework is proposed to provide a solution to the inspection of the newly manufactured gearbox where the fault is identified manually without any scientific technique. To develop this technique, five types of gear conditions are simulated in the experimental setup and the vibration data is acquired. The statistical features extracted from the time domain vibration signals are applied to train the classifiers such as ANN and RBFNN. The accuracy for training and testing of the classifier is obtained by dividing the datasets by the 70% and 30%, respectively. It is found that the ANN would be a better classifier as it can work well even in the presence of noise in the dataset. The trained ANN file is invoked in LabVIEW while acquiring the real-time signal and the extracted features from it are the input to the trained

ANN module and the result of it glows the visual indicator setup developed in the GUI of LabVIEW. The Virtual Instrumentation (VI) is developed with different colour indicators designed to indicate a specific fault and this indicator glows when the vibration signal is acquired from the particular fault. This method proposes an effective technique that could be applied in the industries for the newly manufactured gearbox.

References

1. Smith JD (2003) Gear noise and vibration, 2nd edn. (Revised And Expanded). Marcel Dekker, Inc., New York, Basel
2. Mohanty AR (2015) Machinery condition monitoring: principles and practices. CRC Press, Taylor & Francis Group, Boca Raton, FL
3. Peng ZK, Chu FL (2004) Application of the wavelet transform in machine condition monitoring and fault diagnostics: a review with bibliography. Mech Syst Signal Process 18(2):199–221
4. Hongyu Y, Joseph M, Lin M (2003) Vibration feature extraction techniques for fault diagnosis of rotating machinery: a literature survey. In: Asia Pacific vibration conference, Gold Coast, Australia, pp 1–7
5. Samuel PD, Pines DJ (2005) A review of vibration-based techniques for helicopter transmission diagnostics. J Sound Vib 282(1–2):475–508
6. Bajrić R, Sprečić D, Zuber N (2011) Review of vibration signal processing techniques towards gear pairs damage identification. Int J Eng Technol 11(4):97–101
7. Jardine AKS, Lin D, Banjevic D (2006) A review on machinery diagnostics and prognostics implementing condition-based maintenance. Mech Syst Signal Process 20(7):1483–1510
8. Worden K, Staszewski WJ, Hensman JJ (2011) Natural computing for mechanical systems research: a tutorial overview. Mech Syst Signal Process 25(1):4–111
9. Sharma V, Parey A (2016) A review of gear fault diagnosis using various condition indicators. Procedia Eng 144:253–263
10. Večeř P, Kreidl M, Šmíd R (2005) Condition indicators for gearbox condition monitoring systems. Acta Polytech 45(6):35–43
11. Lebold M, McClintic K, Campbell R, Byington C, Maynard K (2000) Review of vibration analysis methods for gearbox diagnostics and prognostics. In: Proceedings of the 54th meeting of the society for machinery failure prevention technology, Virginia Beach, VA, pp 623–634
12. Alattas MA, Basaleem MO (2007) Statistical analysis of vibration signals for monitoring gear condition. Damascus Univ J 23(2):67–92
13. Demuth H, Beale M (1993) Neural network toolbox for use with MATLAB

Multi-body Dynamics Simulations of Spacecraft Docking by Monte Carlo Technique

V. Sri Pavan Ravi Chand, Vijay Shankar Rai, M. Venkata Ramana, Anoop Kumar Srivastava, Abhinandan Kapoor, B. Lakshmi Narayana, B. P. Nagaraj, and H. N. Suresha Kumar

Abstract Spacecraft docking is one of the key technology demonstrations in space sector. It paves way for several advanced space missions including on-orbit servicing of satellites, space robotics, and development of space station. One of the major subsystems in this mission is mechanisms sub-system. This has several constituents including mechanism for capture, mechanism for integrating two satellites after capture. The success of the docking event is dependent on the proper alignment of the two spacecrafts rings, which in turn influences the dynamics during the docking. From this perspective, the spacecraft docking problem statement can be considered to be determination of the dynamic envelope for successful capture of the two spacecrafts. This involves study on the dynamic behavior of the two spacecrafts during docking under different approach conditions. This paper presents the different studies that have been carried out to understand the dynamic docking envelope of two spacecrafts. The number of simulations required is very high given the large number of permutations and combinations among the approach parameters. Toward this, a simulation-based approach has been adopted for analysis based on Monte Carlo technique for generating random sample space of approach parameters. The studies have been carried out with the objective of optimizing the computational resources required for multi-body dynamics analysis while not compromising upon the accuracy of results. Multi-body dynamics software MSC software ADAMS has been used for carrying out the dynamic simulations. The details of this methodology along with the results obtained from simulations have been presented in the paper.

Keywords Spacecraft docking · Multi-body dynamics · Simulation · Monte Carlo method

V. S. P. R. Chand · V. S. Rai · M. V. Ramana · A. K. Srivastava · A. Kapoor (✉) · B. L. Narayana · B. P. Nagaraj · H. N. S. Kumar
Spacecraft Mechanisms Group, U.R.Rao Satellite Centre, ISRO, Bengaluru 560017, India
e-mail: akapoor@ursc.gov.in

S. Sharma et al. (eds.), *Intelligent Control, Robotics, and Industrial Automation*, Lecture Notes in Electrical Engineering 1066,
https://doi.org/10.1007/978-981-99-4634-1_68

1 Introduction

Spacecraft docking is an important milestone in space endeavors. It enables further progress toward advanced space missions like on-orbit satellite servicing involving space robotics and development of space station as well as commercial space missions like space tourism. It is based on low impact docking concept using two small satellites in a low earth orbit. A low impact docking concept can be adopted as it produces very low docking disturbance. Spacecraft docking refers to making a physical connection between the two spacecrafts under on-orbit conditions, under low velocity using closed-loop guidance with proximity sensors. This connection is between the corresponding mating rings on the two spacecrafts, which are referred to as approach and destination spacecrafts. A basic overview of two satellites before and after docking has been shown in Fig. 1.

The success of docking is dependent on proper alignment of the two spacecrafts rings, which in turn influences the dynamics during the docking. Various studies using multi-body dynamics simulations have been carried out based on Monte Carlo technique (for generation of random sample space of different input parameters) toward estimating the dynamic docking envelope that ensures 100% successful docking. There are efforts in this direction globally that involves simulations and testing to understand and ensure successful docking [1]. The simulations presented in this paper are prerequisite for moving toward integrated docking tests on ground and subsequently planning the on-orbit experiments. This paper presents the details of the analyses that have been carried out along with the reasons for adoption of the corresponding methodology.

2 Multi-body Dynamics Analysis

Multi-body Dynamics (MBD) model of the two spacecrafts with requisite joints has been generated using commercial software MSC ADAMS. The model of both spacecrafts before docking has been shown in Fig. 2. The mass of each spacecraft

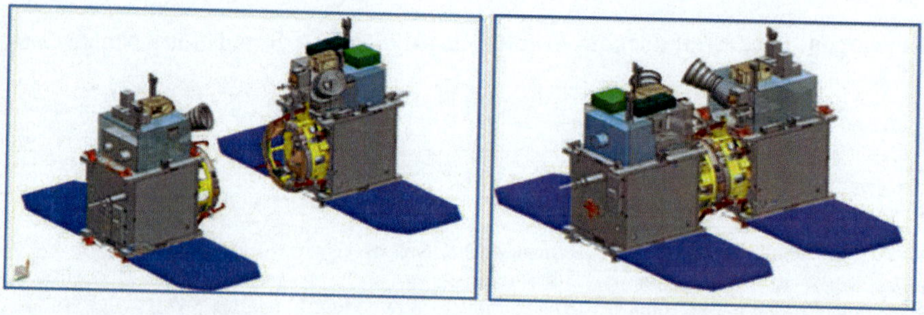

Fig. 1 Spacecraft docking

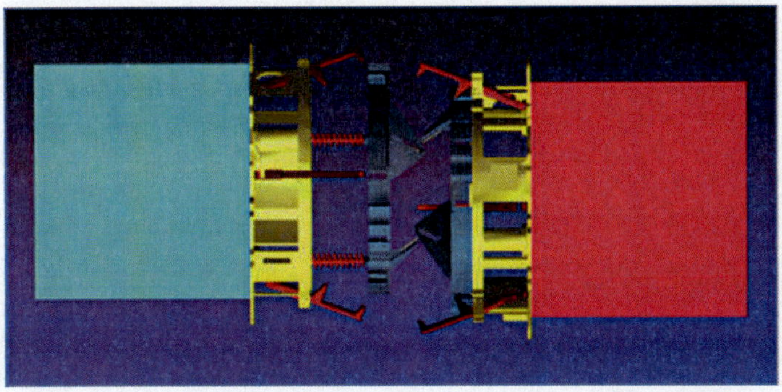

Fig. 2 Multi-body dynamics (MBD) model of spacecraft docking

is assumed to be 200 kg and the corresponding center of mass and inertia properties have been assigned accordingly for simulations.

An important constituent in this process is modeling of the contact between the mating parts during docking. A contact is a near discontinuous event which is challenging to solve numerically for an integrator. Impact method in ADAMS has been used for solving as it is less sensitive to error tolerances. It calculates forces locally at the point of contact based on the input parameters like contact stiffness, depth of penetration, force exponent, and damping. The parameters used in the present study have been given in Table 1. These have been taken in line with the properties specified in literature [2].

The simulations have been carried out initially by considering the condition that the mating rings of both approach and destination spacecrafts are perfectly aligned with the objective of validating the model and also to understand the sensitivity of hardware properties like spring stiffness, preload, center of mass. Subsequently, simulations have been carried out by varying the approach conditions like position and angular variables along with their time derivatives in order to evaluate dynamic behavior of the two spacecrafts during docking.

From the perspective of spacecraft docking, the problem statement that is considered in the present paper is determination of the dynamic envelope for successful capture of the two spacecrafts in terms of 11 input parameters of approach spacecraft with respect to the destination spacecraft. Success is defined by positive engagement of capture levers so as to hold the target spacecraft. Typical values of these 11 input

Table 1 Contact parameters

S. No.	Parameter	Value
1	Damping (N/m/s)	28
2	Stiffness (N/m)	3.5E7
3	Penetration depth (33)	0.001
4	Exponent	1.5

parameters used for present analysis, in roll–pitch–yaw coordinate system, have been mentioned in Table 2.

The values mentioned in Table 2 refer to those that are estimated from the static point of view based on geometric calculations. These are in line with the initial studies carried out to understand response characteristics [3] and impact behavior [4]. These values are then taken as input for dynamic analysis under different approach conditions to arrive at the dynamic envelope for successful docking. It is evident that number of simulations required is very high as there can be large number of permutations and combinations among the approach parameters. This can be comprehended from the fact that in an 11-dimensional vector space, mere consideration of only corner points in positive and negative directions along with zero makes the sample size of the order of 59,049. In this context, a simulation-based approach has been adopted for analysis based on Monte Carlo technique for generating random sample space of approach parameters like position and angular variables along with their time derivatives. The goal is to understand the behavior of the system under variation of individual parameters or their combinations with the objective of optimizing the computational resources required for multi-body dynamics analysis while not compromising upon the accuracy of results.

Monte Carlo technique is a statistical method that is used in analysis of systems that have uncertainty and variability of outcome owing to large sample size involving multiple input parameters. It helps to characterize the uncertainty and variability in estimation of outcome. In addition, it also helps in quantifying the relative contribution of different input parameters toward prediction of uncertainty in outcome. In the present study, Monte Carlo studies with uniform distributions have been used to get the probabilistic estimation of successful capture. Uniform distribution has been selected for analysis as it helps in giving a conservative estimation, which is very essential in critical (Go or No-Go) missions like spacecraft docking. A typical distribution of uniform sample is shown in Fig. 3.

The methodology adopted for the multi-body dynamics analysis carried out in this paper based on Monte Carlo technique has been summarized below:

(a) The system of docking of the two spacecrafts is considered to be a binary outcome problem dependent on 11 input parameters of approach spacecraft. The binary output is either "0" or "1". The formulation of this binary function has been done by taking magnitude of distance between the centers of mating

Table 2 Approach parameters

S. No.	Parameter	Value
1	Approach velocity-yaw axis (mm/s)	[5,15]
2	In-plane linear offsets—roll, pitch axes (mm)	[−35,35]
3	Angular offsets—roll, pitch, yaw (deg)	[−0.5, 0.5]
4	Angular velocities—roll, pitch, yaw (deg/s)	[−0.1, 0.1]
5	In-plane velocities—roll, pitch (mm/s)	[−1,1]

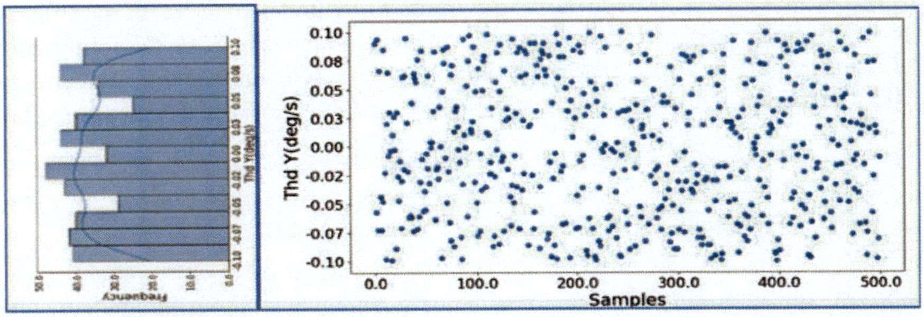

Fig. 3 Typical distribution of uniform sample

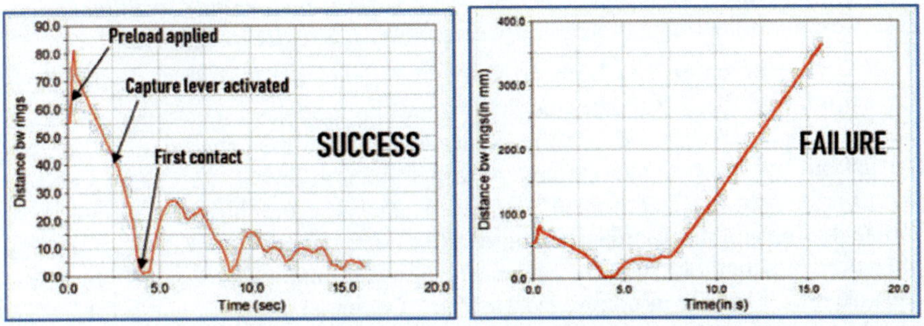

Fig. 4 Typical profiles of binary outcome for successful docking and that of the failed docking

rings of two spacecrafts as the base. Typical profiles of the binary outcome for the case of successful docking and that of the failed docking are shown in Fig. 4.

(b) The destination spacecraft is considered to be stationary and input parameters of approach spacecraft are taken as relative to the destination spacecraft.

(c) The input parameter sample space is generated randomly using Monte Carlo method based on normal distribution.

(d) Sample size is decided as 1000 based on the criteria of computational time to get a reasonably accurate prediction in less time. The process of elimination has been adopted to study the sensitivity of each parameter in sample space. With this, probability of failure is estimated from results of the finite sample experiment.

(e) Subsequently upon identification of a solution that determines the dynamic docking envelope for 100% success, the sample size has been enhanced to 3000 to confirm the validity of the obtained solution.

The profile of binary output for the success and failed cases, as shown in Fig. 4, is understandable. In the former scenario, as the mating rings come into first contact, after the actuation of capture lever, both the rings try to align themselves as can be seen from the cyclic nature of outcome plot. However, in the case of the latter, the rings get separated out after the first contact. This can be seen from the diverging nature of the outcome plot. This phenomenon needs to be understood based on different sets of approach parameter combinations in order to arrive at a solution in the form of input parameter vector space that determines the dynamic docking envelope for 100% success.

3 Different Sets of Simulations Carried Out

The system can be clustered into three major divisions from the point of view of the input approach parameters. These divisions are with regard to

(a) Approach velocity *(S. No. 1 in Table 2)*.
(b) In-plane linear offsets and in-plane velocities *(S. Nos. 2 and 5 in Table 2)*.
(c) Angular offsets and angular velocities *(S. Nos. 3 and 4 in Table 2)*.

This categorization has been arrived at based on the preliminary simulations carried out initially to validate and understand the model, as mentioned in Sect. 2. The docking analysis from satellite point of view involves integration of analyses carried out by different sub-systems at respective unit levels. This perspective has been taken into consideration in the present analysis. For instance, based on the initial studies and discussions, it is understood that the margin of dispersion available on parameters like in-plane linear offsets and in-plane velocities *(S. Nos. 2 and 5 in Table 2)* is lesser in comparison to that of parameters like angular offsets and angular velocities *(S. Nos. 3 and 4 in Table 2)*. Thereby, the sample sets have been chosen accordingly wherein the sub-sizing has been considered in the latter case and enlarging or retention of the range has been considered for the former case. There have been efforts globally with regard to the dynamics modeling and analysis of docking process [5], wherein the spacecraft dynamics have been evaluated using simulations. This simulation-based approach is inevitable in systems like spacecraft docking, which involves several possibilities.

The studies in the present paper have been carried out in a sequential manner so as to identify the sensitive parameters and tune their allowable range accordingly, so as to ensure 100% successful docking. The different sets considered in simulation studies have been shown in Table 3.

Observations and reasoning regarding the Table 3 studies are:

(a) The dynamic envelope for 100% successful docking is a sub-space of static envelope obtained from geometric studies as can be seen from set 1. This is understandable, as in general the dynamic characteristics are nonlinear.
(b) From set 2, it can be said that it may not be feasible to enhance the range of the in-plane velocity parameter.
(c) From set 3, it can be said that the sub-spacing of velocity seems to reduce the failure probability with respect to set 1. Here, it is to be noted that this parameter has been studied for discrete cases and it has been observed that the parameter range of [10–15 mm/s] has more probability of success in comparison to that of [5–10 mm/s] range.

Table 3 Sets of simulations

Set	Parameter sample space (all parameters are taken as per specification *(Table 2)*, except)	Sample size	Number of failures
1	Nil (all parameters are taken as per specification)	1000	3
2	In-plane velocities are taken as [−2 to 2] mm/s *(S. No. 5, Table 2)*	1000	6
3	Approach velocity is taken as [8–12] mm/s *(S. No. 1, Table 2)*	1000	1
4	All parameters are taken as 50% of their respective specification	1000	0
5	Angular offsets are taken as 50% specification *(S. No. 3, Table 2)*	1000	2
6	Angular velocities are taken as 50% specification *(S. No. 4, Table 2)*	1000	4
7	Angular offsets and angular velocities are taken as 50% of their specification *(S. Nos. 3 and 4, Table 2)* (with enhanced sample size of 3000)	1000 3000	0 0
8	Angular offsets and angular velocities are taken as 60% of their specification (S. Nos. 3 and 4, Table 2)	1000	1

(d) A few cases have been studied discretely by taking parameters within 80 and 70% of specification to understand the system trend. It has been observed that these sub-spaces are also having at least one failure in 1000. Accordingly, the next set of simulations has been chosen as set 4, with all the parameters within 50% of their respective specification. It is observed that the set meets the goal of 100% successful docking.

(e) Subsequently, the parameters related to angular offsets and angular velocities have been studied one by one and also in combination for 50% specification. It can be observed, as given in sets 5, 6, and 7, that sub-spacing of one of these parameters alone may not be adequate to avoid failure, while the sub-spacing of the combination ensures 100% successful docking. Thus, a solution, which is more or less feasible from system point of view, of the problem statement has been obtained.

(f) Further studies, in the form of set 7 with enhance sample size (3000) and set 8, have been carried out to confirm the correctness of the solution obtained. It has been observed that the set 8, using angular offsets and angular velocities in combination for 60% specification, could not avoid a failure while the previous set 7, using angular offsets and angular velocities in combination for 50% specification, could ensure 100% successful docking for enhanced sample size of 3000.

Based on Table 3 analysis, the docking envelope for 100% successful docking can be defined in terms of the 11 input approach parameters. This is the sample space as given in Table 4. It is to be noted that this is one of the possible and feasible solutions taking into consideration the system constraints like allowable dispersion, controllability, etc. This approach has novelty in the sense of considering clustering based on parameters within the complete space of vectors.

Table 4 Sample space for 100% successful docking

S. No.	Parameter	Value
1	Approach velocity—yaw axis (mm/s)	[5,15]
2	In-plane linear offsets—roll, pitch axes (mm)	[−35,35]
3	Angular offsets—roll, pitch, yaw (deg)	[−0.25, 0.25]
4	Angular velocities—roll, pitch, yaw (deg/s)	[−0.05, 0.05]
5	In-plane velocities—roll, pitch (mm/s)	[−1,1]

It is found that the probability of successful docking for the uniform distribution-based sample space, under static considerations, of 11 input approach parameters is more than 99.8%. In order to make this probability of successful docking as 100% under dynamic conditions, a new sample space of 11 input approach parameters has been proposed. This solution is defined as the sample space wherein all parameters are taken within the specification taken under static considerations, except for the six angular parameters which are to be modified as 50% of that static specification. This ensures 100% successful docking and it has been further confirmed by increasing the sample size to 3000.

4 Conclusion

This paper brings out a methodology for carrying out multi-body dynamics analysis involving multiple variables based on Monte Carlo simulations. The analysis has been adopted in a logical and sequential manner by way of elimination of options as can be seen from the studies catering to different sets of sample space. The results of eight sets, consisting of 10,000 cases, along with the observations and reasoning obtained from these studies have been presented.

Finally, the paper highlights the need for adoption of Monte Carlo-based simulations for spacecraft docking analysis. This technique helped in getting a reasonably quick estimation of the docking outcome that has numerous possibilities under different input conditions. It also enabled optimum use of computational resources for multi-body dynamics analysis while not compromising upon the accuracy of results. This is not only beneficial but also necessary for analysis of complex systems for advanced space missions.

Acknowledgements The authors would like to thank Shri Alok Kumar Shrivastava, Deputy Director, MSA, and Shri. M. Sankaran, Director, URSC, for their constant support and encouragement.

References

1. Mitchell JD et.al (2008) Integrated docking simulation and testing with the Johnson space center six-degree-of-freedom dynamic test system. AIP Conf Proc 969:709–716
2. Davidson C, Karsten T (2009) Contact in ADAMS. MSC Lunch and Learn Series
3. Rai VS, Soni D, Kumar HNS, Murthy KAK (2018) Dynamic simulation studies for on-orbit spacecraft docking experiment. In: 11thNational conference and exhibition on aerospace and defense related mechanisms (ARMS-2018)
4. Narayana BL, Ranganath R, Nataraju BS (2000) Studies of docking dynamics using Adams. J Spacecraft Technol 10(1):40–51
5. Tchoryk P, Hays AB, Pavlich J, Wassick G, Ritter G, Nardell C, Sypitkowski G (2001) Autonomous satellite docking system. In: Space 2001 conference exposition (AIAA 2001–4527)

Solutions to Diffusion Equations Using Neural Networks

Sampath Routu, Madhughnea Sai Adabala, and G. Gopichand

Abstract In this study, the Forward Euler approach and the neural network approach are tested by the authors to solve the diffusion equation. The continuous rate of change in the quantity of a chemical species present at a specific time point is related by a diffusion equation. A nonlinear network model called the formation diffusion equation employs a number of neurons, each with a unique weight vector and bias. They used finite difference methods to compare the performance to the outcomes. Future research should look at the neural network's performance at various learning rates and iteration counts.

Keywords Eigenpairs · Diffusion equations · Mean-squared error · Neural networks · Numerical methods

1 Introduction

Solving differential equations (DEs) is an important and central problem in physics, finance, and other fields. Therefore, doing so efficiently and accurately is of great interest, and various algorithms such as the one used in Lagaris et al. [1] have been used to solve this problem. Complex DEs typically require high-order algorithms, while simple DEs can be solved using first or second-order algorithms such as Forward Euler Scheme. Time and computational power consumption are the main problems in solving DEs. A coupled model computing several equations for thousands or millions of grid boxes may suffer badly from a poor DE solving algorithm

S. Routu (✉) · M. S. Adabala · G. Gopichand
SCOPE, Vellore Institute of Technology, Vellore, India
e-mail: venkat.saisampath2019@vitstudent.ac.in

M. S. Adabala
e-mail: adabalamadhughnea.sai2019@vitstudent.ac.in

G. Gopichand
e-mail: gopichand.g@vit.ac.in

S. Sharma et al. (eds.), *Intelligent Control, Robotics, and Industrial Automation*, Lecture Notes in Electrical Engineering 1066,
https://doi.org/10.1007/978-981-99-4634-1_69

in terms of accuracy and time consumption. The paper by Lagaris et al. [1] is one of many that suggests that solving DEs can be done with high precision using neural networks (NNs). However, the computational cost may be quite large [2, 3]. The question of interest regarding using NNs is whether they can outperform traditional algorithms in speed and efficiency. This experimentation will investigate comparing the Forward Euler approach with the neural network approach. Furthermore, following the approach in the paper by Knoke and Wick [4], the NN will be used to compute extrema eigenvectors and corresponding eigenvalues. In this project, the Diffusion Equation will be solved (approximated). Physically, this equation can, for example, represent the temperature gradient through a rod, describing how the temperature decays along the rod with time [5]. In Sect. 2, all theory and background information and ideas will be presented, including the metrics used to evaluate the results. Section 3 describes how the algorithms are built and what they produce. In Sect. 4, the methods used to solve the diffusion equation will be compared, and main findings will be presented, and Sect. 5 provides a discussion of the results. Finally, Sect. 6 provides a conclusion and suggestions for future work [6–8].

2 Theory

2.1 The General Problem

The problem to be solved is the simple diffusion equation

$$\frac{\partial \mathrm{u}(x,\, t)}{\partial t} = \frac{\partial^2 \mathrm{u}(x,\, t)}{\partial x^2} \quad t > 0 \quad x \in [0,\, L]. \tag{1}$$

Another way to write this problem is $\mathrm{u}_{xx} = \mathrm{u}_t$. Initial conditions are necessary. Using $L = 1$, the initial condition at $t = 0$ is given by

$$\mathrm{u}(x,\, 0) = \sin(\pi x). \tag{2}$$

Furthermore, Dirichlet boundary conditions are also used, given by

$$\mathrm{u}(0,\, t) = \mathrm{u}(L,\, t) = 0 \quad t \geq 0.$$

As an example, this differential equation and its initial and boundary conditions could represent the temperature of a heated rod. As time progresses, the heat is transported through the rod while the temperature decreases along the way.

2.2 Analytical Solution of the Diffusion Equation

The analytic solution is the benchmark for comparing approximations from the Forward Euler method and the neural network. Through separation of variables, the exact solution can be expressed as

$$\mathrm{u}(x,\ t)\ =\ X(x)T(t). \tag{3}$$

Differentiating this according to (1) and moving some terms, we get

$$\frac{X''(x)}{X(x)} = \frac{T'(t)}{T(t)}. \tag{4}$$

As to the both sides of (4) are not dependant on the same variables, they must both be equal to a constant. For convenience the constant is chosen to be $-\,\omega^2$. This gives the two equations.

$$\begin{aligned} X''(x) &= -\ \omega^2 X(x), \\ T\prime(t) &= -\ \omega 2T(t). \end{aligned} \tag{5}$$

Now, the solution X can appear in three possible ways given by the characteristic equation. In order to satisfy the initial condition (2), $X(x)$ must be on the form

$$X(x)\ =\ B\sin(\omega x)\ +\ C\cos(\omega x).$$

The initial condition then rules $C\ =\ 0,\ \omega\ =\ \pi$. For $T(t)$ in (5), the solution is on the form

$$T(t)\ =\ \mathrm{Ae}^{-\,\omega 2\mathrm{t}}.$$

As we know $\omega\ =\ \pi$, the solution is then:

$$\mathrm{u}(x,\ t)\ =\ X(x)T(t)\ =\ \mathrm{Ae}^{-\pi 2t}B\sin(\pi x).$$

Finally, from the initial condition, we know that $A\ \cdot\ B\ =\ 1$; hence, the exact solution denoted *exact* must be

$$\mathrm{u_{exact}}(x,\ t)\ =\ \mathrm{e}^{-\pi*\pi*t}\sin(\pi x). \tag{6}$$

2.3 Solution Using Explicit Euler Scheme

Now, it is desired to solve the equation with an Euler scheme. To make this possible, Eq. (1) must be discretized in both time and space. As time is only used in first-order derivative, we will use the explicit Forward Euler Scheme, which gives an error proportional to Δt. This scheme is given as

$$\mathrm{u}(x,\, t)t \approx \frac{\mathrm{u}(x,\, t + \Delta t) - \mathrm{u}(x,\, t)}{\Delta t}. \tag{7}$$

For the spatial discretization, a centered difference is used, which has an error proportional to Δx^2, given by

$$\mathrm{u}(x,\, t)x = \frac{\mathrm{u}(x + \Delta x,\, t) - 2\mathrm{u}(x,\, t) + \mathrm{u}(x - \Delta x,\, t)}{\Delta x^2} \tag{8}$$

on a discrete time and space grid, $\mathrm{u}(x,\, t) = \mathrm{u}(x_i,\, t_n)$, $t + \Delta t_n = t_{n+1}$, and so on. Simplifying this notation yields $\mathrm{u}^n{}_i = \mathrm{u}(x_i,\, t_n)$. On a discrete form, Eq. (1) is then

$$\begin{aligned} \mathrm{u}_{\mathrm{xx}} &= \mathrm{u}_t \\ [\mathrm{u}_{\mathrm{xx}}]_i^n &= [\mathrm{u}_t]_i^n \\ \frac{\mathrm{u}_{i+1}^n - 2\mathrm{u}_i^n + \mathrm{u}_{i-1}^n}{\Delta x^2} &= \frac{\mathrm{u}_i^{n+1} - \mathrm{u}_i^n}{\Delta t}. \end{aligned} \tag{9}$$

Solving this for u_i^{n+1} provides the solution u to (1) for each spatial point i:

$$\mathrm{u}_i^{n+1} = \frac{\Delta t}{\Delta x^2}\left(\mathrm{u}_{i+1}^n - 2\mathrm{u}_i^n + \mathrm{u}_{i-1}^n\right) + \mathrm{u}_i^n. \tag{10}$$

(10) is stable for the following grid resolutions:

$$\frac{\Delta t}{\Delta x^2} \le \frac{1}{2}.$$

2.4 Solution Using a Neural Network

Solving the PDE can also be done using a neural network. In this project, the neural network functionality within *TensorFlow* for python3 is used, as this is stable, fast, and simple to use compared to building a neural network from scratch.

In order to solve PDEs with a neural network, we approximate the true function u with a trial function $\Theta(x,\, t)$. Thus, the aim is to calculate Θ as close to the true function u as possible [1]. When aiming to solve Eq. (1), the corresponding equation for the trial function is

$$\Theta(x, t)x = \Theta(x, t)t \quad t > 0 \quad x \in [0, L].$$

The residual of this approximation is then

$$E = \Theta(x, t)x - \Theta(x, t)t. \tag{11}$$

The cost function minimized by the neural network is the sum of E, evaluated at each point in the space and time grid. This is equivalent to reducing the Mean Squared Error. For each iteration, the trial function is updated based on the networks' previous calculated trial function [9]. Choosing the correct "form" of the trial function is important to restrain the residual, and this "form" is based on the order of the PDE and its initial condition. To satisfy both the initial condition and the Dirichlet condition of Eq. (1), the following form is chosen:

$$\Theta(x, t) = (1 - t)I(x) + x(1 - x)tN(x, t, p), \tag{12}$$

where $I(x)$ is the initial condition, $N(x, t, p)$ is the output from the neural network, and p is the weights and biases of the network.

Ideally, the process should work accordingly: for each iteration in the network, the partial derivatives of Θ are calculated according to the new state of the network $N(x, t, p)$, updating the cost along the way. As the cost is minimized, the error term E gets closer to zero, and the trial function $\Theta(x, t)$ approaches the true solution of the PDE. In theory, the cost can practically reach zero if the number of iterations is big enough, but since it is of small interest to run infinite iteration, a *minimum cost value* is chosen to 10^{-3}. Learning rate, the structure of the neural network in terms of the number of hidden layers, and the number of nodes in each layer are also important in minimizing the cost [10].

2.5 *Computing Eigenpairs with Neural Networks*

It is also desired to solve another problem using the neural network, namely computing the eigenvectors Λ and corresponding eigenvalues λ of a symmetric matrix A.

Knoke and Wick [4] provide a neat way to compute Λ_{max} and λ_{max} which is provided. This computation is done by solving the ordinary differential equation

$$\frac{d\Lambda(t)}{dt} = -\Lambda(t) + \mathrm{f}(\Lambda(t)), \quad t \geq 0, \tag{13}$$

where $\Lambda = [\Lambda, \Lambda, \ldots, \Lambda_n]^T$. Here, I is the $n \times n$ identity matrix.

According to [4], when $t \to \infty$, any random non-zero initial Λ will approach Λ_{max} if it is not orthogonal to Λ_{max}. The corresponding eigenvalue λ_{max} is computed by the equation

$$\lambda = \frac{\upsilon^T A \upsilon}{\upsilon^T \upsilon}. \tag{14}$$

In (14), A is a symmetric matrix given by

$$A = \frac{Q^T + Q}{2}, \tag{15}$$

where Q is a random, real matrix.

After finding the largest eigenvalue λ_{max}, the smallest eigenvalue λ_{min} corresponding to the eigenvector Λ_{min} is easily found by substituting A with $-A$ in Eq. (15) [4].

To solve Eq. (13) with a neural network, a trial solution is proposed. Since $v \in R^n$, we choose a trial function $\Psi(x, t)$ dependent both on position x and time t, so that for each time step, the approximated eigenvector is given as $[v(1,t), v(2,t), \ldots, v(n, t)]^T$. Equation (13) can then be rewritten as

$$\frac{\partial \Psi(x, t)}{\partial t} = -\Psi(x, t) + f(\Psi(x, t)) \tag{16}$$

with $t \geq 0$ and $x = 1, 2, \ldots, n$. We defined the trial function as

$$\Psi(x, t) = v_0 + tN(x, t, p),$$

where v_0 is the initial v, chosen at random. The error is then the difference between the two hand sides of Eq. (16).

2.6 *Metrics Used to Evaluated Solutions*

The main metric used to evaluate the performance of the neural network compared to the analytic scheme is the Mean Squared Error (MSE). As the name suggests, this metric quantifies the mean of the squares of the error between the prediction $\hat{x}_i$ and the observed values x_i. Thus,

$$\text{MSE} = \frac{1}{n}\sum_{i=1}^{n}(x_i - \hat{x}_i)^2,$$

where n is the number of samples, MSE value of 0 would predict the observed value, thus the closer to zero the better prediction.

3 Method—Implementation of Algorithms

Solving the problem with the Forward Euler Scheme is done by iterating in time and space over Eq. (10), returning u and x values for a given initial function and total time T. Later, the solutions are plotted, and the MSE from the exact solution is calculated [11].

The neural network solves the diffusion equation using a given number of layers, neurons, total time, and learning rate. A cost function is calculated based on a trial solution, which is then minimized. The network iterates until a given threshold cost value is reached, then returns u and x. After that, the solutions are plotted, and the MSE from the exact solution is calculated.

Computing the eigenpairs is done using the approach described in Sect. 2.5. Functions defining the function f and the eigenvalues are defined and calculated using the symmetric matrix A. Then, the neural network is designed similarly to the one described above, now returning eigenvectors v_{dnn}, a time vector t, and number of completed iterations i. The eigenvectors produced by the network are then used to calculate eigenvalues, which is then plotted with time to verify convergence. Also, *numpy.linalg* is used to calculate reference eigenpairs.

4 Results

In Fig. 1, the solution of the diffusion equation by the Forward Euler Scheme is displayed. Blue lines are solution after 0.02 s, while red lines are solutions after 0.2 s. Inspecting Fig. 1 immediately reveals that the analytic solution of the equation is very good and that smaller spatial steps provide a solution closest to the exact solution. Also, we observe that the solution for the spatial step δx=0.1 is closer to the exact solution at t=0.2 than at t=0.02. For both moments in time, the solution for spatial step δx=0.01 is very hard to distinguish from the exact solution, meaning that it is a very good approximation.

In Fig. 2, the diffusion equation is solved using a neural network with Nt=10, Nx=100 and two hidden layers with 20 neurons each. The learning rate was set to 10^{-3} and the number of iterations was set to 10^3. A 3D visualization of the diffusion equation solved by the neural network can be found in Fig. 3.

Running the Euler scheme, we get the MSE scores provided in Table 1, and running the neural network, we get the MSE scores provided by in Table 2.

The neural network is very slow in solving the differential equation compared to the analytical solution.

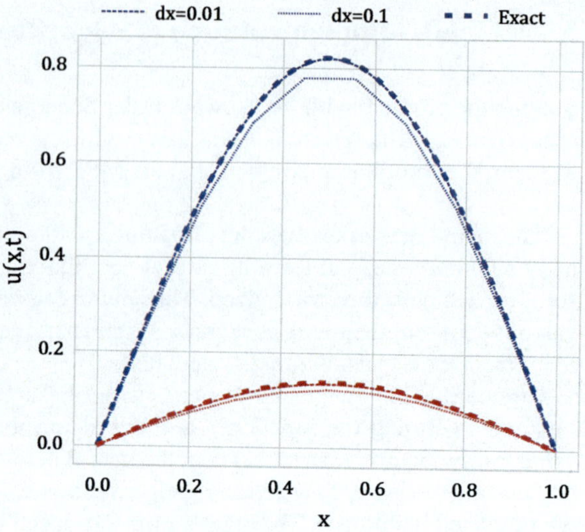

Fig. 1 Solution of the diffusion equation using Forward Euler Scheme. Solutions for δx=0.1 and δx=0.1 are displayed with the exact solution. Blue lines are solutions after 0.02 s. Red lines are solution after 0.2 s

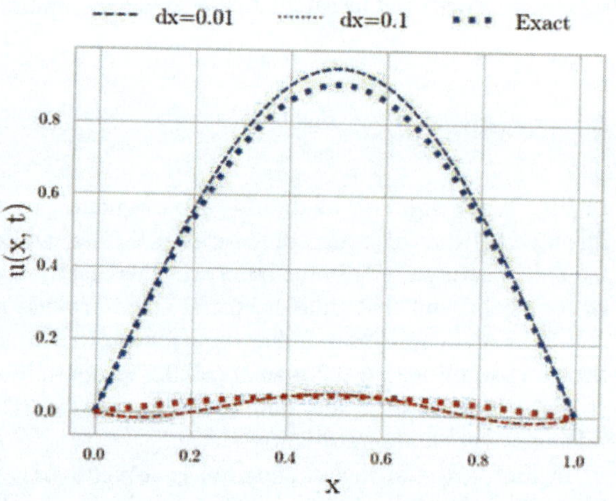

Fig. 2 Solution of the diffusion equation using a neural network. Solutions for δx=0.1 and δx=0.1 are displayed with the exact solution. Blue lines are solutions after 0.02 s. Red lines are solution after 0.2 s

4.1 Eigenvectors and Eigenvalues

Using the approach provided by Knoke and Wick [4], the neural network produced eigenvalues displayed in Figs. 4 and 5.

The target cost value was set to 10^{-3} because the algorithm was very slow for any lower values. As displayed in Fig. 5, the required number of iterations to calculate the $v_{\min}$ eigenvalue was 56,845. It is also evident from Figs. 4 and 5 that the network can in fact calculate the eigenvalues, as almost all of them seem to converge at some value with time.

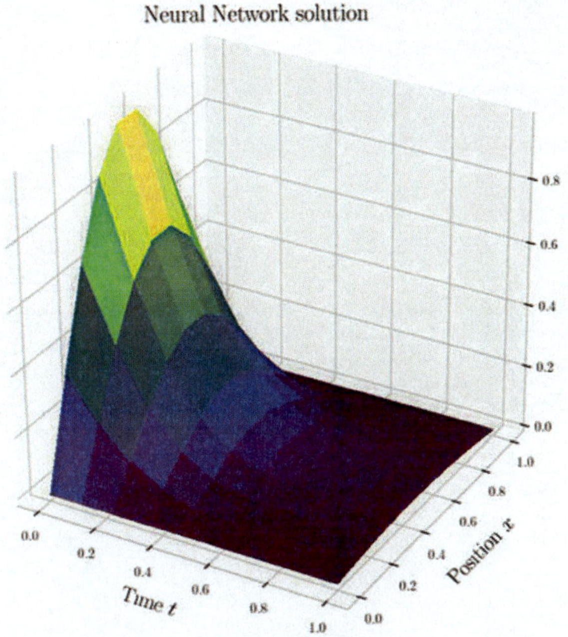

Fig. 3 Solution of the diffusion equation using a neural network. Nt=10, Nx=100, and 10^3 iterations was done

Table 1 MSE from the Euler solution at t=0.2 s and t=0.02 s with different spatial steps

Time (t)	Spatial step (dx)	Mean square error (MSE)
0.02	0.01	2.4336e−06
0.2	0.01	7.7725e−06
0.02	0.1	4.3479e−04
0.2	0.1	1.1393e−04

Table 2 MSE from the neural network solution at t=0.2 s and t=0.02 s with different spatial steps

Time (t)	Spatial step (dx)	Mean square error (MSE)
0.02	0.01	6.7583e−03
0.2	0.01	5.2900e−03
0.02	0.1	7.5181e−03
0.2	0.1	5.1556e−03

5 Discussion

Letting the finite difference approach serve as a benchmark, the MSE is on the order of 10^3 for δx=0.01 and 10^1 for δx=0.1 lower than for the neural network. This is partly because the target MSE value for the neural network was chosen to 10^{-3} due to an issue with time consumption. Allowing the network to iterate to lower MSE values

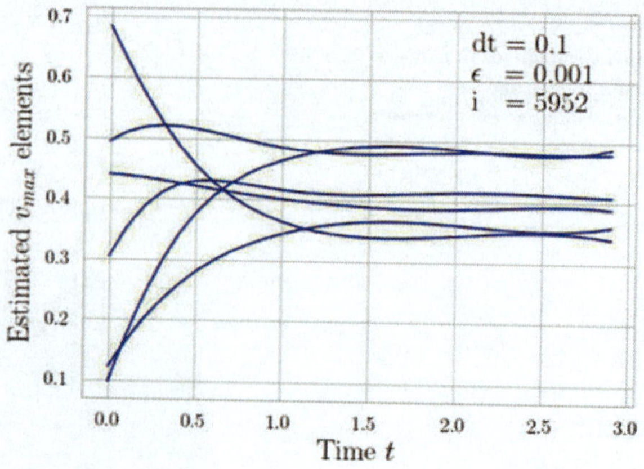

Fig. 4 Time evolution of $v_{\max}$

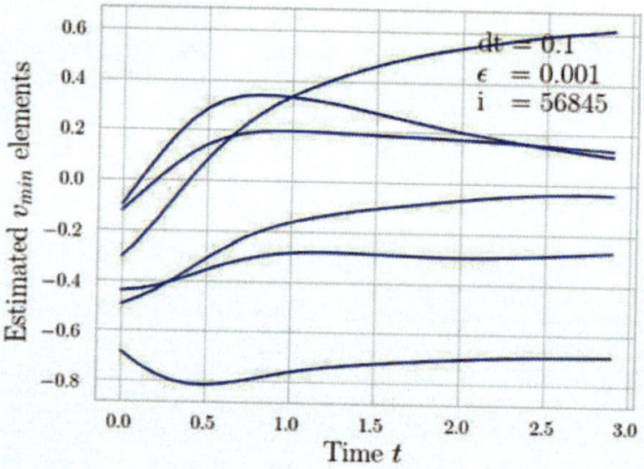

Fig. 5 Time evolution of $v_{\min}$

was possible, but required lots of time to complete, making the code close to useless. Depending on what model the neural network could have been used in, the MSE of the neural network is still quite good. However, in a complex model, the Euler solution would be preferred over the neural network. Keeping the calculation error as small as possible is often crucial for obtaining good result over time. Chiaramonte and Kiener [12] solved the Laplace equation using a neural network, obtaining MSE values of order 10^{-4} [12]. This clearly shows that a neural network could perform better than showcased in this paper with a better algorithm. Also, since the network algorithm is considerably slower compared to the Euler algorithm, it would most likely cause serious problems for complex models or models iterating over large periods of time or space. Despite relatively good agreement with the exact solution in Fig. 2, something appear to be wrong with the algorithm as the solution is negative in some parts of the domain. The figure is still left for illustrational purposes [13]. This neural network is not tuned in any particular way. Since it already provides relatively low MSE, tuning it would improve the performance. Running the network with varying iterations and learning rates would likely reveal a lower MSE. Additionally, optimizing the temporal and spatial steps to the specific problem of interest would most likely improve the time consumption, possibly making it comparable to the Euler approach [14].

6 Conclusion and Future Work

A neural network has been used to solve the diffusion equation. Performance has been compared to the solution obtained by the finite difference method. This study shows that partial differential equations can be solved using a neural network, despite being slower and yielding higher MSE than the Euler approach. Future work would include investigating how the neural network performs for different learning rates and several iterations. The performance depends on both, so researching this could result in a better performance and time consumption trade-off. Making the neural network algorithm faster would be an exciting topic for continuing research, as neural networks already have a broad range of applications that could benefit from the fastest possible algorithm.

References

1. Lagaris IE, Likas A, Fotiadis DI (1998) Artificial neural networks for solving ordinary and partial differential equations. IEEE Trans Neural Networks 9(5):987–1000. https://doi.org/10.1109/72.712178
2. Pasini ML, Perotto S (2023) Hierarchical model reduction driven by machine learning for parametric advection-diffusion-reaction problems in the presence of noisy data. J Sci Comput 94:36-1–36-22. https://doi.org/10.1007/s10915-022-02073-6
3. Xu X, D'Elia M, Glusa C, Foster JT (2022) Machine-learning of nonlocal kernels for anomalous subsurface transport from breakthrough curves. arXiv:2201.11146, arXiv:2201.11146v2, https://doi.org/10.48550/arXiv.2201.11146
4. Knoke T, Wick T (2021) Solving differential equations via artificial neural networks: findings and failures in a model problem. Examples Counterexamples 1:100035. https://doi.org/10.1016/j.exco.2021.100035
5. Ryczko K, Krogel JT, Tamblyn I (2022) Machine learning diffusion Monte Carlo energies. J chem Theory Comput 18(12):7695–7701. https://doi.org/10.1021/acs.jctc.2c00483
6. Mulani AO, Mane PB (2017) Watermarking and cryptography based image authentication on reconfigurable platform. Bull Electr Eng Inf 6(2):181–187. https://doi.org/10.11591/eei.v6i2.651
7. Kulkarni PR, Mulani AO, Mane PB (2017) Robust invisible watermarking for image authentication. In: Emerging trends in electrical, communications and information technologies. Lecture notes in electrical engineering, vol 394. Springer, Singapore, pp 193–200. https://doi.org/10.1007/978-981-10-1540-3_20
8. Mulani AO, Mane PB (2016) Area efficient high speed FPGA based invisible watermarking for image authentication. Indian J Sci Technol 9(39):1–6. https://doi.org/10.17485/ijst/2016/v9i39/101888

9. MacPhee N (2022) Use of machine learning for outlier detection in healthy human brain magnetic resonance imaging (MRI) diffusion tensor (DT) datasets. PhD thesis, McMaster University
10. Guo L, Wu H, Yu X, Zhou T (2022) Monte Carlo fPINNs: deep learning approach for forward and inverse problems involving high dimensional fractional partial differential equations. Comput Methods Appl Mech Eng 400:115523. https://doi.org/10.1016/j.cma.2022.115523
11. Wu W, Wang S, Sun Q (2022) Topological quantum cathode materials for fast charging Li-ion battery identified by machine learning and first principles calculation. Adv Theory Simul 5(3):2100350. https://doi.org/10.1002/adts.202100350
12. Chiaramonte M, Kiener M (2003) Solving differential equations using neural networks. https://cs229.stanford.edu/proj2013/ChiaramonteKiener-SolvingDifferentialEquationsUsingNeuralNetworks.pdf
13. Li C, Yang Y, Liang H, Wu B (2022) Learning quantum drift-diffusion phenomenon by physics-constraint machine learning. IEEE/ACM Trans Networking 30(5):2090–2101. https://doi.org/10.1109/TNET.2022.3158987
14. William P, Badholia A, Verma V, Sharma A, Verma A (2022) Analysis of data aggregation and clustering protocol in wireless sensor networks using machine learning. In: Evolutionary computing and mobile sustainable networks. Lecture notes on data engineering and communications technologies, vol 116. Springer, Singapore, pp 925–939. https://doi.org/10.1007/978-981-16-9605-3_65

SNAVI: A Smart Navigation Assistant for Visually Impaired

Madhu R Seervi and Adwitiya Mukhopadhyay

Abstract Our lives are made easier by automated solutions based on the Internet of Things (IoT). Navigating from one place to another can be challenging for blind people. IoT can increase navigational confidence while simultaneously decreasing reliance on others. The goal is to make the device less bulky and to aid them with two ultrasonic sensors and a flame sensor, which will be cost-effective for blind people who can comfortably travel both indoors and outdoors with minimum sensor use. SNAVI: A Smart Navigation Assistant for the Visually Impaired might be molded into a stick for visually impaired persons so that they receive a notification through their headphones when a barrier is identified in front of them via two ultrasonic sensors, as well as the height of the obstacle. The obstruction might be stationary or moving, and the system can detect fire through a flame IR sensor and send voice alerts. This method helps vision-impaired people travel about with less stress.

Keywords Internet of Things (IoT) · Ultrasonic sensor · Flame IR sensor · Raspberry Pi · Visually impaired · Voice alerts · Fire detection · Obstacle detection

1 Introduction

1.1 General Overview

With the help of the Internet of Things (IoT), almost all electric equipment can be connected to the Internet. The Internet of Things is changing the way we live. It enables us to have a greater understanding of the inner working of objects and the world around us.

The Internet of Things (IoT) is a network of interconnected devices that use the Internet to send and receive data. People are seeking to use IoT to make their jobs easier, so smart homes are growing in popularity. Air conditioning, doorbells, fire

M. R. Seervi · A. Mukhopadhyay (✉)
Department of Computer Science, Amrita School of Computing, Amrita Vishwa Vidyapeetham, Mysuru Campus, Mysuru, India
e-mail: adwitiya@my.amrita.edu

S. Sharma et al. (eds.), *Intelligent Control, Robotics, and Industrial Automation*, Lecture Notes in Electrical Engineering 1066,
https://doi.org/10.1007/978-981-99-4634-1_70

alarms, water heaters, and security alarms, among other devices, can be linked to communicate data with users via a mobile app. IoT devices are divided into two categories: general devices and sensor devices. The main components of a data hub and information exchange are general devices, which may be connected via wire or wireless interfaces. These IoT-based solutions have made our lives easier, so we can use them to make the lives of visually impaired people easier to navigate.

Visually challenged people's mobility is constrained by their environment. Without the assistance of a person, move safely and privately in a metropolitan area or at home. So we decided to assist them by developing an IoT-based device that would make them easier to navigate at public spaces. The ultrasonic sensor detects the moving or static objects on the way. Notify the individual if a fire hazard is in his or her route. There will be a voice alert message provided. These features will aid blind people in navigating their environment, whether at home or on the streets. There has been a lot of study done to improve the lives of blind people easier. There are articles that try to make deaf, dumb, and blind people [4] better by employing numerous sensors.

1.2 Introduction to SNAVI

The IoT device SNAVI: A Smart Navigation Assistant for Visually Impaired has two Ultrasonic sensors for obstacle detection if the obstacle is less than 30 cm in height, ultrasonic sensor 1 sends a voice alert, if the obstacle is greater than 30 cm in height, ultrasonic sensor 2 sends that particular alert and if objects height is between 10 and 30 cm voice alert is sent by the ultrasonic sensor 1. A flame IR sensor for fire detection within the range of 30 cm. With the use of two ultrasonic sensors and a

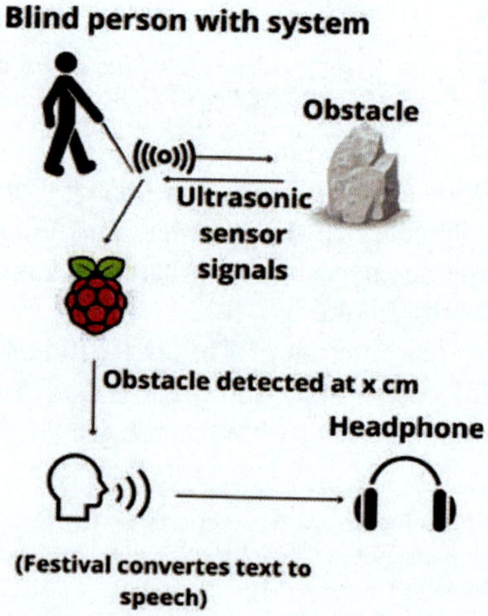

Fig. 1 Overall system architecture

flame sensor, we tried to make this system less bulky compared to other proposed systems so that users may feel light while traveling. The overall system architecture of SNAVI is as shown in Fig. 1. This will be cost-effective for blind individuals who can comfortably walk both indoors and outdoors with little sensor use.

2 Related Works

2.1 Works Related to Image Processing

Some studies have attempted to recognize objects using image processing [3]. This device allows users to easily navigate both indoors and outdoors. It is designed for people who have reduced vision (but are not completely blind). The authors here have used a phone camera to take photos and then categorized them based on their location. They used a computer vision technique to merge IoT and picture processing. In [5], authors have proposed a system that employs a camera to take pictures and collect data from both indoor and outdoor locations, and they used MATLAB and a Raspberry Pi to simulate the process [13]. They have proposed a system that detects and recognizes the faces that have been registered in the system, and if an unknown face is found, a notice is sent to the user's smartphone.

In [15], the authors have discussed eye-wearing devices having a camera, laser sensor, GPS module, and accelerometer. The device recognizes an object, such as a currency or a table, and delivers a notification to the user. The device may send the user's location's latitude and longitude. If a free fall happens, the call is forwarded to the individual who is concerned. The system recognizes the collection of objects that have been registered. As a single-board computer, they used the Raspberry Pi and the Coral model. For the blind, in [16], this method includes two devices: a cap and a stick. The items are identified at a maximum distance of 7 cm. The cap uses deep learning methods to capture the object spotted by the camera module and informs the user through an audio message. The stick uses a vibrator and a buzzer to notify the user, as well as an ultrasonic sensor and an infrared sensor to identify obstacles and a water sensor to detect moisture.

In [22], the authors have proposed a system that recognizes objects and faces, shares positions, and recognizes obstacles. The Raspberry Pi is used as the controller. This system is difficult to use because it includes a sensor and a camera for face recognition, as shown in the image. For facial recognition, they employed the Viola-Jones algorithm. The authors of [6] describe using the k-means clustering method and the backpropagation algorithm to optimize energy usage in wireless sensor networks. In [14], authors have proposed a system for automated irrigation system which sends relevant information like soil moisture, humidity, and dryness to the farmer at a regular intervals of time using ZigBee. In [12], the authors have proposed the device which is developed for blind people as a security device, detecting moments and faces, and sending the captured image to a smartphone. The system suggested by the authors in [21] is based on classification, segmentation, and recognition from the video users' smartphones are sending to them, as well as sending the voice alert with which object has been detected. However, how the video will be taken while the user is walking is not mentioned, and the design and implementation of the system is only partially explained.

2.2 *Works Related to Different Sensors*

An IoT-based system uses an ultrasonic sensor and a gas sensor to determine the distance between obstacles and the presence of gas [18]. The Arduino Mega is the microcontroller. The data has been posted to the ThinkSpeak website. The sending of alert messages via the headset via text-to-speech methods [17], either via the Wi-Fi module or via the Bluetooth module [25], was one thing that was noticed. Some authors have utilized simulations instead of real equipment to get more accurate findings than they would have gotten with genuine hardware. Some rely on direct landmark detection, while others use motion sensors like accelerometers, magnetometers, and gyroscopes.

In [1], the author has proposed a system using three ultrasonic sensors and Arduino to detect the barrier from three sides. Second, the system uses an IR sensor to identify the hole, and when the hole is discovered, an alarm is sent to the user. The next module is the water detection module, which is accomplished with the help of a moisture sensor. The stick also uses an LDR sensor to detect street lights, and when the emergency button is touched, it sends the user's location.

The authors in [7] propose a system that includes three sensors: ultrasonic, wetness, and infrared. When the button is pressed, it transmits the location to the worried individual. When a moisture sensor is detected, a warning is sent to the speaker. They used an ultrasonic sensor to detect obstacles and a Bluetooth module to convey the alarm message to the headset in [8]. This system is simple and useful for blind persons because it does not require the use of several sensors and is straightforward to operate.

The Radio Frequency Identification (RFID)-based mapping [2] system uses RFID tags and RFID cane readers that are interfaced with Bluetooth technology to provide easy and free travel across public spaces. The biggest problem is that it interferes with traffic light frequency, and the expense of installing such a system is prohibitively expensive. In [23], the impediment is detected using an ultrasonic sensor in this system. When an object is recognized, a message is sent to a laptop attached to an Arduino microcontroller, which is then copied to a CSV file. This system can be used as a starting point for study. The goal of [27] and [10] study is to create a system for those who have deafness or are completely blind. In [27], when an impediment is recognized, the technology amplifies the voice and delivers a voice alert, which can assist those who are both deaf and blind. The authors of [11] employed smartphones with Internet connections to transfer ECG signals, and Firebase cloud database and Google cloud storage for data storage and management were used to enable efficient data storage and transmission. In [10], authors have proposed a system that is used for indoor navigation using a GIS system giving them audio guidance using the built-in mobile sensors.

In [24], the authors perform obstacle detection using an Arduino and an ultrasonic sensor. They have provided a detailed explanation of how to connect an ultrasonic sensor to an Arduino as well as the operation of the ultrasonic sensor's receiver and transmitter. In the case of [26], a binary prediction algorithm is employed along with a three-class prediction method for manhole, pitfall, and staircase identification, as well as an ultrasonic sensor to detect obstacles. When a barrier is recognized through GSM, this gadget sends an alternate path.

In [9], authors have proposed a system which is based on two ultrasonic sensors connected to a controller, and the device is essentially a handheld device that the user holds in his hand, and when an obstacle is detected, an alert is sent via earphone and a vibration alert is given. Ultrasonic sensors are placed at angles of 0 and 40°C. In [19], the device uses four ultrasonic sensors at various heights to identify stairs and barriers of various heights. The motion sensor detects moving things in front of the stick, while the proximity sensor, which faces downwards, the buzzer, vibrator, and Wi-Fi module convey the device/location as stated in the study [28]. This research examines how ultrasonic sensors determine object distance in great detail. In [20], authors have proposed a system that uses experimental settings to compare the electromagnetic and ultrasonic systems.

3 System Design

SNAVI uses an ultrasonic sensor and a flame sensor to address two primary issues: obstacle detection and fire detection. The Pi is connected to the headphones, which will get notifications from sensors and guide them through their journey. Two ultrasonic sensors are utilized to detect and alert visually impaired people to both minor and lengthy obstacles.

3.1 Connections and Setup

Before we begin, we must first install the Operating System (OS) on our Raspberry Pi. We load Raspbian OS on a micro-SD card and insert it into the Raspberry Pi, then connect the HDMI cable to the monitor, turn on the Raspberry Pi, and set it up. We create our code in the Python IDE and execute the scripts after everything is set up.

After that, we will begin by connecting the sensor to the Raspberry Pi. To begin with, the ultrasonic sensor: The four pins of an ultrasonic sensor are VCC, Trigger (TRIG), Ground (GND), and ECHO (Echo), as well as an ultrasonic transmitter and receiver. When the Ultrasonic Transmitter collides with obstruction and bounces back, the ultrasonic sound is received, and the distance is determined. We connect

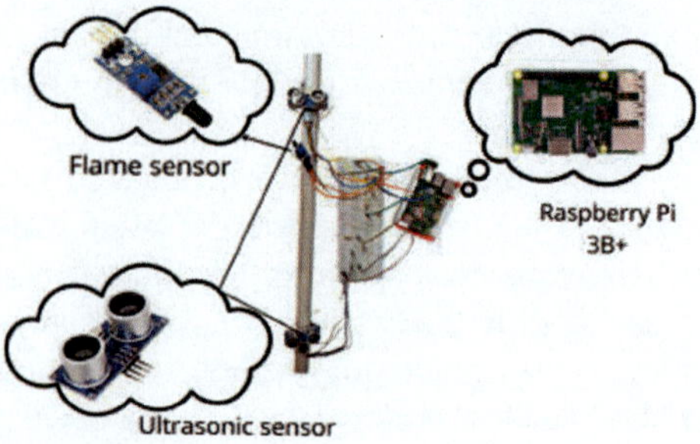

Fig. 2 SNAVI: working system

one of the Raspberry Pi's GPIO pins to the ultrasonic sensor's TRIG, the ultrasonic sensor's VCC to 5V of the Raspberry Pi, and the Raspberry Pi's GND to the ultrasonic sensor's GND. We also connect a 1-Ω and 2-Ω resistor between the echo pin and the ultrasonic sensor's GND pin, and the ECHO pin to the Raspberry Pi's GPIO pin. Connect another ultrasonic sensor at the desired height using the same procedure. Second, the flame sensor must be connected; there are various types of flame sensors; in our system, we used an infrared (IR) flame sensor. It contains three pins: VCC, GND, and digital output (DO), which are connected to the Pi's 5V pin, GND, and DO, respectively. Finally, connect the Bluetooth speaker to the Raspberry Pi. By turning on the Raspberry Pi's Bluetooth setting, we connect the Bluetooth device (which can be any Bluetooth device; for our experiment, we used Bluetooth earbuds). When the connections are made correctly and the code is loaded, the sensor begins to work, and when the fire catch is detected, the sensor's LED illuminates and the alarm message is produced. A setup of SNAVI is shown in Fig. 2 where two ultrasonic sensors and one flame sensor are linked to the Raspberry Pi 3B+. To prevent connections from becoming complicated, use a breadboard.

4 Experimental Design

We aimed to keep the SNAVI as lightweight as possible by employing only a few sensors and providing the finest help at an inexpensive price. When a small barrier is encountered with this system, such as stones or boxes of any height between 10 and 30 cm, the ultrasonic sensor located at the bottom recognizes it and sends an alarm message to the user via headphones. Sending voice alerts to users has a very limited time window or none at all. As soon as an obstacle is found, the user hears a voice alert. If the barrier is higher than 30 cm, the ultrasonic sensor 30 cm above it will detect it and deliver a message indicating the obstacle's height. The high alert voice message is issued twice if the obstruction is within 30 cm of the sensor. When the flame of the fire is detected by the flame IR sensor the user receives voice alert with caution. To remove the need of Bluetooth device separately and make the gadget easier to use, headphones are connected to the Raspberry Pi's 3B+ build in Bluetooth connection

Algorithm 1 Algorithm of Ultrasonic sensor detection

```
TRIG = True
StartTime ← time.time()
StopTime ← time.time()
while ECHO = 0 do
    StartTime ← time.time()
    while ECHO = 1 do
        StopTime ← time.time()
        distance ← StartTime − StopTime
        distanceInCm ← (distance ∗ 34300)/2
        while True do
            if distance ≤ 30 then {alert with distance calculated}
            end if
        end while
    end while
end while
```

When an impediment is encountered, the ultrasonic sensor detects the obstacle in front of it and sends the voice alert. A Python application uses festival to convert a text message to a voice message, which is then sent to the user via headphones. The voice alert is sent as soon as an object is detected. The ultrasonic sensor detects both moving and stationary obstructions. When the obstacle height is greater than 10 cm but less than 30 cm, the ultrasonic sensor located at the bottom will send the voice alert to the user. Both ultrasonic sensors will detect the obstacle if it is higher than 30 cm in height. Because it correctly distinguishes both fixed and moving objects, only the ultrasonic sensor located at the top will send the voice alert when the height of the object is greater than 30cm. The alarm message is sent twice if the obstacle is within a 30cm radius of the sensor. If the flame is detected by the flame sensor, the voice alert is sent twice with high alert caution. To better comprehend the system's flow, look at Fig. 3, which explains the system's flow step by step. The used ultrasonic sensor will have a 4 m range. In addition to using higher-end flame sensors with a wider field of view, SNAVI features a threshold value of 30 cm that can be modified when put into action.

5 Result and Discussion

In SNAVI, we have used the stick for experiment purpose which is approximately 80–90 cm in height. We experimented with a variety of objects ranging in height from less than 30 cm to more than 40 cm, including stones, bottles, furniture, doors, and walls. When an obstruction is identified within a 30 cm distance, the message is transmitted twice with a high warning level.

SNAVI has two ultrasonic sensors that are linked such that when a small barrier and obstacles higher than 30 cm. If the barrier is taller than 30 cm, both ultrasonic sensors will detect the obstacle, and a voice alert is sent by the ultrasonic sensor

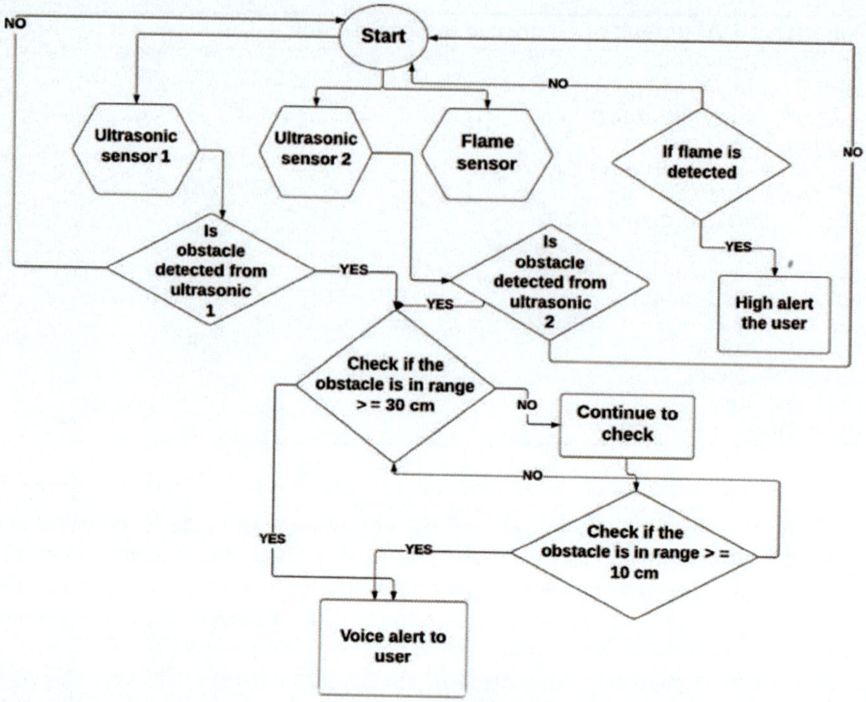

Fig. 3 Flow of the system

Fig. 4 Furniture as obstacle

Fig. 5 Small object as obstacle

Fig. 6 Random obstacle between path

located at 30 cm height indicating the height of the obstacle is a minimum of 30 cm like in Fig. 4. If the smaller object is detected like in Fig. 5, the alert is send by the ultrasonic sensor located at the end of the stick. When the distance between the user and the obstacle is less than 30 cm, the alarm message is issued twice, causing the user to become hyper-alert and avoid colliding like we got twice when the random object came in path of the user in Fig. 6. In the case of fire detection in Fig. 7, user gets an alert twice with the caution statement. Not only static obstacle is detected even moving object is detected by both the ultrasonic sensor and flame sensor. When the fire is detected, the voice alert "Danger caution: Fire is detected" is sent twice. We have used a flame IR sensor for experiment purposes which has a very small range of detecting fire this can be replaced with a higher level of the flame sensor when brought into use. We observed that there is a very small window of time, if any, for sending voice alerts to users. A voice alert is heard by the user as soon as an obstruction is detected.The obstacles detected by both ultrasonic sensors are depicted

Fig. 7 Fire detection

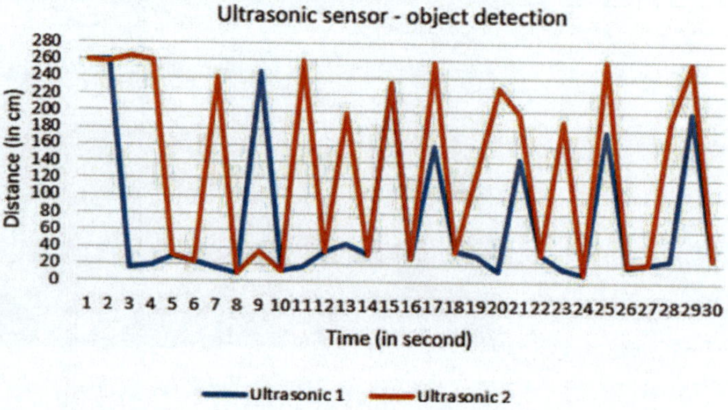

Fig. 8 Analysis of ultrasonic sensors with small objects as obstacle

in Figs. 8 and 9; the ultrasonic sensor's maximum range is 260 cm. Only when the obstacle is within a range of less than or equal to 30 cm do we issue the notifications the range can be set to a higher level for better navigation.

The blue line graph illustrates the first ultrasonic sensor's obstacle detection from the ultrasonic sensor located at the bottom, while the red line graph depicts the second ultrasonic sensor's obstacle detection from above. We can see that at some point, both the ultrasonic sensor and the user received a notification stating that a barrier of a minimum distance of 30 cm had been spotted. The analysis of ultrasonic sensor detection of random objects is shown in Fig. 8, and the detection of furniture and walls is shown in Fig. 9. Because the height of the wall and furniture is more than 30 cm, the graph of furniture and wall detection is given as a bar graph. Both

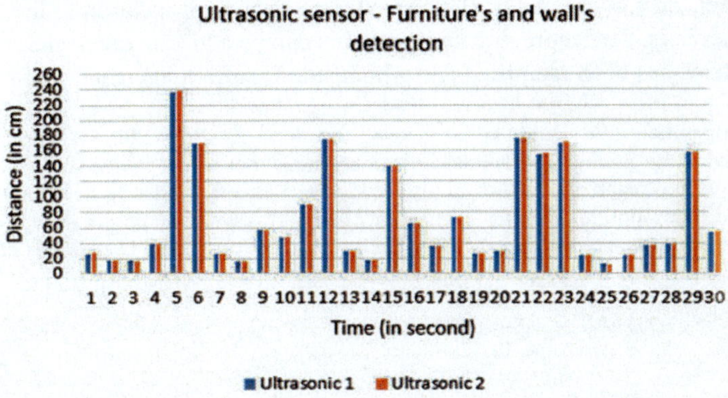

Fig. 9 Analysis of ultrasonic sensors with furniture and walls as obstacle

ultrasonic sensors detect the distance with nearly identical readings. As a result, the bar graph clearly shows the difference in data from both ultrasonic sensors. The x-axis reading is time in seconds, and the y-axis is the distance in cm. The system's efficiency is 85%; however, we may improve it by using a different microcontroller, such as an Arduino Mega, because the Raspberry Pi's OS is unstable, and we require an extra display to access the Pi, which consumes more power and causes delays when several sensors are connected.

A flame sensor detects fire within a 30 cm range the alert is received by the user. We tried out different distances with the lighter and the bonfire. We discovered that every fire test was successful if the fire caught within a 30 cm range and the voice alert was sent as soon as the fire was detected. The message was transmitted twice, with a 2-s interval between each transmission. To make the system more efficient, we can employ a higher-threshold fire sensor and can set the range as per the requirement.

6 Conclusion

The suggested technology identifies the obstruction and provides an alarm message to the visually handicapped. This approach can be utilized by the sight impaired to navigate so that they feel free to do it without hesitation or assistance from others. Because it includes fewer modules and is less sophisticated, this simple system can be inexpensive and lightweight. SNAVI contains two ultrasonic sensors, one of which is at the end of the stick and the other at a height of 30 cm. The alarm message is sent twice when there is less than 30 cm between the user and the barrier, making the user hyperaware and preventing a collision. Both the flame sensor and the ultrasonic sensor may detect moving objects in addition to just static ones. The alarm message "Danger caution: Fire is detected" is issued twice when a fire is discovered. In the future, this system might be enhanced by adding a GPS connection so that if a user is in danger, they only need to click a button to send their location to the registered mobile numbers. Two more ultrasonic sensors can be added to determine which side of the barrier has been identified and which side is safe to navigate.

Acknowledgements We would like to pay our gratitude to our Chancellor, Sri. Mata Amritanandamayi Devi is the guiding light and inspiration behind all our works toward societal benefit. We would also like to thank all the staff at Amrita Vishwa Vidyapeetham who have provided us the support and motivation in the completion of this work. This work would not have been possible without the infrastructure and support provided by the Discovery Labs, Department of Computer Science, Amrita Vishwa Vidyapeetham, Mysuru Campus.

References

1. Agrawal S, Vaval S, Chawla V, Agrawal MN, Namdev MK, Smart blind helping stick using IoT and android
2. Choudhary S, Bhatia V, Ramkumar K (2020) IoT based navigation system for visually impaired people. In: 2020 8th international conference on reliability, infocom technologies and optimization (trends and future directions) (ICRITO). IEEE, pp 521–525
3. Croce D, Giarre L, Pascucci F, Tinnirello I, Galioto GE, Garlisi D, Valvo AL (2019) An indoor and outdoor navigation system for visually impaired people. IEEE Access 7:170406–170418
4. Karmel A, Sharma A, Garg D et al (2019) IoT based assistive device for deaf, dumb and blind people. Procedia Comput Sci 165:259–269
5. Khade S, Dandawate YH (2016) Hardware implementation of obstacle detection for assisting visually impaired people in an unfamiliar environment by using Raspberry Pi. In: International conference on smart trends for information technology and computer communications. Springer, pp 889–895
6. Krishnapriya K, Anand S, Sinha S (2019) A customised approach for reducing energy consumption in wireless sensor network. Int J Innov Technol Explor Eng (IJITEE)
7. Kunta V, Tuniki C, Sairam U (2020) Multi-functional blind stick for visually impaired people. In: 2020 5th international conference on communication and electronics systems (ICCES). IEEE, pp 895–899
8. Mala NS, Thushara SS, Subbiah S (2017) Navigation gadget for visually impaired based on IoT. In: 2017 2nd international conference on computing and communications technologies (ICCCT). IEEE, pp 334–338
9. Mehta U, Alim M, Kumar S (2017) Smart path guidance mobile aid for visually disabled persons. Procedia Comput Sci 105:52–56
10. Mukhopadhyay A, Nagashree M, Amrutha S, Vedavathi S (2021) In-hospital navigation with audio visual guidance for hearing and visually impaired: a GIS assisted venture. In: 2021 Asian conference on innovation in technology (ASIANCON). IEEE, pp 1–6
11. Mukhopadhyay A, Xavier B, Sreekumar S, Suraj M (2018) Real-time ECG monitoring over multi-tiered telemedicine environment using firebase. In: 2018 international conference on advances in computing, communications and informatics (ICACCI), pp 631–637. https://doi.org/10.1109/ICACCI.2018.8554736
12. Othman NA, Aydin I (2017) A new IoT combined body detection of people by using computer vision for security application. In: 2017 9th international conference on computational intelligence and communication networks (CICN). IEEE, pp 108–112
13. Othman NA, Aydin I (2018) A face recognition method in the internet of things for security applications in smart homes and cities. In: 2018 6th international Istanbul smart grids and cities congress and fair (ICSG). IEEE, pp 20–24

14. Parween S, Manjhi P, Sinha S (2018) Design of automated irrigation system using zigbee
15. Rahman MA, Sadi MS (2021) IoT enabled automated object recognition for the visually impaired. In: Computer methods and programs in biomedicine update, p 100015
16. Rahman MW, Tashfia SS, Islam R, Hasan MM, Sultan SI, Mia S, Rahman MM (2021) The architectural design of smart blind assistant using IoT with deep learning paradigm. Internet of Things 13:100344
17. Rodrigo-Salazar L, Gonzalez-Carrasco I, Garcia-Ramirez AR (2021) An IoT-based contribution to improve mobility of the visually impaired in smart cities. Computing 103(6):1233–1254
18. Saquib Z, Murari V, Bhargav SN (2017) Blindar: An invisible eye for the blind people. In: IEEE international conference on recent trends in electronics information communication technology
19. SathyaNarayanan E, Nithin B, Vidhyasagar P et al (2016) IoT based smart walking cane for typhlotic with voice assistance. In: 2016 online international conference on green engineering and technologies (IC-GET). IEEE, pp 1–6
20. Scalise L, Primiani VM, Russo P, Shahu D, Di Mattia V, De Leo A, Cerri G (2012) Experimental investigation of electromagnetic obstacle detection for visually impaired users: a comparison with ultrasonic sensing. IEEE Trans Instrum Measur 61(11):3047–3057
21. Sharma T, Apoorva J, Lakshmanan R, Gogia P, Kondapaka M (2016) Navi: navigation aid for the visually impaired. In: 2016 international conference on computing, communication and automation (ICCCA). IEEE, pp 971–976
22. Sharmila V, Paul NR, Ezhumalai P, Reetha S, Kumar SN (2020) IoT enabled smart assistance system using face detection and recognition for visually challenged people. Mater Today Proc
23. Singh B, Kapoor M (2021) A framework for the generation of obstacle data for the study of obstacle detection by ultrasonic sensors. IEEE Sens J 21(7):9475–9483
24. Sirumalla M (2021) Ultrasonic distance detector using arduino. Available at SSRN 3918137
25. Subbiah S, Ramya S, Krishna GP, Nayagam S (2019) Smart cane for visually impaired based on IoT. In: 2019 3rd international conference on computing and communications technologies (ICCCT). IEEE, pp 50–53
26. Varalakshmi I, Kumarakrishnan S (2019) Navigation system for the visually challenged using internet of things. In: 2019 IEEE international conference on system, computation, automation and networking (ICSCAN). IEEE, pp 1–4
27. Vasanth K, Macharla M, Varatharajan R (2019) A self assistive device for deaf & blind people using IoT. J Med Syst 43(4):1–8
28. Vidhya D, Rebelo DP, D'Silva C, Fernandes LW, Costa C (2016) Obstacle detection using ultrasonic sensors. IJIRST—Int J Innov Res Sci Technol 2(11)

Application of Quantum-Based K-Nearest Neighbors Algorithm in Optical Fiber Classification

H. B. Ramesh and Kaustav Bhowmick

Abstract Due to the various features of fiber-optic cables, the fields of use vary from one type of cable to another. Therefore, identifying the cable is important before its use. Aging or other factors can make cables inaccessible, such as underground cables that can lose their markings or have torn or worn labels. By analyzing the behaviors of a particular fiber type, it is possible to classify using machine learning algorithms such as K-Nearest Neighbors (KNN). Meanwhile, the possibilities of quantum computing are still being explored in terms of advantages and limitations. The idea of making the best use of both machine learning and quantum computing is the foundation for this work. So, the problem statement of the current project was to first implement the classical KNN algorithm and then develop an efficient KNN algorithm with the support of quantum technology. Finally, the classical and quantum-based KNN algorithms are compared in terms of time complexity and prediction accuracy. During the project, it was found that the quantum KNN algorithm can achieve the same accuracy as conventional algorithms, along with reducing the complexity, i.e., classical version has log complexity whereas quantum version has quadratic complexity.

Keywords Optical fiber · Cable classification · Machine learning · Quantum computing · KNN · Quantum KNN

1 Introduction

Fiber-optics have revolutionized the telecommunication industry and enabled efficient data transmission. While both coaxial (coax) and optical fiber are guided mediums for mode of transmission, the optical fibers are highly efficient, lightweight, have high noise immunity and better transmission speeds compared to coax

H. B. Ramesh · K. Bhowmick (✉)
PES University/ECE, Bangalore, India
e-mail: kaustavbhowmick@pes.edu

S. Sharma et al. (eds.), *Intelligent Control, Robotics, and Industrial Automation*, Lecture Notes in Electrical Engineering 1066,
https://doi.org/10.1007/978-981-99-4634-1_71

cables. While each variety of cable vary in its applications and properties, identification of cable plays an important role in cable maintenance and error detection. Demonstrated implementation of the K-Nearest Neighbor (KNN) algorithm [1] for identifying coaxial cables using S-parameters based on frequencies, was proposed. Following a similar approach by enabling quantum computing advantages, in-order to classify fiber-optic cables, definitely pave a way for many research applications. In a previously reported research work, a quantum KNN (Q-KNN) algorithm-based Image classification was demonstrated [2–4]. On a classical computer, therein [2], the authors have extracted feature vectors from images, followed by putting them into a quantum superposition state. Finally, the image is classified by quantum measurement. By considering the above-cited works, classification of optical fibers using KNN, based on Attenuation versus Wavelength characteristics, was performed. The coaxial cable classification based on the conventional KNN algorithm was kept as a benchmark to set up the classification problem for optical fibers. Based on the shortcomings observed such as linearly increasing complexity along with dataset size and requirement of lot of measurements in test set, the proposed Q-KNN would be developed for application in the optical fiber classification problem, to compare the efficiency of the algorithm. The Image classification-based Q-KNN algorithm [5] was considered as a reference for the quantum KNN algorithm for classifying fiber-optic cables.

In this paper, a new approach for optical fiber classification is proposed which is based on quantum machine learning. Various classes of 10 different optical fiber classes are considered and dataset was prepared for training and testing of ML algorithm using attenuation versus wavelength characteristics. The structure of this paper is as follows. Firstly, the scheme and algorithm of the work is explained in Sect. 3. After which the datasets are prepared and verified for coax and optical fiber (Sect. 3.1, 3.2, 3.3), followed it by testing out KNN on coax dataset, then implementing classical KNN for optical fiber (Sect. 3.4). Finally, implementing Q-KNN for fiber-optic dataset (Sect. 4) and concluding results (Sect. 5) based on complexity, accuracy, and future scope (Sect. 6).

2 Related Works

With the fast development of fiber-optic communication technology, it has emerged as an essential platform for various information including voice, data, and pictures. Maintenance of cable includes proper identification of fiber type and its characteristics [6]. Classifying an optical fiber can depend upon the label printed on it or the datasheet provided by the manufacturer. But, cables like underground cables, because of usage over a lot of time or other factors, their label might also vanish or be out of reach [1]. With the help of K-L transform, Zhou et al. proposed Q-KNN algorithm based on K-L transform for image classification with which they were able to prove theoretical advantage of quantum over classical KNN. The application of quantum machine learning [7, 8] in classification is surveyed by Abohashima et al. [3]. They elaborated on the number of technical advances, similarities, and strengths of QML research work. By using a swap test [9] circuit for calculating fidelity between states and using Durr's modified algorithm, they were able to classify quantum states without explicitly describing the classical basis.

Ning Yang explored the simulation of KNN based on quantum information [10] for text classification. By running Grover's search algorithm for a particular number of times, we can iteratively come to the minimum element in a given set. Hence using this in image classification gave big speed up for the work of Dang et al. [5] and Li et al. [11]. In natural language processing (NLP), text classification is one of the basic tasks. It is the process of assigning tags or categories to text based on its content. By using the IBM Qubit toolkit, therein [5], the author simulated circles of the algorithm which are nothing but the swap test circuits composed of several quantum gates, and passed the test. Schuld et al. [12] gave a detailed study of existing ideas and approaches to quantum machine learning (QML). Sun et al. by exploiting the idea of fiber sensor along with phase modulation technology [13] and accomplished fiber-optical cable identification, but have not applied any automated ML algorithms to cable identification, which turns out as a future scope for that work. Bader et al. proposed a work [1] presenting the application of a machine learning-based algorithm to identify a coaxial cable based on its properties. KNN [14] is a non-parametric ML technique for classification and regression-based algorithms. Whereas in quantum computing, an algorithm known as Durr's minimum search algorithm [15], is based upon Grover's unsorted search algorithm to find a minimum element in a set, can be used with dataset even without sorting.

3 Proposed Method

The presented Q-KNN algorithm is an enhanced version of classical KNN algorithm which uses quantum subroutines in-order to achieve much better performance [5]. This algorithm can be viewed as a combination of classical and quantum procedures. Figure 1 presents the overview of the work done for application of Q-KNN in optical fiber classification, adopted from the coax cable work [1] for benchmarking.

As we can see from Fig. 1, the first step was to extract the COAX dataset that was used in paper [1] so that it will be possible to train and test the functionality of the classical KNN algorithm on it. In work [1], spectral measurements of the coaxial cable in terms of the "input port reflection's magnitude" | S11 | with Nano Vector Network Analyzer (VNA) in the evenly spread frequency range between 50 kHz and 900 MHz, were performed. The k-nearest associates (KNN) algorithm is one of the non-parametric ML approaches used for classification or regression and is used here for cable identification.

Once the KNN algorithm was correctly applied for the coax dataset, then the following step was to prepare dataset for optical fiber. With the developed optical fiber dataset, the next step was to divide the entire dataset into 70% training set and 30% testing set. Afterward, the classical KNN and quantum KNN are tested on the

Fig. 1 Flow diagram of current work

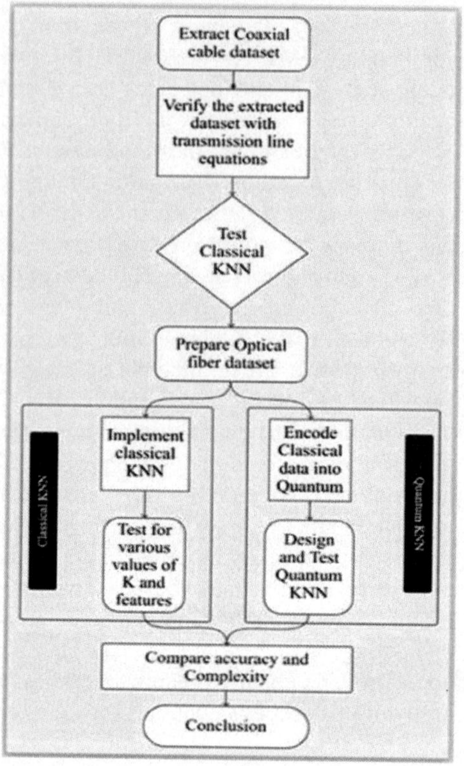

developed dataset. The aforesaid conventional KNN algorithm was verified for its functionality and correct predictions were recorded.

The Q-KNN version of the KNN algorithm was customized to the fiber-optic dataset and applied to the fiber-optic classification. The first step in Q-KNN was to prepare the quantum state by extracting features from the training and test datasets and encoding them into quantum states. Q-KNN uses quantum superposition states to compute similarities in parallel. Using Durr's minimum search algorithm, the nearest neighbors were found. Compared with the existing algorithms, the quantum scheme has a significant increase in efficiency, providing excellent classification performance.

3.1 Coaxial Cable Dataset Preparation

The identification of the coaxial communication cable [1] was based on the scattering parameters of the amplitude S11 against the frequency characteristic. 10 coaxial communication cables, cables of different lengths, sizes, and types of connectors are

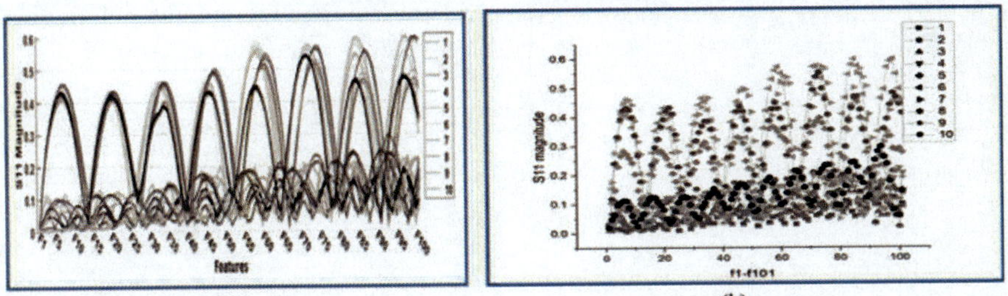

(a) (b)

Fig. 2 S11 Magnitude versus frequency features **a** original plot from paper [1], **b** extracted dataset plot in this work from [1]

taken, and the length, type, and connector of the cable are considered as a unique class. First, we can apply the algorithm to this dataset, and then if the algorithm works correctly the same can be used for the fiber-optic dataset. The dataset of coaxial cable has been prepared by extracting frequency versus s11 magnitude plot to verify correctness of previously implemented work (Fig. 2a and b). Also, dataset was verified by reproducing a cable's feature using RLGC transmission line equations [16].

Figure 2a and b, the frequency versus s11 magnitude plot is plotted. It was verified that the extracted dataset follows similar patterns as it was in the original work from [1].

3.2 *Testing of Classical KNN Algorithm for Coax Dataset*

To implement the KNN algorithm for the COAX dataset, the KNeighborsClassifier function from the **sklearn** library was used. A confusion matrix was plotted for the prediction class versus the true class. Figure 3 shows the confusion matrix plot of KNN outputs. Figure 3a shows that the authors of [1] were able to achieve almost 99% accuracy, and the Fig. 3b shows the output from our prepared dataset.

3.3 *Preparation of Optical Fiber Dataset*

Similar to the data set for coax, a dataset for the optical fibers was required to classify them using the KNN algorithm. Optical fiber identification was based on the attenuation characteristics versus wavelength of different optical fibers (can be seen from Table 1).

Table 1 list the various optical fiber classes that were considered for our present work. The plots for all the different fibers are traced by using WebPlotDigitizer software tool and the dataset is prepared, of which the plot can be seen from Fig. 4.

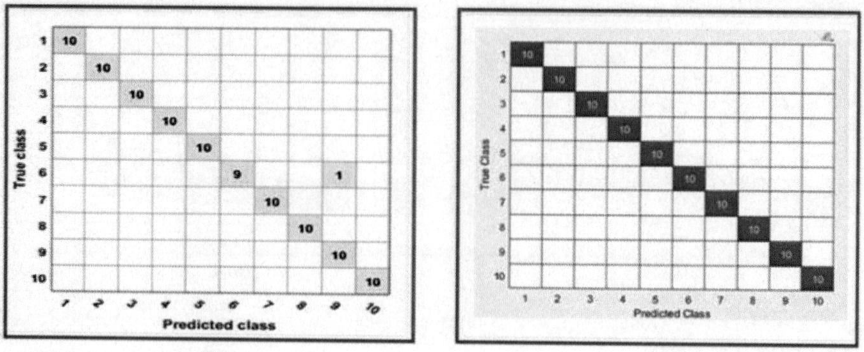

(a) (b)

Fig. 3 **a** Confusion matrix from paper [1] **b** confusion matrix of our implementation

Table 1 Different optical fiber classes used for database generation

Serial No	Class Name
1	25um 0.10NA MMF
2	50um 0.22NA MMF
3	200um 0.22NA MMF
4	smf 28 e
5	smf 28
6	150um 0.39NA MMF
7	200um 0.39NA MMF
8	200um 0.50NA MMF
9	105um 0.22NA MMF
10	105um 0.10NA MMF

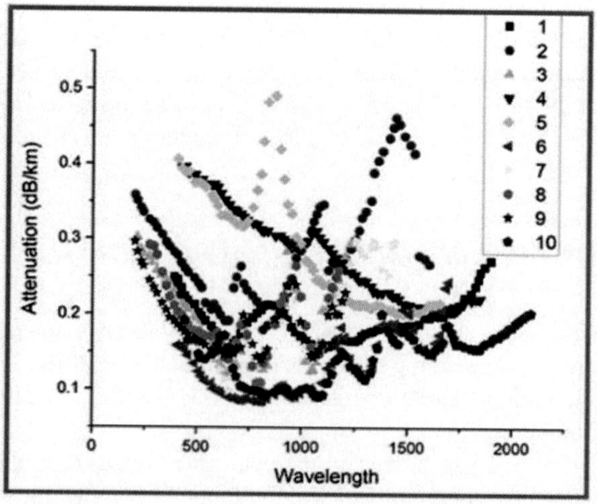

Fig. 4 Attenuation versus wavelength plot for various optical fibers

3.4 *Testing of Classical KNN Algorithm for Optical Fiber Dataset*

The complete optical data set is divided into 70% training set and 30% test set, then using the KNeighborsClassifier function, the KNN algorithm is implemented. From the Fig. 5a, the accuracy of conventional KNN algorithm can be seen in terms of predicted class versus true class confusion matrix. The dataset considered was 400 × 101 out of which 30% of dataset was selected as testing dataset and used to implement KNN. As the dataset is just an interpolated extension of dummy dataset generated using the basic attenuation versus wavelength optical fiber data from Thorlabs.com (https://www.thorlabs.com/), along with a subsequent verification with standard optical equations, a 100% classification accuracy for a K value of 3 can be observed. Whereas from Fig. 5b, it can be seen that the classification accuracy decreases as the value of K increases.

4 Quantum KNN

On a classical computer, a bit can have two different values 0 or 1. It's discrete in nature. In quantum computing, a "quantum bit or qubit is the basic unit of quantum information", and the behavior of its circuit is based on quantum mechanics. Qubits can have the values 0 or 1 or a superposition of 0 and 1. The qubit resides in the Hilbert vector space with the two elements indicated as the base, Indicated as $|0\rangle$ and $|1\rangle$. A pure qubit state is a coherent superposition of the basis states. Hence, a single qubit can be described by a linear combination of $|0\rangle$ and $|1\rangle$, i.e.,

$$|\Psi\rangle = \alpha|0\rangle + \beta|1\rangle \tag{1}$$

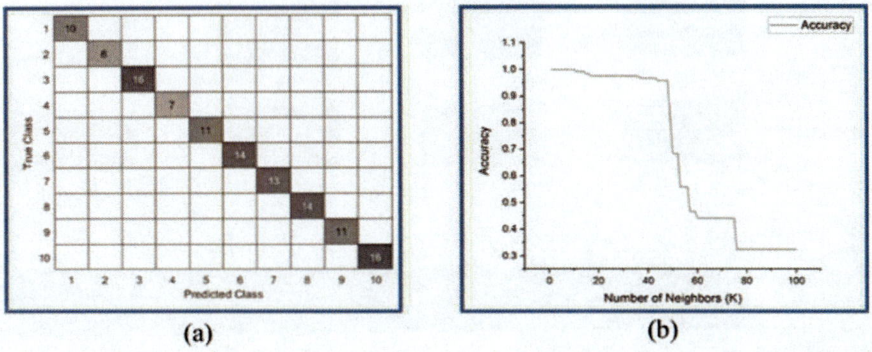

Fig. 5 **a** Confusion matrix of optic dataset of 400 × 101, **b** accuracy plot for various value of K for 400 × 101 dataset

where α and β are the probability amplitudes, that are both complex numbers. Hadamard (H) gate, Pauli-X (X) gate, Controlled Not (CNOT), SWAP, CCNOT, Pauli-Y (Y), etc. are few quantum gates [17] that can modify qubits.

4.1 Storing the Dataset

As the primary step in a KNN algorithm, it was needed to have data stored to operate upon it and predict the output. In a classical way, storing of the data in data and memory registers is needed. In quantum computation, there is no possibility of storing the data [17], so preparation of quantum registers for the values of optical fiber features for various classes from the dataset was required. There are various methods of storing the data in quantum registers [17]. All data provided to the quantum circuit is stored in the qubits of the circuit. This is done by encoding the data. Among various encoding formats, such as binary encoding, angle encoding, amplitude encoding. Using encoding methods, such as AMPLITUDE ENCODING, Fig. 6 which use quantum superposition to reduce the amount of storage required to store the dataset. In amplitude encoding, by modifying wave function $|\psi\rangle$ of quantum bits, which also determines the measurement probabilities, the data is encoded [17]. Data is encoded as quantum state amplitude Eq. (1).

Figure 6 is an example of encoding classical data using amplitude encoding, where first we will normalize the values, and then store them into respective amplitudes of various qubit combinations. With the current qubits limitation of qasm_simulator of Qiskit, 32 qubits are available. So, it is possible to encode a maximum of 2^15 features per class of dataset. Because, 2^15 for training one, 2^15 for testing one, and one ancilla bit for measuring fidelity. This means that for storing 32,768 features, 32,768 classical bits are required. But, in the quantum version, only 15 qubits are

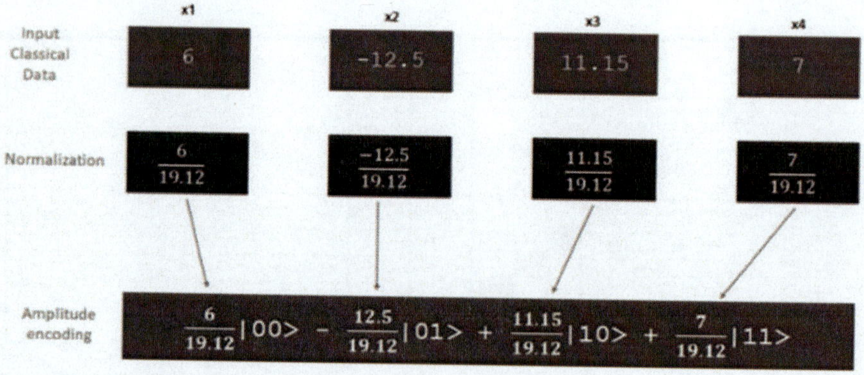

Fig. 6 Example of amplitude encoding highlighting each step

needed. By creating a quantum circuit for log (n) qubits and then calling initialize function, encoding of dataset is performed.

4.2 Calculation of Distance Between Training and Testing Classes

The KNN algorithm classification is based upon the distance metric between training and testing classes. It is a vital step to measure the distance between a testing element with each training element to find the nearest neighbors. In the quantum version, a special circuit known as a swap test circuit was used which helps to calculate fidelity between 2 quantum circuits Fig. 7 [1].

When the auxiliary qubit $|0\rangle$ is passed through first Hadamard gate, it is mapped to: $\frac{1}{\sqrt{2}}(|0\rangle + |1\rangle)$.

For a controlled swap gate, this superposition acts as a control qubit to get:

$$\frac{1}{\sqrt{2}}(|0\rangle\,|\Psi\rangle\,|\varphi\rangle + |1\rangle|\Psi\rangle|\varphi\rangle) \tag{2}$$

After passing through final H gate, ancilla bit becomes:

$$\frac{1}{\sqrt{2}}(|0\rangle\,|\Psi\rangle\,|\varphi\rangle + |\varphi\rangle|\Psi\rangle)\frac{1}{2}(|1\rangle\,|\Psi\rangle\,|\varphi\rangle - |\varphi\rangle|\Psi\rangle) \tag{3}$$

Finally, the output probability of being 0 is defined by:

$$P\,(0) = \frac{1}{2} + \frac{1}{2}\langle\Psi \mid \varphi\rangle^2 \tag{4}$$

And, output probability of being 1 is defined by:

$$P\,(0) = \frac{1}{2} - \frac{1}{2}\langle\Psi \mid \varphi\rangle^2 \tag{5}$$

If the P (0) value is very high, it indicates that the two quantum states have more fidelity, meaning they have high similarities between them, which is exactly what is needed here to classify nearest neighbors in Q-KNN [1].

Fig. 7 Swap test circuit

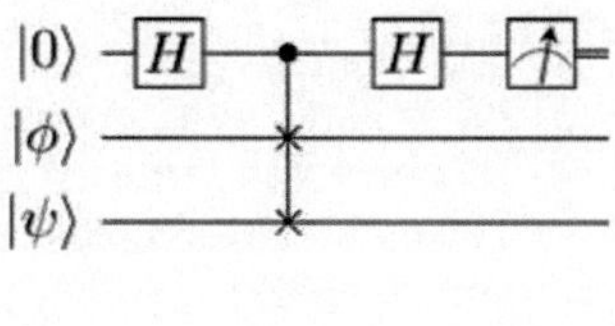

4.3 Majority Voting, Classification, and Prediction

The final step in the KNN algorithm was to perform a majority vote on all the nearest neighbors of the test class. Following the previous swap test-based accuracy measurement procedure, a list of all fidelities between each test class and all training classes was obtained. Now, requirement was to find the K number of maximum fidelities, i.e., nearest neighbors for each class. Based on the majority voting, the output class label can be predicted for the test dataset. To do the majority voting in a classical way, sorting of all the fidelities and then select the K number of classes that have maximum fidelity was needed. Whereas in quantum, the cost of sorting the entire list of values can be avoided by implementing a quantum-based minimum search algorithm.

Quantum computers have an advantage over conventional computers in terms of their excellent database retrieval speed. Grover's algorithm displays this possibility by using amplitude amplification technique. By speeding up an unstructured search problem quadratically, it can be used as an acceleration function in many applications. Durr's algorithm starts by searching a minimum element in a given list by first selecting a threshold index randomly from the given list. Then it follows Grover's searching algorithm [18] for a specified number of times by every time replacing the chosen threshold with the latest obtained minimum value from Grover's search. Finally, after all the iterations, the minimum value in a given list was obtained. For the current work, a modified version was used in which we find the maximum element because elements with maximum fidelity are nearest neighbors to a specified test class [15].

5 Comparison of Accuracy and Complexity of KNN versus Q-KNN

Implementations of both algorithms are performed on the same optical fiber datasets to compare the accuracy and complexity of the traditional KNN algorithm and the quantum KNN algorithm. To verify the correctness of the Q-KNN algorithm, the algorithm is also tested on the Iris dataset. By running KNN and quantum KNN algorithms, the accuracy found was, without any kind of data pre-processing KNN have 100% accuracy for most values of K, so with pre-processing, the accuracy was 100% lower value of K and then decreased gradually as K increases. Whereas quantum KNN didn't have much good accuracy if pre-processing is done, i.e., only 93.33% accuracy, but if pre-processing is not done, it has 100% accuracy for small K values and then decreases gradually with increase in K value.

The detailed comparison between classical and quantum KNN accuracy is provided below in terms of 2 key ways. **Confusion matrix** and **plotting of accuracy** versus K values. In first way of creating a confusion matrix of specific K values for both KNN and quantum KNN, the algorithms are run for different values of K

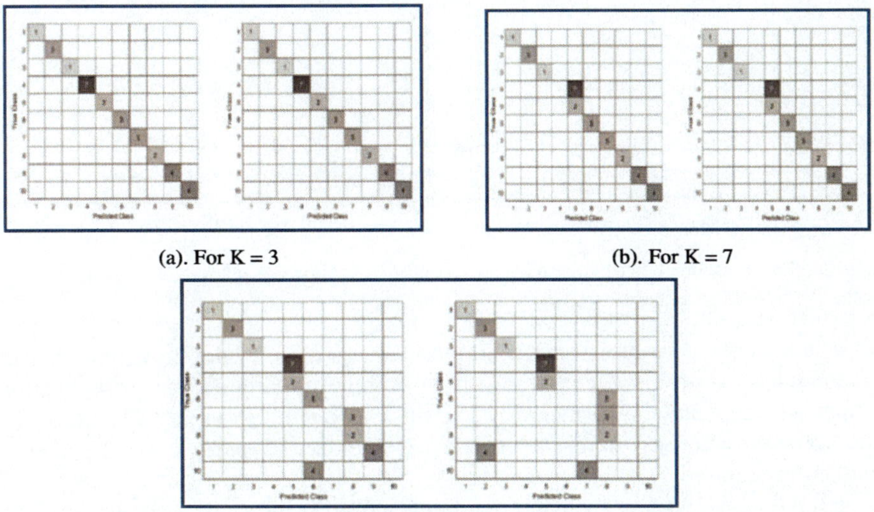

(a). For K = 3

(b). For K = 7

(c). For K = 15

Fig. 8 Confusion matrix of optic dataset of 100 × 8 for classical (left) versus quantum KNN (right)

and their predictions are compared with true class labels. Figure 8 shows that for a particular value of K and dataset size, both classical and quantum KNN algorithms displays similar results in confusion matrix. For a K value of 3, both classical and quantum KNN has 100% classification accuracy (Fig. 8a), as is expected [19], that K values are usually chosen by optimization, where extreme values of K reduce accuracy. For a K value of 7 and 15, both classical and quantum KNN has reduced classification accuracy (Fig. 8b and c) so that the confusion matrix shows similar wrong predictions.

The second way was to plot a graph of various values of K and accuracy. This helps to analyze the performance of both algorithms in an efficient way. The dataset sizes considered were 400 × 8 and 100 × 8. By analyzing both confusion matrix and accuracy plots from Fig. 9, the conclusion drawn was that the quantum KNN was able to achieve accuracies comparable to the classical version of KNN. It was clear that the rate of fall in terms of accuracy was quite high in Q-KNN with respect to KNN. And both the algorithms tend to lose accuracy at very high K values and perform very well at smaller K values.

The classical KNN algorithm and the quantum KNN algorithm were compared in terms of BIG-O notation [20]. It is one of the basic techniques for analyzing the cost of an algorithm. The limiting behavior of a function when its arguments tend to be specific values or infinity is described by a mathematical notation in this technique. O(n) has the linear time complexity, O(logn) has lesser complexity than O(n), and as we go from O(nlogn), O(n!) the complexity increases exponentially.

Compared to the classical KNN algorithm, quantum KNN has reduced complexity in terms of storing the dataset, distance measurement, and majority voting. In terms

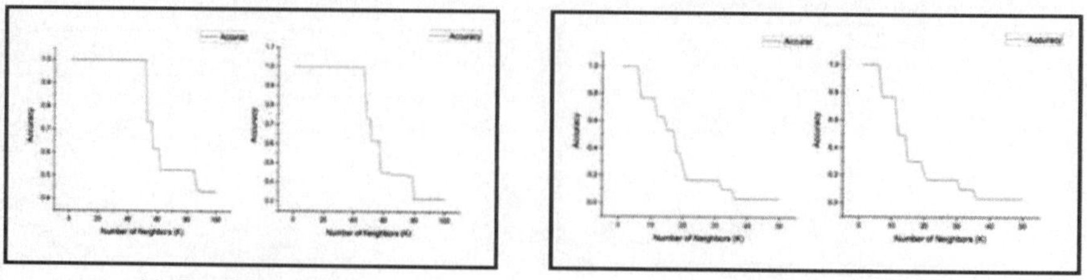

(a). For 400x8 dataset (b). For 100x8 dataset

Fig. 9 Plots of accuracy versus K values for classical (left) versus quantum KNN (right)

of data encoding, the classical KNN algorithm has no reduction in the number of bits required to store the dataset. Storing the entire training dataset for prediction requires a large amount of memory. The more size the dataset has, the worse the storage requirements. Whereas to encode a 2^n number of values, only n number of qubits are required.

For n dataset features, there is a requirement of a for loop n times to compute the distance. This will result in O(n) time complexity. Whereas in the quantum version, since n number of features are encoded into log2(n) qubits, it will greatly reduce the complexity. Only 2 H gates, O (1) complexity for each, and log2(n) of c-swap gates with O (logn) complexity can be used, which makes the overall complexity of O(logn) Which represents the swap test circuit containing CNOT and H gates. Hence, the observation was that the complexity of distance measurement was decreased in quantum version.

In the last step, i.e., majority voting, in terms of classical KNN sorting of the distances was required which need O(n) complexity. A majority vote was then cast based on the number of K-nearest neighbors. So, the total complexity was – O(n*logk). The quantum version uses a search based on Grover's algorithm (Durr's algorithm) to find the element with maximum fidelity with complexity O(sqrt(n)). It then votes for a majority based on the K nearest labels. So, the total complexity was found to be O(sqrt(n)*logk)[5].

Figure 10a, b and c are the plots of various of K for majority voting complexity between classical and quantum versions of KNN, and the observation was clear that the quantum computing was really advantageous as the size of dataset scales up. While the complexity is increasing linearly for classical version, for quantum version, the increase in complexity was very negligible. But, for very small dataset, the complexity was quite comparable with classical KNN.

6 Conclusion and Future Scope

In the present work involving classification of optical fiber classes/types using an efficient quantum KNN (Q-KNN) algorithm, a reduced time complexity compared to the classical counterpart was demonstrated. Specifically, time complexity of

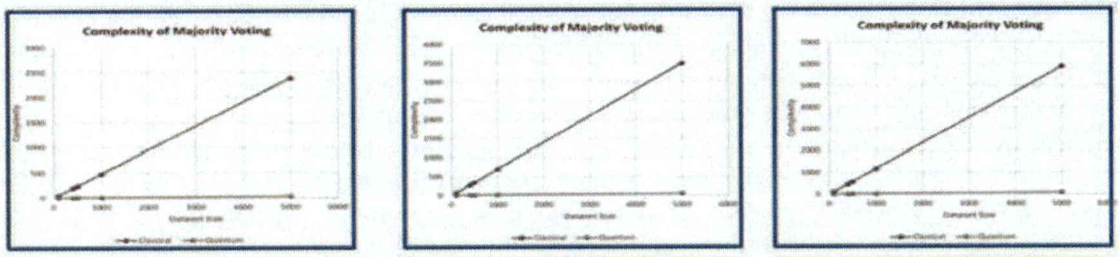

(a) (b) (c)

Fig. 10 Plots for the complexity of classical versus quantum for **a** K = 3, **b** K = 5 and **c** K = 15

O(sqrt(n)*logk) was obtained for Q-KNN as opposed to O(n*logk) in the case of Classical KNN. For accuracies in prediction similar to Classical KNN, with dataset sizes of the scale ~ 400 × 101, smaller datasets in the scale ~ 400 × 8 and 100 × 8 was found to be enough in case of Q-KNN. Such accurate predictions and simple implementations encourages Q-KNN to be a better choice in case of optical fiber classification where measurements can be delicate. However, current limitations in the availability of number of Q-bits posed a hindrance to processing a sufficiently large dataset at one run. With quantum computers and simulation platforms becoming robust in the future, in terms of number of Q-bits, many applications involving huge datasets and big data will be possible.

Classification algorithms and sorting algorithms in their quantum forms show a promise of better efficiency. Applications involving delicate handling and measurement, in case of optical fibers for example, will benefit greatly in the future with the reduction in establishment cost and increased Q-bit capacity in quantum computers to come.

References

1. Bader O, Haddad D, Kallel AY, Amara NEB, Kanoun O (2021) Identification of communication cables based on S-parameters and K-nearest neighbors algorithm. In: 2021 18th International multi-conference on systems, signals and devices (SSD). IEEE, pp 808–811. https://doi.org/10.1109/SSD52085.2021.9429367
2. Zhou N-R, Liu X-X, Chen Y-L, Du N-S (2021) Quantum K-nearest-neighbor image classification algorithm based on K-L Transform. Int J Theor Phys 60:1209–1224. https://doi.org/10.1007/s10773-021-04747-7
3. Abohashima Z, Elhosen M, Houssein EH, Mohamed WM (2020) Classification with quantum machine learning: a survey. arXiv:2006.12270, arXiv:2006.12270v1, https://doi.org/10.48550/arXiv.2006.12270
4. Ruan Y, Xue X, Liu H et al (2017) Quantum algorithm for K-nearest neighbors classification based on the metric of hamming distance. Int J Theor Phys 56:3496–3507. https://doi.org/10.1007/s10773-017-3514-4
5. Dang Y, Jiang N, Hu H et al (2018) Image classification based on quantum K-nearest-neighbor algorithm. Quantum Inf Process 17(239):1–18. https://doi.org/10.1007/s11128-018-2004-9
6. Dutta A (2015) Mode analysis of different step index optical fibers at 1064nm for high power fiber laser and amplifier. 6(3):74–77. Retrieved from https://osf.io/5bja7/download

7. Khan TM, Robles-Kelly A (2020) Machine learning: quantum versus classical. IEEE Access 8:219275–219294. https://doi.org/10.1109/ACCESS.2020.3041719
8. Kopczyk D (2018) Quantum machine learning for data scientists. arXiv:1804.10068, arXiv:1804.10068v1, https://doi.org/10.48550/arXiv.1804.10068
9. Basheer A, Afham A, Goyal SK (2020) Quantum *k*-nearest neighbors algorithm. arXiv:2003.09187, arXiv:2003.09187v3, https://doi.org/10.48550/arXiv.2003.09187
10. Yang N (2019) KNN algorithm simulation based on quantum information. In: Proceedings of the student-faculty research day conference, CSIS, Pace University, pp 1–6
11. Li J, Lin S, Yu K et al (2022) Quantum K-nearest neighbor classification algorithm based on Hamming distance. Quantum Inf Process 21(18):1–17. https://doi.org/10.1007/s11128-021-03361-0
12. Schuld M, Sinayskiy I, Petruccione F (2015) An introduction to quantum machine learning. Contemp Phys 56(2):172–185. https://doi.org/10.1080/00107514.2014.964942
13. Sun Q, Wu Q (2014) Research of cable identification method based on single fiber. In: Proceedings of the 5th international conference on optical communication systems (OPTICS-2014), pp 45–50. https://doi.org/10.5220/0005022500450050
14. Moldagulova A, Sulaiman RB (2017) Using KNN algorithm for classification of textual documents. In: 2017 8th International conference on information technology (ICIT). IEEE, pp 665–671. https://doi.org/10.1109/ICITECH.2017.8079924
15. Durr C, Hoyer P (1996) A quantum algorithm for finding the minimum. arXiv:quant-ph/9607014, arXiv:quant-ph/9607014v2, https://doi.org/10.48550/arXiv.quant-ph/9607014
16. Zhang J, Koledintseva MY, Drewniak JL, Antonini G, Orlandi A (2004) Extracting R, L, G, C parameters of dispersive planar transmission lines from measured S-parameters using a genetic algorithm. In: 2004 International symposium on electromagnetic compatibility (IEEE Cat. No.04CH37559), vol 2. IEEE, pp. 572-576. https://doi.org/10.1109/ISEMC.2004.1349861
17. Cortese JA, Braje TM (2018) Loading classical data into a quantum computer. arXiv:1803.01958, arXiv:1803.01958v1, https://doi.org/10.48550/arXiv.1803.01958
18. Grover LK (1996) A fast quantum mechanical algorithm for database search. In: Proceedings of the twenty-eighth annual ACM symposium on theory of computing (STOC'96), pp 212–219. https://doi.org/10.1145/237814.237866
19. Lubis Z (2020) Optimization of K value at the K-NN algorithm in clustering using the expectation maximization algorithm. IOP Conf Ser Mater Sci Eng 725:012133-1–012133-13. https://doi.org/10.1088/1757-899X/725/1/012133
20. Chivers I, Sleightholme J (2015) An introduction to algorithms and the big O notation. In: Introduction to programming with FORTRAN. Springer, Cham, pp 359–364. https://doi.org/10.1007/978-3-319-17701-4_23

Application of Artificial Neural Network for Successful Prediction of Lower Limb Dynamics and Improvement in the Mathematical Representation of Knee Dynamics in Human Locomotion

Sithara Mary Sunny, K. S. Sivanandan, Arun P. Parameswaran, T. Baiju, and N. Shyamasunder Bhat

Abstract The motivating factor behind this research is the significance and demand for developing relatively affordable yet effective assistive technologies for disabled people. In the presented work, an intelligent model that predicts the lower limb joint angles for an entire gait cycle and a model that modifies the constant terms of the average value-based modeling representation of the knee dynamics are both developed through the application of artificial neural networks (ANN). Hip and knee joint angles were predicted using a model with ground reaction force (GRF) and joint angle being input parameters. The coefficients of correlation and determination values of the model were found to be close to the ideal value, while the mean square error value was determined to be within the tolerance limit. In another developed model, the linear displacement of a human gait cycle was predicted based on the inputs like angular displacement, velocity, and acceleration of the hip and knee joints. The average value-based modeling representation of the knee dynamics was more accurately represented after obtaining the modified values of the constant terms (C_0, C_1, C_2). The resulting models can be used to design and develop assistive technologies for physically disabled people, thereby enabling their reintegration into society and helping them to lead normal lives.

S. M. Sunny · A. P. Parameswaran (✉)
Department of Electrical & Electronics Engineering, Manipal Institute of Technology, Manipal Academy of Higher Education, Manipal, Karnataka 576104, India
e-mail: arun.p@manipal.edu

K. S. Sivanandan
Department of Biomedical Engineering, Manipal Institute of Technology, Manipal Academy of Higher Education, Manipal, Karnataka 576104, India

T. Baiju
Department of Mathematics, Manipal Institute of Technology, Manipal Academy of Higher Education, Manipal, Karnataka 576104, India

N. Shyamasunder Bhat
Department of Orthopaedics, Kasturba Medical College, Manipal Academy of Higher Education, Manipal, Karnataka 576104, India

S. Sharma et al. (eds.), *Intelligent Control, Robotics, and Industrial Automation*, Lecture Notes in Electrical Engineering 1066,
https://doi.org/10.1007/978-981-99-4634-1_72

Keywords Artificial neural network · Joint angle · Ground reaction force · Average value-based model · Variational terms

1 Introduction

Rehabilitation engineering deals with the development of technologically advanced assistive devices to help people with physical disabilities. The objective of these devices is to meet the absolute need of disabled subjects to perform their daily activities by taking minimal external help. The highly customized character of these devices makes their design highly complex and different from regular engineering systems.

Physical disability is qualitative. The design of the assistive device requires a quantitative understanding of the qualitative nature of the disability, which is accomplished through the use of models. Models aid in the indirect analysis of engineering problems without injuring, deforming, or disrupting the already-existing original system. When a direct evaluation is not feasible or preferred, it is a perfect technique to study or evaluate system characteristics. Mathematical models developed are used to understand the basic principles of natural human motions, and knowledge about the principles of human locomotion is vital for rehabilitation engineering [1, 2].

1.1 Artificial Neural Network (ANN) Modeling

Neural networks are an effort to develop a smart system that can mimic the operation of the human brain to some extent. It can be considered a mathematical model of a biological neuron. Artificial neural networks (ANN) are a type of advanced adaptive learning dynamic system and pattern recognition algorithm that, to some extent, resemble the central nervous system (CNS).

Rehabilitation evaluation is the most important step in the rehabilitation treatment procedure. Physiotherapy study focuses on identifying abnormalities in human locomotion. Different techniques are employed to diagnose lower extremity problems, out of which the most prevalent is gait analysis [3–5]. As gait is the outcome of the synchronous movement of the hip, knee, and ankle joints, a problem in one can affect the other, altering the overall gait pattern [6]. ANN is a versatile nonlinear modeling technique that is highly effective in doing the same. Existing clinical data were used as training data for the ANN in [7], which was used to extract the gait characteristics of patients of varying ages.

Closed trajectories generated by plotting one joint versus another joint angle to represent gait kinematics throughout the gait cycle are termed cyclograms. Cyclograms can provide information regarding joint coordination and joint interaction [8]. Using cyclograms and artificial intelligence, the gait is analyzed over time, and the leg movement is predicted. Cyclograms aid in understanding cyclic processes like walking [9]. Another research work presents cyclogram-based methods for evaluating difficulties during bipedal motion and using the same cyclograms in the control systems of prosthetic actuators or other assistive devices [10]. The hypothesis tested in another work was the prediction of cyclogram tracks using an intelligent system. ANN was taught and trained to predict the future location of the lower leg using cyclograms. The purpose of this research was to see how effectively an ANN predicted the angle of a lower limb joint [11].

1.2 *Average Value-Based Technique*

The ordinary linear differential equation with constant coefficients is the most extensively used mathematical model for examining the dynamic response. This average value-based approach can be applied to functions having an average value. Human locomotion is monitored where the knee position is recorded as a sequence of digital values via the optical technique. These values are sorted and tabulated using time as the basic variable. The displacement vs. time relationship is represented digitally in these tables. These data were used to formulate the mathematical equation using the average value-based approach [12]. The characteristics of human locomotion differ from one individual to the next. As a result, to generalize the characteristics, it is explained on an average basis. Conducting such studies on various people regularly improves the average characteristics.

An nth order linear differential equation with constant coefficients between q_i the input function and q_o, the output function is commonly used to express a system's input–output function.

$$q_{\text{out}} = C_m \frac{\mathrm{d}^m q_0}{\mathrm{d}t^m} + C_{m-1} \frac{d^{m-1} q_0}{\mathrm{d}t^{m-1}} + \cdots + C_1 \frac{\mathrm{d}q_0}{dt} + C_0 q_0, \tag{1}$$

where the constants Cs represent the physical parameters of the system. $C_0 q_0$ is the base term, $C_1 \frac{\mathrm{D}q_0}{\mathrm{D}t}$ is the first variational term, and $C_2 \frac{\mathrm{D}^2 q_0}{\mathrm{D}t^2}$ is the second variational term. In other words, the output can be represented as the sum of the base term and variational terms.

The base value contains a coefficient C_0 multiplied by the variable's average value, the first variational term contains a coefficient C_1 multiplied by the variable's first variational part, and the second variational term contains coefficient C_2 multiplied by the variable's second variational part, and this extends up to infinity. To generalize the model, these coefficients are defined as constant values representing the system's physical properties (in the case of an engineering system). In the case of a biological system, the constant terms refer to both the physical parameters and the psychological components of the biological system. The number of variational terms that must be considered is determined by the level of accuracy sought by the researcher through the study [13].

In this paper, we propose a method for predicting lower limb joint angles using GRF as an input parameter, as well as an effort to improve the mathematical representation of knee dynamics developed using an average value-based modeling technique.

2 Methodology

2.1 Data Collection and Processing

Generally, standardized laboratories are often used to conduct quantitative gait analysis. In well-known laboratories, advanced analytical tools, such as motion capture devices and ground reaction force (GRF) measurement plates, are exclusively available for locomotion analysis and rehabilitation engineering objectives. A healthy volunteer was requested to walk a short distance on the force plate as part of the experiment, and a high-resolution video was taken of the activity. As illustrated in Fig. 1, markers were placed at the hip and knee joints to determine the joint kinematics. To collect and tabulate the kinematic features of the hip and knee joints, the recorded video was processed using the suitable software program[14, 15]. Ground reaction forces were determined from the force plate.

The knee joint angle, its angular velocity, and angular acceleration at a 40ms time interval are tabulated, and Figs. 2, 3, and 4 show the variation of the same. The knee joint is flexible during the swing phase, and hence, the lower leg swings freely during the stance phase, the joint becomes stiff to balance the body [16, 17].

3 Development of Models

The study for the prediction of joint angles was performed using the MATLAB Neural Network Toolbox along with the required python coding. The output of the

Fig. 1 Location of markers on the left limb of the healthy volunteer during experimental studies

ANN was compared to the desired output for each input data set to calculate the error. The weights of the ANN were then adjusted to minimize the error, bringing the neural model closer to generating the desired output with each iteration. This error was then fed back to the NN as feedback (backpropagated). The measure of performance is mean squared error (MSE), and it is calculated as the average of the squared difference between the target and predicted values.

$$\text{MSE} = \frac{1}{n}\sum_{t=1}^{n}(A_t - P_t)^2, \quad (2)$$

where MSE is the Mean Square Error, A_t is the target value, P_t is the anticipated value, and n represent the sample size [18].

The coefficient of correlation is the measure of the relation between two variables. Here, r shows the relation between the target and predicted value.

$$r = \frac{\sum_{t=1}^{n}(A_t - \bar{A}_t)(P_t - \bar{P}_t)}{\sqrt{\sum_{t=1}^{n}(A_t - \bar{A}_t)^2 \sum_{t=1}^{n}(P_t - \bar{P}_t)^2}}, \quad (3)$$

where r is coefficient of correlation, A_t is the target value, P_t is the anticipated value, n represent the sample size, $\bar{A}_t$ and $\bar{P}_t$ are the average of target value and predicted value, respectively.

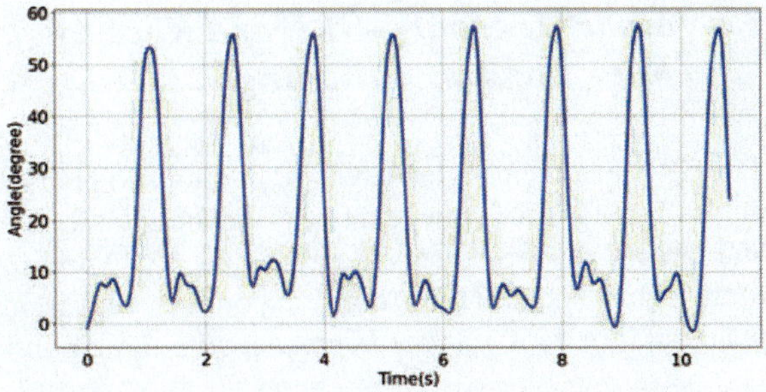

Fig. 2 Variation of knee joint angle with time

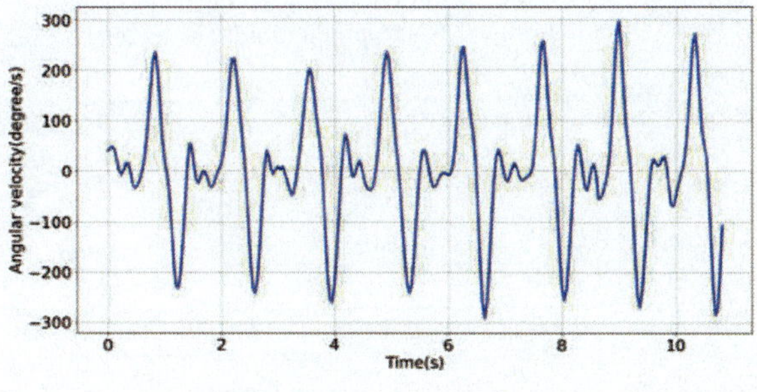

Fig. 3 Variation of knee joint angular velocity with time

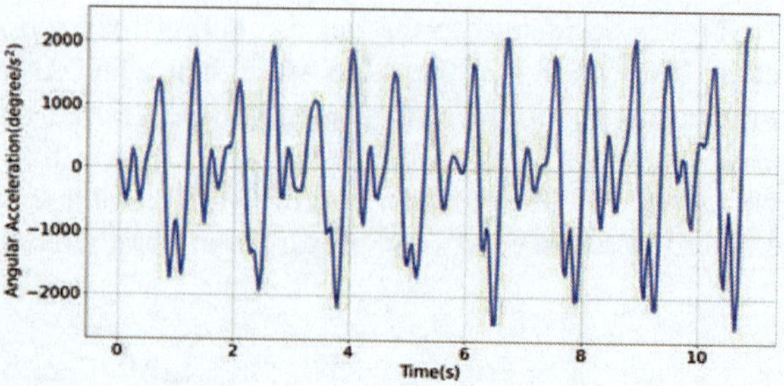

Fig. 4 Variation of knee joint angular acceleration with time

The coefficient of determination is the ability to find the likelihood of future events falling within the predicted outcome.

$$R^2 = 1 - \frac{\sum_{i=1}^{n}(A_t - P_t)^2}{\sum_{i=1}^{n}(P_t)^2}, \tag{4}$$

where R^2 is coefficient of determination, A_t is the target value, P_t is the predicted value, and n represents the sample size.

For accurate ANN models, the coefficient of correlation (r) and coefficient of determination (R^2) must be close to one, and the MSE must be close to zero.

Sections 3.1 and 3.2 describe the models developed in this study.

3.1 *Model-I: Input—GRF and Knee Angle; Output—Future Knee Angle and Hip Angle*

The GRF and knee angle were used as inputs to predict the hip angle and future knee angle (Table 1). First row represents the input I, second row represents the IInd input, and the nth row represents the nth input. The knee angle was also regarded as one of the inputs because the ground response force was only available during the stance phase and the foot was not in contact with the ground during the swing phase. The

Table 1 Data to train the NN to predict joint angles (GRF, joint angle as input)

Input data					Target data	
Ground reaction force (GRF)				Knee Joint angle	Future knee joint angle	Hip joint angle
GRF_1	GRF_2	GRF_3	GRF_4	θ_{k1}	θ_{k2}	θ_{h1}
GRF_2	GRF_3	GRF_4	GRF_5	θ_{k2}	θ_{k3}	θ_{h2}
…	…	…	…	…	…	…
$GRF_{(n-3)}$	$GRF_{(n-2)}$	$GRF_{(n-1)}$	GRF_n	$\theta_{k(n-1)}$	θ_{kn}	$\theta_{h(n-1)}$

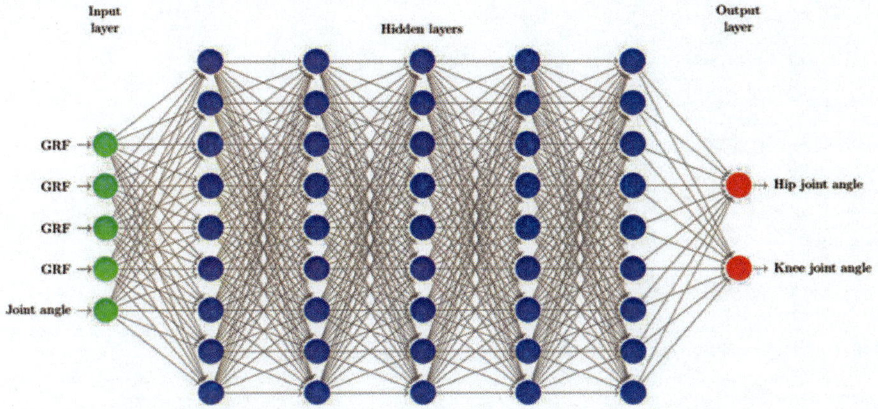

Fig. 5 Neural network structure of the first model

inputs were four GRF values and one knee angle value, with the next knee value and hip angle being the outputs. The activation function was 'tansig,' and there were five hidden layers. The network was trained after which tested with data that had not been utilized in the training process. 85% of data was used for training and 15% was used for testing. Figure 5 is the structure of the neural network which shows the inputs to the net and the outputs which will be predicted.

The regression plot depicts the relation between the output of the network and the targets. The perfect result is represented by the dotted line in each plot: Outputs and targets are equal. The best-fit linear regression line between anticipated values and targets is shown by a solid line. The R-value expresses the relation between the anticipated value and the desired result. The Y-label on these graphs represents the equation between the predicted value and the target value, with the anticipated value as the dependent variable and the target as the independent variable. The target coefficient depicts the ratio between the anticipated value and targets. As a result, it should be close to unity for a high-performance neural network. The error or residue that should be added is the constant term (second term). It should ideally be zero.

3.2 *Model-II: Input—Angular Displacement, Velocity and Acceleration; Output—Linear Displacement*

Figure 6 shows the structure of the neural network with displacement, velocity, and acceleration of the hip and knee joint as input and the linear displacement as the output. Here, the ankle joint is not considered. The inputs are generated using the average value-based modeling method. The average value-based model, which is derived from the nth order linear differential equation and infinite series, was used for this. As a result, the associated input variable is predicted [19].

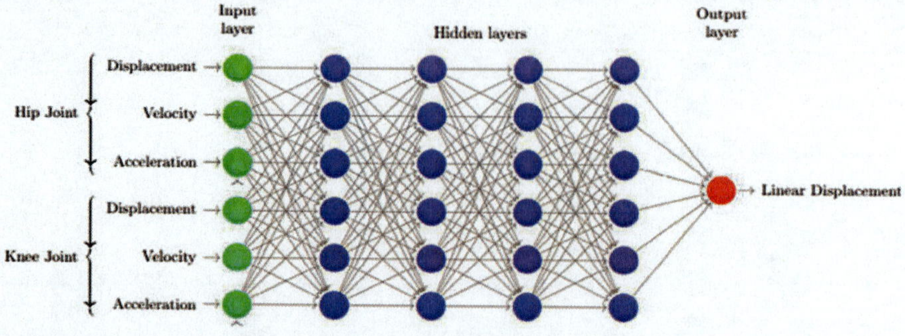

Fig. 6 Neural network structure of the second model

$$C_0 b_{\text{avg}} + C_1\,\delta b_{\text{avg}} + C_2\,\delta^2 b_{\text{avg}} = \text{distance covered} \tag{5}$$

Different sectors were formulated, and different values of inputs were generated which were fed to the network. The initial weights were given as the values of C_0, C_1, and C_2 acquired instead of starting with an arbitrary value of weights, and the terms derived from the average value-based model were given as input to the neural network. This initialization of weights helped to reduce the number of iterations. The activation function used was the 'sigmoid' function. The resulting output was calculated and compared to the actual result. The backpropagation method was used to reduce the error by changing the values of C_0, C_1, and C_2 until the error was low. After each iteration, the contribution of each term to the error was calculated and the value corresponding to the percentage of contribution was added/subtracted from the constant term.

C_0, C_1, and C_2 are the initial weights, which are multiplied by the displacement, velocity, and acceleration. The multiplied values are added in the summing module. The output of the summation module is fed into the activation function module. In this case, the sigmoid function was used as the activation function. By comparing the sigmoid of the cumulative value to the sigmoid of the actual output, the error is computed. The contribution of each constant term to the error is determined. From C_0, C_1, and C_2, the inverse sigmoid of contribution is computed and added/subtracted. The updated values C_0, C_1, and C_2 are multiplied by displacement, velocity, and acceleration, respectively. The procedure is repeated until the error is as minimal as possible.

4 Results and Discussions

As mentioned in Sect. 3.1, the first model was constructed to predict joint angles during the gait cycle using GRF as input. The variation of the target and predicted knee joint angle is shown in Fig. 7, and the variation of the target and predicted

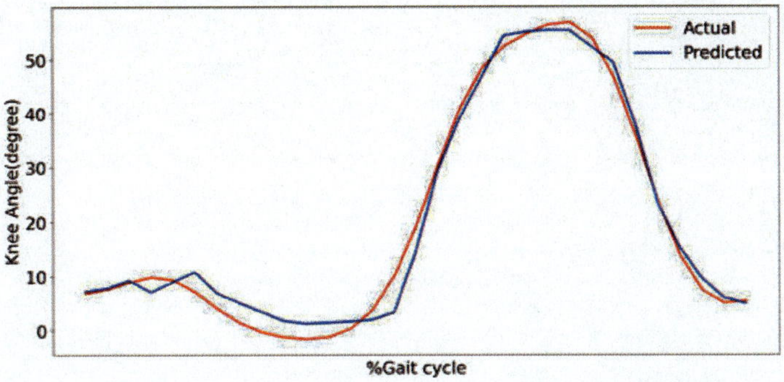

Fig. 7 Variation of actual and predicted values of knee joint angle with GRF and knee joint as input

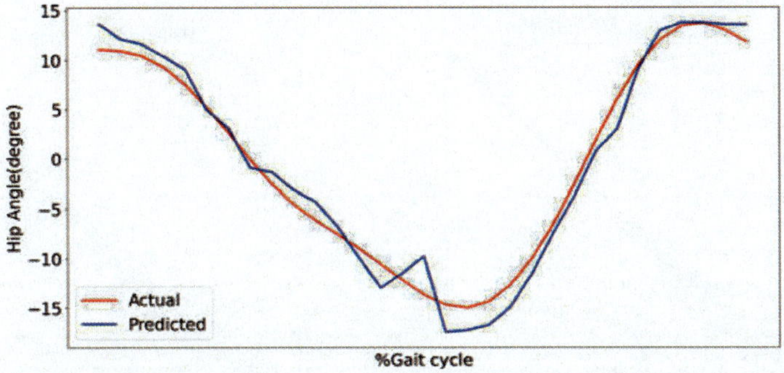

Fig. 8 Variation of actual and predicted values of hip joint angle with GRF and knee joint as input

hip joint angle is shown in Fig. 8. The coefficient of correlation (r) was calculated as 0.99452 for the knee and 0.95174 for the hip joint and the coefficient of determination was calculated as 0.9946 for the knee and 0.927694 for the hip from the network.

Cyclogram is the variation of any two joint angles of the lower limb. Here, in this work, as the hip and knee joints are considered, the variation of hip and knee joints (both actual and predicted) during a gait cycle are plotted and shown in Fig. 9. Figure 10 shows the regression plot.

The model accurately predicted joint kinematics with a low level of inaccuracy. MSE was computed to be 3%, which was considered to be within the tolerance limit. The value of r ranges from 95 to 99%, while the value of R^2 ranges from 93 to 99%. The second model which is mentioned in Sect. 3.2 is used for updating the constant values (C_0, C_1, and C_2) and thereby increasing the accuracy of the average value-based model [19]. The average value-based model represents the dynamics of the knee joint during the locomotion. It is represented as the ordinary differential equation. The constant C_0, C_1, and C_2 of the equation represents the physical and psychological aspects of the subject.

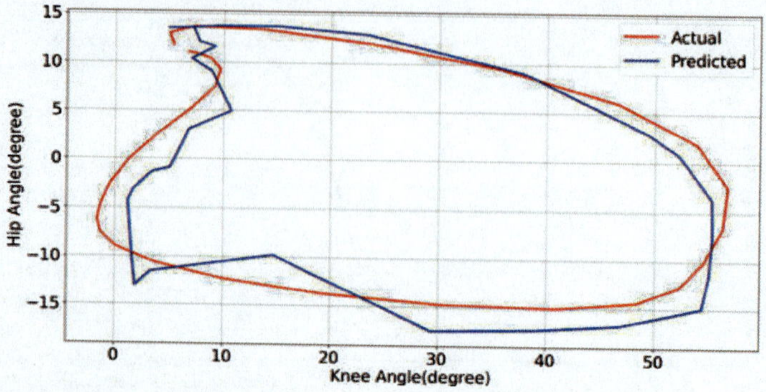

Fig. 9 Cyclogram of target data and predicted data for one gait cycle

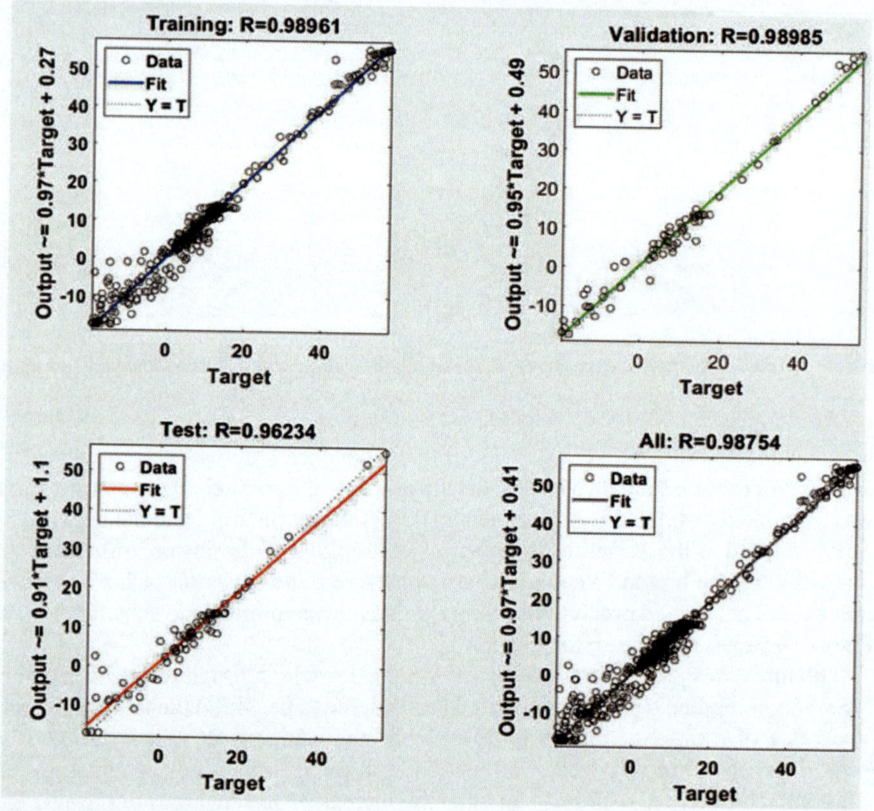

Fig. 10 Regression plot of GRF and joint angle input network

5 Conclusion

This manuscript attempts to construct intelligent models that may be used to examine the dynamics of the human lower limb for a gait cycle. The developed models aid in predicting the motion of the lower limb joints, and this predicting data may be utilized to assess the characteristics of human locomotion. These models can also be used as control algorithms in lower limb prostheses in the field of rehabilitation engineering. As stated in Sect. 4, the first model predicted with low inaccuracy and the values of MSE and r and R^2 are appreciable, and the second model improved the accuracy of the average value-based representation of the knee dynamics.

Acknowledgements The authors would like to express their gratitude for the facilities provided by the 'National Institute of Technology, Calicut, and Manipal Institute of Technology, Manipal, India.' The financial assistance through the intramural grant (MAHE/CDS/PHD/IMF/2019) by Manipal Academy of Higher Education (MAHE) is deeply appreciated.

References

1. Doebelin E (1998) System dynamics: modeling, analysis, simulation, design. CRC Press
2. Doebelin EO, Manik DN (2007) Measurement systems: application and design
3. Zhao H, Wang Z, Qiu S, Shen Y, Zhang L, Tang K, Fortino G (2018) Heading drift reduction for foot-mounted inertial navigation system via multi-sensor fusion and dual-gait analysis. IEEE Sens J 19(19):8514–8521
4. Tanigawa A, Morino S, Aoyama T, Takahashi M (2018) Gait analysis of pregnant patients with lumbopelvic pain using inertial sensor. Gait & Posture 65:176–181
5. Sutherland DH (2002) The evolution of clinical gait analysis: Part II kinematics. Gait & Posture 16(2):159–179
6. Yan SH, Liu YC, Li W, Zhang K (2021) Gait phase detection by using a portable system and artificial neural network. Med Novel Technol Dev 12:1000092
7. Tang K, Luo R, Zhang S (2021) An artificial neural network algorithm for the evaluation of postoperative rehabilitation of patients. J Healthcare Eng 2021
8. Lee HS, Ryu H, Lee SU, Cho Js, You S, Park JH, Jang SH (2021) Analysis of gait characteristics using hip-knee cyclograms in patients with hemiplegic stroke. Sensors 21(22):7685
9. Kutilek P, Farkasova B (2011) Prediction of lower extremities' movement by angle-angle diagrams and neural networks. Acta Bioeng Biomech 13(1)
10. Kutilek P, Viteckova S (2012) Prediction of lower extremity movement by cyclograms. Acta Polytechnica 52(1)
11. Caparelli TB, Naves ELM (2017) Reconstruction of gait biomechanical parameters using cyclograms and artificial neural networks. Res Biomed Eng 33:229–236
12. Schmalz J, Paul D, Shorter K, Schmalz X, Cooper M, Murphy A (2021) Modelling human gait using a nonlinear differential equation. bioRxiv
13. Sujalakshmy V, Sivanandan K, Moideenkutty K (2012) Average value based model for electrical distribution system load dynamics. Int J Electr Power Energy Syst 43(1):1285–1295
14. Wang JJ, Singh S (2003) Video analysis of human dynamics—a survey. Real-Time Imaging 9(5):321–346
15. Whittle MW (1996) Clinical gait analysis: a review. Human Movement Sci 15(3):369–387
16. Tucker MR, Shirota C, Lambercy O, Sulzer JS, Gassert R (2017) Design and characterization of an exoskeleton for perturbing the knee during gait. IEEE Trans Biomed Eng 64(10):2331–2343

17. Kalita B, Narayan J, Dwivedy SK (2021) Development of active lower limb robotic-based orthosis and exoskeleton devices: a systematic review. Int J Soc Robot 13(4):775–793
18. Farooq U, Shabir MW, Javed MA, Imran M (2021) Intelligent energy prediction techniques for fog computing networks. Appl Soft Comput 111:107862
19. Sirish TS (2014) Modeling, analysis and realization of supporting system for afflicted human locomotion. Ph.D. thesis

Pruning and Quantization for Deeper Artificial Intelligence (AI) Model Optimization

Suryabhan Singh, Kirti Sharma, Brijesh Kumar Karna, and Pethuru Raj

Abstract Artificial intelligence (AI) models are being produced and used to solve a variety of business and technical problems. AI model engineering processes, platforms, and products are acquiring special significance across industry verticals. Due to deeper automation, the number of features being used for model generation is numerous and hence the resulting AI models are bulky. Therefore, AI researchers have come out with a number of powerful optimization techniques and tools to compress AI models. This paper explores a suite of pioneering methods for achieving AI model compression. Pruning and quantization techniques are being used together to reduce the size of complex AI model architectures and make them optimized and performant so that they can be easily deployed in the IoT edge devices. We have elaborated on three different methods to compress an AI model and drawn inferences on the same. Further, proper comparisons were conducted between model metrics on the basis of minimal loss in accuracy and precision. To identify the most accurate and compressed model, we utilized pruning techniques and dived deeper by comparing performances using right metrics. We have implemented two types of pruning for the purposes of this paper: weight pruning and unit pruning. These pruning techniques were implemented with 0 to 100 percent sparsity to create a network model that is lightweight and does not significantly affect performance metrics. We have also reduced the size of the model by removing weights and neurons that were used more than once. We have compared pruning and quantization compression methods, which are implemented using the Tensor flow lite library. We have found that Tensor flow lite model is better for edge deployment. Additionally, a practical survey of network compression techniques is also included.

Keywords AI model optimization · Pruning · Quantization · Edge deployment · Model performance

S. Singh (✉) · K. Sharma · B. K. Karna · P. Raj
Edge AI Division, Reliance Jio Platforms Limited, Avana Building, Bangalore 560103, India
e-mail: Suryabhan2.Singh@ril.com

P. Raj
e-mail: pethuru.chelliah@ril.com

S. Sharma et al. (eds.), *Intelligent Control, Robotics, and Industrial Automation*,
Lecture Notes in Electrical Engineering 1066,
https://doi.org/10.1007/978-981-99-4634-1_73

1 Introduction

In the field of computer vision (CV) and natural language processing (NLP), neural networks are being used in innovative and impactful ways. However, the computational resources required for implementing neural networks (NNs) (realized through machine and deep learning (ML/DL) algorithms) are on the higher side. Further on, the energy consumption of these artificial intelligence (AI) models is also high while the heat dissipated by them into our fragile environment is hugely damaging. Additionally, deploying such features-rich neural networks in the Internet of Things (IoT) edge devices is beset with a number of technical challenges and concerns. In this paper, we discuss methodologies to create light weight (compressed) models and compare various models that have been trained on numerous datasets. This paper will focus on two such methodologies to create lightweight and superior models. Quantization and pruning are the two common techniques using which ML algorithms can be made to work well. Quantization is a machine learning technique that involves converting data from floating point 32 bits to a less precise format, such as integer 8 bits, and then using integer 8 bits to do all the convolution operations. In the last stage, the output with lower precision is changed back to the output with higher precision in floating point 32. This process allows models to be represented in a compact manner and allows victories operations to run quickly on a wide range of hardware platforms. Pruning is a technique used to remove unwanted and unnecessary computation from a network. It works in two ways: structured pruning and unstructured pruning (also known as weight pruning and unit pruning (of neurons)). The weight matrix is set to zero in weight pruning or unstructured pruning. This allows the removal of connections. In order to obtain k percent sparsity, arrange the weights in order of their size and set them all to zero to create a weight matrix. In this paper, we have highlighted the different types of compression techniques for neural networks and described how network compression works in real-life applications. The goal of this paper is to highlight the number of ways in which pruning and quantization approaches can be used optimally to create a compressed model with minimal loss and drop-in accuracy (less computations with no drop-in performance).

2 Related Work

The paper includes a practical implementation that shows different ways to process information quickly [1]. There are pruning and quantization techniques to sharply reduce model complexities. By using sparsity percentage and utilizing pruning techniques on neural networks, the accuracy can be greatly improved, the learning parameters can be speed up significantly, and it helps get rid of weights and neurons that aren't needed [2]. We can use magnitude-based, rank-based, unit-based, and weight-based pruning to shrink the network, after which this light weight model can be deployed in mobile and other low-power applications [3]. In another research paper, the author has suggested a "lottery ticket hypothesis", which aids in making the network train faster and produce better results compared with the original by ensuring the presence of sparse subnetworks at the start [4] and [5] came up with the idea of a learning rate to make weight compression work well for winning lottery tickets. Pruning can be done before or after training, and it can also be done over and over again until the required compression is reached. In [6], the authors provide a description of the updated pruning scenario that will help find gaps in the current metrics and benchmarks. Similarly, we can utilize the super mask method that will help "super mask", which is used on randomly initialized and untrained networks to make a better model with higher performance [7]. Quantization has also been used in frameworks like Okeras [8] and Brevitas [9] to develop quantized neural network training. They use the quantized operator format with clipping and quantize-clip-dequantize (QCDQ) format and then introduce a novel higher-level ONNX format called quantized ONNX (QONNX) that presents three new operators—Quant, BipolarQuant, and Trunc—in order to characterize uniform quantization. Hessian-aware-quantization is a method that uses information about the second derivative to automatically choose the relative precision of bits in each layer [10]. Using the 0-bit limit of quantization is one way to find out about structure pruning and quantization [11]. By adding the convolution operation with a low-cost multiplexer to one proposed work, the combination of pruning and quantization techniques will make convolution neural networks less complicated, use less memory, and use less power [12]. Field-programmable gate arrays (FPGAs) are used to build multiple variations of hardware. In this paper, the author defines and elaborates on the quantization process and how it works in hardware, as well as how it can be used effectively on devices. FP32 weights and activations are used to train all of the neural networks. Quantize, or representations with fewer bits, like INT8, are used to reduce not only data transfer but also size and energy used when doing MAC operations.

3 Proposed Solution for the Model Compression

3.1 Methodology

In machine learning, there are multiple methods and algorithms to influence neural networks to work as well as they can. In network optimization, this can be done by removing a few of the connections between neurons and layers. This process is an example of unit/neuron pruning. By doing so and by getting rid of the parameters that aren't needed, we will be able to significantly speed up calculations. This will result in a network structure which will resemble. Every neuron in the large layer is connected to the layer above, which means that we will have to multiply many float values together in this case. This type of network structure is known as "sparse network". The following are some well-known methods to trim a neural network with rectified linear unit (based on weights) or ReLU(xW).

Weight Pruning In neural network weight pruning, the weight matrix is the primary focus, and each individual weight in the weight matrix is set to zero.

- We need to rank the individual weights in the weight matrix "W" according to their magnitude or absolute value—wi, j—to achieve the sparsity of K (in percentage).
- After this, we need to focus on the ranked weight matrix and order it on the basis of magnitude/absolute or norm value (L1-Norm, L2-Norm) to know the individual weights in the weight matrix which have less priority, are unnecessary weights, or are unwanted and set them to zero the smallest K.
- We need to set to zero those weights which do not provide enough information or are redundant, unwanted, and inadequate.
- Take that sparse weight matrix and remove the connection between the neurons according to the steps followed above.

Unit/Neuron Pruning In neural network unit pruning, neurons in the layer are prioritized, and entire columns in the weight matrix are set to zero to delete the output neuron. Remove the connections and neurons as depicted in.

- The columns of weight matrix must be ranked according to their L2-norm to achieve the sparsity of K (in percentage).
- Set the entire weight matrix column to zero.
- Select from those the smallest k value and remove the output neurons with smallest K.

When the value of the sparsity percentage (k percent) is raised, the network will tend to become less dense, and the zeroth association in the corresponding matrix will grow. We used different datasets, such as the MNIST and FMNIST datasets, to test weight pruning and unit pruning for CIFAR along with a comparison on their performances. If while implementing, the desired sparsity is not obtained, the frequency arguments might need to be changed to a lower value. To get the best results from these networks, you may need to train them longer and make hyper-tuning changes when necessary. We suggest making a copy of the original model first, and then pruning the copied model with a lower magnitude.

3.2 AI Model Pruning

Here, we consider two types of such criteria: Consider a network, train it without pruning, and apply pruning and quantization to the trained model on the basis of the percentage sparsity of K. Consider a CNN network, train it without pruning, and then take the trained network and apply pruning and quantization with more training and hyper tuning to achieve a compression ratio. In the first work, we use pruning and quantization on three different types of datasets and compare the results for both weight pruning and unit pruning. For this, the Keras model library was used to implement pruning model exploration. A set of three different datasets has been used to compare parameters such as performance and efficiency of the network in different situations, such as pruning on a trained model, training a model first, applying pruning, and picking a random network. Here, the problem of pruning is broken down into several steps. We suggested pruning on a custom-made assumed network, using a Keras model without pruning and four hidden layers with ReLU-activated neural networks. There are 200, 500, 1000, and 1000 of them, and they all connect to each other. We have thought of the output logit layer as the fifth layer, and 10 is the size given to it. All of these layers are directly connected to the output layer, so we can use them as spare layers for any pruning. The convolution layers, dropout layers, batch normalization layers, and AVG pooling layers have also been left out. This network is used to run the algorithms listed above, and models are trained on it without pruning the network. We used the MNIST, FMNIST, and CIFAR datasets, which have different numbers of classes and input shapes, to build the model architecture. At the time of building architecture, the sparsity percentage is also considered to shrink the model. For building models, an uncompressed Keras is used which has a dense layer with shapes [1000, 1000, 500, 200]. An optimizer updates steps by hand, following the constraints of training with 60,000 samples and validating on 10,000 samples. After 50 epochs of training without pruning with Keras, the following are the performance metrics obtained as given in Table 1. For weight and unit pruning, the sparsity percentage K is [0.25, 50, 60, 70, 80, 90, 95, 97, and 99]. Sparsity in the trained pruning method will not cut away the weight that is causing the SoftMax layer. The way it works is that it takes kernel and bias matrices (for a dense layer) and gives back the unit-pruned versions of each. The k weights matrices will be 2-D, b weight will be a 1-D matrix of the biases of a dense layer, and k sparsity is the percentage of weights to set to zero. They give back kernel weights, a sparse matrix with the same shape as the original weight matrix, and bias weights, a sparse array with the same shape as the original bias array. For weight pruning in the proposed network, we can use the trained model. It will take input arguments as, k weights, b weights, and k sparsity. In the processing function, the network will copy the kernel weights and get ranked indices of the abs for k weight. In k sparsity, it will set the number of indexes to be set to 0. In b weight, it will copy the bias weights and get the ranked indexes of the abs after all the processing is done. Finally, it will return the kernel weights and bias weights. In the case of unit pruning, it will take k weights, b weights, and k sparsity as arguments and run the function processing on them. That means the 2-D weight matrix's indices are set to zero. For b weight, the function processing will copy the bias weights and get ranked by the abs. In this case, sparse cut-off indices, the indexes in the 1-D bias weight matrix are set to zero, which is the same as the indexes of the column that was taken out.

3.3 L2 Norm

The L2 norm is calculated by taking the square root of the sum of the square vector values. Like the L1 norm, the L2 norm is used to fit the regularization method for

Table 1 Performance metrics for the MNIST, FMNIST, and CIFAR datasets

Dataset	Baseline test accuracy	Test accuracy	Test loss	Total pram
MNIST	0.985199	0.985199	0.310332	2,388,710
FMNIST	0.893800	0.893800	0.890166	2,388,710
CIFAR	0.477999	0.477999	3.94429]	4,676,710

machine learning. Here in pruning and quantization, it will also help to find the weight magnitude and rank them by magnitude by calculating the L2 norm. On this weight, L2 norm will be the square of the weight matrix and will give a value of positive magnitude.

3.4 Obtained Results Illustrated in a Pandas Data Frame

K sparsity—percentage of sparsity for network layer (percentage of network made sparse (zero)). Figures 1, 2, and 3 are given to illustrate performance metrics.

FMNIST loss weight—loss using weight pruning.

FMNIST acc unit—accuracy using unit/neuron pruning.

FMNIST loss weight—loss using weight pruning.

FMNIST acc weight—accuracy using weight pruning.

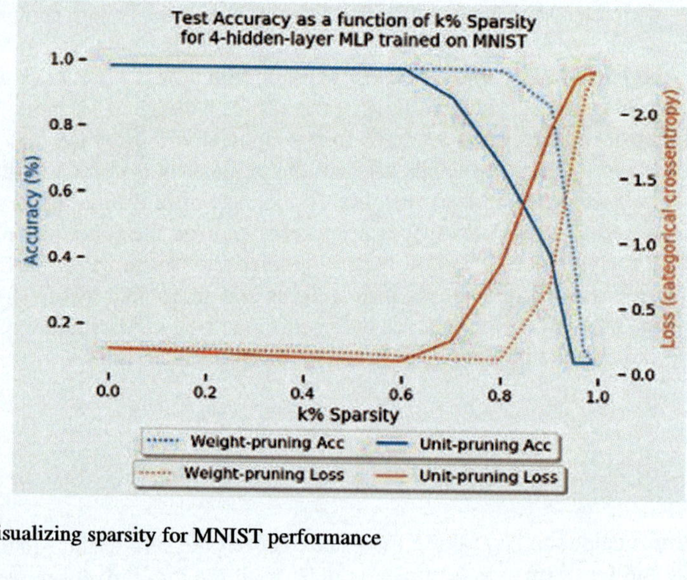

Fig. 1 Visualizing sparsity for MNIST performance

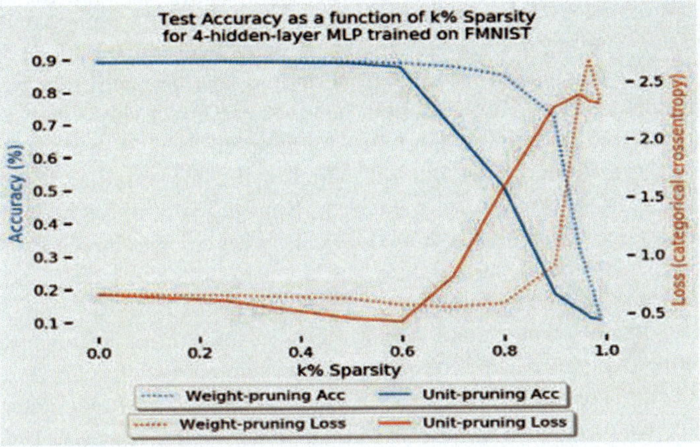

Fig. 2 Sparsity for FMNIST data performance

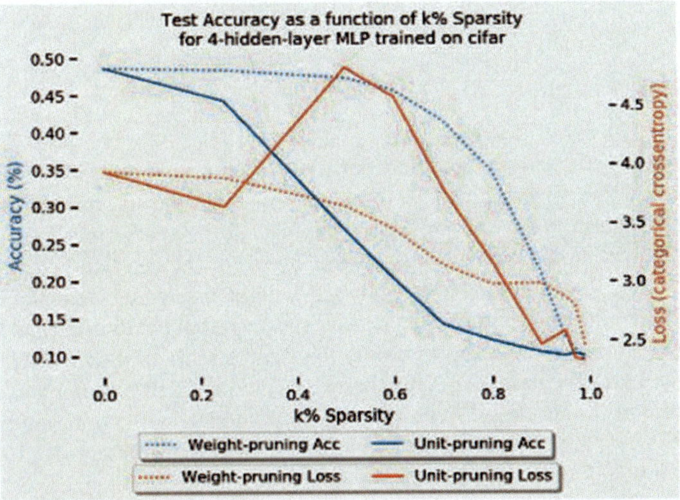

Fig. 3 Visualizing sparsity for CIFAR data performance

FMNIST loss unit—loss using unit/neuron pruning.

3.5 *Performance Evaluation*

Here we can easily identify that on the MNIST dataset, the performance curves for unit and weight pruning differ. Pruning weight matrices of dense matrices does not result in dramatic drops in accuracy or increases in loss until around k = 80. Moreover, the accuracy does not begin to noticeably decrease until k = 90. For unit pruning, accuracy begins to fall earlier, at around k = 70 (with loss beginning to increase around k = 60). However, both methods can effectively remove more than half of the network weights without any dramatic differences in test classification performance. On the FMNIST dataset, which had a lower initial accuracy and a higher initial loss, the same pattern emerges. Similarly, pruning weight matrices of dense matrices does not result in dramatic drops in accuracy or increases in loss until around k = 80 and the accuracy does not begin to noticeably decrease until k = 90. For unit pruning, the differences manifest earlier for FMNIST than in MNIST. Accuracy begins to fall at around k = 60 (with loss beginning to increase around k = 60). Pruning weight matrices of dense matrices do not result in dramatic drops in accuracy or increases in loss until around k = 70 and the accuracy does not begin to noticeably decrease until k = 70. For unit pruning, accuracy begins to fall earlier, around k = 25 (with loss beginning to increase around k = 60). In conclusion, both methods can effectively remove more than half of the network weights without any dramatic differences in test classification performance.

3.6 Compression

We can also test the extent to which our model can be compressed with this new sparsity and if this will in turn speed up execution. We do not have to use the TF model optimization toolkit to shrink our model. All the sparse matrices in this list of weight matrices must be compressed and saved for the last two layers. After the function is run, it will return a compressed weight list, which is a list of weight matrices that are like the input list of matrices, except that the dimensions have been reduced by leaving out columns that only have zeros in them. Essentially, when we compress, we take the base Keras file that we trained without pruning and use it as a starting point for the pruning. After zipping those models with zip functionality, the model size is now around 8,884,279.00 bytes, which is the original model size. After unit pruning with k = 0.5 (50%) sparsity, the compressed model size is 2953777.00 bytes, while for the same sparsity, the compressed model size is 2910981 bytes when using weight pruning. When we think about quantization and combine it with pruning, the size of the model will further decrease and the model will become light weight, as per our goal. Quantization cuts down on computations by reducing the size of the datatype, when we think about pruning and quantization together with the TFLite tool. The zipped, pruned, and quantized TFLite model is 964419.00 bytes, which is about 10× times smaller than the original.

4 Comparative Study among the Pruning Methods

4.1 *Prune Trained Network and Fine-Tune*

This section can be organized as follows: First, we focus on the idea of "on-significance" in function and neural networks and try to find ways to prune the trained neural network. Some of the neural network choices will also be made at random. Post pruning, we can analyze the results and compare them with the results of other pruning method to conclude. The final evaluation of the code will be based on the TensorFlow model, which is the choice made for implementation purposes. Neural networks are also known as function approximations. We can teach them to recognize the representations of input data points, which are fundamental and help them learn parameters as well. Because weights and biases are known as learnable parameters, weight is often called the learnable coefficient function.

4.2 *Gradient Decent*

Gradient descent is known as an optimization technique used to find the local minima of a given differentiable function. It aids in minimizing the given function.

The following steps are used to locate the significant weights.

- Sort the weights based on their magnitudes in descending order and choose the ones that occur at the start of the queue. This scenario, combined with a sparsity level % of weight to be pruned, is what we want to achieve.
- We can set a threshold value and those whose weight magnitude is higher than that threshold would be considered significant. In this scheme, we can get several flavors.
- The threshold to which we are referring should be the lowest weight magnitude in the entire network.

We have some clarity on what could be called a significant weight. Here, we use magnitude-based pruning, which conveys that in pruning we will consider the weight magnitude to prune the model. By pruning, we essentially zero out the significant weights. We test this concept with the MNIST datasets in this paper, but we can extend this work to more datasets as well. We consider shallow networks that are fully connected. The network has got a total of 20,410 trainable parameters. Training this network for 10 epochs gives us a good baseline. Using TensorFlow model optimization, there are two methods for pruning:

- Choose a trained network and prune it with more training.
- Randomly initialize a network and train it with pruning from scratch.

4.3 Choose a Trained Network, Prune It with More Pruning

In Figs. 4 and 5, x-axis blue line shows changes in train value on each epoch and y-axis shows corresponding accuracy.

We begin with the network we have already set up and trained, and prune it from there. In order to maintain the sparsity level, we use the pruning schedule (to be mentioned by the developer) throughout the training process. Before we can prune, we need to recompile the trained model, and we will compile the model in a similar manner as before. It adds a mask that is either 0 or 1. In this case, the pruning model d will not hamper the performance. In order to prune, the pruning scheme must be specified at the time of training. We have set an end-step argument in the pruning schedule to be greater than or equal to the number of epochs we used to train a model. The frequency argument, which is how often pruning is done to achieve good performance and the desired sparsity, must also be considered here.

To understand the power of pruning, we need to dive deeper into its theory.

- Export the pruned and unpruned model (network) and compress them and note their size.
- Apply quantization on both. Quantize and compress them with the quantized version. Note their size and then at last evaluate their performance.

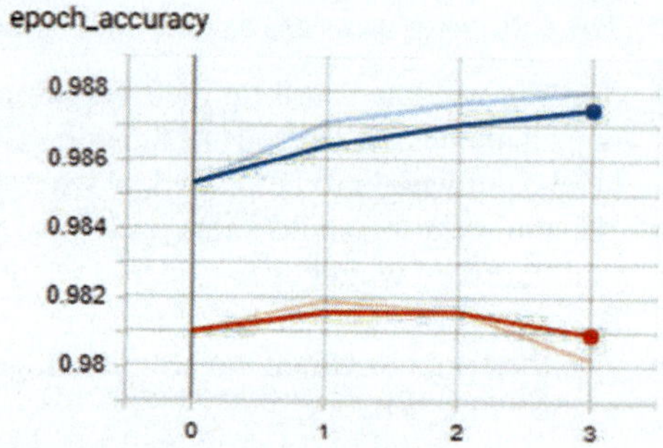

Fig. 4 Resulting graphs obtained for accuracy

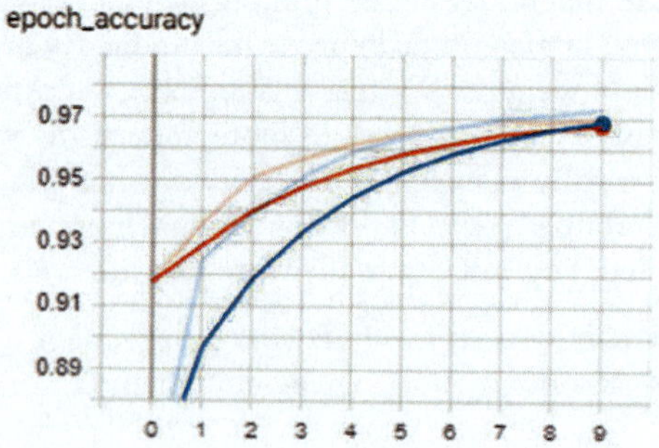

Fig. 5 Model behavior for pruning randomly initialized network: accuracy

The model behavior on pruning for randomly initialized network is illustrated in the graph in Figs. 4 and 5.

4.4 Performance Evaluation

We used the zip file library in order to compress the model in zip format. We also need to use tfmot.sparsity.keras.strip-pruning for serializing the pruned model time. It aids in the removal of the pruning wrapper added to the model by TensorFlow model optimization.

4.5 Quantizing the Models, Compressing Them, and Comparing Performance

We can see this from Tables 2 and 3. Table 2 is giving compression and accuracy achieved using three types of pruning. In Table 3, we have quantized our model using TensorFlow Lite to further compress the model size without hampering the performance. While passing the model to TensorFlow Lite, it needs to be kept in mind that we need to strip the pruning wrappers. When we want to work on bulk data, it is better to load the baseline model that had been serialized earlier and convert them using TFLite.

Table 2 Compression achieved using three types of pruning

Pruning type	Size (bytes)	Validation accuracy (%)
Baseline (no pruning)	78,039	0.980199993
Pruning a trained network	48,338	0.980199993
Pruning (training a network from scratch)	48,883	0.970000029

Table 3 Results obtained after quantization using TensorFlow Lite model

TF Lite model type	Size (bytes)	Validation accuracy (%)
Baseline (no pruning)	17,820	0.9807
Pruning a trained network	13,430	0.9806
Pruning and training a network from scratch	13,224	0.9704

Table 4 Result obtained for MNIST

Datasets	Optimization methods	Test accuracy	Models, compressing
MNIST	Pruning on trained network	0.9851999878883362	0.9721 (sparsity = 0.5%)
FMNIST	Pruning on trained network	0.8938000202178955	0.8824 (when sparsity = 0.5%)
CIFAR	Pruning on trained network	0.4779999852180481	0.4726 (when sparsity = 0.5%)
MNIST	Prune network with more pruning (using *TFLite)	0.98719	0.980001
MNIST	Prune network after training from scratch (using TFLite)	0.9790000	0.9704

* TFLite: TensorFlow optimization kit

5 Results

The performance comparison of each dataset and type of compression can be seen in Table 4. The fire three are implementations for MNIST, FMNIST, and CIFAR datasets and describe the performance of the model without using the TensorFlow toolkit. We observe that the test accuracy and accuracy after compressing the model are very close. These implementations show better compression of the model with a sparsity of 0.5. The last two implementations for MNIST dataset denote the performance of model using the TensorFlow toolkit for two scenarios, the first one is for the scenario "choose a trained network, prune it with more pruning" and other is for "take randomly initialize network, prune it after training from scratch". In these implementations, using the TensorFlow toolkit results in test accuracy and post compression accuracy that are closer to each other but when it comes to a comparison on the basis of size, the first three implementations (without using TFLite) shows a better result compared with the last two implementations (using TFLite).

6 Conclusion

The goal of this paper is to prune and compress AI models in order to make them lighter and faster to deploy when compared with traditional models. Here, pruning and compression techniques are applied to models without hampering their original performance metrics making them suitable for deployment in edge devices that have a limit on computation power and power consumption. With these, we obtain a compression ratio of 3× times for MNIST datasets with sparsity of 50% (excluding the usage of tensor flow optimization kit). When this is combined with pruning (excluding the usage of tensor flow optimization) and quantization techniques (using TFLite), the compression ratio is 10× times. Further on, we are performing pruning and quantization techniques using a TensorFlow optimization tool. As a part of this, two types of practical work have been proposed. In the "take trained network then prune" method, the obtained compression ratio is around 1.6 or 1× times for pruning using the TensorFlow optimization tool. When this pruning method is combined with quantization using TensorFlow optimization, the compression ratio obtained is around 10x.

References

1. Frankle J, Carbin M (2019) The lottery ticket hypothesis: training pruned neural networks. In: 7th International conference on learning representations, pp 1–42
2. Renda A, Frankle J, Carbin M (2020) Comparing rewinding and fine-tuning in neural network pruning. In: 8th International conference on learning representations, pp 1–16
3. Blalock D, Ortiz JJG, Frankle J, Guttag J (2020) What is the state of neural network pruning?In: Proceedings of the 3rd MLSys Conference
4. Zhou H, Lan J, Liu R, Yosinski J (2019) Deconstructing lottery tickets: zeros, signs, and the supermask. In: 33rd Conference on neural information processing systems (NeurIPS 2019), pp 1–11
5. Coelho Jr CN, Kuusela A, Li S, Zhuang H, Ngadiuba J, Aarrestad TK, Loncar V, Pierini M, Pol AA, Summers S (2021) Automatic heterogeneous quantization of deep neural networks for low-latency inference on the edge for particle detectors. Nat Mach Intell 3:675–686
6. Pappalardo A, Umuroglu Y, Blott M, Mitrevski J, Hawks B, Tran N, Loncar V, Summers S, Borras H, Muhizi J, Trahms M, Hsu S-C, Hauck S, Duarte J (2022) QONNX: representing arbitrary-precision quantized neural networks. In: Contribution to 4th workshop on accelerated machine learning (AccML) at HiPEAC 2022 Conference. arXiv:2206.07527, arXiv:2206.07527v3, https://doi.org/10.48550/arXiv.2206.07527
7. Dong Z, Yao Z, Arfeen D, Gholami A, Mahoney MW, Keutzer K (2020) HAWQ-V2: hessian aware trace-weighted quantization of neural networks. In: 34th Conference on neural information processing systems (NeurIPS 2020), pp 1–12
8. Nagel M, Amjad RA, Van Baalen M, Louizos C, Blankevoort T (2020) Up or down? Adaptive rounding for post-training quantization. In: Proceedings of the 37th international conference on machine learning, PMLR, vol 119, pp 7197–7206
9. Hacene GB, Gripon V, Arzel M, Farrugia N, Bengio Y (2020) Quantized guided pruning for efficient hardware implementations of deep neural networks. In: 2020 18th IEEE International new circuits and systems conference (NEWCAS). IEEE, pp 206–209
10. Habi HV, Peretz R, Cohen E, Dikstein L, Dror O, Diamant I, Jennings RH, Netzer A (2021) HPTQ: hardware-friendly post training quantization. arXiv:2109.09113, arXiv:2109.09113v3, https://doi.org/10.48550/arXiv.2109.09113
11. Schaub NJ, Hotaling N (2020) Assessing intelligence in artificial neural networks. arXiv:2006.02909, arXiv:2006.02909v1, https://doi.org/10.48550/arXiv.2006.02909
12. Karbachevsky A, Baskin C, Zheltonozhskii E, Yermolin Y, Gabbay F, Bronstein AM, Mendelson A (2021) Early-stage neural network hardware performance analysis. Sustainability 13(2):717

COVID-19 Disease Classification Using DL Architectures

Devashish Joshi, Ruchi Patel, Ashutosh Joshi, and Deepak Maretha

Abstract Numerous deaths have already occurred as a direct result of poor diagnosis procedures and COVID-19 diagnosis errors. The current work suggests deep learning (DL)-based classification approaches for recognising COVID-19 using patient chest X-rays (CHI). It is intended to utilise a deep learning (DL) assist computer-aided finding structure to divide X-ray descriptions into two groups: negative (0) and positive (1), in order to automatically discover COVID-19. The proposed study includes all three processes, including pre-processing, data collecting, and categorization. During pre-processing, distracting noise at the margins is removed. For feature extraction, U-NET-based models and Gaussian blur are used. On the collected features, two DL (DL)-based classifiers are applied: ResNet50 and Inception V3. The SARS-CoV-2 CT-scan (CTS) dataset is utilised to develop and assess COVID classification models. The proposed method was evaluated using a number of outcomes measuring criteria, including precision, recall, accuracy, and F-score. It was demonstrated that when U-NET-based segmentation and Gaussian blur were combined as features, the ResNet50 classifier outperformed the InceptionV3 classifier in terms of performance. The proposed system, using the ResNet50 classifier, surpassed the state-of-the-art COVID-19 classification on the test dataset, achieving 99% accuracy, 94% precision, 96% recall, and 95% F-score.

Keywords SARS-CoV-2 · CT-scan (CTS) · Image segmentation · DL architectures · Gaussian blur

D. Joshi (✉)
Prosirius Technologies, Indore, Madhya Pradesh, India
e-mail: Joshi.devashish01@gmail.com

R. Patel
Gyan Ganga Institute of Technology and Sciences, Jabalpur, Madhya Pradesh, India

A. Joshi
Tata Consultancy Services, Pune, Maharashtra, India

D. Maretha
Webkorps Services India Private Limited, Indore, Madhya Pradesh, India

S. Sharma et al. (eds.), *Intelligent Control, Robotics, and Industrial Automation*,
Lecture Notes in Electrical Engineering 1066,
https://doi.org/10.1007/978-981-99-4634-1_74

1 Introduction

China's Hubei province's capital, Wuhan, announced the first coronavirus epidemic late last month. This new infection has spread quickly throughout the entire world. As of January 30, 2020 [1], 216 nations had declared the spreading coronavirus an international public health emergency. On February 11, 2020, the WHO named this new coronavirus-related acute lung illness as COVID-19 [2]. Acute respiratory distress syndrome (ARDS), severe interstitial pneumonia, and the ensuing multi-organ malfunction are substantially more likely to affect those who are already more vulnerable to virus-related complications, such as the elderly or those with pre-existing conditions. Acute respiratory distress and a rise in fatalities may result from this. Since this virus is highly contagious and spreading quickly, several governments in the affected nations are prioritising the quarantine of sick persons. These hints, however, are inadequate on their own to justify further investigation. In the majority of cases, a pathogenic test for COVID-19 and a positive chest CTS were discovered in patients who had no symptoms [3].

Radiologists frequently diagnose COVID-19. Therefore, manually diagnosing COVID-19 patients is a time-consuming, difficult, error-prone, and demanding task that requires a highly competent radiologist. Due to a paucity of radiologists, diagnostic tools, and test kits, COVID-19 is challenging to diagnose in underdeveloped nations [4]. Concerns concerning the lack of radiologists with the necessary training for illness diagnosis have also been raised as a result of the high increase of COVID-19 patients. Radiologists look for signs of COVID-19-induced lung deformation on X-rays or CTSs. Due to hospitals being overrun with patients after the COVID-19's rapid spread, there has been a scarcity of doctors to assist in the fight against the virus. False diagnoses of chronic virulent illness are frequently made as a result of poor image quality in X-rays 19 (COVID-19). Ineffective or hazardous treatments are frequently given to patients as a result of incorrect diagnoses. Given the seriousness of this health issue, it is imperative that quicker, more accurate ways of detecting COVID-19 from chest X-ray pictures be created. DL in particular has been found to be more accurate than others in machine learning [3, 5, 6]. To identify COVID-19 disease, the authors of [3, 5] used DL methods on X-ray and CTS pictures. There was a problem with class imbalance even though the suggested models in this research were only tested on a small number of X-ray images, particularly in the positive case images of COVID-19 patients. Additionally, they exclusively employed the CNN model whilst using DL. This problem can be resolved using DL models and algorithms, which have lately shown tremendous growth as a result of better processing capacity and a larger dataset [7]. DL aims to train a multi-hidden-layer machine learning model by a vast dataset in order to obtain extremely precise characteristics in order to increase the precision of classification and forecasting [8, 9].

2 Related Work

In [10], they create a DL algorithm aimed at dataset of chest X-ray images (CHI). An already-trained deep convolutional neural network (CNN) and domain extension transfer learning (DETL) were used in this study. The proposed method was built with the express purpose of identifying cases of COVID-19 in CHI. COVID-19 was identified by means of GRAD-CAM, and the locations wherever the model absorbed extra through classification were displayed using a gradient class activation map. Overall accuracy was 0.14 per cent or 90.13 per cent.

The authors of the paper mentioned in [11] established that image analysis might be used in the main finding and prevention of COVID-19 illness. The team used a database of CHI to study the infection of COVID-19 in people who had pneumonia or lung disease. 5 pre-learned CNN-based models were developed by the researchers to aid in the diagnosis of patients suffering from coronavirus pneumonia (ResNet50, ResNet101, ResNet152, InceptionV3, and InceptionResNetV2) by far, the best results across the board were achieved by the pre-trained ResNet50 model (96.1% accuracy across datasets 1 and 2, 99.5% accuracy across dataset 3, and 99.7% accuracy across dataset 4). In [12], the author introduces a CNN-based m architecture for diseases of the lungs.

Prediction of COVID-19 from CHI using deep transition training was reported by Minaee et al. [13]. For the purpose of identifying COVID-19 disease in a collection of X-ray images from different sources, we used transfer training algorithms such as ResNet18, ResNet50, SqueezeNet, and DenseNet-121. Patients with COVID-19 infections who underwent full-body X-rays were included. After putting the models through their paces, they found that they were 98% sensitive and 90% specific. Whilst CNNs and pre-trained models were used for feature extraction, the results were not as trustworthy. In [14], a CNN model was trained to diagnose 14 different diseases using approximately 100,000 CHI dataset. CNN has also used this to forecast the spread of pneumonia [15]. As a result of a prediction method introduced in [16], it is now much simpler to identify diseased areas in CTSs. Therefore, convolutional neural networks (CNNs) make sense for identifying COVID-19 patients. In [17, 18], researchers analysed 7 different existing DL neural network designs using only 50 images. All but twenty-five of the samples were from people who tested positive for COVID-19. Amongst the 5 pre-learned DL models presented by Wang et al. [19], Xception achieved the best overall performance and accuracy (96.75%). There is a total of 1102 CHI, split evenly between control and COVID-19-infected persons for the learning and evaluation phases. In order to accurately identify and diagnose COVID-19 pneumonia in medical imaging of lung tissue, the development of DL is crucial.

This research proposes an effective classifier that uses CHI to reliably categorise COVID-19 cases as either negative (uninfected) or positive (infected). The proposed work creates a DL-based framework for automated COVID-19 disease categorization, with the goals of maintaining a high level of accuracy whilst decreasing the number of false-negative cases. In most cases, CHI of those diseased with COVID-19 was segmented using image segmentation.

3 Proposed Framework

In this research, a framework is planned for classification of COVID-19 disease on chest CTS images. The key steps are covered in the proposed work as shown in Fig. 1. The procedure includes image pre-processing, feature extraction, and disease classification using the COVID-19 system. During the image processing stage, noise was eliminated by rescaling and shrinking the images as well as by normalising them to a set size and applying filtering algorithms. In order to separate the foreground and background images and locate edges inside each of them, characteristics for doing so were extracted. Thirdly, it is customary for the ResNet50 and InceptionV3 architectures to extract significant data from the input image. The Gaussian blur features and the U-NET image segmentation data were combined so that the classification model can be trained, which then used the combined data to determine the classification from pre-processing to training, validation, and testing, sigmoid functions are used in these designs as a function training technique. The image is then subjected to ResNet50 and Inception to assign it to one of two groups (normal or COVID-19). Figure 1 is a detailed breakdown of the proposed COVID classification system's phases.

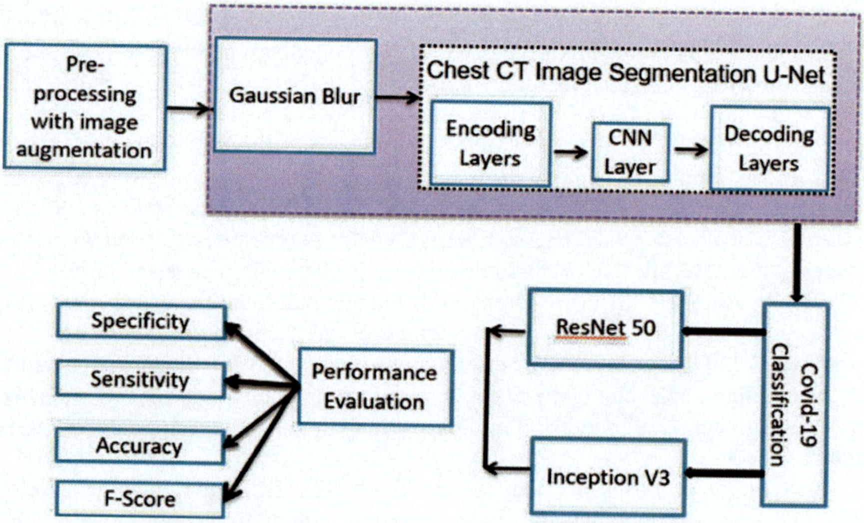

Fig. 1 Proposed framework for COVID-19 classification

3.1 *Phase1: Pre-processing*

Raw CTS images were provided by the SARS-CoV-2 dataset. Images in the dataset come in many different shapes, and this research modifies the ones that don't fit the desired profile by collecting and processing a total of 264 dimensions of images. The original form was at (1000, 209088), and the new one is at (1000, 264, 264, 3). Images are also transformed by standardising and resizing them. Elastic transform, grid distortion, optical distortion, vertical flip, and horizontal flip are just some of the image transformation techniques used in image augmentation.

3.2 *Phase 2: Features Extraction*

There are several techniques for retrieving various features from the images after they have been altered and processed: U-Net architectures for Gaussian blur and image segmentation.

Gaussian Blur When trying to extract contours from an image, blurring it with a Gaussian blur and then removing the high-frequency components are a common technique. The standard for further processing of the retrieved image is for it to have the shape (264, 264, 3). Specifically, a Gaussian blur function from Python's computer vision library (cv2) is used in the investigation. This fuzzy image is what gets used for sorting purposes.

Chest CT Image Segmentation Using U-Net In this study, the popular U-Net architecture is used for the segmentation task, with one each in the encoder and decoder layers. The number of channels in each layer of the three presented models varies. The model contains the elements of an encoder, a decoder, and a convolution block. Position values and information are reduced, and abstract features are discovered by the encoding block, which employs the pooling layer. Precise placement is achieved by employing pixel properties discovered at the decoding layer. The up-sampling process combines these neighbourhood features with the new feature map to preserve some useful data from the original down-sampling approach. Skipping the link between the encoder and decoder levels allows the network to maintain low-level features.

Model is comprised of three encoding layers, one convolution layer, and three decoding layers. Each layer's maximum pooling operation doubles the size of the encoder block, which begins at 64. After that, a 512-dimensional convolution block is used to maximise pooling and perform batch normalisation on the input images. Maximum pooling is a technique for minimising mistakes and conserving texture details in images. The decoder block up-samples and down-samples the image to determine the original 64-by-64 pixel dimension. The decoder layers are responsible for the convolution, transposition, and concatenation of the skip connection feature vectors. At the end of this process, we have five distinct features that can be used separately for classification.

3.3 Phase 3: COVID-19 Classification

COVID uses a binary system of classification, with the two possible outcomes being "positive" and "negative". The framework makes use of two DL classification architectures: ResNet50 and InceptionV3. These classifiers divide the dataset into a training (70%) and test set (30%).

ResNet50 The ResNet50 architecture's input layer receives the featured images and sends them to the convolution block-1, which is comprised of the conv2d, batch normalisation, ReLU activation, and max pooling processes. Methods like these are employed to map and filter images for the purpose of identifying instances of contextual similarity. The 50 weighted layers that make up ResNet50 are all connected to one another in the same way that block-1 is connected to block-2. It internally modifies the sizes of 32-bit, 64-bit, 128-bit, 256-bit, 512-bit, 1024-bit, and 2048-bit forms. Two dense layers followed by dropout layers were added to the model to mitigate any potential for overfitting. The dense layer features ReLU activation, L2 kernel regularisation with a penalty value of 0.01, and dropout with a value of 0.5. The final step is an application of a sigmoid activation function to the output layer. The model is built with hyper-parameters like a learning rate of 0.0001 for the Adam optimizer and a binary cross-entropy loss function. Testing the model with 32 and 40 epoch batches shows that the model overfits and becomes unresponsive after epoch 10.

InceptionV3 InceptionV3 uses convolutions, average pooling, max pooling, concatenations, dropouts, and fully linked layers to feed input images into the symmetric and asymmetric model. The model makes extensive use of batch normalisation, including for the activation inputs. There are 42 weighted layers in this architecture, and the connections between the blocks here follow a specific order. In order to prevent the model from becoming overly specific, we add two dense layers followed by dropout layers. The dense layer employs ReLU activation, L2 kernel regularisation with a penalty value of 0.01, and dropout with a value of 0.5. The final step is an application of a sigmoid activation function to the output layer. The model is built with the aid of hyperparameters like the Adam optimizer's acceleration of training of 0.0001 and the binary cross-entropy loss function. Testing the model with 32 and 40 epoch batches shows that the model overfits and becomes unresponsive after epoch 10.

4 Experimentation

This work [20] makes use of the SARS-CoV-2 CTS dataset, the biggest freely accessible collection of CT images for detecting COVID-19. The dataset includes both CTSs that tested positive for SARS-Cov-2 and CTSs that tested negative for the virus, which are 2481 CTS images. This statistics comes directly from hospital records in the Brazilian metropolis of Sao Paulo. It is hoped that this dataset will encourage the development of AI-enabled methods for determining whether a patient is diseased with the fatal virus by analysing the results of a scan. CT-scan is a COVID and non-COVID dataset with a total of 1737 images: 744 for training and 840 for validation. The proposed COVID classification system was written in Python. In order to train the models and tools for TensorFlow 2.0 (with Keras), we made use of a free and open-source DL framework. DL models were developed and evaluated on Google Collaboratory.

5 Performance Evaluation

The overall efficacy of the proposed COVID detection system is measured by a number of different metrics, including accuracy, precision, recall, and F-score. Calculating the recall ratio involves comparing the number of correct predictions of COVID classes to the total number of reference COVID classes. Precision can be calculated as the fraction of correct predictions for COVID classes relative to the total fraction of correct predictions for COVID classes. Accuracy and recall play together in perfect harmony to create the F1-score. Recall, precision, and F-score are defined in Eqs. (1), (2), and (3), and reference and predicted COVID classes, $S_{\text{reference}}$ and $S_{\text{predicted}}$, are used. It is recommended that accuracy and recall be determined before attempting to compute the F-score. Where y is the label and $p(y)$ is the foreseen probability of the point being green for all N points, the binary cross-entropy loss function is shown in Eq. (4).

$$R = \frac{\left|S_{\text{reference}} \cap S_{\text{predicted}}\right|}{S_{\text{reference}}} \quad (1)$$

$$P = \frac{\left|S_{\text{reference}} \cap S_{\text{predicted}}\right|}{S_{\text{predicted}}} \quad (2)$$

$$F = \frac{2 \times P \times R}{P + R} \quad (3)$$

$$H_p(q) = -\frac{1}{N}\sum_{i=1}^{N} y_i \,.\log(p(y_i)) + (1 - y_i)\,.\log(1 - p(y_i)) \quad (4)$$

Table 1 Performance of proposed system on validation and test data

	ResNet50 (%)		InceptionV3 (%)	
CTS dataset	Validation data	Test data	Validation data	Test data
Precision	96	94	92	93
Recall	92	96	90	91
F-score	94	95	91	92
Accuracy	98	99	90	90

6 Results and Discussions

The proposed system used two classifiers based on feature extraction methods to distinguish between COVID-19 positive and negative classes. Table 1 displays the results of a comparison of the proposed system's performance with that of various classifiers built on the basis of feature extraction. ResNet50 outperforms Inception V3 in terms of accuracy, precision, recall, and F-score. When combined with the ResNet50 classifier, feature fusion in the proposed framework yielded the best results, with an F-score of 95%, the highest precision (94%), the highest recall (96%), and the highest accuracy (99%).

This paper also demonstrates the loss and accuracy of classifiers on epoch-based validation and training data. Validation and training loss for ResNet50 are plotted as a function of epochs in Fig. 2a. Both validation and training loss tend to decrease with increasing epochs. Epoch-by-epoch comparison of validation and training accuracy for ResNet50 is displayed in Fig. 2b. Training accuracy was inconsistent up until epoch 17 but remained constant thereafter. Validation accuracy was consistent across all epochs. Figure 3a displays the Inception V3 validation and training loss over time. The loss decreases with increasing epochs, and until epoch 11 the values for validation and training loss are very close to one another. For Inception V3, Fig. 3b displays a comparison of validation and training accuracy across epochs. Through epoch 17, the gap between validation and training accuracy was relatively small, but after that, it widened significantly. Tabulated in Table 2 are results from a comparison of the proposed framework to other commonly used methods for categorising COVID-19. The outcomes endorse the superiority of the proposed approach over its rivals. The Inception ResNet method [11], which combines two classifier models, also demonstrated enhanced accuracy.

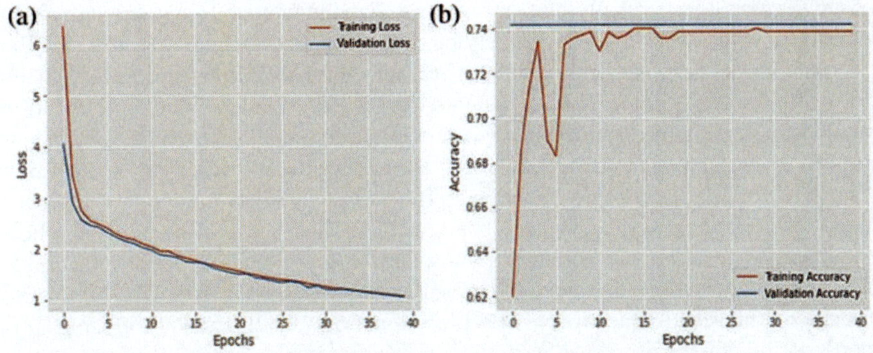

Fig. 2 **a** Assessment of validation and training loss of ResNet50, **b** assessment of validation and training accuracy of ResNet50

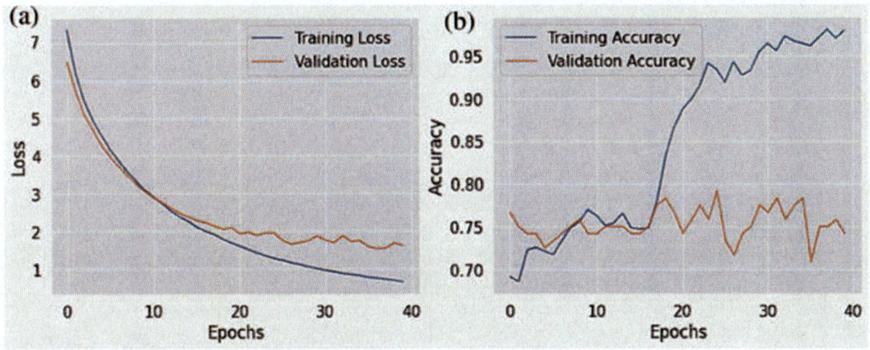

Fig. 3 **a** Assessment of validation and training loss of Inception V3, **b** assessment of validation and training accuracy of Inception V3

Table 2 Comparison of the proposed method and current COVID-19 classification methods

Authors	Method	Accuracy (%)
[11]	InceptionResNet	96
[21]	DL	86.7
[22]	DL	89.5
[23]	SVM	90
Proposed System	Feature fusion on ResNet50	99

7 Conclusion

This work suggests a framework for classifying COVID-19 data. The planned framework consists of three stages: pre-processing, feature extraction, and classification. In this framework, two methods of feature extraction are used: Gaussian blur and image segmentation using U-Net architecture. Each technique is applied separately to a CHI, with Gaussian blur identifying and extracting image contours and the U-Net model identifying and retrieving segmented areas. After features have been extracted, the outcomes of feature fusion are subjected to independent classification methods. The proposed framework uses the DL architectures ResNet50 and InceptionV3. Image analysis utilising these methods reveals two groups, with "0" denoting "negative" and "1" denoting "positive". These classification models are assessed based on their accuracy, precision, recall, and F-score. It was shown that the best classification results were achieved by combining the U-Net model and Gaussian blur feature extraction techniques. The suggested framework using the ResNet50 classifier achieves 99% accuracy, 94% precision, 96% recall, and 95% F-score. It is worth noting that the proposed framework outscored all competing classifiers across the board. It is planned to expand this study to incorporate clinical aspects of COVID for chest X-ray classification. Whilst recent efforts have focussed solely on binary classification, future efforts will incorporate multi-classification for more accurate findings.

References

1. G Pascarella A Strumia C Piliego F Bruno R Buono Del F Costa S Scarlata FE Agrò 2020 COVID-19 diagnosis and management: a comprehensive review J Intern Med 288 2 192 206
2. AM Ayalew AO Salau BT Abeje B Enyew 2022 Detection and classification of COVID-19 disease from X-ray images using convolutional neural networks and histogram of oriented gradients Biomed Signal Process Control 74 103530
3. Sethi R, Mehrotra M, Sethi D, Deep learning based diagnosis recommendation for COVID-19 using chest x-rays images. In: Second international conference on inventive research in computing applications (ICIRCA). IEEE, pp 1–4
4. D Yang C Martinez L Visuña H Khandhar C Bhatt J Carretero 2021 Detection and analysis of COVID-19 in medical images using deep learning techniques Sci Rep 11 19638 1 13
5. H Jiang S Tang W Liu Y Zhang 2021 Deep learning for COVID-19 chest CT (computed tomography) image analysis: a lesson from lung cancer Comput Struct Biotechnol J 9 1391 1399
6. Sadik R, Reza ML, Al Noman A, Al Mamun S, Kaiser MS, Rahman MA (2020) COVID-19 pandemic: a comparative prediction using machine learning. Int J Autom Artif Intell Mach Learn 1(1):1–16
7. Bejnordi BE, Veta M, van Diest PJ, van Ginneken B, Karssemeijer N, Litjens G, van der Laak JAWM, Hermsen M, Manson QF, Balkenhol M, Geessink O et al (2017) Diagnostic assessment of deep learning algorithms for detection of lymph node metastases in women with breast cancer. JAMA 318(22):2199–2210
8. J Amin M Sharif M Yasmin SL Fernandes 2018 Big data analysis for brain tumor detection: deep convolutional neural networks Futur Gener Comput Syst 87 290 297
9. Pastur-Romay LA, Cedrón F, Pazos A, Porto-Pazos AB (2016) Deep artificial neural networks and neuromorphic chips for big data analysis: pharmaceutical and bioinformatics applications. Int J Mol Sci 17(8):1313
10. Basu S, Mitra S, Saha N (2020) Deep learning for screening COVID-19 using chest x-ray images. In: 2020 IEEE Symposium series on computational intelligence (SSCI). IEEE, 2521–2527

11. A Narin C Kaya Z Pamuk 2021 Automatic detection of coronavirus disease (COVID-19) using X-ray images and deep convolutional neural networks Pattern Anal Appl 24 3 1207 1220
12. M Anthimopoulos S Christodoulidis L Ebner A Christe S Mougiakakou 2016 Lung pattern classification for interstitial lung diseases using a deep convolutional neural network IEEE Trans Med Imaging 35 5 1207 1216
13. Minaee S, Kafieh R, Sonka M, Yazdani S, Soufi GJ (2020) Deep-COVID: predicting COVID-19 from chest x-ray images using deep transfer learning. Med Image Anal 65:101794
14. Rajpurkar P et al (2017) CheXNet: radiologist-level pneumonia detection on chest x-rays with deep learning. arXiv:1711.05225 [cs.CV], arXiv:1711.05225v3 [cs.CV], https://doi.org/10.48550/arXiv.1711.05225
15. E Luz 2022 Towards an effective and efficient deep learning model for COVID-19 patterns detection in x-ray images Res Biomed Eng 38 1 149 162
16. Shan F et al (2020) Lung infection quantification of COVID-19 in CT images with deep learning. arXiv:2003.04655 [cs.CV], arXiv:2003.04655v3 [cs.CV], https://doi.org/10.48550/arXiv.2003.04655
17. Hemdan EE-D, Shouman MA, Karar ME (2020) COVIDX-Net: A framework of deep learning classifiers to diagnose COVID-19 in x-ray images. arXiv:2003.11055 [eess.IV], arXiv:2003.11055v1 [eess.IV], https://doi.org/10.48550/arXiv.2003.11055
18. Simonyan K, Zisserman A (2014) Very deep convolutional networks for large-scale image recognition. arXiv:1409.1556 [cs.CV], arXiv:1409.1556v6 [cs.CV], https://doi.org/10.48550/arXiv.1409.1556
19. D Wang J Mo G Zhou L Xu Y Liu 2020 An efficient mixture of deep and machine learning models for COVID-19 diagnosis in chest X-ray images PLoS ONE 15 11 e0242535
20. Soares E, Angelov P, Biaso S, Froes MH, Abe DK (2020) SARS-CoV-2 CT-scan dataset: a large dataset of real patients CTS scans for SARS-CoV-2 identification. In: medRxiv preprint, pp 1–8
21. Xu X et al (2020) A deep learning system to screen coronavirus disease 2019 pneumonia. Engineering 6(10):1122–1129
22. Wang S et al (2021) A deep learning algorithm using CT images to screen for Corona virus disease (COVID-19). Eur Radiol 31(8):6096–6104
23. Barstugan M, Ozkaya U, Ozturk S (2020) Coronavirus (COVID-19) classification using CT images by machine learning methods. arXiv:2003.09424 [cs.CV], arXiv:2003.09424v1 [cs.CV], https://doi.org/10.48550/arXiv.2003.09424

Deep Learning Based Models for Annual Rainfall Forecasting: An Empirical Study on Tamil Nadu

V. Vanitha, M. Sathish Kumar, L. Vishal, and S. Srivatsan

Abstract Rainfall prediction is a complex element in the hydrological cycle. It is critical to know the spell of rainfall in advance to plan strategies in various sectors—especially in agriculture. More than 25% of Tamil Nadu's economy depends on agriculture, which in turn relies on monsoons. Though several statistical models have been implemented to forecast rain, they are not accurate due to chaotic patterns in the rainfall data. To assist the stakeholders of the agricultural sector in attaining abundant yields, this study aims to employ modern technology to forecast annual rainfall in Tamil Nadu. From the data supply portal of Indian Meteorological Department, gridded rainfall data for the period 1901–2017 has been obtained. The data is pre-processed to extract information about rainfall in Tamil Nadu. Two variant models based on long short-term memory (LSTM), namely stacked LSTM and bidirectional LSTM, as well as a gated recurrent unit (GRU), were created, tested on a dataset, and their results were compared. The proposed bidirectional LSTM model has performed better in terms of forecast accuracy than the other proposed models.

Keywords Forecast · LSTM · GRU · Rainfall · Prediction

1 Introduction

Rainfall is a natural phenomenon influencing the manner in which every living being inhabits. It affects the entire ecological system, right from human beings to flora and fauna. It can be beneficial as well as destructive in nature. Knowing the likelihood of rain spell ahead of time will be extremely beneficial to a variety of industries, including agriculture, fishing, tourism, hydroelectric power generation, and planning strategies for floods in towns and river basins, disaster management, and landslides [1–3]. Hence, rainfall prediction has drawn the attention of industries, government,

V. Vanitha (✉) · M. S. Kumar · L. Vishal · S. Srivatsan
Sri Ramachandra Faculty of Engineering and Technology, Sri Ramachandra Institute of Higher Education and Research, Chennai, Tamil Nadu, India
e-mail: vanikkdi@gmail.com

S. Sharma et al. (eds.), *Intelligent Control, Robotics, and Industrial Automation*, Lecture Notes in Electrical Engineering 1066,
https://doi.org/10.1007/978-981-99-4634-1_75

research communities and even general public since their businesses and everyday decisions rely on this. For the agricultural community, rainfall prediction brings in huge benefits—it helps them decide on what crops to plant, harvesting time, soil moisture, fertiliser application, food grain storage, and livestock protection.

In India, agricultural dependency on monsoon is very high and over 70% of the sown farmland relies on the monsoon rain. As a result, the agricultural production that feeds the Indian economy becomes vulnerable to the monsoon. The failure of monsoon will bring undesired effects on the livelihood of farmers and on Indian economy. From 1995 to 2014, almost 3,00,000 people in the farming community [4] committed suicide due to insufficient agricultural revenue. The numbers of 10,677 farmer suicides in 2020 [5] show that the dreadful situation persists. The monsoon in India is unpredictable and changes according to geographical region. In some parts of India like Kerala, it causes flood; whilst Rajasthan and Gujarat are drought-prone states. A delayed monsoon may decrease the yield and increase the price of essential commodities, thus accelerating food inflation.

It is critical to forecast rain in order to boost agricultural output and consequently the economy. Having an accurate and timely method to predict rain, its intensity and duration is vital to plan agricultural activities. Additionally, it contributes significantly in mitigating and minimising the adverse impact of natural phenomena such as landslides, cyclones, flash floods, avalanches, and droughts. Rainfall triggered waterborne diseases can be brought under control using this early warning system. The prediction of weather and rainfall is not something novel, it dates back to the Babylonian era. In the last decade, remarkable viewpoints are unwrapped by several weather prediction models. But, to date, none of them seem to be accurate. Despite advancements and innovations in science and technology, the failure to build a precise forecasting model is due its complexity [6]. The information needed to predict rainfall can't be taken from future depends on collections of past data and models. As rainfall depends on various factors like precipitation, temperature, wind direction, speed, and pressure that are chaotic and random in nature, estimates have a large error band, resulting in unreliability.

Several statistical and deep learning models have been developed on varieties of data like physical data, remote sensing data and combinations of them to build an accurate forecasting model. Despite the recent advancements in hardware, technologies, computational power, and availability of data, prediction of rainfall still remains a concern and needs improvement unceasingly. Despite widespread usage of statistical models like moving average (MA) and its superior variants, they possess several disadvantages such as inclusion of past errors and yield best results only on stationary data. To address the limitations of statistical models, focus on forecasting has been shifted to developing deep learning models. The main advantage is that they do not require parameter calculation and support ease of handling multivariate data. On the negative side, these modes require large sized datasets to achieve good accuracy.

Based on the current requirements and research gap, the primary goal of this study is to forecast annual rainfall of Tamil Nadu. The trend and seasonality factors have been extracted and analysed. Finally, rainfall forecasting models based on long short-term memory, its modifications, and the gated recurrent unit are developed, and their performances are analysed and compared.

2 Related Works

The literature review on rainfall forecasting using machine learning (ML) models showcased those two flavours of ML has been adopted. The classical ML modes like artificial neural networks, support vector machine, linear regression, and deep learning techniques like LSTM and GRU. Few samples of research works conducted on rainfall prediction using machine learning approaches have been summarised in Table 1.

Table 1 Summarization of rainfall forecast

Ref	Country	Dataset	Long/short term	Method	Evaluation metrics	Rainfall measure parameters
[7]	India	Monthly rainfall data for period 1939–1994	Short term	ANN	RMSE	Temperature
[8]	India	Monthly rainfall data	Short term	Multilayer Perceptron	RMSE	Temperature
[9]	Australia	Monthly rainfall data for period 1900–2009	Short term	RNN	RMSE	Historical rainfall data
[10]	India	Monthly rainfall data for period 1871–2010	Short term	ANN	RMSE, mean, variance	Historical rainfall data
[11]	Korea	Hourly data weather data from 2012	Short term	LSTM	RMSE	Temperature, wind, speed, humidity, sea surface pressure
[12]	China	Hourly weather data 2015 and 2016	Short term	LSTM	RMSE, MAE	Wind, temperature, humidity, radiation, and rainfall
[13]	India	Monthly weather data from July 1979 to January 2018	Short term	LSTM ConvNet	RMSE, MAPE	Infrared, microwave and rainfall data
[14]	Sri Lanka	Annual rainfall data from 1981 to 2010	Trend analysis	Linear regression	–	Rainfall

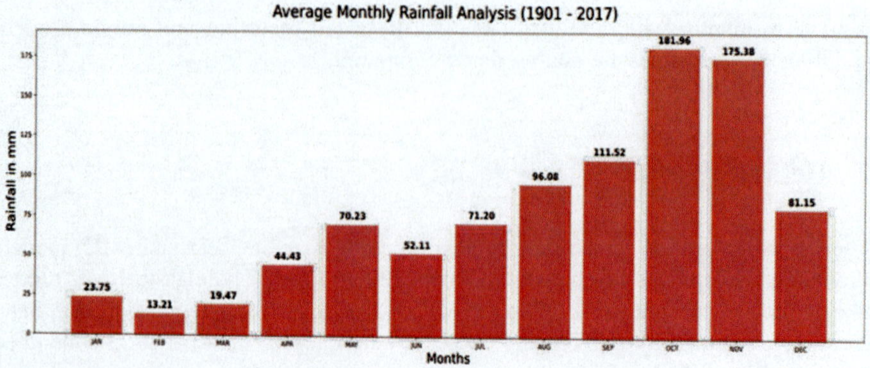

Fig. 1 Average annual rainfall in TN

3 Materials and Methods

3.1 *Study Region and Data Description*

The state of Tamil Nadu (TN) which lies on the southernmost part of India has been considered for this study. In this state, winter runs through the months of Jan and Feb where there is little rain. During the summer period (March—May), the heat is high, and there is scanty rainfall throughout the state. The state receives rain from South-West and North-East monsoon during monsoon and post monsoon periods. The South-West monsoon arrives on June and retreats on November. The North-East monsoon period lies between October and December. Figure 1 displays the average monthly rainfall for the period 1901 to 2017. Analysing 117 years of data, a bell curve (normal distribution) from the months June to December is observed, confirming that monsoons bring rain in June and recedes in December.

The historical rainfall data was obtained from India Meteorological Department (IMD) for entire India for 117 years (period 1901–2017). IMD divides the entire India into 36 meteorological subdivisions and records subdivision-wise hydrological data. The state of Tamil Nadu falls into 31st subdivision along with Puducherry and Karaikal. The monthly, annual, and seasonal (four seasons) rainfall data for the study area have been extracted from the entire dataset and analysed for further processing.

3.2 *Methods*

Long Short-Term Memory A recurrent neural network (RNN) is the key deep learning technique on time-series data to extract temporal correlations hidden in the data [15]. It has one to many hidden states distributed the temporal way and can

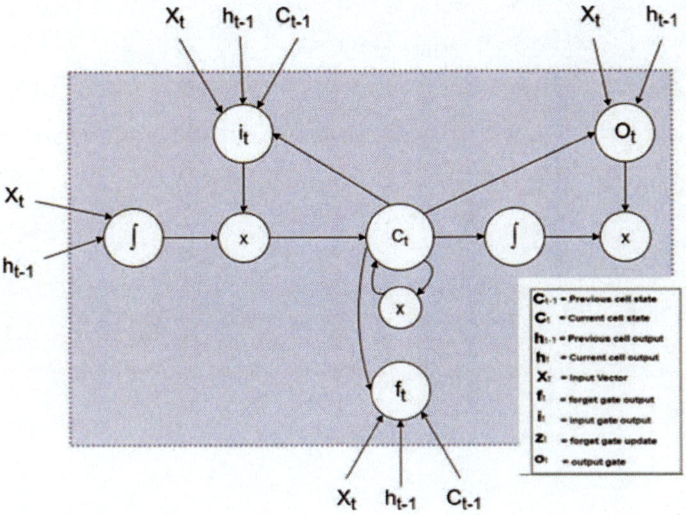

Fig. 2 LSTM cell

forecast the future with better accuracy than traditional methods. The major disadvantage with this method is its inability to overcome the vanishing gradient problem. To address this shortcoming, LSTM was developed to regularise the gradient flow [16]. Through memory cells (LSTM cells), LSTM can learn and uncover long timescale dependencies hidden in the data. The dissection of LSTM cell is shown in Fig. 2.

These dependencies and temporal correlation of the input are captured in the LSTM cell through the series of gates. The three main gates of LSTM, namely forget gate, input gate, and output gate, along with their corresponding activation function control the flow information. The computation at each gate in LSTM cell is shown in the equations [17].

$$i_t = \sigma(W_{\mathrm{xi}} x_t + W_{\mathrm{hi}} h_{t-1} + W_{\mathrm{Ci}} C_{t-1} + b_i) \tag{1}$$

$$f_t = \sigma\left(W_{\mathrm{xf}} x_t + W_{\mathrm{hf}} h_{t-1} + W_{\mathrm{Cf}} C_{t-1} + b_f\right) \tag{2}$$

$$o_t = \sigma(W_{\mathrm{xo}} x_t + W_{\mathrm{ho}} h_{t-1} + W_{\mathrm{Co}} C_t + b_o) \tag{3}$$

$$c_t = \sigma(f_t c_{t-1} + i_t \tanh(W_{\mathrm{xc}} x_t + W_{\mathrm{hc}} h_{t-1}) + b_c) \tag{4}$$

$$h_t = o_t \tanh(c_t) \tag{5}$$

where σ and tanh are non-linear activation functions.

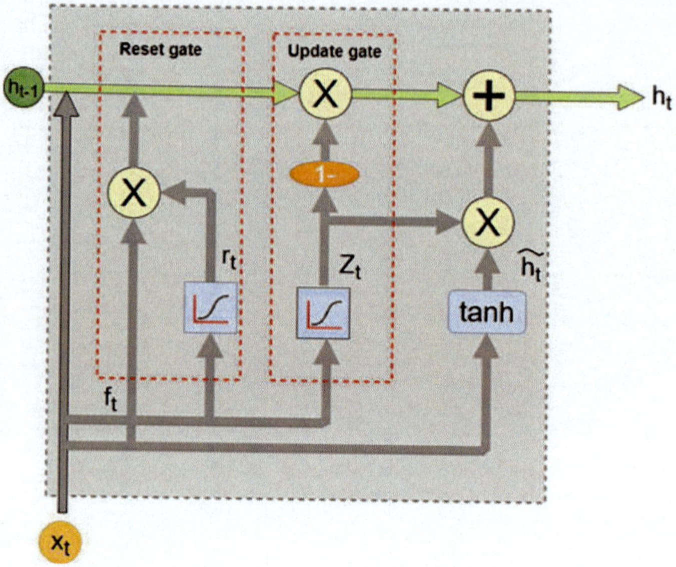

Fig. 3 GRU cell

Gated Recurrent Unit GRU is a simpler version of LSTM as it has fewer number of gates and no internal memory. The gate architecture of GRU is similar to LSTM. The forget gate that decides on what information to be retained and discarded and the input gate that decides on the relevant and important information are combined as a single reset gate. Long-term memory is handled by the update gate, and reset gate acts as a short-term memory unit. It calculates the hidden state (h_t) in two steps: (i) computation of candidate hidden state and (ii) computation of hidden state from candidate hidden state. GRU works well for smaller datasets and is faster to train due to its architecture. The GRU architecture is depicted in Fig. 3.

3.3 Evaluation Metrics

The proposed models are evaluated using root mean square error (RMSE). This metric is frequently employed in forecasting models to assess how well they perform in terms of error. It is calculated using the formula given in Eq. (6).

$$\text{RMSE} = \sqrt{\frac{1}{n}\sum_{i=1}^{N}\left(y_{\text{pred}} - y_{\text{actual}}\right)^2} \tag{6}$$

where N is total samples. y_{pred} is predicted value, and y_{actual} is actual value.

3.4 *Proposed Metrics*

Stacked LSTM It is a simple model with one LSTM hidden layer with 2000 vertically stacked memory units. It is followed by three dense layers, each of which has 500 neurons. A drop out layer with drop out probability of 0.4 is added after the LSTM hidden layer and dense layers to reduce overfitting. This layer will deactivate 40% of the neurons during training to minimise the complexity of the model. The final output dense layer has 1 neuron with linear activation function. As the forecast output value need not to be bounded between 0 and 1, linear activation function is chosen. On trial-and-error basis, Adam optimiser has been chosen. The input time sequence is set to 1 considering the significance of one whole year's rainfall history.

Bidirectional LSTM (BiLSTM) The proposed bidirectional LSTM model is shown in Fig. 4. It has three bidirectional LSTM hidden layers, each with 100 units. It can be viewed as two LSTM models applied on the data. In the first pass, the data is applied from past to future, and the model learns the data in the same order. In the second pass, it learns the reverse order of the data. When data is fed in both directions, the dependency in the data can be learnt more accurately. The ReLU activation function is applied on hidden layers due to its faster computation nature. Batch normalisation layer is added to standardise the input. It is followed by an output dense layer. A drop out layer with drop out probability of 0.5 is added. RMSprop is the chosen optimiser.

Gated Recurrent Unit (GRU) GRU network is built for this study which is shown in Fig. 5. A three-layer GRU architecture with 150 units in each layer is proposed. A dropout layer with dropout probability of 0.5 is included after every GRU hidden layer to reduce overfitting. It is followed by an output dense layer. This model is designed with stochastic gradient descent (SGD) and mean squared error loss function.

4 Results and Discussion

All the proposed models are built using appropriate software and hardware in the Google Colab environment. The dataset is divided into two parts in the ratio 80:20, namely the training and validation set. The train dataset is used to build the proposed models, and the validation dataset is used for testing. The forecasting for all proposed models is based on the 'annual rainfall' attribute. The proposed models are trained for various number of epochs (100, 150, 500, and 600). The training-validation loss curve shown in Fig. 6a, b, c, d, e, f is used to interpret the performance of the models—whether they underfit, overfit, or perfectly fit the data. Underfitting models have high bias, meaning that training loss will not decrease with increase in epochs. This indicates that the model is not able learn from the training data. On the other hand, overfitting indicates high variance. The model can perform well on the training data but poorly on unseen data. This means that the model cannot generalise well.

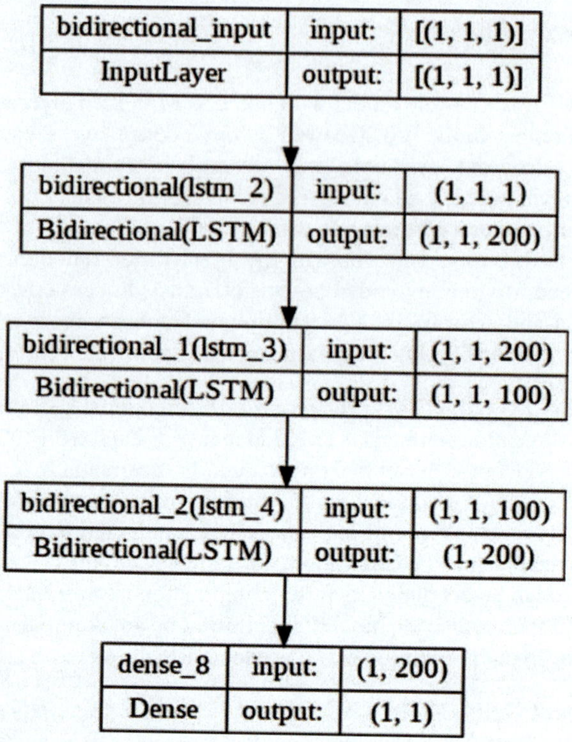

Fig. 4 Proposed bidirectional LSTM

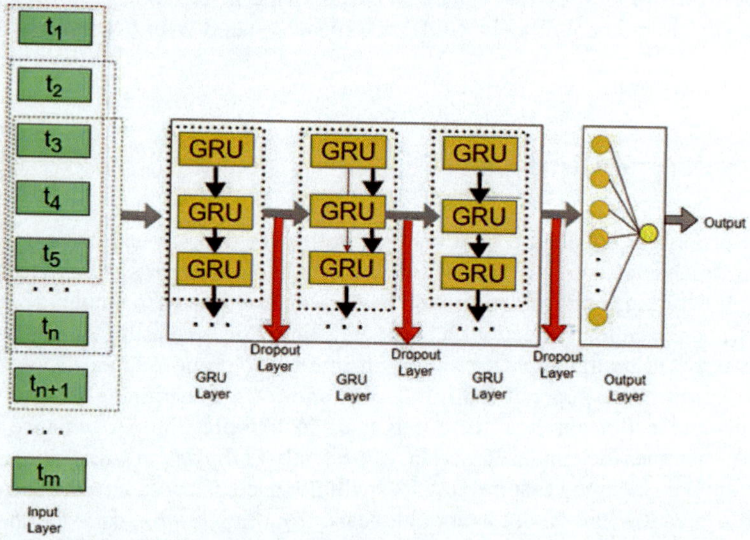

Fig. 5 Proposed GRU model

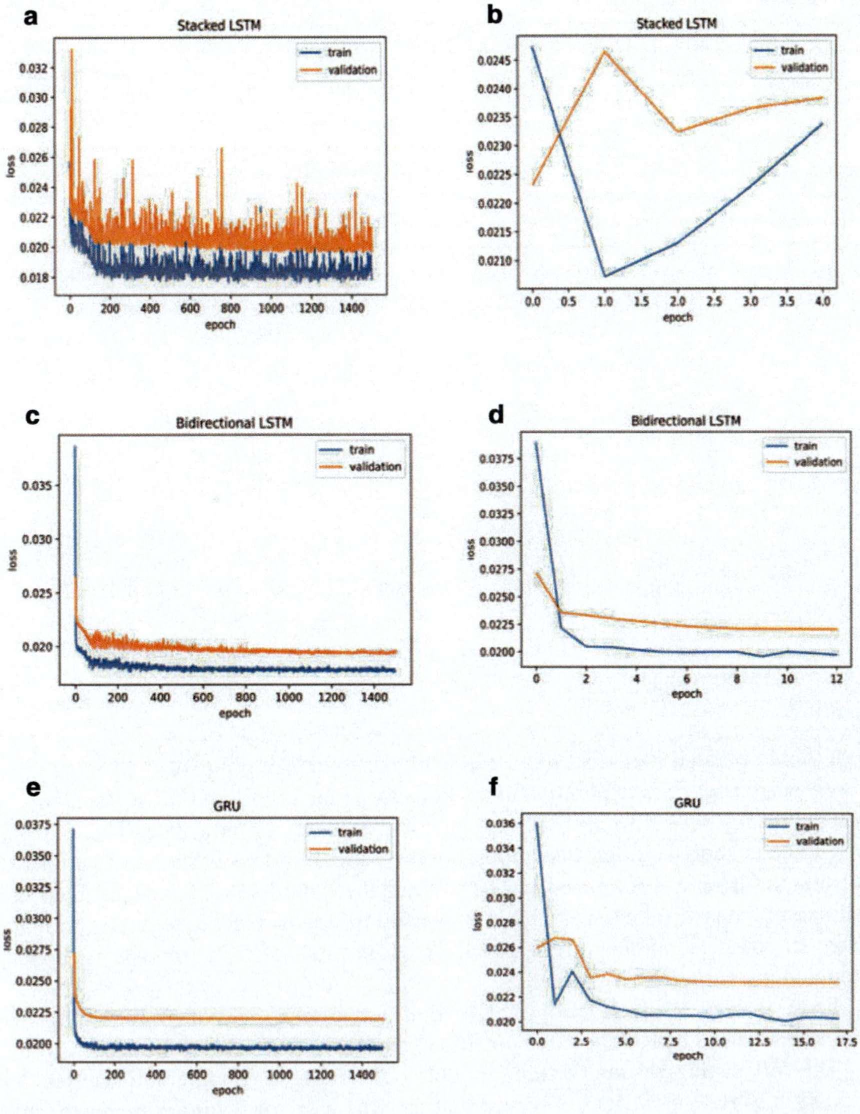

Fig.6 Training-validation loss plot **a** stacked LSTM loss plot, **b** stacked LSTM loss plot (early stop), **c** bidirectional loss plot, **d** bidirectional loss plot (early stop), **e** GRU loss plot, **f** GRU loss plot (early stopping)

Table 2 Performance measures

Proposed models	RMSE (train dataset)	RMSE (test dataset)	MAD	Correlation coefficient
Stacked LSTM	61.74	64.23	14.3	0.826
Bidirectional LSTM	57.03	60.19	11.7	0.915
GRU	68.19	61.76	12.9	0.891

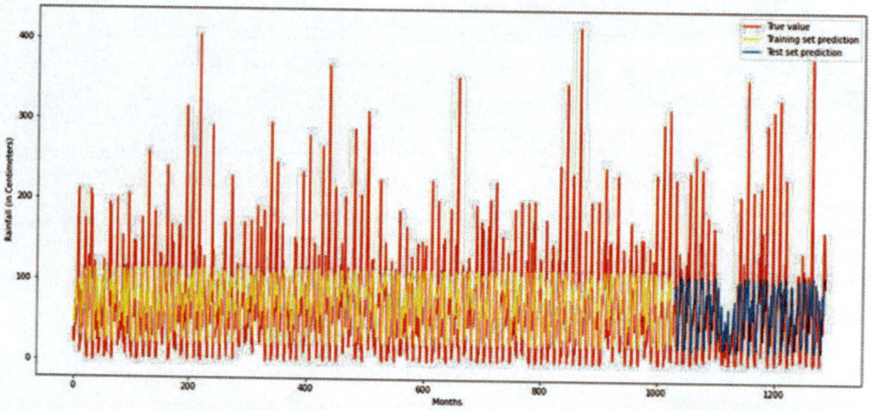

Fig. 7 LSTM forecast plot

LSTM revealed that validation loss is higher than training loss and shoots up at several instances of data points. This indicates that the proposed bidirectional model has overfitting problems and will not able to generalise on new data. The loss plot of GRU has shown similar behaviour, but the validation loss is smaller than that of the first model and is stabilised at 0.025. When these models are tested on an unseen validation dataset, they have exhibited overfitting issues. Hence, regularisation and early stopping techniques are applied. The loss plots of three models with early stopping are shown in Fig. 6b, d, f.

Error measures of the proposed models are calculated both on training and validation data. The RMSE value obtained from the experiment is shown in Table 2.

The forecasting results of three models on training and validation datasets are plotted in Figs. 7, 8, 9. It is observed that the RMSE error is higher than expected, and hence, data needs to be studied and analysed. As the data has high seasonality, in the future work, the data will be smoothened using moving average method.

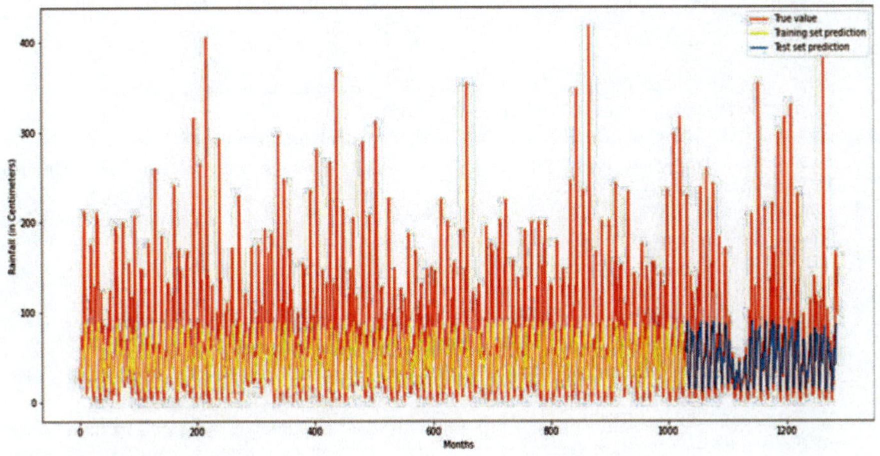

Fig. 8 Bidirectional forecast plot

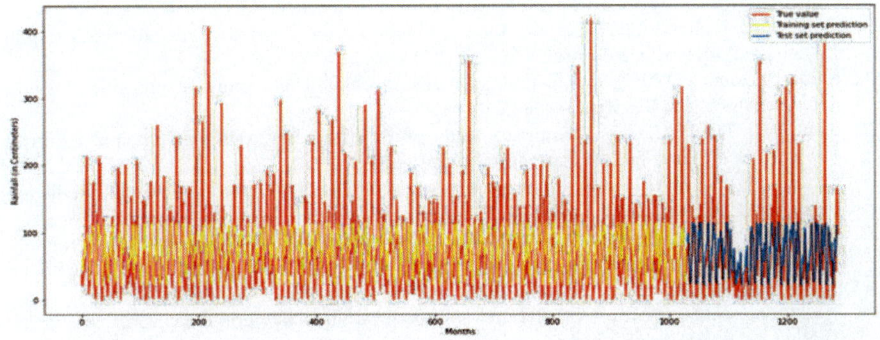

Fig. 9 GRU forecast plot

5 Conclusion

This study proposed two LSTM variant models and a GRU model to forecast annual rainfall in Tamil Nadu. Data sample of 117 years has been used. The results shows that the correlation coefficient is 91% for BiLSTM and 89.1% for GRU. The superiority of BiLSTM is endorsed with low RMSE and MAD mean absolute deviation (MAD) values. The future scope of this study involves taking a new perspective to detect change of point in the rainfall pattern using Pettitt and Alexanderson tests. This will help to analyse and detect any abrupt change in the annual rainfall data. Based on the analysis, we intend to design an appropriate deep learning model to make a short-term and a long-term forecast.

References

1. Trinh TA (2018) The impact of climate change on agriculture: findings from households in Vietnam. Environ Resource Econ 71(4):897–921
2. Hartmann H, Snow JA, Su B, Jiang T (2016) Seasonal predictions of precipitation in the Aksu-Tarim river basin for improved water resources management. Global Planet Change 147:86–96
3. Bui A, Johnson F, Wasko C (2019) The relationship of atmospheric air temperature and dew point temperature to extreme rainfall. Environ Res Lett 14(7):074025
4. Jain A (2017) A study of trends and magnitude of farmer suicides in India. Int J Adv Sch Res Allied Educ 13(2):80–85
5. Bhattacharyya S, Venkatesh P, Aditya KS, Burman RR (2020) The macro and micro point of view of farmer suicides in India. Natl Acad Sci Lett 43(6):489–495
6. Zhao Q, Liu Y, Ma X, Yao W, Yao Y, Li X (2020) An improved rainfall forecasting model based on GNSS observations. IEEE Trans Geosci Remote Sens 58(7):4891–4900
7. Venkatesan C, Raskar SD, Tambe SS, Kulkarni BD, Keshavamurty RN (1997) Prediction of all India summer monsoon rainfall using error-back-propagation neural networks. Meteorol Atmos Phys 62(3):225–240
8. Chattopadhyay S (2006) Anticipation of summer monsoon rainfall over India by artificial neural network with conjugate gradient descent learning. arXiv:nlin/0611010, arXiv:nlin/0611010v1, https://doi.org/10.48550/arXiv.nlin/0611010
9. Abbot J, Marohasy J (2012) Application of artificial neural networks to rainfall forecasting in Queensland, Australia. Adv Atmos Sci 29(4):717–730
10. Singh P, Borah B (2013) Indian summer monsoon rainfall prediction using artificial neural network. Stoch Env Res Risk Assess 27(7):1585–1599
11. Kim H-U, Bae T-S (2017) Preliminary study of deep learning-based precipitation. J Korean Soc Surv Geod Photogramm Cartogr 35(5):423–429
12. Chao Z, Pu F, Yin Y, Han B, Chen X (2018) Research on real-time local rainfall prediction based on MEMS sensors. J Sens 2018(6184713):1–9
13. Aswin S, Geetha P, Vinayakumar R (2018) Deep learning models for the prediction of rainfall. In: 2018 International conference on communication and signal processing (ICCSP). IEEE, pp 657–661
14. Wickramagamage P (2016) Spatial and temporal variation of rainfall trends of Sri Lanka. Theoret Appl Climatol 125(3):427–438
15. Sherstinsky A (2020) Fundamentals of recurrent neural network (RNN) and long short-term memory (LSTM) network. Physica D Nonlinear Phenom 404:132306
16. Schmidhuber J, Hochreiter S et al (1997) Long short-term memory. Neural Comput 9(8):1735–1780
17. Yu Y, Si X, Hu C, Zhang J (2019) A review of recurrent neural networks: LSTM cells and network architectures. Neural Comput 31(7):1235–1270

Analysis of Weather Forecasting and Prediction Using Neural Networks

Manish Choubisa, Manish Dubey, Surendra Kumar Yadav, and Harshita Virwani

Abstract Over the past few decades, weather forecasting has grown in importance as a research area but the traditional methods are still not forecasting the weather conditions accurately. It is now one such domain that attracted many researchers because it is well connected with the human life. In this research study, several weather constraints were collected from open access platform and using linear regression (LR) and deep neural network (DNN), the model is trained for different combinations. The dataset was taken from Kaggle (an online platform for public data). The results achieved using this model are compared and analyzed through mean absolute error and median absolute error between actual and predicted values.

Keywords Linear regression · DNN regressor · OLS · Weather prediction

1 Introduction

The concept of weather forecasting using machine learning can be defined in a simplified manner as a method with a particular approach in which, based on the previous observed pattern or behavior of weather over a given area and timeline, considering *n* number of factors as well as their dependencies, the weather conditions of the future time is predicted. More emphasis is laid upon the reliability of the models used by testing them frequently along with the help of testing datasets and once the

M. Choubisa (✉) · M. Dubey · S. K. Yadav · H. Virwani
Poornima College of Engineering, Jaipur, Rajasthan, India
e-mail: ermanishchoubisa@gmail.com

M. Dubey
e-mail: manish.dubey@poornima.org

S. K. Yadav
e-mail: surendra.yadav@poornima.org

H. Virwani
e-mail: harshita.virwani@poornima.org

S. Sharma et al. (eds.), *Intelligent Control, Robotics, and Industrial Automation*, Lecture Notes in Electrical Engineering 1066,
https://doi.org/10.1007/978-981-99-4634-1_76

model proves itself reliable enough based on the accuracy measure, the model is set ahead to be trained with the training dataset.

For multiple and crucial applications, weather prediction is quite a necessary domain including military utilities, agriculture and production, aviation industry, etc. These are some of the verticals where weather forecasting plays a major role in mitigating the risk.

Precise weather prediction is a tactful task because of the dynamic nature of Earth's atmosphere, at any instance the weather condition over particular demographics can be represented by some variables. Some of these variables are more significant and hence decisive then others, and hence with the help of redundancy techniques priority can be assigned to these variables [1].

Weather forecasting systems must be clever enough to interpret statistical data and establish patterns and rules to research and predict the future based on past data. A numerical weather prediction model might be used to reflect the global importance of forecasting. They used sub grid lakes in global forecasts to reduce forecast errors at high latitudes, particularly in the spring and summer [2].

Although the technologies, such as radars, wireless sensors, and satellites, are more than adequate for dealing with weather circumstances, these advancements and enhancements are the consequence of betterment for the future and under computational methodologies, the better observation. To increase accuracy, numerous academics have worked to discover well-defined forecasting models that include linear, nonlinear, and hybrid methodologies [3, 4].

Recent research has found that hybrid computational models can result in the decomposition of linear and nonlinear forms in some situations, making them a more efficient approach than single or individual techniques. It also has a lower probability of making mistakes. On the other hand, other research argues that hybrid or combined models are ineffective. In this instance, such uncertainties always open the door to new chances to improve model precision. Even with the large penalty of imbalance systems, relying on meteorological forecasts all of the time is not a good idea.

2 Related Work

The majority of the research on weather forecasting applications has been done using machine learning methods. In this part, some of the work is condensed.

From last several years, there have been numerous technological advancements in order to solve weather prediction problem using numerical weather prediction (NWP) and statistical modeling using machine learning techniques which are achieving successful results. The input data used for forecasting is high-dimensional data collected from weather stations, satellites, Doppler radar from different regions. The study by Salman et al. [5] proposes recurrent neural network (RNN) can be carried out in prediction rainfall with good enough accuracy degree.

Singh et al. [6] presented "Weather Forecasting Using Machine Learning Techniques" in the year 2019. This research investigates three machine learning (ML) models for weather prediction: SVM, ANN, and a time series-based RNN. The weather is predicted using an RNN using time series, a linear SVC, and a five-layered neural network. The root mean squared error between the anticipated and actual values is used to examine and compare the results of various models. After analysis of all the models, it is observed that time series RNN is better than SVM and ANN for this problem. Time series RNN had RMS values of 1.41, ANN had value of 3.1, and SVM had value 6.67.

Samya and Rathipriya [7] published "Predictive Analysis for Weather Prediction Using Data Mining with ANN: A Study" in 2016. The primary goal of this research is to examine the various weather forecasting techniques that use data mining and artificial neural networks (ANN). Temperature, rainfall, and wind speed are the most often utilized factors for interpreting weather forecasts. In the future, Big Data and data analytics will be used to achieve 100 percent accuracy. According to the findings, data mining techniques, ANN, and fuzzy logic result in higher accuracy.

Biradar et al. [8] published "Weather Prediction Using Data Mining" in 2017. This method will forecast weather based on variables like temperature, humidity, and wind speed. The fluctuation in historical weather circumstances must be used to anticipate future weather conditions. It is extremely likely that it will match within the next two weeks of the preceding year. This study proposes the use of K-medoids and the Naive Bayes algorithm for weather forecasting with variables such as temperature, humidity, and wind. The prediction can be considered reliable as the forecast would be based on previous record.

Kunjumon et al. [9] used many dimensions for weather survey on numerous approaches for weather prediction. Kareem et al. [10] built weather forecasting prediction models using neural networks, Naive Bayes, random forests, and K-nearest neighbor algorithms. These models categorize the instances of unobserved data into numerous classes, including rain, fog, partly cloudy day, clear day, and cloudy day. Synoptic data from the Kaggle website was used to train and test these models' performance for each method. We have (1796) instances and (8) characteristics for this dataset. The random forest algorithm, when compared with other algorithms, produced the best performance accuracy of 89 percent.

Zofishan and Ghazal [11] discuss a critical review on weather forecasting using data mining applications. They present weather dataset is based on metrological parametric input data collected from the secondary sources, in order to purify the already available dataset according to the implemented techniques and methodologies related to the different fields of data mining. Metrological parametric input dataset based on basic parameters, e.g., temperature, rainfall, humidity, wind, clouds, and snow conditions. Grace and Suganya [12], worked on to predict the rainfall for the Indian dataset using multiple linear regression and provide improved results in terms of accuracy, MSE, and correlation. The general comparison of the present related work is given in Table 1.

Table 1 Hypothesis and limitations of the several contributed methodologies

Author	Title	Year	Techniques employed	Limitations
Salman et al. [5]	Weather forecasting using deep learning techniques	2015	Recurrence neural network (RNN), conditional restricted Boltzmann machine (CRBM), and convolutional network (CN) models	Convolutional neural network (CNN) that offers the accurate representation, classification and prediction on a multitude of time series problems and compared with the shallow approaches when configured and trained properly
Singh et al. [6]	Weather forecasting using machine learning techniques	2019	Support vector machine (SVM), artificial neural network (ANN), and a time series-based recurrent neural network (RNN)	Authors find that time series-based RNN does the best job of predicting the weather but the implementation is more complex
Samya and Rathipriya [7]	Predictive analysis for weather prediction using data mining with ANN: a study	2016	Artificial neural network (ANN)	Authors used the simple artificial neural network as case study
Biradar et al. [8]	Weather prediction using data mining	2017	Decision trees and k-mean clustering	Only a comparison is made by authors which shows that decision trees and k-mean clustering are best suited data mining technique for weather prediction
Kunjumon et al. [9]	Survey on weather forecasting using data mining	2018	Artificial neural network, support vector machine, FP growth algorithm, Hadoop with map reduces, K-medoids algorithm, Naive Bayes algorithm, and decision tree classification algorithm	Not implement any such methods
Kareem et al. [10]	Predicting weather forecasting state based on data mining classification algorithms	2021	Random forest; Naïve Bayes; K-nearest neighbor	Only classification algorithms are used

(continued)

Table 1 (continued)

Author	Title	Year	Techniques employed	Limitations
Zofishan and Ghazal [11]	A critical review on weather forecasting using data mining applications	2021	K-mean clustering, ANN	Hidden layers of ANN are not applied
Grace and Suganya [12]	Machine learning-based rainfall prediction	2020	Multiple linear regression	Only predicts rainfall

3 Methodology

This section introduces the practice being used and the analysis performance. There are various machine learning and deep learning algorithms. But for this study we are using linear regression with OLS method and deep neural network (DNN) regressor for advancement for prediction.

3.1 Linear Regression with OLS Method

Linear regression is primarily used for supervised machine learning. Supervised machine learning tasks are classified into regression and classification. In regression, numerical or continuous values as target are predicted based on features. Linear regression models are extensively used in practical applications because the model which are linearly dependent on unknown parameters are easier to fit than models which are nonlinearly related to the parameters [13].

Generally, linear regression is defined in terms of equation $Y = a + bX$, where an independent variable is defined by X and dependent variable defined by Y. The line slope is defined by b and a represent the intercept.

$$a = \frac{(\sum y)(\sum x^2) - (\sum x)(\sum xy)}{n(\sum x^2) - (\sum x)^2} \tag{1}$$

$$b = \frac{n(\sum xy) - (\sum x)(\sum y)}{n(\sum x^2) - (\sum x)^2} \tag{2}$$

The ordinary least squares regression (OLS) estimates the unidentified attribute in regression model. It chooses the attributes of liner function by the least squares.

3.2 Deep Neural Network (DNN)

When there is a consistency of multiple layers between the input and output layers in an artificial neural network (ANN), then it is recognized as a deep neural network (DNN). Though there can be a variety of neural networks, the base components that can be identified are in fact common neurons, synapses, weights, biases, and functions.

When multiple layers of nodes are put to use in order to derive high-level functions from available input information, such a type of machine learning represents a deep neural network. Hence the transformed creative and abstract component of data elucidates the use of deep neural networks [14].

A contextual hierarchical (layered) organization of neurons (similar to the neurons in the brain) fastened to other neurons is a rather streamlined version of deep neural network. Based on the accrued input, information in the form of a message or a signal is bequeathed to the subsequent other neurons. This leads to formation of a complex network that grasps and learns with feedback mechanisms [15].

3.3 Research Framework

Figure 1 shows the framework which is used for studying weather prediction data model.

The experiment phases used for the analysis are: training phase of linear regression with OLS and DNN regressor models and testing each of these models.

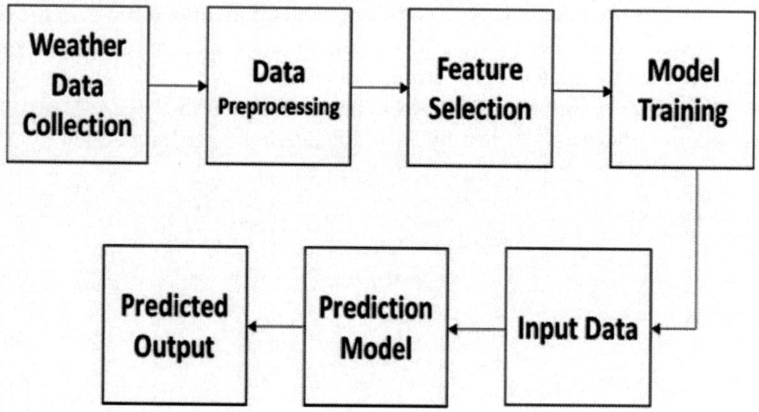

Fig. 1 Framework for data model

3.4 Dataset

The dataset was taken from Kaggle (an online platform for public data) consisting 40 features and 676 instances over a period of two years from May 4, 2016 to March 11, 2018. The weather dataset contains features such as date, max temperature, mean temperature, mean dew point, minimum temperature, max humidity, minimum humidity, precipitation temperature, max dew point in the Jaipur city of Rajasthan, India. Two weather prediction models will be explored in this study which are namely: (i) linear regression with OLS method and (ii) deep neural network (DNN) regressor.

3.5 Linear Regression with OLS Method

One approach to evaluate the linearity between independent variable, which until further notice will be the mean temperature, and the other free factors is to compute the Pearson connection coefficient. To evaluate the connection in this information, we will call the corr() strategy for the Pandas dataframe object.

This will yield the output from most negatively related to most positively related. Features having less absolute value than 0.6 will be removed from the dataset. We have seen that the maxtempm and mintempm variables are useless as for the prediction of meantempm, we will eliminate the nan values using fillna() method.

In order to find linearity between variables, we are using matplotlib pyplot module to graph the linear relationship between the independent variable and dependent variable.

As we know that a linear regression model utilizes statistical test for selection, so to locate the statistically significant features, statsmodels library has been used. After finding highest p-value predictor, compare if it is greater than our selected alpha. After finding the predictor we will drop it from our dataframe.

3.6 DNN Regressor

The removal of the maxtempm and mintempm columns from the dataframe does not affect in predicting the average meantempm and we have separated targets(Y) and features(X).

We have defined a function, i.e., reusable which we can refer as an input function that will call wx_input_fn(). For feeding the data into neural network during the testing and training phases, we will use this function. After defining our input function, we can train our model on training dataset.

4 Performance Evaluation

In order to evaluate the quality of fitted model regressor model score() function is used. This function proves that the given model is able to justify the given explained variance collected from outcome variable.

Also, the difference between the actual values and predicted values can be calculated using mean absolute error (MAE) and median absolute error.

The mean absolute error is average of all absolute error. For computing the absolute error, the formula is:

$$(\Delta x) = |X_i - X| \tag{3}$$

To calculate the MAE:

$$\text{MAE} = \frac{1}{n}\sum_{i=1}^{n}|X_i - X| \tag{4}$$

To measure how the data is spread we use median absolute error or median absolute deviation (MAD).

For a univariate data set $X_1, X_2, \ldots, X_n$, the MAD is defined as the median of the absolute deviations from the data's median.

$$\tilde{X} = \text{median}(X) \tag{5}$$

To calculate the MAD:

$$\text{MAD} = \text{median}\left(\left|X_i - \tilde{X}\right|\right) \tag{6}$$

5 Result and Discussion

In this presented method of neural network, the training datasets are separated in to subsets with testing and training set. Using these methods, the predictor mean temperature is the independent variable. By using linear regression the regressor model's score() function proves that our model is able to justify 95% of the variance collected from the outcome variable, i.e., mean temperature. In the Fig. 2 the variance for different attributes in dataset is displayed using matplotlib pyplot module.

Table 2 gives the experimental result values in terms of explained variance, mean absolute error and median absolute error. The mean_absolute_error() and median_

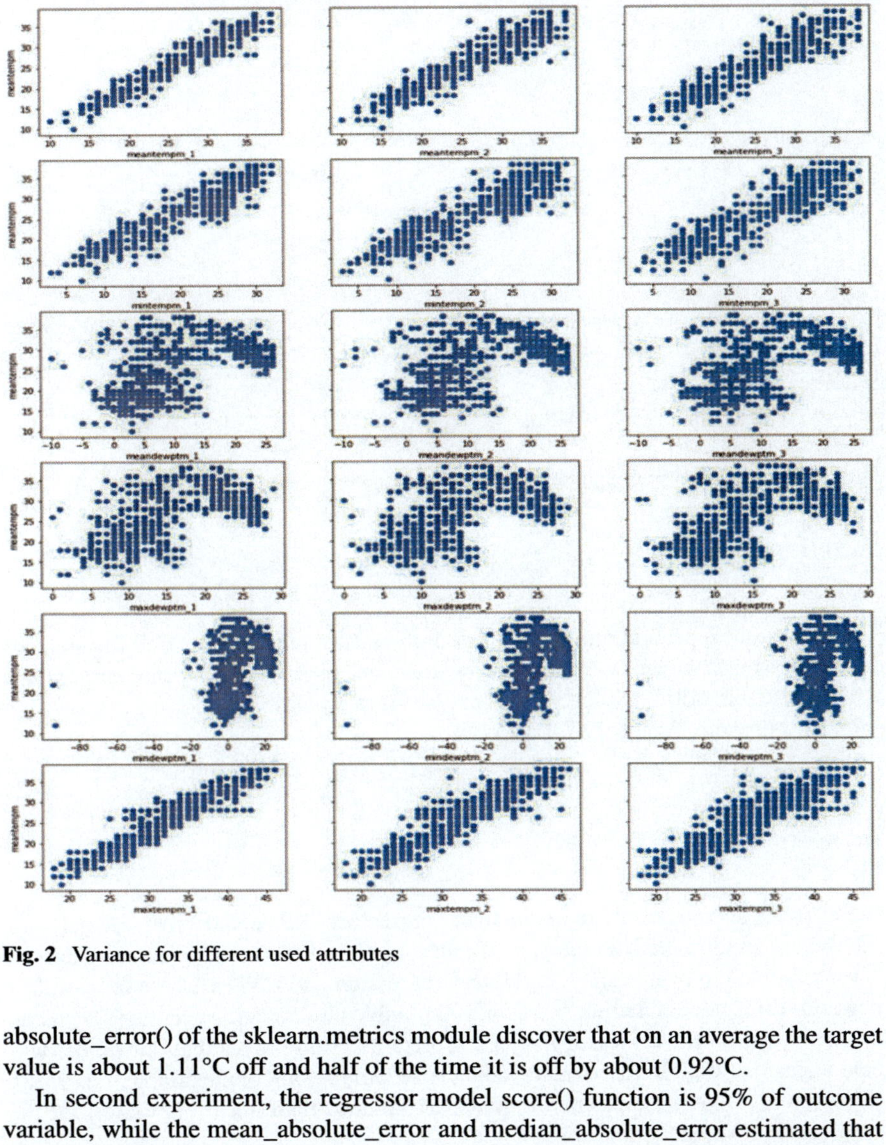

Fig. 2 Variance for different used attributes

absolute_error() of the sklearn.metrics module discover that on an average the target value is about 1.11°C off and half of the time it is off by about 0.92°C.

In second experiment, the regressor model score() function is 95% of outcome variable, while the mean_absolute_error and median_absolute_error estimated that the predicted value is about 1.09°C off and half off the time it is off by 1.10°C. There

Table 2 Experiment result

Model	Explained variance	Mean absolute error	Median absolute error
Linear regression with OLS method	0.93	1.31	0.96
DNN regressor	0.95	1.09	0.90

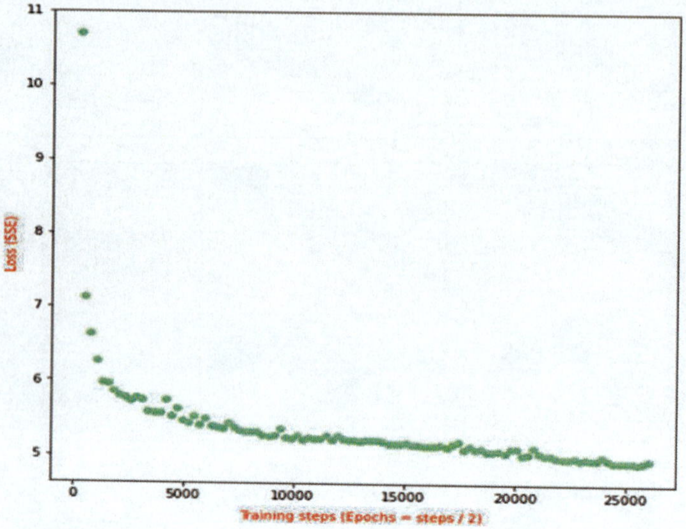

Fig. 3 Training steps and loss

is a collection of evaluation for each iteration and we plot them as training steps of function, so that we can validate that, we have not overtrained our model therefore we will use matplotlib pyplot module to plot simple scatter plot. Figure 3 shows the training steps and loss of the used model.

6 Conclusion

In this paper, deep learning methods linear regression (LR) and deep neural network (DNN) are used to analyze the performance of high-dimensional weather datasets. The experimental results show that linear regression and DNN are able to learn fast in a high-dimensional dataset and achieved notable classification accuracy. By using linear regression the regressor model's score() function proves that our model was able to justify 95% of the variance collected from the outcome variable, i.e., mean temperature. DNN has achieved higher performance than linear regression for the prediction of weather conditions.

References

1. Chandrayan SS, Singh K, Bhoi AK (2022) Atmospheric weather fluctuation prediction using machine learning. In: Cognitive informatics and soft computing. Lecture notes in networks and systems, vol 375. Springer, Singapore, pp 431–443
2. Abrahamsen EB, Brastein OM, Lie B (2018) Machine learning in python for weather forecast based on freely available weather data. In: Proceedings of the 59th conference on simulation and modelling (SIMS 59), pp 169–176
3. Kalaiyarasi P, Kalaiselvi A (2018) Data mining techniques using to weather prediction. Int J Comput Sci Trends Technol (IJCST) 6(3):249–254
4. Ehsan BMA, Begum F, Ilham SJ, Khan RS (2019) Advanced wind speed prediction using convective weather variables through machine learning application. Appl Comput Geosci 1:100002
5. Salman AG, Kanigoro B, Heryadi Y (2015) Weather forecasting using deep learning techniques. In: 2015 International conference on advanced computer science and information systems (ICACSIS). IEEE, pp 281–285
6. Singh S, Kaushik M, Gupta A, Malviya AK (2019) Weather forecasting using machine learning techniques. In: Proceedings of 2nd international conference on advanced computing and software engineering (ICACSE)
7. Samya R, Rathipriya R (2016) Predictive analysis for weather prediction using data mining with ANN: a study. Int J Comput Intell Inform 6(2):150–154
8. Biradar P, Ansari S, Paradkar Y, Lohiya S (2017) Weather prediction using data mining. IJEDR 5(2):211–214. ISSN: 2321–9939
9. Kunjumon C, Nair SS, Rajan SD., Suresh P, Preetha SL (2018) Survey on weather forecasting using data mining. In: 2018 Conference on emerging devices and smart systems (ICEDSS). IEEE, pp 262–264
10. Kareem FQ, Abdulazeez AM, Hasan DA (2021) Predicting weather forecasting state based on data mining classification algorithms. Asian J Res Comput Sci 9(3):13–24
11. Zofishan M, Ghazal F (2021) A critical review on weather forecasting using data mining applications. Adv Res Comput Networking 13:247–278
12. Grace RK, Suganya B (2020) Machine learning based rainfall prediction. In: 2020 6th International conference on advanced computing and communication systems (ICACCS). IEEE, pp 227–229
13. Burton AL (2021) Chapter 104–OLS (linear) regression. In:The encyclopedia of research methods in criminology and criminal justice, vol 2, pp 509–514
14. Yadav SK, Chouhan Y, Choubisa M (2022). Predictive hybrid approach method to detect heart disease. Math Stat Eng Appl 71(1):36–47. https://doi.org/10.17762/msea.v71i1.40
15. Fatima N, Imran AS, Kastrati Z, Daudpota SM, Soomro A (2022) A systematic literature review on text generation using deep neural network models. IEEE Access 10:53490–53503

A Systematic Review on Latest Approaches of Automated Sleep Staging System Using Machine Intelligence Techniques

Santosh Kumar Satapathy, Hari Kishan Kondaveeti, and Debabrata Swain

Abstract Sleep staging plays a vital role in sleep research because sometimes sleep recording errors may cause severe problems like misinterpretations of the changes in characteristics of the sleep stages, medication errors, and, finally, errors in the diagnosis process. Because of these errors in recordings and analysis, the sleep behavior and automated sleep staging system are adopted by different researchers with different methodologies. This study identifies specific challenges with the existing studies and highlights certain points that support the improvement of automated sleep staging-based polysomnography signals. This work provides a comprehensive review of a computerized sleep staging system, which was contributed by the different researchers in the recent research developments using electroencephalogram, electrocardiogram, electromyogram, and combinations of these signals. Our review in this research area shows that single-modal and multi-modal signals are used for sleep staging. Also, we have observed some great points from the existing methodologies: (1) It has been noticed that the 30-s length of the epoch of EEG signals may not be sufficient to extract enough information for discriminating the sleep patterns, but in the other hand that a 10-s and 15-s length epoch is well suitable for sleep staging, (2) due to similar characteristics on N1 and REM sleep stages, most of the traditional classification models misclassified N1 sleep stages as REM stage, which alternatively degrades the sleep stagging performance, (3) consideration of heterogeneous form signal fusions can give the improvement results on sleep staging, and (4) applying deep learning models. The review mentioned above points simultaneously improves automated sleep staging by polysomnography signals. These points

S. K. Satapathy (✉)
Information and Communication Technology, Pandit Deendayal Energy University, Gandhinagar, Gujarat 382007, India
e-mail: Santosh.Satapathy@sot.pdpu.ac.in

H. K. Kondaveeti
School of Computer Science Engineering, VIT-AP University, Vijayawada, Andhra Pradesh 522237, India

D. Swain
Computer Science and Engineering, Pandit Deendayal Energy University, Gandhinagar, Gujarat 382007, India

S. Sharma et al. (eds.), *Intelligent Control, Robotics, and Industrial Automation*, Lecture Notes in Electrical Engineering 1066,
https://doi.org/10.1007/978-981-99-4634-1_77

can help to focus our research work from the traditional feature extraction method to systematic improvements such as automatic feature recognition without explicit features, a proper characterization of the sleep stage's behavior, safety, and reduced cost.

Keywords Sleep stage · Electroencephalogram · Machine learning · Deep learning

1 Introduction

Sleep is a continuous and dynamic process, which has its cyclic structural framework with a time course. There are two main types of sleep stages as observed in human sleep [1]. The sleep cycle covers NREM and REM sleep stages cyclically, and typically each cycle lasts on an average 45–60 min [2, 3]. There are mainly two sleep rules followed by Rechtschaffen and Kales (R&K) and the American Academy of Sleep Medicine (AASM). According to the R&K rule, mainly human sleep is divided into three main sleep stages: wake, rapid eye movement (REM), and non-REM (NREM). Based on the R&K rules, the entire sleep cycle is divided into REM, four sleep stages of NREM, namely N1, N2, N3, and N4, and the wake stage [4–6]. During sleep staging, the standard practice is to divide the sleep records into a 30-s length window, called the epoch. Each epoch represents certain sleep stage, and its representation plot is called a hypnogram, which generally provides the overall overview of sleep architecture and presents the information of sleep stage distribution chronologically. The sleep scoring procedure is characterized and analyzed by the presence of certain waves and events associated with the recorded signals. The dynamic pattern of the wake-NREM-REM sleep cycle provides information about sleep quality. It also helps to determine sleep efficiency, sleep quality, and sleep-related disorders [7]. The sleep quality can be determined based on the presence of different sleep stages from the hypnogram. Generally, the sleep cycles cover two phases, NREM and REM, which alternate in the sleep cycle, from the beginning of NREM Stage 1 to the REM stage. The average period usually lasts 90–110 min for each phase in a sleep cycle. Each sleep stage is characterized differently from the others by their different change characteristics during the sleep time. The typical changes in patterns of the brain in the different stages of sleep are briefly discussed here. However, the same brief discussion can be found in [8].

2 Clinical Importance

Across the world, most sleep centers obtain manual sleep staging practices in the early years. This has a good solution during that period. Still, several drawbacks are observed in the sleep staging process: (1) This approach consumes much time for manual interpretation of the sleep stages' behavior throughout the night. (2) There are also huge variations caught between experts in sleep staging annotations. (3) Laboratory set-up is also too expensive [9, 10].

These certain limitations forced the existence of an automated sleep staging process in the field of sleep research. It helps to reduce the cost and also provides accurate analysis of sleep patterns and less manual interpretation. Common standards are followed and no variations in the sleep scoring. This automated sleep scoring process also opens a new way for early detection of sleep-related diseases and quantifying sleep patterns. This automatic sleep scoring process is vital in diagnosing different sleep-related disorders [11, 12].

3 Visual Scoring Procedures

Before taking any specific diagnosis solutions, it is essential to understand the subject's sleep cycle and irregularities during sleep time. Therefore, sleep experts go for visual inspection of entire recordings of sleep hours. The first rules and regulations for the visual scoring procedure were published by the R&K in 1968. During the N2 sleep stage, subjects entered from slight sleep into a deep sleep, in which the eye movements and muscle activities are completely seized. During this stage, two major patterns have appeared that is sleep spindles and K-complexes as the N2 sleep stage progresses [13]. Afterward, the subjects can move into the Slow Wave Sleep (SWS) stage (N3 and then N4); during this stage, the most dominant patterns are delta wave patterns (0.5–2 Hz) of the EEG signal and no eye movements [14]. All the sleep centers or clinics for sleep staging followed this rule until 2007. The AASM re-edited the sleep standards regarding technical set-ups, annotation process, execution procedure, analytic scoring, and the interpretation of results. In AASM, the total number of sleep stages is reduced from seven to five stages. Till 2018(V2.5) in every year, there are some changes with AASM guidelines. The required data for visual scoring is recorded from the different body parts by placing other electrodes. The fixing of electrodes is set according to the 10–10 and 10–20 standards [15, 16]. It has been noticed that most sleep specialists prefer EEG signals for understanding the sleep behavior of the subjects. There are different combinations of derivations used for recording the sleep data.

4 Automatic Sleep Scoring by Artificial Intelligence

It has been found that there are some disadvantages seen during the manual sleep staging, firstly consuming more time for scoring the entire PSG recordings from the whole night sleep. The other main challenges with subject to manual scoring are the variations and inconsistency on sleep scoring, which mainly happens due to the different levels of the expertise domain experts. Such differences in sleep scoring create ambiguous sleep stages [3, 13, 17–28] similarly, [29–53] result analysis presented in this review work.

4.1 Deep Learning—Knowledge from Raw Data

Zhao et al. [19] proposed a dual-modal and multi-scale deep neural network to classify sleep stages. They used EEG and ECG signals to have three binary classifications: sleep and wake, light and deep sleep, and REM and NREM. Using the MIT-BIH database, they trained and implemented their algorithm to obtain accuracies of 88.80, 97.97, and 98.84%, respectively, for binary classification and 80.40% for four-class types. The MIT-BIH dataset, used for training and testing the proposed model, contains 18-night periods of sleep of 16 healthy adults ranging from age 32 to 56. The proposed model uses nine convolution layers to classify EEG signals and 13 convolutional layers to classify ECG signals. Li et al. [20] offered a convolution neural network called CCN-SE for classifying sleep stages. They have implemented their algorithm on three different datasets. The SHHS consisted of 100 near-normal PSG recordings. While Sleep-EDFx has two cohorts, SC and ST, from which 100 recordings sampled at 100 Hz were chosen, and the Fpz-Cz EEG channel was selected for single-channel sleep staging. The PSG files from the clinic in Hong Kong consisted of 3-channel EEG sampled at 256 Hz. Ten such subjects were chosen from the database. The algorithm's structure consisted of 3 CNN blocks and a SE block. Each of the CNN blocks was followed by batch normalization. The SHHS database obtained an overall accuracy of 86.70%, while the precision, recall, and F1-score are 70.80, 69.60, and 68.50%, respectively. Similarly, for Sleep-EDFx, the accuracy was 81.30%, while the precision, recall, and F1-Score were 70.30, 66.60, and 65.50%, respectively. Li et al. [13] worked on a sleep staging classification based on a single-channel EEG signal. BiSALnet adopts an SPD manifold structure to encapsulate desired feature distribution properties in the proposed architecture. BiSALnet is tested upon two databases, the NPH and the Sleep-EDF databases. The NPH database comprises 15 patient recordings, with classification done every 30 s among five classes: Wake, N1, N2, N3, and REM. The EEG signal recorded was from O1-M2. The Sleep-EDF database consisted of 8 patients' 30 s EEG epochs sampled at 100 Hz. The EEG signal used in Sleep-EDF is Fpz-Cz. The proposed model achieves an accuracy of 80.00% when tested on the NPH dataset, while sensitivity and F1-scores are 76.00% each. The accuracy improves when tested on the Sleep-EDF dataset reaching 91.00%, while sensitivity and F1-score are 75.00 and 77.00%, respectively. Zhang et al. [21] propose a model focusing on latent discriminative features since it has more receptive fields. The hyperbolic parts perform well in detecting the local information spots, and Manifold Learning Block working in cohesion with Hyperbolic Block helps in sleep feature extraction. The database used is the Sleep-EDF database which includes 197 PSG recordings. The EEG signal is divided into two parts based on electrode location. The Sleep Cassettes database consists of 78 healthy Caucasians ranging from age 25 to 101. Of these extensive subjects, eight are chosen at random to experiment. RMH-net achieves an accuracy of 89.00% when tested on the Sleep-EDF database and a Kappa value of 78.00%. Wang et al. [22] propose an automatic sleep staging analysis model using transfer learning and network fusion. The model, Seq-Deepsleepnet, consists of three networks: Seqsleepnet subnetwork, deepsleepnet subnet, and LightBGM classifier. The Seqsleepnet subnetwork consists of a sequence-to-sequence classifier having a bi-directional RNN. Deepsleepnet consists of CNNs, and bi-directional LSTM and LightBGM merge the output of the other two networks as input to the classifier. The Sleep-EDF expanded dataset is used to evaluate the proposed model. For pretraining, MASS dataset is used, which consists of 200 nights of PSG of 97 males ranging from age 19 to 49 and 103 females ranging from age 18 to 38; all recordings have a sampling frequency of 256 Hz. For evaluation, Sleep-EDF expanded database consists of 153 Sleep Cassette files of 20 healthy Caucasians ranging from age 25 to 101; all recordings have a sampling rate of 100 Hz. The model on the Sleep-EDF database gives an accuracy of 87.84%. Jain and Ganesan [23] proposed a model to classify sleep stages of unknown subjects using single- and multiple-channel EEG signals. The classification was performed for both six-class and two-class sets. The algorithm suggested is a mixture of temporal, nonlinear, spectral, time–frequency, and statistical features of EEG signal and RUSBoost (random under sampling and boosting technique). The model was trained using three datasets, namely Sleep-EDF (8 subjects; 4 healthy and 4 having sleep disorders), DREAMS (20 subjects; 16 females and 4 males), and Expanded Sleep-EDF (197 topics; 153 healthy and 44 having sleep disorders). Along with issues from these three datasets, the model was tested using subjects unseen by the model. The model's performance and accuracy were accurate and reliable for unknown problems. The testing subjects from Sleep EDF obtained 92.6 and 97.9% accuracies for six stages and two stages. In contrast, the topics from the expanded version of the same dataset obtained more efficient accuracies of 96.3% for six steps and 99.85 for two sets. In [24], the authors considered two different kinds of networks. The first network contained three hidden layers of 8, 16, 32, or 128 LSTM units. Other important processes were done, like normalization (for scaling the input). This architecture resulted in seven network configurations (Fig. 1).

5 Accumulating Previous Research Gaps and Highlighting Future Directions

It has been noticed that most of the studies used single-channel, which sometimes not captures the hidden changes in the sleep stages' behavior over the individual sleep stages. Finally, there is one more limitation with concern to sleep studies is the lack of datasets. Most of the studies used the same public datasets, which becomes quite tricky regarding analyzing the improvements in the sleep scoring system. Future work in the field of sleep staging mainly focuses on improving the preprocessing techniques, considering multi-modal fusions, and obtaining plans to

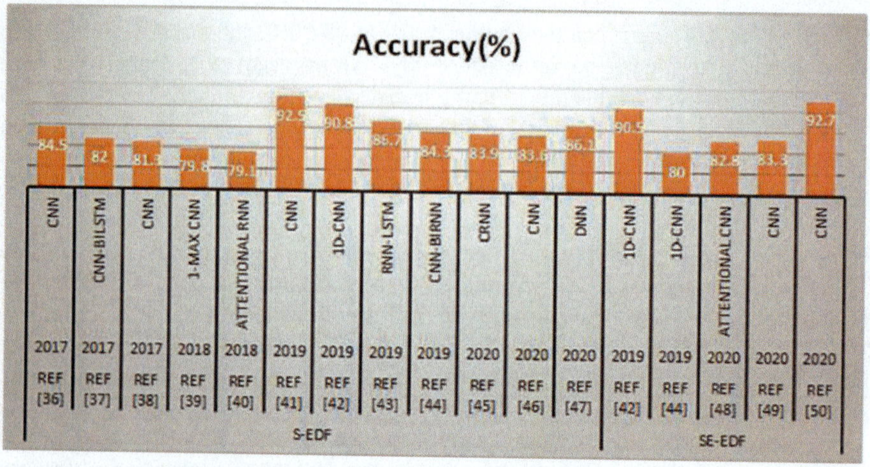

Fig. 1 Sleep staging classification performance using S-EDF and S-EDFX data

manage a large number of samples, and also focuses on some general public health tools for monitoring sleep quality and sleep irregularities during sleep periods (Fig. 2).

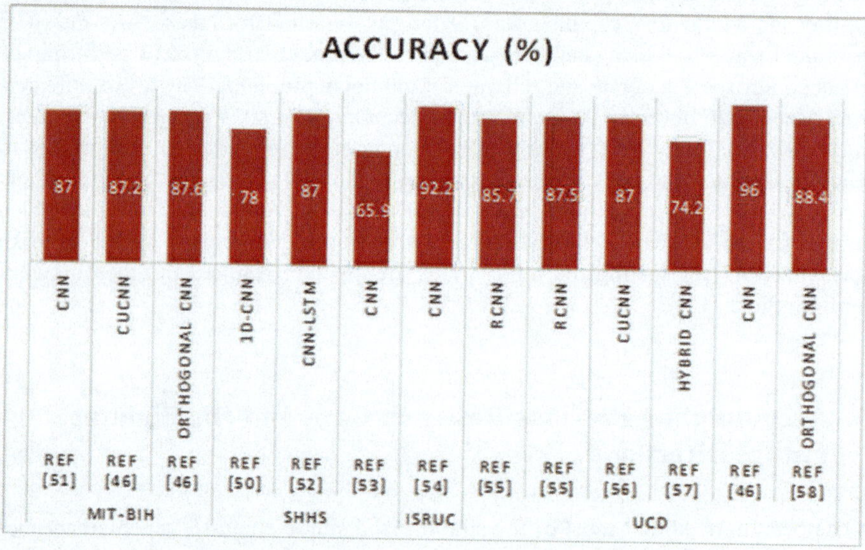

Fig. 2 Sleep staging performance using MIT-BIH, SHHS, ISRUC, and UCD data

6 Conclusion

In this review, the recently contributed methods in sleep scoring systems based on ML and DL have been sincerely exercised, comparing these studies in different aspects. ML methodologies performed well for binary-class classification problems but poorly for multi-class classification problems. The main limitation concerning ML algorithms is the required hand-crafted features for inspecting the changes in the behavior of sleep stages over the individual sleep stages. Another challenging job for ML algorithms is identifying the relevance of the feature subject to the patient's behavior. In ML-based sleep-scoring system, the selection of classification algorithms is also of difficult job. Sometimes it may directly impact the classification performance. Due to these limitations, most researchers adopted the DL approaches, making it more comfortable and accurate to discriminate the changes in sleep behavior over individual sleep stages. The most significant advantage of DL approach features is automatically recognized without explicit features. From this review, we have found that DL models achieved improved results compared to ML models. This review helps to understand the significant improvements achieved and the limitations of sleep staging. Therefore, in this review study, we have pointed out that the future direction of the sleep scoring system should be multi-modal signal fusions. Also, in the future, we should focus on how different DL models will be employed in the real-time diagnosis process for various sleep-related disorders.

References

1. Panossian LA, Avidan AY (2009) Review of sleep disorders. Med Clin North Am 93(2):407–425
2. Satapathy SK, Kondaveeti HK (2021) Automated sleep stage analysis and classification based on different age specified subjects from a single-channel of EEG signal. In: 2021 IEEE Madras section conference (MASCON), Chennai, India. IEEE, pp 1–7
3. Surantha N, Lesmana TF, Isa SM (2021) Sleep stage classification using extreme learning machine and particle swarm optimization for healthcare big data. J Big Data 8(14):1–17
4. Satapathy SK, Ravisankar M, Logannathan D (2020) Automated sleep stage analysis and classification based on different age specified subjects from a dual-channel of EEG signal. In: 2020 IEEE International conference on electronics, computing and communication technologies (CONECCT), Bangalore, India. IEEE, pp 1–6
5. Satapathy SK, Loganathan D (2022) Automated accurate sleep stage classification system using machine learning techniques with EEG signals. In: Kannan SR, Last M, Hong TP, Chen CH (eds) Fuzzy mathematical analysis and advances in computational mathematics. Studies in fuzziness and soft computing, vol 419. Springer, Singapore, pp 137–161
6. Liu C, Tan B, Fu M, Li J, Wang J, Hou F, Yang A (2021) Automatic sleep staging with a single-channel EEG based on ensemble empirical mode decomposition. Phys A Stat Mech Appl 567:125685
7. Satapathy SK, Madhani H, Garg S, Swain D, Rajput N (2022) AutoSleepNet: a multi-signal framework for automated sleep stage classification. In: 2022 IEEE World conference on applied intelligence and computing (AIC), Sonbhadra, India. IEEE, pp 745–750

8. Yücelbaş Ş, Yücelbaş C, Tezel G, Özşen S, Yosunkaya S (2018) Automatic sleep staging based on SVD, VMD, HHT and morphological features of single-lead ECG signal. Expert Syst Appl 102:193–206
9. Widasari ER, Tanno K, Tamura H (2020) Automatic sleep disorders classification using ensemble of bagged tree based on sleep quality features. Electronics 9(3):512
10. Adnane M, Jiang Z, Yan Z (2012) Sleep-wake stages classification and sleep efficiency estimation using single-lead electrocardiogram. Expert Syst Appl 39(1):1401–1413
11. Xiao M, Yan H, Song J, Yang Y, Yang X (2013) Sleep stages classification based on heart rate variability and random forest. Biomed Signal Process Control 8(6):624–633
12. Rahimi A, Safari A, Mohebbi M (2019) Sleep stage classification based on ECG-derived respiration and heart rate variability of single-lead ECG signal. In: 2019 26th National and 4th international Iranian conference on biomedical engineering (ICBME), Tehran, Iran. IEEE, pp 158–163
13. Li Y, Peng C, Zhang Y, Zhang Y, Lo B (2022) Adversarial learning for semi-supervised pediatric sleep staging with single-EEG channel. Methods 204:84–91
14. Rahman MM, Bhuiyan MIH, Hassan AR (2018) Sleep stage classification using single-channel EOG. Comput Biol Med 102:211–220
15. Abdulla S, Diykh M, Laft RL, Saleh K, Deo RC (2019) Sleep EEG signal analysis based on correlation graph similarity coupled with an ensemble extreme machine learning algorithm. Expert Syst Appl 138:112790
16. Koley B, Dey D (2012) An ensemble system for automatic sleep stage classification using single channel EEG signal. Comput Biol Med 42(12):1186–1195
17. Eldele E et al (2021) An attention-based deep learning approach for sleep stage classification with single-channel EEG. IEEE Trans Neural Syst Rehabil Eng 29:809–818
18. Smith A, Anand H, Milosavljevic S, Rentschler KM, Pocivavsek A, Valafar H (2021) Application of machine learning to sleep stage classification. In: 2021 International conference on computational science and computational intelligence (CSCI), Las Vegas, NV, USA. IEEE, pp 349–354
19. Zhao R, Xia Y, Wang Q (2021) Dual-modal and multi-scale deep neural networks for sleep staging using EEG and ECG signals. Biomed Signal Process Control 66:102455
20. Li F, Yan R, Mahini R, Wei L, Wang Z, Mathiak K, Liu R, Cong F (2021) End-to-end sleep staging using convolutional neural network in raw single-channel EEG. Biomed Signal Process Control 63:102203
21. Zhang C, Liu S, Han F, Nie Z, Lo B, Zhang Y (2022) Hybrid manifold-deep convolutional neural network for sleep staging. Methods 202:164–172
22. Wang H, Guo H, Zhang K, Gao L, Zheng J (2022) Automatic sleep staging method of EEG signal based on transfer learning and fusion network. Neurocomputing 488:183–193
23. Jain R, Ganesan RA (2021) Reliable sleep staging of unseen subjects with fusion of multiple EEG features and RUSBoost. Biomed Signal Process Control 70:103061
24. Malafeev A, Laptev D, Bauer S, Omlin X, Wierzbicka A, Wichniak A, Jernajczyk W, Riener R, Buhmann J, Achermann P (2018) Automatic human sleep stage scoring using deep neural networks. Front Neurosci 12(781):1–15
25. Abdollahpour M, Rezaii TY, Farzamnia A, Saad I (2020) Transfer learning convolutional neural network for sleep stage classification using two-stage data fusion framework. IEEE Access 8:180618–180632
26. Phan H et al (2021) Towards more accurate automatic sleep staging via deep transfer learning. IEEE Trans Biomed Eng 68(6):1787–1798
27. Sundar GN, Narmadha D, Jone AAA, Sagayam KM, Dang H, Pomplun M (2021) Automated sleep stage classification in sleep apnoea using convolutional neural networks. Inf Med Unlocked 26:100724
28. Kumar CB et al (2022) SCL-SSC: supervised contrastive learning for sleep stage classification. TechRxiv. Preprint, pp 1–10
29. Smaldone A, Honig JC, Byrne MW (2007) Sleepless in America: inadequate sleep and relationships to health and well-being of our nation's children. Pediatrics 119(Supplement_1):S29–S37

30. Satapathy SK, Loganathan D, Narayanan P, Sharathkumar S (2020) Convolutional neural network for classification of multiple sleep stages from dual-channel EEG signals. In: 2020 IEEE 4th Conference on information and communication technology (CICT), Chennai, India. IEEE, pp 1–16
31. Wei L, Lin Y, Wang J, Ma Y (2017) Time-frequency convolutional neural network for automatic sleep stage classification based on single-channel EEG. In: 2017 IEEE 29th International conference on tools with artificial intelligence (ICTAI), Boston, MA, USA. IEEE, pp 88–95
32. Supratak A, Dong H, Wu C, Guo Y (2017) DeepSleepNet: a model for automatic sleep stage scoring based on raw single-channel EEG. IEEE Trans Neural Syst Rehabil Eng 25(11):1998–2008
33. Vilamala A, Madsen KH, Hansen LK (2017) Deep convolutional neural networks for interpretable analysis of EEG sleep stage scoring. In: 2017 IEEE 27th International workshop on machine learning for signal processing (MLSP), Tokyo, Japan. IEEE, pp 1–6
34. Phan H, Andreotti F, Cooray N, Chen OY; De Vos M (2018) DNN filter bank improves 1-max pooling CNN for single-channel EEG automatic sleep stage classification. In: 2018 40th Annual international conference of the IEEE engineering in medicine and biology society (EMBC), Honolulu, HI, USA. IEEE, pp 453–456
35. Phan H, Andreotti F, Cooray N, Chén OY, De Vos M (2018) Automatic sleep stage classification using single-channel EEG: learning sequential features with attention-based recurrent neural networks. In: 2018 40th Annual international conference of the IEEE engineering in medicine and biology society (EMBC), Honolulu, HI, USA. IEEE, pp 1452–1455
36. Qureshi S, Karrila S, Vanichayobon S (2019) GACNN SleepTuneNet: a genetic algorithm designing the convolutional neural network architecture for optimal classification of sleep stages from a single EEG channel. Turk J Electr Eng Comput Sci 27(6):4203–4219
37. Yildirim O, Baloglu UB, Acharya UR (2019) A deep learning model for automated sleep stages classification using PSG signals. Int J Environ Res Public Health 16(4):599
38. Michielli N, Acharya UR, Molinari F (2019) Cascaded LSTM recurrent neural network for automated sleep stage classification using single-channel EEG signals. Comput Biol Med 106:71–81
39. Mousavi S, Afghah F, Acharya R (2019) SleepEEGNet: automated sleep stage scoring with sequence to sequence deep learning approach. PLoS ONE 14(e0216456):1–15
40. Seo H, Back S, Lee S, Park D, Kim T, Lee K (2020) Intra- and inter-epoch temporal context network (IITNet) using sub-epoch features for automatic sleep scoring on raw single-channel EEG. Biomed Signal Process Control 61:102037
41. Zhang X, Xu M, Li Y, Su M, Xu Z, Wang C, Kang D, Li H, Mu X, Ding X et al (2020) Automated multi-model deep neural network for sleep stage scoring with unfiltered clinical data. Sleep Breath 24:581–590
42. Xu M, Wang X, Zhangt X, Bin G, Jia Z, Chen K (2020) Computation-efficient multi-model deep neural network for sleep stage classification. In: ASSE'20: proceedings of the 2020 Asia service sciences and software engineering conference, Nagoya, Japan. Association for Computing Machinery (ACM), New York, NY, USA, pp 1–8.
43. Zhu T, Luo W, Yu F (2020) Convolution-and attention-based neural network for automated sleep stage classification. Int J Environ Res Public Health 17(11):4152
44. Jadhav P, Rajguru G, Datta D, Mukhopadhyay S (2020) Automatic sleep stage classification using time-frequency images of CWT and transfer learning using convolution neural network. Biocybern Biomed Eng 40(1–2):494–504
45. Fernandez-Blanco E, Rivero D, Pazos A (2020) Convolutional neural networks for sleep stage scoring on a two-channel EEG signal. Soft Comput 24:4067–4079
46. Sors A, Bonnet S, Mirek S, Vercueil L, Payen J-F (2018) A convolutional neural network for sleep stage scoring from raw single-channel EEG. Biomed Signal Process Control 42:107–114
47. Zhang L, Fabbri D, Upender R, Kent D (2019) Automated sleep stage scoring of the sleep heart health study using deep neural networks. Sleep 42(11):1–10
48. Li Q, Li Q, Liu C, Shashikumar SP, Nemati S, Clifford GD (2018) Deep learning in the cross-time frequency domain for sleep staging from a single-lead electrocardiogram. Physiol Meas 39:124005

49. Cui Z, Zheng X, Shao X, Cui L (2018) Automatic sleep stage classification based on convolutional neural network and fine-grained segments. Complexity 2018(9248410):1–13
50. Biswal S, Kulas J, Sun H, Goparaju B, Westover MB, Bianchi MT, Sun J (2017) SLEEPNET: automated sleep staging system via deep learning. arXiv:1707.08262, arXiv:1707.08262v1, https://doi.org/10.48550/arXiv.1707.08262
51. Zhang J, Wu Y (2018) Complex-valued unsupervised convolutional neural networks for sleep stage classification. Comput Methods Programs Biomed 164:181–191
52. Yuan, Jia K, Ma F, Xun G, Wang Y, Su L, Zhang A (2019) A hybrid self-attention deep learning framework for multivariate sleep stage classification. BMC Bioinform 20(Suppl 16):1–10
53. Zhang J, Yao R, Ge W, Gao J (2020) Orthogonal convolutional neural networks for automatic sleep stage classification based on single-channel EEG. Comput Methods Programs Biomed 183:105089

TPredDis: Most Informative Tweet Prediction for Disasters Using Semantic Intelligence and Learning Hybridizations

M. Arulmozhivarman and Gerard Deepak

Abstract Twitter is one of the most common social networking platforms, and millions of tweets are generated every hour. Various non-profit organizations and relief agencies monitor Twitter data to control and help the people in emergency and need. The popularity and accessibility of smartphone to people has made this scenario possible. It allows the user to announce the emergency they are facing in real time. This paper proposes a novel framework incorporating a deep learning architecture transformer to predict the tweets that signify the disaster situation. The disaster dataset based on Twitter is collected, preprocessing the data. The preprocessing steps include tokenization and lemmatization, stop word removal, and word entity recognition. The individual informative tweets are extracted and enriched using semanto sim and Twitter semantic similarity after preprocessing the dataset. Disaster ontology is generated using ontocolab, and it is incorporated with the enriched words to generate metadata for the disaster dataset.

Keywords Transformers · Ontocolab · Metadata · Semantic similarity · Logistic regression · Twitter semantic similarity

1 Introduction

Due to its IN-semantic nature and hybridization of certain machine learning classifiers, the proposed model has the maximum precision, recall, accuracy, and F-measure. It is an IN-semantic machine learning hybridization technique that uses word enrichment using two semantic similarities models, i.e., semanto sim measure

M. Arulmozhivarman
Department of Electronics and Electrical Engineering, SASTRA Deemed University, Thanjavur, Tamil Nadu, India

G. Deepak (✉)
Department of Computer Science and Engineering, Manipal Institute of Technology Bengaluru, Manipal Academy of Higher Education, Manipal, India
e-mail: gerard.deepak.cse.nitt@gmail.com

S. Sharma et al. (eds.), *Intelligent Control, Robotics, and Industrial Automation*, Lecture Notes in Electrical Engineering 1066,
https://doi.org/10.1007/978-981-99-4634-1_78

and Twitter semantic similarity model that has been integrated. The approach also generates metadata, which ensures a large amount of relevant World Wide Web information is incorporated, further classified using the logistic regression. Most importantly, the glowworm optimization to refine the initial recommendable word enriched set to the final recommendable set using the semantic similarity.

The incorporation of disaster ontology and the metadata, which is classified, increases the density of auxiliary knowledge fed into the framework, which is in the domain relevance with various disasters. They provide much higher background knowledge when compared with the word embeddings as used in the previous approaches, and intermediate vocabulary generation is also from the metadata as well as the disaster ontology enriches the density of knowledge into the approach and makes sure the proposed approach performs better than the other baseline models.

The transformers predict the most informative tweet by categorically matching the tweets in the dataset with the rich intermediate vocabulary—the alignment of initial words with the classified instances using Twitter semantic similarity. Results are represented with the specific threshold under the glowworm algorithm, ensuring the optimization is much better. Most relevant entities are furnished by the approach used for predicting the most informative tweets.

Motivation. The disaster dataset based on the tweets used in this research project is preprocessed. The informative words extracted during the preprocessing are then combined with the retrieved words to create an ontology. Twitter and semantic similarity phrases are used to determine the semantic similarity of the disaster dataset. The glowworm optimization method categorizes the generated metadata using the top 50% of occurrences. The resulting intermediate vocabulary is built using Twitter semantic similarity.

Contribution. Individual informative words are extracted during preprocessing. The disaster ontology created using OntoColab is combined with the words retrieved in the previous phase and enhanced with a semantic similarity measure. Semanto sim and Twitter semantic similarity are used to compute semantic similarity. The disaster dataset's metadata is created using the Internet and enhanced terms. To categorize the generated metadata, use the glowworm optimization method. The top 50% of identified metadata occurrences are utilized to construct a comprehensive intermediate vocabulary. Using Twitter semantic similarity, the starting words are aligned with the identified occurrences and integrated into the lexicon. The most informative tweets are created from the vocabulary using the unique transformer architecture.

Organization. The following is the structure of the paper that was published. Section 2 contains related works, while Sect. 3 encircles the proposed framework. Sections 4 and 5 comprise the implementation and performance evaluation sections, respectively. Finally, the paper comes to a close in Sect. 6.

2 Related Work

Dhanya et al. [1] present an approach to collect and analyze the generated requests for help or resource availability in a given geographic area. The resulting data will be analyzed using various machine learning algorithms. Madichetty et al. [2] introduced a method that uses the learned features of various popular phrases such as aid, need, food, and packets, to generate a classifier that can detect tweets containing these terms. The proposed model is performed on various test setups. Pekar et al. [3] proposed a method that can be commonly used to classify words into various classes. This method bases its decision on the distributional similarity of a given the word to the target class. The semantic relatedness of the word to the class can then be used to inform the classification decision. Khosla et al. [4] proposed various neural network-based retrieval models for automatically identifying tweets. We find that these models outperform the previous techniques. Carrillo et al. [5] proposed method to measure the word co-occurrence of tweets based on the corpus. The corpus has become more sophisticated in terms of identifying temporal resolution. This is because the corpus can now detect the collective expression of thoughts through the use of its corpus. Krishnanand et al. [6] developed a method for detecting the sources of a given general nutrient profile on a two-dimensional space using multiple robots. This method is commonly used for monitoring and containing spilled chemicals or nuclear materials. Rudra et al. [7] suggested a method that is based on the concept of disaster-related concepts that evolve in Twitter during disasters. It helps us identify and classify complex and misleading messages. Shah et al. [8] divided the proposed algorithms into three parts: logistic regression, random forest, and K-nearest neighbor. The results were analyzed and compared with each of them. Leena et al. [9] introduced a semantic similarity measure proposed by semanto sim that is computed between the concepts and the query. It is a context-driven recommendation system that takes into account the semantic relatedness of the query. Hettiarachchi et al. [10] introduced a task that will automatically identify if a tweet is informative or not. It will then inform the system about the details of the case, such as the location of the victims and the confirmed death cases. Wålinder et al. [11] hypothesize that both the logistic regression and the random forest models were evaluated using the same set of datasets. The models were compared and evaluated on their accuracy and metadata analysis. Madani et al. [12] proposed a hybrid method that combines fuzzy logic with the information extraction system (IRS) concepts. It achieves its goal by fuzzifying a tweet to classify it as two opinion documents. In [13–17] several ontology-based approaches in favor of the proposed approach have been discussed.

3 Proposed Work

The proposed Fig. 1 depicts the most informative tweet classification and prediction using semantic infused approach. Disaster tweet dataset is subjected to tweet preprocessing for which tokenization lemmatization and NER has been incorporated, so on preprocessing, the individual informative words for the tweets are extracted, and words enrichment is done laterally by using a disaster ontology which is generated by ontocolab from real-world articles and blog concerning the disaster. Once the ontology-driven word enrichment is achieved, the word enrichment is done by using semanto measure and Twitter semantic similarity (TSS). The semantic threshold levels are 0.75 for semanto sim and 0.75 for Twitter semantic similarity to generate the metadata classification features, which is further used to generate metadata. Using these enriched terms, metadata has been generated for disaster using metadata tools such are RDF distiller and open catalis, ggrdl.

RDFa 1.1 is a specification that enables developers to create attributes using HTML or XML. The markup language used to create these attributes is RDF. This package contains the RDFa 1.1 core document, XHTML + RDFa specifications, and HTML + RDFa. Gleaning resource descriptions are defined in the GDRL specification. This markup establishes a method for describing the contents of an XML document in terms of its RDF and XPath relationships. The GRDDL mechanism allows an XML document to declare that all associated data is derived from 'Gleanable Data' or 'Gibable Data' This mechanism is also used to link to an algorithm for extracting data from sources. The Open Catalyst Project is a group of people working together on a research project. The goal is to employ artificial intelligence to predict and discover new catalysts for use in renewable energy storage to mitigate climate change.

Since there is an extensive amount of metadata, they are classified using logistic regression, and the top 50 percent of the metadata instances are classified. Using all the instances, we are building a rich intermediate vocabulary that aligns with

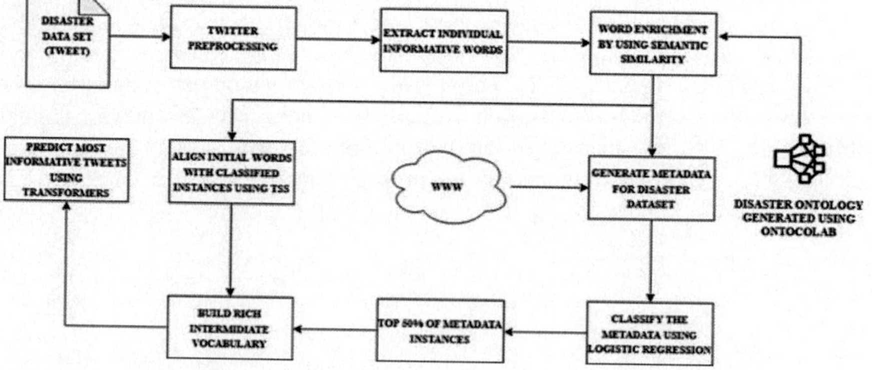

Fig. 1 Proposed TPredDis architecture

the initially enriched words ontology using the Twitter semantic similarity under ant colony optimization. The reason for using metaheuristic optimization like ant colony optimization is predominantly to regulate the large volume of metadata. The initial population alone is insufficient. So using Twitter semantic similarity as the objective function, the initial population has to be refined to get the seasoned and most relevant vocabulary set to build and model the intermediate rich vocabulary set.

The vocabulary set is fed into transformers to find the word to vec and vec to word conversion. Since the intermediate regression vocabulary is extensive in number and tweets are sentenced, this vocabulary is passed into the transformers to recognize the tweet that matches the vocabulary. So each of the tweets that match the vocabulary and the most informative are based on the most amounts of the matching words in the specific tweets. They are rearranged in the ascending order of the information content, and the most informative tweets are predicted.

4 Implementation and Performance Evaluation

4.1 Dataset Preparation

The disasters folder contains over 11,000 tweets that refer to a disaster event. The text was manually classified depending on the context of the tweet. The tweets are associated with the words such as ‘crash’, ‘quarantine’, ‘bush fire’, etc. Disasters on social media inherited the data structure.

A common semantic similarity metric, the Semanto measure is derived from the pointwise mutual information metric. It pairs the terms A and B if there are two terms in a query. Permutations of the three pairings are examined if there are three query phrases. If the query is a single phrase, the semantic similarity between the phrase a and its most closely related semantically relevant word is regarded. The tokenized and stemmed query keywords are the query words. The Semantosim determines how semantically connected two phrases are (a, b). The equation calculates the PMI(a, b) (1). f(a, b) is the probability of the term an occurring in conjunction with b. f(a, b) is the probability of the term b occurring in conjunction with a. f(a), and f(b) are the probabilities of the terms a and b, respectively.

The frequency $\Phi(\theta)$ can be calculated as the mean of amount of time in between the tweets in this sequence, taking into account the term θ and the time stamp series $\{(\theta)\}$ of N tweets comprising θ. By limiting it by 100 tweets per query, the number of tweets in the time stamp series, N, is formally a programmable criterion that relies on the Twitter API. We use $N = 30$. As a result, the velocity of a word is calculated as the mean of the difference between the time stamps of the latest 30 tweets that contain that word.

Algorithm 1: Algorithm for the proposed TPredDis Framework

Input: Disaster dataset, Disaster ontology
Output: Most informative Tweets that best serves the need of disaster management

Start
Step 1: Tweets based disaster dataset used in this research study undergoes preprocessing
Step 2: Upon preprocessing individual informative words are extracted
Step 3: The disaster ontology generated using OntoColab is incorporated with the words extracted from the previous step and they are enriched using semantic similarity measure
Step 4: Compute semantic similarity using Semanto sim and Twitter semantic similarity
Step 5: The metadata for the disaster dataset is generated based on world wide web and enriched words
Step 6: Apply Glowworm optimization algorithm to classify the generated metadata
Start
Use luciferin update rule to modify luciferin value
Setup glowworms location and local decision zone
Compute the fitness of glowworm
For each iteration, x
For every glowworm, j
Use neighborhood range update rule to modify the glowworms' moment
If (Termination criteria met)
Yes then end
No then again compute fitness of glowworm
End
Step 7: The top 50% of the classified metadata instances are used to build a rich intermediate vocabulary
Step 8: The initial words are aligned with the classified instances using Twitter semantic similarity and incorporated into the vocabulary
Step 9: The most informative Tweets are generated from the vocabulary using the novel transformer architecture
End

Similarly, independent of their relative order inside the tweet, we estimate the regularity of co-occurrence of the two terms θ 1 and θ 2, $\Phi(\theta\ 1 \wedge \theta\ 2)$, from the output of tweets that hold together. We choose $\eta = 1/4$ to get a decent scale when defining the TSS between terms θ 1 and θ 2, with η as a scaling factor.

$$\text{Semanto sim}(a, b) = \frac{\text{pmi}(a, b) + f(a, b)\log[f(a, b)]}{[f(a).f(y) + \log[f(a, b)]} \tag{1}$$

$$\text{TSS}(\theta_1, \theta_2) = \left(\frac{\phi(\theta_1 \wedge \theta_2)}{\max(\phi(\theta_1), \phi(\theta_2))}^{\eta}\right) \tag{2}$$

5 Results and Performance Evaluation

In order to compare the performance of the deep red, it has base lined with extracting and summarizing situational information from the Twitter social media during disasters (EITSMD), a stacked convolutional neural network for detecting the resource tweets during a disaster (SCNNDST), Twitter-based disaster management system using data mining (TDMUSD), and spectral clustering + Word2Vec + TF-IDF + word embeddings + random forest. The proposed TPredDis performance is computed by using accuracy, precision, F-measure, recall, and false negative rate as potential metrics. Precision, F-measure, recall, and accuracy compute the significance of the outcomes. In contrast, the false negative rate calculates the value of false negatives furnished by the proposed methodology (Table 1).

$$F - \text{measure} = \frac{2 * \text{Precision} * \text{Recall}}{\text{Precision} + \text{Recall}} \tag{3}$$

$$\text{FNR} = 1 - \text{Recall} \tag{4}$$

$$\text{Precision\%} = \frac{\text{True Number of Positive}}{\text{True Number of Positives} + \text{False Number of Positives}} \tag{5}$$

$$\text{Recall\%} = \frac{\text{True Number of Positive}}{\text{True Number of Positives} + \text{False Number of Negatives}} \tag{6}$$

$$\text{Accuracy\%} = \frac{(\text{Precision} + \text{Recall})}{2} \tag{7}$$

It is very clear that the suggested deep red produces the highest precision, F-measure, recall, accuracy and lowest FNR compared with the baseline models. Figure 2 depicts the precision vs number of recommendations graph, from which

Table 1 Performance evaluation of the TPredDis model in comparison with other baseline models

Search technique	Average precision%	Average recall %	Accuracy %	F-measure%	FNR
EITSMD [1]	50.53	53.45	51.99	51.94	0.47
SCNNDST [2]	74.14	78.38	76.26	76.20	0.22
TDMUSD [3]	78.43	81.18	79.80	79.77	0.19
Spectral Clustering + Word2Vec + TF-IDF + Word Embeddings + random forest	83.28	87.43	85.35	85.30	0.13
Proposed TPredDis	**86.89**	**90.01**	**88.45**	**88.42**	**0.10**

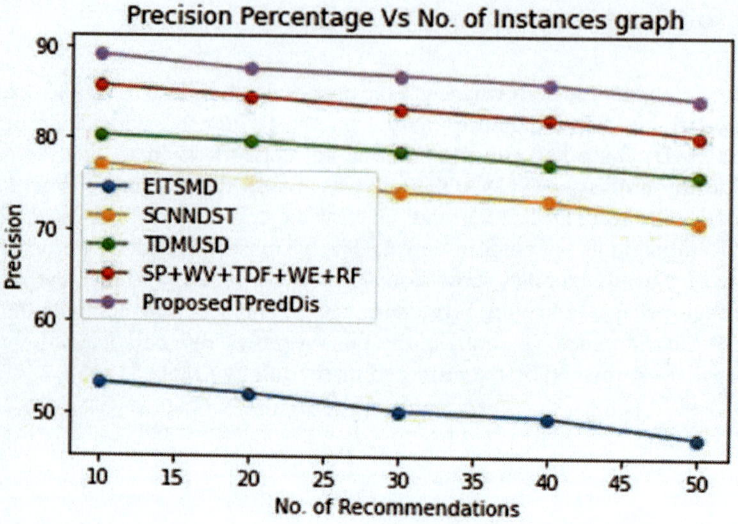

Fig. 2 Graphical representation of precision percentage versus no. of instances

we can deduce that the proposed model has greater precision compared with other baseline compared models. The EITSMD yields 50.53% precision, 53.45% recall, 51.99% accuracy, 51.94% F-measure, and an FNR of 0.47. Similarly, SCNNDST yields 74.14% of precision, 78.38 of recall, 76.26 of accuracy, 76.20 of F-measure, and an FNR of 0.22. TDMUSD yields 78.43% of precision, 81.18 of recall, 79.80 of accuracy, 79.77 of F-measure, and an FNR of 0.19. Compared with Word2Vec plus TF-IDF plus word embeddings plus random forest, a spectral clustering follows a higher precision, recall, accuracy, F-measure, and FNR of 83.28%, 87.43% 85.3%, 85.30%, 0.13, respectively. The suggested TPredDis produces the highest precision of 86.89%%, the highest recall of 90.01%, the highest accuracy of 88.45%, the highest F-measure of 88.42%, and the lowest FNR of 0.10.

The EITSMD has the lowest precision, F-measure, recall and accuracy with the highest FNR mainly because it uses tweets that are not in English. There is a cross-language dependency, and most importantly, they use the low-level lexical and syntactic features to differentiate between situational and non-situational tweets. Initially, the summarization is done, and other content words that are used for the summarization is approached. A domain-independent classifier has been modeled in the proposed paradine, and for classification, they use a traditional SVM model. All the things ensure that their precision, F-measure, recall and accuracy are distinctively low with the highest FNR compared with the other models.

The SCNNDST approach uses stack CNN with crises word embeddings. Most importantly, they use a series of classifiers like deep learning model, SVM, gradient boosting, random forest, K-nearest neighbors, decision tree, and Naive Bayes classification technique. However, the approach is an amalgamation of deep learning with a series of machine learning classifiers. This ensures the computational collaboration is very high, and most importantly, the percentage of precision, F-measure, recall, accuracy is comparatively higher than the previous EITSMD. However, there is always scope for whether precision, F-measure, recall and accuracy can be increased using preferable semantics. Hence, the SCNNDST is also computationally expensive, although it has a significantly low application response time.

The TDMUSD model uses linear rich regression SGD classifier and Naïve-based algorithm for initial filtering, and these three classifiers, when merged with the dataset, significantly has an increase in precision, F-measure, recall and accuracy with a lower FNR. These three machine learning models, when applied together, perform much better than the SCNNDST, which amalgamates much deep learning and machine learning classifiers along with stack CNN and crises word embeddings. TDMUSD has only three classifiers, namely linear regression, Naïve Bayes, and SGD classifiers. The amalgamation of these three in this approach increases precision, F-measure, recall, accuracy and lowers the FNR value.

6 Conclusions

The proposed model has the highest precision, recall, accuracy, and F-measure because it is IN-semantic and hybridizes specific machine learning classifiers. It is an IN-semantic machine learning hybridization technique that uses word enrichment using two semantic similarities models that is semanto sim measure and Twitter semantic similarity model that has been integrated. The approach also generates metadata, which ensures a large amount of relevant World Wide Web information is incorporated, further classified using the logistic regression. Most importantly, the glowworm optimization to refine the initial recommendable word enriched set to the final recommendable set using the semantic similarity.

The incorporation of disaster ontology and the metadata, which is classified, increases the density of auxiliary knowledge fed into the framework, which is in the domain relevance with various disasters. They provide much higher background knowledge when compared with the word embedding's as used in the previous approaches, and intermediate vocabulary generation is also from the metadata as well as the disaster ontology enriches the density of knowledge into the approach and makes sure the proposed approach performs better than the other baseline models.

The transformers predict the most informative tweet by categorically matching the tweets in the dataset with the rich intermediate vocabulary—the alignment of initial words with the classified instances using Twitter semantic similarity. Results are represented with the specific threshold under the glowworm algorithm, ensuring the optimization is much better. Most relevant entities are furnished by the approach used for predicting the most informative tweets.

References

1. Rudra K, Ganguly N, Goyal P, Ghosh S (2018) Extracting and summarizing situational information from the twitter social media during disasters. ACM Trans Web (TWEB) 12(3):1–35
2. Madichetty S (2021) A stacked convolutional neural network for detecting the resource tweets during a disaster. Multimed Tools Appl 80(3):3927–3949
3. Dhanya VG, Jacob MS, Dhanalakshmi R (2021) Twitter-based disaster management system using data mining. In: Computer networks, big data and IoT. Springer, Singapore, pp 193–203
4. Khosla P, Basu M, Ghosh K, Ghosh S (2017) Microblog retrieval for post-disaster relief: applying and comparing neural IR models. arXiv preprint arXiv:1707.06112
5. Carrillo F, Cecchi GA, Sigman M, Slezak DF (2015) Fast distributed dynamics of semantic networks via social media. Comp Intell Neurosci
6. Krishnanand KN, Ghose D (2005) Detection of multiple source locations using a glow-worm metaphor with applications to collective robotics. In: Proceedings 2005 IEEE Swarm Intelligence Symposium, 2005. SIS 2005, pp 84–91. IEEE
7. Shah K, Patel H, Sanghvi D, Shah M (2020) A comparative analysis of logistic regression, random forest and KNN models for the text classification. Augment Human Res 5(1):1–16
8. Giri GL, Deepak G, Vemugopal MSH (2017) A query relevant context driven ontology recommendation system incorporating semantics preservation and semantic ontology matching. Int J Adv Eng Res Develop 4(5)
9. Hettiarachchi H, Ranasinghe T (2020) Infominer at wnut-2020 task 2: transformer-based covid-19 informative tweet extraction. arXiv preprint arXiv:2010.05327
10. Wålinder A, Evaluation of logistic regression and random forest classification based on prediction accuracy and metadata analysis
11. Pekar V, Staab S (2003) Word classification based on combined measures of distributional and semantic similarity. In: 10th Conference of the European Chapter of the Association for Computational Linguistics
12. Madani Y, Erritali M, Bengourram J, Sailhan F (2020) A multilingual fuzzy approach for classifying Twitter data using fuzzy logic and semantic similarity. Neural Comput Appl 32(12):8655–8673
13. Arulmozhivarman M, Deepak G (2021) OWLW: ontology focused user centric architecture for web service recommendation based on LSTM and whale optimization. In: European, Asian, Middle Eastern, North African Conference on Management & Information Systems, Springer, Cham, pp 334–344
14. Deepak G, Priyadarshini JS (2018) Personalized and enhanced hybridized semantic algorithm for web image retrieval incorporating ontology classification, strategic query expansion, and content-based analysis. Comput Electr Eng 72:14–25
15. Deepak G, Teja V, Santhanavijayan A (2020) A novel firefly driven scheme for resume parsing and matching based on entity linking paradigm. J Discrete Math Sci Crypt 23(1):157–165
16. Surya D, Deepak G, Santhanavijayan A (2021) QFRDBF: query facet recommendation using knowledge centric DBSCAN and firefly optimization. In: International Conference on Digital Technologies and Applications, pp 801–811. Springer, Cham
17. Deepak G et al (2021) An ontology-based semantic approach for first aid prediction in aviation services incorporating RBFNN Over a cloud server. In: International Conference on Emerging Trends and Technologies on Intelligent Systems. Springer, Singapore

Histopathological Colorectal Cancer Image Classification by Using Inception V4 CNN Model

Rakesh Patnaik, Premanshu Sekhara Rath, Sasmita Padhy, and Sachikanta Dash

Abstract It is crucial to analyze colorectal cancer histological pictures with objectivity. One of the main causes of mortality globally is colorectal cancer (CRC), sometimes referred to as bowel cancer. Early diagnosis has become crucial for a treatment to be effective. A cutting-edge deep convolutional neural network (CNN) with transfer learning can classify several images of CRC into a variety of categories of artificial intelligence (AI). We have adjusted the Inception V4 model to classify the CRC histopathology images in this study's experiment. We also use transfer learning (TL) and modification techniques for increasing accuracy. Inception V4 net is currently one of the finest CNN designs for identifying CRC histopathology pictures from the National Center for Tumor Diseases (NCT) Biobank an open-source collection, according to the findings of our experiment. Additionally, we are able to achieve 97.7% accuracy using TL on the validation dataset, outperforming all prior values we could find in the literature.

Keywords Deep learning · Colorectal cancer · Machine learning · And CNN · Inception V4

R. Patnaik · P. S. Rath · S. Dash
CSE Department, GIET University, Gunupur, Odisha 765002, India
e-mail: rakesh.patnaik@giet.edu

P. S. Rath
e-mail: premanshusekhararath@giet.edu

S. Dash
e-mail: sachikanta@giet.edu

S. Padhy (✉)
School of Computing Science and Engineering, VIT Bhopal University, Bhopal-Indore Highway, Kothrikalan, Sehore, Madhya Pradesh 466114, India
e-mail: pinky.sasmita@gmail.com

S. Sharma et al. (eds.), *Intelligent Control, Robotics, and Industrial Automation*, Lecture Notes in Electrical Engineering 1066,
https://doi.org/10.1007/978-981-99-4634-1_79

1 Introduction

The third most common cancer globally in terms of incidence (6.1%) and death (9.2%) is colorectal cancer (CRC) [1]. By 2030, the number of new cases and fatalities from CRC is predicted to rise by 60% worldwide [2]. Numerous research' findings have demonstrated that an additional precise categorization of medical imaging can successfully predict the progression of colorectal cancer (CC) [3].

The medical method known as optical colonoscopy is typically employed to make a clinical diagnosis by looking at a number of the location, morphology, and pathological changes of abnormalities on the colonial surface. In order to administer the most suitable therapeutic treatment, this increases the intensity of how accurate the diagnosis was and the capacity to gauge the disease's severity. A thorough visual examination by pathologists with an extreme specialization is currently required for the diagnosis of CRC. Using samples of frozen or formalin-fixed paraffin-embedded (FFPE) tissues that have been H&E-stained, digital whole-slide images (WSIs) are used to make diagnoses. The analysis of WSI has difficulties due to the extremely large picture size (More than 10,000X10, 000 pixels), and histological changes in shape, size, texture, and nuclei staining, which complicate and lengthen the diagnostic process [4]. While the requirements for colonial specimen evaluation in gastrointestinal clinics are rigorous, pathologists must complete lengthy (more than 10 years) training programs [5]. Therefore, it is crucial to provide trustworthy CRC detection and pathological image analysis techniques that can enhance clinical effectiveness and efficiency without unintentional human bias.

Modern artificial intelligence (AI) techniques with state-of-art, including deep learning (DL), are particularly effective at classifying data and making predictions. The deep learning technique often needs enormous amounts of data for training, therefore the more data used to train a model, the more effective it will be. For the manual identification and labeling of histology pictures, however, specialists are required. It might involve significant time and money expenditures. It still takes prospective validation studies to properly identify common biomarkers for therapeutic use, even if the underlying technique automatically pays attention to discriminative information for improved categorization. In summary, the decision-making during the subjective assessment for cancer diagnosis continues to be made by highly qualified pathologists. Deep learning algorithms can help doctors make more accurate projections, but they cannot take the place of a doctor's responsibility. Convolutional neural network (CNN) has been used successfully in WSI analysis for lung [6], breast [7, 8], prostate [9], and skin [10, 11] malignancies, among other diseases. Since Alex Krizevsky initially released the groundbreaking AlexNet in 2012 [12], CNN has made significant contributions to several advances in the field of computer vision technology and has transformed the area of image investigation. CNN, which may be trained on training data and can learn to discriminate between things depending on their attributes, can efficiently handle complex visual tasks.

Here, we created a brand-new automated AI method based on a weaker labeled supervised DL for the very first widespread clinical CRC diagnostic applications. To categorize the histological pictures of CRC, our AI method employs Inception-v4 CNN architecture [13] with weights initialized using TL.

2 Literature Review

The outcomes of many approaches of classifying histopathological image dataset given in Table 1, demonstrate the widespread usage of digital technology in classification of medical pictures now. In 5000 histology photographs, Kather [3] employed a variety of textual descriptors to investigate a multi-class issue of tumor epithelium and simple stroma. He suggested the following four categories: The following methods are used to classify categories: (1) The k-neighbor algorithm (k-NN) (2) try to classify all categories using an Support Vector Machine (SVM) decision function; (3) create decision tree models with the RUSBoost technique, and (4) train the classifiers using tenfold cross-validation without using an explicit stratification approach. The findings showed that SVM, which had an accuracy rate of 87.4% across eight classes, was the best classification technique. Recently, it has been discovered that the CNN classification approach is more accurate for classifying different tumor kinds. [12] Used feature selection and a deep learning method using CNN architecture for detecting pneumonia by the analysis of from chest X-rays image dataset, and they were able to reach an accuracy rate of within 82.1%.

Tsai and Tao [13] also employed a TL strategy with GoogLeNet and attained an accuracy of 90.2%, indicating the viability of doing so to categorize the tumor stroma ratio (TSR). Xu et al. [14] enhanced the AlexNet model's activation characteristics and introduced the ability to classify and segment neurons by examining them at the last hidden layer. An accuracy of 97.5% was obtained using the framework, which was trained on ImageNet, and it effectively converted the features derived from the network into small histopathological image structures for visualization and training. Additionally, Kather [3] updated the classification layer with VGG19, which had the greatest accuracy rate at 98.7%.

2.1 *How Deep Learning Works*

Another important branch of machine learning is called deep learning (DL). It was utilized by Hubel [15] to identify correspondences between neuron systems based on cortical cells. DL uses a combination of hidden layers and several nonlinear processing layers, which are inspired by biological nerve systems, to learn characteristics directly from input. As seen in Fig. 1, Hinton [16, 17] hypothesized that learning features using many hidden layers is helpful for classification.

Table 1 Redirecting summary of literature

Literature	Research objective	Classification technique	Accuracy rate (%)
[12]	Chest X-ray pneumonia diagnosis at radiologist level using machine learning	CNN and feature selection	80.9
[13]	For the purpose of separating and categorizing the stromal and epithelial sections of histopathology pictures, a deep CNN	Two number of convolutional layers, two max-pooling layers, two fully connected layers, and a softmax layer make up the CNN network	84
[2]	Analysis of colorectal cancer histology through textures with Multiple classes	Decision trees, radial-basis function SVM, linear SVM, and one-nearest neighbor	87.4
[14]	Classification of tumor epithelium and stroma using deep CNN-learned image features	Utilizing TL techniques with CNN and GoogLeNet	90.2
[15]	Identifying and classifying TAS in diagnosing breast biopsies using deep CNN	Offer a few new geometrical characteristics for benign biopsies	97.5
[3]	Deep learning-based survival prediction: a retrospective multicenter analysis from CC histology slides	VGG19, AlexNet, SqueezeNet, GoogLeNet, and Resnet50 were the five CNN models whose performance was evaluated	98.7

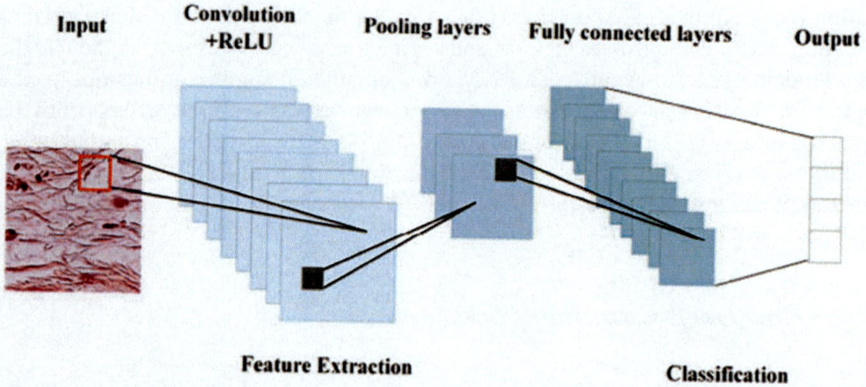

Fig. 1 Architecture of CNN

2.2 CNN Model Networks

AlexNet. The commonly used deep CNN AlexNet [18] may nevertheless compare favorably to other kinds of networks in categorization. The input image is scaled down to 224X224 pixels and forwarded to the network during the training phase of the AlexNet model.

SqueezeNet. With 50x fewer parameters, On ImageNet, SqueezeNet [19], a small CNN architecture, obtains accuracy comparable to AlexNet.

VGGNet. There are two number of convolutional layers, with the ReLU activation function used in each layer, are the first two layers of the VGG architecture. These two layers are followed by one max-pooling layer and numerous fully connected layers, all of which also use the ReLU activation function. A Softmax layer for classification makes up the last layer of the VGGNet model. In VGG-E, the convolution filter's dimensions are also altered to a 3X3 filter with a 2 span.

ResNet. Networks that avoided suffering from a vanishing gradient issue can be helped by residual functions in the ResNet an ultra-deep network framework for residual learning. Unexpectedly, the ResNet framework network's accuracy reaches a saturation point as its depth rises, while adding more layers introduces training mistakes.

Inception. A significant milestone in the creation of CNN classifiers is the Inception model. Its design is intricate (heavily engineered), and it employs a variety of techniques to boost performance in terms of both speed and accuracy.

The popular versions on the Inception model are:

- *Inception V1, V2, and V3*
- *Inception V4 and Inception-ResNet.*

3 Resources and Techniques

3.1 Details of Dataset

The dataset utilized in this study was published in 2019 [3] by Kather. The National Center for Tumor Disease's first dataset (named NCT-CRC-HE-100 K) is made up of 100,000 carefully collected 86 HE-stained human cancer tissue slides from the University Medical Center Mannheim (UMM) pathology collection and the NCT biobank are shown in non-overlapping images. Every picture in the collection is 224 $\times$ 224 pixels and 0.5 Micron per pixel (MPP). Additionally, the author uses the Macenko approach to color-normalize all photos to remove irregularities in the process of staining and to improve the result of classification [17, 20–22]. Table 2 specifies Different tissue classifications' image distributions in the NCT database.

Table 2 Different tissue classifications' image distributions

	ADI	BACK	DEB	LYM	MUC	MUS	NORM	STR	TUM
NCT-CRC-HE-100 K	10.407	10.566	11.512	11557	8896	13,536	8763	10,446	14.317
CRC-VAL-HE-7 K	1.338	847	339	634	035	592	741	421	1.233

3.2 Methodology

CNN has greatly increased the popularity of deep learning in scientific computing, and many businesses employ its techniques to resolve challenging issues. For comparison reasons, we looked at several network designs in this study [12, 18, 19, 23, 24]. Figures 2 and 3 depict the proposed model for training the datasets and testing the datasets, respectively. The parameters and the optimal architecture selection process is shown in Fig. 4.

Model Training. NCT-CRC-HE-100 K pictures with a 150X150 format of pixel and 9 number of distinct tissue classifications were employed. We separated it into a dataset for training purposes that made up 70%, a validation dataset that made up 15%, and a test dataset that made up 15%. We used the ratio of each kind's numbers to determine how many training, validation, and tissue classification tests were necessary because the quantity of the original dataset's types weren't all equal. This allowed us to confirm that the percentage was correct.

Finding the Superior Architecture and Parameters. We examined three training network approach optimizers in the first research study using the NCT-CRC-HE-100 K [3] dataset to evaluate the effectiveness of the network designs used by various

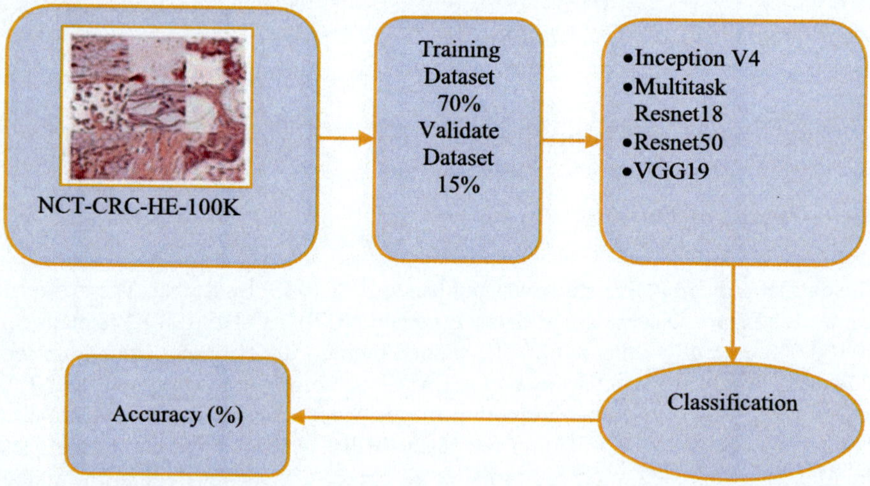

Fig. 2 Proposed model training

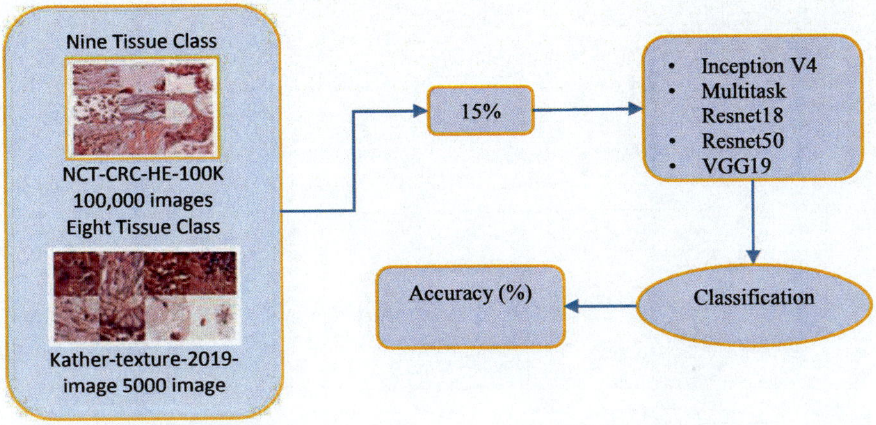

Fig. 3 Proposed model testing

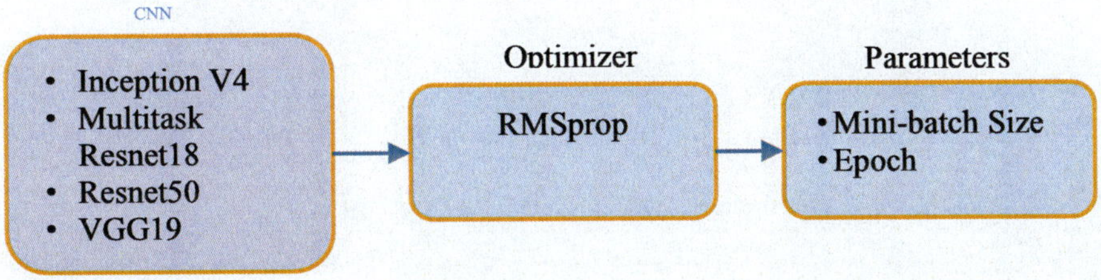

Fig. 4 Selecting optimal architecture and parameters

CNN models: adaptive moment estimation (Adam), root mean square propagation, and stochastic gradient descent with momentum (SGDM).

Model Testing. The first dataset (NCT-CRC-HE-100 K), which had 100,000 picture patches obtained from 86 whole-slide images, was utilized for testing after neural networks were trained on all of the patches.

4 Experimental Result

4.1 Training Inception V4 via TL

The NCT-CRC-HE-100 K open histology dataset of nine tissue types was utilized in this experiment to train the classifier. Kather et al. [3] created these photos, which contain 86 tissue slides stained with hematoxylin and eosin (H&E). The NCT-UMM website was used to get the labels for the histology pictures in the data that was made accessible. Figure 5 includes a list of the nine tissue classes' sample photographs.

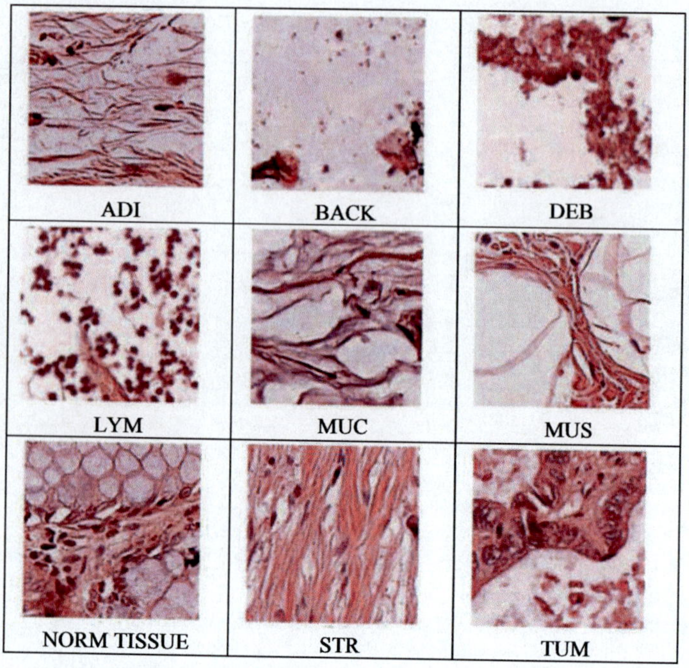

Fig. 5 Examples of the nine tissue classifications that the NCT-CRC-HE-100 K dataset represents

4.2 Evaluation Criteria

The accuracy of the network's performance was assessed in order to compare our findings to those of previous research. To assess our performance, we used the whole validation set of NCT-CRC-HE-7 K 7180 pictures and calculated the average accuracy across the 9 tissue classifications.

4.3 Classification Results

We train our optimized Inception V4 using TL on the NCT biobank dataset and assess the classification performance on a separate collection of 7,180 photos from a range of patients. Using Inception V4 with TL and fine-tuning approaches, we performed a classification of multi-class classifier; diagnostic investigation on the histological pictures of CC. Figure 6 displays the test results of our proposed work of CC histopathology image analysis.

The validation accuracy was 97.7%, whereas the training accuracy was close to 99%. In each of the nine groups of CRC tissues, high accuracy was attained. While

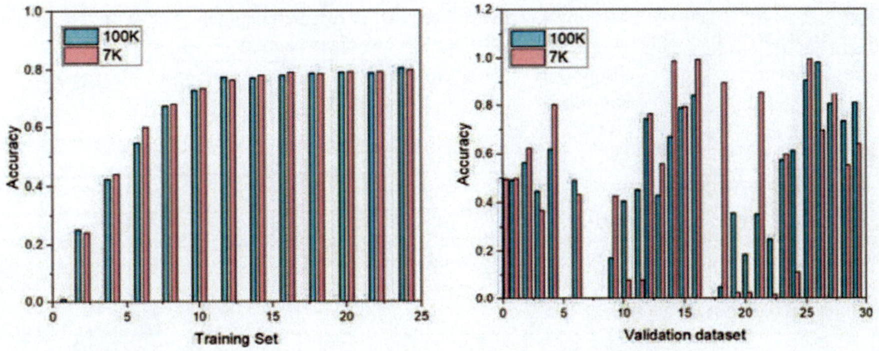

Fig. 6 The CRC histological images' accuracy in training and validation

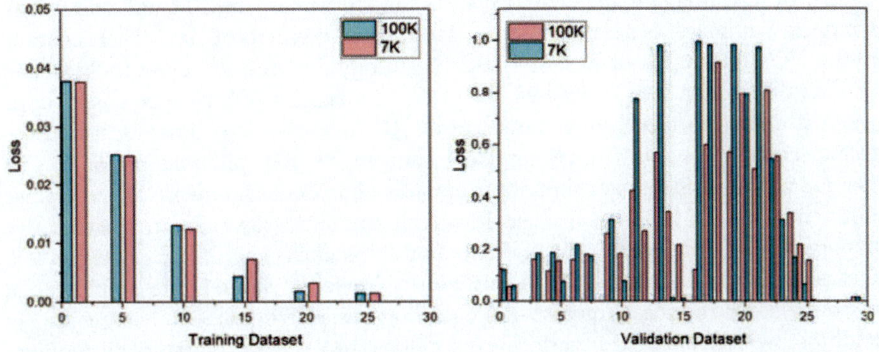

Fig. 7 Loss function value during Inception V4 training and validation using CRC histopathology images

the findings in Fig. 7 demonstrate that throughout the training test, the loss function's value declines quickly and smoothly before converging to a tiny number.

4.4 Comparison of Result

Table 3 compares our findings with those of other research that used the image dataset of NCT-CRC-HE-100 K and subsequently NCT-CRC-HE-7 K 7180 image datasets. We compared the findings from [3, 20] to our classification accuracy. The findings show that, of all the research we could find in the literature about the classification of histological pictures of CC based on the NCT Biobank database, our fine-tuned network had the highest accuracy. The Inception V4 with TL fine-tuned neural network is one of the superb options for categorizing the histopathological pictures of CRC in the current trends of research works.

Table 3 Comparison of our results with those from earlier research using the NCT-CRCHE-100 K and NCT-CRC-HE-7 K databases using CRC multi-class classification

Testing accuracy	(Proposed) fine-tuned inception V4 + TL	Multitask Resnet18 [20]	Resnet50 [20]	VGG19 [3]
NCT-CRC-HE-100 K	**99.9**	98.6	98.8	98.8
NCT-CRC-HE-7 K	**97.7**	95.0	94.2	94.3

5 Conclusion

In this work, we effectively improved the histological scans of CRC's classification accuracy. In our investigation, we used the Inception V4 model to categorize the CRC histopathology pictures. To increase accuracy, we also used TL and fine-tuning methods. Summary: In a series of internal tests, the 9 numbers of class-level accuracy of the NCT-HE-100 K dataset of 1 lakhs histology picture were close to 99%, and in a set of external tests, it was 94.3% [3]. The Inception V4 network has an edge when it comes to identifying traits from CRC histopathology images, according to experimental results and comparisons with those from previous research. Our method was equivalent to or superior to existing approaches for the rapid and precise differentiation of CRC cases from healthy or inflammatory cases. Even more effective than pathologists at testing on a big scale multiple-center data. As far as we can tell, this is the initial AI research for a trustworthy, versatile, and reliable supplemental instrument for routine clinical pathologic diagnosis of initial CRC screening. In addition, we were able to outperform all previous studies in the literature by applying the TL approach. The research work and approach can also be improvised and applied to the histopathological analysis of the types of images in other types of cancers.

References

1. Bray F, Ferlay J, Soerjomataram I, Siegel RL, Torre LA, Jemal A (2018) Global cancer statistics 2018: GLOBOCAN estimates of incidence and mortality worldwide for 36 cancers in 185 countries. CA Cancer J Clin 68(6):394–424
2. Arnold M, Sierra MS, Laversanne M, Soerjomataram I, Jemal A, Bray F (2017) Global patterns and trends in colorectal cancer incidence and mortality. Gut 66(4):683–691
3. Kather JN, Krisam J, Charoentong P, Luedde T, Herpel E, Weis C-A, Gaiser T, Marx A, Valous NA, Ferber D et al (2019) Predicting survival from colorectal cancer histology slides using deep learning: A retrospective multicenter study. PLoS Med 16:e1002730
4. Komura D, Ishikawa S (2018) Machine learning methods for histopathological image analysis. Comput Struct Biotechnol J 16:34–42
5. Black-Schaffer WS, Morrow JS, Prystowsky MB, Steinberg JJ (2016) Training pathology residents to practice 21st century medicine: a proposal. Acad Pathol 3:2374289516665393

6. Campanella G, Hanna MG, Geneslaw L, Miraflor A. Werneck Krauss Silva V, Busam KJ, Brogi E, Reuter VE, Klimstra DS, Fuchs TJ (2019) Clinical-grade computational pathology using weakly supervised deep learning on whole slide images. Nat Med 25(8):1301–1309
7. Bulten W, Pinckaers H, van Boven H, Vink R, de Bel T, van Ginneken B, van der Laak J, Hulsbergen-van de Kaa C, Litjens G (2020) Automated deep-learning system for Gleason grading of prostate cancer using biopsies: a diagnostic study. Lancet Oncol 21(2):233–241
8. Strom P, Kartasalo K, Olsson H (2020) Artificial intelligence for diagnosis and grading of prostate cancer in biopsies: a population-based, diagnostic study. Lancet Oncol 21(2):E70
9. Szegedy C, Vanhoucke V, Ioffe S, Shlens J, Wojna Z (2016) Rethinking the Inception Architecture for Computer Vision. In: 2016 IEEE Conference on Computer Vision and Pattern Recognition (CVPR): 27–30 June 2016, pp 2818–2826
10. Kather JN, Krisam J, Charoentong P, Luedde T, Herpel E, Weis CA et al (2019) Predicting survival from colorectal cancer histology slides using deep learning: A retrospective multicenter study. PLoS Med 16(1):1–22
11. Padhy S, Dash S, Routray S, Ahmad S, Nazeer J, Alam A (2022) IoT-based hybrid ensemble machine learning model for efficient diabetes mellitus prediction. Comp Intell Neurosci
12. Peng T, Boxberg M, Weichert W, Navab N, Marr C (209) Multi-task learning of a deep K-nearest neighbour network for histopathological image classification and retrieval. In: Medical image computing and computer assisted intervention—MICCAI 2019. Springer International Publishing, Cham, pp 676–684
13. Tsai MJ, Tao YH (2019) Machine learning based common radiologist-level pneumonia detection on chest X-rays. In: Proceedings of the 2019 13th International Conference on Signal Processing and Communication Systems (ICSPCS), Gold Coast, Australia, pp 16–18
14. Xu J, Luo X, Wang G, Gilmore H, Madabhushi A (2016) A deep convolutional neural network for segmenting and classifying epithelial and stromal regions in histopathological images. Neurocomp 191:214–223
15. Du Y, Zhang R, Zargari A, Thai TC, Gunderson CC, Moxley KM, Liu H, Zheng B, Qiu Y (2018) Classification of tumor epithelium and stroma by exploiting image features learned by deep convolutional neural networks. Ann Biomed Eng 46:1988–1999. [CrossRef] Xu Y, Jia Z, Wang L-B, Ai Y, Zhang F, Lai M, Chang EI-C (2017) Large scale tissue histopathology image classification, segmentation, and visualization via deep convolutional activation features. BMC Bioinform 18:281 [CrossRef]
16. Dash S, Padhy S, Parija B, Rojashree T, Patro KAK (2022) A simple and fast medical image encryption system using chaos-based shifting techniques. Int J Info Sec Priv (IJISP) 16(1):1–24
17. Shankar TN, Padhy S, Dash S, Teja MB, Yashwant S (2022) Induction of secure data repository in blockchain over IPFS. In: 2022 6th International Conference on Trends in Electronics and Informatics (ICOEI), pp 738–743. IEEE
18. Pranitha G, Rukmini T, Shankar TN, Sah B, Kumar N, Padhy S (2022) Utilization of blockchain in e-voting system. In: 2022 2nd International Conference on Intelligent Technologies (CONIT), pp 1–5. IEEE
19. Padhy S, Shankar TN, Dash S (2022) A comparison among fast point multiplication algorithms in elliptic curve cryptosystem
20. Panda R, Dash S, Padhy S, Das RK (2023) Diabetes mellitus prediction through interactive machine learning approaches. In: Next generation of internet of things. Springer, Singapore, pp 143–152
21. Bejnordi BE, Mullooly M, Pfeiffer RM, Fan S, Vacek PM, Weaver DL, Herschorn S, Brinton LA, Van Ginneken B, Karssemeijer N et al (2018) Using deep convolutional neural networks to identify and classify tumor-associated stroma in diagnostic breast biopsies. Mod Pathol 31:1502–1512
22. Macenko et al (2009) A method for normalizing histology slides for quantitative analysis. In: 2009 IEEE International Symposium on Biomedical Imaging: From Nano to Macro, pp 1107–1110. Available from: https://doi.org/10.1109/ISBI.2009.5193250

23. Dash S, Das RK (2020) An implementation of neural network approach for recognition of handwritten Odia text. In: Lecture Notes in Networks and Systems, pp 94–99. https://doi.org/10.1007/978-981-15-2774-6_12
24. Dash S, Panda R, Padhy S (2021) Blockchain-based intelligent medical IoT healthcare system. SPAST Abst 1(1)

Machine Learning for Prediction of Nutritional Psychology, Fast Food Consumption and Its Impact on Students

Parth P. Rainchwar, Rishikesh S. Mate, Soham M. Wattamwar, Dency R. Pambhar, and Varsha Naik

Abstract The intake of fast food is rapidly accelerating due to the factors specifically the cost-effectiveness and being tasty, but not all individuals are aware of its harmful long-term effects on physical and mental health. On the issue of nutrition and fast food intake this is an unexplored field of research so far. In this paper, we have profoundly analyzed the relationship using Machine Learning models, which is a new approach for nutrition-based analysis. A general questionnaire is prepared dealing with all the factors of nutrition and the immune system. The survey was hosted on an online platform and participants were college students from MIT WPU School of Engineering. Responses were then analyzed implying onto the food-habits and eating behavior using Random Forest, Naive Bayes and Extremely Randomized Trees. According to our understanding and knowledge, this is the earliest research to include all these factors along-with Machine Learning algorithms especially on college students as target audience. The primary objective is to apply association, classification, and regression algorithms in order to predict BMI, sickness, pre-COVID and post-COVID eating schedule and the experiments conducted during this research reveal that this method significantly improves the analysis of real-world data as compared to the traditional statistical approach with a commendable accuracy of 98%.

Keywords Machine learning · BMI · Random forest · Naive Bayes · Fast food intake · Nutrition · Classification · Prediction

1 Introduction

Healthy eating habits and a well-balanced diet plays a critical role in an individual's happy life. A person following a healthy and balanced diet can live a happy life. Being a nutritionist is in high demand these days. Unfortunately, the current generation do not share the similar viewpoint. Modern trends show that people from age 18 to 23 are

P. P. Rainchwar · R. S. Mate (✉) · S. M. Wattamwar · D. R. Pambhar · V. Naik
Dr. Vishvanath Karad's MIT World Peace University, Pune, India
e-mail: rishikeshmate09@gmail.com

S. Sharma et al. (eds.), *Intelligent Control, Robotics, and Industrial Automation*, Lecture Notes in Electrical Engineering 1066,
https://doi.org/10.1007/978-981-99-4634-1_80

likely to be addicted to fast food, which can affect both their physical and emotional health. For the past two decades, studying how people from various age groups choose between food items has been seen as a critical justification for encouraging research in the field of nutritional psychology.

Additional factors, such as location, group, age, tradition, social class, and gender also influences fast food consumption preference. Fast food eating has become a subculture among university students, and it is rising to become a point of concern which in future we might need to worry about. Numbness, tiredness, and tightness are few of the immediate effects of junk food consumption.

As a result, we included a question in our survey asking "if you feel fatigued after eating fast food?". This inspired us to conduct study in this area and use machine learning methods to gain understanding of it [1] conducted a survey of American students from which they came across some critical findings such as 75 percent of individuals aged 20–25 are overweight, and those individuals find it difficult to have weight loss. From the studies done, we understood that fast food intake can also lead to food intolerance and can lead to exhaustion. Which can lead to communicable diseases such as cold, flu and stomach ache. Our results even showed that, even by comprehending the negative consequences of excessive fast food eating, consumers who are addicted to fast food do not alter their fast food consumption. Similarly, eating at irregular intervals disrupts dietary habits, which has an adversarial impact on a person's well-being.

Certain inferences from the survey showed that college crowds prefer to consume fast food on a regular basis which raises their monthly expenditure. Unfortunately, there isn't pertinent research on using machine learning in understanding nutritional psychology [2] demonstrated LR [3], demonstrated NLR [4], demonstrated Random Forest [5], demonstrated Extra Trees Regressor [6], demonstrated Ridge Regression of regression type and [7] demonstrated Decision Tree [8], demonstrated Association FP Growth [9], demonstrated Gaussian NB, and [10] demonstrated Multinomial NB for classification. Taking those machine learning models into consideration we performed a detailed study on how fast food consumption happens and its impact on an individual's mental and physical health. While performing our research we also came across the study performed by [11] who talks about optimization methods from machine learning perspective that is taken in consideration for resource consumption. Inspired from the available observations and research performed in the field of nutritional psychology we performed our own research to understand college youth and their expenditure.

2 Related Work

While looking at the health-related issues faced we came across various studies performed in this field [12] demonstrated a study which focuses on fast food eating habits. Their study showed the relationship of students with respect to their BMI, 21.8% of responders came out to be overweight and 10.59% underweight. The paper was not able to present any further solid insights due to inadequacy of test and clamor data [13] researched on the issues of fast food intake. Their discovery depicted that the specific time of MASHHAD is the most optimum time when the fast food intake rate was high [14] demonstrated the nutritional status and physical activity performed by the girls aged from 20 to 25, major attributes taken in consideration were BMI, height, weight, waist size, and hip circumference. The study came with a conclusion that intake of fast food must be controlled as it has the potential to cause disease associated with liver, arthritis, cardiovascular, and diabetic [15] conducted a research to identify how gender affects the fast food intake in school and university level students. Their findings discovered that there is a positive link between the gender and fast food intake in terms of type, portion size, and frequency of consumption. Their findings showed that commercials and promotions have a huge impact on school and university students, young adults are inclined to prefer fast food over nutritional food even after knowing the bad effects of excessive fast food consumption [16] performed research to identify the rapid changes occurring in non-vegetarian food intake behavior and the factors which are affecting it. A structured survey was conducted by them with multiple sections of questions such as socioeconomic profile, purchasing behavior, and products in the market were taken into consideration [17] demonstrated the study which focused on the change in lifestyle due to pandemics has led to the change in eating habits. In [18], demonstrated the research focusing on how health was affected by the emotional state. Lockdown caused an increase in depression, anxiety, and loneliness which directly affects the immune system and health of an individual [19] demonstrated the study to determine energy drinks intake among young adults and factors affecting consumption like lack of sleep, increase in energy for studying or while driving. The main reason for the consumption of energy drinks is to overcome lack of sleep, mainly during exam period. Their study showed that the caffeine required for stimulating cognitive effects is less than that consumed by consumers in energy drinks which has negative effects on health [20] presented the hypothesis on sugar consumption in the United States and its relationship with psycho-sociology. The finding identified that there is an invariable relationship between standard of living and sugar consumption [21] used two supervised machine learning techniques namely K-means and logistic regression to understand food consumption [22] proposed an experimental analysis of fast food consumption of students from University of Bangladesh stating that overall prevalence of fast food consumption of males came out to be more in proportion to 56:44 [23] identified that weight and obesity are one of the challenging features to most of the human beings, modern machine learning techniques like SVM, RF, ANN can be used for weight predictions.

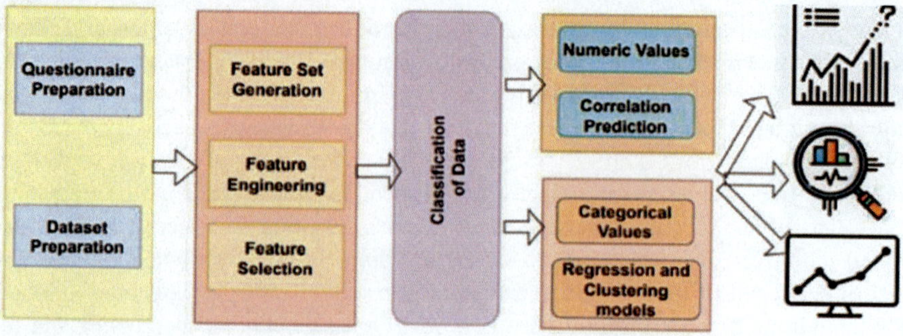

Fig. 1 Implementation flow of research work

3 Methodology

The data for this research was collected through a personally administered survey conducted from the year 2018 till 2020. The survey was conducted for the age group between 18 and 23. A total of more than 150 responses were gathered from individuals belonging to different classes of the society. Figure 1 presents workflow for the gathered data.

Figure 2 presents the survey questionnaire structuring. The survey questionnaire's developed for the study have questions representing 5 different sections such as socioeconomic background, Health, Food branding, Food and Friends. The first section includes the general questions related to Monthly expenditure, daily minimal expenditure, fast food frequency consumption. The second section of the questionnaire is related to health which includes the question focusing on health and mental status. The third section consists of questions related to the food market and impact of branding in the food chain market. The last section consists of food and friends related questions which focuses on the fast food consumption scenario which consists of comfort of eating, spending, and dining out with groups.

4 Experimental Evaluation

The classification model namely decision tree, Gaussian Bayes, multinomial Bayes are implemented to identify the relationship between certain feature variables and target variables with help of which further predictive analysis can be done. Table 1 provides result for below discussed cases.

Following are the resultant cases and observation which were built based upon the data gathered:

(a) Identification of body mass index based upon all the questions asked in the survey. Categorized distribution of BMI is done where BMI less than 18.5 is

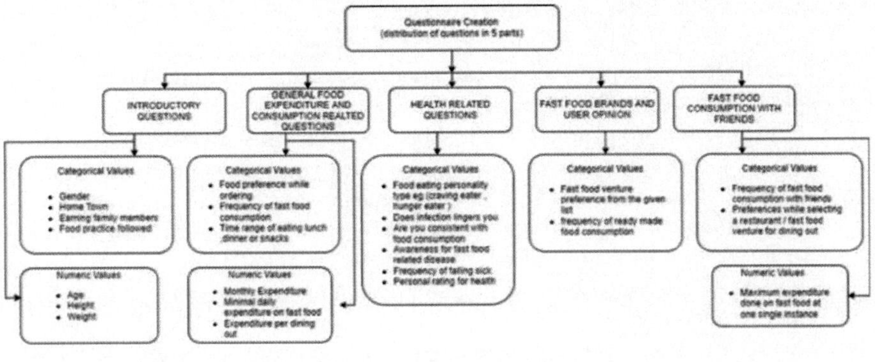

Fig. 2 Survey taxonomy

Table 1 Comparison of classification methods

Cases	Target	Model	Accuracy (%)
Case 1	BMI	Decision Tree Multinomial Bayes	83.26 83.72
Case 2	Frequency of falling sick	Decision Tree Gaussian Bayes	98 97.82
Case 3	Infection to linger	Decision Tree Multinomial Bayes	86.04 90.69
Case 4	Post-COVID and pre-COVID scenario	Decision Tree Multinomial Bayes	79.06 76.74
Case 5	Preference of food quality	Decision Tree Multinomial Bayes	83.72% 79.06%

considered underweight, greater than or equal to 18.5 and less than 25 as normal, greater than or equal to 25 and less than 30 as overweight, and for BMI greater than or equal to 30 as obese.

(b) The frequency of falling sick based upon certain feature variables such as gender of the person, subject's BMI in a category such as underweight, normal, overweight or obese, how much subject rates himself on a personal scale of 1–5 for the question how much he will rate his eating habits and certain choice-based questions as if he is aware of harmful effects of excessive consumption of fast food if he feels fatigued after excessive consumption of fast food if infection tends to linger him. The target variable is the frequency of falling sick where it is divided into categories such as never, rarely, often, sometimes [25].

(c) If infection tends to linger based upon features such as the subject is hostilities or localities, subject's BMI, the subject's eating type category namely vegetarian, non-vegetarian or vegetarian, his daily minimal expenditure, his frequency of fast food consumption, his eating behavior such as hunger eater, influenced eater, craving eater or stress eater, the subject is aware of excessive consumption of fast food if he feels fatigued after excessive food consumption if he prefers food quality or not if he has fast food venture near his home, how much subject rates himself on a personal scale of 1–5 for the question how much he will rate his eating habits.

(d) Identifying that the subject has a healthy food intake post-COVID-19, the feature variables identified for the same are BMI category of the subject such as underweight, normal, overweight or obese, frequency of fast food consumption, and other choice-based questions such as if subject skips his meal post-COVID-19, if subject used to skip meal pre-COVID-19, subject's lunchtime either before noon or in between 12 and 4 pm, subject's dinner time either between 6 and 9 pm or after 9 pm.

(e) Identifying if the user has a preference for food quality while ordering in or dining out from a restaurant where target variable considered are BMI categorized as underweight, normal, overweight or obese, subject's choice for ordering while dinging out if it is only food, only drinks or food and drinks, the maximum amount of spending done by the subject in one order, subject's preference while dinging out for ambiance, budget, and comfort [24, 26].

Table 2 showcases the prominent results evaluated on these target variables.

(a) Identification of body mass index with respect to all the features of our data. For implementation of the model all categorical values are converted to integer using label encoding in order to use all features as dependent variables.

(b) Features for predicting the target having healthy food intake post-COVID-19, the feature variables identified are BMI categorized as underweight, normal, overweight or obese, frequency of fast food consumption and if subject skips his meal post-COVID-19, used to skip meal pre-COVID-19, subject's lunch time, dinner time.

Table 2 Results of target variables

Cases	Target	Model	Accuracy (%)
Case 1	BMI	Linear Regression	92.83
		Random Forest	96.53
		Extremely Randomized Trees	81.22
		Ridge Processor	92.83
		Non-Linear	87.06
Case 2	Pre-COVID and Post-COVID	Linear Regression	–
		Random Forest	89.93
		Ridge Regression	Alpha = 40
Case 3	Food Quality and Preference	Linear Regression	–
		Random Forest	80.97
		Ridge Regression	Alpha = 20

(c) For identifying the target variable where user has a preference for food quality while ordering in or dining out where target variable is as BMI category of the subject such as underweight, normal, overweight or obese, subject's choice for ordering while dining out if the choices are only food, only drinks or food and drinks, maximum amount of spending done in one ordering, preference while dining out for ambience, budget, and comfort.

5 Visualizations

This unique method of representing the data clearly allows an individual to never miss any insight the data is about to reveal. The below visualizations are made upon the responses gathered from MIT WPU school of engineering students whether consumption of excessive fast food affects the health of students.

In the Fig. 3, the visualization depicts the relation between the monthly expenditure and how often a student falls sick the fig shows that the people who are hostilities and have a high monthly expenditure are more likely to fall sick compared to those who have a normal monthly or low monthly expenditure.

In Fig. 4, The above visualization shows the frequency of fast food intake on a daily basis it shows that 4.23% consumed fast food more than once a day, 7.75% ate fast food at most once a day, 21.83% on odd days, 23.94% ate fast food based upon college/ workspace schedule, 40.14% rarely a week.

In Fig. 5, the visualization shows the comparison between frequency of feeling sick, monthly expenditure, and the minimal expenditure eating fast food. The visualization shows that the student who feels sick usually has excessively daily minimal spending on fast food per day.

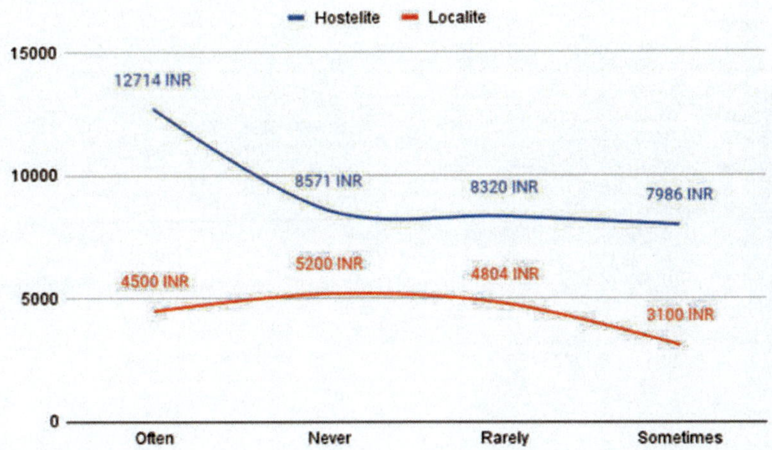

Fig. 3 Categorization of localities and hostilities on how often do they fall sick over their monthly expenditure

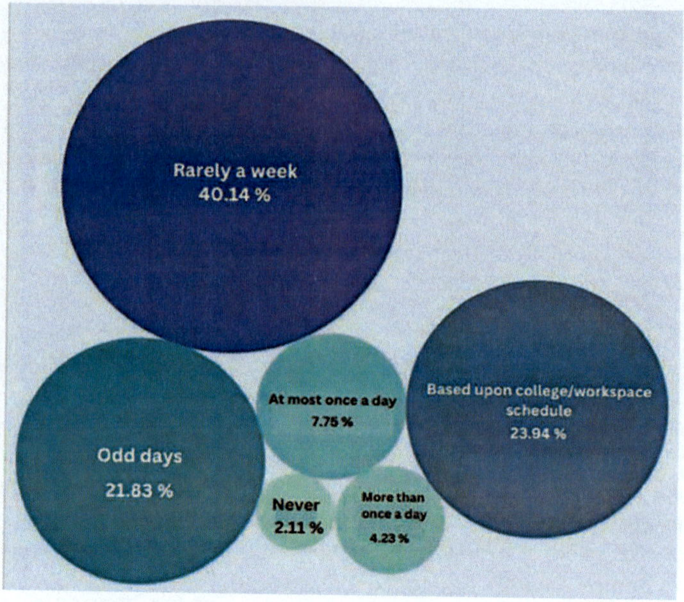

Fig. 4 Frequency consumption of fast food on a daily basis

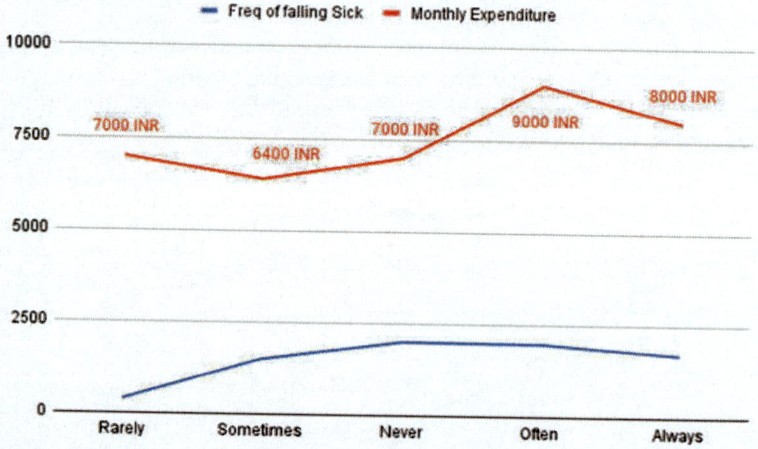

Fig. 5 Showcasing count of people falling sick under each category and their average monthly spending

6 Conclusion

Based on the research done it becomes very clear that there is a huge opportunity for innovation and exploration for machine learning and its application in nutritional psychology and consumer consumption system. From the experiment conducted we got a critical observation that there is a strong relationship between a person's fast food consumption and his locality, students or people living far from their home in hostel prefers to eat fast food in comparatively greater extent as compared to the one living in their hometown with their families. More a person is addicted to fast food consumption, the more are his chances to experience laziness. People with jam packed schedule prefers to consume fast food more frequently as compared to other individuals as they consider it as an escape from their day to day life. 44% individuals who were able to change their fast food consumption habits during the period of pandemic preferred to stay constant to their dietary lifestyle and preferred to control their fast food consumption. Herewith, study must be conducted to determine factors like the opinion of parents, health ministry, nutritionists, and health experts. It is commended to educate and familiarize families, teenagers, and young adults about the long-term adverse effects after intake of fast food.

References

1. Satia JA, Galanko JA, Siega-Riz AM (2004) Eating at fastfood restaurants is associated with dietary intake, demographic, psychosocial and behavioural factors among African Americans in North Carolina. Public Health Nutr 7(8):1089–1096
2. Schneider A, Hommel G, Blettner M (2010) Linear regression analysis: part 14 of a series on evaluation of scientific publications. Deutsches Arzteblatt international 107(44):776–782. https://doi.org/10.3238/arztebl.2010.0776
3. Uyanık GK, Güler N (2013) A study on multiple linear regression analysis. Procedia Soc Behav Sci 106:234–240. https://doi.org/10.1016/j.sbspro.2013.12.027.doi:10.1016/j.sbspro.2013.12.027
4. Breiman L (2001) Random forests. Mach Learn 45:5–32. https://doi.org/10.1023/A:1010933404324
5. Geurts P, Ernst D, Wehenkel L (2006) Extremely randomized trees. Mach Learn 63(1):3–42. https://doi.org/10.1007/s10994-006-6226-1.doi:10.1007/s10994-006-6226-1
6. Hoerl AE, Kennard RW (2000) Ridge regression: biased estimation for nonorthogonal problems. Technomet 42(1):80–86. https://doi.org/10.2307/1271436
7. Quinlan JR (1986) Induction of decision trees. Mach Learn 1:81–106. https://doi.org/10.1007/BF00116251
8. Borgelt C (2005) An implementation of the FP-growth algorithm. In: Proceedings of the 1st International Workshop on Open Source Data Mining: Frequent Pattern Mining Implementations, pp 1–5, Aug
9. Raizada RDS, Lee Y-S (2013) Smoothness without smoothing: why Gaussian Naive Bayes is not naive for multi-subject searchlight studies. PLoS ONE 8(7):e69566. https://doi.org/10.1371/journal.pone.0069566
10. Kibriya AM et al (2004) Multinomial naive bayes for text categorization revisited. In: Australasian Joint Conference on Artificial Intelligence. Springer, Berlin, Heidelberg

11. Sun S, Cao Z, Zhu H, Zhao J (2020) A survey of optimization methods from a machine learning perspective. IEEE Trans Cybernet 50(8):3668–3681. https://doi.org/10.1109/TCYB.2019.2950779
12. Chowdhury MR, Subho MR, Rahman MM, Chaki ISD (2018) Impact of fast food consumption on health: a study on University Students of Bangladesh. In: 21st International Conference of Computer and Information Technology (ICCIT), pp 1–6
13. Saghaian S, Mohammadi H (2018) Factors affecting frequency of fast food consumption. J Food Dist Res 49:22–29
14. Monika S, Chishty S, Verma K (2015) Fast food consumption pattern among postgraduate female student living in hostel of University of Rajasthan, India. Asian J Dairy Food Res34(4). https://doi.org/10.18805/ajdfr.v34i4.6887
15. Kayisoglu S, İçöz A (2014) Effect of gender on fast-food consumption habits of high school and university students in Tekirdag, Turkey. Acta Alimentaria 43:53–60. https://doi.org/10.1556/AAlim.43.2014.1.6
16. Banerjee M, Mishra M (2017) Retail supply chain management practices in India: a business intelligence perspective. J Retail Consum Serv 34:248–259
17. Hassen TB, Bilali HE, Allahyari MS (2020) Impact of COVID-19 on food behavior and consumption in Qatar. Sustain 12(17):6973–6973. https://doi.org/10.3390/su12176973
18. Heidal KB, Colby SE, Mirabella GT, Al- Numair KS, Bertrand B, Gross KH (2012) Cost and calorie analysis of fast food consumption in college students. Food Nutri Sci 3(7):942–946. https://doi.org/10.4236/fns.2012.37124
19. Malinauskas BM, Aeby VG, Overton RF, Carpenter-Aeby T, Barber-Heidal K (2007) A survey of energy drink consumption patterns among college students. Nutrit J 6(1):35–35. https://doi.org/10.1186/1475-2891-6-35
20. Barthes R (2018) Toward a psychosociology of contemporary food consumption. Food Cult. Routledge, 13–20
21. Abdella GM et al (2020) Sustainability assessment and modeling based on supervised machine learning techniques: the case for food consumption. J Cleaner Product 251:119661
22. Goon S, Bipasha MS, Islam S (2014) Fast food consumption and obesity risk among university students of Bangladesh. European J Preventive Med 2(6):99–104. https://doi.org/10.11648/j.ejpm.20140206.14
23. Babajide O et al (2020) A machine learning approach to short-term body weight prediction in a dietary intervention program. In: Krzhizhanovskaya VV et al. (eds) Computational science—ICCS 2020. ICCS 2020. Lecture Notes in Computer Science, vol 12140. Springer, Cham. https://doi.org/10.1007/978-3-030-50423-6_33
24. Jadhav MM (2021) Machine learning based autonomous fire combat turret. Turkish J Comp Math Educ (TURCOMAT) 12(2):2372–2381
25. Mulani AO, Jadhav MM, Seth M (2022) Painless machine learning approach to estimate blood glucose level with non-invasive devices. In: Artificial intelligence, internet of things (IoT) and smart materials for energy applications. CRC Press, pp 83–100
26. Kashid MM, Karande KJ, Mulani AO (2022) IoT-based environmental parameter monitoring using machine learning approach. In: Proceedings of the International Conference on Cognitive and Intelligent Computing, pp 43–51. Springer, Singapore

A Unified System for Crop Yield Prediction, Crop Recommendation, and Crop Disease Detection

Arpitha Varghese and I. Mamatha

Abstract Agriculture is the most important factor in ensuring livelihood. Crop disease growth has recently increased due to catastrophic weather patterns and a dearth of immunogenicity in crops. Machine learning (ML) could be a critical perspective for achieving a realistic and true solution to the crop yield issue. It will assist Indian farmers in predicting crop yield and recommending crops based on environmental conditions. Machine learning and IoT are used in this work to implement crop recommendations, crop yield prediction, and disease detection using deep learning. The hardware system is intended to collect soil and environmental data from the surrounding environment for crop recommendation and yield. A web application is developed for all these three applications.

Keywords Crop yield prediction · Crop disease detection · Machine learning · Deep learning · IoT · ThingSpeak

1 Introduction

Agriculture has quite a long history in India. Despite being the largest financial industry by population and playing an essential function in India's aggregate socioeconomic fabric, current research findings have revealed a ceaseless downward trend in agriculture's participation in Indian economic development. Most farmers relied on deep-rooted field experience with selective plants to forecast a good return in the upcoming harvest season and still didn't receive the crop's valued profit margin as expected. This is usually caused by sparse irrigated agriculture or penurious crop choice. Weather conditions, soil characteristics, infectious diseases, and other characteristics always have an impact on crop production. There have been few works

A. Varghese · I. Mamatha (✉)
Department of Electrical and Electronics Engineering, Amrita School of Engineering, Amrita Vishwa Vidhyapeetham, Bangalore, India
e-mail: i_mamtha@blr.amrita.edu

S. Sharma et al. (eds.), *Intelligent Control, Robotics, and Industrial Automation*, Lecture Notes in Electrical Engineering 1066,
https://doi.org/10.1007/978-981-99-4634-1_81

reported in the literature for crop yield prediction, crop advice, and disease detection. Ferrández-Pastor et al. [1] have concluded that software technology needs to be carefully planned in order to be incorporated into businesses without creating additional issues. The user interfaces and services need to be extremely intuitive and appropriate for the way technicians and farmers operate. Balamurugan et al. [2] used the random forest algorithm to classify and estimate crop yield by considering the season and other weather elements as input. Niketa et al. [3] have demonstrated that crop yield is affected by seasonal climate and machine learning algorithms can effectively model climatic variations and predict yield. They forecast imminent statistics using records from the past years and classified the results using SMO classifiers in Waikato Environment for Knowledge Analysis (WEKA). Ashok et al. [4] proposed an improvised technique using a convolutional neural network (CNN) to detect leaf disease in tomato crops with a 98 percent accuracy level. Ahmad et al. [5] proposed an efficacious approach for thoroughly categorizing crop illness signs using CNN. These do seem to be retention capable, so merged well with the suggested training structure, they enable the swift growth of industrial uses by cutting training times in half. A steady, substantial dataset and the use of several feature extraction approaches have made it easier to get positive experimental results as approached by Pooja et al. [6]. After sorting the qualities of the feature retrieved photos that used a data gain approach, an SVM classifier is utilized to categorize them by Kuricheti et al. [7]. The two specified illnesses have been identified, and a GUI has been developed to display the many steps of the image processing technique. Rekha et al. [8] provide information here on the system's parts and client services. A mobile application developed as a result of this study helps farmers with numerous agriculture practices like fertilization and irrigation. Nandhini et al. [9] provide information here on the system's parts and client services. A mobile application developed as a result of this study helps farmers with numerous agriculture practices like fertilization and irrigation. Nandhini et al. [10] proposed a method where diseased area of the plant is dissected and studied using the k-mean clustering approach. The client has the option to choose how many clusters to use. Devi et al. [11] suggested a solution which uses IoT and image analysis to identify plant illnesses sooner. Comparing the suggested method to the current KNN classifier, the classification accuracy has increased by an average of 24 percent. This method permits remote management of the hill banana field and early disease identification in hill bananas. The current method for detecting plant diseases is simply a naked-eye observation, which necessitates more manpower, properly equipped laboratories, expensive devices, and so on. Inadequate disease detection can also lead to inexperienced pesticide use, which can lead to the development of long-term resistance to diseases, significantly lowering the crop's ability to fight back. Therefore, there is a need for a unified system with a good accuracy rate that can do all these three applications like crop recommendation, crop yield prediction, and crop disease detection together to help our farmers. As a necessary consequence, the objective is to predict crop yield and give crop recommendations by considering various factors affecting the crops including N, P, and K using machine learning and IoT. Second objective is to detect crop disease using deep learning and inform farmers about the cause and what can be done to cure or prevent the disease:

Third objective is to implement a hardware system that can be used to take data from the surrounding using sensors for these applications, develop a web application that has all three applications crop recommendation, crop yield prediction, and disease detection together.

2 Methodology for Crop Yield Prediction and Recommendation

Machine learning is an important guide for estimating crop production, along with recommending the crop that can be grown in that soil to gain more yield. A few ML techniques have been implemented for sets of data for crop recommendation and yield forecast. Crop production is primarily determined by weather conditions and soil quality. Through examining the soil as well as the environment, a specific region's best crop for increased crop production and total combined yield of crops could be forecast. This forecast will be beneficial to farmers. The methodology followed in the proposed work for crop yield prediction and crop advice is depicted in Fig. 1.

The crop recommendation and yield prediction datasets were obtained from various government websites such as data.gov.in. Details about the area, production, crop name, and N, P, K, and Ph values were obtained from these datasets. Rainfall details are collected from Indianwaterportal.org. Temperature and humidity data were obtained from power.larc.NASA.gov. Preprocessing stage is just a technique for converting original data into a dataset without any null set or empty spaces. The information has been gathered from various references and is in raw format, which further tends to make examination extremely difficult. We could perhaps change the data into more of a more readable form by using specific methodologies such as filling in the missing and void values. Correlation among the various factors affecting crop recommendation and yield prediction is shown in Fig. 2a, b, respectively, where

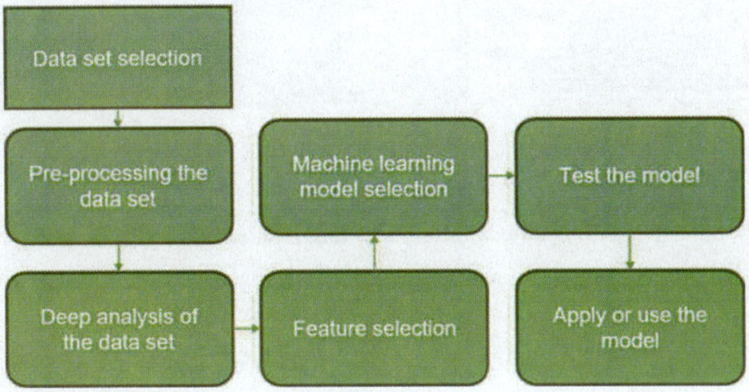

Fig. 1 Methodology for crop advice and prediction

it may be observed that P and K are highly correlated. Similar way, the correlation among the factors affecting crop yield prediction shows a high correlation between soil moisture and humidity and is quite natural. A variety of variables affect crop yield forecasting [12]. And those would be such characteristics that aid in forecasting crop yields year-round. The elements that affect them are climatic conditions such as rainfall and soil characteristics such as Nitrogen (N), Phosphorous (P), Potassium (K), and Ph value. To investigate statistical properties and the events that influence modification, various information visualization methods are implemented. Decision trees [13], Naive Bayes [14], random forest classifier [15], XGBoost [16], SVM [17], and logistic regression [18] are implemented and compared for crop recommendation. Decision tree, SVM, linear regressor [19], and random forest regressor are implemented and compared for crop production forecast. Among the data collected, 80% of data is considered for training, and 20% is used for testing. Accuracy from these methodologies has been compared, which aids to forecast the best approach which is depicted in Fig. 3a for crop recommendation and Fig. 3b for crop yield prediction. Crop recommendation is implemented using the random forest classifier, which achieved a 99 percent accuracy rate for our dataset. Crop yield prediction is implemented using the random forest regressor, which has a 97 percent accuracy rate. The plants which have been used for the study are kidney beans, coconut, tomato, potato, coffee, maize, lentil, jute, apple, chickpea, rice, cotton, black grams, muskmelon, papaya, moth beans, areca nut, arhar, bajra, barley, banana, mango, orange, wheat, ragi, watermelon, grapes, beetroot, bitter gourd, brinjal, cabbage, carrot, cauliflower, cashew nut, colocasia, small millets, coriander, cucumber, drumstick, chilies, ginger, garlic, squash, strawberry, peach, cherry, jackfruit, soybean, urad, and gram.

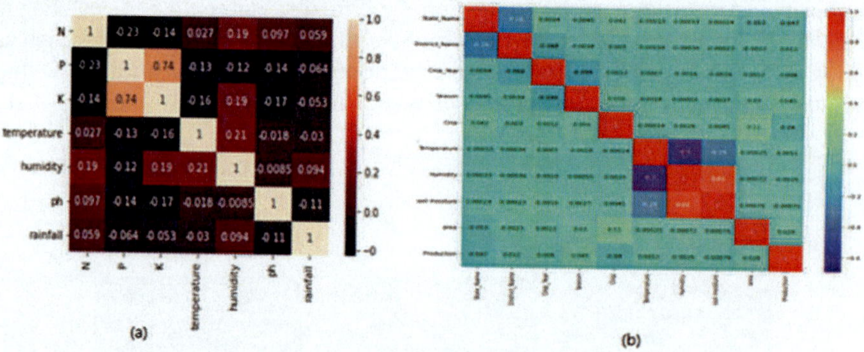

Fig. 2 **a** Correlation between various factors affecting crop recommendation and **b** Crop yield prediction

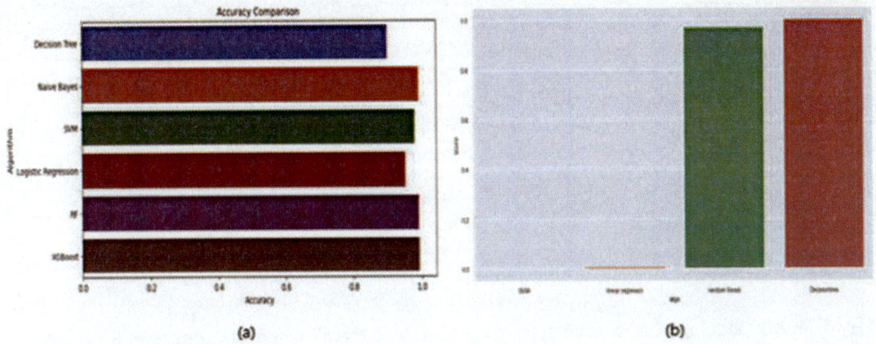

Fig. 3 **a.** Accuracy comparison for crop recommendation and **b.** R2 score comparison for crop yield prediction

3 Methodology for Crop Disease Detection

Deep learning is useful for detecting crop diseases. When a crop disease is detected, the farmer can take the appropriate action to avoid production losses. The various steps followed for crop disease detection using deep learning and the parameters considered for ResNet are shown in Fig. 4a, b respectively. The dataset of crop disease detection consists of around 87 k images which are categorized into 38 different classes. From the datasets, 14 types of unique crops and 26 types of disease classes, and 12 healthy classes are chosen in the present work. The pixel values for images used for disease detection were transformed from (0–255) to (0–1). Data augmentation techniques like cropping, flipping, and rotation are done to give the model a variety of images. The residual network's primary objective is to construct a greater-depth neural network. The whole framework established the notion of residual frames to overcome the issue of said exploding gradient. The ResNet model is explained in detail in [20]. 80% of the data is considered for training and 20% is used for testing from the data collected. The parameter considered for the model is shown in Fig. 8. Crop disease detection is carried out using the ResNet model, which has a 99.2 percent accuracy rate. The types of diseases that have been detected using deep learning algorithms are tomato_late_blight, tomato_early_blight, tomato_septoria_leaf_spot, tomato_curl_virus, tomato_bacterial_spot, tomato_target_spot, tomato_mosaic_virus, tomato_leaf_mold, tomato_spider_mites, orange_haunglongbing, squash_powdery_mildew, corn_northern_leaf_blight, corn_cercospora_leaf_spot, corn_common_rust, grape_black_measles, grape_black_rot, grape_isariposis_leaf_spot, strawberry_leaf_scorch, apple_scab, apple_black_rot, cherry_powdery_mildew, peach_bacterial_spot, potato_late_blight, potato_early_blight, bell_pepper_bacterial_spot.

(a)

Parameters	Value
Batch size	32
Steps per epoch	558
Epoch	10
Validation steps	1
Seed	42
Optimizer	SGD(stochastic gradient descent)
Learning rate	Default
Decay	Default
Momentum	Default

(b)

Fig. 4 **a.** Methodology for disease detection using deep learning and **b.** Parameters for ResNet

4 System Design, Implementation, and Results

The analysis involves the study of cracking down the sophisticated problem into small segments for a finer insight into it. Engineering analysis considers requirements, structures, mechanisms, and system dimensions. Figure 5a is the structure designed and Fig. 5b shows the implementation for crop yield and crop recommendation application. The system designed collects the data from temperature, humidity, and Ph (N, P, K) sensors and then sends the data to ThingSpeak via Wi-Fi. Figure 5a shown below is the block diagram for crop disease detection using IoT and Fig. 5b is the implementation for the same. The output of the system is sent to a Telegram on a smartphone via Wi-Fi.

Python 3.8.5 (Jupyter Notebook) is used to implement ML and deep learning techniques, which include input libraries such as ScikitLearn, Numpy, Keras, and Pandas. The hardware component's script is composed in the Arduino IDE and the sensor data are sent to ThingSpeak, which in turn takes data for input. Figure 6a, b shows the hardware system designed and implemented for crop production prognosis and crop recommendation for collecting data from the environment without a power supply. When the system is powered up, the sensor begins to collect data. The sensor

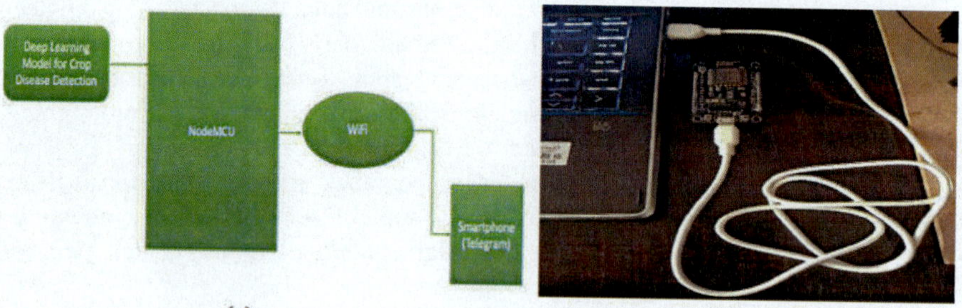

(a) (b)

Fig. 5 **a.** System designed **b.** Implementation for crop disease detection using nodeMCU

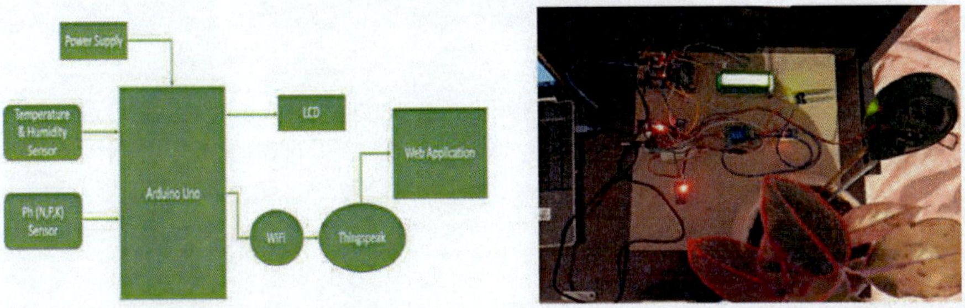

(a) (b)

Fig. 6 **a.** System designed for crop yield prediction and recommendation and **b.** System implemented for crop yield prediction and recommendation

readings are then sent to ThingSpeak. The value sent to ThingSpeak is depicted in Fig. 7i–v. The sensor data sent to ThingSpeak is used as input for the web application. Figure 5 depicts the hardware connection with the power supply turned on. The crop disease detection code is uploaded into nodeMCU, and the name of the disease detected is sent to the user via Wi-Fi using Telegram. Figure 8a shows the home page for a web application for crop advice, disease detection, and production forecast. Figure 8b, c show the crop recommendation page where the input for N, P, and, K is directly taken from ThingSpeak, and the rest of the data is entered by the user and also shows the output for crop recommendation. Based on the data entered, we can see that the system recommended growing kidney beans as shown in Fig. 8d. Figures 9a–c show the input for crop yield prediction, and Fig. 9d shows the output for crop yield prediction. The values for temperature and humidity are directly taken from the sensor. Based on the parameters entered, the system predicts the user will get a yield of 434.512 tons per hectare. Figure 10a shows the front page and input for crop disease detection, and Fig. 10b–d show the output for crop disease detection. The user gets the name of the disease detected, the cause of the disease, and how it can be prevented. Figure 11a shows the input which is given to the system for crop disease detection using nodeMCU and the output of the system as shown in Fig. 11b is communicated to the client via Telegram as shown in Fig. 11c.

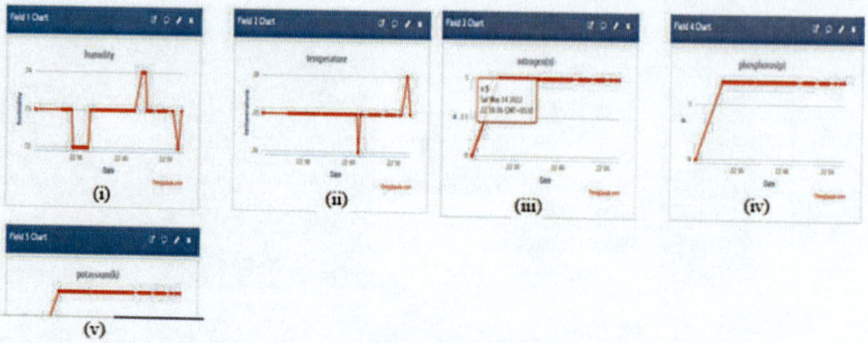

Fig. 7 Screenshot showing values of (i) Temperature (ii) Humidity (iii) Nitrogen (iv) Phosphorus (v) Potassium sent to ThingSpeak

Fig. 8 **a.** Homepage for the web application, **b.** and **c.** Input for crop recommendation **d.** Output for crop recommendation

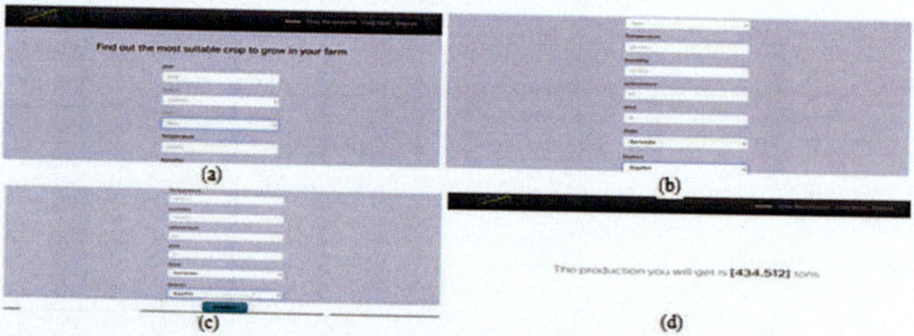

Fig. 9 **(a)–(c).** Input for crop yield prediction **d.** Output for crop yield prediction

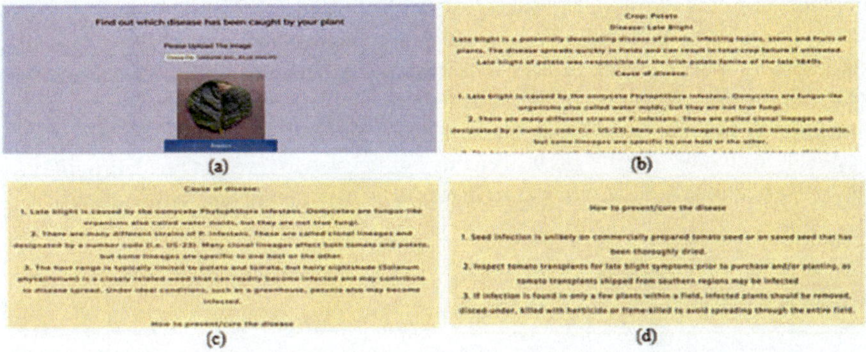

Fig. 10 **a.** Input for crop disease detection and **(b)–(d).** Output for crop disease detection

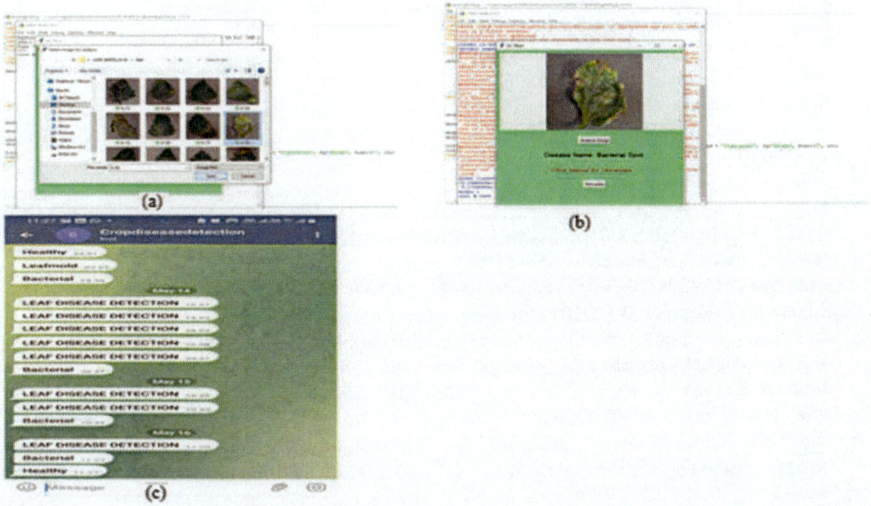

Fig. 11 **a.** Input for crop disease detection using nodeMCU **b.** Output for crop disease detection using nodeMCU **c.** Name of the disease detected sent to the user using Telegram

5 Conclusion

Agriculture is indeed the lifeblood of several countries, including India. Nevertheless, in order to prevent agriculture loss, the utilization of agricultural technologies must be prioritized. The following conclusions can be drawn from the analysis, design, and the first prototype work. The proposed system will assist farmers in estimating yields depending on the climatic parameters and crop area. If yield estimations are undesirable, the producer can use this to decide either to cultivate that specific plant or an alternative plant and detect disease in crops so that the necessary steps can be taken to protect crops from further damage. In this paper, a random decision forest regressor is used to forecast crop yield which gave an R2 score value of 97%, and a random forest classifier to recommend crops which gave an accuracy rate of 99%, as well as the ResNet model for crop disease detection is implemented with a 99.2% accuracy rate. A hardware system for crop yield prediction and crop recommendation to gather and use real-time environmental data and a web application which has all these three applications is developed. This will enable our farmers to forecast yield, provide crop recommendations, and identify crop diseases using a single web application.

References

1. Ferrández-Pastor, Francisco-Javier, Mora-Pascual J, Díaz-Lajara D (2022) Agricultural traceability model based on IoT and blockchain: application in industrial hemp production, 100381. https://doi.org/10.1016/j.jii.2022.100381
2. Priya P, Muthaiah U, Balamurugan M (2015) Int J Eng Sci Res Techn Pred Yield Crop Using Machine Learning Algorithm
3. Gandhi N et al (2016) Rice crop yield forecasting of tropical wet and dry climatic zone of India using data mining techniques, 357–363. https://doi.org/10.1109/ICACA.2016.7887981
4. Ashok S, Kishore G, Rajesh V, Suchitra S, Sophia SSG, Pavithra B (2020) Tomato leaf disease detection using deep learning techniques, 979-9839. https://doi.org/10.1109/ICCES48766.2020.9137986.
5. Ahmad M, Abdullah M, Moon H, Han D (2021) Plant disease detection in imbalanced datasets using efficient convolutional neural networks with stepwise transfer learning, 9:140565–140580. https://doi.org/10.1109/ACCESS.2021.3119655
6. Pooja V, Das R, Kanchana V (2017) Identification of plant leaf diseases using image processing techniques, pp 130–133. https://doi.org/10.1109/TIAR.2017.8273700
7. Kuricheti G, Supriya P (2019) Computer vision based turmeric leaf disease detection and classification: a step to smart agriculture. In: 2019 3rd International Conference on Trends in Electronics and Informatics (ICOEI), pp 545–549
8. Rekha P, Rangan VP, Ramesh MV, Nibi KV (2017) High yield groundnut agronomy: an IoT based precision farming framework, pp 1–5. https://doi.org/10.1109/GHTC.2017.8239287
9. Nandhini SA, Hemalatha SR, Indumathi K (2018) Web-enabled plant disease detection system for agricultural applications using WMSN, pp 725–740. https://doi.org/10.1007/s11277-017-5092-4
10. Reddy JN, Vinod K, Ajai ASR (2019) Analysis of classification algorithms for plant leaf disease detection, pp 1–6. https://doi.org/10.1109/ICECCT.2019.8869090
11. Devi R, Deepika S, Nandhini A, Hemalatha R, Radha S (2019) IoT enabled efficient detection and classification of plant diseases for agricultural applications, pp 447–451. https://doi.org/10.1109/WiSPNET45539.2019.9032727
12. Liliane TN, Charles MS (2020) Agronomy-climate change & food security. In: Chapter, factors affecting yield of crops factors affecting yield of cropsl IntechOpen
13. Rokach L, Maimon O (2005) Decision trees. In: Data mining and knowledge discovery handbook. Springer, Boston, MA, pp 165–192
14. Rish I (2001) An empirical study of the naive Bayes classifier. IJCAI 2001 Works Emp Methods Artif Intell 3(22):41–46
15. Breiman L (2001) Random forests. pp 5–32. https://doi.org/10.1023/A:1010933404324
16. Chen T, Guestrin C (2016) Xgboost: a scalable tree boosting system. pp 785–794. https://doi.org/10.48550/arXiv.1603.02754
17. Wang L (ed) (2005) Support vector machines: theory and applications, vol 177. Springer Science & Business Media

18. Peng CYJ, Lee KL, Ingersoll GM (2021). An introduction to logistic regression analysis and reporting. J Educ Res 96(1):3–14
19. Kumari K, Yadav S (2018) Linear regression analysis study. J Practice Cardiovas Sci 4(1):33
20. He K, Zhang X, Ren S, Sun J (2016) Deep residual learning for image recognition, pp 770–778. https://doi.org/10.1109/CVPR.2016.90

Performance Analysis of ExpressJS and Fastify in NestJS

M. Praagna Prasad and Usha Padma

Abstract A lot of research and resources are available on the back-end development using NodeJS because of its flexibility and ease among develop ers, but in the recent years, there has been a rise in migration of back-end framework to NestJS by developers due to the various shortcomings of NodeJS. In this paper, we have given an insight on the shortcomings of NodeJS compared to NestJS. Fundamental concepts of NestJS are discussed along with the construction of the structure of the application. We have also highlighted features of NestJS including unit testing, validation, and database connection with PostgreSQL. A comparative analysis is drawn between these ExpressJS and Fastify based on latency and throughput varying with respect to the number of concurrent HTTP requests sent to both Express.js and Fastify using Apache Jmeter. Results are drawn in graphical form and inference is discussed. Fastify proved to be 5% faster in performance than Express.

Keywords NestJS · ExpressJS · Fastify · Test-driven development

1 Introduction

Back-end development refers to client-side development, whereas front-end development refers to server-side development. Although they are very distinct from one another, these two phrases are extremely important for web development. To increase the functioning of the website, each side must successfully communicate with and work with the other as a unified entity. Database management, API handling, data validation, and ensuring that everything on the client-side functions properly is all part of back-end development process. Various frameworks, including NodeJS, Django, Rails, Laravel, and Spring, are used for this. Front end refers to the area of a website

M. P. Prasad (✉) · U. Padma
Electronics and Telecommunication Engineering, RV College of Engineering, Bangalore, India
e-mail: praagnaprasad@gmail.com

U. Padma
e-mail: kamakshimb@rvce.edu.in

S. Sharma et al. (eds.), *Intelligent Control, Robotics, and Industrial Automation*,
Lecture Notes in Electrical Engineering 1066,
https://doi.org/10.1007/978-981-99-4634-1_82

where users interact. Everything a user immediately experiences is included. Some of the frameworks used for front-end development include AngularJS, ReactJS, jQuery, and SASS. The work conducted includes comparison between NodeJS and NestJS as well as Express and Fastify in NodeJS. A system is created to migrate the back-end framework to NestJS.

Operations of PUT, GET, DELETE, PATCH, UPDATE are performed on the back-end framework which can be observed using Postman. End-to-end testing, validation were performed on the NestJS program with test cases. A database was created on Postgres and a database connection was made in NestJS. Two projects were created, one with Fastify and one with Express. Both were able to fetch, create, update, and delete data from database and were able to connect to the web browser. They were also able to handle the concurrent HTTP requests that were sent, and comparative analysis was drawn between these two using Apache JMeter.

2 Literature Survey

The various methods of deployment data into different cloud services were presented in [1]. The paper also included different cloud services like Data as a Service, Product as a Service, Software as a Service, and the advantages and disadvantages of different cloud services. For cloud web services, a suggested integrated identity and attribute-based access management system were presented in [2]. It draws attention to how attribute-based access control and authentication work together to increase the security of cloud web services. It examines concepts related to security and cloud computing. The study in [3] designed a methodology for structural analysis of rest APIs, which are expected to become increasingly more significant in the context of cloud computing, the internet of things, and microservices. Rest APIs are a method for implementing distributed systems, and this research established it as such. React Native, Native Script, and Ionic are three frameworks discussed in [4] for the creation of mobile JavaScript apps that were offered for a study of learning for programmers. In contrast to Ionic and React Native, which displayed identical worldwide results with a little upward trend in global mean results for React Native relative to Ionic, the results demonstrated that Native Script was the most challenging to learn, framework with the most negative feedback. A comparison of JavaScript-based native and hybrid mobile applications (Reactive Na tive, NativeScript, and Ionic) was made in [5]. This study demonstrated that React Native achieves the best outcomes across all principles analyzed while still providing advantages for hybrid development as compared to native. Comparisons between the performance of different web development technologies were analyzed in [6], mainly in PHP, Python, and Node.js. From the data presented in this paper, Node.js was more efficient, easy to handle, process, and handled more requests than PHP and Python-Web. The different methods of development of web applications that are dynamic using web frameworks were discussed in [7]. It highlights the features of different web frameworks and discusses their merits and demerits. A complete analysis and assessment of security and security services on node.js platform were studied in [8]. Node.js is a synchronous programming interface for input–output operations. The work has shed light on some security ineffectiveness and pitfalls like the fragility of node applications, dos attacks, error handling. Three open-source tools of NodeJS and Angular were discussed in [9]. Two of these technologies work with NodeJS, a back-end technology that is generally used to manage servers using JavaScript. The two NodeJS tools are nodejsonld and jldc, a command-line utility for NodeJS. The performance of PostgreSQL, MYSQL, and other SQL servers with test conditions was analyzed in [10]. For working with big data, numerous tools and DBMSs have been developed. DBMS examples include Oracle MS SQL Server. Most big data systems require significant computational power to guarantee acceptable performance. It also discusses the TPC-H test and 22 of its queries, whose execution enables you to assess the efficacy of database management systems (DBMS). Node.js's core technology was discussed in [11]. Node.js is a server programming platform that was developed based on the JavaScript runtime environment of the Chrome V8 engine, in contrast to single-threaded PHP and multi-threaded JAVA. By employing its own built-in and defined properties, Node.js makes up for the deficiencies of the backdrop programming language in the conventional sense. In [12] it was demonstrated how to use REST APIs for testing. The cost vs. benefit of including usage examples in the documentation must be considered by REST API developers in order to priorities' documentation efforts.

3 Key Technologies

3.1 NodeJS

Node.js is a JavaScript application that is event-driven. It can execute code asynchronously. It is single-threaded framework which is relatively efficient and easy to use. Also, Node.js provides deadlock-free process under the hood as there are no locks. Node.js handles multiple concurrent connections. Upon each connection, a callback is called. Node.js sleeps when no new connections are made, and this contrasts with common concurrency models in today's market. It does not block any process since almost no function performs I/O operation directly. This non-blocking nature increases scalability of the system. Thus, different scalable systems are developed in Node.js. It presents an event loop as a runtime construct like Ruby's Event Machine. In other system, the behavior is defined through callbacks at the beginning of script. Server is started through a blocking call at the end. Node.js does not use such start-the- event-loop call. After executing the input string, it enters the event loop, and it exits event loop when there are no more callbacks to perform. This event loop is hidden from user like JavaScript browsers. Node.js mainly focuses on HTTP protocol. Thus, it is designed focusing on streaming and low latency.

3.2 NestJS

To begin with the origins of NestJS, it was developed by Kamil Mysliwiec as a method to facilitate development of server-side applications that are large scale. It is more of like an abstraction layer and it mainly accepts JavaScript along with TypeScript. One of the most salient features of Nest is that is meant to resolve one of the most crucial issues faced for developers which is the design structure of the app, and while it has Angular architecture including the dependency injection, it does not only cater to front end.

Nest CLI is the building block of Nest, it is a Command-line Interface tool that helps in automation and app initialization which is considered as the most tedious tasks, and hence, they speed up the web application development process by a large value. To create a project in nest, run the code shown in Fig. 1 on the terminal.

Next step is to pick a packet manager, and either npm or yarn can be chosen. After selecting the yarn, an app will be installed with a provider, controller, and module along with its dependencies. The structure of the app will look like as shown in Fig. 2.

It provides a clean and efficient solution and it is easier when working with Angular. The main advantage is that it provides out-of-box features that might be required at the early stages of the project. It not only helps in building new services and controllers, but it also arranges it in a separate folder connected to the modules that are relevant.

Fig. 1 To run NestJS

```
$ npm i -g @nestjs/cli
$ nest new project-name
```

Fig. 2 Structure of an app on NestJS

```
.
├── README.md
├── nest-cli.json
├── package.json
├── src
│   ├── app.controller.spec.ts
│   ├── app.controller.ts
│   ├── app.module.ts
│   ├── app.service.ts
│   └── main.ts
├── test
│   ├── app.e2e-spec.ts
│   └── jest-e2e.json
├── tsconfig.build.json
├── tsconfig.json
├── tslint.json
└── yarn.lock
```

4 Comparison of NodeJS and NestJS

The system compares current ExpressJS framework with NestJS framework. Community support available for NestJS is higher than NodeJS. Thus, has fewer components to maintain. Robustness and security are higher in NestJS than NodeJS. Third-party libraries can be used in NodeJS as it does not provide similar libraries explicitly. This compromises security of the application. In NestJS, it provides different libraries as it has high community support; thus, security of the application is higher as compared to NodeJS. Structure of NestJS is built similar to AngularJs. It provides high dependency injection with the help of decorators. In case of NodeJS dependency, injection is not applicable. Interceptor support is also higher in NestJS as compared to NodeJS. Route versioning is used in order to maintain backward compatibility of web application. NestJS provides flexible route versioning as compared to NodeJS. NestJS supports TypeScript for development. Variable declaration in TypeScript is statically typed. One has to declare the type of variable before it is used. This feature increases robustness of the application. In case of NodeJS, variable declaration with its type is not supported. NodeJS does not provide build-in software architecture. Thus, developers have come up with different types of architecture patterns. In big application, this architecture is difficult to maintain as the framework does not force developers to use the same architecture in overall system. This increases difficulty to maintain the application as the consistency is low. NestJS provides Model–View–Controller (MVC) architecture as a built-in feature. Thus, it enforces all the developers to use the same architecture. This results in maintained consistency of overall application. This architecture also separates business logic-based code from database interaction-based code, resulting in the ease of maintaining and understanding the code. Interceptor support is also higher in NestJS than c NodeJS framework. Based on this study, it can be concluded that NestJS is better than NodeJS framework.

5 System Design

Two similar systems with identical designs have been created, one with Express in NestJS and another with Fastify in NestJS. Since the work involves comparing the two, the design and its components are discussed below. The system is designed with MVC as shown in Fig. 3. Model part handles all the database-related operations. Controller handles all the incoming HTTP requests and sends requested data from model to client. View handles UI part of the application.

The system has mainly two modules:

- **App Module**—All the database-related configurations are declared here. This is the root module of the application.
- **Student-Record Module**—This module organizes the core feature of the application.

Fig. 3 MVC system design

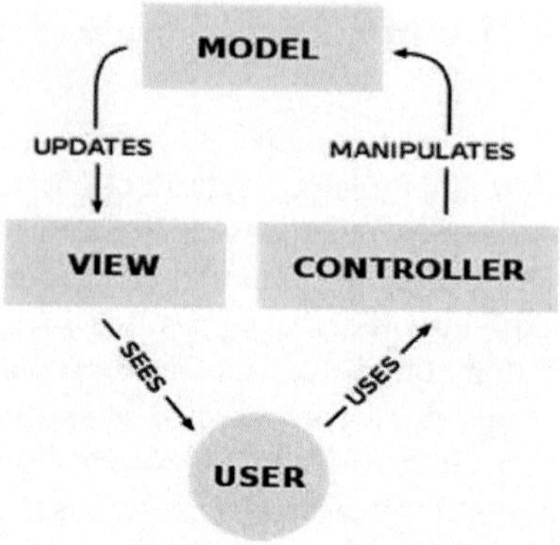

A Student-Record Module is created for this work as a database, and it is shown in Fig. 4 which has the following components:

- **Controller**—receives incoming HTTP requests and calls respective function depending on incoming request method, sends data to client through response.
- **Middleware**—validates or modifies incoming data before it is handled by the handler. Handler is a function, which is called depending on incoming request method.
- **Service**—it is called after validation or modification of incoming data is done by middleware. Services perform business logic of the application. It calls database-related function if it is required by calling the repository.
- **Repository**—this handles all the database-related logics. It receives data from service and manipulates it. It fetches data from database or writes data into the database.

Fig. 4 Master student-record module components

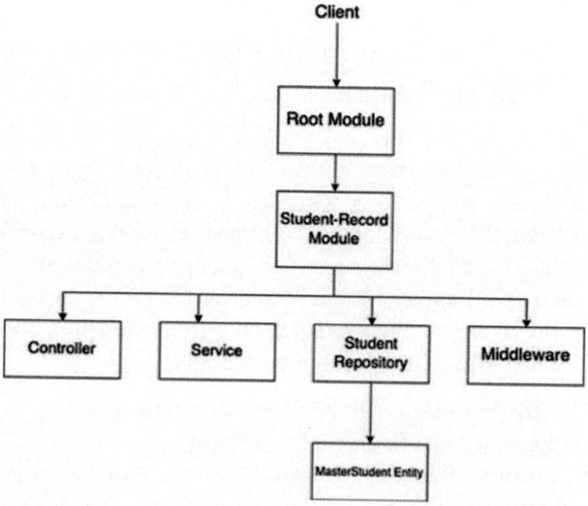

- **Entity**—defines data to be stored in database. It includes some built-in functions like save (), create(), find(), findAll(), update(), etc.
- **Data Transfer Object (Dto)**—these files are classes that define the type of data that needs to be given in the request. These classes are used to validate incoming data. For validation purpose, Validation Pipes are used. These Validation Pipes provided by NestJS validate incoming data with respect to its corresponding dto. If the data are not in the required format, the request is not completed. Thus, robustness is provided.

6 System Requirement Specification

To measure performance of both Express and Fastify, two different programs are created. Both the systems should satisfy below requirements: Functional Requirements.

The systems should be able to handle GET, POST, PATCH, DELETE HTTP requests by all browsers that is sent across. The systems should be able to fetch, create, update, and delete data from database. The systems should be able to validate incoming data through all the HTTP requests. Handling of HTTP requests and database should be identical. If the two systems are met with these four criteria's, the system will be able to run, and this can be viewed on Postman.

7 Implementation

Figure 5 displays node in Express framework. Salient features of Express framework are:

- It enables middleware to be configured to reply to HTTP requests.
- Establishes a routing table that is utilized to carry out various operations.
- Enables dynamic HTML page rendering based on template inputs.

```
const express = require('express')
var bodyParser = require('body-parser')
var paytmRoute = require('./paytmRoute')
const app = express()
app.use('/paytm', paytmRoute)
app.use(bodyParser.urlencoded({ extended: false }))
app.use(bodyParser.json())

require('dotenv').config()

app.listen(process.env.PORT,() => {console.log("Server listening at",process.env.PORT)})
```

Fig. 5 Node with express framework

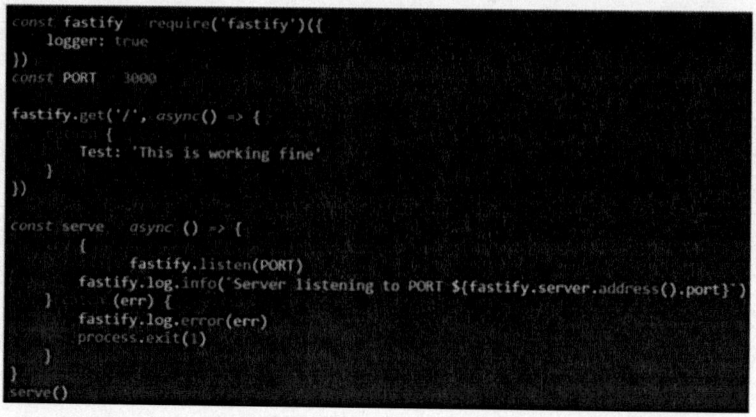

Fig. 6 Node with fastify framework

Let us examine the snippet's functional components displayed in Fig. 6:

- We turned on the Fastify logger first. It is disabled by default.
- Next, we gave our program a port number. However, using process.env. PORT is advised when delivering to production.
- We then developed our first route! Our controller function is synchronous, as you may have noticed. When we develop our controller, we will see more of that.
- We then launch the server on port 3000.

8 Performance and Results

As explained in this paper, two systems were created and developed in NestJS with the same architecture and design. The objective was to take the comparison between Fastify and Express and quantify their performance on NestJS. To do the comparative analysis, evaluation and conclusion were drawn based on different parameters.

The parameters used to do the analysis were average latency, maximum latency, average throughput, and minimum throughput.

Latency—the amount of time it takes for data to appear in a database or data warehouse after an event has occurred is known as data latency.

Throughput—it tracks how inputs and outputs flow throughout the production process. The system was compiled and then executed on VS code with the database connection made on PostgreSQL. To obtain the graphical representation of the results, JMETER was used. Below are graphs for measurement of output for Fastify vs. ExpressJS.

In Fig. 7, average latency is found out for program running on Express and Fastify, when 100 API calls are sent concurrently to both using Apache JMeter.

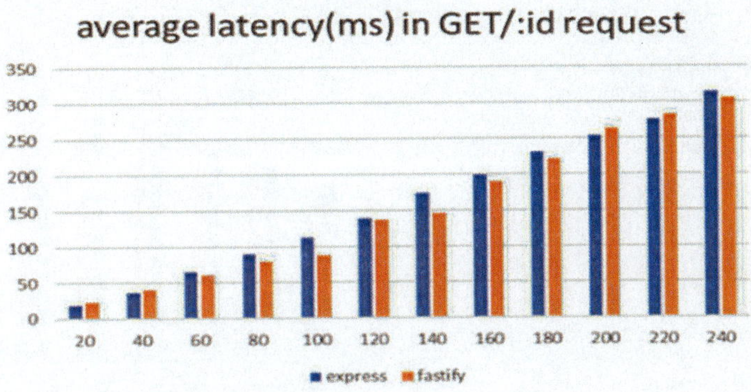

Fig. 7 Depicting average latency in GET request

Table 1 draws a comparison between the average latencies between Express framework and Fastify for a program when 100 API calls were sent concurrently using Apache.

In Fig. 8, maximum latency is found out for program running on Express and Fastify, when 100 API calls are sent concurrently to both using Apache JMeter.

Table 2 draws a comparison between the maximum latencies between Express framework and Fastify for a program when 100 API calls were sent concurrently using Apache Jmeter.

In Fig. 9, average throughput is found out for program running on Express and Fastify, when 100 API calls are sent concurrently to both using Apache JMeter.

Table 3 draws a comparison between the maximum latency between Express framework and Fastify for a program when 100 API calls were sent concurrently, and you can see that Express has more average throughput.

Table 1 Comparison for average latency in ms

Express (ms)	Fastify(ms)
19.19	22.97
36.46	40.05
65.56	60.01
89.62	79.06
112.57	87.28
139.08	136.06
174.53	144.86
198.56	189.33
F230.18	219.71
252.93	262.46
275.01	281.26
313.7	305.87

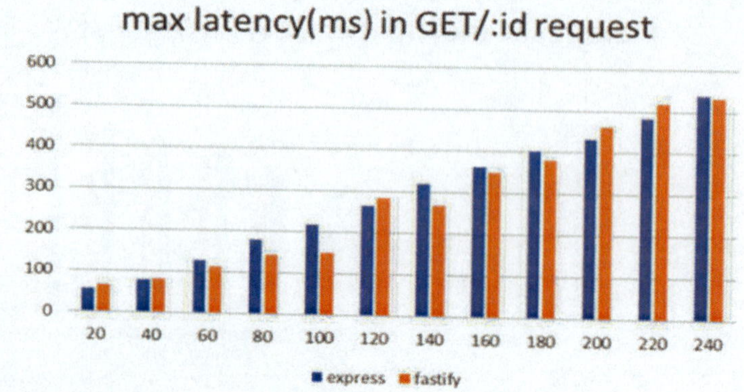

Fig. 8 Maximum latency (in ms) in GET request

Table 2 Comparison table for max latency (ms)

Express (ms)	Fastify(ms)
56.44	65.34
78.69	80.27
129.03	112.81
180.21	144.55
218.09	151.5
266.57	284.07
320.29	267.83
363.07	348.25
402.74	380.68
432.49	462
484.51	519.74
540.01	534.21

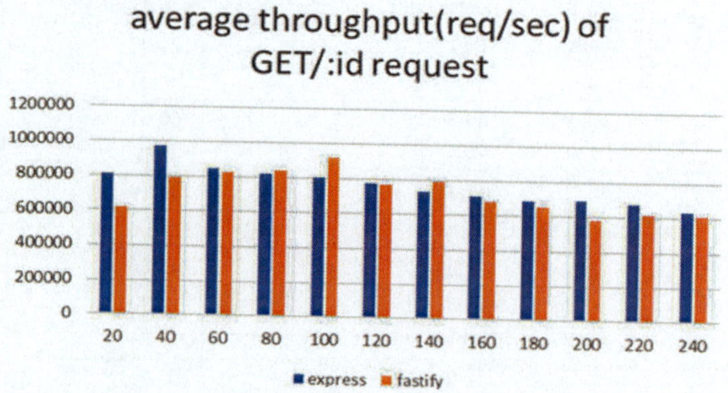

Fig. 9 Average throughput (req/sec) in GET request

Table 3 Depicts a comparison table for average throughput (req/sec)

Express (ms)	Fastify(ms)
809,472	619,904
971,622.4	792,768
847,293.44	822,496
819,910.14	839,139.2
804,013.41	919,945.92
775,441.34	772,276.19
729,691.33	791,621.22
709,333.93	682,426.12
688,085.39	658,181.81
692,834.14	585,958.5
675,798.61	618,365.5
636,271.06	612,390.15

In this Fig. 10, minimum throughput is found out for program running on Express and Fastify, when 100 API calls are sent concurrently to both of them using Apache Jmeter.

Table 4 draws a comparison between the minimum throughputs between Express framework and Fastify for a program when 100 API calls were sent concurrently, and you can see that Express has more minimum throughput.

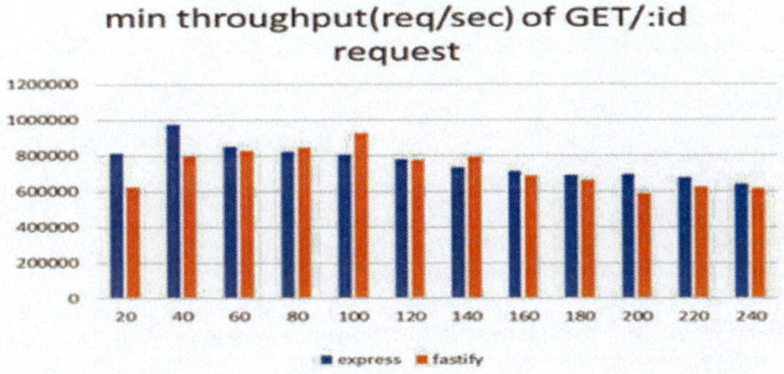

Fig. 10 Minimum throughput (req/sec) in GET request

Table 4 Comparison table for minimum throughput

Express (ms)	Fastify(ms)
809,508.2	619,878.7
971,520.02	792,754.17
847,241.7	822,449.22
819,939.57	839,102.42
804,066.46	919,963.02
775,485.34	772,210.09
729,656.71	791,609.49
709,333.87	682,449.05
688,098.39	658,132.69
692,743.48	585,935.07
675,815.25	618,349.07
636,267.73	612,444.31

9 Conclusion

The paper contains a detailed account about the various technology stacks that were used like NestJS, NodeJS, JavaScript, etc. It also covers in depth about the motivation behind the paper, the problems currently faced by the developers in NodeJS back-end framework. The framework migration from Node to Nest was successful and the code was written in TypeScript which is a subset of JavaScript. Upon the back-end framework migration, two programs were created with essentially the same design and architecture. Unit testing of end-to-end data, validation of data were also performed. Through the help of Postman, the back-end structure of the code and API calls could be witnessed. A comparison between these two programs was performed. One code ran on Express and the other on Fastify. The performance was measured for throughput and latency. Fastify and Express both could handle all the incoming requests with proper incoming data validation. Also, in both the programs, data consistency in database is achieved. Performance comparison between Fastify and Express shows that Fastify is nearly 5% faster than Express.

References

1. Charan NRG, Rao ST, Srinivas PVS, Deploying an application on the cloud. (IJACSA) Int J Adv Comp Sci Appl 6(5):237–242. ISSN No. 2455-2143
2. Dhakal P, Munikar M, Dahal B (2019) One-shot template matching for automatic document data capture. Artificial Intell Transform Bus Soc (AITB) 2019:1–6
3. Haupt F, Leymann F, Scherer V-H, Karolina A (2017) A framework for the structural analysis of REST APIs. In: 2017 IEEE International Conference on Software Architecture (ICSA), pp 251–256

4. Brito H, Santos A, Bernardino J, Gomes A (2019) Learning analysis of mobile JavaScript frameworks. In: 14th Iberian Conference on Information Systems and Technologies (CISTI)
5. Brito H, Santos A, Bernardino J (2019) JavaScript in mobile applications: react native vs ionic vs NativeScript vs native development. Int Res J Eng Tech (IRJET) 6(6):3666–3669
6. Lei K, Ma Y, Tan Z (2020) Performance comparison and evaluation of web development technologies in PHP, Python and Node.js. In: International Conference on Intelligent Computing and Human-Computer Interaction (ICHCI)
7. Okanovic V (2014) Web application development with component frameworks. In: 37th International Convention on Information and Communication Technology, Electronics and Microelectronics (MIPRO)
8. Ojamaa A, Düüna K (2012) Assessing the security of Node.js platform. In: 2012 International Conference for Internet Technology and Secured Transactions
9. Sterling A (2019) Nodejs and angular tools for JSON-LD. In: IEEE 13th International Conference on Semantic Computing (ICSC)
10. Vershinin S, Mustafina AR (2021) Performance analysis of PostgreSQL, MySQL, Microsoft SQL server systems based on TPC-H tests. In: 2021 International Russian Automation Conference (RusAutoCon)
11. Huang X (2020) Research and application of Node.js core technology. In: International Conference on Intelligent Computing and Human-Computer Interaction (ICHCI)
12. Sohan SM, Maurer F, Anslow C, Robillard P (2017) A study of the effectiveness of usage examples in REST API documentation. In: IEEE Symposium on Visual Languages and Human-Centric Computing (VL/HCC)

Index

S. Sharma et al. (eds.), *Intelligent Control, Robotics, and Industrial Automation*, Lecture Notes in Electrical Engineering 1066,
https://doi.org/10.1007/978-981-99-4634-1